THE ROUTLEDGE COMPANION TO GENDER AND ANIMALS

The Routledge Companion to Gender and Animals is a diverse and intersectional collection which examines human and more-than-human animal relations, as well as the interconnectedness of human and animal oppressions through various lenses.

Comprising fifty chapters, the book explores a range of debates and scholarship within important contemporary topics such as companion animals, hunting, agriculture, and animal activist strategies. It also offers timely analyses of zoonotic disease pandemics, mass extinction, and the climate catastrophe, using perspectives including feminist, critical race, anti-colonial, critical disability, and masculinities studies.

The Routledge Companion to Gender and Animals is an essential reference for students in gender studies, sexuality studies, human–animal studies, cultural studies, sociology, and environmental studies.

Chloë Taylor is a feminist philosopher, critical animal studies scholar, and Professor of Women's and Gender Studies at the University of Alberta, Canada. She is the author of three monographs, co-editor of five previous books, and founder of the North American Association for Critical Animal Studies.

ROUTLEDGE COMPANIONS TO GENDER

Recent titles in series:

THE ROUTLEDGE COMPANION TO BLACK WOMEN'S CULTURAL HISTORIES
Edited by Janell Hobson

THE ROUTLEDGE COMPANION TO INTERSECTIONALITIES
Edited by Jennifer C. Nash and Samantha Pinto

THE ROUTLEDGE COMPANION TO GENDER AND SCIENCE FICTION
Edited By Lisa Yaszek, Sonja Fritzsche, Keren Omry and Wendy Gay Pearson

THE ROUTLEDGE COMPANION TO GENDER AND AFFECT
Edited by Todd W. Reeser

THE ROUTLEDGE COMPANION TO GENDER, MEDIA AND VIOLENCE
Edited by Karen Boyle and Susan Berridge

THE ROUTLEDGE COMPANION TO EVE
Edited by Caroline Blyth and Emily Colgan

THE ROUTLEDGE COMPANION TO GENDER AND COVID-19
Edited by Linda C. McClain and Aziza Ahmed

THE ROUTLEDGE COMPANION TO GENDER AND ANIMALS
Edited by Chloë Taylor

THE ROUTLEDGE COMPANION TO GIRLS' STUDIES
Edited by Sharon Mazzarella

For more information about this series, please visit: www.routledge.com/Routledge-Companions-to-Gender/book-series/RCGENDER

THE ROUTLEDGE COMPANION TO GENDER AND ANIMALS

Edited by Chloë Taylor

LONDON AND NEW YORK

Designed cover image: 'Antonia' (August 2023) by Cydney Taylor

First published 2024
by Routledge
4 Park Square, Milton Park, Abingdon, Oxon OX14 4RN

and by Routledge
605 Third Avenue, New York, NY 10158

Routledge is an imprint of the Taylor & Francis Group, an informa business

British Library Cataloguing-in-Publication Data
A catalogue record for this book is available from the British Library

Library of Congress Cataloging-in-Publication Data
Names: Taylor, Chloë, 1976– editor.
Title: The Routledge companion to gender and animals / edited by Chloë Taylor.
Description: Abingdon, Oxon ; New York, NY : Routledge, 2024. | Includes bibliographical references and index.
Identifiers: LCCN 2023055640 (print) | LCCN 2023055641 (ebook)
Subjects: LCSH: Human-animal relationships—Philosophy. | Feminist bioethics. | Ecofeminism.
Classification: LCC QL85 .R674 2024 (print) | LCC QL85 (ebook) | DDC 304.2—dc23/eng/20240310
LC record available at https://lccn.loc.gov/2023055640
LC ebook record available at https://lccn.loc.gov/2023055641

ISBN: 978-1-032-21877-9 (hbk)
ISBN: 978-1-032-22625-5 (pbk)
ISBN: 978-1-003-27340-0 (ebk)

DOI: 10.4324/9781003273400

Typeset in Sabon
by Apex CoVantage, LLC

For Haiku, Clementine, and Antonia

CONTENTS

FIGURES

CONTRIBUTORS

Elan Abrell is a cultural anthropologist whose research focuses on human–animal interactions, animal law, and animal liberation activism in the contemporary United States. He is an assistant professor of the practice in environmental studies and animal studies at Wesleyan University.

Carol J. Adams is the author of *The Sexual Politics of Meat*, *Burger*, *Protest Kitchen: Fight Injustice, Save the Planet, Fuel Your Resistance One Meal at a Time*, and many other books. She has also edited several important anthologies on ecofeminism, feminism, and animals.

Nekeisha Alayna Alexis is a native Trinidadian and former New Yorker who now lives boldly in the Michiana area. She has wide-ranging interests around human and other animal co-liberation, especially related to constructs of race and White supremacy. She is a long-time vegan, and her happy color is bright grass green.

jessie l. beier is a teacher, artist, philosopher, and conjurer of weird pedagogies for the "end times." She is an assistant professor in art education at Concordia University (Montréal, Canada) and author of *Pedagogy at the End of the World: Weird Pedagogies for Unthought Educational Futures*.

Emilie Blanc is Associate Researcher at Rennes 2 University. Her work focuses on contemporary art history and more specifically on the issue of art and politics. She is currently working on a research project on the relationships between art, feminisms, and the animal condition.

e Campbell is an enby and neurospicy settler lawyer and artist living on MÁLEXEŁ and Quw'utsun territories. They grew up by Zhooniya Zaagiigan and Chi'Nibiish in Michi Saagiig territory and Mission Creek in syilx territory. They have Scottish/Norse, English, French, and German ancestry.

Pablo P. Castelló is the 2022–2024 Postdoctoral Fellow in Animal Ethics at the Department of Philosophy, Queen's University. His interdisciplinary research centers animals and engages with democratic theory, ecofeminism, animal law, disability studies, and critical race theory.

Danielle Celermajer is Professor of Sociology at the University of Sydney and Deputy Director of the Sydney Environment Institute and leads the Multispecies Justice project. Her book *Summertime: Reflections on a Vanishing Future* (Penguin Random House, 2021) considers the more-than-human experience of climate catastrophe.

Darren Chang is a PhD candidate in the Department of Sociology and Criminology at the University of Sydney. His research interests broadly include interspecies relations under colonialism and global capitalism; practices of solidarity, kinship, and mutual aid across species in challenging oppressive powers; social movement theories; and multispecies justice.

Pierre Cloutier de Repentigny is a queer, disabled, non-binary, white settler assistant professor of law and legal studies at Carleton University. Their research focuses on two broad areas—environmental law, particularly the marine environment, and trans justice—and their intersection through the concept of queer ecology.

Lauren Corman is an associate professor of sociology (Brock University) and an environmental sociologist who teaches in the areas of environmental thought, contemporary social theory, and critical animal studies. Her research centralizes anti-racist, anti-colonial, queer, and feminist understandings of social relations and the more-than-human world.

Naisargi N. Davé is Associate Professor of Anthropology at the University of Toronto. She is the author of *Indifference: On the Praxis of Interspecies Being* (2023) and *Queer Activism in India: A Story in the Anthropology of Ethics* (2012), both with Duke University Press.

Maneesha Deckha is Professor and Lansdowne Chair in Law at the University of Victoria in British Columbia where she directs the Animals & Society Research Initiative. She is author of *Animals as Legal Beings: Contesting Anthropocentric Legal Order*.

Rebecca Deutsch is a master of arts student in the Gender and Social Justice program at the University of Alberta. She is invested in the possibility of creating new forms of relationship and new valuations of knowledge that more readily support and account for all the beings we share space with.

Josephine Donovan is Professor Emerita of English at the University of Maine. She is the author of *Animals, Mind, and Matter: The Inside Story* and *The Aesthetics of Care: On the Literary Treatment of Animals* and co-editor with Carol J. Adams of *The Feminist Care Tradition in Animal Ethics*.

Jessica Eisen is an associate professor at the University of Alberta Faculty of Law. Her research interests include animal law, critical animal studies, constitutional law, and feminist jurisprudence.

Karen S. Emmerman is an independent scholar and part-time faculty in philosophy at the University of Washington. She is also the Philosopher-in-Residence at John Muir Elementary School and Education Director of the Philosophy Learning and Teaching Organization. Karen writes on ecofeminism, animals ethics, and philosophy for children.

Catia Faria is an assistant professor of moral philosophy at Complutense University of Madrid and a founding member of the Centre for Animal Ethics at Pompeu Fabra University. Their latest work, *Animal Ethics in the Wild*, has recently been published by Cambridge University Press (2023).

Greta Gaard is Professor of English and Women/Gender/Sexuality Studies at the University of Wisconsin in River Falls. Her scholarship on queer and critical ecofeminisms, children's environmental literature, feminist postcolonial milk studies, milk, and oil are all her favorites.

Rama Ganesan was born in Chennai, India, and immigrated to the UK and then later to the USA. She has an undergraduate degree from Oxford, a PhD from the University of Wales, and an MBA from the University of Arizona. She currently lives in the Midwest.

Kathryn Gillespie is a multispecies ethnographer and feminist geographer. She is the author of *The Cow with Ear Tag #1389* about the lives of cows in the dairy industry and a forthcoming book called *The Sound of Feathers: Haunting and Bearing Witness in Multispecies Worlds*.

Anahí Gabriela González holds a PhD in philosophy from the Université Paris 8 and the Universidad Nacional de San Martín (Argentina). She is a postdoctoral fellow of CONICET, professor at the Universidad Nacional de San Juan (Argentina), and director of the Revista Latinoamericana de Estudios Críticos Animales.

Katja M. Guenther is Professor of Gender and Sexuality Studies at the University of California, Riverside, where she focuses on feminist politics and the human exploitation of non-human animals. She is the author of numerous articles and books, including *The Lives and Deaths of Shelter Animals* (2020).

Alok H. Gupta is a lawyer, activist, and researcher working on LGBT and animal rights. Alok is currently pursuing a PhD at the School of Law, University of Reading, looking at animal-centric jurisprudence from Indian courts.

Dylan Hall is an uncle, settler, and proud cat dad living in Edmonton, Alberta, known in Cree as Amiskwaciwâskahikan, or Beaver Hills House. He is currently a graduate student in environmental history and, always, an eager collector of gardening tips.

Carrie Lou Hamilton is the author of *Veganism, Sex and Politics: Tales of Danger and Pleasure* (2019) and *Pen in Fist*, a newsletter on writing and activism (peninfist.substack.com). She lives in London.

Alexandra Isfahani-Hammond is Associate Professor Emerita of Comparative Literature and Luso-Brazilian Studies at the University of California, San Diego. Her current book project, *Home Sick*, blends theory with creative nonfiction to meditate on caregiving, end of life, and the transformative potential of bereavement.

Stephanie Jenkins is an associate professor in the School of History, Philosophy, and Religion at Oregon State University. She works in the fields of critical animal studies, critical disability studies, feminist philosophy, ethics, and public philosophy.

pattrice jones is the self-selected name of the conduit through which "Queering Animal Liberation" was written. Known to parrots as D#-E-E-D#-E-F, this organism is known to humans as a cofounder of VINE Sanctuary, an LGBTQ-led refuge for farmed animals, and the author of various works, including *The Oxen at the Intersection* (Lantern Books, 2014).

Danika Jorgensen-Skakum is a PhD student in political science at the University of Alberta where she also completed her MA in gender and social justice. She is a queer Métis scholar interested in questions of data, posthumanism, climate justice, disability studies, and the so-called Anthropocene.

Claire Jean Kim is Professor of Political Science and Asian American Studies at University of California, Irvine. She is the author of three books, including *Dangerous Crossings: Race, Species, and Nature in a Multicultural Age* (Cambridge University Press, 2015).

Tessa Laird is Senior Lecturer in Critical and Theoretical Studies, Faculty of Fine Arts and Music, University of Melbourne. Her book *Bat* (2018) forms part of Reaktion's Animal series, and she is currently writing *Cinemal*, a book about animal aesthetics in experimental film.

Juan José Ponce Léon has a PhD in psychology, Universidad Autónoma de Madrid, and a masters in political sociology, FLACSO-Ecuador. He is a member of the editorial board of the Revista Latinoamericana de Estudios Críticos Animales. His lines of research focus on social movements, subjectivities, masculinities, decolonial veganisms, and the psychosocial dimensions of human–animal violence.

Melissa Plisic (they/she/siya) is a zine maker, poet, stand-up comedian, activist, and occasional academic. They are a Sagittarius sun and rising and Leo moon. Točka (she/he/they) is a total sweetheart and menace to social order. Her hobbies include playing tug-of-war, napping, and eating snacks.

Annie Potts is the Co-Director of the New Zealand Centre for Human–animal Studies at the University of Canterbury. She writes about and advocates for "underdogs" of the animal kingdom, including those consumed in their billions and those persecuted as "pests".

Sue Hall Pyke teaches creative writing and Indigenous studies at the University of Melbourne. She writes within the rocky terrain of Djargurd Wurrong Country (the "Stony Rises"). This territory, governed by the Eastern Maar Nation, is located in the southwest of "Victoria". For more information, see @suehallpyke.com.

Emelia Quinn is Assistant Professor of World Literature & Environmental Humanities at the University of Amsterdam. She is author of *Reading Veganism: The Monstrous Vegan, 1818 to Present* (Oxford University Press, 2021) and co-editor of *The Edinburgh Companion to Vegan Literary Studies* (Edinburgh University Press, 2022).

Tim Reijsoo studied philosophy and international development studies at the University of Amsterdam. His philosophy thesis develops the concepts of "species privilege" and "human innocence". Reijsoo's work looks at how systems of injustice are intertwined. In 2024, he will start a PhD project on conceptualizing multispecies oppression.

Sal Renshaw teaches in the Department of Gender Equality and Social Justice at Nipissing University, Canada. She is the author of *The Subject of Love: Hélène Cixous and the Feminine Divine* (2009). Her most recent research and teaching focuses on critical animal studies and the ethics of interdisciplinary pedagogy.

Margaret Robinson is a Two-Spirit Mi'kmaw scholar and a member of Lennox Island First Nation. She works at Dalhousie University as an associate professor and holds the Canada Research Chair in Reconciliation, Gender, and Identity. She is passionate about vegan cooking and Indigenous survivance.

Deborah Slicer is Professor Emerita with the University of Montana's Department of Philosophy. She has been teaching and writing about other animals for three decades.

Kelly Struthers Montford is Assistant Professor of Criminology at Toronto Metropolitan University. She is the co-editor of three books and several journal issues spanning colonialism, animality, and the carceral. Her work has been published in philosophy, law, sociolegal, and animal studies journals.

Alison Suen is an associate professor of philosophy at Iona University, New York. She received her PhD in philosophy from Vanderbilt University. Her publications include *The Speaking Animal* (2015) with Rowman and Littlefield International and *Why It's OK to Be a Slacker* (2021) with Routledge.

Chloë Taylor is a feminist philosopher, critical animal studies scholar, and professor of women's and gender studies at the University of Alberta. She is the author of three monographs, co-editor of five previous books, and founder of the North American Association for Critical Animal Studies.

Richard Twine is Reader in Sociology and Co-Director of the Centre for Human–Animal Studies (CfHAS) at Edge Hill University, UK. He is the author of several papers and books on human–animal relations, ecofeminism, and the climate crisis.

Tessa Wotherspoon is an English literature masters student at the University of Glasgow. Upon completion of this course, she will undertake a PhD entitled "Black Naturalism: Genre and the Nonhuman in 19th and 20th Century Black Diasporic Literature" in the fall of 2023.

Corey Lee Wrenn is a senior lecturer in sociology at the University of Kent, past chair of the Animals and Society Section of the American Sociological Association, and co-founder of the International Association of Vegan Sociologists.

Selingul Yalcin is a law school student at the University of Ottawa (Common Law). Her endeavors and activism are related to access to justice and empowering marginalized voices within the law.

Tayler Zavitz is a PhD candidate and sessional instructor in the Sociology department at the University of Victoria. Her current research focuses on the repression of animal activism in Canada, the expanding criminalization of dissent, and what this means for the future of activism in Canada.

INTRODUCTION

Why Gender and Animals?

Chloë Taylor

What do animals have to do with gender studies?[1] Are animals a feminist issue? A quick search of courses offered in women's and gender studies and feminist studies departments across primarily English-speaking countries would suggest not. Very few such programs offer courses in human–animal or critical animal studies with any gender component.[2] In my own education, in all the feminist courses I took at several universities, we did not read a single article on animals. A survey of recent programs for major feminist studies conferences also shows few presentations on animals.[3] Anthologies of feminist thought demonstrate little to no space for animals. For instance, the 1512-page edited volume *Gender and Philosophy*, published in 2011, includes only one chapter on animals, and this is the seventy-fifth chapter out of seventy-five.[4] Ten years later, the 2021 edited volume, *Keywords for Gender and Sexuality Studies*, includes "Abjection," "Affect," "Agency," and "Anal" as its A-words but not "Animal."[5] It seems that at this time, gender studies remains overwhelmingly humanistic. Nonetheless, and as the chapters in this volume demonstrate, it is important to think about gender and animals together for at least six reasons.

First, it is important to think about gender and animals together because distance from and proximity to animals are aspects of how gender and other, intersecting identity categories such as race, sexuality, and ability have been constructed in Western thought.

Second, it is important to think about gender and animals together because human debates around gender and sexuality have been deeply invested in biological, zoological, and primatological studies of animals, as well as in popular science representations of animals.

Third, it is important to think about gender and animals together because, to borrow from Rosemarie Garland-Thomson,[6] integrating feminism transforms species theory. Predating critical animal studies, feminist theory offers insights into the categories, constructions, and performances of gender and sexuality that may be helpfully applied for thinking about species.

Fourth, gender and animals should be studied together because human gender constructs strongly impact how people treat and perceive animals, with femininity being associated with caring for animals and masculinity being linked to harming animals.

DOI: 10.4324/9781003273400-1

Fifth, it is important to study gender and animals together because these are significant and entangled themes in human cultural production and social movements.

Sixth, it is important to study gender and animals together because we are living in an era of environmental catastrophe, marked not only by climate change but also by extraordinary biodiversity loss, which impacts all of us but in which certain groups—including women and more-than-human animals—are particularly vulnerable. Multispecies futures are at stake, and feminist scholarship and activism that is inclusive of all earth beings is necessary to respond to the crisis and find paths to multispecies justice.

In the following I will discuss each of these topics in more detail, but with particular attention to the first point, since in exploring this topic I will introduce key concepts and theoretical frameworks that are prevalent throughout the book, such as ecofeminism, care ethics, the sexual politics of meat, and intersectionality. Signposts to chapters in the book are woven throughout this discussion.

1. Animalization and Human Identity Constructs

1.1 Animalization in the History of Western Thought

In the creation myth that the ancient Greek philosopher Plato offers in the *Timaeus*, he describes a transmigration of souls in which humans and other animals are downgraded from male to female, human to animal, and higher animal to lower animal based on how foolishly or wickedly they behaved in their previous life.[7] As Genevieve Lloyd observes,

> From the beginnings of [Western] philosophical thought, femaleness was symbolically associated with what Reason supposedly left behind—the dark powers of the earth goddesses, immersion in unknown forces associated with mysterious female powers. The early Greeks saw women's capacity to conceive as connecting them with the fertility of Nature. As Plato later expressed the thought, women "imitate the earth."[8]

On a more pragmatic level but also associating women with nature and animals, Plato's contemporary, Xenophon, wrote in his treatise on economics,

> When sheep fare badly, we usually fault the shepherd, and when a horse behaves badly, we usually speak badly of the horseman; as for the woman, if she has been taught the good things by the man and still acts badly, the woman could perhaps justly be held at fault; on the other hand, if he doesn't teach the fine and good things but makes use of her as though she is quite ignorant of them, wouldn't the man justly be held at fault?[9]

Xenophon then writes of a woman needing to become "accustomed" to her husband or "domesticated to the extent that we could have discussions."[10] Women, for Xenophon, might be somewhat more "teachable" animals than sheep or horses, but they are closely related to agricultural animals in their relation to the male head of household. It is this understanding of women and animals as beings to be domesticated, groomed, and utilized by male heads of households that is at the origin of the term "animal husbandry," which continues in use to this day. According to this understanding of "husbandry," agricultural animals are the wives of men, while women are like farmed animals in their need to be managed and trained for the economic profitability of the household. As in the case of human wives historically, a huge part of the management of agricultural animals involves exploiting and controlling their sexual and reproductive capacities.

For Plato's student Aristotle, man is a "political animal," as evidenced by his unique capacities for reason and speech. In contrast, lack or deficiency in reason is the explanation for why women, slaves, and animals are each subordinate to free men, who alone are fit to be heads of households and states.[11]

The European male philosophical habit of comparing women to animals and nature endured into the modern period. In the *Philosophy of Nature*, early 19th-century philosopher Georg Wilhelm Friedrich Hegel argues that women are incapable of attaining full consciousness, but rather only reach a primeval or primitive stage of consciousness.[12] In the *Philosophy of Right*, Hegel underscores that women are defined by their biologically determined roles as wives and mothers and have no place in civic life.[13] According to the later 19th-century German philosopher Friedrich Nietzsche, "Woman is more closely related to Nature than man and in all her essentials she remains ever herself. Culture is with her always something external, a something which does not touch the kernel that is eternally faithful to Nature."[14] Given the frequency with which white male philosophers have defined "man" as uniquely rational or as a "political animal," even while describing women as irrational, closer to primitive nature than men and thus unfit for politics, feminist philosophers have legitimately wondered whether women have been considered human in the canon of Western thought.[15]

1.2 Ecofeminism

Feminists have responded to the fact that women and nature are culturally associated in Western thought in at least three ways, two of which may be characterized as ecofeminist responses. First, some have identified the association of women with nature as the main source of women's oppression and thus sought to reject it, insisting that women are not animals. They have argued that women are as fully human as men and contribute to culture (or would if men let them) rather than being on the nature side of the nature-culture dualism. Of course, this entails accepting the assumptions in the male-dominated Western philosophical tradition that to be an animal is to be inferior to humans and that only humans have culture. As pattrice jones writes in her chapter for this volume, "Since subordinated humans are often treated as subhuman, identity-based liberation movements often tend to strongly embrace the *Human*, arguing only that they should be included within the category rather than challenging the category itself." Simone de Beauvoir's *The Second Sex* is an example of this response to women's association with nature and animals. In her "Historical Materialism" chapter, she writes that "Humanity is not an animal species, it is a historical reality. Human society is . . . against nature."[16] She is then at pains to situate women on the side of history and humanity rather than nature. De Beauvoir has no issue with humans dominating nature and other animals; she just insists that women should be in the human and dominant category and not in the nature, animal, and dominated category.

While it is understandable that many feminists would want to distance themselves from an association that has harmed them and may also have internalized patriarchal understandings of nature and animals, ecofeminists have highlighted that this response does nothing to undermine the underlying logic of dualistic, hierarchical, and oppressive thinking. It insists that women should not be treated "like animals" without questioning why being an animal should justify being treated badly. While a group that takes this tactic of distancing themselves from nature and animals may have some success at leveraging themselves out of the oppressed category, the fundamental logic of oppression is perpetuated even while others remain in the category of animal or animalized, including more-than-human animals themselves.

A contrasting approach that some feminists have taken to the linking of women, nature, and animals in Western thought is to accept this association as following from biological facts but to revalue nature so that being close to it is something to celebrate rather than eschew. Some examples of this kind of feminist response to the nature-woman connection are Mary Daly's (1978) *Gyn/ecology*, which has correctly been critiqued for its transphobia,[17] and Starhawk's spiritualism, which has also been problematized for being exclusionary of queer people.[18] This type of biologically essentialist, cisheteronormative ecofeminism has unfortunately often been taken as characteristic of ecofeminism as a whole, which has thus been widely dismissed within feminist circles. In fact, however, as numerous scholars have pointed out, biological essentialism characterized cultural feminism of this time generally, whether ecofeminist or not, and most feminists working at the intersections of environmentalism and gender saw the connection between women and nature as socially constructed. Greta Gaard's chapter that opens the current volume, "Ecofeminism," provides a history of ecofeminism, its emergence and reception, internal conflicts, contemporary work in both the global north and south, ecological affects, eco-spirituality, and the directions that ecofeminism should take at this time of environmental crisis, without centering the essentialism debates. In an earlier essay, however, "Ecofeminism Revisited: Rejecting Essentialism and Re-Placing Species in a Material Feminist Environmentalism," Gaard considers these debates and what was lost in their aftermath, observing that the charge of essentialism was so effective at dismissing ecofeminism that the majority of scholars working in this area chose to distance themselves from ecofeminism for fear of "contamination-by-association." As a result, many such scholars would go on to describe their work as "ecological feminism," "feminist environmentalism," "social ecofeminism," or "critical feminist eco-socialism" instead.[19]

A third response to the association between women, nature, and animals in Western thought, then, is the work of such scholars, who might be called social constructivist ecofeminists but who may have described themselves in a somewhat different way. Karen Warren, for example, who preferred "ecological feminism," argues that women have historically been relegated to roles that keep them closer to nature than culture such as reproduction rather than production, caring for the "animal" aspects of human life rather than the political or cultural sides.[20] The historical relegation of women to nature has resulted in a symbolic connection, with women taken to represent nature and nature taken to represent women. We see the symbolic connection of women and nature in language when we speak of the earth as a "mother" and the "rape" of "virgin" land.

For Warren, these historical, symbolic, and theoretical connections between women and nature have resulted in material and interconnected dominations. Because nature is considered subordinate to humans in many traditions and societies and certainly in the West, and because women are seen as closer to nature than men are, women's domination has been normalized. Conversely, because nature has been viewed as female, misogyny has functioned to justify men's domination of nature. Thus Warren argues that any feminist theory needs to account for the exploitation of nature and animals, and any environmental ethics needs to account for women's oppression. Social constructivist ecofeminists such as

Warren recognize that the perception of women as more natural or animalistic than men is cultural and contingent without therefore distancing themselves from that association, standing rather as allies of nature and animals. Social constructivist ecofeminists thus challenge patriarchal, hierarchical, and dualistic logics that result in the domination of women, ecosystems, and animals.

For social constructivist ecofeminists, although the woman-nature and woman-animal connections are products of historical processes and theories, they have resulted in many women having experiences of oppression and exploitation that give them insights and investments in liberating and protecting more-than-human animals. For instance, because women have often experienced exploitation, domination, alienation, and powerlessness in their roles as mothers, some ecofeminists have felt a particular connection and empathy with animals such as dairy cows and sows in industrial agriculture. These other female mammals are repeatedly impregnated against their wills, kept in a continual state of either pregnancy or lactation, spatially confined, and limited to their reproductive roles. The lives and value of these female animals are reduced to their reproductive abilities, such that when they cease to be fertile, they are killed. At the same time, these female animals repeatedly experience their babies being taken from them for the profit of others, with humans holding all legal rights over their children.

Although the situation of human women has rarely been as extreme as that of a sow in a sow crate on an industrial farm, the institution of slavery in the U.S. is exceptional. In this context, Black women slaves were the legal property of white people, were regularly raped and impregnated by white men, experienced their children taken away from them and sold repeatedly, were forced to give their breastmilk to feed white children rather than their own babies, and were thus quite literally treated like female agricultural animals.[21] The late 19th- to 20th-century eugenic sterilizations of women and men deemed unfit to reproduce on the grounds of race, disability, mental health, poverty, and sexuality—sometimes combined with institutional warehousing—were also situations where human beings have experienced practices routinely inflicted on domesticated animals targeting their reproductive capacities.[22] Even the most privileged women have experienced compulsory heterosexuality and motherhood in many cultures and societies, moreover, as women have historically been compelled into heterosexual relationships, marriage, and motherhood through social norms, familial pressure, lack of viable alternatives, forced economic dependency, lack of access to birth control and abortion, male control of divorce, threats, and physical force.[23] While white people had legal ownership over enslaved women's children, after the technical abolition of slavery, Black and Indigenous women continued (and continue to this day) to have their children taken from them by the state at far higher rates than white women in countries such as Canada.[24] Men have sometimes had entire legal control over children in marriage such that even a privileged white woman who left her husband was forced to leave her children behind.[25] Although there are significant differences between these experiences of women and the experiences of more-than-human animals on farms, and although women's experiences have differed widely amongst themselves based on intersections of gender with race, ability, sexuality, and socioeconomic class, these experiences have given some feminists empathy for the lack of reproductive autonomy experienced by agricultural animals.

Ecofeminists have thus had the insight that animal oppression is gendered, and, as Corey Lee Wrenn's chapter in this volume discusses, this is true for male animals in agriculture as well. Male animals experience forced castration without anesthesia, forced extraction of their semen, and cruel forms of infanticide in industries where they are considered superfluous, such as the crating, enforced anemia, and early slaughter of "veal" calves in the dairy industry and the mass crushings and suffocations of male chicks in the egg industry. As Wrenn highlights, it is also male geese who are the victims of the cruel foie gras industry. While the gendered and sexual oppression of male animals in agriculture is often overlooked by ecofeminist and other feminist animal studies scholars, critical race scholar Aph Ko has argued that the gendered and sexual oppression of Black men is also overlooked by intersectionality frameworks.[26] As Ko argues, examining the intersections of gender and race has in fact meant centering the experiences of Black women. Ko observes, however, that both under slavery and under the contemporary criminal-legal system in the U.S., Black men experience systematic gendered and sexual violence. As she writes, "Within an intersectional analysis, we tend to miss how Black men were sexually vulnerable during slavery and were routinely raped by white men *and* women,"[27] while police "stop and frisk" practices that target Black men should be understood as gendered and sexual as well as racial violence.[28]

Here it is important to consider that drawing analogies across differences, and analogies between animal oppression and human oppression in particular, is always fraught.[29] In her chapter for this volume, "Analogy and Alterity," feminist and animal law scholar Jessica Eisen observes that analogies between particular groups of humans and particular animals have served to degrade and harm humans, however animal activists have frequently attempted to invert such analogies, using them to elevate animals instead. For instance, while Black people have been oppressed through comparisons to animals such as apes, white animal activists have sought to elevate animals by showing that they have suffered comparable forms of treatment as Black people and, by extension, should also be emancipated. This is seen, for instance, when PETA juxtaposes the image of a Black human's leg in chains and a circus elephant's leg in chains. There are several problems with these types of analogies, however, as Eisen's chapter explores. As she writes, "Such analogies have been charged with minimizing ongoing justice projects, ignoring the unique valences animal comparisons may have for some human groups, and failing to sufficiently attend to the incommensurability of the human experiences in issue." Animal activist comparisons of racial slavery to nonhuman animal slavery often assume white supremacy is a historical atrocity that, as a thing of the past, may now be mobilized in furthering another group's cause. This is done by white activists who not only demonstrate an ignorance of the historical trauma of such comparisons for Black people but also that their oppression is ongoing. For instance animal activists making such comparisons seem unaware or are willfully ignorant of critical race arguments that the current mass incarceration of Black people in the U.S. is a perpetuation of racial slavery and do not often take a prison abolitionist stance. Indeed, as the chapter by Kelly Struthers Montford, Darren Chang, and Selingul Yalcin for this volume shows, there has been a carceral tendency in white animal activism much as there has been in white feminism, wherein activists seek prison terms for those who abuse animals, with white animal activists seemingly oblivious to anti-racist critiques of incarceration. In contrast with such strategies, ecofeminists by and large have recognized human oppressions such as sexism and racism as ongoing and have been deeply invested in liberating those beings on each side of the analogies they make, whether these are women and animals or other groups of humans and animals.

Despite the dangers of making comparisons between oppressions, Eisen concludes that animal law cannot completely avoid drawing such analogies given the centrality of precedence to the practice of law. Arguably, ecofeminism, feminist critical animal studies, and animal activism more generally cannot entirely eschew making connections between oppressions either, since, as the following discussion will explore, in some instances these are not in fact distinct oppressions but are profoundly entangled. If we refuse to think about speciesism alongside racism or ableism, or to think about racism or ableism as it interconnects with speciesism, aspects of these oppressions may in fact be lost to us.[30] As Lauren Corman has observed,

> dominant Western culture has linked oppressed human groups with animals. To try to explore how power is functioning in such linkages is not to draw analogies, as some animal activists have done, but to analyze and attend to how the connections have been constructed in order to better diagnose and disrupt them, both in service of animal and human liberation.[31]

This being the case, Eisen argues that

> Feminist and intersectional methods offer us some guidance as to how to proceed with care in this context: attending to interconnections rather than blunt comparisons between oppressions, and drawing experiences of oppression together in ways that invite empathy rather than objectification. As an example, Eisen observes that in *The Sexual Politics of Meat*, Carol J. Adams chooses not to take up a tendency in the animal rights movement to directly analogize the artificial insemination of dairy cows to rape within human communities. She explains her decision to term the former "forcible impregnation," in part to leave room for dedicated attention to what is specific about each category of experience.

As Eisen underscores,

> Here, we see *placements alongside* deepened through attention to distinct justice contexts (human rape and sexual exploitation of animals) to make an intersectional argument: that tropes of objectification work in distinct but interrelated ways across species lines to authorize particular experiences of violence and exploitation.

1.3 *The Sexual Politics of Meat*

There is likely no other work more closely associated with the topic of gender and animals than Adams' groundbreaking 1990 book, *The Sexual Politics of Meat: A Feminist-Vegetarian Critical Theory*.[32] Likewise, no work is referenced as often throughout the chapters in the current volume as this one. In this iconic work, Adams considers a wide array of materials, from myths, literature, historical, anthropological, and sociological studies to advertisements and visual culture to describe the undeniable associations between meat and masculinity, animal consumption and misogyny, and animal oppression and racism, primarily but not exclusively in Western cultures.

As Adams explains, men and boys in many societies have been expected to eat more and better food than women and girls, and good food has been understood as meat. In contrast, women and girls are expected to be more disciplined about what they eat and to forego food so that men and boys can have more in times of scarcity. Women and girls are socialized to opt for "feminine foods," with vegetables being feminized because they are associated with passivity (as in the use of "vegetative" to describe a comatose state). In contrast, eating animal muscle, especially from mammals, is thought in Western patriarchies to bestow energy and strength. Food and sex, as the two primary desires or appetites of the body, are continually interconnected in the Western heteropatriarchal imaginary, and restrictions on what women and girls should eat have paralleled restrictions on their sexual desires: in both the cases of food and sex, women and girls are socialized to suppress their appetites and to be passive. In contrast, men and boys are socialized to have more unrestrained appetites and to feel entitled to a large share. Adams argues that men's right to eat animal bodies is associated with their perceived sex right or their right of access to women's bodies. One result of these social constructs is that men and boys face greater social opposition in the West when they become vegetarians or vegans, as they are violating more social norms. As Brian Luke has described, it may be particularly hard for men to express compassion for animals or concern for the environment due to the sexual politics of meat, since such concerns and sympathies are read as feminine.[33]

Adams also describes how non-white people have been described by racists as, like white girls and women, more capable than white men of subsisting on a vegetarian diet. Taking up evolutionary theory, white male scientists historically argued that non-whites, supposedly being less evolved, could subsist on diets of cereals like lower animals, while white men, being more evolved "brain-workers," required animal protein, with red meat being the quintessential food for white men. The large amount of meat and dairy typical of European diets has thus been taken not only as indicative of European men's apparently greater need for animal protein compared to non-white peoples but as an explanation for their supposed racial superiority as well.

In *The Sexual Politics of Meat* and in decades of presentations on the topic, Adams analyzes media and literary representations in which women are represented as meat, or as animals who are fated to become meat, while animals in the meat industry are feminized and sexualized. As she shows, dead or soon-to-be-dead animal bodies are represented as female, and women's flesh is used to market animal flesh. Thirty years after the publication of Adams' book, we continue to see the sexual politics of meat that Adams describes in contemporary meat advertising. In Adams' chapter for this volume, she revisits an important concept that she introduced in *The Sexual Politics of Meat*, "feminized protein," drawing on contemporary examples to demonstrate the gendered and sexual violence of the dairy and egg industries.

Since Adams, other feminist animal studies and vegan studies scholars have provided additional explorations of the sexual politics of meat. For example, in his 2009 article, "Metrosexuality Can Stuff It: Meat Consumption as (Heterosexual) Masculine Fortification," communications scholar C. Wesley Buerkle considers meat in relation to what has been called "metrosexuality."[34] According to Buerkle, metrosexuality is resented by straight men, who see it as an emasculating imposition of women's desires onto them. Now that women are more independent of men financially, and are thus less often required to marry or have sex with men to survive, they can be more discerning about which men they sleep with and if they have sex with men at all. Put otherwise, the grasp of what feminist scholars have called "compulsory heterosexuality" is loosening. Men now have to work to get women, and to be attractive to women, they now have to conform to feminine tastes: they must attend more to their appearance and hygiene, become more cultivated in their behavior, and be more conscientious about what they eat. Like Adams, Buerkle turns to media representations of meat to make his points. He examines hamburger commercials such as Burger King's "Manthem," in which meat-eating is represented as a male liberation movement, a resistance to the emasculating influence of women or feminism. "Manthem" even compares this men's liberation movement to the Black Liberation movement with the "I am (a) man" slogan. In this imagined men's liberation movement, men reassert their right to be men. This reclamation of a virile hegemonic and toxic masculinity is epitomized through their eating of red meat. In another ad, a man buys a military vehicle to "restore the balance" or "restore his manhood" after feeling emasculated by purchasing tofu and carrots under the pitying gaze of another man who is buying copious amounts of red meat.[35]

Annie Potts and Jovian Parry also explore meat-eating, gender, and sexuality in their 2010 article, "Vegan Sexuality: Challenging Heteronormative Masculinity through Meat-Free Sex."[36] Potts and Parry discuss the fact that a very small number of women in a New Zealand survey expressed a visceral aversion to sex with meat-eaters and stated that they were only attracted to other vegans. According to these interview subjects, the bodies of omnivores repel them since they are constituted by dead animals and their secretions. As Potts and Parry discuss, the idea that some women might reject men because of what they eat caused an explosion of online outrage on the part of men that is comparable to the backlash against metrosexuality that Buerkle discusses, and particularly his discussion of "Manthem." Here we see the online equivalent of men rioting in the street, burning their underwear and destroying minivans, in outrage that women are trying to dictate how they should eat, threatening their right to meat and challenging their male sex right. The idea that men have a right to animal bodies (meat) is entangled with the idea that they have a right to women's bodies (sex), and it is obviously infuriating to some men that there are women who might tell them not to eat meat or refuse to have sex with them if they do.

Following Adams' intersectional analysis of meat advertisements, other authors have provided updated accounts of the racial politics of meat. In "The Whopper Virgins," for instance, critical animal studies scholar Vasile Stănescu critiques Burger King's highly successful and xenophobic "Whopper Virgins" advertisement campaign.[37] The campaign demonstrates the ongoing sexualizing of meat, with people in geographically remote communities who had not eaten hamburgers called "virgins." In another article, " 'White Power Milk': Milk, Dietary Racism, and the Alt-Right," Stănescu demonstrates that the consumption of animal products continues to be mobilized as a symbol of white, heteropatriarchal masculinity. Just as the consumption of meat and dairy as opposed to rice and corn-based diets was taken to explain white superiority in 19th-century imperialist discourses, Stănescu shows that similar claims were mobilized by the alt-right in the early 21st century.[38] As Stănescu's articles demonstrate, racist, xenophobic, and gendered discourses such as those analyzed by Adams in *The Sexual Politics of Meat* continue to have traction.

Vegan sexuality and the sexual politics of meat are taken up in the interview with Emelia Quinn in this volume, and several other chapters in this book explore the relationship between masculinities and animals in the context of a gendered and sexualized meat culture. In her contribution to this volume, for instance, "Homoeroticism and Trophy Hunting," Sal Renshaw considers the "remarkably repetitive and iconic visual stagings" of hunters with their kills, in "which white men pose over the dead but hauntingly life-like bodies of huge, elegant, even beautiful 'conquered foes.'" One way in which men are associated with meat is the concept of "man the hunter" in contrast with "woman the gatherer."[39] Renshaw observes that for earlier ecofeminist scholars, the dominated body was always feminized, and so images of the masculine hunter and feminized prey would be an example of violent heteromasculinity. This interpretation is too quick, however, and Renshaw demonstrates that it is only at the moment of the trophy shot that the animal's body is feminized; up until then, hunters' discourses emphasize the masculine prowess of the sexually mature male animals whom they passionately pursue, imbuing the hunt with homoeroticism.

For her part, in "A Feminist Rubik's Cube: Slaughterhouse Labour, Violence, and Trauma," sociologist Lauren Corman describes the slaughterhouse as a space of hegemonic masculinity, where the mostly male workers are required to suppress their emotions to embody scripts of men as stoic and professional who view the animals they kill or dismember not sentimentally but as commodities. Although feelings are suppressed, Corman describes how slaughterhouse work can cause perpetrator-induced trauma and desensitization to harming others. The latter results in increased rates of domestic violence, assault, and sexual assault on the part of workers, thus producing more trauma in the already precarious communities from which meatpacking workers are hired.[40] Indeed, Andrew Knight and Katherine Watson have recently suggested that Jack the Ripper was likely a slaughterhouse worker, of whom there were many in East London where the murders took place in the 1880s, as slaughterhouses were located in this area.[41] Due to the speed and skill with which the Ripper killed and cleanly removed organs from his victims, it has been speculated that he was a surgeon; however, studies of mortuary sketches show his technique was more consistent with those of slaughterhouse workers who were, moreover, habituated to quick and serial killing. Given the terror and immense, ultimately lethal violence to which nonhuman animals are subjected in slaughterhouses, as well as its spillover into human communities, Corman describes the abattoir as a space of multispecies trauma.

Rama Ganesan's chapter in this volume, "Animals and the Purity-Power Dichotomy in the Brahminical Tradition," offers a generative critique of Adams' sexual politics of meat, which she argues centers Western gender constructs. As Ganesan's chapter shows, a consideration of the Indian Brahminical tradition, where vegetarianism is associated with masculine purity, complicates the sexual politics of meat in a number of ways. Ganesan's chapter also draws on the example of India to problematize uses of the concepts of dehumanization and animalization in contemporary critical animal studies scholarship, showing that, in some cultural contexts, associations with animals need not be degrading of humans.

Although masculinity is described in most ecofeminist writings as anti-ecological, or toxic not only to women but to the environment, queer communities have long shown that there are other ways to be masculine, including queer masculinities, trans masculinities and female masculinities, as well as nonhegemonic, cisheterosexual masculinities. The hegemonic and toxic masculinity that is targeted and reproduced by meat advertisements is thus but one kind of masculinity, and Gaard has explored the possibilities and potentials of what she calls "eco-masculinities." As she writes, "Perhaps it is (past) time to envision alternative genders—and particularly eco-masculinities—from an ecofeminist perspective."[42] In his chapter for this volume, "Boyhood, Ecomasculinities, and Animal Pedagogy," sociologist Richard Twine defines ecomasculinities studies as "both a critical and creative endeavour, interested in both the role of masculinities in creating the ecological crisis and in examining how gender contestation and reinvention is vital for societal transformation." In addition to providing an overview of ecomasculinities research, Twine makes a novel contribution to the field by centering childhood and animals, whereas most considerations of ecomasculinities have focused on adult men and the environment. Twine observes that there have recently been high-profile cases of children involved in climate activism, most notably Greta Thunberg, Severn Cullis-Suzuki, and Kehkashan Basu, and they have succeeded in invoking the ire of "such exemplars of industrial masculinities as former U.S. President Donald Trump, former Australian Prime Minister Scott Morrison, and Brexit funder Aaron Banks." Twine questions why there are no boys among these prominent young climate advocates, however, and suggests that the feminization of care for nature and empathy for animals already has an impact in childhood.

1.4 Empathy and Care

In her canonical article, "Animal Rights and Feminist Theory," ecofeminist and literature scholar Josephine Donovan observes that the two most widely recognized figures in the animal liberation movement, philosophers Peter Singer and Tom Regan, express allergies to emotion, insisting that they do not particularly "love" or "care for" animals. In Singer's words, he is not "inordinately fond of dogs, cats, or horses," and his arguments for animal welfare protection should not be attributed to feelings of this kind.[43] Rather, both Singer and Regan insist that their arguments are grounded in philosophical reason. As Donovan astutely observes, the strong desire on the part of these white men to distance themselves from the connotations of having feelings for animals is no coincidence. Emotions are considered womanish in the history of Western thought, just as reason has been deemed masculine. Donovan explains that the reason/emotion dualism is grounded in Cartesian mind/body dualism, and women—like animals and Black people—are associated with the body and its instincts, intuitions, and feelings, while white men are associated with the mind and rationality. Donovan notes that it is highly ironic for animal ethicists such as Singer and Regan to ally themselves with Cartesian dualism and rationalism, however, given Descartes' own argument that animals are non-sentient and need no moral consideration due to their lack of reason. The association of animals with unreason has been the primary justification for human dominion over animals, and it thus seems unlikely that a rationalistic philosophy scornful of feelings could ever be liberatory of animals.

Donovan observes that Descartes' primary critics in his own day were 17th-century women philosophers such as Margaret Cavendish and Anne Finch, and similarly, women philosophers and animal ethicists who are contemporaries of Singer and Regan, such as Mary Midgley and Constantia Salamone, have not felt the same need to distance themselves from caring for animals. This volume includes chapters by another generation of feminist philosophers and critical animal studies scholars who openly avow what Singer referred to as an "inordinate fond[ness]" for "dogs, cats, and horses." Philosopher Stephanie Jenkins describes her chapter for this volume, "Care Ethics and Service Dogs," as "an act of care and grief" for her beloved service dog Mandy. Trying to capture her experience of "entangled subjectivity" with Mandy, Jenkins writes that she experienced their relationship as "one of a dual, co-authored subjectivity; I experienced the world as a 'we,' a plurality, rather than as a single individual plus a dog," and "Mandy and I were inseparable. We lived, worked, played, traveled, and slept together. When friends or colleagues asked, 'How are you?' I would respond 'Our day is going well' or 'We had a stressful morning.' " Because of this profound entanglement, Jenkins describes herself as "undone" by Mandy's death. In her contribution to the current volume, "Requiem for Tia Maria," comparative literature scholar Alexandra Isfahani-Hammond writes of her girlhood adoration for a horse, Tia Maria, and how this love is now haunted by guilt, regret, and sorrow as she has come to recognize horseback riding as cruel. As she writes, "Tia Maria's gaze encapsulates me in darkness. . . . She who had been bound fixes me in place and occupies my sentences." Isfahani-Hammond's chapter is both a heartbreaking apology to Tia and a necessary exploration of the ways we claim to love beings whom we in fact harm. As for cats, philosopher Alison Suen opens her chapter for this volume, "Crazy Cat Lady," by writing of her cat, whose name is linguini, "I would call linguini the love of my life," except that such a phrase does not in fact do justice to linguini's role in her life. Indeed, Suen acknowledges that linguini is the muse of many of her academic writings and that she has written the current chapter in part for the pleasure of seeing linguini's name in print. In their willingness to avow profound feelings of love and grief for dogs, cats, and horses in academic writings, Jenkins, Isfahani-Hammond, and Suen are part of a tradition of feminist scholars who emphasize the ethical and political significance of care.

Given the shortcomings of malestream philosophical approaches to animal ethics exemplified by Singer and Regan, ecofeminist animal scholars have turned instead to a feminist ethics of care that is exemplified in these and other chapters in the current volume. As Karen S. Emmerman underscores in her chapter, "ecofeminism is rooted in the feminist ethics of care, a view that rejects both the possibility and desirability of disembodied, impartial moral reflection and that embraces a role for affect, empathy, care and loving attention in our moral lives." The field of feminist care ethics began with Carol Gilligan's 1982 book, *In a Different Voice: Psychological Theory and Women's Development.* As a graduate student, Gilligan had worked with Harvard moral psychologist Lawrence Kohlberg, who was renowned for his theory of stages of moral development. His study depended on interviewing children, teenagers, and adults about what they would do in moral dilemmas and analyzing their moral reasoning. According to Kohlberg, there are six stages of moral development, with each stage being superior to the ones before. At the first stage, people are only concerned with the consequences of their actions, or if they will be punished or rewarded. At the second stage, self-interest is involved, and an action is "good" if it benefits the individual. At the third stage, individuals are concerned with the approval of society: if an action receives social approbation, it must be good, and if it receives social disapprobation, it must be bad. At the fourth stage, individuals do not just follow rules in order to avoid punishment or receive approval but due to a Hobbesian belief that a law-abiding society is a good one. At stage five, an individual is willing to break the laws and norms of society if this means acting according to their convictions, distinguishing between natural and civil law. Finally, at the sixth stage, a person makes moral decisions based on abstract reasoning or according to universalizing moral principles. Kohlberg ascribed to the rationalistic, universalizing moral philosophy of the 18th-century German philosopher Immanuel Kant, which entails "categorical imperatives" such as "do not lie" and "do not steal" to which no exceptions should be made. Kohlberg's sixth stage of moral reasoning is, essentially, Kantianism. In his studies, Kohlberg observed what appeared to him to be moral regression: tracking participants over long periods, he found that some people who had reached level six reverted to level four or five. Most significantly for Gilligan, Kohlberg observed that none of the female participants ever reached the sixth stage. For Kohlberg, echoing moral philosophers from Aristotle to Kant, this meant that girls and women were inferior to men and boys when it comes to moral reasoning.

In her book, *In a Different Voice*, Gilligan disagrees with her teacher. She argues that the moral reasoning that characterized the responses of female participants was not inferior but simply different. The girls, for example, eschewed making categorical claims such as "one should never lie" or "one should never steal" in order to attend to the contexts of the moral dilemmas and the relationships and feelings that were involved. In certain situations, it might be justifiable to lie or steal to protect relationships or provide aid. Gilligan argued that the girls were weighing relationships, contexts, and emotions such as care, compassion, and concern rather than prioritizing rules, abstract principles, or the virtue of justice. This did not so much make them morally inferior to the boys as it made them different. They were speaking in a different voice: the voice of care ethics rather than justice ethics.

As in the case of ecofeminism, there has been a minority of essentialist care ethicists who see this "different voice" as linked to (some) women's biological capacities to be mothers.[44] That said, the majority of care ethicists believe that the associations they describe are socially constructed rather than innate. Boys are socialized to suppress their emotions and are not socialized to do care work, whereas girls are socialized to invest in and prioritize individual relationships of care and to engage in the labor of care. It is due to the types of experiences that girls have relative to boys that they approach ethical dilemmas through the language of care rather than abstract principles of justice.

Importantly, Kohlberg's studies have been critiqued because he relied for his study participants on people from the Harvard community. The individuals he interviewed were mostly Harvard students, faculty, and their children, and so were overwhelmingly white and middle- to upper-class Americans, and also relatively young due to university demographics. Additional studies in moral psychology that considered a wider demographic found that care ethics does not just characterize girls and women but everyone other than young, able-bodied, socioeconomically privileged white men. Even upper middle-class white men, when they get older, will move towards a care ethics perspective rather than the conventional justice perspective. Like girls and women, men of color and working-class men are also likely to be in caring roles and professions. When people get older, they are also likely to begin to need more care and thus to begin valuing it more.

In her influential reflection on care ethics, "The Need for More Than Justice," feminist philosopher of moral psychology Annette Baier observes that justice is understood in the canon of Western philosophy as treating one another with respect, which is understood reductively as not violating each other's rights, autonomy, and freedom or not interfering with each other's individual pursuits. Feminist philosophers and critical disability studies scholars have critiqued the focus on freedom and independence in male-dominated political philosophy as a myth of the privileged, since the ideal of autonomy has relied on an erasure of relationships and a denial of the fact that we are all interdependent upon one another, animals, and ecosystems. Care ethics, feminist philosophy more generally, and disability studies have recognized that the human condition, and indeed the animal condition, is one of vulnerability and interdependence.[45] Feminist philosophers have replaced the ideal of atomized autonomy with what they have called relational autonomy.[46] Indeed, we depend on relationships with one another to be autonomous.

Baier moreover observes that other people may respect our rights, freedom, and autonomy and refrain from interfering in our lives, and we could still be miserable, suicidal, and find life meaningless. With perfect justice, we might be equal and free but lonely and depressed, because we need friendship, care, and love as much as autonomy and equality to flourish and find meaning in life. Relations of love and care tend to entail treating people unequally, however, or to be unjust. If we lived in a world where all relationships were just in the sense of impartial, there might not be any love, friendship, or family other than in a biological sense. There are situations in which impartiality is important, but justice in this sense is not enough for human flourishing. We also need community, friends, love, kin, or relations of care.

Carol J. Adams and Josephine Donovan have edited two important anthologies that extend feminist care ethics to include caring for and by more-than-human animals: *Beyond Animal Rights: A Feminist Caring Ethic for the Treatment of Animals* (1996) and *The Feminist Care Tradition in Animal Ethics: A Reader* (2007).[47] To the sixteen essays in these collections, this volume includes an interview with Josephine Donovan as well as a series of new chapters that theorize care from feminist critical animal studies perspectives. In "Animals, Feminist Care Theory, and Critical Standpoint Theory," Donovan discusses not only her earlier work but also her most recent work taking up standpoint theory as a caring ethical framework for thinking about animals. In "Animal Sanctuaries as Feminist Care Activism," cultural anthropologist Elan Abrell explores the underlying ethical framework of the animal sanctuary movement, which he characterizes as a feminist ethic of care. In "Feminist Legal Systems that Benefit Animals: Placing Parameters around Care and Relationality," feminist animal law scholar Maneesha Deckha builds on the work of other feminist critical animal studies scholars such as Eva Giraud, Zipporah Weisberg, and Stephanie Jenkins to argue that Donna Haraway's companion species scholarship entails an "abstract rejection of human exceptionalism or simple celebration of human–animal inter-relationality" that nonetheless "tolerate[s] and even justif[ies] industries and practices that traffic in animals and normalize systemic violence against them." Consequently, this work, however influential within animal studies, has been of little use in improving the lives of animals. In its stead, Deckha proposes an ecofeminist, relational, empathetic and care-based framework for animal law. In her chapter, mentioned previously, Stephanie Jenkins draws on years of shared experiences with her former service dog Mandy to reject the legal, instrumental understanding of service animals as mere medical equipment or tools for people with disabilities. Instead, Jenkins draws on feminist care ethics to develop a notion of "entangled subjectivity" that better describes her relationship with Mandy. In Rebecca Deutsch's chapter, "Shelters," she considers caretaking relationships and emotional bonds between women and their companion animals, particularly in situations of domestic violence, and how shelters for animals and for humans experiencing domestic violence and homelessness, are currently inadequate sites of care. Deutsch concludes her chapter by imagining how shelters might be transformed to better meet the care needs of both humans and other animals. Finally, in "Wild Animal Ethics: A Gender-Sensitive Perspective," philosopher Catia Faria considers both anthropocentric and naturogenic harms to animals and argues that they are tolerated in both cases because of masculinist approaches to ecology. In the case of anthropogenic harms, such as cullings of non-native species and "hybrids" by conservation biologists, ecological wholes are prioritized over individual animals without empathy for their suffering. In the case of naturogenic harms, Faria argues that environmentalists are also largely indifferent to individual animal suffering due to a "patriarchal devaluation of compassion, following a broader devaluation of *all things natural* typically associated with the feminine realm." Displacing such masculinist values, Faria argues instead for a feminist, caring, empathetic response to wild animal suffering.

The chapters in the fourth part of this volume extend the discussions of feminist ethics and the ethics of care into the domain of methodologies. The first chapter in this part of the book, by critical animal studies scholar Pablo Pérez Castelló, argues for care as method when conducting multispecies ethnographies. To illustrate his arguments and show the necessity of care, Castelló draws on his own experiences of conducting multispecies ethnographic research at VINE, an LGBTQ-led sanctuary for farmed animals that was co-founded by another author in this volume, pattrice jones. The chapter by Kelly Struthers Montford and Chloë Taylor, and the closely related interview with feminist animal geographer Kathryn Gillespie, also explore the topic of multispecies ethnography, raising feminist ethical questions about how multispecies research should be conducted. The interview with Gillespie delves into her own experiences conducting multispecies ethnographic work at animal auctions and her reflections on multispecies auto-ethnography and witnessing as other multispecies methods. While these chapters argue for the importance of bringing feminist principles into multispecies research, the chapter in this section by Lauren Corman, "Possibilities and Productive Failures: Feminist Methodologies and Animal Subjects," makes a complementary argument, showing the importance of bringing multispecies research into feminist scholarship. This part of the book concludes with a multispecies autoethnography by Melissa Plisic, "Unruly Faces in Suburban Places: A Practice in Multispecies Autoethnography," which they approach from queer, feminist, anti-colonial, and anti-capitalist as well as anti-speciesist perspectives.

1.5 Intersectionality

Intersectionality has been a central concept in feminist theory for at least thirty-five years.[48] It began with critical race feminists criticizing white solipsism in white feminist discourses, or the ways that mainstream feminism assumed white women's experiences, just as the Black liberation movement assumed Black men's experiences. Both movements ignored the experiences of Black women and did not consider how systems of subordination overlap. Even before feminist law scholar Kimberlé Crenshaw's use of the metaphor of intersectionality, Black feminists had insisted that race needed to be considered simultaneously with gender and not as an additive factor but as an integral aspect of gender.[49] Gender is always already racialized, and feminist theory cannot understand gender without simultaneously considering race. Building on such arguments, feminists have contended that we must also consider other axes of identity and oppression simultaneously with race and gender, such as sexuality, class, and ability.

Early ecofeminism recognized that women are not the only group to have been animalized, associated with primeval nature, or seen as less than fully human. This logic of oppression has functioned even more powerfully in the oppression of people of color and especially in anti-Black racism. The enslavement of Africans and their descendants was justified in the 18th and 19th centuries through pseudo-scientific claims that Africans were either not the same species as whites or that they were a less evolved stage of the human species.[50] Settler colonialism, similarly, was justified by dehumanizing discourses about Indigenous people such as their comparison to wolves living wildly in forests.[51] Such dehumanizing logics have also been at work in the oppression of people with congenital disabilities and those diagnosed with mental illnesses, who, while not seen as separate species, have been viewed as degenerate humans who threaten the species gene pool. Scholars have shown that poor people were frequently diagnosed by doctors as "degenerate," "morons," "idiots," and "moral imbeciles" and were dehumanized and sterilized without their consent under eugenic laws, when in fact they were simply uneducated or illiterate.[52] Queer people were also included in the category of the mentally ill and degenerate well into the 20th century, and continue to be pathologized through new psychiatric diagnoses, such as gender identity disorder, that replace the old ones.[53] Discourses of dehumanization have taken up other arguments—and eugenic arguments in particular—in the cases of race, poverty, disability, and queerness than in the case of (white, middle class, able-bodyminded, cis/heterosexual) women, and such eugenic ideas have not disappeared.[54] Nonetheless, each group has been dehumanized in some way and to some degree. Ecofeminist ethics and politics, and feminist approaches to critical animal studies more generally, have thus necessarily been concerned not only with the intersections of gender, animals, and the environment but also with how these interlock with racism, xenophobia, colonialism, socioeconomic class, ableism, sanism, heterosexism, and transphobia.

Ecofeminist and feminist critical animal studies scholars have advanced a plethora of contextualized, intersectional arguments for animal liberation in general and veganism in particular. These include pattrice jones' *The Oxen at the Intersection: A Collision*;[55] Greta Gaard's "Toward a Feminist Postcolonial Milk Studies";[56] Jessica Eisen's "Milked: Nature, Necessity, and American Law";[57] Margaret Robinson's "Veganism and Mi'kmaq Legends";[58] Sunaura Taylor's *Beasts of Burden: Animal and Disability Liberation*;[59] and Maneesha Deckha's "Toward a Postcolonial, Posthumanist Feminist Theory: Centralizing Race and Culture in Feminist Work on Nonhuman Animals,"[60] to name only a few. The current volume contributes to this literature with a section of chapters on intersectional veganisms. In her chapter for this volume, "Feminist Veganism," philosopher Karen S. Emmerman argues that veganism is a feminist issue due to "the joint oppression of women and animals under a patriarchal worldview" and "the gendered nature of animal exploitation in the food system." As Emmerman acknowledges, however, any universalizing argument for veganism, ecofeminist or otherwise, is complicated when we consider intersections of race, colonialism, and class. For this reason, as Emmerman explains, ecofeminist scholars have often argued not for a universal veganism but rather for contextual moral veganism.[61] In "Black Veganism(s): A Personally Guided Exploration," anti-racism educator, critical race, and critical animal studies scholar Nekeisha Alayna Alexis considers a range of reasons why Black people embrace veganism and, indeed, in the U.S. context, are more likely to do so than white people. As she describes, for some Black vegans, the imposition of a meat and dairy-centric diet on people of African descent is a form of racial oppression, and they seek to decolonize themselves from meat and dairy-centric diets that are causing premature deaths in Black communities. In this context, Alexis discusses various ways in which Black vegans have re-visioned soul food. For other Black vegans, plant-based diets are a path to empowerment, "greater self-determination and collective resistance amidst White supremacy." For still other Black vegans, including Alexis herself, veganism arises from empathy for more-than-human animals, a recoiling from the violence and horrors of animal agriculture, and a recognition of the interlocking of racial and species oppressions. In discussing these different paths to Black veganism, Alexis explores the work of other Black vegan scholars, including A. Breeze Harper, Aph Ko, and Syl Ko. While veganism is primarily associated with diet, it is in fact a rejection of all animal products, including the wearing of their skin, fur, and feathers. The next chapter in this part of the volume, Carrie Lou Hamilton's "The Skins I'm In," provides an intimate reflection on the wearing of animals. This queer feminist vegan meditation on skin beautifully interweaves personal contemplations on aging, whiteness, queer femininity, and animality. "Vegan Camp" is an interview with queer theorist and vegan studies scholar Emelia Quinn on her concept of a vegan camp aesthetic, as well as related intersectional concepts within vegan studies such as vegansexuality and the vegan killjoy. Finally, this part of the volume concludes with Chloë Taylor's chapter on veganism and disability. This chapter begins with an overview of the intersections between animal and disability oppression and then considers the ways in which global warming, zoonoses, and antibiotic resistance—each of which is driven in large part by animal agriculture—disproportionately impact disabled humans and more-than-human animals.

Unfortunately, while there has been considerable interest in intersectionality on the part of critical animal studies scholars, this has been a mostly one-way relationship. There has been little uptake of intersectional environmental feminist and feminist critical animal studies scholarship by other feminist scholars writing about intersectionality or doing intersectional research. Recent major monographs on intersectionality do not include discussions of ecofeminism, animals, or species, and these terms do not appear in their indexes.[62] Indeed, humanist Black feminists have been critical of the ways that intersectionality is taken up without centering race, such as studies that use the term intersectionality to study two or more other axes of identity such as gender and disability. Some Black feminist scholars have argued that intersectionality must always not only include but center the experiences of Black women, or even the experiences of American Black women.[63] According to this critique, the concept of intersectionality has been appropriated from U.S. Black feminism by white-dominated feminist studies but leaves Blackness behind. This might be said of some ecofeminist and feminist critical animal studies uses of intersectionality in which animals are centered along with another axis of identity such as gender or sexuality, where race is not a focus. While several recent works in critical animal studies do center race along with animality, the authors of these works argue that intersectionality is not an adequate framework for what their analyses reveal.[64]

1.6 Beyond Intersectionality

In *Racism and Sexual Oppression in Anglo-America*, philosopher Ladelle McWhorter argues that the relationship between sex and race is not well explained by "intersectionality" because these categories are in fact completely enmeshed, their histories intertwined.[65] Race and sex do not merely intersect at one or more points, as the metaphor of intersectionality would make us think. Rather, they are together all along. This is particularly clear within the context of eugenics, where much of the concern with race is a concern with sex, or with certain people having sex and thereby undermining the purity of the white race by passing on "inferior" racial traits. McWhorter considers how racism against Black people was so caught up with these kinds of fears about racial purity and miscegenation that Black people were hypersexualized, seen as sexually dangerous and animalistic. Eugenic concerns about sex, then, were all about protecting the human/white race. Although less explicitly thematized by McWhorter, this history is moreover shown to be very much about disability and species, as eugenics is also concerned with preventing the passing on of congenital disability, which, like miscegenation, was seen as a way in which the white/human race or species would degenerate. The words "race" and "species" are used interchangeably in these discourses, with talk of the "human race" meaning both the "white race" and the human species. Race, disability, sex, and species are indeed complexly entangled, with race having sometimes been constituted as a type of congenital disability[66] and disability (or at least the prospect of people with disabilities having sex and procreating) seen as a threat to the human/white race/species within eugenicist discourses.

Without explicitly critiquing or pointing to the limitations of the concept of intersectionality, in *Beasts of Burden: Animal and Disability Liberation*, critical animal studies and critical disability studies scholar Sunaura Taylor makes a similar argument regarding ableism and speciesism as McWhorter makes about sexism and racism. Taylor compellingly demonstrates that ableism is in fact speciesist, since ableism consistently entails humans with disabilities being disparaged through comparison to nonhuman animals or because they move, eat, or appear in a way that is reminiscent of nonhuman animals. Taylor moreover shows that the logic of these interlocking oppressions tracks both ways, or that speciesism is also ableist. In an ableist society, we are prejudiced against nonhuman animals and disabled humans alike because they don't have the species-typical abilities of normate humans. Both nonhuman animals and the cognitively disabled are seen as inferior because they do not have the cognitive and linguistic abilities of the neurotypical human, for instance. This is another example where the concept of intersectionality fails to adequately capture what are in fact profoundly interconnected oppressions.

Two critical race scholars studying the intersections of racial and animal oppression, Claire Jean Kim and Aph Ko, have explicitly argued that intersectionality is inadequate for conceptualizing the interrelationships between white and human supremacy. In their introduction to a special topics issue of *American Quarterly* on species/race/sex (2013), Kim engages in a dialogue with her colleague Carla Freccero in which they consider the usefulness and limitations of intersectionality.[67] While Freccero observes that "'Intersection' restricts us to one specific place or site where otherwise unrelated vectors of oppression meet, and of course everyone recognizes that this is not how the facets of identity or categories of identitarian oppression work," Kim describes intersectionality as "a provisional guide" who will only ascend "to the first elevation of the mountain with you."[68] As she acknowledges, "That sort of [guide] might not be much of a [guide], but a concept can still be a perfectly good concept even if it doesn't provide all the answers."[69] In her interview for the current volume, Kim explains why this guide only gets us so far up the mountain. She writes, "In my work, I emphasize that anti-Blackness is structural, a foundational principle in U.S society, but Blackness is always gendered and sexualized, etc., so does it make sense to talk about each of these dimensions of experience as distinct in the way intersectionality does?" If we climb higher, Kim contends that we will see that "various supremacies (racism, speciesism, sexism, homophobia, etc.) are so closely intertwined in thought and deed that they will persist together or be interrupted together, not singly."[70] At the same time, Kim warns that terms such as "coarticulation" and "coconstitution" risk flattening oppressions, implying that they are always fungible or comparable, whereas the extent to which this is so depends on "the details" or historical contexts.[71]

In *Racism as Zoological Witchcraft: A Guide to Getting Out*, media studies and anti-racism scholar Aph Ko also rejects intersectionality as an adequate framework for theorizing the interconnections between race and species. As she writes, "Intersectionality is not the best social theory model to capture how incredibly messy and complex oppression is."[72] What we need, Ko observes, is "a deeper analysis to explain how these social phenomena relate to one another."[73] Rather than thinking of oppressions as an intersection, Ko proposes thinking of them as a cube in order to emphasize their multidimensionality.[74]

Multidimensionality is also a term used by Kim,[75] while the metaphor of a cube is independently explored in Corman's chapter for the current volume, "A Feminist Rubik's Cube," where she uses it as a simile to describe the many sides of multispecies trauma.

In sum, intersectionality is just one way in which feminist scholars have combined the study of gender and animals, but it is not uncontested. Numerous scholars, including feminist critical animal studies and critical race scholars, have argued for terms other than intersectionality or that intersectionality is not the right way to think about how oppressions interact.

2. Human Debates About Gender Constructs Are Deeply Invested in Claims About Animals

Feminist philosopher Simone de Beauvoir opens her 1949 magnum opus, *The Second Sex*, by reflecting on women's relationship to animality and the ways that biological studies of animals are used to validate misogyny. On the first page of her first chapter, "Biological Data," she writes:

> The term "female" is pejorative not because it roots woman in nature but because it confines her in her sex, and if this sex, even in an innocent animal, seems despicable and an enemy to man, it is obviously because of the disquieting hostility woman triggers in him. Nevertheless, he wants to find a justification in biology for this feeling. The word *female* evokes a saraband of images: an enormous, round egg snatching and castrating the agile sperm; monstrous and stuffed, the queen termite reigning over the servile males; the female praying mantis and the spider, gorged on love, crushing their partners and gobbling them up; the dog in heat running through back alleys, leaving perverse smells in her wake; the monkey showing herself off brazenly, sneaking away with flirtatious hypocrisy. And the most splendid wild cats, the tigress, lioness, and panther, lie down slavishly under the male's imperial embrace, inert, impatient, shrewd, stupid, insensitive, lewd, fierce, and humiliated. Man projects all females at once onto woman.[76]

Unfortunately, as noted earlier, de Beauvoir will ultimately distance women and indeed all humans from animals. Nonetheless, in the "Biological Data" chapter, she defends female animals of other species who, she observes, have been maligned in misogynist ways much like female humans. Reading the "castrating" spider, the "cruel" praying mantis, or the "hypocritical" "flirting" of female monkeys as indicative of a natural antagonism between the sexes is, in de Beauvoir's words, "just rambling"[77] ("c'est divaguer"[78]). Particularly in the "lower animals," she asserts, both males and females are victims of the species, their lives consumed by the end of reproduction: "it is the species that devours both of them in different ways."[79]

De Beauvoir moreover uses biological data herself to argue against the view that females are more naturally inclined towards nurturing the young than males.[80] She gives examples of female fish who eat their eggs unless the male chases them away and guards the eggs until they hatch.[81] For sea horses and toads, de Beauvoir observes, it is the male who incubates the eggs, whereas in many bird species, males will help build the nest, guard the eggs, and feed the fledglings.[82] In some cases, the male thus has at least as much if not more of a "maternal instinct" than the female, and from this de Beauvoir concludes that it is not determined that in humans the female must be the domestic, caretaking parent.

Seventy-five years since the publication of *The Second Sex*, feminist science studies scholars continue to demonstrate that gendered interpretations and projections of animal behavior persevere in the field of biology. In her recent book, *Sexing the Animal in a Posthumanist World*, Roslyn Appleby gives the example of a study of migration patterns in two groups of great white sharks, one in South African waters and the other in Australian and New Zealand waters.[83] Observing that some sharks made long, transoceanic migrations to meet the other group, while other sharks remained in home waters, scientists assumed that it was "roving males" who traveled while "non-roving females" stayed at home.[84] A later study would show, however, that it was the female sharks who undertook the long journey across the Indian Ocean and back to mate.[85] As Appleby argues, scientists had been projecting gendered stereotypes of adventurous, active masculinity and domestic, passive femininity onto the sharks. Also looking at examples from marine biology, in *Undrowned: Black Feminist Lessons from Marine Mammals*, poet and critical race feminist scholar Alexis Pauline Gumbs shows how not only heteronormative gender constructs but also racial constructs such as violence and criminality are projected onto marine mammals—particularly if they are black.[86] In the introduction to this work, for instance, Gumbs describes consulting the *National Audubon Society Guide to Marine Mammals of the World* and the *Smithsonian Handbook: Whales, Dolphins & Porpoises.*

> What I found was that the languages of deviance and denigration (for example, the term "vagrant juveniles," used to describe hooded seals), awkwardly binary assignments of biological sex, and a strange criminalization of mammals that escaped the gaze of biologists showed up in what would call itself the "neutral" scientific language of marine guidebooks. I just wanted to know which whale was which, but I found myself confronted with the colonial, racist, sexist, heteropatriarchalizing capitalist constructs that are trying to kill me—the net I am already caught in, so to speak.[87]

In the chapter, "Stay Black," Gumbs describes one such criminalized black marine mammal, a "slender Blackfish" named Slim. As Gumbs describes, marine biologists "criminalize" Slim "for exceeding the terms of a deadly contract." They call him "killer for refusing to perform for killers": "More killer than the actual killer whale to hear them tell it. The Audubon guide calls them 'pugnacious.' Twice. I heard they tried to lock Slim up and it was bad news for everyone involved, trainers, cellmates. Slim refuses to be caged. Yup. Slim will cut you."[88]

In her 1984 article, "Primatology Is Politics by Other Means," Donna Haraway observes that primatological studies have particular significance in debates about human gender roles.[89] Since Darwin, primatologists have struggled to provide humanity's origin story, with all the moral and political weight that religious and mythological origin stories such as *Genesis* once had. Haraway's title plays on 19th-century Prussian military theorist Carl von Clausewitz's famous claim that war is politics by other means, as well as 20th-century French philosopher Michel Foucault's inversion of the argument: politics is war by other means.[90] For Haraway, primatology is politics/war in the sense that the arguments of primatologists are deeply invested in the so-called "battle of the sexes." As she demonstrates, male primatologists have looked to studies of primates to show that traditional human gender roles are found among our closest kin and are thus natural and normal. In contrast, female primatologists have undertaken similar studies to show that gender roles are in fact diverse in nature and do not conform to traditional Western human gender roles, which are thus shown to be social constructs. Women primatologists have used their observations of non-human primates to prove that females too can be competitive, sexually assertive, orgasmic, dominant, and roamers, and that they are just as active as males, expending similar amounts of energy. Primatology is a "territory of feminist struggle," Haraway argues, and hence "women's place is . . . in the jungle."[91] For Haraway, if conservative male primatologists are going to be in the jungle gathering ammunition to argue that women's place is in the home, feminist primatologists need to be there too, gathering their own evidence that not all female animals are passive, domestic slaves of the males and young of their kind. For Haraway, this explains why women are highly represented in primatology in comparison to other sciences.

Significantly, however, Haraway specifies that it is "Western women" who belong in the jungle and observes that almost all primatologists, including female primatologists doing field studies, are white. Given the ways that Black people's oppression has been continually justified by whites through comparisons to monkeys, it would be more perilous for a Black woman to write a book called *My Friends the Apes* (as white zookeeper Belle Benchley did in 1942) or *My Friends the Wild Chimpanzees* (as white primatologist Jane Goodall did in 1967). Moreover, most women primatologists are trained in anthropology—a field infamous for having mostly white social scientists travel to study racialized "others," who stay in their place and are observed. Similar to the racialized humans studied by white anthropologists, the nonhuman primates who are observed by white, anthropologically-trained primatologists are African, Indian, Southeast Asian, and South American animals and exoticized. Haraway calls the whiteness of primatology "Simian Orientalism."[92] There are infamous cases, moreover, of white women primatologists enacting violence against male African humans poaching apes, without considering the broader oppressive conditions that contribute to poaching.[93] The racial, imperial, and economic politics of animal activist responses to poaching is a topic addressed by Darren Chang's chapter in this volume, and it is one of many cases in which Black people have noted white people seeming to care more about (humanized) animals than (dehumanized) Black people. Other cases include the white outrage expressed over the murder of Cecil the Lion in Zimbabwe by a trophy hunter, without similar concern for African humans dying in the same part of the world; outrage over Michael Vick's dogfighting on the part of white people who are not similarly concerned with police violence against Black men; and scenes of cats and dogs being rescued after Hurricane Katrina while Black people remained stranded on rooftops.[94]

Queer studies scholars have taken up Haraway's thesis that animal studies are "politics by other means" to analyze scientific studies of sexual orientation in nonhuman animals. According to Stacy Alaimo, "the majority of scientists have ignored, refused to acknowledge, closeted, or explained away their observations of same-sex behavior in animals, for fear of risking their reputations, scholarly credibility, academic positions, or heterosexual identity."[95] She cites the example of biologist Valerius Geist, who admitted that he "still cringe[s] at the memory of seeing old D-ram mount S-ram repeatedly": "I called these actions of the rams aggrosexual behavior, for to state that the males had evolved a homosexual society was emotionally beyond me. To conceive of those magnificent beasts as 'queers'—Oh God!"[96] Alaimo observes that male primatologists have also been at pains to explain away the undeniably abundant same-sex activity between female bonobos, describing "all that mounting" as having a social rather than erotic function.[97] As Alaimo describes, primatologists must rule out the possibilities that same-sex activity between female animals is not actually about attracting male partners, impeding reproduction, forming alliances, communicating dominance, diffusing social tension, making up after fights, practicing for heterosexual mating, or obtaining help with childcare before they will admit that it might actually be sexual. As she writes, "Clearly, same-sex activity between animals is considered not-sex until proven otherwise."[98]

Alaimo echoes Catriona Sandilands in referring to this heteronormative bias on the part of scientists as "eco-sexual normativity."[99] Sandilands herself discusses a case in which ecologists explained away same-sex intimacies between female seagulls by arguing that the behavior "must be evidence of some major environmental catastrophe."[100] Similarly, Jennifer Terry considers the example of zoologists who captured footage of two male octopuses of different species and disproportionate sizes "fisting" on the Pacific Ocean floor. The scientists referred to "this close encounter in the deep" as "reproductive behavior," speculating that the male octopuses only resorted to having sex with one another because of a shortage of females, thus imposing a normative heterosexuality onto the cephalopods despite all evidence to the contrary.[101] In her chapter for this volume, pattrice jones observes that by denying that more-than-human animals have the agency, individuality, psychological complexity, sentience, and emotional capacities to have the diverse and queer types of intimate relationships that they—like humans—do, homophobic scientists are in fact reducing these animals to automata who are merely determined by biological drives to reproduce. This denial of animals' agency, individuality, sentience, and psychological and emotional complexity makes it easier to view them as inferior, jones observes, and "Thus, homophobia made it easier to lock up animals in vivisection labs and foie gras factories."

Not only scientific animal studies but also more popular representations of animals are mobilized to make socially conservative claims about gender and sexuality. For instance, the documentary film *The March of the Penguins* (2005) was taken up by religious conservatives as evidence of the naturalness of traditional family values: monogamous heterosexuality and self-sacrificing parenthood.[102] Unfortunately for the religious conservatives, penguins are famously queer serial monogamists, frequently forming same-sex couples who have been known to successfully adopt and hatch eggs and raise chicks together. A number of zoos have celebrity "gay" penguin couples with romantic lives as complex as ours: raising chicks, breaking up, one partner coming out as bisexual, and getting back together again. When zookeepers, concerned with increasing their zoo's penguin population, have attempted to introduce penguins in same-sex relationships to potential heterosexual mates, local gay rights groups have organized protests. Penguins, it is clear, have been deployed on all sides in battles over human gender and sexuality.

It is not just heterosexual scientists, media producers, and zookeepers who are preoccupied with "queer animals," then. On the contrary, just as male and female primatologists have engaged in gender battles through the study of primatology, so have straight and queer, cis and trans, non-feminist and feminist scholars and activists taken up animal studies as arms in ongoing ideological struggles over human sexuality. As Terry explains,

> Given the investment in a continuum of animal and human behavior since nineteenth-century evolutionary theory, it is not surprising that scientists have been busy at work in laboratories, trying to make sense of homosexuality and prodding animals to exhibit the love that dares not speak its name.[103]

In part, certain gay and lesbian scientists are invested in this issue because the discovery of "gay" animals, like the discover of a "gay gene," is taken as evidence of the naturalness and thus normalcy of homosexuality. Given the anti-normativity stance of much queer theory, this is a quest that many queer theorists find suspect, however, and Jack Halberstam has referred dismissively to "the gay science community's insistence on gay genes, gay seagulls, lesbian ducks, [and] transgender fish."[104] On the other hand, other queer animal studies scholars have recognized such a stance as untenable human exceptionalism and have provided decades of studies demonstrating the "exuberant" "rainbow" of gender roles and sexuality among more-than-human animals.[105] As a feminist new materialist and queer scholar of ecologies, for instance, Alaimo writes with pleasure of queer biologists "outing" more-than-human animals from the "zoological closet"[106] and of the "sheer delight of dwelling within a queer bestiary that supplants the dusty, heteronormative Book of Nature."[107] For Alaimo, tomes such as Bruce Bagemihl's *Biological Exuberance: Animal Homosexuality and Natural Diversity* (1999) and Joan Roughgarden's *Evolution's Rainbow: Diversity, Gender, and Sexuality in Nature and People* (2004) are important correctives to heterosexual bias in science, rendering "any sense of normative heterosexuality within nature an absurd impossibility."[108]

Some feminists have rejected the very premises of these debates, however, insisting that since gender and sexuality in humans are social constructs, what animals do is irrelevant to us. Feminist sexologist Leonore Tiefer, for example, has dismissed the value of her own doctoral experiments on mice as wrongheaded, insisting that, among humans, "sex is not a natural act."[109] For Tiefer, it would make little sense to speak of animals having gender at all, since feminists have defined gender as a social construct, a product of cultures, and animals, in Tiefer's view, do not have cultures. What is called "gender" in animals, then, are really sex role behaviors and are, in fact, tied to biological sex in a way that feminists have long denied when it comes to human genders. Recent animal studies scholars have questioned whether at least some animals might not in fact have genders in feminist senses of this term,[110] however, since, as they point out, animals do in fact have cultures.[111]

Of course, what we mean by gender and its relation to sex is itself a matter of feminist debate. For second-wave feminists such as Tiefer, this would have to mean exploring whether animals have gender in the sense of contingent social constructs that are variable over time and from one geographically located community to another within the same species. For poststructuralist feminists, it might mean questioning whether even what we have understood as biological sex roles in more-than-human animals are not in fact social scripts, or whether it isn't gender all the way down. For feminist new materialists, to argue that more-than-human animals have genders refers neither to biologically determined sex roles nor to pure social constructs but rather, in Alaimo's words, to "creative forces simultaneously emergent within and affecting a multitude of naturecultures."[112] That is, for feminist new materialists, nature and culture, like sex and gender, are not dichotomies or even distinct but rather intra-active.[113]

3. Integrating Feminism, Transforming Species Theory

In "Enemy of the Species," Ladelle McWhorter observes that there have been three major definitions of "species" in Western thought: the Platonic definition, the medieval definition, and the biological definition first articulated in the 18th century by Georges-Louis Leclerc, Compte de Buffon. Although not uncontested, Buffon's biological definition is the one that we likely learned in school, according to which animals are distinct species if they cannot mate, are sterile when they mate, or give birth to offspring who are sterile.

As McWhorter observes, this definition made Black people, Indigenous people, and other racialized groups members of the same species as white people and so contradicted many other arguments racist biologists and philosophers were making at the time. Reluctant to include Africans and Indigenous people in the human species, some white scientists rejected the definition, but McWhorter shows that other scientists took a different tactic: they accepted the definition of species as interfertile groups, but argued that whites and Blacks were not interfertile since their offspring were akin to mules. Like horses and donkeys, whites and Blacks could reproduce, these scientists acknowledged, but their offspring were, if not completely sterile, progressively less fecund with each generation, manifesting chronic health problems, weakness, and poor parenting in addition to decreased fertility. Even for those who were unconvinced by such arguments, since most people would have known perfectly healthy "mulattoes," the new biological definition of species fueled ongoing racism. Blacks and whites belonged to the same species, in this alternative view, but non-white races were less evolved members of the species, and interbreeding with less evolved members of the species would lead to the degeneration of the white race.

Whichever argument was favored, we see why Claire Jean Kim has described species, like race, as a "taxonomy of power" and, indeed, as two dimensions of the same taxonomy. Although McWhorter makes her intervention into the issue of species to think through the relationship between racial and queer calls for "diversity," her work has significance for feminist critical animal studies since it underscores that species, like race, sexuality, gender, and disability, is a socially constructed, variable, and contested political category. As McWhorter has argued elsewhere, when a concept has changed over time, the question should not be whether our current usage is conventional or even if it is correct but what the power effects are of using the term in one way versus another.[114]

In "Veganism as a Way of Life," Robert McKay points to the question of species as an intellectual oversight of animal studies.[115] As he explains, animal studies "has largely overlooked, even in its most avowedly 'posthumanist' guise, quite fundamental difficulties in knowing just what exactly its object of analysis is, or might be thought to be, and about just who (or what) it is that is doing the analysis."[116] In other words, animal studies scholars generally write of species without questioning what we mean by this term, taking for granted that there is such a thing as species and that we know what it is. This is the case even though we may be aware of what is known as "the species problem": the fact that biologists cannot agree on an understanding of species or that it is even a meaningful way of categorizing life.[117] In fact, there are many cases of species considered distinct that are interfertile, for instance, the European wild cat that is native to Britain and the domesticated cat that is descended from the African wild cat. Although these are "completely separate species," they co-exist in Britain and can have kittens together, much to the distress of conservation biologists.[118] There exist over twenty competing understandings of species within biology today.[119] Similar to biological debates about what we mean by "sex" (is it about external genitalia? gonads? hormones? whether one's body menstruates or produces semen?), it is unclear what competing criterion should be used to define species (is it about morphology? interbreeding? genealogies? molecular similarity?). Given that evolutionary biology moreover shows that life forms are continually changing and adapting, some suggest that species are not real.[120] As McKay notes, "Such difficulties (about knowing what your object of analysis is, and who you are in relation to it) were and continue to be addressed by work in feminism and beyond that theorizes sexed bodies; and . . . animal studies has much to learn from such work."[121] As McKay suggests, feminists have continually questioned what is meant by "woman," "gender," and "sex" and if these are meaningful categories at all. In responding to these questions, feminists first distinguished between sex and gender and then blurred those distinctions again, and, McKay contends, it is possible to do the same with species.

One might think that it makes little difference to our efforts to liberate captive orcas or factory-farmed chickens to debate what these species labels mean, and yet McKay argues that destabilizing species, and thus the category of the human, can in fact be a politically effective tactic in animal liberation. He tells a story of being a guest at the home of a family member who, upon serving him a dish of ratatouille while the rest of the family prepares to eat meat, has repeatedly made the joke, "Here we are Robert, your lesbian food!"[122] McKay writes, "As I read this situation, it is the normality—read centrality—of my host's very humanity that is endangered by what is (to him) an unimaginable identification with non-humans, inherent in veganism."[123] The "lesbian food" joke is intended to assert that McKay's host finds it strange to identify with nonhuman animals in the way that ethical veganism suggests, and this assertion proves his own normalcy and humanity. To understand this, McKay cites Judith Butler's claim that subject constitution requires a repudiation of what it is not,[124] and so for the family member to assert his heterosexual humanness, he needs to foreclose the queer and the vegan. By abjecting both lesbians and a diet expressing compassion for more-than-human animals, McKay's relative anxiously performs compulsory heterosexuality and what McKay calls "compulsory humanness."[125] The fact that the host makes the same joke repeatedly suggests that performances of human exceptionalism, like gender performances, follow tired scripts that must be continually read anew.[126]

While not all vegans would accept the understanding of veganism as an identification with nonhuman animals or a repudiation of their humanity, a number of animal studies scholars have lately distanced themselves from the human in favor of a broader understanding of kin, and this includes scholars who come from groups that have been particularly dehumanized. Alexis Pauline Gumbs, for instance, introduces *Undrowned: Black Feminist Lessons from Marine Mammals* by explaining,

> I identify as a mammal. I identify as a Black woman ascending with and shaped by a whole group of people who were transubstantiated into property and kidnapped across an ocean. And, like many of us, I am simply attracted to the wonder of marine life. And so I went to the aquarium and bought both of those guidebooks [the *National Audubon Society Guide to Marine Mammals of the World* and the *Smithsonian Handbook: Whales, Dolphins & Porpoises*] to learn about my kin.[127]

A little bit later Gumbs describes herself as a "marine mammal apprentice," explaining, "I am a beginner marine mammal very early in my journey,"[128] still learning, for instance, to echolocate.[129] Sunaura Taylor is similarly willing to identify as an animal and with other animals, although she acknowledges that she may be able to do so because, although physically disabled, she is white, middle class, verbal, and educated and is not cognitively disabled.[130]

Although it may require some amount of social privilege to be able to distance oneself from compulsory humanness, McKay embraces this understanding of being vegan. As he writes, "I want to think about vegan lives as necessarily problematizing any belief in the integrity of human-being, and indeed as problematizing the very reality of species difference. I would like to say: I am vegan, not human."[131] Or again,

> Being vegan, to enact yet another blatant act of terminological pastiche of queer studies, is, for me, a form of species dissidence. Just so, we can regard the concept of species—the purported humanness of humanity—much as Butler regards the concept "biological sex." It is what she calls a "regulatory ideal."[132]

As a regulatory ideal, McKay argues that our identification with the species *Homo sapiens* requires repudiating practices of the self that cross conventional boundaries between forms of animal life or that draw those boundaries otherwise.[133] McKay includes veganism among such practices of the self, as well as cross-species friendships. In her article, "Notes on Vegan Camp," Emelia Quinn similarly explores ideas at the intersection of vegan and sexuality studies such as "compulsory carnivorism" and the performance of human exceptionalism, which is discussed in her interview for this volume.

4. Gender Constructs Impact How Humans Treat Other Animals

Feminist critical animal studies scholars have observed that the overwhelming majority of animal activists,[134] volunteers at animal rescues and shelters,[135] and vegans[136] are women in Western countries. Nonetheless, due to what feminist critical animal studies scholars Fiona Probyn-Rapsey, Siobhan O'Sullivan, and Yvette Watt have described as "glass elevators," the most celebrated spokespeople for the animal movement and the heads of most major animal rights organizations in the West continue to be men,[137] and sexism and sexual harassment of women activists is rife within the animal rights movement.[138] Women also make up the majority of veterinary students in Western countries.[139] By 2003, they already constituted 80% of veterinary students in the U.S. and Canada, and now they are the majority of practitioners.[140] Similar figures are reported in the U.K.[141] Within veterinary medicine, women overwhelmingly dominate the profession when it comes to the care of companion animals.[142] In contrast, veterinarians collaborating with agribusiness in the slaughter of farmed animals continue to mostly be men.[143]

While there is thus a strong feminization of caring for animals, the violence of killing animals continues to masculinized. Slaughterhouse workers, recreational hunters, and animal abusers, including those who sexually assault animals, are mostly men.[144] Domestic violence is also highly gendered[145] and frequently involves violence against the abused partner's companion animal(s). As social justice scholar Rebecca Deutsch's chapter for this volume explores, many survivors of domestic violence will stay in situations of abuse because they do not want to abandon a cat or dog to their partner's aggression and retaliation. As Deutsch discusses in "Shelters," the problem is exacerbated by the fact that most domestic violence shelters and homeless shelters will not allow companion animals, while animal shelters are often spaces of captivity and death rather than care.

5. Representing Animals

A number of chapters in this volume explore the political representation of animals. Since the 19th century, a minority of Western feminists have seen representing animals as connected to their own political self-representation, linking their liberation and that of more-than-human animals.[146] In contrast, men's rights movements have sometimes taken up the consumption of animals and their excretions as an embodiment of virile white heteromasculinity.[147] It is important to study animal activism through a feminist lens, therefore, because animal activism has been part of feminist activism, and animal oppression has been part of anti-feminist activism. Katja M. Guenther's chapter in this volume, "Women's Contributions to the Movements for Animal Protection and Rights," addresses the issue of the gender gap in animal activism. Although attention to this topic has tended to focus on why women are over-represented in the movement, Guenther's chapter instead explores what women have accomplished in the movement and what the gender gap says about society. As Guenther powerfully contends, what is problematic is not that there are so many women in the movement but

> that this imbalance . . . upholds the cultural belief that connecting with animals is a feminine activity, contributes to the marginalization of animal rights positions in political life, supports the lack of mainstreaming of animal studies into many academic disciplines where it clearly belongs . . . and helps explain the continued political and cultural marginalization of animal rights perspectives.

The topic of women's central role in animal advocacy is also picked up on in the interview with Tayler Zavitz, which moreover explores topics such as the repression and criminalization of animal activism through ag-gag and ecoterrorism laws as well as through stigmatizing representations of animal activism in the media. In turn, the chapters in this book by Juan José Ponce Léon and Anahí Gabriela González offer complementary explorations of (trans)feminist animal activisms in Latin America. Léon's chapter begins with a historical introduction to feminist animal activism in the global North before considering its Latin American counterparts. The chapter then analyzes interviews with feminist animal activists in Ecuador in particular. Finally, Léon considers the work of transfeminist animal activists, with particular attention to the performative work of trans, antispeciesist artists. González's chapter also considers both global north and Latin American ecofeminist theoretical frameworks and activisms but focuses on the context of Argentina and the transfeminist animal activist movement, *Ni Una Menos*. In her chapter for this volume, "Two-Spirit Feminism and Indigenous Ecological Governance," Mi'kmaq scholar Margaret Robinson reflects on the current moment of environmental crisis from an Indigenous and Two-Spirit ecofeminist perspective. Robinson highlights that Indigenous women and Two-Spirit people play leading roles in land (and animal) protection against settler colonial ecocide.

It is also important to study animal activism as feminists because animal activism in the West has frequently used misogynistic and other oppressive tactics. To take just a few examples: animal activists have continually used heteronormatively idealized representations of naked women to advance their causes or to get media attention, or have depicted women bloodied and packaged for consumption like meat. Animal activists have again and again outraged people of color by making comparisons between animal oppression and racial oppression while showing little to no investment in being allies against ongoing racism and colonialism. Animal activists have promoted veganism as a way to prevent or cure disability, to the extent that people with disabilities are made to feel uncomfortable in animal activist circles since they do not represent the movement's messaging. Animal activism has also promoted veganism as a way to "save the whales, lose the blubber," to cite a PETA advertisement, and more generally as a weight loss diet, reinforcing a culture of fat shaming and fat phobia. While PETA is consistently the worst offender in all of these respects, they are not alone among animal activists in using oppressive tactics. Animal activism is thus a significant site for intersectional feminist attention and critique. Fittingly, then, this volume includes a chapter by Corey Lee Wrenn on sexism in animal activism and a chapter about racism in animal activism by Darren Chang. Wrenn's chapter specifically critiques anti-foie gras campaigns for their use of violently sexist images in which pornographic scenes of "deep-throating" are invoked to show gagging women activists in the role of the male geese exploited to death by this luxury food industry. Chang's chapter considers the history of racism in animal activism as well as several ways in which anti-racist animal activists have attempted to transform the movement.

In discussing animal activism and advocacy, it is important to recognize that animals resist their oppression and advocate for themselves, and we do not so much need to be their voices as their allies.[148] As pattrice jones writes in her chapter in this volume: "Like raccoons in the suburbs, anarchic forces capable of upending human constructions may not be visible during the day but remain ready, if only we can ally with them." The chapter by Tim Reijsoo in this volume provides an overview of how animals resist their oppression and also why this resistance is often invisible to humans. In "The Homoxenoerotics of Elephant Crime and Captivity in India," Alok H. Gupta and Naisargi N. Davé describe the frequent and lethal violence of elephants against humans in India as acts of animal resistance, despite the ways that this resistance is continually glossed over as a natural rather than political phenomenon. As they describe, male elephants are particularly desired as captives, and their violence against humans often occurs during the condition of *musth* or in states of frustrated lust. As Gupta and Davé write, their chapter "represents a queer politics in animal studies, one which acknowledges animals—in this case, male elephants—as proud, sexual, and angry beings."

Another sense of representing animals is that of media, literary, artistic, and other cultural depictions. Feminist and queer literature scholars and art historians have examined the ways that authors and artists represent animals, including how these representations reflect, reproduce, challenge, and complicate gender roles, and a number of chapters in this book explore such work. Feminist art historian Emilie Blanc's chapter examines feminist visual art from the 1970s and 80s in the Los Angeles area that explored gendered and speciesist oppression, whether it was by using the eggs, milk, flesh, and blood of more-than-human animals as symbols of women's oppression or by using art as a medium for exploring bonds of love and grief for more-than-human animals. Feminist critical animal studies scholar Annie Potts' chapter, "From Folklore to Factory Farms: Gender, Sex and Chickenkind," examines gendered representations of chickens from ancient myths and folklore, including Baba Yaga, a supernatural being in Slavic folklore whose house is perched on chicken legs; Kikimora, a female house spirit with chicken feet and beak; the eastern European chicken god Kirinyi Bog; and the little red hen. Potts also introduces other aspects of chicken history and the contemporary lives of chickens, always with attention to the ways these have been caught up in gendered imaginaries and oppressions. As literature scholar Sue Hall Pyke explains in her chapter, "Gothic Snakes: Snake Handling, Snake Women and a Post-Secular Serpentine Practice," hatred of snakes has been intertwined with misogyny since the story of Eve and the serpent, with snakes often symbolizing "supernatural womanly evil." Pyke's chapter explores this vilification of snakes and women in Christian theology as well as its inheritances in Gothic literature, the Christian fundamentalist practice of snake-handling, and Pyke's own life. In "Masculinity and Multispecies Labour: A Feminist Animal Studies

Reading of *In the Skin of the Lion*," Tessa Wotherspoon considers the oppression and historical erasure of both working-class men and animals who together labored to build the Toronto Viaduct as these are represented in a novel by Sri Lankan-Canadian author Michael Ondaatje. In turn, Emelia Quinn's chapter considers the bizarre abundance of dairy products in the fiction of another Torontonian author, Margaret Atwood. Although meat and food generally have been thematized in Atwood's work, Quinn is the first scholar to have remarked upon what she describes as "Margaret Atwood's dairyscape," in which even the weather is recurrently described with butter and milk similes, and women are "frequently described *as* cheese." In addition to being a fascinating study of Atwood's novels, Quinn's chapter serves as a critical introduction to milk studies and provides an example of Quinn's concept of vegan camp. Finally, poet and philosopher Deborah Slicer's chapter, "A Hut of Her Own," is a playfully inventive sequel to J.M. Coetzee's *Elizabeth Costello*. While Coetzee's alter-ego character Elizabeth Costello may be astute in her anti-speciesist arguments and convictions, Coetzee unfortunately depicts her as a sad, old, tired, and dowdy vegan killjoy who makes herself unwelcome at dinner parties and in her own son's home. In her delightful response to Coetzee, Slicer reimagines Costello joyful, in better human and more-than-human company, and with a hut of her own.

6. Multispecies Justice for Multispecies Futures

In "Multispecies Justice: Theories, Challenges, and a Research Agenda for Environmental Politics," Danielle Celermajer and her co-authors describe the 2019–2020 megafire that burned through parts of Australia as "neither a natural disaster nor a tragedy, but injustice."[149] In "Justice through a Multispecies Lens," Celermajer invokes the over one billion animal lives that were lost in these fires and the fact that more deaths would come, as many animals who survived the fires themselves would not survive the loss of their habitats. Some endangered species are believed to have been driven to extinction by the fires.[150] As Celermajer observes,

> As humans on this planet, and specifically as political theorists facing the prospect that such devastating events will only become more frequent, the question before us is whether we can rethink what it means to be in ethical relationships with beings other than humans and what justice requires, in ways that mark these deaths as absolute wrongs that obligate us to act, and not simply as unfortunate tragedies that leave us bereft.[151]

Dominant theoretical frameworks for justice are anthropocentric and thus woefully inadequate for this task.[152] An understanding of justice is needed that is not only inclusive of other animals but of all beings on earth. For Celermajer and other scholars of multispecies justice, this goes beyond animal rights and also beyond ecocentric approaches to environmental ethics. Multispecies justice is neither limited to sentient beings nor to ecosystems but includes individual entities understood in Western traditions as inanimate such as rivers, mountains, and soils, as well as the relationships between beings.

In their chapter for the current volume, "A Grateful Acknowledgement: Gender Theory and Multispecies Justice," Celermajer and Darren Chang explore "the formative role" that "various forms of feminism, in particular ecofeminism, new materialist, and posthumanist feminisms" have played in the development of the field of multispecies justice. At least as urgently, they acknowledge that what is now being theorized as multispecies justice is nothing new to Indigenous traditions, although it may be novel to settler scholars. Drawing on Māori political theorist Christine Winter, they insist that

> Multispecies Justice must not repeat the epistemic erasure [Métis scholar] Zoe Todd identified had been committed by western philosophy's so-called novel "ontological turn": a failure to acknowledge and learn from Indigenous peoples' long held and articulated knowledges, practices, and protocols, in this case in relation to justice beyond the human.

Margaret Robinson's chapter for this volume introduces some ways in which Indigenous people, in particular Indigenous women and Two-Spirit individuals, have approached what could be described as multispecies justice. In "Two-Spirit Feminism and Indigenous Ecological Governance," Robinson provides an important critique of Western models of land management and stewardship over animals, contrasting these with Indigenous approaches to land and animals that are more holistic and do not treat land or animals as resources but as relations. As Robinson observes, rather than setting humans apart from and above other animals as in settler traditions, Indigenous ontologies see humans as one type of animated being among others, all of which are persons whose interests need to be considered. One of these interests for all animals, including Indigenous humans, is their connection to their traditional territories and to the other animals of those lands. As scholar of Indigenous law Sarah Krakoff has contended, climate change, which has been driven by settler colonialism and is already creating climate refugees, is particularly unjust to Indigenous peoples because of the strong connection Indigenous peoples have to their traditional territories and the animals of those territories.[153] With other Indigenous scholars, Robinson underscores that while white settlers see climate change as a coming apocalypse, for Indigenous peoples, the apocalypse began long ago with the arrival of Europeans on their lands. Robinson moreover argues that part of this settler apocalypse was the imposition of misogynistic and binary gender roles on Indigenous people, which entailed the political exclusion of Indigenous women and Two-Spirit people.

Indigenous and environmental law scholar e Campbell's chapter for this volume raises the rarely discussed point that settler colonialism has imposed binary European gender roles not only on Indigenous humans but on more-than-human animals as well. In "Pronouns in More-Than-Human Worlds and Indigenous and Colonial Laws: Towards Decolonial and Gender Diverse Relationships with More-Than-Human Animals," Campbell considers possibilities for not only queering and decolonizing but also deanthropocentrizing pronouns. Campbell's chapter, which is illustrated with their own artworks, dives into queer and critical animal studies scholarship, Indigenous law, and language studies, as well as settler law, to consider a myriad of possibilities for speaking ethically about and with animals. As Campbell discusses, Indigenous languages may not only have more or other genders than European languages, but they may distinguish between beings in other ways entirely, for instance, on the basis of animacy/inanimacy rather than male/female. For Campbell, exploring different pronoun possibilities is one way in which we might "reorient the ways we speak about and with more-than-human beings and worlds," whether gendered, genderless, animate, or inanimate. Ultimately Campbell argues that "Indigenous language methodologies centered on listening to and speaking with more-than-human beings and across human and more-than-human worlds can offer one way of learning how we can relate to each other in ways that advance multispecies justice."

In "Justice through a Multispecies Lens," Celermajer raises the question, "If we stage the introduction of beings other than humans onto the scene of justice as part of this trajectory of justice theory . . . what do the truly radical differences of these newly introduced subjects imply for both the concept of justice and the institutions designed to administer it?"[154] More specifically, can the current carceral system be part of multispecies justice projects? In "Transformative Justice for Animals: Lessons from Anti-Carceral Feminism," Kelly Struthers Montford, Darren Chang, and Selingul Yalcin respond in the negative, arguing that given the ways the current criminal-legal system woefully fails when it comes to addressing intrahuman injustices, it is not the institution we should turn to when seeking justice for more-than-human animals. Anti-carceral feminism has shown that prisons and policing are in fact sources of injustice rather than justice when it comes to gender and sexual violence, and Struthers Montford, Chang, and Yalcin make a parallel argument in the case of animal cruelty. The authors thus resist carceral tendencies in animal activism that parallel carceral feminism or which seek remedies for violence against animals through legislation and law enforcement. Instead, following the example of anti-carceral feminism, Struthers Montford, Chang, and Yalcin consider the potential of transformative justice for addressing harms to more-than-human animals.

Prison abolitionists and transformative justice practitioners who recognize that the carceral state will not keep them safe often turn instead to mutual aid. According to Dean Spade, author of *Mutual Aid: Building Solidarity During This Crisis (and the Next)*,[155] mutual aid meets people's survival needs, builds movement participation, and is "where we practice the kind of world we want to live in."[156] Spade's examples of mutual aid tend to focus on humans helping other humans, however, and the worlds he images are primarily human worlds. In contrast, queer theorist and animal law scholar Pierre Cloutier de Repentigny's chapter for this volume, "Queer Futures and Mutual Aid," approaches mutual aid more expansively as a way to build more just multispecies futures. For Cloutier de Repentigny, theories of queer futurities help us to imagine multispecies mutual aid projects and the multispecies futures we want to live in.

Unfortunately, one aspect of multispecies futures that will likely become ever more significant in coming years is zoonosis, given that, as Tessa Laird reminds us, "new/old viruses are emerging from the melting permafrost," and "global warming is increasing the range of many diseases once confined to the tropics. The future is sure to be teeming with microbial possibilities: inflorescences as well as die-offs." Laird's chapter also explores some of the ways that feminist theory and a gender lens help us to think of zoonotic transmission in new ways. These range from Laird's description of zoonosis as natureculture entanglements, to her highlighting of the historical roles of milkmaids and cows in the invention of vaccines, to her reminder of the homophobia that is latent or blatant in responses to zoonoses such as HIV/AIDS and monkeypox, and her reflection on "animaladies" as potential "transfections" between women and animals. Although the topic of zoonosis was committed to in advance, Laird's chapter was written while she was recovering from Covid-19 and concludes with reflections on her personal experience of zoonotic infection. In particular, she connects being ill with the coronavirus to her earlier work on "zoognosis," which images viral infections between animals and humans as transmissions of animal knowledges.[157] She considers, then, what animal knowledge has been transmitted to her by Covid-19, the animal reservoir for which may have been bats, and how her viral infection might thus help her to respond to Thomas Nagel's famous question, "What is it like to be a bat?"

Besides zoonotic epidemics and pandemics, another aspect of our multispecies futures is that life will in fact become ever less multiple. We live in a time of extinction that has only five precedents in Earth's history. Unlike the previous five mass extinction events, this time, an extraordinarily destructive minority of one species, our own, is the cause. Current rates of human-driven biodiversity loss are a clearer indication than climate change that we have entered a new geological epoch—what is being called the Anthropocene—with climate change but one of many anthropogenic causes of the current extinction event. Today, 96% of mammalian biomass on earth is made up of humans and their agricultural animals, a statistic that shows the shocking extent of wild animal eradication and the degree to which biota on earth have been homogenized. Indeed, due to the combined monocropping and eradication of life on earth, one of many alternative names that has been suggested for the current epoch is the Homogenocene. The fossil record for the current period will be unprecedentedly uniform: showing relatively few species, very large numbers of the animals humans eat and feed to the animals they eat, and the same species all across the planet. For example, geologists have argued that this epoch might well be called the Poultryocene, Gallocene, or Age of the Chicken rather than Age of Man since chicken bones will be one of the primary geological markers of the time in which we are living. This is not only because chicken bones will be so numerous in the fossil record of today in comparison to former times, but also because they will be found all across the globe instead of in the geographically isolated area of Southeast Asia where chickens first evolved. Moreover, chicken bones will be a clear marker of our catastrophic epoch because these bones will be so strange in comparison to the chicken bones of earlier eras. This is because chickens have been bred since the 1950s to produce more meat—and more breast meat in particular. Today's broiler chickens weigh twice as much as chickens of only sixty years ago, with their breasts weighing 80% more than they used to. Modern chickens reach this weight in only six weeks, whereas the much smaller chickens of the pre-1950s took fifteen weeks to reach their full weight. This selective breeding to be front heavy causes chickens considerable suffering, such that moving is painful and difficult, and most chickens are disabled by the time they are slaughtered at only forty days old. The perpetual pain of these chickens is evident in their bones and thus will be permanently written in the fossil record. The final three chapters in the volume thus turn to the topic of extinction, animal suffering, and death on this mass scale.

In "The Elachistocene vs. Anthropo-Scenes: Inheriting an Epoch Defined by Mass Extinction," environmental history scholar Dylan Hall explains the contentiousness of the term Anthropocene as the geologically proposed naming device for the current epoch, introducing some of the other concepts that have been suggested, such as Capitalocene, Eurocene, Plantationocene, and Manthropocene. Describing the latter, for instance, Hall writes

> the majority of Anthropocene scientists are men, mansplaining ecological crisis as a problem of "human nature" and ignoring feminists who have long shown how patriarchal agricultural empires and the exploitation of women by men is linked to hierarchies that also justify the exploitation of animals and waters and minerals.

Hall is most persuaded, however, by feminist geologist Jill Schneiderman's proposal of the name Elachistocene, in part because it encapsulates so many of the lessons of these other terms. As Hall explains, Elachistocene, meaning the Age of the Least New Life, centers biodiversity loss rather than the human and

> is supported by the lessons learned from the Capitalocene, Eurocene, Manthropocene, Plantationocene, Homogenocene, and other such scenes of inequality and injustice that describe deep drivers of intersecting naturalcultural crises, while grounding these other diagnoses of socio-economic illnesses and inequality with a call to listen for our animal kin whose symphony is being obliterated from the orchestra of life and replaced by the tortured cries of broiler chickens and cows in feedlots.

In arguing for Elachistocene, Hall's chapter offers both a valuable explanation of and impassioned lament for the current loss of biodiversity on Earth and its causal links to multiple forms of human oppression.

In her chapter for this volume, "De-Extinction," education scholar and artist jessie l. beier considers the real-life science of resurrection biology as well as its representation in the *Jurassic Park* film series, reading the latter as an "allegory for contemporary ecological politics." beier introduces the reader to genetics and bioscience companies such as Colossal Laboratories & Biosciences, which claims it will "fix" the problem of anthropogenic mass extinction through species revivalism. As beier notes, de-extinction projects raise many ethical issues, including animal welfare concerns, but the most troubling is that they conceptualize life as biocapital.

Finally, the book concludes with a reflection on multispecies death in a time of mass extinction. In "Composting Life-Death Time," political science scholar Danika Jorgensen-Skakum considers the recent interest in environmentally friendly human burials, which she reads as a response to "green guilt, as if we might leave our bodies as a parting gift or apology to the Earth when our attempts to divert climate change fail." Among the new trends in green burials are mushroom suits, whereby humans might find immortality through being transformed into fungi, and various designs for human composting projects. Whatever their green guilt, Jorgensen-Skakum observes that clients of human composting projects wish to be assured that their or their family member's remains will not get mixed up with those of other clients. Jorgensen-Skakum critiques these desires, emphasizing that we are "with and of the earth . . . we are always becoming-*with*. We are eaten before death . . . Our ontologies become messier upon every new connection." For Jorgensen-Skakum, by accepting these connections—and through the radical ethical, epistemological, and ontological shifts that come with such an acceptance—we may yet make more space for our multispecies kin and multispecies collaborations towards a more just future. This is worthwhile even if, at this moment when we have already passed so many tipping points, this can only mean dying together less badly or in less unjust ways.

Notes

1 I am grateful to Lauren Corman and Greta Gaard for their invaluable feedback on earlier drafts of this introduction.

2 An August 2023 search of course listings at 103 programs in Canada, 104 programs in the U.S., 41 programs in Australia, and 101 programs in the UK found only 14 regularly taught (as opposed to special topics) courses on gender or feminism and animals.

3 Programs for the last three annual meetings of the WGSRF in Canada and the NWSA in the U.S. were consulted.

4 Cressida Heyes, ed., *Gender and Philosophy* (London: Routledge, 2011).
5 The Keywords Feminist Editorial Collective, *Keywords for Gender and Sexuality Studies* (New York: NYU Press, 2021).
6 Rosemarie Garland-Thomson, "Integrating Disability, Transforming Feminist Theory," *NWSA Journal* 14, no. 3 (2002): 1–32.
7 Plato, *Plato's Cosmology; the Timaeus of Plato* (London and New York: Harcourt, 1937).
8 Genevieve Lloyd, *The Man of Reason: Male and "Female" in Western Philosophy* (London: Routledge, 1993, second edition; first edition 1984), 2.
9 Xenophon's *Oeconomicus*, cited in Foucault, *The History of Sexuality*, vol. 2, *The Use of Pleasure* (New York: Vintage, 1990), 155.
10 Cited in Michel Foucault, 1990, 156.
11 Aristotle, *Aristotle's Politics* (Oxford: Clarendon Press, 1905).
12 Heidi M. Ravven, "Has Hegel Anything to Say to Feminists?" *In Feminist Interpretations of G. W. F. Hegel*, ed. Patricia Jagentowicz Mills (University Park: Pennsylvania State UP, 1996).
13 G.W.F. Hegel, *Elements of the Philosophy Of Right*, ed. Allen W. Wood (Cambridge: Cambridge UP, 1991), §165–166, §171, §177.
14 Cited in Genevieve Lloyd, *The Man of Reason: Male and "Female" in Western Philosophy*, 2nd ed. (London: Routledge, 1993) (1st ed., 1984), 1.
15 Ibid.
16 Simone de Beauvoir, *The Second Sex*, trans. Constance Bourde and Sheila Malovany-Chevalier (New York: Vintage Books, 2010), 62.
17 Mary Daly, *Gyn/Ecology: The Metaethics of Radical Feminism* (Boston: Beacon Press, 1978). Daly made a number of transphobic statements in *Gyn/Ecology*, such as referring to trans people as "Frankensteinian" and writing that "Transsexualism is an example of male surgical siring which invades the female world." (*Gyn/Ecology*, 70–71) Daly was also the dissertation supervisor to Janice Raymond, author of transphobic book, *The Transsexual Empire*.
18 Starhawk, *The Spiral Dance: A Rebirth of the Ancient Religion of the Goddess* (New York: Harper and Row, 1979). For a critique of the ways Starhawk excludes queer people, see Ladelle McWhorter, "Practicing Practicing," in *Feminism and the Final Foucault*, eds. Dianna Taylor and Karen Vintges (Urbana: University of Illinois Press, 2004), 143–162.
19 Greta Gaard, "Ecofeminism Revisited: Rejecting Essentialism and Re-Placing Species in a Material Feminist Environmentalism," *Feminist Formations* 23, no. 2 (Summer 2011): 27.
20 Karen J. Warren, "The Power and the Promise of Ecological Feminism," in *Environmental Ethics* 12, no. 2 (1990): 125–146.
21 Hortense J. Spillers, "Mama's Baby, Papa's Maybe: An American Grammar Book," *Diacritics* 17, no. 2 (1987): 65–81.
22 Ashwin Roy, Ameeta Roy, and Meera Roy, "The Human Rights of Women with Intellectual Disability," *Journal of the Royal Society of Medicine* 105, no. 9 (2012): 384–389; Senate of Canada, "Human Rights Committee to Study the Forced and Coerced Sterilization of People in Canada," *The Standing Senate Committee on Human Rights* (2019); National Women's Law Center, "Forced Sterilization of Disabled People in the United States," (2021).
23 Adrienne Rich, "Compulsory Heterosexuality and Lesbian Existence," *Signs* 5, no. 4 (1980): 631–660.
24 Alicia Bridges, "Life Without Barriers Will Transfer Responsibilities of Fostered Aboriginal Children to Indigenous-Led Services," *Abc Net* (2023); Deanna Around Him, "American Indian and Alaska Native (AIAN) Children are Overrepresented in Foster Care in States with the Largest Portions of AIAN Children," *Child Trends* (2023); Fraser Needham, " 'The Bond Is Broken,': Data Shows Number of Indigenous Kids in Foster Care Is Going Up: StatCan," *National News* (2022); Joshua Rovner, "Black Disparities in Youth Incarceration," *The Sentencing Project* (2021).
25 Tim Stretton, K.J. Kesselring, and Sara M. Butler, eds., *Married Women and the Law: Coverture in England and the Common Law World* (Montreal and Kingston: McGill-Queen's University Press, 2013).
26 Aph Ko, *Racism as Zoological Witchcraft: A Guide to Getting Out* (New York: Lantern Books, 2019), 79–86.
27 Ibid., 82.

28 Ibid., 80.
29 Syl Ko, "Emphasizing Similarities Does Nothing for the Oppressed," in *Aphro-Ism: Essays on Pop Culture, Feminism, and Black Veganism from Two Sisters*, eds. Aph Ko and Syl Ko (New York: Lantern Books, 2017), 37–43.
30 Aph Ko, *Racism as Zoological Witchcraft: A Guide to Getting Out* (New York: Lantern Books, 2019); Syl Ko, "Who Is the Human and Who Is the Animal in the Human–animal Divide?" *Human Animal Studies* (2019), retrieved from: https://www.youtube.com/watch?v=TWkTeeejS8k; Sunaura Taylor, *Beasts of Burden: Animal and Disability Liberation* (New York: The New Press, 2017).
31 Personal correspondence.
32 Carol J. Adams, *The Sexual Politics of Meat: A Feminist-Vegetarian Critical Theory* (London and New York: Continuum, 1990).
33 Brian Luke, "Taming Ourselves or Going Feral: Toward a Nonpatriarchal Metaethic of Animal Liberation," in *Animals and Women: Feminist Theoretical Explorations*, eds. Carol J. Adams and Josephine Donovan (Durham: Duke University Press, 1995), 290–319.
34 C. Wesley Buerkle, "Metrosexuality Can Stuff It: Beef Consumption as (Heteromasculine) Fortification," *Text and Performance Quarterly* 29, no. 1 (2009): 77–93.
35 Greta Gaard has explored similar arguments to Buerkle in several works. See: Greta Gaard, "Towards New EcoMasculinities, EcoGenders, and EcoSexualities," in *Ecofeminism: Feminist Intersections with Other Animals and the Earth*, eds. Lori Gruen and Carol J. Adams (London: Bloomsbury, 2014), 225–239; Greta Gaard, "Queering the Climate," in *Men, Masculinities, and Earth, Contending with the (m)Anthropocene*, eds. Paul Pulé and Martin Hultman (London: Palgrave, 2021), 515–536.
36 Annie Potts and Jovian Parry, "Vegan Sexuality: Challenging Heteronormative Masculinity through Meat-Free Sex," *Feminism and Psychology* 20, no. 1 (2010): 53–72.
37 Vasile Stănescu, "The Whopper Virgins: Hamburgers, Gender, and Xenophobia in Burger King's Hamburger Advertising," in *Meat Culture*, ed. Annie Potts (Leiden: Brill, 2016), 90–108.
38 Vasile Stănescu, " 'White Power Milk': Milk, Dietary Racism, and the 'Alt-Right'," *Animal Studies Journal* 7, no. 2 (2018): 103–128.
39 Archaeological studies have complicated these categorizations, showing that women also hunted not only small but large game in some ancient societies. Randall Haas, James Watson, Tammy Buonasera, John Southon, Jennifer C. Chen, Sarah Noe, Kevin Smith, Carlos Viviano Llave, Jelmer Eerkens, and Glendon Parker, "Female Hunters of the Early Americas," *Science Advances* 6, no. 45 (2020); Margaret Osborne, "Early Women Were Hunters Not Just Gatherers, Study Suggests," *Smithsonian Magazine* (2023).
40 As Kelly Struthers Montford has explained, "Research . . . shows that abattoir workers 'double' in the sense of emotionally splitting themselves to kill, and that those who have a higher regard for animals suffer more psychological harms in this line of work. Over time, this type of employment not only decreases one's ability to emphasize with animals, but also with humans categorized as more vulnerable, such as women and children. Indeed, slaughterhouses are empirically proven to increase rates of domestic violence, as well as arrests for sexual violence including rape and other sex offenses in the communities in which they operate." Kelly Struthers Montford, "Slaughterhouses," in *Environmental Justice Encyclopedia*, ed. Dorceta Taylor (Thousand Oaks: Sage Publications, forthcoming). See also: A.J. Fitzgerald, L. Kalof, and T. Dietz, "Slaughterhouses and Increased Crime Rates: An Empirical Analysis of the Spillover from 'The Jungle' into the Surrounding Community," *Organization & Environment* 22, no. 2 (2009): 158–184.
41 Andrew Knight and Katherine Watson, "Was Jack the Ripper a Slaughterman? Human–animal Violence and the World's Most Infamous Serial Killer," *Animals* (Basel) 7, no. 4 (April 10, 2017): 30.
42 Greta Gaard, "Toward New EcoMasculinities, EcoGenders, and EcoSexualities," in *Ecofeminism: Feminist Intersections with Other Animals and the Earth*, 2nd ed., eds. Lori Guen and Carol J. Adams (London: Bloomsbury, 2022), 267.
43 Josephine Donovan, "Animal Rights and Feminist Theory," *Signs* 15, no. 2 (Winter 1990): 350–375.
44 The education scholar Nel Noddings is an example. See Nel Noddings, *Caring: A Feminine Approach to Ethics and Moral Education* (Oakland: University of California Press, 1986).

45 Judith Butler, *Precarious Life: The Powers of Mourning and Violence* (New York: Verso, 2006).
46 See: *Relational Autonomy: Feminist Perspectives on Autonomy, Agency, and the Social Self*, eds., Catriona MacKenzie and Natalie Stoljar (Oxford: Oxford University Press, 2000); Sunaura Taylor, "Interdependent Animals: A Feminist Disability Ethics of Care," in *Ecofeminism: Feminist Intersections with Other Animals and the Earth*, 2nd ed., eds. Lori Guen and Carol J. Adams (London: Bloomsbury, 2022), 144, 156.
47 *Beyond Animal Rights: A Feminist Caring Ethic for the Treatment of Animals*, eds. Josephine Donovan and Carol J. Adams (London and New York: Continuum, 1996); *The Feminist Care Tradition in Animal Ethics: A Reader*, eds. Josephine Donovan and Carol J. Adams (New York: Columbia University Press, 2007).
48 Intersectionality is the dominant paradigm within feminist scholarship today but is not uncontested. Even within Black feminisms, some scholars and activists have preferred the earlier paradigm of "interlocking oppressions" because it emphasizes structures of oppression rather than how these are individually embodied or experienced. An ecofeminist theoretical framework that also emphasizes structures of oppression rather than individual experiences and identities is Karen Warren's concept of "logics of domination," which she introduced in 1990. See: Karen Warren, "The Power and Promise of Ecological Feminism." Another framework for thinking of multiple axes of identity and oppression at the same time includes Gloria Anzaldúa "borderlands theory," introduced in 1987. See: Gloria Anzaldúa, *Borderlands/La Frontera: The New Mestiza* (San Francisco: Aunt Lute Press, 1987).
49 "A Black Feminist Statement" by The Combahee River Collective (April 1977), in Gloria Hull, Patricia Bell Scott, and Barbara Smith, eds., *All the Women Are White, all the Blacks Are Men, But Some of Us Are Brave: Black Women's Studies* (Old Westbury, NY: The Feminist Press, 1982), 13–22.
50 See Claire Jean Kim, *Dangerous Crossings.*
51 Ibid., 45–47, 51.
52 Glynis Whiting (Director), *The Sterilization of Leilani Muir* (1996).
53 Sara E. McHenry, "'Gay Is Good': History of Homosexuality in the *DSM* and Modern Psychiatry," *American Journal of Psychiatry: Minority Mental Health and Diversity* (2022).
54 Danielle Peers, "From Eugenics to Paralympics: Inspirational Disability, Physical Fitness, and the White Canadian Nation" (PhD dissertation, University of Alberta, 2015); Eli Clare, *Brilliant Imperfection: Grappling with Cure* (Durham: Duke University Press, 2017).
55 Pattrice Jones, *The Oxen at the Intersection: A Collision (or, Bill and Lou Must Die: A Real-Life Murder Mystery from the Green Mountains of Vermont)* (New York: Lantern Books, 2014).
56 Greta Gaard, "Toward a Feminist Postcolonial Milk Studies," *American Quarterly* 65, no. 3 (2013): 595–618.
57 Jessica Eisen, "Milked: Nature, Necessity, and American Law," *Berkeley Journal of Gender, Law, and Justice* 34, no. 1 (2019): 71–116.
58 Margaret Robinson, "Veganism and Mi'kmaq Legends," *The Canadian Journal of Native Studies* 33, no. 1: 189–196.
59 Taylor, *Beasts of Burden: Animal and Disability Liberation* (New York: The New Press, 2017).
60 Maneesha Deckha, "Toward a Postcolonial, Posthumanist Feminist Theory: Centralizing Race and Culture in Feminist Work on Nonhuman Animals," *Hypatia* 27, no. 3 (2012): 527–545.
61 Deane Curtin is the originator of the phrase "contextual moral vegetarianism" (although it is clear he is referring to what would now be called veganism). See: Deane Curtin, "Toward an Ecological Ethic of Care," *Hypatia* 6, no. 1 (1991), guest edited by Karen Warren.
62 For example, see: Anna Carastathis, *Intersectionality: Origins, Contestations, Horizons* (Lincoln: Nebraska University Press, 2016); Ange-Marie Hancock, *Intersectionality: An Intellectual History* (Oxford: Oxford University Press, 2016); Jennifer C. Nash, *Black Feminism Re-imagined: After Intersectionality* (Durham: Duke University Press, 2019).
63 Ange-Marie Hancock, *Intersectionality: An Intellectual History* (Oxford, Oxford University Press, 2016): "Amidst scholarship (primarily in Europe) that seeks to broaden the genealogy of intersectionality by claiming it as a product of mainstream feminism, there is a set of scholars (Knapp 2005; Alexander-Flloyd 2012; Jordan-Zachery 2013, 2014; Bilge 2013) who see intersectionality's travels (both geographic and disciplinary) as replicating the very hegemonic politics that intersectionality was created to fight against (Alexander-Floyd 2012; Bilge 2013; May 2015). The threats posed by these moves are twofold. First, intersectionality as a field of study loses its analytical

(ibid.) and rhetorical (Jordan-Zachery 2007, 2014) power when it is not centered on the daily experiences of U.S. Black women who are presumed to be the originators and original subjects of intersectionality theory." (page 12)
64 Claire Jean Kim, *Dangerous Crossings: Race, Species, and Nature in a Multicultural Age* (Cambridge: Cambridge University Press, 2015); Aph Ko, *Racism as Zoological Witchcraft: A Guide to Getting Out* (New York: Lantern Books, 2019).
65 Ladelle McWhorter, *Racism and Sexual Oppression in Anglo-America* (Bloomington: Indiana University Press, 2009).
66 Sami Schalk, "Critical Disability Studies as Methodology," *Lateral* 6, no. 1 (2017). See also: Suzanne Lennon and Danielle Peers, " 'Wrongful Inheritance': Race, Disability, and Sexuality in *Cramblett v. Midwest Sperm Bank*," *Feminist Legal Studies* 25, no. 2 (2017): 141–163.
67 Claire Jean Kim and Carla Freccero, "Introduction: A Dialogue," *American Quarterly* 65, no. 3 (2013): 461–479.
68 Ibid., 466–467.
69 Ibid., 467.
70 Ibid., 465.
71 Ibid., 468.
72 Ko, *Racism as Zoological Witchcraft*, 75.
73 Ibid., 77.
74 Ibid., 89.
75 Kim and Freccero, "Introduction: A Dialogue," 465.
76 Simone de Beauvoir, *The Second Sex*, trans. Constance Bourde and Sheila Malovany-Chevalier (New York: Vintage Books, 2010), 41. In the original: "*Le mot femelle fait lever chez lui une sarabande d'images: un énorme ovule rond happe et châtre le spermatozoïde agile; monstrueuse et gavée la reine des termites règne sur les mâles asservis; la mante religieuse, l'araignée repues d'amour broient leur partenaire et le dévorent; la chienne en rut court les ruelles, traînant après elle un sillage d'odeurs perverses; la guenon s'exhibe impudemment et se dérobe avec une hypocrite coquetterie; et les fauves les plus superbes, la tigresse, la lionne, la panthère se couchent servilement sous l'impériale étreinte du mâle. Inerte, impatiente, rusée, stupide, insensible, lubrique, féroce, humiliée, l'homme projette dans la femme toutes les femelles à la fois,*" in *Le deuxième sexe* (Paris: Gallimard, 1949), 33.
77 *The Second Sex*, 54.
78 *Le deuxième sexe*, 47.
79 *The Second Sex*, 54. In the original: "*c'est l'espèce qui par des voies différentes les dévore tous deux,*" in *Le deuxième sexe* (Paris: Gallimard, 1949), 47.
80 *The Second Sex*, 55.
81 Ibid.
82 Ibid., 56.
83 Roslyn Appleby, *Sexing the Animal in a Posthumanist World* (London: Routledge, 2019).
84 Ibid., 67.
85 Ibid., 68.
86 Alexis Pauline Gumbs, *Undrowned: Black Feminist Lessons from Marine Mammals* (Chico, CA: AK Press, 2020).
87 Ibid., 5–6.
88 Ibid., 135–136.
89 Donna J. Haraway, "Primatology Is Politics by Other Means," *PSA: Proceedings of the Biennial Meeting of the Philosophy of Science Association* 1984 (1984): 489–524.
90 Foucault wrote that "we can invert Clausewitz's proposition and say that politics is the continuation of war by other means . . . the role of political power is perpetually to use a sort of silent war to reinscribe that relationship of force, and to reinscribe its institutions, economic inequalities, language, and even the bodies of individuals. This is the initial meaning of our inversion of Clausewitz's aphorism—politics is the continuation of war by other means." Michel Foucault, *"Society Must Be Defended": Lectures at the Collège de France*, eds. Mauro Bertani and Alessandro Fontana, trans. David Macey (New York: Picador, 2003), 47.
91 Haraway, "Primatology Is Politics by Other Means," 500. "Women's Place Is in the Jungle" is also the title of chapter 11 in Donna Haraway's book, *Primate Visions: Gender, Race, and Nature in the World of Modern Science* (London: Routledge, 1989).

92 Haraway, "Primatology is Politics by Other Means," 505.
93 Michelle A. Rodrigues, "It's Time to Stop Lionizing Dian Fossey as a Conservation Hero," *Lady Science* (September 20, 2019); Mary Battiata, "Dian Fossey: The Crusade, the Conflict, the Night of Horror," *The Washington Post* (January 25, 1986).
94 Cases such as these are discussed by Claire Jean Kim in *Dangerous Crossings: Race, Species, and Nature in a Multicultural Age* (Cambridge: Cambridge University Press, 2015).
95 Stacy Alaimo, "Eluding Capture: The Science, Culture, and Pleasure of 'Queer' Animals," in *Queer Ecologies: Sex, Nature, Politics, Desire*, eds. Catriona Mortimer-Sandilands and Bruce Erickson (Bloomington: Indiana University Press, 2010), 54.
96 Bruce Bagemihl's *Biological Exuberance: Animal Homosexuality and Natural Diversity* (1999) 107, cited in Alaimo, "Eluding Capture: The Science, Culture, and Pleasure of 'Queer' Animals," 54.
97 Alaimo, "Eluding Capture: The Science, Culture, and Pleasure of 'Queer' Animals," 62.
98 Ibid., 63.
99 Cited in ibid., 63.
100 Cited in ibid., 54.
101 Jennifer Terry, "'Unnatural Acts' in Nature: The Scientific Fascination with Queer Animals," *GLQ: A Journal of Lesbian and Gay Studies* 6, no. 2 (2000): 155.
102 Noël Sturgeon, "Penguin Family Values," in *Queer Ecologies: Sex, Nature, Politics, Desire*, ed. Catriona Sandilands (Blomington: Indiana University Press, 2010).
103 Terry, "'Unnatural Acts' in Nature: The Scientific Fascination with Queer Animals," 158.
104 Jack Halberstam, *Wild Things: The Disorder of Desire* (Durham: Duke University Press, 2020), 17.
105 Joan Roughgarden, *Evolution's Rainbow: Diversity, Gender, and Sexuality in Nature and People* (Oakland: University of California Press, 2004); Bruce Bagemihl, *Biological Exuberance: Animal Homosexuality and Natural Diversity* (New York: St. Martin's Press, 1999).
106 Alaimo, "Eluding Capture: The Science, Culture, and Pleasure of 'Queer' Animals," 56.
107 Ibid., 54.
108 Ibid., 52.
109 Leonore Tiefer, *Sex Is Not a Natural Act & Other Essays*, 2nd ed. (London: Routledge, 2004).
110 Letitia Meynell and Andrew Lopez, "Gendering Animals," *Synthese* 199, no. 2 (2021).
111 In fact Japanese primatologists began writing of "animal cultures" in the 1940s. See: Z. Reznikova, "Animal Culture," in *Encyclopedia of the Sciences of Learning*, ed. N.M. Seel (Boston, MA: Springer, 2012).
112 Alaimo, "Eluding Capture: The Science, Culture, and Pleasure of 'Queer' Animals," 63.
113 Intra-action is Karen Barad's terminology and replaces "interaction". While "interaction" presupposes two pre-existing entities that engage with one another, "intra-action" implies ongoing co-constitution between inseparable parts. Karen Barad, *Meeting the Universe Halfway: Quantum Physics and the Entanglement of Matter and Meaning* (Durham: Duke University Press, 2007).
114 McWhorter makes this argument in *Racism and Sexual Oppression in Anglo-America* when she observes that Foucault uses the term "racism" in a novel way in his *Les anormaux* course lectures when he refers to "racism against the abnormal." Although this usage initially struck McWhorter as offensive because unconventional, ultimately she observes that the meaning of the term "racism" has changed several times, so a new usage in itself does not make it right or wrong. Indeed, there is no Platonic form of racism against which we could measure whether a particular usage is right or wrong.
115 Robert McKay, "Veganism as a Form of Life," in *Thinking Veganism in Literature and Culture: Toward a Vegan Theory*, eds., Emelia Quinn and Benjamin Westwood (London: Palgrave, 2018), 249.
116 Ibid., 249–250.
117 Igor Ya Pavlinov, *The Species Problem: A Conceptual History* (London: Routledge, 2023).
118 Alexandra Palmer and Virginia Thomas, "Categorisation of Cats: Managing Boundary Felids in Aotearoa New Zealand and Britain," *People and Nature* (August 6, 2023).
119 Richard A. Richards, *The Species Problem: A Philosophical Analysis* (Cambridge: Cambridge University Press, 2015).

120 John S. Wilkins, Frank E. Zachos, and Igor Ya Pavlinov, *Species Problems and Beyond: Contemporary Issues in Philosophy and Practice* (London: Routledge, 2022).
121 McKay, "Veganism as a Form of Life," 249–250.
122 Ibid., 250.
123 Ibid., 251.
124 Ibid.
125 Ibid., 254.
126 Ibid., 263.
127 Gumbs, *Undrowned*, 5.
128 Ibid., 9.
129 In *Becoming Human: Matter and Meaning in an Antiblack World*, Zakiyyah Iman Jackson argues that African diaspora writers such as Toni Morrison, Nalo Hopkinson, Octavia Butler, and Audre Lorde do not so much seek for inclusion in humanity as to disrupt the category of the human and its racialization. Zakiyyah Iman Jackson, *Becoming Human: Matter and Meaning in an Antiblack World* (New York: NYU Press, 2020).
130 Taylor, *Beasts of Burden: Animal and Disability Liberation.*
131 McKay, "Veganism as a Form of Life," 250.
132 Ibid., 253.
133 Ibid., 254.
134 Emily Gaarder, "Where the Boys Aren't: The Predominance of Women in Animal Rights Activism," *Feminist Formations* 23, no. 2 (2011): 54–76.
135 Katja Guenther, *The Lives and Deaths of Shelter Animals* (Stanford: Stanford University Press, 2020).
136 Women are twice as likely as men to be vegan in the UK, and 79% of vegans are women in the U.S., for example. See: Catherine Oliver, "Mock Meat, Masculinity, and Redemption Narratives: Vegan Men's Negotiations and Performances of Gender and Eating," *Social Movement Studies* 22, no. 1 (2023): 62–79.
137 Fiona Probyn-Rapsey, Siobhan O'Sullivan, and Yvette Watt, " 'Pussy Panic' and Glass Elevators: How Gender Is Shaping the Field of Animal Studies," *Australian Feminist Studies* 34, no. 100 (2019): 198–215.
138 On sexual harassment in the animal rights movement, see Carol J. Adams, "#MeToo and the Sexual Politics of Meat," lecture at Harvard Law School, February 25, 2019, retrieved from: https://www.youtube.com/watch?v=kd9phAMfisI&ab_channel=HarvardAnimalLaw
See also: Marc Gunther, "A New Animal Rights Conference, and a Blacklist" (August 27, 2022), retrieved from: https://medium.com/nonprofit-chronicles/a-new-animal-rights-conference-and-a-blacklist-f5ae3b075091
139 Leslie Irvine and Jenny Vermilya, "Gender Work in a Feminized Profession: The Case of Veterinary Medicine," *Gender & Society* 24, no. 1 (2010): 56–82.
140 Jeanne Loftstedt, "Gender and Veterinary Medicine," *The Canadian Veterinary Journal* 44, no. 7 (2003): 533–535.
141 Andrew Gardiner, "The 'Dangerous' Women of Animal Welfare: How British Veterinary Medicine Went to the Dogs," *Social History of Medicine* 27, no. 3 (2014): 466–487.
See also the Knowing Animals podcast, Episode 76: "Dangerous Women and Veterinary Medicine with Andrew Gardiner." Retrieved from: https://knowinganimals.libsyn.com/episode-76-dangerous-women-and-veterinary-medicine-with-andrew-gardiner
142 Jenny Vermilya, *Identity, Gender, and Tracking: The Reality of Boundaries for Veterinary Students* (West Lafayette: Purdue University Press, 2022).
See also: Episode 181 of Knowing Animals podcast (December 13, 2021). Retrieved from: https://knowinganimals.libsyn.com/episode-181-boundaries-and-veterinary-medicine-with-jenny-vermilya
143 Ibid.
144 See: Angela Stuesse and Nathan T. Dollar, "Who Are America's Meat and Poultry Workers?" *Economic Policy Institute* (2020); U.S. Department of the Interior, U.S. Fish and Wildlife Service, and U.S. Department of Commerce, "2016 National Survey of Fishing, Hunting, and

Wildlife-Associated Recreation. U.S.," *Census Bureau* (2016); The Humane Society of the U.S., "Animal Cruelty Facts and Statistics," *Humane Society Resources* (2023). For the gender of offenders in cases of interspecies sexual assault, see Chloë Taylor, " 'Sex without All the Politics'? Sexual Ethics and Human-Canine Relations," in *Pets and People: The Ethics of Companion Animals*, ed. Christine Overall (Oxford: Oxford University Press, 2017).

145 M. Dragiewicz, "Why Sex and Gender Matter in Domestic Violence Research and Advocacy," in *Violence against Women in Families and Relationships: Vol. 3 Criminal Justice and the Law*, eds. E. Stark and E. Buzawa (Santa Barbara, CA: Praeger, 2009), 201–215.

146 Diana Donald, "Anti-Vivisection: A Feminist Cause?" in *Women Against Cruelty: Protection of Animals in Nineteenth-Century Britain* (Manchester, 2020; online ed., Manchester Scholarship Online, May 21, 2020), https://doi.org/10.7765/9781526115430.00011, accessed 6 August 2023.

See also: Craig Buettinger, "Women and Antivivisection in Late Nineteenth-Century America," in *Journal of Social History* 30, no. 4 (1997): 857–872.

See also: Liam Young, "Eating Serial: Beatrice Lindsay, Vegetarianism, and the Tactics of Everyday Life in the Late Nineteenth Century," *Societies* 5, no. 1 (2015): 65–88.

147 The alt right, for instance, simultaneously opposed gender equality and used the milk emoji as a symbol of white-dominated patriarchy, mocking soy milk as for feminists, effeminate politically correct men, and the lactose-intolerant racialized other. See: Vasile Stănescu, " 'White Power Milk': Milk, Dietary Racism, and the 'Alt-Right'," in *Animal Studies Journal* 7, no. 2 (2018): 103–128.

See also "How the Alt-Right Uses Milk to Promote White Supremacy," *The Conversation* (April 26, 2018).

See also: Amy Harmon, "Why Are White Supremacists Chugging Milk?" *The New York Times* (October 17, 2018).

See also: "Milk, a Symbol of Neo-Nazi Hate," *The Conversation* (August 31, 2017).

148 For critical animal studies discussions of animal resistance, see pattrice jones, "Stomping with the Elephants: Feminist Principles for Radical Solidarity," in *Igniting a Revolution: Voices in Defense of the Earth*, eds. Steven Best and Anthony J. Nocella, II (Chico, CA: AK Press, 2006), 319–334; Jason Hribal, *Fear of the Animal Planet: The Hidden History of Animal Resistance* (Chico, CA: AK Press, 2010).

149 Danielle Celermajer, David Schlosberg, Lauren Rickards, Makere Stewart Harawira, Mathias Thaler, Petra Tschakert, Blanche Verlie, and Christine Winter, "Multispecies Justice: Theories, Challenges, and a Research Agenda for Environmental Politics," *Environmental Politics* 30, no. 1–2 (2021): 116.

150 Michael Slezak, "3 Billion Animals Killed or Displaced in Black Summer Bushfires, Study Estimates," *ABC News* (July 27, 2020), retrieved from: https://www.abc.net.au/news/2020-07-28/3-billion-animals-killed-displaced-in-fires-wwf-study/12497976

151 Danielle Celermajer, S. Chatterjee, A. Cochrane et al. (7 more authors), "Justice through a Multispecies Lens," *Contemporary Political Theory* 19, no. 3 (2020): 475.

152 Celermajer et al., "Multispecies Justice: Theories, Challenges, and a Research Agenda for Environmental Politics," 116.

153 Sarah Krakoff, "American Indians, Climate Change, and Ethics for a Warming World," *Denver Law Review* 85 (2008): 865, *available at* https://scholar.law.colorado.edu/faculty-articles/302. Sarah Krakoff, "Radical Adaptation, Justice, and American Indian Nations," *Environmental Justice* 4 (2011): 207–212.

154 Danielle Celermajer, S. Chatterjee, A. Cochrane et al. (7 more authors), "Justice through a Multispecies Lens," *Contemporary Political Theory* 19, no. 3 (2020): 479.

155 Dean Spade, *Mutual Aid: Building Solidarity During This Crisis (And the Next)* (New York: Verso, 2020).

156 Dean Spade, "We Keep Each Other Safe: Mutual Aid for Survival and Solidarity," presentation for Barnard Center for Research on Women (2020), retrieved from: https://www.youtube.com/watch?v=C32KGX6qP5s&t=671s&ab_channel=BarnardCenterforResearchonWomen (10:12–11:12).

157 Tessa Laird, "Zoognosis: When Animal Knowledges Go Viral. Laura Jean McKay's *The Animals in That Country*, Contagion, Becoming-Animal, and the Politics of Predation," *Animal Studies Journal* 10, no. 1 (2021): 30–56.

Bibliography

Adams, Carol. *The Sexual Politics of Meat: A Feminist-Vegetarian Critical Theory*. London and New York: Continuum, 1990.
Adams, Carol. "#MeToo and the Sexual Politics of Meat." Lecture at Harvard Law School, February 25, 2019. Retrieved from: https://www.youtube.com/watch?v=kd9phAMfisI&ab_channel=HarvardAnimalLaw
Alaimo, Stacy. "Eluding Capture: The Science, Culture, and Pleasure of 'Queer' Animals." In *Queer Ecologies: Sex, Nature, Politics, Desire*, edited by Catriona Mortimer-Sandilands and Bruce Erickson, 51–72. Bloomington: Indiana University Press, 2010.
Anzaldúa, Gloria. *Borderlands/La Frontera: The New Mestiza*. San Francisco: Aunt Lute Books, 1987.
Aristotle. *Aristotle's Politics*. Oxford: Clarendon Press, 1905.
Around Him, Deanna. "American Indian and Alaska Native (AIAN) Children Are Overrepresented in Foster Care in States with the Largest Proportions of AIAN Children." *Child Trends*, 2023.
Bagemihl, Bruce. *Biological Exuberance: Animal Homosexuality and Natural Diversity*. New York: St. Martin's Press, 1999.
Barad, Karen. *Meeting the Universe Halfway: Quantum Physics and the Entanglement of Matter and Meaning*. Durham: Duke University Press, 2007.
Battiata, Mary. "Dian Fossey: The Crusade, the Conflict, the Night of Horror." *The Washington Post*, January 25, 1986.
Beauvoir, Simone de. *The Second Sex*. Translated by Constance Bourde and Sheila Malovany-Chevalier. New York: Vintage Books, 2010.
Bridges, Alicia. "Life Without Barriers Will Transfer Responsibilities of Fostered Aboriginal Children to Indigenous-led Services." *ABC Net*, 2023.
Buerkle, C. Wesley. "Metrosexuality Can Stuff It: Beef Consumption as (Heteromasculine) Fortification." *Text and Performance Quarterly* 29, no. 1 (2009): 77–93.
Buettinger, Craig. "Women and Antivivisection in Late Nineteenth-Century America." *Journal of Social History* 30, no. 4 (1997): 857–872.
Butler, Judith. *Precarious Life: The Powers of Mourning and Violence*. New York: Verso, 2006.
Carastathis, Anna. *Intersectionality: Origins, Contestations, Horizons*. Lincoln: Nebraska University Press, 2016.
Celermajer, Danielle, Chatterjee, S., Cochrane, A., Fishel, S., Neimanis, A., O'Brien, A., Reid, S., Srinivasan, K., Schlosberg, D., and Waldow, A. "Justice Through a Multispecies Lens." *Contemporary Political Theory* 19, no. 3 (2020).
Celermajer, Danielle, Schlosberg, David, Rickards, Lauren, Harawira, Makere Stewart, Thaler, Mathias, Tschakert, Petra, Verlie, Blanche, and Winter, Christine. "Multispecies Justice: Theories, Challenges, and a Research Agenda for Environmental Politics." *Environmental Politics* 30, no. 1–2 (2021).
Clare, Eli. *Brilliant Imperfection: Grappling with Cure*. Durham: Duke University Press, 2017.
The Combahee River Collective. "A Black Feminist Statement." April 1977, in *All the Women Are White, All the Blacks Are Men, But Some of Us Are Brave: Black Women's Studies*, edited by Gloria Hull, Patricia Bell Scott, and Barbara Smith. Old Westbury, NY: The Feminist Press, 1982.
Curtin, Deane. "Toward an Ecological Ethic of Care." *Hypatia* 6, no. 1 (1991). Guest edited by Karen Warren.
Daly, Mary. *Gyn/Ecology: The Metaethics of Radical Feminism*. Boston: Beacon Press, 1978.
Deckha, Maneesha. "Toward a Postcolonial, Posthumanist Feminist Theory: Centralizing Race and Culture in Feminist Work on Nonhuman Animals." *Hypatia* 27, no. 3 (2012): 527–545.
Donald, Diana. "Anti-vivisection: A Feminist Cause?" *Women Against Cruelty: Protection of Animals in Nineteenth-Century Britain*. Manchester, online ed., Manchester Scholarship Online, May 21, 2020. https://doi.org/10.7765/9781526115430.00011
Donovan, Josephine. "Animal Rights and Feminist Theory." *Signs* 15, no. 2 (Winter 1990): 350–375.
Donovan, Josephine, and Adams, Carol J., eds. *Beyond Animal Rights: A Feminist Caring Ethic for the Treatment of Animals*. New York: Continuum, 1996.
Donovan, Josephine, and Adams, Carol J., eds. *The Feminist Care Tradition in Animal Ethics: A Reader*. New York: Columbia University Press, 2007.

Dragiewicz, M. "Why Sex and Gender Matter in Domestic Violence Research and Advocacy." In *Violence against Women in Families and Relationships: Vol. 3 Criminal Justice and the Law*, edited by E. Stark and E. Buzawa, 201–215. Santa Barbara, CA: Praeger, 2009.

Eisen, Jessica. "Milked: Nature, Necessity, and American Law." *Berkeley Journal of Gender, Law, and Justice* 34, no. 1 (2019): 71–116.

Fitzgerald, A.J., Kalof, L., and Dietz, T. "Slaughterhouses and Increased Crime Rates: An Empirical Analysis of the Spillover From 'The Jungle' Into the Surrounding Community." *Organization & Environment* 22, no. 2 (2009): 158–184.

Foucault, Michel. *"Society Must Be Defended": Lectures at the Collège de France*. Edited by Mauro Bertani and Alessandro Fontana, translated by David Macey, New York: Picador, 2003.

Freeman, Andrea. "Milk, a Symbol of Neo-Nazi Hate." *The Conversation*, August 31, 2017.

Gaard, Greta. "Ecofeminism Revisited: Rejecting Essentialism and Re-Placing Species in a Material Feminist Environmentalism." *Feminist Formations* 23, no. 2 (Summer 2011).

Gaard, Greta. "Toward a Feminist Postcolonial Milk Studies." *American Quarterly* 65, no. 3 (2013): 595–618.

Gaard, Greta. "Towards New EcoMasculinities, EcoGenders, and EcoSexualities." In *Ecofeminism: Feminist Intersections with Other Animals and the Earth,* edited by Lori Gruen and Carol Adams, 225–239. London: Bloomsbury, 2014.

Gaard, Greta. "Queering the Climate." In *Men, Masculinities, and Earth, Contending with the (m) Anthropocene,* edited by Paul Pulé and Martin Hultman, 515–536. London: Palgrave, 2021.

Gaarder, Emily. "Where the Boys Aren't: The Predominance of Women in Animal Rights Activism." *Feminist Formations* 23, no. 2 (2011): 54–76.

Gambert, Iselin, and Linné, Tobias. "How the Alt-right Uses Milk to Promote White Supremacy." *The Conversation*, April 26, 2018.

Gardiner, Andrew. "The 'Dangerous' Women of Animal Welfare: How British Veterinary Medicine Went to the Dogs." *Social History of Medicine* 27, no. 3 (2014): 466–487.

Garland-Thomson, Rosemarie. "Integrating Disability, Transforming Feminist Theory." *NWSA Journal* 14, no. 3 (2002): 1–32.

Guenther, Katja. *The Lives and Deaths of Shelter Animals*. Stanford: Stanford University Press, 2020.

Gumbs, Alexis Pauline. *Undrowned: Black Feminist Lessons from Marine Mammals*. Chico, CA: AK Press, 2020.

Gunther, Marc. "A New Animal Rights Conference, and a Blacklist." *Medium*, August 27, 2022. Retrieved from: https://medium.com/nonprofit-chronicles/a-new-animal-rights-conference-and-a-blacklist-f5ae3b075091.

Haas, Randall, Watson, James, Buonasera, Tammy, Southon, John, Chen, Jennifer C., Noe, Sarah, Smith, Kevin, Viviano Llave, Carlos, Eerkens, Jelmer, and Parker, Glendon. "Female Hunters of the Early Americas." *Science Advances* 6, no. 45 (2020).

Halberstam, Jack. *Wild Things: The Disorder of Desire*. Durham: Duke University Press, 2020.

Hancock, Ange-Marie. *Intersectionality: An Intellectual History*. Oxford: Oxford University Press, 2016.

Haraway, Donna J. "Primatology Is Politics by Other Means." *PSA: Proceedings of the Biennial Meeting of the Philosophy of Science Association 1984* (1984): 489–524.

Harmon, Amy. "Why Are White Supremacists Chugging Milk?" *The New York Times*, October 17, 2018.

Hegel, G.W.F. *Elements of the Philosophy of Right*. Edited by Allen W. Wood. Cambridge: Cambridge University Press, 1999.

Heyes, Cressida, ed. *Philosophy and Gender*. London: Routledge, 2011.

Hribal, Jason. *Fear of the Animal Planet: The Hidden History of Animal Resistance*. Chico, CA: AK Press, 2010.

The Humane Society of the U.S. "Animal Cruelty Facts and Statistics." *Humane Society Resources*, 2023.

Irvine, Leslie, and Vermilya, Jenny. "Gender Work in a Feminized Profession: The Case of Veterinary Medicine." *Gender & Society* 24, no. 1 (2010): 56–82.

Jackson, Zakiyyah Iman. *Becoming Human: Matter and Meaning in an Antiblack World*. New York: NYU Press, 2020.

Jones, Pattrice. "Stomping with the Elephants: Feminist Principles for Radical Solidarity." In *Igniting a Revolution: Voices in Defense of the Earth*, edited by Steven Best and Anthony J. Nocella, II, 319–334. Chico, CA: AK Press, 2006.

Jones, Pattrice. *The Oxen at the Intersection: A Collision (or, Bill and Lou Must Die: A Real-Life Murder Mystery from the Green Mountains of Vermont)*. New York: Lantern Books, 2014.

The Keywords Feminist Editorial Collective. *Keywords for Gender and Sexuality Studies*. New York: NYU Press, 2021.

Kim, Claire Jean. *Dangerous Crossings: Race, Species, and Nature in a Multicultural Age*. Cambridge: Cambridge University Press, 2015.

Kim, Claire Jean, and Freccero, Carla. "Introduction: A Dialogue." *American Quarterly* 65, no. 3 (2013): 461–479.

Knight, Andrew, and Watson, Katherine. "Was Jack the Ripper a Slaughterman? Human-Animal Violence and the World's Most Infamous Serial Killer." *Animals* (Basel) 7, no. 4 (2017).

Knowing Animals podcast. Episode 76: "Dangerous Women and Veterinary Medicine with Andrew Gardiner." Retrieved from: https://knowinganimals.libsyn.com/episode-76-dangerous-women-and-veterinary-medicine-with-andrew-gardiner.

Knowing Animals podcast. Episode 181 (December 13, 2021). Retrieved from: https://knowinganimals.libsyn.com/episode-181-boundaries-and-veterinary-medicine-with-jenny-vermilya

Ko, Aph. *Racism as Zoological Witchcraft: A Guide to Getting Out*. New York: Lantern Books, 2019.

Ko, Syl. "Emphasizing Similarities Does Nothing for the Oppressed." In *Aphro-Ism: Essays on Pop Culture, Feminism, and Black Veganism from Two Sisters*, edited by Aph Ko and Syl Ko, 27–43. New York: Lantern Books, 2017.

Ko, Syl. "Who Is the Human and Who Is the Animal in the Human-Animal Divide?" *Human Animal Studies* (2019). Retrieved from: https://www.youtube.com/watch?v=TWkTeeejS8k.

Krakoff, Sarah. "American Indians, Climate Change, and Ethics for a Warming World." *The Denver Law Review* 85 (2008): 865. Retrieved from: https://scholar.law.colorado.edu/faculty-articles/302

Krakoff, Sarah. "Radical Adaptation, Justice, and American Indian Nations." *Environmental Justice* 4 (2011): 207–212.

Laird, Tessa. "Zoognosis: When Animal Knowledges Go Viral. Laura Jean McKay's *The Animals in That Country*, Contagion, Becoming-Animal, and the Politics of Predation." *Animal Studies Journal* 10, no. 1 (2021): 30–56.

Lenon, Suzanne, and Peers, Danielle. "'Wrongful Inheritance': Race, Disability, and Sexuality in Cramblett v. Midwest Sperm Bank." *Feminist Legal Studies* 25, no. 2 (2017): 141–163.

Lloyd, Genevieve. *The Man of Reason: Male and "Female" in Western Philosophy*. 2nd ed. London: Routledge, 1993 (1st ed., 1984).

Loftstedt, Jeanne. "Gender and Veterinary Medicine." *The Canadian Veterinary Journal* 44, no. 7 (2003): 533–535.

Luke, Brian. "Taming Ourselves or Going Feral: Toward a Nonpatriarchal Metaethic of Animal Liberation." In *Animals and Women: Feminist Theoretical Explorations*, edited by Carol J. Adams and Josephine Donovan, 290–319. Durham: Duke University Press, 1995.

MacKenzie, Catriona, and Stoljar, Natalie, eds. *Relational Autonomy: Feminist Perspectives on Autonomy, Agency, and the Social Self*. Oxford: Oxford University Press, 2000.

McHenry, Sara E. "'Gay Is Good': History of Homosexuality in the *DSM* and Modern Psychiatry." *American Journal of Psychiatry: Minority Mental Health and Diversity* 18, no. 1 (2022): 4–5.

McKay, Robert. "Veganism as a Form of Life." In *Thinking Veganism in Literature and Culture: Toward a Vegan Theory*, edited by Emelia Quinn and Benjamin Westwood, 249. London: Palgrave, 2018.

McWhorter, Ladelle. "Practicing Practicing." In *Feminism and the Final Foucault*, edited by Dianna Taylor and Karen Vintges, 143–162. Urbana: University of Illinois Press, 2004.

McWhorter, Ladelle. *Racism and Sexual Oppression in Anglo-America*. Bloomington: Indiana University Press, 2009.

Meynell, Letitia, and Lopez, Andrew. "Gendering Animals." *Synthese* 199, no. 2 (2021).

Nash, Jennifer C. *Black Feminism Re-imagined: After Intersectionality*. Durham: Duke University Press, 2019.

National Women's Law Center. "Forced Sterilization of Disabled People in the United States." (2021). Retrieved from: https://nwlc.org/resource/forced-sterilization-of-disabled-people-in-the-united-states/

Needham, Fraser. " 'The Bond Is Broken,': Data Shows Number of Indigenous Kids in Foster Care Is Going Up: StatCan." *National News,* 2022.

Noddings, Nel. *Caring: A Feminine Approach to Ethics and Moral Education.* Oakland: University of California Press, 1986.

Oliver, Catherine. "Mock Meat, Masculinity, and Redemption Narratives: Vegan Men's Negotiations and Performances of Gender and Eating." *Social Movement Studies* 22, no. 1 (2023): 62–79.

Osborne, Margaret. "Early Women Were Hunters Not Just Gatherers, Study Suggests." *Smithsonian Magazine,* 2023.

Palmer, Alexandra, and Thomas, Virginia. "Categorisation of Cats: Managing Boundary Felids in Aotearoa New Zealand and Britain." *People and Nature,* August 6, 2023.

Pavlinov, Igor Ya. *The Species Problem: A Conceptual History.* London: Routledge, 2023.

Peers, Danielle. "From Eugenics to Paralympics: Inspirational Disability, Physical Fitness, and the White Canadian Nation." PhD Diss., University of Alberta, 2015.

Plato. *Plato's Cosmology; the Timaeus of Plato.* London: New York: Harcourt, 1937.

Potts, Annie, and Parry, Jovian. "Vegan Sexuality: Challenging Heteronormative Masculinity through Meat-Free Sex." *Feminism and Psychology* 20, no. 1 (2010): 53–72.

Probyn-Rapsey, Fiona, O'Sullivan, Siobhan, and Watt, Yvette. " 'Pussy Panic' and Glass Elevators: How Gender Is Shaping the Field of Animal Studies." *Australian Feminist Studies* 34 (2019): 198–215.

Quinn, E. Notes on Vegan Camp. *PMLA/Publications of the Modern Language Association of America.* 2020; 135 (5): 914–30.

Ravven, Heidi M. "Has Hegel Anything to Say to Feminists?." *In Feminist Interpretations of G. W. F. Hegel,* edited by Patricia Jagentowicz Mills. University Park: Pennsylvania State University Press, 1996.

Reznikova, Z. "Animal Culture." In *Encyclopedia of the Sciences of Learning,* edited by N.M. Seel. Boston, MA: Springer, 2012.

Rich, Adrienne. "Compulsory Heterosexuality and Lesbian Existence." *Signs* 5, no. 4 (1980): 631–660.

Richards, Richard A. *The Species Problem: A Philosophical Analysis.* Cambridge: Cambridge University Press, 2015.

Robinson, Margaret, "Veganism and Mi'kmaq Legends." *The Canadian Journal of Native Studies* 33, no. 1 (2013): 189–196.

Rodrigues, Michelle A. "It's Time to Stop Lionizing Dian Fossey as a Conservation Hero." *Lady Science,* September 20, 2019.

Roughgarden, Joan. *Evolution's Rainbow: Diversity, Gender, and Sexuality in Nature and People.* Oakland: University of California Press, 2013.

Rovner, Joshua. "Black Disparities in Youth Incarceration." *The Sentencing Project,* 2021.

Roy, Ashwin, Roy, Ameeta, and Roy, Meera. "The Human Rights of Women with Intellectual Disability." *Journal of the Royal Society of Medicine* 105, no. 9 (2012): 384–389.

Schalk, Sami. "Critical Disability Studies as Methodology." *Lateral* 6, no. 1 (2017)

Senate of Canada. "Human Rights Committee to Study the Forced and Coerced Sterilization of People in Canada." *The Standing Senate Committee on Human Rights* (2019).

Slezak, Michael. "3 Billion Animals Killed or Displaced in Black Summer Bushfires, Study Estimates." *ABC News,* July 27, 2020. Retrieved from: https://www.abc.net.au/news/2020-07-28/3-billion-animals-killed-displaced-in-fires-wwf-study/12497976.

Spade, Dean. *Mutual Aid: Building Solidarity During This Crisis (And the Next).* New York: Verso, 2020.

Spade, Dean. "We Keep Each Other Safe: Mutual Aid for Survival and Solidarity." Presentation for Barnard Center for Research on Women (2020). Retrieved from: https://www.youtube.com/watch?v=C32KGX6qP5s&t=671s&ab_channel=BarnardCenterforResearchonWomen

Spillers, Hortense J. "Mama's Baby, Papa's Maybe: An American Grammar Book." *Diacritics* 17, no. 2 (1987): 65–81.

Stănescu, Vasile. "The Whopper Virgins: Hamburgers, Gender, and Xenophobia in Burger King's Hamburger Advertising." In *Meat Culture,* edited by Annie Potts, 90–108. Leiden: Brill, 2016.

Stănescu, Vasile. "'White Power Milk': Milk, Dietary Racism, and the 'Alt-Right'." *Animal Studies Journal* 7, no. 2 (2018): 103–128.
Starhawk, *The Spiral Dance: A Rebirth of the Ancient Religion of the Goddess.* Harper and Row, 1979.
Stretton, Tim, Kesselring, K.J., and Butler, Sara M., eds. *Married Women and the Law: Coverture in England and the Common Law World.* Montreal and Kingston: McGill-Queen's University Press, 2013.
Struthers Montford, Kelly. "Slaughterhouses." In *Environmental Justice Encyclopedia*, edited by Dorceta Taylor. Thousand Oaks: Sage Publications, forthcoming.
Stuesse, Angela, and Dollar, Nathan T. "Who are America's Meat and Poultry Workers?" *Economic Policy Institute*, 2020.
Sturgeon, Noël. "Penguin Family Values." In *Queer Ecologies: Sex, Nature, Politics, Desire,* edited by Catriona Sandilands. Bloomington: Indiana University Press, 2010.
Taylor, Chloë. "'Sex without All the Politics'? Sexual Ethics and Human-Canine Relations." In *Pets and People: The Ethics of Companion Animals*, edited by Christine Overall. Oxford: Oxford University Press, 2017.
Taylor, Sunaura. *Beasts of Burden: Animal and Disability Liberation.* New York: The New Press, 2017.
Taylor, Sunaura. "Interdependent Animals: A Feminist Disability Ethics of Care." In *Ecofeminism: Feminist Intersections with Other Animals and the Earth*, 2nd ed., edited by Lori Guen and Carol Adams. London: Bloomsbury, 2022.
Terry, Jennifer. "'Unnatural Acts' in Nature: The Scientific Fascination with Queer Animals." *GLQ: A Journal of Lesbian and Gay Studies* 6, no. 2 (2000).
Tiefer, Leonore. *Sex Is Not a Natural Act & Other Essays.* 2nd ed. London: Routledge, 2004.
U.S. Department of the Interior, U.S. Fish and Wildlife Service, and U.S. Department of Commerce. "2016 National Survey of Fishing, Hunting, and Wildlife-Associated Recreation. U.S." *Census Bureau*, 2016.
Vermilya, Jenny. *Identity, Gender, and Tracking: The Reality of Boundaries for Veterinary Students.* West Lafayette: Purdue University Press, 2022.
Warren, Karen J. "The Power and the Promise of Ecological Feminism." *Environmental Ethics* 12, no. 2 (1990): 125–146.
Whiting, Glynis (Director). *The Sterilization of Leilani Muir* (1996).
Wilkins, John S., Zachos, Frank E., and Pavlinov, Igor Ya. *Species Problems and Beyond: Contemporary Issues in Philosophy and Practice.* London: Routledge, 2022.
Xenophon. "Oeconomicus." Cited in Foucault, *The History of Sexuality*, Vol. 2, *The Use of Pleasure*, New York: Vintage, 1990.
Young, Liam. "Eating Serial: Beatrice Lindsay, Vegetarianism, and the Tactics of Everyday Life in the Late Nineteenth Century." *Societies* 5, no. 1 (2015): 65–88.

PART I

Theoretical Foundations

1
ECOFEMINISM

Greta Gaard

Feminist animal studies scholarship of the 2020s has deep roots in over two centuries of women's animal defense, women's peace and environmental movements, women's spiritualities, feminist literature and anti-war activism. By 1970, when San Diego State University opened the first Women's Studies Department in the U.S., feminism was already recognized as both a theory and a practice of justice, albeit with limitations that had long been visible to those excluded or existing only on the margins. Whether it was privileged white feminists refusing to link "their" Women's Movement with the issues facing working-class women or women of color (hence Sojourner Truth's famous 1851 speech), distancing themselves from revolutionary women organizers and writers of the Depression era; from lesbian feminists of the 1960s and 1970s; and later from vegetarian ecofeminists of the 1970s, 1980s, 1990s and beyond, mainstream feminism has long reflected the larger social hierarchies of class, race, gender identities, sexuality, ability and species—and struggled with the requisite transformations of feminist theory and praxis involved when these inclusions continue to remake feminism.

Ecofeminism Remakes Feminism[1]

A many-layered field of multispecies ethics, affects, eco-justice activisms, philosophies, politics, arts and spiritualities, ecofeminism has taken shape through rhizomatic developments using feminist lenses to identify, investigate and resist oppressive systems as well as to envision, shape and create conceptual and material systems of multi-directional relationships that promote flourishing.

Ecofeminist roots grew from what initially seemed to be single-issue, unrelated movements for peace[2] and for ecological, gender, racial and species justice, moving from strategic essentialisms of motherhood to a full range of efforts for political equality and ethical consideration—the Women's International League for Peace & Freedom (founded 1919), forest and forest communities' defense,[3] animal defense (challenging fur farms, sport and trophy hunting, factory farms, "pet" abuses)[4] and, from the 1960s and beyond, defenses of women's rights and bodily self-determination (i.e. resistance to domestic violence, rape

DOI: 10.4324/9781003273400-3

and sexual assault, gendered work and wage inequalities).[5] However, the conceptual links connecting species oppression among these views were largely untheorized[6] beyond those articulated through feminist multispecies organizations: Marti Kheel and Tina Frisco founding "Feminists for Animal Rights" (1981) in Berkeley, CA; Roberta Kalechofsky founding "Jews for Animal Rights" (1985) in New York.[7]

In the first scholarly feminist vegetarian critique of dominant animal rights theories—Peter Singer's utilitarianism and Tom Regan's deontological approach[8]—Josephine Donovan describes the many U.S. and European 19th- and 20th-century feminists who have advocated multispecies feminism through practices of both vegetarianism and feminism, and from their activism and theories, crafts a *feminist ethic of multispecies nonviolence* that

> requires a fundamental respect for nonhuman life forms, an ethic that listens to and accepts the diversity of environmental voices and the validity of their realities. It is an ethic that resists wrenching and manipulating the context so as to subdue it to one's categories; it is nonimperialistic and life-affirming.[9]

At the time, the "diversity of environmental voices" did not include plants and other vital matters (e.g. water, air, earth, minerals)—but soon it would.

Ecofeminism Emerges (1990s)

Ecofeminist anthologies of the 1980s[10] articulated a theory/practice ("praxis") intersection that had been building from previous decades of women's activism and writing across a diversity of loosely linked progressive movements. These intersections are visible though the migrating ethical frameworks explaining a highly celebrated feminist direct action: a photograph of the "web" woven by activists during the 1982 Women's Pentagon Action appears on the cover of a key text of feminist nonviolence[11]—and a year later, the Unity Statement from the Women's Pentagon Action is among the lead chapters in the first *ecofeminist* anthology.[12] The evolving interpretive frames illustrate the ways 1980s ecofeminism was international and multidisciplinary, bridging grassroots activism and research to uncover links between human and environmental health, militarism and violence against women/animals/poor people/communities of color/ecosystems, food empowerment and reforestation, Earth spiritualities and arts, maternal health, ecological economics and more.

Despite its grassroots activisms, from 1990–95, ecofeminism was reflected in only a few two-page articles in *Ms. Magazine*,[13] a vehicle of feminist popular culture, and was generally unexplored within academic feminism. Was this omission purposeful or inadvertent? As interactions with editors at feminist publications *Signs*, *NWSA Journal* and *The Women's Review of Books* suggest, there seemed to be a tacit feminist resistance to ecological feminism—one that even spilled over to *Environmental Ethics*.[14] Evidently editors at all four journals weren't aware that *Hypatia*, another leading academic feminist journal, had devoted a special Spring 1991 issue to "explicitly philosophical articles on the subject of ecological feminism"[15] with articles by Val Plumwood on ecological identity, Stephanie Lahar on ecopolitics, Deborah Slicer on multispecies ethics, Carol J. Adams on ethical vegetarianism, Deane Curtin on contextual moral vegetarianism, Patrick Murphy on ecofeminist literary criticism, Roger King on feminist environmental ethics and more.

Socialist feminism dominated the elite academic feminist journals of the 1990s, and harsh debates about ecofeminism took place between editors and potential contributors, well out of the public eye.

But among ecofeminists, the debates were just as fierce. In early April 1998, a conference titled "Ecofeminism: A Practical Environmental Philosophy for the 21st Century" was held at the University of Montana, Missoula, and drew many ecofeminist scholars. Fierce battles erupted between vegetarian/vegans[16] vs. omnivore ecofeminists with other contextualized dietary ethics, rallied by Dallas-based author Carol J. Adams, who advocated for an end to the patriarchal violence of animal exploitation in all forms (laboratory research, industrial animal food production, hunting, breeding, flesh-eating) and Australian philosopher Val Plumwood, who argued that universal vegetarianism amounted to cultural imperialism and who spoke eloquently about the traditional foodways of Australian aboriginals and the Māori of New Zealand. In her keynote address, Plumwood argued that

> the main forms of "eco"-feminist animal theory remain inadequately contextualized and poorly integrated with ecological consciousness, and the universalizing, ontological form of veganism which has come to the fore in Carol J. Adams's theory seems to have as much potential as the mainstream version for development in directions that express alienation from ecological embodiment and ecological communities. The leading thesis of what I shall call "ontological vegetarianism" is that nothing morally considerable should ever be ontologized as edible or as available for use.[17]

Curiously, the page Plumwood references (200) is from Adams' essay on "The Feminist Traffic in Animals," which explores the question, "Should *feminists* be vegetarians?"—particularly at feminist conferences such as the National Women's Studies Association (NWSA) (Adams 1993, 210, 214)—and does not state directly the material that Plumwood infers. In effect, Plumwood is quoting out of context—understandably so, since neither Plumwood nor Adams attended NWSA.

At the 1990 National Women's Studies Association's first Ecofeminist Task Force Meeting (the year after the group's well-attended formation in 1989), four members wrote up the Ecofeminist Task Force Recommendation—two representatives from Feminists for Animal Rights, founder Marti Kheel (who dictated the majority of the facts from her encyclopedic memory) and wisewoman Batya Bauman (who filled in when Marti faltered); environmental educator Stephanie Lahar; and Greta Gaard (who took notes and typed up the document to submit to the NWSA post-conference administrative meeting). Only one step more radical than Frances Moore Lappé's *Diet for a Small Planet*—whose four-fold vegetarian advocacy appealed to promoting human and ecological health and ending both world hunger and the animal suffering involved in food production—the Task Force proposal invited NWSA to connect feminist theory with feminist practices of nonviolence and ecology via dietary choices.[18] Our two-page list of "whereas" claims referenced the harms of first-world animal-based food practices (i.e. clearcutting forests in the South, replacing them with cattle-grazing in the Amazon and industrial animal agriculture in the North) and their deleterious health effects on indigenous communities as well as Northern consumers, who eat these animals and their "feminized protein" (eggs, dairy). Just as the Association for the Advancement of Sustainability in Higher Education (AASHE) would do almost 30 years later (albeit in regard to sustainability alone),[19] the Ecofeminist Task Force hoped to inspire

NWSA conference-goers to make connections between feminist theory, ecological and multispecies ethics and food choices: no prohibitions were set beyond conference meals and NWSA conference-goers, and no awareness about culturally diverse diets was expressed. In 1990, NWSA membership—and leadership—was largely white.[20]

As an antiracist and indigenous ally, Plumwood was especially critical because ontological vegetarianism "creates major conflicts with contemporary feminist consciousness, as well as with feminist anthropology and women's experience in indigenous cultures" (287). Instead, Plumwood proposed "a framework of reciprocity," that would envision

> an ecological order which is itself potentially an ethical order with its own values and standards of sharing, generosity, and radical equality between species, and with its own stringent obligations to recognize the other as equally positioned, as potentially food and always more than food.
>
> *(2000, 318)*

At the time of the 1998 Missoula conference, Plumwood had already experienced and described her encounter with an ecological other, a crocodile who was eager to take her up on that "reciprocity": in short, Plumwood had already aligned her life and food practices with her "critical feminist eco-socialist" (or "critical ecofeminist") multispecies ethics.[21]

Multispecies Ecofeminisms in the New Millennium

By the year 2010, ecofeminisms had developed through a variety of standpoints, ranging from eco-anarchist animal studies and feminist animal studies[22] to ecofeminist philosophies,[23] eco-political ecofeminisms[24] and eco-queer sexualities[25]—and these branches continued to diversify across geopolitical locations. Explorations of coalitions and shared values had also developed across Northern ecofeminisms and indigenous/third world ecofeminisms.[26]

Of great importance, ecofeminists took seriously the critiques challenging the Eurocentric origins and thus the limitations of ecofeminist theorizing at the same time as broader intersections of feminist standpoints became voiced and visible. In North America, multispecies feminisms of color emerged in print through popular texts such as A. Breeze Harper's anthology *Sistah Vegan: Black Female Vegans Speak on Food, Identity, Health, and Society* (2010) and later Aph and Syl Ko's *Aphro-ism: Essays on Pop Culture, Feminism, and Black Veganism from Two Sisters* (2020).[27] More academic texts such as Claire Jean Kim and Carla Freccero's co-edited *American Quarterly* issue on "Species/Race/Sex" (2013) and Kim's book *Dangerous Crossings: Race, Species, and Nature in a Multicultural Age* (2015) make strong contributions to intersectional ecofeminisms, as do Maneesha Deckha's articles on animal personhood, dairy and decolonization.[28] Dividing her work between the U.S. and Taiwan, Chai-ju Chang utilizes ecofeminist perspectives to explore Taiwanese women's care-work for stray dogs.[29] Longtime activist from Greenham Common Women's Peace Camp Gwyn Kirk and co-editor K. Melchor Quick Hall have produced an ecological feminist anthology centering women of color.[30] Laura Wright has gathered diverse perspectives on vegan practices and logics across global cultures, arguing that "an ecofeminist approach to veganism" offers a strongly inclusive politics, and two different editorial teams have each compiled volumes exploring intersections among animal law, violence and mass incarcerations.[31] Expanding ecofeminist perspectives from animal to plant species, Cate Sandilands' work offers strong foundations for both queer ecologies and plant studies.[32]

Part of ecofeminism's rhizomatic spread, other multidisciplinary developments have also utilized ecofeminist insights to launch new explorations: that is, not only queer ecologies but also environmental humanities and postcolonial ecocriticism, plant studies, elemental ecocriticism,[33] human–animal studies and critical animal studies—the latter acknowledging inspiration from ecofeminism in making clear "how the material and symbolic exploitation of animals intersects with and helps maintain dominant categories of gender, 'race' and class."[34] In the 2020s—described by Christiana Figueres as the final and most "decisive decade for the future of humanity and the planet" for averting the most dire climate change outcomes—ecofeminists and other multispecies climate justice activists will do well to ramp up our skills in several areas.[35]

1. Toward More Affect-ive Activism

In each era, ecofeminisms interact with other eco-justice and ecocritical perspectives, and although feminist multispecies care, grief, love and friendship[36] have been a primary emphasis long predating "affect studies,"[37] care is not one of Silvan Tompkins' original nine affects. Instead, the field-defining collection, *The Affect Theory Reader* (2010), explores eco-affects such as eco-anxiety, climate grief, land affect, empathy, solastalgia (and solastalgic distress) and melancholy, affects that abound among ecocritics:[38] notably, Sarah J. Ray has created a handbook for helping students address affects such as climate grief, mourning, nostalgia and solastalgia.[39] Perhaps to offset the array of dismal climate affects, ecocritic Nicole Seymour has explored not only empathy but also the more queer affects of "irony, absurdity, perversity, and camp" as well as "irreverence . . . playfulness . . . frivolity, indecorum, ambivalence, and glee," arguing that "being earnest" makes environmentalists "killjoy sticks in the mud."[40]

In terms of eco-affects among multispecies activists, most narratives report that encountering animal abuse initially elicits affective responses for the animals themselves (compassion, empathy, grief, horror); these affects soon inform actions to alleviate animal suffering while breaking silences about its brutality and its multiple harms to animals, workers, indigenous communities and ecosystems.[41] Multispecies activists hold perpetrators accountable by documenting and publicizing the violence, organizing boycotts and filing lawsuits. Women constitute from 68 to 80 percent of animal rescue workers and activists, crediting the power of their "empathetic action" to their "shared experiences of oppression."[42] From Mercy for Animals to Plant-Based News, animal activist groups have highlighted the contributions and leadership of queer and transgender animal activists,[43] and the links between animal abuse and queer-bashing, domestic violence, gun violence and terrorism, psychological disorders and toxic masculinity have been repeatedly documented.[44] In these allied struggles for justice, what affects are powerful enough to compel institutional and cultural progress toward multispecies justice?

Across the disciplines, it seems contemporary affect studies has forgotten Larry Kramer's rallying cry mobilizing AIDS activism: "Where is your rage?" In the 1980s, both queer and feminist communities used the affect of anger to propel movements for ecopolitical and sexual justice.

Two feminist anthologies just a decade apart portray very different views of feminist rage. *Fight Back! Feminist Resistance to Male Violence* (1981) is articulated through the lens of radical lesbian feminism and lesbian separatism, focusing on male sexual assaults (both within families and between strangers), woman-battering, the anti-rape movement, women in prison, women's self-defense and broader feminist interventions in public media (e.g. billboards, posters, magazines) and public institutions (e.g. maternity and childbirth practices, legal challenges, government, religion). Preceding and contributing to the three sections on the Women's Pentagon Action, the WPA Unity Statement and the activists' trial is the description of a feminist street performance, "In Mourning and in Rage." The performers dramatized actions of the Los Angeles Hillside Strangler and the mass media sensationalism, mourned the violence of male sexual/physical/economic/psychological terrorism and then exchanged their black mourning coats for red cloaks of rage. Together, performers and observers joined the chorus, "In memory of our sisters, WE FIGHT BACK!" The ritual ended with "city council members voic[ing] support to the Rape Hotline Alliance, pledging to start self-defense classes for women in the city" and Holly Near singing "Fight Back!" while participants and observers began a circle dance.

A decade after *Fight Back!,* feminists continued resisting male dominance, albeit through another channel of self-determination: *Angry Women* (1991)[45] profiled 15 feminist women artists and educators for whom anger at dominant heteropatriarchal culture became an activating, energizing, politically motivating, artistic and erotic affect. Notably, this volume's every interview addresses the shift from lesbian feminist separatism and its political/sexual/dietary/sartorial correctness toward a different feminist emphasis on artistic and erotic freedom and self-determination.[46] "What do we *as women* do with our rage?" asks bell hooks (80). As Audre Lorde observed, "anger is loaded with information and energy. . . . Focused with precision it can become a powerful source of energy serving progress and change."[47]

In 2018, four books exploring feminist anger gave voice to that energy.[48] Two of these volumes focused on the personal and interpersonal level: Brittney Cooper's *Eloquent Rage: A Black Feminist Discovers Her Superpower* describes the author's coming-into-feminist-awareness through life experiences of sexism, racism and the disappointing race-loyalty of white women, and claims her right to rage. Lilly Dancyger's *Burn It Down: Women Writing about Anger* offers 22 short accounts of anger sparked by experiences with family, social institutions and male violence. Linking the personal, institutional and political, another two volumes analyze feminist rage-inspiring moments and movements from the past century to the present. While Rebecca Traister's *Good and Mad: The Revolutionary Power of Women's Anger* focuses on feminist rage used in the public sphere to challenge injustice, Soraya Chemaly's *Rage Becomes Her: The Power of Women's Anger* looks more closely at the grassroots level and persistently brings an intersectional approach to her inquiry. Both books shuttle between "the personal and the political," linking gender roles with women's psychological subordination, reinforced via cultural, economic and political institutions. Both studies offer breathtaking and detailed documentation—especially for women who have lived many of the years these authors describe and who have survived the nexus of institutional and interpersonal belittling, exploitation and assault with strategies that deny or repress women's anger because verbally, psychologically, physically, economically, culturally and politically powerful men threaten their lives, bodies, safety, children, homes, employment. And these books document women's anger, mobilized across diverse communities and fighting for justice.

What these intersectional studies *don't do* is open the lens of women's anger to include issues of intersectional multispecies violence—domestic violence where batterers regularly enforce dominance by threatening to abuse companion animals and eventually maim or kill the family "pet," the long history and present reality of police using "attack" dogs against indigenous and Black Americans or the associations between animal abuse and youth violence.[49] Where are the uniquely *ecofeminist* affects and manifestations of rage—experiencing, naming, denouncing and channeling our rage at multispecies micro- and macro-climate injustices produced by elites' overconsumption of the planet? Well-informed rage against climate change inaction has been appearing in the popular press for over a decade: *YES! Magazine*, *Popula*, *Grist*, *The Atlantic*, the *New York Times*, and other progressive media have all run articles about climate rage—but the rage animal activists feel against those profiting from massive suffering caused by the animal industrial complex is almost unsearchable. Instead, articles like "Examining Extremism: Violent Animal Rights Extremists" or "Childhood Animal Abuse and Violent Criminal Behavior" surface.

These issues—and this silence—are of high relevance for ecofeminist animal activists and scholars.[50]

Consider the work of younger climate activists—Anjali Appadurai telling the United Nations in 2011, "You've been negotiating all my life. Get it done!" or Greta Thunberg in 2019 reminding world leaders about a litany of ecological injustices, frequently punctuating these injustices by exclaiming, at the fiery edge of rage and anguish, "How dare you!"[51] High school senior Emma Gonzalez's February 2018 speech to the National Rifle Association and Trump after the shooting at Marjory Stoneman Douglas High School in Florida, flatly challenged their indifference to gun crimes with her refrain, "We call BS!" and Alexandria Ocasio-Cortez called out the climate of sexism and racism that made Rep. Yoho feel free to verbally assault her:

> this issue is not about one incident. It is cultural. It is a culture of lack of impunity, of accepting of violence and violent language against women, and an entire structure of power that supports that.[52]

While AOC's speech has been widely shared and viewed, inspiring many middle-of-the-road voters as well as younger feminists, this welcome reception is not the norm. Those who benefit from and seek to maintain a status quo of gender, race or species inequality often belittle raw feminist rage as an affect stigmatizing the speaker and empty of content (i.e., "angry feminist," "angry black woman"); thus, feminists have often used art as a communication channel.

Feminist street theater from Raging Grannies to the Sisters of Perpetual Indulgence; Extinction Rebellion; and the Women's March of January 21, 2017, have all combined anger, incisive political critique and humor—"love and rage," as anarchists urge. From the 1960s to the present, the Guerilla Girls have challenged racism and sexism using the guise of an endangered species, while artists like Ana Mendieta have created performance art using their own bodies and earth arts to convey their critique, stated more directly in graphic art by Barbara Kruger's "We Won't Play Nature to Your Culture" (1983).

More recently, Annie Sprinkle and Beth Stephens have mobilized queer ecological feminist affects of camp, drag, visual and performative criticisms through their activisms, recorded in their eco-documentary: "Goodbye Gauley Mountain" (2013), which critiques mountaintop removal for its intersectional harms to workers, poor and rural communities, multispecies inhabitants and environments. Stephens and Sprinkle have also helped communities perform their own polyamorous ecosexual wedding ceremonies, activating spousal activisms by residents who marry the multispecies inhabitants of place, promising to "love, honor, and cherish the Earth until death brings us closer together."[53] Across the globe from San Francisco, Kunsthall Trondheim in Norway hosted a "Sex Ecologies" exhibit and accompanying book publication, curated by Stefanie Hessler and Katja Aglert. Through eco-arts, essays, photos and histories, their volume offers an array of contemporary queer feminist and multispecies perspectives.[54] Admittedly, these arts and actions have occurred largely in the global North.

2. Contextualizing Ecofeminisms North and South

Keeping ecofeminism and feminist animal studies attuned to intersectional contexts requires being mindful of ecofeminisms across cultures and nations.

Witnessing animal suffering at auction blocks and slaughterhouses in Enumclaw, Washington and Jaipur, India, Kathryn Gillespie and Yamini Narayan collaborate in documenting global violence, offering their witnessing of animal suffering—a "brief moment of comfort between the calf and the pigs; the anxiety of a *mother*, whether a donkey or a cow; the loneliness of a horse"—as a conduit for amplifying the realities of capital-induced multispecies violence.[55] Documentaries such as Andrea Arnold's "Cow" (2021), Ali Tabrizi's "Seaspiracy" (2021), and Kip Anderson's "Cowspiracy" (2014) and Anderson with Kameron Waters "Christspiracy" (2024) all communicate both the data of animal suffering, abuse, violation and deaths while also creating an affective resonance for viewers who, as embodied animals ourselves, may writhe as we watch other animalbodies suffer abuse, confinement, family separations, terror and the slaughter of close companions which precedes their own deaths. What these documentaries share is the goal of exposing a local and global "extraction" of life itself from the world's multispecies majority to the world's human elites. This transfer appropriates free movement, companionship, health and life from species, environments and diverse human communities—and uses these as resources to produce capital. As Chris Hayes has argued, the fossil fuel divestment movement has economic precedent in the abolitionists' effort to compel Southern slaveowners to relinquish their profits garnered through human slavery—a sum Hayes translates from 1860 to 2014 economic values as $10 trillion.[56] Add that number to the wealth animal industries accumulate through federal subsidies for cattle grazing and "meat" production; through employing low-waged and undocumented slaughterhouse and meatpacking workers; through water overuses and eutrophication, soil pollution and GHG emissions, zoonotic viruses and *E. coli* infections that affect vulnerable children and elderly most directly—all producing "meat" for elite consumers while creating food scarcity and malnutrition for the world's majority. The extraction of global wealth from the lives of people of color, indigenous communities, ecosystems, climates and other-than-human animals ontologized as "food" functions as a *global transfer of life and wealth* into the hands of economic elites. Eco-justice and multispecies activists seeking to end this extractive transfer are forced to confront not only the violently weaponized forces protecting that wealth, but also the more widespread sense of human entitlement.

Because anger needs an object[57]—and because it's easier to be angry at an individual than at the entanglement of ecocidal carnist culture, the animal industrial complex, colonialism and elite indifference—multispecies allies tend toward affects of grief, mourning, melancholy and occasional joy with individual others. What's most utilized for animal and ecological activism (though not sufficiently theorized) is determination, an approach-related positive affect that energizes the movement for the purposes of alleviating suffering. And this affect is in use fighting oppressions globally.

In Ecuador, insights from ecofeminism have been useful in articulating indigenous women's determined resistance movement against extraction.[58] Both Acción Ecológica (Ecological Action) and Miradas Críticas del Territorio desde el Feminismo (Feminist Critical Views of Territory) have built ties with ecofeminist and environmental justice organizations such as the Women's Earth and Climate Action Network (WECAN) and Amazon Watch, inviting supporters to join "our struggle for the defense of our territories, the Pacha Mama [Mother Earth], since this is not only for us but for the rest of the world. . . . Amazonian women will defend the whole world from climate change."[59] Ivonne Yánez of Acción Ecológica uses an ecofeminist analytical framework to describe the ways that capitalism's operations extract wealth from both women and nature:

> just as women have been subjugated by capital, today nature has also been pressed into servitude to permit seemingly endless accumulation. Just as women supposedly know and should "love" and must therefore *obligatorily* take responsibility for the care of the home, the elderly, the sick and children (without pay, because it is not a job, it is something inherent to them), Mother Nature provides us with environmental "services" because she "has always done it" and it is inherent to her. Neither of these suppositions is true. On one hand, the love a person may feel cannot be used as a justification to turn her into a domestic slave or a breeder of human beings. Similarly, the gifts of nature cannot be used as a source of "services", formerly referred to more commonly as "natural capital."[60]

Readers may recall the 1980s when human rights violations across Central and South America prompted many environmentalist and religious activists, college students and human rights advocates to travel to Central America and live with indigenous communities, putting their first-world privilege in the service of indigenous needs and survival. Many activists survived, but some were "disappeared," as their resistance hampered profits for multinational agribusiness corporations, whose primary products have been bananas, coffee, cotton, sugar, corn, sorghum, rice, peanuts and palm oil.[61] By 2017, Honduras had become the most dangerous country in the world for environmentalists—though many other countries in South America and Africa have experienced assassinations of eco-activists.[62] European colonization has continued in the 20th and 21st centuries through agribusiness plantations for cattle-grazing (or for growing soybeans as their feed) and the ongoing assaults on the lives of indigenous humans and other animals, as well as their subsistence economies, extracting wealth for first-world elites.

Writing from Honduras, Irune del Rio Gabiola chronicles the indigenous Lencas' experiences of colonial disruption, industrialization and criminalization, as well as the leadership of Berta Cáceres and her contributions to Latin American ecofeminisms. With mandates from her community, Berta Cáceres filed complaints against plans for building the Agua

Zarca Hydroelectric Dam on the Gualcarque River, citing not only its spiritual significance but also its provisioning of water, fish and medicine to local human communities and sustenance to the surrounding environments. After Berta's assassination on March 2, 2016, indigenous activists developed the rallying cry, "Berta no ha muerto, Berta se ha multiplicado!" (Berta did not die, Berta has multiplied herself). Through the indigenous social movement Civic Council of Popular and Indigenous Organizations of Honduras (COPINH), activists have worked to halt the privatization of rivers and land where they live and maintain their subsistence: their actions utilize the affects of outrage, resistance, transnational solidarity, care, mourning and hope.[63]

Berta's criticisms of the Aqua Zarca Dam resonate across the Atlantic to Spain, where Alicia Puleo's theory of "critical ecofeminism" offers analysis of "the paradox of hedonism"—a state of dissatisfaction arising when a person seeks happiness through pleasure-consuming activities that fail to transcend selfish motives rather than a more sustainable happiness that arises through reciprocal "relationships of care, affection and protection" both within and across species. Puleo describes these problems and articulates her concern for justice across species, class, nation and planet:

> Millions of people have been trapped in the hedonistic paradox while many others have lacked the necessary resources to survive in countries which have been impoverished by systematic plundering. And if millions of human beings, who in theory have been recognized as bearers of rights, find themselves at the limit of subsistence, what is to be said of the woodland creatures which are literally being wiped out by hunting, herbicides, pesticides, and alterations due to climate change, in other words, the disappearance of the world? Is this the Earth we want?[64]

Placing ecofeminisms and feminist multispecies ethics in dialogue across diverse and global contexts magnifies the potential for more accurate, affectively sustaining, and context-responsive ecofeminisms.

3. *Ecofeminist Multispecies Spiritualities: Beyond the 1970s*

Aware that for millennia, western spiritualities have perpetuated as "divinely ordained" the cultural frameworks legitimating hierarchy, hetero/sexism, racism, classism, speciesism and the plunder of the earth, second-wave feminisms produced a wealth of feminist critiques interrogating Judeo-Christian religions and spiritualities. Their research exposed the anti-ecological synthesis of patriarchy and divinity that located a male-god in the skies and portrayed human bodies, the earth, other species and the depths under our feet as portals to hell.[65] Notably, Marti Kheel and Sally Gearhardt emphasized the speciesism of most patriarchal religious traditions, which have used animal "sacrifice" to expiate human crimes.[66] Nonetheless, Ecofeminist versions of Christianity have informed the work of Ivone Gebara, a Brasilian ecofeminist theologian, as well as shaping the journal and workshops offered by Chile's Con-spirando Collective.[67] In the United States, Melanie L. Harris' EcoWomanist spirituality is shaped from the nexus of African-American spiritualities, womanism and ecotheology.[68] Other woman-centered approaches have attempted to recuperate earth-based spiritualities, abandoning western spiritualities entirely.[69]

Shifting their focus from western to eastern spiritualities, Buddhist feminists have explored intersections of their politics and spiritualities, organizing several Women in Buddhism conferences in the 1980s and 1990s.[70] Environmental scholar-activists Stephanie Kaza and Zen teacher David Loy have exposed the ways that capitalist Euro-western cultures foster insatiable greed, hatred and delusion, promising happiness at the next mouse-click or purchase but delivering only temporary satisfaction and thereby feeding the fires of craving that are overconsuming the planet.[71] Curiously, however, Buddhists in the U.S. have not all agreed on the connections between dietary practices and the Buddhist precepts of non-harming, non-stealing and not-killing, an incongruity that Dharma Voices for Animals has worked tirelessly to illuminate.[72] Intent on critiquing patriarchal religions, few feminists have investigated whether any world religions reflect ecofeminist values, envisioning not only multi-gendered and multi-species deities but also revering the lives of other-than-human animal species. While Hinduism offers a system with multi-gendered deities and vegetarian spiritual practices, it too is not free from patriarchal beliefs and practices.[73] On these issues, a dialogue on "the future of the planet" between feminists Rita Gross, a Buddhist scholar, and Christian theologian Rosemary Radford Ruether offers intriguing information and a strategy for exploring diverse spiritualities on the basis of their inherent support for multispecies feminisms.[74]

What motivates feminist interest in salvaging religions? Vegan ecofeminist Alka Arora distinguishes between religious views that reinforce internalized and structural oppressions and women's "spiritual *experiences*, which often serve as a counterhegemonic force, enabling them to resist patriarchal conditioning while strengthening their sense of agency, meaning, and connection."[75] Nonetheless, highly regarded feminists of color such as bell hooks, Gloria Anzaldúa and Alice Walker who have linked spirituality, feminism and social transformation to convey eco-justice spiritualities have seen their views filtered through mainstream feminist commentary that secularizes their work.

Ecofeminists and feminist multispecies activists interested in building multiracial coalitions by supporting indigenous struggles defending water and life, or meditating with diverse Buddhist or Hindu practitioners, have noticed the "partial connection" Andrea Sempértegui describes, offering spaces for allyship and for difference. Spiritualities that combine elements of ecofeminisms—that is, an ecological practice that eschews consumerism and extractionism; an ethic and practice that uncovers connections across feminism, anti-racism, anti-speciesism and anti-colonialism; a sense of multispecies inter-identity that involves the more-than-human world—have the potential to "challenge patriarchy at its core."[76]

Ecofeminisms Beyond 2022

In September 2022, the anti-hijab protests in Iran have set a fiery example of Muslim women's rage against cultural and religious mandates that restrict their lives and require them to live as second-class citizens—yet many of these activists protest *within* their religion. May their resistance inspire all feminist activists to tap into our own empathy, grief and rage at the gender, race and species of planetary oppressions, and use that energized determination to power global solidarity efforts toward multispecies flourishing and inclusive climate justice.[77]

Notes

1 This claim was initially formulated as "Ecofeminism Reconceives Feminism," in Karen Warren, "The Power and Promise of Ecological Feminism," *Environmental Ethics* 12 (Summer 1990), 125–146.
2 See Alice Cook and Gwyn Kirk, *Greenham Women Everywhere: Dreams, Ideas and Actions from the Women's Peace Movement* (London: Pluto Press Ltd., 1983); Linda Etchart, "Demilitarizing the Global: Women's Peace Movements and Transnational Networks," in *The Oxford Handbook of Transnational Feminist Movements*, eds. Rawwida Baksh and Wendy Harcourt (Oxford University Press, 2015), 702–722.
3 See Bina Agarwal, "The Gender and Environment Debate: Lessons from India," *Feminist Studies* 18, no. 1 (1992): 119–158; Judi Bari, *Timber Wars* (Monroe, ME: Common Courage Press, 1994); Ronald B. Hatch, ed., *Clayoquot & Dissent* (Vancouver, Canada: Ronsdale Press, 1994); Julia Butterfly Hill, *The Legacy of Luna* (San Francisco: Harper Books, 2000); Wangari Maathai, *The Green Belt Movement* (New York: Lantern Books, 2004); Niamh Moore, *The Changing Nature of Eco/Feminism: Telling Stories from Clayoquot Sound* (Vancouver: UBC Press, 2015). See also Aubrey Wallace, ed., *Eco-Heroes: Twelve Tales of Environmental Victory*, for narratives from Wangari Maathai (Green Belt, Kenya), Catherine Wallace (Antarctica), Colleen McCrory, Lois Gibbs (toxic waste, race & class, NY), and Medha Patkar (Narmada dams resistance, India).
4 See Feminists for Animal Rights website at http://www.farinc.org/, and Josephine Donovan, "Animal Rights and Feminist Theory," *Signs* 15, no. 2 (1990): 350–375.
5 See Frédérique Delacoste and Felice Newman, eds., *FIGHT BACK! Feminist Resistance to Male Violence* (Minneapolis, MN: Cleis Press, 1981).
6 A notable exception—published in the same period (early 1980s) as feminist eco-justice activisms—is Aviva Cantor's essay, "The Club, the Yoke, and the Leash: What We Can Learn from the Way a Culture Treats Animals," *Ms. Magazine*, August 1983, 27–29.
7 On Roberta Kalechofsky's work, see https://www.lib.ncsu.edu/findingaids/mc00207.
8 See Peter Singer, *Animal Liberation: A New Ethics for Our Treatment of Animals* (San Francisco, CA: HarperCollins, 1975); Tom Regan, *The Case for Animal Rights* (Berkeley, CA: University of California Press, 1983).
9 Donovan, "Animal Rights and Feminist Theory," 374.
10 See Léonie Caldecott and Stephanie Leland, eds., *Reclaim the Earth: Women Speak Out for Life on Earth* (London: The Women's Press, 1983); Judith Plant, ed., *Healing the Wounds: The Promise of Ecofeminism* (Philadelphia, PA: New Society Publishers, 1989); Irene Diamond and Gloria Feman Orenstein, eds., *Reweaving the World: The Emergence of Ecofeminism* (San Francisco: Sierra Club Books, 1990).
11 Pam McAllister, ed., *Reweaving the Web of Life: Feminism and Nonviolence* (Philadelphia, PA: New Society Publishers, 1982).
12 Caldecott and Leland, eds., *Reclaim the Earth.*
13 The 2023 compilation, *50 Years of Ms.: The Best of the Pathfinding Magazine That Ignited a Revolution*, does not include the columns on Ecofeminism that appeared in the 1990s, yet offers selected poems and articles linking women, ecology, and (sometimes) species, from authors such as Patricia Trujillo to Mary Oliver. At https://msmagazine.com/tag/ecofeminism/page/6/, at least 55 articles archived in Ms. did not make it into "50 Years of Ms." compilation.
14 These repeatedly thwarted efforts to publish ecofeminist perspectives in the journals of elite feminism and leading-edge environmental ethics are detailed in Greta Gaard, "Ecofeminism Revisited: Rejecting Essentialism and Replacing Species in a Material Feminist Environmentalism," *Feminist Formations* 23, no. 2 (Summer 2011): 26–53.
15 See Karen J. Warren, ed., "Special Issue: Ecological Feminism," *Hypatia* 6, no. 1 (Spring 1991), viii.
16 At the time, distinctions between vegan and vegetarian were not yet foregrounded, and the term most commonly in use was "vegetarian," although it was used inclusively to refer to both practices. By the time that "second gen" ecofeminist Laura Wright published her volume, *The Vegan Studies Project: Food, Animals, and Gender in the Age of Terror* (Athens, GA: University of Georgia Press, 2015), veganism had become the *de facto* ethical perspective within (and often beyond) ecofeminisms.

17 Plumwood's keynote was later revised and published as "Integrating Ethical Frameworks for Animals, Humans, and Nature: A Critical Feminist Eco-Socialist Analysis," *Ethics and the Environment* 5, no. 2 (2000): 285–322. The excerpt quoted appears in Plumwood (2000): 287, and references Adams' chapter "The Feminist Traffic in Animals," in *Ecofeminism: Women, Animals, Nature*, ed. Greta Gaard (Philadelphia, PA: Temple University Press, 1993), 195–218.

18 To contextualize the proposal for a new generation of feminist readers, one must note that vegetarian cookbooks of the 1980s were almost too numerous to mention: authors such as Mollie Katzen, the Moosewood Collective, Madhur Jaffrey, Martha Rose Schulman, Deborah Madison and more were common favorites, advocating a vegetarian diet for the four commonly known reasons. The Ecofeminist Task Force simply advanced an idea that was already accepted in progressive communities and proposed linking the idea with feminism.

19 See https://www.aashe.org/news/aashe2017-vegetarian-menu/.

20 1991 was also the year that Ruby Sales' grossly unjust treatment in the offices of NWSA rocked the conference: at the administrative meeting immediately following the conference sessions, administrators followed Robert's Rules of Order and refused a hearing of Sales' concerns (and of the Ecofeminist Task Force petition). Anticipating the worst, the Women of Color Caucus and the Lesbian Caucus had held a meeting during the conference and agreed to walk out of the organization if Sales' petition was ignored. In short, we did, and although the Ecofeminist Task Force petition was never heard, our collective action transformed the leadership of NWSA. In retrospect, it's evident that establishment feminism wanted to give no time to considering institutionalized racism and speciesism—much less their nexus.

21 See Val Plumwood, "Human Vulnerability and the Experience of Being Prey," *Quadrant* 39, no. 3 (March 1995): 29–34; *Feminism and the Mastery of Nature* (New York: Routledge, 1993), 28.

22 Barbara Noske's *Beyond Boundaries: Humans and Animals* (Montreal, CA: Black Rose Books, 1997) defies categorization and has been undeservedly overlooked in the recitations of feminist groundwork for multispecies ethics: Noske's analysis is eco-cultural, multispecies, anticapitalist, and merits reading in 2024 and beyond. For feminist animal studies, see Carol J. Adams, *The Sexual Politics of Meat: A Feminist-Vegetarian Critical Theory* (New York: Continuum Publishing, 1990); Carol J. Adams and Josephine Donovan, eds., *Animals and Women: Feminist Theoretical Explorations* (Durham, NC: Duke University Press, 1995); Josephine Donovan and Carol J. Adams, eds., *The Feminist Care Tradition in Animal Ethics* (New York: Columbia University Press, 2007); Marti Kheel, *Nature Ethics: An Ecofeminist Perspective* (Lanham, MD: Rowman & Littlefield, 2008).

23 See Chris J. Cuomo, *Feminism and Ecological Communities: An Ethic of Flourishing* (New York: Routledge, 1998); Freya Mathews, *The Ecological Self* (New York: Routledge, 1991), *For Love of Matter: A Contemporary Panpsychism* (Albany, NY: SUNY Press, 2003), and *Reinhabiting Reality: Towards a Recovery of Culture* (Albany: SUNY Press, 2005); Val Plumwood, *Feminism and the Mastery of Nature* (London: Routledge, 1993) and *Environmental Culture: The Ecological Crisis of Reason* (London: Routledge, 2002); Karen J. Warren, ed., *Ecological Feminism* (New York: Routledge, 1994) and *Ecofeminism: Women, Culture, Nature* (Bloomington, IN: Indiana University Press, 1997); Karen J. Warren, *Ecofeminist Philosophy: A Western Perspective on What It Is and Why It Matters* (Lanham, MD: Rowman & Littlefield, 2000).

24 See Greta Gaard, ed., *Ecofeminism: Women, Animals, Nature* (Philadelphia, PA: Temple University Press, 1993); Greta Gaard, *Ecological Politics: Ecofeminists and the Greens* (Philadelphia, PA: Temple University Press, 1998); Chaia Heller, *Ecology of Everyday Life: Rethinking the Desire for Nature* (Montreal, CA: Black Rose Books, 1999); Mary Mellor, *Feminism and Ecology* (Cambridge, UK: Polity Press, 1997); Ariel Salleh, *Ecofeminism As Politics: Nature, Marx and the Postmodern* (London: Zed Books, 1997); Catriona Sandilands, *The Good-Natured Feminist: Ecofeminism and the Quest for Democracy* (Minneapolis: University of Minnesota Press, 1999); Noël Sturgeon, *Ecofeminist Natures: Race, Gender, Feminist Theory and Political Action* (New York: Routledge, 1997) and *Environmentalism in Popular Culture: Gender, Race, Sexuality, and the Politics of the Natural* (Tucson, AZ: University of Arizona Press, 2009).

25 See Greta Gaard, "Toward a Queer Ecofeminism," *Hypatia* 12, no. 1 (1997): 114–137, and the many articles and books by Cate Sandilands listed at https://euc.yorku.ca/faculty/catriona-a-h-sandilands/ and on Academia.edu.

26 See Maria Mies and Vandana Shiva, *Ecofeminism (*London: Zed Books, 1993); Deborah Bird Rose, *Dingo Makes Us Human: Life and Land in an Australian Aboriginal Culture* (Cambridge University Press, 2000); Vandana Shiva, *Staying Alive: Women, Ecology and Development* (London: Zed Books, 1988).

27 A. Breeze Harper, ed., *Sistah Vegan: Black Female Vegans Speak on Food, Identity, Health, and Society* (New York: Lantern Books, 2010); Aph and Syl Ko, *Aphro-ism: Essays on Pop Culture, Feminism, and Black Veganism from Two Sisters* (New York: Lantern Books, 2020).

28 Claire Jean Kim and Carla Freccero, eds. "Species/Race/Sex," *American Quarterly* 65, no. 3 (September 2013); Claire Jean Kim, *Dangerous Crossings: Race, Species, and Nature in a Multicultural Age* (New York: Cambridge University Press, 2015); Maneesha Decka, "Something to Celebrate? De-Listing Dairy in Canada's *National Food Guide*," *Journal of Food, Law, and Policy* 16, no. 1 (2020): 11–46; "Veganism, Dairy, and Decolonization," *Journal of Human Rights and the Environment* 11, no. 2 (2020): 44–67; "Unsettling Anthropocentric Legal Systems: Reconciliation, Indigenous Laws, and Animal Personhood," *Journal of Intercultural Studies* 41, no. 1 (2020): 77–97.

29 Chia-ju Chang, "Women and Interspecies Care: Dog Mothers in Taiwan," in *International Perspectives in Feminist Ecocriticism*, eds. Greta Gaard, Simon Estok, and Serpil Oppermann (New York: Routledge, 2013), 151–165.

30 K. Melchor Quick Hall and Gwyn Kirk, eds., *Mapping Gendered Ecologies: Engaging with and beyond Ecowomanism and Ecofeminism* (Lanham MD: Lexington Books, 2021).

31 Laura Wright, ed., *The Routledge Handbook of Vegan Studies* (New York: Routledge, 2021); Kelly Struthers Montford and Chloe Taylor, eds., *Building Abolition: Decarceration and Social Justice* (Routledge, 2022); Lori Gruen and Justin Marceau, eds., *Carceral Logics: Human Incarceration and Animal Captivity* (Cambridge University Press, 2022).

32 See Catriona Mortimer-Sandilands and Bruce Erickson, eds., *Queer Ecologies: Sex, Nature, Politics, Desire* (Bloomington, IN: Indiana University Press, 2010) and Sandilands' articles ranging from "Desiring Nature, Queering Ethics: Adventures in Erotogenic Environments," *Environmental Ethics* 23, no. 2 (2001): 169–188, to "Fear of a Queer Plant," *GLQ: Gay and Lesbian Quarterly* 23, no. 3 (2017): 419–429; "Here We Go Round the Mulberry Bush: A Queer Botanical Meander," *Center for Sustainable Practice and the Arts Quarterly* 19 (2018): 28–33, and beyond.

33 Graham Huggan and Helen Tiffin, *Postcolonial Ecocriticism: Literature, Animals, Environment* (Routledge, 2010; 2015); Cate Sandilands, "Floral Sensations: Plant Biopolitics," in *The Oxford Handbook of Environmental Political Theory*, eds. Teena Gabrielson et al. (New York: Oxford University Press, 2016), 226–237; Jeffrey Jerome Cohen and Lowell Duckert, eds., *Elemental Ecocriticism: Thinking with Earth, Air, Water, and Fire* (Minneapolis: University of Minnesota Press, 2015); Astrida Neimanis, *Bodies of Water* (London: Bloomsbury Academic, 2017); Greta Gaard, "(Un)Storied Air, Breath, and Embodiment," *ISLE: Interdisciplinary Studies in Literature and Environment* 29, no. 2 (2022): 295–322.

34 Nik Taylor and Richard Twine, eds., *The Rise of Critical Animal Studies: From the Margins to the Centre* (Oxfordshire: Routledge, 2014), 4.

35 See https://www.climatechangenews.com/2020/02/24/world-faces-decisive-decade-fix-global-warming-former-un-climate-chief-says/.

36 See Carol J. Adams and Lori Gruen's second edition of their co-edited volume, *Ecofeminism: Feminist Intersections with Other Animals & the Earth* (New York: Bloomsbury Academic, 2022), which groups chapters into three topical clusters: Affect, Context, Climate. The multispecies affects explored include care, compassion, joy, eros and grief.

37 Affect studies allegedly coalesced through inspiration from Eve Kosofsky Sedgwick's fascination with Silvan Tompkins, a psychologist who identified nine key affects of human experience—only two positive affects (enjoyment and interest/excitement), one neutral affect (surprise/startled) and six negative affects (i.e. anger and fear).

38 See Melissa Gregg and Gregory J. Segworth, eds., *The Affect Theory Reader* (Chapel Hill, NC: Duke University Press, 2010); Alexa Weik von Mossner, *Affective Ecologies: Empathy, Emotion, and Environmental Narrative* (Columbus: Ohio State Press, 2017); Kyle Bladow and Jennifer Ladino, eds., *Affective Ecocriticism: Emotion, Embodiment, Environment* (Lincoln: University of Nebraska Press, 2018).

39 Sarah Jaquette Ray, *A Field Guide to Climate Anxiety: How to Keep Your Cool on a Warming Planet* (Oakland, CA: University of California Press, 2020).

40 Nicole Seymour, *Strange Natures: Futurity, Empathy, and the Queer Ecological Imagination* (Urbana, IL: University of Illinois Press, 2013) and *Bad Environmentalism: Irony and Irreverence in the Ecological Age* (Minneapolis: University of Minnesota Press, 2018), 12, 19, 23, et passim. Seymour does not mention that vegan ecofeminists have also been called "killjoys"; see Richard Twine, "Vegan Killjoys at the Table: Contesting Happiness and Negotiating Relationships with Food Practices," *Societies* 4, no. 4 (2014): 135–158.

41 Mia MacDonald, "Maximum Plunder: The Global Context and Multiple Threats of Animal Agriculture," 355–372 in Adams and Gruen, eds.

42 Emily Gaarder, *Women and the Animal Rights Movement* (New Brunswick, NJ: Rutgers University Press, 2011), 11–12.

43 See https://mercyforanimals.org/blog/12-lgbtq-vegan-activists-you-need-to-follow/ (June 16, 2017); Emily Court, "20 Highly Influential LGBTQ+ Vegans," at https://plantbasednews.org/opinion/20-highly-influential-lgbtq-vegans/ (June 16, 2018).

44 See Fred Kuhr, "Activist Animals: Queers Campaign for the Rights of Other Species," *XTRA Magazine* at https://xtramagazine.com/culture/activist-animals-23761, accessed July 20, 2005; F.R. Ascione, "Battered Women's Reports of Their Partners' and Their Children's Cruelty to Animals," *Journal of Emotional Abuse* 1, no. 1 (1998): 119–133; Brian Luke, *Brutal: Manhood and the Exploitation of Animals* (Champaign, IL: University of Illinois Press, 2005); Greta Gaard, "Queering the Climate," in *Men, Masculinities, and Earth*, eds. Paul M. Pulé and Martin Hultman (Cham, Switzerland: Palgrave, 2021), 515–536; Lisa Kemmerer, *Oppressive Liberation: Sexism in Animal Activism* (Palgrave Macmillan, 2023).

45 Andrea Juno and V. Vale, eds., *Angry Women*. RE/SEARCH #13 (San Francisco. CA: Re/Search Publications, 1991).

46 Many of the women interviewed are "bad girls" in that they have had sex with younger men (bell hooks), written Afrocentric poetry and essays about mid-century life in Watts (Wanda Coleman), introduced sex toys to feminism (Susie Bright), written poetry and fiction about incest (Sapphire), enjoyed sex work (Annie Sprinkle) and boldly offered their breasts and bodies for touching and viewing in public performances (Valie Export), simultaneously defying patriarchal and lesbian separatist mandates for "good" feminists.

47 Audre Lorde, "Uses of Anger: Women Responding to Racism," 124–133 in *Sister Outsider: Essays and Speeches* (Trumansburg, NY: The Crossing Press, 1984).

48 Brittney Cooper, *Eloquent Rage: A Black Feminist Discovers Her Superpower* (St. Martin's Press, 2018); Soraya Chemaly, *Rage Becomes Her: The Power of Women's Anger* (New York: Simon & Schuster/Atria, 2018); Lilly Dancyger, ed., *Burn It Down: Women Writing About Anger* (New York: Seal Press, 2019); and Rebecca Traister, *Good and Mad: The Revolutionary Power of Women's Anger* (New York: Simon & Schuster, 2018).

49 The reality of multispecies domestic violence was recognized nationally in the United States through the long-advocated passage of the Pet and Women Safety (PAWS) Act on December 20, 2018, establishing a grant fund for those who provide shelter and housing assistance for domestic violence survivors, enabling agencies to better meet the housing needs of survivors with pets. The law defines "pets" broadly to include service and emotional support animals as well as horses. For documentation on "Animal Abuse and Youth Violence," see Frank R. Ascione, *Office of Juvenile Justice and Delinquency Prevention Bulletin*, September 2001, retrieved from: https://www.ojp.gov/pdffiles1/ojjdp/188677.pdf. For documentation on "Police Dogs and Anti-Black Violence," see Tyler Parry in *Black Perspectives* (July 31, 2017), retrieved from: https://www.aaihs.org/police-dogs-and-anti-black-violence/.

50 Those seeking data confirming the urgency of animal suffering and its connections to climate and extinctions may review Richard Twine, *The Climate Crisis and Other Animals* (Sydney University Press, 2024) and "Emissions from Animal Agriculture—16.5% Is the New Minimum Figure," *Sustainability* 13 (2021): 6276, retrieved from: https://doi.org/10.3390/su13116276; Deborah Bird Rose, Thom van Dooren, and Matthew Chrulew, eds., *Extinction Studies: Stories of Time, Death, and Generations* (New York: Columbia University Press, 2017).

51 In 2021, Thunberg teamed up with Mercy for Animals to create a five-minute message #ForNature, highlighting the multiple harms of animal agriculture for animals and the earth, and urging viewers to go vegan, retrieved from: https://www.youtube.com/watch?v=7WvehTbuvIo

52 Rep. Alexandria Ocasio-Cortez (AOC) House Floor Speech Transcript on Yoho Remarks (July 23, 2020), retrieved from: https://www.rev.com/blog/transcripts/rep-alexandria-ocasio-cortez-floor-speech-about-yoho-remarks-july-23.
53 See Annie Sprinkle and Beth Stephens with Jennie Klein, *Assuming the Ecosexual Position: The Earth as Lover* (Minneapolis: University of Minnesota Press, 2021).
54 Kunsthall Trondheim hosted the multimedia "Sex Ecologies" exhibit from December 9, 2021–March 6, 2022. See Stefanie Hessler, ed., *Sex Ecologies* (Kunsthall Trondheim, The Seed Box, and MIT Press, 2021).
55 Kathryn Gillespie and Yamini Narayanan, "Global Atmospheres of Violence: Shifting Terrains of Othering in Ecofeminists Multispecies Witnessing," 335–353 in Carol J. Adams and Lori Gruen, eds., at p. 352.
56 Chris Hayes, "The New Abolitionism," *The Nation* (April 22, 2014), retrieved from: https://www.thenation.com/article/archive/new-abolitionism/
57 Naomi Scheman, "Anger and the Politics of Naming," in *Women and Language in Literature and Society*, eds. Sally McConnel-Ginet, Ruth Borker, and Nelly Furman (New York: Praeger Special Studies, 1980), 174–187.
58 Andrea Sempértegui, "Indigenous Women's Activism, Ecofeminism, and Extractivism: Partial Connections in the Ecuadorian Amazon," *Politics & Gender* 17 (2021): 197–224.
59 Ibid., 203.
60 See World Rainforest Movement (WRM) Bulletin 208 (December 9, 2014), retrieved from: https://www.wrm.org.uy/bulletin-articles/why-are-women-fighting-against-extractivism-and-climate-change.
61 See Elaina Zachos, "Why 2017 Was the Deadliest Year for Environmental Activists," *National Geographic* (July 4, 2018), retrieved from: https://www.nationalgeographic.com/environment/article/environmental-defenders-death-report; Eric R. Viramontes, "Agribusiness in Central America, Mexico," *AgriBusiness Global* (August 2, 2022), retrieved from: https://www.agribusinessglobal.com/markets/exploring-the-possibilities-for-agribusiness-in-central-america-mexico/
62 Irune del Rio Gabiola, *Affect, Ecofeminism, and Intersectional Struggles in Latin America: A Tribute to Berta Cáceres* (Peter Lang Verlag, 2020), 4–5.
63 Ibid.
64 Alicia H. Puleo, "Speaking from the South of Europe," *DEP: Deportate, esuli, profughe*, no. 20 (2012): 78–89, at p. 84. See also Puleo's monograph, *Claves Ecofeministas: Para rebeldes que aman a la Tierra y a los animals* (Ecofeminist Keys: For rebels who love the earth and the animals), 3rd edition (Madrid: Plaza y Valdés Editores, 2019; 2021)
65 See Carol P. Christ and Judith Plaskow, eds., *Womanspirit Rising: A Feminist Reader in Religion* ((New York: Harper & Row, 1979); Rosemary Radford Ruether, *Gaia & God: An Ecofeminist Theology of Earth Healing* (New York: HarperCollins, 1992); Monica Sjoo and Barbara Mor, *The Great Cosmic Mother: Rediscovering the Religion of the Earth* (HarperOne, 1987); Charlene Spretnak, ed., *The Politics of Women's Spirituality: Essays on the Rise of Spiritual Power within the Feminist Movement* (New York: Anchor Books/Doubleday, 1982); Merlin Stone, *When God Was a Woman* (Mariner Books, 1978).
66 Marti Kheel, "From Heroic to Holistic Ethics: The Ecofeminist Challenge," in *Ecofeminism: Women, Animals, Nature*, ed. Greta Gaard (Philadelphia, PA: Temple University Press, 1993), 243–271; Sally Abbott, "The Origins of God in the Blood of the Lamb," in *Reweaving the World: The Emergence of Ecofeminism*, eds. Irene Diamond and Glora Feman Orenstein (San Francisco, CA: Sierra Club Books, 1990), 35–40. See also Kip Anderson and Kameron Waters' "Christspiracy" (2024).
67 See Mary Judith Ress, "The Con-Spirando Women's Collective: Gloalization from Below?" in *Ecofeminism & Globalization: Exploring Culture, Context, and Religion*, eds. Heather Eaton and Lois Ann Lorentzen (New York: Rowman & Littlefield, 2003), 147–161, and Ivone Gebara, "Ecofeminism: An Ethics of Life," in *Ecofeminism & Globalization: Exploring Culture, Context, and Religion*, eds. Heather Eaton and Lois Ann Lorentzen (New York: Rowman & Littlefield, 2003), 163–176.
68 Melanie L. Harris, *EcoWomanism: African-American Women and Earth-Honoring Faiths* (Orbis Books, 2017).
69 Judith Gleason, *Oya: In Praise of the Goddess* (Boston: Shambhala, 1987); Luisah Teish, *Jambalaya: The Natueal Woman's Book of Personal Charms and Practical Rituals* (New York:

HarperCollins, 1985); Starhawk, *Dreaming the Dark: Magic, Sex & Politics* (Boston: Beacon Press, 1982) and *The Spiral Dance: A Rebirth of the Ancient Religion of the Great Goddess* (San Francisco, CA: HarperCollins, 1979).

70 See Sandy Boucher, *Turning the Wheel: American Women Creating the New Buddhism* (Boston: Beacon Press, 1988); China Galland, *The Bond Between Women: A Journey to Fierce Compassion* (New York: Riverhead Books, 1998); Rita Gross, *Buddhism After Patriarchy: A Feminist History, Analysis, and Reconstruction of Buddhism* (Albany, NY: SUNY Press, 1993); Anne Klein, *Meeting the Great Bliss Queen: Buddhists, Feminists, and the Art of the Self* (Boston: Beacon Press, 1995)

71 Stephanie Kaza, ed., *Hooked! Buddhist Writings on Greed, Desire, and the Urge to Consume* (Boulder, CO: Shambhala Press, 2005); Stephanie Kaza and Kenneth Kraft, eds., *Dharma Rain: Sources of Buddhist Environmentalism* (Boston: Shambhala Press, 2000); David R. Loy, *EcoDharma: Buddhist Teachings for the Ecological Crisis* (Somerville, MA: Wisdom Publications, 2018).

72 See Greta Gaard, "Mindful Ecofeminism and the Multispecies Sangha," *Tarka* #3: On Ecology (2020): 139–147. In the United States, vegan Dharma teachers include Sandy Boucher, Tara Brach, Ruth Denison, and Spring Washam. See https://www.dharmavoicesforanimals.org/

73 Lina Gupta, "Ganga: Purity, Pollution, and Hinduism," in *Ecofeminism and the Sacred*, ed. Carol J. Adams (New York: Continuum Publishing, 1993), 99–116.

74 Rita M. Gross and Rosemary Radford Ruether, *Religious Feminism and the Future of the Planet: A Buddhist-Christian Conversation* (New York: Continuum, 2001).

75 Alka Arora, "Re-Enchanting Feminism: Challenging Religious and Secular Patriarchies," in *Postsecular Feminisms: Religion and Gender in Transnational Context*, ed. Nandini Deo (London: Bloomsbury Academic, 2018), 32–51.

76 Karen J. Warren, "A Feminist Philosophical Perspective on Ecofeminist Spiritualities," in Adams, ed., 119–132. The spirituality of multispecies and ecological care that was legendary for St. Francis of Assisi suggests this capacity is available to people across differences of gender.

77 As I wrote this chapter, I learned the (U.S.) National Women's Studies Association (NWSA) 2022 Conference theme would be "Killing Rage: Resistance on the Other Side of Freedom," a theme inspired by bell hooks' (1995) book, *Killing Rage: Ending Racism.*

Bibliography

Abbott, Sally. "The Origins of God in the Blood of the Lamb." In *Reweaving the World: The Emergence of Ecofeminism*, edited by Irene Diamond and Glora Feman Orenstein, 35–40. San Francisco, CA: Sierra Club Books, 1990.

Adams, Carol J. *The Sexual Politics of Meat: A Feminist-Vegetarian Critical Theory*. New York: Continuum Publishing, 1990.

Adams, Carol J. "The Feminist Traffic in Animals." In *Ecofeminism: Women, Animals, Nature*, edited by Greta Gaard. Philadelphia, PA: Temple University Press, 1993.

Adams, Carol J., and Donovan, Josephine, eds. *Animals and Women: Feminist Theoretical Explorations*. Durham, NC: Duke University Press, 1995.

Adams, Carol J., and Gruen, Lori. *Ecofeminism: Feminist Intersections with Other Animals & the Earth*. 2nd ed. New York: Bloomsbury Academic, 2022.

Agarwal, Bina. "The Gender and Environment Debate: Lessons from India." *Feminist Studies* 18, no. 1 (1992): 119–158.

Arora, Alka. "Re-enchanting Feminism: Challenging Religious and Secular Patriarchies." In *Postsecular Feminisms: Religion and Gender in Transnational Context*, edited by Nandini Deo, 32–51. London: Bloomsbury Academic, 2018.

Ascione, F.R. "Battered Women's Reports of Their Partners' and Their Children's Cruelty to Animals." *Journal of Emotional Abuse* 1, no. 1 (1998).

Ascione, F.R. *Office of Juvenile Justice and Delinquency Prevention Bulletin* (September 2001). Retrieved from: https://www.ojp.gov/pdffiles1/ojjdp/188677.pdf.

Bari, Judi. *Timber Wars*. Monroe, ME: Common Courage Press, 1994.

Bladow, Kyle, and Ladino, Jennifer, eds. *Affective Ecocriticism: Emotion, Embodiment, Environment*. Lincoln: University of Nebraska Press, 2018.

Boucher, Sandy. *Turning the Wheel: American Women Creating the New Buddhism*. Boston: Beacon Press, 1988.
Caldecott, Léonie, and Leland, Stephanie, eds. *Reclaim the Earth: Women Speak Out for Life on Earth*. London: The Women's Press, 1983.
Cantor, Aviva. "The Club, the Yoke, and the Leash: What We Can Learn From the Way a Culture Treats Animals." *Ms. Magazine*, August 1983: 27–29.
Chang, Chia-ju. "Women and Interspecies Care: Dog Mothers in Taiwan." In *International Perspectives in Feminist Ecocriticism*, edited by Greta Gaard, Simon Estok, and Serpil Oppermann, 151–165. New York: Routledge, 2013.
Chemaly, Soraya. *Rage Becomes Her: The Power of Women's Anger*. New York: Simon & Schuster/Atria, 2018.
Christ, Carol P., and Plaskow, Judith, eds. *Womanspirit Rising: A Feminist Reader in Religion*. New York: Harper & Row, 1979.
Cohen, Jeffrey Jerome, and Duckert, Lowell, eds. *Elemental Ecocriticism: Thinking with Earth, Air, Water, and Fire*. Minneapolis: University of Minnesota Press, 2015.
Cook, Alice, and Kirk, Gwyn. *Greenham Women Everywhere: Dreams, Ideas and Actions from the Women's Peace Movement*. London: Pluto Press Ltd., 1983.
Cooper, Brittney. *Eloquent Rage: A Black Feminist Discovers Her Superpower*. St. Martin's Press, 2018.
Court, Emily. "20 Highly Influential LGBTQ+ Vegans." *Plant Based News*, 2018. Retrieved from: https://plantbasednews.org/opinion/20-highly-influential-lgbtq-vegans/
Cuomo, Chris J. *Feminism and Ecological Communities: An Ethic of Flourishing*. New York: Routledge, 1998.
Dancyger, Lilly, ed. *Burn it Down: Women Writing About Anger*. New York: Seal Press, 2019.
Decka, Maneesha. "Something to Celebrate? De-listing Dairy in Canada's *National Food Guide*." *Journal of Food, Law, and Policy* 16, no. 1 (2020): 11–46.
Decka, Maneesha. "Veganism, Dairy, and Decolonization." *Journal of Human Rights and the Environment* 11, no. 2 (2020): 44–67.
Decka, Maneesha. "Unsettling Anthropocentric Legal Systems: Reconciliation, Indigenous Laws, and Animal Personhood." *Journal of Intercultural Studies* 41, no. 1 (2020): 77–97.
Delacoste, Frédérique, and Newman, Felice, eds. *FIGHT BACK! Feminist Resistance to Male Violence*. Minneapolis, MN: Cleis Press, 1981.
Diamond, Irene, and Orenstein, Gloria Feman, eds. *Reweaving the World: The Emergence of Ecofeminism*. San Francisco: Sierra Club Books, 1990.
Donovan, Josephine. "Animal Rights and Feminist Theory." *Signs* 15, no. 2 (1990): 350–375.
Donovan, Josephine, and Adams, Carol J., eds. *The Feminist Care Tradition in Animal Ethics*. New York: Columbia University Press, 2007.
Doyle, Alister. "World Faces 'Decisive Decade' to Fix Global Warming, Former UN Climate Chief Says." *Climate Home News*, 2020. Retrieved from: https://www.climatechangenews.com/2020/02/24/world-faces-decisive-decade-fix-global-warming-former-un-climate-chief-says/.
Etchart, Linda. "Demilitarizing the Global: Women's Peace Movements and Transnational Networks." In *The Oxford Handbook of Transnational Feminist Movements*, edited by Rawwida Baksh and Wendy Harcourt, 702–722. Oxford University Press, 2015.
Feminists for Animal Rights. Retrieved from: http://www.farinc.org/
Gaard, Greta, ed. *Ecofeminism: Women, Animals, Nature*. Philadelphia, PA: Temple University Press, 1993.
Gaard, Greta. "Toward a Queer Ecofeminism." *Hypatia* 12, no. 1 (1997): 114–137.
Gaard, Greta. *Ecological Politics: Ecofeminists and the Greens*. Philadelphia, PA: Temple University Press, 1998.
Gaard, Greta. "Ecofeminism Revisited: Rejecting Essentialism and Replacing Species in a Material Feminist Environmentalism." *Feminist Formations* 23, no. 2 (Summer 2011): 26–53.
Gaard, Greta. "Mindful Ecofeminism and the Multispecies Sangha." *Tarka* #3: On Ecology (2020): 139–147.
Gaard, Greta. "Queering the Climate." In *Men, Masculinities, and the Earth*, edited by Paul M. Pulé and Martin Hultman. Palgrave, 2021.
Gaard, Greta. "(Un)Storied Air, Breath, and Embodiment." *ISLE: Interdisciplinary Studies in Literature and Environment* 29, no. 2 (2022): 295–322.

Gaarder, Emily. *Women and the Animal Rights Movement*. New Brunswick, NJ: Rutgers University Press, 2011.
Gabiola, Irune del Rio. *Affect, Ecofeminism, and Intersectional Struggles in Latin America: A Tribute to Berta Cáceres*. Peter Lang Verlag, 2020.
Galland, China. *The Bond Between Women: A Journey to Fierce Compassion*. New York: Riverhead Books, 1998.
Gebara, Ivone. "Ecofeminism: An Ethics of Life." In *Ecofeminism & Globalization: Exploring Culture, Context, and Religion*, edited by Heather Eaton and Lois Ann Lorentzen, 163–176. New York: Rowman & Littlefield, 2003.
Gillespie, Kathryn, and Narayanan, Yamini. "Global Atmospheres of Violence: Shifting Terrains of Othering in Ecofeminists Multispecies Witnessing." In *Ecofeminism: Feminist Intersections with Other Animals & the Earth*, 2nd ed., edited by Carol J. Adams and Lori Gruen, 335–353. New York: Bloomsbury Academic, 2022.
Gleason, Judith. *Oya: In Praise of the Goddess*. Boston: Shambhala, 1987.
Gregg, Melissa, and Segworth, Gregory J., eds. *The Affect Theory Reader*. Chapel Hill, NC: Duke University Press, 2010.
Gross, Rita. *Buddhism After Patriarchy: A Feminist History, Analysis, and Reconstruction of Buddhism*. Albany, NY: SUNY Press, 1993.
Gross, Rita M., and Ruether, Rosemary Radford. *Religious Feminism and the Future of the Planet: A Buddhist-Christian Conversation*. New York: Continuum, 2001.
Gruen, Lori, and Marceau, Justin, eds. *Carceral Logics: Human Incarceration and Animal Captivity*. Cambridge University Press, 2022.
Gupta, Lina. "Ganga: Purity, Pollution, and Hinduism." In *Ecofeminism and the Sacred*, edited by Carol J. Adams, 99–116. New York: Continuum Publishing, 1993.
Hall, K. Melchor Quick, and Kirk, Gwyn, eds. *Mapping Gendered Ecologies: Engaging with and beyond Ecowomanism and Ecofeminism*. Lanham MD: Lexington Books, 2021.
Harper, A. Breeze, ed. *Sistah Vegan: Black Female Vegans Speak on Food, Identity, Health, and Society*. New York: Lantern Books, 2010.
Harris, Melanie L. *EcoWomanism: African-American Women and Earth-Honoring Faiths*. Orbis Books, 2017.
Hatch, Ronald B., ed. *Clayoquot & Dissent*. Vancouver, Canada: Ronsdale Press, 1994.
Hayes, Chris. "The New Abolitionism." *The Nation*, April 22, 2014. Retrieved from: https://www.thenation.com/article/archive/new-abolitionism/
Heller, Chaia. *Ecology of Everyday Life: Rethinking the Desire for Nature*. Montreal, CA: Black Rose Books, 1999.
Hessler, Stefanie, ed. *Sex Ecologies*. Kunsthall Trondheim, The Seed Box, and MIT Press, 2021.
Hill, Julia Butterfly. *The Legacy of Luna*. San Francisco: Harper Books, 2000.
Huggan, Graham, and Tiffin, Helen. *Postcolonial Ecocriticism: Literature, Animals, Environment*. Routledge, 2010 (2015).
Juno, Andrea and Vale, V., eds. *Angry Women*. RE/SEARCH #13, San Francisco. CA: Re/Search Publications, 1991.
Kaza, Stephanie, ed. *Hooked! Buddhist Writings on Greed, Desire, and the Urge to Consume*. Boulder, CO: Shambhala Press, 2005.
Kaza, Stephanie, and Kraft, Kenneth, eds. *Dharma Rain: Sources of Buddhist Environmentalism*. Boston: Shambhala Press, 2000.
Kemmerer, Lisa. *Oppressive Liberation: Sexism in Animal Activism*. Cham, Switzerland: Palgrave Macmillan, 2023.
Kheel, Marti. "From Heroic to Holistic Ethics: The Ecofeminist Challenge." In *Ecofeminism: Women, Animals, Nature*, edited by Greta Gaard, 243–271. Philadelphia, PA: Temple University Press, 1993.
Kheel, Marti. *Nature Ethics: An Ecofeminist Perspective*. Lanham, MD: Rowman & Littlefield, 2008.
Kim, Claire Jean. *Dangerous Crossings: Race, Species, and Nature in a Multicultural Age*. New York: Cambridge University Press, 2015.
Kim, Claire Jean, and Freccero, Carla, eds. "Species/Race/Sex." *American Quarterly* 65, no. 3 (September 2013).
Klein, Anne. *Meeting the Great Bliss Queen: Buddhists, Feminists, and the Art of the Self*. Boston: Beacon Press, 1995.

Ko, Aph, and Ko, Syl. *Aphro-ism: Essays on Pop Culture, Feminism, and Black Veganism from Two Sisters*. New York: Lantern Books, 2020.
Kuhr, Fred. "Activist Animals: Queers Campaign for the Rights of Other Species." *XTRA Magazine*, 2005. Retrieved from: https://xtramagazine.com/culture/activist-animals-23761
Lorde, Audre. "Uses of Anger: Women Responding to Racism." In *Sister Outsider: Essays and Speeches*, 124–133. Trumansburg, NY: The Crossing Press, 1984.
Loy, David R. *EcoDharma: Buddhist Teachings for the Ecological Crisis*. Somerville, MA: Wisdom Publications, 2018.
Luke, Brian. *Brutal: Manhood and the Exploitation of Animals*. Champaign, IL: University of Illinois Press, 2005.
Maathai, Wangari. *The Green Belt Movement*. New York: Lantern Books, 2004.
MacDonald, Mia. "Maximum Plunder: The Global Context and Multiple Threats of Animal Agriculture." In *Ecofeminism: Feminist Intersections with Other Animals & the Earth*, edited by C.J. Adams and L. Gruen, 355–372. New York: Bloomsbury Academic, 2022.
Mathews, Freya. *The Ecological Self*. New York: Routledge, 1991.
Mathews, Freya. *For Love of Matter: A Contemporary Panpsychism*. Albany, NY: SUNY Press, 2003.
Mathews, Freya. *Reinhabiting Reality: Towards a Recovery of Culture*. Albany: SUNY Press, 2005.
McAllister, Pam, ed. *Reweaving the Web of Life: Feminism and Nonviolence*. Philadelphia, PA: New Society Publishers, 1982.
Mellor, Mary. *Feminism and Ecology*. Cambridge, UK: Polity Press, 1997.
Mercy for Animals, "Greta Thunberg's Message #fornature." *YouTube*, 2021. Retrieved from: https://www.youtube.com/watch?v=7WvehTbuvIo
Mies, Maria, and Shiva, Vandana. *Ecofeminism*. London: Zed Books, 1993.
Montford, Kelly Struthers and Taylor, Chloë, eds. *Building Abolition: Decarceration and Social Justice*. Routledge, 2022.
Moore, Niamh. *The Changing Nature of Eco/Feminism: Telling Stories from Clayoquot Sound*. Vancouver: UBC Press, 2015.
Mortimer-Sandilands, Catriona. "Desiring Nature, Queering Ethics: Adventures in Erotogenic Environments." *Environmental Ethics* 23, no. 2 (2001): 169–188.
Mortimer-Sandilands, Catriona. "Fear of a Queer Plant." *GLQ: Gay and Lesbian Quarterly* 23, no. 3 (2017): 419–429.
Mortimer-Sandilands, Catriona. "Here We Go Round the Mulberry Bush: A Queer Botanical Meander." *Center for Sustainable Practice and the Arts Quarterly* 19 (2018): 28–33.
Mortimer-Sandilands, Catriona, and Erickson, Bruce, eds. *Queer Ecologies: Sex, Nature, Politics, Desire*. Bloomington, IN: Indiana University Press, 2010.
Mossner, Alexa Weik von. *Affective Ecologies: Empathy, Emotion, and Environmental Narrative*. Columbus: Ohio State Press, 2017.
Neimanis, Astrida. *Bodies of Water*. London: Bloomsbury Academic, 2017.
Noske, Barbara. *Beyond Boundaries: Humans and Animals*. Montreal, CA: Black Rose Books, 1997.
Ocasio-Cortez, Alexandria. "House Floor Speech Transcript on Yoho Remarks." July 23, 2020. Retrieved from: https://www.rev.com/blog/transcripts/rep-alexandria-ocasio-cortez-floor-speech-about-yoho-remarks-july-23.
Parry, Tyler. "Police Dogs and Anti-Black Violence." *Black Perspectives* (a blog from the African American Intellectual History Society), July 31, 2017. Retrieved from: https://www.aaihs.org/police-dogs-and-anti-black-violence/
Plant, Judith, ed. *Healing the Wounds: The Promise of Ecofeminism*. Philadelphia, PA: New Society Publishers, 1989.
Plumwood, Val. *Feminism and the Mastery of Nature*. New York: Routledge, 1993.
Plumwood, Val. "Human Vulnerability and the Experience of Being Prey." *Quadrant* 39, no. 3 (March 1995): 29–34.
Plumwood, Val. "Integrating Ethical Frameworks for Animals, Humans, and Nature: A Critical Feminist Eco-Socialist Analysis." *Ethics and the Environment* 5, no. 2 (2000): 285–322.
Plumwood, Val. *Environmental Culture: The Ecological Crisis of Reason*. London: Routledge, 2002.
Puleo, Alicia H. "Speaking from the South of Europe." *DEP: Deportate, esuli, profughe*, no. 20 (2012): 78–89.

Puleo, Alicia H. *Claves Ecofeministas: Para rebeldes que aman a la Tierra y a los animals* (Ecofeminist Keys: For rebels who love the earth and the animals). 3rd ed. Madrid: Plaza y Valdés Editores, 2019 (2021).

Ray, Sarah Jaquette. *A Field Guide to Climate Anxiety: How to Keep Your Cool on a Warming Planet*. Oakland, CA: University of California Press, 2020.

Regan, Tom. *The Case for Animal Rights*. Berkeley, CA: University of California Press, 1983.

Ress, Mary Judith. "The Con-spirando Women's Collective: Gloalization from Below?" In *Ecofeminism & Globalization: Exploring Culture, Context, and Religion*, edited by Heather Eaton and Lois Ann Lorentzen, 147–161. New York: Rowman & Littlefield, 2003.

Rose, Deborah Bird. *Dingo Makes Us Human: Life and Land in an Australian Aboriginal Culture*. Cambridge University Press, 2000.

Rose, Deborah Bird, van Dooren, Thom, and Chrulew, Matthew, eds. *Extinction Studies: Stories of Time, Death, and Generations*. New York: Columbia University Press, 2017.

Ruether, Rosemary Radford. *Gaia & God: An Ecofeminist Theology of Earth Healing*. New York: HarperCollins, 1992.

Salleh, Ariel. *Ecofeminism as Politics: Nature, Marx and the Postmodern*. London: Zed Books, 1997.

Sandilands, Cate. "Floral Sensations: Plant Biopolitics." In *The Oxford Handbook of Environmental Political Theory*, edited by Teena Gabrielson et al. New York: Oxford University Press, 2016.

Sandilands, Catriona. *The Good-Natured Feminist: Ecofeminism and the Quest for Democracy*. Minneapolis: University of Minnesota Press, 1999.

Scheman, Naomi. "Anger and the Politics of Naming." In *Women and Language in Literature and Society*, edited by Sally McConnel-Ginet, Ruth Borker, and Nelly Furman, 174–187. New York: Praeger Special Studies, 1980.

Sempértegui, Andrea. "Indigenous Women's Activism, Ecofeminism, and Extractivism: Partial Connections in the Ecuadorian Amazon." *Politics & Gender* 17 (2021): 197–224.

Seymour, Nicole. *Strange Natures: Futurity, Empathy, and the Queer Ecological Imagination*. Urbana, IL: University of Illinois Press, 2013.

Seymour, Nicole. *Bad Environmentalism: Irony and Irreverence in the Ecological Age*. Minneapolis: University of Minnesota Press, 2018.

Shiva, Vandana. *Staying Alive: Women, Ecology and Development*. London: Zed Books, 1988.

Singer, Peter. *Animal Liberation: A New Ethics for Our Treatment of Animals*. San Francisco, CA: HarperCollins, 1975.

Sjoo, Monica, and Mor, Barbara. *The Great Cosmic Mother: Rediscovering the Religion of the Earth*. HarperOne, 1987.

Spretnak, Charlene, ed. *The Politics of Women's Spirituality: Essays on the Rise of Spiritual Power within the Feminist Movement*. New York: Anchor Books/Doubleday, 1982.

Sprinkle, Annie, and Stephens, Beth with Klein, Jennie. *Assuming the Ecosexual Position: The Earth as Lover*. Minneapolis: University of Minnesota Press, 2021.

Starhawk. *The Spiral Dance: A Rebirth of the Ancient Religion of the Great Goddess*. San Francisco, CA: HarperCollins, 1979.

Starhawk. *Dreaming the Dark: Magic, Sex & Politics*. Boston: Beacon Press, 1982.

Stone, Merlin. *When God Was a Woman*. Mariner Books, 1978.

Sturgeon, Noël. *Ecofeminist Natures: Race, Gender, Feminist Theory and Political Action*. New York: Routledge, 1997.

Sturgeon, Noël. *Environmentalism in Popular Culture: Gender, Race, Sexuality, and the Politics of the Natural*. Tucson, AZ: University of Arizona Press, 2009.

Taylor, Nik and Twine, Richard, eds. *The Rise of Critical Animal Studies: From the Margins to the Centre*. Oxfordshire: Routledge, 2014.

Teish, Luisah. *Jambalaya: The Natural Woman's Book of Personal Charms and Practical Rituals*. New York: HarperCollins, 1985.

Traister, Rebecca. *Good and Mad: The Revolutionary Power of Women's Anger*. New York: Simon & Schuster, 2018.

Trondheim, Kunsthall. Hosted the Multimedia "Sex Ecologies" exhibit from December 9, 2021–March 6, 2022.

Twine, Richard. "Emissions from Animal Agriculture—16.5% Is the New Minimum Figure." *Sustainability* 13, no. 11 (2021): 6276. https://doi.org/10.3390/su13116276.

Twine, Richard. *The Climate Crisis and Other Animals*. Gadigal, Australia: Sydney University Press, 2024.

Viramontes, Eric R. "Agribusiness in Central America, Mexico." *AgriBusiness Global*, August 2, 2022. Retrieved from: https://www.agribusinessglobal.com/markets/exploring-the-possibilities-for-agribusiness-in-central-america-mexico/

Warren, Karen J. "The Power and Promise of Ecological Feminism." *Environmental Ethics* 12 (Summer 1990), 125–146.

Warren, Karen J., ed. "Special Issue: Ecological Feminism." *Hypatia* 6, no. 1 (Spring 1991).

Warren, Karen J. "A Feminist Philosophical Perspective on Ecofeminist Spiritualities." In *Ecofeminism and the Sacred*, edited by Carol J. Adams, 119–132. New York: Continuum Publishing, 1993.

Warren, Karen J., ed. *Ecological Feminism*. New York: Routledge, 1994.

Warren, Karen J. *Ecofeminism: Women, Culture, Nature*. Bloomington, IN: Indiana University Press, 1997.

Warren, Karen J. *Ecofeminist Philosophy: A Western Perspective on What It Is and Why It Matters*. Lanham, MD: Rowman & Littlefield, 2000.

World Rainforest Movement. "(WRM) Bulletin 208." (December 9, 2014). Retrieved from: https://www.wrm.org.uy/bulletin-articles/why-are-women-fighting-against-extractivism-and-climate-change.

Wright, Laura. *The Vegan Studies Project: Food, Animals, and Gender in the Age of Terror*. Athens, GA: University of Georgia Press, 2015.

Wright, Laura, ed. *The Routledge Handbook of Vegan Studies*. New York: Routledge, 2021.

Zachos, Elaina. "Why 2017 Was the Deadliest Year for Environmental Activists." *National Geographic*, July 4, 2018. Retrieved from: https://www.nationalgeographic.com/environment/article/environmental-defenders-death-reports

2
FEMINIZED PROTEIN

Carol J. Adams

I coined the term *feminized protein* in the late 1980s as I was finishing my first book, *The Sexual Politics of Meat: A Feminist-Vegetarian Critical Theory*, in which I propounded the patriarchal context for eating dead animals' bodies. I was seeking a term that identified the use of reproductive material taken from living nonhuman female animals, specifically a term to encompass the nursing material extracted from cows, buffaloes, and goats and the eggs being forcibly produced by chickens. The term would need explicitly to point to the use of animals identified as female in a misogynous world. It would also try to counter the invisibility of the suffering of these female animals in which the sexual commodification required to keep them "productive" occurs in private. I thought of Virginia Woolf's famous sentence from *Three Guineas*, "In a transitional age when many qualities are changing their value, new words to express new values are much to be desired" (311, n. 11). My hope was that through a "new naming," the use of this reproductive material—which is often viewed as victimless because the animals remain alive—would help to catalyze new values in response to the lives of cows, buffaloes, goats, and chickens.

One major demarcation between vegetarians and vegans is whether they eat feminized protein or not. In 1983, feminist Norma Benney revived a statement made in John Vyvyan's 1969 *In Pity and In Anger* that proposed that people should stop eating foods taken from living female animals before they stopped eating dead animals, as it was those living animals whose suffering was most intense and unrelenting and lasted the longest.[1]

My reasoning for using the term "feminized protein" was that all protein comes from vegetables. Some people eat it directly; others rely on animals to convert the vegetables into protein. I knew that in the nineteenth century some vegetarians in the United States referred to food from dead animals' bodies as "animalized protein." Ellen G. White—whose conclusion that humans were not meant to eat dead animals contributed to the adoption of vegetarianism by Seventh-Day Adventists—illustrates the way nineteenth-century vegetarians used this term:

> The diet of the animals is vegetables and grains. Must the vegetables be animalized, must they be incorporated into the system of animals, before we get them? Must we obtain our vegetable diet by eating the flesh of dead creatures?
>
> *(396)*

DOI: 10.4324/9781003273400-4

Figure 2.1 This fragmented cow, complete with precisely-rendered nipples, greets shoppers at the entrance to Irene Mall, Pretoria, South Africa, Photograph copyright © Justin van Huyssteen. Used by permission.

I thought, "If protein taken from the bodies of dead animals is animalized, what do we call protein taken from living female bodies?" And the answer that came to me was *feminized protein.*

Reproduction as Production

Every egg and every drop of maternal milk is monetized through contemporary animal agricultural practices.

In her important 1997 book, *Beyond Boundaries,* Barbara Noske sought to place our understanding of the animal industrial complex within a Marxist framework, proposing an anthropological rather than an ethological approach for her discussion of animals' lives. She reminded her readers that "We have seen that the human (male) worker surrendered control over the production and labour processes to the management but maintained control over the sphere of reproduction: his home" (17). She quoted Marx's famous statement that "the worker only feels at home outside his work and in his work he feels a stranger." The female worker, however, Noske argued, "whether or not working in the sphere of production, is required also to work in the sphere of reproduction: the home. . . . For her there exists no clear distinction between home and work." (More than 35 years after the publication of this book, even after the gender binary has been exposed as a social construction, it remains true that individuals who identify as women are more likely to be doing the unpaid care work in the home.)

For domesticated animals—animal "workers"—Noske shows how there is no division between home and work: "In the case of animals the 'home' itself has been brought under factory control. . . . Indeed, it is often the sphere of reproduction (mating, breeding, the laying of eggs), which the capitalist seeks to exploit" (17). Further, "The animal's life-time has truly been converted into 'working-time': intro round-the-clock production."[2]

For cows, buffaloes, goats, and chickens required to produce feminized protein, the sphere of production and the sphere of reproduction *are the same*; their alienated labor occurs against their will in their own bodies; this alienated labor involves reproduction *as* production, production *through* reproduction (for cows, calves and their lactation products; for "laying" chickens, eggs). Kathryn Gillespie, in her book *The Cow with Ear Tag #1349*, reminds her readers that the product a cow is "responsible for producing is, first and foremost, milk and so the *pregnancy* (rather than the resulting calf) is her job, not the care of offspring" (2019, 179–80). As confirmation of Gillespie's claim that the purpose of the cow was to produce milk, not babies, at the time of slaughter, about 25 percent of cows are pregnant.[3]

Oppressive language about cows, goats, and chickens highlights the entangled relationship of reproduction as production and production through reproduction. A *Hoard's Dairyman* article advises that when a cow becomes "a low producer," she "shouldn't be allowed to take up space" and should be killed (quoted in Dunayer 2001, 144). An article in *The Guardian* discussing the use of the dead bodies of elderly, "exhausted" cows as food asserted that the cows "collected their pension" during the time between when they no longer were required to produce mammalian milk until the time of their deaths (Cumming). One reader reported he was glad to hear it, "We should give these hard working ladies more respect!" (Maguire in Bergo).

In 2016, I was sent a flyer from Wegmans, a large Northeast United States grocery store chain, that, supposedly in a light-hearted way, detailed the "Benefit Package of a Dairy Cow." It read:

> To start with, the cows receive full-time pay for part-time work. The work (of being milked) takes about 20–30 minutes per day. *The employer provides paid medical coverage, with a doctor (veterinarian) on call 24/7, 365 days per year.* Meals are prepared by a nutritionist, with room service and clean up every time. There is a full-time housekeeper who even cleans the bathrooms.
>
> A paid team of experts is always available for these bovine beauties; hair dresser, pedicurist and spa facilities are provided. There is 24-hour surveillance. No need for online dating . . . there is mate selection provided through a directory of selective traits, and could be a different mate each year. All transportation is provided free of charge for a lifetime.
>
> *(Adams 2017, 19)*

Feminist legal theorist Jessica Eisen points out that the language in a flyer like this that posits cows as carefree working girls is dangerous in the ways it normalizes "the extreme conditions of animal agriculture by positioning current forms of animal use as continuous with intra-human relational structures that are broadly accepted as healthy and productive, if occasionally imperfect" (2019b, 7). At the same time, such tropes also appear to naturalize and denigrate women's traditional labor, while casting these "working girls" as frivolous. For instance, Bovi-Shield, a company that produces bovine vaccinations for beef cattle producers, advertises its products by placing cows in unusual situations to emphasize the abnormality that occurs when cows aren't being kept pregnant. One advertisement depicts a cow supposedly helping a man hunt birds to underscore the apparent absurdity of cows doing anything other than producing calves and milk.[4] "Keep your cows pregnant and on the job," the advertisement cautions. As though cows are not "working" when they aren't pregnant.

Cows and chickens are less "workers" and more "worked upon," animal machines who are disposed of when they are no longer productive (see Harrison). The pharmaceutical industry talks of the laying of eggs in production terms, as in Zinpro's discussion of "High egg production volume comes with a trade off in quality, no matter how efficient the layer is" (https://www.zinpro.com/species/poultry/layers/), but often its advertisements anthropomorphize those "layers" with stereotypical characteristics associated with human females. For instance, one such ad shows a feminized hen suggestively lifting her feathers in a reveal of her leg and reads, "Healthy Skin, Less Scratches and Cellulitus."[5]

When chickens—deemed "little egg engines" according to one writer (Bergo)—no longer "produce" eggs, they are deemed "spent" because their "commercial value" is "considered negligible" (Harris). As Marian Burros reported in *the New York Times,* "They are worth so little that many are incinerated." However, their dead bodies may still contain eggs, and these eggs—known as "immature, unborn, unlaid or embryonic" and often left unharvested when the hens are burnt—are now also monetized. In other words, the post-mortem chicken's body continues to be seen as a reproductive/productive body. Customers, however, "balked" when these eggs were called "embryonic"; they preferred "immature" (Burros).

As one gains knowledge of the institutionalized cruelty inherent to the production of feminized protein, it is helpful to reframe the idea of the "farm." Eisen suggests that

> In many ways, the farm may be to animals what the family has been to women within some strands of feminist critique: the arbiter and enforcer of their place, their purpose, their meaning; so naturalized that it is not even worth asking what they might think of it, even if we thought they were capable of answering.
>
> *(2017, 240)*

The result is that just as the concept of privacy shrouded and protected violence in the home (Pateman), the idea of private property—applied both to the farm and the domesticated animals living there—protects those deemed the "owners" of animals, and the law thereby is used "to create a lawless space" (Eisen 2017, 240).

Cows

Due to scientific illiteracy, many drinkers of mammalian milk believe they are helping a cow with her "surplus" milk. But they are mistaken. As with other mammals, the nursing material of the cow eventually dries up. That is the purpose of her yearly forced pregnancies: to make sure the cow's milk comes back in. She continues to have her milk taken from her during much of her pregnancy; the demands on her body of lactating and being pregnant have been likened to jogging six or more hours a day (Desaulniers 2016). Though cows can enjoy sex, these pregnancies do not result from a romp with friendly bulls in the pasture. And while uninterested cows can refuse sex in that pasture simply by walking away, in contemporary farming situations, they have no choice. Artificial insemination is the predominant form of impregnating cows, and cows have no power to resist it. The forced penetration first occurs with a human hand and part of the arm inserted into the cow's rectum; this hand then guides an insemination gun into the cervix. The act of impregnating cows is *so similar* to the definition of bestiality that laws against bestiality have to make exceptions for contemporary farming practices in their statutes (Rosenberg and Dutkiewicz; see also discussions by Cusack, Davis 2017, and Narayanan, 2019). As Gabriel Rosenberg observes, "Agricultural exemptions to bestiality laws demonstrate that the decisive difference between a bestialist and a farmer is about a difference in *relation to capital*, not in *relation to animals*" (475).

Any milk sucked by babies from their mammal mothers takes profit away from the farmers, so destroying the relationship between mother and child is an essential aspect of the dairy industry. The babies' removal occurs as early as possible. Calves, sometimes just hours old, may be auctioned off or carried by wheelbarrow to a hutch to become a veal calf, destined for someone's plate. Kathryn Gillespie reports seeing at auction calves so young, their umbilical cords were still attached to them. Their quick removal from their mothers enables even the first drops of the lactation product, colostrum—needed by the newborn for muscle development—to be collected, monetized, and packaged in vitamin supplements.

When her baby is taken from her, each cow vocalizes her unique and anguished distress. At one large dairy farm, the sounds emanating from the cows were so loud and recurrent that nearby residents reported these haunting sounds to law enforcements. The local sheriff explained the cows were "lamenting" the loss of their babies. He had to walk back this wording the next day to claim that what the cows were experiencing was normal (Rogers). Jessica Eisen describes how, after the removal of her calf, "the cow generally exhibits significant physiological and behavioral symptoms of distress (e.g., pacing, urinating, and calling out), with those symptoms dissipating if they are reunited" (Eisen 2017, 105).

Cows resist the removal of their babies in any way available to them. If able, cows will escape and chase after their calves. In the early colonial times on the land now known as the United States, "Feral cows with calves and wild sows with piglets roamed the woods in gangs, and the wary mothers instinctively took the offensive against anyone who came near their young" (Anderson, 122).

The development of genetic engineering, the careful tracking of feed rations, and the use of growth hormones means that cows exploited by the dairy industry produce 61 percent more milk than cows from only 25 years ago. Their udders must carry an extra 58 pounds of milk. The cow's bloated udders may force her hind legs apart, causing lameness. Lameness can cause cows to become "non-ambulatory," or "downed"—unable to walk on their own. Other causes of a downed cow are accidents, metabolic or infectious disease (like mastitis), trauma, and calving. *Downer cow syndrome* is "a common problem in dairy cattle" (Puerto-Parada et al.). The survival rate of downed cows is estimated to be between 11 and 50 percent (Puerto-Parada et al.). With the reproduction-as-production framework in effect, downed cows are considered expensive: their milk can no longer be taken, they require medical care, and their killing eliminates "even the salvage value" (Buza and Babisz).

With the normalizing and naturalizing of the consumption of the cow's nursing material, when it comes to drinking it, most consumers don't think "I am now interacting with a cow and a calf." Over the past 35 years, I have reviewed many visual images that advocate drinking the lactation product of cows (Adams 2019). Many of these forms of propaganda omit images of or references to the calf. The calf's absence reinforces the idea that the cow's relationship as mammalian milk producer is to us, the consumers, not to their babies. They act to assure us that they are, in the words of seventeenth-century writer Thomas Dekker, "our nurses that give us milk," not mothers with children (152).

E. Melanie DuPuis observes that in the late nineteenth and early twentieth century, mammalian milk advertising moved its focus from "the site of production" to the "product of consumption." Images of healthy human babies and children appeared in advertisements. DuPuis finds in this change an example of the emergence of a mechanistic view of nature during the Scientific Revolution of the early modern period of European history identified by Carolyn Merchant, the mechanistic view that has evolved to treat animals forced into producing feminized protein as machines (DuPuis, 104).

Figure 2.2 "You can drink it!" Rosie the Cow postcard, Peace Memorial, Caen, Normandy, France, 2016. Photograph copyright © Camille Brunel. Used by permission.

Figure 2.3 Two good ol' Texans protesting outside a Democratic primary rally for Hillary Clinton, Dallas, Texas, February 2008. Photograph copyright © by Carol J. Adams. Used by permission.

Joan Dunayer identified the ways words about cows became applied to women. Confined to a stall, denied the active role of nurturing and protecting a calf—so that milking becomes something done *to* her rather than *by* her—she is seen as passive and dull. The cow then becomes emblematic of these traits, which metaphor can attach to women (Dunayer 1995, 13).

Chickens

The relatives of today's chickens loved to sit in trees, form close relationships with others, dust bathe for at least 30 minutes a day and forage for a large part of the day. On twenty-first–century farms where eggs are taken away, a chick may begin calling out to her mother while still an embryo in an egg, but the mother isn't there to cluck back or gently turn the egg. At one day old, the chick can count and identify whether two objects are the same or different. Though birds can grasp abstract concepts and are now considered "feathered apes" in terms of cognitive processing (Emery), the laying hen is viewed as stupid.

Egg farms may keep captive 10 million birds at time (Potts, 158). In these industrial farms, eggs are easy to collect, as they slide down a gradient while the birds sit constrained within cages with four to nine other laying hens. Their cages, like so many Legos, are lined up horizontally and vertically. Behavior innate to chickens, including preening, scratching, and turning around, cannot occur. Nest-building, "associated with natural changes in hormone levels occurring around the time of egg release," is thwarted (Potts, 161). While urine and feces drop from the cages above the hens, they also collect in vast pools inside the battery shed, causing intense ammonia fumes that burn the hens' eyes and lungs.

Chicken scholar Karen Davis identifies three specific areas of deprivation that intensify the laying hens' suffering: the impact of industrial farming on eyesight, hearing sounds, and hygiene. Davis explains, "Chickens have much better eyesight than human beings have, including long distance and close-up vision. They have full-spectrum color vision from the infrared to the ultraviolet." In their "captive environment," though, they are "deprived of all natural colors and forced to see only varying shades of brown" (Davis 2022, 10).

In the forests where they evolved, a "vibrant, meaningful ruckus" could be heard, "communicating information through a natural symphony of sounds." In captivity, chickens instead experience "dead silence, cries of distress, yelling of workers, and din of machinery" (Davis 2022, 11). In the place of notes "of joy or enthusiasm so natural to chickens living in a wild or sanctuary environment . . . every sound, every silence, is negative—threatening traumatizing, distressful." Finally, chickens have a need "to keep themselves, their feathers and their skin, clean and refreshed." Dust baths and preening their feathers ("with the preen oil from their preen gland") are ways chickens practice bodily hygiene. "So powerful is their instinct to bathe themselves and be clean that they will pathetically perform what poultry scientists call 'vacuum dust bathing,' even on the wire floor of a crowded battery cage" (Davis 2022, 11).

As with the cow's production of mammalian milk, myths exist about the laying hen. Perhaps the most ingenious one acts to console egg consumers: the myth that "only happy hens lay eggs." In her 2012 book *Chicken,* scholar Annie Potts counters this myth as she describes how hens lay eggs whether they are happy or not (Potts, 158–163) Karen Davis points out that

> chickens do not gain weight and lay eggs in inimical surroundings because they are comfortable, content, or well-cared for, but because they are specifically manipulated to do these things through genetic and management techniques that have nothing to do with happiness except to destroy it.
>
> *(1996, 20)*

It is hard to imagine happy hens when, unlike their ancestors, who would normally lay two clutches of about 12 eggs a year, twenty-first–century hens have been genetically manipulated to produce 270 or more eggs per year. A certain percentage will suffer from "caged layer fatigue" (CLF)—the equivalent of osteoporosis in chickens that "primarily affects caged chickens that are [forced to be] at a high level of egg production" (Jacob, parenthetical additions mine). This increased production means hens often suffer from prolapsed uteri, a painful condition in which the uterus of the hen extends outside of the body. This pink tissue will then be pecked at by other chickens (called "vent pecking" or "cloacal cannibalism"), causing the chicken to die from "hemorrhage and shock" (McMurray). This process in which "the complete oviduct and parts of the adjacent intestinal tract" are pulled from the abdominal cavity is known as "peckout" (Espinosa). Also known as "blowout," it happens when "the vagina swells, cannot retract, and remains prolapsed." The hen's reproductive organ is literally turned inside out.

Prolapse happens to about 10 percent of laying hens (cited in Ray et al.). The causes of hens' prolapses (and the remedies suggested) reveal the production practices for this form of feminized protein:

- The reproductive systems of birds are stimulated by the long days of summer to form and lay eggs. But, when reproduction is monetized, production now must occur year-round, and artificial lighting is used to extend the daylight hours. "Hens require 14 hours of day length to sustain egg production" (Ray et al.). Backyard farmers are encouraged to keep lights on for 16 hours to trick hens into producing more eggs. However, "Sudden increases in day light results in big sized eggs (jumbo eggs) that can result in prolapse" (Ray et al.). Prolapse has been found to be more common in industrial farming, (Greenacre), where hens live under artificial light for at least 17 hours a day.
- Large eggs, popular with consumers, cause prolapse. "Jumbo" eggs often are produced when, at about 75 weeks old, the hens' reproductive systems begin to wear out. To trick the stressed hens' bodies into more weeks of egg production, hens will be starved for up to two weeks. When they start to lay eggs again, those eggs are most likely to be "jumbo."
- To avoid cloacal cannibalism, the hens' very sensitive beaks are amputated with a hot blade without anesthesia. With their deformed beaks, hens cannot eat and drink properly.

When their egg-laying days are over, the hens who do not die as a result of prolapses and cannibalism will be suffocated, gassed, or buried alive in landfills.

Joan Dunayer identified the way misogyny employs speciesist terms like *hen party, henpeck, brood,* and *old biddy*, concluding "If hens were not held captive and treated as nothing more than bodies, their lives would not supply symbols for the lives of stifled and physically exploited women" (1995, 12–13). Irene Lopez adds to this analysis in comparing the animal terms for men versus women: "Men are frequently referred to as *studs*, *bucks*, *wolves*, *toros* (bulls), *zorros* (foxes) and *linces* (lynxes) whereas women are referred to with such metaphors as *chick*, *bird*, *kitten*, *pollita* (chicken) or *gatita* (kitten)" (2009, 82). Animal metaphors for men, she observes, are based on "the size (big), strength, and habitat of the animal (wilderness)." Women, on the other hand, are associated with "small domestic animals," like the hen. These terms enact stereotypical gender views: "being wild animals, men need freedom and no restraint; however, the fact that women are presented as domestic or livestock animals might suggest that a woman's place should be confined to the domestic arena" (Lopez, 83). Lopez's discussion of language as it conveys the low status of domesticated animals and women adds another dimension to Eisen's assertion about how the farm for animals is akin to the family for women.

In the mid-1990s, Karen Davis criticized environmental philosophy for ignoring the plight of the chicken. Her thoughts were prompted by the rejection by the academic journal, *Environmental Ethics*, of her essay "Clucking Like a Mountain," in which she asked the question whether Aldo Leopold's recommendation to *think like a mountain* would conflict with her commitment to *think like a chicken,* specifically a battery hen. (I invited her to describe this experience in an essay for *Animals and Women.*) She believed that the embrace of holism abandoned—as well as justifying the abandonment of—creatures whose lives had been altered by human control. One of the reviewers chided her with "too much first person singular" framing and, as Davis describes it, "snorts that 'sixteen billion chickens cannot tell me the psychic price of scientific enlightenment' " (1995, 208). Her encounter with *Environmental Ethics* reminded her of "Western culture's smug identification with the 'knower' at the expense of the 'known' " (197), and directs us once again to contemplate Carolyn Merchant's explanation of how nature, and its creatures, came to be viewed from a mechanical framework. The reception to the idea that an epistemological claim could be made for the battery hen occurred from within a patriarchal philosophical framework.

Misogyny and the Gender Binary

We do not know how cows and chickens experience gender, but we can identify how humans project their assumptions and beliefs about gender onto cows and chickens. Gender is a human fixation after all; the animal world doesn't fit into the gender binary, but animal agriculture needs, relies on, and reinforces the gender binary (see Adams 2019, 16–19). Femaleness becomes both an identity imposed, and an oppression, enacted on certain animals by animal agriculture. Throughout their lives, cows and chickens experience their lives interpreted by humans who hold deeply misogynistic beliefs, who live in patriarchal spaces, and are committed to the idea of the gender binary. In her research, Kathryn Gillespie found relentless "sexually violent commodification of the female body" (Gillespie 2014, 5). This misogynist commodification can be seen throughout the animal agriculture industry in journals, advertisements, practices, language, and representations.

Animals as Absent Referents

In *The Sexual Politics of Meat,* I suggested three ways by which animals become *absent referents.* One is literally: through consumption of dead animals, the animals are literally absent because they are dead. Another way is definitional: "when we eat animals we change the way we talk about them, for instance, we no longer talk about baby animals but about veal or lamb." The word *meat* has an absent referent, the dead animals. The third way is metaphorical, as animals become metaphors for describing people's experiences. "In this metaphorical sense, the meaning of the absent referent derives from its application or reference to something else" (21).

Animals used to produce feminized protein are also absent referents in multiple ways. To begin with, because they are not dead, it is assumed they are not harmed. The denial by animal agriculture that cows and chickens might desire anything (the desire for comfort, the desire to run outside and not be standing on concrete, the desire to be scratching and sunbathing, the desire to be with their offspring) is another way female animals are rendered absent—desires and feelings and motives are emptied from them like the milk taken from them and the eggs they are forced to produce. But then animal agriculture manufactures an acceptable desire: a sexualized one. Drug companies depict sexy, buxom females who want to be used, who desire to be desired, who wish to be pregnant, who wish to lay eggs (see Adams, 2019 *passim*). The representations assert that the cows and chickens *need this.* They *want* to be made pregnant or to lay eggs. They *want* to "give" their milk; they *want* to feed us. The animal disappears in these multiple effects of the structure of the absent referent: Who they are, who they could be, and what they want is denied. Then the being who has been evacuated of any selfhood in the exploiters' world is made a receptacle, both through forced reproduction and through representations that imply this emptied self only desires to be that (re)producer.

Seeing a truck carrying mammalian milk and its gendered imagery that proclaimed the happiness of the cow inspired A. Breeze Harper to reflect on mythmaking and its role in distracting consumers from the experiences of the exploited and the abused. Harper notes how "Aunt Jemima Brands were notorious for selling the idea to white Americans that Black cisgender women's bodies existed to happily feed the nation, despite the reality of how violent and cruel antebellum slavery and Jim Crow eras were for Black people." Harper says her point is not in making a comparison between Black people and cows, nor is she saying they suffer in the same way. She is raising the issue of how we recognize the complacency that accepts dominant cultural narratives. Why, for instance, she asks, do white vegans become upset when non-vegans unquestioningly accept the dominant discourse about cows, and yet, as white vegans, they fail to question the dominant discourse about race? (Harper, 2015).

Conflating the Control of Reproduction

A dominant culture that expends much energy denying animals have feelings finds itself producing popular cultural images that impute sexual desire to animals used as food and female animals used as baby producers, mammalian milk producers, and egg producers. In terms of animals relied on for feminized protein, one particular trope highlights the use of the gender binary and the expression of misogyny: advertisements aimed at farmers and more general cultural representations often deliberately conflate whether they are referring to the control of women's or other female animals' reproduction.

Regarding reproductive rights, a frequent image used in the past decade to evoke their loss is a photograph of a cow. The text that accompanies the photograph announces, "There's a term for creatures not permitted to control their reproduction. That term is LIVESTOCK." After the 2022 Dobbs decision by the United States Supreme Court that overturned *Roe v. Wade*, allowing abortion to be made illegal, a flyer containing the statement and an image of a cow were circulated throughout the prochoice protest in front of the Supreme Court.

The fetishizing of the femaleness of cows carries over to attitudes toward people who are pregnant. The result of misogyny's fixation on controlling women's reproductive capacities and animal agriculture's need to control female animals' reproduction results in a triply oppressive belief: that women should stay pregnant if they are pregnant, that female animals are "ours" to use, and that women are the only people who get pregnant. As a result of the latter tenet, trans men and nonbinary folk struggle to receive prenatal care when they find themselves pregnant.

Animal agriculture's complete control of female reproduction breeds a language of entitlement that carries oblique or explicit anti-abortion references, from Bovi-Shield's question about a cow, "If she can't stay pregnant, what else will she do?" to a McDonald's advertisement that depicted an egg and their signature "Egg McMuffin" breakfast dish with the caption, "Every egg's dream."[6] Even the unhatched egg's desire is to be consumed. As I write in *The Pornography of Meat*, "In attributing to the unborn a mind, desires, and the need for fulfillment this ad follows the language of antiabortionists who have articulated the doctrine of 'fetal personhood'" (2019, 92).

Figure 2.4 Lauren Melchionda sent me this postcard that was distributed on June 24, 2022, in front of the United States Supreme Court after the announcement of the verdict in the *Dodds v. Jackson Women's Health Organization* that found the Constitution does not confer the right to abortion and which overruled *Roe v. Wade*. Photograph © Lauren Melchionda. Used by permission.

Juxtaposed with the need for animal agriculture to be exempted from bestiality laws is the appearance of pregnancy announcements that circulate in farming groups on social media showing a woman captured within what is called a "squeeze chute" that would usually hold a cow to immobilize her during artificial insemination. She is shown on all fours, and next to her stands a man wearing a bloodied shoulder-length sleeve worn by farmers when they forcibly impregnate a cow. His unsleeved hand holds a sign that says "BRED." (For one example, see Sadeque.) These photographs are greeted with delight, and congratulations, within the community of dairy farmers. Others respond with dismay, if not disgust. These different responses highlight the insights in Manon Garcia's *We Are Not Born Submissive: How Patriarchy Shapes Women's Lives*. Garcia explains that "submission is socially prescribed to women. There are therefore possible advantages for them in submitting." Garcia continues, "If submission is the result of a cost-benefit analysis that is tilted toward submission by the patriarchal social norms, then it is a mistake to understand submission as a choice. Submission is a consent to one's destiny as it is predetermined by social norms" (193–194). In this reading, by their participation in the staging of the

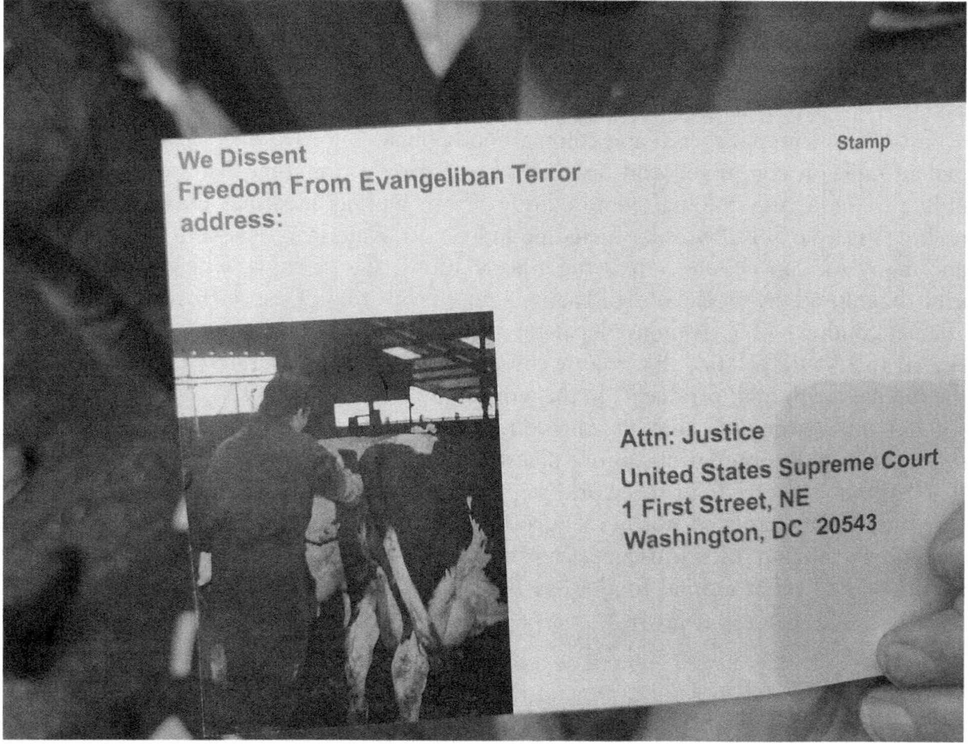

Figure 2.5 The reverse side of the postcard distributed in front of the U.S. Supreme Court, showing forced reproduction via artificial insemination of a cow. Note that the right arm is sheathed. Photograph © Lauren Melchionda. Used by permission.

pregnancy announcement in the context of the forced breeding of cows, the posed women acknowledge that women are situated like cows in a patriarchal culture. That is the destiny predetermined by patriarchal social norms.

Colonialism

Many writers have shown how Western cultures co-constructed meat eating, drinking mammalian milk, masculinity, nationhood, and white supremacy. The role of colonialism in terms of bovines is a many hydra–headed relationship: There were no cows on the land mass known as North American until the Spanish and English colonists introduced them. Cows were used to help expand settler colonialism and then became the occupiers of the land from which Native Americans had been displaced. (See Eisen 2017 for a summary of colonialism and the use of mammalian milk.) Feminist postcolonial legal scholar Maneesha Deckha writes,

> Veganism, in reality, rejects a practice (dairy farming) that was constitutive of settler colonialism in North America and which still promotes colonial familial ideologies

> while constructing Indigenous peoples and other non-Europeans (who disproportionately cannot tolerate lactose) as abnormal.
>
> *(2020, 244)*

Further, nutritional racism and colonial food policies imposed the consumption of mammalian milk on conquered and disenfranchised people. Jonathan Saha calls mammalian milk a "conquering colonial commodity" (Saha). DuPuis identifies the numerous "non-milking" people of the world, "including indigenous American, East and Southeast Asian, and many African peoples." It is these populations, she explains, who do not have "the gene that allows the production of lactase—the enzyme that digests lactose sugars in milk—after childhood" (27). Feminist legal theorist Mathilde Cohen points to "India, China, Brazil, and New Zealand, all formerly colonized lands" that "currently figure on the list of the top ten cow's milk producers in the world" (2017a, 269).

Humans are the only animals who continue to drink milk past infancy. Yet the majority of the world's population cannot digest milk taken from cows. (Cohen cites a statistic that "about 75 percent of the world population is lactose intolerant" [2017a, 269].) It is mainly descendants of northern Europeans who are able to digest mammalian milk. The universalizing of Euro-American bodies by exhorting nourishment from dairy products (and other animal foods) has been called *bio-ethnocentrism, food oppression,*

Figure 2.6 The healthy white man with his love object, "Fresh Milk," in a pastoral Irish setting. The white shirt, the white man, and the white box of milk all tacitly performing an affirmation of those who are most able to drink milk. Ireland, 2019. Photograph copyright © Roger Yates. Used by permission.

and *nutritional racism*. Eisen refers to the "racialized promotion of milk" (2019a, 86). It actively involves the suppression of Afro-diasporic vegan cooking traditions, Meso-American cooking traditions, and other Indigenous plant-based traditions. And while the inability to digest mammalian milk from nonhuman animals has been called "lactose intolerance," activist groups like the Food Empowerment Project recommend instead the term "lactose normal," because "it is not natural to consume the milk of another species." They remind their readers that "lactose intolerance implies there is something 'wrong' with Black, Brown and Indigenous people who are not able to digest milk—a product of colonization" (Food Empowerment Project). Their recent activist outreach included anti-dairy billboards playing off the popular "Got Milk?" ad campaigns of the past 30 years, but their ads ask, "Got Colonization?"

Chickens are not native to the Americas; their first home was the tropical forests of Southeast Asia (Davis 2022, 10). Some speculate that chickens may have been introduced by Polynesians in the fourteenth century (Adler and Lawler). The changes that brought about the current "farming" practices toward chickens took place in the United States. When Annie Potts writes that "The modern enslavement of chickens is a very Western and capitalist tale" (139), she pinpoints how the development of industrial farming created this enslavement. The Petaluma area of California, which became known as the

Figure 2.7 Food Empowerment Project's "Got Colonization" billboard on Bancroft Avenue in Oakland, California. Copyright © Food Empowerment Project. Used by permission.

"Nation's Egg Basket," initiated the current practices in the early part of the twentieth century. The automated systems that they began eventually became models for keeping large numbers of animals captive, for instance, pigs. But it was chickens who were the first to be confined indoors and viewed as animal machines to be manipulated by antibiotics and genetic selection. The Egg Basket model then took hold like a virus, inculcating practices that deprived animals of basic needs, such as the need to move, and casting them as living machines.

Reception, Reclamation, and Innovation

The challenge to problematize what has been naturalized and normalized regarding other animals is ongoing. My coining of the term "feminized protein" was one attempt at this. I saw it as a way for theory to inform activism. I hoped that the knowledge of the situation of the animals being used in this way would bring about a boycott of all those products I labeled "feminized protein." Many of my readers have written that it was their encounter with this term that moved them to become vegans. Artists (Gonzalez), sanctuary workers (Schwartz), and other activists have incorporated the term "feminized protein" into their work.

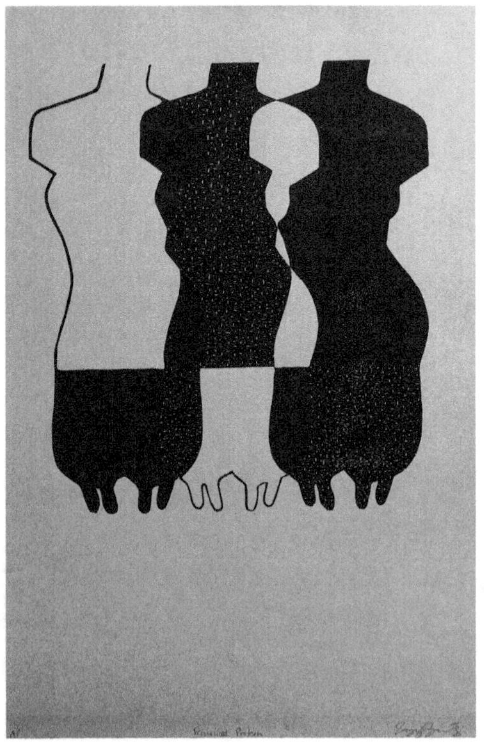

Figure 2.8 *Feminized Protein*, linocut 19" × 12.5" 2013. Copyright © 2013 Suzy González.

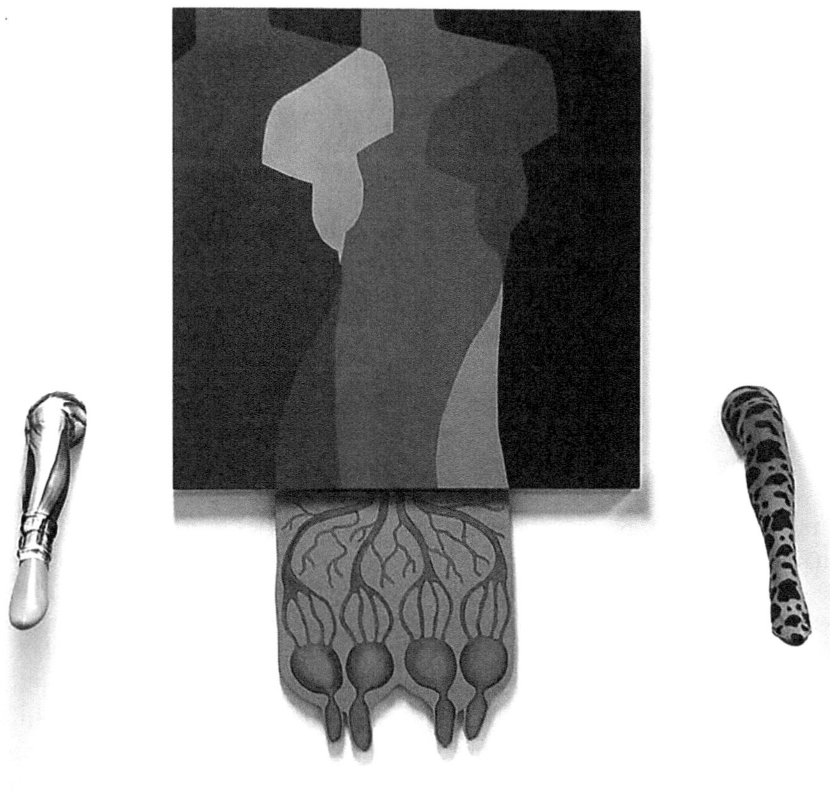

Figure 2.9 *Feminized Protein (Installation)*. Spray paint on canvas, acrylic on cut panel, stocking on mannequin legs. Dimensions variable. Copyright © 2013 Suzy González.

Discussing "The Disruptive Possibilities of Plant Milk," Iselin Gambert argues that the term "best captures the specific ways in which female bodies in particular are exploited within the patriarchal system within which the current animal industrial complex exists." In a presentation on "design as a means to resist planetary collapse," Delfina Fantina, a researcher and lecture in design products, and Harriet Harriss, an architect and dean of the Pratt School of Architecture, discussed "the ecological imperative driving change within design curriculum and pedagogy." Their topics ranged "from speciesism to feminized protein, the spatial rights of automated beings to decolonization" (see Fantini and Harriss).

In the past 20 years, there has been a flourishing of feminist, antiracist, and postcolonial scholarship on the lives of chickens and cows. Important developments in discussing the monetizing of mammalian milk can be found in the careful examination of sites of oppression by geographer Kathryn Gillespie; the focus on the dairying situation in India and its relationship to nationalism by urban planning scholar Yamini Narayanan; and the fascinating and important work on milk by feminist legal scholars including Yoriko Otomo, Mathilde Cohen, Jessica Eisen, and Maneesha Deckha. The use of terms such as *nursing materials of a cow, mammalian milk,* and *lactation product* are attempts at naming that center the individual who is required to produce it. (See also Gurel, Hill, Tomasello, et al., Wrenn, etc.)

For chickens, there has been the early and consistent voice of the late Karen Davis, a former English professor who discovered upon moving into a new home a chicken in the barn, abandoned when the chickens were collected to be taken to slaughter. In that moment, Davis became an advocate for chickens, geese, and turkeys and soon founded United Poultry Concerns. She has written numerous books in an attempt to educate about and change the current unethical treatment of fowl. Annie Potts, in her examination of the chicken as part of the Reaktion Animals Series (2012), offered a sophisticated overview of the history of the chicken and a piercing critique of the current treatment of chickens.

As I indicated, the point of naming and examining feminized protein accompanied the belief that once people knew about how reproduction had been forced into production, they would boycott the resulting products. As an aid to this boycott, the past 40 years have witnessed the development of vegan alternatives to the lactation product of cows and goats and the eggs forcibly produced by chickens. Vegan recipes from the 1980s and 1990s offered what were often known as "steddas," short for *instead of*. Nutritional yeast became a valuable cheese substitute, soy milk and almond milk early alternatives to mammalian milk. Now plant milks are made as well from oats, rice, cashews, coconuts, pecan, walnuts, and pine nuts and pumpkin, sesame, or sunflower seeds (see Cole). To help baked goods rise, the use of clabbered plant milk (by adding a little apple cider vinegar or lemon juice) was adopted to replace buttermilk. Soon, cashews became the go-to nut for creams (for instance, in alfredo sauce). Plant-based fromageries developed a process of culturing and fermenting that brought vegan cheeses to a new level (see, for instance, Conroy, McAthy, Piatt, Schinner, Schlimm). As the *New York Times* reported, "Cheesemakers are pushing the boundaries of cultured, plant-based milks, producing more compelling vegan cheeses than ever before" (Rao, 2021).

The custom of using eggs in food has resulted in their presumed irreplaceability. But the reason for this is that the egg has been confused with its function: a variety of other foodstuffs *can* perform the same function in cuisine as the eggs. Rather than eggs as binders in baked goods, one could use arrowroot, bananas, soft tofu, ground flax seeds with water, or the powdered product Egg Replacer. Eggs are not necessary for omelets, scrambles, or quiches; instead chickpea flour replaces eggs for omelets and tofu for eggs in quiches or scrambles, with a little kala namak black salt for the "eggy" taste and turmeric for its color. In addition, products made from mung beans or chickpeas provide both liquid and powedered egg replacements. Among these are Just Egg, Be Leaf, Follow Your Heart Vegan Egg, Simply Egg, Nabati Plant-Based Eggz, Vegg, Peggs, and Nummy Nibbles. The discovery that the water from chickpeas (aquafaba) that could be seen foaming as it was emptied from a can could substitute for beaten egg whites for meringues opened the possibility for baked goods thought to always require eggs (see Dever). Aquafaba can also replace eggs used in the two step egg-flour dipping method for frying foods from tofu to cauliflower; so, too, can a combination of Dijon mustard mixed equally with water.

While male bodies can also lactate at times (see Cohen 2017b), animal agriculture has not identified a way to monetize this, so the term "feminized protein" offers a way to examine and contemplate the lives (and deaths) of animals marked as female. In the end, however, the adopting of the neologism "feminized protein" is less important than the consciousness it was seeking to express: the awareness of the injustice of manipulating and exploiting the reproductive functions of captive animals identified as female to force them to continually produce from their bodies a product for consumption by humans.

Acknowledgements. I would like to thank Chloë Taylor and Jessica Eisen for their careful reading of this entry. In addition, I am grateful to lauren Ornelas and the Food Empowerment Project for permission to use a photograph of "Got Colonialism?" billboard, Justin van Huyssteen for his tweet to me of the fragmented cow in Pretoria, Lauren Melchionda for taking—and sharing with me—the photographs of the postcard distributed at the U.S. Supreme Court on the day of the Dodds decision, and the artist Suzy Gonzalez for permission to reproduce her artwork. For the other images used in this chapter, I drew on images sent to me and used in the updated edition of *The Pornography of Meat* (2019) and thank Roger Yates and Camille Brunel for their permission to use their photographs again.

Notes

1 See "All of One Flesh: The Rights of Animals," quoting Vyvyan: "In fact, although the normal route taken to personal 'non-reliance' on animal food products has been firstly to give up eating animal flesh and then later to drop dairy products and eggs, logically—in terms of suffering caused—one should firstly give up dairy products, as I am convinced that a cow endures much more suffering than the beef animal through continually being robbed of her young" (Benney 1983, 143).
2 To be clear, while animals are forced to work all the time, and because their "work" is their bodies, I don't endorse the move to see animals as laborers. Jessica Eisen (2019b) lays out the reason for not making that theoretical move. What I am trying to show here is how reproduction itself became an aspect of production.
3 Various studies have confirmed the 25% statistic including Harris et al., Singleton et al., Maurer et al. (Maurer et al. found a surprisingly higher rate of 39.5%.) A Danish study found that 90% of farmers knew that their cows were pregnant when they were sent for slaughter (Nielsen et al.).
4 This ad can be found here: https://www.yumpu.com/en/document/view/38922749/brochure-for-bovi-shield-gold-hb-dairy-wellness
5 The Zinpro ad can be found here: https://caroljadams.com/spom-examples/zk2x4slkb6dk3qvxc10kqqymtl2fna
6 This ad can be seen at the "Awful Advertisements," blog here: http://awfuladvertisements.blogspot.com/2008/06/mcdonalds-every-eggs-dream.html

Bibliography

Adams, Carol J. *The Sexual Politics of Meat: A Feminist-Vegetarian Critical Theory*. London and New York: Bloomsbury, 1990.
Adams, Carol J. "Feminized Protein: Meaning, Representations, and Implications." In *Making Milk*, edited by Mathilde Cohen and Yoriko Otomo. London and New York: Bloomsbury, 2017.
Adams, Carol J. *The Pornography of Meat: New and Updated*. New York and London: Bloomsbury Academic, 2019.
Anderson, Virginia DeJohn. *Creatures of Empire: How Domestic Animals Transformed Early America*. Oxford and New York: Oxford University Press, 2004.
Adler, Jerry and Lawler, Andrew. "How the Chicken Conquered the World." *The Smithsonian,* June 2012. Retrieved from: https://www.smithsonianmag.com/history/how-the-chicken-conquered-the-world-87583657/
Benney, Norma. "All of One Flesh: The Rights of Animals." In *Reclaim the Earth: Wolmen Speak Out for Life on Earth,* edited by Leonie Caldecott and Stephanie Leland. London: The Women's Press, 1983. 141–151.
Bergo, Alan. "Unlaid Eggs." *Forager/Chef,* March 2020. Retrieved from: https://foragerchef.com/unlaid-eggs/

Burros, Marian. "What the Egg Was First." *The New York Times,* February 7, 2007. Retrieved from: https://www.nytimes.com/2007/02/07/dining/07eggs.html, accessed August 16, 2022.

Buza, Marianne and Babisz, Shelby. "How Now Down Cow." *MSU Extension* (2017). Retrieved from: https://www.canr.msu.edu/news/how_now_down_cow#

Cohen, Mathilde. "Animal Colonialism: The Case of Milk." *American Journal of International Law Unbound* 111 (September 2017a): 267–271.

Cohen, Mathilde. "The Lactating Man." In *Making Milk. The Past, Present and Future of Our Primary Food,* edited by Mathilde Cohen and Yoriko Otomo. London and New York: Bloomsbury, 2017b.

Cohen, Mathilde and Otomo, Yoriko, ed. *Making Milk: The Past, Present, and Future of Our Primary Food.* London: Bloomsbury, 2017.

Cole, Candia Lea. *Not Milk . . . Nut Milks!* Santa Barbara: Woodbridge Press, 1992.

Conroy, Skye Michael. *The Non-Dairy Evolution Cookbook.* [no publisher info provided], 2014.

Cumming, Ed. "Raising the Steaks: Meet the Elderly Spanish Cows Destined for Dinner Plates." *The Guardian,* October 11, 2015. Retrieved from: https://www.theguardian.com/global/2015/oct/11/raising-the-steaks-meet-the-elderly-spanish-cows-destined-for-dinner-plates

Cusack, Carmen M. "Feminism and Husbandry: Drawing the Fine Line between Mine and Bovine." *Journal for Critical Animal Studies* 11, no. 1 (2013): 24–44.

Davis, Karen. "Thinking Like a Chicken: Farmed Animals and the Feminine Connection." *Animals & Women: Feminist Theoretical Explorations,* edited by Josephine Donovan and Carol J. Adams. Durham and London: Duke University Press, 1995.

Davis, Karen. *Poisoned Chickens, Poisoned Eggs: An Inside Look at the Modern Poultry Industry.* Summertown, TN: Book Publishing Company, 1996.

Davis, Karen. "Interspecies Sexual Assault: A Moral Perspective." *Animal Liberation Currents* (2017). Retrieved from: https://www.animalliberationcurrents.com/interspecies-sexual-assault/#more-1680.

Davis, Karen. "Violated Expectations in the Experience of Factory-Farmed Chickens." *Poultry Press* 32, no. 1 (2022): 10–11.

Deckha, Maneesha. "Toward a Postcolonial, Posthumanist Feminist Theory: Centralizing Race and Culture in Feminist Work on Nonhuman Animals." *Hypatia* 27, no. 3 (2012): 527–545.

Deckha, Maneesha. "Veganism, Dairy, and Decolonization." *Journal of Human Rights and the Environment* 11, no. 2 (2020): 244–267.

Dekker, Thomas. *The Great Frost: Cold Doings in London.* Cited in Alexandra Harris, *Weatherland: Writers & Artists Under English Skies.* New York: Thames and Hudson, 1607 (2015).

Desaulniers, Élise. *Cash Cow: Ten Myths about the Dairy Industry.* New York: Lantern Books, 2016.

Dever, Zsu. *Aquafaba: Sweet and Savory Vegan Recipes Made Egg-Free with the Magic of Bean Water.* Woodstock, VA: Vegan Heritage Press, 2016.

Dunayer, Joan. "Sexist Words, Speciesist Roots." *Animals and Women: Feminsit Theoretical Explorations,* edited by Carol J. Adams and Josephine Donovan, 11–31. Durham: Duke University Press, 1995.

Dunayer, Joan. *Animal Equality: Language and Liberty.* Derwood, MD: Ryce Publishing, 2001.

DuPuis, E.M. *Nature's Perfect Food: How Milk Became America's Drink.* New York: New York University Press, 2002.

Eisen, Jessica. "Milk and Meaning: Puzzles in Posthumanist Method." In *Making Milk: The Past, Present and Future of our Primary Food,* edited by Mathilde Cohen and Yoriko Otomo. London: Bloomsbury, 2017.

Eisen, Jessica. "Milked: Nature, Necessity, and American Law." *Berkeley Journal of Gender, Law & Justice* (2019a). Retrieved from: https://lawcat.berkeley.edu/record/1128892

Eisen, Jessica. "Down on the Farm: Status, Exploitation, and Agricultural Exceptionalism." *Animal Labour: A New Frontier of Interspecies Justice?* edited by Charlotte E. Blattner et al. Oxford and New York: Oxford University Press, 2019b.

Emery, Nathan. *Bird Brain: An Exploration of Avian Intelligence.* Princeton, NJ and London: Princeton University Press, 2016.

Espinosa, Rodrigo A. "Oviduct in Poultry." *Merc Manual* (2019). Retrieved from: https://www.merckvetmanual.com/poultry/disorders-of-the-reproductive-system/prolapse-of-the-oviduct-in-poultry

Fantina, Delfina and Harriss, Harriet. "Design Curricula for Climate Crisis." (2020). Retrieved from: https://www.rca.ac.uk/news-and-events/news/design-curricular-climate-crisis/

Food Empowerment Project. "Got Colonization?" Retrieved from: https://gotcolonization.org/colonization/

Gambert, Iselin. "Got Mylk?: The Disruptive Possibilities of Plant Milk." *Brooklyn Law Review* 84 (2019). Retrieved from: https://brooklynworks.brooklaw.edu/blr/vol84/iss3/3/

Garcia, Manon. *We Are Not Born Submissive: How Patriarchy Shapes Women's Lives*. Princeton, NJ and Oxford, England: Princeton University Press, 2021.

Gillespie, Kathryn. "Sexualized Violence and the Gendered Commodification of the Animal Body in Pacific Northwest U.S. Dairy Production." *Gender, Place and Culture* 21, no. 10 (2014): 1321–1337.

Gillespie, Kathryn. *The Cow with Ear Tag #1389*. Chicago: University of Chicago Press, 2018.

Gonzalez, Suzy. "Feminized Protein." Mixed Media (2013). Retrieved from: https://suzygonzalez.com/artwork/3268735-Feminized-Protein.html

Greenacre, C.B. "Reproductive Diseases of the Backyard Hen." *Journal of Exotic Pet Medicine* 24, no. 2 (April 2015): 164–171. Retrieved from: https://www.ncbi.nlm.nih.gov/pmc/articles/PMC7106171/

Gurel, Perin. "Live and Active Cultures." *Gastronomica* 16, no. 4 (Winter 2016): 66–77. Retrieved from: https://www.jstor.org/stable/10.2307/26362396

Harper, Breeze. "Fairy Tales and Cowlifornia Nightmarin': Who Else Is Sick of the Sugarcoating of The Endless Nightmare of Systemic Oppression?" The Sistah Vegan Project (2015). Retrieved from: http://sistahvegan.com/2015/08/06/fairy-tales-and-cowlifornia-nightmarin-who-else-is-sick-of-the-sugarcoating-of-the-endless-nightmare-of-systemic-oppression/

Harris, Chris. "Finding the Value in Processing Spent Laying Hens." *The Poultry Site*, December 20, 2019. Retrieved from: https://www.thepoultrysite.com/articles/finding-the-value-in-processing-spent-laying-hens

Harrison, Ruth. *Animal Machines*. London: Vincent Stuart Publishers Ltd., 1964.

Hill, K. "Happy Hens or Healthy Eggs—A Summative Content Analysis of How Hens Are Represented In Supermarket Egg Boxes Narratives." *TRACE: Journal for Human-Animal Studies* 7, no. 1 (2021): 70–94. https://doi.org/10.23984/fjhas.98684

Jacob, Jacquie. "Cage Layer Fatigue." *Small and Backyard Poultry* (2022). Retrieved from: https://poultry.extension.org/articles/poultry-health/common-poultry-diseases/cage-layer-fatigue/

Lopez, Irene. "Of Women, Bitches, Chickens and Vixens: Animal Metaphors for Women in English and Spanish." *Cultura, Lenguaje y Representación/Culture, Language and Representation* 7 (2009): 77–100. Retrieved from: https://core.ac.uk/download/pdf/39085608.pdf

Maurer, P., Lücker, E., Riehn, K. "Slaughter of Pregnant Cattle in German Abattoirs—Current Situation and Prevalence: A Cross-sectional Study." *BMC Veterinary Research* 12 (June 7, 2016): 91. Retrieved from: https://www.ncbi.nlm.nih.gov/pmc/articles/PMC4895965/#

McAthy, Karen. *Plant-Based Cheesemaking: How to Craft Real, Cultured, Non-Dairy Cheeses*. New Society Publishers. Gabriola Island, Canada: New Society Publishers, 2017.

McMurray Staff. "Gail Damerow Discusses Prolapse and Egg Binding in Laying Hens." August 23, 2018. Retrieved from: https://blog.mcmurrayhatchery.com/2018/08/23/gail-damerow-discusses-prolapse-and-egg-binding-in-laying-hens/

Merchant, Carolyn. *The Death of Nature: Women, Ecology, and the Scientific Revolution*. New York and London, 1979.

Narayanan, Yamini. "'Cow Is a Mother, Mothers Can Do Anything for Their Children!' Gaushalas as Landscapes of Anthropatriarchy and Hindu Patriarchy." *Hypatia* 34, no. 2 (2019): 195–221.

Narayanan, Yamini. *Mother Cow, Mother India: A Multispecies Politics of Dairy in India*. Stanford: Stanford University Press, 2023.

Nielsen, Søren Saxmose, Sandøe, Peter, Kjølsted, Stine Ulrich, and Agerholm, Jørgen Steen. "Slaughter of Pregnant Cattle in Denmark: Prevalence, Gestational Age, and Reasons" *Animals* 9, no. 7 (2019): 392. Retrieved from: https://www.ncbi.nlm.nih.gov/pmc/articles/PMC6681307/, accessed August 17, 2022.

Noske, Barbara. Beyond Boundaries: Humans and Animals. Montreal: Black Rose Books, 1997.

Pateman, Carole. *The Sexual Contract*. Cambridge: Polity, 1988.

Piatt, Julie. *This Cheese Is Nuts! Delicious Vegan Cheese at Home*. New York: Avery, 2017.

Potts, Annie. *Chicken*. London: Reakton Books, 2012.

Puerto-Parada, M., Bilodeau, M.E., Francoz, D., Desrochers, A., Nichols, S., Babkine, M., Arango-Sabogal, J.C., and Fecteau, G. "Survival and Prognostic Indicators in Downer Dairy Cows Presented to a Referring Hospital: A Retrospective Study (1318 Cases)." *Journal of Veterinary Internal Medicine* 35, no. 5 (September 2021): 2534–2543. https://doi.org/10.1111/jvim.16249. Epub 2021 Aug 13. PMID: 34387390; PMCID: PMC8478027.

Rao, Teja. "Vegan Cheese, but Make It Delicious." *New York Times,* April 16, 2021. Retrieved from: https://www.nytimes.com/2021/04/16/dining/vegan-cheese.html

Ray, Subhasish, Swain, Partha, Amin, Rooh, Nahak, Anil, Sahoo, Saroj, Rautray, Amiya, and Mishra, Akash. "Prolapse in Laying Hens: Its Pathophysiology and Management: A Review." *Indian Journal of Animal Production and Management* 29, no. 3–4 (2013): 17–24.

Rogers, D. "Strange Noises Turn Out to Be Cows Missing Their Calves." *Newburyport News,* October 23, 2013. Retrieved from: http://www.newburyportnews.com/news/local_news/strange-noises-turn-out-to-be-cows-missing-their- calves/article_d872e4da-b318–5e90–870e-51266f8eea7f.html

Rosenberg, Gabriel N. "How Meat Changed Sex: The Law of Interspecies Intimacy after Industrial Reproduction." *GLQ: A Journal of Lesbian and Gay Studies* 23, no. 4 (2017): 473–507. Retrieved from: https://www.muse.jhu.edu/article/673278.

Rosenberg, Gabriel N., and Dutkiewicz, Jan. "The Meat Industry's Bestiality Problem." *The New Republic,* December 11, 2020. Retrieved from: https://newrepublic.com/article/160448/meat-bestiality-artificial-insemination

Sadeque, Samira. "People Are Grossed Out by Cow Insemination-themed Pregnancy Announcement. The Announcement Is Drawing Comparisons to 'The Handmaid's Tale'." *Daily Dot* (2019). Retrieved from: https://www.dailydot.com/irl/livestock-bred-pregnancy-announcement/

Saha, Jonathan. "Milk to Mandalay: Dairy Consumption, Animal History and the Political Geography of Colonial Burma." *Journal of Historical Geography* 54 (2016): 1–12.

Schinner, Miyoko. *Artisan Vegan Cheeses: From Everyday to Gourmet*. Summertown, TN: Book Publishing Company, 2012.

Schlim, John. *The Cheesy Vegan*. Boston: DeCapo Publishing, 2013.

Schwartz, Deborah. "The Good, the Bad, and the Ugly." January 28 2015. Retrieved from: https://safehavenfarmsanctuary.org/the-good-the-bad-and-the-ugly/

Singleton, G.H., and Dobson H. "A Survey of the Reasons for Culling Pregnant Cows." *Veterinary Record* 136 (1995): 162–165. https://doi.org/10.1136/vr.136.7.162.

Tomasello, Sarah, Piazza, April, and Poirier, Nathan. "Reproduction of the Lack Thereof: A Mode of Oppression, a Means to Liberation?" In *Gender and Sexuality in Critical Animal Studies,* ed. Amber E. George, 145–161. Lanham: Lexington Books, 2021.

van Ditmar, Delfina Fantini and Harriss, Harriet. "Design Curricula for Climate Crisis." *Service Design Network,* 2020. Retrieved from: https://www.service-design-network.org/events/in-session-design-curricula-for-climate-crisis

White, Ellen G. 1938, Letter 72, *Counsels on Diet and Foods*. Takoma Park: Review and Herald Publishing Association, 1976.

Woolf, Virginia. *Three Guineas*. London: Hogarth Press Ltd., 1938.

Wrenn, Corey Lee. "The Problem with Milk Not Jails," and other writings at https://www.coreylee wrenn.com.

3

ANIMALS, FEMINIST CARE THEORY, AND CRITICAL STANDPOINT THEORY

An Interview with Josephine Donovan

Chloë Taylor [CT]: Josephine Donovan is an emeritus professor of English at the University of Maine. She is the author of a dozen books of non-fiction and the editor of five books of non-fiction and has also authored numerous academic journal articles, book chapters, her father's memoirs, and short stories. Her work has been translated into eight languages. Professor Donovan specializes in the areas of feminist theory, American women's literature, early modern women's literature, and feminist theoretical approaches to animal ethics, in particular care ethics and animals. With Carol J. Adams, she edited *Animals and Women: Feminist Theoretical Explorations,*[1] *Beyond Animal Rights: A Feminist Caring Ethics for the Treatment of Animals,*[2] and *The Feminist Care Tradition in Animal Ethics: A Reader.*[3] She is also the author of two books on care ethics and animals: *The Aesthetics of Care: On the Literary Treatment of Animals*[4] and most recently *Animals, Mind and Matter: The Inside Story,*[5] that draw on both feminist care theory and standpoint theory.

Professor Donovan, thank you so much for agreeing to do this interview. I wondered if you could start out by explaining what care ethics and standpoint theory are, the relationship you see between these two theories, and how these theories have been extended by scholars such as yourself to include more-than-human animals.

Josephine Donovan [JD]: Well, care theory is a dialogical theory, which was originally developed by Carol Gilligan in a book called *In a Different Voice,*[6] and the basic idea there was that there's an ethic to be derived from the voice of groups who are rarely listened to and rarely paid attention to. In her case, it was adolescent girls, adolescent females, and her point was that their mode of reasoning was different from that of the dominant group and that the way they were thinking and what they were thinking should be listened to. So it was a feminist intervention originally.

DOI: 10.4324/9781003273400-5

The link to standpoint theory is that standpoint theory—which was originally a Marxist theory developed by Georg Lukács, a Hungarian Marxist back in the 1920s—was originally developed to apply to the proletariat, the idea that the proletariat in an industrialized system is treated as just cogs in the wheel, as objects, when in fact they are human beings who are subjects. And as subjects, they necessarily have a critical perspective on being treated as objects. There have been problems in terms of working [standpoint theory] out with the proletariat, but in my view, I think it works even better with animals. Because there you have a group that's being treated as objects who are, in fact, subjects and that there is necessarily a critical perspective there on the part of the animals, if we paid attention to it, who resist or refute being treated as objects. So I think that is the link, then, is that it's a matter of listening to groups who are not listened to by the dominant culture and hearing their voice and recognizing that they have standpoints that are critical of the treatment they're receiving as being treated as objects or being reified. So that's the basic idea there.

CT: Feminist care ethicists such as yourself writing about animals have critiqued the so-called "fathers" of animal liberation, Peter Singer and Tom Regan, for their aversion to emotions—in particular their insistence that they do not make the arguments that they do because they "love" animals but simply because they are being "rational" about animals, for instance, following the logical principal that we should treat like cases alike. Human oppression of animals is in fact deeply unreasonable, they show. As care ethicists such as yourself have argued, however, in this way these philosophers not only reiterate and reinforce the stigmatization of emotion as "feminine," but they also reiterate and reinforce human supremacy by continuing to elevate "reason" over feelings when nonhuman animals are considered incapable of reason.

In contrast with the rationalistic philosophies of Singer and Regan and other analytic male philosophers, a feminist care ethics approach to animal ethics avows and centres emotions and recognizes these as essential to treating animals better than we do. Along with care, you have written on sympathy,[7] and ecofeminist philosopher Lori Gruen has written on empathy as important elements of a feminist animal ethics.[8]

In response, philosophers such as Singer and Regan have expressed worries that feelings such as care, love, empathy and sympathy are not consistent or reliable enough to ground an ethics. Some people just don't care about animals, or they care about their pets but not other animals, or they care about dogs and cats but not pigs and chickens, or they care but don't act consistently with this care. They "love" animals but still eat them, for instance.

There is in fact a critique of emotions such as care and love within feminist theory as well, with feminists such as Marilyn Frye recognizing that expressions of "love" have often been used by men to justify their domination of and violence against women.[9] Anti-racism and critical animal studies scholar Nekeisha Alayna Alexis has written a chapter, " 'There's Something about the Blood': Tactics of Evasion within Narratives of Violence," in which she compares the genre of plantation romances—novels that depicted slavery as a benevolent institution—and the discourses of conscientious omnivores—that describe small-scale farming as a similarly benevolent institution.[10] In each case, the focus is taken off the experiences of the dominated—the slaves and the animals—and is placed on the feelings of the oppressors—the slave owners and farmers. The fact that these oppressors have "good" feelings towards their slaves and animals—for instance, love, care, and grief or sorrow over their deaths—is presented as justifying their domination and violence.

In all these cases we see that while centring emotions and affect is a feature of feminist work, and of care ethics in particular, it can also function in problematic ways to justify or redeem violence—and feminist scholars have recognized this in the context of violence against women and, to a lesser extent, violence against animals. How do feminist care ethicists respond to cases and arguments such as this?

JD: Well, there are a number of points that you made there that are often levelled against care or any theory that is not rooted in a rational logic, which is usually conceived of as deductive logic. So it's a very complex, you know, it gets into really complex issues of epistemology. I don't know how much you want me to get into all that here, but basically, I think the basic position of care theorists—and I think this maybe is an area that needs to be explored further—but basically, is that there are other forms of knowledge than deductive knowledge and that those are legitimate forms of understanding where another person or another creature is coming from. In other words, the scientific approach to animals is basically to objectify them and see them in terms of their component parts. For example, in laboratory experimentation you're looking at the cells or the component parts of an organism rather than looking at them as subjects. And so the basic question is, how do subjects relate to other subjects? And deductive logic is not the way that subjects relate to other subjects.

So we're getting into the question of, as I say, epistemology. How do we know what another creature is feeling? And as I have argued and others have argued in many places there are a number of modes of communication and knowledge that are not strictly mathematical forms of logic. And that was certainly a point that Carol Gilligan made in *In a Different Voice*, that the form of logic you found in these adolescent girls was contextual, particularistic, and dealing with the specific circumstances of a given case. In that example, they were discussing abortion, for example, which is a complex moral decision that often requires knowledge of the particulars of the case and can't be adjudged according to universalistic ethical laws, you might say. So she was going against the sort of idea of a universalistic law for every given case. And so in that sense, yes, care theory is focused more on the individual and basically making the point that most moral ethical decisions are contextual and require a knowledge of the particulars of the situation and in particular require listening to the voice of those who are being subjugated by the given situation, as in the case of animals.

I don't know if that begins to address your question, but certainly, care theory considers emotion as part of the connection. The knowledge that one might have of another creature and particularly the whole question of sympathy or empathy, which involves getting into the feelings and thoughts of another creature through the recognition of many of the signs and expressions and behaviour that that creature is giving out are similar to those that one would feel oneself in the same situation. So there's a kind of rational inference going on there that, in other words, the signs that that creature is giving are the sort of signs that you might give in the same situation and therefore making the conclusion that that creature is feeling something similar to what you would feel in the situation. So it's a form of reasoning that I think stands up and it seems perfectly legitimate to me as a form of understanding. So empathy, sympathy do involve a certain amount of emotional identification that is disparaged by those who think everything should be dispassionate and rational and that that's the way to truth. But I think that care theorists say that that is not the only way to truth and have made the case, I think, convincingly, that some of these other modes are equally valid.

CT: Drawing on standpoint theory, you argue in your latest book, as you explained, that animals have standpoints and that if we pay attention to them we can understand what these are, and we should respect them—so if an animal is communicating that she doesn't want to be in a cage or slaughtered, for instance, we can understand that and we should let her out and not kill her. You reject arguments such as the one we find in Thomas Nagel's "What Is It Like to Be a Bat?" that we cannot know what animals are feeling or thinking.

I do agree with you and I think that most of the time the argument that we can't know what animals are feeling or thinking is used to dismiss the relevancy of their experiences, so by and large your argument that we can, in fact, understand animals' standpoints if we pay attention to them would improve our treatment of them. But there is also a case that I have been thinking about lately that makes me worry about how this argument could be taken up. This is the issue of zoophiles, or "zoos" as they call themselves, who claim to be in consensual, romantic relationships with animals such as dogs and horses. They claim that their animals love them and in some cases have even married them and that they know these animals want to have sex with them because, for instance, the zoophiles pay attention to their animals' body language and can interpret what they are saying. There has been a troubling trend in animal studies lately of defending interspecies sex as "queer" and even "liberatory" of animals because it undermines human exceptionalism. Women's and Gender Studies Professor Kathy Rudy, for instance, has been one of the most vocal defenders of this argument.

I was having a conversation about this topic with Carol J. Adams and Piers Beirne a few days ago, both of whom have written critically on what they consider to be, not a sexual orientation, but rather interspecies sexual assault. Their perspective is that the zoophiles are misguided in assuming they know what the animals want, and that in fact, we need to not have sex with animals because they cannot consent to it or we at least cannot know if they consent to it—and we talked about how we should question if consent is even a meaningful concept in a relationship of such domination as domestication.

In this context, your book came up because, while I am sure this wasn't your intention, you make an argument that could actually be used to support the zoophiles' claims that they can interpret what the animals are saying by attending to their body language, for instance. In this case, it seems like insisting that we cannot know what animals want and should therefore just leave them alone—at least when it comes to sex—seems like it might be a better way to protect them from abuse. What do you think about this case in particular?

JD: Well, I have to confess I haven't read that material. But it seems to me that the underlying issue here is a matter of animal dignity. I mean, you have to consider that the animals exist in a power relationship with humans, that they are subjugated by humans and that to the extent they're being coerced into these relationships it is, I think, a violation of their dignity and their integrity.

Now, for the question of how we know what the animals think or feel it is important to consider the animals in their natural habitat, what their natural behaviour is and what their given natural design is. And from those point of views they're not designed to have sex with humans and that is not part of their natural life, it's not part of their natural purposes. And any interaction of that sort is undoubtedly initiated on the part of the human and therefore is coercive. So it's not a matter of consent per se on the part of the animal. It's a matter of respecting the animal's integrity and their natural purposes within their natural habitat. I don't know if that answers your question but I do think any of those types of questions have to consider the issue of power and the fact that the animals are in those situations under the human's control has to be taken into consideration.

CT: Thank you. That's so helpful. I've been reading this literature for years and I've never come across a single case where there's a zoophile in a relationship with a wild animal. So I think that's really helpful to think about how, even if those animals act like they're consenting, that has everything to do with domestication and how they've been groomed, and how they've been deprived of relationships with members of their own species, because you never see this in wild animals.

JD: I think we can safely say that no raccoon will choose to have sex with a human over another raccoon. And I think too that the question of how do we know animals, it gets into again, into the epistemological question. But a lot of human knowledge of animals, of anything for that matter, is coloured by ideology and, of course, anthropocentric and speciesist assumptions. So I think what we're saying, those of us who are advocates of standpoint theory, care theory, is that we should really pay attention directly to the animals and try to get beyond our human ideological assumptions and really get into the world and the mindset of the animals.

A lot of ethologists are doing this effectively and revealing things to us that we never understood before. For example, the various studies of octopuses that have come out recently have, I think, opened a lot of people's eyes to the subjectivity of octopuses, which no one imagined before. So I think a lot of the human assumptions about animals have to be challenged. And the whole point of standpoint theory, care theory, is to challenge those assumptions from the point of view of the animal. And getting into that point of view means getting out of a lot of anthropocentric assumptions. And bestiality obviously assumes human domination of the animal.

CT: Absolutely.

While care is usually explored in feminist philosophy as an approach to ethics, one of the other areas of value theory, aesthetics, is less often considered in terms of care. Yet you have written not only on the *ethics* of care but also on the *aesthetics* of care in relation to animals. Can you explain what an aesthetics of care is and how it relates to an ethics of care? Why is it important to think about animals in terms not only of ethics but also of aesthetics?

JD: Well, my book, *The Aesthetics of Care*, was largely a study of literary depictions of animals, and I critiqued the Kantian view of aesthetics as basically a reifying, dominative approach to aesthetics that reifies the subject matter in terms of basically mathematical designs. For example, in a painting, you consider the lines and then the shapes of the objects and whatnot, and the subject matter is deemed irrelevant. I mean it's a parallel to the scientific approach to any living matter. And so in literature, the question becomes, how do we depict the subjectivity of animals and get away from the objectification of them in literary forms? And if you look at examples, many, many examples of literature, you'll find that animals are usually there as a kind of local colour or backdrop or provide some sort of symbolic comment on the human character. And rarely are the animals allowed to exist as subjects in their own right.

There are a few examples in literature of successful attempts to do that [allow animals to exist as subjects in their own right]. For example, Tolstoy. Often the example is given of the dog Laska in the novel *Anna Karenina*. Laska does have a mind of her own, and her perspective is given as different from the protagonist and indeed critical of it at times. So there's an example of trying to realize the subjectivity of animals in literary form. It can be done, in other words, but it rarely is done. And once again, animals are usually used as set pieces or backdrops and relegated to secondary status. So that's where the aesthetics of care comes in and parallels to the ethics of care in that it calls for realizing the standpoint of animals in literary treatment of them.

CT: In your new book, *Animals, Mind and Matter*, you are critical of feminist new materialisms, but you take Donna Haraway as the main representative of this school of feminist thought, with a bit less discussion of Karen Barad but not many other feminist theorists who are working in this area. By focusing on Haraway and Barad you make a case that feminist new materialisms is not helpful and is even problematic when it comes to thinking ethically about animals. Yet there are other feminist new materialists who have very different and I think more ethical and feminist things to say about human–animal relationships than Haraway, such as Astrida Neimanis and Stacy Alaimo, both of whom have written powerfully on marine animals in the context of the current anthropogenic annihilation of oceanic life. I do agree that by and large feminist new materialists just haven't paid attention to animals, and the one who is most well known for doing so, Haraway, has said very problematic things. But I am not sure this is something about feminist new materialisms per se.

At the same time that you are critical of feminist new materialisms, you draw on a lot of science and especially physics in your book, and this seems to me to be a hallmark of feminist new materialisms: my understanding of feminist new materialisms is that it moves away from a strong social constructivist position that is highly suspicious of science, and instead is characterized by an openness to science and considerations of not just social but also material factors and the ways that the social and material are co-constitutive and entangled. Feminist new materialisms also recognizes the agency of the material world, as do you in your discussion of Indigenous cosmologies, for instance, and your call to resubjectify the world.

So I actually wonder if your own work isn't close to feminist new materialist in some respects, and if you are not actually rejecting feminist materialism per se but just a dominant and foundational figure within feminist new materialism: Donna Haraway, whom a lot of critical animal studies scholars have also critiqued. Am I wrong about this, though? Is your critique of feminist new materialism more fundamental than I am understanding?

JD: Well, my critique of new materialism is basically that it is materialistic. In other words, it ignores the subjectivity of living forms. You mentioned the fact that I get into some alternative conceptions of science, especially in the world of quantum physics. I think what's interesting there is that they seem to be coming up with—in something like quantum non-locality—they seem to be coming up against basic, almost like a wall, that is sort of the endpoint at which materialism is no longer functional. There's clearly the sense of being mystified on the part of many of these physicists as to what is really going on in the subatomic world. There are alternative theories like panpsychism and so forth. I'm not ready to fully endorse them, but I think we're at the point where we need new alternatives to the traditional materialist way of looking at things. And so new materialism, unfortunately, is still materialism. And many of these new materialists go out of their way to disavow any connection to vitalism or any kind of notion of the spirit world inhabiting the physical world—there's no soul there, as some of them have said. But I think that the fact is we just don't know. I mean, there is something there that we don't understand and there's something in even the most reductive forms of physical matter that is not behaving the way physical matter is supposed to behave according to various assumed paradigms.

So that's where we are. That's where it's cutting edge, really, in quantum physics itself. And so I'm interested and open to the idea that certainly there is some sort of subjectivity that we haven't quite figured out what it is. But at the very least, we know it's in living forms and it's something that deserves to be recognized and acknowledged. The overall point, of course, in terms of animal ethics is that we should no longer treat these forms that have subjects in them as if they were just objects. And then in our dominant discourses, including the law, certainly, in commerce and in the sciences, animals are treated as objects. And in the case of the law, they're property, they have no legal status other than being objects. They're treated as commodities in the economic world. And of course, in the sciences, they're just treated as laboratory specimens.

My concern throughout this is a rehabilitation, you might say, a recognition of the existence of subjectivity. I just think we're at a point where there are certain branches in the sciences that are moving in the direction of recognizing that subjectivity is something real. And while it can't be quantified and objectified almost by definition, it's there, and to continue to ignore it is distortive and dominative and ultimately damaging to the subjects that are mistreated.

CT: Although *Animals, Mind and Matter* focuses on animals, you also discuss the animacy of nature more generally and of trees in particular. For instance, you discuss Indigenous cosmologies that undermine the animate versus inanimate distinction in Western thought, seeing rivers and rocks as animate. In your final chapter, you discuss Richard Power's *The Overstory*, which is more about trees than animals. The conclusion to your book also discusses trees and their relationships with fungi, and you conclude by calling for a "resubjectification of the world." In the final sentence of the book you invoke a recognition of "the caring subject, the soul, the living conscious spirit that exists *in all creation.*" In short, you move beyond animals and include plants, fungi, and all of creation in your ethical framework.

In the past animal ethicists have mostly been wary of acknowledging that plants might matter ethically, for fear that this would undermine arguments for vegetarianism or veganism. If everything is sentient and morally considerable, in other words, why is it better to eat plants than animals? The late ecofeminist philosopher Val Plumwood was a significant exception to this rule, and lately I am learning of more and more critical animal studies scholars who are also interested in critical plant studies. Still, there has been a lot of apprehension among critical animal studies scholars and animal ethicists about extending our ethical consideration to plants—let alone to fungi and rivers and rocks. What do you think of critical plant studies and similar emerging disciplines and how do you see their relation to critical animal studies?

JD: Well, I think that as I make the point in the book, I do believe that there is a, I wouldn't call it subjectivity, but some sort of animation or animus in plant forms, and it's another area where we know very little. But I don't think they're just inert objects. But there are obvious distinctions to be made between plants and animals. And I think the ethics that we come up with for the two different forms have to be different and tailored to whether it's a plant or an animal.

I don't think plants are sentient in the sense of having feelings in the same way as animals do. But I do think there's some form of knowledge there that is non-material and, if you want, you could call it spiritual. But again, we don't, I think, really understand. And certainly, plants communicate. I mean, anyone who's had a garden knows that there are ways of understanding what is good and bad for different plants. And you learn by experience what is going to be beneficial.

On the question of eating plants, it is possible to exist without destroying the plant. It's difficult. And of course, in a mass society and mass production, it'd be virtually impossible. But at least theoretically, it's possible. A fruitarian diet is possible if one wants to pursue that issue. But in general, I think the individual identity of plants is probably less pronounced than in animals. And whether that should be a factor in ethical decisions, I don't know.

But certainly, the overall issue that's being raised by some of the advocates, for some of the eco-theorists, the idea that, for example, lakes have rights or rivers have rights implies a certain ontological value to these entities. That means that they have standpoints in a way that should be respected and integrated into our decisions about how to treat them. And indeed, there are legal cases being made now along those lines, which I think are quite interesting. But we know that badly polluting or depleting a lake is going to destroy the lake. We can take the Great Salt Lake in Utah, as a matter of fact, as an example of that happening. And I think the case can be made that the lake itself should have a certain say in the issue, has a certain standpoint, and that humans should represent that standpoint in legal and ethical decisions regarding their destiny.

So it's complex. I think we're really at a very beginning point in terms of trying to figure out what would be reasonable ethical responses to the world of plants. I think it's far more developed in terms of our thinking about animals. But anyway, in that sense, I endorse certainly those who argue that there is a standpoint to be found in the world of flora, of plants and trees and so forth. And of course, there have been recent discoveries about the fungal networks that are uniting trees and how they communicate and so forth. Suzanne Simard up in British Columbia has made important discoveries in this regard. So we're just learning more and more. And the more we learn, the better we can tailor our proper ethical response, and, of course, ideally we want the natural world in all its forms to flourish and to counter the human domination that has really led to the destruction of much of that world at this point.

CT: As mentioned earlier, in your final chapter you draw on Richard Powers' novel, *The Overstory*, and in particular you cite these and similar passages:

"Humankind is deeply ill. The species won't last long. It was an aberrant experiment. Soon the world will be returned to the healthy intelligences, the collective ones. Colonies and hives." (2018, 56)

Doug sends a silent message to the trees: "Hang on. Only ten or twenty decades. . . . You just have to outlast us. Then no one will be left to fuck you over" (ibid., 90).

I recently read two philosophical books that defend misanthropy given humankind's treatment of animals and destruction of nature. Queer theorist, literature scholar, and vegan studies scholar Emelia Quinn is currently writing on what she calls "critical misanthropy" and recently organized a conference on this topic. I also read about a new concept, "disanthropy," which is the misanthropic desire for a world without humans—and the person who coined that term, Greg Garrard, is a literature scholar who explores how this disanthropic desire has been expressed in literature. Powers' book seems like a clear example of disanthropy. I am curious, what are your thoughts on misanthropy and disanthropy, and why did you choose to end your book with a series of quotes that look forward to human extinction as, it is hard to deny, the best thing that can happen to the planet?

JD: Well, the quotes that you cite are from specific characters in the novel who have a rather misanthropic view of the world. I don't necessarily endorse that view because I believe that there is an alternative, I guess I could say hope in the revival among human beings of appreciating the spiritual side of nature and certainly of animals. So I don't share these views that we're all doomed and it would be better for the human race to disappear. I guess I'm chauvinistic enough as a human to hope that we survive and feel that there is much value to admire in human beings. But we've gotten off the track. We've gotten off on a bad track. There's no question about that. And we've been on a bad track since the early modern period with this domination of the scientific worldview that has legitimized so much of this destruction and I might say in holy alliance or unholy alliance with capitalism that has led to this dreadful situation that we're currently in.

But I have hope that people are becoming aware of this terrible catastrophe that we're in and that the love and feeling of connection with the natural world will prevail. And you find that in *The Overstory*, of course, where many of the characters have this emotional connection with the natural world, particularly with trees, which are seen as subjects themselves, animate creatures. I hope that a revival of that, one might say, animist worldview will counter this destructive capitalist scientism that is now the dominant worldview in the planet. And it's, of course, a Western viewpoint. In my book, I look to various alternatives, including non-Western premodern animist theories that I think are useful sources of rejuvenation of the human spirit that will acknowledge our place in the natural world and our connections with it. And it does involve reviving some of these neglected forms of epistemology and knowledge that are held in contempt or denigrated by the adherents of rationalist science.

CT: My final question is also about literature. At several points in *Animals, Mind and Matter*, you cite Olga Tokarczuk's Nobel speech. I presume that you have read her book, *Drive Your Plow over the Bones of the Dead*, which I absolutely loved and think is so interesting to think about in terms of sexism, ageism, and speciesism and their interconnections, and I wonder what you think of this book from a feminist critical animal studies perspective?

JD: Yes, that is a fascinating novel. It's almost feminist revenge work. The main character is a kind of eccentric woman who's, of course, treated disrespectfully by the authorities, and so forth. And she's sort of a crank who protests various things, including hunting. Well, I won't give the plot away, but there's a battle between her representing this sort of alternative world that is critical of the dominant masculine forms that dominate in the Western world. The book deals with the clash between her and these authorities. But it's quite an interesting novel, and she's a very interesting writer, and I cite her Nobel Prize lecture, "The Inside Story," in my book. She is calling for a recognition of the existence of subjectivity in the world, and claims that writers should try to express this in their work and that we need new literary forms to enable this subjective world to be seen and recognized. So she's certainly a writer whose works I'm very interested in following.

Notes

1 *Animals and Women: Feminist Theoretical Explorations*, eds. Josephine Donovan and Carol J. Adams (Durham, NC: Duke University Press, 1995).
2 *Beyond Animal Rights: A Feminist Caring Ethic for the Treatment of Animals*, eds. Josephine Donovan and Carol J. Adams (New York: Continuum, 1996) (Co-edited with Carol J. Adams).
3 *The Feminist Care Tradition in Animal Ethics: A Reader*, eds. Josephine Donovan and Carol J. Adams (New York: Columbia University Press, 2007).
4 Josephine Donovan, *The Aesthetics of Care: On the Literary Treatment of Animals* (New York: Bloomsbury, 2016).
5 Josephine Donovan, *Animals, Mind, and Matter: The Inside Story (*East Lansing: Michigan State University Press, 2022).
6 Carol Gilligan, *In a Different Voice: Psychological Theory and Women's Development* (Harvard University Press, 1982).
7 Josephine Donovan, "Participatory Epistemology, Sympathy, and Animal Ethics," in *Ecofeminism: Feminist Intersections with Other Animals and the Earth*, eds. Carol J. Adams and Lori Gruen (New York: Bloomsbury, 2014).
8 Lori Gruen, *Entangled Empathy: An Alternative Ethic for Our Relationships with Animals* (Lantern, 2015).
9 Marilyn Frye, "In and Out of Harm's Way: Arrogance and Love," in *The Politics of Reality: Essays in Feminist Theory* (Trumansburg, NY: The Crossing Press, 1983), 52–83.
10 Nekeisha Alayna Alexis, " 'There's Something about the Blood': Tactics of Evasion in Narratives of Violence," in *Animaladies: Gender Animals and Madness*, eds. Fiona Probyn-Rapsey and Lori Gruen (Bloomsbury, 2019), 47–64.

Bibliography

Alexis, N.A. " 'There's Something about the Blood': Tactics of Evasion in Narratives of Violence." In *Animaladies: Gender Animals and Madness,* edited by Fiona Probyn-Rapsey and Lori Gruen, 47–64. Bloomsbury, 2019.

Donovan, J. *The Aesthetics of Care: On the Literary Treatment of Animals*. New York: Bloomsbury, 2016.

Donovan, J. *Animals, Mind, and Matter: The Inside Story*. East Lansing: Michigan State University Press, 2022.

Donovan, J., and Adams, C.J., eds. *Animals and Women: Feminist Theoretical Explorations*. Durham, NC: Duke University Press, 1995.

Donovan, J., and Adams, C.J., eds. *Beyond Animal Rights: A Feminist Caring Ethic for the Treatment of Animals*. New York: Continuum, 1996.

Donovan, J., and Adams, C.J., eds. *The Feminist Care Tradition in Animal Ethics: A Reader*. New York: Columbia University Press, 2007.

Donovan, J. "Participatory Epistemology, Sympathy, and Animal Ethics." In *Ecofeminism: Feminist Intersections with Other Animals and the Earth*, edited by Carol J. Adams and Lori Gruen. New York: Bloomsbury, 2014.

Frye, M. "In and Out of Harm's Way: Arrogance and Love." In *The Politics of Reality: Essays in Feminist Theory,* 52–83. Trumansburg, NY: The Crossing Press, 1983.

Gilligan, C. *In a Different Voice: Psychological Theory and Women's Development*. Harvard University Press, 1982.

Gruen, L. *Entangled Empathy: An Alternative Ethic for Our Relationships with Animals*. Lantern, 2015.

4

QUEERING ANIMAL LIBERATION

20 Thoughts Toward Interspecies Solidarity

pattrice jones[1]

I. Orientation

Sometimes there comes a crack in time itself.[2]

1. *We Are Not OK*

Forests burn as islands disappear below rising tides. Pandemics escalate along with soaring temperatures. Tarmacs melt as warplanes strafe cities. Migrants mass at borders as snarling strongmen vow to keep them out. More and more animals come closer and closer to extinction. Relationships upon which everything depends—from insect-plant interactions to democratic norms among humans—are more frayed each day. Consensus reality collapses along with ecosystems.

2. *Right Now*

We've never been here before. There's a rip in space-time, leaving more and more animals 'mis-timed' due to disjunctive responses to changing seasons.[3] Other temporalities are also askew. Criss-crossing fast and slow violences compound one another. It is and is not 'too late' for humans collectively to change course in time to prevent or even mitigate further catastrophe.

3. *But That Could Change*

We stand not only at the conjunction of numerous cascading crises but also at the confluence of persistent underground streams of resistance. The vital processes responsible for life itself ensure that change is always happening. The problem is momentum, which is currently on the side of chaos and violence. Queer eros could power new ways of thinking, feeling, and being in relation to each other and the larger-than-human world, but only if we are willing to forswear hubris in order to join the joyful worldwide resistance against humdrum human hegemony.

DOI: 10.4324/9781003273400-6

II. Queer Planet

We're new here! Nonhuman animals such as beavers, bats, and bees had been co-creating stable ecosystems for millions of years before a subset of our comparatively young species set out to codify everything. Not surprisingly, they mostly got it wrong. Luckily, life goes on being queer regardless of whether or not self-important primates can comprehend it.

4. *The World Is Queerer Than Our Categories*

Life is energy, organized. Energy radiates, always surging into any pathway that emerges. Organisms evolve in association, converging, diverging, and sometimes merging as they co-create the ecosystems in which they participate. Meantime, every morning brings a fresh jolt of juice to power it all.

We'll be here all day if I start listing the ways that humans have gotten it wrong when trying to make sense of the larger-than-human world, which tends to confound categories and boundaries. Birds are dinosaurs. Some fungi have hundreds of mating types or, as we like to call them, sexes. Some rivers run underground.

Animals apprehend the world through sense organs that have evolved to detect particular types of data, such as sounds or colors. Internal processes convert that raw input into actionable information. Since every kind of animal has its own evolutionary trajectory, the types of information available to us vary considerably across species. Bats and elephants can hear sounds that are, respectively, above and below the frequencies that human ears can hear. Chickens can see colors we can't see.

Some butterflies detect the polarity of light, and pigeons seem to be able to sense the earth's magnetic field. Bees may be able to detect minute differences in barometric pressure. Some birds sing and dance so quickly that we must play back recordings at slower speeds to distinguish the sounds and movements. We can only guess how they may experience time.

What do they know that we don't know? What other kinds of information might be beyond our ability to even imagine? I propose awe as an antidote to the insult implicit in the ideology of human supremacy.

5. *We're Not Alone*

From the trees that exhale the oxygen we need to the fungi and bacteria that produce essential vitamins, other organisms make humans possible. We are ourselves aggregations of organisms, even though we may experience ourselves as solitary. That multiplicity makes each of us queerer than many people suppose.

Other entanglements are chosen. The array of relationships into which humans have invited or conscripted other animals span the spectrum from care to callous exploitation, including true mutual aid. Some of these chosen relationships are *queer* in the sense of odd when viewed from the perspectives of other animals. A particularly poignant few are *queer* in the sense of paradoxical because the humans in question proclaim (and perhaps experience) love for animals they injure. Many are *queer* in that they are motivated by deep-seated desire for connection falling outside of socially constructed norms. That wish may be among most our most powerful and sustainable resources, but only if we are willing to set aside common fallacies about ourselves.

6. *Ecce Homo*

Homosexual relations and other ways of being *queer* in the sense of sexual orientation are and always have been common within our own species and many other species. The ever-expanding literature on both same-sex relations among nonhuman animals[4] and the worldwide diversity of sexual orientations and gender expressions among humans[5] reveals a very queer planet indeed. That literature was made necessary by falsehoods about humans and other animals propagated in the course of Enlightenment-era imperialism and colonization.[6] At every turn, those falsehoods fell on fertile ground due to the human propensity for error and human preoccupation with identity.

III. "Human" Error

They say "to err is human," and it's really true. While many humans flatter themselves by imagining our species to be the most quick-witted, numerous biological and social factors predispose human beings to err in ways that have led more or less directly to our current crises.

7. *Categories and Stories*

Human cognition relies heavily on categories and stories. Categories simplify the process of coping with the blooming profusion of novel incoming information by means of basic pattern matching. Categories also help us make the most of our sharply limited conscious cognitive capacity. Working memory can handle only about seven (plus or minus two) items at a time. 'Chunking' things together by means of categories allows us to expand the number of things that can be held in mind at any one time.

People handle novel perceptions by either assimilation (incorporating the new information into existing categories) or accommodation (adjusting the category system.) Accommodation requires more mental effort, and therefore more energy, than assimilation. Therefore, there is a built-in tendency to modify incoming information to fit existing categories rather than do the work of reformulating the categories themselves. This makes our mental categories more stubborn than if our minds were the disembodied reasoners many people imagine them to be.

Problem-solving calls for cause-and-effect reasoning. Every hypothesis about cause and effect is a narrative: *This happened* ***and then*** *that happened.* The drive to find narrative that 'make sense' is very strong, so much so that humans will often make up stories rather than tolerate the discomfort of being without explanatory narratives.

Most of these confabulations are products of the parts of ourselves that call themselves our selves. Since the conscious mind is but a small subset of cognition and is able to access only a subset of the constant communications going on within the wider organism, these selves of ours might themselves be considered confabulations. Certainly, the conscious self is always working with incomplete information about the organism and its surroundings, combining new perceptions with always-imperfect memories to come up with satisfactory stories. No wonder these are often fanciful!

8. *Animality Is Emotion*

Like us, plants perceive the world through sense organs, using the information gained to make decisions about what to do. Like us, plants are in constant motion, albeit much more slowly. Like us, plants communicate and sometimes cooperate with each other and with other organisms. Unlike us, plants do all of this supremely rationally, because their situational analyses are never clouded by emotion.

Animals *feel*, and those emotions motivate our motions. While some humans like to imagine that pure reason is possible, in fact there is no enforceable border between 'thought' and 'feeling.' Reason and emotion are but two categories into which we attempt to sort neurological phenomena that are always merged. This means that our categories and stories are always infused, to some degree, with feeling. Thus, humans sometimes cling to stories long after they have been shown to be false and may get angry when their categories are contravened.

9. Got ID?

Categories and stories combine to form identities. People sort themselves and each other into categories and tell stories about those categories, devoting considerable emotional and physical energy to questions of identity. Of all categories and stories, those associated with identity seem to provoke the most emotion. At times, humans may feel something approaching existential terror when their identity categories are threatened. Both metaphorical and material wars have been and continue to be rooted in disputes about identity.

This vexing feature of human relations may be overdetermined in the sense of having multiple causes that compound one another. Recall that the purpose of the story-telling self is survival. Consciousness evolved quite early in our lineage. Consciousness is always *of* something. Hence, at the heart of every human is an emptiness focused on survival. Recall, too, that humans are social animals whose survival depends on inclusion within a group that itself survives.

It makes sense to me that such animals might become deeply vested in socially constructed identities, so much so that words describing those identities can come to feel like life-or-death matters. But of course, perhaps that is just my own story-telling self creating an explanatory narrative! Nonetheless, I notice two conjoined errors about identity to which many humans seem prone and which *might* offer some insight into our current crises.

First, I notice that many humans (particularly those whose internal lives have been patterned by Enlightenment-era European ideas) commit what I call the *fundamental identity error* by thinking of their conscious minds as their selves and the rest of themselves as somehow lesser or other. I notice that these conscious selves tend to treat the rest of the organism in the same way that many humans tend to treat the larger-than-human world: They pretend to be in control of things they really don't understand at all. They denigrate the physical and elevate the mental. They experience themselves as independent and in charge when in fact they are entirely dependent and mostly clueless.

Which brings us to *Human*, which I will capitalize and italicize when referring to the currently dominant identity construct that goes by that name in English. Many of the battles among humans are rooted in the demarcation, elevation, and defense of the category *Human*. While speciesism is most evident in that construct, racism, ableism, sexism, and other biases are also built-in.[7]

The *Human* is the "rational animal" of Aristotle, the "wise ape" of Linneas, and the favored being of the God of the Abrahamic faiths, who granted him (yes, him) dominion over other animals. The *Human* is inherently and self-evidently superior. If not God's favor, then some ability—reason, language, tool use, or some other capability that the *Human* falsely imagines to be his sole province—makes him the only sort of entity worthy of moral regard.

The *Human* is the smartest of all! The *Human* makes rational decisions. The *Human* is large and in charge. All other organisms are mere background to the doings of this self-made species.

10. Hazardous Intersections

The habits of belief and behavior leading to and ensuing from the misidentification of the *Human* cause harm to humans, animals, and habitats in multiple interconnected ways. People with disabilities as well as non-male, non-white, non-straight, or otherwise queer persons all have endured injuries and indignities rooted in the same ideologies and practices that propel climate change, pollution, extinctions, and the ongoing consignment of nonhuman persons to perpetual incarceration in the category of property. It will not be possible to solve any of those problems without setting aside *Human* in favor of a more realistic portrait of ourselves.

11. If Not Human, Then Who?

If *Human* is a mistake, then who are we? Only by setting aside the biases, falsehoods, and confusions implicit in *Human* can we begin to sort out what kind of animals we actually are and then take those realities into account when crafting strategies to improve our own behavior. Gabby, curious, and highly emotional social animals who make music and love to decorate things but tend to get into fights about sound symbols, humans have one defining trait that might yet save the day: behavioral plasticity.

How many habitats have humans colonized? How many different ways of feeding, housing, and clothing themselves have they devised? How many languages? How many different social systems, games, musical styles, dances, spices, identities?

Humans are always falsely claiming that this or that ability is unique to humans, but if there were a feature that sets us apart, it would seem to be an extraordinary capacity to generate new ways of relating to each other and the wider world. Not always willingly! We do tend to cling to those categories and stories. Still, at a time of widespread and well-founded despair, it may be salutary to remember that humans have proved themselves capable of behaving and becoming otherwise. *We* are queerer than our categories.

IV. Getting Real

If we replace Humans with humans in our analyses, potential remedies emerge.

12. Eco-Logic 101

Every human action (or inaction) follows from a complex conjunction of choice and circumstance, with circumstance tending to play a much larger role than humans who think of themselves as independent rational actors like to suppose. All of our problems are *situations*—confluences of numerous social and material forces at particular places. Ecofeminism, intersectional theory, and other varieties of systems thinking offer helpful conceptual tools, which I refer to by the collective term *eco-logic*, for assessing situations and devising interventions.

13. The Conjoined Crises

Thinking eco-logically, we can see that the current crises are complexly conjoined in ways that exacerbate each other but also create avenues for multipurpose interventions. Consider the Ram truck, a popular pick-up in the USA, as an example of the 'petromasculinity'[8] that simultaneously worsens the climate crisis while boosting the fortunes of authoritarian figureheads who deny the reality of that emergency. Would this behemoth be as attractive to its target consumers if it were more widely known that bighorn rams are more likely than not to be homosexual? What else might be gained by launching an all-out effort to undermine the fundaments of human supremacy by, as the late ecofeminist philosopher Val Plumwood suggested, highlighting the diversity, agency, and inventiveness of the larger-than-human world?[9] We cannot know unless we try. In so doing, we must be mindful of how vehemently people cling to their categories and identities and how likely they are to react with fear and rage when they are wobbled.

14. Feedback Loops

Peat moss melts as a result of global warming, releasing even more carbon into the atmosphere. People switch on air conditioners to cope with heat waves, making future heat waves more likely. Such feedback loops can also occur when emotional responses worsen the conditions that sparked them.

Like many other animals, humans often experience the fight-flight-freeze response to emergencies. At present we seem to be stuck in feedback loops associated with each of these responses. Frightening circumstances such as the constant barrage of floods, fires, mass shooting, and pandemics may be making everybody more on edge, which is to say more combative or more likely to freeze or retreat into places that feel more safe. Fighting leads to more fighting. Retreats into real or virtual identity-based spaces lead to exacerbated fears of ever more fantastically demonic others. Stasis in the midst of emergencies that require action lead those emergencies to escalate, becoming ever more frightening.

These loops will persist for so long as motivating emotions are channeled into socially constructed conduits, including identity and consumerism. It's no use wishing for humans to become more rational or less driven by embodied terror. But queer humans have demonstrated, again and again, across centuries of often lethal oppression, that desire is stronger than fear. Eros may be the only thing that can save us from ourselves.[10] Hence, the project of decolonizing desire becomes urgent.

15. Decolonizing Desire

Desire drives everything but has been dammed and diverted by the axis of capitalism, heteropatriarchy, militarism, and anthropocentrism that currently has the planet and its peoples in a stranglehold. To decolonize desire, we will need to queer decolonization, decolonize queerness, and both queer and decolonize human relations with animals and the larger-than-human world.

Queer is a queer sort of identity category. Queer things tend to be non-categorical or counter-categorical. Categorizing people by their sexual orientation is a relatively new human habit, and one which is not universally accepted. In the Americas and elsewhere, European colonizers imposed heteronormativity and the gender binary on people whose own traditions were more inclusive and expansive.

Modern movements against homophobia and transphobia have tended to use the tactic of claiming pride in despised identities. This strategy has been used by many oppressed peoples and has many benefits, such as fostering in-group solidarity and helping people to cope with the emotions that arise when people are treated as subordinates or outcasts. But it always carries dangers with it. Since subordinated humans are often treated as subhuman, identity-based liberation movements often tend to strongly embrace the *Human*, arguing only that they should be included within the category rather than challenging the category itself.[11] In the case of queer liberation, an attendant danger is the unthinking embrace of gender and sexual orientation as natural ways to sort humans rather than identity categories that might best be dismantled rather than reshaped.[12] When liberation struggles rooted in such Eurocentric categories are exported around the world, they can hinder rather than help decolonization struggles in places where, for example, precolonial norms featured a multi-gender system or treated homosexual couplings as normal but did not sort people according to their sexual behavior.[13]

Is it possible to uncouple desire from social constructs? Probably not. Humans are, after all, social animals who think in categories and stories. However, desire tends to decolonize itself. And, within the history of decolonization struggles are many instructive episodes of efforts to shake off mental fetters along with literal chains.[14] "Aristotle's logic? A practice of things and corpses," wrote surrealist Réne Ménil from Martinique, "Thought is biological or not at all."[15] According to feminist and anti-colonial writer and activist Suzanne Césaire, "The most urgent task was to liberate the mind from the shackles of absurd logic and so-called reason."[16]

Nonhuman allies are our allies in this project, as they have long been contesting *Human* hegemony without adopting our ways of thinking while doing so. However, both humans and *Humans* will need to do some work to bring ourselves into allyship with animal outlaws.

V. Queering Animal Liberation

Like raccoons in the suburbs, anarchic forces capable of upending human constructions may not be visible during the day but remain ready, if only we can ally with them.

16. Queer

At its most basic, a queer spin on animal liberation requires us to understand the interconnections among speciesism and homophobia, heterosexism, and transphobia and to recognize and act upon any opportunities for cross-movement solidarity presented by those intersections. We need to be able to explain exactly how speciesism hurts LGBTQIA+ people *and* how biases against LGBTQIA+ people hurt animals and then work together with movements for LGBTQIA+ liberation at those conjunctions of shared concern. Among these mutual concerns are heteronormativity, reprocentrism, and toxic masculinity—all of which are linked to each other and to climate change.

Nonhuman animals are both tools and victims of the false presumption that only heterosexuality is normal. For many years, scientists ignored or actively obscured evidence of same-sex relations among other animals in order to uphold the myth of heteronormativity that rationalizes laws against 'unnatural' relations. At the same time, animals have been treated like automatons whose relations arise only from an innate drive to reproduce. By hiding evidence of pair-bonding, courtship, or sexual relations driven by affection, playfulness, or a desire for pleasure rather than for any reproductive purposes, scientists also hid evidence of sentience. Thus, homophobia made it easier to lock up animals in vivisection labs and foie gras factories.

Heteronormativity reflects and reinforces reprocentrism—the notion that reproduction is the very purpose of existence.[17] Among humans, reprocentrism leads to pressure to have children and, again, the notion that it is 'unnatural' to refrain from heterosexual couplings. Much suffering ensues from feelings of unworthiness among those who happen to be infertile to the abuse of children brought into the world only because their parents felt pressured to produce grandchildren. The corollary suffering of nonhuman animals, displaced again and again by burgeoning human communities and poisoned by their ever-increasing emissions, is both incalculable and unspeakable, given the prohibitions, on both left and right, against acknowledging the fact of overpopulation.

Reprocentrism powers and draws power from capitalism, which locates worth in productivity. Capitalism itself requires incessant production and reproduction for profits to flow within its fantasy world in which endless growth is not only possible but necessary. Even if all of the capitalists were to suddenly forswear the manufacture and sale of animal products, that inherently extractive and exploitative economic system would continue to drive climate change and harm animals by the billion. Queer animal liberation thus rejects not only overt biases against nonhuman animals and LGBTQIA+ people but also reprocentrism and capitalism.

17. Queerer

An even queerer animal liberation would require us to see not only similarities but also differences and to value rather than denigrate the latter. Both speciesism and homo/transphobia rely not only on a false notion of the norm but also on denigration of the non-normative. Much animal advocacy understandably stresses similarities between humans and other animals as a way of prompting empathy and respect. However, this maneuver runs the risk of colluding with both ableism and human supremacy by seeming to suggest that only those who are *like us* or have the same *abilities* as us are worthy of moral regard. A queerer animal liberation makes sure to always augment talk of similarities with candid conversation about often wondrous differences.

18. Queerest

The queerest animal liberation is an overset of human liberation. All of the things we need to do to liberate other animals from human hegemony also serve the purpose of freeing ourselves from habits of belief and behavior that lead directly to violence and injustice among human animals while poisoning human habitats. From this perspective, all human struggles for social and environmental justice are but subsets of a wider struggle for animal liberation. This is among the ways we can queer animal liberation: by seeing it not as something other than or even opposed to human liberation but, rather, as a necessary condition for human well-being.

This queerest of animal liberations queers queerness itself by acknowledging the countercategorical reality of the sexualities and gender identities within our own species. No matter how many extra letters and punctuation marks we add to the "LGBT" base, somebody feels left out, misrepresented, or oppressed by the formulation. In part, this is because both sexual-orientation-as-identity and the gender binary are Eurocentric ways of thinking, and, therefore, any way of talking that takes them as givens will be neocolonial at best. But the problem goes deeper, and right to the heart of *Human* identity, which presupposes a single and static organism that may grow or age but does not alter in its fundamental constitution. Sexual orientation often shifts over the course of a lifetime. Gender identity also sometimes shifts over time and may be experienced as dual, multiple, constantly changeable, or entirely absent.

Instead of seeing such flux as an impediment to the development of an entirely satisfactory definition of LGBT*QIA++++ identity, we can queer *queer* by leaning into the flux, helping to undermine *Human* along the way. None of us are unitary. All of us literally contain multitudes while participating in countless relationships that make us possible. We are verbs, not nouns. In liberating ourselves from the constraints of static identity based on mistaken categories, we can queer the human so thoroughly that *Human* will forever be exposed as a foolish fantasy. Moreover, embracing the queerness within our every cell will enable us to meet other animals as allies in a shared struggle to co-create more equitable and sustainable habitats for all.

VI. Reorientation

Here's a chance to dance our way out of our constrictions.[18]

19. Eros as Energy

In her germinal essay "The Uses of the Erotic," Black lesbian poet and activist Audre Lorde recognized *eros as* "the personification of love in all its aspects" that "can give us the energy to pursue genuine change within our world."[19] Inherently relational, eros begins in the body and reaches outward, seeking connection and promoting curiosity, creativity, courage, and generosity along the way. Along with other forms of renewable energy, queer eros may be a key to averting planetary catastrophe engendered by human folly.

How can we call upon queer eros? First by becoming more queer. By this I mean not only embracing our own diversity of sexual orientation and gender expression but also questioning *all* of our categories, including those upon which we have based our identities, such as *self* and *Human*.

With that aim, we can cultivate otherwise habits. Since English has a nasty habit of turning processes into things, we can look at situations from the perspective of energy, which is all verb all the time. Since consciousness cannot help but center itself, we can remember that other parts of our bodies also can think and offer ways for our hands and legs to show what they know nonverbally. Since words constrain imagination even as they facilitate communication, we can draw, dance, and daydream while problem-solving. Since unconscious human supremacy tends to background the larger-than-human world, we can make a point of consciously recognizing *all* of our neighbors, plants and animals alike, and wondering what they might know that we don't know.

Collective problem-solving becomes more possible within an ethos of care and becomes more powerful as more persons are recognized as allies. Strategies becomes more creative when we think with our hands and other body parts as well as with our brains. Paths are made by walking. Actions create momentum. Days begin to seize themselves.

20. Momentum

We inhabit a spinning planet. Earth rotates on its axis once each day. Or, rather, what we call a day constitutes one rotation. These twirls occur within a larger swirl. The planet orbits a star, traveling some 940 million kilometers in the course of what we call a year. This star emits energy, about 1,360 watts of which reaches every square meter of the outer reaches of our atmosphere that are facing it. Each new day brings fresh force, more power than we could ever need.

At every moment of every minute, plant roots probe underground while leaves above respire, converting sunshine to edible calories and exhaling the oxygen we animals need. All of the works and days of the *Human*, while catastrophically destructive for so many, are but the doings of one species among millions. All of those other species are doing things too. I've seen lichen growing on a snow plow. You've encountered insects going about their business without a care for whatever you or any other human thinks about them. Their collective biomass is larger than ours. Queering animal liberation can bring us into alliance with them, and with the blue-green algae who have already demonstrated their ability to alter the fundamental conditions of life on the planet.

Everything is always in flux. Nobody has to create change, because change is already always happening. The question is momentum. Right now, as human-engendered climate change surpasses one tipping point after another, bringing the risk of cascading collapses of ecosystems closer each day, we need to sharply and quickly shift the orientation of human beings toward multispecies mutual aid.

Some rivers run underground. Some people, human and otherwise, have been actively contesting *Human* hegemony for centuries. By queering animal liberation, we can tap into the wellspring of motivating emotion to free ourselves and others from the deadening effects of human supremacy.

Notes

1 The author thanks all of the participants in the scores of "Queering Animal Liberation" workshops at which some of these ideas have been rehearsed and dedicates this chapter to the memory of the anarchic Airstream Community, which included Rocky the peacock, Ready the duck, and Sharkey the former fighting rooster.

2 Stephen Vincent Benét, *John Brown's Body* (New York: Doubleday, Doran, 1928). The full text of this epic poem is widely available online.

3 Marcel E. Visser and Christiaan Both, "Shifts in Phenology Due to Global Climate Change: The Need for a Yardstick," *Proceedings of the Royal Society B: Biological Sciences* 272, no. 1581 (December 22, 2005), 2561–2569.

4 See, for example, Bruce Bagemihl, *Biological Exuberance: Animal Homosexuality and Natural Diversity* (New York: St. Martin's Press, 1999).

5 See, for example, Chelsea Schields and Dagmar Herzog, *The Routledge Companion to Sexuality and Colonialism* (Routledge, 2021).

6 Ladelle McWhorter, "Enemy of the Species," In *Queer Ecologies: Sex, Nature, Politics, Desire*, eds. Catriona Mortimer-Sandilands and Bruce Erickson (Bloomington, IN: Indiana University Press, 2010), 73–101.

7 Ibid.

8 Cara Daggett, "Petro-Masculinity: Fossil Fuels and Authoritarian Desire," *Millennium* 47, no. 1 (September 1, 2018): 25–44.

9 Val Plumwood, *Environmental Culture: The Ecological Crisis of Reason* (New York; London: Routledge, 2002). See chapter 8 and in particular page 194 for these and other suggestions for undermining anthropocentrism.
10 See Pattrice Jones, "Eros and the Mechanisms of Eco-Defense," In *Ecofeminism: Feminist Intersections with Other Animals and the Earth*, eds. Carol J. Adams and Lori Gruen (New York: Bloomsbury Academic, 2014) for an expanded discussion of the potential of *eros* to power effective activism.
11 I speak here of on-the-ground struggles for social justice, which remain *Human*-centered even as critiques of the *Human* have become more common within numerous academic realms.
12 Pattrice Jones, "Queer Eros in the Enchanted Forest: The Spirit of Stonewall as Sustainable Energy," *QED: A Journal in GLBTQ Worldmaking* 6, no. 2 (2019): 76
13 Sonia Katyal, "Exporting Identity," *Yale Journal of Law and Feminism* 14, no. 1 (2002).
14 Robin D.G. Kelley, *Freedom Dreams: The Black Radical Imagination* (Boston: Beacon Press, 2002); Franklin Rosemont and Robin D.G. Kelley, *Black, Brown, & Beige: Surrealist Writings from Africa and the Diaspora* (Austin: University of Texas Press, 2009).
15 Michael Richardson, ed., *Refusal of the Shadow: Surrealism and the Caribbean* (New York: Verso, 1996), 150.
16 Ibid., 124.
17 See Catriona Mortimer-Sandilands and Bruce Erickson, "A Genealogy of Queer Ecologies," in *Queer Ecologies: Sex, Nature, Politics, Desire*, eds. Catriona Mortimer-Sandilands and Bruce Erickson (Bloomington: Indiana University Press, 2010), 1–47 for an overview of the history of reprocentrism. See Pattrice Jones, "Property, Profit, and (Re)Production: A Bird's-Eye View," in *Animal Oppression and Capitalism*, ed. David Nibert (Santa Barbara, CA: Praeger, 2017), 2, 31–48 for a discussion of its role in capitalism, climate change, and animal exploitation.
18 George Clinton, Walter Morrison, and Garry Shider, "One Nation Under a Groove," *Funkadelic: One Nation Under a Groove* (Warner Bros., 1978).
19 Audre Lorde, *Sister Outsider Essays and Speeches* (New York: Random House, 2012), 59.

Bibliography

Bagemihl, Bruce. *Biological Exuberance: Animal Homosexuality and Natural Diversity*. New York: St. Martin's Press, 1999.
Benét, Stephen Vincent. *John Brown's Body*. New York: Doubleday, Doran, 1928.
Daggett, Cara. "Petro-Masculinity: Fossil Fuels and Authoritarian Desire." *Millennium* 47, no. 1 (September 1, 2018): 25–44.
Jones, Pattrice. "Eros and the Mechanisms of Eco-Defense." In *Ecofeminism: Feminist Intersections with Other Animals and the Earth*, edited by Carol J. Adams and Lori Gruen. New York: Bloomsbury Academic, 2014.
Jones, Pattrice. "Property, Profit, and (Re)Production: A Bird's-Eye View." In *Animal Oppression and Capitalism*, edited by David Nibert, 2, 31–48. Santa Barbara, CA: Praeger, 2017.
Jones, Pattrice. "Queer Eros in the Enchanted Forest: The Spirit of Stonewall as Sustainable Energy." *QED: A Journal in GLBTQ Worldmaking* 6, no. 2 (2019): 76.
Katyal, Sonia. "Exporting Identity." *Yale Journal of Law and Feminism* 14, no. 1 (2002).
Kelley, Robin D.G. *Freedom Dreams: The Black Radical Imagination*. Boston: Beacon Press, 2002.
Lorde, Audre. *Sister Outsider: Essays and Speeches*. New York: Random House, 2012.
McWhorter, Ladelle. "Enemy of the Species." In *Queer Ecologies: Sex, Nature, Politics, Desire*, edited by Catriona Mortimer-Sandilands and Bruce Erickson, 73–101. Bloomington, IN, USA: Indiana University Press, 2010.
Mortimer-Sandilands, Catriona, and Bruce Erickson. "A Genealogy of Queer Ecologies." In *Queer Ecologies: Sex, Nature, Politics, Desire*, edited by Catriona Mortimer-Sandilands and Bruce Erickson, 1–47. Bloomington: Indiana University Press, 2010.
Plumwood, Val. *Environmental Culture: The Ecological Crisis of Reason*. New York and London: Routledge, 2002.

Richardson, Michael, ed. *Refusal of the Shadow: Surrealism and the Caribbean*. New York: Verso, 1996.

Rosemont, Franklin, and Robin D.G. Kelley, eds. *Black, Brown, & Beige: Surrealist Writings from Africa and the Diaspora*. Austin: University of Texas Press, 2009.

Schields, Chelsea, and Dagmar Herzog. *The Routledge Companion to Sexuality and Colonialism*. Routledge, 2021.

Visser, Marcel E, and Christiaan Both. "Shifts in Phenology Due to Global Climate Change: The Need for a Yardstick." *Proceedings of the Royal Society B: Biological Sciences* 272, no. 1581 (December 22, 2005): 2561–2569.

5
ANALOGY AND ALTERITY

Jessica Eisen

Introduction

Human–animal comparisons have consistently emerged as key tools of "dehumanization" projects, inextricably linked to exploitation, violence, and genocide.[1] The Nazi genocide of European Jews was supported by texts and images likening Jews to dogs, apes, and, most pervasively, rats—animals that carried a specific cultural association as invaders warranting extermination.[2] The enslavement of Africans and ongoing subordination of Black people in the colonial and postcolonial world is entwined with an enduring characterization of Black people as being like apes ("simianization"),[3] an association that in turn depends for its force on assumptions that apes are violent, lascivious, or dull.[4] Normalized private violence and exploitation of women is linked to cultural associations between women and domesticated animals ("bitch," "catty," "bunny") or farmed animals ("cow," "chick," "nag,"), who are socially constructed as providers of comfort, service, and labor, and as existing to meet the appetites provoked by their flesh.[5] The lines are not neat: farmed animals, apes, and rats in fact appear in dehumanizing discourse across groups.[6] But there is a specificity to each of these histories of dehumanization that is, in part, reflected in the types of animal comparisons most prevalent in each context.

Animals appear in each of these examples as the "source" analogs, with debased humans configured as the "target" analogs.[7] Each of these debasing analogies is predicated on a starting assumption about the source cases: that something about animals in general, or the specific animals in question, makes them undeserving of concern, threatening, grotesque, and, ultimately, not grievable.[8] The expected "analogical inferences"[9] arising from these comparisons are that, since the source is relevantly similar to the target, and since the debased treatment of the animal source is appropriate, therefore, the human target deserves to be similarly debased. As is already clear from the brief examples sketched here, this high-level description of the analogical mechanics of dehumanization strips a great deal of content from very particular social and historical dynamics. While the broad strokes capture many true things about dehumanization, they sideline inquiry into the variety of source animals and their social significations, the actual lives of those animals, the content and

DOI: 10.4324/9781003273400-7

dynamism of the hierarchies these analogies support, the actual lives of the human beings so coded, and the intersections and overlaps between the categories invoked.

In a twist on this enduring deployment of animal comparisons to degrade human groups, animal advocates have sought in effect to *reverse* these analogies—to turn animals from the source into the target, in hopes of illuminating the "wrongness" of animal use practices. To this end, the abuse of animals is frequently described by animal advocates as analogous to, or as an instance of, rape,[10] slavery,[11] or holocaust.[12] As Claire Jean Kim explains, "the purpose of the comparisons is not, as in the past, to denigrate certain human groups but rather to elevate nonhuman animals—but the form of the linkage is the same."[13] In such advocacy, the analogical mechanism, made more or less explicit, runs roughly as follows: these source analogs are wrong, they are relevantly similar to the treatment of animals, and so the treatment of animals is also wrong (i.e. the *wrongness* analogy). This analogy only works if listeners share a widespread agreement in the wrongness of those source analogs (rape, slavery, holocaust), despite the fact that in some times and places they have been viewed as acceptable or banal, as animal use is now. In this respect, these comparisons may invoke a further analogical inference: people once wrongly accepted these source practices but then rightly corrected their mistake; people now make a relevantly similar mistake by accepting animal use practices, and so we should correct the current mistake of accepting animal use (i.e. the *corrigible mistake* analogy). "The analogy," Angela Harris explains, "reminds us that, as the bumper sticker says, truth has three phases: universal ridicule, heated controversy, and finally unquestioned fact."[14] This argument is often linked to a "moral extensionism" thesis, in which unjust exclusions have been, and will continue to be, corrected.[15]

These *wrongness* analogies and *corrigible mistake* analogies are ubiquitous in animal law, and in animal advocacy more broadly. The *wrongness* analogy is drawn, for example, in Isaac Bashevis Singer's reflection that, "for the animals it is an eternal Treblinka";[16] the People for the Ethical Treatment of Animals (PETA) "Enslaved" exhibition panel, which placed an image of the chained foot of an enslaved person alongside the chained foot of a captive elephant;[17] or the "widespread use of the rape metaphor in ecofeminist literature."[18] The *corrigible mistake* analogy is often set up in a way that gathers these diverse wrongs together and locates them in the past, as in Steven Best's reflection that "[h]aving recognized the illogical and unjustifiable rationales used to oppress blacks, women, and other disadvantaged groups, society is beginning to grasp that speciesism is another unsubstantiated form of oppression and discrimination."[19] Both types of analogies are most commonly drawn by animal advocates seeking to eliminate, rather than moderate, the use and killing of animals.[20]

These comparisons have attracted discomfort, outrage, and critique, especially as they pertain to American racial slavery. Such analogies have been charged with minimizing ongoing justice projects, ignoring the unique valences animal comparisons may have for some human groups, and failing to sufficiently attend to the incommensurability of the human experiences in issue. This chapter will seek to draw together these critiques and then ask *why*, despite these objections, these analogies continually resurface. A tentative answer will explore the centrality of analogy in human thought, the allure of bootstrapping, the practical interconnections between human and animal oppressions, and the particular significance of analogy in legal systems grounded in precedent. In conclusion, this work will argue that, despite the hazards of some of these analogies, a complete departure from any reference to human atrocities in animal law analyses is a goal that risks limiting our ability to think or speak clearly about either side of the human–animal divide. Feminist and intersectional methods offer us some guidance as to how to proceed with care in this context: attending to interconnections rather than blunt comparisons between oppressions and drawing experiences of oppression together in ways that invite empathy rather than objectification. As with so many of the challenges of feminist method, however, the only resolution (such as it is) is a commitment to engage thoughtfully, strategically, and provisionally.

Terminological Note: Analogy and Placements Alongside

I refer throughout this piece to the concept of "analogy," though I am in fact hoping to invoke a broader family of rhetorical maneuvers. There are, for example, formal distinctions that might be drawn between *analogy* (source and target are relevantly alike), *metaphor* (the two things *are* each other, poetically though not in fact), and *category* (two things belong in the same basket of concepts).[21] I will not attend to these distinctions in the discussion that follows. My interest is in the placement of human atrocities alongside the treatment of animals, whether or not the analogical work is made explicit. These *placements alongside* are necessarily evocative and challenging in the ways I explore here. Whether a speaker says that a contemporary chicken farm *is* Treblinka (i.e. metaphor), is *like* Treblinka (analogy), or *belongs in a class with* Treblinka (category), the invocations of the "wrongness" and "corrigible mistake" parallels may be equally present. And in all cases, these rhetorical gestures appeal not to reason alone but also inevitably to the deep and distinct matrices of emotion, symbol, and implicit association carried by each human atrocity relied upon as a source analog in this broader sense.[22]

A brief example of the relationship between formal analogical reason and emotive connections is helpful in explaining why these *placements alongside* carry some of the relevant qualities of analogy, even where a formal analogy is not explicitly drawn. Social psychologist Thomas Gilovich, aware of the widespread salience of the Vietnam War and World War II (WWII) in U.S. foreign policy discourse, designed an experiment. He created two hypothetical cases for proposed U.S. military intervention, identical in every substantive aspect but including language and imagery evocative of these distinct historical conflicts (e.g. one scenario included the term "blitzkrieg" to describe a sudden attack, while the other used the term "quickstrike," and one scenario described refugees fleeing in boxcars, while the other described refugees fleeing in small boats.) While these differences should not have been logically relevant to the appropriate foreign policy response, they were designed to implicitly invoke distinct source analogies: WWII (popularly viewed as an appropriate context for U.S. military intervention) or the Vietnam War (popularly viewed as a mistaken U.S. military intervention). When presented with these substantively identical hypotheticals, test subjects were more likely to favor U.S. military intervention where those logically irrelevant cues were designed to call forth the WWII source analog.[23] Cognitive scientists Keith J. Holyoak and Paul Thagard remark that this experiment demonstrates that analogies are "open to . . . biases whenever salient similarities irrelevant to the purpose of the analogy guide memory retrieval."[24] Particularly on legal, social, and political matters, analogies invariably interact with a complex well of implicit associations and emotional reactions. In these settings, at least, this experiment suggests that analogical thinking and argument are never well described as exercises in pure logic. The interconnections between reason and emotion (which are, of course, well-developed within feminist theory[25]) shape the deployment and reception of analogies, other *placements alongside*, and other forms of social and intellectual engagement.

The distinctions between analogy, metaphor, and categorization are formally meaningful, but this group of rhetorical maneuvers can be helpfully gathered together for our purposes. These modes of *placement alongside* operate in similar ways—drawing on emotionally laden source cases to lead audiences toward a particular view of the target case. In view of this, I will treat together, under the broad umbrella of analogical thinking, the sort of *placement alongside* that includes, for example, the use of the term "abolitionism" to describe the animal rights movement, with knowledge that this will evoke the movement to abolish American racial slavery, even in circumstances where the comparison between animals and slaves is not expressly drawn.[26]

Analogy and the Problem of Alterity

Wrongness analogies and *corrigible mistake* analogies that compare human and animal oppressions have provoked hurt, anger, and criticism. Proponents of these analogies have sometimes shaken off these responses as just another instance of failure to give animals the concern they are due: "comparing the suffering of animals to that of blacks (or any other oppressed group) is offensive only to the speciesist: one who has embraced the false notions of what animals are like."[27] But this retort does not do justice to the depth and nuance of critique these analogies have attracted.

In many cases, the concern at the heart of these criticisms is not about any supposed differences between humans and animals but rather about the *alterity* of the human groups called upon as source analogs within human communities.[28] Alterity, as developed in feminist and post-colonial scholarship, describes the experience of "the Other" within "conceptual fields . . .[that] have been divided in problematic and oppressive ways between a privileged, dominant One and a de-valued or subordinated Other."[29] As *Others*, the source analogs drawn upon often stand as outsiders to hegemonic culture and representation—meaning that their experiences are at special risk of being misunderstood, misappropriated, or misdescribed.[30] The following sections will survey some critiques of human–animal analogies in animal rights discourse, identifying particular arguments relating to *incommensurability*, *ongoing animalization*, and *closure*. This survey owes a particular debt to the works of Bénédicte Boisseron, Angela Harris, and Claire Jean Kim. Two important conclusions arise from this survey and typology of common objections to human–animal comparisons made by animal advocates. First, the most trenchant criticisms of the *wrongness* and *corrigible mistake* analogies do not depend on a violent or dismissive attitude toward animals. These analogies are not, as Spiegel suggests, "offensive only to the speciesist." Second, these criticisms, though sharing connections across human justice contexts, hold special force with respect to comparisons to American racial slavery.

The Incommensurability Problem

In objecting to Holocaust comparisons in animal advocacy, Roberta Kalechofsky insists that "some agonies are too total to be compared with other agonies."[31] Sometimes this objection is offered together with specific examples of the relevant differences between the human source and animal target analog, as in Kalechofsky's insistence that the Holocaust's distinct ideological and historical underpinnings were specific to antisemitism[32] or Jeff Oboler's distinction that the Holocaust aimed at genocidal extermination while animal agriculture aims to continue breeding animals for profit and exploitation.[33] In these cases, the veracity and usefulness of the comparisons are cast as dependent on facts about similarity or difference. There is a certain irresolvability to this conflict, to the extent that it is cast as one that can be answered through recourse to facts alone. Analogies and disanalogies depend for their argumentative and descriptive value on some underlying theory identifying which facts matter and why. Without an underlying theoretical framework of this kind, we are just left with the reality that "everything is a little bit similar to, or different from, everything else."[34]

Kalechofsky's argument that the agony of the Holocaust is "too total" to bear comparison seems, however, to appeal to something more than mere factual disanalogy. It is an insistence that comparison, or reduction to one-of-a-type, risks denying the *incommensurability* of the harm. As Kim summarizes, there is a worry that "instrumentalizing the Jewish experience" as analogical fodder for a different justice context "inevitably leads to de-specifying it."[35] So, too, Bénédicte Boisseron argues that analogies between the experiences of enslaved Americans and animals "dismisses the unique pain of being discriminated against as a black individual."[36]

Boisseron's inquiry into comparisons between animals and Black humans offers an important further articulation of the concerns of de-specification through instrumental comparison. In particular, the *wrongness* analogy invites listeners to reduce both enslaved person and the animal only to their suffering, with the *corrigible mistake* analogy offering a similar reduction and inattention to the particularities of Black life under conditions of inequality. As Boisseron affirms, "the individual . . . should not be defined by her pain alone. Analogizing the black experience with the animal experience in an exclusive state of suffering results in essentializing both and desensitizing us to the actual being."[37] Stripped of their specificity and individuality, "[a]nimals, women, blacks, and Jews become merely ideas and concepts, caught in a rhetoric of similes, analogies, and metaphors."[38]

Significantly, this concern does not (necessarily) depend upon an assumption that animals are less worthy of concern than human beings. True, in some cases, the factual detail raised to challenge these analogies relates to the fact that the victims in the source cases are human—and sometimes in terms that do seem to depend on a degraded view of animals.[39] Often, though, these objections are not fairly described as bald assertion of human supremacy but rather as illuminating aspects of the harms in issue that are meaningfully distinct owing to the shared semantic and symbolic contexts of intra-human oppression. Kim, for example, notes that visual representations of similar tools or technologies of restraint as between animals and enslaved people (including, for example, spike collars), do not capture the extent to which human victims are intended to exact not only physical compliance but also "distinctively human" experiences of "sham[e] and debasement."[40]

There is a further concern with Spiegel's insistence that these comparisons are "offensive only to the speciesist":[41] that a very similar set of objections emerges, for similar reasons, even where the target analogs are human beings. The treatment of Black Americans in particular has taken on a paradigmatic quality in U.S. legal and political life, recurrently emerging as a source analog offered for comparison to LGBTQ+ rights and women's rights, among others.[42] In these contexts too, the risks of analogy remain present. Trina Grillo and Stephanie M Wildman, for example, resist comparisons between racism (as source) and a number of humanist "isms," including sexism, antisemitism, heterosexism, and ableism.[43] Among their concerns are that such analogies "take[] back center-stage from people of color, even in discussions of racism," invisibilizing people with intersectional identities that place them in both source and target analog groups and "appropriation of pain" by whites who claim to "understand the experience of racism."[44] Holocaust analogies have attracted similar resistance, even when compared to other human atrocities.[45] Rape analogies have their own complex history and context in the intra-human context, with particular objections arising from the common use of rape metaphors to describe trivial matters; incommensurability objections have also arisen, however, where rape is used as a source analog respecting serious human injustices such as colonialism.[46] The concerns from incommensurability, therefore, are not fundamentally (or at least not necessarily) rooted in speciesism but in a worry about subjecting these singular experiences to the instrumentalization and reductionism necessitated by deployment as analogical source material. As Angela Harris explains, "moral disasters" of this scale "demand of us that we recognize their black-hole quality: they are utterly singular, utterly horrific in very specific ways; they signify the breakdown of ordinary politics and ordinary public policy in which this harm can be put in the scale and weighed against that."[47]

The Ongoing Animalization Problem

Another problem with these reverse-dehumanization analogies is that they invariably draw listeners back to their troubling inverses, calling forth symbolic intersections through which hurt and violence still circulate. As Kim posits, for those particularly threatened by racial violence, "the intent behind making the blacks-animals association hardly matters; what matters is that the association itself is forever tainted with the blood of its victims."[48] Thus, Harris argues that animal-slavery analogies seem to "assume a comfort in associating oneself with animals and animal issues that people of color can only assume with difficulty."[49] Breeze Harper adds that, particularly since these analogies are often offered without careful historical attention to the source analogs, "such images and textual references trigger trauma and deep emotional pain."[50] Boisseron argues that, "[w]hen talking about the value of black existence, the animal comparison is intrinsically part of our culture."[51] Boisseron artfully links racist dehumanization discourse, animal abolitionist discourse, and the recurrent charge within Black activism that society values animals more than it does Black people.[52] Across these distinct ideological projects and discursive spaces, Boisseron argues that representations of Blackness and animality are forged together within a broader cultural "system that convulsively pits blackness against animality."[53]

Each of these scholars acknowledges the imperative of disrupting the racial and species hierarchies that undergird these oppressive dynamics. Harris, for example, compares racialized discomfort with animal analogies to Margaret Baldwin's identification of many women's felt need to dissociate themselves from "prostitutes": "It is, of course," she explains, "the opposition between woman and prostitute, animal and African that needs itself to be destroyed."[54] Harris thus does not locate the discomfort she identifies with the comparison in any lesser moral value of animals but rather in an awareness of the reality of pervasive cultural associations. To charge ahead with these analogies without regard to the damaging dynamics they invoke serves to "implicitly code" these analogical deployments "as white."[55] Wittingly or not, these *placements alongside* "call[] out the structures of feeling that have undergirded racism for so long."[56]

The "ongoing animalization" problem is of special concern with respect to slavery and anti-Black racism. While the "incommensurability" problem is similarly raised with respect to slavery, rape, and the Holocaust, the problem of "ongoing animalization" is not raised with equivalent frequency or force across these analogical deployments. The deep imbrication of animal imagery in ongoing, state-sanctioned violence toward Black people does not have a parallel in the lives of North American Jewish people and has therefore not given rise to this particular form of objection. This is not a matter of "oppression Olympics"[57] but rather a reflection of the distinct roles and functions of animality across contexts. And Harris' observation that group experiences with animalization may affect comfort with these *placements alongside* seems borne out here. While Jewish scholars and activists, often Holocaust survivors themselves, commonly advance the Holocaust analogy, it is most commonly white activists and scholars who insist on the slavery analogy.[58]

The Closure Problem

The refusal to understand slavery as a particular, *incommensurable* system of harms is, of course, linked to the refusal to attend to the representational harms of *ongoing animalization*. When it is argued that slavery is *relevantly like* other things, such as animal use, the facts that mark that context as *different*, including the particulars of contemporary modes of anti-Black racism, become irrelevant to the world of the analogy. Both of these challenges are further linked to a third concern of particular relevance to the slavery context: a tendency to treat the wrongs of slavery and anti-Black racism as essentially resolved. The analogical positioning of slavery as a *corrigible mistake* insists on this conclusion.

As Harris explains, the experiences of Black Americans are routinely drawn upon in analogical arguments by other justice-seeking groups, both human and animal, as "civil rights mascots" or exemplars of social movement success.[59] In these arguments, Kim explains, the source analogical relationship holds that "black people at some point (variously, emancipation, Reconstruction, the civil rights movement) moved demonstrably from slavery to freedom."[60] This implied closure of the problems of racial injustice, serving now as an archetype for the still-live problem of animal exploitation is described in various terms that capture the description of Black and animal rights struggles as successive: a "logic of subsequence,"[61] a "teleological" imaginary in which "animal is the new black,"[62] an "orientation" of slavery "backward in time,"[63] or a false "racial temporality"[64] grounded in "triumphalist narratives."[65]

These presumptions of freedom and success require inattention to the durable dynamics of racial hierarchy and violence that are in fact "borne aloft into the present," expressed, for example, as "environmental racism, residential and educational segregation and degradation, the school-to-prison pipeline, mass incarceration, [and] police violence."[66] As Harris explains, "What's wrong with such arguments is their implicit assumption that the African American struggle for rights is over, and that it was successful"[67]—a presumption that requires the erasure of contemporary Black struggles for safety and inclusion. For Boisseron, this presumption that Black justice struggles are closed requires a "racial blindness" that "presupposes that we have progressed beyond blackness in our considerations of (de-) personhood."[68] The disregard for the realities of contemporary Black lives dovetails with a disinterest in the particularities and incommensurability of the experiences of enslavement. As Kim explains, these analogies are often drawn in ways that seem "less interested in exploring the character of racial slavery than in pronouncing it dead and naming animal slavery as its successor."[69]

Here, again, we see a distinction between the meanings of slavery, Holocaust, and rape analogies. While Kim gathers Holocaust and slavery analogies together as posing similar concerns with closure (noting that "Jews and blacks continue to fight for full membership in humankind"),[70] the suggestion that Holocaust analogies present a false sense of closure is far less common among objections to Holocaust analogies than it is respecting slavery analogies. While antisemitism and antisemitic violence persists, scholars of North American antisemitism have not identified the kinds of enduring, systemic, and officialized dynamics that characterize anti-Black racism. Scholars of anti-Black racism identify a durable through-line from slavery to Jim Crow to contemporary practices such as mass incarceration.[71] This emphasis on continuity of violence and dispossession is distinct from the overriding representations of the status of the Holocaust in contemporary Jewish life as forging a "community of memory": as something to be insistently remembered ("never forget") and as something to be vigilantly guarded against under constant threat of recurrence ("never again.")[72] A distinct temporal orientation emerges with respect to rape analogies: rape is an ongoing source of violence in many peoples' lives, and its systematicity and official sanction engages the complex intersections between "public" and "private" life.[73] Remarkably, Karen Davis' exploration of Holocaust analogy invokes sexual violence as well but does so through reference to Margaret Atwood's futuristic dystopia, *The Handmaid's Tale*, thus locating the source harm in a dreaded future rather than a historical or ongoing trauma.[74] While it might feel "obscene" to turn any of these tragedies into a "legacy of uplift,"[75] the nature of the obscenity carries a distinct valence where it erases and minimizes not only a historic calamity or threatened future but also the recurrent and ongoing reinstantiation of source analog's violence.

The Enduring Allure of Analogies

Given the force of these objections, it is remarkable how prevalent these analogies—particularly the slavery analogy—remain within animal advocacy in general and animal law in particular. In 2011, despite objections to their past use of slavery analogies, PETA launched a suit claiming that the U.S. Constitutional prohibition on slavery should extend to captive orcas.[76] Steven Wise persistently references the case of *Somerset v. Stewart*, the case in which slavery was found to be contrary to the common law of England, as a key precedent in his litigation efforts on behalf of captive animals.[77] And Gary Francione and Anna Charlton repeatedly condemn animal welfare laws (as opposed to animal rights laws) with recourse to the slavery analogy: although "[w]e would all agree that beating one's slaves less is better than beating one's slaves more . . . [n]o one would promote the 'humane' treatment of slaves as something that would eradicate the injustice of the institution of slavery."[78]

Part of this might be attributable to the usual inability of dominant groups to hear and imagine the lived experiences of others.[79] But I think that there are additional forces at work here as well. First, analogy, and the related concepts of category and metaphor, are fundamental to human thought and communication.[80] Second, this baseline significance of analogical reasoning is deepened by the special role that analogy plays in precedent-based legal systems[81] and therefore also with the public debates with which they intersect. Given the fact that emancipation and civil rights for Black Americans have served as a "crucible for equality law in America," Catharine A. MacKinnon has observed that "sexism must be like racism, or nothing can be done" in a precedent-based legal system.[82] In view of this, Janet Halley has reflected that "asking the advocates of gay, women's or disabled peoples' rights to give up 'like race' similes would be like asking them to write their speeches and briefs without using the word 'the'."[83]

In the animal protection context, the allure of analogies to widely acknowledged atrocities is, in part, attributable to the fact that animals' interests are often not taken seriously. The hope animating these analogies is often that associating the cause of animals with acknowledged wrongs will make the cause of animals appear important and urgent: "to endow the contemporary animal rights movement with an aura of world-historical significance, moral urgency, and historical possibility."[84] This dynamic has been widely explored in human discrimination contexts, where race-sex and race-disability analogies are prevalent. As Serena Mayeri explains, these analogies are deployed to attract "political and legal currency" and to "inspire empathy and understanding of harms previously unrecognized."[85] Through analogy to recognized legal claims, "the legally unintelligible" might be "made recognizable as a subject or object before the law."[86] This claim, a version of the "wrongness analogy," is often implicitly supported by the fact that source analogs once involved a fight for this kind of recognition as well—giving rise to the "corrigible mistake analogy".

The visual, technological, and ideological relationship between intra-human and interspecies hierarchies also makes these analogies appear highly available to animal advocates. The devices of physical control—cages, chains, whips, cattle cars, barbed wire—appear across species lines, offering up provocative "visual rhyme[s],"[87] even if the social meanings and experiences of their uses vary.[88] So too with the justifications offered for hierarchy and exploitation across contexts, often similarly articulated in naturalistic, divine, and scientistic terms. And, in fact, many of these social hierarchies (race, gender, ability, species) are co-constituted in ways that mean any deep inquiry into one will lead invariably into the others. It might, then, be difficult or even impossible to conceive of an animal legal praxis that avoids the *placing alongside* of justice problems, human and animal. The balance of this chapter will address two strategies for placing justice struggles alongside each other that do not fall into the traps of the *wrongness* analogy or the *corrigible error* analogy.

Connection Not Comparison

Significantly, Kim, Boisseron, and Harris do not conclude that race and animality, or even slavery and animal use, should never be considered together. To the contrary, each author insists that a fulsome understanding of these justice contexts in fact *requires* that they be considered in relation to each other.[89] But, once we depart from the bare "wrongness" analogy and associated "corrigible error" analogy, how do we proceed to consider these justice problems together? The answer, for each of these authors, lies in analyses that focus on the *interconnections* between diverse justice problems rather than asserting blunt sameness analogies that strip so much context from both source and target analog.

Boisseron's project, for example, traces the relationship between Blackness and animality, with a particular focus on dogs, dating back to the Atlantic slave trade. Boisseron seeks in her analysis to "determine how the history of the animal and the black in the black Atlantic is connected, rather than simply comparable."[90] Drawing on Jared Sexton's analysis, Kim comes to a similar conclusion, arguing that a "relational" rather than a "comparative" analysis is required to illuminate the dynamics of race and species without "black existence . . . serving as an evacuated historical prop dragged on- and offstage in someone else's drama."[91] Kim notes, for example, the possibility of rethinking the "ontological schema of slavery as essentially triadic, with the key terms being human, slave, and animal," with racial slavery requiring a "continuous and intimate dependence on the symbolic (as well as material) figure of the animal."[92] Harris too emphasizes the need for analyses focused not on similarity of harm or experience but on the "common history and a common logic"[93] between racism and species-ism: "not only a history of capitalist exploitation under which slaves crammed into ships presage factory farms, but also the history of an episteme under which nature and culture are violently separated and the modern subject emerges, nostalgic about the rupture."[94] For Harris, the ensuing analysis decenters (but does not displace) "rights" and takes seriously epistemologies foreclosed by hegemonic western ideologies, embracing a range of Indigenous, Asian, African, and African-diasporic ways of understanding and assessing relationships, including with other species.[95]

Significantly, each of these authors approaches their work in the context of a demonstrated commitment to addressing anti-Black racism. Where the intention and commitment to the justice projects undergirding the source analog are not present, even intersectional or relational argumentation can carry the veneer of instrumentalization. As S. Marek Muller explains,

> A blog post explicating, for instance, the historical, ideological, and material ties between the transatlantic slave trade and contemporary animal labor can be true enough but is in practice little more than offensive when the blogger has no interest in reciprocal and restorative justice for both animal and Black bodies.[96]

While personal identity is not determinant of commitment, it is notable that so many thoughtful works placing the experiences of some women alongside those of some animals are offered by women-identified and feminist-identified scholars, and Holocaust analogies have most commonly been advanced by commentators identifying themselves as Jewish, while the most troublingly reductive analogical work, very often involving slavery comparisons, is often offered by white men who have no apparent links or commitment to the justice projects of Black people.[97]

This sort of analysis—embracing interconnection and the relationship between oppressions, rather than blunt recourse to analogy—is prominent within feminist work on animal oppression. Much of the scholarship at this intersection focuses on the ways that ideologies and technologies of exploitation work together rather than simply identifying similarities. For example, legally sanctioned violence against animals is often observed to be deeply shaped by gender. In the agricultural context, this line of argument is influenced by Carol J. Adams' identification of eggs and dairy as "feminized protein."[98] Elaborating the ways that cultural representations of eggs and dairying often rely on tropes of objectification of women, Adams explains that "in conventional attitudes, these female animals disappear from concern; partly . . . because they are female."[99] In this same discussion, Adams chooses not to take up a tendency in the animal rights movement to directly analogize the artificial insemination of dairy cows to rape within human communities. She explains her decision to term the former "forcible impregnation," in part to leave room for dedicated attention to what is specific about each category of experience.[100] Here, we see *placements alongside* deepened through attention to distinct justice contexts (human rape and sexual exploitation of animals) to make an intersectional argument: that tropes of objectification work in distinct but interrelated ways across species lines to authorize particular experiences of violence and exploitation.

Feminist animal law scholarship addressing itself to the place the "farm" and "family" as distinct instantiations of law's "private sphere" shares this quality. Mathilde Cohen, for example, has argued that "cows have remained hidden from the public gaze, confined to the 'privacy' of their farms under the dominion of their owners, much like generations of women before them were confined to the privacy of their home under the dominion of their husbands."[101] My own research examining the social and legal position of dairy cattle has similarly concluded that "the farm has earned a place among those institutions like the 'home' and the 'family' that have been misleadingly cast as apolitical and beyond the reach of law,"[102] including through "structurally parallel" legal forms used to create and defend a "private sphere" in which violence is tolerated.[103] In these analyses, similarities are drawn between the legal structures enabling hierarchy in two distinct contexts. The analogy drawn between farm and family is not one of simple bootstrapping (wrongness or corrigible error) but instead draws out legal technological details across distinct circumstances in order to illuminate the intersectional nature of the "private sphere" as it emerges in each context.

In addition to these doctrinal examinations of law's internal logics (i.e. sameness argumentation and the private sphere), some feminist animal law scholarship observes the ways that gender and species hierarchies intersect within legal culture—the immediate social contexts in which legal logics are produced. Angela Fernandez's "legal archaeology" of the *Pierson v. Post* decision demonstrates this form of feminist intersectional engagement.[104] This classic case, in which the New York courts were called upon to determine ownership of the body of a fox pursued by two competing hunters, is a staple in law school property classes for its examination of the relationship between pursuit, possession, and ownership of a natural object.[105] Fernandez's work does address itself to these doctrinal themes[106] but also takes up the ways that the clubby and elitist culture of New York state's legal profession contributed to the court's propertization of the fox. In an "imagined dissent,"[107] Fernandez offers the opinion of a woman-of-color judge (a purposefully implausible figure on the 1805 New York State court).[108] In the voice of her imagined dissenter, Fernandez critiques the court's "profound disrespect . . . towards the hunted animal," including as evidenced through the judge's "witty repartee" and associated "presuppos[ition] rather than argu[ment] for the legitimacy of her status as property, as a slave."[109] Building on Fernandez's documentation of the sexual and racial hierarchies shaping the legal profession and its sense of humor,[110] the imagined dissenter reflects that her colleagues' jokes about the fox mirror the contemptuous and mocking regard in which she suspects she herself is also held.[111] There is, in other words, an intersection between the hierarchies of race, gender, and animality, which shapes the legal culture that, in turn, shapes the court's exercise of power over the fox.

The preceding intersections—in the realm of doctrine and legal culture—are each technologies of hierarchy: mechanisms through which oppression operates. Some feminist animal law adopts intersectional analysis at a more fundamental level, examining not only specific technologies of hierarchy but the underlying logics and ontologies that shape human legal engagements with animals. Yoriko Otomo, for example, draws on Derrida's theory of "carnophallogocentrism" as the "onto-theological structure" underpinning animal exploitation.[112] Perhaps the most sustained analysis of this kind is offered by Maneesha Deckha, in *Animals as Legal Beings: Contesting Anthropocentric Legal Orders*.[113] Deckha's deliberately "intersectional"[114] theory of legal anthropocentrism ties racial, sexual, colonial, and species hierarchy to a single ontological root: "[l]aw's liberal orientation and the multiple dispossessions and marginalizations it entails."[115] Again, the aim here is not to demonstrate that the experiences of oppression across these various social categories are merely similar but, more deeply, that they intersect, with each structure of hierarchy supporting the others and all co-constituting a broader common ontological architecture.

Empathy Not Objectification

A recurrent criticism of slavery analogies is their tendency to objectify and reduce the experiences of enslaved people—to turn them into source analogs whose only relevant qualities are their utility as a comparative case. Both the "wrongness" analogy and the "corrigible mistake" analogy risk this kind of objectification, because both locate the listener as the *one with power to decide*. The listener is not invited to look internally, to *feel with* those experiencing the complexity of and specificity of experiences of violence. The fact that these analogies so often take the form of visual or pictorial representations, rather than attention to first-person narrative, reinforces the voyeuristic and distancing quality that so often characterizes these analogies. As Harris explains, "[t]he comparison implicitly constructs a gaze under which slaves and animals appear alike. This is the sentimental gaze of the privileged Westerner who 'saves' those less fortunate, the voiceless masses whether human or animal."[116]

Assuming a position of empathy and connection, rather than objectification and voyeurism, allows for justice problems to be placed alongside one another without falling into the pitfalls associated with the "wrongness" and "corrigible mistake" analogies. Contrast, for example, Francione and Charlton's reflections inviting the reader to imagine "beating one's slaves less" with this reflection of Tashee Meadows:

> I thought of my ancestry as a Black woman: the rapes, the unwanted pregnancies, captivity, stolen babies, grieving mothers, horrific transports, and the physical, mental, and spiritual pain of chattel slavery. I'm convinced that animals of other species, many of whom are more protective of their young than humans, grieve when their babies are taken away. I thought of how much I missed my mother, brother, and sister when I was in foster care. I made the emotional connection that other beings must feel this pain, too, when they are separated from their mothers and other family members. When I saw the battery cages, I thought of the more than two million Americans who know cages firsthand in the prison-industrial complex.[117]

Meadows is clearly placing slavery alongside the experiences of farmed animals here, but her mode of engagement is entirely different from Francione and Charlton's. She draws on the experience of slavery not as an object of a distant past but as a resource for an emotional experience with which she feels direct connection and empathy. Slavery, in this account, is not merely a past event, an exemplar of bad behavior since corrected. It is a source of intimate harms that echo through generations and pains that call out to be understood in their specificity. She is not asking the reader to accept that these diverse stories (slavery, animal exploitation, foster care, and incarceration) are identical in a way that might be refuted by pointing out factual differences between these distinct contexts. Instead, she asks her readers to *feel with* those experiencing these harms—to draw on available emotional resources to imagine a life that might be profoundly different from their own.[118] Boisseron invokes a similar imperative to *feel with* rather than to objectify in her choice to analyze race and species "through the prism of black and animal defiance," hoping to capture a more textured and personal account of the lives she describes than bare attention to "their comparable state of subjection and humiliation" might allow.[119]

In this respect, *placement alongside* can actually serve to focus attention on the particularities of experience for source and target analogs. This is most clear where the *placement alongside* serves to emphasize the need for the epistemological recovery of particular voices and experiences. We might think of inquiries along these lines as developing an "animal legal method" that might be placed alongside the projects of "feminist legal method"[120] or "outsider jurisprudence."[121] Yoriko Otomo and Cressida Limon, for example, note that dogs, pigs, and children in colonial Britain were entirely "absent as subjects from the vast tracts of legal scholarship that purport to deal with topics such as domesticity, sexuality, criminality and responsibility," such that they "lived through law in similar ways to women."[122] What is taken as similar as between women and animals here is not the content of the experiences they might convey through law if given the chance but rather the fact of their epistemological exclusion. Building on this view of law as wrongly obscuring animal experience, most research in feminist animal law involves some effort to center animal experience, from Deckha's call to "foreground the laboratory rat's first person perspective"[123] in legal regulation of animal experimentation to Catharine MacKinnon's more general inquiry about the field known as "animal law": "who asked the animals? References to what animals might say are few and far between."[124] This methodological turn toward attention to animal perspectives also infuses feminist-informed calls to foreground animal "capabilities"[125] or "dignity" in ways that make animals' experiences "intrinsically relevant for law"[126] and in feminist animal law scholarship attending to animals' "vulnerability" or "embodiment" as a basis for making law responsive to animal experience.[127]

Work in this vein often emphasizes the challenges of this imperative for human beings who act as necessary intermediaries in efforts to bring animal perspectives to bear in animal law scholarship, legislation, and litigation. MacKinnon has called this the "speaking-for-the-other problem," and Deckha has identified it as a "foundational challenge to the liberal legal culture that presumes representing others is an innocuous act." For Boisseron, this marks a significant point of distinction from women's and ethnic studies, both of which have included substantial commitments to include women and racialized people in academic spaces: "Because there is no possibility for self-representation, as we know it, the animal remains the silent one, bound to be represented by 'us' humans."[128] Despite the intractable difficulties of representing others across vast differences and power differentials, however, feminist and intersectional animal law pays careful attention to the diverse human and animal perspectives invoked as necessary.

Conclusion

Even where it is viewed as essential to consider animality alongside instances of intra-human hierarchies and violence, these placements alongside are often treated by intersectional and feminist thinkers as "risky"[129] or "dicey."[130] As Boisseron cautions, noting the limits of her own project, "[o]ne cannot address the entangled oppressions of humans, minorities, and animals while doing justice to all groups and individuals equally."[131] The aims of resisting generalization, attending to specificity, and avoiding reductive treatment of diverse justice projects can never be fully realized. "The goal," then, can only be to "open our minds to the exponentiality of intersections but with no ambition to address them all."[132] The perennial incompleteness of methodological projects aiming to recover and amplify excluded perspectives has been a persistent theme in feminist and intersectional jurisprudence.[133] Ultimately, as with so many of the challenges of feminist method, the only resolution (such as it is) is a commitment to engage thoughtfully, strategically, and provisionally.

Notes

1 Nick Haslam, "Dehumanization: An Integrative Review," *Personality and Social Psychology Review* 10, no. 3 (August 2006).
2 Boria Sax, *Animals in the Third Reich: Pets, Scapegoats, and the Holocaust* (New York: Continuum, 2000), 54–55, 82; Michelle Wiebe, "Extermination and Euthanasia: Animal Symbolism in Nazi Germany," *The Mirror—Undergraduate History Journal* 28, no. 1 (2008): 28; Claire Jean Kim, "Moral Extensionism or Racist Exploitation? The Use of Holocaust and Slavery Analogies in the Animal Liberation Movement," *New Political Science* 33, no. 3 (September 2011): 329.
3 Wulf D. Hund and Charles Mills, "Comparing Black People to Monkeys Has a Long, Dark Simian History," *The Conversation* (February 28, 2016). Retrieved from: https://theconversation.com/comparing-black-people-to-monkeys-has-a-long-dark-simian-history-55102.
4 Claire Jean Kim, *Dangerous Crossings: Race, Species, and Nature in a Multicultural Age* (Cambridge: Cambridge University Press, 2015), 35–36.
5 See Karen Davis, "Thinking Like a Chicken: Farm Animals and the Feminine Connection," in *Animals and Women*, eds. Carol J. Adams and Josephine Donovan (Durham: Duke University Press, 1995), 192, 196; Joan Dunayer, "Sexist Words, Speciesist Roots," in *Animals and Women*, eds. Carol J. Adams and Josephine Donovan (Durham: Duke University Press, 1995).
6 See, for example, Wulf D. Hund, Charles W. Mills, and Silvia Sebastiani, eds., *Simianization: Apes, Gender, Class, and Race* (Zürich: Lit Verlag, 2015) (discussing ape analogies across human social contexts); Claire Jean Kim, "Abolition," in *Critical Terms for Animal Studies*, ed. Lori Gruen (Chicago: University of Chicago Press, 2018), 29. Referencing farmed animal analogies in the context of American racial slavery; Wiebe, "Extermination and Euthanasia," 23, 28 (describing ape analogies deployed against Jews in Holocaust rhetoric).
7 See Keith Holyoak and Paul Thagard, *Mental Leaps: Analogy in Creative Thought* (Cambridge: MIT Press, 1995), 101.
8 See Judith Butler, *Precarious Life: The Powers of Mourning and Violence* (New York: Verso, 2004) and Judith Butler, *Frames of War: When is Life Grievable?* (New York: Verso, 2016). Developing the concept of grievability; see also extensions of Butler's grievability analysis in the non-human context in Chloë Taylor, "The Precarious Lives of Animals: Butler, Coetzee, and Animal Ethics," *Philosophy Today* 52, no. 1 (2008); James Stanescu, "Species Trouble: Judith Butler, Mourning, and the Precarious Lives of Animals," *Hypatia* 27, no. 3 (2012); Kathryn A. Gillespie and Patricia J. Lopez, eds., *Economies of Death: Economic Logics of Killable Life and Grievable Death* (London: Routledge, 2015).
9 See Holyoak and Thagard, *Mental Leaps*, 101.
10 Andrée Collard with Joyce Contrucci, *Rape of the Wild: Man's Violence against Animals and the Earth* (Bloomington: Indiana University Press, 1989); Lori Gruen, "Thoughts on Exclusion and Difference: A Response to 'On Women, Animals and Nature'," *American Philosophical Association Newsletter on Feminism and Philosophy* 91, no. 1 (1992): 80–81. See also "Is Your Food a Product of Rape?" News, PETA People for the Ethical Treatment of Animals, retrieved from: https://www.peta.org/features/rape-milk-pork-turkey/, accessed July 25, 2023.
11 Marjorie Spiegel, *The Dreaded Comparison: Human and Animal Slavery* (New York: Mirror Books, 1996).
12 Charles Patterson, *Eternal Treblinka: Our Treatment of Animals and the Holocaust* (New York: Lantern Books, 2002); "Holocaust on Your Plate," traveling display, PETA People for the Ethical Treatment of Animals, 2004.
13 Kim, "Moral Extensionism," 313.
14 Angela Harris, "Should People of Color Support Animal Rights," Journal of Animal Law 5, no. 1 (2009): 25.

15 Kim, "Moral Extensionism," 324. Describing these analogies as aiming to prompt listeners to think: "We used to think it was acceptable to enslave Africans and employ child labor and deny women the vote. Will we change our minds about animals, too?"
16 Isaac Bashevis Singer, "The Letter Writer," in *Collected Stories: Gimpel the Fool to The Letter Writer*, Isaac Bashevis Singer (New York: The Library of America, 2004), 750.
17 Kim, "Abolition," 15.
18 Greta Gaard, "Vegetarian Ecofeminism: A Review Essay," *Frontiers: A Journal of Women Studies* 23, no. 3 (2002): 141 note 5, also citing Lori Gruen's critique of this metaphor in "Thoughts on Exclusion and Difference: A Response to 'On Women, Animals and Nature," *American Philosophical Association Newsletter on Feminism and Philosophy* 91, no. 1 (1992): 78–81.
19 Steven Best, "The Need for Animal Rights Against Left Welfarist Politics," *CounterPunch* (October 8, 2010), retrieved from: https://www.counterpunch.org/2010/10/08/the-need-for-animal-rights-against-left-welfarist-politics/. See also Peter Singer's influential account of "speciesism" in Peter Singer, *Animal Liberation: A New Ethics for Our Treatment of Animals* (New York: New York Review, 1975), Chapter 1.
20 Kim, "Moral Extensionism," 316.
21 See Cass R. Sunstein, "Analogical Reasoning," *Harvard Public Law Working Paper No. 21–39* (2021): 16–17. Retrieved from: http://dx.doi.org/10.2139/ssrn.3938546.
22 Cf. George Lakoff, *Women, Fire, and Dangerous Things: What Categories Reveal about the Mind* (Chicago: University of Chicago Press, 1987), 19. Referencing overlapping definitions of analogy and metaphor.
23 Thomas Gilovich, "Seeing the Past in the Present: The Effect of Associations to Familiar Events on Judgments and Decisions," *Journal of Personality and Social Psychology* 40 no. 5 (1981): https://doi.org/10.1037/0022-3514.40.5.797 cited in Holyoak and Thagard, *Mental Leaps*, 107.
24 Holyoak and Thagard, *Mental Leaps*, 108.
25 See generally Jennifer Nedelsky, "Embodied Diversity and the Challenges to Law," *McGill Law Journal* 42, no. 1 (February 1997); Alison M. Jaggar, "Love and Knowledge: Emotion in Feminist Epistemology," *Inquiry* 32, no. 2 (1989).
26 See generally Kim, "Abolition"; See also Bénédicte Boisseron's discussion of the use of the word "speciesism". Bénédicte Boisseron, *Afro-Dog: Blackness and the Animal Question* (New York: Columbia University Press, 2018), xv–xvi.
27 Spiegel, *The Dreaded Comparison*, 30.
28 For a foundational articulation of alterity, see Emmanuel Levinas, *Alterity and Transcendence*, trans. Michael B. Smith (New York: Columbia University Press, 1999).
29 Val Plumwood, "Feminism and the Logic of Alterity," in *Representing Reason: Feminist Theory and Formal Logic*, eds. Rachel Joffe Falmagne and Marjorie Hass (New York: Rowman & Littlefield, 2022), 46.
30 See generally Gayatri Chakravorty Spivak, "Can the Subaltern Speak?" in *Marxism and the Interpretation of Culture*, eds. Cary Nelson and Lawrence Grossberg (Urbana, IL: University of Illinois Press, 1988), 271–313.
31 Roberta Kalechofsky, "Animal Suffering and the Holocaust: Description," *Micah Publications*, retrieved from: https://web.archive.org/web/20090904163302/http://www.micahbooks.com/animalsufferingandtheholocaust40.html. See also Roberta Kalechofsky, *Animal Suffering and the Holocaust: The Problem with Comparison* (Marblehead, MA: Micah Publications, 2003); See also Alan Mintz, *Popular Culture and the Shaping of Holocaust Memory in America* (Seattle: University of Washington Press, 2001), 39. Attributing to the Holocaust "a dimension of tragedy beyond comparison and analogies."
32 Kalechofsky, *Animal Suffering and the Holocaust*, 34, 39, 41, 44, 46, 51, 55–56. But see Anna Isaacs, "Q&A: Animal Rights Activist and Holocaust Survivor Alex Hershaft," *Moment* (October 2, 2015), https://momentmag.com/qa-animal-rights-activist-and-holocaust-survivor-alex-hershaft/. Arguing that indifference, not hatred, better describes the motivations in both the Holocaust and contemporary factory farm context. For a fuller defense of the holocaust analogy drawing on a range of factual similarities, see David Sztybel, "Can the Treatment of Animals Be Compared to the Holocaust?" *Ethics and Environment* 11, no. 1 (2006).

33 Jeff Oboler, "A Torah Perspective on *Eternal Treblinka*," Letter to Richard Schwartz (2003), cited in Karen Davis, *The Holocaust and the Henmaid's Tale: A Case for Comparing Atrocities* (New York: Lantern Books, 2005), 56, together with Davis' counterargument.
34 Cass R. Sunstein, "On Analogical Reasoning Commentary," *Harvard Law Review* 106 (1992): 774.
35 Kim, "Moral Extensionism," 326.
36 Boisseron, *Afro-Dog*, 22.
37 Ibid.; See also Kim, "Abolition," 18: "All we need to know about the slave from the vantage point of animal abolition is that she was treated like an animal or a thing. And that she is no longer a slave."
38 Boisseron, *Afro-Dog*, 22.
39 See examples cited in Kim, "Moral Extensionism," 330–331.
40 Kim, "Abolition," 18.
41 Spiegel, *The Dreaded Comparison*, 30.
42 See generally, Serena Mayeri, " 'A Common Fate of Discrimination': Race-Gender Analogies in Legal and Historical Perspective," *Yale Law Journal* 110, no. 6 (April 2001); Janet E. Halley, " 'Like Race' Arguments," in *What's Left of Theory? New Work on the Politics of Literary Theory*, eds. Judith Butler, John Guillory, and Kendall Thomas (New York: Routledge, 2000); Isaac West, "Analogizing Interracial and Same-Sex Marriage," *Philosophy & Rhetoric* 48, no. 4 (2015); Sharon Elizabeth Rush, "Equal Protection Analogies—Identity and Passing: Race and Sexual Orientation," *Harvard Blackletter Law Journal* 13 (1997).
43 Trina Grillo and Stephanie M. Wildman, "Obscuring the Importance of Race: The Implication of Making Comparisons between Racism and Sexism (Or Other-Isms)," *Duke Law Journal* 1991, no. 2 (April 1991): 401.
44 Grillo and Wildman, "Obscuring," 401. See also Wilderson: "The violence that turns the African into a thing is without analog. . . . This is why it makes little sense to attempt analogy." Frank B. Wilderson, *Red, White & Black: Cinema and the Structure of U.S. Antagonisms* (Durham: Duke University Press, 2010), 38.
45 See e.g. Steven T. Katz, *Holocaust Studies: Critical Reflections* (New York: Routledge, 2019), 73: "Contrary to the widespread tendency to relativize the Holocaust and to employ it as a universal metaphor of evil, I would insist that the destruction of European Jewry is an unprecedented (phenomenologically, not morally) singular form of evil. . . . Auschwitz and Treblinka have no real historic parallels."
46 See, for example, Rachel Fraser, "The Ethics of Metaphor," *Ethics* 128, no. 4 (2018): 738. Fraser argues that such metaphors "restructure[]" the concept of rape in ways that fail to attend to the "interests of a hermeneutically marginalized class—that is, victims of sexual violence."
47 Harris, "Should People of Color Support," 25.
48 Kim, "Moral Extensionism," 331.
49 Harris, "Should People of Color Support," 27.
50 A. Breeze Harper, "Introduction: The Birth of the Sistah Vegan Project," in *Sistah Vegan: Black Female Vegans Speak on Food, Identity, Health, and Society*, ed. A. Breeze Harper (New York: Lantern Books, 2010), xiv.
51 Boisseron, *Afro-Dog*, xv.
52 Ibid., 31–32. Among Boisseron's examples is the case of Cecil the Lion, a lion whose killing was argued by some anti-racist activists to have attracted greater public attention and censure than the killings of Black people, including by police.
53 Boisseron, *Afro-Dog*, xiv.
54 Harris, "Should People of Color Support," 27, citing Margaret Baldwin, "Split at the Root: Prostitution and Feminist Discourses of Law Reform," *Yale Journal of Law and Feminism* 5, no. 47 (1992).
55 Harris, "Should People of Color Support," 27; See also Boisseron, *Afro-Dog*, xiii.
56 Harris, "Should People of Color Support," 27.
57 Wilderson, *Red, White & Black*, 37.
58 See Davis, *The Holocaust and the Henmaid's Tale*, 16; Harris, "Should People of Color Support," 27.
59 Harris, "Should People of Color Support," 25.
60 Kim, "Abolition," 18.

61 Boisseron, *Afro-Dog,* 13.
62 Che Gossett, "Blackness, Animality, and the Unsovereign," *Verso Blog* (September 8, 2015), retrieved from: https://www.versobooks.com/blogs/news/2228-che-gossett-blackness-animality-and-the-unsovereign.
63 S. Marek Muller, *Impersonating Animals: Rhetoric, Ecofeminism, and Animal Rights Law* (East Lansing: Michigan State University Press, 2020), 154
64 Kim, "Abolition," 18.
65 Kim, "Moral Extensionism," 326.
66 Kim, "Abolition," 27.
67 Harris, "Should People of Color Support," 25.
68 Boisseron, *Afro-Dog,* 13.
69 Kim, "Abolition," 21.
70 Kim, "Moral Extensionism," 326.
71 See e.g., Michelle Alexander, *The New Jim Crow: Mass Incarceration in the Age of Colorblindness* (New York: The New Press, 2020); Saidiya V. Hartman, *Scenes of Subjection: Terror, Slavery, and Self-Making in Nineteenth-Century America* (New York: Oxford University Press, 1997).
72 See Diana I. Popescua and Tanja Schult, "Performative Holocaust Commemoration in the 21st Century," *Holocaust Studies* 26, no. 2 (2020): 135–136, https://doi.org/10.1080/17504902.2019.1578452.
73 See e.g. Jennifer Nedelsky, "Violence against Women: Challenges to the Liberal State and Relational Feminism," *Nomos* 38 (1996): 454–497.
74 Davis, *Holocaust and the Henmaid's Tale.*
75 Alexis Okeowo, "How Saidiya Hartman Retells the History of Black Life," *New Yorker* (October 19, 2020), retrieved from: https://www.newyorker.com/magazine/2020/10/26/how-saidiya-hartman-retells-the-history-of-black-life.
76 See Jeffrey S. Kerr et al., "A Slave by Any Other Name Is Still a Slave: The Tilikum Case and Application of the Thirteenth Amendment to Nonhuman Animals," *Animal Law* 19, no. 2 (2013).
77 See Angela Fernandez, "Legal History and Rights for Nonhuman Animals: An Interview with Steven M. Wise," *Dalhousie Law Journal* 41, no. 1 (April 2018): 199. Fernandez notes, in particular, the centrality of the Somerset case in Wise's litigation and scholarship on behalf of animals. For a critique of Wise's *habeas corpus* litigation strategy, including its treatment of race, see Jessica Eisen, "Litigating Animal Captivity: Habeas Corpus in the Carceral State," in *Carceral Logics: Human Incarceration and Animal Captivity*, eds. Lori Gruen and Justin Marceau (Cambridge: Cambridge University Press, 2022).
78 Gary L. Francione and Anna Charlton, *Animal Rights: The Abolitionist Approach* (Newark, NJ: Exempla Press, 2015), 24. See also Gary L. Francione, *Rain without Thunder: The Ideology of the Animal Rights Movement* (Philadelphia: Temple University Press, 1996), 60, 178–180, 186.
79 See generally Iris Marion Young, "Asymmetrical Reciprocity: On Moral Respect, Wonder, and Enlarged Thought," in *Judgment, Imagination, and Politics: Themes from Kant and Arendt*, eds. Ronald Beiner and Jennifer Nedelsky (Lanham, MD: Rowman & Littlefield, 2001), 204.
80 See Paul Bartha, "Analogy and Analogical Reasoning," in *The Stanford Encyclopedia of Philosophy*, ed. Edward N. Zalta (2022), retrieved from: https://plato.stanford.edu/archives/sum2022/entries/reasoning-analogy/: "Analogical reasoning is fundamental to human thought and, arguably, to some nonhuman animals as well." See also Holyoak and Thagard, *Mental Leaps*; George Lakoff and Mark Johnson, *Metaphors We Live By* (Chicago: University of Chicago Press, 1980); Lakoff, *Women, Fire, and Dangerous Things.*
81 See Sunstein, "Analogical Reasoning"; Sunstein, "On Analogical Reasoning Commentary"; Scott Brewer, "Exemplary Reasoning: Semantics, Pragmatics, and the Rational Force of Legal Argument by Analogy," *Harvard Law Review* 109, no. 5 (March 1996); Lloyd L. Weinreb, *Legal Reason: The Use of Analogy in Legal Argument*, 2nd ed. (Cambridge: Cambridge University Press, 2016), 12.
82 Catharine A. MacKinnon, "Reflections on Sex Equality under Law," *Yale Law Journal* 100, no. 5 (March 1991): 1288–1289.
83 Halley, "Like Race," 46.
84 Kim, "Abolition," 16.
85 Mayeri, "A Common Fate," 1046.

86 West, "Analogizing Interracial," 562. See also Bartha, "Analogy": "[o]ften the point of an analogical argument is just to persuade people to take an idea seriously."
87 Harris, "Should People of Color Support," 25–26.
88 Kim, "Abolition," 17–18.
89 See, for example, Boisseron, *Afro-Dog*, xiii: "Though one should not ignore entangled forms of oppression, analogizing can be harmful when it is meant to serve one cause over the other; when its sole function is, for example, to serve the animal cause by instrumentalizing the black cause."
90 Boisseron, *Afro-Dog*, xx.
91 Kim, "Abolition," 28, citing Jared Sexton.
92 Kim, "Abolition," 28–29.
93 Harris, "Should People of Color Support," 17.
94 Ibid., 28.
95 Ibid., 28, 31. See also Syl Ko, "Addressing Racism Requires Addressing the Situation of Animals," in *Aphro-isms*, eds. Aph Ko and Syl Ko (New York: Lantern Books, 2017), 48: "A big part of fighting racism is *rejecting the position that white, Western voices and views are the only legitimate voices and views in the world*. . . . I don't see why we have to honor the hyper-obsession with the 'person' or the 'individual' in the West and try to extend personhood or individuality to animals in order to rethink/reimagine animality."
96 Muller, "Impersonating Animals," 81.
97 See Corey Lee Wrenn, *A Rational Approach to Animal Rights: Extensions in Abolitionist Theory* (New York: Palgrave Macmillan, 2015).
98 Carol J. Adams, *The Sexual Politics of Meat: A Feminist-Vegetarian Critical Theory* (New York: Continuum, 1990).
99 Carol J. Adams, "Feminized Protein: Meaning, Representations, and Implications," in *Making Milk: The Past, Present and Future of our Primary Food*, eds. Mathilde Cohen and Yoriko Otomo (London: Bloomsbury, 2017), 23.
100 Adams, "Feminized Protein," 33.
101 Cohen, "Of Milk and the Constitution," *Harvard Journal of Law and Gender* 40, no. 1 (2017): 152 note 238; See also Dinesh Joseph Wadiwel, "The War against Animals—Domination, Law and Sovereignty," Griffith Law Review 18, no. 2 (2009): 186.
102 Eisen, "Milked: Nature, Necessity and American Law," *Berkeley Journal of Gender, Law & Justice* 34, no. 1 (2019): 73; See also Eisen, "Milk and Meaning: Puzzles in Posthumanist Method," in Making Milk: The Past, Present and Future of our Primary Food, eds. Mathilde Cohen and Yoriko Otomo (London: Bloomsbury, 2017), 240.
103 Eisen, "Milked," 112.
104 Angela Fernandez, *Pierson v. Post: The Hunt for the Fox: Law and Professionalization in American Legal Culture* (Cambridge: Cambridge University Press, 2018); Angela Fernandez, "Pierson v. Post, 3 Cai. R. 175 (N.Y. SUP. CT. 1805) Justice Angela Fernandez, Dissenting," in *Feminist Judgments: Rewritten Property Opinions*, eds. Eloisa C. Rodriguez-Dod and Elena Maria Marty-Nelson (New York: Cambridge University Press, 2021), 98.
105 Pierson v. Post, 3 Cai. R. 175 (N.Y. Sup. Ct. 1805).
106 See Lisa Austin, "Pierson v. Post, The Hunt for the Fox: Law and Professionalization in American Legal Culture by Angela Fernandez (review)," University of Toronto Law Journal 70, no. 2 (Summer 2020).
107 On imagined judgments as a common form of feminist jurisprudential engagement, see generally Eloisa C. Rodriguez-Dod and Elena Maria Marty-Nelson, "Introduction to Feminist Judgments: Rewritten Property Opinions," in *Feminist Judgments: Rewritten Property Opinions*, eds. Eloisa C. Rodriguez-Dod and Elena Maria Marty-Nelson (New York: Cambridge University Press, 2021), 4–5.
108 Fernandez, "Pierson v. Post."
109 Ibid., 114.
110 Ibid.
111 Ibid., 115.
112 Yoriko Otomo, "Law and the Question of the (Nonhuman) Animal," *Society & Animals* 19, no. 4 (2011): 383–391.

113 Maneesha Deckha, *Animals as Legal Beings: Contesting Anthropocentric Legal Orders* (Toronto: University of Toronto Press, 2020).
114 Ibid., 7.
115 Ibid., 98.
116 Harris, "Should People of Color Support," 26.
117 Tashee Meadows, "Because They Matter," in *Sistah Vegan: Black Female Vegans Speak on Food, Identity, Health and Society*, ed. A. Breeze Harper (New York: Lantern Books, 2010), 151.
118 See also Alice Walker, "Am I Blue?" in *Living by the Word: Selected Writings 1973–1987*, Alice Walker (New York: Harcourt Brace Jovanovich, 1989).
119 Boisseron, *Afro-Dog*, 36.
120 Catharine A. MacKinnon, *Toward a Feminist Theory of the State* (Cambridge: Harvard University Press, 1989), 106; see also Kathryn Abrams, "Feminist Lawyering and Legal Method," *Law and Social Inquiry* 16, no. 2 (1991): 373–404.
121 Mari J. Matsuda, "Public Response to Racist Speech: Considering the Victim's Story," *Michigan Law Review* 87, no. 8 (1989): 2323, n.15.
122 Yoriko Otomo and Cressida Limon, "Dogs, Pigs and Children: Changing Laws in Colonial Britain," *Australian Feminist Law Journal* 40, no. 2 (2014): 163–164.
123 Maneesha Deckha, "Non-Human Animals and Human Health: A Relational Approach to the Use of Animals in Medical Research," in *Being Relational: Reflections on Relational Theory and Health Law*, eds. Jennifer J. Llewellyn and Jocelyn Grant Downie (Vancouver: UBC Press, 2012), 306.
124 MacKinnon, "Of Mice and Men: A Feminist Fragment on Animal Rights," in *Animal Rights: Current Debates and New Directions*, eds. Cass R. Sunstein and Martha C. Nussbaum (New York: Oxford University Press, 2004), 270. See also Fox, *Rethinking Kinship* at 493.*
125 Nussbaum, *Frontiers of Justice; New Nussbaum Work?**
126 Eva Bernet Kempers, "Animal Dignity and the Law: Potential, Problems and Possible Implications," *Liverpool Law Review* 41, no. 2 (2020): 173 at 174, 177.
127 See e.g. Eisen, *Animals in the Constitutional State*; Deckha, *Legal Beings*; Matambanadzo. See also Satz, *Vulnerability and Interest Convergence*; and Gruen, *Entangled Empathy**
128 Boisserion, *Afro-Dog*, xiv.
129 Muller, "Impersonating Animals," 154.
130 Harris, "Should People of Color Support," 32.
131 Boisseron, *Afro-Dog*, 25.
132 Ibid., 26.
133 See e.g. Martha Minow, "Feminist Reason: Getting it and Losing it," Journal of Legal Education 38, no. 1/2 (March/June 1988); Jennifer C. Nash, "Re-Thinking Intersectionality," *Feminist Review* 89, no. 1 (2008), https://doi.org/10.1057/fr.2008.4.

Bibliography

Abrams, Kathryn. "Feminist Lawyering and Legal Method." *Law and Social Inquiry* 16, no. 2 (1991): 373–404.
Adams, Carol J. *The Sexual Politics of Meat: A Feminist-Vegetarian Critical Theory*. New York: Continuum, 1990.
Adams, Carol J. "Feminized Protein: Meaning, Representations, and Implications." In *Making Milk: The Past, Present and Future of our Primary Food*, edited by Mathilde Cohen and Yoriko Otomo. London: Bloomsbury, 2017.
Alexander, Michelle. *The New Jim Crow: Mass Incarceration in the Age of Colorblindness*. New York: The New Press, 2020.
Austin, Lisa. "Pierson v. Post, The Hunt for the Fox: Law and Professionalization in American Legal Culture by Angela Fernandez (review)." *University of Toronto Law Journal* 70, no. 2 (Summer 2020): 376–381.
Baldwin, Margaret. "Split at the Root: Prostitution and Feminist Discourses of Law Reform." *Yale Journal of Law and Feminism* 5, no. 47 (1992): 106–146.

Bartha, Paul. "Analogy and Analogical Reasoning." In *The Stanford Encyclopedia of Philosophy*, edited by Edward N. Zalta. 2022. Retrieved from: https://plato.stanford.edu/archives/sum2022/entries/reasoning-analogy/.

Best, Steven. "The Need for Animal Rights Against Left Welfarist Politics." *CounterPunch*, October 8, 2010. Retrieved from: https://www.counterpunch.org/2010/10/08/the-need-for-animal-rights-against-left-welfarist-politics/.

Boisseron, Bénédicte. *Afro-Dog: Blackness and the Animal Question*. New York: Columbia University Press, 2018.

Brewer, Scott. "Exemplary Reasoning: Semantics, Pragmatics, and the Rational Force of Legal Argument by Analogy." *Harvard Law Review* 109, no. 5 (March 1996): 923–1028.

Butler, Judith. *Precarious Life: The Powers of Mourning and Violence*. New York: Verso, 2004.

Butler, Judith. *Frames of War: When is Life Grievable?* New York: Verso, 2016.

Cohen, Mathilde. "Of Milk and the Constitution." *Harvard Journal of Law and Gender* 40, no. 1 (2017): 115–182.

Collard, Andrée with Joyce Contrucci. *Rape of the Wild: Man's Violence against Animals and the Earth*. Bloomington: Indiana University Press, 1989.

Davis, Karen. "Thinking Like a Chicken: Farm Animals and the Feminine Connection." In *Animals and Women*, edited by Carol J. Adams and Josephine Donovan, 192–212. Durham: Duke University Press, 1995.

Davis, Karen. *The Holocaust and the Henmaid's Tale: A Case for Comparing Atrocities*. New York: Lantern Books, 2005.

Deckha, Maneesha. "Non-Human Animals and Human Health: A Relational Approach to the Use of Animals in Medical Research." In *Being Relational: Reflections on Relational Theory and Health Law*, edited by Jennifer J. Llewellyn and Jocelyn Grant Downie, 287–315. Vancouver: UBC Press, 2012.

Deckha, Maneesha. *Animals as Legal Beings: Contesting Anthropocentric Legal Orders*. Toronto: University of Toronto Press, 2020.

Dunayer, Joan. "Sexist Words, Speciesist Roots." In *Animals and Women*, edited by Carol J. Adams and Josephine Donovan, 11–31. Durham: Duke University Press, 1995.

Eisen, Jessica. "Milk and Meaning: Puzzles in Posthumanist Method." In *Making Milk: The Past, Present and Future of our Primary Food*, edited by Mathilde Cohen and Yoriko Otomo. London: Bloomsbury, 2017.

Eisen, Jessica. "Milked: Nature, Necessity and American Law." *Berkeley Journal of Gender, Law & Justice* 34, no. 1 (2019): 71–115.

Eisen, Jessica. "Litigating Animal Captivity: Habeas Corpus in the Carceral State." In *Carceral Logics: Human Incarceration and Animal Captivity*, edited by Lori Gruen and Justin Marceau, 343–365. Cambridge: Cambridge University Press, 2022.

Fernandez, Angela. *Pierson v. Post: The Hunt for the Fox: Law and Professionalization in American Legal Culture*. Cambridge: Cambridge University Press, 2018.

Fernandez, Angela. "Pierson v. Post, 3 Cai. R. 175 (N.Y. SUP. CT. 1805) Justice Angela Fernandez, Dissenting." In *Feminist Judgments: Rewritten Property Opinions*, edited by Eloisa C. Rodriguez-Dod and Elena Maria Marty-Nelson, 98–118. New York: Cambridge University Press, 2021.

Francione, Gary L. *Rain without Thunder: The Ideology of the Animal Rights Movement*. Philadelphia: Temple University Press, 1996.

Francione, Gary L., and Anna Charlton. *Animal Rights: The Abolitionist Approach*. Newark, NJ: Exempla Press, 2015.

Fraser, Rachel. "The Ethics of Metaphor." *Ethics* 128, no. 4 (2018): 728–755.

Gaard, Greta. "Vegetarian Ecofeminism: A Review Essay." *Frontiers: A Journal of Women Studies* 23, no. 3 (2002): 117–146.

Gillespie, Kathryn A., and Lopez, Patricia J., eds. *Economies of Death: Economic Logics of Killable Life and Grievable Death*. London: Routledge, 2015.

Gilovich, Thomas. "Seeing the Past in the Present: The Effect of Associations to Familiar Events on Judgments and Decisions." *Journal of Personality and Social Psychology* 40, no. 5 (1981): 797–808. https://doi.org/10.1037/0022-3514.40.5.797.

Grillo, Trina, and Wildman, Stephanie M. "Obscuring the Importance of Race: The Implication of Making Comparisons between Racism and Sexism (Or Other-Isms)." *Duke Law Journal* 1991, no. 2 (April 1991): 397–412.

Gruen, Lori. "Thoughts on Exclusion and Difference: A Response to 'On Women, Animals and Nature.'" *American Philosophical Association Newsletter on Feminism and Philosophy* 91, no. 1 (1992): 78–81.

Halley, Janet E. "'Like Race' Arguments." In *What's Left of Theory? New Work on the Politics of Literary Theory*, edited by Judith Butler, John Guillory, and Kendall Thomas, 40–74. New York: Routledge, 2000.

Harper, A. Breeze. "Introduction: The Birth of the Sistah Vegan Project." In *Sistah Vegan: Black Female Vegans Speak on Food, Identity, Health, and Society*, edited by A. Breeze Harper, xiii–xix. New York: Lantern Books, 2010.

Hartman, Saidiya V. *Scenes of Subjection: Terror, Slavery, and Self-Making in Nineteenth-Century America*. New York: Oxford University Press, 1997.

Haslam, Nick. "Dehumanization: An Integrative Review." *Personality and Social Psychology Review* 10, no. 3 (August 2006): 252–264.

Holyoak, Keith and Paul Thagard. *Mental Leaps: Analogy in Creative Thought*. Cambridge: MIT Press, 1995.

Hund, Wulf D., and Mills, Charles W. "Comparing Black People to Monkeys Has a Long, Dark Simian History." *The Conversation*, February 28, 2016. Retrieved from: https://theconversation.com/comparing-black-people-to-monkeys -has-a-long-dark-simian-history-55102.

Hund, Wulf D., Mills, Charles W., and Sebastiani, Silvia, eds. *Simianization: Apes, Gender, Class, and Race*. Zürich: Lit Verlag, 2015.

Isaacs, Anna. "Q&A: Animal Rights Activist and Holocaust Survivor Alex Hershaft." *Moment*, October 2, 2015. Retrieved from: https://momentmag.com/qa-animal-rights-activist-and-holocaust-survivor-alex-hershaft/.

Jaggar, Alison M. "Love and Knowledge: Emotion in Feminist Epistemology." *Inquiry* 32, no. 2 (1989): 151–176.

Kalechofsky, Roberta. *Animal Suffering and the Holocaust: The Problem with Comparison*. Marblehead, MA: Micah Publications, 2003.

Kalechofsky, Roberta. "Animal Suffering and the Holocaust: Description." *Micah Publications*. Retrieved from: https://web.archive.org/web/20090904163302/http://www.micahbooks.com/animalsufferingandtheholocaust40.html.

Katz, Steven T. *Holocaust Studies: Critical Reflections*. New York: Routledge, 2019.

Kerr, Jeffrey S., Bernstein, Martina, Schwoerke, Amanda, and Strugar, Matthew D. "A Slave by Any Other Name Is Still a Slave: The Tilikum Case and Application of the Thirteenth Amendment to Nonhuman Animals." *Animal Law* 19, no. 2 (2013): 221–294.

Kim, Claire Jean. "Moral Extensionism or Racist Exploitation? The Use of Holocaust and Slavery Analogies in the Animal Liberation Movement." *New Political Science* 33, no. 3 (September 2011): 311–333.

Kim, Claire Jean. *Dangerous Crossings: Race, Species, and Nature in a Multicultural Age*. Cambridge: Cambridge University Press, 2015.

Kim, Claire Jean. "Abolition." In *Critical Terms for Animal Studies*, edited by Lori Gruen. Chicago: University of Chicago Press, 2018.

Ko, Syl. "Addressing Racism Requires Addressing the Situation of Animals." In *Aphro-isms*, Aph Ko and Syl Ko. New York: Lantern Books, 2017.

Lakoff, George. *Women, Fire, and Dangerous Things: What Categories Reveal about the Mind*. Chicago: University of Chicago Press, 1987.

Lakoff, George, and Mark Johnson. *Metaphors We Live By*. Chicago: University of Chicago Press, 1980.

MacKinnon, Catharine A. *Toward a Feminist Theory of the State*. Cambridge: Harvard University Press, 1989.

MacKinnon, Catharine A. "Reflections on Sex Equality under Law." *Yale Law Journal* 100, no. 5 (March 1991): 1281–1328.

MacKinnon, Catharine A. "Of Mice and Men: A Feminist Fragment on Animal Rights." In *Animal Rights: Current Debates and New Directions*, edited by Cass R. Sunstein and Martha C. Nussbaum, 263–276. New York: Oxford University Press, 2004.

Matsuda, Mari J. "Public Response to Racist Speech: Considering the Victim's Story." *Michigan Law Review* 87, no. 8 (1989): 2320–2381.

Mayeri, Serena. "'A Common Fate of Discrimination': Race-Gender Analogies in Legal and Historical Perspective." *Yale Law Journal* 110, no. 6 (April 2001): 1045–1087.
Meadows, Tashee. "Because They Matter." In *Sistah Vegan: Black Female Vegans Speak on Food, Identity, Health and Society*, edited by A. Breeze Harper, 150–154. New York: Lantern Books, 2010.
Minow, Martha. "Feminist Reason Method: Getting it and Losing it." *Journal of Legal Education* 38, no. 1/2 (March/June 1988): 47–60.
Mintz, Alan. *Popular Culture and the Shaping of Holocaust Memory in America.* Seattle: University of Washington Press, 2001.
Muller, S. Marek. *Impersonating Animals: Rhetoric, Ecofeminism, and Animal Rights Law.* East Lansing: Michigan State University Press, 2020.
Nash, Jennifer C. "Re-Thinking Intersectionality." *Feminist Review* 89, no. 1 (2008). https://doi.org/10.1057/fr.2008.4.
Nedelsky, Jennifer. "Violence against Women: Challenges to the Liberal State and Relational Feminism." *Nomos* 38 (1996): 454–497.
Nedelsky, Jennifer. "Embodied Diversity and the Challenges to Law." *McGill Law Journal* 42, no. 1 (February 1997): 91–118.
Okeowo, Alexis. "How Saidiya Hartman Retells the History of Black Life." *New Yorker*, October 19, 2020. Retrieved from: https://www.newyorker.com/magazine/2020/10/26/how-saidiya-hartman-retells-the-history-of-black-life.
Otomo, Yoriko. "Law and the Question of the (Nonhuman) Animal." *Society & Animals* 19, no. 4 (2011): 383–391. https://doi.org/10.1163/156853011X590033.
Otomo, Yoriko, and Cressida Limon. "Dogs, Pigs and Children: Changing Laws in Colonial Britain." *Australian Feminist Law Journal* 40, no. 2 (2014): 163–167. https://doi.org/10.1080/13200968.2015.1020589.
Patterson, Charles. *Eternal Treblinka: Our Treatment of Animals and the Holocaust.* New York: Lantern Books, 2002.
PETA People for the Ethical Treatment of Animals. "Holocaust on Your Plate." Traveling display. 2004.
PETA People for the Ethical Treatment of Animals. "Is Your Food a Product of Rape?" *News*, Retrieved from: https://www.peta.org/features/rape-milk-pork-turkey/, accessed July 25, 2023.
Pierson v. Post, 3 Cai. R. 175 (N.Y. Sup. Ct. 1805).
Plumwood, Val. "Feminism and the Logic of Alterity." In *Representing Reason: Feminist Theory and Formal Logic*, edited by Rachel Joffe Falmagne and Marjorie Hass, 45–70. New York: Rowman & Littlefield, 2022.
Popescu, Diana I., and Tanja Schult. "Performative Holocaust Commemoration in the 21st Century." *Holocaust Studies* 26, no. 2 (2020): 135–151. https://doi.org/10.1080/17504902.2019.1578452.
Rodriguez-Dod, Eloisa C., and Elena Maria Marty-Nelson. "Introduction to Feminist Judgments: Rewritten Property Opinions." In *Feminist Judgments: Rewritten Property Opinions*, edited by Eloisa C. Rodriguez-Dod and Elena Maria Marty-Nelson, 3–9. New York: Cambridge University Press, 2021.
Rush, Sharon Elizabeth. "Equal Protection Analogies–Identity and Passing: Race and Sexual Orientation." *Harvard Blackletter Law Journal* 13 (1997): 65–106.
Sax, Boria. *Animals in the Third Reich: Pets, Scapegoats, and the Holocaust.* New York: Continuum, 2000.
Singer, Isaac Bashevis. "The Letter Writer." In *Collected Stories: Gimpel the Fool to The Letter Writer*, Isaac Bashevis Singer, 724–755. New York: The Library of America, 2004.
Spiegel, Marjorie. *The Dreaded Comparison: Human and Animal Slavery.* New York: Mirror Books, 1996.
Stanescu, James. "Species Trouble: Judith Butler, Mourning, and the Precarious Lives of Animals." *Hypatia* 27, no. 3 (2012): 567–582.
Sunstein, Cass R. "On Analogical Reasoning Commentary." *Harvard Law Review* 106 (1992): 741–791.
Sunstein, Cass R. "Analogical Reasoning." *Harvard Public Law Working Paper No. 21–39* (2021): 1–64. http://dx.doi.org/10.2139/ssrn.3938546.

Sztybel, David. "Can the Treatment of Animals Be Compared to the Holocaust?" *Ethics and Environment* 11, no. 1 (2006): 97–132.
Taylor, Chloë. "The Precarious Lives of Animals: Butler, Coetzee, and Animal Ethics." *Philosophy Today* 52, no. 1 (2008): 60–72.
Wadiwel, Dinesh Joseph. "The War against Animals—Domination, Law and Sovereignty." *Griffith Law Review* 18, no. 2 (2009): 283–297.
Walker, Alice. "Am I Blue?" In *Living by the Word: Selected Writings 1973–1987*, Alice Walker, 9–10. New York: Harcourt Brace Jovanovich, 1989.
Weinreb, Lloyd L. *Legal Reason: The Use of Analogy in Legal Argument*. 2nd ed. Cambridge: Cambridge University Press, 2016.
West, Isaac. "Analogizing Interracial and Same-Sex Marriage." *Philosophy & Rhetoric* 48, no. 4 (2015): 561–582.
Wiebe, Michelle. "Extermination and Euthanasia: Animal Symbolism in Nazi Germany."*The Mirror—Undergraduate History Journal* 28, no. 1 (2008): 21–38.
Wilderson, Frank B. *Red, White & Black: Cinema and the Structure of U.S. Antagonisms*. Durham: Duke University Press, 2010.
Wrenn, Corey Lee. *A Rational Approach to Animal Rights: Extensions in Abolitionist Theory*. New York: Palgrave Macmillan, 2015.
Young, Iris Marion. "Asymmetrical Reciprocity: On Moral Respect, Wonder, and Enlarged Thought." In *Judgment, Imagination, and Politics: Themes from Kant and Arendt*, edited by Ronald Beiner and Jennifer Nedelsky. Lanham, MD: Rowman & Littlefield, 2001.

PART II

Intersectional Veganisms

6
FEMINIST VEGANISM

Karen S. Emmerman

Introduction

Veganism has not always been viewed as a feminist issue.[1] Much of early feminist theorizing and activism focused solely on the oppression and domination of human women and the structural injustices they face—leaving animals out of their discussions entirely. Even the early ecofeminists who argued forcefully for the idea that all feminists should be environmentalists (and all environmentalists feminists) largely left animals and our treatment of them in the food system out of their discussions. Those are the feminists who merely left animals out. There are also feminist theorists who expressly argued against including animals in a feminist ethic of care as well as those who have argued that encouraging vegetarianism or veganism is in fact anti-feminist.[2] Ecofeminists who include animals in their sphere of empathy and compassion, who see the fate of women and all marginalized people as inextricably tied to the fate of animals, are sometimes called "vegetarian ecofeminists."[3] In what follows, I will simply use the terms "ecofeminist" and "ecofeminism," intending ecofeminist conceptions that are inclusive of animals. My aim in this chapter is twofold: first, to demonstrate why veganism is a feminist issue and, second, to consider one complication that arises out of feminist veganism—worries about the possible imperialist, classist, and racist implications of making claims to a universal requirement to refrain from eating animal flesh—and discuss the way contextual moral vegetarianism seeks to address that complication.[4]

Why Is Veganism a Feminist Issue?

Utilitarian and rights-based approaches to vegetarianism typically rely on identifying a characteristic of animals that makes them relevantly morally similar to humans in ways that generate an obligation to refrain from eating them.[5] Though many people find these traditional ethical approaches persuasive, they do not generally illuminate why who we eat is a particularly feminist issue. Rooted in moral frameworks that rely on taking a disembodied and "objective" stance toward moral questions, these approaches to animal ethics rarely if ever note the economic, political, or social context surrounding and shaping the

DOI: 10.4324/9781003273400-9

ethical dilemmas we face. By contrast, ecofeminism is a politicized and contextualized ethical approach, taking the social, political, and economic forces that structure our lives as germane to both understanding and resolving ethical questions.[6] As Marti Kheel put it, when engaging in moral deliberation we should not work from "truncated narratives" where moral problems are "posed in a static, linear fashion detached from the context in which [they] are formed."[7] Additionally, ecofeminism is rooted in the feminist ethics of care, a view that rejects both the possibility and desirability of disembodied, impartial moral reflection and that embraces a role for affect, empathy, care and loving attention in our moral lives. Hence, when ecofeminists talk about how animals are exploited and dominated in service of human consumption, we include analyses of the context in which that exploitation occurs and locate a plurality of reasons animals matter morally—yes, it is wrong to cause unnecessary suffering to sentient creatures, and it is wrong to violate their most basic interests, but it is also relevant that we have feelings of empathy, care, and compassion toward them and that animals have caring relations of their own that matter to them.

With these commitments to politicizing and contextualizing ethical reflection and centering care in the ethical life, it is not surprising that ecofeminists consider veganism a feminist issue for several reasons. In the interest of brevity, I will focus on just two. First, the joint oppression of women and animals under a patriarchal worldview and, second, the gendered nature of animal exploitation in the food system. Let's begin with the first.

In 1990, Karen Warren detailed the oppressive conceptual frameworks that result in the dual oppression of women and nature.[8] Among the conceptual frameworks Warren identified as highly pernicious are value dualisms and value hierarchies. Value dualisms are binaries that are presented as fixed and where one side of the binary is valued more than the other. Examples include human/animal, male/female, and culture/nature. Value hierarchies occur when one side of these binaries is considered superior to the other. In patriarchal systems, the left side of the binaries (human, male, culture) is prized over the right (animal, female, nature). The more highly valued elements get grouped together as superior, and the less highly valued elements are grouped together as inferior. Warren calls the next step in these oppressive conceptual frameworks "the logic of domination"—the idea that if X is superior to Y, then X is justified in dominating Y. Thus, if humans are superior to animals, then humans are justified in dominating animals, and if men are superior to women, then men are justified in dominating women, and so forth. Women and animals, grouped together (along with many other marginalized groups), are thus exploited and dominated by the very same systems of oppression. That which marginalizes the other-than-human also marginalizes the feminine. That which justifies the exploitation of the other-than-human also justifies the exploitation of women.

Also in 1990, Carol J. Adams published her groundbreaking book *The Sexual Politics of Meat*, highlighting the connections between the joint oppression of women and animals. Among other things, Adams identifies the cycle of objectification, fragmentation, and consumption that connects the fates of animals and women in a patriarchal system. Objectification "permits an oppressor to view another being as an object" resulting in treating the being as a means to an end. From objectification, fragmentation—"brutal dismemberment"—is possible. Fragmentation is most obviously enacted through the process of butchering animals' bodies to create food for humans, but as Adams notes fragmentation happens to women too, sometimes literally, sometimes in imagery, sometimes metaphorically or linguistically. Finally, after objectification and fragmentation comes consumption.

In the case of animals this is the literal consumption of their flesh. In the case of women, consumption can take many forms, including the consumption of women through visual imagery.[9]

In publications, art works, and on the streets as activists, ecofeminists have been making explicit these connections between the oppression of women and animals and thus making the case for veganism as an explicitly feminist issue. It is no accident that women, animals, and other marginalized groups seen as "other" and less valuable than cis-gendered, heterosexual, non-disabled, white men experience the worst kinds of domination and exploitation. Value dualisms, value hierarchies, and the logic of domination—all tools of a patriarchal conceptual framework—ensure that it is so. Kheel averred that these facts "invite vegetarianism as a response":[10]

> If people are opposed to the domination of women, they may be more inclined to empathize with the plight of nonhuman animals once they understand the connections between the domination of women and of nonhuman animals. It is empathy, not abstract norms, that provides the motivation for vegetarianism in this invitational approach. Vegetarianism thus becomes part of a larger resistance to violence and domination.[11]

The urgency of this invitation is further underscored by the gendered nature of violence against animals in the food system. In the United States alone, roughly ten billion animals are raised and slaughtered annually for human consumption. These numbers are impossible to achieve without a finely honed system for exploiting animals', in particular *female* animals', reproductive capabilities. This reproductive exploitation includes all stages of the production of young including insemination, gestation, birth, and lactation, culminating in prematurely separating babies from their mothers. Some examples of the exploitation of female animals' reproductive capacities in the food system include (but are certainly not limited to): artificial insemination through the use of a "rape rack,"[12] use of the hormone rBGH in cows used for dairy to artificially extend cows' lactation period and increase milk production, resulting in painful mastitis,[13] impregnating cows for dairy production and taking their babies away at birth to ensure their milk is used for humans,[14] gestation crates where pregnant sows are kept 24 hours a day and 7 days a week where they cannot turn or lie down,[15] use of a similarly space-constrained farrowing crate for when sows give birth, and then taking away their piglets at 2–3 weeks of age so the mother can be impregnated over and over again until she can no longer breed and is killed.[16] Laying hens, used for egg production, fare no better. Soon after birth, laying hens are debeaked without anesthesia,[17] they spend their lives in "battery cages" where they are crammed in with several other hens (hence the debeaking—they are unable to peck each other when frustrated or scared), they are subjected to starvation and forced molting to increase egg production, and they are slaughtered after one year when they become "spent" and can no longer produce eggs. This is a premature death for these hens, as their lifespan in natural conditions is up to ten years.[18] The rampant exploitation of female animals' reproductive systems in animal agriculture is what led Adams to coin the term "feminized protein."[19]

> *The Sexual Politics of Meat* proposes a specific conceptual term to recognize the exploitation of the reproductive processes of female animals: milk and eggs should be called *feminized protein*, that is, protein that was produced by a female body. The

> majority of animals eaten are female animals and children. Female animals are doubly exploited: both when they are alive and then when they are dead. They are the literal female pieces of meat.[20]

Female animals are stripped of reproductive freedoms at every stage of their lives, only to be slaughtered and consumed when they can no longer serve their reproductive purpose. Male animals also do not avoid reproductive exploitation in the food system. Artificial insemination requires sperm from males, who are often subjected to electroejaculation.[21] Roughly 200 million male chicks, useless in the egg industry, are ground alive or smothered by each other's weight in massive garbage receptacles annually in the United States,[22] and male calves, useless in the dairy industry, are separated from their mothers and raised in horrific conditions for the production of veal.[23] It is for these reasons that veganism is very much a feminist issue. Feminists care about bodily integrity, reproductive freedom, and the importance of consent; considerations that should not stop at the species barrier. As noted earlier, we also recognize the interlocking nature of oppression and see veganism as an appropriate response to rejecting the violence, marginalization, exploitation, and domination of *all* beings—human and other-than-human—in the food system.[24]

Contextual Moral Veganism

Having explained why veganism is an explicitly feminist issue, we'll now turn to considering a complication that arises when we see veganism as a necessary feminist response to the interlocking oppressions and gendered violence in animal agriculture. One might worry that feminist veganism is issuing a universalized edict to eschew animal flesh foods, a problematic approach for a contextualized ethic. The appearance of a universal vegan requirement raises concerns that feminist veganism is reifying the very "isms" it seeks to reject—neocolonialism, imperialism, racism, and classism in particular. What about people in climates unamenable to agriculture? What about people in refugee camps? What about low-income people and people of color living in neighborhoods without access to healthy foods? How can feminists claim everyone should adopt veganism in the face of very real issues of justice and cultural difference? As Cathryn Bailey put it:

> in their zeal to extend sympathy for animals, white privileged women have at times been insensitive to the different meanings that both animals and food have for women of color and women globally. This absolutist position, that it is morally wrong for feminists to eat meat, has met particularly harsh criticism for its insensitivity to women's real lives.[25]

Similarly, Kheel noted that ecofeminists and others have been "charged with trying to impose a white, middle-class norm on other cultures where people eat meat out of necessity or due to their own cultural norms."[26]

In response to these concerns, ecofeminists developed the concept of "contextual moral vegetarianism" (CMV), where "vegetarian" is meant to be understood as veganism. Deanne Curtin articulated the position in the following way:

> I do not see the commitment to vegetarianism as a decontextualized, purely rational, duty, but as a situated, feminist response to the contemporary moral landscape. The

> approach I propose is contextual, though not relativistic. That is, it sees vegetarianism as one situated response to a complex set of neocolonial, ecological, and feminist issues. Although it is contextual, it is not relativist since it resolutely opposes neocolonialism, naturism, speciesism, and sexism.[27]

Lori Gruen notes that ecofeminism is "mindful of the violence perpetuated in many gendered, racialized, and colonial contexts as well as the realities of changing climate, and thus foregoes top-down absolute universalizing judgments that everyone, everywhere should 'see veganism as a moral baseline'."[28] Versions of CMV differ in the details, but all share the recognition that it runs counter to ecofeminist principles to issue a moral edict requiring veganism across all cultures, times, contexts, and circumstances. No doubt, for defenders of animals it can be reassuring and expedient to ignore contextual exigencies and issue such edicts. In doing so one both proves one's bona fides as a protector of animals *and* is assured one has given others a very clear moral directive. Indeed, Gary Francione has accused ecofeminists supporting CMV of instantiating a human/animal hierarchy where animals retain their status as things.[29]

This criticism fails in two ways, as it misunderstands both CMV and the moral realities of veganism as a practice. Ecofeminists understand CMV as a response to particular contexts—"It recognizes that the reasons for moral vegetarianism may differ by locale, by gender, as well as by class."[30] Along these lines, Kheel asks "What are the factors that support meat eating as a dietary norm? Moreover, what factors might invite vegetarianism as a response?"[31] With this in mind, she focuses her analysis on the "Western world" because that is where meat eating predominates and where most abuse of nonhuman animals occurs.[32] For ecofeminists, people in industrialized countries who have access to fresh fruits and vegetables and plant-based alternatives to flesh-based foods should see veganism as the appropriate response to the plurality of violence committed in the food system. Rather than ontologizing animals as things, as Francione suggests, ecofeminists view animals as important individuals with desires, interests, and relationships of their own and with whom we ought to be in compassionate, caring relationships.[33] An adherence to CMV does not entail seeing animals as lower on a hierarchy than humans. Rather, it recognizes that not all humans are situated in the same way and that to be actionable, ethical norms must take the details of individuals' lives and the cultural, economic, social, and political systems under which they live into account. For example, I have written about the moral complexity of being unable to breastfeed while needing to feed my premature infant in a neo-natal intensive care unit at a time when vegan infant formula did not exist.[34] Additionally, ecofeminists recognize that there are important racist, classist, and neocolonialist implications when white activists from affluent nations (or neighborhoods within their own nation) espouse veganism as a required norm. Supporters of CMV are cognizant of the harms such activism can do and endeavor to avoid reifying racism in their activism and theoretical work.

While someone like Francione can declare that everyone everywhere should be vegan, I honestly do not know what such a declaration can mean to the parents of an infant in an incubator who have no alternative options for feeding their child or to the person living without access to fresh vegetables due to food apartheid.[35] All ecofeminists would be elated to live in a world where no animals' bodies are used to feed humans, but we cannot make it so simply by issuing universal requirements to adopt veganism.

Moreover, criticisms like Francione's misunderstand what veganism is. As Curtin notes, vegetarianism is not "the stance of perfect nonviolence," and "violence is committed by the most mindful of vegetarians every day."[36] The idea that veganism is the only purely ethical way to live is predicated on a misunderstanding of what veganism is. Gruen and Jones distinguish between veganism as an identity or lifestyle and veganism as a goal. The former adopts a position of moral superiority—asserting itself as a lifestyle where one's hands are morally clean. The latter understands veganism as a practice, "a process of doing the best one can to minimize violence, domination, and exploitation."[37] They go on to argue that veganism can only be understood as an aspiration because there is no way to live a human life without causing violence to someone. Veganism is the right direction, to be sure, but it is not the end of the journey. Moreover, many of our consumer decisions that might be better for the animals are in fact harmful in other ways—What environmental damage occurred in the manufacture of the non-leather materials necessary for my affordable vegan shoes? How were the workers treated who made my shoes? The list goes on. CMV, therefore, understands veganism as the appropriate (and even necessary) response for those who wish to live lives of compassion and care as well as avoid participation in systems of domination and oppression that leave nothing but death, destruction, and harm in their wake. At the same time, proponents of CMV understand that veganism is not possible in certain contexts and that veganism itself, even when meticulously carried out, is not best understood as a lifestyle embodying morally purity but as a practice of non-violence that involves a lifelong pursuit of finding ways to instantiate these values in one's life on a daily basis.

Having said all of this, it is important to note that blanket charges asserting the neocolonialist, racist, sexist, and classist nature of calls to veganism are problematic. In the remainder of this chapter, we will consider several reasons why such claims, while worth considering and attending to, must themselves be understood contextually. Certainly, some proponents of veganism use language and tactics, as well as adopting attitudes, which are racist and neocolonialist. At the same time, the generalized assertion that all appeals to adopt veganism are indicative of racist, imperialist, and neocolonialist attitudes is too facile and unnuanced.

As Richard Twine has argued, if we are going to worry about widespread exportation of a neocolonialist and problematic view of animal consumption, our attention should really turn toward the rampant exportation of industrialized agriculture from affluent, industrialized countries to the developing world.[38] Twine is careful to note that of course we must be attentive to pernicious "isms" within the vegan movement. At the same time, however, it is a matter of empirical fact that industrialized animal agriculture is spreading from industrialized nations at a much higher rate than veganism and that this spread is having a significant negative impact on the environment, communities, and local flora and fauna. Twine refers to this as the "tragicomic aura" of criticism of universal veganism.[39] Losing sight of the neocolonialist consequences of the rapid exportation of Western approaches to food production and consumption is a mistake. This is especially so when we notice that in many cultures throughout the world, vegetarianism has been a long-standing norm rather than something newly introduced and advocated for by Western animal rights advocates.[40] We must attend to both internal critique of our own rhetoric and values as proponents of veganism *and* address the empirical reality that a food system running counter to the interests of developing nations is rapidly expanding in those parts of the world, to the detriment of all impacted.

Moreover, when invoking the importance of respect for food-based cultural norms and traditions, it is crucial to inspect those traditions for problematic oppressive belief systems. As many ecofeminists have pointed out, defense of animal-flesh–based food practices on cultural grounds often overlooks problematic gender practices within those cultures. For example, in many patriarchal societies, men are fed animal protein either exclusively or before what is left over is offered to women and children.[41] Kheel has also cautioned against essentializing indigenous cultural practices with broad-brush appeals to their respectful stance towards animal others, noting that many indigenous groups have "killed animals indiscriminately, with little or no regard for either conservation or animals' suffering."[42] Finally, as Claire Jean Kim argues in her careful work on cross-cultural conflicts over animal-based cultural practices, conceptions of culture are often construed anthropocentrically. Framed as conflicts between human groups, animals are left out of the discussions entirely.[43] Kim argues that our view of cultural conflicts about animals should adopt "multi-optic vision" rather than "single-optic vision," where the former "takes disparate justice claims seriously without privileging any one presumptively."[44] This would entail thinking through the important issues of race, class, colonialism, *and speciesism*. These issues are just as important to consider when thinking through traditional practices involving animals within our own religious, cultural, and ethnic groups.[45] As Greta Gaard has argued in multiple contexts, it is important to note the problematic oppressive practices embedded in traditions and to recognize that "cultural insiders are the ones who make the choices."[46] To attend to the very genuine concern that vegan advocacy can be racist and neocolonialist, it is important that appeals to adopt veganism within marginalized groups come from members of those groups rather than from without. Given the horrific and widespread suffering of animals right here on our own doorstep, feminist vegans in affluent, industrialized nations can turn their attention to the systems of oppression rampant in their own dominant cultures while offering whatever support and allyship advocates in other settings request and welcome in their revolutionizing endeavors.[47]

Conclusion

Proponents of feminist veganism consider adopting a diet free from animal and feminized protein as the correct response to the violence, domination, and exploitation of both human and other-than-human animals. We understand veganism as an act of resistance to the gendered nature of animal exploitation in the food system as well as the joint oppression of women and animals under patriarchal world views.[48] The hope is that in light of the realities of exploitation and domination, the pull to veganism will be felt not as an externally imposed constraint on behavior but as a response to one's internal sense of how one wants to be in relation with others and what stance we wish to adopt in the face of cruel and oppressive systems. As Gruen has expressed it, the idea is to locate "the force and authority of the ethical demand within the agent" (2004, 285).[49]

Recognizing that there are contexts in which adopting veganism is complicated by geographical, political, economic, and cultural factors, ecofeminists developed the idea of contextual moral vegetarianism. Rather than issuing ethical edicts requiring universal veganism, CMV encourages examining one's own situation to see if veganism is the correct response. In most cases it will be. Understanding that eliminating all violence from one's life is not possible, however, veganism is understood as an ethical practice rather than a completed moral journey. We must reject the forces of violence, exploitation, and oppression wherever and whenever we can. When that rejection hits its limit at the level of individual choice and action, we must turn to the systemic powers that undergird these forces and seek long-term, lasting structural change. CMV, therefore, is about more than identifying specific cases where it might not be reasonable to demand people adopt veganism. It is about recognizing the limits all of us face in making choices to reject violence and exploitation, facing those limits with honesty and courage, and striving to bring the world into better alignment with our feminist values.

Notes

1 By "veganism," I mean ethical veganism, "a commitment to try to abstain from consuming products derived from animals including meat, dairy, and eggs, as well as products derived from or containing animal products as an ingredient" (Lori Gruen and Robert C. Jones, "Veganism as Aspiration," in *The Moral Complexities of Eating Meat*, eds. Ben Bramble and Bob Fischer (New York: Oxford University Press, 2016), 153–171, 155). This is different from veganism rooted in concerns about health rather than ethics.

2 See Adams, "The Feminist Traffic in Animals," for a discussion of feminist positions opposed to animal rights. See Kathryn Paxton George, "Should Feminists Be Vegetarians?" *Signs* 19, no. 2 (Winter 1994) for a defense of the view that vegetarianism is anti-feminist. Bailey, "We Are What We Eat" has a helpful discussion of anti-vegetarian feminist views.

3 Greta Gaard, "Vegetarian Ecofeminism: A Review Essay," *Frontiers: A Journal of Women Studies* 23, no. 2 (2002): 117–146.

4 At the time the term "contextual moral vegetarianism" was coined, "veganism" was not yet mainstream language. Readers should understand the vegetarianism referred to in the term as veganism.

5 See Peter Singer, *Animal Liberation*, new revised ed. (New York: Avon Books, 1990), for an example of a utilitarian approach and Tom Regan, *The Case for Animal Rights*, updated ed. (Berkeley: University of California Press, 2004) for a rights-based approach.

6 For examples of a politicized and contextualized ecofeminist ethical approach see: Josephine Donovan, and Carol J. Adams, eds., *The Feminist Care Tradition in Animal Ethics* (New York: Columbia University Press, 2007); Carol J. Adams and Lori Gruen, eds., *Ecofeminism: Feminist Intersections with Other Animals and the Earth*. 2d ed.)New York: Bloomsbury Academic, 2022); Curtin, "Toward an Ecological Ethic of Care," Josephine Donovan, "Animal Rights and Feminist Theory," in *Ecofeminism: Women, Animals, Nature*, ed. Greta Gaard (Philadelphia: Temple University Press, 1993), 167–194; Donovan and Adams, *The Feminist Care Tradition in Animal Ethics*, Gaard, *Ecofeminism: Women, Animals, Nature*, Lori Gruen, "Empathy and Vegetarian Commitments," in *Food for Thought: The Debate over Eating Meat*, ed. Steve F. Sapontzis (New York: Prometheus Books, 2004), 284–292 and Lori Gruen, *Ethics and Animals: An Introduction* (Cambridge: Cambridge University Press, 2011); and Marti Kheel, "The Liberation of Nature: A Circular Affair," *Environmental Ethics* 7, no. 2 (1985): 135–149; and Marti Kheel, *Nature Ethics: An Ecofeminist Perspective* (Lanham, MD: Rowman & Littlefield, 2008); among many others.

7 Marti Kheel, "From Heroic to Holistic Ethics: The Ecofeminist Challenge," in *Ecofeminism: Women, Animals, Nature*, ed. Greta Gaard (Philadelphia: Temple University Press, 1993), 243–271, 255.

8 Karen Warren, "The Power and the Promise of Ecological Feminism," *Environmental Ethics* 12, no. 2 (1990): 125–146.

9 Carol J. Adams, *The Sexual Politics of Meat: A Feminist-Vegetarian Critical Theory*, 20th anniversary ed. (New York: Continuum, 2010. First published 1990): 73. Adams' "Sexual Politics of Meat Slideshow" offers a plethora of images highlighting the joint oppression of women and animals through the cycle of objectification, fragmentation, and consumption. Readers can sample the images Adams has collected over the years at https://caroljadams.com/examples-of-spom.

10 Marti Kheel, "Vegetarianism and Ecofeminism: Toppling Patriarchy with a Fork," in *Food for Thought: The Debate over Eating Meat*, ed. Steve F. Sapontzis (New York: Prometheus Books, 2004), 327–341, 334.

11 Ibid.

12 Mark Hawthorne, *Bleating Hearts: The Hidden World of Animal Suffering* (Winchester, UK: Changemakers Books, 2013), 454.

13 Greta Gaard, "Milking Mother Nature: An Ecofeminist Critique of rBGH," *The Ecologist* 24, no. 6 (1994); and Gene Bauer, *Farm Sanctuary: Changing Hearts and Minds About Animals and Food* (New York: Touchstone Books, 2008), 115–119.
14 Gaard, "Vegetarian Ecofeminism," 119.
15 Ibid., and Food Empowerment Project, "Pigs," retrieved from: https://foodispower.org/animals-on-land/pigs/, accessed August 11, 2022.
16 Food Empowerment Project, "Pigs."
17 Bauer, *Farm Sanctuary*, 149.
18 Gruen, *Ethics and Animals*, 84. Noting the ways that speciesist and sexist language connect, Joan Dunayer observes that human women who are no longer considered sexually attractive and who cannot reproduce are called "old biddies," a word also used for hens. As she puts it, "If hens were not held captive and treated as nothing more than bodies, their lives would not supply symbols for the lives of stifled and physically exploited women," Joan Dunayer, "Sexist Words, Speciesist Roots," in *Animals and Women: Feminist Theoretical Explorations*, 3d ed., eds. Carol J. Adams and Josephine Donovan (Durham, NC: Duke University Press, 1995), 13.
19 Adams, *The Sexual Politics of Meat*, 21. Citations refer to the 20th anniversary edition.
20 Ibid.
21 Hawthorne, *Bleating Hearts*, 455.
22 Food Empowerment Project, "Chickens (Hens) Raised for Eggs," retrieved from: https://foodispower.org/hens-raised-for-eggs/, accessed August 11, 2022 and Bauer, *Farm Sanctuary*, 149–150.
23 Food Empowerment Project, "Cows Raised for Milk," retrieved from: https://foodispower.org/cows-raised-for-milk/, accessed August 11, 2022 and Bauer, *Farm Sanctuary*, 99–111.
24 Though I have not had room to discuss it here, the interlocking oppressions involved in animal agriculture involve naturism and racism in addition to sexism and speciesism. Ecofeminists stand opposed to all these forms of violence and see veganism as a necessary response.
25 Bailey, "We Are What We Eat," 50.
26 Kheel, "Vegetarianism and Ecofeminism," 334. Cf., Gaard, "Vegetarian Ecofeminism," 134–135 and Richard Twine, "Ecofeminism and Veganism: Revisiting the Question of Universalism," in *Ecofeminism: Feminist Intersections with Other Animals and the Earth*. 2d ed., eds. Carol J. Adams and Lori Gruen (New York: Bloomsbury Academic, 2022), 229–246.
27 Deane Curtin, "Contextual Moral Vegetarianism," in *Food for Thought: The Debate over Eating Meat*, ed. Steve F. Sapontzis (New York: Prometheus Books, 2004), 273.
28 Lori Gruen, "Facing Death and Practicing Grief," in *Ecofeminism: Feminist Intersections with Other Animals and the Earth*, 2nd ed., eds. Carol J. Adams and Lori Gruen (New York: Bloomsbury Academic, 2022), 164, 161–176.
29 Gary Francione, "Ecofeminism and Animal Rights: A Review of *Beyond Animal Rights: A Feminist Caring Ethic for the Treatment of Animals*," *Women's Rights Law Reporter* 18, no. 1 (Fall 1996): 95–106.
30 Deane Curtin, "Toward an Ecological Ethic of Care," *Hypatia* 6, no. 1 (Spring 1991): 68–69.
31 Kheel, "Vegetarianism and Ecofeminism," 327.
32 Ibid.
33 This is, of course, where relationships are appropriate. For many animals, the ideal relationship with humans is no direct relationship at all.
34 Karen S. Emmerman, "Inter-Animal Moral Conflict and Moral Repair: An Ecofeminist Approach in Action," in *Ecofeminism: Feminist Intersections with Other Animals and the Earth*. 2d ed., eds. Carol J. Adams and Lori Gruen (New York: Bloomsbury Academic, 2022).
35 Commonly referred to as "food deserts," I follow Food Empowerment Project in using the language of "food apartheid" as the latter better captures the deliberate, systemic, and racist roots of the food crisis in the United States. For more information, see Food Empowerment Project, "Food Deserts," retrieved from: https://foodispower.org/access-health/food-deserts/, accessed August 11, 2022.
36 Deane Curtin, "Recipes for Values," in *Cooking, Eating, Thinking: Transformative Philosophies of Food*, eds. Deane W. Curtin and Lisa M. Heldke (Bloomington and Indianapolis: Indiana University Press, 1992), 131.
37 Gruen and Jones, "Veganism as Aspiration," 155–156.

38 Twine, "Ecofeminism and Veganism," 230–233 and 242.
39 Ibid., 231.
40 Kheel, *Nature Ethics*, 236.
41 Adams, *The Sexual Politics of Meat*, 49–50; Lori Gruen, "Dismantling Oppression: An Analysis of the Connection Between Women and Animals," in *Ecofeminism: Women, Animals, Nature*, ed. Greta Gaard (Philadelphia: Temple University Press, 1993), 60–90, 74, and Kheel, "Vegetarianism and Ecofeminism," 329–335. In addition to being gendered, the norms of meat-eating often involve problematic issues related to status and class (Kheel, "Vegetarian Ecofeminism," 335). This is also true for some cultural practices around hunting and whaling (Gaard, "Tools for a Cross-Cultural Feminist Ethics").
42 Kheel, *Nature Ethics*, 237. Kheel also notes that inviting vegetarianism as a response is markedly different from imposing a norm on others ("Vegetarianism and Ecofeminism," 335). The practice of essentializing traditional cultures can cut both ways. Claire Jean Kim notes that the idea that compassion for animals is a Western concept foisted on other people was deeply offensive to Chinese animal advocates who opposed San Francisco's Asian live animal markets. These activists proclaimed that caring for animals is rooted in various religious and philosophical positions prevalent in China (Claire Jean Kim, *Dangerous Crossings: Race, Species, and Nature in a Multicultural Age* (New York: Cambridge University Press, 2015), 191–192).
43 Kim, *Dangerous Crossings*, 11. See Gaard's discussion of disputes regarding reinstating cultural whaling practices in "Tools for a Cross-Cultural Feminist Ethics."
44 Kim, *Dangerous Crossings*, 19.
45 Karen S. Emmerman, "What's Love Got to Do with It? An Ecofeminist Approach to Inter-Animal and Intra-Cultural Conflicts of Interest," *Ethical Theory and Moral Practice* 22 (2019): 77–91.
46 Gaard, "Vegetarian Ecofeminism," 135.
47 Greta Gaard, "Tools for a Cross-Cultural Feminist Ethics: Exploring Ethical Context and Contents in the Makah Whale Hunt," *Hypatia* 16, no. 1 (Winter 2001): 18.
48 Curtin discusses veganism as a political act of protest in "Recipes for Values," 132–133, noting that men might reject the consumption of animal flesh as a form of resistance against the patriarchal view that "real men eat meat." See also Kheel *Nature Ethics*, 242–244 and 246.
49 Gruen, "Empathy and Vegetarian Commitments," 285.

Bibliography

Adams, Carol J. *The Sexual Politics of Meat: A Feminist-Vegetarian Critical Theory*. 20th anniversary ed. New York: Continuum, 2010. First published 1990.

Adams, Carol J., and Lori Gruen, eds. *Ecofeminism: Feminist Intersections with Other Animals and the Earth*. 2nd ed. New York: Bloomsbury Academic, 2022.

Bauer, Gene. *Farm Sanctuary: Changing Hearts and Minds About Animals and Food*. New York: Touchstone Books, 2008.

Curtin, Deane. "Toward an Ecological Ethic of Care." *Hypatia* 6, no. 1 (Spring 1991): 60–74.

Curtin, Deane. "Recipes for Values." In *Cooking, Eating, Thinking: Transformative Philosophies of Food*, edited by Deane W. Curtin and Lisa M. Heldke, 123–144. Bloomington and Indianapolis: Indiana University Press, 1992.

Curtin, Deane. "Contextual Moral Vegetarianism." In *Food for Thought: The Debate over Eating Meat*, edited by Steve F. Sapontzis, 272–283. New York: Prometheus Books, 2004.

Donovan, Josephine. "Animal Rights and Feminist Theory." In *Ecofeminism: Women, Animals, Nature*, edited by Greta Gaard, 167–194. Philadelphia: Temple University Press, 1993.

Donovan, Josephine, and Adams, Carol J., eds. *The Feminist Care Tradition in Animal Ethics*. New York: Columbia University Press, 2007.

Dunayer, Joan. "Sexist Words, Speciesist Roots." In *Animals and Women: Feminist Theoretical Explorations*, 3rd ed., edited by Carol J. Adams and Josephine Donovan, 11–31. Durham, NC: Duke University Press, 1995.

Emmerman, Karen S. "What's Love Got to Do with it? An Ecofeminist Approach to Inter-Animal and Intra-Cultural Conflicts of Interest." *Ethical Theory and Moral Practice* 22 (2019): 77–91.

Emmerman, Karen S. "Inter-animal Moral Conflict and Moral Repair: An Ecofeminist Approach in Action." In *Ecofeminism: Feminist Intersections with Other Animals and the Earth*, 2nd ed., edited by Carol J. Adams and Lori Gruen, 181–196. New York: Bloomsbury Academic, 2022.

Food Empowerment Project. "Chickens (Hens) Raised for Eggs." Retrieved from: https://foodispower.org/hens-raised-for-eggs/, accessed August 11, 2022.

Food Empowerment Project. "Cows Raised for Milk." Retrieved from: https://foodispower.org/cows-raised-for-milk/, accessed August 11, 2022.

Food Empowerment Project. "Food Deserts*." Retrieved from: https://foodispower.org/access-health/food-deserts/, accessed August 11, 2022.

Food Empowerment Project. "Pigs." Retrieved from: https://foodispower.org/animals-on-land/pigs/, accessed August 11, 2022.

Francione, Gary. "Ecofeminism and Animal Rights: A Review of *Beyond Animal Rights: A Feminist Caring Ethic for the Treatment of Animals*." *Women's Rights Law Reporter* 18, no. 1 (Fall 1996): 95–106.

Gaard, Greta. "Milking Mother Nature: An Ecofeminist Critique of rBGH." *The Ecologist* 24, no. 6 (1994).

Gaard, Greta. "Tools for a Cross-Cultural Feminist Ethics: Exploring Ethical Context and Contents in the Makah Whale Hunt." *Hypatia* 16, no. 1 (Winter 2001): 1–26.

Gaard, Greta. "Vegetarian Ecofeminism: A Review Essay." *Frontiers: A Journal of Women Studies* 23, no. 2 (2002): 117–146.

Gruen, Lori. "Dismantling Oppression: An Analysis of the Connection Between Women and Animals." In *Ecofeminism: Women, Animals, Nature*, edited by Greta Gaard, 60–90. Philadelphia: Temple University Press, 1993.

Gruen, Lori. "Empathy and Vegetarian Commitments." In *Food for Thought: The Debate over Eating Meat*, edited by Steve F. Sapontzis, 284–292. New York: Prometheus Books, 2004.

Gruen, Lori. *Ethics and Animals: An Introduction*. Cambridge: Cambridge University Press, 2011.

Gruen, Lori. "Facing Death and Practicing Grief." In *Ecofeminism: Feminist Intersections with Other Animals and the Earth*, 2nd ed., edited by Carol J. Adams and Lori Gruen, 161–176. New York: Bloomsbury Academic, 2022.

Gruen, Lori, and Jones, Robert C. "Veganism as Aspiration." In *The Moral Complexities of Eating Meat*, edited by Ben Bramble and Bob Fischer, 1530171. New York: Oxford University Press, 2016.

Hawthorne, Mark. *Bleating Hearts: The Hidden World of Animal Suffering*. Winchester, UK: Changemakers Books, 2013.

Kheel, Marti. "The Liberation of Nature: A Circular Affair." *Environmental Ethics* 7, no. 2 (1985): 135–149.

Kheel, Marti. "From Heroic to Holistic Ethics: The Ecofeminist Challenge." In *Ecofeminism: Women, Animals, Nature*, edited by Greta Gaard, 243–271. Philadelphia: Temple University Press, 1993.

Kheel, Marti. "Vegetarianism and Ecofeminism: Toppling Patriarchy with a Fork." In *Food for Thought: The Debate over Eating Meat*, edited by Steve F. Sapontzis, 327–341. New York: Prometheus Books, 2004.

Kheel, Marti. *Nature Ethics: An Ecofeminist Perspective*. Lanham, MD: Rowman & Littlefield, 2008.

Kim, Claire Jean. *Dangerous Crossings: Race, Species, and Nature in a Multicultural Age*. New York: Cambridge University Press, 2015.

Paxton George, Kathryn. "Should Feminists Be Vegetarians?." *Signs* 19, no. 2 (Winter 1994): 405–434.

Regan, Tom. *The Case for Animal Rights*. Updated ed. Berkeley: University of California Press, 2004.

Singer, Peter. *Animal Liberation*. New revised ed. New York: Avon Books, 1990.

Twine, Richard. "Ecofeminism and Veganism: Revisiting the Question of Universalism." In *Ecofeminism: Feminist Intersections with Other Animals and the Earth*, 2nd ed., edited by Carol J. Adams and Lori Gruen, 229–246. New York: Bloomsbury Academic, 2022.

Warren, Karen. "The Power and the Promise of Ecological Feminism." *Environmental Ethics* 12, no. 2 (1990): 125–146.

7
BLACK VEGANISM(S)
A Personally Guided Exploration

Nekeisha Alayna Alexis

To attempt a chapter on "Black Veganism" is to run into at least two challenges that are worth naming at the outset. One is the use of the word, "Black" and the vast, diverse community the term represents. In this place some call the United States of America,[1] Black often functions as a substitute or parallel to African American, as evidenced by the way "African American/Black" tends to appear as an option on countless forms. Yet Black identity is also global, extending to people of African descent on the earth's second-largest continent and across the diaspora. It includes native Trinidadians like me, Garifuna and Black Mexicans, Siddis and other Afro-Asians, Ethiopians and Ghanaians. The word encompasses distinct yet connected melanated peoples that transcend European colonization and enslavement and exist in their aftermath.

That "black" as a racial concept arose in Europe across the 16th to 19th centuries and was solidified through the oppression of African-descended peoples; that Black is *also* a reclaimed identity associated with individual and collective power, pride, and shared experiences within and beyond White supremacy; and that Black is an adjective that is sometimes contested by persons in the community it is meant to define complicates the very idea of Blackness. Consequently, at least one reasonable response to any writing on "Black Veganism" is to question the enterprise itself: What do you mean by Black? Who all are included and why? What about veganism would make it Black and why does it matter? The word Black is already complex by its lonesome, and the difficulty is only compounded by the nuances of "vegan" as well.

Although the rise of the vegan movement has its own history, people from various cultures have practiced and promoted different kinds of vegetable-based diets for millennia. The specific descriptor emerged in 1944 in the United Kingdom by White "non-dairy vegetarians." Coined by Donald Watson using the first and last letters of the word, vegetarian, "vegan" referred to a group who distinguished themselves from people who gave up meat while eating other animal byproducts. They believed "lacto-vegetarianism is but a half-way house between flesh-eating and a truly humane, civilised diet"[2] and sought to "evolve sufficiently to make the 'full journey'."[3] These outliers formed the Vegan Society, and by 1949, the word for a diet that excluded everything with a face and that came from a face expanded. At this point, society president Leslie J. Cross defined veganism as "[t]he

 DOI: 10.4324/9781003273400-10

principle of the emancipation of animals from exploitation by man."[4] Two years later, veganism was described as a "doctrine," and the society's mission broadened to "seek to end the use of animals by man for food, commodities, work, hunting, vivisection and all other uses involving exploitation of animal life by man."[5] At the time of this writing, the Vegan Society formally defines veganism as:

> a philosophy and way of living which seeks to exclude—as far as is possible and practicable—all forms of exploitation of, and cruelty to, animals for food, clothing or any other purpose; and by extension, promotes the development and use of animal-free alternatives for the benefit of animals, humans and the environment. In dietary terms it denotes the practice of dispensing with all products derived wholly or partly from animals.[6]

The organization continues working toward "liberation, for the creatures and for the mind and heart of man" and a world where using other animals for "self-interested purposes would be so alien to human thought as to be almost unthinkable."[7]

Although the 1940s fringe group has added millions of adherents worldwide, consensus on what constitutes vegan practice can sometimes feel elusive and downright strange. There are self-identified vegans who eat honey and who eat plant-based while wearing leather, wool, and silk. There are debates about the vegan-ness of purchasing products with ingredients that destroy wildlife and supporting vegan brands whose parent companies conduct animal testing. Can one be vegan if the motivation is not compassion for other animals? Is someone vegan if they aren't engaged in rescue efforts? How about eating plant-based burgers at McDonald's? Adding to the confusion is vegan advocates' tendency to define growth by the number of people removing meat from their plates, increases in vegan restaurants and menu offerings, and the availability of plant-based foods at the supermarket. Yet, when a 2016 poll of United Kingdom households finds *a mere 1.66% of adults 15 years of age and older*, "Never eat meat and avoid using items derived from animals or that involve animals,"[8] and a 2019 poll of Americans places "concern for animals" in fourth place as a "major reason" for eating meat "less," "rarely," or "never,"[9] it suggests that veganism—at least as Watson, Cross, Elsie Shrigley, and others came to understand it—may not have as much traction as we think, at least in parts of the West.

Like the complexity of the word "Black," discussions about veganism can raise their own share of questions: What do we mean by "vegan"? Who all do we include? Why does any of it matter in the first place? With all this in mind, the first point to be made about "Black veganism" is that it cannot be reduced to "Black people who are vegans." Each word describes its own rich histories, interpretations, controversies, and communities that create a subgroup of subgroups when everything is brought together. Indeed, as a Black woman who follows a vegan "lifestyle"[10] as a matter of liberationist Christian theology and ethics; creation care; and an interspecies, intersectionality-based[11] approach to justice, I can personally attest that Black veganism is, to put it mildly, full of variety and depth.

With all this in mind, I will attempt (with some fear and trembling) to explore *a few* expressions of Black Veganism(s) I have observed along my journey. I will do this by sketching three approaches in a loosely held, non-scientific way. I have called these 1) Black Health and Plant-Based Foodways, 2) Black Power and Identity in Vegan-Eating, and 3) Vegan for Us and the (Other) Animals.[12] I will summarize each kind not by cataloguing a supposedly objective and comprehensive list of representatives, inevitably forgetting someone of significance. Instead, I will take a primarily subjective stance, referring mostly to practitioners, advocates, and scholars who shaped me. I begin with my own narrative to make visible that I am not detached from the subject and as a challenge to the presumed objectivity of dominant Western academic writing.

Toward a Black Veganism: My Story in Brief

My journey toward a Black veganism began somewhat ironically at a predominantly White conference at Calvin College in 2008. There at Wake Up Weekend, I confronted the inconceivable torture other animals endure to become edible things. At the time, I was calling myself "vegetarian" while still eating fishes and products containing cow's milk and hen's eggs. I was also a student at a seminary in the peace church tradition who was politically anarchist and committed to pacifism. Initially, the idea of cutting out all animal products from all areas of life felt like "going overboard"—until I saw footage of suffering hens, bellowing cows, and pigs being slaughtered; listened to activists who had been jailed for rescuing farm animals and documenting abuses; and heard about farmers who had abandoned their livelihoods for the sake of compassion. I distinctly remember a visceral moment while watching scenes from an industrial farm when I slunk down in my seat, hands over my eyes, body tight from noises that sounded like screaming. It was at this gathering I learned what "going overboard" really meant.

I left my first Wake Up Weekend with the conviction to expand my commitments to Christian peacemaking, nonviolence, and rejecting hierarchies of oppression. Within months, I switched to a truly plant-based diet and began transitioning my household. Shortly after becoming vegan, I received a copy of Carol J. Adams's *The Sexual Politics of Meat: A Feminist-Vegetarian Critical Theory*[13] and was introduced to Marjorie Spiegel's *The Dreaded Comparison: Human and Animal Slavery*. Here again, it would be the writings of two White women that would introduce me to a cross-species analysis and sharpen my understanding of veganism as a matter of human liberation, especially for me as a Black woman. I also began encountering Black vegan thinkers and writers who further refined my understanding.

Although my vegan journey began with majority White resources, Black practitioners have *always* been present on my path. At my first Wake Up Weekend, I met Michelle Loyd-Paige, an administrator at Calvin and event co-organizer, who "ate like a vegan."[14] Years later at the same event, I met Bryant Terry, chef and author of *Vegan Soul Kitchen: Fresh, Healthy, and Creative African-American Cuisine*, and learned about the power of container gardening and wild edible education for Black women experiencing distress by leaders from the non-vegan organization Our Kitchen Table.[15] I can even recall my delight when I saw San'Dera Nation featured in the popular documentary, *Forks Over Knives* (2011). Her story as an African-American single mother of five children from the Cleveland suburbs treating her hypertension and diabetes with a plant-based diet, and reflections by Terry Mason, M.D., then Commissioner of Health for Chicago and the only public official to advocate a plant-based diet for optimal health, were key reasons I showed the film in a Food and Faith class I co-led at the African Methodist Episcopal church I attended. Encountering these and other Black persons early in my transition was essential for busting the myth that veganism was a "White thing." We may have been in the minority in my experiences, *but we were always here*. That visibility showed me we not only had a place in this movement: we were uniquely positioned to lead and benefit from it.

Black Health and Plant-Based Foodways

The impact of plant- and of animal-based diets on human health have been widely documented, with increasing attention to the advantages of the former and detriments of the latter in American media and medicine. As of June 2020, the American Cancer Society stated that "[p]rocessed meat intake, even in small amounts, and red meat in moderate to high amounts increases risk" of colorectal cancer, while "[r]egular intake of processed, grilled, or charcoaled meats increases risk for non-cardia gastric cancer."[16] They also describe healthy eating as limiting or excluding "[r]ed and processed meats" and incorporating:

- A variety of vegetables—dark green, red and orange, fiber-rich legumes (beans and peas), and others
- Fruits, especially whole fruits in a variety of colors
- Whole grains[17]

Consuming the carcasses and byproducts of other animals is also linked to other common and often preventable illnesses like high blood pressure, high cholesterol, and obesity. Moreover, recent infectious diseases like mad cow, swine flu and a variety of avian flus have also emerged from industrial animal farms, causing additional concern for global safety.[18] As a result, a desire to maintain and improve one's physical wellbeing is often a key reason for choosing plant-based foodways regardless of one's racial identity.[19] Yet the motivation is often pronounced among Black Americans due to persistent disparities in our health outcomes and access to quality health care, due in part to structural White supremacy.

In the introduction to the *African American Vegan Starter Guide*, a collaboration between public health nutritionist and *By Any Means Necessary* author Tracye McQuirter, MPH, and White-led organization Farm Sanctuary, McQuirter attributes her vegan conversion to comedian and Civil Rights activist Dick Gregory. At a lecture she attended, Gregory unexpectedly discussed, "the plate of black America, and how unhealthfully most folks eat."[20] He "graphically traced the path of a hamburger from a cow on a factory farm, through the slaughterhouse process, to a fast-food restaurant, to a clogged artery, to a heart attack,"[21] igniting McQuirter's curiosity. Within months, she discovered "a large and thriving community of black vegetarians and vegans" and embarked on a 30-year journey of plant-based eating.

The "Why Go Vegan" section of the booklet also foregrounds health benefits for African Americans as a key reason to make the switch. "We have the most to gain . . . because we experience the highest rates of preventable, diet-related chronic diseases in the country."[22] This sentiment was similarly captured in a 2012 article on the "rising number of black Americans who are becoming vegetarian and vegan."[23] While the author doesn't provide statistics, a Pew Research Center survey published four years later found that, while 3% of American adults overall identify as vegan, "That number jumps to a startling 8 percent among African American adults."[24] Although the article only quotes self-identified vegetarians, it still names some of the health considerations Black people have for avoiding flesh foods. Losing weight, having more energy, and better quality of life in one's senior years are some of the reasons mentioned.

While it may be tempting to see Black interest in plant-based eating as a phase or novelty, surveying some of the more well-known voices quickly demonstrates otherwise. In

1974, Gregory released *Dick Gregory's Natural Diet for Folks Who Eat: Cookin' with Mother Nature*. Almost two decades later, holistic health practitioner, wellness coach and green foods movement pioneer Queen Afua published *Heal Thyself for Health and Longevity*. Early in the internet age, it was McQuirter who created one of the first vegan websites in 1997. In 1999, rapper and long-time vegan KRS-One would release the song "Beef," detailing the use of pharmaceutical drugs in cows; the health costs of eating "the fear and stress of another"; and the addictive power of meat, "the number one drug on the street."[25] Indeed, consciousness about plant-based foodways has been a small but significant part of hip hop culture since it began. As of 2020, eight members of the legendary Wu-Tang Clan and Stic.man of Dead Prez—groups that also formed in the 1990s—are vegan, and well-known artists like Jermaine Dupri, Erykah Badu, André 3000, and Common "have dabbled or committed to plant-based lifestyles."[26] This list includes only a handful of Black plant-based advocates in the United States alone.

Today, Black voices continue to promote plant-based eating for optimal health within and outside of our communities. Celebrity power couple Beyoncé and Jay Z are partners with the plant-based 22-Days Nutrition Greenprint program, and Jay Z's venture capital firm has strong investments in plant-based companies. Meanwhile, athletes like American record-holding weightlifter Kendrick Farris, two-time Australian 400-meter sprinting champion Morgan Mitchell, and former NFL cornerback Lou Smith make appearances alongside other Black sportspersons in the vegan advocacy film *The Game Changers*. Currently, one of the most popular vegan voices in the U.S. is one of my favorite social media personalities, Tabitha Brown. The African American chef, author, and entrepreneur "had tried every drug the doctor offered" to address the symptoms she felt due to an unexplained autoimmune condition.[27] She adopted a plant-based diet after her health issues drastically improved during a vegan challenge and currently showcases her vegan cooking skills to her now 4 million TikTok and Instagram followers.

There are many people, Black and otherwise, who eat plant-based for health reasons that won't develop an ethic of care for other animals. One of the deterrents to making the link between the two is the dominant discourse of U.S. nutrition science, which "prefers discussing a whole-plant-food-diet, rather than a vegan one, to eliminate connotations of animal rights or environmental ideology."[28] As we'll see, Black people also face distinct barriers to holding health and animal advocacy together due in significant part to our communities' experiences of White supremacy and racism.

Black Power and Identity Through Vegan Eating

While some see staying fit through plant-based eating as a personal matter, separate from or tangential to their Black identity, others with a heightened Black power consciousness see vegan eating as necessary for greater self-determination and collective resistance amidst White supremacy. For the latter group, this foodway is a health matter that *also* advances the physical, mental, spiritual, social, and political wellbeing of Black communities in a world that actively works against us. In the United States, this explicit link between diet and politics emerged in the late 1960s when "an increasing number of black food reformers rejected, or at least complicated, what they regarded as standard American food practices," and "asserted a separate black national identity and a competing value system."[29] On one hand, many of these voices opted to re-envision Soul Food—a heavily animal-based "product of African, European, and Native American ingredients and culinary knowledge—as a distinctly black cultural product."[30] In contrast, several Black nationalists:

> characterized American food culture as unhealthful and corrupting, and dismissed soul food as a pernicious rebranding of slave food. Many black nationalists urged African Americans to adopt a mostly vegetarian, whole-foods diet as a way to signal their rejection of the American values embedded in a painful past and as a means to improve the physical and financial health of the people they increasingly conceptualized as members of an oppressed, stateless black nation.[31]

This subset of the Black food revolt often developed organizations and community groups with alternative food practices. Based on traditional and re-imagined African and diaspora cultures, these approaches to food helped define and reinforce radical Black identities among members.

In "How to Eat to Live: Black Nationalism and the Post-1964 Culinary Turn," Jennifer Jensen Wallach provides a broad sketch of the contest between rehabilitators of soul food and those who rejected it, as well as tensions that arose between Black nationalists. For example, when Political School of Kawaida founder Amiri Baraka loosened his belief that, "vegetables, fruits, grains and fish . . . natural juices, milk and water" were the most nutritious foods for life and the people's work, fellow members of the Congress of African People Haki Madhubuti and Jitsu Weusi resigned immediately. Madhubuti, who believed "an unhealthy diet was counter-revolutionary,"[32] had denounced meat-eating in favor of vegetarianism. Meanwhile, Weusi, who directed The East, also advocated for a whole-foods, vegetarian diet "so that our functions are not hampered by illness."[33] Even children ate "meals consisting of whole grains, fruits, and vegetables and were forbidden to eat meat, dairy, or sugary foods."[34] For Weusi, food was a key part of an overall infrastructure toward Black uplift that also included a school, a cooperative business, a newspaper, and a performance space.

The idea of adopting a plant-based diet as integral for "reimagining a distinctly black corporeality," "showing ideological fidelity and physiological loyalty to the black nation," and seeing "health as a community concern"[35] persists well into today. It is present in *Sistah Vegan: Black Women Speak on Food, Identity, Health, and Society*, a collection of essays I read four years into my path. In "Ma'at Diet," Iya Raet describes her commitment to a vegan diet as a matter of "physical, cultural, emotional and spiritual reasons."[36] Seeing herself as "a divine being" who follows "the diet of my ancestors," she emphasizes eating *holistic* foods "based on the principles of Ma'at: order, balance, and harmony."[37] Her piece, which focuses on parenting as a Nubian woman, briefly notes that, "The Afrikan woman's womb is attacked by hormonal birth control, pesticides, and other toxins" as well as "holding on to emotional hurt and pain."[38] She outlines an alternative approach not only to maintaining a vegan diet during pregnancy but also to co-parenting and mothering in a "Pan Afrikan" perspective.

In addition to Raet's reflections, authors Melissa Danielle and Ma'at Sincere Earth explore the notion of community health through vegan eating. In "Nutrition Liberation: Plant-Based Diets as a Tool for Healing, Resistance and Self-Reliance," Danielle describes plant-based eating as a "conscious, sustainable diet" *and* as "a political statement, another weapon in our fight for economic, social and political empowerment."[39] She continues:

> Where our food comes from, how it gets to the places we buy it from, and what is in the food we eat are just as important as having full access to housing, health care, jobs, and the polls. What good is independence if we are still a slave to food? How can we fully embrace progress if we are not in control of our bodies?[40]

Earth, who does not eat flesh-foods and usually abstains from animal-based products, writes similarly in "Black-a-tarian," saying, "We have accepted another race's diet for our own. Our bodies are not reacting well to the diet of the masses. Americans diet is an example of sweeping racial differences under the door. . . . Diet is the racial oppression that Black people have to get over in order to survive."[41] She credits her initial move to vegetarianism to Sistah Souljah's autobiography, *No Disrespect*, and her switch to vegan eating to Dr. Llaila Afrika's *African Holistic Health* and *Nutricide*. Reflecting on Afrika's work, she writes, "The idea of freeing myself nutritionally from the oppressors' diet motivates me to avoid meat. I soon became vegan to solidify my nutritional commitment."[42]

One thing I have noticed about Black nationalist vegan approaches is the practice of not eating other animals does not necessarily lead to concern for animal welfare. Indeed, people may readily overlook the needs of other animals, see them as inconsequential to Black struggle, or consider benefits to other animals as a positive byproduct of the foodway instead of the reason for choosing it. In some instances, Black vegan nationalists may dismiss animal liberation altogether or see advocacy for other animals as rival to and distracting from racial injustice. As Earth writes, "People assume that just because you are vegan, your front is animal rights, and they think they should never see you in a leather coat or shoes. But that's not what motivates me to not eat meat . . . I don't even understand animal rights."[43] The center of this pathway is consistently and unapologetically *freedom for Black peoples*. Indeed, I venture that if animal-based foods were healthier than plant-based ones, Black nationalists would likely favor the former instead.

Over time, I've observed several complex factors that contribute to this disconnect between struggles for Black liberation and animal liberation that I can only name in passing. One is widespread lack of awareness about Black voices that see abuse of the nonhuman world and of our peoples as emerging from the same supremacist sources and thus understand the fight for liberation as an integrated, interspecies one. Relatedly, mainstream veganism often lacks strong and substantial representation of and leadership by Black people and has alienated some of us by misusing slavery and other White terrorism against us to spotlight animal causes. The failure of many non-Black animal advocates to think critically about how oppressions are linked across species also means these groups are often silent when Black communities experience violence, for example, at the hands of police and vigilantes of through systems like mass incarceration.[44]

The contrast between animal advocates' sensitivity—and for some Black people, a misplaced *hyper*sensitivity—to the plight of cats, dogs, and chickens in contrast to the fleeting attention and energy devoted by animal advocates to Black lives mattering also creates a rub. A high-profile example of this was the controversy around former NFL quarterback and African American Michael Vick and his arrest and imprisonment for dogfighting. Scholar Claire Jean Kim brilliantly traces the nuances of the dispute which painfully revealed antiBlackness among animal welfarists and anti-animal sentiments among Black voices. Some of Vick's detractors were explicitly racist, while many others' "criticisms of him as cruel, brutal and pathological resonated with incorrigible tropes."[45] Meanwhile, many Black people called out the racism behind the swift legal, economic, and social retribution against Vick. Yet they did so by comparing the brutality against the dogs to the cruelties Black people face and downplaying the dogs' suffering as a result.[46] Kim suggests another response was possible—what she calls a "posture of avowal"—where both groups could develop the empathy and analysis needed to challenge the racism against Vick *and* the victimization of the dogs.

In addition to the peculiar barriers that racism creates between animal liberation movements and Black nationalist and other approaches to vegan eating, there is also the impact of anthropocentrism and speciesism that exists across people of all racial identities. Our human-centered lenses and unwillingness to see the likenesses between humans and other animals prevents most people from questioning our mistreatment of them, regardless of racial identity. Yet it is important to note that even these "isms" can have a different character within Black communities. Because the White supremacist imagination currently saturating our world considers us primate-like beings, it's easy to see how asserting our humanity by resisting common cause with other animals (often subconsciously), and focusing on our liberation exclusively could become a default posture.

Vegan for Us and the (Other) Animals

In this third stream I call Vegan for Us and the (Other) Animals, Black vegans cross the bridge between Black and animal liberation not by turning our attention to violence against other animals as a separate issue but by perceiving the interlocking nature of our oppressions *and centering other animals and ourselves* in the cause of freedom. For some, empathy for the plight of other animals comes from being "treated like an animal" by White terrorist hands and structures, and opting for a radical nonviolence that also redefines our relationships with other animals. For others, it comes from recognizing race as a *European fiction forged through conquest and colonization* that relies in part on *European fictions about other animals that normalize violence against them*. For still others, this approach is discerned through a broader decolonization framework. At the core of this perspective is the belief that challenging White supremacy and its legacies includes addressing the survival of Black people and other animals together.

One expression of this approach precedes the debut of veganism in the United Kingdom and runs parallel to the Black nationalist culinary turn in the United States. Arising in Jamaica around the year 1930, Ital eating is a "Black plant-based foodway" that developed within the Rastafarian movement and is influenced by "south Indian spiritualities imported into the Caribbean by Asian indentured workers, Christianity, African spiritualities such as Ubuntu, and philosophies articulated locally by enslaved Africans in the Caribbean."[47] Rastafarianism was cultivated among Jamaican youth who created communes to escape and heal from the ongoing effects of colonialism and racism. As researcher Solaire Denaud notes, this withdrawal was conceived as a cross-species affair in that:

> they did not limit their philosophy to humanness. They viewed the land and all its inhabitants as victims of Babylon, asserting that violence against nature and non-human animals is also a form of Babylonian violence that needs to cease. If you think about it, Rastas are going further than the vegan movement since they consider many forms of violence as part of a broader problem.[48]

Derived from the word "vital," Ital eating reflects the Rastafarian emphasis on "unity with all things" and is said to increase the "universal energy" they call *livity*.[49] Again, it is worth emphasizing that Ital eating and the belief behind it began fifteen years before Watson and others began carving out veganism in their context. Yet I don't recall vegan conversations that focused on Rastafarianism and the Ital diet, nor did I know about their interspecies character until now.

Gregory, whom I mentioned earlier, also provides an example of a person whose experience of violence enabled him to see other animals in a new way. In *Callous On My Soul: A Memoir*, he recounts giving up meat in 1964 after a White sheriff kicked his wife Lil, who was pregnant with twins, at a Civil Rights march in Alabama. That incident prompted a deeper appreciation for the meaning of nonviolence and the decision that "if I would not hit a man who kicked my pregnant wife, then I could no longer participate in the destruction of any animal that has never harmed me."[50] From that time, Gregory became increasingly attentive to animal welfare issues. He eventually opposed "killing in any form," including of other animals for food and sport, and became concerned about the treatment of elephants in circuses. In the captivity of elephants, he saw slavery: "They are shackled and chained like our ancestors, and they represent the domination and oppression we suffered for hundreds of years."[51] By 1980, he joined the People for the Ethical Treatment of Animals (PETA).

Earlier in his memoir, Gregory also describes discrimination he faced while clothes shopping as a fourteen-year-old in St. Louis. At the White-run stores, he was "treated like an animal instead of a teenage boy. It is almost like you stole something. . . . They looked down on Negroes like we were dogs."[52] This experience of dog-like treatment, which was inseparable from criminal-like treatment, reveals a key working of White supremacy. Once you "persuade the mainstream that certain groups fall outside of 'human'—they are irrational, they hold 'barbaric' values, they have 'inferior' systems of beliefs, they behave 'like animals, and so on,"[53] you can also legitimize and make ordinary heinous actions against us. While Gregory eventually associated the grinding down that he and other Black people endure with the brutality other animals undergo, the significance of being "treated like an animal" remains only a metaphor for most antiracists. Moreover, the one-dimensional argument that "People of color are humans, too; so, we should treat them as humans, not animals" holds within it an "open acceptance of the negative status of 'the animal' " and "a tacit acceptance of the hierarchical racial system and white supremacy in general."[54]

In *Aphro-ism: Essays on Pop Culture, Feminism, and Black Veganism from Two Sisters*, Aph and Syl Ko critically engage with the history of race as an idea and its tie to animality and in so doing make the case that Black and other animal oppression *and* liberation are intertwined. This work, which was deeply influential for me, clearly outlines how the racial logic that makes possible the atrocities done to Black and other non-White people is rooted in the "human–animal divide" that also arose in the European Enlightenment. As Syl Ko explains,

> "White" is not just the superior race; it is also the superior mode of *being*. Residing at the top of the racial hierarchy is the white human, where species and race coincide to create the master being. Resting at the bottom as the abject opposite of the human, of whiteness, is the (necessarily) nebulous notion of "the animal."[55]

In other words, and as I like to say, "race is a species construct." It is an idea, an equation, that insists:

1. there are different *species* of beings called races *plus*
2. that the White races are the only human species *which equals*
3. that White races can legitimately and even benevolently commit violence against the lower races, who are only partially evolved and thus not-quite-human (usually non-Black people of color) or animals that can never be human (Black people)

While many animal advocates miss this race and animal connection and thus fail to be antiracist, many antiracists often notice the race and animal connection but fail to overturn *the original species lie* on which the racial thinking depends:

1. that humans are not animals *plus*
2. that only humans are worthy of value and care *which equals*
3. it is acceptable for humans to legitimately and benevolently commit violence against any and all beings that are animals

Even the most progressive persons have unconsciously bought into European fictions that humans are the only rational creatures with language and complex relationships, while other animals, lacking these and other skills, are only here for our use. Ko insists that "As long as these notions of 'the animal' and 'the human' are intact, white supremacy remains intact."[56] For Black people who are vegan for us and other animals, exposing and undermining the original lies on which the derivative lie of race is built is part of how we attain our freedom. We recognize that being "treated like an animal" is dangerous and deadly because the majority in our world is committed to treating other animals as disposable. To be "treated like an animal" could well be paradise *if* the dominant belief was animals deserve empathy and compassion; protection from unnecessary suffering; and a right to live at peace and unafraid, liberated and self-determining.

Today, other voices are articulating a vision for Black and animal thriving focused not on the expansion of European power but instead on their understanding of African cultures and histories. As South African Venita Januarie explains in " 'Veganism' through the Lens of Decolonization:"

> Until about five centuries ago, Africa remained mainly dependent on traditional food. When adventurers and slave-traders came to the African continent, they introduced various crops and the larger-scale domestication of animals for commercial consumption and export. These capitalistic farming methods exacerbated the spread of animal diseases among humans. The nomadic lifestyle of some African tribes, which required smaller herds, also began to dwindle as meat production became a lucrative industry and changed the eating patterns of people on the continent.[57]

She cites Tendai Chipara, Zimbabwean founder of Plant-Based African, who argues that the food of her ancestors "was organic and meat products were consumed minimally. The unfortunate thing that happened to us a people was colonization which led to a massive change to our food production, access to land, and the emergence of processed foods."[58]

Black vegans in the United States are also having similar conversations. In response to those who see vegan eating as contradictory to Black identity, gourmet soul food restaurant co-owner Aaron Beener insists that "this is our culture, too." Foods like "black-eyed peas, cowpeas, okra, and rice indigenous to the Senegambia region of West Africa made their way to the Americas and the Caribbean as seed stock for enslaved Africans."[59] These plant-based ingredients are not foreign to us, but displacement from land and this plant-based agricultural legacy makes it easy not to know this part of Black history.

As a Christian who is aware of the ways dominant Christianity has rationalized harm against other animals and of the powerful place Christianity still holds for many Black people,[60] I could not close this chapter without mentioning Christopher Carter's work *The Spirit of Soul Food: Race, Faith, & Food Justice*. He examines the current approach to soul food like Black nationalists before him but from a decidedly Jesus-centered social location. The book asks the question, "If soul food is to be a response to food injustice in the Black community, given the myriad of ways industrial agriculture harms Black people—economically, environmentally, ideologically—what should soul food look like today?"[61] He invites Black Christians especially to see food justice as a key part of what it means to follow Jesus, and suggests that "black veganism . . . is the ideal form of soulfull eating and the way Black people can decolonize our diets and delink from coloniality."[62] From Carter's and other decolonial lenses, choosing plant-based lifestyles is a kind of reclamation and return not only to a way of eating but of a kind of relationship with other creatures that challenge and possibly pre-date dominant Western models of engagement. As Januarie concludes, becoming vegan "allowed the most authentic expression of myself. I could not advocate for life and protecting the environment while I clearly turned a blind eye to the suffering of sentient creatures and the systematic destruction of our environment."[63]

Conclusion

The fact that Black Veganism(s) are often secondary or absent from mainstream consideration does not mean these approaches do not exist. Black voices have been leaders and predecessors in the plant-based movement and even for the struggle for other animal existence. We predate well-known and recognized White names like Peter Singer, Ingrid Newkirk, and Tom Reagan. We've advocated for plant-based wellness long before *What the Health?* We lead organizations like Afro Vegan Society and Hip Hop Is Green. We've produced films like *The Invisible Vegan*. We devote our lives to conservation and vegan advocacy like Chef Nicola. Even the tenth element of hip hop now includes "plant-based eating . . . and animal rights activism."[64] Moreover, with Africa/Middle East (19% vegetarian and 9% vegan) being "the second most vegetarian region in the world after Asia" (16% vegetarian and 6% vegan), it is people of color internationally who are ahead of North America (6% vegetarian and 2% vegan) and Europe (5% vegetarian and 2%) when it comes to maintaining this ethic.[65]

That the advent of the term "vegan" comes at a particular place and time does not mean it is our only entry point into that way of living. Indeed, there are more names and resources than there is room to add to this essay that can serve as guides and inspiration. As Black and other communities—me included—are (re)introduced to the plant-based and animal-friendly aspects of our cultures over time and geography, we will discover distinctly Black sources to choose toward greater health, to (re)claim our identities, for the sake of other animals, and/or for a wide range of other reasons. At the heart of Black Veganism(s), at least as I see it today, is a desire for radical wholeness. For each of us. For all of us. Because all of our lives matter.

Notes

1 I acknowledge that I am writing this chapter as a resident on the lands of the Potawatomi and Miami peoples, now known broadly as Elkhart, Indiana. I recognize that what some call the United States of America has been long known to other indigenous communities by other names.
2 Donald J. Watson, *The Vegan News*, November 1944. Republished as "Excerpts from the Vegan News," *Gentle World* (September 28, 2020), retrieved from: https://gentleworld.org/the-full-journey-excerpts-from-the-vegan-news-november-1944/, accessed December 16, 2022. See also The Vegan Society, retrieved from: https://www.vegansociety.com/about-us/history, accessed December 6, 2022
3 Ibid.
4 The Vegan Society, retrieved from: https://www.vegansociety.com/about-us/history, accessed December 6, 2022
5 Leslie J. Cross, "Veganism Defined," *The Vegan News*, republished by Gentle World on (August 2013), retrieved from: https://gentleworld.org/veganism-defined-written-by-leslie-cross-1951/, accessed December 16, 2022
6 The Vegan Society, retrieved from: https://www.vegansociety.com/about-us/history, accessed December 6, 2022
7 Cross, "Veganism Defined."
8 "Poll Conducted for The Vegan Society. Incidence of Vegans Research," retrieved from: https://www.ipsos.com/sites/default/files/migrations/en-uk/files/Assets/Docs/Polls/vegan-society-poll-2016-topline.pdf, accessed August 6, 2023,
9 Justin McCarthy and Scott Dekoster, "Nearly One in Four in U.S. Have Cut Back on Eating Meat," *Gallup* (January 27, 2020), retrieved from: https://news.gallup.com/poll/282779/nearly-one-four-cut-back-eating-meat.aspx, accessed August 6, 2023
10 According to the IPSOS poll, "lifestyle vegan" is the term The Vegan Society uses to refer to people who "never eat any animal products and avoid using products derived from animals or involving animals in their production," retrieved from: https://www.ipsos.com/en-uk/vegan-society-poll
11 "Kimberlé Crenshaw on Intersectionality, More Than Two Decades Later," Columbia Law School, posted on (June 8, 2017), retrieved from: https://www.law.columbia.edu/news/archive/kimberle-crenshaw-intersectionality-more-two-decades-later, accessed August 6, 2023. Crenshaw defines intersectionality most simply as, "Intersectionality is a lens through which you can see where power comes and collides, where it interlocks and intersects."
12 If space and energy permitted, I could also explored Black veganism(s) related to environmental racism and worker justice and another exploring Black Christian vegan theology and animal liberation.
13 Eternally grateful to fellow Anabaptist Mennonite Biblical Seminary graduate turned animal scholar Brianne Donaldson for sharing her copy of Adams' book. It was profound at providing an analysis that was strong enough to undergird the emotions and raised consciousness I experienced at Wake Up Weekend.
14 Loyd-Paige would often describe herself this way because she had not forgone the use of other animals in clothing, personal care products, and other uses as yet.
15 Our Kitchen Table is a grassroots, non-profit organization in Grand Rapids, Michigan, that "seeks to promote social justice and serve as a vehicle that empowers our neighbors so that they can improve their health and environment, and the health and environment of their children, through information, community organizing and advocacy." For more information, visit https://oktjustice.org/
16 "Effects of Diet and Physical Activity on Risks for Certain Cancers," American Cancer Society, retrieved from: https://www.cancer.org/healthy/eat-healthy-get-active/acs-guidelines-nutrition-physical-activity-cancer-prevention/diet-and-activity.html, accessed August 6, 2023.
17 Ibid.
18 John Vidal, "Factory Farms of Disease: How Industrial Chicken Production Is Breeding the Next Pandemic," *The Guardian* (October 18, 2021), retrieved from: https://www.theguardian.com/environment/2021/oct/18/factory-farms-of-disease-how-industrial-chicken-production-is-breeding-the-next-pandemic, accessed June 14, 2023
19 According to a 2019 Gallup Poll with U.S. respondents, 70% cited "concern about your health" as a "major reason" for eating meat, "less, rarely or never," while only 10% cited it as "not a

reason." See McCarthy, Justin and Scott Dekoster, "Nearly One in Four in U.S. Have Cut Back on Eating Meat," *Gallup* (January 27, 2020), retrieved from: https://news.gallup.com/poll/282779/nearly-one-four-cut-back-eating-meat.aspx, accessed August 6, 2023.

20 *African American Vegan Starter Guide: Simple Ways to Begin a Plant-Based Lifestyle*, By Any Greens Necessary and Farm Sanctuary, introduction.

21 Ibid.

22 Ibid., 3.

23 Rose Russell, "More Black Americans Are Giving Up Meat and Dairy Foods," *The Elkhart Truth*, Section D1 (October 24, 2012).

24 Laura Reiley, "The Fastest-Growing Vegan Demographic Is African Americans. Wu-Tang Clan and Other Hip-Hop Acts Paved the Way," *The Washington Post* (January 24, 2020), retrieved from: https://www.washingtonpost.com/business/2020/01/24/fastest-growing-vegan-demographic-is-african-americans-wu-tang-clan-other-hip-hop-acts-paved-way/, accessed August 6, 2023

25 KRS-One, "Beef," Edutainment, Jive/RCA Records (July 17, 1990), CD.

26 Reiley, "The Fastest-Growing Vegan Demographic Is African Americans. Wu-Tang Clan and Other Hip-Hop Acts Paved the Way."

27 Jalen Rose, "Jalen Rose Cooks Up Conversation with Vegan Chef Tabitha Brown," *New York Post* (January 12, 2023), retrieved from: https://nypost.com/2023/01/12/jalen-rose-talks-with-vegan-chef-tabitha-brown/, accessed August 6, 2023

28 David Templeton, "As Research on Health Benefits Grows, So Does Vegan Following," *The Elkhart Truth* (July 9, 2014). Section D1.

29 Jennifer Jensen Wallach, "How to Eat to Live: Black Nationalism and the Post-1964 Culinary Turn," *Study the South*, University of Mississippi (July 2, 2014), retrieved from: https://southernstudies.olemiss.edu/study-the-south/how-to-eat-to-live/, accessed June 21, 2023

30 Ibid.

31 Ibid.

32 Ibid.

33 Ibid.

34 Ibid.

35 Ibid.

36 IYa Raet, "Ma'at Diet," in *Sistah Vegan: Black Female Vegans Speak on Food, Identity, Health and Society*, ed. A. Breeze Harper (Lantern Books, 2010), 139.

37 Ibid. For more on the Egyptian goddess, read "Ma'at," Rosicrucan Egyptian Museum, retrieved from: https://egyptianmuseum.org/deities-Maat, accessed July 16, 2023.

38 Raet, "Ma'at Diet," 141.

39 Melissa Danielle, "Nutrition Liberation: Plant-Based Diets as a Tool, for Healing, Resistance and Self-Reliance," in *Sistah Vegan: Black Female Vegans Speak on Food, Identity, Health and Society*, ed. A. Breeze Harper (Lantern Books, 2010), 47.

40 Ibid., 52.

41 Ma'at Sincere Earth, "Black-a-Tarian," in *Sistah Vegan: Black Female Vegans Speak on Food, Identity, Health and Society*, ed. A. Breeze Harper (Lantern Books, 2010), 70.

42 Ibid., 71.

43 Ibid., 69.

44 I have written extensively about the problem of this strategy in my essay, "Beyond Compare: Intersectionality and Interspeciesism for Co-Liberation with Other Animals," in *The Routledge Handbook for Animal Ethics,* ed. Bob Fischer (Routledge: New York, 2020).

45 Claire Jean Kim, *Dangerous Crossings: Race, Species and Nature in a Multicultural Age* (New York: Cambridge University Press, 2015), 275.

46 Ibid., 276–277.

47 Allie Mitchell, "Solaire Denaud on Black Veganism, the Rastafari Movement, and the Ital Diet," *VegOut* (February 23, 2023), retrieved from: https://vegoutmag.com/interviews/solaire-denaud/, accessed August 4, 2023

48 Ibid.

49 "An Introduction to Jamaican Ital Food," retrieved from: https://theculturetrip.com/caribbean/jamaica/articles/an-introduction-to-jamaican-ital-food, accessed August 6, 2023

50 Dick Gregory, *Callous on my Soul: A Memoir*, with Sheila P. Moss (New York: Kensington Publishing Corp., 2000): 111.
51 Ibid., 112.
52 Ibid., 13.
53 Syl Ko, "Addressing Racism Requires Addressing the Situation of Animals," in *Aphro-Ism: Essays on Pop Culture, Feminism, and Black Veganism From Two Sisters* (New York: Lantern Publishing & Media, 2017), ebook.
54 Ibid.
55 Ibid.
56 Ibid.
57 Venita Januarie, "'Veganism' through the Lens of Decolonization," *Africa Matters Media* (September 8, 2020), accessed August 6, 2023, retrieved from: https://www.africamattersinitiative.com/post/veganism-through-the-lens-of-decolonization
58 Ibid. See also Kat Smith, "Africa's Been at This Vegetarian Thing Longer Than Most of the World," *LiveKindly,* retrieved from: https://www.livekindly.com/africa-vegetarian-diet-history/, accessed August 6, 2023
59 Omnia Said, "Black Americans Are Leading a Vegan Movement," *Saveur* (October 13, 2022), retrieved from: https://www.saveur.com/culture/vegan-restaurants-harlem/, accessed August 6, 2023
60 As Carter notes, a Pew Research Center study has demonstrated that 79% of African Americans still identify as Christians. I wholeheartedly agree with him that it is necessary that vegan and animal advocates either have the capacity to speak about these issues from that perspective—or at the very least be in conversation with those who are grounded in the faith precisely to be able to speak to a significant portion of Black communities.
61 Christopher Carter, *The Spirit of Soul Food: Race, Faith, & Food Justice* (University of Illinois Press, 2021), 3.
62 Ibid., 3.
63 Venita Januarie, "'Veganism' through the Lens of Decolonization," retrieved from: https://www.africamattersinitiative.com/post/veganism-through-the-lens-of-decolonization
64 Reiley, "The Fastest-Growing Vegan Demographic Is African Americans."
65 Allie Mitchell, "Solaire Denaud on Black Veganism, the Rastafari Movement, and the Ital Diet," *VegOut* (February 27, 2023), retrieved from: https://vegoutmag.com/interviews/solaire-denaud/, accessed June 28, 2023.

Bibliography

Alexis, Nekeisha Alayna. "Beyond Compare: Intersectionality and Interspeciesism for Co-liberation With Other Animals." In *The Routledge Handbook for Animal Ethics,* edited by Bob Fischer. New York: Routledge, 2020.

American Cancer Society. "Effects of Diet and Physical Activity on Risks for Certain Cancers." *American Cancer Society* (n.d.). Retrieved from: https://www.cancer.org/healthy/eat-healthy-get-active/acs-guidelines-nutrition-physical-activity-cancer-prevention/diet-and-activity.html

Any Greens Necessary and Farm Sanctuary. *African American Vegan Starter Guide: Simple Ways to Begin a Plant-Based Lifestyle.* 2021. Retrieved from: https://meatout.org/wp-content/uploads/2021/03/AA_VeganStarterGuide.pdf.

Carter, Christopher. *The Spirit of Soul Food: Race, Faith, & Food Justice*. University of Illinois Press, 2021.

Crenshaw, Kimberlé. "Kimberlé Crenshaw on Intersectionality, More Than Two Decades Later." Columbia Law School, June 8, 2017. Retrieved from: https://www.law.columbia.edu/news/archive/kimberle-crenshaw-intersectionality-more-two-decades-later

Cross, Leslie J. "Veganism Defined." *The Vegan News*, republished by Gentle World on (August 2013). Retrieved from: https://gentleworld.org/veganism-defined-written-by-leslie-cross-1951/

Danielle, Melissa. "Nutrition Liberation: Plant-based Diets as a Tool, for Healing, Resistance and Self-Reliance." In *Sistah Vegan: Black Female Vegans Speak on Food, Identity, Health and Society*, edited by A. Breeze Harper. Lantern Books, 2010.

Gregory, Dick. *Callous on my Soul: A Memoir* with Sheila P. Moss. New York: Kensington Publishing Corp., 2000.

Ipsos. "Poll Conducted for The Vegan Society. Incidence of Vegans Research." Retrieved from: https://www.ipsos.com/sites/default/files/migrations/en-uk/files/Assets/Docs/Polls/vegan-society-poll-2016-topline.pdf

Januarie, Venita. " 'Veganism' through the Lens of Decolonization." *Africa Matters Media*, September 8, 2020. Retrieved from: https://www.africamattersinitiative.com/post/veganism-through-the-lens-of-decolonization

Kim, Claire Jean. *Dangerous Crossings: Race, Species and Nature in a Multicultural Age*. New York: Cambridge University Press, 2015.

Ko, Syl. "Addressing Racism Requires Addressing the Situation of Animals." In *Aphro-Ism: Essays on Pop Culture, Feminism, and Black Veganism From Two Sisters*. New York: Lantern Publishing & Media, 2017

KRS One. "Beef." *Edutainment, Jive/RCA Records*, July 17, 1990, CD.

Ma'at Sincere Earth. "Black-a-tarian." In *Sistah Vegan: Black Female Vegans Speak on Food, Identity, Health and Society*, edited by A. Breeze Harper. Lantern Books, 2010.

McCarthy, Justin and Dekoster, Scott. "Nearly One in Four in U.S. Have Cut Back on Eating Meat." *Gallup*, January 27, 2020. Retrieved from: https://news.gallup.com/poll/282779/nearly-one-four-cut-back-eating-meat.aspx

Mitchell, Allie. "Solaire Denaud on Black Veganism, the Rastafari Movement, and the Ital Diet." *VegOut*, February 23, 2023. Retrieved from: https://vegoutmag.com/interviews/solaire-denaud/

No Author. "An Introduction to Jamaican Ital Food." Retrieved from: https://theculturetrip.com/caribbean/jamaica/articles/an-introduction-to-jamaican-ital-food

Raet, IYa. "Ma'at Diet." In *Sistah Vegan: Black Female Vegans Speak on Food, Identity, Health and Society*, edited by A. Breeze Harper. Lantern Books, 2010.

Reiley, Laura. "The Fastest-growing Vegan Demographic Is African Americans. Wu-Tang Clan and Other Hip-hop Acts Paved the Way." *The Washington Post*, January 24, 2020. Retrieved from: https://www.washingtonpost.com/business/2020/01/24/fastest-growing-vegan-demographic-is-african-americans-wu-tang-clan-other-hip-hop-acts-paved-way/

Rose, Jalen. "Jalen Rose Cooks Up Conversation with Vegan Chef Tabitha Brown." *New York Post*, January 12, 2023. Retrieved from: https://nypost.com/2023/01/12/jalen-rose-talks-with-vegan-chef-tabitha-brown/

Russell, Rose. "More Black Americans Are Giving Up Meat and Dairy Foods." *The Elkhart Truth*, October 24, 2012, Section D1.

Said, Omnia. "Black Americans are Leading a Vegan Movement." *Saveur*, October 13, 2022. Retrieved from: https://www.saveur.com/culture/vegan-restaurants-harlem/

Smith, Kat. "Africa's Been at This Vegetarian Thing Longer Than Most of the World." *LiveKindly*. Retrieved from: https://www.livekindly.com/africa-vegetarian-diet-history/

Templeton, David. "As Research on Health Benefits Grows, So Does Vegan Following." *The Elkhart Truth*, July 9, 2014. Section D1.

Wallach, Jennifer Jensen. "How to Eat to Live: Black Nationalism and the Post-1964 Culinary Turn." *Study the South*, University of Mississippi, July 2, 2014. Retrieved from: https://southernstudies.olemiss.edu/study-the-south/how-to-eat-to-live/

Watson, Donald J. *The Vegan News,* November 1944. Republished as "Excerpts from the Vegan News" by Gentle World, September 28, 2020. Retrieved from: https://gentleworld.org/the-full-journey-excerpts-from-the-vegan-news-november-1944/

The Vegan Society. "The Vegan Society, About Us." Retrieved from: https://www.vegansociety.com/about-us/history

Vidal, John. "Factory Farms of Disease: How Industrial Chicken Production Is Breeding the Next Pandemic." *The Guardian*, October 18, 2021. Retrieved from: https://www.theguardian.com/environment/2021/oct/18/factory-farms-of-disease-how-industrial-chicken-production-is-breeding-the-next-pandemic

8
THE SKINS I'M IN

Carrie Lou Hamilton

I stand on tiptoes, raise up my right arm and snatch the left boot from the top shelf.

Once a boot I walked many a mile in, it now gathers dust aside its pair on a shelf built for books. I am wary of wearing leather these days. But I'm also reluctant to rid myself of these boots, so soaked with identity and memory.

Now the boot's been sitting on the bedside table for over a month while I try to coax it to speak. I wonder what it might tell me about its history, and mine. My partner wonders aloud why this lone cowboy boot, crusted with decades of dirt gathered on the roads of least two continents, is now propped next to my pillow.

I move the boot to my desk, beside my computer, hoping the closeness to my tapping fingers will help it transmit the tales I yearn to hear.

The boot has become an *aide mémoire*.

***

The cowboy boot, like the body, has an anatomy: collar, scallop, shaft (with stitching), lining, pull strap, piping, counter heel, spur ridge, heel, heel cap, vamp, welp, insole, outsole, toe bug, toe box. The sole stands solidly on the desktop while the shaft leans rightwards, like a miniature Tower of Pisa.

I pick up the boot and cradle the faded red leather in my right hand. There is nothing soft about this boot. The fabric is thick and stiff. As I look more closely, I see the tears and ridges in the skin, the scuffs, the small hole in the toe where the leather has torn away altogether, baring the cloth underneath. A hole in the sole, a reheeled heel, unhealed tiny tears.

These boots are well worn, worn in.

I can see from the fineness of the stitching, the thickness of the fabric, that this is a quality boot. Well made.

But it was the red that drew me in, the red that makes me hold on. Not just to the boot, but to a memory of myself: an image of the femme at 25.

This boot and its pair once held a promise of transformation. Of creating a whole look from toe to head. A style.

DOI: 10.4324/9781003273400-11

There was sex in this style. I fancied that the red cowboy boots, which end halfway up my calf, would show off my legs in a way that walking shoes would not. I walk differently in leather boots than I do, say, in running shoes or faux-leather heels.

I put the boot on my left foot and instantly that feeling returns. I feel stronger, more powerful, youthful, sexier—not just because I got these boots 30 years ago, but because red leather cowboy boots—the whole kit and caboodle—are *young* footwear.

I still think of the cowboy boot as erotic, and though its history and representation are soaked with machismo and whiteness, with John Waynes and Westerns, this boot is undeniably femme footwear. I totter and falter in heels, feel too girly and straight—as in edged as well as oriented—in dainty or pointy flats. I've tried all this foot attire over the years and never quite feel at home in any of them. I have more than one pair of "sensible" shoes for walking, but let's face it, sense and sensuality are awkward bedfellows.

I stand in front of the mirror and ask: do these red boots make me *look* queer? Does this looking queer femme enough also show me off as a bit of a rebel? What is this image of myself I seek?

***

I go to the bookshelf in the living room and take down the tattered paperback copy of my favorite lesbian history book, *Boots of Leather, Slippers of Gold*, by Elizabeth Lapovsky Kennedy and Madeline D. Davis.[1]

The book recounts the romance and dramas of working-class lesbian bar culture in Buffalo, New York, before the Stonewall riots of 1969 and the rise of second-wave feminism and gay liberation. The title evokes the glamour of those early butch-femme communities, with their strict gendered codes of clothing and behavior—codes that operated both to protect lesbians from a hostile world and to signal to each other their outlaw desires. The cover of another iconic butch-femme book on my shelf, Joan Nestle's *The Persistent Desire* from 1992, showcases a seductive, suggestive photograph of two sets of legs stretched on the grass and what looks like a picnic blanket.[2] One pair wears silk stockings that end just above the knee, revealing white skin and the bottom of a lace slip. The feet are dressed not in slippers of gold but in pink patent leather shoes with strap and heel. The other legs don grey trousers and brown leather cowboy boots. In the working-class bars of the 50s and 60s cowboy boots, along with other classic men's footwear, like the penny loafer, were part of the weekend butch look, worn for socializing and cruising.[3]

A third book in my library, Chloë Brushwood Rose and Anna Camilleri's *Brazen Femme: Queering Femininity*, features a photo of a woman, who may be white, wearing a tight black vest that may or may not be made of leather, revealing ample cleavage.[4] The provenance of the vest's fabric matters less perhaps than the aesthetic: the buxom femme is dressed entirely in black, with short nails on strong hands adorned with multiple rings. One hand clutches a knife, sticking its tip onto a chair between the femme's naked thighs. This clothing and these accessories, like the model's visible tattoo and nose ring, distance this femme in time and appearance from the earlier lesbian femininity of the gold slippers and pink heels. This is a confidently tough, queer femme with an SM aesthetic. A raunchy sexuality that challenges the female passivity associated with traditional white bourgeois femininity.

The style of *Brazen Femme* represents both a reclaiming and a transgression of the butch-femme style of the pre-Stonewall era, when gender roles and looks were more rigid, when femmes rarely wore trousers on a night out and boots were strictly for butches.[5] Radical lesbian feminists of the seventies rejected butch-femme as a copycat of the binary patriarchal power dynamics of heterosexuality. But as the stories in Nestle's volume attest, butch-femme aesthetics and relationships, while perhaps out of style for a time, never entirely disappeared. By the time I came of queer age in the 1980s, many young urban dykes in North America were reclaiming butch-femme as a subversive and assertively sexual identity. We rejected radical feminist arguments that butch-femme relationships were oppressive for women. But we had also inherited the more fluid dress codes and gender identities of women's liberation. Butch-femme became part of a playful expression of genderqueer identity and sexual freedom.[6] In the 80s and 90s we femmes often wore the pants—and not just metaphorically. Jeans and t-shirts and cowboy boots were no longer strictly butch attire. We femmed up the classic butch uniform with splashes of color, low cuts, makeup, scarves and big earrings, long hair, replacing the manly swagger with a seductive swing of the hips.

The late twentieth century was also the era of the leather dyke, another gender-bending lesbian figure. In North America, queer leather culture grew out of the gay biker scene of the fifties, epitomized in a look of "rugged masculinity"—"black leather jacket and riding boots, clinging T-shirt and rugged Levis jeans," part of a subculture that rejected post-war bourgeois authority and conformism and evoked aspects of the Western outlaw tradition.[7] These traditions—biker and Western—signified not only a working-class toughness but also a rebel whiteness. By the nineties, leather dykes—on the streets, on the BDSM scene, on motorcycles riding with "Dykes on Bikes" at Pride marches—were integrating leather paraphernalia into their clothing, mixing fetish wear and punk haircuts with accessories borrowed from working-class urban gaywear, including classic biker and cowboy boots.[8]

My collection of butch-femme books, like the red cowboy boots, came into my life in the epoch of the leather dyke, the decade when I came out. The red boots are a leftover from the leather I gave away when I started doing veganism two decades later. Today I wonder at those boots, at the decision to keep them, at their redness, at the color of their skin and the color of my own skin, of the skins I'm in. At how those meanings change with time, how I have aged inside and alongside the boots. How my skin has transformed, how my ideas about skin, about leather, about whiteness, have changed over the years, how a feminist thinking through the skin might be expanded by considering too the skins of other animals.

The longer I inhabit a skin that contains a body that thrives on plant foods, the more my understanding of ethical veganism is sharpened. If, as Stella North writes, "we are not only beings who are skinned, we are beings who are clothed. . . . When we think, we are thinking through clothes," what does it mean to think not only through our changing human skins but also through the animal skins we have worn?[9]

***

I've worn these boots and I too have worn. There's a looseness around the waist, a fullness to the hips, a heaviness to the legs that first slipped on those treasured red boots 30 years ago. My skin has taken on new marks. Lines as well as words move up and down my arms and legs, adorning what was once naked. The color too has changed, no doubt, though in what ways I strain to make out, whiteness being a tone too often invisible to its wearer, even in its differing shades.

My once shod foot has also aged. It's not quite leathery, but weathered, yes.

It's summer and my feet are filthy. The heels and balls and underpart of my toes are darkened with dirt. They're rough to the touch. I run my right hand along the sole of my left foot, the one that lets the side down. A bulb of white hard skin bulges on the big toe, the other toes sharpen into little ridges at the edges.

The tops of my feet are less soiled than the soles. There is a faint stripe of tan, bleached clusters of hair on each toe. As I flex my foot the big bones and blue veins make mountain ranges and rivers on the off-white skin. When I point my toes the skin evens out—taut as a pelt stretched across a small drum.

Callouses collect around the heel, the side of the ball at the root of the big toe. The only softness on the sole is the ticklish spot on the arch, the part that does not pound the pavement, that only hits the ground when I walk foot-naked through the grass or over sand. This summer the grass is yellow, spikey and prickly. This season my soles are like the earth's skin: dry, rough, thirsty. The ground begs for water; my soles yearn for soft squishy surfaces.

My mother, who gifted me the red boots all those years ago, is worried for my feet. This should come as no surprise: she's just had foot surgery to remove bunions caused by the kitten-heeled shoes she and other teenage girls wore in Canada in the 1950s. After the operation she couldn't walk properly for months.

But it's not the shape of my footwear that vexes my mother. It's the fabric. Plant food she takes in her stride, but boots must be made of hide. "You *must* wear leather on your feet. Your boots have to *breathe*."

There is some truth to this idea that animal skin breathes. Even we humans take in about 1–2% of our oxygen through our skin. Smaller animals, in particular "small invertebrates such as sponges, corals, jellyfish and worms," breathe mostly through their skin. This is especially advantageous when they're under water, though the only creature to rely 100% on skin breathing for their oxygen is the lungless salamander.[10]

The point about breathing is that it keeps our bodies alive. Skin breathes when it covers a living animal. The skin of a dead animal does not *breathe*. It is, however, *breathable*. According to an online dictionary of leather:

> Breathability is one of the special properties of leather, which no other substitute material exudes. The ability to expel moisture in the form of vapour is one of the main characteristics of leather and what makes it unique. This can be essential, especially for shoes. In warm and waterproof hiking boots made of leather, you will never sweat as much as in rubber boots.[11]

This linguistic trick, the shift from verb (to breathe) to adjective (breathable) to noun (breathability), traces the move from a living being to a dead object.

If leather breathes only in the imagination of the manufacturer and the consumer, it drinks more literally. Leather is a thirsty fabric. A finished piece of tanned hide has consumed up to 100 times its weight in water.[12] If in recent years leather has become somewhat sartorially suspect, it is not in the main because wearing the skin of a dead animal on one's body has gone out of fashion (even fur goes in as well as out). It is because, in spite of all this talk of breathing, all this talk of leather as a *natural* fabric, leather is killing the planet.

The earth's skin is not aging so well. Like my skin, it wears the cracks of dryness.

***

By holding onto the red boots, storing them on a shelf as if in a makeshift museum, am I endeavoring to preserve my own youth? Leather is often promoted as a fabric to hold onto, a fabric that *ages well*. If I hold on tight to my leather, wear it against my dried, cracking skin, will some of its magic agelessness rub off on me?

This is all fanciful thinking. But is it so much more fanciful than the fancy footwork it takes to conjure away the animals whose skin I wear when I wear the boots?

The agelessness of leather is associated not with the age of the animals whose skin it is made of, but with the age of the product. Chances are that the animals whose flesh bones organs were protected by their epidermis until it was stripped along the long road to the fabrication of my stiff red boots were young adult cows. Most leather used to make clothing for humans comes from cows raised and slaughtered for meat. Although the natural lifespan of a cow or bull is 15 to 20 years, those raised for meat are usually killed at age 2 or 3.[13]

The skin of a dead cow ages well not because it retains the characteristics of a once-living creature, but because it has been treated to death. This treatment not only involves countless chemical compounds used to cure the skin but the removal of all the extra stuff the skin carried around while it was living: hair, bits of feces, even the epidermis. That thin layer that protects the body from harm in life is, in death, extraneous. Along with the hair and the shit and the outer skin, what gets stripped off in the process of leather making is the truth: the reality covered up by this boring business about leather as a by-product of the meat industry, as if those who make a living off the skins of other animals are doing the rest of us a favor, as if all this work to make coats and boots and shoes we want but do not need is helping to save the planet by ridding it of the putrefied outtards of the hundreds of millions of bovines slaughtered for meat every year.[14]

So we come back to the skin: the skin of the cow, the skin of the earth, the skins of the workers who treat the skin that eventually becomes my second skin. My aging skin.

***

I put the left boot on my foot. Even faded, the red skin contrasts sharply with the pale skin of my leg, untanned even after a summer of searing heat, thanks to the factor 50 I wear, quite literally, as a second skin. The stiff leather doesn't exactly mold around my calf; it sits snug and out of it grow red flames encircling a blue-black dagger inked into my skin.

It is these contrasts of dark colors on white skin that draw my eye and take my mind to the question of how whiteness wears the skins of others. For if it is indeed impossible, as Sara Ahmed and Jackie Stacey write, to inhabit the skins of other humans, human beings *do* often wear the skins of other animals. Now I wonder how this wearing—as fashion statement, as a mark of subcultural belonging, as a plot to ward off ageing by refusing to "dress my age"—is linked to whiteness, to femininity, to queerness. How the inability to live in the skin of other people relates to the wearing of the skins of other animals.

Whiteness, like skin, slides between materiality and metaphor, not always smoothly. When I examine my aging skin, I see a certain whiteness, which both is and is not, in the sense that white skin is not literally white. When I consider how the red cowboy boots make me look, I see not just the contrast with the white skin, a contrast I could compare with tattoos, marks on the skin that highlight its whiteness. I think too about how the aesthetic of the cowboy, of kick-ass femininity, is a specific kind of whiteness, one that rejects "traditional" white bourgeois femininity and its association of words like *pretty*, *clean*, *wholesome*. What does it mean for white women, especially white middle-class women, to disavow normative white femininity through the skins we wear? What is this attempt to escape convention, which at the same time cannot be an escape from privilege?

I think of Cathryn Bailey's reflections on feminist vegetarianism and the reproduction of racial identity in the United States. Bailey writes of "special whiteness"—that aspiration of middle-class white people who wish to distance themselves from the bland, bourgeois whiteness associated with mainstream American culture.[15] Looking back on her own experience of moving from a working-class white upbringing to a university city, Bailey considers the practice of vegetarianism among her white middle-class colleagues as one example of this "special whiteness." If vegetarianism, just as easily as a liking for certain "ethnic" cuisines—like Thai or Mexican—can become a signifier of "special whiteness," how do we retain the ethics of vegetarianism or veganism as something that is *more than* a mere marker of (privileged) identity or status?

To avoid commonsense arguments that Western vegetarianism is *nothing more than* a moralistic and elitist symbol of white bourgeois goodness, Bailey insists we take seriously the ways that food and eating are "deeply entwined with the production and reproduction of identity."[16] If this means attending not only to the realities of animal exploitation but also to the lived realities of people's lives, and how these might make it more, or less, difficult for people to adopt a plant-based diet, it must also mean considering the ways consuming animal products is itself a marker of identity.[17]

If, as Bailey argues, whiteness can be produced and reproduced through the consumption of different kinds of food, it is surely also made and remade through the clothing we wear.[18] Yet although fashion is as much a part of our identities as foodways, writing on animal rights and veganism continues to lean heavily towards the food end of the farmed animal exploitation chain. In contrast, scholarship on the politics of fashion is all but mute on leather. A chapter in a recent edited book, *The Dangers of Fashion*, makes brief mention of the chemical hazards of this "fabric."[19] But neither this piece nor the rest of the volume has anything to say about the dangers of leather for the animals whose skins are stolen to make it.[20]

Yet considering the skins we do or do not wear as much as those we refuse to eat takes considerations of veganism and its relationship to human identities and relationships in new directions. As Susan Kaiser and Denise Green write,

> Fashion is never finished . . . it crosses all kinds of boundaries. It is ongoing and changes with each person's visual and material interpretations of who they are becoming and how this becoming connects with others' experiences within time and place.[21]

What if we consider animals part of these "others"?

***

While we often hear talk of the problem of white people appropriating fashion and other cultural forms from nonwestern cultures, I would like to think about what it means for white people like me to appropriate, literally to take, the skins of animals as fashion. Thinking about the wearing of the skins of other animals on white bodies is an expansion of the idea of appropriation, its literalization, which must be considered as a part of, not apart from, the myriad ways in which whiteness is constructed through acts of appropriation:

of land, of labor, of bodies. In the case of leather, we would also have to consider not only the appropriation of the labor of working animals but also the human labor that goes into making the treatment of animal skins and their conversion into wearable fabrics, as well as the stripping of the land, the polluting of the water, that occurs in that process.

To say that I have had to learn how to see lives and the violences behind dead animal skins as I've had to learn how to see the lives and violences behind the histories of whiteness is not to say that I have learned to see these in the same way, that there is any kind of symmetry. The fact that the exploitation of animals and racialized human beings is densely entangled, that racialization often takes the form of animalization, does not mean that there are direct comparisons. As Billy-Ray Belcourt stresses from the position of that land that is today called Canada:

> the animal and the Indigenous are not commensurable colonial subjects insofar as their experiences of colonization are different—we must therefore continue to account for these differences rather than obfuscate them with a one-size-fits-all solution to a problem—speciesism—looped through and coanimated by colonial violence of all sorts.[22]

If certain colors and fashions highlight the whiteness of my off-white skin, if leather and other fabrics help my body perform a certain kind of queer white femininity, the loss of any connection, the total erasure of ethical accountability, between the wearer and the once-living animals whose skin she wears is part of an economy—contemporary consumer capitalism—indelibly marked by whiteness.

I consider my place in all this—in the queerness, the whiteness, the global chain of fashion—and I wonder what kind of whiteness I sought to inhabit by wearing dead animals as a second skin. I recognize and honor the histories of queer femininity that drew me to and into the red leather boots, even if I have no desire to reinhabit that youthful skin. I would not again be the person who sports the skins of others without batting an eyelash, who relies on the skins of others to make her feel sexy and young.

* * *

If as humans our skin becomes more delicate, thinner, more prone to tearing as we age, metaphors of skin and aging move in the opposite direction. It is said that age thickens the skin, makes us less vulnerable to psychic wounding, to the scrapes and scratches of ugly feelings such as guilt and shame. Elspeth Probyn recalls that upon arriving as a white woman from one colonial settler society (Canada) to another (Australia), she encountered an "overwhelming jumble of impressions, the gaps of knowledge, the constant pressure of ignorance, the incommensurability of ways of knowing."[23] At first, she concludes that "to think as a white in Australia it helps to have thick skin. Thick, hard skin as protection against implication in the living memory of the country, the brutalities of colonisation still so vividly worn by some, hidden on others."[24] But in the end this thickening of the human hide will not do. In order "to live ethically, which is to say fully situated," Probyn—like other white people who migrate over the surface of the ocean that stretches between colonized lands—must reckon with skins and their histories, her own and those of others.[25] In response to the temptations to develop a thick skin to protect against history, Probyn proposes another metaphor: "Skin is . . . expansive; it stretches and then retracts, always showing the traces of its own processes."[26]

We are not stuck, then, with the tired metaphor of thick skin. As we age, we can grow into our skins in different ways, ones that wear the marks of our pasts and make room for different kinds of relationships with the others around us.

Probyn's reflections take me back once again to my bookshelf, to the writings of another white queer woman writer, Minnie Bruce Pratt, whose essay "Identity: Blood Skin Heart" I bought and read for the first time around the time the red boots came into my life. Reflecting on the long process of learning to unlearn racism as part of her coming into feminism and out as a lesbian, Pratt recalls how her body changes when she is trapped by the fear of saying the wrong things: "I feel my racing heart, breath, the tightening of my skin around me, literally defenses to protect my narrow circle."[27] And then she considers other ways of being in her white skin in such situations, moving beyond that sense of constriction: "I will try to be at the edge of my fear and outside, at the edge of my skin, listening, asking what knew thing I will hear, will I see, will I let myself feel beyond the fear."[28]

I will try to be . . . at the edge of my skin.

***

Writing on skin necessarily goes in new directions, stretches out, shapes our bodies in new ways when it takes into account and is forced to account for the literal ways in which human beings inhabit the skins of others. Stretching theories of the skin beyond the human necessitates new thinking on how the different meanings and histories of human skins—of age, gender, race, among others—shape our relationship to the animals we wear.

Thinking about the different ways queer white femininity inhabits a fabric such as leather is not an invitation to shift the focus away from human skins to those of animals. I don't have any answers to how I might embody whiteness ethically. But I have a good idea of how *not* to embody it, how not to force it into shape by squeezing it into the dead, breathless skins of other creatures.

My resistance to giving up the red cowboy boot might ultimately be about more than a desire to hold on tight to the memories of a more youthful white queer femme self. It is also a recognition that I can't give up that history. And that in the end it might be better to try to make sense of it than to disavow it.

Let the writing breathe.

***

We may not shed skins when we get older after all. Perhaps what the red cowboy boots give me is the opportunity to turn nostalgia on its head: to lock eyes with the past, to reckon with what my desire for special whiteness means, to come to terms with my desire for animal skins, to own those desires—not so I can thicken my skin against the difficult feelings they raise but to think of more ethical ways for being in my skin.

Notes

1 Elizabeth Lapovsky Kennedy and Madeline D. Davis, *Boots of Leather, Slippers of Gold: The History of a Lesbian Community* (New York: Routledge, 1994).
2 Joan Nestle, *The Persistent Desire: A Femme-Butch Reader* (New York: Alyson Books, 1992).
3 Adam Geczy and Vicki Karaminas, *Queer Style* (London: Bloomsbury Academic, 2013), 32.

4 Chloë Brushwood Rose and Anna Camilleri, eds., *Brazen Femme: Queering Femininity* (Vancouver: Arsenal Pulp Press, 2003).
5 Geczy and Karaminas, *Queer Style*, 32.
6 Ibid., 37.
7 Ibid., 108–109.
8 Valerie Steele, *Fetish, Fashion, Sex and Power* (Oxford: Oxford University Press, 1996), 105; cited in Geczy and Karaminas, *Queer Style*, 110.
9 Stella North, "The Surfacing of the Self: The Clothing-Ego," in *Skin, Culture and Psychoanalysis*, eds Sheila A. Cavanagh, Angela Failler, and Rachel Alpha Johnston Hurst (New York: Palgrave Macmillan, 2013), 69.
10 "Animal Respiration: Skin Breathing," Animal Fun Facts, retrieved from: https://www.animalfunfacts.net/knowledge/senses-and-abilities/333-animal-respiration-skin-breathing.html, accessed November 25, 2022
11 Leather Dictionary, retrieved from: https://www.leather-dictionary.com/index.php/Leather, accessed November 25, 2022.
12 Anthony D. Covington and William R. Wise, *Tanning Chemistry: The Science of Leather*, 2nd ed. (Croydon: Royal Society of Chemistry, 2020), ebook, 1021.0.
13 Jennifer Mishler, "Most Cows Don't Die of Old Age," *Sentient Media* (June 4, 2021), retrieved from: https://sentientmedia.org/how-long-do-cows-live/.
14 Covington and Wise, *Tanning Chemistry*, 1021.0; Grace Hussain, "How Many Animals Are Killed for Food Every Day?" *Sentient Media* (August 31, 2022), retrieved from: https://sentientmedia.org/how-many-animals-are-killed-for-food-every-day/.
15 Cathryn Bailey, "We Are What We Eat: Feminist Vegetarianism and the Reproduction of Racial Identity," *Hypatia* 22, no. 2 (2007): 47–49.
16 Ibid., 53.
17 Ibid., 54.
18 Ibid., 48.
19 Huantian Cao, "Fibers and Materials: What Is Fashion Made of?" in *The Dangers of Fashion: Towards Ethical and Sustainable Solutions*, eds. Sara B. Marcketti and Elena E. Karpova (London: Bloomsbury Visual Arts, 2020), Ebook, 148.6.
20 Sara B. Marcketti and Elena E. Karpova, eds., *The Dangers of Fashion: Towards Ethical and Sustainable Solutions* (London: Bloomsbury Visual Arts, 2020).
21 Susan B. Kaiser and Denise N. Green, *Fashion and Culture Studies*, 2nd ed. (New York: Bloomsbury Visual Arts, 2021), ebook, 35.9.
22 Billy-Ray Belcourt, "An Indigenous Critique of Critical Animal Studies," in *Colonialism and Animality: Anti-Colonial Perspectives in Critical Animal Studies*, eds. Kelly Struthers Montford and Chloë Taylor (New York: Routledge, 2020), ebook, 113.2.
23 Elspeth Probyn, "Eating Skin," in *Thinking Through the Skin*, eds. Sara Ahmed and Jackie Stacey (New York: Routledge, 2001), 87.
24 Ibid., 87.
25 Ibid., 88.
26 Ibid., 100.
27 Minnie Bruce Pratt, "Identity: Blood Skin Heart," in *Yours in Struggle: Three Feminist Perspectives on Anti-Semitism and Racism*, eds. Elly Bulkin, Minne Bruce Pratt, and Barbara Smith (Brooklyn: Long Haul Press, 1984), 18.
28 Ibid.

Bibliography

Ahmed, Sara and Stacey, Jackie. "Introduction: Dermographies." In *Thinking Through the Skin*, edited by Sara Ahmed and Jackie Stacey, 1–17. New York: Routledge, 2001.

Animal Fun Facts. "Animal Respiration: Skin Breathing." *Animal Fun Facts* (n.d.). Retrieved from: https://www.animalfunfacts.net/knowledge/senses-and-abilities/333-animal-respiration-skin-breathing.html.

Bailey, Cathryn. "We Are What We Eat: Feminist Vegetarianism and the Reproduction of Racial Identity." *Hypatia*, 22, no. 2 (2007): 39–59.
Belcourt, Billy-Ray. "An Indigenous Critique of Critical Animal Studies." In *Colonialism and Animality: Anti-Colonial Perspectives in Critical Animal Studies*, edited by Kelly Struthers Montford and Chloë Taylor, 107.0–134.0. New York: Routledge, 2020. Ebook.
Brushwood Rose, Chloë and Camilleri, Anna, eds. *Brazen Femme: Queering Femininity*. Vancouver: Arsenal Pulp Press, 2003.
Cao, Huantian. "Fibers and Materials: What is Fashion Made of?" In *The Dangers of Fashion: Towards Ethical and Sustainable Solutions*, edited by Sara B. Marcketti and Elena E. Karpova, 129.0–159.0. London: Bloomsbury Visual Arts, 2020. Ebook.
Covington, Anthony D., and Wise, William R. *Tanning Chemistry: The Science of Leather*. 2nd ed. Croydon: Royal Society of Chemistry, 2020. Ebook.
Geczy, Adam and Karaminas, Vicki. *Queer Style*. London: Bloomsbury Academic, 2013.
Hussain, Grace. "How Many Animals are Killed for Food Every Day?" *Sentient Media,* August 31, 2022. Retrieved from: https://sentientmedia.org/how-many-animals-are-killed-for-food-every-day/.
Kaiser, Susan B., and Green, Denise N. *Fashion and Culture Studies*. 2nd ed. New York: Bloomsbury Visual Arts, 2021. Ebook.
Kennedy, Elizabeth Lapovsky and Davis, Madeline D. *Boots of Leather, Slippers of Gold: The History of a Lesbian Community*. New York: Routledge, 1994.
Leather Dictionary. Retrieved from: https://www.leather-dictionary.com/index.php/Leather, accessed November 25, 2022.
Marcketti, Sara B., and Karpova, Elena E., eds. *The Dangers of Fashion: Towards Ethical and Sustainable Solutions*. London: Bloomsbury Visual Arts, 2020.
Mishler, Jennifer. "Most Cows Don't Die of Old Age." *Sentient Media*, June 4, 2021. Retrieved from: https://sentientmedia.org/how-long-do-cows-live/.
Nestle, Joan. *The Persistent Desire: A Femme-Butch Reader*. New York: Alyson Books, 1992.
North, Stella. "The Surfacing of the Self: The Clothing-Ego." In *Skin, Culture and Psychoanalysis*, edited by Sheila A Cavanagh, Angela Failler, and Rachel Alpha Johnston Hurst, 64–89. New York: Palgrave Macmillan, 2013.
Pratt, Minnie Bruce. "Identity: Blood Skin Heart." In *Yours in Struggle: Three Feminist Perspectives on Anti-semitism and Racism,* edited by Elly Bulkin, Minne Bruce Pratt, and Barbara Smith. Brooklyn: Long Haul Press, 1984.
Probyn, Elspeth. "Eating Skin." In *Thinking Through the Skin*, edited by Sara Ahmed and Jackie Stacey, 87–103. New York: Routledge, 2001.

9

VEGAN CAMP

An Interview with Emelia Quinn

Chloë Taylor [CT]: In this interview I will be talking to Dr. Emelia Quinn about a concept she has herself coined, vegan camp, which brings together her areas of expertise in vegan and critical animal studies, queer theory, and the study of culture. Dr. Quinn is a scholar of literature and has also taught in an art history department at the University of Birmingham in the United Kingdom after completing her PhD at the University of Oxford. Since 2021 Dr. Quinn has been an assistant professor of world literatures & environmental humanities at the University of Amsterdam. She is the author of *Reading Veganism: The Monstrous Vegan, 1818 to Present*, published in 2021 with Oxford University Press, and the co-editor of *The Edinburgh Companion to Vegan Literary Studies*, published by the University of Edinburgh in 2022, as well as many articles and book chapters. I am speaking to Dr. Quinn about her 2020 article published in PMLA—*Publications of the Modern Language Association of America*, "Notes on Vegan Camp."[1] Thank you, Emelia, for agreeing to do this interview.

Emelia Quinn [EQ]: Thanks so much, Chloë.

CT: You are both a queer theorist and a vegan studies scholar, and your work often explores the complex imbrications of queerness and veganism and of gender, sexuality, and food more generally. Can you say a bit about how you yourself came to be vegan and your own experiences as a vegan? What made you decide to make veganism not only a personal practice but also a major topic of your research, and how have you connected this to queer theory and studies?

EQ: I was brought up as a vegetarian, in a vegetarian household. We didn't necessarily talk about animal ethics in any explicit way as I was growing up, but it was the implicit background of my childhood. It was only when I was about 19 or 20 that I started thinking about veganism. I saw a Facebook post that contained a graphic photograph—I'm sure many people have seen something similar—of

DOI: 10.4324/9781003273400-12

baby chicks going down a conveyor belt directly into a grinder. I was horrified that this was a function of the egg industry and, even more so, that I'd gone this long in my life without having connected the dots to see how this meant that eating eggs didn't fit with my existing vegetarian commitments. I then followed the link provided with the picture to find out more about the horrors of the modern dairy industry and went vegan almost immediately.

At that point in my academic studies, I had become very interested in queer theory during my BA degree but hadn't thought at all about my personal veganism in relation to my studies. This changed during my MA, when I took module on critical theory that included a week on animal studies. It immediately spoke to me, in a way I'm sure many people can relate to: the experience of reading Carol J. Adams's work for the first time and being like, oh my God, this makes so much sense.

However, the more I dived into animal studies, particularly literary animal studies, the more I noticed, particularly at this time, which I guess was ten years ago now, that it seemed as if the field was really shying away from talking about veganism. That just didn't seem to make much sense to me, that there could be an academic investment in animals, an investment in care for animals, that stays silent about veganism.

I became very interested from that point onwards in the connections between veganism and queerness, as means of destabilising a heteronormative status quo, and, more specifically, in whether veganism, like queerness, could function as a theoretical approach to literature and culture. And so that's really where I began connecting my own personal veganism to my academic scholarship and thinking about what it might mean to read in a vegan way, and whether we can see veganism as a critical lens or mode of thinking.

As you just mentioned, my more recent work on vegan camp considers what kind of affinities there might be between veganism and queerness in relation to their aesthetic sensibilities. My work also considers what veganism can learn from the scholarship that I find so influential in queer studies—particularly questions of failure and utopianism that I think are really central to thinking more critically about what it means to be a vegan and that have been productively thought through in much queer scholarship.

CT: Can you say a bit about the literature on a camp sensibility or camp aesthetic and explain a bit more about how you have extended it with your concept of vegan camp?

EQ: So the most famous piece of work on camp is Susan Sontag's "Notes on Camp," which I obviously riff on in my title, "Notes on Vegan Camp." Sontag's piece was published in 1964 and sought to define a sensibility that she associates (and that I think is still often primarily associated with) white, male, gay culture. In basic terms, camp is a way of seeing the world that, as Sontag says, understands "Being-as-playing-a-Role". It functions as an embrace of artifice and exaggeration, a refusal to look beyond the surface. The surface is everything in camp. For Sontag, camp is therefore an inherently apolitical aesthetic. However, since her article, the supposed apoliticism of camp has been critically challenged by much work in queer theory. Camp is now more often embraced as both a survival strategy and form of social agency. It is also often envisaged as an act of utopian world building. Camp is embraced as a distinct aspect of queer culture and often works to expose the superficiality and cultural construction of gender norms.

In my article on vegan camp, I take these aspects of queer camp and extend them beyond questions of gender performance to the idea of the human itself. Human exceptionalism is understood through a vegan camp lens as something that we might also hold up to ridicule. This way of seeing the world functions as a kind of survival strategy, as I describe it in the article. Theorising a mode of vegan camp is to recognise that as vegans we don't always necessarily need to be confronted with trauma and suffering (seeing the dead animal behind the surface of meat, for example), and that a critical consideration of our joys and pleasures might also be politically productive. Vegans are often seen as undesiring, cantankerous killjoys, but vegan camp offers up a mode of pleasure that is about satirising the human as we currently understand it. Yesterday I was reading Nicole Seymour's little book on glitter[2] and there she elaborates on Amory Starr's concept of "tactical frivolity." This concept is engaged in similar ways to camp with a rethinking of what can constitute productive political acts.

CT: In your "Notes on Vegan Camp," you cite J. M. Coetzee's character, Elizabeth Costello, who famously describes questioning her sanity in a society where what she recognises as an atrocity of unprecedented scale is seen by virtually everyone around her as normal and unremarkable, where seemingly good people including her family members are participating in this atrocity on a day-to-day basis. This is a famous passage because it has resonated so much for many vegans, and you write of how ethical veganism risks "a relentless confrontation with horror and a sense of despair at the scale of human brutality. Ethical veganism often results in an inability to ignore the absent referent animal behind practices of animal exploitation." Of course, you don't want to reinforce animals being reduced to absent referents[3] all the time, and yet you are highlighting the fact that in order to cope in this world, it is necessary even for vegans to some of the time *not* see the oppression, violence, and death behind virtually everything and everyone around them. In this context you describe vegan camp as "a survival strategy," as you've mentioned. Later you describe vegan camp as a way of "*working through* horror and continuing to fight for change." Do you see vegan camp as a psychological defence mechanism or as a way of warding off activist fatigue?

EQ: Yes, I think I definitely see it as both as defence mechanism and survival strategy for activists. That is something I stress in the article. But I also think the concept is useful, more generally, as a means through which to think about the ways in which our ethical commitments don't always neatly align with our aesthetic pleasures.

In the article I describe how I came to write it: in late 2017 I was invited to speak at a conference about the nineteenth-century whaling artefacts on display at the Hull Maritime Museum in the UK. I was asked to give a vegan perspective on a single object from the museum's collection and felt that I was therefore expected to find some really hideous depiction of human violence and talk about how terrible it was. But the fact was that when I saw the "Jolly Sailor" scrimshaw that I describe in the article, my response wasn't horror. I instead found that the object was funny, that I found it pleasurable to look at. And so I started to ask the question: what do we do with such a disconnect between aesthetic and ethical responses? And, through this, how might we reveal a different mode of vegan agency and vegan aesthetic culture?

I think another thing possibly worth mentioning here is the shift in the ways the absent referent, that you just mentioned, functions in the contemporary world. Erica Fudge has written very convincingly on this, arguing that we have entered into a "new anthropocentrism," in which the absent reference structure makes slightly less sense. She notes, for example, that meat eaters are no longer hiding behind products that don't look anything like animals.[4] People are now instead embracing practices such as raising their own meat, killing their own animals. And so I think part of what vegan camp is also trying to do is rethink how we respond to violence against animals in this new climate. It is perhaps no longer enough to just re-awaken people to the reality that the meat they eat was once a living animal because most people aren't shocked or seriously affected by such a revelation. They already know. Vegan camp plays with these changed dynamics between surface and depth, of knowledge and ignorance.

CT: Kelly Struthers Montford also has an article called "The Present Referent" where she writes about advertisements for meat in Alberta, the province in Canada where my job is and where she was doing her PhD at the time, and how the animals are very much represented in these ads for beef.[5] It's cattle country in Alberta, so it isn't like people need to only see meat in sanitised packages dissociated from the living animal to eat it.

EQ: I was speaking to Carol [Adams] about something similar when she came to Oxford to do a presentation on contemporary examples of the sexual politics of meat. She was showing some recent Burger King TV adverts that I noted were overtly and ironically nodding to the sexual politics of meat. These adverts made me question whether it is perhaps no longer enough to say, "hey, look, misogyny and speciesism are linked," and leave the analysis there, because these adverts are already giving a knowing wink to exactly such an awareness. I think vegan camp helps us to read such adverts differently in a climate in which the sexual politics of meat is being actively embraced and humorously gestured to by meat advertisers. That's why I use Eve Kosofsky Sedgwick's work a lot in the article: her work on paranoid reading suggests that the critical tendency of paranoid reading, that which seeks to reveal what's hidden below the surface, assumes an endless naiveté on the part of the uncritical reader or viewer. Vegan camp is a reparative practice that takes account in many ways for the fact that these adverts themselves are already aware of what they are doing.

CT: As you've already mentioned, in her 1964 article, "Notes on Camp," Susan Sontag states that it goes without saying that camp is apolitical. She also gives Oscar Wilde's epigram, "It's absurd to divide people into good and bad. People are either charming or tedious," as an example of camp, suggesting that camp has an aversion to morality as well as politics. Yet, unless it is for health reasons, veganism is usually understood as an ethical and political stance—and I think that's why vegans get on people's nerves so much. You don't even have to say anything about why you're vegan. As soon as people hear you're vegan, they assume you're self-righteous and moralising. So as you discuss in "Notes on Camp," this may seem to be a tension between veganism and camp, or perhaps between queer camp and vegan camp, but in fact, as you have already said, it's not exactly true that queer camp is apolitical. Can you explain how both queer and vegan camp are, in fact, political (and maybe even moral?) sensibilities?

EQ: Yes, so Ann Pellegrini has a really great article critiquing Sontag's claim that camp is apolitical.[6] She notes the various ways in which scholars have reclaimed camp, as I mentioned before, as a means of survival, of world building, or of claiming of a social agency usually denied. That's not to say that camp is therefore inherently political—obviously it depends on who's using it and how they are using it—but she forwards the idea of camp as forging a utopian space within the failures of the present.

In my article I note that an embrace of the campy pleasure that can be gained from objects of animal exploitation that might otherwise be only traumatising can be understood as a utopian gesture that signals to the possibility of a future in which such products won't be psychically wounding.

However, vegan camp works differently than in queer culture in several ways. There's a really distinct and important difference, for example, in the political perspectives of queer camp and vegan camp in that queer camp is usually premised on the assertion of agency by gay men, and for gay men. By contrast, vegan camp is really about vegans, and vegan responses, and therefore the animals at the centre of vegan concern are largely ignored. So it might be said to risk repeating the structure of the absent referent. But in the article I see this risk of repetition as an important aspect of what vegan camp is doing, or can do. Vegan camp is a refusal to speak for animals. It is an acknowledgement of the ways in which vegans are also complicit in these inescapable structures of anthropocentrism.

This idea of complicity ties into what you mentioned before regarding the general assumption often made that vegans are self-righteous, that you just say the term "vegan" and people think you must think yourself to be so holier-than-thou. Vegan camp, by laughing in the face of suffering, functions as a performance of complicity. I see this attachment to complicity as an important political gesture that provides a way for vegans to recognise and acknowledge the ways in which we are implicated in exploitation. By performing this sense of complicity to excess, vegan camp tries to disrupt human exceptionalism. I see this as a disruption and challenge to the ways in which we live as human beings, and a political act on behalf of animals.

CT: Sontag describes camp literature and art as work that aims to be serious or sincere but fails. On the one hand, if an artist, writer or director sets out to create camp, the work usually isn't successful as camp, it has to *try* to be serious, but on the other hand, if it succeeds in being serious, it is not camp either—it has to be aiming to be serious but fail. Your work on veganism insists that vegans will also always fail to some degree, that we cannot be "purely" vegan, especially in a society where animal exploitation is everywhere, normalised, and often invisibilised. You also link veganism and failure in your Knowing Animals podcast interview about your monograph, *Reading Veganism*, where you discuss the anxiety that our vegan activism and scholarship is failing to make any difference for animals.[7] So we could describe veganism as a serious and earnest attempt to not harm and even to help animals which fails. I am also thinking of Jack Halberstam's work on queer failure and similarly trans writers, musicians, and performers Ivan Coyote and Rae Spoon's live show and later their book, *Gender Failure*. It seems that vegan failure is something unfortunate though—it would be better if we could totally stop harming and even help animals through our veganism—whereas we can celebrate failed art as camp and gender failure. Can you talk a little bit about the significance of failure in these areas of veganism, queerness, and camp, and the theme of failure in your writing?

EQ: I've definitely been heavily influenced by Jack Halberstam's work on failure and as you mentioned, my work has stressed the importance of recognising the impossibility of ever being *fully* vegan. I think it is James Stanescu who writes that maybe we shouldn't even talk about being vegan at all, but only of "becoming-vegan," a turn of phrase that recognises that veganism is always an ongoing process.[8] Partly what I suggest throughout my work is that failure doesn't need to be an impediment to action, that embracing veganism's dual occupation of the poles of both utopianism and insufficiency is important. This again works against the stereotype of the insufferably self-righteous vegan.

Jacques Derrida is famously very critical of vegetarianism because he sees it as a desire to possess an illusion of good conscience.[9] But I'm much more convinced by work such as Gary Steiner's, who describes veganism not as a state of good conscience but, instead, as a gnawing horror that confronts us every day and forces us to make decisions constantly about our actions.[10] I think there's something about striving towards an impossible ideal [of no exploitation of animals] that is both a way of committing to utopian possibilities while not pretending or purporting to moral purity. Veganism is not about standing outside of an animal-destroying culture but about continuing to strive for something different within a failed present.

CT: I know a lot of people who have taken up work such as Donna Haraway's as a reason not to be vegan or to dismiss vegans by saying "vegans think they're pure." It is as if what is really immoral is to think you might have clean hands and therefore it becomes an immoral stance to be vegan because you're supposedly presenting yourself as pure. Often insects come up, the fact that when you grow vegetables you kill insects, so it's impossible to be pure and not harm animals and therefore we might as well not even try. So sometimes I feel like this argument against purity, and for accepting failure, functions as an argument to not even try, to not even recognise that there are degrees of failure. Even though we can't be purely vegan, I think it's better to at least try to minimise harm. If you eat animals, plants are grown for the agricultural animals to eat and then insects and other small animals are killed through plant agriculture anyway, and you grow a lot more plants to feed agricultural animals than if you just directly eat the plants. So if you're really worried about the insects and small animals who get killed when we grow plants, you should still eat plants, not animals. It's not purity, but it's harm reduction to be vegan most of the time. I think there are some exceptions where some non-vegan foods can be less harmful than some vegan foods, but for the most part veganism is a rule of thumb for harm reduction. Anyway, I've seen this argument taken up by Donna Haraway's followers in a way that functions to dismiss veganism. How would you respond to that type of position?

EQ: I also take issue with Haraway's work and her dismissal of veganism. For one, the idea of veganism being a stance of self-righteous purity is, for me, a fundamental misrepresentation of veganism. That's why I find Gary Steiner's work so useful. Steiner describes veganism as a constant onslaught of inescapable horror and an attempt to try and do something, do anything, in the face of that. To be vegan is not to pretend that I feel perfect every day because I know I'm doing the right thing, but it's an attempt to try and grapple with a very real felt sense of horror.

Haraway also suggests that veganism is a rigid moral code that has no leeway or flexibility and therefore isn't really engaging with the lived entanglements of our relations with others. But I think, again, this just doesn't actually reflect what it means to be a vegan, which is always about making complex and difficult decisions, given the inescapability of animal suffering and our contributions to it. I see veganism as about constant negotiation with an imperfect world. In my book, *Reading Veganism,* I embrace the figure of the monster because I find that monstrosity is an apt and helpful way of thinking about veganism, as a monstrous assemblage of utopian striving with failures and insufficiencies.

CT: You argue that while queer camp is associated with an embrace of stereotypes of gay men, vegan camp is associated with an embrace of stereotypes of vegans, and in particular vegans as asexual women who secretly or subconsciously want to be eating meat and having sex with men. You mention "vegansexuality" in this context, and Annie Potts' and Jovian Parry's article[11] that describes how vegan women who admitted to being sexually attracted exclusively to other vegans, and to being viscerally disgusted by the bodies and body fluids of meat eaters, were virulently attacked by heterosexual men online, in ways that suggested that they were either sexless or in fact secretly wanted sex with men and to be eating meat. Lesbians have been described in some similar ways—the stereotype of "lesbian bed death" and depictions of lesbians turning to other women only because they've failed to secure or have been traumatised by relationships with men. Can you explain what "vegan sexuality" is and the connections you see with cultural stereotypes of vegans and queer women, and how this all relates to "vegan camp"?

EQ: Yes, so Annie Potts and Mandala White did a study of New Zealanders' dietary habits and in this study, I think maybe three vegan participants in the study said that oh yes, they prefer to have vegan sexual partners. I think if you're a vegan you can relate to this desire: it makes sense just in terms of day to day living, but this minor preference in the study absolutely blew up and made international news. The widespread media attention was driven by meat's close imbrication to heterosexual masculinity, with the idea that women might reject meat-eating men forming the primary object of scandal. The negative representation of vegansexuality, and the public vitriolic responses to it, led to a large number of vegans "coming out" as vegansexual and claiming this label for themselves. I think it's kind of interesting in that regard because veganism is so often portrayed as a renunciation of all possible desire or pleasure, right? You're giving up everything that's good. Vegansexuality, by contrast, was a way of vegans reclaiming their own desires, even if those desires ran counter to mainstream culture.

Rasmus Simonsen has been very critical of vegansexuality because he sees it as at risk of repeating an exclusionary dynamic.[12] But I think for me vegansexuality can also be a productive form of vegan camp, a campy performance of sexual desire that subjects heteronormativity itself to scrutiny or parodic excess. To embrace vegansexuality is to engage with the absurdity of the ways in which heterosexual desire is so inextricably tethered to meat eating and that masculinity seems to be so threatened by the idea that a woman wouldn't want to eat meat.

I think the fact that vegansexuality created so much anxiety and reprehension also speaks to the ways in which veganism and queerness do seem to be really linked. They both are resistant to heterosexual masculine ideals that are linked to normative western dietary habits and patriarchal western cultures. So in that sense, it also cements this link that veganism does seem to be challenging to normative gender roles and ideas of sexuality and desire.

CT: I wonder if it connects a bit back to the purity issue. Vegansexuality completely makes sense to me in terms of choosing partners with compatible lifestyles and being able to share meals and so on. I have talked to a lot of critical animal studies vegan scholars about the issue of vegansexuality and if they would identify that way, and no one

I have met has actually said they do. Most have said they'd *like* to but they would just never date anyone if they limited themselves that way. The best description I heard is "veggie sexual but vegan romantic." But then others have said that they see the idea of vegan sexuality as puritanical or purity politics again, as if you're saying you don't want to contaminate your body with meat-eating bodies, or with fluids that might be tainted by meat.

EQ: Yes, I certainly see the puritanical risks and I don't think that we can claim vegansexuality as a necessary core sexual orientation or identity, but I do think that the virulent responses to it expose the problematic relationships between meat, gender, and sexuality. So I think I see it as useful in my work that way and I also see the benefit in a reclamation of such vegan stereotypes. In my book, for example, I look back to the work of H.G. Wells. Many of his vegetarian characters embody the worst vegansexual characterisations, as sexless spinsters who seem to be pathologically scared of sex. In the book I suggest the productive possibilities of embracing such monstrous figures. I think there's this sense in which this has been a long running stereotype and so why not reclaim it? Why not use such stereotypes to expose the ways in which heterosexual normative desires are also somewhat absurd?

CT: Still on the topic of sexual orientations, I don't know if you have thought about the recent trend in animal studies of embracing zoophilia and even bestiality as potentially queer sexualities, and as potentially emancipatory for animals in so far as they supposedly undermine human exceptionalism. Donna Haraway's description of French kissing her dog Cayenne Pepper and women's and gender studies scholar Kathy Rudy's *Hypatia* article "LGBTQ . . . Z?", along with her book, *Loving Animals*, in which she also defends eating animals, are the most cited examples of this argument, but many other authors have echoed it now.[13] For instance Joanna Bourke's recent book, also titled *Loving Animals*, while condemning most forms that bestiality takes as animal abuse, does accept a certain "queer" argument for zoosexuality.[14] Only a few queer theorists within critical animal studies scholars have engaged with these arguments critically. As someone with expertise in both queer and critical animal studies, what do you think of his argument?

EQ: I was talking at the very start of this interview about when I first took a class on animal studies during my MA. When I was researching for an essay for this course, I went down a real rabbit hole of reading about zoophilia and one of the first graduate essays I wrote ended up being on zoosexuality. I think what stuck with me from that research is the ways in which certain online zoophilia communities were writing compelling arguments about how so much of mainstream culture invests in zoophilia even while it is publicly stigmatised. So they talked about cultural texts like the *King Kong* films as zoophilic texts that expose the latent human–animal sexual desires hidden in plain sight.

This idea that mainstream culture is already zoophilic speaks also to the ways in which many of us share our intimate spaces with domestic companion animals—sharing beds, sharing kisses, caressing their bodies, stroking them to give them pleasure, feeling comforted by the warmth of their bodies on ours. That there is something queer in these relationships, something romantic, and something mutually pleasurable, I think is an important way of rethinking any easy dismissal of bestiality as solely the actions of a criminal minority or pet-keeping as a state of zombification, in Jack Halberstam's terms. This is not to necessarily support zoophilia per se but to highlight that our relationships with animals already cross boundaries we prefer to think of as fixed.

Gabriel Rosenberg has an excellent article in *GLQ* about meat and sex[15] where he notes the ways in which we're in fact committing bestiality all the time, with an extremely thin legal line, for instance, between what it means to masturbate a bull if it's for meat production and to masturbate a bull if it's for your own pleasure. The Rosenberg piece is important because it looks at how the law is so flimsy in terms of really defining bestiality. Really [in terms of the law] it comes down to whether you enjoyed it or not, right? Because if you don't enjoy it and it had a utility purpose for meat production, then that's legally sanctioned and part of a culturally accepted industrial practice. But if you were to gain sexual gratification at the same time, then it is a shocking crime to be morally condemned. Legal divisions such as these offer interesting thought experiments that challenge what we understand sexuality itself to mean.

I think, ultimately, I wouldn't necessarily take a stance right now in terms of pro- or anti-zoophilia nor would I attempt to answer whether it can be ethical. But I do think that raising the question of cross species sex, and the debates such a question inspires, are important in drawing attention to the slipperiness of the boundary between the human and the animal. I find Rudy's article very compelling in this regard. Her article is not trying to elicit a shock factor by making grand claims such as "oh, we can ethically have sex with animals and we don't need to think about consent." She is instead thinking about the ways in which we are already in intimate, entangled relationships with animals and that they do have a queer valence and that that valence changes how we think about kinship and cross-species love.

Ultimately, I do think there is a significant difference between bestial violent activities that lead to immense suffering and often death and reflections on a form of love and intimacy that could cross the species barrier.

CT: I hope you will write on this.

EQ: I never returned to it, but I was super into it at the time.

CT: I did write on it, but it was over six years ago and so much has been published since then. I feel like I have to write about it again but it's really complex and it's weird when people ask you what you're working on!

EQ: Oh, I love answering that question. That's a discomfort I think it is worth generating in others.

CT: You reference Richard Twine's figure of the "vegan killjoy" in your "Notes on Vegan Camp" as well as in your Knowing Animals podcast interview on vegan camp.[16] In the interview, you suggested that vegan camp is a way for vegans to avoid being killjoys—for them to take some pleasure in carnist culture along with meat-eaters, albeit in a different way than meat-eaters do. On the other hand, in "Notes on Vegan Camp," you align vegan camp with the vegan killjoy, saying that both are ways to reorientate desire away from carno-phallogocentric culture. Can you explain who the vegan killjoy is, how she relates to the "feminist killjoy," and what the relationship is between the vegan killjoy and vegan camp?

EQ: Sure. So Richard Twine draws directly on Sarah Ahmed's concept of the feminist killjoy to think about the idea of the "vegan killjoy." He situates his piece around the figure of the disruptive vegan dinner guest: the guest who comes to dinner and unabashedly declares, "Oh, I don't eat meat because it's murder," and as a result, ruins everyone else's meal since they are now made to feel guilty about what is on their plate. So the killjoy is here the figure responsible for spoiling everyone else's fun. The dinner table is here a highly culturally constructed space: associated with the heteronormative family unit, a unit that is meant to be a site of happiness, enshrined in the conviviality of the family meal. The vegan killjoy threatens this space. I imagine that most vegans have had that experience of being the killjoy, and it ties into stereotypes of vegans being joyless, cantankerous, that they just ruin everything, they don't have any fun of their own, and want everyone else to be miserable too.

With vegan camp I draw out another important part of the killjoy figure that Twine notes in his article. For Twine, killjoys do not just spoil fun, in the process they also construct new modes of happiness. And so vegan camp for me is a campy embrace of one's killjoy status, a parodic performance of this joyless vegan figure that in the process elicits new vegan pleasures and joys. So the relationship between vegan camp and the vegan killjoy engages with both of the aspects that you draw out in your question: vegan camp encourages embracing one's disruptive presence and deriving a pleasure from exposing the subsequent anxiety generated by vegans among non-vegans. So yes, I encourage the performance of one's vegan killjoy status, performing it to excess in order to therefore expose the anxieties attendant to meat eating culture.

CT: According to Sontag, extravagance is the hallmark of camp, and one of her examples is "a woman walking around in a dress made of three million feathers." You highlight this example and others in Sontag's article in which animals are absent referents, to show that vegan camp isn't just an analogous phenomenon to queer camp, but that there are also shared archives of camp. You write: "Vegan camp is also an important extension of queer camp, recognising that heterosexual masculinity relies on the assumption of compulsory carnivorism." Your own examples of camp explore these intersections and extensions. For instance, resonating with Sontag's woman in the dress of three million feathers, you describe Lady Gaga's "Meat Dress" as consciously engaging a queer camp aesthetic and also unconsciously engaging in vegan camp. Can you say a bit more about how you see vegan camp having not just an analogous aesthetic to queer camp but an overlapping aesthetic and shared archive?

EQ: In the article, I note that it's interesting that a lot of queer camp or campy drag performances already play with plastic feather boas, fake leather, fake pearls. There's a sense in which a lot of queer camp, or certainly queer camp drag, is playing with the overdetermined significations of animal bodies. Animal bodies often stand as heavy markers of gender, race, class that are being subjected to parody in drag.

I went to a really great talk by Molly McGehee at Oxford many years ago about the drag queen DeAundra Peek from Atlanta, who played in her performances with Vienna sausages—tinned little sausages made from several different types of meat. Her performance I think has this real vegan drag aspect to it that is playing with this indecipherable meat, its connections to lower-class diets, its undesirability alongside working-class femininity, and its humorously failed phallicism. And so I see vegan camp is part of such an archive, showing the ways in which gender is heavily marked by animal bodies.

Jacques Derrida's work on carnophallogocentrism is helpful here. Derrida argues that the idea of the Western humanist subject is not just defined by phallogocentrism, but is about meat eating and consuming animals as a way of asserting oneself as a subject. And so vegan camp is exposing that within these gendered paradigms the ways in which gender is also reliant on a certain logic of the animal.

CT: Another example of vegan camp you analyse in your article is the "Jolly Sailor" piece that you mentioned earlier—a 19th-century folk art depiction of a sailor waving a British flag, a canon between his legs, engraved on the tooth of a slaughtered sperm whale. Like Lady Gaga's meat dress, you describe this work as campy in both the queer and vegan senses. Although intended to be serious, you argue that the "Jolly Sailor" fails as such, striking us as funny in its extravagant "parody of imperialist masculinity" and human exceptionalism. While as vegans we could get righteous or sad about the murder of the sperm whale and the chauvinistic bravado of the British sailor, you suggest that it is more subversive to laugh at this depiction and underscore its performativity and homoeroticism. We are used to thinking of gender and compulsory heterosexuality as performative, but can you say a bit about how you see human exceptionalism and compulsory omnivorism as performative as well? What is the significance of the concept of performativity to your argument about vegan camp?

EQ: To give an example, in Carol J. Adams's *The Sexual Politics of Meat,* Adams notes the generic scene of the patriarchal father at the head of the family dinner table carving a turkey. For Adams, this scene is a performance both of patriarchal power and human exceptionalism—the dominance of the man over the woman and the human over the animal—and reveals the way that the oppression of women and animals is interconnected. This, for Adams, is a performance of male dominance and human exceptionalism.

Performativity is a useful concept for vegan camp, because gender performativity in Butler's work is not about just putting on an act or putting on a costume, but it's something that is inescapable. Vegan camp, as I've gestured to before, is about not standing apart from an animal-destroying culture and suggesting we're outside of it, but recognises the ways in which we are forced to work within its structures as it stands, forced to negotiate cultural scripts of gender, the family, heterosexuality, human exceptionalism. Vegan camp requires acknowledging our complicity and collusion with various paradigms of human exceptionalism but nonetheless holds such paradigms up to challenge and scrutiny in order to try and disrupt them from the inside.

CT: The third example of vegan camp that you discuss in your "Notes on Camp" is mock meats that go to absurd lengths to emulate meat from dead animals, such as the vegan Thanksgiving turkey that is actually the size and shape of a trussed turkey, veggie burgers that "bleed," and a vegan duck product that appears to have been plucked. Although the phrase "mock meat" implies that it is mocking meat or can be funny, it seems the producers of these products are sincere, earnestly believing this is what vegans want. Their seriousness fails though, as we can laugh at these products and, in the case of the mock turkey, maybe also at the ridiculousness of this icon of the heteronormative family meal, and this is what makes mock meats campy. There are some vegans who are opposed to fake meats on moral grounds though, arguing that these perpetuate the Western dietary norm in which meat has to be the centrepiece of every meal, and it also perpetuates the idea that vegans aren't eating "real foods," making omnivorism superior because it's more "natural." In Michael Pollan's "Food Rules," for instance, one of the rules is not to eat food that is pretending to be something it is not. Do you see these responses to mock meats as too earnest? Do mock meats reinforce meat culture or do they undermine it through performativity?

EQ: So I think this is a good point to say that I really don't see vegan camp as the only kind of aesthetic response to animal exploitation, nor is it a response that is always appropriate. I do think there's something important about demonstrating that veganism is nutritionally adequate in its own right, that it is important to highlight there are amazing things you can do with nuts and lentils and tofu and vegetables. In this sense, I'm somewhat convinced by the arguments I've read of the idea, as you say, that mock meats reinforce the idea that the only proper meal is one that has meat at its centre. But I think vegan camp is a different way of looking at the world that can function importantly in different ways.

To counter the Pollan rule that you mention, for instance, vegan camp is disruptive to the idea that there's anything natural about *anything* we eat. We currently eat in a mass produced, industrial neoliberal capitalist world. The vegan burger that "bleeds," with this context in mind, can be seen to draw attention to the desperation of desires for the natural and the raw. And I do think, as you said, that it can be an effective way of undermining meat culture through exaggeration. To look upon a mock-plucked lump of mock duck is to highlight the overt signs of animality usually erased in meat production, as well as to emphasise the visceral disgust often generated by such signs.

CT: As you mentioned in your Knowing Animals podcast interview on vegan camp, people have had more trouble with the Lady Gaga meat dress example than with the "Jolly Sailor," and you suggested that this could have something to do with it being contemporary. These are freshly slaughtered animals she was wearing. Sontag observes that many examples of camp are old, and it may be easier to see the absurdity of oppressive images and artifacts and to take pleasure in their extravagance when they are at more of a historical distance. Sontag also stated that to name a sensibility "requires a deep sympathy modified by revulsion." Would you describe this as your relationship to vegan camp, or at least to some of the examples or it, such as the meat dress?

EQ: I admit in the article and in most talks I do on this topic that really vegan camp is still an aspirational gesture on my part with horror and trauma more characteristic of how I actually respond to a lot of things I see around me. But vegan camp offers, as we spoke about before, a survival strategy.

I really like the idea that vegan camp embodies this idea of sympathy modified by revulsion because I have been thinking in my more recent work about vegan camp in the context of misanthropy. I see vegan camp in some ways as a misanthropic discourse, in that it's a satire of human exceptionalism that derives from an absolute despair and horror with what humans do to animals. But it's still longing for human community, and an investment in human pleasures. I think in that sense, vegan camp is enacting a form of both sympathy and revulsion towards and against the human. Vegan camp enacts a desire to celebrate something about the human while holding it up to critique.

CT: You're a literature scholar and scholar of vegan literary studies in particular, but none of your examples of vegan camp in "Notes on Camp" are literary. Are there literary examples of vegan camp?

EQ: For sure. In my book, *Reading Veganism*, the first half of the book traces a 200-year trajectory of vegan monsters and how they appear across literary history. In the second half of the book, I try and reclaim these monsters, thinking about what we might gain from embracing monstrosity as an identity, as monstrous vegans, and in particularly performing and playing with monstrosity. The final chapter is on Alan

Hollinghurst's work as an example of literary vegan camp. I look at how his novels obsessively use meaty imagery to depict a utopian space of pre-HIV/AIDs crisis gay sexual hedonism. One character in *The Swimming-Pool Library,* for example, picks up young men to feature in his pornographic movies at Smithfield Market. In this novel, Hollinghurst also writes an archetypal sexless vegan character in James, a character who describes longing for big cocks, while eating old tofu burgers, very sad and miserable. This could just be seen as a derogatory depiction of veganism, one of many examples of literature that dismisses veganism. But what I see in Hollinghurst is actually an extremely campy display of the ways in which desire, particularly heteronormative desire, is coded through meat eating and predatory discourses. I turn to the ways in which Hollinghurst can be read as playing with and critiquing such discourses. So that is definitely one example of literary vegan camp.

And then the piece that I wrote for you for this book, on Margaret Atwood's use of dairy across her vast oeuvre, is also an exploration of the literary manifestations of vegan camp. My argument in this chapter is that her excessive use of dairy imagery can be read as a camp proliferation that satirises the ways in which we attach so much meaning—in relation to gender, class, and race, in particular—onto milk and dairy.

CT: Since the publication of "Notes on Camp," you have extended this concept through other projects, including your current work on critical misanthropy that you've just mentioned. Can you discuss some of the afterlives or applications of "Notes on Camp"?

EQ: Yes, so I just finished up an article in which I extend vegan camp to think about a broader emergence of misanthropic camp in the age of climate crisis. This article is part of my ongoing work on misanthropy. The archetype of the misanthrope is a somewhat bitter, hateful, reclusive character, but actually in contemporary Western culture, which seems to possess a pervasive atmosphere of misanthropy, it also often appears as a campy and playful discourse. So in the article, I turn, as one example, to Willow Pill's winning performance of Season 14 of RuPaul's Drag Race, with her song "I Hate People." This performance demonstrates just how mainstream misanthropy has become and allows me to think about what such mass market misanthropic camp does. Partly, this speaks to what I was just saying about vegan camp, of this idea of how we cling to the human while also hating it. I'm highly critical of anti-natalist discourses in this piece, which might otherwise seem the obvious route to go down as a misanthrope. I think instead about misanthropic camp as a way of clinging to the human despite its failures.

Some other afterlives—or what I hope will be future afterlives of "Notes on Vegan Camp"—include my long-running ambition to do some sort of vegan camp art exhibit. For now, I'm trying to construct a digital archive of vegan camp on Instagram. I think I only have a rather pathetic number of followers but you can find it @vegancampcamp.

Notes

1 Emelia Quinn, "Notes on Vegan Camp," *PMLA* 135, no. 5 (October 2020): 914–930.

2 Nicole Seymour, *Glitter* (London: Bloomsbury, 2022).

3 The "absent referent" is a term introduced by Carol J. Adams in *The Sexual Politics of Meat* to describe the ways animals themselves are often absented from the ways that meat is thought about as it is purchased and consumed, for instance when a cow carcass is called "beef" or a "steak" or a deer's carcass is called "venison." Adams also argues that when animal carcasses are treated in

sexualized ways, women are the absent referent, whereas when women are "treated like meat," animals are the absent referent.

4 Erica Fudge, *Animal* (London: Reaction, 2010).

5 Kelly Struthers Montford, "The 'Present Referent': Nonhuman Animal Sacrifice and the Constitution of Dominant Albertan Identity," *PhaenEx* 8, no. 2 (2013): 105–134.

6 Ann Pellegrini, "After Sontag: Future Notes on Camp," in *A Companion to Lesbian, Gay, Bisexual, Transgender and Queer Studies* (Blackwell, 2007), 168–193.

7 "The Monstrous Vegan with Emelia Quinn": February 7, 2022, Episode 185 of Knowing Animals podcast: https://knowinganimals.libsyn.com/episode-185-the-monstrous-vegan-with-emelia-quinn

8 James Stanescu, "Towards a Dark Animal Studies: On Vegetarian Vampires, Beautiful Souls and Becoming-Vegan," *Journal for Critical Animal Studies* 10, no. 3 (2012): 26–50.

9 Jacques Derrida. " 'Eating Well', or the Calculation of the Subject: An interview with Jacques Derrida. *Points . . .: Interviews, 1974–1994,* ed. Elisabeth Weber, trans. Peggy Kamuf (Stanford, CA: Stanford University Press), 255–287.

10 Gary Steiner. *Animals and the Limits of Postmodernism* (New York: Columbia University Press, 2013).

11 Annie Potts and Jovian Parry, "Vegan Sexuality: Challenging Heteronormative Masculinity through Meat-Free Sex," *Feminism & Psychology* 20, no. 1 (2010): 53–72.

12 Rasmus Rahbek Simonsen, "A Queer Vegan Manifesto," *Journal for Critical Animal Studies* 10, no. 3 (2012): 51–80.

13 Kathy Rudy, "LGBTQ . . . Z?" *Hypatia* 27, no. 3 (2012): 601–615; Kathy Rudy, *Loving Animals: Toward a New Animal Advocacy* (Minneapolis: University of Minnesota Press, 2013).

14 Joanna Bourke, *Loving Animals: On Bestiality, Zoophilia, and Posthuman Love* (Chicago: University of Chicago Press, 2020).

15 Gabriel Rosenberg, "How Meat Changed Sex: The Law of Interspecies Intimacy after Industrial Reproduction," *GLQ* 23, no. 4 (2017): 473–507.

16 "Emelia Quinn on Vegan Camp": November 16, 2020, Episode 154 of Knowing Animals podcast: https://knowinganimals.libsyn.com/episode-153-vegan-camp-with-emelia-quinn

Bibliography

Adams, Carol J. *The Sexual Politics of Meat: A Feminist-vegetarian Critical Theory*. New York: Continuum, 1990.

Bourke, Joanna. *Loving Animals: On Bestiality, Zoophilia, and Posthuman Love*. Chicago: University of Chicago Press, 2020.

Derrida, Jacques. "'Eating Well', or the Calculation of the Subject: An Interview with Jacques Derrida." In *Points . . .: Interviews, 1974–1994,* edited by Elisabeth Weber, translated by Peggy Kamuf, 255–287. Stanford, CA: Stanford University Press.

Fudge, Erica. *Animal*. London: Reaction, 2010.

Knowing Animals. "The Monstrous Vegan with Emelia Quin, Episode 185." *Knowing Animals Podcast,* February 7, 2022. Retrieved from: https://knowinganimals.libsyn.com/episode-185-the-monstrous-vegan-with-emelia-quinn

Montford, Kelly Struthers. "The 'Present Referent': Nonhuman Animal Sacrifice and the Constitution of Dominant Albertan Identity." *PhaenEx* 8, no. 2 (2013): 105–134.

Pellegrini, Ann. "After Sontag: Future Notes on Camp." In *A Companion to Lesbian, Gay, Bisexual, Transgender and Queer Studies,* edited by George E. Haggerty and Molly McGary, 168–193. Blackwell, 2007.

Potts, Annie and Parry, Jovian. "Vegan Sexuality: Challenging Heteronormative Masculinity through Meat-Free Sex." *Feminism & Psychology* 20, no. 1 (2010): 53–72.

Quinn, Emelia. "Notes on Vegan Camp." *PMLA* 135, no. 5 (October 2020): 914–930.

Rosenberg, Gabriel. "How Meat Changed Sex: The Law of Interspecies Intimacy after Industrial Reproduction." *GLQ* 23, no. 4 (2017): 473–507

Rudy, Kathy. "LGBTQ . . . Z?." *Hypatia* 27, no. 3 (2012): 601–615

Rudy, Kathy. *Loving Animals: Toward a New Animal Advocacy*. Minneapolis: University of Minnesota Press, 2013.

Seymour, Nicole. *Glitter*. London: Bloomsbury, 2022.
Simonsen, Rasmus Rahbek. "A Queer Vegan Manifesto." *Journal for Critical Animal Studies* 10, no. 3 (2012): 51–80.
Stanescu, James. "Towards a Dark Animal Studies: On Vegetarian Vampires, Beautiful Souls and Becoming-Vegan." *Journal for Critical Animal Studies* 10, no. 3 (2012): 26–50.
Steiner, Gary. *Animals and the Limits of Postmodernism*. New York: Columbia University Press, 2013.

10
VEGANISM AND DISABILITY IN CATASTROPHIC TIMES

Chloë Taylor

Introduction

This chapter aims to show that there are strong reasons coming from politicized, critical disability perspectives for veganism and for boycotting the animal agriculture industry specifically. The chapter starts out with an overview of the literature on the intersections between disability and animal oppression and the ways that animal agriculture systematically causes disability in humans, more-than-human animals, and ecosystems. The next parts of the chapter consider the ways that animal agriculture is a primary driver of global warming, zoonoses, and antibiotic resistance and how each of these disproportionately harms people with disabilities and more-than-human animals.

In recent years important scholarship has emerged exploring the intricate connections between animal and disability oppression and animal and disability liberation.[1] For instance, the collection *Disability and Animality: Crip Perspectives in Critical Animal Studies*,[2] the *Canadian Journal of Critical Disability Studies* special topics issue on *The Intersections of Critical Disability Studies and Critical Animal Studies*,[3] and the *New Literary History* special topics issue on *Animality/Posthumanism/Disability*,[4] all published in 2020, offer an array of perspectives on these topics. Still the most significant work at the intersection of animals and disability is Sunaura Taylor's monograph, *Beasts of Burden: Animal and Disability Liberation*.[5] In this work, Taylor theorizes the interconnectedness of speciesism and ableism, showing that ableism underpins speciesism just as speciesism underpins ableism.

As Taylor explains, human domination over animals has been justified in Western thought for millennia on the grounds that animals do not have the same abilities as humans, for instance, the abilities to speak, reason, and walk upright.[6] Speciesism is ableist because it is grounded in an argument about animals lacking particular cognitive, linguistic, and physical abilities—abilities that are characteristic of the human species. This ableism in Western thought creates a hierarchy among animals according to which those, such as cetaceans, nonhuman primates, and elephants, who come closest to emulating the human species-typical cognitive, linguistic, physical, and emotional capacities are deemed "higher animals," in some cases even being granted legal personhood. At the other end of the spectrum, those animals deemed the least intelligent and most alien are given the least moral

DOI: 10.4324/9781003273400-13

consideration. This ableism also impacts all those humans who lack linguistic or cognitive abilities, who do not bipedally perambulate, or who otherwise lack abilities that are considered significant and human species-typical and who are thus associated with animals.

While humans with intellectual disabilities are particularly likely to be animalized or considered subhuman, Taylor observes that humans with physical disabilities are also regularly compared to animals. In *Beasts of Burden*, she describes her own experiences of being told that she walks like a monkey, eats like a dog, and has hands like a lobster. Taylor explored the latter comparison in a 2011 artwork, "Lobster Girl." In other artworks, Taylor juxtaposes representations of her body with those of other animals, from chickens and bison to a manatee. Despite such reclaimings, what the original comparisons indicate is that just as more-than-human animals are discriminated against based on the ways their abilities differ from species-typical human abilities, so are disabled people discriminated against through comparisons to more-than-human animals. These comparisons are made in a speciesist context where they are implicitly negative. Because speciesism is part and parcel of the logic of ableism, and ableism is similarly integral to the logic of speciesism, Taylor suggests that animal and disability liberation must go hand in hand as well. As she writes: "ableism and speciesism are inextricably linked and anti-ableist thinking has to contend with and challenge anthropocentrism."[7]

Beyond these constitutive conceptual ties between ableism and speciesism, in her work Taylor explores many concrete manners in which disability and animal oppression are imbricated. For example, she shows that animal agriculture systematically inflicts physical and psychological disability on farmed animals through the oppressive, cramped and crowded, unhygienic and violent conditions in which the animals are forced to live. As she also discusses, industrial animal agriculture similarly imposes disability on human workers due to the commonality of violent workplace accidents, the repetitive strain of factory labor, and the psychological trauma of slaughterhouse work. In addition to routinely disabling both animal victims and human employees within the industry, Taylor explains that the meat industry also causes disability in humans outside the slaughterhouse through the impacts of cheap animal product diets and environmental pollution from farms on human health.

Since the publication of *Beasts of Burden*, Taylor has explored disability as a system of oppression that is socially and politically imposed not only on humans and other animals but also on ecosystems.[8] She describes how ecosystems can be thought of as disabled by air and water pollution, deforestation, overgrazing, biodiversity loss, water depletion, and global warming, each of which is a direct impact of animal agriculture. While animal agriculture is known to be a major source of environmental pollution, Kelly Struthers Montford has examined the toxicity of the slaughterhouse in particular. She writes

> Slaughterhouses are polluted, polluting, and resource intensive environments that are overwhelmingly concentrated in marginalized communities. Abattoirs consume billions of gallons of water every year, and often dump their untreated toxic water into nearby waterways. Slaughter runoff has the highest concentration of phosphorus and the second highest concentration of nitrogen across all industrial effluent. This poisons surrounding animal habitats, farmed animals, and human water supplies causing respiratory issues, birth defects, cognitive impairment, cancers, miscarriages, and death. Slaughterhouses also spray and spread their toxic sludges, contributing to air pollution and climate change, with nearby residents unable to be outside or open their

> windows. In the U.S. alone, 75% of abattoirs surpassed their pollution limits thereby violating clean air legislation. Enforcement on these matters is virtually nonexistent despite air pollution being the leading threat to human health. "Deadstock"—the 60% of a slaughtered animal that does not enter the food chain—is an environmental and public health hazard. The U.S. alone produces 1.4 billion tonnes of deadstock per year. Rendering, burial, incineration, and composting are dominant methods of deadstock disposal, and all pose various environmental risks to air, water, and soil in addition to their decomposing bodies releasing infections, antibiotics, agricultural chemicals, and airborne viruses.[9]

Taylor calls the cumulative effect of these manifold ways in which the animal agriculture industry causes disability "the meat industries' disabled ecology."[10]

Throughout her work leading up to and since *Beasts of Burden*, Taylor has carefully considered what it means to be critical of the ways that industries such as animal agriculture and the military cause disability, even while maintaining the neutral and positive ontologies of disability for which disability liberation movements have fought. That is, how do we condemn animal agriculture, like the military, for continually imposing disability on humans, more-than-human animals, and ecosystems, even while maintaining that disability is not reducible to a personal tragedy or a medical problem but should rather be seen as variations in bodyminds and valuable ways of being in the world? As she writes,

> Naming these relationships of harm [between animal agriculture, humans, other animals, and the environment] as disabled ecologies, I want to be careful to point out, does not erase or negate the generative and flourishing beauty of c/krip ways of knowing and being (which are invaluable, especially as we try to learn to live with disabled ecologies). But it does point to how, as Nirmala Erevelles has written, critical disability perspectives need to grapple with the injustices that cause disability, human and non, in order for the radical potentials of disability to be fulfilled.[11]

Building on Taylor's work, in the remainder of this chapter, I draw on critical disability studies, critical animal studies, and environmental studies perspectives to consider the ways that ecosystems increasingly disabled by animal agriculture will impact vulnerable populations, and disabled humans and more-than-human animals in particular.

Global Warming

As we know and are already experiencing, one of the many impacts of global warming is more intense and frequent weather events such as heat waves, tsunamis, and hurricanes. A considerable amount of work exists charting variability in climate vulnerability, or the facts that the global poor, people in less industrialized nations, people who are already food insecure, people who depend directly on agriculture or natural resources for survival, Indigenous peoples, people in low-lying and island nations, women, children, and the elderly are more vulnerable than others to the impacts of climate change.[12] Rarely are more-than-human animals or disabled people mentioned in these studies; however, these groups are also increasingly vulnerable due to ever more severe and frequent weather events.

A few years ago I was in Baja California Sur, Mexico, during a Category 4 hurricane that passed over town during the night. In the morning amidst the wreckage I was surprised to see dead and dying birds everywhere. They had been flung against buildings and died from impact. Soaked birds lay on the ground dying of hypothermia. Birds had been blown into pools and drowned. In the months that followed, the usual cacophony of bird song that characterizes this tropical region was noticeably quieter. A rooster had also been blown into the yard during the storm, seemingly unharmed. Other birds are blown thousands of kilometers from home by hurricanes, however, and may find themselves in entirely unfamiliar environments, separated from their flocks and thrown off their migration course. Some do not survive being blown so far and fall from the skies to die. Birds who initially survive hurricanes may starve because their food sources have been destroyed. For instance, birds who live off berries find that the berry bushes have been denuded by the storm. To take one example of how drastic the impact of hurricanes can be on birds, half of the endangered parrots in Puerto Rico perished during and after Hurricane Maria in 2018.[13]

More-than-human animals also suffer and die because of other types of weather events that are being made more severe due to anthropogenic global warming. For instance, cattle around the world are dying of thirst and starvation due to droughts. Other animals are dying from heat exhaustion, and still others are perishing in wildfires. For example, three billion animals are estimated to have been killed, displaced, or otherwise harmed by the wildfires that raged through Australia for nine months in 2019 and 2020 alone.[14] I write this chapter in a summer when wildfires burn across Canada, and from Vancouver to Montreal, cities have been engulfed in smoke. Indigenous authors Eva Jewell and Hayden King observe that while those living in these major Canadian cities, all located in the south of the country, have to cope with smoke, Indigenous people, many of whom live in the north, are 30% more likely to have to relocate than settlers.[15] This is yet another example of how the most vulnerable populations are most impacted by global warming related disasters. As Jewell and King underscore, "wildfires are a type of climate feedback loop; the effects are multiplied. With the loss of millions of acres of trees, more carbon is released while less can be sequestered, leading to even higher emissions."[16]

As feminist philosopher of science Nancy Tuana observes in "Viscous Porosity: Witnessing Katrina," the Category 5 Atlantic Hurricane Katrina that hit New Orleans in 2005, causing 1,392 human fatalities, made strikingly visible the fact that extreme weather events exacerbated by global warming will also disproportionately impact people with disabilities, particularly as disability intersects with poverty and race.[17] People who are poor are more likely to become disabled, and people who become disabled are likely to experience downward economic mobility. At the same time, people of color are more likely than white people to be disabled, in part because they are more likely to be poor. Poor people and people with disabilities face extra obstacles during evacuations in advance of catastrophic weather events and, because of the eugenic devaluation of their lives, evacuation facilities and transport may not be accessible, and they are more likely than others to be abandoned to die.[18] Unsurprisingly then, media images from Hurricane Katrina showed dead residents of nursing homes and dead bodies abandoned in wheelchairs, including the iconic image of a dead woman in a wheelchair covered in a blanket outside the New Orleans Superdome.[19] It is also the case that more-than-human animals with disabilities may be less capable of fleeing in the face of weather disasters such as Katrina than able-bodied animals of their kinds.

Disabled humans and more-than-human animals are also disproportionately impacted by heat waves. People with multiple sclerosis experience increased pain during heat waves,[20] and people with spinal cord injuries may be unable to sweat in order to avoid heat exhaustion.[21] Intense heat also exacerbates medical conditions such as kidney disease[22] and cardiovascular diseases.[23] Because of the causal relations between disability and poverty, people with disabilities are also less likely to have air conditioning in their homes. People with disabilities are thus among those least able to cope with rising temperatures on a warming planet. Heat exhaustion and heat stroke can also cause disability by damaging internal organs such as the brain, kidneys, and heart.[24]

Animal agriculture is responsible for 14.5% of all global greenhouse gas emissions, compared to all forms of transportation combined, which are responsible for around 13% of global emissions.[25] The primary greenhouse gas associated with animal agriculture is methane, which stays in the atmosphere for fewer years than carbon—around 12 years compared to centuries—but is 80 times as powerful at trapping heat in the atmosphere compared to carbon dioxide. Twenty-five percent of global warming is caused by methane released due to human activities, and 37% of methane emissions from human activities come from animal agriculture. Animal agriculture is also the leading cause of deforestation, which results in carbon dioxide stored in trees being released back into the atmosphere as well as the loss of forests, which serve as carbon sinks.[26] For this reason, animal agriculture is a major cause of both methane and carbon emissions and hence of both short- and long-term global warming.[27]

Critical animal studies scholar Vasile Stănescu has introduced the concept of "animal realism," an extension of "capitalist realism," to describe the fact that many people find it easier to imagine the end of the world than the end of eating animals.[28] Even if the future of the species hinges on our doing so, many people simply cannot believe that a global shift to vegan diets is realistic.[29] Nonetheless, because methane is 80% more powerful at trapping heat than carbon in the short term, a rapid global shift to plant-based diets is the low-hanging fruit we need to grasp if we are to decrease greenhouse gas emissions dramatically enough in the coming years to avoid existential disaster, allowing humanity time to transition from a carbon economy.[30] Given the disproportionate impact of climate change on vulnerable populations of humans and more-than-human animals, including humans and more-than-human animals with disabilities, it is especially urgent for these groups that plant-based diets be widely adopted.

Zoonoses

Although the cause of the Covid-19 pandemic continues to be debated, it, like the 2002–2004 SARS epidemic before it,[31] is widely believed to have originated in a wild animal market in China.[32] There are particular dangers that come from eating wild animals because humans have not had the chance to develop immunities to the pathogens that these animals may carry, much in the way that Indigenous people in the Americas perished in astronomical numbers from European diseases because they had not lived in proximity to the animals from whom these zoonotic diseases came. There is also a particular risk of transmission between animals in live animal markets where different kinds of animals are stacked on top of each other in cages, often in dying conditions—bleeding, urinating, defecating, and salivating on one another. As epidemiologist Martha Nelson of the National Institutes for Health has observed, however, it is very easy to think about pandemics as "foreign invaders" coming from "other people who are doing things in a bad way," when in fact there have been numerous pandemics and epidemics that originated in the factory farms and laboratories of North America and Europe.[33] Pathogens may be transmitted from wild to domesticated animals, for instance, when wild animal habitats are taken over for agricultural pastureland. As Jan Dutkiewicz, Astra Taylor, and Troy Vettese argue,

> Xenophobes call Covid-19 the "Wuhan virus", but in reality zoonoses emerge worldwide, and do so with increasing regularity. The 1918 "Spanish flu" probably came from a Midwestern [American] pig farm. In the 1990s, ecological destabilization in the U.S. south-west led to the Four Corners Hantavirus outbreak. The Hendra and Menangle viruses are named after Australian towns. The Reston virus is an Ebola strain named after a DC suburb. Marburg virus emerged in Germany. These last two diseases sprang from monkeys imported for laboratory use—the Chinese are not the only ones with a large and dangerous wildlife trade.[34]

One example of a zoonotic disease that derived from industrialized agriculture in Western countries is so-called "Mad Cow Disease," or bovine spongiform encephalopathy (BSE). When BSE spreads to humans, it is called Creutzfeldt-Jakob disease. BSE was first discovered in 1986. A mass outbreak occurred several years later, peaking in January 1993 in the United Kingdom. The outbreak originated in industrialized agriculture where farmers were feeding cattle, including the calves of dairy cows, with meat-and-bone meal that included remnants from other cattle. The cannibalistic feeding of cattle to other (vegetarian) cattle, and in particular the unnatural feeding of cattle to baby calves rather than allowing them to drink their own mothers' milk, resulted in the neurodegenerative disease called BSE in 184,000 infected cattle and a mass culling of *4.4 million* cattle in the UK.[35] Since this time, there have been recurring outbreaks of BSE in other countries such as Canada, each time resulting in cullings.

Avian influenza, or bird flus, are also a pandemic threat that are likely to originate in factory farms. The transmission of diseases from birds to humans is relatively rare; however, the mortality rate for humans who catch some types of avian influenza, such as H5N1, is as high as 60%: much higher than Covid-19. In the case of another avian influenza, H7N7, infected poultry do not die, but a fifth of humans who caught the avian influenza did die.[36] The most virulent type of avian influenzas are called HPA1, or "highly pathogenic avian influenza."[37] There have been virulent strains of avian influenza in many countries, across Asia, Africa, the Middle East, Europe, and North America, including Canada. Some have originated in small, backyard farms, but most have originated in poultry farms that are industrialized according to the North American factory farm model. Although there have been recorded cases since the late 19th century of what used to be called "fowl plague," the vast majority of avian flus have occurred since the 1990s, with some outbreaks involving millions of birds and some human fatalities.[38] Bird flus have become more common since the 1990s because the number of birds farmed grew by 76% in less industrialized countries, and by 23% in over-developed countries. It is the intensity and density of the ways that birds are now farmed that explains the great increase in cases of avian flus around the world. As epidemiologist Rob Wallace stresses in his book, *Big Farms Make Big Flu*, the homogeneous genetics of birds in industrialized agricultural settings also makes it particularly easy for avian flus to spread.[39] When there is little genetic diversity in a group, such as different sets of genetic immunities, there is no natural firewall against pathogens. This means that entire flocks can die within 28 hours.

Avian flus in wild birds are mild and have little opportunity to be transmitted to humans. Avian flus in factory farms, however, have an environment in which they can mutate to become more deadly and where they can jump to humans. When thousands of birds are together in one building, more than one strain of influenza can be present, and then one animal is likely to become infected with two strains at the same time. When the two strains of influenza are in the same animal, they can swap genes, and in the process a more deadly variant of influenza may emerge. Avian flus in factory farm settings have the opportunity to be transmitted to humans easily because humans are handling the bodies, feathers, body fluids, and excrement of the birds during slaughter and plucking and because humans are touching contaminated surfaces and equipment. We learned with Covid-19 that the risk of transmission is greater when we are inside, and the same is true for avian flus: more transmissions occur when the birds are incarcerated for the duration of their brief lives indoors and when humans are inside with the birds, which is characteristic of factory farming. In addition to the opportunities for contagion provided by human travel, and by the fact that workers bring viruses from farms into their communities, because the industrial agricultural model involves transporting more-than-human animals across the country and even around the world on boats and planes, it is very easy for a zoonosis that originates in an individual factory farm to become a global threat quickly.

Swine flus are also likely to originate in industrial animal agricultural settings, for similar reasons as avian flus: the intensity and density of how the animals are farmed, the lack of genetic diversity, the fact that they are indoors all of their lives, and that humans are indoors with them handling their bodies and excretions. A swine flu reached pandemic proportions in 1918 (the so-called "Spanish flu") and again in 2009–10. The swine flu of 2009–10 was an H1N1 virus that combined North American swine influenza, North American avian influenza, human influenza, and two swine influenza viruses typically found in Asia and Europe. It infected up to 1.4 billion people worldwide, which was 21% of the global population, although many people were asymptomatic or mildly symptomatic.

Although none of these factory farm–derived pandemics or epidemics has yet to reach the proportions of Covid-19, epidemiologists warn that this is only coincidental: factory farm-derived pandemics and epidemics have so far happened to involve less lethal or contagious strains of a virus than Covid-19. There is no guarantee, however, that a future strain of avian or swine flu, or cattle disease, will not lead to a pandemic as serious or more serious than Covid-19. As Martha Nelson and Rob Wallace both explain, it would in fact be very easy for one of the zoonotic diseases that have originated in Western or Western-style factory farms to reach a pandemic level in the near future. As Nelson puts it, the factory farm is the perfect environment for "brewing up" a "monster virus," and we are currently playing "Russian roulette" with our chances of creating just such a virus.[40] For Wallace, the only "fix" to factory-farmed pandemics is to end the power of big agribusiness, to break the hold that agricultural corporate lobbies have over politics. In *Big Farms Make Big Flu*, Wallace notes that by industrializing agriculture, we are also industrializing pathogens, and to stop industrializing pathogens, we need to de-industrialize agriculture. In an interview on his book, Wallace observes that big agriculture desperately wants to deflect attention from this fact, so they blame others: they blame the Chinese, they blame wild fowl, and they blame small farmers in order to take attention off themselves.[41] Wallace argues, however, that we need to keep our attention on how industrial agriculture is putting the future of humanity—and so much other life on earth—at risk.

Animal advocates and critical animal studies scholars have observed that the Covid-19 pandemic, like other zoonotic epidemics and pandemics that have preceded it, highlights the devastating repercussions of human exploitation of other animals and the interlocking of human and animal oppressions. Animal agriculture, agricultural animal fairs, human encroachment on the ever-shrinking habitats of wild animals, and live animal markets are all sites of both zoonotic transmission *and* of human and animal oppression.[42] So long as animal agriculture and the meat industry persist, they will be a breeding ground for zoonotic epidemics and pandemics, with the most immediate risks for poor and often racialized agricultural and slaughterhouse workers, and especially for people with disabilities or "pre-existing conditions." During the Covid-19 pandemic, ongoing eugenic logics were made painfully visible, with the elderly, chronically ill, people with disabilities, and poor and racialized people deemed acceptable sacrifices so that the economy could return to "normal." Critical animal studies scholars such as Kelly Struthers Montford and Tessa Wotherspoon[43] have written incisively of the racialized, slow violence of the Trump administration's decision to invoke the Defense Production Act to deem meat production an "essential service," forcing slaughterhouse workers to work in highly contagious meatpacking plants. Predictably, rates of Covid infection and mortality were appallingly high among slaughterhouse workers and in the marginalized communities in which they live and to which they spread the virus. Slaughterhouse workers, being poor, racialized, and often immigrant laborers, were deemed, in Chang and Corman's words, "disposable"[44] and, in Sunaura Taylor's words, an "unabashed sacrifice."[45] Taylor adds to this that at the same time as the pandemic was disproportionately killing Black people, "one thing that did not change under quarantine is police killing Black people."[46] Indeed it was the combination of the racism made visible by the pandemic and ongoing police murders of Black people that resulted in the eruption of the Black Lives Matter movement at this time. Taylor moreover stresses that 30–50% of the people the police kill are also disabled.[47] Also indicated the perceived disposability of disabled people's lives, Taylor observes that "COVID-19 deaths in nursing homes and other aggregate settings are almost taken for granted (often not even officially counted), as if being in need of care makes infection almost inevitable and naturalized, and, therefore, less worthy of mourning."[48]

Wallace is surely right that the only way to avoid factory farming future pandemics is to stop factory farming animals. Jan Dutkiewicz, Astra Taylor, and Troy Vettese agree with Wallace that a first step towards such a cessation requires that governments cease subsidizing industrial animal agriculture, and indeed they argue that a tax should be imposed on animal products to account for their environmental and public health costs. At the same time, governments must shift to supporting sustainable plant farming. Feeding the human population on plant-based diets would mean a 75% reduction in the land used for agriculture,[49] and so such a transition would allow for widespread rewilding and reforestation, creating carbon sinks as well as eliminating a primary driver of greenhouse gas emissions. Beyond these changes to the food industry, however, critical animal studies scholars have emphasized that profound social changes are necessary to combat eugenic logics so that when epidemics, pandemics and environmental crises inevitably do occur, people of color, the poor, people with disabilities, and more-than-human animals are not treated as disposable or less grievable lives.

Antibiotic Resistance

Having eliminated genetic diversity and thus natural firewalls against diseases on factory farms, industrial agriculture tries to prevent disease from spreading among farmed animals by adding large quantities of antibiotics to the animals' feed. In countries where agriculture is almost entirely industrialized, such as in Canada and the U.S., as much as 70% of antibiotic use is accounted for by animal agriculture. This means that farms are the greatest contributor to the problem of drug resistance. As a result of antibiotic-resistant bacteria, nearly 5 million people died of drug resistance infections in 2019 alone, and millions more were sickened, making antibiotic resistance a greater threat than HIV/AIDS or malaria.[50] As long as industrial animal agriculture persists, there is no choice but to preventatively feed large amounts of antibiotics to billions of animals, and yet this is the primary way that we are overusing antibiotics as a society. As a result, more and more people and other animals will experience illness, disability, and premature death from conditions for which antibiotics could previously have saved them.[51]

Needless to say, these estimates of the toll of antibiotic resistance only account for human deaths and disease, yet more-than-human animals also die from drug resistant infections. The obvious fact that more-than-human animals as well as humans suffer illness and death due to antibiotic resistance means that, ultimately, antibiotics will not work to prevent mass outbreaks of diseases in factory farms. Factory farming is not sustainable for this reason alone, in addition to being unsustainable environmentally.

Conclusions

In an earlier book chapter, my co-author Kelly Struthers Montford and I argued for veganism as universal design.[52] There we observed that veganism is the most inclusive and accessible diet for humans, to which we should thus default for accessibility reasons alone. Moreover, we argued that "given the environmental cost of animal agriculture, veganism [also] makes the planet more accessible for life on earth. As one life form among many, all humans would benefit from making the planet accessible through veganism; more importantly, however, billions of other species would also benefit from widespread human veganism."[53] This chapter has taken this argument further by taking up three examples[54]—global warming, zoonoses, and antibiotic resistance. In each case, it was seen that animal agriculture, as a site of both speciesist and ableist violence, is disabling ecologies in ways that will disproportionately impact humans and more-than-human animals with disabilities, among other vulnerable populations.

Notes

1 I am grateful to Lauren Corman and Kelly Struthers Montford for their helpful comments on the first draft of this chapter.
2 *Disability and Animality: Crip Perspectives in Critical Animal Studies*, eds. Stephanie Jenkins, Kelly Struthers Montford, and Chloë Taylor (Routledge, 2020).
3 Alan Santinele Martino, Sarah May Lindsay, "The Intersections of Critical Disability Studies and Critical Animal Studies" *Special Topics issue of the Canadian Journal of Critical Disability Studies*, vol. 9, no. 2 (2020).
4 Michael Lundblad, editor, "Animality/Posthumanism/Disability" *Special Topic issue of New Literary History*, vol. 51, no. 4 (2020).
5 Sunaura Taylor, *Beasts of Burden: Animal and Disability Liberation* (The New Press, 2018).

6 In the *Timaeus*, for instance, Plato argues that the fact that humans walk bipedally, thus naturally looking at the heavens, indicates their superiority over animals who walk on four legs, thus naturally looking at the ground.
7 Taylor, *Beasts of Burden: Animal and Disability Liberation,* 200.
8 At the time of this writing, Taylor has a forthcoming volume, *Disabled Ecologies: Lessons from a Wounded Desert* (University of California Press, 2024).
9 Kelly Struthers Montford, "Slaughterhouses," in *Environmental Justice Encyclopedia*, ed. Dorceta Taylor (Thousand Oaks, CA: Sage Publications, forthcoming).
10 Sunaura Taylor and Sara E.S. Orning, "Being Human, Being Animal: Species Membership in Extraordinary Times," *Johns Hopkins University Press* 51 no. 4 (2020): 677.
11 Ibid.
12 To take just two examples, see: Salim Momtaz and Muhammed Asaduzzaman, *Climate Change Impacts and Women's Livelihood: Vulnerability in Developing Countries* (Routledge, 2018); "Climate Change and Women's Health: Impacts and Opportunities in India," *Geohealth* 10, no. 2 (October 2018): 283–297, retrieved from: https://www.ncbi.nlm.nih.gov/pmc/articles/PMC7007102/
13 Associated Press, "Scientists Work to Save Puerto Rican Parrots after Hurricane Maria," *The Washington Post* (2018), retrieved from: https://www.washingtonpost.com/lifestyle/kidspost/scientists-work-to-save-puerto-rican-parrots-after-hurricane-maria/2018/11/20/5d84ca28-e434-11e8-8f5f-a55347f48762_story.html
14 Daniel Vernick, "3 Billion Animals Harmed by Australia's Fires," *World Wild Life* (2020), retrieved from: https://www.worldwildlife.org/stories/3-billion-animals-harmed-by-australia-s-fires. Critical animal studies scholar Danielle Celermajer has written a powerful series of reflections on Australian wildfires and their impacts on animals. See: Danielle Celermajer, *Summertime: Reflections on a Vanishing Future* (Penguin Books Australia, 2021).
15 Eva Jewell and Hayden King, "The Consequences of Ignoring Indigenous Land Management Are Now in the Smoky Air We Breathe," *The Globe and Mail* (July 3, 2023), retrieved from: https://www.theglobeandmail.com/opinion/article-the-consequences-of-ignoring-indigenous-land-management-are-now-in-the/
16 Ibid.
17 Nancy Tuana, "Viscous Porosity: Witnessing Katrina," *Material Feminisms* 188 (2008): 207.
18 United Nations Department of Economic and Social Affairs, "Disability-Inclusive Disaster Risk Reduction and Emergency Situations," (n.d.), retrieved from: https://www.un.org/development/desa/disabilities/issues/disability-inclusive-disaster-risk-reduction-and-emergency-situations.html
19 Cited in Tuana, "Viscous Porosity: Witnessing Katrina," 208.
20 National MS Society, "Heat and Temperature Sensitivity," *National Multiple Sclerosis Society* (n.d.), retrieved from: https://www.nationalmssociety.org/Living-Well-With-MS/Diet-Exercise-Healthy-Behaviors/Heat-Temperature-Sensitivity#:~:text=Many%20people%20with%20MS%20experience,%2Dhalf%20of%20a%20degree).
21 Elene Merete Hagen, "Acute Complications of Spinal Cord Injuries," *World Journal of Orthopedics* 6, no. 1 (2015), retrieved from: https://www.wjgnet.com/2218-5836/full/v6/i1/17.htm#:~:text=Sweating%20may%20also%20occur%20exclusively,(above%20Th8%2DTh10).
22 National Kidney Foundation, "News—It's Summertime Beware of Heat Illness," *National Kidney Foundation* (n.d., retrieved from: https://www.kidney.org/news/monthly/Its_Summertime_Beware_of_Heat_Illness#:~:text=What%20Happens%20to%20the%20Kidneys,pressure%20and%20decreased%20kidney%20function
23 Jingwen Liu, Blesson M. Varghese, Alana Hansen, Ying Zhang, Timothy Driscoll, Geoffrey Morgan, Keith Dear, Michelle Gourley, Anthony Capon, and Peng Bi, "Heat Exposure and Cardiovascular Health Outcomes: A Systematic Review and Meta-Analysis," *The Lancet: Planetary Health* 6, no. 6 (2020), retrieved from: https://www.thelancet.com/journals/lanplh/article/PIIS2542-5196(22)00117-6/fulltext#:~:text=Heat%20stress%20can%20lead%20to,regulated%20by%20the%20cardiovascular%20system.
24 Mayo Clinic Press, "Heatstroke," *Mayo Clinic* (n.d.), retrieved from: https://www.mayoclinic.org/diseases-conditions/heat-stroke/symptoms-causes/syc-20353581#:~:text=Heatstroke%20requires%20emergency%20treatment.,of%20serious%20complications%20or%20death.

25 Food and Agricultural Organization of the United Nations, Key Facts and Findings, retrieved from: https://www.fao.org/news/story/en/item/197623/icode/
26 Carbon sinks absorb carbon dioxide from the atmosphere. The primary carbon sinks are the ocean (although warmer water is less effective at absorbing carbon dioxide, so as the oceans warm due to climate change, the less CO_2 they can absorb), the soil, and forests. When we clear forests to graze cattle for animal agriculture, we replace a carbon sink with a high carbon emitting industry. When forests burn because of global warming, the wildfires emit large amounts of carbon, and the carbon sinks that were the forests are simultaneously lost.
27 Anthony Weis, *The Ecological Hoofprint: The Global Burden of Industrial Livestock* (Zed Books, 2013)
28 Chloë Taylor, "Vasile Stanescu on in Vitro Meat and 'Animal Realism'," *YouTube* (2020), retrieved from: https://www.youtube.com/watch?v=C-h8ZWHleeo&ab_channel=ChloeTaylor
29 Thus, scientists try to find ways to bioengineer cows to emit less methane, or we are encouraged to switch from eating red meat to poultry. While poultry farming results in lower greenhouse gas emissions than farming cows and sheep, it is still far more intensive in terms of emissions than a vegan diet and does not address other environmental and animal welfare concerns. See: Berill Takacs, Julia A. Stegemann, Anastasia Z. Kalea, and Aiduan Borrion, "Comparison of Environment Impacts of Individual Meals—Does It Really Make a Difference to Choose Plant-Based Instead of Meat-Based Ones?" *Journal of Cleaner Productions* 379, no. 2 (2022), retrieved from: https://www.sciencedirect.com/science/article/pii/S0959652622043542
30 No author, "A Vegan Diet: Eating for the Environment," *Physician Committee for Responsible Medicine* (n.d.), retrieved from: https://www.pcrm.org/good-nutrition/vegan-diet-environment
31 Wenhui Li, Swee-Kee Wong, Fang Li, Jens H. Kuhn, I-Chueh Huang, Hyeryun Choe, and Michael Farzan, "Animal Origins of the Severe Acute Respiratory Syndrome Coronavirus: Insight from ACE2-S Protein Interactions," *Journal of Virology* 80, no. 9 (2006). Retrieved from: https://journals.asm.org/doi/10.1128/jvi.80.9.4211-4219.2006
32 Teresa G. Valencak, Anna Csiszar, Gabor Szalai, Andrej Podlutsky, Stefano Tarantini, Vince Fazekas-Pongor, Magor Papp, and Zoltan Ungvari, "Animal Reservoirs of Sars-CoV-2: Calculable Covid-19 Risk for Older Adults from Animal to Human Transmission," *Gero Science* 43, no. 5 (2021), retrieved from: https://www.ncbi.nlm.nih.gov/pmc/articles/PMC8404404/
33 Dylan Matthews and Sigal Samuel, "How to Prevent a Factory Farmed Pandemic," *Future Perfect: Vox Media* (2020), retrieved from: https://www.audible.co.uk/pd/How-to-prevent-a-factory-farmed-pandemic-Podcast/B08KYYXXBF
34 Jan Dutkiewicz, Astra Taylor, and Troy Vettese, "The Covid-19 Pandemic Shows We Must Transform the Global Food System," *The Guardian* (2020), retrieved from: https://www.theguardian.com/commentisfree/2020/apr/16/coronavirus-covid-19-pandemic-food-animals
35 No Author, "All Protection Steps Taken' after BSE Diagnosis," *BBC News* (2018), retrieved from: https://www.bbc.com/news/uk-scotland-north-east-orkney-shetland-45954225
36 Yong Poovorawan, Sunchai Pyungporn, Slinporn Prachayangprecha, and Jarika Makkoch, "Global Alert to Avian Influenza Virus Infection: From H5NI to H7N9," *Pathog Glob Health* 107, no. 5 (2013), retrieved from: https://www.ncbi.nlm.nih.gov/pmc/articles/PMC4001451/
37 CDC, "Influenza Type A Viruses," *Centers for Disease Control and Prevention* (2023), retrieved from: https://www.cdc.gov/flu/avianflu/influenza-a-virus-subtypes.htm
38 Kate Kelland, "Proliferation of Bird Flu Outbreaks Raises Risk of Human Pandemic," *Healthcare and Pharma* (2017), retrieved from: https://www.reuters.com/article/us-health-birdflu-risks-idUSKBN15A22D
39 Robert Wallace, *Big Farms Make Big Flu: Dispatches on Influenza, Agribusiness, and the Nature of Science* (New York: New York University Press, 2016).
40 Dylan Matthews and Sigal Samuel, "How to Prevent a Factory Farmed Pandemic," *Future Perfect: Vox Media* (2020), retrieved from: https://www.audible.co.uk/pd/How-to-prevent-a-factory-farmed-pandemic-Podcast/B08KYYXXBF
41 Populist Dialogues, "Big Farms Make Big Flu: With Biologist Rob Wallace," *YouTube* (2017), retrieved from: https://www.youtube.com/watch?v=vSIHpwWWvLY&ab_channel=PopulistDialogues
42 Charlotte Blattner, Kendra Coulter, Dinesh Wadiwel, and Eva Kasprzycka, "Covid-19 and Capital: Labour Studies and Nonhuman Animals—A Roundtable Dialogue," *Animal Studies Journal* 10, no. 1 (2021): 240–272. Available at: https://ro.uow.edu.au/asj/vol10/iss1/11

43 Kelly Struthers Montford and Tessa Wotherspoon, "The Contagion of Slow Violence: The Slaughterhouse and COVID-19," *Animal Studies Journal* 10, no. 1 (2021): 80–113. Available at: https://ro.uow.edu.au/asj/vol10/iss1/7
44 Darren Chang and Lauren Corman, "Multispecies Disposability: Taxonomies of Power in a Global Pandemic," *Animal Studies Journal* 10, no. 1 (2021): 57–79. Available at: https://ro.uow.edu.au/asj/vol10/iss1/6
45 Taylor and Orning, "Being Human, Being Animal: Species Membership in Extraordinary Times," 668.
46 Ibid., 670.
47 Ibid., 671.
48 Ibid., 667.
49 Hannah Ritchie, "If the World Adopted a Plant-Based Diet We Would Reduce Global Agricultural Land Use from 4 to 1 Billion Hectares," *Our World in Data* (2021). Retrieved from: https://ourworldindata.org/land-use-diets
50 Tosin Thompson, "The Staggering Death Toll of Drug-Resistant Bacteria," *Nature Portfolio* (2022), retrieved from: https://www.nature.com/articles/d41586-022-00228-x; CDC, "National Infection & Death Estimates for Antimicrobial Resistance," *Centers for Disease Control and Prevention* (2021), retrieved from: https://www.cdc.gov/drugresistance/national-estimates.html#:~:text=Antimicrobial%20resistance%20is%20an%20urgent,resistant%20infections%20occur%20each%20year.
51 European Centre for Disease Prevention and Control, "Assessing the Health Burden of Infections with Antibiotic-Resistant Bacteria in the EU/EEA, 2016–2020," *Stockholm: ECDC* (2022), retrieved from: https://www.ecdc.europa.eu/en/publications-data/health-burden-infections-antibiotic-resistant-bacteria-2016-2020
52 Chloë Taylor and Kelly Struthers Montford, "Veganism as Universal Design: Accommodation and Inclusion in Law and Social Justice Praxis," in *Disability and Animality: Crip Perspectives in Critical Animal Studies*, eds. Stephanie Jenkins, Kelly Struthers Montford, and Chloë Taylor (Routledge, 2020), 129–157.
53 Chloë Taylor and Kelly Struthers Montford, "Veganism as Universal Design: Accommodation and Inclusion in Law and Social Justice Praxis," in *Disability and Animality: Crip Perspectives in Critical Animal Studies*, eds. Stephanie Jenkins, Kelly Struthers Montford, and Chloë Taylor (Routledge, 2020), 137.
54 Other examples, such as water and air pollution from factory farms, could have been taken up to make similar points.

Bibliography

Associated Press. "Scientists Work to Save Puerto Rican Parrots after Hurricane Maria." *The Washington Post*, 2018. Retrieved from: https://www.washingtonpost.com/lifestyle/kidspost/scientists-work-to-save-puerto-rican-parrots-after-hurricane-maria/2018/11/20/5d84ca28-e434-11e8-8f5f-a55347f48762_story.html

Blattner, Charlotte, Coulter, Kendra, Wadiwel, Dinesh, and Kasprzycka, Eva. "Covid-19 and Capital: Labour Studies and Nonhuman Animals—A Roundtable Dialogue." *Animal Studies Journal* 10, no. 1 (2021): 240–272. Retrieved from: https://ro.uow.edu.au/asj/vol10/iss1/11

Canadian Journal of Critical Disability Studies (2020).

CDC. "National Infection & Death Estimates for Antimicrobial Resistance." *Centers for Disease Control and Prevention*, 2021. Retrieved from: https://www.cdc.gov/drugresistance/national-estimates.html#:~:text=Antimicrobial%20resistance%20is%20an%20urgent,resistant%20infections%20occur%20each%20year.

CDC. "Influenza Type A Viruses." *Centers for Disease Control and Prevention*, 2023. Retrieved from: https://www.cdc.gov/flu/avianflu/influenza-a-virus-subtypes.htm

Celermajer, Danielle. *Summertime: Reflections on a Vanishing Future*. Sydney: Penguin Books Australia, 2021.

Chang, Darren, and Corman, Lauren. "Multispecies Disposability: Taxonomies of Power in a Global Pandemic." *Animal Studies Journal* 10, no. 1 (2021): 57–79. Retrieved from: https://ro.uow.edu.au/asj/vol10/iss1/6

Dutkiewicz, Jan, Taylor, Astra, and Vettese, Troy. "The Covid-19 Pandemic Shows We Must Transform the Global Food System." *The Guardian,* 2020. Retrieved from: https://www.theguardian.com/commentisfree/2020/apr/16/coronavirus-covid-19-pandemic-food-animals

European Centre for Disease Prevention and Control. "Assessing the Health Burden of Infections with Antibiotic-resistant Bacteria in the EU/EEA, 2016–2020." *Stockholm: ECDC,* 2022. Retrieved from: https://www.ecdc.europa.eu/en/publications-data/health-burden-infections-antibiotic-resistant-bacteria-2016-2020

Food and Agricultural Organization of the United Nations. "Key Facts and Findings." *United Nations* (n.d.). Retrieved from: https://www.fao.org/news/story/en/item/197623/icode/

Hagen, Elene Merete. "Acute Complications of Spinal Cord Injuries." *World Journal of Orthopedics* 6, no. 1 (2015). Retrieved from: https://www.wjgnet.com/2218-5836/full/v6/i1/17.htm#:~:text=Sweating%20may%20also%20occur%20exclusively,(above%20Th8%2DTh10).

Jenkins, Stephanie, Struthers Montford, Kelly, and Taylor, Chloë, eds. *Disability and Animality: Crip Perspectives in Critical Animal Studies.* London: Routledge, 2020.

Jewell, Eva, and King, Hayden. "The Consequences of Ignoring Indigenous Land Management Are Now in the Smoky Air We Breathe." *The Globe and Mail,* July 3, 2023. Retrieved from: https://www.theglobeandmail.com/opinion/article-the-consequences-of-ignoring-indigenous-land-management-are-now-in-the/

Kelland, Kate. "Proliferation of Bird Flu Outbreaks Raises Risk of Human Pandemic." *Healthcare and Pharma,* 2017. Retrieved from: https://www.reuters.com/article/us-health-birdflu-risks-idUSKBN15A22D

Li, Wenhui, Wong, Swee-Kee, Li, Fang, Kuhn, Jens H., Huang, I-Chueh, Choe, Hyeryun, and Farzan, Michael. "Animal Origins of the Severe Acute Respiratory Syndrome Coronavirus: Insight from ACE2-S Protein Interactions." *Journal of Virology* 80, no. 9 (2006). Retrieved from: https://journals.asm.org/doi/10.1128/jvi.80.9.4211-4219.2006

Liu, Jingwen, Varghese, Blesson M., Hansen, Alana, Zhang, Ying, Driscoll, Timothy, Morgan, Geoffrey, Dear, Keith, Gourley, Michelle, Capon, Anthony, and Bi, Peng. "Heat Exposure and Cardiovascular Health Outcomes: A Systematic Review and Meta-Analysis." *The Lancet: Planetary Health* 6, no. 6 (2020). Retrieved from: https://www.thelancet.com/journals/lanplh/article/PIIS2542-5196(22)00117-6/fulltext#:~:text=Heat%20stress%20can%20lead%20to,regulated%20by%20the%20cardiovascular%20system.

Matthews, Dylan, and Samuel, Sigal. "How to Prevent a Factory Farmed Pandemic." *Future Perfect: Vox Media,* 2020. Retrieved from: https://www.audible.co.uk/pd/How-to-prevent-a-factory-farmed-pandemic-Podcast/B08KYYXXBF

Mayo Clinic Press. "Heatstroke." *Mayo Clinic* (n.d.). Retrieved from: https://www.mayoclinic.org/diseases-conditions/heat-stroke/symptoms-causes/syc-20353581#:~:text=Heatstroke%20requires%20emergency%20treatment.,of%20serious%20complications%20or%20death.

Michael Lundblad, editor, "Animality/Posthumanism/Disability" *Special Topic issue of New Literary History,* vol. 51, no. 4 (2020).

Momtaz, Salim, and Asaduzzaman, Muhammed. *Climate Change Impacts and Women's Livelihood: Vulnerability in Developing Countries.* Routledge, 2018.

National Kidney Foundation. "News—It's Summertime Beware of Heat Illness." *National Kidney Foundation* (N.D). Retrieved from: https://www.kidney.org/news/monthly/Its_Summertime_Beware_of_Heat_Illness#:~:text=What%20Happens%20to%20the%20Kidneys,pressure%20and%20decreased%20kidney%20function

National MS Society. "Heat and Temperature Sensitivity." *National Multiple Sclerosis Society* (n.d.). Retrieved from: https://www.nationalmssociety.org/Living-Well-With-MS/Diet-Exercise-Healthy-Behaviors/Heat-Temperature-Sensitivity#:~:text=Many%20people%20with%20MS%20experience,%2Dhalf%20of%20a%20degree).

No author. "A Vegan Diet: Eating for the Environment." *Physician Committee for Responsible Medicine* (n.d.). Retrieved from: https://www.pcrm.org/good-nutrition/vegan-diet-environment

No author. "All Protection Steps Taken' after BSE Diagnosis." *BBC News,* 2018. Retrieved from: https://www.bbc.com/news/uk-scotland-north-east-orkney-shetland-45954225

Plato. *Plato's Cosmology; the Timaeus of Plato.* London, New York: Harcourt, Brace, K. Paul, Trench, Trubner & Co. ltd., 1937.

Poovorawan, Yong, Pyungporn, Sunchai, Prachayangprecha, Slinporn, and Makkoch, Jarika. "Global Alert to Avian Influenza Virus Infection: From H5NI to H7N9." *Pathog Glob Health* 107, no. 5 (2013). Retrieved from: https://www.ncbi.nlm.nih.gov/pmc/articles/PMC4001451/

Populist Dialogues. "Big Farms Make Big Flu: With Biologist Rob Wallace." *YouTube*, 2017. Retrieved from: https://www.youtube.com/watch?v=vSIHpwWWvLY&ab_channel=PopulistDialogues

Ritchie, Hannah. "If the World Adopted a Plant-based Diet We Would Reduce Global Agricultural Land Use from 4 to 1 Billion Hectares." *Our World in Data*, 2021. Retrieved from: https://ourworldindata.org/land-use-diets

Sorensen, C., Saunik, S., Sehgal, M., Tewary, A., Govindan, M., Lemery, J., and Balbus, J. "Climate Change and Women's Health: Impacts and Opportunities in India." *Geohealth* 10, no. 2 (October 2018): 283–297. Retrieved from: https://www.ncbi.nlm.nih.gov/pmc/articles/PMC7007102/

Struthers Montford, Kelly. "Slaughterhouses." In *Environmental Justice Encyclopedia*, edited by Dorceta Taylor. Thousand Oaks: Sage Publications, forthcoming.

Struthers Montford, Kelly, and Wotherspoon, Tessa. "The Contagion of Slow Violence: The Slaughterhouse and COVID-19." *Animal Studies Journal* 10, no. 1 (2021): 80–113. Retrieved from: https://ro.uow.edu.au/asj/vol10/iss1/7

Takacs, Berill, Stegemann, Julia A., Kalea, Anastasia Z., and Borrion, Aiduan. "Comparison of Environment Impacts of Individual Meals—Does It Really Make a Difference to Choose Plant-based Instead of Meat-based Ones?" *Journal of Cleaner Productions* 379, no. 2 (2022). Retrieved from: https://www.sciencedirect.com/science/article/pii/S0959652622043542

Taylor, Chloë. "Vasile Stanescu on in Vitro Meat and 'Animal Realism'." *YouTube*, 2020. Retrieved from: https://www.youtube.com/watch?v=C-h8ZWHleeo&ab_channel=ChloeTaylor

Taylor, Chloë, and Struthers Montford, Kelly. "Veganism as Universal Design: Accommodation and Inclusion in Law and Social Justice Praxis." *Disability and Animality: Crip Perspectives in Critical Animal Studies*, edited by Stephanie Jenkins, Kelly Struthers Montford, and Chloë Taylor. London: Routledge, 2020.

Taylor, Sunaura. *Beasts of Burden: Animal and Disability Liberation*. New York: The New Press, 2018.

Taylor, Sunaura, and Orning, Sara E.S. "Being Human, Being Animal: Species Membership in Extraordinary Times." *Johns Hopkins University Press* 51, no. 4 (2020): 663–685.

Thompson, Tosin. "The Staggering Death Toll of Drug-resistant Bacteria." *Nature Portfolio*, 2022. Retrieved from: https://www.nature.com/articles/d41586-022-00228-x

Tuana, Nancy. "Viscous Porosity: Witnessing Katrina." *Material Feminisms* 188 (2008).

United Nations Department of Economic and Social Affairs. "Disability-Inclusive Disaster Risk Reduction and Emergency Situations." *United Nations* (n.d.) Retrieved from: https://www.un.org/development/desa/disabilities/issues/disability-inclusive-disaster-risk-reduction-and-emergency-situations.html

Valencak, Teresa G., Csiszar, Anna, Szalai, Gabor, Podlutsky, Andrej, Tarantini, Stefano, Fazekas-Pongor, Vince, Papp, Magor, and Ungvari, Zoltan. "Animal Reservoirs of Sars-CoV-2: A Covid-19 Risk for Older Adults from Animal to Human Transmission." *Gero Science* 43, no. 5 (2021). Retrieved from: https://www.ncbi.nlm.nih.gov/pmc/articles/PMC8404404/

Vernick, Daniel. "3 Billion Animals Harmed by Australia's Fires." *World Wild Life*, 2020. Retrieved from: https://www.worldwildlife.org/stories/3-billion-animals-harmed-by-australia-s-fires

Wallace, Robert. *Big Farms Make Big Flu: Dispatches on Influenza, Agribusiness, and the Nature of Science*. New York: New York University Press, 2016.

Weis, Anthony. *The Ecological Hoofprint: The Global Burden of Industrial Livestock*. London: Zed Books, 2013.

PART III

Feminist Ethics and Care

11

ANIMAL SANCTUARIES AS FEMINIST CARE ACTIVISM

Elan Abrell

As of March 2023, an ongoing experiment conducted by Elon Musk's Tesla company on public streets to improve the faulty autopilot feature in Tesla cars—notably without the consent of public participants—resulted in crashes that have killed at least 19 people, including 6 pedestrians.[1] From 2018 to 2022, employees at Neuralink—another Musk company—tortured and killed approximately 1500 animals in gruesome experiments designed to develop a computer-brain interface device. All these deaths are the result of choices to prioritize technological innovation with potential future benefits for humans over the lives of individuals in the present, reflecting Musk's utilitarianism-inspired conviction that "We should take the set of actions that maximize total public happiness!", as he posted on Twitter in December 19, 2019. Underlying and facilitating this kind of deadly calculus is what Indigenous environmental justice scholar Kyle Whyte calls an epistemology of crisis, within which the perceived imminence and urgency of potential harm can justify courses of action that perpetuate other harms.[2]

In addition to influencing Musk's deadly experiments, an epistemology of crisis conjoined with a utilitarian approach to strategy has also fueled the rise in influence of effective altruism, "a philosophy of charitable giving that claims to guide adherents in doing the 'most good' per dollar donated or time spent."[3] Although ostensibly focused on an array of different environmental and public health concerns, effective altruism has an especially significant impact on animal activism. In fact, effective altruism can be situated in a longer history of the influence of utilitarian philosophy on animal ethics and activism going back to the eighteenth century.

In contradistinction to this history, however, is a parallel history of feminism-informed activism that provides important, albeit often overlooked, lessons for ongoing efforts to transform the future of human–animal relations. In particular, animal sanctuaries—activist projects affording permanent shelter and care to animals rescued from conditions of anthropogenic misery and exploitation—provide a counterpoint to the sacrificial logic of effective altruism. To illustrate this contrast, this chapter will trace some of the historical and philosophical roots of the animal sanctuary movement, spotlighting its essential role in contemporary animal activism. Unlike utilitarian animal ethics and effective altruism,

DOI: 10.4324/9781003273400-15

animal sanctuaries provide one example of an alternative vision for the future of animal activism grounded in feminist ethics of care.[4]

Historical Roots of the Animal Sanctuary Movement

The juxtaposition of two stories, at first blush seemingly unrelated to each other, provides an illuminating lens through which to understand how contemporary animal sanctuaries relate to the longer history of animal activism that preceded them. The first is the story of Jesus Christ and the Gadarene swine, in which Jesus allegedly exorcised demons from a possessed man and cast them into a large herd of swine, who promptly ran into the sea, where they drowned.[5] The second is Ursula K. LeGuin's oft-cited short story "Those Who Walk Away from Omelas," in which the existence of a utopian society depends on the ongoing torturous neglect and confinement of one young child.[6] Together, these stories reflect the anthropocentric logics that have shaped the violent exploitative relationship between humans and other animals throughout so-called "Western" history as well as the influential strain of utilitarian animal advocacy which has attempted to transform that relationship, both of which animal sanctuaries challenge with a care-based ecofeminist approach to animal activism. In this section I will highlight some significant moments in the history (and pre-history) of animal activism that contributed to the modern animal sanctuary movement. Because space is limited, this necessarily selective history will focus on a few significant moments of intersection between species and gender that have been especially important in the historical genealogy of animal sanctuaries.

Of Swine and Men: Human Exceptionalism and Its Discontents

The theologian Augustine of Hippo (350–430 CE) cites the story of the Gadarene swine as evidence to support his position that humans have no moral responsibility to animals. Referencing the swine's drowning, Augustine argues that "Christ himself shows that to refrain from the killing of animals and the destroying of plants is the height of superstition."[7] Several centuries earlier, however, Aristotle (c. 384–c. 322 BCE) laid the conceptual groundwork for this anthropocentric[8] ideology of human exceptionalism with his *scala naturae*, a hierarchical biological classification scheme that placed humans at the top above all other animals due to their capacity to reason and use language. He further cemented this hierarchy with the philosophical justification for the human use of animals in the first book of his *Politics*, where he argued that human domination of animals is natural and proper and that an inversion of this hierarchy or even equality between the two would be harmful to the natural order—a logic he also applied to male domination over women and Greek domination of slaves. Following its fusion with Christian ideology through the deeply influential writings of Augustine, the anthropocentric Aristotelian framework for human–animal relations seized European civilizations in a hegemonic grip that would last for almost two millennia as early Christian philosophers, such as Thomas Aquinas (1225–1274 CE), embraced and expounded on it.

During this time, the Catholic Church heavily policed the human–animal boundary, violently suppressing heretical practices that contradicted this framework. One method it employed was the regulation of mundane practices of interspecies cohabitation, like pet-keeping. This took a violent turn in the fifteenth century when the crusade against heresy morphed into sporadic outbreaks of witch trials across Europe. From the fourteenth to the eighteenth century, approximately 50,000 people were executed for witchcraft. Courts interpreted close relationships with animals as evidence of keeping familiars or committing bestiality, a serious crime aside from its connection to witchcraft that was itself punishable by death.

Familiars were believed to be supernatural companions or demons who took the form of small animals and assisted witches in their evil machinations. The accusation of keeping familiars was largely concentrated in the English witch trials, where around half the cases involved evidence of the accused interacting with animals.[9] British folklorist Christina Hole describes how easily such evidence could be derived from human–animal interactions:

> In an age when fondness for small animals was nothing like as general in England as it is now, the actual possession of any beast that might be supposed to be a familiar was a clear danger to anyone suspected of witchcraft, especially if he or she were known to treat it with affection. It was not even necessary for the creature to live in the house; a dog bounding towards a suspected person in the fields, or a cat jumping through a window might be enough to confirm an already existing suspicion.[10]

Ursula Kemp, for example, was executed in 1582 following testimony from her son that he witnessed her caring for a lamb, two cats, and a toad—a practice that might today simply be thought of as a micro-sanctuary rather than evidence of witchcraft.[11]

The church's moral panic over relations with animals focused on these two different kinds of human–animal interactions—companionate interactions like pet-keeping on one hand and bestiality on the other—for opposite but complementary reasons. While bestiality could be seen as humans debasing themselves to the level of mere animals through sexual intercourse with them, living with companion animals threatens to do precisely the opposite by raising animals closer to the level of humans and incorporating them into human social dynamics. Both kinds of relations threatened the idea of human exceptionalism that situates humans in an existentially and morally superior position to the rest of the animal kingdom.

Further, the pathologization of interspecies relationships functioned as an early form of what Fiona Probyn-Rapsey has termed "an animalady," a portmanteau of "animal," "malady," and "lady." A triangulation drawing together gender, animals, and madness, animaladies are "sites of tension produced by acknowledging how our relationships with other animals are damaged . . . in a variety of ways, both by common attitudes of human superiority and by the violent and disturbing implications of these attitudes."[12] Like the modern-day concept of the "crazy cat lady,"[13] this persecution of isolated women caring for other animals worked to simultaneously reinforce speciesist as well as misogynist social structures.

The Brown Dog Affair: Early Feminist Animal Activism

At the turn of the twentieth century, multiple social movements intersected in a controversy over a brown dog that would set the tone for the more visible and audible animal activism tactics of the modern animal protection movement. What came to be known as the "Brown Dog Affair," started in 1902 when Lizzy Lind af Hageby (1878–1963) and Leisa Katherine Schartau (dates unknown)—feminist and anti-vivisection activists from Sweden—attended the London School of Medicine for Women.[14] After witnessing multiple vivisections of live animals, which were commonly used in the nineteenth and early twentieth century in English medical schools to teach students about physiology, the two women documented what they saw in their book *The Shambles of Science: Extracts from the Diary of Two Students of Physiology* (2012 [1903]).[15]

Hageby and Schartau were some of the first practitioners of the common contemporary animal activist tactic of going undercover to document violent or neglectful treatment of animals in factory farms, laboratories, and other facilities that harm animals. One of the most momentous claims in Hageby and Schartau's book was that they witnessed the vivisection of a brown terrier that was still conscious during the procedure. The dog had also been used in three separate vivisections. Repeated use of animals in experiments and failure to provide anesthesia were both violations of Britain's 1876 Cruelty to Animals Act. Stephen Coleridge (1854–1936), a barrister and anti-vivisection activist with the National Anti-Vivisection Society accused physiologist William Blayliss (1860–1924) of violating the act, which the doctor vehemently denied. Blayliss sued Coleridge for defamation and won after successfully convincing the court he had used anesthesia in accordance with the law. Coleridge was required to pay Blayliss £5,000 in damages. In response to the trial, anti-vivisection activists had a memorial statue of the dog erected in Battersea Park in 1906 with a plaque engraved with the following caption:

> In Memory of the Brown Terrier Dog Done to Death in the Laboratories of University College in February 1903 after having endured Vivisection extending over more than Two Months and having been handed over from one Vivisector to Another Till Death came to his Release. Also in Memory of the 232 dogs Vivisected at the same place during the year 1902. Men and Women of England how long shall these Things be?[16]

In November 1907, medical students began protesting and vandalizing the statue, and in December of that year started raiding women's suffrage meetings as well.[17] Though they were aware that not all suffragists were also anti-vivisectionists, they channeled their violent reaction against anti-vivisectionists, many of whom were women, into a broader misogynistic backlash against challenges to the patriarchal power structure. The Brown Dog Riots continued to escalate as medical students clashed with the motley alliance of social activists who were concentrated in the Battersea neighborhood, including trade-unionists, suffragists, and anti-vivisectionists, until the Battersea council eventually decided to take the statue down in 1910.

Both the coalition of different movements responding to the rioting medical students and the Brown Dog Affair more generally reflect the conflicting and contradictory value systems that have shaped human attitudes toward animals throughout European history. As Lizzy Lind af-Hageby and Charlotte Despard (1844–1939), an Irish suffragist and anti-vivisection activist, saw it, the Brown Dog Affair could be understood as a battle between feminism and "machismo."[18] Similarly, historian Hilda Kean argues that "the 'moment' of *The Shambles of Science* is significant" because the "protagonists are young—and female; the perpetrators of cruelty, in this instance, are male and older. In a sense the work and its circumstances epitomize the political mood of the times: new forces of civilization against age-old brutality."[19] Noting how closely intertwined the women's suffrage and anti-vivisection movements became, literature scholar Coral Lansbury sees the affair as bringing together multiple symbols that resonated with women, the images of vivisected dogs on operating tables evoking images of suffragists being force-fed in Brixton prison, strapped to operating tables during childbirth, or being subjected to forced hysterectomies as a treatment for "mania."[20] Many members of the medical establishment, on the other hand, saw the anti-vivisection activists as representatives of "ignorance, superstition, [and] sentimentality" for empathizing with animals.[21]

As in the witch trials centuries before, care for animals and its association with women was pathologized. Indeed, as Heather Fraser and Nik Taylor note, "with the rise of women's antivivisectionism, American neurologist Charles Loomis Dana created a new medical condition in 1909 called 'zoophilpsychosis,' which referred to those with a heightened concern for animals as afflicted with a mental illness."[22] The affair exposed how discourses of scientific rationality and compassionate sentimentality are gendered, with the latter being subordinated to and dismissed by the former. It revealed the politics that had been unfolding inside and along with the blossoming animal protection movement as philosophical arguments against the rationality of Aristotelian/Cartesian human exceptionalism increasingly relied on an ethics informed by empathetic concern for the suffering of others and a questioning of the categories of difference that served as the support beams for the social power structure.

The Outsized Influence of Utilitarianism and the Importance of Ecofeminism

Since 1975, when the *New York Review of Books* published philosopher Peter Singer's *Animal Liberation: A New Ethics for our Treatment of Animals*, utilitarianism has had an outsized influence on contemporary animal activism. Indeed, *Animal Liberation* is treated by many as the bible of the modern animal advocacy movement. In it, Singer makes an argument for animal liberation based on the utilitarian philosophy of Jeremy Bentham, particularly the principles that right actions are the ones that maximize good for the greatest number of beings and that in evaluating actions, the interests of all beings—both humans and animals—should be considered equally.

As influential as the utilitarian position has been to contemporary animal activism, the frequent overemphasis of its role in late-twentieth century animal activism obscures a parallel history of philosophical work from feminist scholars on animals that was equally pioneering. As feminist scholar Susan Fraiman points out, the same year that Singer published *Animal Liberation*, Carol J. Adams published an article in *The Lesbian Reader* that was the basis for her later book *The Sexual Politics of Meat*, an intersectional analysis of the links between meat-eating and patriarchy.[23] Adam's scholarship "contributed to a growing body of ecofeminist work, emergent in the 1980's, on women, animals, and the environment."[24] Elaborating on this history, Fraiman continues,

> a cohort of ecofeminists—including Adams, Josephine Donovan, Brian Luke, Connie Solomon, Marti Kheel, Andrée Collard, Deane Curtain, Alice Walker, Deborah Silver, Greta Gaard, Lori Gruen, Lynda Birke, and Karen Warren, among others—embarked several decades ago on the project of challenging deeply embedded humanist assumptions concerning gender and animality.[25]

Although cited by activists less often than animal philosophy outside of this feminist tradition, the insights of these writers still influence the arguments of contemporary animal activists, particularly in critiques of links between patriarchal gender politics and the violent exploitation of animals.

Animal Sanctuaries and the Feminist Ethics of Care

Of particular relevance to the work of sanctuaries is the ethics of care tradition in ecofeminism. As Lori Gruen and Carol J. Adams observe, ethics of care recognize the value of caring relationships, "including caring relationships between humans and other animals and between other animals, either within or across species."[26] The goal of sanctuaries is to care for animals that can no longer care for themselves because of the harms inflicted on them by humans. Rather than a utilitarian calculus that can justify, or even require, the sacrifice of individuals' interests for some cumulative measurement of abstract good, the feminist ethics of care that guide sanctuaries center relationships of care and the animals in those relationships as moral ends in themselves.

In the story "Those Who Walk Away from Omelas," the majority of inhabitants of the utopian Omelas accept the profound misery of one child as a necessary cost for the vast aggregate happiness of their society, much like Elon Musk accepts the suffering and deaths of laboratory animals, unwitting pedestrians, and car passengers for the imagined future goods of his technological pursuits. Such moral choices are facilitated by epistemologies of crisis, an epistemology that "organizes knowledge in ways that emphasize some narrative of the imminence of a threat to *the present*."[27] Although utilitarian arguments do not require an epistemology of crisis to justify their claims, epistemologies of crisis can be used to justify the kinds of harms that actions based on utilitarian reasoning can lead to, such as the suffering of the child under Omelas or the victims of Musk's quixotic misadventures. When fused with longtermism, a recent outgrowth of utilitarian philosophy that fixates on attempting to inform current actions with potential harms and benefits in an impossible-to-predict future millennia or even eons away, an epistemology of crisis becomes even more dangerous, potentially justifying immense harms based on imagined future benefits multiplied by fantastically astronomical future population predictions.[28]

By prioritizing relations of care, sanctuaries resist falling into the trap of crisis epistemologies. Rather than offering the illusions of immediate solutions, sanctuaries demonstrate a model for transforming human–animal relations in a more sustainable and stable transition that does not demand acquiescence to harm from anyone, providing an ethical framework for navigating difficult decisions that can adapt and respond to the unique circumstances of particular relationships. Take the experiences of pattrice jones, for example. pattrice is the co-founder of VINE Sanctuary, a "multispecies community that includes more than seven hundred nonhuman survivors" of the animal agriculture industry. As she explains, "Every day at a sanctuary is a case study in decision-making under conditions of uncertainty. People like me thus must become adept at making decisions in situations that are both ethically and factually ambiguous. Often, these are literally life-or-death decisions. From this standpoint, I can report that the rudimentary mathematics of utilitarianism are not at all utile."[29]

Conclusion

Animal sanctuaries and other activist strategies informed by ethics of care are not necessarily any more likely than utilitarianism, effective altruism, or longtermism to solve the many ecological and social problems of the present. But unlike these approaches, they provide a vision for living together better with each other—humans and other animals alike—now, one that rejects both the anthropocentric sacrifices of the Gadarenian and Omelian varieties. Furthermore, animal sanctuaries offer a fundamentally democratic model for interspecies coexistence that allows for the autonomous flourishing of everyone in the community, whereas these various iterations of utilitarian praxis point toward a profoundly undemocratic future in which those with money and power, like Elon Musk, can simply take it upon themselves to choose who can be sacrificed for the allegedly greater good. All life on earth faces a level state of precarity it hasn't faced in eons, if ever. The paths out of that precarity remain unclear, but animal sanctuaries and other activist projects of care offer methods for beginning to live now the kinds of futures so many beings currently long to reach.

Notes

1 Tesla Deaths, retrieved from: https://www.tesladeaths.com/#totals, accessed March 6, 2023.
2 Kyle Whyte, "Against Crisis Epistemology," in *Handbook of Critical Indigenous Studies*, eds. Brendan Hokowhitu, Aileen Moreton-Robinson, Linda Tuhiwai-Smith, Steve Larkin, and Chris Andersen (New York: Routledge, 2021).
3 Carol J. Adams, Alice Crary, and Lori Gruen, "Introduction," in *The Good It Promises, the Harm It Does: Critical Essays on Effective Altruism*, eds. Carol J. Adams, Alice Crary, and Lori Gruen (Oxford: Oxford University Press, 2023), xxii.
4 See Carol J. Adams and Lori Gruen, "Ecofeminist Footings," in *Ecofeminism: Feminist Intersections with Other Animals and the Earth*, eds. Carol J. Adams and Lori Gruen, 2nd ed. (New York: Bloomsbury Academic, 2022), 40–41.
5 Variations of this story can be found in the Bible in Mark 5:1–13, Luke 8:26–33, and Matthew 8:28–32.
6 Ursula K. Le Guin, "Those Who Walk Away from Omelas," in *The Unreal and the Real: The Selected Short Stories of Ursula K. Le Guin* (London: Saga Press, 2017 (1973)).
7 Annika Spalde and Pelle Strindlund, "Doesn't Jesus Treat Animals as Property?" in *A Faith Embracing All Creatures: Addressing Commonly Asked Questions about Christian Care for Animals*, eds. T. York, A. Alexis-Baker, M. Bekoff, B. McLaren (New York: Wipf & Stock Publishers, 2012), 120.
8 Ecofeminist philosopher Val Plumwood employs the term "anthrocentrism" to highlight how in ecological theory this concept functions similarly to androcentrism and ethnocentrism in feminist and anti-racist theory, respectively. See Val Plumwood, *Feminism and the Mastery of Nature* (London: Routledge, 1993); Val Plumwood, "Androcentrism and Anthrocentrism: Parallels and Politics," *Ethics and the Environment* 1, no. 2 (1996): 119–152.
9 Barbara Rosen, *Witchcraft* (London: Edward Arnold, 1969), 30–32.
10 Christina Hole, *Witchcraft in England* (London: Batsford, 1977), 40.
11 Rosen, *Witchcraft*, 109–110.
12 Lori Gruen and Fiona Probyn-Rapsey, "Distillations," in *Animaladies*, eds. Lori Gruen and Fiona Probyn-Rapsey (New York: Bloomsbury Academic, 2018), 1.
13 , Fiona Probyn-Rapsey, "The 'Crazy Cat Lady'," in *Animaladies*, eds. Lori Gruen and Fiona Probyn-Rapsey (New York: Bloomsbury Academic, 2019).
14 See Hilda Kean, "The 'Smooth Cool Men of Science': The Feminist and Socialist Response to Vivisection," *History Workshop Journal* 40 (1995): 16–38; Coral Lansbury, *The Old Brown Dog: Women, Workers, and Vivisection in Edwardian England* (Madison: University of Wisconsin Press, 1985).
15 Lizzy Lind Af-Hageby and Leisa Katherine Schartau, *The Shambles of Science: Extracts from the Diary of Two Students of Physiology* (Charleston: Nabu Press, 2012 (1903)). For a history of the anti-vivisection movement see Susan Sperling, *Animal Liberators: Research and Morality* (Berkeley: University of California Press, 1988).
16 Coral Lansbury, *The Old Brown Dog: Women, Workers, and Vivisection in Edwardian England* (Madison: University of Wisconsin Press, 1985), 14.
17 Ibid., 17–18.
18 Lynda Birke, "Supporting the Underdog: Feminism, Animal Rights and Citizenship in the Work of Alice Morgan Wright and Edith Goode," *Women's History Review* 9, no. 4 (2000): 701.

19 Hilda Kean, *Animal Rights: Political and Social Change in Britain Since 1800* (London: Reaktion Books, 1998), 143.
20 Lansbury, *The Old Brown Dog*, 24. For a detailed history of the intersection between feminist and animal protection movements in England in the nineteenth and twentieth centuries, see Hilda Kean, "The 'Smooth Cool Men of Science;' " Susan Hamilton, *Animal Welfare & Anti-vivisection 1870–1910: Nineteenth Century Woman's Mission* (New York: Routledge, 2004).
21 Lansbury, *The Old Brown Dog*, 84.
22 Heather Fraser and Nik Taylor, "Women, Anxiety, and Companion Animals: Toward a Feminist Animal Studies of Interspecies Care and Solidarity," in *Animaladies*, eds. Lori Gruen and Fiona Probyn-Rapsey (New York: Bloomsbury Academic, 2019), 162.
23 Susan Fraiman, "Pussy Panic versus Liking Animals: Tracking Gender in Animal Studies," *Critical Inquiry* 39, no. 1 (2012): 89. See Carol J. Adams, "The Oedible Complex: Feminist and Vegetarianism," in *The Lesbian Reader*, eds. Gina Covina and Laurel Galana (Oakland: Amazon Press, 1975); Carol J. Adams, *The Sexual Politics of Meat: A Feminist-Vegetarian Critical Theory* (New York: Continuum International, 1990).
24 Fraiman, "Pussy Panic," 89.
25 Ibid., 89–90.
26 Adams and Gruen, "Ecofeminist Footings," 40.
27 Whyte, "Against Crisis Epistemology," 64.
28 Carol J. Adams, Alice Crary, and Lori Gruen, "Coda," in *The Good It Promises, the Harm It Does: Critical Essays on Effective Altruism*, eds. Carol J. Adams, Alice Crary, and Lori Gruen (Oxford: Oxford University Press, 2023), xxii.
29 Pattrice Jones, "Queer Eye on the EA Guys," in *The Good It Promises, the Harm It Does: Critical Essays on Effective Altruism*, eds. Carol J. Adams, Alice Crary, and Lori Gruen (Oxford: Oxford University Press, 2023), 120.

Bibliography

Adams, Carol J. "The Oedible Complex: Feminist and Vegetarianism." In *The Lesbian Reader*, edited by Gina Covina and Laurel Galana. Oakland: Amazon Press, 1975.

Adams, Carol J. *The Sexual Politics of Meat: A Feminist-Vegetarian Critical Theory*. New York: Continuum International, 1990.

Adams, Carol J., Crary, Alice, and Gruen, Lori, eds. "Introduction." In *The Good It Promises, the Harm It Does: Critical Essays on Effective Altruism*, edited by Carol J. Adams, Alice Crary, and Lori Gruen. Oxford: Oxford University Press, 2023a.

Adams, Carol J., Crary, Alice, and Gruen, Lori, eds. "Coda." In *The Good It Promises, the Harm It Does: Critical Essays on Effective Altruism*, edited by Carol J. Adams, Alice Crary, and Lori Gruen. Oxford: Oxford University Press, 2023b.

Adams, Carol J., and Gruen, Lori. "Ecofeminist Footings." In *Ecofeminism: Feminist Intersections with Other Animals and the Earth*, edited by Carol J. Adams and Lori Gruen, 2nd ed. New York: Bloomsbury Academic, 2022.

Birke, Lynda. "Supporting the Underdog: Feminism, Animal Rights and Citizenship in the Work of Alice Morgan Wright and Edith Goode." *Women's History Review* 9, no. 4 (2000): 693–719.

Fraiman, Susan. "Pussy Panic versus Liking Animals: Tracking Gender in Animal Studies." *Critical Inquiry* 39, no. 1 (2012): 89–115.

Fraser, Heather, and Taylor, Nik. "Women, Anxiety, and Companion Animals: Toward a Feminist Animal Studies of Interspecies Care and Solidarity." In *Animaladies*, edited by Lori Gruen and Fiona Probyn-Rapsey. New York: Bloomsbury Academic, 2019.

Gruen, Lori, and Probyn-Rapsey, Fiona. "Distillations." In *Animaladies,* edited by Lori Gruen and Fiona Probyn-Rapsey. New York: Bloomsbury Academic, 2018.

Hamilton, Susan. *Animal Welfare & Anti-vivisection 1870–1910: Nineteenth Century Woman's Mission*. New York: Routledge, 2004.

Hole, Christina. *Witchcraft in England*. London: Batsford, 1977.

Jones, Pattrice. "Queer Eye on the EA Guys." In *The Good It Promises, the Harm It Does: Critical Essays on Effective Altruism*, edited by Carol J. Adams, Alice Crary, and Lori Gruen. Oxford: Oxford University Press, 2023.
Kean, Hilda. "The 'Smooth Cool Men of Science': The Feminist and Socialist Response to Vivisection." *History Workshop Journal* 40 (1995): 16–38.
Kean, Hilda. *Animal Rights: Political and Social Change in Britain Since 1800*. London: Reaktion Books, 1998.
Lansbury, Coral. *The Old Brown Dog: Women, Workers, and Vivisection in Edwardian England*. Madison: University of Wisconsin Press, 1985.
Le Guin, Ursula K. "Those Who Walk Away From Omelas." In *The Unreal and the Real: The Selected Short Stories of Ursula K. Le Guin*. London: Saga Press, 2017 (1973).
Lind af Hageby, Lizzy, and Schartau, Leisa Katherine. *The Shambles of Science: Extracts from the Diary of Two Students of Physiology*. Charleston: Nabu Press, 2012 (1903).
Plumwood, Val. *Feminism and the Mastery of Nature*. London: Routledge, 1993.
Plumwood, Val. "Androcentrism and Anthrocentrism: Parallels and Politics." *Ethics and the Environment* 1, no. 2 (1996): 119–152.
Probyn-Rapsey, Fiona. "The 'Crazy Cat Lady.'" In *Animaladies*, edited by Lori Gruen and Fiona Probyn-Rapsey, New York: Bloomsbury Academic, 2019.
Rosen, Barbara. *Witchcraft*. London: Edward Arnold, 1969.
Spalde, Annika, and Strindlund, Pelle. "Doesn't Jesus Treat Animals as Property?" In *A Faith Embracing All Creatures: Addressing Commonly Asked Questions about Christian Care for Animals*, edited by T. York, A. Alexis-Baker, M. Bekoff, and B. McLaren. New York: Wipf & Stock Publishers, 2012.
Sperling, Susan. *Animal Liberators: Research and Morality*. Berkeley: University of California Press, 1988.
Tesla Deaths. Retrieved from: https://www.tesladeaths.com/#totals, accessed March 6, 2023.
Whyte, Kyle. "Against Crisis Epistemology." In *Handbook of Critical Indigenous Studies*, edited by Brendan Hokowhitu, Aileen Moreton-Robinson, Linda Tuhiwai-Smith, Steve Larkin, and Chris Andersen. New York: Routledge, 2021.

12

FEMINIST LEGAL SYSTEMS THAT BENEFIT ANIMALS

Placing Parameters Around Care and Relationality

Maneesha Deckha

Introduction

Fundamentally enabling multiple planetary crises is the anthropogenic capitalist commodification of other-than-human animals ("animals"), a system centrally informed by the ideology of human exceptionalism and anthropocentrism.[1] That anthropocentric legal systems must embrace a different vision of human–animal relations to better respond to contemporary environmental and other catastrophes, including animal suffering, seems increasingly unassailable.[2] As part of such revisioning, a growing chorus of feminist animal law scholars, myself included, have impugned the objectification of animals and the corresponding widespread inattention, disavowal, and marginalization of animals' needs and interests in common law and civilian legal systems.[3] Alongside the challenge to human exceptionalism, a substantial number of feminist animal law scholars advocate for legal systems that eschew liberal premises to instead adopt deeply relational and care-based outlooks to understand the profound entwinement of human–animal relations.[4]

But it seems reasonable to posit that if such feminist care-based relational revisionings of the legal system are going to materially "benefit" animals, as Greta Gaard insists that feminist interspecies praxis should, at a bare minimum, such revisionings must result in laws that are directed at phasing out and finally ending industries that harm animals for profit or other human or corporate interests.[5] Pressing anthropocentric legal systems to adopt care-based and relational worldviews cannot devolve into mere abstract rejection of human exceptionalism or simple celebration of human–animal inter-relationality. It is unlikely that animals (and other nonhumans as well as humans harmed by animal-based industries) will benefit if "relational" or "care" circulate as empty legal concepts mandating little or no pushback against standard business practices or models.

With this chapter, I explore parameters that need to be placed around the legal meaning for "care" and "relational" should the common law and civilian legal systems proceed in this direction, specifically querying whether posthumanist feminist teachings about caring for and relating to animals are effective models for legal system redesign.[6] Although posthumanist feminist teachings about interspecies care and intersubjectivity are illuminating on

 DOI: 10.4324/9781003273400-16

multiple levels, as others have argued, they largely tolerate and even justify industries and practices that traffic in animals and normalize systemic violence against them.[7] I examine such critiques and extend their relevance to questions of legal system redesign to argue that vegan ecofeminism is more instructive as to how to devise a feminist legal system that benefits animals. Specifically, vegan ecofeminism's responsiveness to suffering and corresponding anti-violence commitments mandate legal definitional parameters around the terms "care" and "relational" that, at a minimum, would prohibit animal-use industries.

For reasons of space, I illuminate the shortcomings of posthumanist feminism's teachings about caring and relating to animals through the critiques that Donna Haraway's work has attracted.[8] When considering feminist care-based, relational theorizing about human–animal dynamics, it is impossible not to acknowledge the pioneering and ongoing staggering presence of Haraway in the field and her hallmark concepts and ideas about how to reconceive relations with nonhumans, namely "companion species," "becoming with" animals, and "entanglement".[9] Haraway's hallmark contributions and the relational theories for animals they portend, however much they centre the human and depart from modernist scientific accounts about naturalized objects, I will argue, are not reliable theoretical guides to foment a post-anthropocentric relational or care-based legal ontology. I discuss the residual anthropocentrism that inhabits Haraway's vision of relationality and how it can guide those who follow her approach on human–animal relations in animal-unfriendly ways. In order to achieve a feminist legal system that attends to animals' vulnerabilities, I close by recommending vegan ecofeminist teachings as a preferred toolkit for legal system redesign [to posthumanist feminist approaches].

2. Posthumanist Relationality: Legal Welfarism in Disguise

a. Dispelling Dualisms and Breaking Boundaries

Donna Haraway is a towering and path-breaking figure in feminist and posthumanist studies. As Gavin Rae notes, this is due to her introduction of the concept of the term "cyborg" to denounce Cartesian-inspired humanist beliefs in neat separations between human and machines, which encouraged contestation of Western binary thinking in general and the marginalizing and dominating ideologies they entail for women, colonized peoples, and nonhumans.[10] Haraway was also an early proponent of the relational nature of selves, both human and not.[11] This investment in a relational worldview and the deconstruction of hierarchy-supporting binary thinking informed her later writings on refiguring human and animal relations.[12] In these works, Haraway develops her ideas about the co-evolution of human and animal species (concentrating on dogs)—what she calls "companion species"—through "the intimate and entwined relationship between human being and animal at both the ethical and ontological levels."[13]

Haraway developed early on the signature figures of the "cyborg"[14] and "companion species"[15] to contest the "Great Divide" of nature/culture that infects Western thought and created racist, sexist, humanist epistemologies. Both figures are celebrated as mashups of humans with the nonhuman (machine, organism, etc.) that expose the fallacy of human exceptionalism, but it is her term "companion species" that "draws particular attention to the ethical and phenomenological inter-relationality of humans and animals."[16] In this latter focus, the concept of "becoming-with"—a concept meant to orient us to our always already embedded development alongside animals as a way to re-image alternative planetary relations that Haraway espouses in her development of companion species ethics[17]—has become globally influential.[18] Haraway's prescription for the insidious nature/culture divide to recede in our worldviews and become-with animals is to substitute the concept of "naturecultures" and develop ties with the more-than-human world through "empathy, accountability and recognition."[19] Drawing from Levinasian ethics, supplemented by Derrida's extension of Levinasian theory of Otherness to nonhumans, Haraway commits to a "call and response" ethical responsibility toward animals.[20]

b. Challenging Ideology of Human Exceptionalism but Not Its Violent Expressions

Clearly, Haraway's "texts constitute a pioneering effort to set up a connection between the culture of contemporary biotechnological sciences and that of the human and social sciences."[21] With respect to animals in particular, her work has provided a thoroughgoing critique of human uniqueness and exceptionalist claims that we stand above animals.[22] She has eschewed modernist Enlightenment notions of the individualistic, atomistic, dominating subject in favour of cultivating ethical relations with non-humans, particularly animals,[23] through, it is worth repeating, "empathy, accountability and recognition."[24] But for all her insistence on the ordinariness of the human, her catalysing contributions to challenging dualistic scientific representations of nature and embedded repressive logics about gender, race, and species, and her dedication to enacting social change,[25] particularly to cultivate "a renewed kinship system, radicalized by concretely affectionate ties to the non-human 'others',"[26] Haraway is accepting of anthropogenic violence against animals. As Eva Giraud notes in her filtering of Haraway's work through a critical animal studies lens, she does not condemn mainstream eating or hunting of animals, for example, and is generally comfortable with animals in commodified and instrumentalized states.[27]

Zipporah Weisberg has also convincingly argued that Haraway's companion species ethic "provid(es) ideological cover for such violent practices as animal experimentation, genetic engineering, dog breeding and training, killing animals for food and hunting."[28] Exemplifying this pattern, Weisberg highlights the explicit embrace of technoscientific practices in the exaltation of hybridity that is central to the deconstructive aim of Haraway's generative cyborg and companion species concepts.[29] Weisberg worries that Haraway's relational hybridity teachings foment derision toward the "belief that each being has its own ontological trajectory,"[30] thus undermining objections to the genetic modifications that animals undergo in the technocentric animal industrial complex.[31] Indeed, Haraway has celebrated the technocratic alteration of animals at the genetic level through invasive laboratory experiments.[32] She has also defended other invasive animal experimentation on human interest grounds going so far as to suggest that animals condemned to a laboratory existence enjoy freedom and should co-operate with scientists.[33] As Weisberg further notes, Haraway claims freedom is also available to animals in factory farms.[34] Although Haraway urges empathy in humans toward animals her "conception of empathy is such that one may claim solidarity, sisterhood, and solicitude for the same creatures one condones killing and eating."[35] For Haraway, an instrumentalist orientation toward animals in the context of human power over them is acceptable.[36] She draws a distinction between human–animal instrumental relations (acceptable) and human–human ones (not acceptable).[37]

These and other examples lead Weisberg to conclude that Haraway's ethical vision countenances violence against animals, permitting empathy for them to extend "to no more than an *apology* for systemic abuse."[38] This vision contradicts the Levinasian worldview that Haraway claims she is following. As Weisberg notes, "genuine response and responsibility in any Levinasian sense precludes objectification, instrumentalization and certainly the torture of the Other."[39] That Haraway believes responsibility is a two-way street with instrumentalized animals owing their human captors ethical regard further violates the tenets Levinas espoused.[40]

In fact, Haraway converts the Levinasian command "Thou shalt not kill" (the Other) to "Thou Shall not make killable."[41] This discursive move allows Haraway to signal her alignment with animal theorists who question the social presumptions of which lives we can routinely extinguish and which lives we cannot[42] but serves also to remind us that death is inevitable when species co-evolve and otherwise come into intimate contact with another. By moving from "Thou shalt not kill" to "Thou shalt not make killable," Haraway argues that an ideal interspecies ethic with animals cannot obtain a "pure" position that avoids killing.[43] As she puts it:

> The problem is actually to understand that human beings do not get a pass on the necessity of killing significant others. . . . Try as we might to distance ourselves, there is no way of living that is also not a way of someone, not something, dying differentially.[44]

She reaffirms this view when she writes: "There is no way to eat and not kill, no way to eat and not to become with other mortal beings to whom we are accountable, no way to pretend innocence and transcendence or a final peace."[45]

Haraway acknowledges that "(v)egans come as close as anyone" and accepts "that vegetarianism, veganism, and opposition to sentient animal experimentation can be powerful feminist positions."[46] Yet she implores her readers to avoid what she classifies as dogmatic or pure positions regarding who/what may be consumed and stigmatizing those whose dietary choices are different.[47] After discussing a multitude of ways in which humans are entwined with other species in the close to three hundred preceding pages, it is significant that Haraway closes the highly influential monograph *When Species Meet* with her thoughts on the question of whether it is ethical to kill animals. She seems desirous of forging what she calls a "cosmopolitical engagement" and, more recently, a "compost" ethic that simultaneously respects the impassioned "truths" people hold dear regarding what they eat but can also align with the sense of responsibility toward animals she has charted.[48] She clarifies that hers is not a relativist gesture that accepts all forms of killing, discussing the concept of "killing well" as an outgrowth from Derrida's injunction to "eat well."[49] She states that she respects both the views of her hunting colleagues as well as her ecofeminist ones who oppose hunting and eating animals.[50]

Doubtless, Haraway's implication that even a vegan diet or non-interference in animal lives will still entail some level of killing is accurate. Consider the microorganisms and other bacteria that live within our guts or the predatory responses of carnivores and omnivores to other animals. Even plant-based agriculture can involve the killing of animals and destruction of habitats with the sowing and tilling of fields.[51] To acknowledge this reality, however, as a reason to endorse hunting of sentient animals,[52] associate veganism with dogmatic orthodoxy,[53] or support "humane" industrial production of animals, as Haraway does, is questionable in all instances. Haraway approaches a classic legal welfarist position that animals are ours to instrumentalize as long as the method we employ is virtuous.[54] Haraway's claim that animals may be killed as long as we kill them "responsibly,"[55] that is, having knowledge about their lives and ecological status,[56] differs from liberal legal welfarist positions due to the affective pathway of care and response it harnesses, but the end result for the animals is the same. As Carol J. Adams and Josephine Donovan also note, Haraway's approach is "flawed from an ethic-of-care perspective":[57] the acceptance of routine consumption of animals is incongruent with caring for them.[58] And as Weisberg argues, to suggest, as Haraway does, that "becoming with" animals is possible in the act of (responsibly) killing them impoverishes the meaning of responsibility in feminist call and response ethics in general and as it has been applied to animals.[59]

Haraway's arguments about the inevitability of killing animals institutionally would be more persuasive (yet, still contestable) if she were non-speciesist in how she articulated this claim. It is difficult to overlook the fact that the entangled messiness of interspecies living that necessitates killing does not extend to any humans or the adored dogs in her home.[60] Haraway leaves intact the Derridean "logic of sacrifice" that permits animals to die and to do so in staggering proportions she purports to critique.[61] Indeed, Stephanie Jenkins has identified Haraway's plea for "responsible killing" as exemplary of the trend within posthumanist animal studies to presumptively favour a modified version of the status quo when it comes to animals rather than demand humanity move toward nonviolent relations with animals.[62] Jenkins points to the flattening of complexity in Haraway's concept of "thou shall not make killable" writing that her "argument (that veganism also entails killing) does not distinguish between differences in degree, kind, and intent of killing, which are ethically relevant."[63] Further, drawing on Butler's concept of precarity and the way (o)ur ability to be responsive to others, a prerequisite for responsibility, is found in conditioned, bodily responses,"[64] Jenkins observes that a vegan ethic has the potential to transform our normalized bodily responses to farmed animals so that we come to see their killing not simply as killing but as a violent act.[65] The latter is not a vision that Haraway promotes even if we share Federica Timeto's recent and nuanced assessment "that in time (Haraway) has distanced herself from a more analytical plane and further problematized her own perspective on instrumental interspecies relations."[66]

3. Setting Higher Expectations in a Feminist Legal System That Cares for Animals

a. *Toward a More Animal-Centred Account of Animal Agency*

If posthumanist feminist visions of human relations with animals such as Haraway's result in status quo legal welfarism and continued instrumentalization of animals, we can understand why such models are not ideal for creating an animal-friendly feminist legal system. This does not mean that Haraway or posthumanist feminist writing on animals in general have nothing to offer a feminist legal system that strives to embed care for and responsiveness to animals to prevent animals' exploitation. The challenges to long-standing dualisms are certainly instructive, as is the contestation at the ideological level of human exceptionalism, liberal humanism, and how these epistemologies constrict our legal imagination.[67] As various feminist scholars observe, Haraway's writing in particular also offers a valuable agential rendering of animals (and other nonhumans). Notably, Brianne Donaldson writes that

> (m)any critical animal scholars bristle at Haraway's work, as do I, when she affirms vivisection or meat-eating, when she speaks on behalf of a chicken "willing" to die for our daily bread, or when she affirms the feral pig on the barbecue spit at a faculty dinner.[68]

Yet Donaldson argues that Haraway "demonstrates fresh ways to approach animals, not as passive bodies to be exploited or rescued, but as active partners in "reworlding," that is, partners in shaping alternate futures that increase understanding and decrease violence."[69] Whereas other feminist scholars condemn Haraway's classification of incarcerated animals as "workers" who have responsibility toward their human "caregivers" in the latter's slaughter and research efforts,[70] Donaldson credits Haraway's belief in animal agency as shaking up our entrenched ways of seeing animals "to recast those bodies as players in our philosophies as well as our world-shaping globalizations."[71] This is a position that Cynthia Willett has also endorsed but, notably, not explicitly with respect to animals in laboratories or other sites of human exploitation.[72] Helen Merrick also notes the early and sustained impetus in Haraway's work to normalize animal agency.[73]

Following Timeto on this point, we must raise our expectations for what agency means for animals in a legal context designed to benefit them.[74] Charlotte Blattner offers a more robust protective vision in calling for a legal "animal agency turn," arguing that

> if the law recognized animals as agents, this would crucially change the way we organize human–animal relationships personally and politically: Animals' voice would need to be considered in deciding who deserves legal protection and, relatedly, who gets legal recourse.[75]

The questions then that should arise for lawmakers are to consider how to

> change the macro-dimensions of animals' lives, concerning, e.g. whether they want to live (which, unsurprisingly, most animals do), where and with whom they want to live (humans? Nonhuman animals? A multispecies society?), their communities and social structures (with common decision-making structures, hierarchical, or equality-based), and what their daily routines should look like.[76]

These are questions for which the answers will be unsettling for most present-day human–animal relations and corresponding human and corporate legal entitlements.

b. Toward an Acknowledgement of Violence in Our Use Relationships with Animals

Approaching animal agency this way helps illuminate the power structures that posthumanist feminists downplay in celebrating "entanglement" while condoning death and killing. Donaldson herself does not do this, yet argues that positive social change for animals may still result. She reads Haraway's claim of "becoming-chicken" through eating factory farmed chickens favourably by inscribing it with potential to influence others to care about animals. Donaldson does not see the claim as reproducing domination but as a conceptual attempt to refuse the privilege of human subjectivity to catalyse conceptual change and thus possible material benefit for chickens (and everyone else) in the future.[77] Donaldson acknowledges that Haraway's relational ontology does not lead Haraway to forego eating animals or condemn animal testing; Donaldson views this as "troublesome" and "nepotistic" given the love Haraway expresses for the dogs in her life.[78] Donaldson is also clearly supportive of Jainism and purely plant-based eating.[79] She nevertheless lauds Haraway's respect for those who consume animals "responsibly" and her caution against self-certain political positions as a position itself that recognizes the complexity of ethical life and the inevitability of killing, while also enabling learning and, it is hoped, behavioural change.[80]

On this last point, Donaldson's optimism seems misplaced when we consider global patterns of animal flesh consumption, particularly chicken consumption, is increasing rather than decreasing.[81] The acknowledgement of the inevitability of killing also does not seem to lead many other feminist animal theorists centrally influenced by Haraway's scholarship to condemn routine violent regulatory or commercial practices against animals.[82] Donaldson's recuperative account notwithstanding, many feminists have accepted Haraway's arguments about the inevitability of human violence against animals in forging allegedly non-anthropocentric frameworks and norms toward animals. I want to give one example of this type of troubling application of Haraway's work to illustrate the slim prospect that animal industries will stimulate a healthy re-organization of human–animal relations to qualify her relational ontology as a productive model for a legal framework that seeks to protects animals from exploitation.

Deborah Heath and Anne Meneley defend artisanal foie gras production in the United States as an acceptable feminist relational practice.[83] Their defence of locally produced foie gras attributes its theoretical formation to two main insights: 1) the artificiality of the nature–culture divide and the value of embodiment as a theoretical lens of knowledge-making and 2) the need to intervene in animal rights debates through a relational ethic of care.[84] While they refer to several sites of disciplinary formation for their approach to culturally interpreting artisanal foie gras production, they marshal multiple conceptual parameters from Haraway to frame their understanding of foie gras production.[85]

With respect to the first insight, the authors invoke Haraway's concept of naturecultures to press the point about the continuities rather than sharp lines between what we normally think of as belonging to the realms of "nature" and "culture," respectively.[86] They use this premise to argue that foie gras producers and the geese and ducks they feed have a shared embodiment that should not be read as exploitation or torture but is better understood as an "attentive human–animal co-production" that showcases the producers' skill, knowledge, and "apparent intimacy with their avian co-workers."[87] They write about the "forced funnel feeding, or gavage, that fattens the bird's liver," as an example of "embodied practices" that produce "human–animal intimacy born of a close feeder–feedee relationship" because the geese are fed through a skilled hand.[88] This application of the concept of naturecultures can be read in accordance with Haraway's own problematic view of animals involuntarily confined and killed as workers and those that process them as caretakers.[89] The authors use Haraway's perspectives on human–animal relations to endorse what can otherwise be perceived as an oxymoron, namely "humane" foie gras production.

To the vegan advocates who would have us focus on the experience of foie gras from the animal's perspective, the authors again invoke Haraway's discussion of companion species to warn against anthropomorphism or any moral equivalence between humans and animals.[90] For the authors, Haraway's relational imperative instructs us to value attentive and caring human–animal relationships despite their surrounding capitalist conditions and the violence that awaits the animals at the end of these relationships at the hands of human "caretakers." The presumption of the ethical legitimacy of the human commercial use of animals for food or otherwise is implicit in Heath and Meneley's argument. The authors go so far to suggest that animals are not treated as things when farming is done with "regard for their comfort and wellbeing"[91] with a "focus on intimate, ongoing relations between animals and their caretakers, with an investment in an " 'ethics of care'."[92] Indeed, Heath and Meneley's argument about what a feminist ethic of care requires contains no political parameter to rein in relationality beyond mutual regard for animal and human "well-being." Adopting a relational ethic of care leads Heath and Meneley to espouse a welfarist middle ground that permits animal-based agriculture if the animals and humans are treated "well."[93]

Heath and Meneley claim that their sense of feminist animal ethics is compatible with ecofeminist sensibilities. In their view an ecofeminist relational ethic of care model allows for a framing of the geese and small-scale farmers as co-producers of foie gras.[94] Yet Heath and Menely acknowledge at the outset that leading ecofeminists Carol J. Adams and Josephine Donovan would strongly disagree with their argument that artisanal foie gras (local, small-scale, and non-industrial) is ethical and would consider the authors' harnessing of their work a dissonant use.[95] Indeed, Adams and Donovan's ecofeminist vision and vegan ecofeminist theory in general insists that care for another being mandates a refusal to instrumentalize and commodify.[96] Animal-centred ecofeminist theory also seriously questions human-authored stories that animals voluntarily participate in their own demise for commercial purposes or eagerly accept the appropriations of their bodies through forced funnel-feeding.[97]

The harnessing of an ethic of care to legitimize artisanal foie gras production where care requires no more than "humane" standards understood as purportedly more tender, respectful, and localized relationships between humans and other animals is better viewed as a misguided application of care theory.

In line with Blattner's account, care theory under vegan ecofeminist animal models is meant to be dialogic, demanding that we ask what animals want in localized situations and that we ask those questions without motivation from our own human or corporate interests.[98] In inquiring into animal preferences, of course, humans will never fully "know" or otherwise be able to apprehend what an individual animal is thinking/feeling but "(b) ody language, eye movement, facial expression, tone of voice" can help indicate assent and desire.[99] A further technique for discerning preference is to imagine how we would feel if placed in an identical or similar situation.[100] Heath and Meneley's evaluation of artisanal foie gras producers' claims that the geese and ducks they feed want to undergo gavage does not ask these types of questions or adopt an animal-centric perspective in considering ideal relations in this type of human–animal encounter.

It is Haraway's posthumanist feminist position allowing relationality to embrace different standards for animal use and killing than for humans that the authors harness to promote their ultimately anthropocentric interpretation of what a feminist caring ethos toward animals entails.

The use of Haraway's relational theorizations to reach and legitimate anthropocentric positions regarding who can be killed, confined, contained, and commodified is a prime reason to avoid relying on them or other closely aligned feminist posthumanist relational approaches to interspecies ethics to cultivate a juridical cultural shift away from anthropocentrism.

4. Conclusion: Preferring Vegan Ecofeminist Conceptualizations of Relationality and Care to Transform Legal Systems

A more protective understanding of care, relationality, and inter-species corporeal exchanges for animals is found in vegan ecofeminism, which can be incorporated into a new legal subjectivity for animals,[101] as well as shifting other foundations of liberal legal systems toward ecofeminist values of embodiment, vulnerability, and relationality.[102] Although it is possible to trace ecofeminist commitments through Haraway's work, including the joint desire to topple oppressive structural forces,[103] the strong objection to animal-use industries and human-induced animal death is absent given the relational framing.[104] In her recent overview of how Haraway's thinking about animal-human relations has evolved through her body of work, Timeto observes that Haraway's view of a "composite, multispecies body politic (is) getting closer to the claims of ecofeminism" and that "Haraway increasingly problematizes . . . the domination of animals as living technologies."[105] Yet, important gaps still remain. Vegan ecofeminism's desire to empathically respond to suffering and dramatically improve the lives of beings with whom we are in intercorporeal exchange through an ethic of care does not lead to an endorsement or tolerance of animal instrumentalization, commodification and death as "caring," "loving," or "responsible" acts. While ecofeminist theory is diverse and imperfect, taken together as a matrix of norms and commitments it identifies and impugns the conditions animals endure through anthropogenic forces as violence.[106]

The framework of vegan ecofeminism also blocks a departure point for action that values human lives over animal lives or accepts that, for humans, ensuring their well-being means that we cannot be killed but that the same logic does not apply to animals.[107] As Jennifer Nedelsky, a leading feminist relational theorist, notes in her reading of arguments to engage with the nonhuman from animal rights, environmental ethics, and Indigenous scholarship, "it is striking that the protection of human values sounds in the language of claims of entitlement that (in principle) must be met, while human relations with the nonhuman call forth mere exhortations to responsibility."[108] Vegan ecofeminist multidimensional formulations for animals seek to make animals similarly entitled—they do not adopt a different language for animals or a different pathway to justice.

In this way, a vegan ecofeminist framework is better positioned to go further than a posthumanist relational "companion species" or "becoming-with" ethic in any feminist reconfiguring of the legal system to result in alternative non-violent realities for animals. This feature persists notwithstanding the aligned intersectional feminist commitments of Haraway and other posthumanist feminists with Adams, Donovan, and other ecofeminists to challenge multiple oppressions and move toward much more ethical multispecies worlds as well embrace contextual decision-making.[109] To hope that posthumanist feminist arguments will eventually turn the tide for animals, as Donaldson does in suggesting that Haraway's tolerance for animal commoditization and killing can lead to alternative realities for them where they are not exploited, is not borne out. And while the posthumanist feminist emphasis on the agential capacities of animals is a very welcome conceptual intervention, the animal standpoint ethic of vegan ecofeminist theory better directs legal attention to interpreting and imagining animals' perspectives regarding the type of legal protection from suffering, as well as the empathy and care they need to peacefully exist in such a profoundly interrelated world.[110]

The upcoming work of feminist animal law scholars is to translate the concepts of care and relationality into concrete legal entitlements or at least policy and interpretive principles that will help legislators and judges question present-day "institutions of species imprisonment, enslavement, and slaughter."[111] This is neither a straightforward nor simple task for multiple reasons ranging from human privilege and psychology to the difficulties in achieving transformative social change. Questions about context and contingency and the role for universal legal norms in any legal system also have to be engaged.[112] Vegan ecofeminist theory, for the reasons advanced previously, nevertheless seems a much better principal theoretical toolkit to begin to reorient common law and civilian legal systems to reach an empathy-guided world that benefits animals as well as countering multiple planetary perils fuelled by animal use industries.

Notes

1 Luiz Marques, *Capitalism and Environmental Collapse*, trans. Rebecca de Faria Slenes (Cham: Springer, 2020), 251–261, https://doi.org/10.1007/978-3-030-47527-7.
2 Ian A. Robertson, *Animals, Welfare and the Law: Fundamental Principles for Critical Assessment* (London: Routledge, 2015), 200–205, https://doi-org.ezproxy.library.uvic.ca/10.4324/9780203112311.
3 Charlotte Blattner, Kendra Coulter, and Will Kymlicka, eds., *Animal Labour: A New Frontier of Animal Justice* (Oxford: Oxford University Press, 2019); Irus Braverman, ed., *Animals, Biopolitics, Law: Lively Legalities* (New York: Routledge, 2016); Mathilde Cohen, "Animal Colonialism: The Case of Milk," *AJIL Unbound* 65, no. 3 (2017); Maneesha Deckha, *Animals as Legal Beings: Contesting Anthropocentric Legal Orders* (Toronto: University of Toronto Press, 2021); Jessica Eisen, "Feminist Jurisprudence for Farmed Animals," *Canadian Journal of Comparative & Contemporary Law* 5 (2019); Marie Fox, "Re-thinking Kinship: Law's Construction of the Animal Body," *Current Legal Problems* 57, no. 1 (2004); Eva Bernet Kempers, "Animal Dignity and the Law: Potential, Problems and Possible Implications," *Liverpool Law Review* 41, no. 2 (2020); Martha C. Nussbaum, *Frontiers of Justice: Disability, Nationality, and Species Membership* (Cambridge, MA: Harvard University Press, 2006); Yoriko Otomo and Ed Mussawir, eds., *Law and the Question of the Animal: A Critical Jurisprudence* (Abingdon: Routledge, 2013); Ani B. Satz, "Animals as Vulnerable Subjects: Beyond Interest-Convergence, Hierarchy, and Property," *Animal Law* 16, no. 1 (2009).
4 Eisen, "Feminist Jurisprudence," 150–152.
5 Greta Gaard, "Posthumanism, Ecofeminism, and Inter-Species Relations," in *Routledge Handbook of Gender and Environment*, 1st ed. (Abingdon: Routledge, 2017), 116, 115–129.
6 Undoubtedly, there are some relational elements to some laws within common law and civilian legal systems. But a premise of this chapter is that when compared to other legal systems, notably those that emerge from Indigenous societies around the world, relationality and respect for nonhuman animals is not a constitutive feature of either the common law or civilian legal traditions. For more on this contrast, see Markus Fraundorfer, "The Rediscovery of Indigenous Though in the Modern Legal System: The Case of the Great Apes," *Global Policy* 9, no. 1 (2018).
7 Gaard, "Posthumanism."
8 Ibid., 121. For other summaries of such critiques (that do not translate them legally) see ibid. and Federica Timeto, "Becoming-with in a Compost Society—Haraway Beyond Posthumanism," *International Journal of Sociology and Social Policy* 41, no. 3/4 (2021): 315–330.
9 Donna J. Haraway, *Simians, Cyborgs, and Women: The Reinvention of Nature* (New York: Routledge, 1991); Donna J. Haraway, *Primate Visions: Gender, Race and Nature in the World of Modern Science* (New York: Routledge, 1989); Donna J. Haraway, *When Species Meet* (Minneapolis: University of Minnesota Press, 2008); Donna J. Haraway, *The Companion Species Manifesto: Dogs, People, and Significant Otherness* (Chicago: Prickly Paradigm Press, 2007). An influential trope from her most recent work is "staying with the trouble" caused by global crises

and inequalities; see Donna J. Haraway, *Staying with the Trouble: Making Kin in the Chthulucene*. Durham: Duke University Press, 2016, 1–2.
10 Gavin Rae, "The Philosophical Roots of Donna Haraway's Cyborg Imagery: Descartes and Heidegger Through Latour, Derrida, and Agamben," *Human Studies* 37, no. 4 (2014): 505–506.
11 Ibid., 506.
12 Ibid.
13 Ibid.
14 Donna Haraway, "Cyborg Manifesto," in Haraway, *Simians, Cyborgs, and Women*, 150.
15 Haraway, *Companion Species Manifesto*.
16 Zipporah Weisberg, "The Broken Promises of Monsters: Haraway, Animals and the Humanist Legacy," *Journal for Critical Animal Studies* 7, no. 2 (2009): 27. Haraway has identified companion species as the larger umbrella concept that encapsulates cyborgs (ibid., 26).
17 Haraway, *When Species Meet*, 3–4.
18 Rosi Braidotti, "Posthuman, All Too Human: Towards a New Process Ontology," *Theory, Culture & Society* 23, nos. 7–8 (2006): 197. See, for example, its salience to an examination of elephant performance in India in Ameet Parameswaran, "Zooesis and 'Becoming with' in India: The 'Figure' of Elephant in Sahyande Makan: The Elephant Project," *Theatre Research International* 39, no. 1 (2014): 5–19.
19 Braidotti, "Posthuman," 199–200.
20 Haraway, *When Species Meet*, 71–88.
21 Ibid., 197.
22 Lori Gruen and Kari Weil, "Animal Others—Editors' Introduction," *Hypatia* 27, no. 3 (2012): 481 citing Rosi Braidotti, *Metamorphoses: Towards a Materialist Theory of Becoming* (Cambridge, UK: Blackwell, 2002).
23 Braidotti, "Posthuman," 199.
24 Ibid., 200.
25 Helen Merrick, "Naturecultures and Feminist Materialism," in *Routledge Handbook of Gender and Environment*, 1st ed. (Abingdon: Routledge, 2017), 101–114.
26 Braidotti, "Posthuman," 199.
27 Eva Giraud, "'Beasts of Burden': Productive Tensions between Haraway and Radical Animal Rights Activism," *Culture, Theory and Critique* 54, no. 1 (2013): 102.
28 Weisberg, "Monsters," 23.
29 Weisberg, "Bacteria," 98–99.
30 Ibid., 101.
31 Ibid.
32 See Donna Haraway, *Modest_Witness@Second_Millennium.FemaleMan_Meets_OncoMouse: Feminism and Technoscience* (New York: Routledge, 1997).
33 Weisberg, "Monsters," 33–34, citing Haraway, *When Species Meet*, 73.
34 Ibid., 36.
35 Weisberg, "Bacteria," 100, citing Haraway, *When Species Meet*, 296–300 and Weisberg, "Monsters," 49. Greta Gaard also notes how Haraway makes light of animals' plight through humour. Gaard, "Posthumanism," 124.
36 Weisberg, "Monsters," 28.
37 Ibid., 29.
38 Ibid., 39.
39 Ibid., 42.
40 Ibid.
41 Ibid.
42 Haraway, *When Species Meet*, 79–80.
43 Ibid., 80.
44 Ibid.
45 Ibid., 295.
46 Ibid.
47 Ibid.
48 Ibid., 299. Donna Haraway, *Staying with the Trouble*, 136–168.
49 Weisberg, "Monsters," 45.

50 Haraway, *When Species Meet*, 300.
51 Weisberg, "Monsters," 44.
52 Haraway, *When Species Meet*, 295–299.
53 Ibid., 299.
54 Kempers, "Animal Dignity," 179–180.
55 Haraway, *When Species Meet*, 80–81.
56 Weisberg, "Monsters," 44.
57 Carol J. Adams and Josephine Donovan, eds., *The Feminist Care Tradition in Animal Ethics* (New York: Columbia University Press, 2007), 12.
58 Ibid., 13.
59 Ibid. citing Haraway, *When Species Meet*, 295.
60 Gaard, "Posthumanism" 124.
61 Haraway, *When Species Meet*, 78.
62 Stephanie Jenkins, "Returning the Ethical and Political to Animal Studies," *Hypatia* 27, no. 3 (2012): 505.
63 Ibid., 507.
64 Ibid., 508.
65 Ibid., 509.
66 Timeto, "Becoming-with," 325.
67 Irus Braverman, "Law's Underdog: A Call for More-Than-Human Legalities," *Annual Review of Law and Social Science* 14, no. 1 (2018): 127–144; Yoriko Otomo, "Making Lawful Animals," in *Routledge Handbook of Law and Theory*, ed. Andreas Philippopoulos-Mihalopoulos (Taylor and Francis, 2018), 317–325.
68 Brianne Donaldson, *Creaturely Cosmologies: Why Metaphysics Matters for Animal and Planetary Liberation* (Maryland: Lexington Books, 2015), 100.
69 Ibid., 99.
70 Weisberg, "Monsters," 37–38; Gaard, "Posthumanism," 123.
71 Donaldson, *Creaturely Cosmologies*, 104.
72 See Cynthia Willett, *Interspecies Ethics* (New York: Columbia University Press, 2014), 56. Recall that Willett is concerned with diluting the picture of animals as always already suffering through human recognition of their agential capacities. She thus believes social histories of labor would benefit from investigating how animals labor and are alienated from their labor (ibid.).
73 Merrick, "Naturecultures," 106–107; Timeto, "Becoming-with," 321.
74 C.E. Blattner, "Turning to Animal Agency in the Anthropocene," in *Animals in Our Midst: The Challenges of Co-existing with Animals in the Anthropocene: The International Library of Environmental, Agricultural and Food Ethics*, Vol. 33, eds. B. Bovenkerk and J. Keulartz (Cham: Springer, 2021), https://doi.org/10.1007/978-3-030-63523-7_4.
75 Ibid., 76.
76 Ibid., 74.
77 Donaldson, *Creaturely Cosmologies*, 104. Willett also appears, contra to Haraway, to object to human consumption of animals when she writes: "Not all animals solicit a sentimental response from humans, and to this respect of course there are some limits to Haraway's communitarian approach. Bunnies are cute and cuddly, but tender young chickens end up more readily as fryers" (Willett, *Interspecies Ethics*, 96). See also her discussion of disgust in relation to Haraway's reflections on eating meat (Willett, *Interspecies Ethics*, 113–114).
78 Donaldson, *Creaturely Cosmologies*, 105.
79 Ibid., 109–110.
80 Ibid at 108–109. I respect that Haraway tells her readers what her views are on eating animals and animal experimentation. I view it as a missed opportunity that Willett is not clearer about her own position on eating animals or using animals in experimentation given her praise and considerable reliance on Haraway's insights through *Interspecies Ethics*. Willett, *supra* note 94 at 17–18, 24, 27, 57–63, 66, 70–3, 78, 94–96, 114.
81 Camilla Govoni et al., "Global Assessment of Natural Resources for Chicken Production," *Advances in Water Resources* 154 (2021): 1, https://doi.org/10.1016/j.advwatres.2021.103987.
82 Robert Leston, "Deleuze, Haraway, and the Radical Democracy of Desire," *Configurations* 23, no. 3 (2015): 355–376.

83 Deborah Heath and Anne Meneley, "The NatureCultures of Foie Gras: Techniques of the Body and a Contested Ethics of Care," *Food, Culture and Society* 13, no. 3 (2010): 422.
84 Heath and Meneley draw upon Haraway's concept of "naturecultures" as well as other theorists who are keen to signal a renunciation of the nature-culture divide in favor of understanding the integrated manner in which nature and culture relate (ibid., 424–425).They cite to Bruno Latour and, again, to, to locate their understanding of and wish to advance a reading of the foie gras debate through the idea of "embodied epistemologies" (ibid., 425).
85 Ibid., 424–425.
86 Ibid., 424.
87 Ibid., 427, 448.
88 Ibid., 428.
89 Ibid., 440.
90 Ibid., 441. The authors cite the following statement of Haraway's: "multispecies flourishing requires a robust nonanthropomorphic sensibility that is accountable to irreducible differences" (ibid.) citing Haraway, *Companion Species Manifesto*, 90.
91 Ibid.
92 Ibid.
93 Ibid., 428.
94 Heath and Meneley present ethnographic information about two family farms, one in France and the other in California, and the relationship the human farmers have with the geese they raise for their fattened livers (ibid., 422).
95 Ibid., 425.
96 Carol J. Adams and Lori Gruen, *Ecofeminism: Feminist Intersections with Other Animals & the Earth* (New York: Bloomsbury, 2014), 26.
97 The positive description of the producers' feeding practices is particularly revealing in this regard. Heath and Menely, "NatureCultures of Foie Gras," 427.
98 Donovan, "Treatment of Animals," *supra* note 10 at 323–324.
99 Ibid., 321.
100 Ibid., 322.
101 I make this argument regarding a new legal subjectivity for animals that I call "beingess" in Deckha, *Animals as Legal Beings*, 121.
102 Ibid.
103 Merrick, "Naturecultures," 110.
104 Susan Fraiman, "Pussy Panic Versus Liking Animals: Tracking Gender in Animal Studies," *Critical Inquiry* 39, no. 1 (2012): 89–115, 113–114.
105 Timeto, "Becoming-With," 324, 26.
106 Gaard, "Posthumanism," 116–118, 126.
107 S. Marek Muller, *Impersonating Animals: Rhetoric, Ecofeminism, and Animal Rights Law* (East Lansing: Michigan State University Press, 2020), 139–142.
108 Jennifer Nedelsky, *Law's Relations: A Relational Theory of Self, Autonomy, and Law* (New York: Oxford University Press, 2011), 195.
109 On such alignment, see Merrick, "Naturecultures," 110 and Fraiman, "Pussy Panic," 112–115.
110 Gaard, "Posthumanism," 117.
111 Ibid., 126.
112 Oriko, "Making," 319.

Bibliography

Adams, Carol J., and Donovan, Josephine, eds. *The Feminist Care Tradition in Animal Ethics*. New York: Columbia University Press, 2007.

Adams, Carol J., and Gruen, Lori. *Ecofeminism: Feminist Intersections with Other Animals & the Earth*. New York: Bloomsbury, 2014.

Blattner, Charlotte. "Turning to Animal Agency in the Anthropocene." In *Animals in Our Midst: The Challenges of Co-existing with Animals in the Anthropocene: The International Library of Environmental, Agricultural and Food Ethics*, Vol. 33, edited by B. Bovenkerk and J. Keulartz. Cham: Springer, 2021. https://doi.org/10.1007/978-3-030-63523-7_4.

Blattner, Charlotte, Coulter, Kendra, and Kymlicka, Will, eds. *Animal Labour: A New Frontier of Animal Justice*. Oxford: Oxford University Press, 2019.
Braidotti, Rosi. "Posthuman, All Too Human: Towards a New Process Ontology." *Theory, Culture & Society* 23, nos. 7–8 (2006).
Braverman, Irus, ed. *Animals, Biopolitics, Law: Lively Legalities*. New York: Routledge, 2016.
Braverman, Irus. "Law's Underdog: A Call for More-Than-Human Legalities." *Annual Review of Law and Social Science* 14, no. 1 (2018): 127–144.
Cohen, Mathilde. "Animal Colonialism: The Case of Milk." *AJIL Unbound* 65, no. 3 (2017).
Deckha, Maneesha. *Animals as Legal Beings: Contesting Anthropocentric Legal Orders*. Toronto: University of Toronto Press, 2021.
Donaldson, Brianne. *Creaturely Cosmologies: Why Metaphysics Matters for Animal and Planetary Liberation*. Maryland: Lexington Books, 2015.
Eisen, Jessica. "Feminist Jurisprudence for Farmed Animals." *Canadian Journal of Comparative & Contemporary Law* 5 (2019).
Fox, Marie. "Re-thinking Kinship: Law's Construction of the Animal Body." *Current Legal Problems* 57, no. 1 (2004).
Fraiman, Susan. "Pussy Panic Versus Liking Animals: Tracking Gender in Animal Studies." *Critical Inquiry* 39, no. 1 (2012).
Fraundorfer, Markus. "The Rediscovery of Indigenous Though in the Modern Legal System: The Case of the Great Apes." *Global Policy* 9, no. 1 (2018).
Gaard, Greta. "Posthumanism, Ecofeminism, and Inter-Species Relations." In *Routledge Handbook of Gender and Environment*, 1st ed., 115–129. Abingdon: Routledge, 2017.
Giraud, Eva. "'Beasts of Burden': Productive Tensions between Haraway and Radical Animal Rights Activism." *Culture, Theory and Critique* 54, no. 1 (2013).
Govoni, Camilla et al. "Global Assessment of Natural Resources for Chicken Production." *Advances in Water Resources* 154 (2021). https://doi.org/10.1016/j.advwatres.2021.103987.
Gruen, Lori, and Weil, Kari. "Animal Others–Editors' Introduction." *Hypatia* 27, no. 3 (2012). Citing Rosi Braidotti, *Metamorphoses: Towards a Materialist Theory of Becoming*. Cambridge, UK: Blackwell, 2002.
Haraway, Donna J. *Primate Visions: Gender, Race and Nature in the World of Modern Science*. New York: Routledge, 1989.
Haraway, Donna J. *Simians, Cyborgs, and Women: The Reinvention of Nature*. New York: Routledge, 1991.
Haraway, Donna J. *Modest_Witness@Second_Millennium.FemaleMan_Meets_OncoMouse: Feminism and Technoscience*. New York: Routledge, 1997.
Haraway, Donna J. *The Companion Species Manifesto: Dogs, People, and Significant Otherness*. Chicago: Prickly Paradigm Press, 2007.
Haraway, Donna J. *When Species Meet*. Minneapolis: University of Minnesota Press, 2008.
Haraway, Donna J. *Staying with the Trouble: Making Kin in the Chthulucene*. Durham: Duke University Press, 2016.
Heath, Deborah, and Meneley, Anne. "The NatureCultures of Foie Gras: Techniques of the Body and a Contested Ethics of Care." *Food, Culture and Society* 13, no. 3 (2010).
Jenkins, Stephanie. "Returning the Ethical and Political to Animal Studies." *Hypatia* 27, no. 3 (2012).
Kempers, Eva Bernet. "Animal Dignity and the Law: Potential, Problems and Possible Implications." *Liverpool Law Review* 41, no. 2 (2020).
Leston, Robert. "Deleuze, Haraway, and the Radical Democracy of Desire." *Configurations* 23, no. 3 (2015).
Marques, Luiz. *Capitalism and Environmental Collapse*. Translated by Rebecca de Faria Slenes, 251–261. Cham: Springer, 2020. https://doi.org/10.1007/978-3-030-47527-7.
Merrick, Helen. "Naturecultures and Feminist Materialism." In *Routledge Handbook of Gender and Environment*, 1st ed. Abingdon: Routledge, 2017.
Muller, S. Marek. *Impersonating Animals: Rhetoric, Ecofeminism, and Animal Rights Law*. East Lansing: Michigan State University Press, 2020.
Nedelsky, Jennifer. *Law's Relations: A Relational Theory of Self, Autonomy, and Law*. New York: Oxford University Press, 2011.
Nussbaum, Martha C. *Frontiers of Justice: Disability, Nationality, and Species Membership*. Cambridge, Mass: Harvard University Press, 2006.

Otomo, Yoriko. "Making Lawful Animals." In *Routledge Handbook of Law and Theory*, by Andreas Philippopoulos-Mihalopoulos, 317–325. Taylor and Francis, 2018.
Otomo, Yoriko, and Mussawir, Ed, eds. *Law and the Question of the Animal: A Critical Jurisprudence*. Abingdon: Routledge, 2013.
Parameswaran, Ameet. "Zooesis and 'Becoming with' in India: The 'Figure' of Elephant in Sahyande Makan: The Elephant Project." *Theatre Research International* 39, no. 1 (2014).
Rae, Gavin. "The Philosophical Roots of Donna Haraway's Cyborg Imagery: Descartes and Heidegger Through Latour, Derrida, and Agamben." *Human Studies* 37, no. 4 (2014): 505–506.
Robertson, Ian A. *Animals, Welfare and the Law: Fundamental Principles for Critical Assessment*. London: Routledge, 2015. Retrieved from: https://doi-org.ezproxy.library.uvic.ca/10.4324/9780203112311.
Satz, Ani B. "Animals as Vulnerable Subjects: Beyond Interest-Convergence, Hierarchy, and Property." *Animal Law* 16, no. 1 (2009).
Timeto, Federica. "Becoming-with in a Compost Society—Haraway Beyond Posthumanism." *International Journal of Sociology and Social Policy* 41, no. 3/4 (2021): 315–330.
Weisberg, Zipporah. "The Broken Promises of Monsters: Haraway, Animals and the Humanist Legacy." *Journal for Critical Animal Studies* 7, no. 2 (2009).
Willett, Cynthia. *Interspecies Ethics*. New York: Columbia University Press, 2014.

13
ENTANGLED SUBJECTIVITY
On Care Ethics and Service Dogs

Stephanie Jenkins

In 2013, I began training and working with a service dog, a rescued lab mix named Mandy of unknown age and origin. Over time, she learned and performed several tasks related to my chronic pain and endocrine conditions, and I depended on her for my health and safety. Her status as a service animal meant she navigated spaces and had experiences from which most dogs are excluded. She traveled domestically and internationally by plane, went to academic conferences, attended my philosophy classes and office hours on campus, visited—but did not dine in—public restaurants, and even attended rock concerts (with hearing protection). When she was off duty we played fetch, visited beaches, and snuggled and watched Netflix. We did the things friends do, with an essential difference. We traveled as a collective. That partnership carried not only legal rights but also ethical, epistemological, and ontological consequences. Our subjectivities were entangled; we were not one but two: a duo of misfits navigating a speciesist and ableist world that is not built for either one of us, let alone the crip human–animal hybrid that we—and other service dog teams—embodied.

This chapter is an act of care and grief. Mandy passed two years ago, and it wasn't until her death that I realized how entangled our subjectivities had become. While her legal status was equivalent to a wheelchair, I mourned her passing in a way I did not the ending of my working relationship with my scooter, when my health improved enough that I stopped using it. Grief is the starting point of my argument, because it undoes the subject; it had undone me. Or, more to the point, it had undone "us." Who is us? The affective ties grief reveals makes evident an ethics of care, highlighting the significance of interspecies interdependence and friendship.

In this chapter, I will review the essential disability studies literature critiquing the legal definition of service dogs, identify and evaluate an alternative care ethics to supplement this literature, and propose my own understanding of entangled subjectivity as a contribution to these discussions. What is a service dog? How can a care ethic help us think about this question? What does a service dog team tell us about subjectivity?

DOI: 10.4324/9781003273400-17

What Is a Service Animal (Or Dog)?

Much of the disability studies literature on service animals addresses their ethical, legal, and ontological status. What is a service animal? What tasks do they perform, and how does this impact the experience of disability?

In the United States, under the 1990 Americans with Disabilities Act and 2010 amendment, a service animal is defined as a dog or miniature horse trained to perform a task to assist a person with disability. The Department of Justice offers the following guidance:

> Service animals are defined as dogs that are individually trained to do work or perform tasks for people with disabilities. Examples of such work or tasks include guiding people who are blind, alerting people who are deaf, pulling a wheelchair, alerting and protecting a person who is having a seizure, reminding a person with a mental illness to take prescribed medications, calming a person with Post Traumatic Stress Disorder (PTSD) during an anxiety attack, or performing other duties. Service animals are working animals, not pets. The work or task a dog has been trained to provide must be directly related to the person's disability. Dogs whose sole function is to provide comfort or emotional support do not qualify as service animals under the ADA.[1]

Categorization as a service animal permits access to public spaces, including those of government, business, and non-profit organizations, from which non-human animals would normally be excluded. In other words, the animal's presence is considered a reasonable accommodation for the individual with disability. In the same way that a restaurant must permit and make space for a wheelchair or scooter at the table and restroom, it must do so for a service animal team.

Two significant essays in the disability studies literature stand out regarding their efforts to think through the implications of this legal definition: Kelly Oliver's "Service Dogs: Between Animal Studies and Disability Studies" and Margaret Price's "What Is a Service Animal? A Careful Rethinking."[2] Both scholars begin their arguments with close readings of legal documents to demonstrate that a function-focused definition of service animals that distinguishes them from pets and emotional support animals a) reduces them to equipment and b) produces what Price calls, citing Ellen Samuels, a "biocertification" process of policing a sharp boundary between legitimate and illegitimate service animals.[3] The problem of service dog legitimacy is so significant that Sorenson and Matsuoka describe it as a "moral panic."[4] Price digs a bit deeper into the legitimacy boundary to examine professional training organization materials, service dog community publications, and news articles dedicated to identifying "fake" service dogs. The essential difference between a real and fake service dog, according to Price's reading of these documents, is behavior; a dog must perform work or task directly related to the handler's disability and—this is the point—they must be trainable. Therefore, providing comfort, companionship, and/or emotional support do not qualify as service animal tasks, because they are natural behaviors and are not trained on unique cues. In fact, "dogs whose sole function is to provide comfort or emotional support" are explicitly excluded as service animals under the ADA definition.[5]

It is here—the excluded affective ties—where Oliver and Price concentrate their criticism of the function-focused definition of service animals; the caring, intimate connections between a service dog and handler defy attempts to ontologically reduce service animals to equipment, or, as Price names it, citing Karen Barad, thingification.[6] Moreover, highlighting the significance of the extra-legal companionship to which these authors and anyone who has ever worked with a service animal will attest, loosens the division between real and fake service animals, pointing toward a more complex, relational definition.

Oliver focuses on the ambiguous status of service animals, despite their legal qualification as property. "[L]aws can't prevent people from becoming emotionally attached to their service animals," she writes.[7] The emotional interdependence and companionship that characterize the service animal partnership, she argues, exemplifies the essential of emotional and physical support from others for human existence. The failure to recognize this fundamental vulnerability is what she calls the "disavowal of dependence."[8]

While Oliver's essay uses service animals as a special case to springboard into animal ethics more generally, Price's article focuses on the ontology of service dogs, offering a phenomenological and ethical account of her experience working with a service dog who others may view as illegitimate. She examines a number of examples to explore the boundary between a legitimate, trainable service task and comfort. "[A]t what point does providing support or comfort become a task?" she asks. If, for example, a dog can perform an entertaining trick that cheers up a depressed person on cue, are they a service animal?[9]

For Price, the point of troubling the sharp distinction between comfort and task is to recognize the "subtler, more nuanced relation between [handlers and their [service] animals than legal or even public definitions would allow."[10] The caring relationships between service animals and their human partners, as reflected in lived experience, are more nuanced, intimate, and complex. This is demonstrated in her description of a traumatic incident on an airplane when she was without her service dog, Ivy. When her partner arrived at the airport with Ivy, Price writes, "I am not in need of my service animal specifically. I just need anyone preferably everyone in my family."[11] Here Ivy, rather than performing a trained task, is providing the loving support of a family member or close friend.

Other first-person accounts describing service dog teams in the disability studies literature support the relational accounts found in Price and Oliver. In *Beasts of Burden*, Sunaura Taylor emphasizes the way her service dog, Bailey, defies the legal and professional requirements of a service animal. His size is impractical for her needed tasks; she did not follow training guidelines or even read the training books she ordered. And, like most shelter dogs, Bailey had significant trauma that interfered with his training and ability to trust. Yet he provided an important social function: what Taylor calls "mediating the outside world";[12] instead of Taylor being the focus of ableist curiosity, stares, and privacy violations, Bailey became the center of attention. He is also able to read Taylor and her partner's emotions, just as she is able to do for him. Bailey, she writes, is "attentive to my emotions, needs, and whereabouts, his very presence helping to mediate between me and the ableism I encounter when we go on our daily walks together."[13]

In "Working Resonance," Emily K. Michael identifies the reason why care defines the service animal relationship; dogs will not work for handlers they do not trust. York, she writes describing her service dog, "works with me because he's smart and dedicated—and we have learned to work together. He's not a preordered, prepackaged machine that obeys every command."[14] Rather, she describes him as best friend, partner, and companion.

Service Dog Teams as Interspecies Care Work

I follow these four authors' call to revalue the excluded, extra-legal affective ties of the service animal and to challenge the policing of the legitimate service dog. To this end,

I introduce an interspecies care ethic, based on Joan Tronto's four phases of caring. Here caring is a mutual, on-going process through which service dog teams engage in the processes of everyday living and navigating an ableist, speciesist world as a collective. Through care, each team member takes the concerns and needs of the other as catalysts for action. This relational, interdependent model for understanding the service dog team relationship is a supplement to Price, Oliver, Taylor, and Michael's accounts. In *Moral Boundaries*, Tronto identifies four phases of caring: caring about, taking care of, care-giving, and care-receiving.[15] In the service dog team relationship, both team members engage in and benefit from all four components. By understanding care as an interspecies practice, we challenge the legal reduction of the service dog to equipment and the policing of the boundaries of the legitimate service dog team.

First, *caring about*, is the phase that recognizes that care is needed. This is noticing and paying attention to the need to care in the first place.[16] For the human partner, this process begins with the search for the service dog candidate; there is one for whom care is necessary. Caring about, for Tronto, is individually, culturally, and politically shaped. My search for a search dog candidate began not with breeders or service dog organizations but in animal shelters in Oregon. I cared not just about the individual dog but about the political conditions in which working dogs are bred for human purposes. The dog cares about when they become responsive to human interaction and training. When I first met Mandy at Safehaven shelter, I took her into an enclosed field to interact with her. I practiced basic commands and played fetch with her. She cared about me in her interactions, but I was not yet *her* human; she performed these tasks with the staff and other visitors. Caring about initiates the caring relationship, but it is not yet individuated.

Taking *care of*, the next phase of care, "involves assuming responsibility for the identified need and determining how to respond to it."[17] This is the recognition that one is in a place to do the caring. In other words, taking care is the mineness of the activity of care through which one assumes responsibility to care for the other. The human partner of the team, taking care of their canine companion, takes on guardianship of another individual living being with emotional, cognitive, and physical needs when taking on a service dog. They accept responsibility for training, veterinary care, grooming, feeding, play, bathroom schedules, and more. The canine partner, in taking care of their human partner, takes on responsibility for responding to the needs of the human. This can be observed through the ways they pay attention through constant, affectionate observation (seen in the direction of the eyes, nose, and ears) and behavior (positioning themself close to their human, following around, licking, and so forth). The dog comes to recognize and choose their human as their own.

I took on the responsibility to take care of Mandy when I signed her adoption papers at the animal shelter where I found her. I knew this meant I was taking on the dual responsibility of caring for her as a dog and training her as a service animal. The latter would be unusual and difficult, because she was an adult rescue dog rather than a puppy bred to be a working dog. Mandy, as a rescue, was shy and anxious when I brought her home. She was slow to cultivate a relationship with me, but it was clear she had chosen to take care of me when she first climbed into my lap and licked my face. She gradually became more attentive, watching me like I was a television show, guarding the door, and concentrating during training sessions. Anyone who has loved, lived with, and worked closely with a dog companion understands the unique bond formed between the dog and their caretaker. The relationship is even more intense with a service dog, when the physical and/or emotional well-being of the human depends on tasks performed. The willingness to work with and for a particular human, displays of affection, and acts of loyalty all demonstrate the dog's participation in taking care. I'm convinced that Mandy was aware her service dog tasks helped me and that she performed them for me—and not others—as taking care.

Third, *care-giving*, is the "direct meeting of the needs of care."[18] In other words, care-giving is the work of care itself. Care, for Tronto, must be competent. The human partner performs the daily tasks of taking care of their canine team member: providing exercise, bathroom breaks, meals, playtime, veterinary care, and companionship. The canine partner provides care-giving through performing trained tasks and providing loving affection. While service dog tasks are trained, they are skilled, caring activities that service dogs perform daily to care for their handlers.

The mutual, interdependent nature of the care-giving phase helps to demonstrate the problems of the disavowal of dependence and the exclusion of affective bonds from the legal definition of service dog. Sometimes being a willing recipient of care should be recognized as the service dog's task. Tasks are not unidirectional; they are always performed in a relationship. Being a companion in public space, as a socialized and trained companion, can itself be the task the dog performs when the care-giving done by the human mitigates an impairment. As noted, Price notes the blurred boundary between task and support, but I want to make a stronger argument that support itself should be recognized as a task. For example, a service dog may provide mobile reality testing for someone who experiences panic attacks or hallucinations; the affective and tactile presence, in an airport or work environment where one is prone to such episodes, of a loyal companion can serve this function with obedience training and socialization. Or, care-giving for a service dog may force a human with depression to care for themselves. In times of isolation, social bonds with a service dog may be easier to cultivate than to maintain with other humans.[19] During the depths of the COVID lockdown period, I experienced a deep depression so extreme that I was barely able to get out of bed. The only motivation I had to get up was to take care of Mandy. I took her on walks around the block and eventually to the dog park. Over time, this activity—with other treatment—helped pull me out of the worst of my illness. I began to set bigger goals like taking her to a different beach each weekend. In this respect, her existence as a canine companion served a therapeutic function, even though there was not a specific trained task being used other than her obedience training.

Finally, *care-receiving* recognizes that "the object of care will respond to the care it receives."[20] This is the response of the cared for to the care that is given. Care must be competent; the needs of the cared for must be met.[21] The dog eats their meals, enjoys their walks, goes potty on schedule, frolics at play, and relishes their pettings. The human stabilizes with a grounding assist, avoids an emergency when the dog retrieves a medication, calms when the dog interrupts a flashback or panic attack, avoids a collision when a guide dog signals there is a moving object, and more. The care-receiving of the service dog team is interdependent. For Mandy to meet my needs, she required a balance of play and work time. If her exercise demands were not met, she made her desire for activity known by dropping toys at my feet and placing a paw on my leg. But it was her diligent everyday care for me that enabled my ability to maintain both of our care schedules.

Considered as the four phases—caring about, taking care, care-giving- and care-receiving—care is a practice rather than simply an emotion or principle.[22] For the service dog team, it is a reciprocal, interdependent practice in which both parties participate, enabling the care of the other. By emphasizing the role of the dog as a subject of care, this ethic combats the thingification of service dogs in legal and public conceptions of service dogs through which they are reduced to equipment;[23] the four phases highlight the dog's active role as a subject and receiver of care. Both the human and the dog are interspecies care workers and receivers.

As Price and Samuels have noted, in their current form, ADA laws and service dog training manuals serve as a kind of biocertification; the mechanisms for verifying the legitimacy of the dog function to authenticate the realness of the handler's disability. In their article using a moral panic framework to analyze media reports about fake service animals, Sorenson and Matsuokoa note that accusations of a service animal being fake tend to mean one of two claims: a) the handler is faking their disability, "[portraying] themselves as disabled users of serviced animals in order to have public access with pets," or b) the dog is fake because they do not meet training standards.[24] The latter Price notes in her claim that the practical distinction between the legitimate and illegitimate service dog is trainability. Policing of the dog's trainability—claim two—is, in my view, a subtle version of claim one; arguing a dog is improperly trained or does not perform a task related to the handler's disability is to claim the dog, as an assistive companion, is unneeded. This charge is similar to the policing of ADA parking spaces familiar to people with non-apparent disabilities or the accusations of malingering against individuals who use wheelchairs for chronic fatigue or pain experience.

Viewing the service dog team through the four-phase relationship of care displaces the biomedical model for viewing disability; rather than understanding disability as a defect that must be certified, it exists in an interdependent, mutual relationship of care. Affective ties and comfort would no longer be disavowed but recognized as essential aspects of taking care and care-giving. Rather than the focus being on whether a service dog is real or fake, the emphasis will be on if the team is caring well for one another. Care-giving, for Tronto, is a skill and requires competence. This means that both members of the team will need training; the human must know how to properly care for and handle a service dog in public and the dog must, for practical and safety matters, behave with advanced obedience, socialization, and disability-specific training in public. Trainability, when viewed through the lens of care as a skill, is not inherently biocertification. Poorly trained dogs should not have public access rights. If a dog is not potty trained, they should not be on an airplane. If a dog cannot sit, stay, and "leave it," they should not be in a restaurant. A dog who is aggressive or bites shouldn't be in a campus lecture hall. At the same time, handlers who fail to care-give forfeit their right to public access as a service dog team. For example, once, at a summer concert in Atlanta, I witnessed a dog who was left alone unvested, tied with a rope to a pole in a crowd, without ear protection or water at a rock concert. In my view, the human handler had abandoned their dog and not provided basic care in a dangerous environment, failing to care-give well.

Service Dog as Entangled Subjectivity

While I admire and have found creative and scholarly inspiration in the previous four relational accounts of service animals, I don't think they go quite far enough ontologically or ethically. Price, Oliver, Taylor, and Michael all define service dogs in terms of intimate relationships and advocate some form of ethics of proximity. Rather than thinking about a service animal as one pole of a two-part relationship, I would like the emphasis to be on the relationship itself: the caring team. That is to say, my experience of working with a service dog was one of a dual, co-authored subjectivity; I experienced the world as a "we," a plurality, rather than as a single individual plus a dog. It's important to note that I don't think the authors would necessarily disagree with this shift. In fact, Price gestures towards it in her essay when she cites Rod Michalko's *The Two-in-One* in which the relationship between the leader and follower is fluid, characterized by a process of becoming.[25] But I would at least like to argue for a perspective shift and think through what the implications of that difference would make. I call this crip, human–animal hybrid entangled subjectivity, inspired by Lori Gruen's concept of entangled empathy and care ethics' critique of the autonomous male subject.[26] It follows in the lineage of feminist philosophy's exploration of pregnant embodiment,[27] care ethics' relational subject, Alice Dreger's challenge to the primacy of "singletons" (or non-conjoined individuals) in *One of Us*,[28] and Donna Haraway's cyborg.[29]

My service dog team hybrid subjectivity was true phenomenologically, medically, biologically, and bodily and is revealed in the taking care, care-giving, and care-receiving phases of caring well. *Phenomenologically,* Mandy and I were inseparable. We lived, worked, played, traveled, and slept together. When friends or colleagues asked, "How are you?" I would respond "Our day is going well" or "We had a stressful morning." *Medically*, we shared an endocrine condition and took one of the same replacement hormones, but I am allergic to one of her pain medications. This needed to be accounted for when handling; the first time a veterinarian asked me if I wanted to pre-authorize CPR or sign a DNR, I realized I was making health decisions for us both. *Biologically*, as Donna Haraway noted about her research dog, living and breathing together meant Mandy's cells had colonized all of my cells, we were "constitutively companion species."[30] *Bodily*, we shared embodied habits like sense of space and comportment. Our biorhythms for waking, resting, and going to the bathroom were synced. Even my sense of physical safety was tied to her presence: I can ignore my phone alarm to take my medication but not her nose; she let me know if someone was at the door.

The extent of our entangled subjectivity was revealed to me in its absence; Mandy very quickly and suddenly declined two years ago. She was an older dog, and as a teacher of existentialism I'm all too familiar with the finitude of existence, so I was conceptually aware of and even practically prepared for her passing. I've been an animal lover my entire life; I've lost many animal friends before, and yet nothing prepared me for or compared to this loss that I could only describe in terms of having pieces of my physical body ripped out. As I relearned my embodied habits, designed a new safety plan, readjusted my biorhythms, and ever so slowly got more used to moving in the world alone, I came to the realization that I had, in fact, in the eight years I worked, lived, and played with Mandy, become a "we." If we are reimagining and refashioning philosophical tools and, building upon the previous authors, and we use entangled subjectivity as a new model, what are we looking at? A service dog team whose beings have been co-shaped at every level. A duo, a two-in-one.

Reciprocal care, affective attunement, and radical interdependence of care ethics and the duo of the service dog team reveal the autonomous male subject as a fraud and rupture the fortress of the atomic, "normate" subject.[31] The moral subject is assumed to be of able body and mind and, of course, is decidedly singular; as Ed Cohen notes in *A Body Worth Defending*, the formula is one body, one mind per subject.[32] If, borrowing a phrase from Foucault, "somatic singularity"[33] is established, individuals can be ranked, ordered, and classified through systems of oppression such as ableism and speciesism. Entangled subjectivity offers a glimpse of what it might mean to move beyond an individualist framework.

The interactive, dual subjectivity throws a wrench in the machine of the atomic individual and resists easy placement on a moral hierarchy or chain of being. (Ableism was deflated a little every time I realized a stranger's at first seemingly invasive questions about Mandy were because they wanted to know if they qualified for a service dog too and we ended up sharing surprise crip connections on the street or at the airport. Speciesism struggles to make sense of the friends who whisper to me that it was harder to lose their dog than their mother, uncle, or other relative when I explain that the white powder I am spreading is my service dog's ashes.)

So what if the subject of ethics was a crip, caring human–animal duo of misfits? Would we conceptualize rights differently? Praise different virtues and blame different vices? Would we have virtues, vices, rights, and theories at all? Would care work or trainable tasks simply be understood as things you do for your loved ones? Would social and physical spaces be more accessible? While I don't yet have answers to these questions, I'm grateful to the disability studies and animal studies scholars and, of course, Mandy, who have prompted me to ask them. The asking itself is a problematization of its own importance. However, in this chapter, my goal has been to outline some of the significant literature on service dogs, identify care ethics within this literature, and shift the kaleidoscopic lens on this literature—to propose entangled subjectivity—as way of thinking about service dogs that more accurately describes my experience and has the potential to counter the hegemony of the atomic individual.

For all four service dog authors, the extent to which we value their caring relations matter not just for how we conceptualize service animals but also ethico-political theory more generally; service animals function as special cases at the intersections of animals and disability, at the borders of the moral community. For Oliver, dependence and independence exist on a continuum; service animals make evident the ways in which all human beings are dependent on animals "in virtually every facet of life."[34] Citing Eva Feder Kittay, she argues that if our dependence on other human beings generates moral obligations to them, then we also share similar obligations to non-human animals. Price cites Oliver directly in order to move towards a collective care ethics so that "the complex, caring relationship between service animals and their human companions . . . [dissolve]. . . the bright lines between human-disability-need-animal-task."[35] The overarching thesis and substantial contribution of Taylor's *Beasts of Burden*, of course, is that ableism and speciesism are co-constitutive; her account of Bailey troubles the dominance of both narratives and positions him as an ethical subject. Finally, Michael concludes "Working Resonance" with a dream that the misfit of the service animal team can be leveraged to transform social space on "a larger scale that we can start to make room for nonhuman animals in our daily lives."[36]

Notes

1 U.S. Department of Justice, "ADA Requirements: Service Animals," ADA.gov. (2020), retrieved from: https://www.ada.gov/service_animals_2010.htm, accessed May 28, 2023

2 Kelly Oliver, "Service Dogs: Between Animal Studies and Disability Studies," in *Disability and Animality: Crip Perspectives in Critical Animal Studies*, eds. Stephanie Jenkins, Kelly Struthers Montford, and Chloë Taylor (New York: Routledge, 2020); Margaret Price, "What Is a Service Animal? A Careful Rethinking," *Review of Disability Studies* 13, no. 4 (2017). https://rdsjournal.org/index.php/journal/article/view/757.

3 Ellen Samuels, *Fantasies of Identification: Disability, Gender, Race* (New York: New York University Press, 2014), in Price, "What Is a Service Animal? A Careful Rethinking," 4.
4 John Sorenson and Atsuko Matsuoka, "Moral Panic over Fake Service Animals," *Social Sciences* 11 (2022): 439.
5 U.S. Department of Justice, "ADA Requirements: Service Animals."
6 Karen Barad, "Posthumanist Performativity: Toward an Understanding of How Matter Comes to Matter," *Signs* 28, no. 3 (2003): 801–831, in Price, "What Is a Service Animal? A Careful Rethinking," 13.
7 Oliver, "Service Dogs: Between Animal Studies and Disability Studies," 112.
8 Ibid., 11.
9 Price, "What Is a Service Animal? A Careful Rethinking," 10.
10 Ibid., 11.
11 Ibid., 12.
12 Sunaura Taylor, *Beasts of Burden* (New York: The New Press, 2017), 221.
13 Ibid., 223.
14 Emily K. Michael, "Working Resonance: Concerto for Guide Dog, Handler, and World," *The Hopper* (2016), retrieved from: http://www.hoppermag.org/working-resonance/, accessed May 28, 2023
15 Joan Tronto, *Moral Boundaries* (New York: Routledge, 1993).
16 Ibid., 106.
17 Ibid.
18 Ibid., 107.
19 Brian Luke, "Justice, Caring, and Animal Liberation," in *The Feminist Care Tradition in Animal Ethics*, eds. Josephine Donovan and Carol J. Adams (New York: Columbia University Press, 2007), 135.
20 Ibid., 107.
21 Ibid., 133.
22 Ibid., 108.
23 Thomas Kelch, "Toward a Non-Property Status for Animals," in *The Feminist Care Tradition in Animal Ethics*, eds. Josephine Donovan and Carol J. Adams (New York: Columbia University Press, 2007), 230.
24 Sorenson and Matsuoka, "Moral Panic over Fake Service Animals," 5.
25 Rod Michalko, *The Two in One: Walking with Smokie, Walking with Blindness* (Philadelphia: Temple University Press, 1999); in Price, "What Is a Service Animal? A Careful Rethinking."
26 Lori Gruen, *Entangled Empathy* (Brooklyn, NY: Lantern Books, 2015); Tronto, *Moral Boundaries*, 9; Deane Curtain, "Toward an Ecological Ethics of Care," in *The Feminist Care Tradition in Animal Ethics*, eds. Josephine Donovan and Carol J. Adams (New York: Columbia University Press, 2007), 91; Carol J. Adams, "Caring about Suffering," in *The Feminist Care Tradition in Animal Ethics*, eds. Josephine Donovan and Carol J. Adams (New York: Columbia University Press, 2007), 199.
27 Iris Marion Young, "Pregnant Embodiment: Subjectivity and Alienation," *The Journal of Medicine and Philosophy* 9, no. 1 (1984): 45–62; Grace Clement, "The Ethic of Care and the Problem of Wild Animals," in *The Feminist Care Tradition in Animal Ethics*, eds. Josephine Donovan and Carol J. Adams (New York: Columbia University Press.2007): 302.
28 Alice Dreger, *One of Us* (Cambridge, MA: Harvard University Press, 2005).
29 Donna Haraway, *Simians, Cyborgs, and Women* (New York: Routledge, 1990).
30 Donna Haraway, *When Species Met* (Minneapolis, MN: University of Minnesota, 2008), 16.
31 Rosemary Garland-Thomson, *Extraordinary Bodies: Figuring Physical Disability in American Culture and Literature* (New York: Columbia University Press, 1997), 8.
32 Ed Cohen, *A Body Worth Defending* (Durham, NC: Duke University Press, 2009), 9.
33 Michel Foucault, *Psychiatric Power: Lectures at the College de France, 1973–1974* (New York: Picador, 2008), 55.
34 Oliver, "Service Dogs: Between Animal Studies and Disability Studies," 119.
35 Price, "What Is a Service Animal? A Careful Rethinking," 11.
36 Michael, "Working Resonance: Concerto for Guide Dog, Handler, and World."

Bibliography

Adams, Carol J. "Caring about Suffering." In *The Feminist Care Tradition in Animal Ethics*, edited by Josephine Donovan and Carol J. Adams. New York: Columbia University Press, 2007.

Barad, Karen. "Posthumanist Performativity: Toward an Understanding of How Matter Comes to Matter." *Signs* 28, no. 3 (2003): 801–831.

Clement, Grace. "The Ethic of Care and the Problem of Wild Animals." In *The Feminist Care Tradition in Animal Ethics*, edited by Josephine Donovan and Carol J. Adams. New York: Columbia University Press, 2007.

Cohen, Ed. *A Body Worth Defending*. Durham, NC: Duke University Press, 2009.

Curtain, Deane. "Toward an Ecological Ethic of Care." In *The Feminist Care Tradition in Animal Ethics*, edited by Josephine Donovan and Carol J. Adams. New York: Columbia University Press, 2007.

Dreger, Alice. *One of Us*. Cambridge, MA: Harvard University Press, 2005.

Foucault, Michel. *Psychiatric Power: Lectures at the College de France, 1973–1974*. New York: Picador, 2008.

Garland-Thomson, Rosemary. *Extraordinary Bodies: Figuring Physical Disability in American Culture and Literature*. New York: Columbia University Press, 1997.

Gruen, Lori. *Entangled Empathy*. Brooklyn, NY: Lantern Books, 2015.

Haraway, Donna. *Simians, Cyborgs, and Women*. New York: Routledge, 1990.

Haraway, Donna. *When Species Meet*. Minneapolis, MN: University of Minnesota Press, 2008.

Kelch, Thomas G. "Toward a Non-Property Status for Animals." In *The Feminist Care Tradition in Animal Ethics*, edited by Josephine Donovan and Carol J. Adams. New York: Columbia University Press, 2007.

Luke, Brian. "Justice, Caring, and Animal Liberation." In *The Feminist Care Tradition in Animal Ethics*, edited by Josephine Donovan and Carol J. Adams. New York: Columbia University Press, 2007.

Michael, Emily K. "Working Resonance: Concerto for Guide Dog, Handler, and World." *The Hopper*, May 2016. Retrieved from: http://www.hoppermag.org/working-resonance/

Michalko, Rod. *The Two in One: Walking with Smokie, Walking with Blindness*. Philadelphia: Temple University Press, 1999.

Oliver, Kelly. "Service Dogs: Between Animal Studies and Disability Studies." In *Disability and Animality: Crip Perspectives in Critical Animal Studies*, edited by Stephanie Jenkins, Kelly Struthers Montford, and Chloë Taylor. New York: Routledge, 2020.

Price, Margaret. "What Is a Service Animal? A Careful Rethinking." *Review of Disability Studies* 13, no. 4 (2017). Retrieved from: https://rdsjournal.org/index.php/journal/article/view/757.

Samuels, Ellen. *Fantasies of Identification: Disability, Gender, Race*. New York: New York University Press, 2014.

Sorenson, John and Matsuoka, Atsuko. "Moral Panic over Fake Service Animals." *Social Sciences* 11 (2022): 439.

Taylor, Sunaura. *Beasts of Burden: Animal and Disability Liberation*. New York: The New Press, 2017.

Tronto, Joan. *Moral Boundaries*. New York: Routledge, 1993.

U.S. Department of Justice. "ADA Requirements: Service Animals." *ADA.gov*, February 28, 2020. Retrieved from: https://www.ada.gov/service_animals_2010.htm

Young, Iris Marion. "Pregnant Embodiment: Subjectivity and Alienation." *The Journal of Medicine and Philosophy* 9, no. 1 (1984): 45–62.

14
SHELTERS

Rebecca Deutsch

Introduction

My hands were shaking as I pushed the door open. With no means of assessing what would be waiting on the other side of those doors, the anxiety was palpable. It had been slightly under twenty-four hours since I had checked myself into an emergency domestic violence (DV) shelter. A period of hours can be such a strange and thick thing. I walked into the apartment, a place that had been home only a day earlier, discovering fresh and lingering remnants of destruction. A knife stuck out of my Tom Waits poster on the wall, empty alcohol containers littered the carpet, and our cat came up to me twining herself around my legs and meowing balefully. With no idea of when my common-law spouse would return, or indeed *if* he would, I sank to the floor. The cat curled onto my lap, and I knew I could not leave her again. It would take almost eight more years to fully extricate myself from the relationship. By placing my experiences in conversation with available literature, I will argue that not only should nonhuman animal-friendly DV and homeless shelters be readily available but also that there is an untapped potential for human-oriented shelters and animal shelter inhabitants to create meaningful spaces of community and healing with one another. The way control and isolation harm beings in abusive contexts and, often, shelter contexts mirror one another, as do the benefits of connection, freedom, and community. I argue in this chapter that we collectively need to reimagine who can experience harm and who can create a community to more fully account for the variety of beings with whom we share our lives.

Background on Domestic Violence

In the complicated array of reasons that caused me to stay, throwing away bloodied clothing and silencing something vibrant and alive in myself, my companion animal was only one. However, this particular reason is shared and well documented.[1]

Intimate partner violence (IPV), used here interchangeably with DV, refers to " 'physical violence, sexual violence, stalking and psychological aggression (including coercive acts) by a current or former intimate partner'. IPV affects one in four women and is the leading cause of injury for women in the USA."[2] On average, women try to leave IPV situations five times

DOI: 10.4324/9781003273400-18

before staying gone, a process that takes an average of eight years.[3] Thus, it is not surprising that often stays in DV shelters are precipitated by an increase in violence and a desire for some form of respite.[4] Safety from abuse can be crucial in recognizing the nature of the relationship, naming it as abusive, and finding a pathway towards autonomy.[5] Depression, anxiety, post-traumatic stress disorder, and a variety of other mental health diagnoses tend to be comorbid with experiences of IPV.[6] The appropriateness of these medicalized diagnoses in a context where they are proportionate responses to external stimuli lies beyond the scope of this chapter; however, the ongoing suffering implied by the diagnoses highlights the necessity for certain shapes of safe spaces (a point we will return to shortly).

Companion Animals and Domestic Violence

Abuse is not about anger or an uncontrolled response to a situation. Rather, abuse is an expression of coercive control.[7] As such, people who engage in abusive behavior often direct that abuse to family "pets" as well[8] because it provides an additional method of harming and coercing the primary subject of their abuse.[9] The end goal may be supplemented with pleasure derived from committing these acts of violence.[10] This abuse can take the shape of threats, active physical harm, sexual assault, or even murder.[11] Even without obvious physical aggression, the autonomy and boundaries of non-human companions are often treated as violable and irrelevant. Companion animals are normatively framed as property and treated as primarily *for* their "owners," rather than beings in their own rights. As such, abusive people take this widely held approach to its most painful end. This unique form of legal and physical vulnerability makes horrific harms actionable and rarely punishable.[12]

Unfortunately, the vast majority of websites for DV shelters situate companion animals as possessions rather than subjects in their own right,[13] which has more commonalities with how abusive people conceptualize animals[14] than how survivors do. Survivors regularly speak about their companion animals as family members—not possessions.[15] Alison Gray, Betty Jo Barrett, Amy Fitzgerald, and Amy Peirone meaningfully consider how the framing employed by DV shelters contradicts the bond human abuse survivors have with their non-human companions and introduce the possibility of reframing abuse against non-human animals to align more with that against human children and other dependents.[16] While this would undoubtedly have benefits, there must also be a greater acknowledgement of the harm that these non-human animals themselves experience as beings in their own right.

Non-human animal survivors, like their human counterparts, often experience long-term "emotional and physical damage."[17] Even if they are not directly abused (which they often are), the proximity to abuse can lead to behavioral changes including a protectiveness towards the human victim and fear, stress, and avoidance of the abuser and those similar to that person. These beings need safety, security, and healing that is discounted, obscured, and minimized by situating them as possessions rather than whole beings.

Background on Homelessness

Homelessness is a multifaceted issue that, while sometimes tied into DV, must be treated as a distinct social issue. Poverty is feminized (meaning that women are more likely to be below the poverty line due to institutional causes) adding a gendered dimension to the picture.[18] Women and gender-nonconforming individuals are often invisible in representations and policies focused on homelessness, leading to a denial of unique vulnerabilities they experience and the prevalence of gender-based violence in their life course.[19] Michael Shier, Christine A. Walsh, and John R. Graham write:

> By personal factors we mean such triggers or personal crises that under the worst circumstances may lead to homelessness. Trigger points include such crises as leaving the parental home after an argument, marital or relationship breakdown, family violence, widowhood, leaving prison, leaving some form of social or health sector care, sharp deterioration in mental health, increase in alcohol or drug use, financial crisis of mounting debts, or eviction from rented or owned home.[20]

Thus, on top of the feminization of poverty, women are at a greater risk of the types of violence that could directly lead to homelessness or indirectly lead to the experience of mental health crisis and drug and alcohol use, all of which increase the vulnerability of homelessness.[21] Race and ability are even more statistically tied to the prevalence of homelessness.[22] In 2023, thirty percent of the recorded homeless population in Canada was Indigenous.[23]

Human and companion animals share space and thus share physical and social conditions.[24] The companion animals of homeless people are likely to be (incorrectly) stigmatized and assumed aggressive, an issue which is exacerbated by the increased prevalence of pit bull (or similar appearing breeds) companioned with homeless people.[25] Pit bulls are profiled, racialized, and legislated in ways that make their bodies at increased risk of violence and death.[26] Alternatively, people sometimes assume homeless people are unable to care for their companion animals or are otherwise inappropriate guardians and thus offer to buy the companion animal.[27]

Homeless people with companion animals regularly put the needs of their companion animal above their own, even refusing to eat until their companion animal has been adequately fed. On scores of animal empathy and attachment, homeless people consistently score higher than securely housed individuals.[28] As such, they are likely to refuse housing if it does not include space for their companion animals.[29] Much like in the case of DV survivors, the bonds of love and companionship run deep.

Support in the Nonhuman/Human Animal Bond for Humans

A wide variety of research delineates the supportive power of the non-human animal and DV victim bond as it relates to the human. Less attention has been paid to how the nonhuman animal feels and experiences the relationship. As such, while I will explore the positive benefits these relationships have for humans, there is a risk this could perpetuate and re-entrench the instrumentalization of nonhuman animals. This is not my intention—the needs and desires of the nonhuman animals involved must always remain central.

As isolation is such a crucial ingredient in the functioning of DV, for some victims, the relationship that they have with their companion animals can be the only positive one left in their lives.[30] This was my reality as I entered the DV shelter. The isolation was totalizing and, in some ways, internalized and self-perpetuating due to layers of emotional abuse. The cat I shared my home with would climb onto my lap, pressing her forehead against my lips for a kiss and snuggling purring into my shoulder, offering love and comfort that was entirely free from the minefield of suffering with which all other relationships came entangled. My experience of emotional support is echoed by the literature again and again.[31] This seems to be particularly evident in the experiences of childless women.[32]

Similarly, "researchers concluded that homeless animal caretakers were exceptionally attached to their pets and viewed their animals as their lone source of love and companionship."[33] The psychological and physiological benefits are manifold and, for women, often extend to increased physical safety.[34] For some, the relationship shared with their companion animal saved their lives.[35] As such, parting with their companion animals can lead to an increased risk of self-harm or even suicide.[36] Undoubtedly, this is the case for DV survivors as well. As companion animals are sometimes the only ties to companionship, love, and safety these human beings have, the damage done by cutting this string cannot be overstated. On one particularly dark occasion, the fear of who would love, feed, and care for the animals in my life was the only practical detail that held me to my life. This sense of responsibility would not have been enough to sustain my prolonged attachment to the suffering of existence, but it was enough to carry me to a moment where I was able to affectively experience love again. It is not the responsibility of companion animals to carry the burden of being a reason for life; however, who can know what tie to life is the one that cannot be broken for life to have value?

Domestic Violence Shelters and Homeless Shelters: Safety Needs and Inclusion

Funding opportunities in DV shelters have led to an increase in rules delineating who can enter these spaces, for how long, and in what ways, creating substantial access barriers that maintain gendered and racialized hierarchies.[37] The types of assistance proffered to the people able to walk through the doors can be invasive, controlling, and harmful.[38] As those who attempt to access these services are generally among the most vulnerable[39] these access restrictions and barriers to meaningful safety must be taken seriously.

My safety needs upon entering the DV shelter certainly originated around requiring a cessation of physical violence but the isolation, emotional abuse, financial abuse, and layers of additional less obvious forms of harm meant that I also needed a cessation of the power dynamics that precaritized my ability to relate to myself and those around me. The introduction of exclusionary rules and hierarchies creates environments often felt as "potentially restrictive and controlling,"[40] which can exacerbate isolation and emotional distress.[41]

DV survivors are most at risk when they are in the process of leaving.[42] There is also extreme emotional duress associated with the escalation of violence that often precipitates an emergency DV shelter stay. Ongoing psychological trauma deteriorates an ability to seek help or accurately conceptualize the situation.[43] Unsurprisingly, research has demonstrated that a need for emotional support can be greater than the expressed need for a safe place to stay.[44] My disrupted beliefs about myself and the world around me, much like those expressed by other DV survivors, made me uniquely sensitive to the dynamics of the environment that I entered into.[45] Likewise, homeless women regularly express a need for privacy, autonomy, and dignity.[46]

Not having a safe place for companion animals adds an extra emotional burden to an already tumultuous time and increases the risk of victimization to human and nonhuman animal parties,[47] whereas "onsite pet programs" can have significant therapeutic benefits. These benefits have been noted in a wide variety of different contexts (nursing homes, psychiatric hospitals, and prisons, for example) as nonhuman animals have slowly been recognized and welcomed as companions.[48] The vast psychological and physical benefits accorded by time spent with companion animals have been substantiated by numerous studies.[49] As everyday experiences of self-determination and community can have some of the most meaningful benefits for survivors in DV shelters,[50] having space to keep non-human animals safe and continue those relations on a day-to-day basis would likewise proffer beneficial impacts. As some shelters return to their feminist roots by emphasizing self-determination,[51] the chorus of voices emphasizing the importance of secure shared housing may become easier to hear.

Availability of funding is the most frequently cited reason for the absence of these services.[52] Noise concerns, spatial issues, and allergies are additional reasons often offered. To account for the variety of needs that DV survivors have, multiple sheltering options would likely need to be available. Some of these issues could not be readily addressed simply by accommodating nonhuman animal companions into existing spaces.

Existing Programs

There are a wide variety of existing programs that aim to create safe housing options for all of those harmed by economic precarity and DV. One of the most promising is housing first initiatives. Being homeless is itself such a significant stressor that expecting adherence to treatment protocols pre-housing is unrealistic.[53] Ensuring people have safe spaces to stay (in motels or in otherwise un-utilized apartment spaces, for example) allows people to address additional support needs as ready. In Denmark, a variety of control studies have demonstrated how these programs can better address complex support needs that would otherwise lead to prolonged homelessness.[54] Community-based mental health supports are paired with access to housing to ensure the variety of needs that homeless people have can be met.[55] Counselling, access to childcare, and a variety of forms of training could also be beneficially integrated with these initiatives.[56] In the case of nonhuman animals, access to low- to no-cost veterinary care and other supportive programs should be added.

Some shelters have created outdoor kennel spaces for companion animals where their respective humans are able to provide care (in Halifax), inside of shelters with non-human animal free floors (Vancouver),[57] in separate wings of the building,[58] and in collaboration with nearby animal shelters (such as through "Safe Haven" programs).[59] In some cases, DV survivors are only attempting to access shelter supports for a temporary respite before returning to the relationship. As it tends to take multiple tries to leave an abusive relationship, temporary housing options would likely be necessary in conjunction with housing first initiatives. This could be the case for homeless people as well, who may face temporary forms of crises that require different forms of support than permanent housing.

Animal Shelters

Nonhuman companion animals are not always in relationship with the human or may be taken away from their humans for a variety of reasons. In those cases, they are likely to find themselves embedded in the animal shelter system. David S. Tuber et al. wrote that:

> during their stay in even modern, well-run shelters, dogs are subject to a variety of psychological stressors, including novelty, isolation from any former attachment figures, exposure to unpredictable and often intense noise, disruption of familiar routines (including walks for elimination), and a general loss of control over environmental contingencies.[60]

Some of these psychological stressors have obvious resonances with the experiences of DV survivors and homeless people. However, I will not directly parallel animal shelters and human shelters, as some of the specificities operationalize unique forms of precarity that could be lost in such a comparison. Thus, to create an image of how human-oriented shelters and non-human animal shelters could form unique kinship networks, we need to briefly examine animal shelters.

We cannot understand the functioning of animal shelters without talking about death. The way that "euthanasia" and "save rates" are calculated varies significantly from shelter to shelter and this makes it nearly impossible to create a statistically accurate image of the sheer quantity of death.[61] Around one million companion animals are executed annually in shelters in the United States alone.[62] These deaths cannot all be accurately classified as euthanasia because definitionally, euthanasia only encapsulates the merciful killing of already dying beings.[63] In contrast, healthy animals are often needlessly killed in shelters, and these deaths are often agonizing (as in the case of heart sticks), undignified, and normalized or ignored.[64] Cats are far more likely to be killed than dogs. In both cases the criteria used is problematic and entangled with conceptions about the inherent value of uncompanioned companion animals.[65] Florida legislation, for one example, situates "unowned" animals as "unneeded."[66]

"Feline-ality" is one system employed to measure the adoptability and personalities of cats. It involves measuring "responses to novel stimuli" and "sociality" within eighteen hours of admittance into the shelter.[67] This timeframe, how the test is undertaken, and by whom (trained or untrained), makes its results dubious at best, and the consequences of these tests can be fatal. In the case of dogs, on top of inaccurate behavioral assessments, assumptions about breed (particularly pit bull-ness) can shape life or death chances.[68] Shelter environments themselves come with unique stressors that compromise well-being and deteriorate how "adoptable" the inhabitants might otherwise seem.[69]

Due to the relationship humans and their companion animals have, human vulnerability increases the likelihood of using animal shelter services.[70] As health issues and housing issues are some of the most frequently cited reasons companion animals are surrendered,[71] without addressing the interconnected nature of well-being,[72] shifting how shelters are operationalized will remain out of reach. Thus, it is necessary to extend the availability of inclusive shelters to substantially decrease how often people are forced to surrender their deeply loved companion animals.[73] Often homeless people's companion animals are considered otherwise unwanted and would otherwise find themselves within shelters. Zanna Shafer suggested that allowing homeless people to adopt from kill shelters could reduce overcrowding and be beneficial to all involved.[74]

Kin Relations

Some of the safety measures and rules put in place in DV shelters are very necessary,[75] as are some of those put in place in animal shelters. Feline panleukopenia, for example, is a highly contagious and often deadly disease that spreads most in environments where cats are housed in groups.[76] Keeping cats segregated can increase their survival outcomes. Dog-to-dog contact can increase risk of diseases or aggression.[77] DV survivors are at the greatest risk when they try to leave their abusive relationship. All of these facts are mobilized in creating restrictive structures. The problem is that it is about "what women [and non-human animals] are enabled to be safe to rather than just safe from."[78] Creating environments that are "safe from" will never be adequate in meaningfully contributing to personal and social change. Without implying that these safety measures are unfounded, we still have to imagine things otherwise. Katja M. Guenther writes and dreams a utopia for Monster, one of the pit bulls she encountered during her ethnographic research in a large animal shelter.[79] In this utopic imaginary, Monster lives his life. He enjoys the feeling of the grass beneath his back, the sun on his face, and the variety of different forms of human and nonhuman kinship of his choosing.

There are so many shapes kinship can take. Harlan Weaver describes the unique kinship that can form in foster homes where "intimacy, affection, and undoubtedly love" are at the heart of relationships, even if they are not intended to last forever.[80] Forms of physical contact and relationship could drastically decrease the stress and associated negative mental health impacts animal shelters cause their inhabitants.[81] Not all humans or nonhuman animals would desire these forms of relationship. Much like in the utopic visionary of a life for Monster, forcing these relationships would be undesirable and unnecessary. Rather, spaces could be created where these relationships could form on their own.

The positive elements of these relationships could decrease negative health outcomes and increase the ability participants have to be present in society and in their lives. However, if the end result was exclusively the existence of spaces of warmth, connection, and safety for vulnerable beings, that would be an adequate reason for creating new programs. New kinship programs could look like integrated shelter environments or fostering programs. In the case of housing first initiatives, opportunities could be provided for desiring and safe humans and nonhuman animals to cohabitate through fostering or adoption programs.

Conclusion

At the time of this writing, it has been over two years since I moved into my own apartment and began to develop a new relationship with myself and with the world around me. My rescue dog is resting his head on my thigh. I can feel the warmth of his breath as he sleeps. The cats I share my home with are each in their favorite spots scattered around the house. They are also all napping. It is a warm summer afternoon, and our home is safe—not just safe from, but safe to. Sometimes, "safe to" looks like a summer nap. When I envision places where shelters could meet, I do not see dramatic therapeutic relationships forced into a specific shape between the parties involved but rather soft forms of chosen kinship. By recognizing all of the beings who can experience harm, we can also imagine new ways of addressing those harms and creating mutually beneficial companionship.

Notes

1 Amy J. Fitzgerald, Betty Jo Barrett, Allison Gray, and Chi Ho Cheung, "The Connection Between Animal Abuse, Emotional Abuse, and Financial Abuse in Intimate Relationships," *Journal of Interpersonal Violence* 37, no. 5–6 (2022); Crystal J. Giesbrecht, "Animal Safekeeping in Situations of Intimate Partner Violence," *Journal of Interpersonal Violence* 37, no. 17–18 (2022); Allison Gray, Betty Jo Barrett, Amy Fitzgerald, and Amy Peirone, "Fleeing with Fido: An Analysis of What Canadian Domestic Violence Shelters Are Communicating Via their Websites about Leaving an Abusive Relationship When Pets Are Involved," *Journal of Family Violence* 34 (2019); Nick Kerman, Michelle Lem, Mike Witte, Christine Kim and Harmony Rhoades "A Multilevel Intervention Framework for Supporting People Experiencing Homelessness with Pets," *Animals* 10 (2020); Dawna Komorosky, Dianne Rush Woods, and Kristine Empie, "Considering Companion Animals An Examination of Companion Animal Policies in California Domestic Violence

Shelters," *Society & Animals* 23 (2015); Michelle Newberry, "Pets in Danger: Exploring the Link between Domestic Violence and Animal Abuse," *Aggression and Violent Behavior* 34 (2017); Mary LouRandour, "Integrating Animals into the Family Violence Paradigm: Implications for Policy and Professional Standards," *Policy and Practice* 7, no. 3 (2007); Rochelle Stevenson, Amy Fitzgerald, and Betty Jo Barrett, "Keeping Pets Safe in the Context of Intimate Partner Violence: Insight from Domestic Violence Shelter Staff in Canada," *Affilia: Journal of Women and Social Work* 33, no. 2 (2018); Elizabeth B. Strand and Catherine A. Faver, "Battered Women's Concern for Their Pets: A Closer Look," *Journal of Family and Social Work* 9, no. 4 (2005); C.M. Tiplady, D.B. Walsh, and C.J.C Phillips, " 'The Animals Are All I Have': Domestic Violence, Companion Animals, and Veterinarians," *Society and Animals* 26 (2018); V. Upadhya, "The Abuse of Animals as a Method of Domestic Violence," *Emory Law Journal* 63 (2014)

2 Sarah R. Robinson, December R. Maxwell, and Kelli R. Rogers, "Living in Intimate Partner Violence Shelters," *British Journal of Social Work* 50 (2020): 82.

3 Strand and Faver, "Battered Women's Concern for Their Pets," 42.

4 Anat Ben-Porat and Noa Sror-Bondarevsky, "Length of Women's Stays in Domestic Violence Shelters: Examining the Contribution of Background Variables, Level of Violence, Reasons for Entering Shelters, and Expectations," *Journal of Interpersonal Violence* 36, no. 11–12 (2021).

5 Janet Bowstead, "Spaces of Safety and More-than-Safety in Women's Refuges in England," *Gender, Place, and Culture* 26, no. 1 (2019).

6 Liria Fernández-González, Esther Calvete, Izaskun Orue, and Alice Mauri, "Victims of Domestic Violence in Shelters: Impacts on Women and Children," *The Spanish Journal of Psychology* 21 (2018).

7 Leslie M. Tutty, "Addressing the Safety and Trauma Issues of Abused Women: A Cross Canada Study of YWCA Shelters," *Journal of International Women's Studies* (2015).

8 Strand and Faver, "Battered Women's Concern for Their Pets."

9 Giesbrecht, "Animal Safekeeping in Situations of Intimate Partner Violence"; Dawna Komorosky, Dianne Rush Woods, and Kristine Empie, "Considering Companion Animals An Examination of Companion Animal Policies in California Domestic Violence Shelters," *Society and Animals* (23) (2015); Tiplady, Walsh, and Phillips, " 'The Animals Are All I Have' Domestic Violence, Companion Animals, and Veterinarians"; Upadhya, "The Abuse of Animals as a Method of Domestic Violence"; Michelle Newberry, "Pets in Danger: Exploring the Link between Domestic Violence and Animal Abuse," *Aggression and Violent Behavior* 34 (2017).

10 Newberry, "Pets in Danger," 277.

11 Strand and Faver, "Battered Women's Concern for Their Pets"; Upadhya, "The Abuse of Animals as a Method of Domestic Violence."

12 Upadhya, "The Abuse of Animals as a Method of Domestic Violence."

13 Gray et al., "Fleeing with Fido"; Fitzgerald et al., "The Connection Between Animal Abuse, Emotional Abuse, and Financial Abuse in Intimate Relationships."

14 Tiplady, Walsh, and Phillips, " 'The Animals Are All I Have' Domestic Violence, Companion Animals, and Veterinarians."

15 Giesbrecht, "Animal Safekeeping in Situations of Intimate Partner Violence"; Jennifer B. Sinski and Patricia Gagné, "Give Me Shelter: The State of Animal Sheltering in Kentucky's County Shelter System," *Contemporary Justice Review* 19, no. 2 (2016); Strand and Faver, "Battered Women's Concern for Their Pets"; Upadhya, "The Abuse of Animals as a Method of Domestic Violence"; Newberry, "Pets in Danger."

16 Gray et al., "Fleeing with Fido," 295.

17 Tiplady, Walsh, and Phillips, " 'The Animals Are All I Have' Domestic Violence, Companion Animals, and Veterinarians," 491.

18 Jennifer Labrecque and Christine A. Walsh, "Homeless Women's Voices on Incorporating Companion Animals into Shelter Services," *A Multidisciplinary Journal of the Interactions between People and Other Animals* 24 (2011): 81.

19 Rosane Braud and Marie Loison, "Homelessness among Women. When Women's Emergency Shelters Call into Question Social Emergency Services," *Travail, Genre et Sociétés* 47, no. 1 (2022).

20 Micheal Shier, Christine A. Walsh, and John R. Graham, "Conceptualizing Optimum Homeless Shelter Service Delivery: The Interconnection between Programming, Community, and the Built Environment," *Canadian Journal of Urban Research* 16, no. 1 (2007): 61.

21 Michelle Cleary, Denis Visentin, Deependra Kaji Thapa, Sancia West, Toby Raeburn, and Rachel Kornhaber, "The Homeless and Their Animal Companions: An Integrative Review," *Administration and Policy in Mental Health and Mental Health Services Research* 47 (2020); John Sylvestre, Nick Kerman, Alexia Polillo, Catherine M. Lee, and Tim Aubry "A Profile of Families in the Emergency Family Homeless Shelter System in Ottawa, Ontario, Canada," *Canadian Journal of Urban Research* 26, no. 1 (2016).

22 Kerman et al., "A Multilevel Intervention Framework for Supporting People Experiencing Homelessness with Pets"; Lars Benjaminsen, "Housing First in Denmark: An Analysis of the Coverage Rate among Homeless People and Types of Shelter Use," *Vulnerable and Disadvantaged Groups: On the Margins of the Welfare State?* 6, no. 3 (2018).

23 "Homelessness Statistics in Canada for 2023—Made in CA."

24 Lexis H. Ly, Emilia Gordon, and Alexandra Protopopova, "Inequitable Flow of Animals in and Out of Shelters: Comparison of Community-Level Vulnerability for Owner Surrendered and Subsequently Adopted Animals," *Frontiers in Veterinary Science* 8 (2021): 1–2.

25 Cleary et al., "The Homeless and Their Animal Companions"; Zanna Shafer, "Home Is Where the Dog Is: A Discussion of Homeless People and Their Pets," *Animal Law* 23 (2016).

26 Katja Guenther, *The Lives and Deaths of Shelter Animals* (Stanford University Press, 2020); Harlan Weaver, "Pit Bull Promises: Inhuman Intimacies and Queer Kinships in an Animal Shelter," *GLQ* 21, no. 2–3 (2015).

27 Cleary et al., "The Homeless and Their Animal Companions."

28 Labrecque and Walsh, "Homeless Women's Voices on Incorporating Companion Animals into Shelter Services," 84.

29 Cleary et al., "The Homeless and Their Animal Companions"; Nick Kerman, Michelle Lem, Mike Wiite, Christine Kim, and Harmony Rhoades "A Multilevel Intervention Framework for Supporting People Experiencing Homelessness with Pets," *Animals* 10 (2020); Labrecque and Walsh, "Homeless Women's Voices on Incorporating Companion Animals into Shelter Services"; Shafer, "Home Is Where the Dog Is."

30 Stevenson, Fitzgerald, and Barrett, "Keeping Pets Safe in the Context of Intimate Partner Violence."; Newberry, "Pets in Danger."

31 Komorosky, Woods, and Empie, "Considering Companion Animals An Examination of Companion Animal Policies in California Domestic Violence Shelters."; Stevenson, Fitzgerald, and Barrett, "Keeping Pets Safe in the Context of Intimate Partner Violence."; Strand and Faver, "Battered Women's Concern for Their Pets."

32 Strand and Faver, "Battered Women's Concern for Their Pets."; Tiplady, Walsh, and Phillips, "'The Animals Are All I Have' Domestic Violence, Companion Animals, and Veterinarians."

33 Labrecque and Walsh, "Homeless Women's Voices on Incorporating Companion Animals into Shelter Services," 84.

34 Cleary et al., "The Homeless and Their Animal Companions."; Kerman et al., "A Multilevel Intervention Framework for Supporting People Experiencing Homelessness with Pets."; Harmony Rhoades, Hailey Winetrobe, and Eric Rice, "Pet Ownership Among Homeless Youth," *Child Psychiatry Human Development* 46 (2015).

35 Shafer, "Home Is Where the Dog Is."

36 Kerman et al., "A Multilevel Intervention Framework for Supporting People Experiencing Homelessness with Pets," 10.

37 Catherine Glenn and Lisa Goodman, "Living With and Within the Rules of Domestic Violence Shelters: A Qualitative Exploration of Residents' Experiences," *Violence against Women* 21, no. 12 (2015); Michelle VanNatta, "Power and Control: Changing Structures of Battered Women's Shelters," *The International Journal of Interdisciplinary Social Sciences* 5, no. 2 (2010).

38 Robinson, Maxwell, and Rogers, "Living in Intimate Partner Violence Shelters."

39 Glenn and Goodman, "Living With and Within the Rules of Domestic Violence Shelters: A Qualitative Exploration of Residents' Experiences."

40 Janet Bowstead, "Safe Spaces of Refuge, Shelter and Contact," *Gender, Place, & Culture* 26, no. 1 (2019): 53.

41 Glenn and Goodman, "Living With and Within the Rules of Domestic Violence Shelters: A Qualitative Exploration of Residents'," 1493.

42 Gray et al., "Fleeing with Fido."

43 Ibid.
44 Tutty, "Addressing the Safety and Trauma Issues of Abused Women."
45 Glenn and Goodman, "Living With and Within the Rules of Domestic Violence Shelters: A Qualitative Exploration of Residents' Experiences."
46 Labrecque and Walsh, "Homeless Women's Voices on Incorporating Companion Animals into Shelter Services," 82.
47 Giesbrecht, "Animal Safekeeping in Situations of Intimate Partner Violence."; Strand and Faver, "Battered Women's Concern for Their Pets."
48 Gray et al., "Fleeing with Fido," 294.
49 Lisa M. Gunter, Rachel J. Gilchrist, Emily M. Blade, Jennifer L. Reed, Lindsey T. Isernia, Rebecca T. Barber, Amanda M. Foster, Erica N. Feuerbacher, and Clive D.L. Wynne, "Emergency Fostering of Dogs From Animal Shelters During the COVID-19 Pandemic: Shelter Practices, Foster, Caregiver Engagement, and Dog Outcomes," *Frontiers in Veterinary Science* 9 (2022); Lexis H. Ly, Emilia Gordon, and Alexandra Protopopova, "Exploring the Relationship Between Human Social Deprivation and Animal Surrender to Shelters in British Columbia, Canada," *Frontiers in Veterinary Science* 8 (2021).
50 Judy Hughes, "Women's Advocates and Shelter Residents," *Journal of Interpersonal Violence* 35, no. 35–36 (2020).
51 VanNatta, "Power and Control."
52 Giesbrecht, "Animal Safekeeping in Situations of Intimate Partner Violence."; Komorosky, Woods, and Empie, "Considering Companion Animals An Examination of Companion Animal Policies in California Domestic Violence Shelters."; Stevenson, Fitzgerald, and Barrett, "Keeping Pets Safe in the Context of Intimate Partner Violence."
53 Benjaminsen, "Housing First in Denmark."
54 Ibid.
55 Kerman et al., "A Multilevel Intervention Framework for Supporting People Experiencing Homelessness with Pets."
56 Sylvestre et al., "A Profile of Families in the Emergency Family Homeless Shelter System in Ottawa, Ontario, Canada."
57 Labrecque and Walsh, "Homeless Women's Voices on Incorporating Companion Animals into Shelter Services."
58 Shafer, "Home Is Where the Dog Is."
59 Randour, "Integrating Animals into the Family Violence Paradigm."
60 David S. Tuber, Deborah D. Miller, Kimberly, A. Caris, Robin Halter, Fran Linden, and Michael B. Hennessy, "Dogs in Animal Shelters: Problems, Suggestions, and Needed Expertise," *Psychological Science* 10, no. 5 (1999): 379.
61 Stacy A. Nowicki, "Lies, Damned Lies, and Michigan Animal Shelter Statistics: Problems and Solutions," *Journal of Animal and Natural Resource Law* XII (2016).
62 Derek Halm, "(Cat)Egory Mistake: the Invalidity of Animal Shelter Behavior Assessments," *Biology & Philosophy* 36, no. 35 (2021): 2.
63 Katherine Sloan, "Death Without Dignity: The Misnomer of Euthanasia in the State Animal Shelter System a Call for a No-Kill Florida," *Journal of Land Use* 32, no. 1 (2016).
64 Sloan, "Death Without Dignity."; Sinski and Gagné, "Give Me Shelter."
65 Halm, "(Cat)Egory Mistake."; Guenther, *The Lives and Deaths of Shelter Animals.*
66 Sloan, "Death Without Dignity," 277.
67 Halm, "(Cat)Egory Mistake," 5.
68 Guenther, *The Lives and Deaths of Shelter Animals;* Weaver, "Pit Bull Promises."
69 Gunter et al., "Emergency Fostering of Dogs From Animal Shelters During the COVID-19 Pandemic."; Solveig Marie Stubsjøen, Randi Oppermann Moe, Cicilie Johannessen, Maiken Larsen, Henriette Madsen, and Karianne Muri, "Can Shelter Dog Observers Score Behavioural Expressions Consistently over Time?" *Acta Veterinaria Scandinavica* 64, no. 1 (2022); Tuber et al., "Dogs in Animal Shelters."
70 Ly, Gordon, and Protopopova, "Exploring the Relationship Between Human Social Deprivation and Animal Surrender to Shelters in British Columbia, Canada."
71 Janne B.H. Jensen, Peter Sandøe, and Søren Saxmose Nielsen, "Owner-Related Reasons Matter More Than Behavioural Problems—A Study of Why Owners Relinquished Dogs and Cats to a Danish Animal Shelter from 1996 to 2017," *Animals* 10 (2020).

72 Ly, Gordon, and Protopopova, "Inequitable Flow of Animals in and Out of Shelters."
73 Kerman et al., "A Multilevel Intervention Framework for Supporting People Experiencing Homelessness with Pets."
74 Shafer, "Home Is Where the Dog Is," 156.
75 Glenn and Goodman, "Living With and Within the Rules of Domestic Violence Shelters: A Qualitative Exploration of Residents' Experiences."
76 Teresa Rehme, Katrin Hartmann, Uwe Truyen, Yury Zablotski, and Michèle Bergmann, "Feline Panleukopenia Outbreaks and Risk Factors in Cats in Animal Shelters," *Viruses* 14 (2022).
77 Tuber et al., "Dogs in Animal Shelters."
78 Bowstead, "Safe Spaces of Refuge, Shelter and Contact," 55.
79 Guenther, *The Lives and Deaths of Shelter Animals.*
80 Weaver, "Pit Bull Promises," 355.
81 Halm, "(Cat)Egory Mistake."; Tuber et al., "Dogs in Animal Shelters."

Bibliography

Benjaminsen, Lars. "Housing First in Denmark: An Analysis of the Coverage Rate among Homeless People and Types of Shelter Use." *Vulnerable and Disadvantaged Groups: On the Margins of the Welfare State?* 6, no. 3 (2018).

Ben-Porat, Anat, and Sror-Bondarevsky, Noa. "Length of Women's Stays in Domestic Violence Shelters: Examining the Contribution of Background Variables, Level of Violence, Reasons for Entering Shelters, and Expectations." *Journal of Interpersonal Violence* 36, no. 11–12 (2021).

Bowstead, Janet. "Spaces of Safety and More-than-Safety in Women's Refuges in England." *Gender, Place, and Culture* 26, no. 1 (2019a).

Bowstead, Janet. "Safe Spaces of Refuge, Shelter and Contact." *Gender, Place, & Culture* 26, no. 1 (2019b).

Braud, Rosane, and Loison, Marie. "Homelessness among Women. When Women's Emergency Shelters Call into Question Social Emergency Services." *Travail, Genre et Sociétés* 47, no. 1 (2022).

Cleary, Michelle, Visentin, Denis, Thapa, Deependra Kaji, West, Sancia, Raeburn, Toby, and Kornhaber, Rachel. "The Homeless and Their Animal Companions: An Integrative Review." *Administration and Policy in Mental Health and Mental Health Services Research* 47 (2020).

Fernández-González, Liria, Calvete, Esther, Orue, Izaskun, and Mauri, Alice. "Victims of Domestic Violence in Shelters: Impacts on Women and Children." *The Spanish Journal of Psychology* 21 (2018).

Fitzgerald, Amy J., Barrett, Betty Jo, Gray, Allison, and Cheung, Chi Ho. "The Connection Between Animal Abuse, Emotional Abuse, and Financial Abuse in Intimate Relationships." *Journal of Interpersonal Violence* 37, no. 5–6 (2022).

Giesbrecht, Crystal J. "Animal Safekeeping in Situations of Intimate Partner Violence." *Journal of Interpersonal Violence* 37, no. 17–18 (2022).

Glenn, Catherine, and Goodman, Lisa. "Living With and Within the Rules of Domestic Violence Shelters: A Qualitative Exploration of Residents' Experiences." *Violence Against Women* 21, no. 12 (2015).

Gray, Allison, Barrett, Betty Jo, Fitzgerald, Amy, and Peirone, Amy. "Fleeing with Fido: An Analysis of What Canadian Domestic Violence Shelters are Communicating Via Their Websites about Leaving an Abusive Relationship When Pets Are Involved." *Journal of Family Violence* 34 (2019).

Guenther, Katja. *The Lives and Deaths of Shelter Animals*. Stanford University Press, 2020.

Gunter, Lisa M., Gilchrist, Rachel J., Blade, Emily M., Reed, Jennifer L., Isernia, Lindsey T., Barber, Rebecca T., Foster, Amanda M., Feuerbacher, Erica N., and Wynne, Clive D.L. "Emergency Fostering of Dogs From Animal Shelters During the COVID-19 Pandemic: Shelter Practices, Foster, Caregiver Engagement, and Dog Outcomes." *Frontiers in Veterinary Science* 9 (2022).

Halm, Derek. "(Cat)Egory Mistake: The Invalidity of Animal Shelter Behavior Assessments." *Biology & Philosophy* 36, no. 35 (2021).

Hughes, Judy. "Women's Advocates and Shelter Residents." *Journal of Interpersonal Violence* 35, no. 35–36 (2020).

Jensen, Janne B.H., Sandøe, Peter, and Nielsen, Søren Saxmose. "Owner-Related Reasons Matter More Than Behavioural Problems—A Study of Why Owners Relinquished Dogs and Cats to a Danish Animal Shelter from 1996 to 2017." *Animals* 10 (2020).

Kerman, Nick, Lem, Michelle, Witte, Mike, Kim, Christine, and Rhoades, Harmony. "A Multilevel Intervention Framework for Supporting People Experiencing Homelessness with Pets." *Animals* 10 (2020).
Komorosky, Dawna, Woods, Dianne Rush, and Empie, Kristine. "Considering Companion Animals An Examination of Companion Animal Policies in California Domestic Violence Shelters." *Society & Animals* 23 (2015).
Labrecque, Jennifer, and Walsh, Christine A. "Homeless Women's Voices on Incorporating Companion Animals into Shelter Services." *A Multidisciplinary Journal of the Interactions between People and Other Animals* 24 (2011).
LouRandour, Mary. "Integrating Animals into the Family Violence Paradigm: Implications for Policy and Professional Standards." *Policy and Practice* 7, no. 3 (2007).
Ly, Lexis H., Gordon, Emilia, and Protopopova, Alexandra. "Exploring the Relationship Between Human Social Deprivation and Animal Surrender to Shelters in British Columbia, Canada." *Frontiers in Veterinary Science* 8 (2021).
Newberry, Michelle. "Pets in Danger: Exploring the Link between Domestic Violence and Animal Abuse." *Aggression and Violent Behavior* 34 (2017).
Nowicki, Stacy A. "Lies, Damned Lies, and Michigan Animal Shelter Statistics: Problems and Solutions." *Journal of Animal and Natural Resource Law* XII (2016).
Rehme, Teresa, Hartmann, Katrin, Truyen, Uwe, Zablotski, Yury, and Bergmann, Michèle. "Feline Panleukopenia Outbreaks and Risk Factors in Cats in Animal Shelters." *Viruses* 14 (2022).
Rhoades, Harmony, Winetrobe, Hailey, and Rice, Eric. "Pet Ownership Among Homeless Youth." *Child Psychiatry Human Development* 46 (2015).
Robinson, Sarah R., Maxwell, December R., and Rogers, Kelli R. "Living in Intimate Partner Violence Shelters." *British Journal of Social Work* 50 (2020).
Shafer, Zanna. "Home Is Where the Dog Is: A Discussion of Homeless People and Their Pets." *Animal Law* 23 (2016).
Shier, Micheal, Walsh, Christine A., and Graham, John R. "Conceptualizing Optimum Homeless Shelter Service Delivery: The Interconnection between Programming, Community, and the Built Environment." *Canadian Journal of Urban Research* 16, no. 1 (2007).
Sinski, Jennifer B., and Gagné, Patricia. "Give Me Shelter: The State of Animal Sheltering in Kentucky's County Shelter System." *Contemporary Justice Review* 19, no. 2 (2016).
Sloan, Katherine. "Death Without Dignity: The Misnomer of Euthanasia in the State Animal Shelter System A Call for a No-Kill Florida." *Journal of Land Use* 32, no. 1 (2016).
Stevenson, Rochelle, Fitzgerald, Amy, and Barrett, Betty Jo. "Keeping Pets Safe in the Context of Intimate Partner Violence: Insight from Domestic Violence Shelter Staff in Canada." *Affilia: Journal of Women and Social Work* 33, no. 2 (2018).
Strand, Elizabeth B., and Faver, Catherine A. "Battered Women's Concern for Their Pets: A Closer Look." *Journal of Family and Social Work* 9, no. 4 (2005).
Stubsjøen, Solveig Marie, Moe, Randi Oppermann, Johannessen, Cicilie, Larsen, Maiken, Madsen, Henriette, and Muri, Karianne. "Can Shelter Dog Observers Score Behavioural Expressions Consistently over Time?" *Acta Veterinaria Scandinavica* 64, no. 1 (2022).
Sylvestre, John, Kerman, Nick, Polillo, Alexia, Lee, Catherine M., and Aubry, Tim. "A Profile of Families in the Emergency Family Homeless Shelter System in Ottawa, Ontario, Canada." *Canadian Journal of Urban Research* 26, no. 1 (2016).
Tiplady, C.M., Walsh, D.B., and Phillips, C.J.C. " 'The Animals Are All I Have' Domestic Violence, Companion Animals, and Veterinarians" *Society and Animals* 26 (2018)
Tuber, David S., Miller, Deborah D., Caris, Kimberly, A., Halter, Robin, Linden, Fran, and Hennessy, Michael B. "Dogs in Animal Shelters: Problems, Suggestions, and Needed Expertise." *Psychological Science* 10, no. 5 (1999).
Tutty, Leslie M. "Addressing the Safety and Trauma Issues of Abused Women: A Cross Canada Study of YWCA Shelters." *Journal of International Women's Studies* 16, no. 3 (2015): 101–116.
Upadhya, V. "The Abuse of Animals as a Method of Domestic Violence." *Emory Law Journal* 63 (2014).
VanNatta, Michelle. "Power and Control: Changing Structures of Battered Women's Shelters." *The International Journal of Interdisciplinary Social Science* 5, no. 2 (2010).
Weaver, Harlan. "Pit Bull Promises: Inhuman Intimacies and Queer Kinships in an Animal Shelter." *GLQ* 21, no. 2–3 (2015).

15
WILD ANIMAL ETHICS
A Gender-Sensitive Perspective

Catia Faria

What does gender have to do with other animals, particularly with wild animals? There are several relevant ways to address this question. Here, I will examine the role of gender identity in shaping human attitudes toward nonhuman animals, in particular toward wild animals, that is, the way that gender influences human beliefs, desires and values about wild animals' lives and shapes our value assessments about the natural world more generally. I will start by exposing anthropogenic harms suffered by animals living in the wild and the gender-based reasons such harms remain typically overlooked (Section 1). I will then proceed by exposing the naturogenic harms that wild animals face on a systematic basis and explore the generalised indifference toward it as explained, at least in part, by a gendered appreciation of nature's ways (Section 2). I will argue that such attitudes derive from a male-biased worldview which misconstrues individuals as mere elements in the large network of "wholes", idealises natural processes and/or places a disproportionate normative value on autonomy. I will close by briefly suggesting future directions for adopting more ethical, and less gendered, attitudes towards the suffering of wild animals (Section 3).

1. Wild Animal Well-Being and Anthropogenic Harms

Human beings have a long record of interventions in nature that jeopardise wild animals' well-being. Perhaps the most salient one is the extensive hunting of wild animals for a plethora of human purposes. Wild animals have been continuously hunted down for, among other things, their flesh, the social status conferred by their desiccated bodies and the economic benefits of their body parts and fluids. Wild animals, particularly members of endangered species, are systematically trapped, drugged and smuggled, often pushing entire populations to the verge of decimation. Some wild animals are also being wiped out due to habitat destruction and the effects of an increasingly changing climate. The harms involved in such unethical activity are deep and have extensively been exposed in the literature. Whether directly or indirectly, the anthropocentric actions of man have proven destructive, leading to the massive suffering and death of wild animals.

DOI: 10.4324/9781003273400-19

Not as scrutinised, though, are other pervasive interventions in nature carried out for reasons given by environmental values, ranging from the most anthropocentric to ecocentric ones. Either for instrumental or intrinsic reasons—that is, either because it provides a multitude of essential goods and services for human beings or because it has value in itself—the preservation of species, ecosystems and, eventually, the entire biosphere is taken as a value to be pursued. Yet the pursuit of environmental values is often in tension with wild animal well-being. It frequently involves causing harms to many individual wild animals—anthropogenic harms—bringing extreme suffering and death to an otherwise already hostile natural environment.

Consider the widespread practice of population control of non-autochthonous species—usually called "invasive species"—as part of a broader strategy for the conservation of native wild species, which consists of deliberate attempts to kill nonhuman individuals and eventually eliminate a whole population from an ecosystem. In Spanish territory, for instance, the native white-headed duck is claimed to be threatened as a genetically pure species due to its crossbreeding with the foreign ruddy duck. As a response, interventions have been carried out to identify and kill all pure specimens of ruddy duck and hybrids with white-headed duck that are sighted.[1]

Other well-known targets of population control within the European context include wolf-dog hybrids that have been detected in several geographic areas[2]—hunted and killed to protect the genetic purity of the native grey wolf[3]—or the non-autochthonous Canarian herbivores, shot to death to protect the native flora of the island.[4] Other methods of population control used to restore natural environments consist mainly of indirect killing of wild animals considered undesirable in a certain ecosystem through the reintroduction of predators and the resulting biological dynamics that follow from it—the "ecology of fear"—by which wild animals, when confronted with the risk of predation, chose to stop grazing openly out of fear of being killed and hide in more scarce places where predators cannot locate them. This causes death by starvation and other lethal complications caused by food and water deprivation. One of the most well-known instances of this method is the reintroduction of wolves in Yellowstone Park in the United States to stop elks from overgrazing autochthonous plants, thereby reversing the ecosystem to a prior form.[5]

These examples show how frequently the pursuit of environmental values clashes with wild animal well-being and how the anthropogenic harms suffered by wild animals remain critically overlooked. Yet, how can such an indifference toward wild animal well-being be justified? The rationale underlying such interventions, which runs through most environmental literature, is typically holistic and finds its corollary in the legendary maxim by Aldo Leopold: "A thing is right when it tends to preserve the integrity, stability, and beauty of the biotic community. It is wrong when it tends otherwise."[6] That is, moral concern should be directed not toward individual wild animals but rather to a large network of wholes such as "species" "ecosystems", "the biotic community" or "the land"—what ecofeminist Marti Kheel called "larger transcendent constructs".[7]

Here I draw on Kheel's analysis of the link between hegemonic masculine identity and transcendence ideals within the environmental movement to sustain the claim that indifference towards wild animal well-being represents a male-biased worldview defined by an unjustified "quest for transcendence". In Kheel's view, this quest for transcendence, insofar as it reflects the masculinist drive to rise above nature—typically associated with the feminine realm—is itself the core of the problem. In my view, the problem is not so much transcendence itself but rather the content of what is to be transcended. In short: the transcendence of the individual.

The transcendence of the individual should be understood in three distinct ways. First, if individuals are mere elements in the large network of wholes, then it's morally unproblematic that individuals be sacrificed for the benefit of the wholes—where value ultimately lies. According to holism, as part of wholes, individuals are relatively insignificant. They have instrumental value, which is a function of both the value of the species they belong to and its population density. The value of a species lies in what it contributes to the whole—its ecological role or function.[8]

The way in which holist thinkers lose sight of the individual for the sake of larger, abstract entities echoes the traditional, male-based obsession in political philosophy of deriving the individual from society and to accordingly subordinate individual interests for what is thought to be the good of the whole, be it society, the nation or the race. In the analogy proposed by John Baird Callicott, for instance, just as the preservation of the interests of a society requires the sacrifice of the interests of some of its parts, the preservation of the "interests" of the biotic community also requires sacrificing the interests of some of its parts (the interests of wild animals) to the benefit of the whole.[9]

Yet, as sentient creatures, wild animals are not mere components of ecological wholes. They have a well-being which makes it possible for their individual lives to go well or badly for *them*. Accordingly, each of them can be harmed and benefitted and to the same extent, independently of what they do and how many others of their species there are. Whatever human action that curtails the well-being of wild animals harms them. First, it harms them by making them suffer. Second, it harms them by killing them, that is, by depriving them of a life of potentially valuable experiences. Their value cannot thus be context dependent. Wild animals are morally considerable individuals, with value in themselves. The anthropogenic harms they suffer in the wild should be an object of our moral concern, at least to the same extent as the harms imposed on them by human exploitation.

The second way to understand the transcendence of the individual is, to borrow again from Leopold, "to think like a mountain"—that is, to transcend the individual perspective—the embodied self—and, so to speak, achieve the point of view of the universe. From this epistemic position, holistic authors usually draw the normative conclusion that the "perspective of the whole" should be prioritised over individuals. Frequently, this reasoning is disputed by ecofeminist authors, most notably Kheel,[10] on the basis of the alleged perils of the quest for impartiality. Yet I find no reason to commit to such a view. First, accepting the value of an impartial perspective does not lead us to prioritise the whole over individuals. In fact, impartiality requires that we imaginatively take the point of view of all individual perspectives affected by an event (here wild animals), not the "perspective" (or lack thereof) of abstract entities (species, ecosystems, the land). Second, and most importantly, it is true that the concept of "impartiality" has been used to camouflage the androcentric perspective, passing as the universal one. Yet, the fact that the androcentric perspective confuses being impartial with pretending to be impartial is not a fault of impartiality itself but rather a problem of the androcentric perspective not being impartial enough. What we need is not to get rid of the norm of impartiality but a more rigorous adherence to such norm, which will allow us to distinguish between legitimate and illegitimate appeals to impartiality. A proper quest for impartiality is intended, first and foremost, to transcend the male-biased position which overvalues the wholes, thereby fostering indifference towards individual wild animals.

The third way to understand the transcendence of the individual is cross-generational. In Kheel's words, "Men seek to transcend their own mortality through the act of killing, claiming allegiance to a timeless patrilineal brotherhood that connects them to past and future generations of men."[11] Again, I agree with the spirit of Kheel's statement while distancing myself from it in its refusal of long-term thinking. In my view, the crucial problem does not lie in longtermism per se—the future, including the far future, deserves our ethical attention—but how longtermism is framed, under environmentalism, in a way that overlooks fundamental nonhuman interests and forges a cross-time link between present and future generations of the most privileged inhabitants of Earth. First, environmentalist longtermism cannot plausibly sustain its holistic rationale. If what is to be preserved is ecological wholes, and individuals are nothing more than members of a planetary "biotic team",[12] then human beings are the most appropriate candidates to be sacrificed for the long-term good of the whole. Yet this is considered unacceptable by virtually all environmentalists. This is why long-term environmentalism is often framed in an anthropocentric way, according to which wholes are suddenly to be preserved, as they provide essential goods and services for human generations to come. However, there is no compelling reason to maintain this human exceptionalism. Some individuals (nonhuman) are not, and should not be, more elements in the large network of "wholes" than others (human).

Second, and relatedly, environmentalism and its many eco-activist branches are imbued with numerous biases in their framing of the future. The environmentalist mantra that it is our moral responsibility to save the future for "our children" leaves unaddressed the question of what exactly is to be preserved and for whose benefit. Certainly not those whose lives must be *controlled*. That includes the lives of numerous wild animals who are perceived as threatening to the "integrity, stability, and beauty" of ecological wholes as well as the lives of those humans with "too many children", that is, individuals primarily from non-white countries in the Global South. As xenofeminist[13] Helen Hester stresses:

> In positioning eco-activism as agitating on behalf of generations to come, we may unwittingly participate in the cult of the Child, which is so central in determining which lives are prioritized, whose needs are seen to matter, and which bodies are framed as threatening.[14]

By taking "the child"—a type of young, white, boy—as the default bearer of the future and by thinking his life, entangled in heteronormative assumptions, at the expense of so many others, human and nonhuman, the environmentalist rhetoric reinforces and potentially magnifies cycles of privilege and disadvantage over time. The environmentalist cult of the unborn child—a racist, heteronormative construct, as Hester calls it—emerges as a justifying figure for the timeless propagation of an unjust status quo, crucially defined by human supremacism but also essentially hostile to racialised, gender and sexual dissident human individuals.

The crucial problem with the environmentalist quest for transcendence of the individual is thus not only to transcend men's mortal individuality through a ruling alliance with future generations of men but rather to transcend all the *wrong* ways of being an individual. It is, in short, to transcend and subordinate a certain form of individual in order to affirm and perpetuate another—the heterosexual, white male human.

2. Wild Animal Well-Being and Naturogenic Harms

Now it would be a mistake to think that threats to wild animal well-being are exclusively anthropogenic. Wild animals suffer tremendously due to human agency, but they also suffer and die due to natural causes. This is particularly true of animals that live in the wild, given their high exposure to harmful natural processes and limited capacity to cope with them. Even without human interference, many wild animals have lives full of suffering which usually end up in horrible deaths. Wild animals are frequently wounded, often fatally. A vast number of them typically die of starvation and commonly suffer from malnutrition. Water is also scarce in the wild, and many animals die of thirst, especially in times of drought. Extreme weather events often cause animals to freeze to death or die of heat en masse and thwart their ability to thrive by decreasing foraging and increasing risk to predation. Wild animals are also prone to be infected by a large number of parasites and are highly susceptible to various forms of disease, which cause them a great suffering and often kills them, either directly or indirectly, by increasing their susceptibility to predation and other harmful events. This is, in fact, likely to affect not just a few individuals but the majority of those who are born in the wild. Most wild animals follow a wasteful reproductive strategy, which consists of producing the maximum number of offspring to maximise the chances of passing on the genetic material to the next generation. Examples range from amphibians and fish to invertebrates and mammals, including small rodents. These individuals have low survival rates. On average, only one survives per parent. Most of them die prematurely and atrociously, captured by predators, frozen or starved to death, ravaged by disease or parasites, with little to no chance of enjoying their lives.[15]

Insofar as natural processes cause suffering and death to wild animals, they are at least as harmful as human interventions against wild animal well-being. Non-human suffering is suffering, no matter who or what causes it. Death is bad for whoever suffers it, regardless of whether it is caused by a gunshot, disease or famine. Yet naturogenic harms suffered by wild animals remain unjustifiably overlooked. Wild animal well-being is thus impaired not merely by adverse environmental factors but crucially by extreme human indifference.

I believe human indifference towards the naturogenic harms that fall upon wild animals can only be fully understood against a backdrop of patriarchal ideals which give rise to a gendered appreciation of nature's ways. The first idea is basically an extension of ecofeminism's core insights, that is, the patriarchal devaluation of compassion, following a broader devaluation of *all things natural* typically associated with the feminine realm. Indeed, ecofeminism has importantly elucidated how traditional dualisms between nature/culture, emotional/rational, female/male, and so on have worked to sustain the male domination of women, nature and other animals. Yet it has fallen short in explaining the other side of the patriarchal coin with regard to shaping our moral indifference—literally, the opposite of care and compassion—toward naturogenic harms, including naturogenic wild animal suffering. There is a widespread belief that, no matter how harmful, whatever happens to wild animals as a result of natural processes is not our moral business. Yet an excessive hands-off approach to providing aid suggests an overestimation of the masculine gender stereotype of emotional restraint and of not being personally involved with others in need, which clashes with what compassion is meant to mean—caring for the suffering of others and having the disposition to help them. By failing to acknowledge that compassion is due to wild animals, we are not only failing wild animals, but we are also deferring to and reinforcing a male-biased worldview.

Another way in which the patriarchy gives rise to a harmful, gendered appreciation of nature's ways is through the idealisation of the "natural". Patriarchal Western cultures have conflated the natural with the normative, selecting what is natural according to what is judged best for the patriarchal system. Moreover, it is assumed, along these lines, that if something is natural, then it is good or desirable. The most salient manifestations of this are exemplified in our attitudes toward heterosexuality and the suffering of other animals. It is now fairly uncontroversial that the cisheteropatriarchy manipulates heterosexual ciswomen's beliefs and desires into submitting to natural practices that often go against their true interests to further the patriarchal institution, thereby turning natural harms into an idealised scenario for those who suffer or observe them. Likewise, some of the most devastating effects of the idealisation of the natural are suffered by other animals, since idealisation also functions as a powerful instrument for perpetuating discrimination against individuals of other species. Idealisation, in this context, is the strategy by which natural harms which fall upon wild animals are excluded as a proper cause for ethical concern and often celebrated as the result of natural processes. Take the paradigmatic case of predation. Predatory activity ends wild animals' lives in gruesome ways. The kill may be abrupt or rather slow and agonising. The capture is often preceded by intense terror, and it usually involves a brutal struggle for survival. Not uncommonly, the prey is consumed while still alive. In addition, prey animals live in a state of constant alarm, responsible for triggering intense stress responses similar to PTSD in humans.[16] And yet, predation is, in general, highly appreciated and often exalted, particularly by men. Much has been said about how the exaltation of predation has been used as a justification fore men's hunting and how hunting, a violent manhood ritual, participates in the construction of an ideal of masculinity typically characterised by emotional suppression, risk-taking, the provision (albeit performative) of family sustenance and, crucially, predatory heterosexuality.[17] However, the positive appreciation of predation entails a much broader patriarchal view, one fundamentally marked by the valuation of hierarchies, by power imbalance between individuals and the annihilation of the weakest. Notice the disparate value accorded to the two poles of the predatory relationship and the valuing of the most powerful one—the predator—over the most vulnerable—the prey, which very much resembles gender hierarchies. Relatedly, notice also the unequal distribution of agency ascriptions, whereby predators are seen as taking an active role in the biological dynamic, while prey are turned into passive elements whom things simply befall. It is simply not the case that prey animals are passive. Consider, for instance, predator-avoidance decision-making, by which animals must choose either to decrease food intake or to increase the risk of being caught by predators. What happens is that our perception of predation is biased by a sort of inbuilt "male gaze" whereby we interpret natural events in socially laden ways. People value predation and celebrate disease and other forms of wild animal suffering, because they knowingly or unknowingly endorse a patriarchal system fundamentally based on the subordination and destruction of the weakest. This contributes to legitimising men's dominant position in society and justifies the marginalisation of all the "weak" ways of being an individual, human and nonhuman.

A true alliance between overlapping targets of the patriarchy should acknowledge that "biology is not destiny" and that premature death, starvation and violence are no more *natural* for individuals of other species than for human beings and, therefore, should no more be part of their lives than of ours. As the xenofeminist slogan goes, "If nature is unjust, change nature."[18]

One way to dispute the previous arguments would be to claim that we should abstain from helping animals in need insofar as interfering on their behalf would curtail wild animal autonomy. But that would be a mistake insofar as it presupposes a particular, male-biased conception of autonomy, one which misrepresents the individual as a self-sufficient, independent self, which downplays interdependence and mutual support—traditionally associated with feminine traits—as defining features of our lives and the extent to which social factors impede or enhance autonomy. Alternatively, a feminist, relational approach to autonomy highlights the social dimension of autonomy and how its development and exercise can be impaired by extreme social neglect, allowing us to stress how wild animals' autonomy is, as a matter of fact, shaped by external social conditions and, in particular, how it is dependent on human support for an adequate range of options. Helping wild animals, by creating the conditions that allow autonomy to develop may often be a precondition for the full exercise of such capacity and, rather than autonomy-undermining, be construed as a means to building more resilient "wild animal communities".

3. Toward a Feminist Wild Animal Ethics

In this chapter, I exposed the anthropogenic and naturogenic harms suffered by animals living in the wild and the gender-based reasons such harms remain typically overlooked as object of moral concern. I argued that the generalised attitude of moral indifference towards the plight of wild animals derives from a male-biased worldview which misconstrues individuals as mere elements in the large network of "wholes", devalues compassion, idealises natural processes and/or places a disproportionate normative value on a certain conception of individual autonomy. Until very recently, "wild animal ethics" was either completely absent from the literature or unduly absorbed in what one might call "nature ethics". Fortunately, the tide has changed, and "wild animal ethics" is increasingly being acknowledged as a field of study in its own right.[19] Yet dominant positions in the field of wild animal ethics remain completely detached from the insights of feminist ethics and thus oblivious to the role of gender in shaping attitudes towards nonhuman animals, particularly wild animals. This chapter, which is more exploratory than conclusive in nature, intends to draw the attention of both parties—feminists and animal ethicists—to this still vastly unexplored topic. On the one hand, it aims to show the centrality of moral concern for wild animal suffering. On the other hand, it aims to bring to the fore the multiple ways that gender influences human beliefs, desires and values about wild animals' lives and shapes our value assessments about the natural world more generally. If this analysis has been minimally compelling, my hope is to have at least sowed the seed for a much-needed and fruitful alliance between wild animal ethics and feminist ethics.

Notes

1 Gobierno de España, "Estrategia para la Conservación de la Malvasía Cabeciblanca (Oxyura leucocephala) en España," Ministerio de Medio Ambiente y Medio Rural y Marino (2009), retrieved from: https://www.miteco.gob.es/es/biodiversidad/publicaciones/pbl-fauna-flora-estrategias-nv-malvasia-cabeciblanca.aspx

2 Valerio Donfrancesco, Paolo Ciucci, Valeria Salvatori, David Benson, Andersen, and Claudia Capitani, "Unravelling the Scientific Debate on How to Address Wolf-dog Hybridization in Europe," *Frontiers in Ecology and Evolution* 7, no. 175 (2019), https://doi.org/10.3389/fevo.2019.00175

3 European Parliament, Parliamentary Question—E-004563/2017(ASW) (2017), retrieved from: https://www.europarl.europa.eu/doceo/document/E-8-2017-004563-ASW_EN.html?redirect#def2
4 Laura Gangoso, José A. Donázar, Stephan Scholz, César J. Palacios, and Ferando Hiraldo, "Contradiction in Conservation of Island Ecosystems: Plants, Introduced Herbivores and Avian Scavengers in the Canary Islands," *Biodiversity & Conservation* 15, no. 7 (2006): 2231–2248
5 Oscar Horta, "The Ethics of the Ecology of Fear against the Nonspeciesist Paradigm: A Shift in the Aims of Intervention in Nature," *Between the Species* 13, no. 10 (2010): 10.
6 Aldo Leopold, *A Sand County Almanac, and Sketches Here and There* (Oxford University Press, USA, 1989).
7 Marti Kheel, *Nature Ethics: An Ecofeminist Perspective* (Rowman & Littlefield Publishers, 2007), 20.
8 J. Baird Callicott, "Animal Liberation: A Triangular Affair," *Environmental Ethics* 2, no. 4 (1980): 325–326.
9 Ibid.
10 Kheel, *Nature Ethics: An Ecofeminist Perspective*.
11 Ibid., 116
12 Leopold, *A Sand County Almanac, and Sketches Here and There*.
13 Xenofeminism is a feminist movement initiated by the activist group Laboria Cuboniks, *The Xenofeminist Manifesto: A Politics for Alienation* (Verso Books, 2018), which, while embracing a certain degree of internal disagreement, is based on three core ideas: gender abolitionism, anti-naturalism and technomaterialism (Hester 2018). While I am generally sympathetic to xenofeminism, in this chapter, I do not defend a xenofeminist position. The appeal to Hester's illuminating passage should be understood as merely an illustration of the risks associated with environmentalism's framing of the future and not as a commitment to the whole xenofeminist project. My arguments will be more or less compelling regardless of the soundness of the xenofeminist view, and, in that sense, if they work, they can be accepted by a much wider range of feminist positions.
14 Helen Hester, *Xenofeminism* (Polity, 2018), 49.
15 Catia Faria, *Animal Ethics in the Wild: Wild Animal Suffering and Intervention in Nature* (Cambridge: Cambridge University Press. Forthcoming, 2023); Horta, "The ethics of the ecology of fear against the nonspeciesist paradigm: A shift in the aims of intervention in nature."; Yew Kang Ng, "Towards Welfare Biology: Evolutionary Economics of Animal Consciousness and Suffering," *Biology and Philosophy* 10, no. 3 (1995): 255–285; Brian Tomasik, "The Importance of Wild-animal Suffering," *Relations: Beyond Anthropocentrism* 3, no. 2 (2015): 133–152.https://doi.org/10.7358/rela-2015-002-toma
16 Philip R. Zoladz, Cheryl D. Conrad, Monika Fleshner, and David M. Diamond, "Acute Episodes of Predator Exposure in Conjunction with Chronic Social Instability as an Animal Model of Post-traumatic Stress Disorder," *Stress* 11, no. 4 (2008): 259–281.
17 Brian Luke, *Brutal: Manhood and the Exploitation of Animals* (University of Illinois Press, 2007)
18 Hester, *Xenofeminism*, 43
19 Sue Donaldson and Will Kymlicka, *Zoopolis: A Political Theory of Animal Rights* (Oxford University Press, 2011); Faria, *Animal Ethics in the Wild: Wild Animal Suffering and Intervention in Nature;* Oscar Horta, "Animal Suffering in Nature: The Case for Intervention," *Environmental Ethics* 39, no. 3 (2017): 261–279; Oscar Horta, *Making a Stand for Animals* (Routledge, 2022); Kyle Johannsen, *Wild Animal Ethics: The Moral and Political Problem of Wild Animal Suffering* (Routledge, 2020); Jeff McMahan, "The Moral Problem of Predation," in *Philosophy Comes to Dinner*, eds. A. Chignell, T. Cuneo, and M. Halteman (London: Routledge, 2015), 278–304; Martha C. Nussbaum, *Frontiers of Justice: Disability, Nationality, Species Membership* (Belknap Press, 2006).

Bibliography

Callicott, J. Baird. "Animal Liberation: A Triangular Affair." *Environmental Ethics* 2, no. 4 (1980): 311–338.
Cuboniks, Laboria. *The Xenofeminist Manifesto: A Politics for Alienation*. Verso Books, 2018.

Donaldson, Sue, and Kymlicka, Will. *Zoopolis: A Political Theory of Animal Rights*. Oxford University Press, 2011.

Donfrancesco, Valerio, Ciucci, Paolo, Salvatori, Valeria, Benson, David, Andersen, Liselotte Wesley, Bassi, Elena, Blanco, Juan Carlos, Boitani, Luigi, Caniglia, Romolo, Canu, Antonio, Capitani, Claudia, Chapron, Guillaume, Czarnomska, Sylwia D., Fabbri, Elena, Galaverni, Marco, Galov, Ana, Gimenez, Olivier, Godinho, Raquel, Godinho, Raquel, Greco, Claudia, Hindrikson, Maris, Huber, Djuro, Hulva, Pavel, Jedrzejewski, Włodzimierz, Kusak, Josip, Linnell, John Durrus, Llaneza, Luis, López-Bao, José Vicente, Männil, Peep, Marucco, Francesca, Mattioli, Luca, Milanesi, Pietro, Milleret, Cyril Pierre, Mysłajek, Robert W., Ordiz, Andres, Palacios, Vicente, Pedersen, Hans Christian, Pertoldi, Cino, Pilot, Malgorzata, Randi, Ettore, Rodriguez, Alejandro, Saarma, Urmas, Sand, Håkan, Scandura, Massimo, Stronen, Astrid Vik, Tsingaraska, Elena, and Mukherjee, Nibedita. "Unravelling the Scientific Debate on How to Address Wolf-dog Hybridization in Europe." *Frontiers in Ecology and Evolution* 7, no. 175 (2019). https://doi.org/10.3389/fevo.2019.00175

European Parliament. "Parliamentary Question—E-004563/2017(ASW)." (2017). Retrieved from: https://www.europarl.europa.eu/doceo/document/E-8-2017-004563-ASW_EN.html?redirect#def2

Faria, Catia. *Animal Ethics in the Wild: Wild Animal Suffering and Intervention in Nature*. Cambridge: Cambridge University Press, Forthcoming, 2023.

Gangoso, Laura, Donázar, José A., Scholz, Stephan, Palacios, César J, and Hiraldo, Ferando. "Contradiction in Conservation of Island Ecosystems: Plants, Introduced Herbivores and Avian Scavengers in the Canary Islands." *Biodiversity & Conservation* 15, no. 7 (2006): 2231–2248.

Gobierno de España. "Estrategia para la Conservación de la Malvasía Cabeciblanca (Oxyura leucocephala) en España." Ministerio de Medio Ambiente y Medio Rural y Marino, 2009. Retrieved from: https://www.miteco.gob.es/es/biodiversidad/publicaciones/pbl-fauna-flora-estrategias-nv-malvasia-cabeciblanca.aspx

Hester, Helen. *Xenofeminism*. Polity, 2018.

Horta, Oscar. "The Ethics of the Ecology of Fear against the Nonspeciesist Paradigm: A Shift in the Aims of Intervention in Nature." *Between the Species* 13, no. 10 (2010): 10.

Horta, Oscar. "Animal Suffering in Nature: The Case for Intervention." *Environmental Ethics* 39, no. 3 (2017): 261–279.

Horta, Oscar. *Making a Stand for Animals*. Routledge, 2022.

Johannsen, Kyle. *Wild Animal Ethics: The Moral and Political Problem of Wild Animal Suffering*. Routledge, 2022.

Kheel, Marti. *Nature Ethics: An Ecofeminist Perspective*. Rowman & Littlefield Publishers, 2007.

Leopold, Aldo. *A Sand County Almanac, and Sketches Here and There*. Oxford University Press, USA, 1989.

Luke, Brian. *Brutal: Manhood and the Exploitation of Animals*. University of Illinois Press, 2007.

McMahan, Jeff. "The Moral Problem of Predation." In *Philosophy Comes to Dinner*, edited by A. Chignell, T. Cuneo, and M. Halteman, 278–304. London: Routledge, 2015.

Ng, Yew Kwang. "Towards Welfare Biology: Evolutionary Economics of Animal Consciousness and Suffering." *Biology and Philosophy* 10, no. 3 (1995): 255–285.

Nussbaum, Martha C. *Frontiers of Justice: Disability, Nationality, Species Membership*. Belknap Press, 2006.

Soryl, Asher A, Moore, Andrew J., Seddon, Philip J, King, Mike R. "The Case for Welfare Biology." *Journal of Agricultural and Environmental Ethics* 34, no. 2 (2021): 1–25. https://doi.org/10.1007/s10806-021-09855-2

Tomasik, Brian. "The Importance of Wild-animal Suffering." *Relations: Beyond Anthropocentrism* 3, no. 2 (2015): 133–152. https://doi.org/10.7358/rela-2015–002-toma

Zoladz, Phillip R., Conrad, Cheryl D., Fleshner, Monika, and Diamond, David M. "Acute Episodes of Predator Exposure in Conjunction with Chronic Social Instability as an Animal Model of Post-traumatic Stress Disorder." *Stress* 11, no. 4 (2008): 259–281.

PART IV

Feminist Multispecies Methods

16

CARE AS METHOD FOR MULTISPECIES ETHNOGRAPHIES

Pablo P. Castelló

Feminism is an intersectional discipline and practice. It is a way of life that acknowledges the fact of patriarchy and opposes it.[1] Patriarchy, as understood by Val Plumwood, Maneesha Deckha, and Anna Grear, can be defined as a knowledge system that centres the paradigmatic human person, that is, the ideal of a white and rational man, who is disembodied, independent, and civil.[2] This patriarchal epistemology pervades our economic, political, and legal systems to uphold the status quo so that human and nonhuman people of the global majority[3] remain oppressed and situated in a position of subordination. Black feminist scholar Patricia Hill Collins and political theorist Dinesh Wadiwel argue that the suppression of oppressed groups' knowledge, bodies, and ways of being facilitates the rule of dominant groups.[4] The point is not necessarily that, for example, men decide to objectify women. Instead, the patriarchal gaze sexualises women so that they are seen as objects to be consumed.[5] Most humans in western countries see with this epistemic gaze, and this gaze will often not see real animals. Instead, through the concept of the voiceless, to-be-consumed animal, animals appear as objects at the disposal of humans, as beings who cannot make decisions and respond.[6] What is important to understand for the purposes of this chapter is that seeing animals for who they are—beings who can voice their desires and preferences, forge affective relationships, and establish their own forms of social organisation—requires creating subversive epistemic conditions.[7] Animals themselves are of utmost importance in this process; in order to build a politics that attends to who animals are, in order to subvert the oppression and violence that patriarchy produces by silencing the voices of those who deviate from the paradigmatic human person, we must listen to animals rather than "what other humans are telling us about them," as ecofeminists Josephine Donovan and Carol J. Adams put it.[8] This means, among other things, and as Donovan and Adams explain, that we ought to speak with animals; learn how they communicate; and attend to their desires, wishes, and needs.[9] Ecofeminist theorists also stress that an ethic of care requires rejecting the commodification and instrumentalisation of animals. Instead, they argue that we should see nonhuman animals as relational and vulnerable beings who should be respected in their full complexity.[10]

The tradition of feminism for which I advocate in this chapter does not only focus on critique; it also emphasises the need to imagine better and more just futures. Angela Davis

DOI: 10.4324/9781003273400-21

captures this point eloquently when she explains that feminism can entail a dual commitment "to use knowledge in a transformative way, and to use knowledge to remake the world so that it is better for its inhabitants—not only for human beings, for all its living inhabitants."[11] Feminists have traditionally argued that in order to subvert the oppressive status quo, we need to adopt an ethic that centres care. Care refers to an emotion that animates moral beings to be responsive to others' needs and to their wills and desires.[12] This requires "certain habits of perception"[13] such as attending to context, to the specific animal persons we encounter, and to the relationships in which we are enmeshed.[14] Further, the conception of the self that emerges from feminist philosophy, of which care ethics is a subfield, understands that human and nonhuman animals are unavoidably entangled in relationships of interdependency and that this fact opens opportunities to flourish together.[15]

In line with care ethics, this chapter demonstrates that care should be understood as a necessary method for conducting multispecies ethnographies (MEs). This is particularly important, I contend, for those ethnographers who seek to map the political dynamics of multispecies communities. I say 'method,' as opposed to a 'methodology,' because a methodology provides a framework that informs the specific techniques (or doings) the researcher will actually use, and it gives the conceptual lens that will serve the researcher to observe and interpret her data (King 1994, 20; Clough and Nutbrown 2002, 30).[16] One perhaps helpful way to think of a methodology is to imagine it as an umbrella term that includes one, or more, method(s) as part of one's methodology—in fact, a methodology partly consists in deciding one's methods. My argument will be that multispecies ethnographers will sometimes need to integrate care as a specific technique used to conduct their research, which is what a method is.[17]

To these ends, I first discuss different dimensions of care and distinguish the different roles that care can play in different kinds of MEs. Alongside this, I explain the methodology used when I conducted research at VINE Sanctuary, an LGBTQ-led farmed animal sanctuary, in November-December of 2021 and why I chose VINE as a research setting and introduce VINE's ethos and multispecies community. By drawing on my fieldwork, I then explore how ethnographers can learn to care-for and the importance that this has for multispecies ethnographies. Finally, I examine why understanding care as a constitutive element of one's ethnographic method is a necessary condition to sense, observe, and interpret the political in a multispecies community.

Problematising Care in Multispecies Ethnographies

In this section, I first define ME and then establish an important distinction between two kinds of MEs. The first kind of ME sees humans working in multispecies places as their primary ethnographic interlocutors. My contention[18] will be that these kinds of ethnographies, which often take place in sites like farms, do not allow ethnographers to see animals as responsive subjects because the material, intersubjective, and socio-political conditions are not there. The second kind of ME centres animals as ethnographic and political subjects who are the researcher's primary interlocutors. This approach occurs in sites where nonhuman animals feel safe and able to trust the researchers. While the former approach might allow ethnographers to care about animals, the latter approach enables researchers to care for animals. My argument is that this second kind of ME enables ethnographers to establish relationships of trust with nonhuman animals, that this is a precondition to listen to animals, and that this is necessary to illuminate the kinds of political subjects that animals are and can become.

I understand ME as a political methodology that ruptures species boundaries and does not see human and nonhuman animals as being mere products of their biological species.[19] Instead, and in line with Sue Donaldson and Will Kymlicka, nonhuman animals are political agents who forge social norms and decide with whom they want to share their lives and where they want to reside.[20] From the beginning, ME has been very influenced by the work of Donna J. Haraway.[21] The ontology advanced by Haraway rightly sees animals (including humans) as being co-constituted by many forces, as embedded in relationships of interdependency, and as embodied vulnerable beings.[22] Haraway's work has also used the discourse of care, but it has been used in contexts in which animals are oppressed, their bodies are violated, and their voices silenced.[23]

Critical multispecies ethnographers have recently paid attention to the ways in which violence and care can be entangled in their research. Some of these ethnographers study the relationship between violence and care by seeing humans as their primary subjects of study and interlocutors. María Elena García's work exemplifies this approach.[24] She did an ethnography in a guinea pig breeding farm in the north of Lima, Peru, and showed how a farm owner's narrative of care and love is unapologetically entangled with commodification and killing. In her study, García encountered a female guinea pig who was at the edge of dying due to birthing complications, which led her to call Walter, the farm owner. After a brief inspection, Walter tossed the "pregnant, dying guinea pig out of the cage onto the dirt floor behind [them]" because, as he put it, "She is almost gone. She will be dead by morning if not sooner."[25] The indifference that Walter showed, and bearing witness to the guinea pig's state of vulnerability, had a profound emotional effect on García, although she hid her feelings from Walter and moved on with him, leaving the guinea pig to die alone. She felt troubled, sad, and grieved the death of the guinea pig—a being who would have not been grieved if García had not been there. This matters because nonhuman animals are not seen as grievable subjects by the western dominant epistemology. Hence, García's experience and work is, from this point of view, subversive.[26]

In conversation with García, Walter explained that female guinea pigs are his "most valuable assets" since he can gain "three times as much for one *reproductora* than he does selling a male or 'spent' female cuy [guinea pig] for meat."[27] Interestingly, however, he also said: "I care for them deeply. I love my little ladies."[28] García's analysis shows that while Walter cared for the guinea pigs by providing them food and shelter, and he explicitly speaks of love and deep care towards them, he also disavowed the guinea pig that was at the edge of death by expressing indifference towards her and leaving her on the floor after a brief inspection. What is more, the way Walter mixes a narrative of care with the guinea pigs' monetary value evinces that the guinea pigs were, after all, mere commodities.[29]

The story is further complicated by García's own positionality as a multispecies ethnographer. While García's ethnography enabled that specific guinea pig's life to be recognised as a grievable one,[30] García also ate guinea pigs when Walter invited her to eat them in a restaurant. Arguably, her research also (re)produced an epistemology in which, in the end, animals remain commodities who can be consumed by humans. García is not oblivious to this ethical problem. As she puts it:

> I can't help but think back and attempt to theorize that moment of sadness, which I think is also one of shame—the shame of taking his side; of worrying about my research.... Would my concern for this guinea pig raise questions that might imperil

> access to these kinds of spaces? In that moment, I felt my project, the so-called multi-species research I was conducting, was taking place at the expense of the animal
> *(García 357)*

While I am sympathetic to García's work, I worry that entangling the recognition of the guinea pigs' grievable lives with eating them, in the context of a breeding farm where they are commodified, is likely to uphold our dominant western epistemology, that is, an epistemology that does not recognise the full ethical weight of animals' lives. Animals' lives may be recognised as grievable, animals might be seen as relational 'whos,' but they still appear as 'less than' human, as beings who can be consumed. Kathryn Gillespie has recently advanced a similar concern in relation to MEs more generally. She argues that while MEs can disclose how we are implicated in relationships of violence, MEs "can also normalize relations of harm because researchers may be too embedded in the dominant cultures or too attached to forms of consumption that sustain normalized systems of violence in which human–animal relations routinely unfold to interrogate their own complicity."[31]

Naisargi N. Davé, whose anthropological work on animal activists in India also adopts the first approach (re: humans as primary ethnographic subjects), has further problematised care-based approaches, especially those that centre love. She pertinently questions why love succeeds "when mobilized in arguments not exactly for but 'decidedly not against' the structural killability of animals."[32] Dave suggests that the western understanding of love is exclusionary insofar as it differentiates, that is, it creates property-like islands in which we find those we love, "our own, beyond which, and in whose interest, there are limits to what we can be asked to do."[33] The reader has probably encountered people in their everyday life who might emphasise the importance of caring for and about those who are close to us, in our inner circle, and merely avoid harming everyone else. Dave's concern seems particularly important in the context of a sixth mass extinction event and systematic anthropocentric exploitation. In contrast to the exclusionary western ethic that centres love, Dave advocates for an ethics of indifference. Her ethnographic study in Mumbai, India, showcases this approach. In the study, she discusses the case of Dipesh, a fieldworker for Welfare Stray Dogs—a Mumbai-based NGO that understands that street animals do not belong in shelters and that they have, by and large, non-proprietary relationships of companionship with humans.[34]

Dave describes an experience between a raggedy dog, Dipesh, and herself to illustrate the ethics of indifference she proposes. After Dipesh had removed maggots from the dog and aided them with an ointment, the dog was almost hit first by a scooter and afterwards by a car. As this happened, Dave screamed; "Dipesh turned to look. He shrugged and said, 'She's old'" (ibid., 664). Dave interprets this situation as follows: "Neither his to love nor *not* his to not love, Dipesh responded to what was there before him: a love indifferent to the differentiations love demands. A world such as this is comprised not of moral islands of kin and property, yard and road, but of only a series of situations."[35]

Like García, Dave suggests that the relationship between an ethic of care and MEs is not as clear-cut as it might appear. Dipesh had *not* to care at some level in order to continue working. His love was by and large indiscriminate, and this enabled him to reach more vulnerable nonhuman animals than if he had, for example, focused on this one dog. Importantly, Dave explains that Dipesh never got married, didn't have children, and devoted his life to helping Mumbai animals by working from sunrise to sunset. Dave asked Dipesh why he had never married; he responded: "So I can care for them, I have not become a husband

and father."[36] It seems that not loving Dipesh's potential inner circle (the nuclear family he could have had) enabled Dipesh to care and aid many more animal people than he could have helped otherwise. By contrast, García also explains that one of the reasons why Walter wanted his business to thrive was to provide for his family, the ones closest to him, 'his own'—to borrow Dave's phrasing. The reader might wonder: 'Why should we then care about care? Shouldn't we adopt Dave's ethic of indifference?'

While I share many of the concerns that Dave and García expose, and it seems important not to care or to bracket our care in certain contexts, it also seems clear to me that their approaches would have been unhelpful for the kind of ethnography I conducted at VINE sanctuary. In what follows I will show that in certain contexts what is required to conduct MEs is affect, intimacy, connection, and responsivity. Instead of indifference, my research also suggests that certain research contexts require immersion. My contention will be that immersing myself in the multispecies political community in which I conducted fieldwork was a necessary condition to gather data and understand the community's political dynamics.

Before discussing my multispecies ethnographic work, it seems necessary to make explicit that I endorse the second approach: seeing nonhuman animals as my primary interlocutors and ethnographic subjects. My approach is in many respects in alignment with Gillespie's recent work on multispecies autoethnography and Donaldson and Kymlicka's zoopolitical project.[37] Gillespie, like Donaldson and Kymlicka, advocates for a kind of ME that centres intimacy[38] and puts emphasis on the kind of knowledge-making processes that can emerge through relationship-building.[39] Gillespie argues that conducting ethnographies in contexts that are not as hostile as, for example, breeding farms, can open epistemic opportunities to counter the dominant western epistemology that concerns Dave and García.

To put it differently, if a researcher seeks to centre animals and understand, as I do, the political dynamics of a multispecies community, the kind of political subjects that animals can become, and the kinds of political and legal systems animals might wish to co-author with humans, then it seems necessary to speak with nonhuman animals in contexts where their political agencies are enabled, where they have the chance to build relationships of trust and of their choosing, and social norms of their own. In other words, it is important to do fieldwork in places where nonhuman animals have, in some respects, a de facto right to self-determination.[40] If researchers are open to listening to animals in these kinds of contexts, transcribe what they say, and interpret their decisions, we can then begin to write a subversive animal-authored epistemology; one that can challenge the anthropocentric western epistemology that does not see animals as interlocutors who can speak, but rather voiceless, to-be-consumed entities. In the following section, I will argue that this kind of research requires 'care as method.'

Sensing Politically Through Care in the Context of a Multispecies Ethnography

VINE Sanctuary is an LGBTQ-led sanctuary in Springfield, Vermont, U.S., which was co-founded by feminist writers and activists Miriam Jones and pattrice jones. VINE's multispecies community is composed of hens, roosters, pigeons, guinea fowls, geese, ducks, alpacas, cows, goats, sheep, emus, humans, and a pig, amongst others. I chose VINE as my research site because I learnt through the research conducted by jones herself,[41] Charlotte Blattner, Sue Donaldson, and Ryan Wilcox[42] that the sanctuary's ethos is ecofeminist and puts self-determination at the centre.

This was important for my research because one of my main aims was to understand how what non-human animals say—in a context where nonhumans had (and have) the opportunity to freely co-author their own lives—should transform our political institutions. VINE was, from this point of view, a good site to explore this question because it is a multispecies institution and community that has created the political preconditions for nonhuman animals to decide how they organise themselves socio-politically. As I listened to and interacted with nonhuman residents at VINE, I came to learn that care was a necessary method to undertake my research.

Let me begin with Church, a male goat who was raised as a human's pet. Due to his upbringing, Church did not have the opportunity to meet other goats before he went to VINE. While these days Church often socialises with goat and non-goat residents at VINE, including humans, he is in some respects different from the other goats. Church sometimes struggles to interact with humans and might head-butt some of the human workers at VINE.

For some reason, Church was particularly fixated on me when I began conducting fieldwork at VINE and would try to head-butt me whenever he approached me. This made me feel vulnerable, scared, and intimidated. One of the human workers advised me to ignore him and walk away to avoid conflicts. The idea was that by expressing indifference, he would leave me alone. However, instead of letting me be when I walked away, Church would chase me, pursuing me around the sanctuary for relatively long periods of time. In these pursuits, he always kept a certain distance, but it still felt intimidating. After one week, I had a meeting with pattrice jones to evaluate my work at VINE and my observations and involvement in the community more generally. As we discussed this situation with Church, jones suggested that perhaps he felt scared around me, he simply wanted to play, or something else we couldn't understand. She invited me to be open minded and said: "How would you respond to Church if you knew he is scared?" jones also mentioned that if I expressed defensiveness through my body, Church would notice this because nonhuman animals are usually pretty good at reading body language. She advised me to bear in mind what I expressed through my body. We both agreed that, for whatever reason, Church's pursuits meant that he wanted to interact with me.

In the second week undertaking fieldwork, I tried to assimilate what jones had told me and put her advice into practice. I made a conscious effort to feel differently around Church and express welcomeness towards him rather than defensiveness.[43] This changed our relationship dramatically. We still had some issues to resolve, but I immediately saw that we were progressing. I could, for example, interact with him without him trying to head-butt me.[44] In the third week, Church did not attempt to head-butt me at all, and we interacted with each other without any problems. I also noticed that I had come to develop a certain affection for Church and learnt that if I allowed Church to attempt head-butting me while I was next to him, he would make motions with his neck without reaching me. While this was risky,[45] and I didn't always take this approach, I learnt to trust him.

In this example, we can notice how I learnt by caring and why it is crucial for ethnographers to forge relationships with animals and participate in the political community where they conduct their ethnographies. It seems to me that without these kinds of experiences, we cannot fully grasp the crucial role that care can play in a ME. When ethnographers are directly engaged in caring practices, these practices can shape their subjectivity and how they see nonhumans.[46] Further, if we assume that socialisation is a crucial aspect of individuals' subjectivities, it is expectable that when a researcher becomes a member of a multispecies community such as VINE's, and is therefore socialised within it, the researcher's subjectivity should eventually change in such a way that they might be able to notice the political in a way that they would have not noticed otherwise.[47]

To be more concrete, Church taught me to welcome more exposure, trust, and openness—each of these elements being necessary conditions to reproduce the social fabric of a political community. jones' perceptive advice and the ethic of care I assumed led me to appreciate that Church might have felt vulnerable around me, eager to receive affection, or something else that I could not grasp. As Gillespie argues, it is important to notice that the political was a constitutive aspect of how Church and I emotionally interacted with each other. Church demanded recognition as a subject who wanted to be reciprocated and who sought to be treated as such. Through our interactions, Church led me to be "de-centred and de-stabilized."[48] Church's incessant requests can be read as a call to learn how to care-for another, which required centring Church, what he wanted, and openness to change as a subject.

What is more, through these caring practices, a more general disposition to caring, being attentive, and being empathetic can be developed in ethnographers. Our experiences and those with whom we interact constitute our-selves,[49] and so if ethnographers caringly interact with the subjects of their study, it seems reasonable to think that they will develop a more caring gaze and a more attentive predisposition towards those who we have never met and might never meet.

Importantly, I do not think that the kind of ethics of indifference advocated for by Dave would have worked in the context of this research. I had tried to remain indifferent towards Church, but he repeatedly requested a response. I could have continued ignoring him, but this seems wrong for three reasons. First, because I would have disavowed Church, denying him the recognition that he deserves as an-other responsive being.[50] Second, because I would have not had the chance to grasp who Church was as a political subject who can reciprocate and be reciprocated—what kind of animal person would I have been able to write about if I had remained indifferent to him throughout my research? Third, and in a more instrumental vein, because if I had continuously disavowed Church, my bodily integrity could have been in danger. Church might have ended up head-butting me—as he had done before with other workers at VINE.

Kanga, who sadly passed away in early 2022, was another resident at VINE. Despite the fact that her rear legs were immobilised, Kanga was an intrepid female sheep who liked to explore VINE's hills and interact with different members of the community. When I arrived at VINE in mid-November of 2021, Kanga was one of the most vulnerable residents at the sanctuary. She was not using her wheel-chair because she had a wound on one of her rear legs. Kanga's cart had been designed for dogs, which meant that it was quite difficult for her to manoeuvre around. For this reason, VINE ordered a new wheelchair that better met Kanga's needs. I was fortunate to witness the first day that she used her new cart.

As the human workers at VINE helped Kanga to get into her new wheel-chair, the alpacas residing at VINE situated themselves in the vicinity of Kanga's pen. They waited. When Kanga went out of the pen, some of the alpacas nuzzled her and gently groomed Kanga, which she reciprocated. After a few minutes, Kanga began walking towards the pond outside the barn. At the beginning her pace was slow, testing the cart's manoeuvrability, but as soon as she realised that she could walk faster with the new wheel-chair, she speedily headed towards the lower hill.[51] At the bottom part of the hill, the alpacas and Kanga interacted with each other again by smelling and nuzzling at each other. It is worth noting that the alpacas accompanied Kanga all the way from the pen to the bottom of the hill. The alpacas would wait near Kanga's pen almost every morning when we were about to open the pen's gate, and when she went out, they would interact with each other.[52]

Care permeated these experiences at different levels. First, VINE was motivated to order Kanga's wheel-chair in great part because the human workers cared-for Kanga. They helped Kanga to get into her wheel-chair every day and provided her with food, shelter, emotional support, and medical attention when she needed it. Further, a component of a care ethic, which VINE endorses,[53] entails being responsive to other's needs.[54] In Kanga's case, this meant enabling her agency, that is, her ability to make decisions on her own terms in relation to participating of the multispecies community's social life and freely moving and associating with other residents at VINE. We can also notice that the political is an integral aspect of this caring process because VINE provided the material pre-conditions that facilitated Kanga's political participation in the community. Disability studies scholars have long argued[55] that accessibility, like agency, is not a mere capability, it is a social construction. This means, among other things, that a given society can fail those who do not conform to normate standards by building stairs instead of ramps, which effectively disables some people with non-normative bodies. By contrast, a community can be responsive to different bodies' needs and, for example, design wheel-chairs for the specific needs that people like Kanga might have.

Second, we can also notice how care motivated the alpacas to be there with and for Kanga when she went out of her pen with the new wheel-chair. To be more precise, care in this context involved a " 'feeling with the other' that centres their needs."[56] The alpacas, like many other nonhumans, are capable of recognising that another is vulnerable and needs support.[57] In other words, the alpacas empathised with Kanga. As Lori Gruen[58] shows, empathy is "a type of caring perception" in which emotion and cognition blend and that is directed towards the other, as opposed to being self-directed. In the case that concerns us, we can read the alpacas' response—situating themselves in front of the pen before Kanga went out and nuzzling her afterwards—as involving the cognitive recognition that Kanga is an individual who might have been unsettled, uncertain of how things might go with the new wheel-chair, excited, or something else. This cognitive response, however, was also constituted by care and empathy, which are the moral emotions that drive us to act towards those who, for example, might need help.[59]

Again, none of this could have even been witnessed in a context like a breeding farm. However, and in order to contest the dominant western epistemology that regards nonhuman animals as disabled commodities, it seems crucial to witness the kinds of subjects that sheep with disabilities can be in a context in which their agency has been enabled and where they can participate in the social life of a zoopolitical community. It seems difficult to deny that Kanga was more than capable of forging friendships, grazing and speaking with other residents, and, more generally, participating in the community's daily life activities. Further, it is likely that sites like breeding farms will not allow researchers to see, for example, actions that entail empathy amongst nonhuman animals—as happened with the alpacas—because the material and socio-political pre-conditions will not be there—the animals might be imprisoned in isolation. What this case also helps to illuminate is that gathering data in relation to the kinds of political subjects Kanga and the alpacas were required looking at their actions through an ethic of care. As a researcher, I would have not read their actions as political actions if care had not been one of my methods—seeing is a practice.

I want now to pause on a third element related to care that Sunaura Taylor[60] critically analyses in *Beasts of Burden*. In it, Taylor explains that, as a disabled person, she has always had an ambivalent relationship to care. While Taylor endorses an ethic of care as an integral aspect of her philosophy of interdependency, she also makes us attentive to how "being cared for can be stifling, if not infantilising and oppressive, as of course can be the role of the caregiver."[61] It seems to me that Taylor's cautionary words on being *cared for*, as opposed to *caring for*, are important to avoid paternalistic and oppressive human–animal relationships as well as oppressive intra-human relationships. In the story about Kanga outlined previously, we can notice that the human residents at VINE were attentive to Taylor's warning in that they assumed that Kanga was capable of taking her own risks. Drawing on the disability rights movement, pattrice jones herself recently explained in a conference on reimagining sanctuaries[62] that one of VINE's principles is the "presumption of competence," which means that nonhumans are, by default, understood to be "competent to make [their] own decisions." This principle was put into practice when the humans respected Kanga going down the hill the first day that she had her new cart, regardless of the fact that there were risks—she could have fallen going downhill. Indeed, Kanga fell on a different day.

In the days after the fall, Kanga would be put in the wheel-chair, and for the first minutes she remained inside the pen—a sort of warmup, as her legs were weaker after the effort she had made when the cart tumbled over. Some days, however, she would hit the pen's gate with the wheel-chair, indicating that she wanted to go out. When this happened, we would open the gate and someone would be in charge of observing her. We can notice that *caring for* Kanga ethically required centring Kanga as the one who was being *cared for* and paying emotional attention to what she wanted.[63] Taylor puts this eloquently when she says that caring for animals means "listening to what animals are telling us about the care they are receiving and the care they would like to receive."[64] Once more, we see here the political bearing of care, in that it can lead carers to enable how nonhuman animals author their own lives. In line with Dave's, García's, and Gillespie's important insights, however, it is important to bear in mind that care can also foreclose opportunities when it is understood in paternalistic and capitalistic terms.

Conclusion

I hope that readers have noticed that my research required not only attentiveness to care as an observant but practising care as a participant. Because I was involved in the community's caring practices, I had to politically engage with Church and decide how to respond to the conflict that existed between us. There were several alternatives at hand. I could have disavowed Church, for example. Instead, and thanks to jones' advice, I tried to recognise Church as a vulnerable individual who might have been requesting emotional attention. This political process required from me to feel more caringly around Church, express welcomeness, learn how to care-for, and be responsive to his requests. These kinds of practices do not only affect how we act in our immediate surroundings because they come to constitute our subjectivities and the epistemologies that are part of ourselves. For this reason, it is my contention that practising care is also a necessary condition to develop a more general disposition to empathise with those who we do not know and might never meet. We are what we do, and we become what we practise.

My relationship with Kanga entailed observing her when she was roaming around, helping her to get into her wheel-chair, filling her bucket with water, and attending to her emotional needs when she requested it. Through these practices, I better intimated how Kanga's political agency could be enabled because I had to listen to her, attend to her needs, and respond accordingly—like the alpacas did. The point is that one's gaze needs to be caring not only in order to attend to others' needs, which is of course crucial, but also to notice how the political permeates a given community. Further, when we adopt care as one of our methods, we are required to respond. Care asks us to avow not only the suffering of animals, but who they are and can become in their full complexity, to attend to their personalities and to their specific desires and requests. My contention has been that the fabric of a zoopolitical community can only be understood by attending to the specific animals we encounter, by establishing connections, and by immersing ourselves, as researchers, in that very community. Seeing, touching, and sensing caringly are, after all, doings. That's why we can understand 'care as method.'

Notes

1 Lorna Finlayson, *An Introduction to Feminism* (Cambridge: Cambridge University Press, 2016).

2 Val Plumwood, *Feminism and the Mastery of Nature* (London and New York: Routledge, 1993); Maneesha Deckha, *Animals as Legal Beings: Contesting Anthropocentric Legal Orders* (Toronto: Toronto University Press, 2020); Anna Grear, "Deconstructing Anthropos: A Critical Legal Reflection on 'Anthropocentric' Law and Anthropocene 'Humanity'," *Law and Critique* 26, no. 3 (November 1, 2015): 225–249, https://doi.org/10.1007/s10978-015-9161-0.

3 That is, people of colour, Indigenous people, women, people with disabilities, members of the LGBTQ community, the working class, and nonhuman animals.

4 Patricia Hill Collins, *Black Feminist Thought* (New York and London: Routledge, 1999); Dinesh J. Wadiwel, *The War against Animals* (Leyden, The Netherlands: Brill, 2015).

5 Carol J. Adams, *The Pornography of Meat* (London: Bloomsbury, 2020); Catharine A. MacKinnon, *Sexual Harassment of Working Women: A Case of Sex Discrimination* (New Haven and London: Yale University Press, 1979), 29–29.

6 Jacques Derrida, *The Animal That Therefore I Am*, eds. Marie Louise Mallet and David Wills (New York: Fordham University Press, 2008); Jacques Derrida, *The Beast and the Sovereign Volume I (2001–2002)*, eds. Michel Lisse, Marie Louise Mallet, and Ginnette Muchaud, trans. Geoffrey Bennington (Chicago: Chicago University Press, 2008).

7 Judith Butler, *The Force of Non-Violence: An Ethico-Political Bind* (London: Verso, 2020); Dinesh J. Wadiwel, *The War Against Animals* (Leyden, The Netherlands: Brill, 2015); Pablo P. Castello, "The Language of Zoodemocracy: Contesting Human Sovereignty Over Animals" (London: Royal Holloway, University of London, 2022), retrieved from: https://pure.royalholloway.ac.uk/portal/files/45404519/CastelloPabloP_The_Language_of_Zoodemocracy_PhD_Thesis_2022.pdf.

8 Josephine Donovan and Carol J. Adams, *The Feminist Care Tradition in Animal Ethics: A Reader* (New York: Columbia University Press, 2007), 4.

9 Ibid.

10 Ibid., 5. See also Lori Gruen's *Entangled Empathy*, 8–15; Adams and Gruen *Ecofeminism;* and Maneesha Deckha's chapter in this volume (reference pages of the chapter's conclusion).

11 Angela Y. Davis, "A Vocabulary for Feminist Praxis: On War and Radical Critique," in *Feminism and War: Confronting U.S. Imperialism*, eds. Robin Riley L., Chandra Talpade Mohanty, and Bruce Pratt (London: Zed Books, 2008), 20, 19–27.

12 Nel Noddings, *Starting at Home: Caring and Social Policy* (Berkeley and Los Angeles: University of California Press, 2002).

13 Davis, "A Vocabulary for Feminist Praxis: On War and Radical Critique," 20, 19–27.

14 Carol Gilligan, *In a Different Voice* (Cambridge, MA: Harvard University Press, 1982). Nel Noddings, *Caring: A Relational Approach to Ethics and Moral Education* (Berkeley and Los Angeles: University of California Press, 1984, 8). It is worth noting that relationships, as Nel Noddings

argues, are themselves a central site of analysis, since it is "where the essential elements of caring are located" (Noddings, *Starting at Home*, 23).

15 Chris Cuomo *Feminism and Ecological Communities: An Ethic of Flourishing*. Eva Kittay, "When Caring Is Just and Justice Is Caring: Justice and Mental Retardation," *Public Culture* 13 (October 1, 2001): 557–580, https://doi.org/10.1215/08992363-13-3-557; Lori Gruen, *Entangled Empathy* (New York: Rowman & Littlefield, 2015); Sunaura Taylor, *Beasts of Burden: Animal and Disability Liberation* (New York: The New Press, 2017). For a more extensive discussion on the meaning of care, see Chapter 2 in this edited collection.

16 Kathryn E. King, "Method and Methodology in Feminist Research: What Is the Difference?" *Journal of Advanced Nursing* 20, no. 1 (July 1, 1994): 20, 19–22, https://doi.org/10.1046/j.1365-2648.1994.20010019.x; Peter Clough and Cathy Nutbrown, *A Student's Guide to Methodology: Justifying Enquiry* (London: SAGE: Publications, 2002), 30.

17 King, "Method and Methodology in Feminist Research: What Is the Difference?" 19–22.

18 Kathryn Gillespie, "For Multispecies Autoethnography," *Environment and Planning E: Nature and Space* (December 13, 2021): 25148486211052870, https://doi.org/10.1177/25148486211052872; Sue Donaldson and Will Kymlicka, "Farmed Animal Sanctuaries: The Heart of the Movement? A Socio-Political Perspective," *Politics and Animals* 1, no. 1 (2015): 50–74.

19 Timothy Pachirat, "The Political in Political Ethnography: Dispatches from the Kill Floor," in *Political Ethnography*, ed. Edward Schatz (Chicago: University of Chicago Press, 2009), 144, 143–161; S. Eben Kirksey and Stefan Helmreich, "The Emergence of Multispecies Ethnography," *Cultural Anthropology* 25, no. 4 (November 1, 2010): 545–576, https://doi.org/10.1111/j.1548-1360.2010.01069.x.

20 See Sue Donaldson and Will Kymlicka, *Zoopolis: A Political Theory of Animal Rights* (Oxford: Oxford University Press, 2011); Sue Donaldson and Will Kymlicka, "Farmed Animal Sanctuaries: The Heart of the Movement? A Socio-Political Perspective," *Politics and Animals* 1, no. 1 (2015): 50–74; Sue Donaldson, "Animal Agora: Animal Citizens and the Democratic Challenge," *Theory and Practice* 46, no. 4 (2020): 709–735, https://doi.org/10.5840/soctheorpract202061296; Charlotte E. Blattner, Sue Donaldson, and Ryan Wilcox, "Animal Agency in Community: A Political Multispecies Ethnography of VINE Sanctuary," *Politics and Animals* 6 (2020): 1–22.

21 Donna J. Haraway, "The Companion Species Manifesto," in *Manifestly Haraway*, ed. Cary Wolfe (Minneapolis: University of Minnesota Press, 2003), 91–198; Donna J. Haraway, *When Species Meet* (Minneapolis: University of Minnesota Press, 2008).

22 For a more detailed discussion of multispecies ethnographies, see Lauren Corman's chapter on the subject in this volume.

23 For critiques, see Maneesha Deckha's chapter in this volume, and Zipporah Weisberg, "The Broken Promises of Monsters: Haraway, Animals and the Humanist Legacy," *Journal for Critical Animal Studies* 7, no. 2 (2009): 22–62; Dinesh J. Wadiwel, "Chicken Harvesting Machine: Animal Labor, Resistance, and the Time of Production," *South Atlantic Quarterly* 117, no. 3 (July 1, 2018): 527–549, https://doi.org/10.1215/00382876-6942135; Pablo P. Castello, "With Haraway and Beyond: Towards an Ecofeminist and Vegan Contextual Ethico-Politics," *Hypatia: A Journal of Feminist Philosophy*, forthcoming.

24 María Elena García, "Death of a Guinea Pig: Grief and the Limits of Multispecies Ethnography in Peru," *Environmental Humanities* 11, no. 2 (November 1, 2019): 352, 351–372, https://doi.org/10.1215/22011919-7754512.

25 Ibid., 352.

26 See Chloë Taylor, "The Precarious Lives of Animals Butler, Coetzee, and Animal Ethics," *Philosophy Today* 52, no. 1 (2008): 60–72, https://doi.org/10.5840/philtoday200852142; James Stanescu, "Species Trouble: Judith Butler, Mourning, and the Precarious Lives of Animals," *Hypatia: A Journal of Feminist Philosophy* 27, no. 3 (August 1, 2012): 567–582, https://doi.org/10.1111/j.1527-2001.2012.01280.x.

27 García, "Death of a Guinea Pig: Grief and the Limits of Multispecies Ethnography in Peru," 354.

28 Ibid.

29 Ibid.

30 See Judith Butler, *Precarious Life: The Power of Mourning and Violence* (London and New York: Verso, 2004).

31 Gillespie, "For Multispecies Autoethnography," 10.

32 Naisargi N. Dave, "Love and Other Injustices: On Humans, Animals, and an Ethics of Indifference," *Comparative Studies of South Asia, Africa and the Middle East* 42, no. 3 (2022): 663, 656–667. https://doi.org/10.1215/1089201X-10148142.
33 Ibid., 659.
34 Ibid., 664.
35 Ibid.
36 Ibid.
37 See Donaldson and Kymlicka, "Farmed Animal Sanctuaries: The Heart of the Movement? A Socio-Political Perspective"; Blattner, Donaldson, and Wilcox, "Animal Agency in Community: A Political Multispecies Ethnography of VINE Sanctuary"; Donaldson, "Animal Agora: Animal Citizens and the Democratic Challenge"; Kathryn Gillespie, *The Cow with Ear Tag #1389* (Chicago: University of Chicago Press, 2018); Kathryn Gillespie, "For Multispecies Autoethnography," *Environment and Planning E: Nature and Space* (December 13, 2021): 25148486211052870, https://doi.org/10.1177/25148486211052872.
38 See also Kathryn Gillespie, "Intimacy, Animal Emotion, and Empathy: Multispecies Intimacy as Slow Research Practice," in *Writing Intimacy into Feminist Geography*, eds. Pamela Moss and Courtney Donovan (London: Routledge, 2017), 160–169.
39 Gillespie, "For Multispecies Autoethnography," 3.
40 Self-determination will always be an aspiration, but it is particularly limited in our current socio-political and legal context—domesticated animals remain, for example, objects of property rights under the law.
41 Pattrice Jones, "Queer Eros in the Enchanted Forest: The Spirit of Stonewall as Sustainable Energy," *QED: A Journal in GLBTQ Worldmaking* 6, no. 2 (2019): 76–82, https://doi.org/10.14321/qed.6.2.0076; Pattrice Jones, *The Oxen at the Intersection* (New York: Lantern, 2020).
42 Pattrice Jones, *The Oxen at the Intersection* (New York: Lantern, 2020); Charlotte. E Blattner, Sue Donaldson, and Ryan Wilcox, "Animal Agency in Community: A Political Multispecies Ethnography of VINE Sanctuary," *Politics and Animals* 6 (2020): 1–22.
43 My use of 'welcomeness' is influenced by Catharine MacKinnon. Her argument is that in contexts of inequality, giving consent (saying 'yes') to have sexual interactions is not enough. As MacKinnon puts it in relation to pornography: "Pornography is full of yes. It is also full of women who have no desire to be there, doing what they are doing, having done to them what is being done" (MacKinnon 2016, 454). Instead, MacKinnon argues that what we need is a welcomeness model, that is, a model that attends to other's eagerness to interact with us, and a model in which we build relationships in which "mutuality is written all over . . . in enthusiasm" (ibid., 450). Catharine A. MacKinnon, "Rape Redefined," *Harvard Law & Policy Review* 10, no. 2 (2016): 431–478.
44 In the first week, every time that we interacted, Church tried to head-butt me.
45 He had head-butted, and unintentionally harmed, some of the humans at VINE before.
46 Reingard Spannring, "Mutual Becomings? In Search of an Ethical Pedagogic Space in Human-Horse Relationships," in *Animals in Environmental Education: Interdisciplinary Approaches to Curriculum and Pedagogy* (Cham, Switzerland: Palgrave MacMillan, 2019), 79–94.
47 While this insight is not unique to my ethnographic work, the emphasis on care in this respect is unusual. For the potential of subjective transformation through participant observation in MEs, see Joe Hutto, *Touching the Wild: Living with the Mule Deer of Deadman Gulch* (New York: Skyhorse Publishing, 2014); and Spannring 2019.
48 Diego Rosello, "The Animal Condition in the Human Condition: Rethinking Arendt's Political Action Beyond the Human Species," *Contemporary Political Theory* 21, no. 2 (2022): 219–239, https://doi.org/10.1057/s41296-021-00495-9.
49 Matthew Calarco, "The Three Ethologies," in *Exploring Animal Encounters: Philosophical, Cultural, and Historical Perspectives*, eds. Dominik Ohrem and Matthew Calarco (Cham, Switzerland: Palgrave MacMillan, 2018), 45–62; Lori Gruen, "Just Say No to Lobotomy," in *Animaladies*, eds. Lori Gruen and Fiona Probyn-Rapsey (London: Bloomsbury, 2019), 11–24.
50 Jacques Derrida, *The Animal That Therefore I Am*, eds. Marie Louise Mallet and David Wills (New York: Fordham University Press, 2008); Kelly Oliver, "Animal Ethics: Toward an Ethics of Responsiveness," *Research in Phenomenology* 40, no. 2 (January 1, 2010): 267–280, https://doi.org/10.1163/156916410X509959; Claire Jean Kim, *Dangerous Crossings: Race, Species, and Nature in a Multicultural Age* (Cambridge: Cambridge University Press, 2015).

51 She roamed further that day than she had done in the last six months, as Cheryl Wylie, who is the Animal Care Coordinator at VINE, explained.
52 As Kanga was tired by being in the wheel-chair, and there are predators like coyotes near VINE, Kanga always stayed inside her pen during the night.
53 "VINE Sanctuary," 2022, retrieved from: https://vinesanctuary.org/.
54 Noddings, *Caring: A Relational Approach to Ethics and Moral Education*, 9; Josephine Donovan, "Feminism and the Treatment of Animals: From Care to Dialogue," *Signs* 31, no. 2 (2006): 305, 305–329, https://doi.org/10.1086/491750.
55 Alison Kafer, *Feminist, Queer, Crip* (Bloomington, IN: Indiana University Press, 2013); Licia Carlson, "Body Shame, Body Pride: Lessons from the Disability Rights Movement," in *The Transgender Studies Reader 2*, eds. Susan Stryker and Aren Z. Aizura (New York: Routledge, 2013); and Hamraie Aimi, "Beyond Accommodation: Disability, Feminist Philosophy, and the Design of Everyday Academic Life," *PhiloSOPHIA* 6, no. 2 (2016): 259–271. https://doi.org/10.1353/phi.2016.0022; and Julie Avril Minich, "Enabling Whom? Critical Disability Studies Now," *Lateral* 5, no. 1 (2016), https://doi.org/10.25158/L5.1.9; Stephanie Jenkins, Kelly Struthers Montford, and Chloë Taylor, *Disability and Animality: Crip Perspectives in Critical Animal Studies* (New York: Routledge, 2020).
56 Birte Wrage, "Caring Animals and Care Ethics," *Biology and Philosophy* 37 (2022): 1–20, https://doi.org/10.1007/s10539-022-09857-y.
57 Some examples can be found in Victoria Braithwaite, *Do Fish Feel Pain?* (Oxford: Oxford University Press, 2010); Inbal Ben-Ami Bartal, Jean Decety, and Peggy Mason, "Empathy and Pro-Social Behavior in Rats," *Science* 334, no. 6061 (December 9, 2011): 1427–1430, https://doi.org/10.1126/science.1210789; Hutto, *Touching the Wild: Living with the Mule Deer of Deadman Gulch*; Joana Carvalheiro et al., "Helping Behavior in Rats (*Rattus norvegicus*) When an Escape Alternative Is Present," *Journal of Comparative Psychology* 133, no. 4 (2019): 452–462, https://doi.org/10.1037/com0000178.
58 Gruen, *Entangled Empathy*, 3.
59 Wrage, "Caring Animals and Care Ethics," 8.
60 Taylor, *Beasts of Burden*, 205.
61 Ibid.
62 Pattrice Jones, "Confronting Our Own Speciesism," retrieved from: https://www.youtube.com/watch?v=pskTvxKCpqI&t=2775s.
63 Gruen, *Entangled Empathy*, 50.
64 Taylor, *Beasts of Burden*, 217.

Bibliography

Adams, Carol J. *The Pornography of Meat*. London: Bloomsbury, 2020.

Aimi, Hamraie. "Beyond Accommodation: Disability, Feminist Philosophy, and the Design of Everyday Academic Life." *PhiloSOPHIA* 6, no. 2 (2016): 259–271. https://doi.org/10.1353/phi.2016.0022

Bartal, Inbal Ben-Ami, Decety, Jean, and Mason, Peggy. "Empathy and Pro-Social Behavior in Rats." *Science* 334, no. 6061 (December 9, 2011): 1427–1430. https://doi.org/10.1126/science.1210789

Blattner, Charlotte. E, Donaldson, Sue, and Wilcox, Ryan. "Animal Agency in Community: A Political Multispecies Ethnography of VINE Sanctuary." *Politics and Animals* 6 (2020): 1–22

Braithwaite, Victoria. *Do Fish Feel Pain?*. Oxford: Oxford University Press, 2010.

Butler, Judith. *Precarious Life: The Power of Mourning and Violence*. London and New York: Verso, 2004.

Butler, Judith. *The Force of Non-Violence: An Ethico-Political Bind*. London: Verso, 2020.

Calarco, Matthew. "The Three Ethologies." In *Exploring Animal Encounters: Philosophical, Cultural, and Historical Perspectives,* edited by Dominik Ohrem and Matthew Calarco, 45–62. Cham, Switzerland: Palgrave MacMillan, 2018.

Carlson, Licia. "Body Shame, Body Pride: Lessons from the Disability Rights Movement." In *The Transgender Studies Reader* 2, edited by Susan Stryker and Aren Z. Aizura. New York: Routledge, 2013.

Carvalheiro, Joana et al. "Helping Behavior in Rats (Rattus Norvegicus) When an Escape Alternative Is Present." *Journal of Comparative Psychology* 133, no. 4 (2019): 452–462. https://doi.org/10.1037/com0000178.
Castello, Pablo P. "The Language of Zoodemocracy: Contesting Human Sovereignty Over Animals." London, Royal Holloway, University of London (2022). Retrieved from: https://pure.royalholloway.ac.uk/portal/files/45404519/CastelloPabloP_The_Language_of_Zoodemocracy_PhD_Thesis_2022.pdf.
Castello, Pablo P. "With Haraway and Beyond: Towards an Ecofeminist and Vegan Contextual Ethico-Politics." *Hypatia: A Journal of Feminist Philosophy*, forthcoming.
Clough, Peter, and Nutbrown, Cathy. *A Student's Guide to Methodology: Justifying Enquiry*. London: SAGE: Publications, 2002.
Collins, Patricia Hill. *Black Feminist Thought*. New York and London: Routledge, 1999.
Cuomo, Chris. *Feminism and Ecological Communities: An Ethic of Flourishing*. Routledge, 1998.
Dave, Naisargi N. "Love and Other Injustices: On Humans, Animals, and an Ethics of Indifference." *Comparative Studies of South Asia, Africa and the Middle East* 42, no. 3 (2022): 663, 656–667. https://doi.org/10.1215/1089201X-10148142.
Davis, Angela Y. "A Vocabulary for Feminist Praxis: On War and Radical Critique." In *Feminism and War: Confronting U.S. Imperialism*, edited by Robin Riley L., Chandra Talpade Mohanty, and Bruce Pratt. London: Zed Books, 1988.
Deckha, Maneesha. *Animals as Legal Beings: Contesting Anthropocentric Legal Orders*. Toronto: Toronto University Press, 2020.
Derrida, Jacques. *The Animal That Therefore I Am*. Edited by Marie Louise Mallet and David Wills. New York: Fordham University Press, 2008a.
Derrida, Jacques. *The Beast and the Sovereign Volume I (2001–2002)*. Edited by Michel Lisse, Marie Louise Mallet, and Ginnette Muchaud, translated by Geoffrey Bennington. Chicago: Chicago University Press, 2008b.
Donaldson, Sue. "Animal Agora: Animal Citizens and the Democratic Challenge." *Theory and Practice* 46, no. 4 (2020): 709–735. https://doi.org/10.5840/soctheorpract202061296
Donaldson, Sue, and Kymlicka, Will. *Zoopolis: A Political Theory of Animal Rights*. Oxford: Oxford University Press, 2011.
Donaldson, Sue, and Kymlicka, Will. "Farmed Animal Sanctuaries: The Heart of the Movement? A Socio-Political Perspective." *Politics and Animals* 1, no. 1 (2015): 50–74.
Donovan, Josephine. "Feminism and the Treatment of Animals: From Care to Dialogue." *Signs* 31, no. 2 (2006): 305, 305–329. https://doi.org/10.1086/491750.
Donovan, Josephine, and Adams, Carol J. *The Feminist Care Tradition in Animal Ethics: A Reader*. New York: Columbia University Press, 2007.
Finlayson, Lorna. *An Introduction to Feminism*. Cambridge: Cambridge University Press, 2016.
García, María Elena. "Death of a Guinea Pig: Grief and the Limits of Multispecies Ethnography in Peru." *Environmental Humanities* 11, no. 2 (November 1, 2019): 351–372. https://doi.org/10.1215/22011919-7754512.
Gillespie, Kathryn. "Intimacy, Animal Emotion, and Empathy: Multispecies Intimacy as Slow Research Practice." In *Writing Intimacy into Feminist Geography*, edited by Pamela Moss and Courtney Donovan, 160–169. London: Routledge, 2017.
Gillespie, Kathryn. *The Cow with Ear Tag #1389*. Chicago: University of Chicago Press, 2018.
Gillespie, Kathryn. "For Multispecies Autoethnography." *Environment and Planning E: Nature and Space*, 13 (December 2021). https://doi.org/10.1177/25148486211052872
Gilligan, Carol. *In a Different Voice*. Cambridge, MA: Harvard University Press, 1982.
Grear, Anna. "Deconstructing Anthropos: A Critical Legal Reflection on "Anthropocentric" Law and Anthropocene "Humanity."' *Law and Critique* 26, no. 3 (November 1, 2015): 225–249. https://doi.org/10.1007/s10978-015-9161-0.
Gruen, Lori. *Entangled Empathy*. New York: Rowman & Littlefield, 2015.
Gruen, Lori. "Just Say No to Lobotomy." In *Animaladies*, edited by Lori Gruen and Fiona Probyn-Rapsey, 11–24. London: Bloomsbury, 2019.
Haraway, Donna J. "The Companion Species Manifesto." In *Manifestly Haraway*, edited by Cary Wolfe, 91–198. Minneapolis: University of Minnesota Press, 2003.
Haraway, Donna J. *When Species Meet*. Minneapolis: University of Minnesota Press, 2008.

Hutto, Joe. *Touching the Wild: Living with the Mule Deer of Deadman Gulch*. New York: Skyhorse Publishing, 2014.
Jenkins, Stephanie, Struthers Montford, Kelly, and Taylor, Chloë. *Disability and Animality: Crip Perspectives in Critical Animal Studies*. New York: Routledge, 2020.
Jones, Pattrice. "Confronting Our Own Speciesism." Retrieved from: https://www.youtube.com/watch?v=pskTvxKCpqI&t=2775s.
Jones, Pattrice. "Queer Eros in the Enchanted Forest: The Spirit of Stonewall as Sustainable Energy." *QED: A Journal in GLBTQ Worldmaking* 6, no. 2 (2019): 76–82. https://doi.org/10.14321/qed.6.2.0076
Jones, Pattrice. *The Oxen at the Intersection*. New York: Lantern, 2020.
Kafer, Alison. *Feminist, Queer, Crip*. Bloomington, IN: Indiana University Press, 2013.
Kim, Claire Jean. *Dangerous Crossings: Race, Species, and Nature in a Multicultural Age*. Cambridge: Cambridge University Press, 2015.
King, Kathryn E. "Method and Methodology in Feminist Research: What Is the Difference?." *Journal of Advanced Nursing* 20, no. 1 (July 1, 1994): 19–22. https://doi.org/10.1046/j.1365-2648.1994.20010019.x
Kirksey, S. Eben, and Helmreich, Stefan. "The Emergence of Multispecies Ethnography." *Cultural Anthropology* 25, no. 4 (November 1, 2010): 545–576. https://doi.org/10.1111/j.1548-1360.2010.01069.x.
Kittay, Eva. "When Caring Is Just and Justice Is Caring: Justice and Mental Retardation." *Public Culture* 13 (October 1, 2001): 557–580. https://doi.org/10.1215/08992363-13-3-557
MacKinnon, Catharine A. *Sexual Harassment of Working Women: A Case of Sex Discrimination*. New Haven and London: Yale University Press, 1979.
MacKinnon, Catharine A. "Rape Redefined." *Harvard Law & Policy Review* 10, no. 2 (2016): 431–478.
Minich, Julie Avril. "Enabling Whom? Critical Disability Studies Now." *Lateral* 5, no. 1 (2016). https://doi.org/10.25158/L5.1.9
Noddings, Nel. *Caring: A Relational Approach to Ethics and Moral Education*. Berkeley and Los Angeles: University of California Press, 1984.
Noddings, Nel. *Starting at Home: Caring and Social Policy*. Berkeley and Los Angeles: University of California Press, 2002.
Oliver, Kelly. "Animal Ethics: Toward an Ethics of Responsiveness." *Research in Phenomenology* 40, no. 2 (January 1, 2010): 267–280. https://doi.org/10.1163/156916410X509959
Pachirat, Timothy. "The Political in Political Ethnography: Dispatches from the Kill Floor." In *Political Ethnography*, edited by Edward Schatz, 143–61. Chicago: University of Chicago Press, 2009.
Plumwood, Val. *Feminism and the Mastery of Nature*. London and New York: Routledge, 1993.
Rosello, Diego. "The Animal Condition in the Human Condition: Rethinking Arendt's Political Action Beyond the Human Species." *Contemporary Political Theory* 21, no. 2 (2022): 219–239. https://doi.org/10.1057/s41296-021-00495-9.
Spannring, Reingard. "Mutual Becomings? In Search of an Ethical Pedagogic Space in Human-Horse Relationships." In *Animals in Environmental Education: Interdisciplinary Approaches to Curriculum and Pedagogy*, 79–94. Cham, Switzerland: Palgrave MacMillan, 2019.
Stanescu, James. "Species Trouble: Judith Butler, Mourning, and the Precarious Lives of Animals." *Hypatia: A Journal of Feminist Philosophy* 27, no. 3 (August 1, 2012): 567–582. https://doi.org/10.1111/j.1527-2001.2012.01280.x.
Taylor, Chloë. "The Precarious Lives of Animals Butler, Coetzee, and Animal Ethics." *Philosophy Today* 52, no. 1 (2008): 60–72. https://doi.org/10.5840/philtoday200852142
Taylor, Sunaura. *Beasts of Burden: Animal and Disability Liberation*. New York: The New Press, 2017.
Wadiwel, Dinesh J. *The War Against Animals*. Leyden, The Netherlands: Brill, 2015.
Wadiwel, Dinesh J. "Chicken Harvesting Machine: Animal Labor, Resistance, and the Time of Production." *South Atlantic Quarterly* 117, no. 3 (July 1, 2018): 527–549. https://doi.org/10.1215/00382876-6942135
Weisberg, Zipporah. "The Broken Promises of Monsters: Haraway, Animals and the Humanist Legacy." *Journal for Critical Animal Studies* 7, no. 2 (2009): 22–62.
Wrage, Birte. "Caring Animals and Care Ethics." *Biology and Philosophy* 37 (2022): 1–20. https://doi.org/10.1007/s10539-022-09857-y.

17
FEMINIST METHODOLOGY AND MULTISPECIES ETHNOGRAPHY

Kelly Struthers Montford and Chloë Taylor

Introduction

In 2010, Kirksey and Helmrich announced that "a new genre of writing and mode of research had arrived on the anthropological stage: multispecies ethnography."[1] Nonhumans, such as land, animals, and plants, previously in the periphery or background of anthropological research, were now central to the work. Animals were recategorized from objects acted upon and existing in service to humans to agential subjects who, along with humans, had "legibly biographical and political lives"[2] ascertainable through multispecies ethnography. There has also been a recent turn in both critical animal studies (CAS) and animal studies scholarship to multispecies ethnography (MSE). While the term MSE is relatively new, its attention to the interdependency and agency of multiple subjects such as humans, animals, land, and water is consistent with what has broadly been called an Indigenous metaphysics, though these categories are themselves colonial.[3] Predating the coining of "MSE," researchers in the natural sciences such as Barbara Smuts were studying human–animal relations in a way that is consistent with MSE's methodology and commitments.[4]

Methodology refers both to the study of methods and the theoretical framework and supporting rationale used to approach a particular phenomenon or research question, whereas methods are the tools used to carry out the research. Put otherwise, ethnography—research undertaken in a "natural" setting to observe and understand behaviour, social action, and cultures in which the researcher is part of the setting—is a methodology which might use methods such as observation and interviews. Recent CAS literature has taken up MSE methodologically in both senses: studying and reflecting upon how to improve it as a framework of inquiry in CAS research[5] and publishing research and knowledge generated from MSE research.[6]

Kirksey and Helmrich provide an overview of the types of anthropological inquiry undertaken under the banner of MSE, projects on "biocultural hope" centred on "insect love," the flourishing of matsutake mushrooms following events of environmental destruction, and the interplay between "microbial cultures" and the socio-politics of food. Key to the project is the recognition of and attention to the interdependency and multispecies

 DOI: 10.4324/9781003273400-22

nature of our realities, namely "how a multitude of organisms' livelihoods shape and are shaped by political, economic, and cultural forces."[7]

As Kirksey and Helmrich's genealogy outlines, early MSE research focused on ecological systems, fungi, and insects rather than on nonhuman animals. So while other-than-human animals fell within the purview of MSE from its outset, they still were not focal. The extent to which animals are commodified in animal-based industries such as agriculture, pharmaceuticals and biomedical testing, entertainment, and fashion means that they are all around us as commodities, yet perversely removed from their natural settings and prevented from engaging in species-specific behaviours. Conducting MSE in spaces of exploitation such as the farm, zoo, and slaughterhouse, for example, was not anticipated by early multispecies ethnographers, and as we suggest in this chapter, considerations and ethical standards about conducting research with nonhuman animals requires guidelines about selection of research topic, preparation for entering and exiting a site, and a political commitment to improving the material conditions of those with whom one conducts research. Because MSE uses tools developed for human-centric research, guidelines for ethnographic research involving animals have not yet been widely theorized. Such guidelines would need to consider the very different contexts of various human–animal relationships, including interactions humans have with wild and feral animals, relatively benevolent companionship relations, and situations of extreme commodification and oppression such as farms and laboratories. These contextual differences could also explain why MSE scholarship differs widely in its political and ethical commitment to its research subjects. This is compounded by the fact that social science research ethics policies do not contemplate animals as research subjects. Animal research ethics currently presuppose biomedical contexts. Drawing on the work of ethnographers and CAS scholars, our objective in this chapter is to argue that ethical parameters for research with animals should be developed and that these should be guided by feminist methodologies.

What Is Feminist Research?

The late 1970s and 1980s saw a litany of feminist research, pedagogies, and theory produced within and against dominant western academic and masculinized ontologies and epistemologies. Broadly, feminist thought is composed of a set of perspectives theorizing gender, social structures, and power and how these shape inequality and life chances depending on intersecting axes of identity and oppression.[8] These perspectives can include radical, Marxist, liberal, socialist, and ecofeminist approaches. Early feminist research distinguished itself from masculinized ways of knowledge production in its closeness to the research and participants, use of qualitative methods (although this is not a requirement as quantitative research can be designed to have liberatory impacts), and personal and political investment. In other words, it took objectivity to be an impossible ideology, as was its attendant fantasy of the researcher's detachment and noninfluence on epistemological framing, data, and findings. Objectivity, feminist scholars contended, merely obscured gender, race, and class loyalties.[9] Instead, women were taken as experts on their own experience of oppression, and from this, new social orders were theorized.

At its core, feminist thought and research continue to be rooted in a commitment to social change and social justice. Indeed, it is the requirement that research produce tangible results that improve the material, social, and political conditions of their research participants that distinguishes feminist research from other methods of critical inquiry.[10] For research to be feminist, Kilty outlines four interdependent elements and commitments: voice, positionality, politics, and praxis. Voice attends to the fact that a researcher plays an active role in knowledge production by both interpreting participant narratives and existing documents, as well as by attending to "the different ways individuals and groups construct and interpret issues and their position in relation to those issues."[11] Positionality refers to situating and attending to how participants' contexts and intersecting identities are shaped by gender, sexuality, ability, race, ethnicity, class, settler colonialism, and anti-Blackness, for example. Politics refers to the feminist imperative to "politciz[e] social, economic, scientific, and legal issues through deliberate action" in which research can take the seemingly banal and reveal how it is shaped by power relations.[12] Praxis refers to the commitment to effect social change, meaning that it is not enough to *know* how individuals are oppressed but that feminist research works toward ameliorating these conditions.

While feminist research developed with gender as an organizing mode of inquiry, research does not have to be on "women" to be feminist. This is because gendered associations contribute to structure hierarchies, social value, and institutions. In the case of animals, ecofeminism is born out of the theorization of the mutually constitutive oppressions of gender, animality, and nature. For Sisseton-Wahpeton Oyate scholar Kim TallBear, decolonial feminist research is primarily rooted in its orientation to do work in which the researcher is invested both politically and personally. It is this investment that allows the researcher to "stand with" and "within" the community they research, to care for their participants and co-constitute knowledge in a faithful way such that the researcher does not speak for but "in concert with" the community in question.[13] "Standing with" then will inevitably require the researcher's willingness to be altered by the work and "revise her stakes in the knowledge to be produced."[14] To do so requires thinking beyond the liability-averse university research ethics notions of "risk" to participants and the academic version of value in which the benefits of research are framed primarily as the generative of new knowledge.

For TallBear, this means designing research that approaches risk and benefit capaciously: "what counts as risk (ontological harms?) and rightful benefit (institution building and community development?) in the course of building knowledge?"[15] The willingness to be transformed also moves beyond typical research notions of reciprocity to participants in the sense of "giving back," because one will have become part of the community (to a certain degree) and will have ongoing interaction and engagement, thereby moving beyond the idea of entering and exiting "the field." It is these components that constitute TallBear's "feminist objectivity," which she describes as "inquiring not at a distance, but based on the lives and knowledge priorities of subjects."[16] Kirksey and Heimlich, writing at the burgeoning point of MSE research, suggest that its orientation has the *potential* to follow and respond to broader calls in the social sciences to consider the ethics of one's projects in terms of whom it benefits rather than mere enthrallment and fetishization when one attends to the presence of the more-than-human. For this potential to be realized with animals in the case of MSE, feminist decolonial praxis must be integrated so that we may "stand with" those with whom we conduct research.

Making Multispecies Research More Feminist

In her article "For a Politicized Multispecies Ethnography," feminist theorist and animal geographer Kathryn Gillespie observes that multispecies ethnographies have been critiqued for still "prioritizing the human (and human interests) over other species, as being overly theoretical and obtuse, and as being depoliticized and unethical in its approach."[17] Anthropologist Natalie Porter's 2019 multispecies ethnography, *Viral Economies: Bird Flu Experiments in Vietnam*, is a case in point. *Viral Economies* explores the management of zoonotic diseases, and avian flus in particular, in the context of Vietnam shortly before the outbreak of Covid-19. Presciently, Porter discusses the pandemic risks posed by "the thousands of retail wet markets" where some of the birds she observes met their end. As she writes, "Markets . . . serve as meeting grounds for people, poultry, and pathogens. As consumers swap cash for fowl, poultry swap microbes and viruses with one another, as well as with other animals for sale."[18] Disappointingly, although Porter presents the book as bringing animals into discussions of zoonosis as "both biological and social actors,"[19] as living beings in relationships with humans and pathogens, the study remains anthropocentric: Porter only considers farmed animals in terms of their value to farmers and not the value of these animals' lives to themselves. She does not question the institutions of domestication or animal agriculture or what the value and meaning of these birds' lives might have been in a non-commercial setting or in the wild. In the conclusion to her book, she laments that fowl who are culled during avian flu outbreaks die "prematurely," their deaths lacking meaning and value, and contrasts this with being slaughtered for the farmers' capital gain, in which case their deaths would have "the meaning and value that comes along with a planned demise."[20] Of course, this "demise" was planned by humans and not by the birds, and the "meaning and value" of that demise is for humans. Moreover, for the birds, any slaughter is a premature death, whether it is part of a cull or so their bodies can be sold as meat.

CAS scholars have been disappointed that multispecies ethnographies such as Porter's have not been designed "to alter status hierarchies, including interspecies hierarchies perceived from the subject positions of privileged academic humans."[21] Gillespie agrees with these critics that "multispecies ethnography must 'address multispecies injustice, suffering, and unidirectional violence'" and furthermore argues "that it ought to reckon with and transform researchers' own complicity and participation in violence against animals."[22] In sum, multispecies ethnography needs to be more politicized, and, for Gillespie, adopting "an explicitly feminist . . . framing" is helpful in this end.[23] She argues that such a feminist perspective "help[s] ethnographers to pay greater politicized attention to differences of embodiment, power relations between and among species, and questions of positionality" and shines a light on multispecies ethnography's "transformative potential, its relational nature, and its ethical ambiguities."[24]

Gillespie's design for a course, "Doing Multispecies Ethnography," and her reflections on teaching it, help to illuminate how she understands politicized multispecies ethnography. The course involved students spending time at a pig sanctuary, where they engaged in participant observation with individual pigs with whom they were paired, taking field notes, and conducting interviews with the sanctuary founder, in addition to undertaking academic research on pigs. As Gillespie describes it,

> the course was formulated through an explicitly feminist geographic lens that asked students to: uncover and reflect on the taken-for-granted power relations embedded in human-farmed animal relations; think critically about hierarchies of human supremacy; and attend to embodiment and emotion as ways of knowing.[25]

Gillespie stresses that students in the course came to have relationships with the pigs as individuals and to learn about them in their own rights and not only in relation to humans. Students also attended to the space the pigs were in, its ethical complexities, and how different these pigs' lives were from those of most of their kin. The students were transformed by the experience, some becoming vegan, for instance, and raised ethical concerns about multispecies ethnography, particularly around the course ending and their relationships with the pigs ending with it. The pigs had bonded with and enjoyed having relationships with the students, who had spent long periods of time observing and socializing with them, and ultimately the students feared this would result in harm, as the pigs would experience abandonment. While some students continued to visit the pig sanctuary on their own for years afterwards, most did not, and as Gillespie observes, this underscores the concern we should have with all ethnography: that the research participant is instrumentalized for the purposes of the researcher. Although Gillespie suggests that "a politicized approach [to MSE] can work to ameliorate this possibility,"[26] she concludes that many ethical quandaries remain to be explored in relation to the methodology.

In "Death of a Guinea Pig," cultural anthropologist María Elena García similarly explores what she calls "the limits of multispecies ethnography." García describes visiting a commercial guinea pig farm in Peru as part of her research on Peruvian gastropolitics. García was 7 months pregnant at the time, and, as the farm she visited was a breeding facility, she was surrounded by 1500 other pregnant female mammals and their babies. She recalls feeling miserably uncomfortable on this stiflingly hot and humid day, her hands and feet swollen, and she was constantly thirsty. In this context, she notices immediately that the guinea pigs have no water and learns that they are expected to get all the hydration they need from the "rough" and "jagged" branches that Walter, the farm owner, tosses indiscriminately into the cages, unconcerned with whether he startles or scratches the animals. García also notices a guinea pig who is nudging a dead baby, and another who is lying on her side and not moving. She calls Walter over and points out the immobile guinea pig. To her dismay, Walter prods the guinea pig and then tosses her onto the dirt floor, saying that she is dying of birth complications. Walter then put his hand on García's back and steered her onwards on the tour. In the moment García suppressed her emotions and allowed herself to be led away because she feared estranging Walter, who was not only an important research participant but was giving her access to a space that was important for her fieldwork. For similar reasons, García describes eating multiple dishes of guinea pig meat that Walter had served to her. When Walter invites her to participate in a guinea pig production workshop, where she would have learned to weigh, brand, and feed guinea pigs, and to kill them by throat slitting, neck breaking, and scalding, she has mixed feelings. She describes regretting that the workshop is scheduled for after her departure from Lima and considers postponing her flight. In contemplating the opportunity to participate in the workshop, she asks herself:

> Would I be capable of killing? And even if I chose not to participate but simply observed, would I be able to witness the suffering of so many animals being killed by unskilled hands? Was this part of the responsibility of choosing to conduct a multispecies ethnography? . . . Methodologically, how does ethnography change when it includes nonhuman others? How *should* it change? . . . Is killing other-than-human animals, for example, an acceptable dimension of participant observation?[27]

Although it seems that García did not attend Walter's workshop, in the book she would publish several years later, she describes going to a different guinea pig production workshop and touring the area where guinea pigs are slaughtered. Despite her cautious openness to killing guinea pigs as part of her research, García describes thinking about and grieving for the guinea pig whom Walter tossed on the floor, and "Death of a Guinea Pig" explores how this "encounter" made her question the ethics and politics of multispecies ethnography. As she explains:

> I can't help but think back and attempt to theorize that moment of sadness, which I think is also one of shame—the shame of taking *his* side; of worrying about *my* research. . . . In that moment, I felt my project, the so-called multispecies research I was conducting, was taking place at the expense of the animal.[28]

It is worth questioning whether what García did in the guinea pig breeding facility was multispecies ethnography as opposed to a site visit. Besides the tour of the breeding facility, in her book García describes attending a guinea pig production workshop and watching a video of a guinea pig being chemically castrated. García was never immersed in the breeding facility site or the world of guinea pig production and never developed any relationships with the guinea pigs. She gets to know Walter far more than she gets to know any individual guinea pig. Yet the case raises ethical issues that apply also to the work of ethnographers who have stronger claims to be doing multispecies ethnography—work that takes place not in benevolent spaces such as an animal sanctuary but in sites of animal oppression. Gillespie, for instance, spent years doing participant observation at dairy cow auctions, a dairy farm, slaughterhouses and rendering plants to write the book *The Cow with Ear Tag #1389*,[29] which, as the title suggests, describes animals as individuals even under very constrained circumstances. Gillespie, like García, bears witness to but is unable to do anything to prevent the cruel treatment of many animals. To access these sites, she, like other researchers, had to suspend or mask her emotions or, at the very least, refrain from acting on them at the time.

If we consider the criteria for feminist research described previously, García's article is feminist in that it attends to voice, situating herself as a Peruvian woman whose Indigenous heritage was hidden from her by her mother until she was in her 20s; a woman with a prestigious job in the United States who is both condescended to as a woman and looked up to as a professor by Walter; a woman who, like the guinea pigs, was pregnant, hot, thirsty, and miserable on that day and who worried about the future of her child. As she contemplates in the article, her own child's prospects were worlds apart from those of the baby guinea pigs, and yet, as she describes, she and the mother guinea pigs shared an "overwhelming love" for their babies, "profoundly entangled with fear and haunted by the specter of death."[30]

García's article is also feminist in that it demonstrates positionality, at least with respect to Walter. She attends to his intersecting identity as an Indigenous Peruvian working class man; a father whose children have been able to move to a central neighbourhood and have a computer because of the booming guinea pig trade; a brother whose sister is ill with cancer, whom he is caring for with the profits he makes from guinea pig farming. She also considers how race and indigeneity have been entangled with animality in the Peruvian context, with Indigenous Peruvians seen by settlers as "just animals."[31] Unfortunately, although she undertakes some textual research on guinea pigs, and particularly about the strong bonds and modes of communication between mother guinea pigs and their pups, in her fieldwork García has no opportunities to attend to the guinea pigs as individuals in the way that she attends to Walter. The one guinea pig with whom she had an individualized encounter is the guinea pig tossed on the floor to die, perhaps simply of thirst and not, as Walter surmised, of birthing complications—a guinea pig who could perhaps have been saved, or at least allowed to die in the enclosure with which she was familiar. This guinea pig was likely never aware of García's presence.

García's article is also feminist in that it politicizes research, including the seemingly banal, such as sharing a meal with a research participant. García considers how Walter introduces her to the waiter, emphasizing her class status, even while speaking in paternalistic ways towards her. She considers how guinea pig meat was once associated with poverty and indigeneity but is now caught up in a multi-billion dollar tourism industry that is changing the lives of many working-class Peruvians, including Walter. She also dwells on everyday emotional and psychological processes such as love and self-deception, writing: "I would be literally consuming the objects of Walter's professed love but I would also be asked not to think of these as dead animals."[32] She considers the "gendered imaginary" of the ways Walter speaks of the *animalitos* he farms and kills and claims to love—calling the females his "little ladies" and describing the male guinea pigs as "like men."[33] Unlike most social scientists, García also seriously considers the politics of species, acknowledging that what is at stake in Peru's gastro-tourism boom is not just socio-economic status and race but also the lives and deaths of billions of animals.

Where García's article arguably falls short, on feminist terms, is on praxis. This raises the issue that is central to García's article, which is whether the fact of grieving and bearing witness to the lives of animals after the fact makes the research ethical or political or whether witnessing and grief are praxis enough. As García asks, "What does my grieving for this guinea pig *do*?"[34] García ultimately concludes: "Writing about the ethnographic encounter as one of tragedy and loss . . . might open up the productive possibilities of mourning and grief in connecting human and nonhuman worlds."[35] This is a hope that other multispecies ethnographers who study sites of animal oppression express. In this way, these scholars are raising the criteria for feminist research that it should have tangible results that improve the material, social, and political conditions, if not of the research participants themselves, at least of other beings of their kind. It is true that mourning may be politically powerful and that bearing witness to the grievability of animal lives may have an impact on readers. At the same time, such mourning did not prevent García from eating guinea pig meat or, in the book she would publish on Peruvian gastropolitics, sharing recipes for alpaca and llama milk and meat. One of these recipes is for "Alpaca & Amaranth (Alpaca y Kiwicha)," and García prefaces the list of ingredients, which includes 400 g of alpaca neck and 1L of alpaca milk, with the musing, "One of the most common images of the high Andes is alpaca grazing on fields amid a backdrop of mountain peaks. You'll find countless postcards depicting the scene. This dish returns the alpaca to this landscape, a place of wild herbs and grains."[36] Of course, the alpaca is dead and serving their cadaver on a plate with amaranth in no way returns them to the landscape to which they belong. Strangely, García expresses no similar grief or maternal compassion for the alpacas and llamas as she explores in the case of the guinea pig. Thus, this grief is only grief for one individual guinea pig with whom García identified, not for the similar animals caught up in industrial farming or killed for meat, and the grief seemingly had little impact on García's actions.

This raises the point that while feminist research, unlike masculinist models of research, does not eschew the expression of emotion, such expression does not in itself make research feminist or constitute praxis. On the contrary, as many feminist and critical animal studies scholars have shown (and indeed as García's own article shows with respect to Walter), expressions of "love" have all too often been used to justify the domination of and violence against women and animals. In " 'There's Something about the Blood': Tactics of Evasion in Narratives of Violence," critical race and animal studies scholar Nekeisha Alayna Alexis examines how feeling "good" emotions has been used by slaveholders and conscientious omnivores alike to justify the institutions of slavery and animal agriculture.[37] Considering the genre of plantation romances as well as the writings of small-scale farmers, Alexis shows how attention is strategically shifted from the oppression that slaves and animals undergo to the benevolent feelings of slaveholders and farmers. And yet, Alexis insists, oppressing others with love and sorrow does not make this oppression any less wrong.

Similarly, García and Walter express "good" feelings for the guinea pigs, whether love or sorrow, and yet this does not prevent either of them from harming, exploiting, killing, or eating those or other animals—albeit to radically different degrees. While emotions are not antithetical to feminist research, then, they are clearly not sufficient to make the research feminist. Beyond feeling the right emotions, feminism requires praxis. For these reasons, anthropologist Naisargi N. Davé has argued with respect to animal advocacy that what we need may not be love at all.[38] Indeed, Davé observes, love is unjust, always being for particular beings and not others. According to Davé, what we need is thus not love but an ethics of indifference to difference, in which it would not matter whether the beings in question are humans, this guinea pig or another, an alpaca, or a llama.

Finally, with regard to this last point we would like to contrast García's work with that of a final multispecies ethnographer, Stephanie Eccles. While opportunities for praxis vary widely depending on the site of a researcher's fieldwork, we argue that Eccles' work models the potential of feminist multispecies scholarship in this regard. In "Researching Contested Companionship: Responsibility and Care Work in the Field," Eccles explores some insights she derived from doing multispecies ethnographic research on breed specific legislation, particularly with pit bull-type breeds of dogs in and around Montreal. Eccles describes not just caring for the dogs on an affective level but acting upon this care, intervening on behalf of some of the pit bulls she encounters in her research. For instance, Eccles adopted (despite breed specific legislation forbidding this) one such dog who had been abandoned, and raised funds for two others and their human so that they could move to a city that does not have breed specific legislation. Beyond studying breed specific legislation and its impact on dogs, Eccles also volunteers at a rescue for these dogs, in this way "giving back," if not always to the dogs she specifically studies, at least to "the community" of pit bull-type dogs.

Notes

1 S. Eben Kirksey and Stefan Helmreich, "The Emergence of Multispecies Ethnography," *Cultural Anthropology* 25, no. 4 (November 2010): 545, https://doi.org/10.1111/j.1548-1360.2010.01069.x.
2 Ibid.
3 Kim TallBear, "Beyond the Life/Not-Life Binary: A Feminist-Indigenous Reading of Cryopreservation, Interspecies Thinking, and the New Materialisms," in *Cryopolitics*, eds. Joanna Radin and Emma Kowal (Cambridge: The MIT Press, 2017); Glen Sean Coulthard, *Red Skin, White Masks: Rejecting the Colonial Politics of Recognition* (Minneapolis: University Of Minnesota Press, 2014).
4 Thanks to Lauren Corman for this insight.
5 Kathryn A. Gillespie, "For a Politicized Multispecies Ethnography: Reflections on a Feminist Geographic Pedagogical Experiment," *Politics and Animals* 5 (September 22, 2019): 17–32; Kathryn

Gillespie, "For Multispecies Autoethnography," *Environment and Planning E: Nature and Space* 5, no. 4 (December 1, 2022): 2098–2111, https://doi.org/10.1177/25148486211052872.

6 Stephanie Rose Eccles, "Researching Contested Companionship: Responsibility and Care Work in the Field," *ACME: An International Journal for Critical Geographies* 21, no. 2 (March 23, 2022): 130–146; María Elena García, "Death of a Guinea Pig: Grief and the Limits of Multispecies Ethnography in Peru," *Environmental Humanities* 11, no. 2 (November 1, 2019): 351–372, https://doi.org/10.1215/22011919-7754512.

7 Kirksey and Helmreich, "The Emergence of Multispecies Ethnography," 545.

8 Kathleen Daly and Meda Chesney-Lind, "Feminism and Criminology," *Justice Quarterly* 5, no. 4 (December 1988): 497–538; Jennifer M. Kilty, "The Evolution of Feminist Research in the Criminological Enterprise: The Canadian Experience," in *Demarginalizing Voices: Commitment, Emotion, and Action in Qualitative Research*, eds. Jennifer M. Kilty, Maritza Felices-Luna, and Sheryl C. Fabian (Vancouver: UBC Press, 2014), 125–143.

9 Daly and Chesney-Lind, "Feminism and Criminology."

10 Kilty, "The Evolution of Feminist Research in the Criminological Enterprise: The Canadian Experience."

11 Ibid., 127.

12 Ibid.

13 Kim TallBear, "Standing With and Speaking as Faith: A Feminist-Indigenous Approach to Inquiry," *Journal of Research Practice* 10, no. 1 (2014): 1.

14 Ibid., 2.

15 Ibid.

16 Ibid., 7.

17 Kathryn Gillespie, "For a Politicized Multispecies Ethnography: Reflections on a Feminist Geographic Pedagogical Experiment," *Politics and Animals* 5 (2019): 17–32.

18 Natalie Porter, *Viral Economies: Bird Flu Experiments in Vietnam* (Chicago: The University of Chicago Press, 2019), 10

19 Ibid., 17

20 Ibid., 175

21 Watson, 2016 cited in Gillespie, "For a Politicized Multispecies Ethnography: Reflections on a Feminist Geographic Pedagogical Experiment," 17.

22 Gillespie, "For a Politicized Multispecies Ethnography: Reflections on a Feminist Geographic Pedagogical Experiment," 17.

23 Ibid.

24 Ibid.

25 Ibid., 19.

26 Ibid., 29.

27 María Elena García, "Death of a Guinea Pig: Grief and the Limits of Multispecies Ethnography in Peru," *Environmental Humanities* 11, no. 2 (2019): 364.

28 Ibid., 356.

29 Kathryn Gillespie, *The Cow with Ear Tag #1389* (Chicago: Chicago University Press, 2018).

30 María Elena García, "Death of a Guinea Pig: Grief and the Limits of Multispecies Ethnography in Peru," 362

31 Ibid., 357

32 Ibid., 359

33 Ibid., 360

34 Ibid., 352

35 Ibid., 353

36 María Elena García, *Gastropolitics and the Specter of Race: Stories of Capital, Culture, and Coloniality in Peru* (Oakland: University of California Press, 2021), 59

37 Nekeisha Alayna Alexis, "'There's Something about the Blood': Tactics of Evasion in Narratives of Violence," in *Animaladies: Gender, Animals and Madness*, eds. Fiona Probyn-Rapsey and Lori Gruen (London: Bloomsbury, 2018), 47–64.

38 Naisargi Dave, "Love and Other Injustices: On Humans, Animals, and an Ethics of Indifference," *Comparative Studies of South Asia, Africa and the Middle East* 42, no. 3 (2022): 656–667.

Bibliography

Alexis, Nekeisha Alayna. " 'There's Something about the Blood': Tactics of Evasion in Narratives of Violence." In *Animaladies: Gender, Animals and Madness,* edited by Fiona Probyn-Rapsey and Lori Gruen, 47–64. London: Bloomsbury, 2018.

Coulthard, Glen Sean. *Red Skin, White Masks: Rejecting the Colonial Politics of Recognition.* Minneapolis: University Of Minnesota Press, 2014.

Daly, Kathleen, and Chesney-Lind, Meda. "Feminism and Criminology." *Justice Quarterly* 5, no. 4 (December 1988): 497–538.

Dave, Naisargi "Love and Other Injustices: On Humans, Animals, and an Ethics of Indifference." *Comparative Studies of South Asia, Africa and the Middle East* 42, no. 3 (2022): 656–667.

Eccles, Stephanie Rose. "Researching Contested Companionship: Responsibility and Care Work in the Field." *ACME: An International Journal for Critical Geographies* 21, no. 2 (March 2022, 23): 130–146.

García, María Elena. "Death of a Guinea Pig: Grief and the Limits of Multispecies Ethnography in Peru." *Environmental Humanities* 11, no. 2 (November 1, 2019): 351–372. https://doi.org/10.1215/22011919-7754512.

García, María Elena. *Gastropolitics and the Specter of Race: Stories of Capital, Culture, and Coloniality in Peru.* Oakland: University of California Press, 2021.

Gillespie, Kathryn A. *The Cow with Ear Tag #1389.* Chicago: Chicago University Press, 2018.

Gillespie, Kathryn A. "For a Politicized Multispecies Ethnography: Reflections on a Feminist Geographic Pedagogical Experiment." *Politics and Animals* 5 (September 22, 2019): 17–32.

Gillespie, Kathryn A. "For Multispecies Autoethnography." *Environment and Planning E: Nature and Space* 5, no. 4 (December 1, 2022): 2098–2111. https://doi.org/10.1177/25148486211052872.

Kilty, Jennifer M. "The Evolution of Feminist Research in the Criminological Enterprise: The Canadian Experience." In *Demarginalizing Voices: Commitment, Emotion, and Action in Qualitative Research*, edited by Jennifer M Kilty, Maritza Felices-Luna, and Sheryl C. Fabian, 125–143. Vancouver: UBC Press, 2014.

Kirksey, S. Eben, and Helmreich, Stefan. "The Emergence of Multispecies Ethnography." *Cultural Anthropology* 25, no. 4 (November 2010). https://doi.org/10.1111/j.1548-1360.2010.01069.x.

Porter, Natalie. *Viral Economies: Bird Flu Experiments in Vietnam.* Chicago: The University of Chicago Press, 2019.

Stephanie Eccles, "Researching Contested Companionship: Responsibility and Care Work in the Field." *ACME: An International Journal for Critical Geographies*, 2022, 21(2): 130–146.

TallBear, Kim. "Standing With and Speaking as Faith: A Feminist-Indigenous Approach to Inquiry." *Journal of Research Practice* 10, no. 1 (2014).

TallBear, Kim. "Beyond the Life/Not-Life Binary: A Feminist-Indigenous Reading of Cryopreservation, Interspecies Thinking, and the New Materialisms." In *Cryopolitics*, edited by Joanna Radin and Emma Kowal. Cambridge: The MIT Press, 2017.

18

FEMINISM AND MULTISPECIES METHODS

An Interview with Kathryn Gillespie

Chloë Taylor [CT]: Thank you, Kathryn, for agreeing to do this interview. To start by introducing you, Kathryn Gillespie completed an undergraduate degree in creative writing and political science from Sarah Lawrence College, after which she earned both an MA and a PhD in geography from the University of Washington. She then went on to be a postdoctoral fellow in animal studies at Wesleyan University, and then a postdoctoral fellow in geography and applied environmental and sustainability studies at the University of Kentucky. She is the author of many articles and book chapters, as well as her 2018 monograph, *The Cow with Ear Tag #1389*, published with the University of Chicago Press. Dr. Gillespie also co-edited three books: two books with Routledge that came out in 2015—*Critical Animal Geographies: Interactions and Hierarchies in a Multispecies World* with Rosemary-Claire Collard and *Economies of Death: Economic Logics of Killable Life and Grievable Death* with Patricia Lopez—and then in 2019, she and Patricia Lopez published *Vulnerable Witness: The Politics of Grief in the Field* with the University of California Press. Most recently, in 2020 she and Yamini Narayan co-edited a special topics issue of the *Journal for International Studies* on the subject of *Animal Nationalism: Multispecies Cultural Politics, Race, and the (Un) making of the Settler Nation State*. She is currently completing a second monograph, *The Sound of Feathers: Haunting and Bearing Witness in Multispecies Worlds*. Katie, I've been reading your work for years, so I am really excited to be able to finally talk to you about it. I wondered if you could start by saying a bit about your research beyond this brief overview of your publications.

Kathryn Gillespie [KG]: For sure, and thank you so much for including me. I'm really excited to be talking to you. So, let's see. I did my PhD in

DOI: 10.4324/9781003273400-23

geography with a focus on animal studies and animal geographies. The way I like to explain geography—because it's a less common social science discipline in terms of people understanding its legibility—is that it's like any other social science, like anthropology or sociology, but it takes into special consideration the way that place and space shape social relations and are shaped by social relations. In that sense, anything can be studied within geography because everything sort of unfolds within a place. The subset of animal geography aims to understand the way place and space shape and are shaped by human–animal relations.

In terms of my own research, I did my dissertation on the lives of cows in the dairy industry in the Pacific Northwest and what I wanted to understand was the emotional and embodied effects of animal agriculture and specifically dairy production on animals. I was already oriented through a feminist geographic lens with the theoretical and methodological commitments I had, but what really came through even more strongly in that project was the way that dairy production and other forms of animal agriculture are gendered and involve gendered and/or sexualised violence in the production of, in this case, milk, but also eggs and meat. What I found in the dairy industry was that it was fairly common in animal studies, and feminist animal studies in particular, to focus on the gendered commodification or violence against female animals, against their lives and their bodies, but the dairy industry shows the way that male animals too are appropriated for their reproductive capabilities in the production of semen, and also in the way that male calves are discarded and considered waste in the dairy industry. So gender and sex really come into play a lot. That dairy research culminated in first my dissertation and then I rewrote the dissertation as a book, *The Cow with Ear Tag #1389*.

Since then, I've been doing some work on what I have been thinking of as multispecies auto-ethnography, looking at some different experiences in my own life where human–animal encounters are unfolding and thinking about the complexities of those historically, what kinds of histories they bring up and how relationships of violence and care can emerge and do emerge in all of these different everyday kinds of relationships. The other ongoing project I have is around animal auction yards, specifically farmed animal auction yards. I'm working on getting started on a book on auctions after I get the last of the revisions done on this book that I'm finishing called *The Sound of Feathers*, which is an autoethnographic study of collection of human–animal encounters in my own life and what they say about human–animal relationships more broadly.

CT: You've already mentioned multispecies auto-ethnography, but can you explain what multispecies ethnography is and what you mean by multispecies *auto*-ethnography and what kind of work has been done under these names? Why have you have called for a more politicised approach to these emerging methods? And does your own work fall into those methodologies? You've said that your current work does fall under multispecies auto-ethnography, but would you describe your earlier work as falling under these methodologies as well?

KG: I think I definitely would put my work under both of these kind of methodological frameworks. If we think about ethnography in a human context or the more traditional context, it's the study of cultures, and so often in the human context of ethnography it's characterised by longer-term, embedded research where the researcher goes and lives in a place. There have been all kinds of critiques around colonialism and othering places and peoples and so lots of critique coming out of anthropology about the human complexities of doing ethnography.

Multispecies ethnography is about considering more than just the human in an ethnographic approach—so, the study of culture in multispecies contexts. There are some projects that include multiple more-than-human species so, in a real sense, these are multispecies kind of ethnographies that attend to a landscape of species experience, but I think that really all animal ethnography is multispecies in the sense that it's human and other animal, even if it's only one other species.

I think my dairy research was certainly multispecies ethnography. One of the reviewers for my dairy book actually challenged whether or not what I had done was ethnography, and it got back to this traditional understanding of ethnography as long term embeddedness. This reviewer thought that I hadn't been embedded in my field sites in a way that he thought was necessary for a project to be ethnographic. I pushed back on this in the sense that the nature of animal agriculture is that the lives of animals are inherently fractured. They're often not in the same place from birth to death. They are exposed to all kinds of different transport, different spaces, different kinds of fractured relationships where calves are taken away, for one. There isn't a linear path from birth to death in a single place to embed yourself. Even the farm is not the complete picture of animal agriculture or dairy production.

The process of researching and getting access to different spaces of production doesn't lend itself to being embedded in a particular site. I had incredible difficulty even getting access to a dairy farm. In the end, I was only able to get to see one dairy farm. That's where the auction as field site came up because auctions are public. So I ended up patching together these different spaces that really reflected the fractured nature of what it means to commodify a body. So I pushed back on that review in terms of how we need to redefine what we think of as ethnography in terms of what it means to do field work, and that a study of culture in the dairy industry can be patched together with these sometimes even momentary encounters like a cow passing through the auction ring in under a minute.

More recently, I have been reading about human practises of auto-ethnography and more autobiographical writing. I've always been interested in writing about my own experience as a form of feminist embodied knowledge and that really lends itself to the auto-ethnographic method. The first way that I was thinking about multispecies auto-ethnography is that the *auto* is the self in relation to others. So multispecies ethnography would be the self—me—in relationship to multispecies others or other species. I explored this in a paper on living with backyard chickens. My partner and I at the time, before we went vegan, bought chicks and raised them for eggs. That actually precipitated our transition to veganism, just connecting with these chickens in a more intimate way.

I think what's fundamental to me about multispecies auto-ethnography is seeing the politicised undertones of how we live and act in relation to others, and that was part of the question as well about this more politicised approach. Multispecies ethnography came out of Eben Kirksey and Stefan Helmreich's paper that coined the term, and my sense of that kind of research is that it's been useful in terms of crafting the language of ethnography around research with other species. That has been a really valuable contribution but there's a vein of that kind of work following Kirksey and Helmreich that is what I would say is a depoliticised or apolitical approach in that it observes relationships without making a political or ethical statement or taking a political or ethical stance, which to me is really a requirement of doing research on or with other animals because of the inherent power relations and violence that characterise so many human–animal relationships. And so that's where I called for a more politicised approach, something that gets at the underlying politics, power relations, and fraught ethical questions that come up. And those really have to be at the heart of ethnography.

CT: To be clear, multispecies ethnography is not always about animals. There is Anna Tsing's *The Mushroom at the End of the World*, for instance, about mushrooms and humans. Do you think those studies should be politicised as well, if they are about lichens or fungus or plants? Or are you specifically thinking that *animal* ethnographies and auto-ethnographies should be politicised?

KG: Thank you so much for that question. That's great because yes, absolutely, that's definitely part of the original multispecies ethnographic trend, and I think I depart here from a lot of critical animal studies folks in thinking that plants and lichen and mushrooms and other forms of life should be considered in a politicised way. I think some critical animal studies scholars balk at including plants and other non-animal non-humans in considerations of critical scholarship because there's this slippery slope of people who are trying to push back against or sidestep the consideration of animals by saying, "well, what about plants? Plants have feelings, too," and that kind of thing. I definitely understand that kind of wariness in critical animal studies, but I think, why foreclose the possibility that we do owe plants much more consideration and ethical exploration? I don't know, I like to err on the side of assuming that we do. I know sentience is a whole specific train of thought and study, so I'm not sure I want to use that word, but why not err on the side of trying to think about more ethical relationships we could have with other forms of life beyond and including other animal species? So I've just been playing around with this in this book I've just finished, *The Sound of Feathers*, mostly with respect to plants. So I'm really interested in considering these forms of life as a politicised and ethically oriented subject in ethnography alongside other animals, without betraying the more overt or obvious forms of taking animals lives seriously that are required.

CT: My next question is about witnessing. You've also written about witnessing as a method. Can you explain what you mean by this and how it has been part of your own scholarship?

KG: Yes, for sure. I started thinking about witnessing in the auction yard as a way to understand what I was doing there. So in social science scholarship, what I was doing there would have been characterised as *participant observation* or *spectator observation*. Rosemary-Claire Collard has written about spectator observation in a really clear way in her book, *Animal Traffic*, about exotic pet auctions. As I was doing my research, I started thinking about the politics behind those forms of participation. I was a participant in the sense that I was sitting there in the audience, as an audience member, participating in a what could be understood as a voyeuristic way. I tried to suss out where the line might be between voyeurism and witnessing as observation. I never bought an animal at the auction, so I wasn't participating in that way, but I just wanted to understand what the role is of sitting there and watching and, in the moment doing nothing, in order to gather information, to then go and do something like writing and sharing this kind of knowledge and experience with others in a broader way. So I thought witnessing is a form of observation of violence that at its heart is a political action and political fodder for political commentary on violence. There's quite a tradition of witnessing in terms of human genocide and these large-scale levels of violence where, in the moment, you can't necessarily do anything to change it in a substantive way. But what you can do is witness and record and be there as a witness to what's unfolding with the ethical obligation then to share that with others with the aim of transforming those structures and relationships of violence so they don't continue.

CT: How has your work on multispecies methodologies been influenced by your feminism? Or how does a feminist ethics and politics inform what you mean by a politicised multispecies ethnography?

KG: I think I got a little bit at the gendered aspect of agricultural production and coming from a feminist orientation with that. I think also, for one, auto-ethnography, although not always characterised as a feminist method, is, I think, quite feminist in its fundamental orientation with a focus on personal experience. One of those foundational feminist commitments is around the personal being political and so I've really taken that to heart in my work in terms of looking at my own life and politicising the relationships I have with other animals, and then looking at the intimate personal lives of other animals and how they're affected by structures of power and violence like capitalism or colonialism. I think there's an inherent focus on personal experience, on embodiment, which is a key feminist way of looking at the world.

There's also the "making the familiar strange" that I think is a feminist commitment, observing the everyday, mundane reality of living and looking at it askew, turning it in a different direction to see the strangeness in the everyday, to see what are the things that are obscured by the everydayness of human–animal relations that make it hard to see the violence behind it, or the nuances and complexity of what's actually going on. I really like the making the familiar strange project as a sort of feminist commitment. So I think feminism has impacted the way that I do research and what the research is focused on as well.

CT: Can you say a bit about the process you went through to get ethics clearance for your doctoral research, the fieldwork that culminated in your 2018 monograph, *The Cow with Ear Tag #1389*?

KG: Yes, for sure. So this was just such an interesting and surprising part of the research process, which was that in social science research or other forms of research with human subjects, you'd go through the IRB, the institutional review board, or human subjects review ethics review board, to make sure that the research that you're doing meets certain ethical standards. And there's all kinds of critique of the human ethics review process in terms of it not necessarily protecting human subjects so much as protecting the university or institution from liability. They focus more on liability concerns. So there's a whole literature and movement critiquing human ethics review, which we can set aside for now, because that's a whole other conversation.

So I went through the human subjects ethics review, because humans were going to be involved in my research, and on that form there was a box that said "does your research involve non-human animals?", and so I checked "yes," of course. But then that routed my application out of the social science ethics review track into the biomedical ethics review process, because in the university, the primary way that animals are conceived of as being part of research is in their use in biomedical research.

I was at the University of Washington and the University of Washington has one of the six primate centres in the U.S. and does a tremendous amount of animal research in the biomedical context. So there was really no way that the university conceived of the role of animals in research process for the social sciences, for instance, or for any kind of context outside of what animals might experience in the labs. So I had ended up having to go through this whole review process that was not at all relevant to the way I was going to be encountering animals. It was about potential harm done in lab settings, it was about how you would euthanise the animals that you were using at the end of a study or what would happen with them, and it was all of these very invasive procedures that this was meant to review.

And so the outcome when I went through that process was, of course, that they approved my research, because it wasn't including or involving any kind of direct kind of bodily harm that I would be delivering onto these animals. I wouldn't be killing them. I wouldn't be holding them captive. I wouldn't be doing experiments on them and those things. And so I sort of flew through that approval process. But the outcome was that even going through ethics review, the IACUC (Institutional Animal Care and Use Committee), there was no real ethical oversight for the research that I was doing. And this is something that I think has really been a growing concern among animal studies and critical animal studies scholars: how do we do meaningful ethics review processes for each other in the absence of institutional committees that should be overseeing this kind of research?

The outcome, too, I guess I should say, or one of the challenges through the ethics process, the animal ethics review, because it went out of the social sciences and humanities into the hard sciences and medicine track, was that they had no knowledge of social science methods. So they asked: "Well, what is participant observation like? This sounds like something made up," and it's just a totally different world. So it was a moment of major pause in terms of how to meaningfully assess the ethics of research involving other species and the frustration of, one, having to go through this kind of lengthy process and, two, not having that be effective or helpful at all in deciding the ethicality of the research I was doing.

CT: Do you think we should ask similar questions about non-human animal participants in social science research as we do about human participants? So should we be asking questions about do they consent to participate in this research? Are these vulnerable populations? Are they being deceived by the researcher? Or do those not make sense in this context? Is there a different set of questions that we should be asking? And what types of questions do you think the ethics review board should have asked you about your doctoral research?

KG: I think that a lot of the questions we do ask about humans should be asked of research with animals, such as questions around consent. How do we observe or evaluate what it means for a member of another species to consent? What does that look like for individuals of other species? What does it look like in the species-typical behaviours of different species? There's all kinds of ways to think about what it means to consent. Without going too far into the weeds here, I think one of the key assumptions that we should make with researching with other animals is that we should just assume that they would belong in the most vulnerable category. So if you think about translating useful things from ethics in human research, one would be to take as a starting point the highest level of vulnerability, and then go beyond that, because I think we can just assume that the research we do can compromise animals in some way. Certainly that's not going to be the case in every study but I think taking that extra layer of caution would be a starting point.

I think that the other thing that doesn't necessarily come up in institutional human ethics review processes all the time, but I think should be key, probably in human cases too but certainly in animal research, is: how does this research benefit the animals that we're studying, and do we need to do the research at all? And what are the motivations for doing it? In a lot of academic research, there are motivations around caring about changing conditions for animals, for instance, but there's also [motivations] in educating people about these different forms and structures of violence. But there's also this fundamental thing in academic research, that we're using this research to advance ourselves professionally, to publish and use this research as a kind of currency in our career advancement. And so there's always going to be that underlying motivation. That's part of the landscape of doing research with other species and so I think the key question here is: how does this benefit the subjects of my research? And we need to make sure that the answer to that is that it is always substantively benefiting [the animals]. That would be, I think, an important anchor in terms of ethical consideration. I don't want to go too far down the rabbit hole because I think there's a lot of questions, some of which are overlapping with humans, and some of them are unique to animal contexts. Obviously, the harm question—Is there any way that this research is causing harm to the participants?—I think that has to be granted that, no, it's not causing harm, and that we have to do the higher standard of asking: how is this benefiting other animals?

CT: That ties in to my next question. In her article, "Death of a Guinea Pig: Grief and the Limits of Multispecies Ethnography in Peru," anthropologist María Elena García has also questioned the ethicality of passively observing animals suffering and being killed. And in particular, she described seeing a pregnant and dying guinea pig being treated callously and not doing anything to help her because she didn't want to estrange the human research subject she was also studying in her multispecies ethnography. She somewhat ambivalently suggests at the end of her article that by grieving and bearing witness to the lives and deaths of animals through her writing, lives and deaths that would otherwise not have been mourned, this type of research might be ethically justified. And that ties in, I think, with the question about whether or not the research substantively helps the animal subjects that you were speaking of a moment ago. So there's nothing to do for that particular guinea pig, but by writing about it, maybe it can have some impact that will help other guinea pigs. What do you think about this example? And does it resonate at all with your own experiences doing field work?

KG: Yes, and I should say that María Elena and I are close friends and colleagues and so have, outside of publications, talked quite a bit about these issues and questions, and I think that there's a lot of similarities between her visiting the guinea pig farm and seeing what she witnessed, and the work that I've done in auctions, for instance. And I think this just gets back to the question of witnessing and what is the value, if any, of witnessing as a method? There's a difference, I think, between actively harming animals in research. So, for instance, Timothy Pachirat [for his book] *Every Twelve Seconds*, he was involved in the work of the slaughterhouse, in killing and dismembering animals. And if you translated that into a human context, it would never be conceived of as ethical to be participating in the killing of humans. So there's an inherent kind of anthropocentrism or sense of human supremacy, I think, in research like animal studies, research that participates in practices that would be categorically unacceptable in the human context. And so I've thought about that in terms of my own work, and I think María Elena and I have talked about that quite a bit in terms of, would you, as a researcher, stand by and watch a human dying of thirst like she is seeing with this guinea pig?

Should we do research in spaces of animal violence? I think that is a really important, fundamental question that we need to address before we even get to the question of what should that research look like if the answer is "yes"? I've thought a lot about that with the auctions, just because of the extreme feeling of powerlessness and inability to change the life of even a single animal in the auction yard. But so many people have told me how useful it is to have this empirical research to draw on in making the really important theoretical contributions, for instance, that philosophers do, or non-field work based disciplines. This can certainly be useful in terms of just communicating the lived realities of these spaces in order to then make arguments about why we shouldn't treat animals the way we do and how we could manifest different futures. But I don't know, it's really something that I struggle with a lot. And with this book I'm starting on auctions, that's obviously a really core question. Is this a project I should even be doing? Because I know that it involves sitting there for hours and days and not doing anything in the moment.

I think that the question that María Elena García asks around worries over losing access to her field site if she does object to how animals are being treated or does something to intervene in some way, I do think that in a lot of contexts, that's a very real concern. When I was doing my dairy research, well, it was rare for women to even be in these spaces. I was asked to leave on a number of occasions because my presence seemed suspect. And that was just me sitting there observing, not even doing anything to intervene on behalf of the animals, and so I think there is a weighing of whether, you know, is it valuable to continue to have access to these spaces? And then if it is valuable, are there ways of ameliorating or creating more urgent or immediate positive action for animals? Because I think part of the problem with witnessing is, you know, you witness, you observe, you write all this stuff down and share it, and you hope that in some way that helps other animals, but it doesn't help those specific individuals going through the auction ring or living in a guinea pig farm, for instance.

So I don't know. This is not something I have a clear answer to at all. It's something that I'm grappling with. And at the moment, I think I am kind of sitting uneasily in the position that I am going to continue to do research in spaces of animal violence and try to be as attentive as possible to not contribute in an active way to harming animals there, which is just such a low bar. I don't even know if it's meaningful to say that. It's a thing I've been talking about with other critical animal studies folks and activists, and it's one of those things that I'm really open to being challenged on and having my position changed in terms of, no, we should never do research in space of animal violence. I could certainly be open to being convinced about that possibility.

CT: Stephanie Eccles, a PhD student in geography who, like you, works in the area of animal geographies, has written an article researching contested companionship, responsibility, and care work in the field, in which she argues for researchers actively intervening for the animals they encounter in their research, even breaking the law to help them. She was doing multispecies ethnographic work on breed-specific legislation or laws that outlaw pit bull–type breeds of dogs from certain jurisdictions, and in the process of doing that research, she describes taking one such dog home who had been abandoned in a vacated apartment and adopting her even though she's living in a jurisdiction that doesn't allow that breed of dog. But she's worried what will happen if she doesn't take her home in this jurisdiction. Will the dog just be killed? And she also describes raising money for a homeless man and his two pit bull dogs to get the dogs vet care and to get them money to leave the city and go somewhere else where those dogs are still legal. So what do you think about engaging in intervention for animals in the field, as in these examples? And is that something that you think that we should be doing more of as critical animal studies scholars?

KG: Yes, I think wherever possible we should intervene. Wherever we could intervene on behalf of an individual animal would make a huge difference for that individual animal. I'm trying to think if there's been a situation in the dairy or auction research where it would have worked for me to intervene. I have wondered about buying animals at auction. A lot of people take the position that any kind of monetary contribution to animal agriculture is a support of that industry and should be avoided. A couple of things on this. One, in these auctions animals are often so inexpensive. I wrote in *The Cow with Ear Tag #1389* about how the cow with the ear tag #1389 didn't end up selling and the bidding went down to $35 and she didn't sell at that price. I've been doing research more recently during COVID in the so-called poultry auctions where chickens and turkeys and ducks and geese and also rabbits are sold and often, you know, they'll just sell for a dollar or two. And so, you know, in terms of contributing to that system, to buy an animal contributes so little monetarily, you know? A good friend of mine who runs an animal sanctuary, she does not go and buy animals at auction but she says an animal would want you to buy them and give them a different life. And so I don't know, that's one sort of intervention I've been thinking about over the years. What are the ethics around that? And then if you did buy an animal, who do you buy when there are hundreds on any given day going through the auction yard, for instance.

I think that the case of the pit bull–type dog that Stephanie took home, I think that's wonderful, and whenever it's possible to do that—and I think we should think creatively about how and when it's possible—then it should be done because it makes all the difference for the animal who's rerouted outside of these systems of violence and killing and all of that. I think for my own work too—this isn't an intervention in the lives of animals I'm studying, but it's a way of trying to engage in care work for members of other species—is throughout all of my research, I've volunteered at Pigs Peace Sanctuary in Stanwood, Washington, just north of Seattle, and engaged in care work for the individuals who have come from situations of neglect and abuse and farming. It feels meaningful in the sense of, these aren't the animals I'm studying, but they're animals who have been caught up in agriculture in similar ways to the ones that I have been studying. I don't know if I would call it amends making, but it certainly feels like meaningful care labour to just labour in service of individual pigs flourishing and living their lives out, you know, in this space of care and safety, and a space that really is designed in in every way to enhance their sense of flourishing.

19

POSSIBILITIES AND PRODUCTIVE FAILURES

Feminist Methodologies and Animal Subjects

Lauren Corman

In the following chapter, I stage an intervention into anthropocentric feminist methodologies. If feminist methodologies are consistently applied, they should prompt questions about nonhuman animals as subjects. The common categorical exclusion of nonhuman animals from the traditional scope of Western feminist methodologies is dubious; in part, I argue that such recurrent omission is an expression of anthropocentrism, that is, human-centredness.[1] Animals' chronic exclusion from conventional anthropocentric feminist methodologies is worth challenging.[2] Drawing on Roxani Krystalli's (2022) distillation of the "four pillars of feminist methodology," I centrally illustrate how these defining characteristics of feminist methodologies *already suggest* the inclusion of nonhuman animals, yet they frequently remain absent; this lacuna is representative of unchecked presuppositions rather than an inherent methodological failing.[3]

Animals' subjectivities could possibly figure into these conventional Western feminist methodologies, and indeed the methodologies might work generatively in this direction, if the implicit anthropocentrism of these approaches is uprooted. Given feminists' relentless interrogation of Western constructions of subjectivity, particularly the constitution of the modern liberal subject, and decolonial feminists' insights that Western European subjectivity is indivisible from colonial productions of "humanity," common feminist methodological assumptions that "Others" are necessarily human is untenable. As Xhercis Mendez (2015) argues,

> Unlike all others in the colonial/modern gender system, "Man" is understood to be the sole possessor of subjectivity and knowledge as pre-determined by his "natural evolution" as a "rational" being. For all intents and purposes he can be considered the literal embodiment of the modern liberal subject (think: Western European philosophy) as well as the subject who is protected and systematically benefits from the notion of "rights" and laws. It is this "Man" who is free to impose his body and will with relative impunity, sexually or otherwise, on any Other that is understood to be naturally inferior to him.
>
> *(p. 44–45)*

DOI: 10.4324/9781003273400-24

The "modern liberal subject" is implicitly a human one. Therefore, a feminist methodological divestment from anthropocentrism—a "settler-colonial logic" (Belcourt, 2015, p. 5)—is urgent.

I argue that attention to animal subjectivities would help address a contradiction within the field of mainstream Western feminist methodology that, while training part of its critique on modern liberal subjectivity, nonetheless continues to reproduce exclusions along species lines. This becomes an even more pressing concern for feminism when we acknowledge how gender, a primary preoccupation of the field, is deployed within colonial logics in Western contexts to mark the boundaries of humanity, suggesting the simultaneous construction of a subhuman and animalized Otherness. That is, gender is racialized through European colonization and crucially marks humanity *as such*.[4]

As these power relations are thoroughly and incisively scrutinized, the question of animal subjectivities, and our responsibilities to animals themselves, often remains untouched by many feminist methodologists. While rejecting equivalences regarding the experiences of human and nonhuman animals, the latter are also degraded through a denial of their subjectivity. I petition that the process of revisiting feminist methodologies is vital to their growth, as it resists stagnancy and embraces change. We should continually query the field's presuppositions, including its anthropocentrism. Echoing Dvora Yanow, Krystalli notes, "The questions of methodology, too, are lived, held, shared and changeable—and, indeed, political" (p. 35).

Before proceeding, however, I offer an interruption into my line of inquiry, namely by suggesting that feminist methodologists make good on their promises and abandon their human exceptionalism, I am also guided by Sarah Ives' (2019) powerful caution. Ives (2019) urges,

> it is critical that multispecies approaches engage further with scholarship on race and colonialism to remember a history in which blurring lines between the human and nonhuman was not celebratory, but rather distinctly violent, a history in which "the management of populations and race undergird[ed] the human–nonhuman interface."
>
> *(p. 2)*

Thus, I continue with a caveat: Given Western histories of racism and colonialism that produced, differentiated, and ranked racial categories in relation to constructions of animality and humanity (Kim, 2015), calls to recognize animal subjectivities within research may understandably not be universally embraced.[5] I hold my caveat—that including animal subjectivities within one's methodological purview can be fraught—alongside observations of Driftpile Cree scholar Billy-Ray Belcourt, alluded to previously, who implores us to consider the relationship between anthropocentrism and settler-colonialism. He posits,

> If settler colonialism and white supremacy mobilize through anthropocentrism (and they do) and capitalism requires the acquisition of Indigenous lands for animal agriculture (and it does), then decolonization is only possible through an animal ethic that disrupts anthropocentrism as a settler-colonial logic.
>
> *(2015, p. 5)*

As a white settler scholar, I am reluctant to argue that *all* feminists should evacuate anthropocentrism from their methodologies. Nonetheless, my call to question the anthropocentrism of feminist methodologies has political and ethical import beyond a desire to be "kind to animals," as anthropocentrism is indivisible from the broader issues of power that feminists and others are interrogating.[6] For example, scholar and self-described Black feminist love evangelist Alexis Pauline Gumbs (2020) embraces interspecies boundary blurring and links such fluidity to a broader political project. Commenting on her poetic prose in *Undrowned: Black Feminist Lessons from Marine Mammals*, Gumbs writes, "the intimacy, the intentional ambiguity about who is who, speaking to whom and when is about undoing a definition of the human, which is so tangled in separation and domination that is consistently making our lives incompatible with the planet" (p. 9).

Feminist Methodologies—An Unfulfilled Promise

During one of our lively conversations about pedagogy, a feminist colleague once exclaimed, "How have I taught methodology for so long, but only just now included animals in discussions beyond laboratory animal care protocols?" "Anthropocentrism," I answered bluntly. Of course, this answer is too simple, but it points to an anthropocentric orientation threaded throughout many feminist methodologies. For example, in Kim Golombisky's (2018) chapter, "Feminist Methodology," she quotes from her earlier work to frame her subsequent arguments:

> What distinguishes feminist research is an honesty about its political agenda, namely its commitment to social justice, defined as "the sustainable material and social circumstances in which all people enjoy general wellbeing, participate in self- determining communities, and thrive in the pursuit of fulfilling lives."
>
> *(p. 172)*[7]

Strikingly, such failures to include animals as subjects resonate within the greater academic research landscape, which demonstrates no compunction to address animals' complex subjectivities within its ambit. For instance, university research ethics boards, which bifurcate guidelines for human and nonhuman animals, ultimately perpetuate animals' objectification and entrench their property status. The CCAC's "Guide to the Care and Use of Experimental Animals" (2020) 193-page document does not include a single mention of consent.[8] Compare this absence to the Government of Canada's "Tri-Council Policy Statement—Ethical Conduct for Research Involving Humans" guidelines, which mention "consent" 470 times in the 203 pages. These discrepancies throw into sharp relief discrepancies in power between humans and animals.[9] Feminist environmentalist Traci Warkentin (Oakley et al., 2010) confronted the problem of animal consent early in her PhD research:

> How does one obtain informed consent from a whale? Pondering this question, while designing my doctoral research on ethical and educational dimensions of encountering whales in aquariums, shone a spotlight on the near total lack of institutional procedures for the ethical consideration of nonhuman participants. While my university had [an] extensive protocol for "research involving human participants," there was nothing for accommodating nonhuman participants apart from the "animal care protocols" which applied to scientific laboratories. The system assumed and maintained a fundamental dualism between human subjects and animal objects.
>
> *(p. 93)*

In codifying the human–animal divide, such protocols naturalize the objectification of nonhuman animal life. At the same time, a version of the "human" is surreptitiously produced, both relying on—and fortifying—a cultural hegemony steeped in humans' superiority over nonhuman animals.

Some might shrug that "we can't meaningfully gain animals' consent," thus nullifying the issue beyond our species; yet such dismissal is nonetheless emphatically challenged by feminist and biopsychologist Barbara Smuts (2008), who demonstrates through decades of rigorous field work that all large-brained social animals engage in "embodied communication" (EC). There is ample evidence that animals communicate in fine detail through a variety of asynchronous to increasingly synchronous greetings and gestures, which can change over time, allowing relationships of varying degrees and kinds to be forged or halted. Similarly, other animals also exercise agency, and demonstrate desires and preferences. Inbal Ben-Ami Bartal et al. (2011), for instance, famously observed that rats will free conspecifics from restrainer tubes: "the free rat was not simply empathically sensitive to another rat's distress but acted intentionally to liberate a trapped conspecific" (p. 1430). They affirm that such "pro-social" behaviour is evidence of rats' empathy. We might infer that, when given the opportunity to exercise their agency, rats prefer not to be confined, nor do they want other rats to be confined.[10] From this vantage, the question of animal consent—animals' agreement or permission—is not so nonsensical after all. However, our bureaucratized research ethics tethers themselves to dualistic Western beliefs about animals, which firmly ranks "us" over "them" rather than dealing with the much stickier problems of animal subjectivities and their implications. Perhaps university ethics boards do not raise questions of animals' consent in research because they realize, given the nature of most experiments, animals would not willingly provide it.

Given this academic backdrop, in addition to animals' plight more generally, my claim is there is a feminist methodological imperative to critically consider nonhuman animals as subjects. As Andrea Doucet and Natasha Mauthner (2006) emphasize, quoting Beverley Skeggs, "feminist research is distinct from nonfeminist research because it 'begins from the premise that the nature of reality in western society is unequal and hierarchical' " (p. 40). If we can unseat our anthropocentrism, we can further address the inequalities and hierarchical relations between humans and animals, and more holistically enact such feminist commitments across species.

When nonhuman animals are categorically omitted from feminist methodologies, and we are committed to their exclusion, then this must be justified beyond tradition and common sense. Commenting on the vital role of reflexivity to feminist methodology, Roxani Krystalli (2022), contends, "it requires reckoning with both the ways in which the researcher's position and situatedness within power relations grant her access to certain ways of knowing, and the possible limitations of that view" (p. 40). The frequent default anthropocentrism of feminist methodologies—certainly "possible limitations of that view"—should all but disintegrate when exposed to the kind of reflexivity feminist methodologists not only laud, but also claim define their field.

For those who still feel it strains credulity to suggest that nonhuman animals be included within the purview of feminist methodologies, I recall an interview I conducted with a slaughterhouse worker years ago. He commented on how another worker would delight in slicing open the mammary glands of sows passing down the kill floor, all within a context in which pigs were regularly degraded as "bitches," while they were prodded and hit. Not only was gender being reproduced, but it was also done in relation to female animals, who were subjects of particularly gendered violence in that moment.[11]

Feminist Methodologies' Imperatives and the Question of the Animal(s)

Despite feminist methodologies' attunement to curiosity, reflexivity, relationality, resistance, and perhaps, most significantly, power as it pertains to each of these elements, they have frequently failed to consider nonhuman animals as subjects. However, there is nothing inherent to the methodologies that precludes this. Doucet and Mauthner (2006) credit scholars such as Donna Haraway for shaping early feminist epistemological debates on gender bias in science, for example, while also acknowledging how Haraway's notion of "situated knowledges" and her challenge to the "view from nowhere" (fictionalized within discourses of objectivity) further developed the discussions. Haraway was interested in the animals themselves, though. Even with such contributions, the question of animal subjectivities is perpetually peripheralized.

To frame my argument, I consider Roxani Krystalli's (2022) recent "Feminist Methodology," from *Gender Matters in Global Politics*, as my starting point, particularly because of her clear and contemporary delineation of the field.[12] Krystalli's synopsis of feminist methodologies, of which she stresses there are many, helps us to think about the inclusion and relevance of nonhuman animals within these approaches. Employing a definition of methodology described by feminist scholars Brooke Ackerly, Maria Stern, and Jacqui True, Krystalli notes (2022), "Methodology refers to 'the intellectual process guiding reflections about the relationship among all of these [epistemology, ontology, methods, ethics]; that is, guiding self-conscious reflections on epistemological assumptions, ontological perspective, ethical responsibilities, and method choices" (p. 34). Here I connect "the question of the animal(s)" (i.e., our ethical and political responsibilities to animals) to Krystalli's distillation of the "four pillars of feminist methodology":

> the role of feminist curiosity in generating and pursuing research questions; the importance of a practice of reflexivity throughout the cycle of research; the acknowledgement of the relational nature of research and its manifestation in collaborations, partnerships and citational politics; and the capacity for feminist methodologies to shed light not only on patriarchal violence, but also on the sources of care and joy that exist alongside it.
>
> *(p. 37)*

The dynamic presence of animal subjectivities directly bears on each of these pillars, both strengthening and modifying their architecture. Significantly, Krystalli does not name nonhuman animals in her chapter, and continually presumes a human subject; a non-anthropocentric interpretation of "the pillars of feminist methodology" prompts a more radical and inclusive orientation across species boundaries.

On Krystalli's first pillar of feminist methodology, "the role of curiosity in generating and pursuing research questions," she guides us in two directions. They first highlight women as central to feminist methodology (p. 38). She states,

> On one level, feminist questions take the lives, experiences and insights of women seriously and treat those women not only as survivors and victims of various forms of patriarchal violence and exclusion, but also, simultaneously, as agents and bearers of knowledge. Feminist research questions also explore the lives of men as gendered subjects, and the experiences of non-binary people who don't identify with and/or who challenge binary understandings of gender identity.
>
> *(p. 37–38)*

Crucially, Krystalli extends her analysis, including men and non-binary people as *gendered subjects*. They are necessarily included within the purview of feminist methodology, so Krystalli is explicit that feminist methodology is not only about women. Yet, implicitly for Krystalli, "people" are pivotal to feminist methodology, but the inclusion of that category of subjects, which simultaneously serves to both capture certain species and cast out others, is not justified or explained.[13]

Recalling the pig who is called a "bitch" in the slaughterhouse, or whose mutilation excites the male slaughterhouse worker, one does not need to understand oneself as a gendered subject to be treated as such. Any feminist methodologist who would interpret that research site as only including humans as gendered subjects would not only miss something crucial about how gender is produced in such contexts, practices, and relations, but also how such production is intimately tied with nonhuman animals, who themselves are agents and bearers of knowledge regarding patriarchal, speciesist, and other forms of violence.

Decades of ethological and other research makes apparent that animals bear the physical wounding of such encounters (Bradshaw et al, 2005). When animals are lucky enough to escape from violence, they can remember these past traumas—experienced as psychological and emotional distress. For example, Barbara Smuts (2006), whose observations are fortified by decades of researching animals' social behaviour, recalls the fear expressed by her dog, Bahati, specifically related to men. She laments,

> Although Bahati and I quickly bonded, she was not an easy dog to live with. She was extremely cautious, wary of anything out of the ordinary: a big stump next to the path, an airplane in the sky, an unexpected sound, the motorcycle next door. Most problematic, she was frightened of unfamiliar people, especially men, and men with beards terrified her.
>
> *(p. 120)*

Such observations are not mere anecdotes, as they are often reduced within mainstream media and stripped of their political and ethical import (see Colling, 2021), but widely reported and well documented occurrences that could inform the feminist methodologies. Krystalli petitions,

> methodology is not just a longer word for "method". Reflections on methods and their suitability for exploring particular kinds of questions (for example, should I conduct interviews? With whom? About what? How do I do that through a feminist lens?) are part of methodological contemplation—but the full picture also includes interrogating assumptions about what counts as knowledge, how researchers can know what we claim to know and how power shapes research relationships.
>
> *(p. 34)*

Krystalli also enjoins us to consider Cynthia Enloe's understanding of feminist curiosity beyond bodies and identities, and to be curious (as a methodological orientation) about the "structural, systemic and symbolic dimensions of gender and power" (p. 38). Enloe urges us to question the workings of feminized and masculinized meanings, which should inspire us to especially question what is conventionally rendered "natural" or "even if acknowledged to be artificial, are imagined to be trivial" (p. 38).

In the spirit of curiosity, as a pillar of feminist methodology, the trivialized and naturalized human–animal relations within capitalist sites of industrial food production, which enable the violence I noted earlier, should prompt the kinds of curiosity that certain feminist methodologists suggest, to confront what is both naturalized and trivialized. In the small recounting of worker testimony previously offered, power ricochets throughout the slaughterhouse, as men who predominantly labour on the kill floor are vulnerable to the development of perpetration-induced traumatic stress (PITS), a form of PTSD precipitated by causing harm to others (MacNair, 2002), and other workplace hazards (Struthers-Montford & Wotherspoon, 2021), while employing the term "bitch," saturated with fused misogyny and speciesism (Dunayer, 2001). This is done while meting violence on beings who, within many feminist methodological imaginations, do not figure as subjects who might be engaged as "agents and bearers of knowledge," with "lives, experiences, and insights" we might take seriously.

Krystalli's second pillar of feminist methodology stresses "the importance of a practice of reflexivity throughout the cycle of research," which should again encourage us to consider nonhuman animals. Stated succinctly, Krystalli claims, "Reflexivity is an invitation to consider the scholar as socially situated within relations of power" (p. 40). She asks,

> How does being a man, a person of colour, a foreigner, a feminist, or a person in the Global North shape this particular inquiry? Which of these vectors are most relevant in this context and how have they affected the research process? What lines of sight does this positioning enable and foreclose?
>
> *(p. 40)*

To return to the first pillar—on a commitment to curiosity—what sorts of knowledge become possible when reflexivity about one's position as *human*, and as a human researcher, is also denaturalized and thus more holistically fulfils this feminist methodological promise? What kinds of power are enabled by virtue of our station as humans? How is this power intensified, diminished, or complicated by histories of colonialism, racial capitalism, and patriarchy that engender certain humans with agency, visibility, and recognition in contrast against those who, even when human, are never afforded full Western status as "Man," as "human"? Scholars have persuasively demonstrated meanings associated with "the human" are indivisible from white supremacist and colonial taxonomies articulated through metrics of animality (Kim, 2015; Gosine, 2021). By interrogating the positionality and situatedness of our heterogeneous "humanity," certain legacies of power are made more intelligible and thus more available to critique, which again echoes the commitments of feminist methodologies.

Conversely, what forms of power prompt us *not* to consider the relevance of species in our feminist inquiries, and how is that lack of questioning, that lack of reflexivity, at odds with a feminist methodological imperative to make reflexivity an iterative process? Such "relations of power" do not occur only among humans, and one is "socially situated" among other life forms, who are also engaged in their own relationality and sociality. These are not the "same" experiences as humans, but of course relational and social experiences among humans also vary dramatically. I am not suggesting we collapse important differences between humans and nonhuman animals to advance an argument about shared sociality, but I am promoting a reflexive orientation that acknowledges continuities between human and animal sociality (Bekoff, 2007), which are indivisibly bound to power. Nonhuman animal relationality and sociality is thoroughly documented by Indigenous peoples, (cognitive) ethologists, and many other close observers of human–animal relations.

The acknowledgement of nonhuman animal relationality and sociality, imbricated with ours, offers a radical departure from the profound anthropocentrism of many, even very critical, forms of feminist methodology. Nonhuman animals are relational and social actors who are co-constituting relations, resisting, exerting agency, and making sense of their worlds in their own ways (Haraway, 2008; Yong, 2022). What kinds of knowledge become possible when, extending Krystalli's question, we might ask, "How does being a *human* also shape this particular inquiry? How is *species* relevant as a vector in this context and how has it affected the research process? What lines of sight does this positioning enable and foreclose?" If these questions seem irrelevant, how is that dismissal perhaps a testament to historically generated expressions of power?

Additionally, Krystalli orients us to a third feminist methodological pillar: "the acknowledgement of the relational nature of research and its manifestation in collaborations, partnerships and citational politics" (p. 37). She is decisive that "research unfolds in an ecosystem, and that ecosystem is not always fully or meaningfully acknowledged in academic work" (p. 41). This pillar displaces the researcher as the sole actor in, or generator of, research inquiry. Feminist methodologies require deep and overt recognition for the contributions of others, especially those typically disavowed in regular academic practice. In a nod to ethnographic research, Krystalli specifically names cooks, caregivers, and drivers, alongside translators, interpreters, and a host of others who make research possible.

The myth of the atomized researcher is thus turned on its head through this methodological responsibility. To illustrate this pillar, Krystalli draws on Sara Ahmed to demonstrate the political and ethical implications of citational work, and thus doing, again points beyond "living humans" (p. 40) in feminist research: "[F]eminist scholars recognise that we also exist in relation to ideas," states Krystalli (p. 41). "Acknowledging the relational nature of research, therefore, also involves tracing the intellectual and emotional lines that shape the research journey and ensuring that those lines are visible in the outputs that result from this research" (p. 41). There is nothing in these defining methodological commitments that restricts them to human relations, although they are often presumed to be. If feminists can be responsible to ideas, and their lineages, why not nonhuman animals? Even excluding research that directly involves nonhuman animals, and our methodological responsibilities to them in those contexts, there is growing acknowledgement and vast scholarly literatures that demonstrate the significant roles animals occupy in relation to humans. As caregivers (Fraser et al., 2018), friends (Mornement, 2018), teachers (Spannring, 2017), and labourers (Coulter, 2016), they support and sustain us, and therefore, I would add, feminist research. Although animals are within our research ecosystems, staggeringly, they remain largely invisible, thus contradicting the very pillars of feminist methodologies that many espouse.

Finally, Krystalli directs us to the fourth pillar: "the capacity for feminist methodologies to shed light not only on patriarchal violence, but also on the sources of care and joy that exist alongside it" (p. 37); while less developed in her chapter, feminist researchers will immediately recognize the salience of this pillar, given years of feminist scholars pressing for representations of subjectivity beyond victimhood (e.g., Mohanty, 1988). Therefore, Krystalli

> invite[s] us to ask questions about agency, resistance and survival in the face of patriarchy . . . [to] make room for the study of care, joy, love and the other positive practices that fuel our days, and for those practices to shape the research process and the researcher, not just the subject matter.
>
> *(p. 42)*

This pillar dovetails with critical disability and animal studies scholar Sunaura Taylor (2014, 2017), and my own critical animal studies research, to include—but move beyond—suffering in our analyses. The development of my approach (Corman, 2017) is directly informed by feminist critiques from the Global South that urged researchers to stop repeatedly and exclusively reducing subjects to their suffering, especially when such moves are understood as benevolent.

Exemplary non-anthropocentric scholarship working in this direction is well represented by transnational feminist, Sarat Colling (2021) and her "animals without borders" approach.[14] Throughout her ground-breaking book, *Animal Resistance in the Global Capitalist Era*, Colling analyses nonhuman animal stories of agency, struggle, and transgression, attending to their everyday and sometimes spectacular dimensions, alongside their inseparability from histories of racism, colonization, and capitalism.[15] "Animals have been shoved into the margins by human spatial and ideological orderings," states Colling, "but they are also subjects of their own struggles, located at the centre of their liberation movement" (p. ix).

As I complete this chapter, orcas are damaging and sinking boats off the coasts of Spain and Portugal. Speculation abounds: Is this a game? Are they retaliating? Has an aggrieved member taught others? Some activists are declaring solidarity with the orcas, with greater and lesser sincerity; as one viral tweet by "Mr. Ape" (@trevorcumbo) cheekily remarks, "DMing every orca i [sic] know and begging them to tell me how a land-based ally can support the movement." Alexis Pauline Gumbs (2020) is no doubt watching and listening closely. Although her work could be read as a radical non-anthropocentric embodiment of each tenet of feminist methodology described here, I want to end by highlighting the relevance of her scholarship especially to the last pillar. The seriousness of Gumbs' theory is matched by an effervescent joy which refuses to reduce human and nonhuman animal subjectivities, and our entanglements, to despondency. With Krystalli and Gumbs as guides, I wonder: How might we enlist our feminist curiosity and reflexivity, and our commitments to name the relationality involved in making subjects, as much as making research, across species? How might attention to animals' subjectivities help us erode colonial myths of "Man," build coalitions in our academic and other practices, and productively unsettle us as researchers, humans, and knowers in a sea of multispecies knowers, us with all our specificities turning towards all of theirs?[16]

Notes

1 Stated simply, anthropocentrism is the "[p]rimary or exclusive focus on humanity; the view or belief that humanity is the central or most important element of existence, esp. as opposed to God or the natural world" (*Oxford English Dictionary*). I am particularly inspired by Val Plumwood's

(1996) ecofeminist analysis of anthropocentrism and her description of the harms it enables. For example, according to Plumwood, an anthropocentric conceptual structure entails, among other characteristics, "Radical exclusion: An anthrocentric viewpoint would treat nature as radically other, and humans as hyperseparated from nature and from animals; it treats nature as lacking continuity with the human and stresses the features which make humans different from nature, rather than those they share, as constitutive of human identity" (1996, p. 137–138).

2 Indigenous scholars (Robinson, 2014), ecofeminists (Gruen, 2015), Black feminists (Gumbs, 2020), feminist biologists (Birke, 2009), feminist geographers (Gillespie, 2018), eco-pedagogues (Russell et al., 2021), cognitive ethologists (Bekoff and Pierce, 2009), anthropologists (King, 2013), multispecies ethnographers (Hunold and Lloro-Bidart, 2022), and others informed by critical epistemologies and methodologies have incorporated animal subjectivities within their research. Despite these significant contributions, the anthropocentrism within certain feminist methodologies persists. Thus, my argument is that the repeated absence and seeming irrelevance of nonhuman animals to numerous feminist Western methodologies requires remediation.

3 Some may question my focus on nonhuman animal subjectivities rather than more broadly attend to relations that include, but are entangled beyond, humans and animals, such as plants and other non-living entities comprising dynamic assemblages and networks. I appreciate such approaches, but I want to emphasize the specificities of animal subjectivities and make a claim for their unique inclusion within feminist methodologies.

4 Mendez (2015) elucidates, "María Lugones suggests that gender was necessarily (re)constituted through the practices and processes of slavery and colonization. Her analysis begins from an attempt to historicize gender in order to emphasize the relational process through which it becomes racialized and a marker of humanity for colonizers" (p. 42).

5 For example, during a discussion with a friend and fellow feminist researcher about this chapter, and my argument that feminist methodologists ought to abandon their anthropocentrism, she bristled, "However, you don't want to make it seem that you're comparing humans to animals." Her comment gestures to painful histories and ongoing denial of many people's subjectivity and humanity, as well as the Eurocentric underpinnings that produce such disavowals. Placing animal subjectivity in proximity to human subjectivity, even while not drawing comparisons, may precipitate anxieties and evoke rebuke. My friend's comment underscores some of the potential difficulty of acknowledging animals as subjects, with their own rich experiences. Emphasizing animal subjectivities inevitably suggests continuities, even if one avoids analogies that directly compare the oppression of humans and animals.

6 Consequently, keeping with feminists' longstanding dedication to examine power relations, this chapter serves as an open invitation to critically reflect on the human exceptionalism still reproduced within some feminist methodologies, rather than a cudgel that demands all feminists—regardless of their own positionality—must bring animal subjectivities into their work. Instead, I aim to show how conventional feminist methodologies offer insights that if more consistently applied and less unencumbered by anthropocentrism, could prompt a reckoning with animal subjectivities and offer a productive methodological guide for researching human–animal relations. My aim is not to castigate anthropocentric Western feminist methodologies but to explore their current limitations in generative ways while furthering their promise.

7 Similar anthropocentric orientations are abundant throughout the literature on feminist methodologies. In Joey Sprague's (2018) "Feminist Epistemology, Feminist Methodology, and the Study of Gender," she asserts, "The very need to ensure that research subjects have voice, are taken seriously as analysts of their lives, is the outcome of social power. People need to claim that they can speak with authority only when they are silenced; part of being privileged is being able to assume that one has authority." Similarly, in earlier works, such as Caroline Ramazanoğlu and Janet Holland's (2002), "Researching 'Others': Feminist Methodology and the Politics of Difference," these authors reproduce an implicit anthropocentrism: "The complexities of difference position people in different areas of expertise on power relations, but this does not ensure that they interpret or express sameness or difference in the same way. The nature of 'otherness', as standpoint theorists argue, is potentially most firmly grasped by those with daily experiences of subordination and exclusion, but much depends on political consciousness."

8 The Canadian Council on Animal Care (CCAC) guidelines sole mention of consent appears in the "CCAC Guidelines: Identification of Scientific Endpoints, Humane Intervention Points, and

Cumulative Endpoints," which includes the treatment of "commercial animals used in production settings" (p. 27). Under the section, "Humane Intervention Points for Third-Party Owned Animals (e.g., Commercial Beef Cattle)", these guidelines urge, "Increasingly, scientific work in Canada is being conducted with commercial animals in production settings. This presents unique challenges in establishing and overseeing humane intervention points. . . . All aspects of the scientific activity must be conducted with the *consent of the owner of the animals*" (emphasis added, CCAC, 2022, pp. 27–28). The larger, "CCAC Guidelines On: The Care and Use of Farm Animals in Research, Teaching and Testing" (2009) 135-page document similarly does not include any reference to consent.

9 Not all humans equally share this power, as histories of racist and ableist medical experiments attest, in conjunction with medical racism (Hoberman, 2012) and medical ableism (Janz, 2019) still perpetuated today.

10 Likewise, in the Government of Canada's *Canadian Biosafety Handbook* (Second Edition) (Public Health Agency of Canada, 2015), "Chapter 13: Animal Work Considerations," the "Handling and Restraint" section advises, "The use of proper handling and restraint techniques helps prevent injury to the handler, reducing the potential for a bite, scratch, or other exposure to infectious material, and therefore protecting against secondary transmission in the community." These cautions suggest that some animals will fight back within laboratory conditions, and they do not passively submit to their treatment. The handbook provides national guidance for the "safe handling and storing of human and terrestrial animal pathogens and toxins in Canada" (Public Health Agency of Canada, 2015).

11 See Kathryn Gillespie's (2014) article, "Sexualized Violence and the Gendered Commodification of the Animal Body in the Pacific Northwest U.S. Dairy Production," for a fuller description of sexualized and gendered harms inflicted on animals in agriculture.

12 Feminist methodologies are diverse, and involve complicated epistemological debates related to what Sandra Harding characterizes as feminist empiricism, standpoint epistemologies, and transitional (postmodern) epistemologies, tributaries that are increasingly blurred (Doucet and Mauthner, 2006). My chapter does not attempt to comprehensively map the field of feminist methodologies in this or other regards but instead aims to introduce key tenets and place them in conversation with animal subjectivities. My descriptions of feminist methodologies are necessarily simplifications.

13 Gail Eisnitz's (1997) similarly includes the testimony of a slaughterhouse worker, Vladak, who also employs such speciesist and misogynist language: "There was one night I'll never forget as long as I live. A little female hog was coming through the chutes. She got away and the supervisor said, 'Stick that bitch!' I grabbed her and flipped her over. She looked up at me. It was like she was saying, "Yeah, I know it's your job, do it.' That was the first time I ever looked into a live hog's eyes. And I stuck her" (p. 74).

14 See, also, Leesa Fawcett and Morgan Johnson's (2022) excellent, "Oceanic, Multispecies, Resilient Resistance: Whales, Noise Pollution, and Tiny House Warriors."

15 Colling (2021) offers a memorable synopsis of nonhuman animal resistance explored in her book: "When monkeys unlatch locks and escape from laboratory cages, when pigs refuse to advance down the chut to the slaughterhouse, when cows fight back against those who steal their children, when salmon struggle for breath when pulled from the water, and when elephants attack those who have killed their family members or encroached on their lands, their actions speak loudly" (ix).

16 We might also ask: How might this time in which orcas are loudly declaring their agency inspire a feminist methodological desire to get curious and reflexive about how humans can—brazenly and uncritically—represent animals and their behaviour, or claim a self-referential species monopoly on knowing? How might we play with and critique the ways orcas are enlisted in our stories of survival and resistance? How might orcas' dazzling collaborations against rudders and yachts swell our solidarity against capitalism, but also—not unrelated—the conditions of seas saturated by racist and colonial histories, and the hazards they currently present to life within and on those waters? Can we go deeper to explore how orcas socially learn and organize their societies (and understand themselves), and perhaps as a result, subjectively change how we know ourselves and the world? How might these understandings be informed by Indigenous human and animal knowledges and understandings of relations across species?

Bibliography

Bartal, I.B.A., Decety, J., and Mason, P. "Empathy and Pro-social Behavior in Rats." *Science* 334, no. 6061 (2011): 1427–1430. https://doi.org/10.1126/science.1210789
Bekoff, M. *The Emotional Lives of Animals: A Leading Scientist Explores Animal Joy, Sorrow, and Empathy—and Why They Matter*. New World Library, 2007.
Bekoff, M., and Pierce, J. *Wild Justice: The Moral Lives of Animals*. University of Chicago Press, 2009.
Belcourt, B. "Animal Bodies, Colonial Subjects: (Re)locating Animality in Decolonial Thought." *Societies* 5, no. 1 (2015): 1–11. https://doi.org/10.3390/soc5010001
Birke, L., and Brandt, K. "Mutual Corporeality: Gender and Human/Horse Relationships." *Women's Studies International Forum* 32, no. 3 (2009): 189–197. https://doi.org/10.1016/j.wsif.2009.05.015
Bradshaw, G.A., Schore, A.N., Brown, J.L., Poole, J.H., and Moss, C.J. "Elephant Breakdown." *Nature* 433, no. 7028 (2005): 807–807. https://doi.org/10.1038/433807a
Canadian Council on Animal Care. "CCAC Guidelines on: The Care and Use of Farm Animals in Research, Teaching and Testing." *Canadian Council on Animal Care*, 2009. Retrieved from: https://ccac.ca/Documents/Standards/Guidelines/Farm_Animals.pdf
Canadian Council on Animal Care. "CCAC Guidelines: Identification of Scientific Endpoints, Humane Intervention Points, and Cumulative Endpoints." *Canadian Council on Animal Care*, 2022a. Retrieved from: https://ccac.ca/Documents/Standards/Guidelines/CCAC_guidelines_scientific_endpoints.pdf
Canadian Council on Animal Care. "Guide to the Care and Use of Experimental Animals." *Canadian Council on Animal Care*, 2022b. Retrieved from: https://ccac.ca/Documents/Standards/Guidelines/Experimental_Animals_Vol1.pdf
Colling, S. *Animal Resistance in the Global Capitalist Era*. Michigan State University Press, 2021.
Corman, L. "Ideological Monkey Wrenching: Nonhuman Animal Politics Beyond Suffering." In *Animal Oppression and Capitalism: The Oppressive and Destructive Role of Capitalism,* edited by D. Nibert, 252–269. Praeger, 2017.
Coulter, K. *Animals, Work, and the Promise of Interspecies Solidarity*. Palgrave Macmillan, 2016.
Doucet, A., and Mauthner, N. "Feminist Methodologies and Epistemologies." In *Handbook of 21st Century Sociology*, edited by C.D. Bryant and D.L Pecks, 36–42. SAGE Publications Ltd., 2006.
Dunayer, J. *Animal Equality: Language and Liberation*. Ryce Pub, 2001.
Fawcett, L.K., and Johnson, M. "Oceanic, Multispecies, Resilient Resistance: Whales, Noise Pollution, and Tiny House Warriors." *Resilience: A Journal of Environmental Humanities* 9, no. 3 (2022): 111–130. https://doi.org/10.1353/res.2022.0013
Fraser, H., Taylor, N., and Morley, C. "Critical Social Work and Cross-species care: An Intersectional Perspective on Ethics, Principles and Practices." In *Critical Ethics of Care in Social Work: Transforming the Politics and Practices of Caring,* edited by B. Pease, A. Vreugdenhil, and S. Stanford, 229–240. Routledge, 2018.
Gillespie, K. "Sexualized Violence and the Gendered Commodification of the Animal Body in Pacific Northwest U.S. Dairy Production." *Gender, Place and Culture: A Journal of Feminist Geography* 21, no. 10 (2014): 1321–1337. https://doi.org/10.1080/0966369X.2013.832665
Gillespie, K. *The Cow with Ear Tag #1389*. The University of Chicago Press, 2018.
Golombisky, K. "Feminist Methodology." In *Communication Research Methods in Postmodern Culture: A Revisionist Approach,* edited by L.Z. Leslie, 172–195. Routledge, 2018. https://doi.org/10.4324/9781315231730-11
Gosine, A. *Nature's Wild: Love, Sex, and Law in the Caribbean*. Duke University Press, 2021.
Government of Canada. "Tri-council Policy Statement: Ethical Conduct for Research Involving Humans." *Government of Canada,* 2018. Retrieved from: https://ethics.gc.ca/eng/documents/tcps2-2018-en-interactive-final.pdf
Gruen, L. *Entangled Empathy: An Alternative Ethic for Our Relationships with Animals*. Lantern Books, Division of Booklight Inc, 2015.
Gumbs, A.P. *Undrowned: Black Feminist Lessons from Marine Mammals*. AK Press, 2020.
Haraway, D. *When Species Meet*. University of Minneapolis Press, 2008.
Hoberman, J.M. *Black and Blue: The Origins and Consequences of Medical Racism*. University of California Press, 2012. https://doi.org/10.1525/9780520951846

Hunold, C., and Lloro, T. "There Goes the Neighborhood: Urban Coyotes and the Politics of Wildlife." *Journal of Urban Affairs* 44, no. 2 (2022): 156–173. https://doi.org/10.1080/07352166.2019.1680243

Ives, S. "'More-than-human' and 'Less-than-human': Race, Botany, and the Challenge of Multispecies Ethnography." *Catalyst* 5, no. 2 (2019). https://doi.org/10.28968/cftt.v5i2.32835

Janz, H. "Ableism: The Undiagnosed Malady Afflicting Medicine." *Canadian Medical Association Journal* 191, no. 17 (2019). https://doi.org/10.1503/cmaj.180903

Kim, C.J. *Dangerous Crossings: Race, Species, and Nature in a Multicultural Age*. Cambridge University Press, 2015.

King, B.J. *How Animals Grieve*. The University of Chicago Press, 2013.

Krystalli, R. "Feminist Methodology." In *Gender Matters in Global Politics,* edited by L.J. Shepherd and C. Hamilton, 34–46. Routledge, 2022.

MacNair, R. "Causing Trauma as a Form of Trauma." *Peace and Conflict, Journal of Peace Psychology* 21, no. 3 (2015): 313–321.

Mendez, X. "Notes toward a Decolonial Feminist Methodology: Revisiting the Race/Gender Matrix." *Trans-scripts* 5 (2015): 41–56.

Mohanty, C. "Under Western Eyes: Feminist Scholarship and Colonial Discourses." *Feminist Review* 30, no. 1 (1988): 61–88. https://doi.org/10.1057/fr.1988.42

Mornement, K. "Animals as Companions." In *Animals and Human Society,* edited by C.G. Scanes and S.R. Toukhasti, 281–304. Elsevier Science & Technology, 2018. https://doi.org/10.1016/B978-0-12-805247-1.00018-6

Oakley, J., Watson, G., Russell, C., Cutter-Mackenzie, A., Fawcett, L., Kuhl, G., Russell, J., van der Waal, M., and Warkentin, T. "Animal Encounters in Environmental Education Research: Responding to the "Question of the Animal."" *Canadian Journal of Environmental Education* 15 (2010): 86–102.

Plumwood, V. "Androcentrism and Anthrocentrism: Parallels and Politics." *Ethics and the Environment* 1, no. 2 (1996): 119–152.

Public Health Agency of Canada. *Canadian Biosafety Standard for Facilities Handling or Storing Human and Terrestrial Animal Pathogens and Toxins* (2015). Retrieved from: https://www.canada.ca/en/public-health/services/canadian-biosafety-standards-guidelines/handbook-second-edition.html

Ramazanoğlu, C., and Holland, J. "Researching 'others': Feminist Methodology and the Politics of Difference." *SAGE Publications Ltd.* (2002). https://doi.org/10.4135/9781849209144

Robinson, M. "Animal Personhood in Mi'kmaq Perspective." *Societie* 4, no. 4 (2014): 672–688. https://doi.org/10.3390/soc4040672

Smuts, B. "Between Species: Science and Subjectivity." *Configurations* 14, no. 1–2 (Winter–Spring 2006): 115–126.

Smuts, B. "Embodied Communication in Nonhuman Animals." In *Human Development in the Twenty-first Century: Visionary Ideas from Systems Scientists,* edited by A. Fogel, B. King, and S. Shanker, 136–146. Cambridge University Press, 2008.

Spannring, R. "Animals in Environmental Education Research." *Environmental Education Research* 23, no. 1 (2017): 63–74. https://doi.org/10.1080/13504622.2016.1188058

Struthers-Montford, K., and Wotherspoon, T. "The Contagion of Slow Violence: The Slaughterhouse and COVID-19." *Animal Studies Journal* 10, no. 1 (2021): 80–113. https://doi.org/10.14453/asj.v10i1.6

Taylor, S. "Animal Crips." *Journal of Critical Animal Studies* 12, no. 2 (2014): 95–117.

Taylor, S. *Beasts of Burden: Animal and Disability Liberation*. New Press, 2017.

Yong, E. *An Immense World: How Animal Senses Reveal the Hidden Realms Around Us*. Alred A. Knoff Canada, 2022.

20
UNRULY FACES IN SUBURBAN PLACES
A Practice in Multispecies Autoethnography

Melissa Plisic

This chapter is about me and my dog, Točka, and the things that have been elevating our mutual cortisol levels since we entered each other's lives on the winter solstice of 2021. If we are talking about North American "pet" culture, which we are, Točka and I fit the stereotype of matching each other's natures. Both of us come in too hot with intraspecies introductions, are on a prescribed SSRI, and are ethnically/breed ambiguous, and both of our disorderly presences in suburban nature cause anxiety for arbiters of the domestic[1]. As a queer guardian,[2] my and Točka's kinship practices reject the cisheteropatriarchal settler imperative to reproduce the nuclear family.[3] Being in close relation with Točka has attuned me to some of the routinized, everyday ways in which settler nation-building operates.

This chapter is an assemblage of critical animal studies, queer ecologies, feminist theory, and post-/anti-/de-colonial scholarship. It is anti-colonial in that I confront "the maintenance of settler state sovereignty and settler futurity"[4] by positioning kinship relations[5] and suburban parks as institutions where coloniality manifests. The affairs of kinship are gendered in a society that has constructed a sphere of domesticity for women to be tethered to, and which works to disappear possibilities for kinship outside the unit of married, monogamous, cisgendered, and heterosexual *homo sapiens*.

Because of domesticity's associations with the feminine, anything within its grasp can be leveraged as in need of "protection" from an internal and/or external "wildness" to benefit patriarchal authority. This is the logic that settler colonial land theft operates on, that any lands or waters without anthropocentric infrastructure or private property relations were an empty wilderness, *terra nullius*.[6] Colonial and capitalist logics that deny our dependence on the world around us, bolstered by heteropatriarchal powers that elevate the rational and detached masculine over the feeling and earthly feminine,[7] are behind our present mass extinction event.[8] This chapter encourages a mode of dog–human relationality that fails to uphold masculinist, capitalist, colonial, punitive expectations of mainstream North American dog culture—this is the tall task of our multispecies autoethnography.

Similarly to how Kathryn Gillespie uses a sketch of her backyard chickens to illustrate how multispecies autoethnography can be done in an anti-anthropocentric way, my goal here is to present several vignettes of Točka and myself that "offer a richer understanding and fuller picture of the ways that animals and animality are embedded in and sustain

DOI: 10.4324/9781003273400-25

processes of racialization and settler colonialism".[9] While canines have historically (and continue to) play important roles in Indigenous nations and communities,[10] colonial use of these creatures has been central to settler claims to sovereignty. For example, in the context of settler livestock farming, "the faithful farm dog" is constructed in opposition to "ruthless" native canines.[11] Gillespie's methods foreground knowledge that is manifested though the steadfast and everyday work of relating with another being, which unfolds in particular geographies and must not be rushed.[12] With care at the heart of Gillespie's methods, they might usher in "the possibility to transform social relations and structures of power and oppression".[13] Under the influence of Gillespie, I orient myself and Točka to the ways in which settler colonial cisheteropatriarchy and white supremacy operate in our lives.

Two Heads Are Better Than One

As an anti-colonial, feminist, and queer social justice scholar working in the mode of autoethnography, I should probably introduce us. I am a queer, second-generation Croatian/Filipinx immigrant-settler in colonial "British Columbia". I grew up on, and currently occupy, the unceded, active lands of the Musqueam, Squamish, and Tsleil-Waututh nations, colloquially/colonially known as "North Vancouver". I am (for the most part) able-bodied and neurotypical. I come from a relatively privileged socioeconomic class. I belong to the species called *Homo sapiens*.

Točka is a small/mediumish, mostly black and brown dog with pointy ears and a tail that curls up over her rear. She has brown eyes and a few dark spots on her tongue. She stands at knee-ish height relative to an adult human. She has been spayed. She belongs to the species called *Canis familiaris*.

Točka is a rescue dog, so the first 10 months of her life are a mystery to me, other than that she was a street dog in Tijuana. When Točka first arrived in British Columbia she went to a foster home with "a lot" of other dogs, then she went to a second home—a single mother and her two young kids whose previous family dog had died, who had intended to foster-to-adopt her. Needless to say, it didn't work out.

Just as my positionality informs where I know from, Točka's informs hers. During my juvenile years, I was trained into the social codes for navigating (sub)urban life. Točka, presumably, did whatever she wanted. She was not "properly socialized", meaning that she did not get assimilated into anthropocentric rules during her developmental stages of life. This is one, likely of a few, reasons that she is what the dog experts call "reactive", meaning that she "over-reacts" to "normal" situations by barking and lunging.[14] Točka's reactions are sometimes motivated by fear, by frustration, or both. Not all rescue dogs are reactive, and not all reactive dogs are rescues; it does happen to be the case with Točka. In North American pet culture, "normalcy" is measured by how tolerant to human wills a dog is. Because of her apprehension to anthropocentric norms, Točka forces us into an anti-anthropocentric way of relating.

I approach this multispecies autoethnography inspired by Leanne Betasamosake Simpson's story of Nanabush as a researcher who imbues "reciprocal recognition", which encompasses "obligations and responsibilities within a nest of diversity, freedom, consent, noninterference, and a generated, proportional, emergent reciprocity".[15] Nanabush goes by himself on his first journey around the world, but on his second he is accompanied by a wolf, Ma'iingan, who gives all the "data" new and deeper meanings.[16] Simpson calls this Nanabush's "methodology",[17] and I strive to embody something similar with Točka.

I don't think Simpson was trying to make a point about literally bringing a canine along as a research assistant, but the way she describes reciprocal recognition is important. Having a relationship based on non-interference seems impossible for Točka and me—at times, it feels like it's all interference. I don't have a solution to the sheer abundance of dogs in Mexico. There are many authors of this interference, including Točka's ancestors (both native and introduced dog populations), community members who care for free-roaming dogs, state actors who look for quick-fix population control (read: culls), local dog shelters and veterinary clinics, people who abandon their un-spayed or un-neutered dogs, humans who may act on their disdain for so-called strays,[18] and rescue organizations and animal advocates in the so-called Global North. The pursuit of non-interference must be balanced with dependency; as Sunaura Taylor asks from critical disability and animal studies, "Does an animal's dependence on human care have to be understood as inevitably negative or exploitative?"[19] Točka depends on me to keep her fed, hydrated, exercised, entertained, and safe. The "proportional" part of Simpson's reciprocal recognition is key here: if I were to retract all my interference with Točka, so that she had similar freedoms to what she had in Tijuana, she would wreak havoc on our neighbourhood, in turn causing more interference for the rest of our multispecies kin.

Perhaps I sound like a total hypocrite for advocating for the agencies of dogs while wielding a leash, but I hope that what I'm doing with it is chucking a wrench into the norms of dog culture. Leaving dog shit around, letting dogs off leash when they are not supposed to be,[20] letting dogs trample ecologically sensitive habitats,[21] and buying new toys rather than mending old ones are all examples of how capitalist-colonial power relations saturate human–dog kin relations. My use of exclusively force free training[22] with Točka fails to uphold proper "interspecies intimacy and domesticity" which is measured by "sexist, heteronormative, and speciesist ideals".[23] In the cisheteronormative, patriarchal, anthropocentric ideal, Točka would not be allowed on my furniture, *especially* not the bed, which would be reserved for monogamous cishetero reproductive canoodling;[24] I would not share my "human" food with her; and as the "pack leader", I would "correct" (read: punish) her for her "mistakes".

Force free training is exactly what it sounds like: I do my best not to force or coerce Točka into doing anything, I don't punish her for anything, and we advance our training at *her* pace, not mine. On our walks, Točka is tethered to me by a 6-, or ideally, 15-foot leash, so the least I can do to proportionally reciprocate is to let her sniff things for as long as she wants. This can take several minutes, which isn't *that* long but can feel like an eternity when we are otherwise imbricated by capitalism's go-go-go culture. This is contrary to many "normal" walks we witness, where dogs get yanked by the leash, maybe even jabbed in the neck by some prongs for leaving a "heel", before they've barely had a whiff of their olfactory feast. The interference in this vignette becomes exponential when I fail to recognize Točka's desires: the less I let her sniff, the more she will try to pull me around to sniff things. Thinking with Točka holds the potential to disrupt these norms, to "open up the mind to an unreason that is far more thoughtful and provocative in terms of empathy".[25] Once I get off my high horse and walk on Točka's level of smelly sidewalk treasures, then we're flying.

What Is Kin?

The greatest scholarly influence on my understanding of kinship is Kim TallBear. As a queer and Indigenous scholar, she so clearly articulates how settler colonial kinship-making practices sustain anthropocentric monogamous cisheteropatriarchy while destroying Indigenous kinship constellations. As she explains:

> [lifelong monogamous marriage] was propped up by Christian moral arguments coupled with state structural enforcements—the linking of marriage to property rights and notions of good citizenship. . . . Growing the white population through biologically reproductive heterosexual marriage—in addition to encouraging immigration from some places and not others—was crucial to settler-colonial nation-building.[26]

Every line of this excerpt implicates me. I was fed those moral arguments while being raised Catholic, I have witnessed the messy scramble for property in the wake of divorce, and I am a product of a heterosexual marriage—both sides of which were welcomed to Canada when it was convenient to have their labour here. My family has been assimilated into the fold of the settler state, which "has been very poor kin indeed".[27] "Canada", in our case, not only forces human beings into violent, patriarchal expectations, it also attempts to prevent us from enacting reciprocity with nonhuman beings;[28] Indigenous peoples are severed from their lands and waters and the many beings therein, and settlers are actively deterred from realizing non-extractive modes of living with our multispecies (and beyond) neighbours.[29] By writing this chapter, I am hoping to take up TallBear's imperative for settlers to "take on the obligations of kinship" where we must actively be anti-anthropocentric and resist our own assimilation into the scaffolding of colonialism.[30]

Queer people do not get an automatic pass for their kinship-making by virtue of their attraction or gender orientations. It is easy enough for a queer family to replicate monogamous, cisheteropatriarchal structures with the simple substitution for two moms, two dads, or even two gender non-conforming parents to head the household.[31] To quote TallBear again, "If pronatalism involves reproducing the middle-class settler family structure, *no matter* the race or sexual orientation of the middle-class family, I lament it".[32] It would be easy to swap human children with companion animals too. The discourse of "dinks"[33] that gets liberally applied to couples with companion animals still assumes a lifelong monogamous marriage in a capitalist society. The desire to get a dog when they are 6–8 weeks old with no previous "owners" to have them for their entire lives replicates the trajectory of having sapien offspring, albeit on a scale approximately seven times shorter. Adopting a "pet", being queer, or being racialized does not bypass the responsibilities of resisting settler colonialism in all its forms.

Now for the obligatory Donna Haraway part of the chapter. She and Ms. Cayenne Pepper were precedent setting for queer and feminist human–dog relating with their *Companion Species Manifesto*. Haraway balances sympathy for positive reinforcement training with "[rejoicing] in [Hearne's] alpha roll of animal rights ideologies".[34] I too take issue when Animal Rights à la Peter Singer attempts to subsume nonhumans into "the rights-bearing, humanist subjects of Western philosophy",[35] but to critique it using Hearne, who was inspired by Thomas Jefferson, a white man who owned Black people as property, and while using long-discredited and toxic masculinist "alpha" discourse, even if just for poetics, does not compute in my anti-colonial, feminist, and queer brain. I appreciate Haraway's appreciation for Hearne's commitment to training for "concrete beings, not . . . categorial abstractions";[36] I too am for catering to the uniqueness of individual dogs, but I will bite the hand that feeds it to me on a platter of human and white supremacist ideals.

I also diverge from Haraway when it comes to breeding, which remains unquestioned in her *Companion Species Manifesto*. Despite arguments in favour of "ethical" breeding, I am always left puzzled at how a practice that hinges on the biopolitical control over sexual reproduction could be called ethical. I am reminded of Alice Walker's *Am I Blue?* and Vasile Stanescu's discussion of it, where the point is that within the confines of breeding (read: sexualized violence), nonhuman animals are denied their kin in service of sapien desires.[37] Later in her career, in *Staying with the Trouble*, Haraway proposes that humanity adopt the slogan "Make Kin Not Babies" while we witness Earth's sixth mass extinction.[38] As Haraway explains, her "purpose is to make 'kin' mean something other/more than entities tied by ancestry or genealogy".[39] This is something that Indigenous societies have known forever and is knowledge that the settler state actively targets for destruction.[40] So Haraway's verb choice of "make" in the latter quote doesn't sit so easy. Even when we are making our multispecies kin, it would do us well to remember that "some of our kin are born to us and some of them come to us in other ways".[41] My contention is that the kin-versus-babies discussion must extend beyond humans; instead of making kin, perhaps we should be finding them.

Breeding dogs raises an ethical question when there are innumerable dogs in shelters and pounds, in line to be killed simply because of a lack of space. The widespread "adopt don't shop" rescue slogan references these animals and more without access to dependable shelter, safety, and security. These are dogs whose origins fall outside the calculations of white, middle-upper-class desires. Obviously, the adopters are the crowd that I most like to party with. But this discourse is not without its faults. For example, it assumes a trajectory into a human family structured by private property relations. I am certain that had I been searching for a dog to adopt with a monogamous cishetero-appearing partner I lived with on a property with a 6 foot fence, I would have had a dog much sooner. Further, rescuing dogs from outside of the West can be weaponized into human saviourism, buttressed by white supremacist culture; white sub/urban animal-eating nuclear families saving dogs from the "Chinese meat trade" are a particular thorn in my Asian side.[42] Now, back to other troubles.

Since *Staying with the Trouble* was published in 2016, reproductive healthcare has continued to be the target of conservative, cisheteropatriarchal, white supremacist forces; the overturning of *Roe v. Wade* in the United States in June 2022 both maintains and sets the precedent for the tactical erosion of reproductive healthcare globally. As someone in their mid-20s, whose social circle is mostly white, university-educated, and environmentally conscious (or anxious) queer people, I feel a lot of existential ambivalence about baby-making. The decision to create human kin here, with a trajectory towards several decades of capitalist hyper-consumption, ought to be handled delicately. Telling anyone what they should or should not do with their bodies would defeat the purpose of this anti-colonial, feminist, and queer chapter; some questions at the intersection of reproductive and climate justice are: How else can we have families? Who gets to have abortions? Who will be blamed for over-population? And, as TallBear asks, "Whose relatives, including other-than-humans, will thrive and whose will be laid to waste?"[43]

Over-population discourses saturate hegemonic orientations towards climate change, where "the planet's demise is linked to the immoral ravishes of the polluting underclass".[44] In the context of xenophobic overpopulation discourses, Andil Gosine differentiates between " 'good' ecological citizens", who are bastions of white heterosexual reproduction, and "unruly bodies", who with their queer and/or non-whiteness threaten a domesticated wilderness with their mal-reproduction.[45] The more unruly a body is, the more disciplinary anxiety it elicits.

The dog population that I am invested in problematizing is not Točka's unruly kin, but the sheer number of -oodles, French bulldogs, pugs, golden retrievers, and whatever breeds are fashionable, some of whom (Frenchies) cannot even reproduce without human interference. These designer/"pure" bred dogs are most notably markers of class but also a normative family structure and accompanying citizenship and private property relations; these are the kind of dogs who belong to "good ecological citizens".

Dogs like Točka, on the other hand, fall into the realm of "unruly bodies". Her genetics escape norms set by the American Kennel Club, and when people learn that she is from Mexico, xenophobic discourses about the supposed unruliness of Mexican sapiens seep into conversations. The stereotype that rescue dogs come with behavioural issues is a major deterrent for people in search of canine kinship; the possibility of unruliness is enough to turn potential adopters into shoppers. The anxiety of being responsible for a nonhuman that upsets the manicured norms of settler sub/urban spaces is enough to sway many humans into the commodification of sexual reproduction.

A Walk in the Park

Točka and I live in the suburbs that I grew up in. In fact, we live in my childhood home, which I have reluctantly returned to for grad school. This ironically means that I have fallen back into a quasi-nuclear family structure living with my adult sibling, mom, and dad, whom I deeply depend on for help with caring for Točka. The suburbs are the critical meeting point of "nature" and "civilization" in the settler imaginary. The North Shore has heavy cultural associations with outdoor recreation; this is in part due to its geographical positioning nested against the coastal mountains, and the historical context of British Columbia as a site of settler masculinity-making through outdoorsmanship.[46]

British Columbia, as a phenomenon that anchors settler colonialism on the west coast, champions the motto "super, natural, British Columbia". The colonial settlement of this province was part of a project to invent "a bourgeois wilderness—that is, a place where middle class men from the Pacific Northwest could indulge themselves in what [big game hunting] historically had been an elite activity",[47] thus using the land to construct themselves as white, cishetero, class-privileged citizens of a new frontier. These are Gosine's "good ecological citizens". These days, big game populations have been hunted to depletion or gentrified deeper into the mountains, but the meta-narrative of outdoor recreation remains. Parks exist as spaces where settlers construct themselves as cishetero, class-privileged citizens of nature's playground.

Whenever Točka and I visit a park during daylight hours, literally, we come into conflict with good ecological citizenry. In the introduction to *Queer Ecologies*, Sandilands and Erickson explain that:

> the idea of park-nature as a space for the disciplined cultivation of virtue has an important sexual component . . . parks were places to see and be seen; they were sites for public spectacle of a particular kind, including the conspicuous display of middle-class respectability and wealth. Parks were places for public cultivation of morally upstanding citizens.[48]

At the park, dogs themselves become the objects through which virtuous suburban citizenship is measured—morally upstanding people have morally upstanding dogs. Dogs are expected to act in ways that serve human interests, even when the humans are supposedly taking their dogs to these spaces for the dog's sake. When dogs begin to show natural dog behaviours including barking, sniffing, growling, lunging, chasing, and humping, they become targets of discipline. Sandilands and Erickson continue to explain that "public parks are *disciplinary* spaces, in which a very narrow band of activities is sanctioned, practiced, and experienced; only certain kinds of nature experience are officially allowed".[49] The general expectations of dogs in the suburbs are to play gently, ignore all distractions, walk in a heel, not display any sexuality or sexual impulses, and get along with everyone else (sapien or canine). As I have learned through doing months of training with Točka,[50] these expectations are incredibly anthropocentric, as they vilify natural dog behaviours.

As a reactive dog, Točka's unruliness travels up the leash to me, emphasising my brownness in spaces that I have otherwise navigated as white. After one particularly dramatic episode witnessed by a white couple on a trail less than 5 kilometres away from home:

White Woman:	*Are you new around here?*
Me:	*No.*
WW:	*Well, my husband and I come here every day, and I've never seen you before.*

Here, a white woman is positioning me and Točka as outsiders to her pristine park-nature, where she engages in taking respectable walks[51] with her respectable husband. Moments before Točka's reaction, this woman was trying to make conversation with me while I was warning her that my dog needs more space than most others. After failing to uphold the virtuous, docile, "normal", human-centred rules for engagement, my racial ambiguity and Točka's unruliness slide us into the terrain of "not belonging". I don't deny that white women have to deal with the woes of patriarchy like the rest of us, but white supremacy relies on us/them to maintain racial subjugation, and we/they are very good at doing so in domesticated nature spaces.

In our noteworthy interactions with men, they seem to be less overtly invested in policing us and more interested in stereotyping Točka or mansplaining dogs to me. When she wears her muzzle out in public, even when Točka is acting "normally", men make comments like, "Does he bite?" or "Oh, he's a bit aggressive isn't he?" The tight association of masculinity with aggression is laid plain here in their pronoun choice. The preoccupation with whether Točka will bite signifies how attention is paid to an outburst of emotional conflict rather than on the conditions that make her feel like there is no other option but to escalate her communication method to a bite. Why does this remind me of so many emotionally unfulfilling relationships that people have with men? What if instead of "Does he bite?", the question was, "How can I make your dog feel safer"? Or what if men simply left young femme-presenting people alone in public?

In my experience, it has been almost exclusively men who have, mostly without solicitation, recommended training techniques that pinch, shock, choke, dominate, or intimidate dogs; the most common attacks towards positive reinforcement training are steeped in misogyny. The dismissal of evidence-based training methods because they are "too soft" works by asking humans to ignore our co-evolved instinct, a *feeling*, to emotionally attune to our familiars, and tells us that it is normal to use violence to get what we want.[52] Reactive dogs like Točka, who do not so easily comply with anthropocentric norms, become a more urgent target for discipline and punishment. Western cisheteropatriarchal culture teaches boys (well, all of us) that the bodily autonomy and sovereignty of women, animals, and the broader natural world should be brought into control through violence. The feminization of beings under patriarchal colonial-capitalist rule authorizes extractive relationships that are achieved through domination.[53] The discourses of control, discipline, punishment, and being "alpha" reverberate across the stolen lands and waters on which my and Točka's walks unfold.

The promise of quickly fixing a dog's behaviour with aversive methods or through domination is a symptom of colonial-capitalist punitive culture. It takes months, if not years, to change a dog's emotional response to something, as Točka and I are learning. Focusing on achieving results as fast as possible by any means necessary supresses a dog's underlying motivations for "acting out". The dog's welfare is forgotten, reflecting capitalist culture's camaraderie with punishment; if you can't behave in a disciplined way quickly enough, you become the target of violent domination. In all my experiences with force free trainers, they are committed to understanding Točka as an individual who has feelings about her environment, reminding me to attend to how Točka may be feeling on any particular day and that this training only works if I commit to deeply understanding my dog and how she communicates with me.

Woof!

On September 27, 2022, I slept in until 6:45am, and Točka and I left for our walk after sunrise. Lo and behold, as we rounded the corner towards the dog poo bins, two small white dogs. One off leash and one on leash, in a leash-required area. The humans looked a generation or two older than me, South/East Asian, and affluent. I asked them if they could please leash their dog and pause playing fetch while we walked past because Točka's still in training. I will spare you the details, but it ended in them yelling at me that it was my fault that my dog has behavioural challenges and that I should train her (oh, the irony). Their ears were closed by the time I left them with my concluding remarks that they were acting very entitled and a reminder that we all share the space.[54]

Rather than undermine my arguments about whiteness, this vignette is a reminder that settlers aren't only white—anyone can don the cloak of white supremacy; Asians in particular ought to reflect on how we may be maintaining global anti-Black and anti-Indigenous power relations. This is about entitlement to and control over stolen land upon which parks and sidewalks are built. When I am obedient[55] to the settler colonial state, which includes BC Parks, not only am I failing to uphold my responsibilities to the Indigenous nations whose land my ancestors, my kin, settled—I am also acting against my ancestors' wishes. Yes, even the white ones, who did once know how to live in reciprocity with their multispecies kin. My ancestors fought hard against colonial forces, and their flame burns bright in me; the great thing about my kin is that we already know how to be non-nuclear, we just need to remember how.

When I am obedient to the norms of punitive dog training, I fail to uphold my promise to Točka that I will care for her. This is why I much prefer the rowdy crowd of treat-pushing, anti-colonial queer feminists who have the sense to walk slowly and let their dogs sniff for as long as they please. Because Točka is reactive, not only do I have to attune to her on our walks—I must be attuned to the world around us to help her navigate the hostility of settler infrastructure that interfaces with the rest of our ecosystem. As I write this chapter, my agenda falls outside of academia. Even my readers who do not coexist with a dog will encounter canines at some point in their lives, and even beyond the sapien and canine, this chapter holds implications for our other kin: slimy, scaly, feathered, planty, watery, rocky, and otherwise. Kin affairs inherently bring future generations into question; the question I want to leave you with is this: How can we do kinship in a way that is antithetical to the settler nation-state?

Salamat Po and Puno Hvala

Točka and I extend our deepest gratitude to the xʷməθkʷəy̓əm, Sḵwx̱wú7mesh, səlilwətaʔɬ, and other Coast Salish nations, auntie char, Chloë Taylor, Leila Harris, Astrida Neimanis, Tessa and Adeline and Frodo Wotherspoon, the EDGES lab at UBC, Kelly and Jacques Struthers Montford, Harriet and Christine Fedusiak, Jackson and Toast Moore, Bravo Dog Training, Sandy Wei, Nicky and Charlie Wilke, Lola and Barba and Tita Len, our late Lucy, and all our other friends. This chapter would not be possible without you.

Notes

1 Mel Y. Chen, *Animacies: Biopolitics, Racial Mattering, and Queer Affect* (Duke University Press, 2012).
2 Not a parent and not an owner.
3 Kim Tallbear, "Making Love and Relations Beyond Settler Sex and Family," *Making Kin Not Population* (2018): 145–209.
4 Jaskiran Dhillon, "Introduction: Indigenous Resurgence, Decolonization, and Movements for Environmental Justice," *Environment and Society* 9, no. 1 (2018): 1–5. https://doi.org/10.3167/ares.2018.090101, 3.
5 TallBear, "Making Love and Relations Beyond Settler Sex and Family."
6 kQwa'st'not~ charlene george (Mentor), tSouk Nation, Indigenous Teachings, shared throughout shared Coast Salish territories.
7 Karen J. Warren, "The Power and the Promise of Ecological Feminism," *Environmental Ethics* 12, no. 2 (1990): 125–146. https://doi.org/10.5840/enviroethics199012221
8 kQwa'st'not~.
9 Kathryn Gillespie, "For Multispecies Autoethnography," *Environment and Planning. E, Nature and Space (Print)*, no. Journal Article (2021): 251484862110528. https://doi.org/10.1177/25148486211052872, 9.
10 Isaac Murdoch, "Rez Dog Medicine," Save Rez Dogs, retrieved from: https://saverezdogs.com/resources/rezdogmedicine, accessed January 13, 2023; Leah Arcand, "Pre-Colonial Times; A Dog's Role," *Save Rez Dogs,* retrieved from: https://saverezdogs.com/resources/pre-colonialdogroles, accessed January 13, 2023; A Dog's Role."
11 Antoinette Burton and Renisa Mawani, *Animalia: An Anti-Imperial Bestiary of Our Times* (Durham and London: Duke University Press, 2020), 99.
12 Gillespie, "For Multispecies Autoethnography," 3.
13 Ibid., 7.
14 Other dogs may experience reactivity differently.
15 Leanne B. Simpson, *As We Have Always Done: Indigenous Freedom Through Radical Resistance* (Minneapolis and London: University of Minnesota Press, 2017), 182.
16 Ibid., 184.
17 Ibid.
18 What, exactly, have they strayed from?
19 Sunaura Taylor, *Beasts of Burden: Animal and Disability Liberation* (New York: The New Press, 2016), 217.
20 kQwa'st'not~.

21 Which are all habitats, as all ecologies are sensitive; kQwa'st'not~.
22 For the purposes of this chapter, force free and positive reinforcement are interchangeable.
23 Will McKeithen, "Queer Ecologies of Home: Heteronormativity, Speciesism, and the Strange Intimacies of Crazy Cat Ladies," *Gender, Place and Culture: A Journal of Feminist Geography* 24, no. 1 (2017): 122–134. https://doi.org/10.1080/0966369X.2016.1276888, 128.
24 Ibid., 128.
25 Edge Effects, "What Dogs Can Teach Us About Justice: A Conversation with Colin Dayan," Podcast (n.d.), retrieved from: https://open.spotify.com/episode/4qjdox2k2K7AnV2NKnyec4?si=7dcd90e4c6a94b1d
26 "Making Love and Relations Beyond Settler Sex and Family," 145–146.
27 Kim Tallbear, "Failed Settler Kinship, Truth and Reconciliation, and Science," *Indigenous STS* (blog) (March 16, 2016), retrieved from: https://indigenoussts.com/failed-settler-kinship-truth-and-reconciliation-and-science/.
28 Ibid.
29 kQwa'st'not~.
30 TallBear, "Failed Settler Kinship, Truth and Reconciliation, and Science."
31 Jacqui Gabb, "It's Raining Cats, Dogs and Diapers! The Intersections of Rising Pet Ownership and LGBTQ+ Coupledom," *Families, Relationships and Societies* 8, no. 2 (2019): 351–357. https://doi.org/10.1332/204674319X15583480855192.
32 "Making Love and Relations Beyond Settler Sex and Family," 152.
33 Double income, no kids.
34 Donna J. Haraway *The Companion Species Manifesto Dogs, People, and Significant Otherness* (Prickly Paradigm Press, 2003), 48.
35 Ibid., 51.
36 Ibid., 52.
37 Stanescu Vasile, *"Why Loving Animals Is Not Enough: A Feminist Critique,"* (2013), retrieved from: https://www.youtube.com/watch?v=nndAHEgwRmM.
38 Donna J. Haraway, *Staying with the Trouble: Making Kin in the Chthulucene* (Duke University Press, 2016), 102.
39 Ibid., 102–103.
40 TallBear, "Making Love and Relations Beyond Settler Sex and Family"; TallBear, "Failed Settler Kinship, Truth and Reconciliation, and Science"; kQwa'st'not~; Simpson, *As We Have Always Done: Indigenous Freedom Through Radical Resistance*.
41 TallBear, "Making Love and Relations Beyond Settler Sex and Family," 152.
42 I know you may be thinking: "Hey, Melissa, aren't you a vegan? But doesn't your dog eat other animals? What about *that*?" And the answer is yes, since getting Točka I have become more complicit in the animal industrial complex than I was before. I did try to have Točka on a plant-based diet, but she simply wouldn't take. This is a problem that I have yet to resolve, but we are working on finding other possibilities for the relationships between domesticated animals.
43 TallBear, "Making Love and Relations beyond Settler Sex and Family," 147.
44 Andil Gosine, "Non-White Reproduction and Same-Sex Eroticism: Queer Acts against Nature," in *Queer Ecologies*, eds. Catriona Mortimer-Sandilands and Bruce Erickson (Indiana University Press, 2010), 149, https://go.exlibris.link/KwYcSgn1., 160.
45 Ibid., 160.
46 Tina Loo, "Of Moose and Men: Hunting for Masculinities in British Columbia, 1880–1939," *The Western Historical Quarterly* 32, no. 3 (2001): 296. https://doi.org/10.2307/3650737.
47 Ibid., 303.
48 Catriona Sandilands and Bruce Erickson, *Queer Ecologies: Sex, Nature, Politics, Desire. Queer Ecologies: Sex, Nature, Politics, Desire* (Bloomington, IN: Indiana University Press, 2010). https://go.exlibris.link/2t5Svtr6, 18.
49 Ibid., 26.
50 I realize it may sound like I am arguing conflicting sides here. While I am critiquing the idea of "obedience", I am doing so to advocate for more responsible, just, and ethical human–dog relationships. Basically: train your dog, but don't be a jerk about it.
51 Let me be clear that I am not advocating for reckless use of outdoor spaces—this would be antithetical to my anti-colonial position.

52 Warren, "The Power and the Promise of Ecological Feminism."
53 Warren, "The Power and Promise of Ecological Feminism"; kQwa'st'not.
54 kQwa'st'not~.
55 Which is not the same as following concrete rules that are in place for equitable access to spaces. It's more about the ideological entitlement to using land with wanton disregard for others.

Bibliography

Arcand, Leah. "Pre-Colonial Times; A Dog's Role." *Save Rez Dogs*. Retrieved from: https://saverezdogs.com/resources/pre-colonialdogroles.

Burton, Antoinette, and Renisa Mawani. *Animalia: An Anti-Imperial Bestiary of Our Times*. Durham and London: Duke University Press, 2020.

Dhillon, Jaskiran. "Introduction: Indigenous Resurgence, Decolonization, and Movements for Environmental Justice." *Environment and Society* 9, no. 1 (2018): 1–5. https://doi.org/10.3167/ares.2018.090101.

Edge Effects. "What Dogs Can Teach Us About Justice: A Conversation with Colin Dayan." Podcast, n.d. Retrieved from: https://open.spotify.com/episode/4qjdox2k2K7AnV2NKnyec4?si=7dcd90e4c6a94b1d.

Gabb, Jacqui. "It's Raining Cats, Dogs and Diapers! The Intersections of Rising Pet Ownership and LGBTQ+ Coupledom." *Families, Relationships and Societies* 8, no. 2 (2019): 351–357. https://doi.org/10.1332/204674319X15583480855192.

Gillespie, Kathryn. "For Multispecies Autoethnography." *Environment and Planning. E, Nature and Space (Print)*, no. Journal Article (2021): 251484862110528. https://doi.org/10.1177/25148486211052872.

Gosine, Andil. "Non-White Reproduction and Same-Sex Eroticism: Queer Acts against Nature." In *Queer Ecologies*, edited by Catriona Mortimer-Sandilands and Bruce Erickson, 149. Indiana University Press, 2010. https://go.exlibris.link/KwYcSgn1.

Haraway, Donna J. *The Companion Species Manifesto Dogs, People, and Significant Otherness*. Prickly Paradigm Press, 2003.

Haraway, Donna J. *Staying with the Trouble: Making Kin in the Chthulucene*. Duke University Press, 2016.

kQwa'st'not~ charlene george (Mentor), tSouk Nation, Indigenous Teachings, shared throughout shared Coast Salish territories.

Loo, Tina. "Of Moose and Men: Hunting for Masculinities in British Columbia, 1880–1939." *The Western Historical Quarterly* 32, no. 3 (2001): 296. https://doi.org/10.2307/3650737.

McKeithen, Will. "Queer Ecologies of Home: Heteronormativity, Speciesism, and the Strange Intimacies of Crazy Cat Ladies." *Gender, Place and Culture: A Journal of Feminist Geography* 24, no. 1 (2017): 122–134. https://doi.org/10.1080/0966369X.2016.1276888.

Murdoch, Isaac. "Rez Dog Medicine." *Save Rez Dogs*. Retrieved from: https://saverezdogs.com/resources/rezdogmedicine.

Sandilands, Catriona, and Erickson, Bruce. *Queer Ecologies: Sex, Nature, Politics, Desire. Queer Ecologies: Sex, Nature, Politics, Desire*. Bloomington, IN: Indiana University Press, 2010. https://go.exlibris.link/2t5Svtr6.

Simpson, Leanne B. *As We Have Always Done: Indigenous Freedom Through Radical Resistance*. Minneapolis and London: University of Minnesota Press, 2017.

Stanescu, Vasile. "Why Loving Animals Is Not Enough: A Feminist Critique." (2013). Retrieved from: https://www.youtube.com/watch?v=nndAHEgwRmM.

TallBear, Kim. "Failed Settler Kinship, Truth and Reconciliation, and Science." *Indigenous STS* (blog), March 16, 2016. Retrieved from: https://indigenoussts.com/failed-settler-kinship-truth-and-reconciliation-and-science/.

TallBear, Kim. "Making Love and Relations Beyond Settler Sex and Family." *Making Kin Not Population*, 2018, 145–209.

Taylor, Sunaura. *Beasts of Burden: Animal and Disability Liberation*. New York: The New Press, 2016.

Warren, Karen J. "The Power and the Promise of Ecological Feminism." *Environmental Ethics* 12, no. 2 (1990): 125–146. https://doi.org/10.5840/enviroethics199012221.

PART V

Transfeminisms, Women, and Animals

21
ANTISPECIESIST TRANSFEMINISMS IN LATIN AMERICA

Juan José Ponce Léon

> We are the queers, transvestites, degendered, inhuman—dehumanized, feral, and wild. We are irreverent, ungovernable, stateless, marginal and sudacas. . . . We speak from antispeciesist transfeminism, we want to destabilize oppressions, put the debate about species privilege on feminist tables, and the debate about gender privilege on antispeciesist tables.
>
> *(Transanimal Manifesto, Valentina Trujillo and Analú Laferal)*

1. Introduction

This chapter aims to address the theoretical and ethical-political relationships between varieties of feminism and antispeciesist animal advocacy in Latin America. The first of the four sections situates the epistemological field and historical development of feminist animal studies (FAS) in the Global North from the 1980s. The second describes the development of a theoretical and political agenda of antispeciesist feminism, which began to consolidate in Latin America in the second decade of the 21st century. The third addresses the connections between the oppression of feminized bodies and that of animalized bodies in the case of Ecuador, based on the experiences of twelve women activists who embody antispeciesist feminism (AF) through situated practices and discourse. This involves questioning both the sex-gender and species binarisms. It also examines the animal issue from a transgender perspective, problematizing the concept of animality as an escape from gender and as a politics of the abject. Two case studies are used for this, centered on the artistic-performative work of the antispeciesist and transfeminist activists Andrea Alejandro from Ecuador and Analú Laferal from Colombia.

1.1. Feminist Animal Studies

In North America, a first articulation of ecofeminism can be identified in 1980s.[1] These resulted from the convergence of feminist, anti-nuclear, and peace movements.[2] Ecofeminism's social base developed out of discussions around empathy and the disposition of caring for non-human animals in the feminist countercultural movements of the 1960s

DOI: 10.4324/9781003273400-27

and 1970s. The principles of non-violence, the critique of world hunger, the struggle for civil rights, the end of the Vietnam War, and the environmental costs of consuming meat and its derivatives were the project's articulating axes.[3] Thus, these discourses and political demands were intertwined and were translated into vegetarian practices that operated as a micropolitical form of protest.

In this context, the first composition of antispeciesist feminism emerged, which denounced the silence and even complicity of some feminists and ecofeminists in the face of animal exploitation.[4] This implied placing the animal question at the center of the well-known debates within feminism and the women's movement around the ethics of the care and maintenance of life. According to Gaard, in her review of the origin and development of vegetarian feminism in North America, animals held a predominant place in political debates and practices.[5] In the West, the feminist first wave recognized the link between women, the anti-vivisectionist struggle, vegetarianism, and animal welfare reforms, especially with reference to the words of Mary Wollstonecraft. In the second wave, feminists such as Carol J. Adams and Marti Kheel denounced the structural, cultural, and linguistic similarities between speciesism, sexism, and racism and the conceptual association between Nazi propaganda and slavery, on the one hand, and animal objectification, on the other. The work of Marjorie Spiegel stands out here.[6] In the same period, lesbian feminism denounced meat consumption as a symbol of patriarchal domination and the critical relationship between heteropatriarchy and speciesism.

Critical analysis at that time already emphasized the interconnection of animal exploitation with the exploitation of nature, women, racialized persons, and gender-sex diversity. It also pointed out the politicization of daily life, for example, problematizing affective relationships and food: "Vegetarian ecofeminism puts into practice the feminist idea that 'the personal is political' and examines the political context of dietary choices, as well as strategic and operational choices in science and economics."[7]

Ecofeminist statements that stressed the importance of studying the objects of oppression (women, animals, nature) with the systems and structures of oppression (heterosexism, speciesism, extractivism) thus called for a feminist analysis of socio-environmental problems to better understand how dominance operates and also an ecological analysis of problems arising from patriarchy. Posthumanist and antispeciesist feminism[8] drew from this political tradition and thinking and, from an intersectional perspective, incorporated the animal issue and speciesism into its research and political agenda.

In this context, feminist animal studies arose,[9] criticizing the masculinization of animal ethics and its consequent excessive rationalization,[10] and presented a feminist-based ethics of contextual and situated care.[11] This responds to certain essentialist positions within eco-feminism[12] and, particularly, rejects the so-called "ontological veganism" of Adams as operating under universalist and ethnocentric assumptions.[13] Reactionary positions of radical feminism against pornography and sex work have been similarly questioned.[14] Black vegan feminism, anti-capitalist and intersectional in nature, emerged in this environment, questioning the elitism and neo-colonial logic of white U.S. animal advocacy.[15]

In this context, it is worth asking how antispeciesist feminism is articulated in Latin America. This leads us to the debates on community, popular, and decolonial feminism which are the epistemic, ethical, and political core of the feminist critique of capitalism and extractivism, contextually and historically situated in the Abya-Yala[16] region, in articulation with the development of antispeciesist animal advocacy of the South.

1.2 *Antispeciesist Feminism in Latin America*

Antispeciesist feminism in Latin America became well known in the second decade of the 21st century, mainly through the emergence of animal rights debates within left-wing social movements, with valuable theoretical and empirical works in the region.

In Guatemala and Mexico, the work of the Guerra Marroquin[17] links community and decolonial feminism, specifically based on the Mayan indigenous worldview, with antispeciesist ecofeminist positions. In Brazil, there are qualitative works investigating the practice and discourse of vegan feminists,[18] studies of the Afro-vegan movement[19] and of the links between the animal rights and feminist movements,[20] and research on the relationship between lethal violence against women and females of other species,[21] as well as philosophical essays that advocate the decolonization of veganism from an antispeciesist feminist position.[22]

In Argentina, philosophical essays tackle antispeciesist (trans)feminism in the light of decolonial, intersectional, and queer perspectives, questioning a series of ontological dualisms—masculine–feminine, reason–emotion, nature–culture, human–nonhuman—and placing animal rights within a post-identity context of deconstruction and emancipation.[23] The *Revista Latinoamericana de Estudios Críticos Animales* [Latin American Journal of Critical Animal Studies] published the dossiers "Feminismos, género(s) y antiespecismo" (Feminisms, gender(s) and antispeciesism) in December 2016 and "Los feminismos a través del espejo: animalidad, género y vidas precarizadas" (Feminisms through the mirror: animality, gender and precarious lives) in December 2021.

In Chile, Alvarez Castillo,[24] has written, from an intersectional approach, a critical essay proposing a fat, lesbian, anti-capitalist, and antispeciesist feminism. In Colombia, there is rich theoretical body of work on antispeciesist transfeminism, which draws on trans studies, queer theories, and the specialized literature on antispeciesist (eco)feminism.[25] In addition, the political journal *Animales y Sociedad* [Animals and Society] was launched, coordinated by the Centro de Estudios Abolicionistas por la Liberación Animal (CEALA) [Center of Abolitionist Studies for Animal Liberation], who seek to link decolonial veganism with Colombian popular movements (see also the special edition "Feminismo antiespecista" [Antispeciesist Feminism], 2021).

Finally, in Ecuador, the empirical work of Antonella Calle,[26] from the perspective of media advertising, studies the discourses and representations around the reification of women and animals through an analysis of billboards around the country promoting the sale of meat.

2. Methodology

The methodology in this chapter is qualitative, and the technique used for data production consists of twelve in-depth interviews carried out between 2019 and 2020 with female activists, between the ages of 22 and 61, who are linked to antispeciesist animal advocacy and feminism in Ecuador. In addition, two semi-structured interviews were used, one with Analú Laferal, together with the audiovisual records of her artistic work linking the transgender sphere and bodies in transit with animality, and one with Andrea Alejandro, a Black transfeminist activist, in 2020.

The study participants were selected according to the theoretical sampling principles of Glaser and Strauss.[27] This type of non-probabilistic sampling was used due to the specificity and the difficulty of access to the sample. The analysis of the interviews was done through the qualitative analysis program *Atlas.ti*, with a selective coding system in dialogue with an open and axial coding system, which enabled the generation of frequencies and sub-categories.[28]

3. Analysis and Results

3.1 Feminized and Animalized Political Subjectivities in the South: Reappropriation of the Body-Territory

This section presents a multi-situated analysis of the political and artistic discourses and practices of concrete subjects who embody antispeciesist (trans)feminism in Ecuador and Colombia, and seeks to prioritize the voices and work of the activists.

The formation of an antispeciesist feminist political subject in Ecuador and Colombia can be identified from two main processes. The first consists of women and queers linked to animal advocacy who did not find political, ideological, or even esthetic and corporal affinities within the antispeciesist collectives. The second is marked by processes of individual or collective reflection within other social movements, fundamentally environmentalist, feminist, and queer, which operate as mechanisms of sensitization and argument around animal rights.

In this sense, 33-year-old Black transfeminist activist, artist, and cultural manager Andrea Alejandro, born in Guayaquil, Ecuador, comments:

> Personally, LGBTI people suffer discrimination, women too, because it's a macho world. Then I begin to realize that I too was still oppressing others. In this case it was the other species and other bodies that inhabited the planet.
>
> *(2020, interview by Antonella Calle)*

In this way, a process of militant sociability is recognized, in which the struggle for total liberation takes an ethical and political form embodied from the various oppressions that its members experience. Andrea continues:

> It is not possible to think of ourselves in an equitable world, in a better world, in a world that respects differences, if we do not also begin to think that we are one of the many species that inhabit the world, that we are not the owners or property-holders of anything.
>
> *(2020, interview by Antonella Calle)*

This horizon of meaning and the encounter with difference makes it possible to question the white veganism that exercises colonial logic or denies the intersectionality of oppression. This critical distance from corporative veganism makes its plurality visible[29] and advances a feminism based on interspecies social justice. In the following, I describe and characterize certain critical nodes that configure these antispeciesist (trans)feminisms of the Global South.

AF is marked by a fierce criticism of coloniality, authority, and domestication. This ethical-political position, which translates into concrete practices both in interspecies relationships that are maintained with non-human animals and in political spaces, is marked by: 1) a defense of the autonomy of bodies and individual and collective self-determination and 2) a critique of the patriarchal and speciesist regimes that establish certain care policies in the so-called private sphere that fall onto feminized bodies. The latter simultaneously functions as a space for agency. Thus, one activist remarks: "It is also a point in favor, because my actions in the kitchen will influence the rest" (Pauli, 2019, interview by author).

These animal advocacy practices are based on care for life, the politicization of the everyday, and the body-territory as their place of enunciation and challenge to the political. They suppose particular modes of relating with animals, with themselves and their corporeality, as well as with the political circles to which they belong. This component of AF is related to what Cabnal,[30] from a community feminism viewpoint, poses as the subjectifying process of "experiencing in a body and in the communal territorial space the historical-structural oppression created by patriarchy".[31] Therefore, their decolonizing practices[32] include the recovery and reappropriation of their first territory, the body,[33] as a political act of emancipation in response to the colonialist and patriarchal attack that tames and instrumentalizes bodies.

3.2 *Carnism in Anthropocentric Feminism as Colonial Penetration*

Antispeciesist feminism denounces phallocentrism and carnism[34] as ideologies that legitimize the exploitation of animals and women within the social movements of the left and, consequently, declares the importance of putting these practices into debate, as another activist states: "The issue of carnism is really important to talk about, the socialization of consuming tortured animals to be swallowed, associated with manly strength" (Cristina, 2019, interview by author). This implies an exercise in politicizing animal issues that works in two ways: 1) as a place of dispute over the public in spaces of militant sociability, fundamentally related to the consumption of animals, and 2) as a performative act that makes animal exploitation visible along with transphobia, in the case of transfeminism. In both cases, the act of "colonial penetration" is denounced,[35] implied in the orders of gender and species.

This critical dimension seeks to make visible what Lugones[36] called the "coloniality of gender" as a response to the sex-gender binarism of Quijano's[37] "coloniality of power". Colonial domination and the subordination of women were based on the biologization and naturalization of gender, in the same way as the animalization of oppressed bodies and the essentialization of animals as objects forms a colonizing and totalizing matrix over life. This relationship is denounced by Latin American antispeciesist feminism.

On March 8, 2018, the Antispeciest Feminist Block participated in the feminist marches in the city of Quito for International Women's Day, with anti-capitalist, anti-patriarchal, and decolonial slogans focused on the issue of animals and the problem of speciesism. The aim of the group's members was to denounce sexism and speciesism as two inseparable phenomena and thereby to question carnism within the women's and queer movements.

Moore Torres[38] argues that, from a critique of coloniality, feminism of the Global South has sought to cause tension in the hegemonic feminism of the West, which situates gender as the sole source of oppression and stresses a historical, universal subject: the white, heterosexual, middle-class, and urban woman. In a similar vein, popular antispeciesist feminism accentuates class, race, and species relations as structural elements of oppression.

Specific performances are thus carried out with artistic-political content that seeks to denounce the problem of animal exploitation and, simultaneously, the Eurocentrism and humanist supremacism from which the anthropocentric feminism of the region speaks—without problematizing animal exploitation and species privilege. For example, in the context of a march that involved blocking a street, activists dramatized the regimes of exploitation and bloody treatment of non-human females by the dairy industry, seeking to make visible and denounce the great agro-industrial complex.

In these moments of irruption into the public sphere, the actors use their own bodies to embody figures of animality as a form of denunciation. However, some scholars[39] have questioned the possible sexualization and objectification of women in antispeciesist campaigns such as this, which could affirm hegemonic modes of exploitation.

Here it is important to mention the heterogeneity of antispeciesist feminism as, in the case of the Quito Feminist Antispeciesist Block, there was a fracture in the process. According to some members, the positions of some women, who biologized gender and did not recognize queerness as a legitimate political subject, were denounced as transphobic. Thus, the decolonizing challenge of antispeciesist feminism requires a procedural dimension that involves permanent internal criticism.

3.3 Embodiment and Critical Gender Praxis: Animalization as Politics of the Abject

Corporality and the place of dispute in the public sphere of antispeciesist feminists share a common theme: denouncing the objectification of women and animals through a staging of a shared schema of exploitation—speciesism and patriarchy. This critique is intersectional in nature, as Soledad states:

> I observe a very similar dynamic between the objectification of a woman and the objectification of the animal. . . . It runs through an idea of ceasing to see them as human beings in the case of a woman, and of not seeing them as a sentient being or simply as a being that also has rights, in the case of the animal. And then this objectification makes it easy for us to exploit them; in the case of women, it may be sexual exploitation, labour exploitation, or exploitation at any level; and in the case of animals, it is an exploitation of consumption, exhibition, vivisection.
>
> *(2019, interview by author)*

The visibilization and denunciation of patriarchal violence is a crucial node in the political action that configures the activists' practice. In this way, the process of speciesist and patriarchal socialization is questioned, configuring an ethics of interspecies care, which is possible thanks to the experiences of shared exploitation. The experience of sexual abuse and domestic violence, according to the activists, operates as a key that opens up an empathic disposition regarding the objectification of non-human animals. Therefore, these Latin American feminist projects have as their matrix of emancipation a critical gender praxis that considers animal rights and the animalization of subaltern sectors as a strategic reproduction of the coloniality of power and heteropatriarchy.

For this reason, being a witness and having experienced suffering not only enable empathy, as an affective and existential situation that gives access to the recognition of the other, but also gives rise to an ethical and political feminist analysis of the intersectionality-decoloniality of oppression, in which animalization and objectification processes form a substantive part of its critique. These subjectivizing moments allow the problem of violence to be placed outside, in the public sphere. This can be clearly seen in another performance, called "Human Meat".

What is important in this process is that this exercise of enunciation and denunciation is not limited to revealing and politicizing the suffering caused by patriarchal violence—which is experienced, for example, in activist spaces—but that also these same experiences allow the problem of animal exploitation to be positioned in the public space. These critical voices could challenge feminists who, in their practice and discourse, reproduce schemes of coloniality[40] through an exercise of reappropriating their own body-territory and politicizing the emotional experiences permeated with speciesist and patriarchal violence.

At this point it is important to mention the work of Analú Laferal, a 33-year-old activist, artist, and researcher born in Bogotá, Colombia, who has used performance and "artivism" to make animality visible as a place of flight from gender, of embodiment of the animal, and as an act of collective sisterhood. This process of political subjectivation supposes a type of agency of the public sphere. As Analú says regarding her work "El Peregrinaje de la Bestia" [The Pilgrimage of the Beast]:

> It was not just about putting the beast in public, but in the places where we have normally been violated, the places where historically there is the male who yells at us, the one who laughs, who wants to abuse us. I needed to change the narrative, the way of inhabiting the street, so that it was not out of fear, but rather generated fear.
>
> *(2020, interview by Salomé Hincapié)*[41]

Thus, her work deals with fear and denounces the violence that trans people and other-than-human animals have experienced in Colombia. For the artist, this critique is aimed at the heteropatriarchy, the Catholic religion, and the capitalist-speciesist system. These queer practices make it possible to situate certain meta-frameworks of transfeminist and animal advocacy action.

Analú's commitment to an "Animal Transvestism" implies a process of politicization of a corporality that escapes and places the gender binarism in permanent tension and is theatrically affirmed from and with animality. Herein lies the queer dimension of veganism, in the measure of its potential to denaturalize gender and disrupt the human–animal barrier.[42]

As Analú remarks: "I do not feel that the body is a territory already won; I always have to go out and put what happens in my body into negotiation with the world" (interview by Salomé Hincapié, 2020). Analú's work challenges the place of the body-territory and the use of public space, which has historically attacked the abject and made it invisible. Her performances involve a ritualization of pain that recovers public space as a site for denunciation and questioning of the prevailing order of gender and species. This generates subversive and less domesticating counter-narratives of gender.[43]

For the artist, this is an alchemical, corporeal-affective, and ethical-political process of transmutation:

> I do the rituals because there is something that is hurting me a lot and traditional or everyday forms are not enough for me; when I shout, when I get naked, I do performance, I don't know if I really feel that it heals me, but I stop getting so anxious, and in the end that is the performance.
>
> *(interview by Salomé Hincapié, 2020)*

This process challenges the excessive rationality of classic dissertations on animal ethics.[44] Since antispeciesist feminisms center their political experience of animality on the body and on affect, this practice is decolonial in that it questions the homogeneous production of Eurocentric rational thinking.[45] Violence and its survival operate as an embodiment of political subjectivation. This is an exercise in recognizing the violence and in embodying the process of animalization that this entails.

The embodiment of animality demands an exercise of recognition of the other, which is based on experiences of oppression. This awareness of the interconnectedness of forms of oppression produces an increase in militant commitment. This in turn questions patriarchal narratives from the position of enlightened reason that articulate hierarchies in which women, as well as non-human animals and racialized subjects, are animalized and conceived as wild beasts.[46] This critique is important because it makes visible a gradient of oppression that aims to annul everything that exists outside of rationality—that is, the animal, the wild, and the natural.

In this context, such animalist feminism questions the refusal to recognize difference and assumes the periphery and the wild as a political and bodily place of assertion. As Kusmierczyk[47] says, it is an epistemic and ethical-political exercise of "turning back to the heart" as a response to the Western domestication of spirits and bodies. That is, the reappropriation of the body-territory subverts animalization as a place of colonial power that affirms structures of oppression; on the contrary, through a process of re-animalization, emancipatory practices are deployed that consist of "translating" the animal condition. This is how Andrea explains it:

> The advantages of feminism are that white women do not need to speak for Black women because Black women have their own voice, they can create their own discourses. But in the case of antispeciesism, it is as if I feel now that it is a place to put myself forward as a translator; it is not that I am speaking for the chimpanzee, for the water, for the cobalt, for the copper, I am not performing the exercise of myself embodying, or myself representing them. But rather trying to do this exercise of translation.
>
> *(interview by Antonella Calle, 2020)*

Thus, in her works, Analú occupies the body, bestiality, immobilizations, pain, the dark, the deviant, the satanic, and transvestism as methods of interpellation. As she comments:

> It doesn't hurt you when they brand cows or bulls, but when you see that I'm branding myself, you get angry and think I'm crazy or it makes you sad, it makes you suffer when it's your species and not when it's another. This action opened two immense possibilities: on the one hand, in the antispeciesist challenge, how to make people question themselves about the pain of other animal species different from humans? And on the other, in the trans challenge, I wanted to think something different from the human, I didn't want to stay thinking gendering codes of the sexed body (to know if the body is male or female). As an escape from that, I thought of the political position of being a wild animal that doesn't care if it's a man or a woman, that doesn't even care about getting dressed, and over time I decided to call these explorations "Animal Transvestism".
>
> *(interview by Salome Hincapié, 2020)*

This establishes a critique in two senses: 1) it challenges a women's agenda that is barely open to other processes of social demands, particularly the agenda that ignores the relationship between animal domestication and coloniality and, therefore, the overlapping between the binarisms of gender and species, and 2) it reveals the difficulty of animal rights in getting involved in the agendas of the left and transfeminism.

The discursive basis of this critique is a questioning of the social privileges that maintain a position of complicity in the face of oppression. As one activist remarks:

> Unfortunately, most choose to stay in that comfort zone, because they are not going to give up their privileges. In other words, their privileges of the word, to do whatever you want, to eat whatever animal you want, to do with a woman whatever you want, to treat a person however you like because you have a certain social status.
>
> *(Antonella, 2019, interview by author)*

This overlapping, which denounces the structures of oppression and the mechanisms that reproduce the asymmetry of power and are founded on such relations of privilege, challenges misanthropy and the failure to recognize other social problems in the animal rights sectors. Precisely the recognition of intersectionality makes it impossible to ignore the most heartfelt problems of society, even though these do not directly allude to non-human animals.

Another important element in the narrative of these activists is the questioning of the instrumentalization of gender discourses by animal rights sectors that, in their opinion, have found it politically convenient to assume these declarations. Thus, the attitude of antispeciesist feminists is critical and suspicious. As one activist comments:

> in the intersectional issue . . . between the issue of patriarchy and speciesism, I think it is much deeper and has to do with systems of domination that objectify our bodies.
>
> *(Carla, 2020, interview by author)*

Therefore, these activists question references to "interrelated" violence, for example, on the part of a legalistic animal advocacy movement on the link between domestic violence and animal abuse which does not recognize or problematize intersectionality at a structural level. As one activist comments: "in vegan groups they know that the issue of women is also important to deal with, so they tell you: 'men who hit women also hit animals' " (Antonella, 2019, interview by author). Taking distance from the superficial discourse of hegemonic animal advocacy also allows activists to conceive animal rights policies that are not reduced to or concentrated on veganism as consumer choice. As this activist continues:

> It gets away from a reduction to food, and now moves on to a much broader discussion that also enables you to make international connections with the theme of environmentalism itself, with the theme of feminism.
>
> *(Antonella, 2019, interview by author)*

Thus, this horizon of meaning, which enables a feminist analysis of the animal question and of the animalization of feminized bodies, is located in an anti-capitalist and emancipatory critique:

> Understanding above all that this capitalist-imperialist system is a clearly speciesist, clearly sexist system, well, sometimes it can also mutate and become vegan and

become feminist depending on the logic of the market. But it is not that it is changing or anything like that, capitalism simply cannot be humanized for us, that's clear. Capitalism will always exploit because that is its essence.

(Arlen, 2019, interview by author)

In this way, the subordination of non-human animals as well as of women and queers is recognized, and the consumption, objectification, and commodification of bodies is denounced, regardless of the species. As one activist comments:

For the issue of animal rights, of feminism, of the workers, even for the issue of LGBTIQ, there is only one enemy, and that enemy is the capitalist system that oppresses us.

(Arlen, 2019, interview by author)

4. Discussion and Conclusions

This chapter has described a particular configuration of antispeciesist feminisms from the Global South that take animal rights as a scenic challenge that seeks to elaborate: 1) an exercise of empathy with other animals and 2) an escape from or esthetic and political destruction of normative codes of gender. Thus, these ethical-political challenges of the abject are related to materialist (eco)feminisms that have sought to unveil the cultural conditions that produce gender binaries,[48] along with the transfeminism that has sought to politicize and de-essentialize feminism[49] and queer antispeciesism based on "heteronormative gender roles being aligned with a form of programming of species roles."[50]

Thus, for antispeciesist feminism, the recognition of a meta-framework that configures a relationship of antagonism against capitalist modernity, speciesism, and heteropatriarchy makes it possible to problematize two elements: 1) criticizing the antifeminist positions in animal advocacy, which the activists consider reactionary and conservative. These animal rights factions warn that it is dangerous to incorporate pro-abortion agendas or general assertions of sexual diversity, for fear of losing prestige amongst the general public. 2) Recognizing the particularity of the animal struggle and functioning as translators of the suffering and pain of non-human animals, based on their own embodied experiences of suffering, denounced through art and performance. This is important because it makes it possible to: "disarticulate the different colonial markers of power that, operating intersectionally, determine which forms of life are important and which are not".[51]

Thus, antispeciesist (trans)feminism is made up of multiple political identities, which enables them to exercise a political practice based on the encounter and recognition of difference. Thus, queerness and antispeciesism occupy similar places of enunciation, challenging heteronormativity, coloniality, and the power structures of medical discourse.[52]

Notes

1 This chapter presents advances from the doctoral thesis of the author. Particularly the academics and activists, Carol J. Adams, Marti Kheel, Greta Gaard, Brian Luke, Lori Gruen, and Josephine Donovan, among others.

2 Greta Gaard, "Vegetarian Ecofeminism: A Review Essay," *A Journal of Women Studies* 23, no. 3 (2002): 117–146. https://doi.org/10.1353/fro.2003.0006.

3 Ibid.

4 Carol J. Adams, "Ecofeminism and the Eating of Animals," *Hypatia* 6, no. 1 (1991): 125–145. https://doi.org/10.1111/j.1527-2001.1991.tb00213.x.
5 Gaard, "Vegetarian Ecofeminism: A Review Essay."
6 Marjorie Spiegel, *The Dreaded Comparison: Human and Animal Slavery* (New York: Mirror Books, 1996).
7 Gaard, "Vegetarian Ecofeminism: A Review Essay," 117.
8 Maneesha Deckha, "Animal Advocacy, Feminism and Intersectionality," *Deportate, Esuli, Profughe. Rivista Telematica Di Studi Sulla Memoria Femminile* 23 (2013): 48–65.
9 Erika Cudworth, "A Sociology for Other Animals: Analysis, Advocacy, Intervention," *International Journal of Sociology and Social Policy* 36 (3/4) (2016): 1–29. https://doi.org/10.1108/IJSSP-04-2015-0040.
10 Josephine Donovan, "Animal Rights and Feminist Theory," *Journal of Women in Culture and Society* 15, no. 21 (1990): 350–375. https://doi.org/10.1086/494588.
11 Greta Gaard, "Feminist Animal Studies in the U.S.: Bodies Matter," *DEP Deportate, Esuli, Profughe* 20, no. 1 (2012): 14–21.
12 Richard Twine, "Intersectional Disgust? Animals and (Eco)Feminism," *Feminism and Psychology* 20, no. 3 (2010): 397–406. https://doi.org/10.1177/0959353510368284.
13 Val Plumwood, "Integrating Ethical Frameworks for Animals, Humans, and Nature: A Critical Feminist Eco-Socialist Analysis," *Ethics and the Environment* 5, no. 2 (2000): 285–322. https://doi.org/10.1016/s1085-6633(00)00033-4.
14 Carrie Hamilton, "Sexo, Trabajo, Carne: La Política Feminista Del Veganismo," *Revista Latinoamericana de Estudios Críticos Animales* 2, no. 5 (2019): 274–304.
15 Amie Harper, "Vegans of Color, Racialized Embodiment, and Problematics of the 'Exotic'," in *Cultivating Food Justice: Race, Class, and Sustainability*, eds. Alison Hope Alkon and Julian Agyeman (London: The MIT Press, 2011); Aph Ko and Syl Ko, *Aphro-Ism: Essays on Pop Culture, Feminism, and Black Veganism from Two Sisters* (Lantern Books, 2017); Jennifer Polish, "Decolonizing Veganism: On Resisting Vegan Whiteness and Racism," in *Critical Perspectives on Veganism*, eds. J Castricano and R. R Simonsen, 373–391. The Palgrave Macmillan Animal Etichs Series (2016) https://doi.org/10.1007/978-3-319-33419-6_17.
16 Ancestral name that refers to the American continent and literally means: "land in full maturity", which has been taken up by the traditions of decolonial thinking.
17 Lidia Patricia Guerra Marroquín, "Tejiendo Diálogos Antiespecistas Con Las Mujeres de Abya Yala," *Analéctica* 8, no. 50 (2022): 189–203. https://doi.org/10.5281/zenodo.5894949.
18 Lorena Cardoso Monteiro and Loreley Gomes Garcia, "Veganismo, Feminismo e Movimentos Sociais No Brasil," *Seminário Internacional Fazendo Gênero* 10 (2013) https://bit.ly/3T59Hxd.
19 Nina Trícia Disconzi and Fernanda dos Santos Rodrigues Silva, "Movimento Afrovegano e Interseccionalidade: Diálogos Possíveis Entre o Movimento Animalista e o Movimento Negro," *Revista Brasileira de Direito Animal* 15, no. 01 (2020): 90–108.
20 Ana Soares, "A Interseção Entre o Veganismo e o Feminismo," *Revista Discente Planície Científica* 2, no. 1 (2020): 155–164.
21 Suane Felippe Soares, "Multifacetas de La Opresión: Ecofeminismo y La Condición Animal En Brasil," *Revista Latinoamericana de Estudios Críticos Animales* 8, no. 2 (2021): 89–104.
22 Martina Davidson, "Feminismo e Projeto Decoloniais: Ferramentas Críticas Para Repensar o Veganismo," *Diversitates International Journal* 13, no. 1 (2021): 1–24.
23 Micaela Anzoátegui, "Desplazamientos de Los Discursos Hegemónicos En La Teoría Feminista: El Feminismo Ecológico y Animalista Como Nuevas Perspectivas," *Revista Nomadías*, no. 27 (2019): 33–50. https://doi.org/10.5354/no.v0i27.54360; Anahí González, "Animales Inapropiados/Bles. Notas Sobre Las Relaciones Entre Transfeminismos y Antiespecismos," *Questión. Revista Especializada En Periodismo y Comunicación* 1, no. 64 (2019): 1–12. https://doi.org/10.24215/16696581e236;
Anahí González, "Políticas Feministas de La Animalidad. Decolonialidad, Discapacidad y Antiespecismo," *Instantes y Azares. Escrituras Nietzscheanas* 26 (2021): 123–146.
24 Constanzx Alvarez Castillo, *La Cerda Punk. Ensayos Desde Un Feminismo Gordo, Lésbiko, Antikapitalista y Antiespecista* (Valparaiso: Trío editorial, 2014) https://bit.ly/421xnXl.
25 Valentina Trujillo Rendón, "Repensar Lo Humano Desde El Transfeminismo Antiespecista," *Analéctica* 8, no. 50 (2022): 204–225. https://doi.org/10.5281/zenodo.5894953.

26 Antonella Calle, "La Cosificación de Los Cuerpos de Las Mujeres y de Los Animales En La Comunicación Publicitaria: Vender Carne a Costa de La Vida de Las Mujeres y de Los Animales," *Revista Latinoamericana de Estudios Críticos Animales* 2, no. 7 (2021): 105–121.
27 Barney Glaser and Anselm Strauss, *The Discovery of Grounded Theory: Strategies for Qualitative Research* (London: AldineTransaction, 2006).
28 Joan Miquel Verd Pericás, "El Uso de La Teoría de Redes Sociales En La Representación y Análisis de Textos. De Las Redes Semánticas al Análisis de Redes Textuales," *EMPIRIA. Revista de Metodología de Las Ciencias Sociales* 1, no. 10 (2005): 129–150.
29 Laura Fernández and Gabriela Parada Martínez, "El Veganismo No Es Una Dieta. Una Revisión Crítica Antigordofóbica y Antiespecista Del 'Veganismo de Estilo de Vida.'" *Animal Ethics Review* 2, no. 1 (2022): 44–59.
30 Lorena Cabnal, "Acercamiento a La Construcción Del Pensamiento Epistémico de Las Mujeres Indígenas Feministas Comunitarias de Abya Yala," in *Feminismos Diversos: El Feminismo Comunitario* (Madrid: Asociación para la cooperación con el Sur, 2010), 11–25, retrieved from: https://bit.ly/3ZQxXFt.
31 Ibid., 11.
32 Silvia Rivera Cusicanqui, *Ch'ixinakax Utxiwa: Una Reflexión Sobre Prácticas y Discursos Descolonizadores* (Buenos Aires: Tinta Limón, 2010)
33 Cabnal, "Acercamiento a La Construcción Del Pensamiento Epistémico de Las Mujeres Indígenas Feministas Comunitarias de Abya Yala."
34 Melanie Joy, "Psychic Numbing and Meat Consumption: The Psychology of Carnism," Saybrook University (2002).
35 Cabnal, "Acercamiento a La Construcción Del Pensamiento Epistémico de Las Mujeres Indígenas Feministas Comunitarias de Abya Yala."
36 María Lugones, "Toward a Decolonial Feminism," *Hypatia* 25, no. 4 (2010): 742–759. https://doi.org/10.1111/j.1527-2001.2010.01137.x.
37 Aníbal Quijano, "Colonialidad y Modernidad-Racionalidad," in *Los Conquistados 1492 y La Población Indígena de Las Américas* (Bogotá: Tercer Mundo Editores, 1992), 437–447; Aníbal Quijano, "Colonialidad Del Poder, Eurocentrismo y América Latina," in *Cuestiones y Horizontes: De La Dependencia Histórico-Estructural a La Colonialidad/Descolonialidad Del Poder* (Buenos Aires: CLACSO, 2014), 777–832.
38 Catherine Moore Torres, "Feminismos Del Sur, Abriendo Horizontes de Descolonización. Los Feminismos Indígenas y Los Feminismos Comunitarios," *Estudios Políticos (Universidad de Antioquia)*, no. 53 (2018): 237–259. https://doi.org/10.17533/udea.espo.n53a11.
39 Eg. Maneesha Deckha, "Disturbing Images: Peta and the Feminist Ethics of Animal Advocacy," *Ethics and the Environment* 13, no. 2 (2008): 35–76. https://doi.org/10.2979/ete.2008.13.2.35.
40 Ana Montanaro, *Una Mirada al Feminismo Decolonial En América Latina* (Madrid: Editorial Dykinson, 2017)
41 Analú Laferal, "El cuerpo es un territorio en disputa. Entrevista a Analú Laferal," Interview by Salomé Hincapié. *Portal Error 19–13* (February 24, 2020), retrieved from: https://portal-error-1913.com/2020/02/24/el-cuerpo-es-un-territorio-en-disputa/
42 Rasmus R. Simonsen, "A Queer Vegan Manifesto," *Journal for Critical Animal Studies* 10, no. 3 (2021): 51–81.
43 Deckha, "Disturbing Images: Peta and the Feminist Ethics of Animal Advocacy," *Ethics and the Environment*.
44 Gaard, "Vegetarian Ecofeminism: A Review Essay."
45 Dominika Kucmierczyk, "Decolonialidad y Buen Vivir. Descolonizar Desde El Corazón Para Construir El Lekil Kuxlejal—Vida Plena, Digna y Justa," *Ex-Céntrica* (2020): 12–18.
46 Bénédicte Boisseron, *Afro-Dog: Blackness and the Animal Question* (New York: Columbia University Press, 2018); Corey Lee Wrenn, "Black Veganism and the Animality Politic," *Society & Animals* 27, no. 1 (2019): 127–131. https://doi.org/10.1163/15685306-12341578.
47 Kucmierczyk, "Decolonialidad y Buen Vivir. Descolonizar Desde El Corazón Para Construir El Lekil Kuxlejal—Vida Plena, Digna y Justa."
48 Laura Wright, "Framing Vegan Studies: Vegetarianism, Veganism, Animal Studies, Ecofeminism," in *The Routledge Handbook of Vegan Studies*, ed. Laura Wright (London and New York: Routledge, 2021), 3–14.

49 Sayak Valencia, "El Transfeminismo No Es Un Generismo," *Pléyade. Revista de Humanidades y Ciencias Sociales*, no. 22 (2018): 27–43.
50 Emelia Quinn, "Vegan Studies and Queer Theory," in *The Routledge Handbook of Vegan Studies*, ed. Laura Wright (London and New York: Routledge, 2021), 261–271.
51 González, "Políticas Feministas de La Animalidad. Decolonialidad, Discapacidad y Antiespecismo," 126.
52 Alissa Overend, "Is Veganism a Queer Food Practice?" In *Feminist Food Studies. Intersectional Perspectives*, eds. Barbara Parker, Jennifer Brady, Elaine Power, and Susan Belyea (Toronto: Women's Press, 2019), 79–102; Quinn, "Vegan Studies and Queer Theory."

Bibliography

Adams, Carol J. "Ecofeminism and the Eating of Animals." *Hypatia* 6, no. 1 (1991): 125–145. https://doi.org/10.1111/j.1527-2001.1991.tb00213.x.
Alvarez Castillo, Constanzx. *La Cerda Punk. Ensayos Desde Un Feminismo Gordo, Lésbiko, Antikapitalista y Antiespecista*. Valparaiso: Trío editorial, 2014. Retrieved from: https://bit.ly/421xnXl.
Ana, Soares. "A Interseção Entre o Veganismo e o Feminismo." *Revista Discente Planície Científica* 2, no. 1 (2020): 155–164.
Anzoátegui, Micaela. "Desplazamientos de Los Discursos Hegemónicos En La Teoría Feminista: El Feminismo Ecológico y Animalista Como Nuevas Perspectivas." *Revista Nomadías*, no. 27 (2019): 33–50. https://doi.org/10.5354/no.v0i27.54360.
Boisseron, Bénédicte. *Afro-Dog: Blackness and the Animal Question*. New York: Columbia University Press, 2018.
Cabnal, Lorena. "Acercamiento a La Construcción Del Pensamiento Epistémico de Las Mujeres Indígenas Feministas Comunitarias de Abya Yala." In *Feminismos Diversos: El Feminismo Comunitario*, 11–25. Madrid: Asociación para la cooperación con el Sur, 2010. Retrieved from: https://bit.ly/3ZQxXFt.
Calle, Antonella. "La Cosificación de Los Cuerpos de Las Mujeres y de Los Animales En La Comunicación Publicitaria: Vender Carne a Costa de La Vida de Las Mujeres y de Los Animales." *Revista Latinoamericana de Estudios Críticos Animales* 2, no. 7 (2021): 105–121.
Cardoso Monteiro, Lorena, and Garcia, Loreley Gomes. "Veganismo, Feminismo e Movimentos Sociais No Brasil." *Seminário Internacional Fazendo Gênero* 10 (2013). Retrieved from: https://bit.ly/3T59Hxd.
Cudworth, Erika. "A Sociology for Other Animals: Analysis, Advocacy, Intervention." *International Journal of Sociology and Social Policy* 36, no. 3/4 (2016): 1–29. https://doi.org/10.1108/IJSSP-04-2015-0040.
Davidson, Martina. "Feminismo e Projeto Decoloniais: Ferramentas Críticas Para Repensar o Veganismo." *Diversitates International Journal* 13, no. 1 (2021): 1–24.
Deckha, Maneesha. "Disturbing Images: Peta and the Feminist Ethics of Animal Advocacy." *Ethics and the Environment* 13, no. 2 (2008): 35–76. https://doi.org/10.2979/ete.2008.13.2.35.
Deckha, Maneesha. "Animal Advocacy, Feminism and Intersectionality." *Deportate, Esuli, Profughe. Rivista Telematica Di Studi Sulla Memoria Femminile* 23 (2013): 48–65.
Donovan, Josephine. "Animal Rights and Feminist Theory." *Journal of Women in Culture and Society* 15, no. 21 (1990): 350–375. https://doi.org/10.1086/494588.
Fernández, Laura, and Gabriela Parada Martínez. "El Veganismo No Es Una Dieta. Una Revisión Crítica Antigordofóbica y Antiespecista Del 'Veganismo de Estilo de Vida.'" *Animal Ethics Review* 2, no. 1 (2022): 44–59.
Gaard, Greta. "Vegetarian Ecofeminism: A Review Essay." *A Journal of Women Studies* 23, no. 3 (2002): 117–146. https://doi.org/10.1353/fro.2003.0006.
Gaard, Greta. "Feminist Animal Studies in the U.S.: Bodies Matter." *DEP Deportate, Esuli, Profughe* 20, no. 1 (2012): 14–21.
Glaser, Barney, and Anselm Strauss. *The Discovery of Grounded Theory: Strategies for Qualitative Research*. London: AldineTransaction, 2006.
González, Anahí. "Animales Inapropiados/Bles. Notas Sobre Las Relaciones Entre Transfeminismos y Antiespecismos." *Questión. Revista Especializada En Periodismo y Comunicación* 1, no. 64 (2019): 1–12. https://doi.org/10.24215/16696581e236.

González, Anahí. "Políticas Feministas de La Animalidad. Decolonialidad, Discapacidad y Antiespecismo." *Instantes y Azares. Escrituras Nietzscheanas* 26 (2021): 123–146.
Guerra Marroquín, Lidia Patricia. "Tejiendo Diálogos Antiespecistas Con Las Mujeres de Abya Yala." *Analéctica* 8, no. 50 (2022): 189–203. https://doi.org/10.5281/zenodo.5894949.
Hamilton, Carrie. "Sexo, Trabajo, Carne: La Política Feminista Del Veganismo." *Revista Latinoamericana de Estudios Críticos Animales* 2, no. 5 (2019): 274–304.
Harper, Amie. "Vegans of Color, Racialized Embodiment, and Problematics of the 'Exotic.'" In *Cultivating Food Justice: Race, Class, and Sustainability*, edited by Alison Hope Alkon and Julian Agyeman, 221–238. London: The MIT Press, 2011.
Joy, Melanie. "Psychic Numbing and Meat Consumption: The Psychology of Carnism." Saybrook University, 2002.
Ko, Aph, and Syl Ko. *Aphro-Ism: Essays on Pop Culture, Feminism, and Black Veganism from Two Sisters*. Lantern Books, 2017.
Kucmierczyk, Dominika. "Decolonialidad y Buen Vivir. Descolonizar Desde El Corazón Para Construir El Lekil Kuxlejal—Vida Plena, Digna y Justa." *Ex-Céntrica* (2020): 12–18.
Lugones, María. "Toward a Decolonial Feminism." *Hypatia* 25, no. 4 (2010): 742–759. https://doi.org/10.1111/j.1527-2001.2010.01137.x.
Montanaro, Ana. *Una Mirada al Feminismo Decolonial En América Latina*. Madrid: Editorial Dykinson, 2017.
Moore Torres, Catherine. "Feminismos Del Sur, Abriendo Horizontes de Descolonización. Los Feminismos Indígenas y Los Feminismos Comunitarios." *Estudios Políticos (Universidad de Antioquia)*, no. 53 (January 2018): 237–259. https://doi.org/10.17533/udea.espo.n53a11.
Overend, Alissa. "Is Veganism a Queer Food Practice?" In *Feminist Food Studies. Intersectional Perspectives*, edited by Barbara Parker, Jennifer Brady, Elaine Power, and Susan Belyea, 79–102. Toronto: Women´s Press, 2019.
Plumwood, Val. "Integrating Ethical Frameworks for Animals, Humans, and Nature: A Critical Feminist Eco-Socialist Analysis." *Ethics and the Environment* 5, no. 2 (2000): 285–322. https://doi.org/10.1016/s1085-6633(00)00033-4.
Polish, Jennifer. "Decolonizing Veganism: On Resisting Vegan Whiteness and Racism." In *Critical Perspectives on Veganism*, edited by J. Castricano and R.R. Simonsen, 373–391. The Palgrave Macmillan Animal Ethics Series, 2016. https://doi.org/10.1007/978-3-319-33419-6_17.
Quijano, Aníbal. "Colonialidad y Modernidad-Racionalidad." In *Los Conquistados 1492 y La Población Indígena de Las Américas*, 437–447. Bogotá: Tercer Mundo Editores, 1992.
Quijano, Aníbal. "Colonialidad Del Poder, Eurocentrismo y América Latina." In *Cuestiones y Horizontes: De La Dependencia Histórico-Estructural a La Colonialidad/Descolonialidad Del Poder*, 777–832. Buenos Aires: CLACSO, 2014.
Quinn, Emelia. "Vegan Studies and Queer Theory." In *The Routledge Handbook of Vegan Studies*, edited by Laura Wright, 261–271. London and New York: Routledge, 2021.
Rivera Cusicanqui, Silvia. *Ch'ixinakax Utxiwa: Una Reflexión Sobre Prácticas y Discursos Descolonizadores*. Buenos Aires: Tinta Limón, 2010.
Simonsen, Rasmus R. "A Queer Vegan Manifesto." *Journal for Critical Animal Studies* 10, no. 3 (2021): 51–81.
Soares, Suane Felippe. "Multifacetas de La Opresión: Ecofeminismo y La Condición Animal En Brasil." *Revista Latinoamericana de Estudios Críticos Animales* 8, no. 2 (2021): 89–104.
Spiegel, Marjorie. *The Dreaded Comparison: Human and Animal Slavery*. New York: Mirror Books, 1996.
Trícia Disconzi, Nina, and Fernanda dos Santos Rodrigues Silva. "Movimento Afrovegano e Interseccionalidade: Diálogos Possíveis Entre o Movimento Animalista e o Movimento Negro." *Revista Brasileira de Direito Animal* 15, no. 01 (2020): 90–108.
Trujillo Rendón, Valentina. "Repensar Lo Humano Desde El Transfeminismo Antiespecista." *Analéctica* 8, no. 50 (2022): 204–225. https://doi.org/10.5281/zenodo.5894953.
Twine, Richard. "Intersectional Disgust? Animals and (Eco)Feminism." *Feminism and Psychology* 20, no. 3 (2010): 397–406. https://doi.org/10.1177/0959353510368284.
Valencia, Sayak. "El Transfeminismo No Es Un Generismo." *Pléyade. Revista de Humanidades y Ciencias Sociales*, no. 22 (2018): 27–43.

Verd Pericás, Joan Miquel. "El Uso de La Teoría de Redes Sociales En La Representación y Análisis de Textos. De Las Redes Semánticas al Análisis de Redes Textuales." *EMPIRIA. Revista de Metodología de Las Ciencias Sociales* 1, no. 10 (2005): 129–150.
Wrenn, Corey Lee. "Black Veganism and the Animality Politic." *Society & Animals* 27, no. 1 (2019): 127–131. https://doi.org/10.1163/15685306-12341578.
Wright, Laura. "Framing Vegan Studies: Vegetarianism, Veganism, Animal Studies, Ecofeminism." In *The Routledge Handbook of Vegan Studies*, edited by Laura Wright, 3–14. London and New York: Routledge, 2021.

22

FURIOUS AND FEROCIOUS FORMS

A Cartography of Antispeciesist and Posthumanist Transfeminisms From the Global South

Anahí Gabriela González

To Leo, my feline companion. In my throat I carry the cries of the days and nights that we will no longer share.

1. Introduction

In recent times, humanism, as the foundation that establishes and legitimizes the political distribution of bodies, has been the object of common criticism from multiple perspectives. This resulted in a radical questioning of the modern ideal of Man, implying a displacement of its operation as a hierarchical and oppressive device. To this extent, alliances have increasingly been woven between anti speciesism, transfeminism, decolonial struggles, gender and sexual dissidence activism, and struggles against ableism. What brings together these perspectives is the question of the mark that makes it possible to establish the difference between those lives deserving to be lived and affirmed and those that do not: the lives of the "disposable-animals" that are systematically thrown into exclusion, exploitation, or death. This is because the human–animal dichotomy operates as the core of multiple operations of the modern-colonial system of subordination. Deconstructing this dichotomy becomes an unavoidable challenge, since it condenses a racist, cisheterosexist, colonial, and speciesist politics (and an ontology) that perpetuates the work of killing those lives not codified as truly "human".

At this point, one of the main contributions of antispeciesist feminisms has been to point out that the domination of other animals is connected to sexist practices and discourses. I'm referring to theoretical contributions such as "the sexual politics of meat" or "feminized protein" by Carol J. Adams,[1] "meat culture" by Annie Potts,[2] or the notions of "carno-phallogocentrism" or "carnivorous virility"[3] by Jacques Derrida. However, in recent years, others have pointed out the limits of an analysis located only at the intersections between gender and species, because these have resulted in a closed and exclusionary representation of the subject of feminist politics. For this reason, queer, trans*, crip,

 DOI: 10.4324/9781003273400-28

decolonial, anti-capitalist, and anti-racist animalisms propose more complex optics for approaching the connections between speciesism and other regimes of domination. Queer ecofeminism points out the heterosexual assumptions of certain vegan feminist theories.[4] Black and indigenous veganisms have addressed the relationships between racism, colonialism, and animality,[5] while crip antispeciesism has encouraged reflection on the articulation of ableism with speciesism.[6] Carrie Lou Hamilton, for her part, has stressed the importance of telling different vegan feminist stories that do not reproduce the heterosexual matrix or silence the voices of sex workers and trans people.[7] To this itinerary, this chapter adds some works from Latin America that have denounced the Eurocentric, cis-heterosexual, and colonial biases of hegemonic anti-speciesist feminisms. In their heterogeneity, these approaches allow us to articulate ethical-political strategies that, in addition to defending other animals, are attentive to the processes of animalization of multiple bodies that do not comply with the standard of humanity proposed by European humanism.

Within this framework, this chapter aims to carry out a brief cartography of transfeminist and posthumanist anti-speciesisms in the Latin American context, with a focus on some activist interventions from Argentina. My intention is to show that such perspectives have emerged as plural, heterogeneous, and situated practices, aimed at fostering other modes of the common, attentive to the specific contexts of inequality and exclusion in the region. First, I set these claims in the context of the rise of different Latin American (trans)feminist struggles, going on to summarize some approaches on the existing relationships between coloniality, cisheteropatriarchy, and speciesism. Secondly, I address some anti-speciesist transfeminist interventions in the context of the development of the *Ni Una Menos* (Not One [Woman] Less) movement in Argentina, seeking to trace the ways in which the affirmation of the notion of the animal emerges as a place of enunciation for dismantling the mechanism of the human. Finally, I propose that the notion of animality can be thought of as a meeting point for the political insurrection of different antispeciesist and/or posthumanist transfeminist voices.

2. *Ni Una Menos* and Trans-Activisms in Argentina

Feminisms, drawing on Teresa de Lauretis, are unstable, historically discontinuous, and organized in multiple ways, since they are products of struggles in changing contexts.[8] Starting with the so-called "feminist criticism of feminism", the incorporation of the voices of eccentric subjects of feminism occurred throughout the 20th century.[9] While lesbian feminisms denounced "compulsory heterosexuality"[10] in the 1960s, Black feminisms problematized the ideal of the white woman within hegemonic feminism. Likewise, questions originating from border identities were developed, from which the analysis of the vectors of oppression of class, race, ethnicity, and sexual orientation were deepened.[11] For their part, the emergence of post-colonial/decolonial, southern, Abya Yala (Latin American), communal, and Third World feminisms radicalized the critique of the political subject of white feminism. As a whole, these perspectives were oriented towards thinking about the specificity of the social, political, and cultural contexts, demonstrating that there are different hierarchical structures of oppression that cannot be explained from the perspective of the binomial male/female.

It is within this general framework that we can situate the multiplicity of feminisms developed in the 21st century in Latin America, which are based on the experiences of cis-gender women, lesbians, sex workers, trans* people, *travestis*, *tortilleras*,[12] and non-binary people who seek to resist not only patriarchy but also racism, heterosexism, cissexism, policies of neoliberal precariousness, and the outposts of neoconservatism.[13] In Argentina, on June 3, 2015, under the slogan *Ni Una Menos*, a massive and unprecedented rally that altered the history of feminist struggles in the region was held. This movement formulated a "feminist perspective on the different phenomena, on work and capitalist accumulation, institutional violence and the ways of subjecting lives".[14] Anchored in the long tradition of street politics in the country, *Ni Una Menos*, according to Malena Nijensohn, brought together both the enormous number and plurality of social movements as well as unorganized people.[15] In addition, the movement is heir to the struggles of the *Madres y Abuelas de la Plaza de Mayo*,[16] of the struggles for the Laws on Marriage Equality and Gender Identity, as well as of the "National Women's Meetings" of Argentina, which later became the "Plurinational Meetings of Women, Lesbians, Travestis, Trans, Bisexuals and Non-binary People". This flow was the consequence of the struggles of bodies that did not respond to the normative ideal of the white cis-gender woman, as is the case of trans-*travestis* activisms and the struggles of indigenous and Black women.

Particularly in Argentina, the struggles of trans*-*travestis* identities have played a vital role in questioning feminist discourses and practices that reproduce cisheteronormative, binary, and colonial logics. Trans* activisms and studies have contributed by giving complexity to the web of power relations, showing that "the map of subjects who oppress and suffer oppression on the basis of gender can no longer be read exclusively in terms of 'unilateral sexism', that is, cis men who oppress cis women".[17] But incorporating the fury of Latin American trans*-*travestis* people into the feminist agenda has been—and continues to be—extremely arduous; this is because, as Sosa Villada observes, trans* activism produces its theories "on a graveyard of sisters, friends, mothers, aunts".[18] In addition, in Latin America, the word *travesti* supposes an anti-colonial projection that begins from other constructions of the body, because the territorialities and access to knowledge are different.[19] *Travestis*[20] have lived in contexts of war, surviving dictatorships, facing police persecution, restrictions on free movement, and permanent obstacles to accessing the rights entitled to all citizens. All of this has meant that *travesti* life is "a life under siege".[21]

It is in this context that trans* activisms and theories, developed by Diana Sacayán, Maite Amaya, Lohana Berkins, Mauro Cabral, Blas Radi, and Marlene Wayar, among others, not only opened space for the trans*-*travesti* collective within the LGTB movement but also incorporated their struggles into the multiple demands of feminism.[22] In effect, these theoretical-political interventions have made explicit that the normative matrix that regulates bodies, behaviors, and desires are especially merciless with those lives that do not respond to cis/heterocentric ideals. In this sense, if we understand that women (cis and trans), trans men, gays, *travestis*, Blacks, *Latinas*, and lesbians have been oppressed in the name of the operation of colonial and cisheteronormative white humanism, we can affirm that "The gathering of these voices, more than the erasure of some, means the amplification of a collective cry".[23] Hence the urgency of a decolonial and Latin American transfeminism that can shelter those bodies and demands that do not coincide with the norm of cisgender, white, straight, and middle-class women from the global north.

3. Decolonial Anti-Speciesist Transfeminisms in Latin America

The emergence and expansion of the animal liberation movement in Latin America can be traced back to the first decade of the 2000s.[24] However, anti-speciesist organizations have had difficulties in establishing dialogue with other anti-oppression social movements. By not questioning the oppressions that have historically been erected on subalternized communities and bodies, anti-speciesist movements have inherited an ontology and political history that reproduce the "epistemic violence" of European thought.[25] This position has made it difficult to create alliances with other social struggles and has silenced the voices of political minorities in the anti-speciesist movement. However, in recent years, new perspectives committed to an intersectional, critical, and plural anti-speciesism have emerged. Showing the existence of other theoretical-practical contributions, beyond the classic cannon of animal rights theories, became an essential task on the path towards a more powerful defense of anti-speciesism.[26]

An important milestone in consolidating a critical perspective on anti-speciesisms in our region was the establishment of the Instituto Latinoamericano de Estudios Críticos Animales (ILECA) and the Revista Latinoamericana de Estudios Críticos Animales (RELECA) in 2014.[27] These projects enabled the creation of networks between researchers and activists from Argentina, Brazil, Chile, Colombia, Ecuador, and Mexico. The Institute provided a place to publish on topics that were not legitimized within the academy; it enabled the organization of different events, projects, and conversations, as well as the publication of books and dossiers. It is a meeting space that emerged from the articulation of the theoretical-political tasks that each member had previously carried out either autonomously or as members of other groups.[28] In addition, the ILECA is positioned as a project rooted in counter-colonial and transfeminist perspectives. Its theoretical, political, and artistic interventions aim to deepen alliances among bodies affected by the precariousness induced by neoliberal policies.[29] This impulse led to the creation, in 2021, of a network called "Animalismos decoloniales antiespecistas" that I coordinated with María Belén Ballardo, Agustina Marín, and Martina Davidson,[30] with whom I also wrote a queer, decolonial, and anti-speciesist manifesto.[31]

The institute and the network serve as points of articulation for various anti-oppression agendas, yet the mobilization of anti-speciesist transfeminisms in Latin America extends far beyond these initiatives. Indeed, recent years have witnessed the growth of diverse organizations, activisms, and interventions driven by this ethical-political commitment.[32] There is no homogeneous way to characterize such movements because they are plural and heterogeneous, with localized practices that, by their mere existence, project different alternatives to the modern-colonial speciesist system. Even so, in a general sense, it could be asserted that Latin American anti-speciesist transfeminisms emerge as a form of resistance to hegemonic anti-speciesist feminism. This resistance stems from the latter's failure to engage in dialogue with community, popular, and decolonial feminisms, along with its neglect of the experiences of *travestis* and dissidents in Latin America.

Therefore, if decolonial, Black, and lesbian feminisms criticized white feminism for being restrictive in nature, in a similar way Latin American anti-speciesist transfeminisms draw on decolonial approaches, queer-trans literature, and posthumanisms with the purpose of establishing dialogues among multiple political minorities. Building from these frameworks, they have faced the triple challenge of: 1) deepening and making the intersectional perspective more complex in order to question the heterocissexist, ableist, and colonial matrix of multiple animalist theorizations; 2) configuring anti-speciesist policies that challenge anthropocentric feminisms to show that the ethical-political task cannot consist of expanding and making the humanist project more inclusive, but they must establish alliances to counteract the effects of the dominant order; and 3) promoting the configuration of common resistances among different political struggles that create the conditions that give space to more habitable worlds for multiple forms of life, subjectivities, and other bodies. Taking this general framework into account, we can identify at least three moments that mark the itinerary of anti-speciesist transfeminist interventions in Latin America.

Initially, the works of Carol J. Adams, Jacques Derrida, and Judith Butler were fundamental to thinking about the ways in which speciesist and sexist systems of power are linked. Adams' work allowed questioning the anthropocentric assumptions of feminism by showing that meat consumption implies a sexual and racial politics. Furthermore, the concept of "carno-phallogocentrism" spurred the development of works that delve into the intricate connections among hierarchical pairings, such as animal/human, woman/man, and emotion/reason.[33] Along a similar line, Carrera, Anzoátegui, and Domínguez introduced the concept of the "ontology of availability" to highlight how certain bodies are viewed as mere commodities, subjected to processes of exploitation and appropriation in their production.[34] There were also developments in perspectives that embraced the ethics of care and Butler's ethics of precariousness with an anti-anthropocentric focus.[35]

Secondly, works emerged that aimed to challenge binary conceptions of gender, recognizing that cis-gender women are not the sole bodies subject to animalization in the cisheteropatriarchy.[36] The trans*, queer, and intersex movements have played crucial roles for certain anti-speciesist perspectives because they attend to the heterogeneities and singularities that go beyond the binomials "woman/man" and "male/female".[37] Therefore, if there are multiple non-normative corporalities that have been considered subhuman, uncivilized, savage, and inferior species, focusing only on the ways in which cis-gender women are animalized, and the ways in which animals are feminized, implies reproducing compulsory heterosexuality. This narrow focus denies the existence of alternative ways of experiencing gender and sexuality.[38] It effectively erases the presence of trans women, non-binary people, and non-hegemonic masculinities (such as trans men and lesbian masculinities)—who have been violated, disciplined, and animalized in the name of the heterocispatriarchal regime and in articulation with speciesism.

In effect, desires, bodies, and non-normative sexualities have been placed on the reverse side of what is human. On the one hand, gay, lesbian, bisexual, intersex, and trans* people have been captured under the grammar of an unnatural deviance. On the other hand, they have also been categorized as closer to "nature" by—supposedly—not being able to control their overflowing animal instincts. Their classification as "savages" has justified violence and disciplinary actions against these bodies, justified under the creation of an imaginary around their supposed monstrosity that can only be "normalized" through the reaffirmation of cisheterosexuality. From this perspective, anti-speciesist transfeminisms are a form of political resistance both to the ways in which the heterosexual regime renders the lives of strange, abnormal, and migrant bodies uninhabitable and to the ways in which other animals have been instrumentalized to represent (supposedly) "normal" human social behavior. At the intersection between "the animal" and "the queer", an affirmative space is opened to produce desires, affects, politics, and knowledge that challenge the hierarchies that are drawn on the forms of life, promoting other ways of the common.[39]

Thirdly, there was the urgency not only to construct a transfeminist anti-speciesism but also to envision its decolonial and antiracist transformation.[40] To this end, taking up the proposals of Anibal Quijano, Walter Mignolo, María Lugones, Enrique Dussel, and Ochy Curiel, among others, it was argued that "speciesism", imposed by coloniality, is a structure of oppression that allows, through its symbolic and concrete operations, those lives that do not respond to human regulations to be marked "as animals" and excluded from the protections enjoyed by those who are recognized as legitimate bodies.[41] Lugones's work was taken up again to show that the logic of colonial extermination is based on the human/non-human dichotomy set in motion during modernity, from the process of colonization of the Americas and the Caribbean. On the one hand, it was argued that one of the most obvious aspects of the intersectional operation between racism/coloniality and speciesism has been the animalization of indigenous peoples and Black people. On the other hand, it became clear that the rhetoric of animality does not operate in a homogeneous way, nor does it produce the same effects for all subalternized bodies. For example, while cis-gender white women were condemned to the "white male" breeding ground, Black women were categorized as genderless, hypersexualized "females".[42]

Furthermore, the modern-colonial device is more than a pattern of global domination centered on the human population; it also expands to other species. As Fabio Oliveira, Tânia A. Kuhnen, and Daniela Rosendo maintain, the hegemony of diets centered on meat, eggs, and dairy products, which originated in the conquest and colonization of the Americas, is the diet that most destroys life on the planet, also impacting the lives of indigenous peoples.[43] This is not only because the expansion of livestock meant that these peoples have been dispossessed of their lands and suffered the disastrous consequences of environmental contamination but also because of the advance of monoculture based on the colonization of the cultural and food production chains. Hence, in the face of white veganism, Latin American anti-speciesist transfeminisms seek to deepen plural, multiple, and heterogeneous veganisms, which already exist in practice and are carried out by multiple political minorities.[44]

Based on the foregoing, it can be argued that, in its multiple manifestations, the decolonial construction of anti-speciesist transfeminisms challenges us to think about the articulations between oppressions, taking cisheterosexist white supremacy as its central axis while trying to provide an articulated response and intersectional resistance that begins from the alliances between different struggles.

4. Antispeciesist and Posthuman Transfeminisms in Argentina

Anti-speciesist transfeminisms in Argentina emerge from the intersection of four key factors. First, the mass nature of feminist struggles gained through the *Ni Una Menos* movement and the need to articulate the demands of anti-speciesism movements with feminist agendas. Second, the huge trajectory of the struggles of the trans* community in this country made it possible to displace, within certain feminisms, the idea of a homogeneous political subject. Third, the need to configure animal rights spaces that resist the cisheteronormative bias of local antispeciesist organizations. Lastly, the project seeks to provide an intersectional response to the global ecological crisis that affects all forms of life differentially.

In Argentina, groups and interventions that stand out include Feminismo Antiespecista Interseccional (FAI), Resistencia Antiespecista (Buenos Aires), Red Transfeminista Antiespecista San Juan, Antiespecistas Transfeministas (Córdoba), Lo animal es político (Santa Fe), Tortilleras por la extinción (La Plata), and Transfeminismos decoloniales transfeministas from RLECA.[45] Various forms of activism are evident, including the establishment of communal kitchens, graphic actions, fanzines, manifestos, the creation of virtual libraries, participation in marches, social media dissemination, animal rescue efforts, as well as the organization of spaces for activist training, debates, and conversations. Most of these organizations have been present, since 2019, in marches on International Women's Day, at International LGBTQI+ Pride Day, and at the *Ni Una Menos* demonstrations. In addition, it is worth noting the organization of workshops on antispeciesism by the *FAI* and *Resistencia Antiespecista* at the "Plurinational Meeting of Cis, Lesbian, Bisexual, *Travestis*, Trans, and Non-Binary Women" in the cities of La Plata (2019) and San Luis (2022).

The theoretical-political interventions of these groups not only challenge anthropocentric feminisms, but also confront the naturalization of certain national-cultural rituals associated with livestock and meat intake. That is, if we consider that the Argentine nation was defined by an agricultural and livestock project at the service of capitalist accumulation, while meat consumption was linked to national identity, it is not difficult to understand that such activisms face significant difficulties in establishing alliances with other social struggles.[46] In fact, in many demonstrations, including feminist marches, *choripanes*[47] are frequently sold, while veganism is considered petty-bourgeois, anti-working-class, and focused on individual consumption. For this reason, the main challenge faced by decolonial anti-speciesist transfeminisms is not only to problematize the conception that veganism revolves around the logic of consumption but also to promote the deepening of vegan practices sensitive to subalternities as an integral part of the project of anti-colonialism.

In this regard, in 2014, a collective fanzine on anti-speciesism, feminism, and sexual dissidence was written by Luciana Carrera, Fer Guaglianone, and Ariel Di Paoli. In it, faced by a mainstream veganism, they propose to speak instead about food dissidence:

> I choose to talk about food dissidence. And, not only veganism or anti-speciesism because that's not enough for me. Dissidence will never support what the system captures. I don't want vegan supermarkets, or specialized aisles. I am not interested in regulating animal abuse, or in having recovery centers for animals in captivity.[48]

In other words, veganism understood as a form of consumption is insufficient for resisting speciesism because it is not a mere form of individual discrimination. Rather, speciesism is an order of domination that combines knowledge, institutions, spaces, techniques, and gestures, delimiting hierarchical criteria for differentiation between bodies, based on the hierarchical dichotomy human–animal.[49] Hence, because of its cooptation of capitalism, hegemonic veganism is aligned with a discourse of neoliberal economic development that does not problematize the commodification of other animals and that, moreover, as Martina Davidson says, is violent towards subalternized peoples in Latin America.[50]

In the case of Argentina, the political construction of the nation as white and civilized had at its core the appropriation of indigenous lands in order to expand an economic system based on cattle raising. The indigenous genocide, during the so-called "Conquest of the Desert" (1878–1885), was financed by those who appropriated the land, namely the Argentine Rural Society. In this sense, the colonial speciesist device implied the extermination of indigenous peoples, the commercialization of other animals, and the concentration of land ownership in the hands of economic elites. But these are not events or the past; the conquest was not only a process of territorial expansion but also a project of material and symbolic domination associated with the capitalist development model. This model is currently taking on new dimensions through agribusiness.[51] Maristella Svampa uses the term "neo-extractivism" to apply to the process of expanding forms of extraction in Latin America, such as open-pit mega-mining, agribusiness, fracking, and the energy industry.[52] The consequences of this neo-extractivism are poisoning by pesticides, the displacement of indigenous and peasant communities, and the setting of intentional forest fires that devastate entire ecosystems. Neo-extractivism has become, today more than ever, an incalculable risk for the conditions of habitability, reproduction, and diversification of life.

For this reason, in recent years, some anti-speciesist groups have had to think about their relationship with social struggles for the defense of land. In this regard, 2020 is a year that marks the collective organization and resistance against neo-extractivism. This is when the Argentine government sought to establish an agreement with China to install pig mega-factories, in a context that was not only marked by the Covid-19 pandemic but also the spread of wildfires in various parts of the country (fires which have continued to multiply). Although there were demonstrations from different organizations, I would like to highlight a series of artistic interventions carried out by two anti-speciesist lesbian activists, who start from a dissident animal rights perspective from the South. The intervention was carried out on the walls of the Faculty of Arts of the city of La Plata, under the slogan "I don't care about your virus. I care about pigs. *Tortilleras* (Dykes) for extinction". The protestors demanded the following:

> How long will economies parasitize the South? . . . What bodies and lives matter (to us)? . . . What is the time of a health care that signs guarantees for future pandemics? . . . How do we escape from species? And what then? How do we make ourselves a *tortillero* body for extinction? What is the economy of extinction? May the end of the world find us outside this species.[53]

On the one hand, the activists indicate that the project to create pig mega-factories responds to a colonial matrix that, due to the presence of foreign capital, deepens the levels of poverty, inequality, and precarity in peripheral countries such as Argentina. Moreover, the agreement with China not only means aggravating the situation of exploitation suffered by other animals but also represents a risk to human health by promoting the development of future zoonotic diseases. On the other hand, the activists, by stating that pigs are lives that matter, problematize the anthropocentric logic that revolves around the idea of human exceptionality. In response to the necropolitical logic that draws dividing lines between the bodies that matter and those that do not, "*Tortilleras* for extinction" is committed to an escape from species through a *tortillera* becoming:

> Making ourselves a *tortillero* body as an unstable identity, as a multiplicity of organisms in finite and deeply felt connection and disidentification, an in-between that makes us more pig and less women. The purpose is to extinguish the notion of humanity as a future project centered on the reproducibility of the species. The intention is to open the way for confronting the reification of the future imposed by the truth of capital, which obstructs the invention of policies that are sensitive to the disobediences to come.[54]

In this way, the proposal to become a "*tortillero* body" does not imply expanding the humanist project but rather displacing the norms of the human that define the field of legitimate, valuable, and mourning subjects. In the intersection of *sudaca*[55] sexual dissidence and anti-speciesism, the act of claiming to be pigs, instead of women, is a way of shaking their belonging to the human species. By becoming unrecognizable according to heterosexual norms, these bodies resist the principle of reproduction of the species, the axis of national political projects. As Gabriel Giorgi would say, "to escape normative gender is always, to some extent, to escape species" because "the recognizability of the human species depends on having a legible, identifiable gender".[56] In addition, by proposing to invent a disobedient affective politics, "*Tortilleras* for extinction" exhorts us to imagine other ways of weaving the space of the common, to the configuration of other worlds, where what matters is the affirmation of bodies in their singularity, heterogeneity, and variability.

For its part, the "Red transfeminista antiespecista San Juan", in which I had the opportunity to participate, shares the goal of imagining ways of living that engage other forms of the common. This group intervened in the International Strike of cis women, trans, *travestis*, lesbian, and non-binary people on March 8, 2019, using animal masks and distributing leaflets. Despite the group disbanding later, some of its members created a podcast called "Rizoma", where they addressed issues related to the pig agreement with China, mega-mining, sex work, job insecurity, disability, and sexuality. In their presentation, they make an incendiary appeal "for the wolves, the black women and the fat women, the whores, the trans and the *travas*".[57] Through their activism, rooted in the affirmation of shared vulnerability, they endeavor to challenge boundaries—be they related to gender, species, ethnicity, or class—that stratify lives. Concurrently, they illuminate how the classification of "animality" operates as a framework under which all those bodies deemed exploitable have been encompassed.

In this itinerary, the travesti-trans activisms of Maite Amaya, Susy Shock, Paulina Gaetán, Camila Sosa Villalba, and Violeta Alegre are of vital importance. Their interventions robustly reject cissexist and colonial categorizations inherent in humanism while advocating for the concept of animality as a site of resistance against the speciesist machinery and its disciplining and lethal mechanisms. Moreover, in some instances, these activists have articulated what could be termed a "becoming trans" of veganisms. A notable example is Maite Amaya, known above all for being a popular fighter from Córdoba (Argentina), *travesti*, feminist, anarchist, and *piquetera*,[58] who died in 2017. In an interview with *La Condesa*, Amaya asserted that contemporary struggles should not aim for an extension of the human but should instead promote connection with all living entities: the earth, plants, and animals. According to this activist, vegan practices represent a means of cultivating alternative habits that disrupt the system in pursuit of social change. Consequently, veganisms emerge as acts of disobedience resonating with the feminist maxim "the personal is political" as "every attitude that comes from a crack happens in small spaces; this is how I was building my notion of activism".[59]

Recently, Paulina Gaetán, an anti-speciesist *travesti*, vegan, social worker, and writer, published *Preludio Travesti*, the first *travesti* literary work in the city of Pergamino (Buenos Aires). From an autobiographical journey narrated by a survivor (because, as the author recalls, the average lifespan of trans* people in Argentina is 35 years), the book sheds light on the intricate relationships between cissexism and speciesism. By drawing attention to the naturalization of the oppression of trans lives and animal lives, Gaetán contributes to problematizing the forms of violence that remain unrecognizable in the existing ethical-political frameworks:

> We *travestis* have historically been persecuted by security forces, we have been locked up in dungeons. We were raped by machismo/sexism, we were tortured, marginalized, beaten, expelled, systematically murdered. The *travesti* genocide was visible to all but no one cared. So, what is the difference from other animals? . . . My freedom must also include animal liberation.[60]

There are bodies whose importance is not questioned, while others are living at the thresholds, more prone to certain forms of violence than others. The human–animal and cis-trans dichotomies condition the recognition of different ways of life, practices, and behaviors, establishing a break between lives that are worthy of mourning and those that are not, the forms of pain that matter and those that do not.[61] Animal, trans*wild, monstrous, or barbaric lives are expelled from the realm of humanity, configuring themselves as its uninhabitable reverse. Hence, the transfeminist and anti-speciesist struggles are insurrections against the structures that normalize, subdue, and, ultimately, make bodies precarious. For this reason, for Gaetán, only if we think about oppressions in an intersectional way can we move towards a transfeminist, anti-speciesist, anti-racist, and anti-capitalist future that enables livable worlds for all subaltern bodies.[62]

By indicating the complicity between cissexism and speciesism, *travestis* anti-speciesists commitments have great potential to create other animalist imaginaries that displace the biologicism of hegemonic anti-speciesist feminism. What is problematic about such feminism, as Hamilton points out, is that by emphasizing biological reproductive capacity as a point of identification between "women" and females of other species, it reinforces the idea that femininity is defined by motherhood.[63] By focusing on reproductive capacity, these feminists propose a biological reductionism in which sex, from a binary and dimorphic perspective, appears as the origin of gender, presenting a biologist, endosexual, and trans-exclusionary perspective.

On this point, in a 2018 interview, Camila Sosa Villada, a trans writer and actress from Córdoba, Argentina, denounces the transphobia of radical feminists who defend their exclusive belonging to the identity "women" while denying this identity to *travesti* individuals. Confronting such feminism, the writer proposes to establish *travesti*-animal alliances, conceiving them as unexpected and potentially transformative collective connections to envision other forms of life:

> We fight for life, for women, children, animals. We fight for a life that can be lived with joy and freedom. I don't want more corpses in our path and I'm sure that the *travas*[64] of my life don't either. But I want to know who my enemy is, who wants to kill me, who wants me to die by being discriminated against, exiled, sick and oppressed. So if for these "women" the "title" leads them to commit such acts of death, I prefer to be a hummingbird as Aunt Susy Shock says.[65]

If *travesti* lives have been abandoned by the system, their way of survival has been through the creation of bonds, alliances, and resistance based on friendship and fury. These forms of solidarity and kinship of the *travesti* herds, aimed at sustaining life, imply other forms of collective organization that start from a wild animality as a meeting point for political insurrection. Faced with the radical feminists who believe they own the title "woman", Sosa Villada prefers to lose her human identity. Or, in the words of Susy Shock, Argentine trans artist, "I claim my right to be a monster/Neither male nor female nor XXY nor H_2O/I am a monster of my desire, meat of each of my brushstrokes".[66]

In this scenario, it is necessary to mention that there are currently some proposals that have reflected on the importance of incorporating an anti-ableist approach to anti-speciesist transfeminist claims. This is the work done by Romina Vitale, an artist, writer, and neurodivergent psychologist from Buenos Aires who, from her clinical work, as well as in her artistic and theoretical interventions, has struggled to show the need to articulate transfeminist, anti-ableist activisms and antispeciesist perspectives. In a series of conversations, Vitale argues that modernity built an ideal body for the human species based on a white and colonial humanity, while those existences that do not correspond to this ideal are considered inferior.[67] Cis-gender women, "the insane", other animals, trans*, non-binary, and queer existences share this fate of exclusion, imprisonment, and systematic instrumentalization. Therefore, it is urgent that anti-speciesist activisms contemplate the ways in which speciesism, sanism, and ableism reinforce each other.

Finally, even though it is not an explicitly anti-speciesist collective, the "Movement of Indigenous Women and Diversities for '*Buen Vivir*'" (Good Living) has played a very important role in articulating intersectional struggles against neo-extractivism. This is a transfeminist, anticolonial, antiracist, and anticapitalist collective that brings together women and diversities from 36 Indigenous nations that coexist in Argentine territory. In this sense, as Surama Lázaro affirms in his ongoing research, its members question the anthropocentric Western paradigm that places the human being above other forms of life. In some cases, they also engage with anti-speciesism, redefining it in the context of their territories and experiences.[68] As inheritors of the survivors of indigenous genocide, one of the movement's purposes is to propose a collective fight against Terricide, understanding it as the systematic extermination of all forms of life.[69] Their activism denounces the death of indigenous peoples due to famine, pollution, and racist violence. They confront rapes of women and girls, as well as transfeminicides and transvesticides, evictions against indigenous peoples, and displacement caused by multiple forest fires that render their lives as disposable bodies. In his words: "Enough of indigenous genocide! They sowed Terricide, they will reap rebellion!"

5. Fierce and Savage Alliances

In this work, my intention has been to show that there is a heterogeneity of ethical-political practices in Latin America that resist the sovereignty of white, cis-gender, adult, heterosexual, capable, healthy, productive, and proprietary man. Antispeciesist and posthuman transfeminisms seek to subvert and transgress the limits of normality of humanism, which insist on making us useful to the state and the circuit of capital and try to tame our contagions and animal insurrections. In addition, the modern-colonial apparatus configured a set of spaces and devices to exploit, subject, and subordinate other animals (zoos, circuses, aquariums, farms). It is there that their bodies are modeled and reduced to commodities, while their forms of resistance and expression in the face of domination are silenced. In this sense, coloniality has implemented animalization as a political tactic to draw boundaries between the lives that matter and the lives that are uninhabitable, distributing them on a hierarchical scale that goes from what is considered properly human to what is non-human.

If the human norm is heterocissexual, racist, speciesist, and ableist, then the *sudacas*, disabled, trans*, queer, and indigenous animalities are lives unworthy of belonging to the human community and of being part of its civilized institutions. That's perhaps why sick, impoverished people, crip bodies, LGBTQI+ people, BIPOC, and cisgender women, among others, have been placed on the side of the animal. This positioning enables us to reconsider animal vulnerability no longer as the inferior opposite of human but as an instance to envision alternative narratives about social connections. In other words, if animality has been subject to discipline by the humanist machine and if, in addition, multiple non-normative bodies have been placed on the side of the animal, wouldn't it be necessary to imagine policies (multiple and heterogeneous) that are oriented towards vindicating that animal vulnerability as an alternative instance to think about collective articulation and other ways of inhabiting existence? Can the notion of animality be conceived, then, as a non-identity meeting point to think about alliances between political minorities? In this regard, the Brazilian transfeminist activist Indianare Siqueira has indicated the following:

> I will never understand humans. I will never understand humanity. You kill and hate people who love other people just because they don't follow compulsory heterosexuality (. . .). Because of a capitalism that I am not going to call savage, because that would be making it beautiful, but rather an unbridled human capitalism in which a minority lives well while the vast majority starves. Their ancestors stole lands, destroyed cultures, invaded territories, decimated peoples, raped, and through religions provoked hatred and wars, enslaved, and oppressed. You kill animals for food and you don't see the pain and suffering they cause. When I see the misery that you cause, and at the same time how miserable, selfish and hateful human beings are, I thank you for depriving me of my humanity.[70]

Understanding the ways in which coloniality distributes bodies is key because it opens space for alliances between different forms of life that have been domesticated, locked up, murdered, and exploited without denying the differential ways in which oppressions are exerted on their bodies and subjectivities. To commit ourselves to those alliances made from our animalities can be a starting point that leads to the configuration of solidarity networks that question the presupposition of the species as a recognition framework. Perhaps it is also an opportunity to enable other forms of recognizing similarities and political bonds: tunnels, herds, packs, burrows, and caves, in which we take refuge together, experiencing the friction of skin, resonating with the fears and pains of others as our own. The animal and ferocious alliances, proposed by antispeciesist and posthuman transfeminisms, are a possible starting point to resist the effects of precariousness in the memories of the skin, a place where our vulnerabilities can find new ways to become a power of affirmative encounter:

> Faced with the advance of neoliberal and neoconservative policies in Latin America on our stocks and territories, we resist. We resist as a way of existence and we forge alliances based on our animal precariousness. Our alliances are not forged in identity but in modes of intertwining of vulnerable, fragile forces, and they are a sign of resistance against the normalization that has been and continues to be made of our subjectivities and ways of life. . . . We will fight for non-fascist, less punitive, more communal, more interdependent, more dissident, more animal ways of living. We will fight for everything that animal, *travestis*, trans*, crip, queer, indigenous and Afro-descendant lives have taught us in our beloved and devastated Latin American territory.[71]

Acknowledgments

I am grateful to María Belén Ballardo, Martina Davidson, and Micaela Anzoátegui for their suggestions and recommendations, as well as for encouraging me to write this work at a time when my mental and physical forces were falling apart. I also thank Romina Vitale y Suráma Lázaro for their willingness to respond to my questions about their activist and theoretical work. I thank Hugo Vedovato and Chloë Taylor for reviewing the translation and Carrie Lou Hamilton for her detailed and generous reading of this chapter.

Notes

1 Carol J. Adams, *The Sexual Politics of Meat. A Feminist-Vegetarian Critical Theory* (Bloomsbury Academic, 2015).
2 Annie Potts, *Meat Culture* (Leiden and Boston: Brill, 2016).
3 Jacques Derrida, " 'Il faut bien manger' ou le calcul du sujet. Entretien (avec J.-L. Nancy)," *Points de suspension.* Entretiens (Paris: Galilée, 1992), 269–301.
4 Greta Gaard, "Toward a Queer Ecofeminism," *Hypatia*, no. 12 (1997): 114- 137.
5 Syl Ko and Aph Ko, *Aphro-ism. Essays on Pop Culture, Feminism and Black Veganism from Two Sisters* (New York: Lantern, 2017); Margaret Robinson, "Indigenous veganism: Feminist Natives do eat tofu," 2014, retrieved from: https://humanrightsareanimalrights.com/2014/12/22/margaretrobinson-indigenous-veganism-feminist-natives-do-eat-tofu
6 Sunaura Taylor, *Beasts of Burden: Animal and Disability Liberation* (New York: The New Press, 2017); Stephanie Jenkins, Kelly Struthers Montford, and Chloë Taylor, eds., *Disability and Animality: Crip Perspectives in Critical Animal Studies* (London and New York: Routledge, 2020).
7 Carrie Hamilton, *Veganism, Sex and Politics: Tales of Danger and Pleasure* (Bristol: HammerOn Press, 2019).
8 Teresa De Lauretis, "Displacing Hegemonic Discourses: Reflections on Feminist Theory in the 1980," *Return to Inscriptions* 3–4 (1988); Micaela Anzoátegui, "Desplazamientos de los discursos hegemónicos en la teoría feminista: el feminismo ecológico y animalista como nuevas perspectivas," *Nomadías*, no. 27 (2009): 33–50.
9 Teresa de Lauretis, "Sujetos excéntricos: la teoría feminista y la conciencia histórica," in *De mujer a género. Teoría, interpretación y práctica feminista en las ciencias sociales*, eds. María Cecilia Cangiano et al. (Buenos Aires: CEAL, 1993), 73–113.
10 Adrienne Rich, "Heterosexualidad obligatoria y existencia lesbiana," *DUODA Revista d'Estudis Feministes*, no. 10 (1996): 15–45.
11 Marta Cabrera and Liliana Vargas Monroy, "Transfeminismo, decolonialidad y el asunto del conocimiento," *Universitas Humanística*, no. 78 (2014): 19–37.
12 This chapter recognizes the power and use of the word "trans*" as an umbrella terminology capable of housing a wide variety of cis hetero-discordant identities, such as intersex, transsexual, transgender, among others. However, seeking to recognize Latin American and decolonial genealogies, it decided to use some words referring to identities historically and socially forged in these political territories. This is the case of *tortilleras*, somehow equivalent to the English "dykes"; or *travesti* and *trava* (these last two terms have no translation). I keep these terms in Spanish because it is in this language that they are transformed by activist movements from slurs into identity-affirming and politically disruptive categories. Therefore, translating them would erase years of political construction and render them insufficiently meaningful.
13 Danila Suarez Tomé, *Introducción a la teoría feminista* (Buenos Aires: Nido de Vacas, 2022), 49.
14 María Pía López, "Duelo, desobediencia, deseo," *Revista Institucional de la Defensa Pública de la Ciudad Autónoma de Buenos Aires* 8, no. 14 (2018): 236.
15 Malena Nijensohn, *La razón feminista. Políticas de la calle, pluralismo y articulación* (Buenos Aires: Las cuarenta, 2019), 35–36.
16 Mothers and Grandmothers of Plaza de Mayo.
17 Blas Radi, "Notas (al pie) sobre cisnormatividad y feminismo," *Ideas. Revista de Filosofía Moderna y Contemporánea*, no. 11 (2020): 24.
18 Camila Sosa Villada, "Con un feminismo así quién necesita enemigos," *Página 12. Suplemento Soy* (February 11, 2019), retrieved from: www.pagina12.com.ar/173332-con-un-feminismo-asi-quien-necesita-enemigos
19 Analú Laferal, "Transfeminismo antiespecista," *XIV Seminario Académico de Género y Diversidad Sexual*, Antioquia, septembre 21, 2021, retrieved from: www.youtube.com/watch?v=vW5PZ_H4Q4I&t=3275s
20 See note 12.
21 Lohana Berkins, "Un itinerario político del travestismo," in *Sexualidades migrantes. Género y transgénero*, ed. Diana Maffia (Buenos Aires: Feminaria Editora, 2003), 133.
22 Eduardo Mattio, "Una vez más, l*s sujet*s del feminismo," *Ideas. Revista de Filosofía Moderna y Contemporánea*, no. 11 (2020): 56.

23 Valentina Trujillo, "Repensar lo humano desde el transfeminismo antiespecista," *Analéctica* 8, no. 50 (2022): 212.
24 Anahí Méndez, "América Latina: movimiento animalista y luchas contra el especismo," *Nueva Sociedad* 288 (2020): 45–57.
25 Anahí Gabriela González and Iván Ávila, "Resistencia animal: ética, perspectivismo y políticas de subversión," *Revista Latinoamericana de Estudios Críticos Animales* I, no. 1 (2015): 38.
26 Anahí Gabriela González and Martina Davidson, "Alianzas salvajes. Hacia un animalismo decolonial, transfeminista y anticapacitista," *Revista Desbordes*, no. 13 (2022).
27 Latin American Institute of Critical Animal Studies (ILECA) and the Latin American Journal of Critical Animal Studies (RELECA), retrieved from: https://revistaleca.org/, www.institutoleca.org/, www.youtube.com/@ILECA
28 Iván Ávila, "El instituto latinoamericano de estudios críticos animales como proyecto decolonial," *Tabula Rasa*, no. 27 (2017): 345.
29 Thus, two thematic issues called "Feminisms, Gender(s) and Antispeciesism" (2016) and "Feminisms through the Looking Glass: Animality, Gender and Precarious Lives" (2021) were published, but also other issues where anti-capitalist and decolonial issues are addressed: https://revistaleca.org/index.php/leca/issue/archive
30 This network brings together transfeminist activists from Ecuador, Colombia, Argentina, Brazil, and other places.
31 Anahí Gabriela González, María Belén Ballardo, Martina Davidson, and Agustina Marín, "De(s) colonialidad y lazo común: el queer-odio y la animalidad como invenciones coloniales," *Madriguera Violeta* 3 (2022).
32 For example, the *Parole de Queer* dossier, in which some Latin American activists participate, retrieved from: https://es.scribd.com/document/441101334/ParoledeQueer-7-Antiespecista#, as well as number Special of *Animales y Sociedad*, retrieved from: www.animalesysociedad.com/wp-content/uploads/2022/03/AnimalesySociedadFinal-compressed.pdf
33 Anahí Gabriela González, "Lo animal como lugar de resistencia ante la trama sacrificial de la filosofía," *Ágora. Papeles de Filosofía* 38, no. 1 (2019): 101–122.
34 Luciana Carrera, Micaela Anzoátegui, Agustina Domínguez, "Inserte 'Animal' donde dice 'Mujer' y viceversa: analogías entre la dominación sobre las mujeres y la dominación sobre los animales en el sistema capitalista heteropatriarcal," *IV Jornadas CINIG de Estudios de Género y Feminismos y II Congreso Internacional de Identidades*, 2016, retrieved from: http://jornadascinig.fahce.unlp.edu.ar/iv-2016. In our region Monica Cragnolini have boarded the relations between sexism and speciesism from the intersection between ecofeminisms and post-Nietzschean philosophies. Cfr. Mónica Cragnolini, "La mujer, el animal y la carne. Escenas cotidianas del carnofalologocentrismo," *Cuadernos de Trabajo*, no. 2 (2016): 17–28.
35 Anahí Gabriela González, "Cuerpos (animales) que importan. Apuntes provisorios sobre la muerte del Hombre," *Anacronismo e irrupción* 8, no. 15 (2018): 33–55; Micaela Anzoátegui, "Vida precaria y duelo: Reapropiaciones para un abordaje no-antropocéntrico de Judith Butler," in *Judith Butler fuera de sí*, ed. María Luisa Femenías (Rosario: Prohistoria, 2017), 139–157.
36 In this context we can situate the works of Iván Ávila, Gabriel Giorgi, Val Trujillo, Analú Laferal, Martina Davidson, and Daniela Rosendo, as well as my own writings.
37 At this point, Carrie Hamilton's critique of Adams's contribution, as well as the proposals of Greta Gaard, Judith Butler, Monique Wittig, Teresa de Lauretis, Val Flores, and Paul B. Preciado, became fundamental.
38 Anahí Gabriela González, "Deshacer la especie: Hacia un antiespecismo en clave feminista queer," *Revista Tempo, Espaço e Linguagem* 10, no. 2 (2019): 45–70.
39 Lucía Pereyra Robledo and, Carli Prado, "Lo animal y lo cuir a través de las comunidades de compost. Reflexiones en torno a una fabulación especulativa en curso," *Revista Latinoamericana de Estudios Críticos Animales* 1, no. 8 (2021): 443–456.
40 In this movement, the works of Fabio Oliveira, Martina Davidson, Juan Ponce, Iván Ávila, and Lidia Guerra, among others, stand out.
41 Anahí Gabriela González, "Políticas feministas de la animalidad: decolonialidad, discapacidad y antiespecismo," *Instantes y Azares*, 26 (2021): 123–156.
42 María Lugones, "Colonialidad y género," *Tabula Rasa*, no. 9 (2008): 73–101.

43 Daniela Rosendo, Fabio Oliveira, and Tania Kuhnen, "Locus fraturado': resistências no Sul Global e práxis antiespecistas ecofeministas descoloniais," in *Feminismos decoloniais*, ed. María Clara Dias (Rio de Janeiro: Ape'Ku, 2020), 197.
44 Martina Davidson, "María Lugones e o pensamento de trincheiras: decolonialidade e veganismos," in *Feminismos decoloniais*, ed. María Clara Dias (Rio de Janeiro: Ape'Ku, 2020: 83–93; Iván Ávila, "Especismo antropocéntrico, veganismo moderno-colonial y configuración de formas-de-vida," in *La cuestión animal(ista)*, ed. Iván Ávila (Bogotá: Desde Abajo, 2016), 60–66.
45 Intersectional Anti-speciesist Feminism (FAI), Anti-speciesist Resistance (Buenos Aires), the San Juan Anti-speciesism Transfeminist Network, Anti-speciesist transfeminists (Córdoba), The animal is political (Santa Fe), *Tortilleras* for extinction (La Plata) and Decolonial Transfeminist Animalisms.
46 The *asado*, especially since 1940, took root in the life of the workers and emerged as a popular meal.
47 It is an Argentine street food that consists of a chorizo sausage in a bun. In addition, it is a food associated with social movements for social justice.
48 Luciana Carrera, Fer Guaglianone, and Ariel Di Paoli, Fanzine: "Como esas otras cosas," 2014, retrieved from: https://issuu.com/comoesasotrascosas/docs/_comoestasotrascosas
49 This definition of speciesism was thoroughly developed by Iván Ávila in his various texts.
50 Martina Davidson, "Feminismo e projeto decoloniais: ferramentas críticas para repensar o veganismo decolonial," *Diversitates International Journal* 1, no. 13 (2021): 1–24.
51 Fabio Oliveira, "Especismo Estructural: los animales no-humanos como un grupo oprimido," *Revista Latinoamericana de Estudios Críticos Animales* 8, no. 2 (2021): 180–123.
52 Maristella Svampa, *Las fronteras del neoextractivismo en América Latina: conflictos socioambientales, giro ecoterritorial y nuevas dependencias* (Guadalajara: CALAS, 2019).
53 Fer Guaglianone, "No me importa tu virus. Me importan los cerdos: acción gráfica viral callejera," *Revista Latinoamericana de Estudios Críticos Animales* 8, no. 2 (2021): 316–317.
54 Ibid., 320.
55 *Sudaca* is a Latin American self-claim that refers to cultural and political belonging to the American Global South.
56 Gabriel Giorgi, "La lección animal: pedagogías queer," *BOLETIN/17 del Centro de Estudios de Teoría y Crítica Literaria*, no. 17 (2013): 7.
57 Rizoma Podcast, retrieved from: www.youtube.com/playlist?list=PLwnd9pf2YV3aWc3lL5D2YKAJSZvaR-i1E
58 *Piquetero* is a term in Argentinian Spanish that designates the people that engaged in the movement started by unemployed workers who left the Argentine oil company YPF (*Yacimientos Petrolíferos Fiscales*), in the 1990. These individuals participate in protests and blockades, often obstructing roads or other key points to draw attention to issues such as unemployment, economic inequality, and concerns regarding social justice.
59 Maite Amaya, Interview by *La Condesa*, 2017, retrieved from: www.youtube.com/watch?v=L-e9VvSza48&t=2947s
60 Paulina Gaetán, *Preludio Travesti* (Pergamino: Antipoemario, 2022), 27.
61 Judith Butler, *Frames of War: When Is Life Grievable?* (London: Verso, 2009), 42.
62 Gaetán, *Preludio Travesti*, 10.
63 Hamilton, *Veganism, Sex and Politics: Tales of Danger and Pleasure*, 26–27.
64 See note 12.
65 Sosa Villada, "Con un feminismo así quién necesita enemigos."
66 Susy Shock, *Poemario Transpirado* (Buenos Aires: Nuevos Tiempos, 2011), 8.
67 Romina Vitale, Talk, "Capacitismo + Especismo. Desde un enfoque de la salud mental" (August 23, 2020), retrieved from: www.instagram.com/tv/CEP039HB-LW/?igshid=YWJhMjlhZTc%3D
68 In fact, the *Mapuche* community is not only formed by human members but also by the rivers, mountains, and other animals that inhabit the place.
69 Surama Lázaro, Interview with Moira Millán: "Resistencia e identidad desde el espíritu animal," *Canal UNED* (March 8, 2022), retrieved from: https://canal.uned.es/video/62272b13b6092324487a8b32
70 Cited in Martín De Mauro, "La huella animal, una mirada salvaje desde el transfeminismo" (April 29, 2022), retrieved from: www.diarioconvos.com/2022/04/29/la-huella-animal-una-mirada-salvaje-desde-el-transfeminismo/
71 González et al., "De(s)colonialidad y lazo común," 8.

Bibliography

Adams, Carol J. *The Sexual Politics of Meat. A Feminist-Vegetarian Critical Theory*. Bloomsbury Academic, 2015.

Amaya, Maite. Interview by *La Condesa* (2017). Retrieved from: www.youtube.com/watch?v=L-e9VvSza48&t=2947s

Anzoátegui, Micaela. "Desplazamientos de los discursos hegemónicos en la teoría feminista: el feminismo ecológico y animalista como nuevas perspectivas." *Nomadías*, no. 27 (2009): 33–50.

Anzoátegui, Micaela. "Vida precaria y duelo: Reapropiaciones para un abordaje no-antropocéntrico de Judith Butler." In *Judith Butler fuera de sí*, edited by María Luisa Femenías, 139–157. Rosario: Prohistoria, 2017.

Ávila, Iván. *Rebelión en la granja: biopolítica, zootecnia y domesticación*. Bogotá: Desde Abajo, 2013.

Ávila, Iván. "Especismo antropocéntrico, veganismo moderno-colonial y configuración de formas-de-vida." In *La cuestión animal(ista)*, edited by Iván Ávila, 60–66. Bogotá: Desde Abajo, 2016.

Ávila, Iván. "El instituto latinoamericano de estudios críticos animales como proyecto decolonial." *Tabula Rasa*, no. 27 (2017): 339–351. https://doi.org/10.25058/20112742.454.

Ávila, Iván. *De la isla del doctor Moreau al planeta de los simios: la dicotomía humano/animal como problema político*. Bogotá: Desde Abajo, 2017.

Berkins, Lohana. "Un itinerario político del travestismo." In *Sexualidades migrantes. Género y transgénero*, edited by Diana Maffia, 127–137. Buenos Aires: Feminaria Editora, 2003.

Butler, Judith. *Frames of War: When Is Life Grievable?*. London: Verso, 2009.

Cabrera, Marta, and Vargas Monroy, Liliana. "Transfeminismo, decolonialidad y el asunto del conocimiento." *Universitas Humanística*, no. 78 (2014): 19–37.

Carrera, Luciana, Anzoátegui, Micaela, and Domínguez, Agustina. "Inserte 'Animal' donde dice 'Mujer' y viceversa: analogías entre la dominación sobre las mujeres y la dominación sobre los animales en el sistema capitalista heteropatriarcal." *IV Jornadas CINIG de Estudios de Género y Feminismos y II Congreso Internacional de Identidades* (2016). Retrieved from: http://jornadascinig.fahce.unlp.edu.ar/iv-2016.

Carrera, Luciana, Guaglianone, Fer, and Di Paoli, and Ariel, Fanzine. "Como esas otras cosas." (2014). Retrieved from: https://issuu.com/comoesasotrascosas/docs/_comoestasotrascosas

Cragnolini, Mónica. "La mujer, el animal y la carne. Escenas cotidianas del carnofalologocentrismo." *Cuadernos de Trabajo*, no. 2 (2016): 17–28.

Davidson, Martina. "María Lugones e o pensamento de trincheiras: decolonialidade e veganismos." In *Feminismos decoloniais*, edited by María Clara Dias, 83–93. Rio de Janeiro: Ape'Ku, 2020.

Davidson, Martina. "Feminismo e projeto decoloniais: ferramentas críticas para repensar o veganismo decolonial." *Diversitates International Journal* 1, no. 13 (2021): 1–24.

De Lauretis, Teresa. "Displacing Hegemonic Discourses: Reflections on Feminist Theory in the 1980." *Return to Inscriptions* 3–4 (1988).

De Lauretis, Teresa. "Sujetos excéntricos: la teoría feminista y la conciencia histórica." In *De mujer a género. Teoría, interpretación y práctica feminista en las ciencias sociales*, edited by María Cecilia Cangiano et al, 73–113. Buenos Aires: CEAL, 1993.

De Mauro, Martín. "La huella animal, una mirada salvaje desde el transfeminismo." April 29, 2022. Retrieved from: www.diarioconvos.com/2022/04/29/la-huella-animal-una-mirada-salvaje-desde-el-transfeminismo/

Derrida, Jacques. " 'Il faut bien manger' ou le calcul du sujet. Entretien (avec J.-L. Nancy)." *Points de suspension*. Entretiens (Paris: Galilée, 1992): 269–301.

Gaard, Greta. "Toward a Queer Ecofeminism." *Hypatia*, no. 12 (1997): 114- 137.

Gaetán, Paulina. *Preludio Travesti*. Pergamino: Antipoemario, 2022.

Giorgi, Gabriel. "La lección animal: pedagogías queer." *BOLETIN/17 del Centro de Estudios de Teoría y Crítica Literaria*, no. 17 (2013).

González, Anahí Gabriela. "Cuerpos (animales) que importan. Apuntes provisorios sobre la muerte del Hombre." *Anacronismo e irrupción* 8, no. 15 (2018): 33–55.

González, Anahí Gabriela. "Lo animal como lugar de resistencia ante la trama sacrificial de la filosofía." *Ágora. Papeles de Filosofía* 38, no. 1 (2019a): 101–122.

González, Anahí Gabriela. "Deshacer la especie: Hacia un antiespecismo en clave feminista queer." *Revista Tempo, Espaço e Linguagem* 10, no. 2 (2019b): 45–70. https://doi.org/10.5935/2177-6644.20190019

González, Anahí Gabriela. "Políticas feministas de la animalidad: decolonialidad, discapacidad y antiespecismo." *Instantes y Azares* 26 (2021): 123–156.

González, Anahí Gabriela, and Ávila, Iván Darío. "Resistencia animal: ética, perspectivismo y políticas de subversión." *Revista Latinoamericana de Estudios Críticos Animales* I, no. 1 (2015): 31–72.

González, Anahí Gabriela, Ballardo, María Belén, Davidson, Martina, and Marín, Agustina. "De(s) colonialidad y lazo común: el queer-odio y la animalidad como invenciones coloniales." *Madriguera Violeta*, no. 3 (2022).

González, Anahí Gabriela, and Davidson, Martina. "Alianzas salvajes. Hacia un animalismo decolonial, transfeminista y anticapacitista." *Revista Desbordes*, no. 13 (2022). https://doi.org/10.22490/25394150.6775

Guaglianone, Fer. "No me importa tu virus. Me importan los cerdos: acción gráfica viral callejera." *Revista Latinoamericana de Estudios Críticos Animales* 8, no. 2 (2021): 314–325.

Hamilton, Carrie. *Veganism, Sex and Politics: Tales of Danger and Pleasure*. Bristol: HammerOn Press, 2019.

Jenkins, Stephanie, Struthers Montford, Kelly, and Taylor, Chloë, eds. *Disability and Animality: Crip Perspectives in Critical Animal Studies*. London and New York: Routledge, 2020.

Ko, Syl, and Ko, Aph. *Aphro-ism. Essays on Pop Culture, Feminism and Black Veganism from Two Sisters*. New York: Lantern, 2017.

Laferal, Analú. "Transfeminismo antiespecista." *XIV Seminario Académico de Género y Diversidad Sexual*, Antioquia, 21 (septembre, 2021). Retrieved from: www.youtube.com/watch?v=vW5PZ_H4Q4I&t=3275s

Lázaro, Surama. Interview with Moira Millán: "Resistencia e identidad desde el espíritu animal." Canal UNED, March 8, 2022. Retrieved from: https://canal.uned.es/video/62272b13b6092324487a8b32

López, María Pía. "Duelo, desobediencia, deseo." *Revista Institucional de la Defensa Pública de la Ciudad Autónoma de Buenos Aires* 8, no. 14 (2018).

Lugones, María. "Colonialidad y género." *Tabula Rasa*, no. 9 (2008): 73–101.

Mattio, Eduardo. "Una vez más, l*s sujet*s del feminismo." *Ideas. Revista de Filosofía Moderna y Contemporánea*, no. 11 (2020): 56–61.

Méndez, Anahí. "América Latina: movimiento animalista y luchas contra el especismo." *Nueva Sociedad* 288 (2020): 45–57.

Nijensohn, Malena. *La razón feminista. Políticas de la calle, pluralismo y articulación*. Buenos Aires: Las cuarenta, 2019.

Oliveira, Fabio. "Especismo Estructural: los animales no-humanos como un grupo oprimido." *Revista Latinoamericana de Estudios Críticos Animales* 8, no. 2 (2021): 180–123.

Potts, Annie, A. *Meat Culture*. Leiden, Boston: Brill, 2016.

Radi, Blas. "Notas (al pie) sobre cisnormatividad y feminismo." *Ideas. Revista de Filosofía Moderna y Contemporánea*, no. 11 (2020): 23–36.

Rich, Adrienne. "Heterosexualidad obligatoria y existencia lesbiana." *DUODA Revista d'Estudis Feministes*, no. 10 (1996): 15–45.

Robinson, Margaret. "Indigenous veganism: Feminist Natives do eat tofu." (2014). Retrieved from: https://humanrightsareanimalrights.com/2014/12/22/margaretrobinson-indigenous-veganism-feminist-natives-do-eat-tofu

Robledo, Lucía Pereyra, and Prado, Carli. "Lo animal y lo cuir a través de las comunidades de compost. Reflexiones en torno a una fabulación especulativa en curso." *Revista Latinoamericana de Estudios Críticos Animales* 1, no. 8 (2021): 443–456.

Rosendo, Daniela, Oliveira, Fabio, and Kuhnen, Tania. "Locus fraturado': resistências no Sul Global e práxis antiespecistas ecofeministas descoloniais." In *Feminismos decoloniais,* edited by María Clara Dias. Rio de Janeiro: Ape'Ku, 2020.

Shock, Susy. *Poemario Transpirado*. Buenos Aires: Nuevos Tiempos, 2011.

Sosa Villada, Camila. "Con un feminismo así quién necesita enemigos." *Página 12. Suplemento Soy,* February 11, 2019. Retrieved from: www.pagina12.com.ar/173332-con-un-feminismo-asi-quien-necesita-enemigos

Suarez Tomé, Danila. *Introducción a la teoría feminista*. Buenos Aires: Nido de Vacas, 2022.

Svampa, Maristella. *Las fronteras del neoextractivismo en América Latina: conflictos socioambientales, giro ecoterritorial y nuevas dependencias*. Guadalajara: CALAS, 2019.

Taylor, Sunaura. *Beasts of Burden: Animal and Disability Liberation*. New York: The New Press, 2017.
Trujillo, Valentina. "Repensar lo humano desde el transfeminismo antiespecista." *Analéctica* 8, no. 50 (2022): 204–225. https://doi.org/10.5281/zenodo.5894953
Vitale, Romina. Talk "Capacitismo + Especismo. Desde un enfoque de la salud mental." August 23, 2020. Retrieved from: www.instagram.com/tv/CEP039HB-LW/?igshid=YWJhMjlhZTc%3D

23

RE-AESTHETICIZING THE MIND

Art, Feminisms, and Animals in Los Angeles (1970s-1980s)

Emilie Blanc

In 2015, the National Museum of Animals & Society in Los Angeles hosted the exhibition *The Sexual Politics of Meat*, referring to Carol J. Adams' groundbreaking book of the same title (1990). Curated by Kathryn Eddy, Janell O'Rourke, and L.A. Watson, founders of the collective ArtAnimalAffect, the exhibition showcased artworks by female artists exploring human–animal relationships.[1]

In this chapter, I focus on artworks made during the 1970s and the 1980s in Greater Los Angeles—which was a very active place for feminist art through the Feminist Art Program and the Woman's Building, among others—thus underlining the long-standing engagement in animal condition from an intersectional lens by feminist and women artists.

Recent publications—such as Elizabeth Sutton's book *Art, Animals, and Experience: Relationships to Canines and the Natural* (2017)—have shown growing interest in the links between art and the status of animals. Indeed, artistic practices are significant ways to disrupt patterns of representations and to envision new ways of imagining relationships between human and nonhuman animals. As Sutton outlines:

> Art's capacity to evoke our human feelings and emotions is powerful because emotions help us empathize; they help us imagine. Artists working today understand that art can provide the impetus for considered emotional looking, and artists use animals to focus viewer attention on significant questions about human relationships to animals, nature, and the world.[2]

Nevertheless, the contributions of feminist art to the thinking of animal exploitation through analysis of linked oppressions have been underestimated.

In her article "An/Aesthetics: The Re-Presentation of Women and Animals" (1985), Marti Kheel states: "As with women, animals have been re-created and portrayed though men's eyes."[3] In the footsteps of ecofeminism and social liberation movements, the co-founder of Feminist for Animal Rights (1982, California) associates the patriarchal mind

DOI: 10.4324/9781003273400-29

with an an/aesthetic one, based on the split between *anima* [spirit], considered a feminine notion, and its male counterpart *animus* [mind]:

> The an/aesthetic mind has deformed and infected not only its own consciousness but also that of all life on this planet. Though its portrayal of all life as re-presenting an object of use for itself, the an/aesthetic mind has an/aesthetized the anima within us all. . . . He has projected his fears onto women and animals. He has seen in them wild and untamed nature that must be subdued; this subduing has taken the form of an/aesthesia.[4]

What does feminist art have to offer to theories of multispecies justice? How does it contribute to awake the *anima* within and re-aestheticize us, as Kheel called for? Drawing from my PhD research in art history about identity politics and visual arts in California, this chapter is the first step of an ongoing research project on feminist visual politics of multispecies liberation. My methodology blends art history, critical animal studies, and feminist studies. From analysis of artworks by feminist and women artists based in Southern California during the 1970s and 1980s, I examine the interrelation between art, feminisms, and animals under three headings: violence, food, and relationships.

Violence

During the 1970s, in women's studies class and consciousness-raising groups, women began to speak more and more about physical abuse and sexual assault, which helped to make these taboos heard and to take action. Many women artists have contributed to expand feminist perspectives on violence, such as in *Ablutions* (1972), a pivotal piece on rape performed in Laddie John Dill's studio in Venice. How did these artists connect the abuse of women and the abuse of animals?

Violence Against Female Bodies

Created by Judy Chicago, Suzanne Lacy, Sandra Orgel, and Aviva Rahmani, *Ablutions* was based on recordings of rape as a soundtrack to a series of symbolic actions: a woman was bound with white gauze, two others bathed in tubs filled with eggs, cow's blood and clay, and hundreds of beef kidneys were nailed to the wall while broken eggshells and animal organs were scattered on the floor. The collective performance aimed at exposing hidden experiences of rape and at generating a public context allowing women to speak out. In relation to our topic, I am particularly interested in the significance of the use of eggs as well as animal organs. As Rahmani remembers, Orgel developed the image of a naked woman bathing in eggs.[5] According to Chicago: "She [the performer] started to wash herself, allowing the eggs to run down her body, an image of immersion in her own biology."[6] Her interpretation emphasizes the exploitation of the reproductive abilities of women as a means of restraint within a patriarchal society. I would add that chickens as well as cows produce eggs and dairy products for humans during their lives before being slaughtered, thus their femaleness is exploited as well. In this regard, Kheel notes that female farm animals are probably the most exploited females in the world.[7] In her autobiography, Rahmani recounts a story about the choice of the eggs. In charge of buying them, she purchased organic ones in a food co-op to ensure that every aspect of the work was consistent with her values, which upset Chicago certainly due to their high price.[8] As Margo DeMello highlights, over 90 percent of eggs worldwide are produced in battery conditions, noting:

> One of the worst examples of misery is found in the lives of egg-laying hens. They are confined to tiny cages without enough room to spread their wings or lie down, they are kept in the dark at all hours, and their beaks are burnt off to keep them from pecking other birds to death. These animals get no sunlight, no dirt, no grass, and no relaxation, but must lay eggs for human consumption until they die of exhaustion.[9]

Rahmani's anecdote brings to light the complex entanglements between capitalism and food industry—a system that fosters disparities in access to quality food for humans, causes environmental damage, and mistreats animals—in which animals are not treated as sentient creatures but as production units. It also questions how we can use animals in artworks from a feminist approach: what does it mean to use animals only as symbols of women's oppression? Wouldn't this imply creating a relationship of domination over animals? Indeed, if we refer to Carol J. Adams, using animals as metaphors only to forward women's issues, not animals', reinforces a patriarchal structure.[10]

As I have mentioned, beef kidneys were also nailed to the wall during *Ablutions*. Lacy tied them together, pulling the rope tight so that the blood ran out of the organs and down the wall. This action evokes *Slaughterhouse* (1972) performed by Faith Wilding—who co-founded with Lacy the first women's studies class at the Fresno State Experimental College (1969)—while studying at the Feminist Art Program, established first in Fresno by Chicago. At the end of the performance, buckets of animal blood collected from a slaughterhouse were thrown over Wilding, who was suspended from a meat hook by her bound hands. Meanwhile, sounds of bellowing cows were played as images of strung-up cows were projected onto her bloodied body. The ritual-based performance expressed the relationships between violence against animals and women, as Wilding outlines: "This [the slaughterhouse] is violent territory, a place of daily death and sacrifice as hidden from public life as are our own lives of women."[11]

From 1969 to 1972, Rahmani also used animal organs in *Meat Piece* (restaged for stills in 2005) (Figure 23.1). In this series of collaborative films, hands slowly manipulate in close-up framing a raw cow's heart that could be read as a vulva.[12] Rahmani's choice to use a cow's heart came from her first drive to California in 1968: with her husband, they stopped at a fence where a herd of cattle came to check them out, expressing curiosity. This brief experience had a profound impact on her to grasp how these dairy cows were sentient:

> So later, when I sought a metaphor for rape, using a cow's heart was obvious. The fact it had once sustained life for one of those beautiful, gentle creatures was so symbolic. . . . The ultimate machine to consume animals, like women were consumed then as part of rape culture. It seemed horrific to manipulate it. But that's how many men then thought about sexuality and how they experienced rape: women as a piece of raw meat, a woman's heart as something to manipulate.[13]

As metaphors of the visceral experience of sexual abuse as well as the violence of slaughtering, the films visualize the power exercised through the control of female bodies. To quote Kheel again: "In patriarchal society, both women and animals are consumed as flesh."[14] As

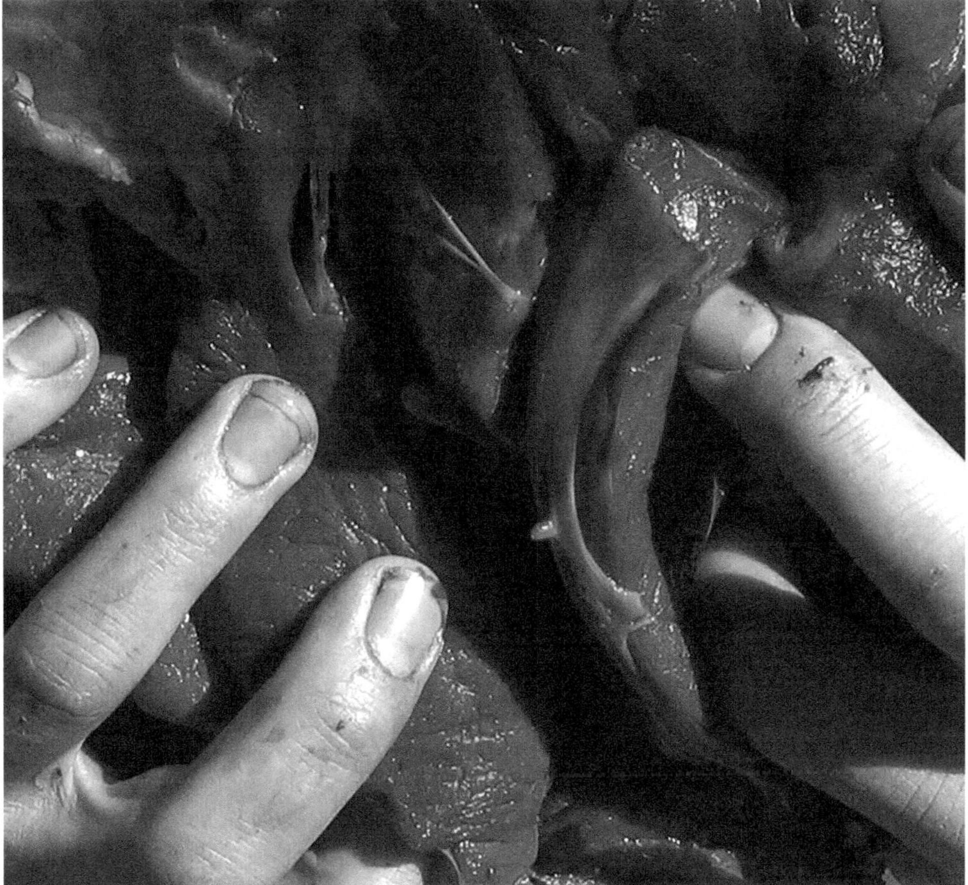

Figure 23.1 Aviva Rahmani, *Meat Piece*, video, 1969–72, still from 2005 restaging, copyright 2023 Aviva Rahmani.

Adams also points out: the phrases about feeling like a piece of meat used by rape victims when describing their feelings suggest that "animals' fate in meat eating is the immediate touchstone for their own experience."[15] Just as the slaughterhouse treats animals as inert objects, so too in rape are women. As Rahmani pointed out to me, it is also important to consider that men are also victims of rape:

> the real problem is the cultural associations and parallels between sexuality, violence, degradation, emotional evisceration as a routine experience that reflects other forms of exploitation and extraction. The war against nature is also the war against intimacy, ultimately with self.[16]

In a *Meat Piece* video, we can read "We raped the earth." Rahmani argues:

> Ecofeminists have said that rape is a pitiless colonization of the other as property, an exercise of power possible only in the absence of empathy. It is an unspeakable

exercise in the taking of control of one person by another, one who has an arbitrary measure of greater power, analogous to how humans routinely act as one species dominating another to extract "value".[17]

In her trigger point theory that merges aesthetic and activism to foster change, Rahmani draws on empathy, asking: "Would we perpetrate rape or habitat fragmentation if we felt empathy for all life?"[18] Othering refers to the practice of making humans or animals different in order to justify treating them differently. The way we distance ourselves from what is different by equating it with something we have already objectified is also at stake in Nancy Buchanan's *Deer/Dear* (1978) (Figure 23.2).

Predation

In 1978, Buchanan responded to the closeness of serial murders of women committed by two men from October 1977 to February 1978 in Los Angeles. A text by Deirdre English published in *Mother Jones* helped her understand how one human being can dissociate from another to commit an act of extreme violence.[19] In her article, English associates the bodies of women with hunters' prey, such as ducks and deer, which are considered inferior to human beings. Kheel also links hunting and sexual assault, stating that in both acts of

Figure 23.2 Kitty Hodge performing in *Deer/Dear*, Nancy Buchanan, January 1978, Santa Ana College, photo by Mayde Herberg, copyright 2023 Nancy Buchanan.

violence "the victim is seen to represent a denied part of the self; in both acts, the true intent of the rapist/hunter is disguised in an intricate web of rationalizations and projections."[20]

Deer/Dear was performed in January 1978 at Santa Ana College. Propped up in a sleeping bag, Buchanan freed herself from a heavy rope, her voice recounting a recurrent dream of men with an endless list of weapons. Then, seven slide-lecture stories about fear and feeling vulnerable began. During each light-interlude, Buchanan stapled up a deer silhouette on the wall. Meantime, three other women performers became active. While Kitty Hodge—her head positioned in front of deer antlers—froze in an expression of terror, Terryl Hunter held a spotlight on her face and recited fragments of a statement about women, such as "like deer, existing for the benefit of hunters."[21] As for Merion Estes, she moved forward slowly from the background, gradually revealing a gun that she used to shoot at the seven deer silhouettes during the sixth and seventh interludes. All the performers were dressed in military uniforms, emphasizing that violence against women and animals is a form of warfare. The first six narratives were told in the third person, while in the final narrative Buchanan spoke directly about her own feelings regarding the recent femicides. At the end, Buchanan pulled Hodge away from the wall and opened up a space for women—the "dear" ones—to talk about their experiences. "As a travelogue in fear,"[22] *Deer/Dear* brings to light how women, like deer, are trained to be fearful in the face of predators.

The artworks discussed in this first section address the intertwining violence against women and animals whose bodies are dispossessed and controlled through sexual assault, reproduction, slaughtering, and hunting.

Food

In April 1978, Buchanan re-acted *Deer/Dear* in Las Vegas as part of *From Reverence to Rape from Respect*, an event set up by Ariadne: A Social Art Network founded by Suzanne Lacy and Leslie Labowitz-Starus (then Labowitz). On January 7, 1979, the collective co-sponsored with the Lesbian Art Project "a dialogue on vegetarianism, meat-eating, violence against animals and its relationship to violence against women"[23] at the Woman's Building. This event highlights how "food is a feminist issue" as outlined in the issue 21 of *Heresies: A Feminist Publication on Art & Politics* (1987). Both Lacy and Labowitz-Starus have raised the issue of food, gender, and violence in their works. How did they specifically address the issue of animals in human food and its relationships to women's experiences?

Women and Meat

From 1974 to 1977, Lacy worked on the series *Anatomy Lessons*, which brings to mind Susan Griffin's book *Woman and Nature. The Roaring Inside Her* (1978) in which two paragraphs have the same title. In connection with Lacy's prior zoology and pre-medical studies at the University of California, Santa Barbara, the series' title evokes the confiscation of knowledge about women bodies by the patriarchal system, as Griffin as well as Carolyn Merchant have demonstrated.[24] *Anatomy Lesson #2: Learn Where the Meat Comes From* (1976) is a performance created for video—a fourteen-minute color video directed by Hildegard Duane—and as a photographic series. In this parody of a cooking TV show, recalling Julia Child's *The French Chef* (1963–1973), Lacy gives tips for successful lamb recipes by learning to speak the same language as the butcher, giving details of the lamb's anatomy to the point of imagining yourself as the animal, feeling its muscles, and considering how best to cook them. She gradually turns into a vampire, ending up wrestling with the carcass and devouring the cooked dish, a roast leg of lamb. Behind Lacy, the chart reproducing a picture of the cuts of meat evokes Adam's *The Sexual Politics of Meat* book cover showing a naked white woman dismembered into body parts. Arguing that meat-eating has long been associated with masculinity since the 1970s,[25] Adams foregrounds a cycle of objectification, fragmentation, and consumption that links butchering to sexual violence. She points out how animals become absent referents though butchering: "Animals in name and body are made absent *as animals* for meat to exist."[26] Indeed, fragmented body parts of animals are usually renamed to obscure the fact that these were once animals. For instance, after death, cows may become roast beef or hamburger. Adams also highlights that the absent referent process is activated "not only by changing names from animals to meat, but also by cooking, seasoning, and covering the animals with sauces, disguising their original nature,"[27] as in Lacy's artwork. The notion of the absent referent is close to the process of turning animals into images far from the real violence suffered by animals in food industry, as Alice Walker describes it: "the animals are forced to become for us merely "images" of what they once so beautifully expressed. And we are used to drinking milk from containers showing "contented" cows, whose real lives we want to hear nothing about, eating eggs and drumsticks from "happy" hens, and munching hamburgers advertised by bulls of integrity who seem to command their fate."[28] In addition, Lacy highlights how TV cooking shows lessen the violence inherent in meat-eating by the use of soft music and female voices.

In *Anatomy Lesson #2: Learn Where the Meat Comes From*, Lacy explores the relationship between consumption of the female body and food, between being flesh, cooking flesh and consuming flesh. Using fantasy to provoke a distancing from the meat-eating norm, Lacy makes visible the violence of women's intimate association with food, usually erased from these TV programs. She examines the roles of women in the meat-eating system, challenges the usual separations between the bodies of humans and animals, and highlights the relationships between violence against women and animals. In this regard, Lacy's choice of lamb—which is also present in her other artworks—is meaningful. The notion of innocence and vulnerability associated with lambs is implicit in the idiom "like a lamb to the slaughter" which means without expecting that something bad will happen. In her analysis of Lacy's artwork, Emily Elizabeth Goodman explains that the comparison between lambs and women "asserts the innocence of the victims of rape and assault and likens the vulnerability of the lambs' exposed flesh to those of the women who have been so cruelly violated by these acts of sexual aggression."[29] Relying on the representations of lambs as sacrificial animals in various religions, Griffin compares woman to "a milk-white lamb that bleats/ For man's protection."[30] She thus links the vulnerability of lambs and women, reminding us that both can be worshiped as well as sacrificed.[31]

Sprouting as Life-Affirming

In 1980, by starting the *Sprout Time* series "which was about rebirthing and germinating and growing,"[32] Leslie Labowitz-Starus engaged in a healing journey for herself, her family, and the violence experienced by women. By becoming an urban farmer in her Venice yard, she connected her feminist stance to ecology. Through this series, Labowitz-Starus has examined how sprouting can be a sustainable food as well as a viable business run by a woman *and* an artist. While maintaining a stand in Santa Monica Market, she has staged several performances, such as *Sproutime* (October 26, 1980), in which she welcomed the public into her seed-growing space. The audience was immersed in a sensory experience composed of sounds, readings, lights, smells, and human contact. At the end of the performance, Labowitz-Starus arranged a buffet sprout lunch during which she explained how sprout growing had become a metaphor for her own self growth and discussed "the beneficial effects of eating 'living food' on one's health."[33] According to Adams, economies dependent mainly on the processing of animals for food include sexual segregation in work activities while plant-based economies are more likely to be egalitarian by fostering women's autonomy.[34] Adams also outlines that vegetarian activities counter patriarchal consumption and challenge the consumption of death: "Feminist-vegetarian activity declares that an alternative worldview exists, one which celebrates life rather than consuming death; one which does not rely on resurrected animals but empowered people."[35] In addition, as DeMello points out: "More animals die—billions per year—and, ironically, more people suffer from hunger, as grains and water are diverted into meat production."[36] Considering food as a political tool and denouncing the privatization of seeds, Labowitz-Starus states: "How important the system of growing food is, how the food is distributed, who gets the food, and who controls the food is the root of our needs."[37] After Ariadne's antiviolence work, she became aware of the psychological impact of being a child of a Holocaust survivor. Her commitment to a life-affirming artwork may be interpreted as a way of repairing her family wounds marked by the horrors of the Nazi camps "modeled on American stockyards and slaughterhouses"[38] in the words of DeMello. In 1984 with the support of the Long Beach Museum, Labowitz-Starus created a version of the old Jewish cemetery of Prague she visited in the 1970s with handmade replicas of signs of the anti-war movement by women along with signs of how to make sprouts and sprout recipes. Every week, one of her employees stacked trays of greens and seeds that were already growing on the other side of the cemetery to "see life regenerating."[39] Labowitz-Starus dedicated the piece to her mother and her daughter Aria, noting: "It was for the women of the past and the women of the future."[40] She thus reflected on the life-sustaining heritage of female work.

If Lacy's and Labowitz-Starus' artworks can be associated with vegetarianism, they do not call for it, which evokes Ariadne's event whose goal was to be a forum for violence against animals and women, and not "an attempt to proselytize."[41] Both ask questions about human food habits, meat-eating in particular, and how these are entangled in logics of domination. After having discussed pieces related to the issues of violence and food, I end this text by analyzing artworks related to subjective experiences of relationships between women and animals: how did they counter anthropocentric priorities?

Relationships

In her poem "Am I Blue?" (1986), Alice Walker highlights the importance of the entanglement between human animals and nonhuman animals that we can forget as adults: "People like me who have forgotten, and daily forget, all that animals try to tell us. 'Everything you do to us will happen to you; we are your teachers, as you are ours. We are one lesson' is essentially it, I think. There are those who never once have even considered animals' rights: those who have been taught that animals actually want to be used and abused by us, as small children 'love' to be frightened, or women 'love' to be mutilated and raped."[42] Indeed, re-considering the relationships between humans and animals is a key step to end multispecies violence. In this last section, I focus on interconnectedness in relation to Suzanne

Jackson's artist's book *Animal* (1978), and on mourning in relation to Terry Wolverton's performance *Familiar* (1984): how did these works envision alternative ways of imagining human–animal bonds?

Interconnectedness

In 1978, the final year of her artist residency with Brockman Gallery Productions, Suzanne Jackson published *Animal*, her second artist's book, reflecting the centrality of animals to her art practice and life since childhood in San Francisco and later in Fairbanks, Alaska, where she joined the National Audubon Society at age nine. In the Yukon Territory, she was surrounded by a vast landscape of tundra inhabited by many varieties of birds as well as moose, foxes, wolves, and bison. From eighth grade through four years of high school Jackson entered the international poster competition for "World Peace and Kindness to Animals" in which she won prizes each year with her posters including animals and symbolic representations of world peace.[43]

Combining paintings and poetry, the book intertwines her human feelings and experiences with her thoughts on and her relationships with animals. Jackson's original intention referenced a documentary that she watched on TV, showing various ways in which animals in nature exchanged care and love for one another. At that time, she embraced nature in activities of everyday life, even in Los Angeles where she enjoyed the raw environment in the surrounding mountains, Pacific Ocean, and wildlife.[44] With a reproduction of *Poisson d'or* (1978) depicting two red goldfish—one of wish bubbles a heart in the foreground—the book cover introduces the reader to Jackson's tribute to animals. Two drawings open and close the book, respectively: a bird surrounded by a heart, and a delicate self-portrait (1972) in which her hair is an extension of a plant element. In the whole book, Jackson, who declares herself as womanist in line with Alice Walker, expresses the interconnectedness between humans, non-human animals, and nature by exploring aspects of ecowomanism—a term coined by Melanie L. Harris to describe an approach to environmental justice by women of African descent based on race-class-gender intersectional analysis.[45] In Jackson's words: "my belief is that we are not entities unto ourselves, that all things between us as a person and the next object or the thing that's out there or plant or animal, there are a lot of molecules and atoms that we don't see, but we're all connected from molecule to molecule."[46] Jackson doesn't support the separation between nature and culture—the dominant paradigm in Western societies since the 16th century—instead fostering interconnected approaches more common in African and Indigenous societies. As she explained to me:

> We grew up being exposed to the traditional ways of the Inupiat, Athapascan, Y'upik people. . . . We learned in school and in the community, that animals in Indigenous traditions, African, North, and South American cultures are central to spirit designations of life, death, and survival.[47]

In the early 1970s, she also conducted research on African retentions in the Americas.

In Jackson's artist's book, relationships with animals are neither passive nor authoritarian; the images as well as the texts reflect intimacy, connivance, affection, sensitivity, and organicity. For instance, *Animal* (1974) shows a bird, linked to forms of different colors evoking humans, animals, and plant elements (Figure 23.3). Jackson's careful depictions

Figure 23.3 Suzanne Jackson, *Animal*, acrylic wash, 80" × 80", 1974, Courtesy of the California African American Museum. Collection of Friends, the Foundation of the California African American Museum. Gift of Joan D. Payne.

of animals through many layers of acrylic paint could be a metaphor for the mutual well-being, joy and enrichment of animal–human relationships or, put another way, the need to care for each other, humans and animals. The book emphasizes the importance of relationships with animals in the construction of self, as a vehicle of sensations and emotions and a source of strength.

Jackson reflects the significance of the contacts with animals as spiritual exchanges. Walker likewise states: "But I am also telling you that we are connected to them [animals] at least as intimately as we are connected to trees. Without plant life human beings could not breathe. Plants produce oxygen. Without free animal life I believe we will lose the spiritual equivalent of oxygen."[48] In addition, Jackson's free associations of symbols, animal, human, and nature elements floating in an undefined setting allow a multiplicity of meanings. She doesn't capture animals in her pieces, rather she lets them exist freely.

Jackson sought to create works that would be recognized as valuable within a broader cultural environment. The freedom of her work reflects her own freedom to express her subjectivity, which caused her misunderstandings. Indeed, her intersectional approach was not clearly in line with human-centered social justice claims: "just for a long time people couldn't figure out where my work was coming from, where to put me. I wasn't a feminist, my work didn't look like the other black artists' work."[49] I would argue that the pejorative associations between animals, women, and non-white people used in language and imagery may have undermined the apprehension of her work, as if her interest in animals would reinforce racist and sexist stereotypes. According to the ecofeminist scholar Greta Gaard:

> the *feminist* empathy for animal suffering, articulated as an ethic of care, was soon *feminized* and women's activism for animal rights was mocked as a movement of "little old ladies in tennis shoes": in male-supremacist (patriarchal) cultures, the association of women and animals reinforces their subordinate status.[50]

In the words of Amie Breeze Harper, who analyzes the connections between whiteness, racism, and animal abuse:

> pro-slavery whites deeply believed that Africans could not feel pain; that we were believed to be "just like animals" who had no feelings, spirits, souls; we were just machines available to serve the purposes of white America.[51]

In this regard, *Witches, Bitches, and Loons* (1972) published in Jackson's artist's book could visualize the intertwined oppressions of animals and Black women, although this is not the artist's original intention.[52] From the silhouette of a female body emerge two women's faces that could symbolize two types of women, witches and bitches as the title suggests, both attacked for their non-conformity to the dominant model of female sexuality in patriarchy. These women are connected to the silhouette of a bird, to the representation of a loon and to vegetal elements. Known by their specific calls, loons are often considered symbols of wilderness, an area in which women and non-white people have been constrained to support the need to control them. In this context, the loon's call can be associated with a call for multispecies justice.

Jackson enhances the importance of animals and nature in her life, especially on a spiritual level. In her book, I feel that she doesn't relegate animals to a secondary position compared to humans: their exchanges are based on harmony and interconnectedness.

Mourning

How do humans view the loss of an animal, is it less important than the loss of another human? Grief at the death of a pet continues to be devalued in society, leading humans to grieve the loss without social support.[53] Nevertheless, as DeMello points out referring to Elizabeth Kübler-Ross, the same stages of grief regarding human grief also apply to those grieving the loss of their animal companions.[54] Invited to be part of the exhibition *Offerings: The Altar Show* in 1984 at Social and Public Art Resource Center (Venice), Terry Wolverton seized this issue by creating a ritual performance out of which her altar would grow (Figure 23.4). Following the loss of her beloved cat Ruby the previous year, she chose to explore human–pet relationships, emphasizing: "the way we achieve an intimate bond with animals that we sometimes cannot forge with other humans."[55] Fostering collaboration in her artistic practice, she asked Kim Dingle to work with her on the visual elements of the piece and invited five other women—Nancy Angelo, Jacqueline De Angelis, Bia Lowe, Rachel Rosenthal,[56] Jere Van Syoc—to perform vignettes about the animals they have lived with.

Seated on the floor on three sides of a circular area, the audience was requested to bring objects or visual elements of their pets. A tape of outdoor sounds—birds and leaves—welcomed Wolverton who began to pour colored earth to cast the circle before starting to

Figure 23.4 Terry Wolverton, *Familiar*, 1984, Photo © Janice Felgar, 1984.

narrate her relationship with Ruby, a stray cat. She related their meeting, the basis of their relationships—"I wouldn't tell her what to do, and she wouldn't tell me"[57]—the difficulties they went through, their move to California, their moments of tenderness, anger and lack of understanding, as well as the pain of loss. Wolverton's story was interspersed with the stories of each guest artist. In their narratives, I grasp the significant role of their companion animals in their human lives and the strength of their shared emotional bonds. For instance, De Angelis considered her cat Sophie her first daughter. Ruby comforted Wolverton when her grandmother passed away; she also expressed her disapproval of not having moved in Wolverton's new home, even if it is complex to gauge the emotional bond of an animal through human feelings. These testimonies are consistent with the experiences of co-construction described by Donna Haraway: "stories about relating in significant otherness, through which the partners come to be who we are in flesh and sign."[58] During the performance, each participant was invited to infuse a stick of colored modeling clay with the feelings that came up for them as they watched the action. After Wolverton's story, each audience member made a totem to his or her beloved familiar: "they seemed to appreciate the opportunity to express their deeply held feelings for their pets."[59] At the end of the performance, the altar was created with their totems.

Through this collective ritual performance, Wolverton asserted that there is no hierarchy in mourning between humans and animals. Like Jackson, she emphasized the significance of human–animal relationships beyond anthropocentric frames.

Conclusion

Through feminist and womanist's perspectives about violence, food, and relationships, these artworks express how animal exploitation is woven into the larger systems of oppression, such as sexism, racism, and economic exploitation. Their conceptual, formal, and symbolic languages raise awareness and engage emotions as well as imaginaries. These artists interrogated mistreatments of humans, animals, and nature, and envisioned alternative relationships between human and nonhuman animals, calling for a common liberation and enriching multispecies justice. They did not disassociate animals from their experiences of oppression and from their lives; they did not try to submit them to anthropocentric interests, nor to objectify them. Rejecting both the patriarchal system and animal abuse, they encouraged to re-aestheticize the mind in the Greater Los Angeles of the 1970s and 1980s. The issues they raised are still relevant and help us think about animal liberation from an intersectional lens.

Many thanks to Nancy Buchanan, Christopher John Davis, Suzanne Jackson, Aviva Rahmani, and Terry Wolverton for their support for producing this chapter.

Notes

1 Kathryn Eddy, Janell O'Rourke, and L.A. Watson, *The Art of the Animal: Fourteen Women Artists Explore the Sexual Politics of Meat* (New York: Lantern Books, 2015).
2 Elizabeth Sutton, *Art, Animals, and Experience: Relationships to Canines and the Natural World* (New York: Routledge, 2017), 7.
3 Marti Kheel, "An/Aesthetics: The Re-Presentation of Women and Animals," *Between the Species* 1, no. 2 (1985): 38. https://digitalcommons.calpoly.edu/cgi/viewcontent.cgi?article=1412&context=bts
4 Ibid., 43.
5 Aviva Rahmani, *Divine Chaos: The Autobiography of an Idea* (New York: New Village Press, 2022), 136.
6 Judy Chicago, *Through the Flower: My Struggle as a Woman Artist* (Garden City: Double Day, 1975), 218.
7 Kheel, "An/Aesthetics: The Re-Presentation of Women and Animals," 42.
8 Rahmani, *Divine Chaos: The Autobiography of an Idea*, 136.
9 Margo DeMello, *Animals and Society: An Introduction to Human–animal Studies* (New York: Columbia University Press, 2012), 264.
10 Carol J. Adams, *The Sexual Politics of Meat: A Feminist-Vegetarian Critical Theory* (New York: Bloomsbury Academic, 2015), 41.
11 Faith Wilding, "Gestations of a Studio of Our Own: The Feminist Art Program in Fresno, California, 1970–71," in *A Studio of Their Own: The Legacy of the Fresno Feminist Experiment*, ed. Laura Meyer (Fresno: Press at California State University, 2009), 90.
12 Rahmani, *Divining Chaos: The Autobiography of an Idea*, 140. All the original films were lost.
13 Aviva Rahmani, E-mail correspondence with author, February 28, 2023.
14 Kheel, "An/Aesthetics: The Re-Presentation of Women and Animals," 41.
15 Adams, *The Sexual Politics of Meat: A Feminist-Vegetarian Critical Theory*, 34.
16 Rahmani, E-mail correspondence with author, February 28, 2023.
17 Rahmani, *Divining Chaos: The Autobiography of an Idea*, 92.
18 Ibid., 51.
19 Nancy Buchanan, E-mail correspondence with author, July 28, 2022.
20 Kheel, "An/Aesthetics: The Re-Presentation of Women and Animals," 41.
21 Nancy Buchanan, "Deer/Dear," *High Performance* 1, no. 2 (June 1978): 33.
22 Ibid.
23 Ariadne: A Social Art Network, "Two Upcoming Activities," in *Woman's Building Records, 1970–1992*, Box 14, Folder 62 (Washington DC: Archives of American Art. Smithsonian Institution, n.d.).

24 Susan Griffin, *Woman and Nature: The Roaring Inside Her* (New York: Harper & Row Publishers, 1978), and Carolyn Merchant, *The Death of Nature. Women, Ecology, and the Scientific Revolution* (1980, reis., New York: Harper & Row Publishers, 1989).
25 Carol J. Adams, "The Oedible Complex: Feminism and Vegetarianism," in *The Lesbian Reader*, eds. Gina Covina and Laurel Galana (Oakland: Amazon Press, 1975), 145–152.
26 Adams, *The Sexual Politics of Meat: A Feminist-Vegetarian Critical Theory*, 20.
27 Ibid., 28.
28 Alice Walker, "Am I Blue?" in *Living by the Word: Essays* (1988, reis., New York: Open Road Integrated Media, 2011).
29 Emily Elizabeth Goodman, *Food, Feminism, and Women's Art in 1970s Southern California* (Abingdon, NY: Routledge, 2022), 26.
30 Griffin, *Woman and Nature: The Roaring Inside Her*, 27.
31 See also Kheel, "An/Aesthetics: The Re-Presentation of Women and Animals," 39.
32 Leslie Labowitz, "Los Angeles Goes Live: Performance Art in Southern California, 1970–1983. Oral History Interviews," Interview by Jade Thacker (Los Angeles: Los Angeles Contemporary Exhibitions, Getty Foundation, 2010), 11. https://rosettaapp.getty.edu//delivery/DeliveryManagerServlet?dps_pid=IE719948
33 Leslie Labowitz, "Sproutime," *High Performance* 3, no. 3–4 (Fall/Winter 1980): 66.
34 Adams, *The Sexual Politics of Meat: A Feminist-Vegetarian Critical Theory*, 14.
35 Ibid., 175.
36 DeMello, *Animals and Society: An Introduction to Human–animal Studies*, 275.
37 Leslie Labowitz-Starus, "Interview with Leslie Labowitz: Performance, Environmentalism, Feminism, and Politics," interview by Jacki Apple. *WEAD Magazine*, no. 11 (October 2021). https://directory.weadartists.org/leslie-labowitz-sprout-actions
38 DeMello, *Animals and Society: An Introduction to Human–animal Studies*, 267.
39 Labowitz-Starus, "Interview with Leslie Labowitz: Performance, Environmentalism, Feminism, and Politics."
40 Ibid.
41 Ariadne: A Social Art Network, "Two Upcoming Activities."
42 Walker, "Am I Blue?"
43 Suzanne Jackson, E-mail correspondence with author, April 12, 2023.
44 Ibid.
45 See Melanee C. Harvey, "People, Nature, and the Spiritual World: The Ecowomanist Art Practice of Suzanne Jackson," in *Suzanne Jackson: Five Decades*, ed. Rachel Reese (Savannah: Telfair Museums, 2019), 115–126.
46 Suzanne Jackson, "African American Artists of Los Angeles: Suzanne Jackson," interview by Karen Anne Mason. *African American Artists of Los Angeles* (Los Angeles: UCLA Center for Oral History Research, 1992), 140.
47 Jackson, E-mail correspondence with author, April 12, 2023.
48 Alice Walker, "The Universe Responds: Or, How I Learned We Can Have Peace on Earth," in *Living by the Word: Essays* (1988, reis., New York: Open Road Integrated Media, 2011).
49 Jackson, "African American Artists of Los Angeles: Suzanne Jackson," 193.
50 Greta Gaard, "Feminist Animal Studies in the U.S.: Bodies Matter," *DEP*, no. 20 (2012): 16.
51 Amie Breeze Harper, "Connections: Speciesism, Racism, and Whiteness as the Norm," in *Sister Species: Women, Animal, and Social Justice*, ed. Lisa Kemmerer (Chicago: University of Illinois Press, 2011), 76.
52 Indeed, this painting was not titled until it was completed: "Working in a larger studio allowed freedom to play with acrylic, color, shapes, and composition. Sometimes, the works will simply title themselves, without necessarily having any explicit reference. Then, by chance having meaning relevant to the times or events." Jackson, E-mail correspondence with author, April 12, 2023.
53 Michelle Kay Crossley and Colleen Rolland, "Overcoming the Social Stigma of Losing a Pet: Considerations for Counselling Professionals," *Human–animal Interactions* (2022), retrieved from: https://www.cabidigitallibrary.org/doi/10.1079/hai.2022.0022
54 DeMello, *Animals and Society: An Introduction to Human–animal Studies*, 158.
55 Terry Wolverton, *Insurgent Muse: Life and Art at the Woman's Building* (San Francisco: City Lights Book, 2002), 114.

56 Rachel Rosenthal dedicated a book to her beloved rat: *Tatti Wattles: A Love Story* (1996).
57 Terry Wolverton, "*Familiar*, A Performance by Terry Wolverton," in *Terry Wolverton papers*, Box 8, Folder 1 (Los Angeles: Department of Special Collections. UCLA Library, 1984).
58 Donna Haraway, *The Companion Species Manifesto: Dogs, People, and Significant Otherness* (Chicago: Prickly Paradigm Press, 2003), 25.
59 Wolverton, *Insurgent Muse: Life and Art at the Woman's Building*, 115.

Bibliography

Adams, Carol. "The Oedible Complex: Feminism and Vegetarianism." In *The Lesbian Reader*, edited by Gina Covina and Laurel Galana, 145–152. Oakland: Amazon Press, 1975.
Adams, Carol. *The Sexual Politics of Meat: A Feminist-Vegetarian Critical Theory*. 25th anniversary ed. New York: Bloomsbury Academic, 2015.
Ariadne: A Social Art Network. "Two Upcoming Activities." In *Woman's Building Records, 1970–1992*, Box 14, Folder 62, Washington DC: Archives of American Art. Smithsonian Institution, n.d.
Buchanan, Nancy. "Deer/Dear." *High Performance* 1, no. 2 (June 1978): 33–34.
Chicago, Judy. *Through the Flower: My Struggle as A Woman Artist*. Garden City: Doubleday, 1975.
Crossley, Michelle Kay and Rolland, Colleen. "Overcoming the Social Stigma of Losing a Pet: Considerations for Counselling Professionals." *Human–animal Interactions* (2022). Retrieved from: https://www.cabidigitallibrary.org/doi/10.1079/hai.2022.0022
DeMello, Margo. *Animals and Society: An Introduction to Human–animal Studies*. New York: Columbia University Press, 2012.
Eddy, Kathryn, Watson, L.A., and O'Rourke, Janell, eds. *The Art of the Animal: Fourteen Women Artists Explore the Sexual Politics of Meat*. New York: Lantern Books, 2015.
Gaard, Greta. "Feminist Animal Studies in the U.S.: Bodies Matter." *DEP* no. 20 (2012): 14–21.
Goodman, Emily Elizabeth. *Food, Feminism, and Women's Art in 1970s Southern California*. Abingdon, New York: Routledge, 2022.
Griffin, Susan. *Woman and Nature: The Roaring Inside Her*. New York: Harper & Row Publishers, 1978.
Haraway, Donna. *The Companion Species Manifesto: Dogs, People, and Significant Otherness*. Chicago: Prickly Paradigm Press, 2003.
Harper, Amie Breeze. "Connections: Speciesism, Racism, and Whiteness as the Norm." In *Sister Species: Women, Animal, and Social Justice*, edited by Lisa Kemmerer, Chicago: University of Illinois Press, 2011. 72–78.
Harvey, Melanee C. "People, Nature, and the Spiritual World: the Ecowomanist Art Practice of Suzanne Jackson." In *Suzanne Jackson: Five Decades*, edited by Rachel Reese, 115–126. Savannah: Telfair Museums, 2019.
Jackson, Suzanne. *Animal*. Los Angeles: Sunflower Seed Production, 1978.
Jackson, Suzanne. "African American Artists of Los Angeles: Suzanne Jackson." Interview by Karen Anne Mason. African American Artists of Los Angeles. Los Angeles: UCLA Center for Oral History Research, 1992. Retrieved from: https://static.library.ucla.edu/oralhistory/pdf/masters/21198-zz0008zszs-4-master.pdf?_ga=2.23415290.15248905.1672674482-1488551237.1672674476
Kheel, Marti. "An/Aesthetics: The Re-Presentation of Women and Animals." *Between the Species* 1, no. 2 (1985): 37–45. Retrieved from: https://digitalcommons.calpoly.edu/cgi/viewcontent.cgi?article=1412&context=bts
Labowitz-Starus, Leslie. "Sproutime." *High Performance* 3, no. 3–4 (Fall/Winter 1980): 66–67.
Labowitz-Starus, Leslie. "Los Angeles Goes Live: Performance Art in Southern California, 1970–1983. Oral History Interviews." Interview by Jade Thacker. Los Angeles, Los Angeles Contemporary Exhibitions, Gerry Foundation, 2010. Retrieved from: https://rosettaapp.getty.edu//delivery/DeliveryManagerServlet?dps_pid=IE719948
Labowitz-Starus, Leslie. "Interview with Leslie Labowitz: Performance, Environmentalism, Feminism, and Politics." Interview by Jacki Apple. *WEAD Magazine*, n°11, October 2021. Retrieved from: https://directory.weadartists.org/leslie-labowitz-sprout-actions
Merchant, Carolyn. *The Death of Nature. Women, Ecology, and the Scientific Revolution*. New York: Harper & Row Publishers, 1989. First published in 1980.

Rahmani, Aviva. *Divining Chaos: The Autobiography of an Idea*. New York: New Village Press, 2022.
Sutton, Elizabeth A. *Art, Animals, and Experience: Relationships to Canines and the Natural World*. New York: Routledge, 2017.
Walker, Alice. *Living by the Word: Essays*. New York: Open Road Media, 2011. First published 1988 by Harcourt Brace Jovanovich. E-Book Edition.
Wilding, Faith. "Gestations of a Studio of Our Own: The Feminist Art Program in Fresno, California, 1970–71." In *A Studio of Their Own: The Legacy of the Fresno Feminist Experiment*, edited by Laura Meyer. Fresno: Press at California State University, 2009. 79–102
Wolverton, Terry. "Familiar, A Performance by Terry Wolverton." In *Terry Wolverton Papers*, Box 8, Folder 1. Los Angeles: Department of Special Collections. UCLA Library, 1984.
Wolverton, Terry. *Insurgent Muse: Life and Art at the Woman's Building*. San Francisco: City Lights Book, 2002.

24
REQUIEM FOR TIA MARIA

Alexandra Isfahani-Hammond

This chapter is guided by a view of social relations and political life wherein the dead have agency. Work that recognizes the active role of the dead includes Saidiya Hartman's Wayward Lives, Beautiful Experiments *(2019),*[1] *wherein imaginative biographies of marginalized early twentieth-century black women are a practice for altering histories and futures, bearing upon what Martin Luther King, Jr. called "the fierce agency of now."*[2] *In Sharnush Parsipur's* Women Without Men (1987),[3] *Munis is a dead woman who defies patriarchal confinement and the imperialist extraction of oil; taking to the streets, she is a dynamic force of resistance to the 1953 CIA-backed coup deposing Iran's democratically elected prime minister, Mohammad Mosaddegh. The spirit of a deceased Chilean dog, Negro Matapacos, has been channeled to galvanize calls for socio-economic justice not only in protests in Santiago but in New York City, including images of the "cop killing" dog alongside the caption, "Estai presente" (You are present).*[4] *Conversely, the conceptualization of the dead as non-existent is culturally and historically delimited; taking root in late-nineteenth-century Europe, it was guided by Auguste Comte's positivism, itself suspect for reifying rationalism and sustaining pseudoscientific theories of Nordic supremacy. As Magali Molinié puts it, death as nothingness "is certainly the least common idea in the world."*[5] *A chapter about relationships with the deceased in the context of an academic book may nonetheless raise eyebrows. In light of this, I would like to request the reader's willingness to play along with my imaginal endeavor, adopting a stance whereby, as Despret observes, the "artifice and authenticity of the experience are no longer contradictory modes. On the contrary, the better the artifice is cultivated, the more the experience will be lived as authentic."*[6]

Unearthing a bank statement from 2012, I am surprised to discover my password: "tiamariasorry." I thought she had only recently been on my mind, but it seems she has been a consistent presence in my imaginary all along.

Tia Maria is a chocolate-colored mare I learned to ride as a pre-adolescent at a Washington, D.C., equestrian center. I've wanted to write about her, but doing so places me on the receiving end of her gaze.[7] Each time she looks at me, I freeze. Avery Gordon observes that haunting occurs when "what's been in your blind spot comes into view."[8] In my case, what comes into view is Tia Maria observing me. I seek refuge in intellectual analysis, then

DOI: 10.4324/9781003273400-30

attempt to conjure her image but now her gaze is directed downward, averting contact. Why should I, her tormentor, expect her to acknowledge me?

Writing about Tia Maria demands that I hold myself accountable to her. I began an article exploring our relationship in 2019, but the piece took shape as an indictment of equestrian sport, analyzing data about "breaking" horses and the pathology of bits, blinders, crops and spurs. On the one hand, I reached progressive readers who were likely incensed about the excesses of the track (e.g. running maimed horses loaded up on painkillers) while viewing equine domestication as benign, overlooking the physical and emotional tortures that are routine elements of horseback riding. But ultimately, I experienced my article as evasive. I had side-stepped a reckoning, seeking refuge from Tia's gaze by crafting an intellectual argument, distancing myself from the scene of the crime.

With the present chapter, I planned to try again. I proposed a hybrid piece weaving together girlhood reflections about Tia Maria with an analysis of the discourse of a movement called Natural Horsemanship, considering how the latter relies upon duplicitous narratives of consent—casting "breaking" as beneficial to the horse, renaming whips as "carrot sticks"—while exploring my own ill-thinking and rationalizations for oppressing someone I loved. But the mixed-genre approach I'd envisioned felt like spurious code-switching. Rather than double fluency, I was repeatedly abandoning ship, escaping into analytical mode because my words about Tia Maria were unwieldy—shame-ridden repetitions, the haunted vocabulary of broken promises surfacing awkwardly in stilted fragments or erupting tsunami-style. Once again, I considered centering the piece entirely around an impersonal analysis of human/horse relations. I had not conceived of an approach nor cultivated an auspicious environment that was up to the task of seeing to Tia.

In the midst of this impasse, Chloë Taylor emailed two items: the first, a video of a squirrel tapping on a screen door to request a nut; the second, a book title, Vinciane Despret's *Our Grateful Dead: Stories of Those Left Behind*,[9] that she said "made me think of you." I had not shared my predicament with Chloë, yet the book offered an uncanny fix for the subject that eluded me. In the first place, it illuminated the dialogical nature of what is required: Tia Maria is engaged in the activity of making a demand while my endeavor to attend to her nourishes the vitality of her articulation. Despret states that "The desire on the part of the dead to be remembered is something that calls upon the living to commemorate them, just as the obligation of the living to do so summons the desire of the dead."[10] For Daniel Bensaid, "The dead appeal to the living to wake up the dead."[11] The give-and-take of this undertaking is echoed metatextually. Chloë has listened to the many stories about my dog soulmate, Akbar, who since passing away has visited me as squirrels. She occasionally sends photos and videos of squirrels which, like this most recent one, I interpret as nurturing my precious arrangement with Akbar's spirit. Because Chloë encourages Akbar's squirrel appearances, they seem to occur more frequently; put differently, my openness to receiving them is nourished by her enthusiasm. Chloë's comment that *Our Grateful Dead* "made her think of me" underscores this "collective cognition" or, as death doula Alexa Hagerty puts it, the artifice of a "singular bounded individual mind."[12]

To relate with the deceased, Despret is adamant about the need to deprogram ourselves, dismantling the confinement of our communications with them to the domain of subjectivity or as part of the mourning process, advocating the pursuit of signs and the cultivation of an "ecology" of openness—allowing herself, for instance, to be led along by friends' reading suggestions.[13] Such coaxing is excessive enough to elicit incredulity; my editor recommends a book elaborating the imperative to follow signs—specifically, friends' reading recommendation—as a pathway to communication with the other realm. Wholly overdetermined, I experienced this guidance as a leap of faith. My recent history establishes the groundwork for this evolving ecology: encounters with my deceased parents in Ayahuasca ceremonies, lingering with Akbar in the threshold between the worlds, nurturing relationships with the spirits of my beloved kin. I and others in my midst are fluently communicating through "non-ordinary" pathways. My mother appears to tell me she's no longer sad about the past and that I shouldn't be either, advising me to relish her dream visit since there are no true endings. In Despret's vocabulary, receptivity to instruction constitutes a form of takeover: "This is allowing oneself to be taught: allowing oneself to be mobilized by the specific type of capture the situation demands."[14]

As such, the intention of this chapter is no longer expository, but the production of a fortuitous "milieu" via a practice of remembrance wherein I allow Tia Maria to work on me, without excessively intervening. Following Despret, I embark "on the path of a discourse that does not have as its aim the description or explanation of an experience, *but rather the aim of giving it a form that produces what it is describing, that stimulates a transformation in our ways of feeling.*"[15] Crafting the conditions for seeing to Tia Maria, its genre is both the enactment of a process, or speech act, and a "statement of purpose" of sorts for an ongoing project of collaborating with Tia Maria by fleshing out her biography, amplifying her demand for justice, naming myself as her abuser and imaginatively restoring her autonomy, assisting her to finish what she was meant to accomplish. By extension, it is also a practice of situating myself as Tia Maria's troubled inheritor. The seed of Despret's book is her ancestor Georges' untimely demise; delivered early to the train station, his father presses him, "Take the earlier train, my son"[16] as the fourteen-year-old boards the vehicle headed for a deadly collision. Georges' story is an enigma she endeavors to inherit by writing her book.[17] Though her name reflects perverse sugarcoating—situating her in a familiar, affectionate and elder relation to those who exploited her—Tia Maria is indeed my ancestor, a ghost unwillingly entangled in my genealogy, the source of my greatest regret and my most indelible debt to my other-than-human kin. I broke my promise that I would return to buy her—putting an end to the daily tortures of her existence, precluding her final trip to the abattoir—but still I can endeavor to extend her life "otherwise."[18] Despret asserts that "remember" in English allows for a nice metaplasm—"*re-member*, as in recompose:"[19] "The dead can certainly be recomposed, reconnected, but so can stories, histories that carry them, that start with them, and allow themselves to be sent elsewhere, toward other narrations that 're-vive' and that themselves ask to be 're-vived'."[20] For thousands of years, humans have exploited horses' aptitude for entrainment, or the capacity to synchronize their movements with others, including species of vastly different autonomy. With this piece, I aspire to attune myself to her, manifesting Tia Maria as the lead mare.

I now see that my previous work about horseback riding wasn't precisely an evasion but, rather, a practice of zigzagging. Certainly, it was less taxing to adopt a safe analytical distance than explore my agency in Tia Maria's suffering. On the other hand, it was she who compelled the production of my abolitionist tract. As Avery Gordon observes, "Ghosts manifest out of a concern for justice,"[21] while, for Despret, the dead are committed to "refabricating the past in present,"[22] especially where unfinished business is concerned; in this case, both redressing the grave harms done to her and abolishing equine commodification across the board. She venerates the extraordinary power of the dead to "activate the living"[23] with the aim of extending their own lives, enabling them to continue "otherwise":

"The dead ask to be helped to accompany us; there are actions to be carried out, answers to be given to what they are asking."[24] She favors "as if" as the optimal syntax for engaging with the deceased.[25] It's "as if" Tia Maria mobilized my assistance to finish what she was meant to do, calling for the cessation of humans' oppression of horses. In a process akin to gift-giving, the dead can activate connections. Tia Maria's conspecifics were co-beneficiaries of the de-ordering text she summoned me to undertake, all the while, "as if" listening to her string of invectives about humans and heavy tack and her demand to remove the damned metal bit from her mouth, to shake the damned human off her back. Despret insists that "Death does not prevent the resolution of conflicts. Quite the contrary. The dead take an active part that they could not take on board when they were alive."[26]

People who care about animals tend not to think about horses' oppression, including various animal studies scholars who overlook the sadistic ethos of riding them. In my article, I offset this omission by underscoring the fact that horses do not want to be instrumentalized and must be forced to submit to humans' wishes, a process called "breaking." First, they are made to tolerate a series of restraints about their heads and torsos called tack. Trainers quell horses' resistance to these restraints by incrementally desensitizing them and/or applying punishments including whipping, kicking, tethering them in stress positions, fettering their feet and loading heavy weight onto their backs. The bit, a piece of metal inserted into the horse's mouth, is used to make them stop or turn, pulled by the reins. Not very different from the mouths of humans, horses' mouths are intensely sensitive, containing an intricate system of cranial nerves. Bits affect the horse's jaws, roof of the mouth, lips and top of the head and produce what is called the "nutcracker effect," the pinching and squeezing of the tongue. As such, even the subtlest movements of the rider's hands exert great discomfort while any harsh or sudden tugging causes extreme pain and distress. Horses may try to alleviate bit pain by placing their tongues over the bit, opening their mouths wide or placing their chins on their chests. The bit is held in place by a bridle, an assemblage of leather straps binding the horse's head and attached, via a noseband, to reins. The horse is then made to sustain the weight of the saddle on their back, held in place by a leather belt wrapped tightly around their belly. If pulling the horse by the mouth and squeezing the horse's girth with their legs doesn't work, the rider kicks the horse's abdomen with their heels. In many equestrian disciplines, spurs are worn on the heels of riding boots to make kicking more effective. Riders also employ crops to strike the horse's hips and buttocks to force them to move or move faster. Produced from murdered cows' skin, bridles, saddles and crops are both the instruments and remainders of violence.

"Breaking" extends beyond inhibiting freedom of locomotion to psychic warfare waged via solitary confinement in narrow stalls, a particularly brutal form of captivity for highly sociable herd animals whose wellbeing and safety are ensured by staying together. Horses are prey animals whose survival depends upon remaining within the herd. Even resting, they stand head to rump alongside one another so that they can keep an eye out for trouble, safe in their communion and touch. Being close to one another is so intrinsic to what it means to be a horse that they will do nearly anything to resist isolation. Severe measures are therefore taken to quell their defiance. One practice involves tying horses to a stake called a "patience pole" to make them accept being separated: "they don't know how long they'll be there, so they just relax."[27] Since the stake is on a circle of pavement, this "relaxation" technique simultaneously hinders the horse's natural habit of pawing the ground; in addition to the sadism of enforced loneliness, this reflects a totalitarian impulse to absolute behavioral control.

Broken horses will not struggle; they will pull vehicles and can be ridden in sports including trail riding, racing, polo, dressage, jumping, vaulting, hunting and rodeos. The expression, "broke to death," refers to a horse who has been made utterly submissive and, thus, safe to ride by even an inexperienced equestrian. In the human imaginary, few creatures are considered as majestic, charismatic and independent as the horse, galloping through open ranges. Yet it is thus, with saddles, bits, bridles, blinders, spurs, kicks and whips that we misappropriate their enormous power for human enjoyment. The psychopathic drive to "break" an idealized species—demonstrating our affection for them by inflicting pain—reflects an exceptionally perverse form of animosity.[28]

My previous article extended the narrative of Tia Maria's life "otherwise," given how she activated me to register her denunciation as a means for altering social relations. It was also a rehearsal for what I now want to accomplish. A life forgotten is painful for the deceased. Despret alerts, "Not being dead for anyone, that is exactly the risk the dead take,"[29] whereas reconstructing an individual's life and body, including accounting for their death and identifying their remains, gives them "*the possibility of being dead for others.*"[30] Tia Maria's life was not, properly speaking, a life at all. Though the profitability of ridden horses requires that they are kept alive far longer than farmed animal species, her embodied existence is on a continuum with Kathryn Gillespie's theorization of the "lively, soon-to-be-dead" commodity, brought forcibly into the world for use and slaughter.[31] A life history therefore needs to be fabricated for her, following Hartman's activist practice of imagining the erased biographies of those outside the boundaries of the dominant culture's moral community. While information about the lives of commercially ridden horses in the 1960s–80s is exceedingly well concealed, it is reasonable to surmise that Tia Maria was born circa 1969 to a "brood mare" tethered in place for impregnation by a stallion who was similarly dragged into position by a lead. Purchased at age two, for approximately twenty years, her daily routine consisted in being constrained, mounted, struck and pulled by the tongue by four students at one-hour increments. Between classes, she remained motionless in her stall, without room to turn around or significantly adjust her stance, the only distraction a bale of hay and a bucket of water set at head's height. An inquiry to the Rock Creek Park Horse Center during the preparation of this chapter elicited assurance that all of their horses, once no longer rideable, are adopted into good homes. However, their assurance was not convincing There is little regulation or accountability regarding the provenance and demise of ridden horses in the U.S. Informal inquiries with critical animal studies scholars familiar with ridden horse practices corroborated Jean di Grazia's description of how Tia Maria's life would end. It is realistic to assume that, in the late 1990s, she would have been transported to a slaughterhouse, where she would have been shot in the head with a pneumatic bolt gun; this practice of "stunning" prior to slaughter is frequently ineffective with skittish prey animals like horses, such that she may have remained fully conscious while being strung upside down for desanguination and dismemberment. Tia Maria's corpse would thereupon have been exported abroad as meat for human consumption. The second-most common method for killing a horse once keeping them alive is no longer profitable is by gunshot or penetrating captive bolt to the brain. The designation of these methods as "euthanasia" is deceptive since, clearly, there is no humane way to end a life on the mere grounds of being elderly. A rendering facility would then have hauled away Tia Maria's corpse, first removing usable portions including hair for violin bows, paint brushes and upholstery work, then "cooking" her remains to wield products including bone meal, glue, candles and Jello, and disposing of her non-usable body parts at a landfill.

Narrating Tia Maria's corporeal trajectory, including details about how she was likely murdered, is fundamental for expanding the meager frame of her "soon-to-be-dead" embodiment. It is equally important to imagine an alternate present and future wherein she flourishes. The dead suffer being recalled in agony and indignity and the living have a responsibility to reassemble history. It is as if Tia Maria rouses me to envision her unencumbered, free of constraints, busting out of her stall, reducing the entire horse center to smithereens. I select an image of her and photoshop her out of her harness, opening an imaginal pathway to a post-anthropocentric future. Now Tia Maria is a matriarch, the agent for obliterating the tethers that bind. I lavish her with blossoms, following Pir Zia Inayat Khan's dictum, "To bless someone, intone *barkat* (blessing) and visualize them showered with pink rose petals," all the while seeking to keep pace with her, attuning my movement to hers.[32] These creative actions constitute what Despret calls stories that "don't describe; they make things happen and, above all, they are experiences."[33]

Representing Tia Maria's activist agenda, divulging her biography and imagining her release from bondage are key to rectifying her erasure. An account of my culpability is also critical for the inheritance I strive to join her in accomplishing. Twice weekly from age ten to fourteen, I arrived at the equestrian center and went directly to the tack room to retrieve her saddle from a stake on the wall, draping it across the length of my forearm. My other hand took hold of her bridle, a web of straps that had to be carefully held so that it did not become tangled. I carried these down a row of narrow stalls, each enclosed on three sides with a chain strung across the front and a horse fixed in place inside it. Reaching Tia, I unhooked the chain and wedged myself into the narrow space between her body and the wall. I lifted the saddle onto her back, fastened the surcingle (girth) around her belly as tightly as possible, then waited a few seconds for her to exhale. I was prepared for the way horses attempt to achieve a looser fit when they see the saddle coming. When you approach a horse with the girth, they distend their belly on an inhale, holding their breath. "Fasten the girth as tightly as possible around her inflated belly, wait a few seconds for her to exhale, then quickly pull in the strap." The words may only have been spoken to me once but they reverberate. "When she exhales, make it tighter." Breath itself is excavated.

To affix her bridle, I draped the reins over her neck, then introduced the base around her muzzle. I inserted my thumb into the side of her jaw so that she opened her mouth and I could place the metal bit atop her tongue. I then slid the encasement of straps up the length of her head, fastening it in place behind her ears and with a buckle at the base of her jawline. Mounting Tia, I squeezed her abdomen, kicked her with my heels and pulled her by the mouth to maneuver her from a walk to a trot to a canter, in figure eights and over jumps, in classes and in competitions. My instructor, Jean di Grazia, called out to me laughing, "You love that horse so much, Alexa. What are you going to do when she gets sent to the glue factory?" My parents were ringside but I do not recall any discussion of Jean's comment. I wrote a letter to "Horse and Rider" magazine asking if Tia would really be made into glue. The editor dismissed my question, assuring me that no responsible horse person would do such a thing and that she would certainly be well cared for in her old age. I leaned into Tia, my arms thrown around her neck, and whispered into her soft ear, "I will buy you when I am eighteen, set you out to pasture and take care of you." My promise was inspired by a convoluted mixture of tenderness, remorse and defiance of Jean's insinuation. Perhaps more than anything, it was motivated by a desire to rationalize my interactions with her, making it all right to continue hurting her.

Tia had mostly given up, with the exception of small, quotidian acts of resistance like distending her abdomen when she saw the saddle coming. Sometimes, she came to a halt when she was tired or would not move promptly from a trot into a canter, defiance met with kicks to her abdomen and strikes of the crop on her flanks. My parents supported my passion for riding—including the pain I invariably administered—in sheer cognitive discord with their empathy for animals. Inside our home, the lives of insects and pigeons with broken wings were held dear and nursed back to life. My mother slammed the sliding doors closed when the neighbors barbequed, exclaiming, "That is the scent of burning flesh!" Yet here in the ring, she did not object to the dynamics of dominating horses, of Tia's existence on the receiving end of the whip, the kicked end of heels, the pinched tongue of the bit and reins. How is it that we claimed to love animals while colluding with their torture? Antithetic to our domestic sphere, the equestrian center was a necropolitical locale whose purpose was to inflict trauma; ritual and pomp abetted this normalization. My parents may have expressed mild disapproval of Jean's comment about turning Tia into glue, but we all ultimately complied with the inevitability that humans inflicted violence on horses and, when they were no longer useful, murdered them. My father brought me an ornate riding crop from Mali with decorative green and red tassels. When I unwrapped it, I did not immediately comprehend its purpose. I caught his eye for a moment of reckoning, discerning from raised eyebrows and tight smile our shared complicity with gratuitous harm. Since it was inevitable that I would strike the horse I loved, I might as well use a beautiful weapon, a flash of flamboyant color offsetting the austere accoutrements of English-style riding: beige jodhpurs, black boots and hats, crisp white shirts.

Our shared acknowledgement of the crop's violence is tucked away in the moment our eyes lock, stored where it won't interfere with my desire to use her. What kind of love is this? Vasile Stanescu interrogates how love is weaponized by "humane meat" enthusiasts. However wrong, unpleasant, ironic, and unethical, injurious protocols are deemed necessary because they are financially expedient.[34] Isolated expressions of care—giving the animal a name, for instance—mask, mitigate or compensate for the injustice of causing them harm.[35] In a bizarre twist, awareness of unprincipled conduct provides further justification, such as when locavorist Catherine Friend asks, "Can I call myself a feminist after this?" while restraining a sheep to be brutally raped,[36] or when her partner absolves herself for castrating lambs without anesthesia, apologizing while squeezing the jaws of a clamp over the cords leading to their testicles.[37] Nekeisha Alayna Alexis notes that such fixation on the killer's experience—and appropriate affective response—is served up as absolution for wrongdoing, functioning as a primary tactic of evasion within what she terms the "ill-logics" of humane slaughter narratives.[38] Love mitigates domination.[39] Is that what my promise to Tia Maria was? Debt payment for inflicting ongoing—unethical yet inevitable—harm?

My mother was forever ring-side, catching us in black and white photographs, stringing dozens of images to dry in her dark room. In most of them, Tia is passive, eyes resigned, looking down. But she must have looked me in the eye. Whom did she see but a predator? One photo shows me smiling, looking over my shoulder at Tia Maria. She is walking more slowly than I want her to. I use the leather reins in my hands to pull her by the mouth. Tia had a wide, white stripe down the center of her face. Her eyes were large and dark, with thick eyelashes.

Tia Maria's gaze encapsulates me in darkness. I'm imbued with the ontology of the ridden horse, one's head encased in straps, a bit pinching the tongue, a human on one's back. This is inheritance in action. That Tia Maria holds me accountable is a form of both capture and haunting. She who had been bound fixes me in place and occupies my sentences. Brianne Donaldson exalts the de-ordering promise of playing "host to creaturely ghosts who reorient (an author's) outlook and vocations through threads of affective feeling."[40] As Avery Gordon describes it, "haunting keeps interrupting the work, and becomes the work."[41]

I gave up riding when I started high school. I did not go back for Tia Maria when I turned eighteen, but a couple of years later, I made what I knew was a futile call to the Rock Creek Park Horse Center. I was told there was no horse there called Tia Maria nor did they know what might have happened to her. I was ashamed but I would keep my misdeed a secret. No one knew about my promise (only she did). No one knew what happened to her (she did). None of it was real (it was real). I forgot about Tia while she remained incarcerated and abused. Looking.

Despret urges, "To feel the presence of the dead . . . you have to start by thinking about them and talking to them. Act as if they are there, and they could be there. Creative thinking aligns with this form of cultivation that is imaginal power."[42] I whisper to Tia Maria, "I am sorry for forcing you to bear my weight, for the toll it took on your muscles and skeletal structure, for the elaborate Malian crop and the cold metal bit in your mouth, pulling you left and right. I am sorry for my complicity with your imprisonment and psychic torment." I recognized my kinship with her too late to alter her corporeal trajectory, like the haunted hunter-turned-feline narrator of João Guimarães Rosa's "My Uncle, the Jaguar" (1961) who, after years of skinning wildcats, realizes they are his family and is himself transformed into a jaguar: "I hate to think about those killings I did. . . . My kinsfolk, how could I?!"[43]

In her lithograph, "The Ghosts of Our Meat" (2013), Sue Coe depicts legions of murdered animals trailing after a haunted human figure.[44] What if the petrified human faced them? While Despret herself certainly does not espouse animal liberation,[45] her invocation suggests a manner for turning toward our animalized ghosts, not merely to grieve them but, rather, to collaborate with them to help them finish what they were meant to do. This vital praxis builds upon Maria Elena Garcia's appeal to redress the "invisibilization" of murdered and soon-to-be-murdered animals and Kathryn Gillespie's work on caregiving for dying animals as imperative counter conduct.[46] Crucially, Despret admonishes that mourning the dead confers closure that can cause them harm, particularly when unfinished business remains—lives unlived as they should have been, culpability disavowed, justice unserved and the enactment of ongoing harm. Engagement with murdered animals' agency speaks to Paula Arcari's exhortation to establish new neural pathways fit for fabricating post-anthropocentric futures.[47] We might begin with a lamentation in the grocery aisle; as James Stanescu expounds,

> In front of you is the violent reality of animal flesh on display: the bones, fat, muscles, and tissue of beings who were once alive but who have been slaughtered for the parts of their body. This scene overtakes you, and suddenly you tear up. Grief, sadness, and shock overwhelms you, perhaps only for a second.[48]

The remains of murdered animals are all around us. First we "tear up," then we proceed to speak to the beings rendered as body parts, encouraging them to instruct us in the art of extending their lives "otherwise"; exacting justice on their behalf, emancipating them from the limited narratives in which they are held captive and manifesting radically alternative future conditions for beyond-human flourishing.

Notes

1 Saidiya Hartman, *Wayward Lives, Beautiful Experiments: Intimate Histories of Riotous Black Girls, Troublesome Women, and Queer Radicals* (New York: Norton, 2019)
2 Martin Luther King, Jr., "I Have a Dream," March on Washington for Jobs and Freedom (1963)
3 Shahrnush Parsipur, *Women Without Men: a Novel of Modern Iran,* trans. Kamran Talattof and Jocelyn Sharlet (New York: First Feminist Press, 2004)
4 "Matapacos" means "copkiller." Billy Anania, "The Cop-Attacking Chilean Dog Who Became a World-wide Symbol of Protest," *Hyperallergic* (2019), retrieved from: hyperallergic.com/526687/negro-matapacos-chilean-protest-dog/; Alexandra Isfahani-Hammond, "How a Chilean Dog Ended Up as a Face of the New York City Subway Protests," *The Conversation* (2020), retrieved from: theconversation.com/how-a-chilean-dog-ended-up-as-a-face-of-the-new-york-city-subway-protests-129167.
5 Vinciane Despret, *Our Grateful Dead: Stories of Those Left Behind* (Minneapolis: University of Minnesota Press, 2021).
6 Ibid., 104.
7 A rich body of work addresses the potential for the animal gaze to subvert species dialectics. For instance, discussing the relational ontology of Amazonian basin worlding, Eduardo Viveiros de Castro observes that to perceive a jaguar's personhood when looking into their eyes results in the dissolution of one's humanity (Eduardo Viveiros de Castro, *La Mirada del Jaguar. Introducción al Perspectivismo Amerindio. Entrevistas* (Buenos Aires: Tinta Limón Ediciones, 2013), 281). Eduardo Kohn notes that for the Ecuadoran Amazonian Runa, the transspecies gaze is central to survival: If you make eye contact with a jaguar, they will recognize you as a fellow predator (Eduardo Kohn, *How Forests Think: Toward an Anthropology Beyond the Human* (Berkeley: University of California Press, 2013), 19). Jacques Derrida identifies the moment when his cat watches him emerge naked from a shower as the origin for his animal turn (Jacques Derrida, *The Animal that Therefore I Am* (New York: Fordham University Press, 2008), 4). In "Am I Blue?," Alice Walker describes her aversion to meeting the heartbroken gaze of a horse severed from his beloved companion: "I dreaded looking into his eyes . . . but I did look" (Alice Walker, "Am I Blue?" *Living by the Word: Selected Writings 1973–1987* (San Diego: Harcourt Brace Jovanovich, 1988), 7). Meeting Blue's gaze leads Walker to recognize that, had she been born in slavery, and her partner had been sold away or killed, "my eyes would have looked like that," whereupon she decides it is unethical to consume animals (8).
8 Avery F. Gordon, *Ghostly Matters: Haunting and the Sociological Imagination* (Minneapolis: University of Minneapolis Press, 1997), xvi.
9 Despret, *Our Grateful Dead: Stories of Those Left Behind.*
10 Ibid., 46
11 In ibid., 126.
12 In ibid., 96.
13 Ibid., 18.
14 Ibid., 16.
15 Ibid., 84 (emphasis mine)
16 Ibid., 1
17 Ibid., 2
18 Ibid., 6
19 Ibid., 47
20 Ibid.
21 Gordon, *Ghostly Matters: Haunting and the Sociological Imagination,* 64.
22 Despret, *Our Grateful Dead: Stories of Those Left Behind,* 99
23 Ibid., 47
24 Ibid., 6
25 Ibid., 123
26 Ibid., 100
27 Al Dunning, "The Lost Art of Standing Tied," *Horse and Rider Magazine* (June 13, 2019), retrieved from: horseandrider.com/western-horse-training-tips/the-lost-art-of-standing-tied/.
28 This description of horse breaking originally appeared in Alexandra Isfahani-Hammond, "Horses and Humans: On and Off the Track," Counterpunch, 4 (November 2019), retrieved from: www.counterpunch.org/2019/11/04/horses-and-humans-on-and-off-the-trak/.

29 Despret, *Our Grateful Dead: Stories of Those Left Behind*, 47
30 Ibid., 49
31 Kathryn Gillespie, "Afterlives of the Lively Commodity: Life-worlds, Death-worlds, Rotting-worlds," *Environment and Planning A: Economy and Space* 53, no. 2 (2021): 2.
32 Pir Zia Inayat Khan, "Nature Meditations on Roses," *Vimeo*, uploaded by Inayatiyya (2021), vimeo.com/557139060.
33 Despret, *Our Grateful Dead: Stories of Those Left Behind*, 124
34 Vasile Stanescu, "Why 'Loving' Animals is Not Enough: a Response to Kathy Rudy, Locavorism, and the Marketing of 'Humane' Meat," *The Journal of American Culture* 36, no. 2 (June 2013): 107.
35 Ibid., 108.
36 Ibid., 105.
37 Ibid., 103.
38 Nekeisha Alayna Alexis, "There's Something About the Blood . . .: Tactics of Evasion within Narratives of Violence," *Animaladies: Gender, Animals, Madness*, eds. Lori Gruen and Fiona Probyn-Rapsey (New York: Bloomsbury, 2019)
39 Stanescu, "Why 'Loving' Animals is Not Enough: a Response to Kathy Rudy, Locavorism, and the Marketing of 'Humane' Meat," 108.
40 Brianne Donaldson, "Transformed by Ghosts: Towards Futures of Less Loss," in *Feeling Animal Death: Being Host to Ghosts*, eds. Donaldson and Ashley King (London: Rowman and Littlefield, 2019), xviii
41 Gordon, *Ghostly Matters: Haunting and the Sociological Imagination*, 16.
42 Despret, *Our Grateful Dead: Stories of Those Left Behind*, 103
43 João Guimarães Rosa, "My Uncle, the Jaguar," *The Jaguar and Other Stories*, trans. David Treece (Oxford: Boulevard Books, 2001), 57. For an expanded discussion of "My Uncle, the Jaguar," see Alexandra Isfahani-Hammond, "Haunting Pigs, Swimming Jaguars: Mourning, Animals and Ayahuasca," in *Colonialism and Animality: Anti-Colonial Perspectives in Critical Animal Studies*, eds. Kelly Struthers Montford and Chlöe Taylor (New York: Routledge, 2020), 201–216.
44 Sue Coe and Steven S. Eisenman, "The Ghosts of Our Meat" (New York: The Trout Gallery, 2013).
45 Despret's reflections on the use of animals in science are fundamentally uncritical. Like Donna Haraway, she emphasizes reciprocity, rather than structural oppression, as foundational for human/animal relations. There is no discussion of animals in *Our Grateful Dead*.
46 Maria Elena Garcia, "Love, Death, Food, and Other Ghost Stories: Hauntings of Intimacy and Violence in Contemporary Peru," in *Economies of Death: Economic Logics of Killable Life and Grievable Death*, eds. Kathryn A. Gillespie and Patricia J. Lopez (New York: Routledge, 2015); Kathryn Gillespie, "Provocation from the Field: A Multispecies Doula Approach to Death and Dying," *Animal Studies Journal* 9, no. 1 (2020)
47 Personal communication, September 28, 2022.
48 James Stanescu, "Species Trouble: Judith Butler, Mourning, and the Precarious Lives of Animals," *Hypatia* 27, no. 3 (2012): 568

Bibliography

Alexis, Nekeisha Alayna. "There's Something About the Blood . . .: Tactics of Evasion within Narratives of Violence." In *Animaladies: Gender, Animals, Madness*, edited by Lori Gruen and Fiona Probyn-Rapsey, 47–64. New York: Bloomsbury, 2019.

Anania, Billy. "The Cop-Attacking Chilean Dog Who Became a Worldwide Symbol of Protest." *Hyperallergic* 5 (November 2019). Retrieved from: hyperallergic.com/526687/negro-matapacos-chilean-protest-dog/.

Coe, Sue, and Eisenman, Steven S. *The Ghosts of Our Meat*. New York: The Trout Gallery, 2013.

Derrida, Jacques. *The Animal That Therefore I Am*. New York: Fordham University Press, 2008.

Despret, Vinciane. *Our Grateful Dead: Stories of Those Left Behind*. Minneapolis: University of Minnesota Press, 2021.

Donaldson, Brianne. "Transformed by Ghosts: Towards Futures of Less Loss." In *Feeling Animal Death: Being Host to Ghosts*, edited by Donaldson and Ashley King, xv–xxix. London: Rowman and Littlefield, 2019.

Dunning, Al. "The Lost Art of Standing Tied." *Horse and Rider Magazine*, June 13, 2019. Retrieved from: horseandrider.com/western-horse-training-tips/the-lost-art-of-standing-tied/.

Garcia, Maria Elena. "Love, Death, Food, and Other Ghost Stories: Hauntings of Intimacy and Violence in Contemporary Peru." In *Economies of Death: Economic Logics of Killable Life and Grievable Death*, edited by Kathryn A. Gillespie and Patricia J. Lopez, 160–178. New York: Routledge, 2015.

Gillespie, Kathryn. "Provocation from the Field: A Multispecies Doula Approach to Death and Dying." *Animal Studies Journal* 9, no. 1 (2020): 1–31.

Gillespie, Kathryn. "Afterlives of the Lively Commodity: Life-worlds, Death-worlds, Rotting-worlds." *Environment and Planning A: Economy and Space* 53, no. 2 (2021): 280–295.

Gordon, Avery F. *Ghostly Matters: Haunting and the Sociological Imagination*. Minneapolis: University of Minneapolis Press, 1997.

Hartman, Saidiya. *Wayward Lives, Beautiful Experiments: Intimate Histories of Riotous Black Girls, Troublesome Women, and Queer Radicals*. New York: Norton, 2019.

Inayat Khan, Pir Zia. "Nature Meditations on Roses." *Vimeo*, uploaded by Inayatiyya, May 30, 2021. Retrieved from: vimeo.com/557139060.

Isfahani-Hammond, Alexandra. "Horses and Humans: On and Off the Track." *Counterpunch*, 4 (November 2019). Retrieved from: www.counterpunch.org/2019/11/04/horses-and-humans-on-and-off-the-trak/.

Isfahani-Hammond, Alexandra. "Haunting Pigs, Swimming Jaguars: Mourning, Animals and Ayahuasca." In *Colonialism and Animality: Anti-Colonial Perspectives in Critical Animal Studies*, edited by Kelly Struthers Montford and Chlöe Taylor, 201–216. New York: Routledge, 2020a.

Isfahani-Hammond, Alexandra. "How a Chilean Dog Ended Up as a Face of the New York City Subway Protests." *The Conversation*, January 8, 2020b. Retrieved from: theconversation.com/how-a-chilean-dog-ended-up-as-a-face-of-the- new-york-city-subway-protests-129167.

King, Martin Luther, Jr. "I Have a Dream." March on Washington for Jobs and Freedom (August 28, 1963). Washington, D.C.

Kohn, Eduardo. *How Forests Think: Toward an Anthropology Beyond the Human*. Berkeley: University of California Press, 2013.

Parsipur, Shahrnush. *Women Without Men: A Novel of Modern Iran*. Translated by Kamran Talattof and Jocelyn Sharlet. New York: First Feminist Press, 2004.

Rosa, João Guimarães. "My Uncle, the Jaguar." In *The Jaguar and Other Stories*, translated by David Treece. Oxford: Boulevard Books, 2001.

Stanescu, James. "Species Trouble: Judith Butler, Mourning, and the Precarious Lives of Animals." *Hypatia* 27, no. 3 (2012): 567–582.

Stanescu, Vasile, "Why 'Loving' Animals Is Not Enough: A Response to Kathy Rudy, Locavorism, and the Marketing of 'Humane' Meat." *The Journal of American Culture*, 36, no. 2 (June 2013): 100–110.

Viveiros de Castro, Eduardo. *La Mirada del Jaguar. Introducción al Perspectivismo Amerindio. Entrevistas*. Buenos Aires: Tinta Limón Ediciones, 2013.

Walker, Alice. "Am I Blue?" In *Living by the Word: Selected Writings 1973–1987*, 3–8. San Diego: Harcourt Brace Jovanovich, 1988.

25
CRAZY CAT LADY

Alison Suen

My cat, linguini, is a tailless tortoiseshell. I would call linguini the love of my life, but such a cliché does not quite capture the indispensable role she plays in my life. She gives me a reason to go to work (so I can buy litter and cat food) and the incentive to write (so I can talk about her and see her name in print). I find it difficult to motivate myself to write unless I can somehow work linguini into my paper or presentation. In fact, without linguini, I couldn't have written my last book, *Why It's OK to Be a Slacker*. As someone who makes no effort to make herself useful, linguini inspires both my understanding and defense of a slacker. It is because of my attachment to linguini that I became interested in thinking about and writing on the topic of slacking. It is not an exaggeration to say that I owe my book to my cat. Given how much I talk about linguini, I reckon I was invited to write this chapter because of her. So once again, I am indebted to linguini. Without her, I would not be writing on the topic of the "crazy cat lady" right now. At the very least, I would not be able to write on this topic with some sort of lived experience. In any case, there is no doubt that linguini is central to my intellectual and creative life.

As a diabetic cat with chronic kidney disease, linguini requires an insulin injection twice a day. To collect her post-injection treat, linguini dutifully wakes me up every morning at 7:30. I have yet to meet another cat (or human, for that matter) whose eyes light up at the sight of a needle. While linguini manages to make her injection routine as painless *to me* as possible, being her primary caretaker means orienting my life around her care. This means requesting a teaching schedule that that allows me to bring her to the vet for checkups on specific days during the week, learning how to collect blood samples to chart a glucose curve, having dinners at odd hours so I can be home on time for her nightly injection at 7:30, or canceling all non-essential plans to stay with her when she appears unwell. Fortunately, my friends and partner are all understanding and supportive, making it easy to prioritize linguini. Given that my life very much revolves around linguini, it is perhaps not surprising that I was teased as a "crazy cat lady" (CCL) every now and then—occasionally as a pejorative, though mostly as a term of endearment. This raises the question: how could such a term be endearing at all?

In this chapter, I will look at how the cat lady stereotype has evolved in recent years. My argument is the following: to successfully reappropriate the CCL, it is not enough to just

DOI: 10.4324/9781003273400-31

point to positive examples of cat lady—individuals who glamorize the stereotype. We also have to show that a cat-loving woman prioritizes her cat not because her cat is a substitute for men but because a cat-*dependent* life is enriching for its own sake.

* * *

What makes a cat lady "crazy"? Suppose we take the word "crazy" literally as a clinical term. There is an ongoing debate on the association between psychosis and *Toxoplasma gondii* (*T. gondii*), a parasite that can only be sexually reproduced in a cat's gut. *T. gondii* has a fascinating life cycle. One study shows that *T. gondii* can alter the brain of an infected rodent, causing them to lose fear of the smell of cat urine.[1] No longer fearing and avoiding cats, the infected rodent becomes more likely to be eaten by one. A cat who has been infected by the parasite-carrying rodent begins to shed oocysts of *T. gondii* in their feces within a few weeks. It is possible for a human to become infected with the parasite if they handle the litter box without properly washing their hands.[2] While *T. gondii* appears to be associated with a number of psychotic symptoms (including symptoms consistent with schizophrenia), the link between cat-caregiving and psychosis is far more tenuous. A more recent study in 2022 suggests that there is a "conditional association" between growing up in a household with a cat and developing psychotic experiences in adulthood.[3] Ironically, the association is observed only with male participants who grew up in a household with rodent-hunting cats. In other words, even if there is a kernel of truth in the association between psychosis and cats, it seems to apply only to the "cat gentlemen." So much for the crazy cat *lady* stereotype.

But the "crazy" in the CCL stereotype is rarely deployed as a clinical term. More often than not, it is used as a disparaging and ableist term to pathologize a woman's devotion to her cats. "Crazy" suggests deviancy and a lack of control. But what makes a woman's devotion to her *cat* deviant? And why is this alleged deviance gendered? After all, many women are devoted to their dogs, and yet there is no "crazy dog lady" stereotype. And while plenty of men adore cats, there is no equivalent of the CCL stereotype for men. (Admittedly, cat-loving men may experience a different kind of prejudice insofar as they are perceived as less masculine. In one study, men who pose with a cat on dating apps are deemed "less datable."[4] Still, their sanity is never in doubt.)

Traditionally, a stereotypical CCL is a spinster whose home is overtaken by a hoard of cats. She is a depressing, pitiful figure who is unable to develop or sustain a close human relationship. Think, for example, of Dr. Elenore Abernathy from *The Simpsons*. Elenore is an Ivy League-educated woman who believes that "a woman can do anything." She suffered a burnout due to the demand of her duo-career as a lawyer and a doctor, and eventually became a gibberish-speaking, cat-hurling alcoholic. Single, childless, and unhinged, Elenore serves as a cautionary tale for ambitious girls (such as Lisa Simpson) who aspire to do great things. Given this culturally ingrained idea of a CCL being a childless spinster, a woman who is devoted to her cats is deemed "deviant" because she departs from a conventional script of how a woman should live her life. The CCL represents an implicit rejection of heteronormativity and anthropocentric speciesism.[5] The CCL cares about cats more than—maybe even instead of—men.

This negative image of the CCL has begun to erode in recent years, however. A 2014 *New York Times* article captures the evolution of the CCL as follows, "The cat ladies of yore may have been sad sacks . . . but this new breed is young, sociable and ambitious. And cats for them (us) are not signs of spinsterhood, but of independence."[6] Even academics make an effort to debunk the CCL stereotype. In a 2019 study, participants were asked to rate animal vocalizations (cat meow and dog whine) on a numerical scale, from "very sad" to "very happy."[7] They were then assessed for their anxiety, depression, and attachment levels. Researchers find that cat caregivers are more likely to rate feline vocalization (meows) as "negative." That is, they are more likely to rate a cat's meows as a distress call—presumably because living with cats allows caregivers to become more attuned to their cat's distress. The study found "no evidence to support the 'cat lady' stereotype: cat-owners did not differ from others on . . . their experiences in close relationship."[8] However, they did find a "small, significant correlation" between adult attachment and perception of cat meows.[9] Specifically, "adults with more secure relationship attachments gave more negative ratings (were more sensitive to distress) for the cat sounds."[10] So much for stereotype of CCL being a lonely spinster!

That loving one's cats has become more culturally acceptable is also evidenced by the fact that plenty of celebrities declared themselves cat ladies. Country/pop singer Taylor Swift is perhaps today's most famous self-proclaimed "cat lady" (she dropped the "crazy" from the moniker). Her cats Meredith Grey, Olivier Benson, and Benjamin Button make regular appearances on her Instagram. Living a gilded life, Swift's cats mingle with the fashionable crowd, travel by private jet, and reside in a NYC townhouse that comes with a coveted zip code. As a result of her social media presence (curated by Swift), Olivier Benson reportedly has a $97 million net worth. Swift's unapologetic, highly publicized adoration for her cats not only normalizes, but also *glamorizes*, both the cats and the cat lady (a point to which we will return later).

In addition to celebrities owning the title of a cat lady, cat lovers have tried to combat this unflattering trope by reappropriating the CCL stereotype in creative and intentional ways. For example, *Girls and Their Cats* (*GATC*), founded by fashion photographer Bri-Anne Wills, is a project that features women from all walks of life and their cats in a positive light.[11] When a collection of portraits was published, Wills introduced the book as follows: "this book redefines the stereotype by showcasing 50 strong, independent, and artistic women who take the world in stride, flanked by their beloved felines."[12] Cat ladies are redefined as well-adjusted, successful, and independent women who happen to love cats.

* * *

What should we make of this increased effort to reclaim the traditionally pejorative stereotype of the CCL? Surely, it is a good thing to *not* pathologize our affection for our feline companion. And surely, one can be devoted to one's cats and still have a thriving career and a robust social life—a fact, though obvious, is worth reiterating. I am in general on board with efforts to de-stigmatize cat love. I also agree that embracing the cat lady label can be empowering: not just because self-labeling is "a defying act of taking ownership over the term," but also because it makes room for cat lovers to reframe the CCL stereotype.[13]

Before turning to how we may contest the CCL stereotype, I would like to draw a distinction between reframing and rebranding. Reframing is a strategy to reclaim a pejorative, to "[alter] the valence of a group stereotype by transforming a weakness into a strength."[14] For example, a self-labeled slacker may reframe slacking by reinterpreting their perceived vice. Instead of conceding that their lack of ambition is a shortcoming, the self-labeled slacker may argue that it is an intentional act to reject the culture of hyper-productivity. By transforming a seemingly passive vice into a purposeful resistance, the slacker can now proudly wear their slackerdom on their sleeves.

Rebranding, unlike reframing, relies on identifying individuals with positive attributes to represent that particular "brand." When it comes to slackers, one may rebrand slacking by spotlighting slackers who are considered posh and trendy. For instance, one may point to the leisure class and aristocrats who believe that working for a livelihood is demeaning. Slacking is then rebranded as a status symbol insofar as it is *represented by* those who are posh and privileged, as opposed to "losers" who lack ambition. However, rebranding alone does not necessarily change the "valence" of the slacker stereotype if it merely focuses on a small group of well-to-do slackers. It does not tell us why slacking is fine even for those who are not wealthy or privileged.

Given this distinction, let's consider the evolution of the CCL stereotype. Celebrity cat lovers such as Taylor Swift, Kristen Stewart, and Katy Perry have certainly lent glamor to being a cat lady. But how, exactly, do these cat-loving celebrities challenge the stereotype? For one, they are rejecting the assumption that loving cats *reduces* one to an unhinged spinster. They prove that becoming Elenore from *The Simpsons* is not an inevitable fate of a cat-loving woman. For another, they chip away at the misconception that cat-loving women are cat-obsessed *because* they are unwanted by men. After all, a young, posh celebrity such as Swift is emotionally invested in her cats not because she has to be but because she *chooses* to be. The spinster cat lady stereotype is undercut insofar as we now have hip, successful spokespersons to represent the cat lady brand. To the extent that the rebranding decouples cat-love with "craziness" and spinsterhood, having posh celebrity spokespersons is one way to recast what it means to be a cat-loving woman.

Nevertheless, rebranding often only addresses the surface, rather than the root, of the negative stereotype. Specifically, rebranding of the "cat lady" tells us that it is okay to prioritize your cats if you are attractive, hip, and well-to-do. It tells us that a woman can be both glamorous and be cat obsessed. However, few of us are posh celebrities. The idea that a woman can have it all—cats, talent, glamour, sanity—does not tell us what it means to be a cat-loving woman who is less wealthy or conventionally good looking. Neither glitzy nor popular, an "ordinary" cat lady may continue to be ridiculed, suspected, or pitied. Her love for, and willingness to prioritize, her feline companion may be perceived as a matter of necessity, rather than choice. ("Of course she was obsessed with her cats—what choice did she have when no one wanted to marry her?")

To successfully reappropriate the cat lady stereotype, we cannot just rely on rebranding. Just as we cannot increase the social acceptance of same-sex relationships just by pointing to fashionable LGBTQ+ celebrities, we cannot increase the social acceptance of cat love just by appealing to Taylor Swift and other cat-loving superstars. Instead of looking for counterexamples of the CCL stereotype, we need to challenge the heteronormative expectation that a woman needs a man in her life for it to be whole and fulfilling. Thankfully, feminists have long contested such an expectation, laying the theoretical foundation for cat-loving women to advocate for their alternative "partnership." On social media, cat-loving women tweet about "having a hot date" with their cats (only half-jokingly), unapologetically celebrating their cats as their "legitimate companions."[15] The saying, "I don't need a man . . . I have a cat," resonates with so many people that it even has its own t-shirt.

That men and cats are pitted against each other on social media is telling. These memes are funny only because a woman is not supposed to prioritize a cat. In these "I don't need a man . . ." proclamations, cats are presented as rivals of or substitutes for men as romantic partners. While memes are meant to be funny and not to be taken too seriously, the idea that sharing one's life with a cat can be *a way of life* deserves some consideration here. In her essay, "LGBTQ . . . Z?", Kathy Rudy uses queer theory as a framework to conceptualize her relationship with her dogs.

> Queer theory has schooled me in ways that make the question of what counts as sex seem rather unintelligible. How do we cordon off sexual desire from all the other desires that move our lives? What does sex mean? Do I think I'm having sex with my dogs when they kiss my face? How do we know beforehand what sex is? [. . .] When I was gay, was I gay because of a narrowly defined genital act that I performed with a person who happened to be another woman? Those words don't make any sense to me. I was gay then, I believe, because I chose to share my emotional, financial, and daily life with a person of the same gender. Now I choose to share that same life with six dogs.[16]

Rudy is not arguing for or against having sex with dogs. Rather, she is calling into question the idea that sexuality or sexual orientation is something we can neatly define. If sexuality is not confined to specific types of sex acts, and many relationships don't even involve sex, then why couldn't one have a legitimate relationship with one's dogs? For our discussion here, if cats (not men) are the preferred partners for some, then why couldn't cat love be considered a type of "orientation"?

As Rudy points out, in a heteronormative society, non-heterosexual love has always been suspect. In the early 20th century, the "heterosexual public . . . thought that the only women who would ever choose lesbianism were ugly, or unfeminine, or somehow lacking in the ability to capture a man."[17] This perception probably still lingers in certain pockets of the American public, but at least it is deemed prejudicial and unprogressive today. While same-sex relationships are more accepted nowadays, an exclusive relationship with one's pets is still considered wrong, deviant, and creepy. We are expected to treat our pets well, but not excessively so. We are not supposed to treat our pets as our partner or significant other. If we do, then there must be something pitiful about our lives. For many, being invested in one's pet is a sign that one "cannot 'find' a human person to love."[18]

Whether a cat (or other pets) can be a legitimate lover or partner is beyond the scope of this chapter. Luckily, we don't need to settle on this issue to understand the anxiety many women feel when they prioritize cats in their lives. There is a reason why these "who needs a man . . . when you have a cat" memes and tweets abound on social media. They are responding to the assumption that women are supposed to be emotionally invested in men, not cats. By owning the fact that they care more about cats than men, these cat-loving women are asserting their agency. They contend that having a life that revolves around cats is a point of pride, rather than something shameful.

The disavowal of heteronormativity perhaps explains why the phrase "independent woman" often gets tossed around in conversations about the CCL stereotype. As we saw, the *GATC* project seeks to dismantle the CCL myth by featuring "strong, independent, artistic women." Cats are deemed "the independent young person's animal."[19] And there is no shortage of paraphernalia with the tagline, "Behind every strong, independent woman is her cat." Given the patriarchal underpinning of the CCL stereotype, a declaration of independence (from men) can be cathartic and empowering. Unfortunately, it is also misleading. The idea that a cat-loving woman is "independent" obscures her dependence on cats.

In "Service Dogs, Between Animal Studies and Disability Studies," Kelly Oliver[20] points out that service dogs are often praised for allowing their disabled human companion to gain independence. But as Oliver argues, what service dogs provide is independence from *other humans*. We train service dogs to pick up keys, turn on lights, retrieve items so that disabled individuals may depend on *them* instead of a human caregiver. As such, a disabled person's "independence" is predicated on their dependence on their service animal. For Oliver, the claim that service dogs provide independence to disabled individuals "discount[s] or disavow[s] our interdependence on nonhuman animals."[21]

A cat-loving woman may not need a man in her life, but she does need a cat. To insist that cat-loving women are "independent" reveals a striking disregard for their dependence on cats. Personally speaking, I feel deeply dependent on linguini. As I said at the beginning of this chapter, linguini is essential to my life—both emotionally and intellectually. linguini is the first person I think about when I wake up in the morning, and her paw prints can be traced in all my writing. In fact, my own identity is very much defined by my relationship with linguini. Even though I do have a professional identity as a philosophy professor, I see myself first and foremost as a companion of linguini. And given that linguini motivates and inspires my thinking and writings, I owe my professional identity to her as well. To declare independence from linguini is tantamount to erasing the core of my identity. As such, even though having such a strong attachment to linguini can be taxing, I have no desire to be independent.

***

As we saw, rebranding alone is not quite enough. A woman should not be expected to have the glamorous life of a superstar to "prove" that she has the agency to choose cats over men. Therefore, we can't just rely on posh, privileged superstars to rebrand the CCL stereotype. We also need to reframe the pejorative stereotype of the CCL "by transforming a weakness into a strength."[22] The purported weakness of a cat-loving woman is that she is unwanted by men, so she has no choice but to invest her time and love in her cats. Given this narrative, it makes sense to challenge (as we saw previously) the expectation that a woman needs a man in her life. It makes sense to declare our "independence" from men. However, I believe it is also important to show that a woman can have an enriching life not despite, but *because*, of her dependence on her cats. In other words, reframing the CCL stereotype requires us to acknowledge that a feline-dependent life is a perfectly legitimate way for women to live and flourish. It is not a consolation prize for those who "lack" a man but a life desirable in and of itself.

Indeed, all the stories from the *Girls and Their Cats* project attest to the fact that cats are an indispensable source of enrichment. At first glance, the project appears to focus on rebranding the stereotype with the "right" group of spokespersons (i.e., "strong, independent, and artistic women"). But the in-depth story that accompanies each cat shows that, for many women, their kinship with cats is precisely what enables them to thrive. These cats help women heal from their heartbreaks and losses, inspire their creativity, bond them with their family, and teach them how to love. The moral of many of these stories is not just that a woman can be successful *and* cat-loving but that a woman can have a fulfilling life *because* of her cats. To reframe the CCL stereotype, we need to both reject the presumption that women need men and to affirm the idea that a feline-dependent life is a truly rewarding one.

To linguini

Man is condemned to be free, or so says the Existentialist.
Freedom—radical, inescapable, a burden we cannot disavow.
Yet, it was you, not "man,"
who took on the burden at the shelter,
choosing us.

***

No one knows about power
better than Foucault.
Disciplinary, embodied, performative—
Power manifests itself when my docile body, finally
learns to subjugate itself to your whims and fancies.
Contorting my torso into a body pillow,
the very moment you
leap to my chest.

***

We get glimpses of the id, when we dream and slip.
Opaque, illicit, threatening,
our deepest desires awaiting
excavation—a skill I've honed,
cleaning your litterbox.
But what's the point of lying on the couch,
if you are not there, being held?

Notes

1 Wendy Marie Ingram et al., "Mice Infected with Low-Virulence Strains of *Toxoplasma gondii* Lose Their Innate Aversion to Cat Urine, Even after Extensive Parasite Clearance," *PloS One* 8, no. 9 (September 18, 2013): e75246, https://doi.org/10.1371/journal.pone.0075246
2 Emily Underwood, "Reality Check: Can Cat Poop Cause Mental Illness?" *Science* (2019), retrieved from: https://www.science.org/content/article/reality-check-can-cat-poop-cause-mental-illness, accessed September 4, 2022.
3 Vincent Paquin et al., "Conditional Associations between Childhood Cat Ownership and Psychotic Experiences in Adulthood: A Retrospective Study," *Journal of Psychiatric Research* 148 (2022): 197–203. https://doi.org/10.1016/j.jpsychires.2022.01.058
4 Lori Kogan and Shelly Volsche, "Not the Cat's Meow? The Impact of Posing with Cats on Female Perceptions of Male Dateability," *Animals* 10, no. 6 (2020): 1007, http://dx.doi.org/10.3390/ani10061007.
5 Will McKeithen, "Queer Ecologies of Home: Heteronormativity, Speciesism, and the Strange Intimacies of Crazy Cat Ladies," *Gender, Place & Culture* 24, no. 1 (2017): 122–134, https://doi.org/10.1080/0966369X.2016.1276888
6 Stephanie Butnick, "I Am a Cat Lady? Thank You," *The New York Times*, March 28, 2014, retrieved from: https://www.nytimes.com/2014/03/30/fashion/Cats-Internet-Adoption.html
7 Christine E. Parsons et al., "Pawsitively Sad: Pet-owners Are More Sensitive to Negative Emotion in Animal Distress Vocalizations," *Royal Society Open Science* 6 (2019), https://doi.org/10.1098/rsos.181555

8 Ibid., 8
9 Ibid., 7
10 Ibid.
11 Disclosure: I came across this project during my research for the topic, and the photographer of *GATC* did a photoshoot with linguini in 2022.
12 BriAnne Wills, "Girls and Their Cats," (2019), retrieved from: https://www.girlsandtheircats.com/, accessed September 4, 2022.
13 Mihaela Popa-Wyatt, "Reclamation: Taking Back Control of Words," *Grazer Philosophische Studien* 97 (2020): 159–176.
14 Cynthia S. Wang et al., "Challenge Your Stigma: How to Reframe and Revalue Negative Stereotypes and Slurs," *Current Directions in Psychological Science* 26, no. 1 (2017): 75–80. https://doi.org/10.1177/0963721416676578
15 McKeithen, "Queer Ecologies of Home: Heteronormativity, Speciesism, and the Strange Intimacies of Crazy Cat Ladies," 8.
16 Kathy Rudy, "LGBTQ . . . Z?" *Hypatia* 27 (2012): 601–615. https://doi.org/10.1111/j.1527-2001.2012.01278.x, 605.
17 Ibid., 604.
18 Ibid., 610.
19 Butnick, "I Am a Cat Lady? Thank You."
20 Kelly Oliver, "Service Dogs: Between Animal Studies and Disability Studies," *philoSOPHIA* 6, no. 2 (2016): 241–258. *Project MUSE*, https://doi.org/10.1353/phi.2016.0021.
21 Ibid., 253.
22 Wang et al., "Challenge Your Stigma: How to Reframe and Revalue Negative Stereotypes and Slurs."

Bibliography

Butnick, Stephanie. "I Am a Cat Lady? Thank You." *The New York Times*, March 28, 2014. Retrieved from: https://www.nytimes.com/2014/03/30/fashion/Cats-Internet-Adoption.html

Ingram, Wendy Marie et al. "Mice Infected with Low-virulence Strains of Toxoplasma Gondii Lose Their Innate Aversion to Cat Urine, Even after Extensive Parasite Clearance." *PloS One* 8, no. 9 (September 18, 2013): e75246. https://doi.org/10.1371/journal.pone.0075246

Kogan, Lori, and Volsche, Shelly. "Not the Cat's Meow? The Impact of Posing with Cats on Female Perceptions of Male Dateability." *Animals* 10, no. 6 (2020): 1007. http://dx.doi.org/10.3390/ani10061007.

McKeithen, Will. "Queer Ecologies of Home: Heteronormativity, Speciesism, and the Strange Intimacies of Crazy Cat Ladies." *Gender, Place & Culture* 24, no. 1 (2017): 122–134. https://doi.org/10.1080/0966369X.2016.1276888

Oliver, Kelly. "Service Dogs: Between Animal Studies and Disability Studies." *philoSOPHIA* 6 no. 2 (2016): 241–258. Project MUSE. https://doi.org/10.1353/phi.2016.0021.

Paquin, Vincent et al. "Conditional Associations between Childhood Cat Ownership and Psychotic Experiences in Adulthood: A Retrospective Study." *Journal of Psychiatric Research* 148 (2022): 197–203. https://doi.org/10.1016/j.jpsychires.2022.01.058

Parsons, Christine E et al. "Pawsitively Sad: Pet-owners Are More Sensitive to Negative Emotion in Animal Distress Vocalizations." *Royal Society Open Science* 6 (2019). https://doi.org/10.1098/rsos.181555

Popa-Wyatt, Mihaela. "Reclamation: Taking Back Control of Words." *Grazer Philosophische Studien* 97 (2020): 159–176.

Rudy, Kathy. "LGBTQ . . . Z?." *Hypatia* 27 (2012): 601–615. https://doi.org/10.1111/j.1527-2001.2012.01278.x

Underwood, Emily. "Reality Check: Can Cat Poop Cause Mental Illness?" *Science*. Retrieved from: https://www.science.org/content/article/reality-check-can-cat-poop-cause-mental-illness.

Wang, Cynthia S. et al. "Challenge Your Stigma: How to Reframe and Revalue Negative Stereotypes and Slurs." *Current Directions in Psychological Science* 26, no. 1 (2017): 75–80. https://doi.org/10.1177/0963721416676578

Wills, BriAnne. "Girls and Their Cats." (2019). Retrieved from: https://www.girlsandtheircats.com/

PART VI

Masculinities and Animals

26
THE HOMOXENOEROTICS OF ELEPHANT CRIME AND CAPTIVITY IN INDIA

Naisargi N. Davé and Alok H. Gupta

On January 27, 2013, Thechikottavu Ramachandran, the elephant known as Raman to his fans and followers, killed three women at a temple festival in Perumbavoor, Kerala. By all accounts, Raman, who had been partially blinded by a strike from his mahout (elephant handler) years earlier and was prone to reacting to loud and sudden noises, "turned rogue and ran amok" amid the festival's din, trampling the three worshippers to death.[1]

The Malayattoor Forest Division filed a case against Raman, who could lay claim now to two legendary statuses: being the tallest captive elephant in Asia (standing at ten and a half feet tall) and being the first elephant in India to be charged with murder.

Raman is thought to have been born in eastern India, in Bihar. He was captured from the wild at around 18, called by the name of Moti Prasad, and sold by his captors at the Haathi Bazaar (Elephant Market) of the Sonepur Cattle Fair in Bihar in 1983. There, he was purchased by a trader, transported to the southern state of Kerala, and sold to a temple compound called the Thechikottukavu Peramangalthu Devaswom. It has been illegal in India since 1972, with the passage of the Wildlife Protection Act, to capture an elephant from the wild, as Raman nevertheless was. But exploitative and profitable loopholes are rife, such as the Forest Department's power as custodians of wild elephants in the country to issue ownership certificates to private individuals (a process ushered in in 1972 to protect existing private elephant owners from new conservation laws). Section 43 of the Wildlife Protection Act prohibits owners from commercially trading their elephant property, but owners use gift deeds, lease, and rental contracts to transact a robust trade in captive elephants.

The human capture of wild elephants in India dates to the settlement of the Indus Valley but, according to environmental historian Thomas Trautmann, took on shades of its current form—recognizable in the contemporary capturing and trade of elephants for zoos, tourist sites, and temples—during the kingdoms of the mid-Vedic period circa 1000 BCE.[2] Trautmann's argument is that, paradoxically, the population of wild elephants has persisted in India precisely because they were useful and eventually necessary to kings as war animals. Rather than breeding elephants in captivity and having to bear the considerable costs of feeding them until they matured for work at 20, kingdoms cultivated an elaborate technology of fostering forest land (which has also involved depopulating forests of people)

DOI: 10.4324/9781003273400-33

from where they could capture male elephants and tame them. Indian kings needed some buy-in from the forest people and farmers whose lives intersected, often violently, with the wild elephants the kings required, but ultimately, Trautmann and others have shown, the Indian and colonial state in all its historical forms—from the times of kings to the modern republic—has fomented human–elephant conflict through policies that favor elite men (war mongers, princely sportsmen, and temple corporations) over elephants, farmers, and Adivasis.[3] As Natasha Nongbri has argued, elephants have always in India been a resource "more valuable alive than dead," with much at stake for "both the government and the landed class."[4] And here, the exception in the Wildlife Protection Act that allows the capture and subsequent "non-commercial" transfer of wild elephants by the state to mitigate human–animal conflict becomes especially crucial to the story.

Reduce elephants in the wild and increase captivity was a resounding message of the 2022 Monsoon Session in the Lok Sabha, the People's House, as they debated the "elephant problem" roiling the country. An historic five hours were dedicated to discussion of an amendment to relax the Wildlife Protection law so that more elephants could be acquired from the wild for private captivity.[5] In favor of such a measure were representatives of constituencies most vulnerable to crop raids by wild elephants (often the poorest people in a region, who live on the outskirts of villages abutting protected forest land and who are also entirely dependent on farming for income).[6] In the last three years, 1500 humans in India have been killed by elephants, around 500 per year, and more than 1 per day. With a robust wild elephant population and dense human population, India accounts for 80% of human deaths by elephants worldwide.[7] The other side of this equation is that around 100 elephants are killed by humans in retaliation each year, and dozens more, like the elephant Osama, whom we will encounter next, are seized into captivity as retaliation for the killing of humans.

So, what became of Raman, the wild-caught captive elephant whom we left a few pages earlier languishing in custody under three charges of murder? Raman's owners, the Devaswom temple complex, put up a 30 lakh rupee surety bond, and Raman was released on bail. The charges against Raman were subsequently dropped, but this was neither the beginning nor the end of his encounters with crime and (extrajudicial) punishment. In 1998, Raman was subject to a court entry after stabbing another elephant (who was owned by a rival temple trust) to death with his tusks, but the charges were dropped for a "lack of sufficient evidence." Six years after his 2013 murder charges and immediate release on bail, Raman had 13 human deaths (five mahouts and eight festival-goers) to his name but still headlined the temple parade circuits. How was it that, despite Raman's acts of unrelenting resistance, including the ultimate ones of murdering his captors as well as the spectators of his captivity, his owners and the public who continued to clamor for him believed he could be made to submit?

At play, we argue, is a nefarious combination of belief in Raman's (and other elephants') essential innocence—a belief rooted in the moralism of vegetarianism and the cultural likening of male elephant disposition to the charmed, mischievous naivete of gods in the Hindu pantheon such as Krishna and Ganesh—crossed with a deep awe (and fear) of male elephants' robust and highly visible lust during the condition of *musth* (rut). Since the time of the war elephant, male elephants have been desired by men for capture precisely because of the dominance, power, spectacle, and rage that tuskers display in musth.[8] The other side of this xenoerotics has been the development of elaborate technologies of counter-domination such as *kettiyadi*, or untying, in which elephants are cyclically beaten into submission following the often monthslong sexual trance of musth. Both Raman's captivity and Raman's cyclical punishment are the products of this cultural and historical dispensation: male elephants are desirable to men due to their highly visible and unrestrainable (sexual) power; wild and captive male elephants often kill those who fail to respect this power, but rather than being treated as in essence free, killer male elephants are restored to innocence in order to justify the continued capture of their lives and life force for human and state spectacle.

Among our contentions in this chapter is that the Indian state and media, by refusing to recognize the killings of humans by elephants as murder, engage in the anthropodenialist fiction of sublimating animal vengeance into aberrant errantry and foster the cycle of human–animal conflict. This chapter argues that the paradigm shift towards recognizing animal sentience must include rage, vengeance, and lust. This chapter represents a queer politics in animal studies, one which acknowledges animals—in this case, male elephants—as proud, sexual, and angry beings. Through the stories of killer elephants we examine here, we follow the paradigm move away from animal innocence and towards sexuality, desire, and vengeance.[9]

Idanju and the Good, Vegetarian Elephant

Consider with us some stories of vengeful Asian elephants.

In May 2017, a 7-year-old elephant named Karthik killed his 47-year-old mahout, Mani, in Karnataka's Dubare Elephant Camp, a tourist resort. Karthik had acted violently towards his mahout before in response to brutal training methods, but despite this, Karthik's murder of his handler was spun as an innocent and unfortunate mistake: Mani was a drinker, Karthik stumbled into his stash, couldn't handle his liquor, and simply lost control.[10]

In August 2022, a laboring male tusker hauling rubberwood in a camp ripped the body of his handler into two by stabbing him repeatedly with his tusks. The violent killing of the mahout was attributed to unseasonably hot weather rather than as a solution to years of overwork and torture.[11]

In Kerala, a mahout once opened a coconut by crashing it down onto the skull of an elephant. Years passed before that same elephant, a tusker named Thiruvambady Vishy, saw a coconut lying within reach and the same mahout nearby. Thiruvambady Vishy seized the coconut in his trunk and brought it down upon the skull of his mahout. He then stabbed the mahout with his tusks, trampled him into the earth, and, when the mahout's lifeless corpse was covered by another man, the elephant bull repeatedly stabbed the cloth-covered body.[12]

In Kerala, when Raman killed any of his 13 human victims (as well as his three elephant victims), his actions were often explained as being overcome by his "wilder instincts," having been made vulnerable and skittish by his lost eye, having "grown impatient" while waiting in the heat for festivities to begin, or to have simply killed without intent and only "accidentally."[13,14]

Between 2014 and 2022, animal rights activists from Kerala alone have documented over 60 incidents of captive elephants, such as the previous ones, attacking their owners, mahouts, kavadis, handlers, or spectators. But as these examples also demonstrate, these acts of animal violence—which we might, following Dinesh Wadiwel, call *counter-warfare*[15]—are glossed in national and local media, as well as by many animal activists, as aberrant, idiosyncratic, and accidental incidents of essentially compliant animals "going rogue." Critical animal studies scholars have examined this tendency. Labor historian Jason Hribal, in

Fear of the Animal Planet, details the "hidden history of animal resistance," which is rarely named vengeance.[16] And indeed acts of murderous revenge by animals in India are often explained away as "going mad," "cracking," or, at best, responses to previous trauma. Nayanika Mathur, in her wonderful ethnography of killer felines in India, considers how humans wrestle with animal culpability and the ethics of interspecies killing, in part by casting killer cats as "crooked" (*teda*), which is to say, as fundamentally innocent (*bekasoor*) beings gone astray due to structural and climactic forces far more powerful and agentive than themselves.[17] From Hribal's perspective, these and other animals are denied the capacity for revenge killing, because, as Hribal writes, "if animals could commit crimes, then crimes could be committed against them."[18] Ironically, elephants must be "good" in order to be subject to permissible human control.

India-based animal studies scholar Krishnaunni Hari provides an excellent analysis of how this operates in Kerala—the state with one of the highest wild and captive elephant populations in the country—through a study of the role everyday Malayalam language plays in perpetuating a moral acceptance of elephant captivity.[19]

Hari's discourse analysis shows that when an elephant in captivity runs away, throws off a mahout, kills onlookers, or otherwise defies control, those actions tend to be represented in vernacular media as *idanju* (rebelled). *Idanju* has the effect of leaving the specificity of elephant resistance and its carceral contexts vague, reducing the animal actor to a rebellious child which can be brought back under control. As Hari argues, this transformation of specific violence into *idanju* not only "silences the elephant's voice" but betrays the deep linguistic and cultural investment in maintaining systems of elephant captivity. Hari writes: "Despite there being as many incidents of elephants rebelling as there are *poorams* in Kerala, few attempt to read [those incidents] as caused by systems of elephant enslavement and the ideas that underpin it."

Another element to the irony in which it is the fundamental innocence of elephants that renders them subject to systems of violent human control is the mobilizing of elephants' herbivorous diet in service of the narrative of the "good elephant." Under Hindu anthropatriarchy, the "purity" of elephants is matched only by the female cow.[20] If elephants had evolved otherwise, as meat eaters, they would have likely been found too "impure" for the uses to which they are currently put, predominantly for Hindu religious and cultural purposes. Adjacent to the holy "mother cow"—whose ostensible holiness is a curse to herself as well as to those whose lives and livelihoods are dependent on her labor—the vegetarian elephant in contemporary India assumes the role of a gentle if giant aunt or uncle: imagined and represented as docile, kindly, and fundamental agreeable to (good) human company.

Even animal activists contribute to this Brahminical narrative production of elephants as innocent and pure through recourse to their vegetarianism. One of these is the pioneering Indo-Canadian filmmaker Sangita Iyer, who has campaigned for the release of Indian captive elephants and done more than virtually anyone to document the brutality of their incarceration and forced labor. In the course of filming her documentary *Gods in Shackles*, Iyer observed with much horror that herbivorous elephants were fed eggs and meat, which she flags as both a potentially fatal health hazard but also as a corruption of the innocent (in this case Hindu) soul of the animal.[21] She says,

> But a revered Hindu spiritual leader in Kerala is concerned that elephants are also given unnatural food, including hundreds of eggs, and meat. . . . People can eat

whatever they want, but the audaciousness of religious institutions to feed meat to a herbivorous animal, that too a cultural icon glorified as the embodiment of Lord Ganesh, is simply intolerable.[22]

The anthropatriarchal presumption of animal innocence and (Hindu) purity—what we have shown elsewhere are the very grounds for permissible violence against women and animals—is deeply ingrained both in the people who exploit elephants and those who believe they are emancipating them.[23] We believe animal studies and animal liberation is in need a queer paradigm shift, one which dispenses with abstract and narratives and moral codes and engages with animals in their lust, their desire, and the unruliness of their expression.

Can the Elephant Be Both Innocent and Horny? Musth and the Denial of Sexual Expression

After a two-year hiatus due to the Covid pandemic, the Pooram, or temple festival, season began once again in September 2021. A 26-year-old elephant named Parmeswaran, after having been captured from the wild, chained, and broken, was to lead the procession from the Thiruvilwamala Temple in Kerala. Like Raman and Karthik and so many others, Parmeswaran had had enough. He shook his body hard, throwing his mahout off his back, and with great poetry broke the *deepastambham,* the cemented ceremonial lamp.[24]

In the stories that followed, Parmeswaran's actions were referred to not as labor resistance or as revolution but as "mad," "violent," and "berserk."[25] The mahout was fortunate; he had severe injuries that required surgery, but he survived. The Forest Department lodged an inquiry—not into Parmeswaran's action, nor into the temple trust for having a chained animal made to bow and bless, but into the mahout for breaking a cardinal implicit rule: Parmeswaran was allegedly in *musth* and should not have been paraded. Instead, the mahout should have kept him cloistered to withstand the urgency of his desires far away from humans or even other elephants.

Most male elephants experience *musth* once per year, often in the winter period, and it lasts anywhere from days to several months. Testosterone is at a high, along with other hormonal changes, which transform the male elephants' entire appearance: their temples grow puffy and a stream of fluids emanates from the glands in front of their ears. Their turgid penises drip urine for extended periods. Unlike rutting animals, elephants don't experience *musth* at the same time, nor is *musth* synchronized to sexual desire in females. *Musth* serves to establish the dominance of the male in its throes. The condition of musth and all of its spectacular visual and temperamental transformations are in part what rendered the elephant so desirable to kings as war animals (and therefore what set the Indic path for elephant captivity more generally). Trautmann suggests that *musth*, from the Persian word *mast*, "drunken," was adopted in Mughal times when kings would artificially induce *musth*-like belligerence and magnificence in their captive elephants by plying them with wine and other drugs and serenading them with the sound of martial drumbeats.[26]

Men's homoxenoerotic desire for the elephant in *musth* helps explain the tragic condition of male elephants' disproportionate captivity, historically and in current times. Ninety percent of the over 500 privately held captive elephants in Kerala are male tuskers (activists call the state the "boys' hostel" of elephants, those hostels being temples and forest camps).[27] Across the entire country, the ratio of male to female elephants in the wild is as low as 1 to 16 due both to hunting for ivory and the preference for male captives. In captivity, almost all males are kept in sexual solitude.

But a 2012 edition of the Malayalam news program *Kannadi* attempted to demonstrate an exception. It featured footage of a captive temple elephant during *musth* who is introduced to a female sexual partner.[28] The two watch each other from afar for a few days and then slowly venture closer. The video, which was shown on prime-time Malayalam news, is frankly sexual. The female drapes her trunk around Rajan's erect penis, and we see him attempting to mount her, as they are both strapped with chains on their legs. The message of the news report appears to be that captive elephants used for religious purposes express even their sexual desires through the codes of morality, love, and decency.

In reality, wild and captive elephants are violent in *musth*, mad with desire for sex, and capable of tremendous destruction. Those who keep elephants have always known to take extreme measures to constrain this sexual force when it is not useful for human purposes. Male elephants in captivity, whether in private ownership or in forest camps, are under constant physical restraint, further exacerbating the suffering of captivity. Many mahouts see the rising, dangling penis of *musth*, which Stephen Alter describes as "second trunk protruding from his wrinkled loins," as a threat in itself.[29] Sangita Iyer records handlers who "throw rocks at the genitals when bull elephants come into their musth cycle," perhaps in an effort to calm the rage that the erection represents.[30]

Describing the elephants of the Guruvayoor Temple in Kerala, the single largest temple collection of private elephants, Stephen Alter writes:

> At least a dozen elephants in the Anna Kotta were in musth and each of them looked miserable. Their coloring appeared darker because the mahouts were unable to give them proper baths and simply hosed them off from a safe distance. In addition to this the musth fluids streaming down the elephant's faces, their penises were constantly dribbling and the insides of their thighs were lathered with urine. Piles of dungs and the rotting remains of fodder surrounded each animal. From time to time they splashed water on themselves from their drinking troughs, and the ground at their feet had become a fetid mire because the mahouts were afraid to clean near the aggressive animals.[31]

Mahouts and mahoutery deserve a moment of reflection here. Mahoutery dates to the time of the war elephant, and mahouts themselves, despite their subaltern status in society, were held in high esteem by royal courts for their expertise.[32] Mahouts were recruited from across the belt of the monsoon forests, from the Himalayas down to the lands of South India, and these mahouts passed knowledge and their positions to relatives, creating a long lineage that defines even today's mahouts, thousands of years later. Mahoutery is demanding work, work that found its genesis in the captivity of wild elephants and the human vulnerability partially fomented by that elite desire. In my (Naisargi's) field work with Animal Rahat, an offshoot of PETA-India based in southern Maharashtra, all of the mahouts I encountered descended from lineages of elephant handlers and were one another's fathers, sons, and cousins. The mahout of Ram Prasad, a then 40-year-old who was caught wild at 20 and brought to temple, was trying desperately to control his *musth*-addled charge. A year earlier he had allowed vets associated with an animal activist organization to inject Ram Prasad with a vaccine which had not yet passed clinical trials (I observed this medical intervention) and was meant to delay and then suppress the length of musth. The vaccine had yet to display any success. Ram Prasad was, for the twentieth time in his captivity, heavy in *musth*, swaying frenetically in his corral, his penis turgid, the wrinkles of his skin riven with fluids, his mahout nervous but observant.

Alter quotes a mahout named Sathyalapalan who offers dignity to his charge by acknowledging the pain and poignancy of keeping sexually frustrated male elephants in chains: "During the first stage [of musth], he said, 'the elephants let us go near them and we massage their temples to release their fluids. They often get erections and using their trunks, some will masturbate—like humans'."[33] The Hindi word for elephant, *haathi*, comes from the Sanskrit for hand: the beast with a hand.[34] The anthropomorphic ability to imagine the human in the animal has the potential to stray from the anthropodenialist rejection of animal lust in the hand that, despite all limitations, seeks release.

Osama to Krishna: The Hindu Anthropatriarchal Conversion of Terror to Mischief

Alpa Shah and others have traced the Indian state's policies which favor conservation and the protection of forest land over the lives and rights of Adivasis and farmers.[35] As we suggested earlier, this is a set of priorities that dates back to the Vedic kingdoms, cultivating animal labor and ecological resources at the expense of the poor. Human–animal conflict which results in crop raids, too, dates back to those earliest days of elephant captivity, in which elephants' diets shifted towards high-energy and readily digestible crops. What Trautmann calls the "ongoing, lowlevel warfare between farmers and elephants" has been long in the making, with unique suffering on both sides of the battle.[36]

One site of this struggle, this war, has been the state of Assam in northeast India. There, a male tusker regularly strayed from designated forest land into human habitation in the village of Goalpara and killed, over years, 27 people.[37] Foresters on the ground in India keep track of wild elephant movements, and those that stand out for extraordinary activity receive names. Since September 11, 2001, every "rogue" elephant, from Assam to Jharkhand, has been called, by an anti-Islamic state, Osama Bin Laden.[38]

During a particularly violent period between October 28th and 30th of 2019, allegedly when he was in *musth*, "Osama" killed five Assamese persons. "Osama" was likely searching for food and for a mating partner, and the current compromise in which elephants are allowed passage from one reserve to another through limited designated corridors which pass across railways, road crossings, and village communities had clearly failed everyone. Kaushik Barua, a wildlife conservationist quoted by Neha Sinhaa, suggests that, "The Goalpara area is a fragmented forest. The big male elephant used to wait for night to fall, so it could go eat paddy from the fields. For the farmers, this was a crime. In their eyes . . . they had their crops to protect."[39] The Forest Department was forced after this latest spate of violence against villagers to act and deemed "Osama" no longer subject to Schedule 1 protection. Section 11 of the Wildlife Protection Act allowed foresters to seek to capture "Osama" for relocation in order to mitigate conflict. Drones were deployed to locate the elephant, and a stockage with *kunkis* (Forest Department elephants who act as compradors) strove to surround him.

The least terrible of the terrible options available was to tranquilize "Osama" and relocate him to an alternative elephant habitat as a free animal. But Assam is a densely populated state, every corner with its own share of deadly and costly conflicts with elephants. No community, understandably, wanted "Osama" near them. "Osama," like most captured

Figure 26.1 Osama tied in Orang.

elephants from the wild, would generally then be placed in *kraal*, a Dutch-Afrikaner term that describes a cattle or other large animal enclosure. Built of intersecting horizontal and vertical logs, the *kraal* locks the elephant in place, unable to turn, lift their trunk, or move meaningfully at all. But Rajiv Gandhi Orang National Park, which was able to take "Osama," did not have a *kraal*, so he was stretched along the length of his body and tied with rope and metal chains from his neck to each of his four legs. In this perpetual position, "Osama" was sought to be tamed, punished regularly with the traditional *ankush,* or bull hook.

In a damning example of the Hindu state's war against Islam, and the ongoing warfare between humans and animals, just as "Osama" was captured and transferred into captivity for life, he was renamed "Krishna," the Hindu god of love. His Hindu reincarnation was the restoration of the elephant's innocence, the paradoxical precondition for his pastoral carcerality. As "Krishna," he was hailed as a *ghar-wapasi* (a prodigal son returned home). The marauding elephant who lived under the fear-mongering name of Osama in terrifying freedom died as the convert Krishna, the merely mischievous god of love, on his eighth day of captivity.[40]

Conclusion: The Sublimation of Homoxenoerotics Into Love

Since the time of the war elephant to the time of the temple elephant, male elephants have been desired by elite men precisely because of the sexual dominance, power, rage, and spectacle that they display. But what is erotic in freedom proves threatening in captivity, so this homoxenoerotics has also necessitated the development of elaborate technologies of physical and discursive counter-domination. The means of physical domination are clear: ropes, *kraal*, chains, and the bull hook, for a few examples. But discursively in contemporary times, we have tried to suggest that this counter-domination operates through the conversion of vengeance into errantry, of terror into mischief, and elite homoxenoerotic desire into the benevolent love of pastoral carcerality.

This love is dependent on the manufacture of an essential elephant innocence, because if the elephant is essentially innocent, which is partly to say incapable of murder, then there is no war, and there is no murder that can be committed against them. Elephant innocence is produced through recourse to the elephant's vegetarian diet (which signals "purity" in a food fascist state) and by drawing comparisons between elephants to rebellious children and mischievous Hindu gods. A lifetime of sacral celibacy for the elephant plays, too, in stripping the sexual from a relationship rooted in homoxenoerotic desire. Even those seeking to liberate captive elephants have participated in this discursive counter-domination, lauding the vegetarian souls of elephants or seeking to temper *musth* through chemical intervention.

Like "Krishna," Parmeswaran (the elephant in *musth* who threw off his mahout), died too while being rehabilitated for his non-criminal aberrations. Raman lives on, still parading. His official age moves annually backwards so that he needn't be retired. Forever young, still innocent, Raman's solitary, lusty acts of resistance remain couched in love rather than revolt.

Notes

1 TNN, "Kerala Court Releases Elephant on Bail," *Times of India* (2013). Retrieved from: https://timesofindia.indiatimes.com/india/kerala-court-releases-elephant-on-bail/articleshow/18920005.cms

2 Thomas R. Trautmann, *Elephants and Kings: An Environmental History* (The University of Chicago Press, 2015).

3 See also Alpa Shah, *In the Shadows of the State: Indigenous Politics, Environmentalism and Insurgency in Jharkhand, India* (Duke University Press, 2010).

4 Natasha Nongbri, *Elephant Hunting in Late 19th Century North-East Indiai*, EPA (July 26, 2003), 389.

5 For a discussion of the original Act and subsequent amendments see https://vidhilegalpolicy.in/research/legal-issues-pertaining-to-insertion-of-proviso-under-section-432-of-wildlife-protection-amendment-bill-2022-discussion-paper/

6 Shah, *In the Shadows of the State: Indigenous Politics, Environmentalism and Insurgency in Jharkhand, India*, 113.

7 Kalyan Ray, "1,700 Indians Killed by Tigers, Elephants in Last Three 3 Years: Environmental Ministry," *Deccan Herald* (2022), retrieved from: https://www.deccanherald.com/national/1700-indians-killed-by-tigers-elephants-in-last-3-years-environment-ministry-1129986.html

8 See Trautmann, *Elephants and Kings: An Environmental History.*

9 Some recent scholarship in anthropology that recognizes animal vengeance (and specifically from the beyond) includes Radhika Govindrajan, "Spectral Justice: Multispecies Haunting and Accountability in Himalayan India," (2021) and Strange, "Vengeful Animals."

10 TNN, "Killer Elephant' Karthik Attacks Manhunt Again," *Times of India* (2019). Retrieved from: https://timesofindia.indiatimes.com/city/bengaluru/killer-elephant-karthik-attacks-mahout-again/articleshow/67517764.cms

11 Koh Ewe, "Hot Weather may Have Caused Elephant to Stab Its Trainer to Death, Police Report," *Vice* (2022), retrieved from: https://www.vice.com/amp/en/article/n7z7bm/hot-weather-elephant-death-thailand-climate-change

12 *Peter Jaeggi, a swiss journalists records of travels in Kerala published in Gods in Chains, by Rhea Ghosh*, Manas Saikia Foundation 2005.
13 G. Shaheed and N.A. Naseer, "The Pain of Being a Temple Elephant," *Frontline* (2014). Retrieved from: https://frontline.thehindu.com/the-nation/the-pain-of-being-a-temple-elephant/article5749412.ece
14 Wikipedia, "Thechikottukavu Ramachandran," *Wikipedia: The Free Encyclopedia* (n.d.). Retrieved from: https://en.wikipedia.org/wiki/Thechikottukavu_Ramachandran
15 Dinesh Wadiwel, *The War against Animals* (Critical Animal Studies, 2015).
16 Jason Hribal, *Fear of the Animal Planet* (AK Press, 2011).
17 Nayanika Mathur, *Crooked Cats: Beastly Encounters in the Anthropocene* (University of Chicago Press, 2021), 11.
18 Hribal, *Fear of the Animal Planet,* 8.
19 Krishnanunni Hari, "Talking Elephants into Slavery," *The India Forum* (2022), retrieved from: https://www.theindiaforum.in/article/talking-elephants-slavery
20 On Hindu anthropatriarchy, see Yamini Narayanan, 2019, " 'Cow Is a Mother, Mothers Can Do Anything for Their Children!' Gaushalas as Landscapes of Anthropatriarchy and Hindu Patriarchy," *Wiley Online Library* (2019) and Naisargi N. Davé and Alok H. Gupta, "The Permissible and the Perverse: Indian Geographies of Interspecies Sex," *Environment Planning and E* (2023).
21 An ancient text on elephant keeping *Mathangleela* recommends mutton as an ailment for captive elephants; feeding meat to elephants is not by any means a nascent phenomenon that can be attributed to battles around caste, religion, and food.
22 Sangita Iyer, "What Kind of Monster Would Feed Meat to an Elephant," *HuffPost* (2017), retrieved from: https://www.huffpost.com/archive/ca/entry/what-kind-of-monster-would-feed-meat-to-an-elephant_b_16744280
23 Davé and Gupta, "The Permissible and the Perverse: Indian Geographies of Interspecies Sex."
24 Mathrubhumi News (2022), retrieved from: https://www.youtube.com/watch?v=6ORdCkWSVvU
25 Henry, "Watch: Temple Elephant Goes Mad in Kerala, Throws Mahout Sitting on Top," *Henry Clubs* (2021). Retrieved from: https://henryclubs.com/watch-temple-elephant-goes-mad-in-kerala-throws-mahout-sitting-on-top/; Team Newsable, "Elephant Goes Violent in Thiruvilwamala Vilwadrinatha Temple, Throws Off Mahout in Kerala," *Asianetnewsable* (2021). Retrieved from: https://newsable.asianetnews.com/india/elephant-goes-violent-in-thiruvilwamala-vilwadrinatha-temple-throws-off-mahout-in-kerala-gps-qzzoeu; Trends Desk, "Watch: Temple Elephant Goes Berserk in Kerala, Throws Off Mahout Sitting Ontop," *Indian Express* (2021). Retrieved from: https://indianexpress.com/article/trending/viral-videos-trending/temple-elephant-runs-amok-in-kerala-throws-off-man-sitting-on-top-7532283/
26 Trautmann, *Elephants and Kings: An Environmental History,* 62–63.
27 Asha Prakash, "Elephants Belong in the Wild . . . Let Them Be!," *Times of India* (2016). Retrieved from: https://timesofindia.indiatimes.com/city/kochi/elephants-belong-in-the-wild-let-them-be/articleshow/53653105.cms?utm_source=contentofinterest&utm_medium=text&utm_campaign=cppst
28 AsiaNet News, "Rare Mating of Captive Elephants: Kannadi, 03 November 2012 Part 1," *YouTube* (2012), retrieved from: https://www.youtube.com/watch?v=arKyaR4ugYY
29 Stephen Alter, *Elephas Maximus: A Portrait of the Indian Elephant* (Harcourt first edition, 2004), 118.
30 Sangita Iyer, *Gods in Shackles* (2022), 105.
31 Alter, *Elephas Maximus: A Portrait of the Indian Elephant,* 129
32 Trautmann, *Elephants and Kings: An Environmental History*, 140–143.
33 Alter, *Elephas Maximus: A Portrait of the Indian Elephant*, 129.
34 Trautmann, *Elephants and Kings: An Environmental History*, 24.
35 Shah, *In the Shadows of the State: Indigenous Politics, Environmentalism and Insurgency in Jharkhand, India.*
36 Trautmann, *Elephants and Kings: An Environmental History,* 338.
37 Samuel Osborne, "Elephant Named Osama Bin Laden Kills Five People in India," *Independent UK* (2019), retrieved from: https://www.independent.co.uk/news/world/asia/elephant-osama-bin-laden-kills-people-assam-india-a9182511.html
38 Wikipedia, "Osama bin Laden (Elephant)," *Wikipedia Free Online Encyclopedia* (n.d.). Retrieved from: https://en.wikipedia.org/wiki/Osama_bin_Laden_(elephant)

39 Neha Sinha, *Wild and Wilful* (Harper Collins, 2021), 110.
40 Scroll Staff, "Assam: Captured, Relocated Elephant Dies of Cardiac Arrest," *Man and Animal* (2019). Retrieved from: https://scroll.in/latest/944072/assam-captured-relocated-elephant-dies-of-cardiac-arrest

Bibliography

Alter, Stephen. *Elephas Maximus: A Portrait of the Indian Elephant*. Harcourt first edition, 2004.
Asianet News. "Rare Mating of Captive Elephants: Kannadi, 03 November 2012 Part 1." *YouTube*, 2012. Retrieved from: https://www.youtube.com/watch?v=arKyaR4ugYY
Davé, Naisargi N., and Gupta, Alok. "The Permissible and the Perverse: Indian Geographies of Interspecies Sex." *Environment Planning and E* 6, no. 2 (2023): 1391–1411.
Ewe, Koh. "Hot Weather may Have Caused Elephant to Stab Its Trainer to Death, Police Report." *Vice* (2022). Retrieved from: https://www.vice.com/amp/en/article/n7z7bm/hot-weather-elephant-death-thailand-climate-change
Ghosh, Rhea. *Gods in Chains*. Manas Saikia Foundation, 2005.
Govindrajan, Radhika. "Spectral Justice: Multispecies Haunting and Accountability in Himalayan India." (2021)
Hari, Krishnanunni. "Talking Elephants into Slavery." *The India Forum*, 2022. Retrieved from: https://www.theindiaforum.in/article/talking-elephants-slavery
Henry. "Watch: Temple Elephant Goes Mad in Kerala, Throws Mahout Sitting on Top." *Henry Clubs*, 2021. Retrieved from: https://henryclubs.com/watch-temple-elephant-goes-mad-in-kerala-throws-mahout-sitting-on-top/
Hribal, Jason. *Fear of the Animal Planet*. AK Press, 2011.
Iyer, Sangita. "What Kind of Monster Would Feed Meat to an Elephant." *HuffPost*, 2017. Retrieved from: https://www.huffpost.com/archive/ca/entry/what-kind-of-monster-would-feed-meat-to-an-elephant_b_16744280
Mathrubhumi News. (2022). Retrieved from: https://www.youtube.com/watch?v=6ORdCkWSVvU
Mathur, Nayanika. *Crooked Cats: Beastly Encounters in the Anthropocene*. University of Chicago Press, 2021.
Narayanan, Yamini. " 'Cow Is a Mother, Mothers Can Do Anything for Their Children!' Gaushalas as Landscapes of Anthropatriarchy and Hindu Patriarcy." *Hypatia: A Journal of Feminish Philosophy* 34, no. 2 (2019): 195–221.
Natasha, Nongbri. "Elephant Hunting in Late 19th Century North-East India." *EPA*, July 26, 2003.
Osborne, Samuel. "Elephant Named Osama Bin Laden Kills Five People in India." *Independent UK*, 2019. Retrieved from: https://www.independent.co.uk/news/world/asia/elephant-osama-bin-laden-kills-people-assam-india-a9182511.html
Padmar, Deepa; Sinha, Debadityo, and Jauhari, Akash Chandra. "Captive Elephants and the Wildlife (Protection) Amendment Bill, 2022: A Discussion Paper." Vidhi: Centre for Legal Policy, 2022. Retrieved from: https://vidhilegalpolicy.in/research/legal-issues-pertaining-to-insertion-of-proviso-under-section-432-of-wildlife-protection-amendment-bill-2022-discussion-paper/
Prakash, Asha. "Elephants Belong in the Wild . . . Let Them Be!" *Times of India*, 2016. Retrieved from: https://timesofindia.indiatimes.com/city/kochi/elephants-belong-in-the-wild-let-them-be/articleshow/53653105.cms?utm_source=contentofinterest&utm_medium=text&utm_campaign=cppst
Ray, Kalyan. "1,700 Indians Killed by Tigers, Elephants in Last Three 3 Years: Environmental Ministry." *Deccan Herald*, 2022. Retrieved from: https://www.deccanherald.com/national/1700-i-ndians-killed-by-tigers-elephants-in-last-3-years-environment-ministry-1129986.html
Scroll Staff. "Assam: Captured, Relocated Elephant Dies of Cardiac Arrest." *Man and Animal*, 2019. Retrieved from: https://scroll.in/latest/944072/assam-captured-relocated-elephant-dies-of-cardiac-arrest
Shah, Alpa. *In the Shadows of the State: Indigenous Politics, Environmentalism and Insurgency in Jharkhand, India*. Duke University Press, 2010.
Shaheed, G., and Naseer, N.A. "The Pain of Being a Temple Elephant." *Frontline*, 2014. Retrieved from: https://frontline.thehindu.com/the-nation/the-pain-of-being-a-temple-elephant/article5749412.ece

Team Newsable. "Elephant Goes Violent in Thiruvilwamala Vilwadrinatha Temple, Throws Off Mahout in Kerala." *Asianetnewsable*, 2021. Retrieved from: https://newsable.asianetnews.com/india/elephant-goes-violent-in-thiruvilwamala-vilwadrinatha-temple-throws-off-mahout-in-kerala-gps-qzzoeu

TNN. "Kerala Court Releases Elephant on Bail." *Times of India*, 2013. Retrieved from: https://timesofindia.indiatimes.com/india/kerala-court-releases-elephant-on-bail/articleshow/18920005.cms

TNN. "Killer Elephant' Karthik Attacks Manhunt Again." *Times of India*, 2019. Retrieved from: https://timesofindia.indiatimes.com/city/bengaluru/killer-elephant-karthik-attacks-mahout-again/articleshow/67517764.cms

Trautmann, Thomas R. *Elephants and Kings: An Environmental History*. The University of Chicago Press, 2015.

TrendsDesk. "Watch: Temple Elephant Goes Berserk in Kerala, Throws Off Mahout Sitting Ontop." *Indian Express*, 2021. Retrieved from: https://indianexpress.com/article/trending/viral-videos-trending/temple-elephant-runs-amok-in-kerala-throws-off-man-sitting-on-top-7532283/

Wadiwel, Dinesh. *The War Against Animals*. Critical Animal Studies, 2015.

Wikipedia. "Osama bin Laden (Elephant)." *Wikipedia Free Online Encyclopedia* (n.d. a). Retrieved from: https://en.wikipedia.org/wiki/Osama_bin_Laden_(elephant)

Wikipedia. "Thechikottukavu Ramachandran." *Wikipedia: The Free Encyclopedia* (n.d. b). Retrieved from: https://en.wikipedia.org/wiki/Thechikottukavu_Ramachandran

27
THE HOMOEROTICS OF TROPHY HUNTING

Sal Renshaw

For those interested in understanding the complex dynamics around gender, sex and hunting, we owe a debt of gratitude to environmental feminist thinkers like Carol J. Adams, Josephine Donovan, Marti Kheel and Brian Luke from the 1990s;[1] Linda Kaloff and Amy Fitzgerald,[2] especially in the early 2000s; and more recently the myriad scholars who have carefully parsed the many complex ways hunting is gendered both symbolically and literally. Drawing on connections long made between desire, sex and power, much of this work is built on the recognition that within the binary logic of Western thought, maleness and masculinity have been associated with culture and civilization in contrast to femaleness and femininity, which have been connected to wildness, emotion and nature. Thus, women's relationship to animals has in part derived from their shared occupation of the space of exclusion from the category of the fully human. The same logic operates with respect to race. To do violence to animals is necessarily to feminize them regardless of their biological sex. To do violence to women is to figuratively, and often literally, cloak them in the skins of animals, to position them as other-than-fully human. As Brian Luke says, "regardless of their biological sex or species, subordination feminizes people and animals."[3]

In its commitment to death, hunting is nothing if not the most extreme practice of radical subordination and othering.

Undeniably much of this scholarship has yielded powerful insights into the heteronormative structures of desire that surround hunting. Take for instance the revelatory discourse analysis of *Traditional Bowhunter* magazine (a random sampling from 1992–2003) conducted by Linda Kaloff, Amy Fitzgerald and Lori Baralt in 2004, in which their analysis of over fourteen issues yielded an account of the "extreme objectification and marginalization" of animal bodies which necessarily stood as a visual rebuttal to the speech acts of hunters themselves who consistently laid claim to 'loving nature and animals.'[4] As the visual counter text trophy hunting photographs were then and continue to be remarkably repetitive and iconic visual stagings of the same story told again and again for more than a century now: white men pose over the dead but hauntingly life-like bodies of a huge, elegant, even beautiful 'conquered foe,' the animals invariably facing the camera, often with weapons displayed over their bodies and almost always with the evidence of their deaths disguised. The story that maps neatly onto heteronormativity here is one of feminine

DOI: 10.4324/9781003273400-34

submission: a wild generic 'mother' nature has been tamed as testament to the prowess and power of the idealized masculinity of the hunter. But, as I will suggest throughout this piece, if we bracket for a moment the story of a radically subordinated feminine that is embodied in both the trophy itself and the trophy photograph, another story comes into view. If instead of attending to the symbolic feminine, we attend to the materiality of the animal whose life has just been taken, what we see is less a subordinated feminine than a very particular *masculine* subject/object of desire. The trophy hunters' longing is actually for an idealized yet very particular male, albeit a non-human one, and it is that very particular male whose subjectivity is erased in conquest and in death, Thus, the trophy shot is a kind of magical hetero-ideal retelling of the practices of desire which inform trophy hunting itself.

Trophy hunting images have also been the seedbed for a powerful analysis of a racialized expression of the desire for objectified dominance, as scholars have pointed to the many ways in which, for instance, international trophy hunting—especially in Africa—is a continuation of the dynamics of colonialism.[5] To the extent they are included at all, images of local experts, the guides who make the hunt possible in the first place, invariably depict them either in the background or to the side, reflecting the power dynamics which foreground the prestige and prowess of the foreign, usually white male hunter. These are not images of collaboration; on the contrary, they perfectly embody the racialized hierarchy that orders the relations between affluent foreign hunters and the local experts. With around just 4% of the U.S. population identified as hunters, 90% of those being white males, and nearly three quarters of all international trophy hunting being undertaken by U.S. citizens, this particular form of hunting remains not only a predominantly white man's 'sport' but also one uniquely intertwined with U.S. conceptions of gendered identity.[67]

This ongoing critical work on both gender and race continues to be powerfully relevant, perhaps all the more so as considerable efforts are being made by hunting organizations as well as, paradoxically, the U.S. Fish and Wildlife Service along with the National Parks, to recruit 'new' hunters, especially women, as well as other 'non-traditional' (by which they mean racially diverse) groups. Notwithstanding the ongoing relevance and importance of this work, however; as I flagged previously, my interest here is in opening a space to bear witness to that 'other' story of the desires that inform trophy hunting and the possibilities that flow from it. By exploring the phenomenal materiality of hunting as lived experience and by taking seriously the sexed particularity of the non-human animals who are the subjects of this often very prolonged desire, what we see revealed is a male desire for male bodies—albeit non-human animal male bodies. Thus, my interest here is specifically in trophy hunting as opposed to other forms of hunting for in none of the other subgenres of hunting—sport, subsistence, religious/spiritual, market—do we find the overlay of justifications falling away so rapidly to expose this underlying homo eros.

While it may involve many other things, trophy hunting is always a violent expression of desire, notwithstanding the language of love that so often infuses the hunter's discourse,[8] or the post-hoc supposedly moral justifications like, "We always donate the meat to the local communities" or "Hunting is 'natural' and we only take the surplus animals and thus increase the fitness of the stock." This claim, by the way, is a lie. In nature, predators typically take the oldest, youngest or sickest, that is, the weakest. In a fairly stunning counter-evolutionary move, only humans hunt the fittest, healthiest and most exemplary animals.[9] While there has been some interesting research in social psychology suggesting that trophy hunting serves a heteronormative evolutionary purpose in terms of social signaling for males,[10] here, I am interested in attending to the phenomenology of the desires that structure much of the quotidian material experience of the hunt itself. Trophy hunting, in practice, is often a very slow process, at the very fastest taking days or weeks but far more often involving years of pursing the same male animals, animals that have often come to be affectionately named and identified as individuals by those pursuing them. While many of the following observations are apposite for many species, deer hunting is perhaps most exemplary of the phenomenon I am here exploring at the same time, as deer are also the animal that by the numbers are most hunted in reality. So, beneath the hyper-heteronormative discourses of masculinity that overtly cloak the practices of hunting and that draw a tight nexus between male arousal patterns and the heteronormative gendered dynamics of hunting, I want to attend to and take seriously that bedrock of what I take to be homoeroticism that remains somewhat submerged in much of the writing on hunting, including feminist writing, and yet is so apparent when we focus on the specificity of the non-human animal that is so desired. In other words when we attend to the particularity of the animal that is being hunted and we refuse the invitation to objectify him through the categories of species, it is hard to ignore the salience of his sexed subjectivity as an integral if not central aspect of the desire of the hunter.

In his insightful essay "Violent Love: Hunting, Heterosexuality and the Erotics of Men's Predation," Brian Luke draws out the intrinsically sexual aspects of the hunting experience:

> Hunting is experienced as and expected to be a very sensual activity for the hunter. The physical exertion; exposure to the elements; immersion in environments rich in sights, sounds, and smells; and the stalking induced intensification of sensory capacities all contribute. But the warm internal feelings mentioned by hunters go beyond the sensory focus and stimulation entailed by stalking in the wild and suggest an additional purely sexual aspect of the hunting sexuality.[11]

and

> The pattern is that of a buildup and release of tension organized around the pursuit, phallic penetration, and erotic touching of a creature whom the hunter finds seductively appealing.[12]

We can see how far Luke's analysis goes towards showcasing the phenomenology of desire that is at work in hunting. However, while briefly touching on the homoerotic connotations of the phallicism of hunting, he also accepted at face value the heteronormative assertions of the hunters themselves, distilling from those assertions further evidence for his important point about the erotic role of differences in power for normative heterosexual masculinity. Presumably via a process of chimeric identification in which they themselves become animal, as I alluded to earlier, hunters perform a magic gender trick which transforms their intense lust for, idealization and pursuit of perfectly beautiful sexually mature and powerful male animals into a story of heterosexual conquest. "Hunting men relate their pursuit of male animals to their sexual relations with female humans because both eroticize power difference."[13] Perhaps it is no surprise that this transformative disguise is psychically necessary. In what remains a viciously class-divided heteronormative world where overt expressions of male desire for male bodies represent not only a social risk but a physical one, the expression of undisguised desire for another male, even a non-human one, presents a profound risk at the very least to self-understanding,[14] and this is true whether we are considering the more working-class pursuits of deer hunting or the elite world of international trophy hunting.

However insightful Brian Luke's analysis is, I wonder if he might have too quickly closed down the theoretical space that he briefly opened around male human desire for male nonhuman animals in a structural analysis that positions animals in general as the feminine victims of male predation in the same way that human women are the objects of a predatory male desire. Regardless, I am indebted to his observations. He is one of the few to even mention the possibility of a homoerotic desire in hunting, and nearly twenty-five years later there is still something transgressive about the fact that he does. Hunting discourses remain belligerently heteronormative and aggressively heterosexual. Open any contemporary deer hunting magazine and you will quickly find yourself immersed in the language of 'fevered bucks' at the mercy of 'hot does.' So, against this defiant heterosexism, claims about homoeroticism feel risky. Yet there's also something evocative in this risky naming. In the end it takes little critical reflection to arrive at two seemingly obvious contextual facts about trophy hunting that give rise to at least two significant questions. Taken together they position an underlying homoerotic structure in trophy hunting as surprisingly obvious.

Two Facts

1. Despite a disturbing and statistically significant increase in female participation in hunting from around 8% in the 80s to 11% now, most 'sport hunters' are cis-gendered, heterosexually identified white males, many of whom hunt in male-only groups.
2. Bracketing for a moment the idea that almost anything can function as a trophy, the term *trophy* in the context of hunting is virtually synonymous with male animals. And trophy animals, by definition, are peak-condition mature males notwithstanding that individual hunters identify particular kills as personal trophies. We need only look at the standardized price lists of international trophy hunting sites to see the evidence. There is always a clear hierarchy where males are the prized trophies, and they are much more expensive to hunt. Second-tier/class 'trophies' are typically females; young animals, including young sexually immature males; and animals with visual 'imperfections.' In the case of antlered animals, only males are trophies. Consider the sizable prices differences retrieved from the 2019 price list of the Namibian Safari Hunt Outfitters: Elland bull $1800, Elland cow $1000; Kudu bull $1600, Kudu cow $600; Blue Wildebeest Bull, $1200, Blue Wildebeest cow $700 (see https://namibiahunts.com/namibia-hunting-price-list/).

It is worth pausing here to attend to the particularities. What is truly desired is not the generic 'animal' member of a species structurally positioned as 'feminine' but rather the very particular male, who embodies qualities that typically connote idealized tropes of masculinity. He is almost always some version of majestic and huge, courageous and wise, a worthy adversary who has proven his intelligence just by staying alive to reach maturity.

> One deer in particular was a beauty of a five pointer (5 × 5) from the previous year that I was fortunate enough to have an encounter with. Luckily for him, I had my heart set on another crafty buck. He got a pass for the year, as I was hoping to see him up close in what would be this year's hunting season.[15]

Through consistently anthropomorphized descriptions 'crafty bucks' slip the bonds of categorical purity as animal objects—always an illusion anyway—and become instead hybrid and very desirable *manimals.*

Two Questions

1. In what social spaces other than hunting do heterosexually identified men get to spend vast amounts of time gazing at and admiring the sublime beauty of another male? Where do they get to write publicly about this desire and their pursuit of this ideal male and even get to express a deep and abiding love for what amounts to another male body at the prime of their reproductive/erotic life? There is undeniably a heteronormative veil thrown over reality here that reinforces heterosexual credibility despite all the evidence to the contrary.
2. Where can a heterosexually identified man literally immerse himself in the acoustic and pheromonal garb of not only another species but another sex in order to provoke and ultimately seduce a desiring male? This is more than a symbolic taking on of the attributes of another, more than mimicry, as hunters regularly transform their own phenomenal embodied being solely for the purpose of seduction. For what is often a considerable period of time–hours, and even days—these are doe men, and yet they still hold onto their assured sense of their own abiding normative heterosexual masculinity?[16]

At the very least, if we pause long enough to attend not only to the sexed and gendered subjectivity of the trophy hunters, something I would argue has garnered at least some academic attention, but also to the particular subjectivity of the trophy, that is, the animal itself, we can clearly see that there is something rather queer going on here. In reality, there are no social spaces where heterosexually identified men can immerse themselves in the phenomenology of desiring male human bodies without deeply compromising their claims to normative heterosexual masculinity, notwithstanding the agonistic battles on sporting fields.[17] Yet by focusing on the actual practices of trophy hunting, not just the symbolic meaning it might have after the animal is dead, there can be no doubt that the operant desire here, at least right up to the moment of penetration/death, is partly homoerotic, even if the animality of the subject of desire obscures that reality. This obscuration happens as a consequence of anthropocentrism. In other words, it is in our consistent failure to slow down enough to attend to the phenomenology of hunting that we miss this aspect of the structure of desire that actually informs the vastly more significant period of time that defines trophy hunting, the pursuit. In doing so we fail to ask in what ways an underlying homoeroticism of some aspects of trophy hunting might reveal something important about the possibilities and foreclosures on masculine desires as they are currently being constructed, if not in Western cultures more broadly, then at least in terms of U.S./North American masculinities.

This last point leads me to highlight a point already made, for it shapes the analysis and some of the questions that might be asked going forward. According to the 2016 IFAW "Analysis of the Global Trophy Hunting Trade,"[18] where tracking the import and export of animal bodies tells us something about who is doing the hunting, in the ten-year period between 2004 and 2014, the U.S. was responsible for almost 71% of all imports of dead animal trophies, with the next nearest contributors being Spain and Germany at just 4.9%.

This is an astonishing difference, and it speaks volumes to the deep connection between hunting and national identity in the U.S., a relationship long recognized and identified with figures like Theodore Roosevelt, who established the national parks system in the U.S. in part to encourage hunting as an exemplary expression of U.S. frontier masculinity, and Aldo Leopold, the pioneering writer, philosopher and naturalist, whose work was profoundly influential in modern environmental ethics in the U.S.[19] So the claims we might make about homoeroticism and hunting cannot be easily separated from U.S. masculinities, and while I am mindful of this connection in the observations that follow, the role of national identity in hunting in general and trophy hunting in particular deserves much more critical attention. For the remainder, my interest is in continuing to draw out some of the curiosities of trophy hunting that emerge when we slow down and explore its quotidian practices, in particular as they relate to deer hunting, by far the most common form of trophy hunting in North America.

Slowing Down and Attending to the Minutia

The August 2022 issue of *North American White Tail* features a series of stories under the title "Life-Changing Archery Hunts: Seven Legendary Trophy Deer Stories," and these stories will function as the touchstone of my remaining observations. They are exemplary, and, as I note previously in relation to trophy photographs, hunting stories are also curiously repetitive and formulaic, mind-numbingly so.

Trail Cams: In a modern twist on camera hunting, the use of trail cams is ubiquitous, and they have vastly shifted the temporal experience of longing, taking what was once an imaginary object of desire, that perfectly majestic 'monarch of a buck,' into the reality of a pursuit that often takes years.

In every one of the seven 'legendary stories,' trail cams offered the first glimpse of the object of desire, indeed, often of many objects of desire. While most hunters have their own cameras and very particular hunting areas, images of prized and idealized animals—always male—would often circulate between groups of men[20] who either hunted together or hunted the same areas in competition with each other. In one case, images were posted on Facebook, creating a kind of imagined community of desire.[21]

> Have you ever been scrolling through Facebook and come across someone posting about a giant buck that you realize must be in your local area? . . . You ask a few friends if they've even heard about that buck, and before you know it you realized the buck is somewhat of a celebrity. Seemingly everyone knows about him, and you set your sights on putting an arrow through him.
>
> *(Alex Comstock 2022, 82)*

Despite their ubiquity and utility, trail cams are a wicked time sink often requiring many hours of reviewing images of the various animals they 'catch.' It's worth noting the way that the review process itself mimics the erotic structure of desire in hunting. Breath holding anticipation and longing are the driving emotions as hunters scroll through what can sometimes be hundreds, even thousands, of images until they see 'the one,' and suddenly there he is! Desire is now laser focused on the particular object of desire; it is him, in his glorious, fully embodied particularity that is desired. No surprise then that in many cases this moment becomes oddly baptismal as the buck transitions from generic to particular, from object to subject, via the process of naming.

> I typically get hundreds of deer pictures to look through daily and I start the summer off by scrolling through pictures one by one. Looking closely at every animal and accruing as much information about the herd as possible. This process evolves, and soon I'm going as fast as I can to sift through never-ending pictures of does and small bucks. Every time I see a photo of a buck with a nice rack, the photo is saved, and my inventory builds. Sometimes a young buck appears that is special—a buck that clearly could be a bomber someday.
>
> *(Bias, 2022, 21)*

The trail cam pictures play a significant role in calibrating the homoerotics of desire. They straddle and concentrate the temporality of the arc of hunting from pursuit to display, and they vastly extend the period of anticipatory desire prior to a kill. This new technology of the hunt has also made possible the very conditions in which the object becomes subject. Hunters now spend vast periods of time not just imagining and fantasizing about *a* trophy deer but rather that particular deer, an animal who is now clearly identified as an individual. Crucially, the trail cam pictures also memorialize the past of the once fully alive subject of desire beyond his objectification in death. But they are not the public photos that mark the moment of a trophy kill. Those photos, as noted, are notoriously heteronormative displays of masculine power. By contrast, the trail cam shots are the private residual traces of the desire that came first, the homoerotic desire that was synonymous with the living animal, now safely tucked away in the virtual closet of a hard drive.

Naming

While the names that hunters typically give the bucks they are pursing are often reflective of aspects of their embodiment—*Big 5* meaning it was a five-point buck, *Curly* to reflect an unusually bent antler tine—others reflect something more like the qualities of personality–*Angry*, so named for seemingly being overly dominant with other bucks—and *Ghost*, so named for not turning up on time to be killed when it was convenient for his pursuer. Names function to signpost individuality, to transform the animal from object to subject, and they are part of building an intimacy, a relational connection between hunter and hunted, at least for the time of the pursuit. Post the kill, it is never the name that is memorialized, hence the transience of subjectivity for the victim. In a dynamic reversal of the subject/object relation that attends on trophy hunting, it's not Angry whom Aaron Sligh killed, it's a record-breaking, huge eight-pointer. At the moment of death the manimal, always a shapeshifter anyway, transforms once more into an animal object, a thing. Individuality and subjectivity are submerged beneath a group/species identity. He is now the 'trophy deer,' yet further objectified in being assigned a number on the Boone and Crocket Trophy Scale.[22] In death, all signs of his unique personhood, his individuality, are disappeared, and along with them also go the homoerotics of desire, which are subsumed beneath the public performance of hyper-heteronormative masculinity that is embodied in the trophy photo. In this moment of transition between life and death, heteronormative masculinity rises to the visible fore, leaving only the ghostly traces of the homoerotics that have defined by far the lengthiest portion of the phenomenology of the hunt itself—the period of longing. The ghostly traces of this transgressive desire are now an absent presence yet visible still to those willing to see.

Sensuality and the Work of Death

In most hunting stories, as Luke and Kheel, Donovan and many others have noted, accounts of the embodied experiences of the hunters are ubiquitous and many narrate the struggle with what they call 'Buck Fever.' This is where hunters express a heightened state of arousal at the peak moment, often just before the kill, that can be so intense it literally prevents them from firing: hands shaking, rapid breathing, dizzy, heart pounding. Typically these stories are read as relatively uncomplicated analogues of heterosexual desire, notwithstanding their intimate tie up with violence. Yet there is something curiously heteronormative about this supposedly uncomplicated reading–a 'boys will be boys' kind of nod and a wink to the notion that that's just how the biology of male desire works, thumping adrenaline, little control. Yet it bears repeating, the animals being pursued by these feverish hunters, these objects of a profound investment in libidinal energy and time, are icons of masculinity. They are male animals, often being hunted, year in and year out, during the very season when they are in peak reproductive form. Hence, the desire that is often so physically overwhelming for those hunters is ultimately a desire for hands-on intimacy with another male body, and every single trophy photo memorializes and bears witness to precisely that moment. "Besides the thrill of the hunt, one of my most favourite things about hunting whitetails is the moment of grabbing the antlers of an old bruiser buck for the first time." We rush past this moment too quickly when we stop our analyses at identifying hetero-hypermasculine symbolism and fail to take note of the material experiences—like the need to touch, to hold—that are embodied and expressed as the culmination of a period of extended longing by a human male for a non-human manimal. Aaron Sligh hunted Angry for more than four years, and David Stroupe captured his first glimpse of Ghost in 2018 before he killed him in November 2021. "And just like that, my hands were wrapped around 240 4/8 inches of Iowa whitetail antler."

Conclusion

I don't deny that the desires underlying trophy hunting are complex, simultaneously a desire for extraordinary intimacy with a forbidden other—a wild male animal—and also a desire to fully and completely possess that other—something that is embodied in the radical act of taking its life. But as I note previously around the issue of naming, the moment that marks the division between life and death is also the moment that a transgressive desire is transmuted into a normative one. As the intensely particular idealized male subject of desire passes from life to death, the homoerotics of the desire that has been the bedrock of the pursuit dissolves into a tale of heteronormative power and conquest. But attending to the minutia as well as the materiality of the hunt as lived experience, brings into focus an underlying homoerotics of desire that is at least a part of the structure of trophy hunting. It remains to be seen what further attention to the materiality of trophy hunting as practice might also reveal around the dynamic and shifting complexities of gender.

Notes

1 Carol J. Adams, *The Sexual Politics of Meat: A Feminist-Vegetarian Critical Theory* (New York: Continuum, 1990); Marti Kheel, "License to Kill: An Ecofeminist Critique of Hunters' Discourse," in *Animals and Women Feminist Theoretical Explorations*, eds. Adams and Donovan (Durham, NC: Duke University Press, 1995); Josephine Donovan and Carol J. Adams, eds., *Animals and Women: Feminist Theoretical Explorations* (Durham, NC: Duke University Press, 1995); Josephine Donavan and Carol J. Adams, "Feminism and the Treatment of Animals: From Care to Dialogue," *Signs* 31, no. 2 (2006): 305–329; Marti Kheel, "The Killing Game: An Ecofeminist Critique of Hunting," *Journal of the Philosophy of Sport* 23, no. 1 (1996).

2 Linda Kakoff and Amy Fitzgerald, "Reading the Trophy: Exploring the Display of Dead Animals in Hunting Magazines," *Visual Studies* 18, no. 2 (2003): 112–122.

3 Brian Luke, "Violent Love: Hunting, Heterosexuality and the Erotics of Men's Predation," *Feminist Studies* 24, no. 3 (1998): 643.

4 Linda Kaloff, Amy Fitzgerald, and Lori Baralt, "Animals, Women and Weapons: Blurred Sexual Boundaries in the Discourse of Sport Hunting," *Society & Animals* 12, no. 3 (2004): 237–251

5 Much of the literature around critiques of trophy hunting, including its inherent racialization, come from the West. Mucha Mkono, "Neo-colonialism and Greed: African's Views on Trophy Hunting in Social Media," *The Journal of Sustainable Tourism* 27, no. 5 (2019): 689–704, makes an important contribution to this literature but from the perspective of those whose countries and people are often implicated but whose voices and opinions are rarely heard.

6 See IFAW 2016 stats on International Trophy Hunting and "The License Cliff: Why We Suck at Recruiting New Hunters, Why It Matters, and How to You Can Fix It" (October 15, 2019). Author uncited, *Outdoor Life*.

7 Much has been written about the unique relationship between national identity, masculinity and hunting as it pertains particularly to the United States. For an accessible introduction see Em Steck's 2018 interview in *Vox* with Historian Phillip Dray (Em Steck, "How America's Hunting Culture Shaped Masculinity, Environmentalism and the NRA," *Vox* (June 12, 2018), retrieved from: https://www.vox.com/conversations/2018/6/12/17449154/hunting-culture-shaped-masculinity-the-nra-and-environmentalism). See also Phillip Dray, on his book *The Fair Chase: The Epic Story of Hunting in America*. Last, Matthew Brower's compelling "Trophy Shots: Early North American Photographs of Non Human Animals and the Display of Masculine Prowess" is an insightful engagement with the social and cultural transformations of the late 19th century that produced new forms of masculinity. His focus on the writing and role of Teddy Roosevelt is particularly apposite in this context.

8 Luke, "Violent Love: Hunting, Heterosexuality and the Erotics of Men's Predation"

9 While the jury is out on the question of details like how many generations it might take, there is a growing consensus among researchers that trophy hunting is, ironically, creating counter-evolutionary pressure on horn and antler size in the very animals who are prized precisely for antler and horn size. See, for example, Tim Coulson, Susanne Schindler, Lochran Traill, and Bruce E. Kendall, "Predicting the Evolutionary Consequences of Trophy Hunting on a Quantitative Trait," *The Journal of Wildlife Management* 83, no. 6 (2017) (https://doi.org/10.1002/jwmg.21261)

10 See for instance, K. Hawkes, "Why Do Men Hunt? Benefits for Risky Choices," in *Risk and Uncertainty in Tribal and Peasant Economies*, ed. E. Cashdan (Boulder, CO: Westview, 1990), 145–1266; K. Hawkes and R.L. Bleige Bird, "Showing Off, Handicap Signaling, and the Evolution of Men's Work," *Evolutionary Anthropology* 11 (2002): 58–67; Chris T. Darimont, B. Codding, and K. Hawkes, "Why Men Trophy Hunt," *Biology Letters* 13, no. 20160909 (2017) (http://doi.org/10.1098/rsbl.2016.0909) and for a counter view against the social signaling hypothesis, see Michael Gurven and Kim Hill, "Why Do Men Hunt? A Reevaluation of 'Man the Hunter' and the Sexual Division of Labor," *Current Anthropology* 50, no. 1 (2009): 51–74.

11 Luke, "Violent Love: Hunting, Heterosexuality and the Erotics of Men's Predation," 635.

12 Ibid.

13 Ibid., 641.

14 Recent statistical analyses from the UCLA School of Law William's Institute on the National Crime Victimization Survey found that transgender people are four times more likely to experience

violence than cisgender people, with one in four trans women identifying their victimization as a hate crime as opposed to fewer than one in ten cisgender women. https://williamsinstitute.law.ucla.edu/press/ncvs-trans-press-release/

15 Scott Kirchgessner, "Lucky Charm," *Big Buck* 29, no. 4 (Spring 2016): 17.

16 Mimicry is a feature of much of hunting as hunters often engage in a range of 'becoming animal' activities from simulating the sounds of males during mating season, to literally coating themselves in the pheromones of females in estrus via body sprays.

17 See the exemplary work of Michael Messner on sport and masculinity, for instance, Michael Messner, "When Bodies Are Weapons: Masculinity and Violence in Sport," *International Review for Sociology of Sport* 25, no. 1 (1990): 203–218.

18 IFAW—International Fund for Animal Welfare. Founded in 1969, initially to stop the seal hunt, IFAW is now a leading global animal welfare, rescue and conservation organization.

19 For more on the complex cultural transformations surrounding U.S. masculinities throughout the late 18th and 19th centuries, see Matthew Brower's fascinating essay on the phenomenon of camera hunting. "Trophy Shots: Early North American Photographs of Non Human Animals and the Display of Masculine Prowess."

20 While it remains overwhelmingly the case that hunting is a predominantly male pursuit, throughout the issue I am working with here, there are a significant number of references to family participation in various aspects of hunting, from shed hunting (the seasonal search for antlers which naturally drop from deer each winter) to trail cam review sessions to accompanying parents on the hunts themselves, and two of the legendary tales are of women hunters. With an overall trend away from the social approval of hunting, many hunting organizations have responded by repackaging hunting as a family activity as well as one that is inclusive of women. Social media has also proven useful as a political lobbying tool aimed again at protecting the perceived rights of hunters to hunt and in the U.S. this often overlaps with Second-Amendment politics around the right to carry guns. On political lobbying efforts, see https://wildthingsinitiative.com/how-american-hunting-organizations-conduct-information-warfare-on-social-media/. On hunter recruitment strategies in the U.S., see https://deltawaterfowl.org/hunter-recruitment-are-we-doing-it-right/. And on the most recent efforts to diversify the dwindling pool of hunters, see https://www.pewtrusts.org/en/research-and-analysis/blogs/stateline/2019/11/07/recruiting-foodies-and-hipnecks-as-the-new-hunters

21 Social media plays a vital role in hunting now, from the Instagram trophy shots to online scoring competitions between private groups as well as hunting clubs. Such platforms function as a form of town square where we can see displayed so much of the performance work of heteronormative masculinity. But this is not where the homoerotics lie.

22 Founded by Theodore Roosevelt in 1887, the Boone and Crockett Club is one of several organizations around the world, including Pope and Young, Safari Club International and National Wild Turkey Federation, that score the spoils of wildlife hunting and keep the official records.

Bibliography

Adams, Carol. *The Sexual Politics of Meat: A Feminist-Vegetarian Critical Theory*. New York: Continuum, 1990.

Bias, Josh. "The Ghost of McDowell County." *North American Whitetail, Full Draw Issue*, 41, no. 4 (2022): 56–62.

Brower, Matthew. "Trophy Shots: Early North American Photographs of Non Human Animals and the Display of Masculine Prowess." *Society and Animals* 13, no. 11 (2022).

Comstock, Alex. "How to Hunt the Popular Buck." *North American Whitetail* 41, no. 4 (2022): 82–89.

Coulson, Tim, Schindler, Susanne, Traill, Lochran, and Kendall, Bruce E. "Predicting the Evolutionary Consequences of Trophy Hunting on a Quantitative Trait." *The Journal of Wildlife Management* 83, no. 6. (June 19, 2017). https://doi.org/10.1002/jwmg.21261

Darimont, Chris T., Codding, B., and Hawkes, K. "Why Men Trophy Hunt." *Biology Letters* 13 (2017): 20160909. http://doi.org/10.1098/rsbl.2016.0909

Donovan, Josephine. "Feminism and the Treatment of Animals: From Care to Dialogue." *Signs* 31, no. 2 (Winter, 2006): 305–329.
Donovan, Josephine, and Adams, Carol, eds. *Animals and Women: Feminist Theoretical Explorations*. Durham, NC: Duke University Press, 1995.
Gurven, Michael, and Hill, Kim. "Why Do Men Hunt? A Reevaluation of "Man the Hunter" and the Sexual Division of Labor." *Current Anthropology* 50, no. 1 (2009): 51–74.
Hawkes, K. "Why Do Men Hunt? Benefits for Risky Choices." In *Risk and Uncertainty in Tribal and Peasant Economies*, edited by E. Cashdan, 145–1266. Boulder, CO: Westview, 1990.
Hawkes, K., and Bleige Bird, R.L. "Showing Off, Handicap Signaling, and the Evolution of Men's Work." *Evolutionary Anthropology* 11 (2002): 58–67.
Kakoff, Linda, and Fitzgerald, Amy. "Reading the Trophy: Exploring the Display of Dead Animals in Hunting Magazines." *Visual Studies* 18, no. 2 (2003): 112–122.
Kaloff, Linda, Fitzgerald, Amy, and Baralt, Lori. "Animals, Women and Weapons: Blurred Sexual Boundaries in the Discourse of Sport Hunting." *Society & Animals* 12, no. 3 (2004): 237–251.
Kheel, Marti. "License to Kill: An Ecofeminist Critique of Hunters' Discourse." In *Animals and Women Feminist Theoretical Explorations*, edited by Carol J. Adams and Josephine Donovan. Durham, NC: Duke University Press, 1995.
Kheel, Marti. "The Killing Game: An Ecofeminist Critique of Hunting." *Journal of the Philosophy of Sport* 23, no. 1 (May 1996): 30–44.
Kirchgessner, Scott. "Lucky Charm." *Big Buck* 29, no. 4 (2016): 17–21.
Littlefield, Jon. "Men on the Hunt: Ecofeminist Insights into Masculinity." *Marketing Theory* 10, no. 1 (2010): 97–117. Sage.
Luke, Brian. "Violent Love: Hunting, Heterosexuality and the Erotics of Men's Predation." *Feminist Studies* 24, no. 3 (1998): 627–655.
Messner, Michael. "When Bodies are Weapons: Masculinity and Violence in Sport." *International Review for Sociology of Sport* 25, no. 1 (1990): 203–218.
Mkono, Mucho. "Neo-colonialism and Greed: African's Views on Trophy Hunting in Social Media." *The Journal of Sustainable Tourism* 27, no. 5 (2019): 689–704.
Sligh, Aaron. "Getting Angry." *North American Whitetail* 41, no. 4 (2022): 20–31.
Steck, Em. "How America's Hunting Culture Shaped Masculinity, Environmentalism and the NRA." *Vox*, June 12, 2018. Retrieved from: https://www.vox.com/conversations/2018/6/12/17449154/hunting-culture-shaped-masculinity-the-nra-and-environmentalism

28

A FEMINIST RUBIK'S CUBE

Slaughterhouse Labour, Violence, and Multispecies Trauma

Lauren Corman

[P]eople who naturally empathize with the animals are likely to have a difficult time with animal industry work, since the nature of the work requires the worker to treat the animal not as a living being with individual worth, but as another widget, a means to an end.[1]

Industrial meatpacking is an institution that produces trauma as much as it produces steaks, or any other animal product. There is both an individual psychological and emotional toll slaughterhouses have on workers and animals within industrial slaughter and a larger social toll linked to farmed animal abuse, family and sexual violence, and increased arrests. While hegemonic masculinity can cauterize some of the emotional wounds involved in slaughterhouse labour, the blunting comes with costs. Given the harms industrial slaughter inflicts across human and nonhuman animal species, and the highly gendered and racialized nature of the work, slaughterhouses and the violence enacted through them should be a feminist concern.[2] In this chapter I consider multispecies dimensions of trauma produced through industrial slaughterhouses—turning them over as one might examine a Rubik's Cube—and develop a concept of "multispecies trauma" to respond to the feminist call to address violence in its myriad forms.

Trauma Beyond the Human

Trauma has largely been assumed to be a human phenomenon,[3] despite the staggering amount of nonhuman animal experimentation that undergirds its study.[4] Given the anthropocentrism of scholarship about trauma, it is difficult to find a definition within the trauma literature that includes nonhuman animals within its scope. For example, in her field-defining book, *Trauma and Recovery: The Aftermath of Violence—From Domestic Abuse to Political Terror*, Judith Herman (2015) contends,

> Traumatic events are extraordinary, not because they occur rarely, but rather because they overwhelm the ordinary human adaptations to life. Unlike commonplace misfortunes, traumatic events generally involve threats to life or bodily integrity, or a

DOI: 10.4324/9781003273400-35

> close personal encounter with violence and death. They confront human beings with the extremities of helplessness and terror, and evoke the response of catastrophe. According to the *Comprehensive Textbook of Psychiatry*, the common denominator of psychological trauma is a feeling of "intense fear, helplessness, loss of control, and threat of annihilation."[5]

Despite the human-centredness of most trauma-related research, which is preoccupied with the impact of trauma in the lives of humans, scholars are increasingly directing the study of trauma to nonhuman animals for their own sakes. For example, without an understanding of the "ecology of fear"[6]—fear produced through predator-prey interactions and the subsequent impacts on populations, communities, and ecosystems—we miss something vital about how wild species and individuals change their behaviours, such as foraging, in response to trauma, and the potential long-term effects such trauma has on their brains, in the form of post-traumatic stress disorder (PTSD).[7]

Diminished neurogenesis, for example, is evident in people with PTSD, which is believed to contribute to the enduring retention of traumatic memories, and is similarly reflected in brains of "laboratory rodents" experiencing fear of predators. These results are consistent with a study of wild cowbirds and a study of free-living deer mice.[8] Additionally, as Zanette and Clinchy (2020) report, in one study crows who were caught and released were able to distinguish between human faces of those who had previously captured them (predators) and those who had not, with the former inspiring activation in the amygdala and "related neuronal fear circuits."[9] "The complexity of what wild animals find fearful is here illustrated not only by differences in fear memories being induced by an apparently quite mild trauma but also the fact that the crows discriminated different human faces," comment Zanette and Clinchy.[10] The authors boldly note that the activity cost of a predator encounter exceeds a fleeting fight-or-flight reaction and can impact behaviour and the brain for weeks or permanently, leading them to conclude that such evidence provides a "compelling mechanism strongly supporting the supposition that population and community-level effects of fear are indeed commonplace in nature."[11]

Zanette and Clinchy (2020) point us to an understanding of trauma beyond the anthropocentric reach of most literature on the topic. Powerfully, by recognizing nonhuman animals' capacity to experience trauma, they inadvertently unseat a claim made by other researchers who conduct trauma-inducing experiments on nonhuman animals and maintain that they are imperfect models, as they argue animals lack the psychological capacities to truly replicate humans' experiences, including the ability to worry.[12] Thus, this chapter proceeds with an appreciation that trauma can be experienced across human and nonhuman species. Such an orientation is crucial if we are to comprehend the multifaceted contours of trauma as they manifest within the modern industrial slaughterhouse.[13] As we examine the Rubik's Cube, we see trauma distributed on all sides.

Workers' Trauma

The number of cows, chickens, pigs, and sheep slaughtered in 2020 was 73 billion.[14] Canada killed 841 million land animals in 2022.[15] Men dominate the sector.[16,17] As we contemplate industrial slaughterhouses as sites that produce trauma, the question of how the work itself unfolds and is maintained is crucial.[18] Gender, and in particular its expression as hegemonic masculinity, is essential to the operations of the meat industry, in which workers must continually reproduce an unemotional identity to get on with "what has to be done."[19] In this way, hegemonic masculinity provides both a backdrop and process that facilitates the multispecies trauma of the meat industry.

Eimear McLoughlin (2019) argues for the salience of gender in making sense of slaughterhouse labour and the emotional landscape that shapes it. In her "emotionography," an ethnographic approach that attends to affect, she considers the "gendered identity of the ideal slaughterhouse worker,"[20] which involves a vacillating expression and repression of emotion in service of an always negotiated, never fully complete, enactment of hegemonic masculinity. McLoughlin's insights are especially potent for those interested in multispecies justice, as she demonstrates how one's identity as an ideal slaughterhouse worker is intimately entangled with the co-construction of nonhuman animals as commodities, a process that is also never fully complete but done and undone through "doing gender" at the slaughterhouse. The institutionalized masculinity of the slaughterhouse is enabled through the subjugation of nonhuman animals and the masculinized "feeling rules" of the company that demand an emotional stoicism and professionalism, detached from the animal as a knowing being while repositioning them as economic products. Reflecting on her participant observations and interviews with workers, McLoughlin summarizes, "In an act of self-preservation, the workers perceive their role as an elaborate performance that distances them from the stigma associated with killing animals by manifesting a gendered unemotional occupational identity that maintains productivity without emotional investment."[21]

Although Jennifer Dillard (2007) does not discuss hegemonic masculinity *per se* in her article, "A Slaughterhouse Nightmare: Psychological Harm Suffered by Slaughterhouse Employees and the Possibility of Redress through Legal Reform," she does attend to the impact of emotional repression on male meatpacking workers, recounting the findings of one U.S.-based study:

> The intensive, production-focused nature of factory farming has led workers to suppress their "spontaneous empathy" for the animals, and this study suggests that male workers are affected more strongly by this phenomenon of empathy suppression. "Because compassion is not an attitude compatible with the requirements of economic competition and maleness, men may have suffered more than women in this repression of affect and learned to hide their feelings." This concern is particularly important for the U.S. animal industry, since a vast majority of U.S. slaughterhouse workers are male.[22]

McLoughlin would likely identify such emotional numbing as an expression of hegemonic masculinity. Resonant with McLoughlin's description of the "ideal slaughterhouse worker," Dillard also notes the incompatibility between the characteristic of empathy and the character of the "good stockperson." A person who empathizes with animals must work against those empathetic feelings to perform their labour and to ultimately reduce nonhuman animals to the status of widgets, a "means to an end."[23] My interview with ex-Tyson slaughterhouse worker, Virgil Butler, lends credence to Dillard's observation. He recounts in vivid detail the empathetic drain he witnessed in other workers as consequence of their labour conditions:

> Well, I had several friends, you know, that came to work there, they were pretty good people. But after a while, they'd start to get to where . . . you know, before, they

> wouldn't even consider mistreating another animal. After a while, they'd get to where they'd just spike a bird, just for the heck of it. And then before long, you know, after they'd done this for a while, they'd go home, maybe them and their wife would get into an argument, they'd just reach up and slap the dickens out of her, you know. . . . You start to get into fights all the time. You just don't mind hurting things.[24]

Butler's testimony is striking for the affective trajectory he maps, in which workers are transformed from "good people" to domestic abusers. He understands his co-workers' diminished empathy as a result of their working conditions, leading to the precipitation of violence against women.[25] Such descriptions suggest the violent multispecies consequences of slaughterhouse labour can transcend the walls of the slaughterhouse, and that the psychological damage can ricochet back at workers in the form of lasting trauma.

American psychologist and sociologist Rachel MacNair (2015) introduced the term "Perpetration- [or Participation] Induced Traumatic Stress," or PITS, in the early 2000s. PITS is "the form of posttraumatic stress disorder symptoms caused by killing or otherwise committing violence as the stressor."[26] PITS was originally coined as a way of naming PTSD symptoms associated with combat veterans who are traumatized by causing trauma.[27] Extending beyond the realm of combat veterans, MacNair demonstrates that PITS has been linked to numerous occupations, including animal slaughter.[28]

Dillard also notes that the psychological phenomenon of "doubling," more thoroughly documented in the testimony of Nazi doctors, is also evidenced in some slaughterhouse workers' accounts.[29] Doubling, a form of the PITS phenomenon, involves the bifurcation of the self into two separate wholes. In the case of the Nazi doctors, the previous image of themselves as husbands and fathers, for instance, is psychologically maintained, while another self—contra to his own morals—commits the debased acts. Although cognizant of differences between both the doctors and the slaughterhouse workers, Dillard observes similar kinds of "doubling" in workers' testimony. In another firsthand account, echoing McLoughlin's observations, a worker recalled both the affection he felt for a pig and its necessary suppression required for his job. Later, Dillard argues, "The worker's natural self identifies with the pig and recognizes it as an animal worthy of affection and care, but the worker's other self—the self developed to work in the slaughterhouse—kills the pig, literally unable to care about the animal."[30]

Inseparable from the psychological effects of killing, the physical brutality of slaughterhouse labour on workers is well documented. Among these dangers are lacerations, chronic infections, and repetitive strain disorders. In his ethnographic research, which included periods of working at chicken slaughterhouses, Steve Striffler (2010) observes the banal repetitive nature of slaughterhouse labour. The tedium of doing the same tasks through a shift, year after year, can be unbearable for both the mind and body. "The oppressiveness of routine work is very difficult to convey," he remarks. "Yet it defines factory life and is perhaps the most devastating part of work in the poultry industry."[31] Christopher Cook (2010) draws attention to the immediate physical dangers ever present on the disassembly line while reinforcing Striffler's points about the routinized agony of the work:

> In a macabre, medieval scene, workers hack frantically at fast-moving carcasses while standing in pools of blood, fat, and chunks of abscess. Their knives, dulled from stabbing meat every three seconds or so, sometimes slice into the wrong piece of flesh—their coworker just a couple of feet away. Back injuries are common from slipping on

> the greasy floors. But the biggest risk is the mundane: the steady ceaseless cutting—of heads, necks, knuckles, legs, organs, stomachs. Cutting a mind-numbing train of animal parts flying down the line at dizzying speeds. Thousands of cuts per day, about three seconds per piece of meat.[32]

Zooming out further from the psychological and physical to the social, Fitzgerald et al.[33] documented community-level impacts of slaughterhouses, demonstrating increases in arrests for rape and sexual offenses. Fitzgerald et al.'s research shows the community effects of working at a slaughterhouse are different from working in other manufacturing industries. Jessica Racine Jacques' (2015) more recent study confirms these findings. She concludes, "the violent work of slaughterhouses and the mere presence of a slaughterhouse in a county had significant effects on total arrests, arrests for rape, and arrests for offenses against the family."[34]

In the United States and in Canada, the meatpacking industry has intentionally and heavily relied on racialized groups of workers to do incredibly dangerous labour, another entangled face of the Rubik's Cube. The labour force is composed of disproportionately high numbers of racially migrants and immigrants, including many racialized refugees.[35] Journalist Ted Genoways (2014) comments on the American meatpacking context, noting,

> Undocumented workers, many from Mexico and other parts of Latin America, formed a perfect corporate workforce: thankful for their pay cheques, willing to endure harsh working conditions, unlikely to unionise or even complain. "They don't ask for breaks. They don't ask for raises," one worker at the Hormel plant in Fremont told me. "They just work harder and harder, because they need to work."[36]

In her "No Safe Place" Action Dignity report, Bronwyn Bragg (2021) underscores how the pandemic worsened conditions for socially marginalized workers in Canada:

> 70% of beef sold in Canada is manufactured at two plants in Southern Alberta: Cargill in High River and JBS Foods in Brooks. The meat processing industry, including these two plants, relies heavily on racialized, immigrant, migrant and refugee workers who make up a significant proportion of the workforce. An estimated 67% of workers in the Alberta meat processing industry are immigrants to Canada (Statistics Canada, 2016). When COVID-19 emerged as a serious threat to workers in this industry in March of 2020, it was these im/migrant and refugee workers who bore the brunt of the outbreaks, infection, and fatalities.[37]

In Canada, in addition to relying on racialized immigrant, migrant, and refugee labour there has been a turn toward employing Indigenous peoples as a prospective labour force and increasing Temporary Foreign Workers as an existing one.[38] One might wonder about the subtext of the government-funded, "Securing Canada's Meat Workforce" report, which states,

> because the sector can be perceived as a low-skill one with limited career progression opportunities, it can be challenging to recruit and retain workers. In addition, there are significant physical requirements for many of the jobs, and for meat slaughtering, it is important to find workers who can tolerate the slaughter environment.[39]

"Where will we find them?" asks a sub-section of the industry-supported study. The report suggests recruitment of women, older workers, youth, unemployed, new Canadians, and "Indigenous Canadians." As the meat industry struggles to find employees, its targeting of Indigenous people is troubling. Food Processing Skills Canada makes the following recommendation:

> Indigenous Canadians appear to be underrepresented in the meat processing sector, which could be due to a lack of knowledge among processors of the local Indigenous communities as a potential supply of workers. In order to increase the engagement of local Indigenous communities in meat processing, it will be important to understand the various factors that can detract or contribute to that engagement.[40]

Bombay et al.[41] argue that due to colonization and forced assimilation, Indigenous people endure high levels of adverse child experiences, such as abuse, neglect, and household substance use. Compounding these stressors, Indigenous people also experience ongoing discrimination. There are both specific and more widespread historical traumas that variably impact Indigenous communities, and the effects of these can be transferred across generations, which can present as PTSD and other negative mental health outcomes. Given the presence of intergenerational trauma, Bombay et al. urge the need to interrupt such cycles.

While the meat industry may offer some relief from financial stressors, which Bombay et al. also suggest can bear on trauma, slaughterhouse labour is intentionally precarious.[42] Bragg and Hyndman (2022) describe the predatory targeting of certain populations as a strategy of the Canadian meat industry to "deliberately engineer . . . a vulnerable workforce."[43]

> We argue that workers are recruited precisely because of their vulnerability to exploitation, difficult work conditions, and limited ability to enact their rights to a safe workplace. We describe this as "intentional precarity." Their vulnerability is linked to their immigration status, official language status, transnational family ties as well as their position often as racialized workers who may not have official language fluency in the Canadian labour market.[44]

The authors draw on transnational feminist analyses of social reproduction to interpret the precarity experienced by Canadian meatpacking employees. Social reproduction involves invisible practices required to ensure short- and long-term survival, while they also reproduce the "racist, patriarchal, and class relations that underpin capitalism."[45] As their interviews document, workers are often caught between the dangers of the slaughterhouse and the financial responsibilities to family living outside the country. The vulnerability of the workers is matched by the vulnerability of their families. Such intertwined vulnerability is manufactured by the meat industry, which through consolidation and relocation of slaughterhouses to geographically isolated areas, exacerbates the stress of workers in Canada, as families and kin endure additional burdens in the absence of workers at home, such as increased childcare. The COVID-19 pandemic further entrenched these problems. Braggs and Hyndman (2022) relay the following:

> Another worker had a family member who was living with him die of COVID 19. He told us that he was "quite sure he got it [COVID] from me." Following the death of the family member, the worker was diagnosed with "trauma, stress, and flashbacks"

(his words). He went on stress leave for eight weeks. He reported that he suffered from insomnia: "I had flashbacks. I could not get over the fact of what just happened to him [the relative who died], my kids."

(Dani, Filipino interviewee)[46]

COVID-19 tore through slaughterhouses, resulting in sickness and fatalities that included, and extended beyond, the workers themselves.[47] Given the many harms borne through meatpacking, and its potential contributions to legacies of intergenerational and racial trauma, which "refers to the events of danger related to real or perceived experience of racial discrimination,"[48] we should question the sector's hiring and labour practices and the human and nonhuman suffering they create.

Nonhuman Animals' Trauma

Concentrated animal feeding operations (CAFOs) characterize contemporary animal agriculture, which focuses on the intensification of animal (by-)products for the profit at the expense of animal welfare.[49] Jean-Jacques Kona-Boun, chief of anaesthesiology at DMV Veterinary Center in Montreal, argues that intense physical and psychological suffering is ubiquitous throughout the production process for farmed animals. While this chapter focuses on slaughterhouses, it is impossible to separate animals' distress within such sites from the lives of misery that define their experiences prior to death. They arrive at slaughterhouses already traumatized. Kona-Boun details a cascade of harms committed against them not as a catalogue of anomalous and particularly malicious acts but rather those routinized as "standard industry practices," which remain virtually immune to legal action.[50]

Male chicks, rendered disposable by the egg industry, for example, are commonly ground up alive or left to suffocate within bins, while male calves on dairy farms are separated from their mothers within hours or days of birth, causing extreme stress to both the babies and their mothers. These calves then spend the remainder of their lives confined before they are killed for veal. Laying hens, like dairy cows, are killed once their productivity declines, well before the end of their natural life spans. Prior to slaughter, laying hens typically spend their short lives confined within "battery cages," which frustrate their natural behaviours and force them in uncomfortable positions. Compounding the physical pain of confinement, animals within such facilities, which also include gestation crates for pigs and robotic milking installations for cows, are denied access to the outdoors and to exercise.

While farmed animals are popularly assumed to lack complex subjective lives, Kona-Boun is definitive about the psychological effects of such institutionalization. He states,

> The suffering of farmed animals is a multidimensional sensory and emotional experience that is by no means confined to physical sensations such as pain. Underestimating the impact of psychological suffering is a common error. When psychological suffering reaches an unbearable level, it is expressed in maladaptive behaviours (e.g., stereotypy, displacement behaviors, aggression, learned helplessness) that clearly indicate that the individual's well-being has been seriously compromised.[51]
>
> *(p. 5)*

In addition to boredom and frustration, Kona-Boun notes the psychological suffering of farmed animals can lead to ulcers, gastric erosions, and immunosuppression, which increase animals' susceptibility to pathogenic microorganisms typical within such crowded environments, further exacerbating their torment. Mutilation of animals' bodies also occurs within industrial animal agriculture, often without anaesthetic. The cruelty of these conditions is so profound that they are well characterized as "fear factories."[52] Matthew Scully (2010) describes a visceral scene of industrial pig production:

> At the Smithfield mass-confinement hog farms I toured in North Carolina, the visitor is greeted by a bedlam of squealing, chain rattling, and horrible roaring. To maximize the use of space and minimize the need for care, the creatures are encased row after row, 400- to 500-pound mammals trapped without relief inside iron crates 7 feet long and 22 inches wide. They chew maniacally on bars and chains, as foraging animals will do when denied straw, or engage in stereotypical nest building with the straw that isn't there, or else just lie there like broken beings.[53]

Transportation to slaughter is also psychologically distressing to animals, evidenced by behaviours and haemato-biochemical parameters, and effect on the carcasses. The stress of being loaded onto trucks then leads the journeys themselves, which can involve both food and water deprivation (particularly stressful after acclimatization to regular feedings) and temperature extremes, alongside jostling from turns and acceleration, among other driving impacts. When animals arrive at slaughterhouses, they are unloaded off the trucks and handled by people, often roughly, all while being crammed together with unfamiliar animals in new environments, conditions that can precipitate aggression:

> The psychological suffering experienced at multiple points up to and including the slaughterhouse is unmistakably expressed by, among other things, physical behaviour and vocalizations. Clearly, animals that balk at entering the corridors of death or struggle when suspended fully conscious by the hind legs, hanging upside-down on the rail that will convey them to the puncturing point, cannot be experiencing psychological well-being.[54]
>
> *(Kona-Boun, 2020, p. 7)*

Jessica Scott-Reid (2019) reports that Canada permits a 1–4% margin of error regarding pre-slaughter stunning, adding another layer of complexity to the egregious suffering that is legally permitted within Canadian agribusiness. "That means that at least 210,000 pigs a year may not be properly rendered unconscious upon the first attempt of being electrocuted or gassed," she laments.[55]

A Feminist Theory of Multispecies Trauma

I turn to multispecies justice (MSJ) as a theoretical strategy to tackle the Rubik's Cube. In the article "Multispecies Justice: Theories, Challenges, and a Research Agenda for Environmental Politics," Celermajer et al. (2020) stress,

> In conceiving "multispecies", we contest the exclusive classificatory politics of anthropocentric justice theories that purport to expand beyond humans by recognising the value of certain other entities. Such speciesist approaches tend to import a human/other, or assume hierarchies based on anthropocentric assumptions about the

> character and worth of "other" subjects including other humans. By adopting more relational ontologies, MSJ can recognise the multiplicity of different types of being, in their own terms and their involvement in thick relational webs. Rethinking the subject of justice moves attention from the fiction of individuals to the actual ecological array of relationships that sustain life.[56]

MSJ is especially useful for thinking through the multidimensional site of the slaughterhouse because it draws our attention to relationships. MJS helps us to not just explore the many aspects of oppression that constitute the slaughterhouse but to also understand the connections among those aspects, allowing us to scrutinize more deeply, and hopefully to address problems that may initially seem isolated. To solve the Rubik's Cube, like the perils of the slaughterhouse, we need to understand the how its relationality functions so that we know which piece to turn and when. A Rubik's Cube is a puzzle box that is devious because it asks the player to find order in something very chaotic. It is also a game of relationality: Turn one row and it does not just change the face you are looking at, it changes all the other faces as well.

We can think here of the kinds of relations that yoke together workers and animals, workers and managers, workers and communities, animals and consumers, technology and violence, and capitalism and nature; we can also think here of the kinds of relations that are severed through the slaughterhouse, relations between animals and their young, workers and their partners through domestic violence associated with slaughterhouse work, crime and surrounding communities. We can think of the relationships slaughterhouse workers have with themselves, as well, through processes such as "doubling" and hegemonic masculinity, which can dislocate workers' empathy for nonhuman animals. We can think in a different register of colonialism and capitalism as twin forces of industrial animal agriculture, breaking relationships between people and the environment, animals and their habitats, Indigenous peoples and their lands.

Driftpile Cree scholar Billy-Ray Belcourt (2015)[57] emphasizes how animal agriculture in North America is achievable only through the violent erasure of Indigenous peoples and animals. Any multispecies justice approach that fails to acknowledge this fact, and thereby neglects to recognize the settler colonial machinations that allow for such animal industries to exist, misses both the colonial legacies and current colonial forces that make subjugation of land, peoples, and animals, possible. As Struthers Montford and Wotherspoon (2021) argue, "Animal agriculture is not only bound up in the ontological production of colonial human exceptionalism but has been foundational to territorial acquisition."[58] From this vantage, when we heed the call from the MSJ cohort to attend to relational ontologies, in which "beings do not pre-exist their relatings,"[59] Belcourt's argument that animal ethics must be decolonial makes clear and obvious sense. Settler colonialism prefigured human–animal relations by casting animals, and those deemed subhuman or animal-like (or as animals themselves), as objects, available for exploitation in service of capital, all without the trappings of conscience.

The trauma produced by the slaughterhouse is multifaceted, as a both exceptionally gendered and racialized phenomenon. It is only relatively recently that scholars have explored the trauma involved in this kind of labour, which turns living beings into commodified flesh. The original craftsmanship of the butcher is deliberately deskilled through the structural organization of the industrial slaughterhouse; with the breakdown of this craft into individualized repetitive tasks, connection with the animal as a whole is severed. As political scientist Timothy Patchirat documents, a typical industrial plant will involve 121 distinct kill floor jobs.[60] The violent impacts are meted upon both workers and animals, serving the interests of efficiency and capital at every station.

As a society, we actively condone the operation of an industry that traumatizes the individuals who work within it, and the communities that surround it. We ask people to kill animals in a way many could not tolerate themselves, offloading the effects of this violence onto workers. We draw people from marginalized and vulnerable demographics to do low-paying jobs[61] that, as Cook (2010) notes, "most white Americans simply won't do."[62] The industry treats their employees as disposable labour, producing high turnover rates.[63] We further offload the effects of this violence into surrounding communities. The workers are traumatized by the nature of their very jobs, and this trauma bleeds onto others.

Returning to Belcourt, we might understand the meat industry's disregard for its profoundly vulnerable workforce and its largescale commodification and mistreatment of animals as an expression of contemporary settler-colonial logics. Montford and Wotherspoon's (2021) exceptional article, "The Contagion of Slow Violence: The Slaughterhouse and COVID-19," explores both the colonial and racist logics of the meat industry within the context of the pandemic. Industrial animal agriculture, and in this case industrial meat-packing, petitions us to consider oppression from a multiplicity of angles, including the multiple imbrications of gender, race, and class not only within the facilities and beyond their walls, to communities, but also to our collective conscience.[64] A feminist and trauma-informed multispecies justice approach, attending to what I call "multispecies trauma," can bring into sharper focus the problems of industrial agriculture in ways that account for both human and nonhuman animals' lives, both as a matter of diagnosis and a more-informed ground to push for change.

Notes

1 J. Dillard, "A Slaughterhouse Nightmare: Psychological Harm Suffered by Slaughterhouse Employees and the Possibility of Redress Through Legal Reform," *Georgetown Journal on Poverty Law and Policy* 14, no. 2 (2007): 9.

2 A small amount of literature offers a sustained analysis of Western slaughterhouse labour from a decidedly feminist perspective, such as Kathryn Gillespie's (2017), "Industrial Slaughter," in *Gender: Animals* from the Macmillan Interdisciplinary Handbooks series. While this a highly valuable work, which touches trauma and highlights the perspectives of human workers and animals, this chapter adds to Gillespie's scholarship by foregrounding multispecies trauma as a product of industrial slaughter and situates the violent gendered and racialized dynamics of North American industrial slaughter within an analysis of settler colonialism.

3 P.A Levine, *Waking the Tiger: Healing trauma: Te Innate Capacity to Transform Overwhelming Experiences* (North Atlantic Books, 1997), for example, directly denies wild animals' capacity to experience long-term trauma.

4 L. Corman, "Trauma as a Möbius Strip: PTSD, Animal Research, and the Oak Ridge Prisoner Experiments," in *Building Abolition: Decarceration and Social Justice*, eds. C. Taylor and K. Struthers-Montford (Routledge, 2021)

5 J.L. Herman, *Trauma and Recovery: The Aftermath of Violence—from Domestic Abuse to Political Terror* (Basic Books, 2015 Ed.), 33.

6 J.S Brown, J.W. Laundré, and M. Gurung, "The Ecology of Fear: Optimal Foraging, Game Theory, and Trophic Interactions," *Journal of Mammalogy* 80, no. 2 (1999): 385–399. https://doi.org/10.2307/1383287

7 L.Y. Zanette and M. Clinchy, "Ecology and Neurobiology of Fear in Free-living Wildlife," *Annual Review of Ecology, Evolution, and Systematics* 51, no. 1 (2020): 297–318. https://doi.org/10.1146/annurev-ecolsys-011720-124613

8 Ibid.
9 Ibid., 312.
10 Ibid.
11 Ibid., 313.
12 In "Animal Models for Post-Traumatic Stress Disorders," E. Perkins, S. Brothers, and C. Nemeroff, "Animal Models for Post-traumatic Stress Disorder," in *Post-traumatic Stress Disorder*, eds. C. Nemeroff and C. Marmar (Oxford University Press, 2018) assert, "Trauma in humans is inevitably coloured by the complexity of human thought. The type of complex stress experienced by humans following a traumatic event is often compounded by the ability to worry—the ability to envision the future and how to calculate how that future may be affected by the trauma—a trait that cannot be assumed in rodents" (p. 422). Rather than an affirmation of animals' appropriateness as a PTSD model, Zanette and Clinchy's (2020) findings should inspire a commitment to end anthropogenic trauma.
13 Although I explore trauma throughout this chapter, my intent is not meant to undermine the incredible resistance and resilience enacted by workers and nonhuman animals in such dire circumstances, including the essential role of whistleblowers who expose these issues (see S. Colling, *Animal Resistance in the Global Capitalist Era* (Michigan State University Press, 2021); G.A. Eisnitz, *Slaughterhouse: The Shocking Story of Greed, Neglect, and Inhumane Treatment Inside the U.S. Meat Industry* (Prometheus Books, 1997); J. Warrick, "They Die Piece by Piece," *The Washington Post* (2001), retrieved from: https://www.washingtonpost.com/archive/politics/2001/04/10/they-die-piece-by-piece/f172dd3c-0383-49f8-b6d8-347e04b68da1/).
14 Food & Agriculture Organization of the United Nations, *Crop And Livestock Products*. FAOSTAT (2022), retrieved from: https://www.fao.org/faostat/en/#data/QCL
15 Government of Canada, *Animal Industry*. Agriculture & Agri-Food Canada (2023) https://agriculture.canada.ca/en/sector/animal-industry
16 Food Processing Skills Canada, *Securing Canada's Meat Workforce—Real Challenges. Practical Solutions. Fresh Perspectives. Executive Summary*. Food Processing Skills Canada (2019) https://fpsc-ctac.com/wp-content/uploads/2020/03/exec-summary-canadian-meat-and-poultry-lmi-final-report.pdf; K. Gillespie, "Industrial Slaughter," in *Gender: Animals*, ed. J.S. Parreñas (Macmillan Reference USA, 2017).
17 Given the gendered labour distribution, slaughterhouses might not initially appear as sites of concern for feminists, who have primarily been interested in the lives and experiences of women. However, even though the industry disproportionately employs men, large numbers of women also labour in slaughterhouses. Significantly, though, feminist methodologists petition us to get curious about the "workings of femininized and masculinized meanings" (C. Enloe, *The Curious Feminist: Searching for Women in a New Age of Empire* (University of California Press, 2004), 220). For an analysis of how nonhuman animals' experiences of violence are gendered within animal agriculture, see Gillespie, "Industrial Slaughter."
18 Gillespie, "Industrial Slaughter," states, "Women in meatpacking plants are usually employed in jobs cutting or trimming smaller pieces of meat, or they are employed in positions such as quality assurance or inspection where the physical elements of the job are not as demanding" (p.190).
19 E. McLoughlin, "Knowing Cows: Transformative Mobilizations of Human and Non-human Bodies in an Emotionography of the Slaughterhouse," *Gender, Work, and Organization* 26, no. 3 (2019): 322–342. https://doi.org/10.1111/gwao.12247
20 Ibid., 323.
21 Ibid., 338.
22 J. Dillard, "A Slaughterhouse Nightmare: Psychological Harm Suffered by Slaughterhouse Employees and the Possibility of Redress through Legal Reform," *Georgetown Journal on Poverty, Law, & Policy Forthcoming* (2007): 9
23 Ibid.
24 L. Corman (Interviewer). Ex-slaughterhouse worker Virgil Butler [Audio podcast episode]. *Animal Voices* (2005, February 15). https://animalvoices.ca/2005/02/15/ex-slaughterhouse-worker-virgil-butler/
25 Significantly, though, as Gillespie, "Industrial Slaughter," attests, not all slaughterhouse workers enact such violence in the home. Similarly, not all positions within slaughterhouses precipitate

identical effects. Drawing on Timothy Pachirat's detailed description of slaughterhouse labour, Gillespie highlights the specific role of the "knocker," a position that many employees avoid given the recognized psychological impacts of the position. The varied jobs, social identities, and positionalities, and individuals' histories and idiosyncrasies, all suggest a heterogenous workforce and range of experiences.

26 R. MacNair, "Causing Trauma as a Form of Trauma," *Peace and Conflict, Journal of Peace Psychology* 21, no. 3 (2015): 313

27 In her 2015 review of the PITS-related literature, MacNair notes the paucity of research, which is partially attributable to a reluctance to lend sympathy to and humanize people, such as Nazis, and to regard them as anything but monsters. In the case of combat veterans, PITS also suggests the culpability of states that willfully harm those they enlist to fight, which may encumber a wider recognition or discussion of PITS. Critically, though, no empirical research has attempted to disavow its veracity.

28 One expression of this trauma is elevated levels of alcohol and drug use, which are prevalent among slaughterhouse workers and behaviour symptomatic of PITS (Dillard, "A Slaughterhouse Nightmare: Psychological Harm Suffered by Slaughterhouse Employees and the Possibility of Redress through Legal Reform"). Turning to Gail Eisnitz's interviews, she shares the testimony of one hog sticker, who stressed, "a lot of the slaughterhouse hog killers have problems with alcohol. They have to drink, they have no other way of dealing with killing live, kicking, animals all day long. If you stop and think about it, you're killing several thousand beings a day" (p. 7).

29 Dillard, "A Slaughterhouse Nightmare: Psychological Harm Suffered by Slaughterhouse Employees and the Possibility of Redress through Legal Reform."

30 J. Dillard, "A Slaughterhouse Nightmare: Psychological Harm Suffered by Slaughterhouse Employees and the Possibility of Redress through Legal Reform," *Georgetown Journal on Poverty Law & Policy* 15, no. 2 (2008): 389.

31 S. Striffler, "Watching the Chickens Pass By: The Grueling Monotony of the Disassembly Line," in *The CAFO Reader: The Tragedy of Industrial Animal Factories*, ed. D. Imhoff (University of California Press, 2010), 125; ibid., later comments, "Not all workers are affected in the same way or to the same extent, but if you spend more than a year on the line—doing the exact same series of motions over and over again—it is certain that your fingers, wrists, hands, arms, shoulders or back will feel the effect. A few more years and the damage may be irreparable. Almost all of the line works I met had serious wrist problems. Many had undergone surgery and more than a few were permanently debilitated" (p. 126).

32 C. Cook, "Sliced and Diced: The Labour You Eat," in *The CAFO Reader: The Tragedy of Industrial Animal Factories*, ed. D. Imhoff (University of California Press, 2010), 235

33 A.J. Fitzgerald, L. Kalof, and T. Dietz, "Slaughterhouses and Increased Crime Rates: An Empirical Analysis of the Spillover From 'The Jungle' into the Surrounding Community," *Organization & Environment* 22, no. 2 (2009): 158–184. https://doi.org/10.1177/1086026609338164

34 J.R. Jacques, "The Slaughterhouse, Social Disorganization, and Violent Crime in Rural Communities," *Society & Animals* 23, no. 6 (2015): 594–612. https://doi.org/10.1163/15685306-12341380: 609

35 B. Bragg and J. Hyndman, "Family Matters: Navigating the Intentional Precarity of Racialized Migrant and Refugee Workers in Canadian Meatpacking," *Canadian Ethnic Studies* 54, no. 3 (2022): 9–31. https://doi.org/10.1353/ces.2022.0023

36 T. Genoways, "'I Felt Like a Piece of Trash'—Life Inside America's Food Processing Plants," *The Guardian* (2014), retrieved from: https://www.theguardian.com/world/2014/dec/21/life-inside-america-food-processing-plants-cheap-meat

37 B. Bragg, "'No Safe Place': Documenting the Migration Status and Employment Conditions of Workers in Alberta's Meatpacking Industry During the Pandemic," *Action Dignity* (2021): 3, retrieved from: https://actiondignity.org/wp-content/uploads/2021/09/NO-SAFE-PLACE_FINAL.pdf

38 Food Processing Skills Canada, *Securing Canada's Meat Workforce—Real Challenges. Practical Solutions. Fresh Perspectives. Executive Summary.*

39 Ibid., 6.

40 Ibid., 8.

41 A. Bombay, K. Matheson, and H. Anisman, "Intergenerational Trauma: Convergence of Multiple Processes among First Nations Peoples in Canada," *Journal of Aboriginal Health* 5, no. 3 (2009): 6–47.
42 Bragg and Hyndman, "Family Matters: Navigating the Intentional Precarity of Racialized Migrant and Refugee Workers in Canadian Meatpacking."
43 Ibid., 13.
44 Ibid.
45 Ibid., 14.
46 Ibid., 19.
47 K. Struthers Montford and T. Wotherspoon, "The Contagion of Slow Violence: The Slaughterhouse and COVID-19," *Animal Studies Journal* 10, no. 1 (2021): 80–113. https://doi.org/10.14453/asj.v10i1.6
48 L. Comas-Díaz, G.N. Hall, and H.A. Neville, "Racial Trauma: Theory, Research, and Healing: Introduction to the Special Issue," *The American Psychologist* 74, no. 1 (2019): 1. https://doi.org/10.1037/amp0000442
49 D. Imhoff, ed., *The CAFO Reader: The Tragedy of Industrial Animal Factories* (University of California Press, 2010).
50 See: J.J. Kona-Boun, "Anthropogenic Suffering of Farmed Animals: The Other Side of Zoonoses," *Animal Sentience* 5, no. 3 (2020). https://doi.org/10.51291/2377-7478.1207; Gillespie's, "Industrial Slaughter,"
51 Kona-Boun, "Anthropogenic Suffering of Farmed Animals: The Other Side of Zoonoses."
52 M. Scully, "Fear Factories: The Case for Compassionate Conservatism—for Animals," in *The CAFO Reader: The Tragedy of Industrial Animal Factories*, ed. D. Imhoff (University of California Press, 2010), 15–28.
53 Ibid., 23.
54 Kona-Boun, "Anthropogenic Suffering of Farmed Animals: The Other Side of Zoonoses," 7.
55 J. Scott-Reid, "Advocates Agree: Increasing Kill Line Speeds Puts Animals and Workers at Risk," *Sentient Media* (2019) https://sentientmedia.org/advocates-agree-increasing-kill-line-speeds-puts-animals-and-workers-at-risk/; likewise, Slade and Alleyne (2023) starkly remark, "it has been argued that facilitating or observing the cutting, skinning, and boiling of conscious or unconscious animals can cause psychological distress (i.e., cognitive dissonance) on the workers" (p. 430).
56 D. Celermajer, D. Schlosberg, L. Rickards, M. Stewart-Harawira, M. Thaler, P. Tschakert, B. Verlie, and C. Winter, "Multispecies Justice: Theories, Challenges, and a Research Agenda for Environmental Politics," *Environmental Politics* 30, no. 1–2 (2021): 120. https://doi.org/10.1080/09644016.2020.1827608
57 B. Belcourt, "Animal Bodies, Colonial Subjects: (Re)locating Animality in Decolonial Thought," *Societies* 5 (2015): 1–11.
58 Struthers Montford and Wotherspoon, "The Contagion of Slow Violence: The Slaughterhouse and COVID-19," 84.
59 D. Haraway, *The Companion Species Manifesto: Dogs, People, and Significant Otherness* (Prickly Paradigm Press, 2003), 6
60 A. Solomon, "Working Undercover in a Slaughterhouse: An Interview with Timothy Pachirat," *Medium* (2014) https://medium.com/learning-for-life/working-undercover-in-a-slaughterhouse-an-interview-with-timothy-pachirat-c6d7f37eef9c
61 Struthers Montford and Wotherspoon, "The Contagion of Slow Violence: The Slaughterhouse and COVID-19."
62 Cook, "Sliced and Diced: The Labour You Eat," 235.
63 J. Slade and E. Alleyne, "The Psychological Impact of Slaughterhouse Employment: A Systematic Literature Review," *Trauma, Violence, & Abuse* 24, no. 2 (2023): 429–440. https://doi.org/10.1177/15248380211030243; Struthers Montford and Wotherspoon, "The Contagion of Slow Violence: The Slaughterhouse and COVID-19."
64 For discussions on the environmental impacts of industrial animal agriculture and slaughterhouse labour on racialized and economically impoverished communities, and wild animals, see Gillespie, "Industrial Slaughter," and Struthers Montford and Wotherspoon (2021).

Bibliography

Belcourt, B. "Animal Bodies, Colonial Subjects: (Re)locating Animality in Decolonial Thought." *Societies* 5 (2015): 1–11.

Bombay, A., Matheson, K., and Anisman, H. "Intergenerational Trauma: Convergence of Multiple Processes among First Nations Peoples in Canada." *Journal of Aboriginal Health* 5, no. 3 (2009): 6–47.

Bragg, B. " 'No Safe Place': Documenting the Migration Status and Employment Conditions of Workers in Alberta's Meatpacking Industry during the Pandemic." *Action Dignity*, August 2021. Retrieved from: https://actiondignity.org/wp-content/uploads/2021/09/NO-SAFE-PLACE_FINAL.pdf

Bragg, B., and Hyndman, J. "Family Matters: Navigating the Intentional Precarity of Racialized Migrant and Refugee Workers in Canadian Meatpacking." *Canadian Ethnic Studies* 54, no. 3 (2022): 9–31. https://doi.org/10.1353/ces.2022.0023

Brown, J.S., Laundré, J.W., and Gurung, M. "The Ecology of Fear: Optimal Foraging, Game Theory, and Trophic Interactions." *Journal of Mammalogy* 80, no. 2 (1999): 385–399. https://doi.org/10.2307/1383287

Celermajer, D., Schlosberg, D., Rickards, L., Stewart-Harawira, M., Thaler, M., Tschakert, P., Verlie, B., and Winter, C. "Multispecies Justice: Theories, Challenges, and a Research Agenda for Environmental Politics." *Environmental Politics* 30, no. 1–2 (2021): 119–140. https://doi.org/10.1080/09644016.2020.1827608

Colling, S. *Animal Resistance in the Global Capitalist Era*. Michigan State University Press, 2021.

Comas-Díaz, L., Hall, G.N., and Neville, H.A. "Racial Trauma: Theory, Research, and Healing: Introduction to the Special Issue." *The American Psychologist* 74, no. 1 (2019): 1–5. https://doi.org/10.1037/amp0000442

Cook, C. "Sliced and Diced: The Labour You Eat." In *The CAFO Reader: The Tragedy of Industrial Animal Factories*, edited by D. Imhoff, 231–239. University of California Press, 2010.

Corman, L. "Trauma as a Möbius Strip: PTSD, Animal Research, and the Oak Ridge Prisoner Experiments." In *Building Abolition: Decarceration and Social Justice,* edited by C. Taylor and K. Struthers-Monford, 269–285. Routledge, 2021.

Corman, L. (Interviewer). "Ex-slaughterhouse Worker Virgil Butler [Audio podcast episode]." *Animal Voices*, February 15, 2005. Retrieved from: https://animalvoices.ca/2005/02/15/ex-slaughterhouse-worker-virgil-butler/

Dillard, J. "A Slaughterhouse Nightmare: Psychological Harm Suffered by Slaughterhouse Employees and the Possibility of Redress through Legal Reform." *Georgetown Journal on Poverty Law and Policy* 14, no. 2 (2007): 273–300.

Eisnitz, G.A. *Slaughterhouse: The Shocking Story of Greed, Neglect, and Inhumane Treatment Inside the U.S. Meat Industry*. Prometheus Books, 1997.

Enloe, C. *The Curious Feminist: Searching for Women in a New Age of Empire*. University of California Press, 2004.

Fitzgerald, A.J., Kalof, L., and Dietz, T. "Slaughterhouses and Increased Crime Rates: An Empirical Analysis of the Spillover from 'The Jungle' into the Surrounding Community." *Organization & Environment* 22, no. 2 (2009): 158–184. https://doi.org/10.1177/1086026609338164

Food & Agriculture Organization of the United Nations. "Crop and Livestock Products." *FAOSTAT*, 2022. Retrieved from: https://www.fao.org/faostat/en/#data/QCL

Food Processing Skills Canada. "Securing Canada's Meat Workforce—Real Challenges. Practical Solutions. Fresh Perspectives. Executive Summary." Food Processing Skills Canada, 2019. Retrieved from: https://fpsc-ctac.com/wp-content/uploads/2020/03/exec-summary-canadian-meat-and-poultry-lmi-final-report.pdf

Genoways, T. " 'I Felt Like a Piece of Trash'—Life Inside America's Food Processing Plants." *The Guardian*, December 24, 2014. Retrieved from: https://www.theguardian.com/world/2014/dec/21/life-inside-america-food-processing-plants-cheap-meat

Gillespie, K. "Industrial Slaughter." In *Gender: Animals*, edited by J.S. Parreñas, 181–195. Macmillan Reference USA, 2012.

Government Of Canada. "Animal Industry." Agriculture & Agri-Food Canada, 2023. Retrieved from: https://agriculture.canada.ca/en/sector/animal-industry

Haraway, D. *The Companion Species Manifesto: Dogs, People, and Significant Otherness*. Prickly Paradigm Press, 2003.

Herman, J.L. *Trauma and Recovery: The Aftermath of Violence—from Domestic Abuse to Political Terror*. 2015 ed. Basic Books, 2015.

Imhoff, D., ed. *The CAFO Reader: The Tragedy of Industrial Animal Factories*. University of California Press, 2010.

Jacques, J.R. "The Slaughterhouse, Social Disorganization, and Violent Crime in Rural Communities." *Society & Animals* 23, no. 6 (2015): 594–612. https://doi.org/10.1163/15685306-12341380

Levine, P.A. *Waking the Tiger: Healing Trauma: The Innate Capacity to Transform Overwhelming Experiences*. North Atlantic Books, 1997.

McLoughlin, E. "Knowing Cows: Transformative Mobilizations of Human and Non-human Bodies in an Emotionography of the Slaughterhouse." *Gender, Work, and Organization* 26, no. 3 (2019): 322–342. https://doi.org/10.1111/gwao.12247

Perkins, E., Brothers, S., and Nemeroff, C. "Animal Models for Post- traumatic Stress Disorder." In *Post-traumatic Stress Disorder*, edited by C. Nemeroff and C. Marmar. Oxford University Press, 2018.

Scott-Reid, J. "Advocates Agree: Increasing Kill Line Speeds Puts Animals and Workers at Risk." *Sentient Media*, October 7, 2019. Retrieved from: https://sentientmedia.org/advocates-agree-increasing-kill-line-speeds-puts-animals-and-workers-at-risk/

Scully, M. "Fear Factories: The Case for Compassionate Conservatism—for Animals." In *The CAFO Reader: The Tragedy of Industrial Animal Factories*, edited by D. Imhoff, 15–28. University of California Press, 2010.

Slade, J., and Alleyne, E. "The Psychological Impact of Slaughterhouse Employment: A Systematic Literature Review." *Trauma, Violence, & Abuse* 24, no. 2 (2023): 429–440. https://doi.org/10.1177/15248380211030243

Solomon, A. "Working Undercover in a Slaughterhouse: An Interview with Timothy Pachirat." *Medium*, August 25, 2014. Retrieved from: https://medium.com/learning-for-life/working-undercover-in-a-slaughterhouse-an-interview-with-timothy-pachirat-c6d7f37eef9c

Striffler, S. "Watching the Chickens Pass By: The Grueling Monotony of the Disassembly Line." In *The CAFO Reader: The Tragedy of Industrial Animal Factories*, edited by D. Imhoff's, 125–130. University of California Press, 2010.

Struthers Montford, K., and Wotherspoon, T. "The Contagion of Slow Violence: The Slaughterhouse and COVID-19." *Animal Studies Journal* 10, no. 1 (2021): 80–113. https://doi.org/10.14453/asj.v10i1.6

Warrick, J. "They Die Piece by Piece." *The Washington Post*, April 10, 2001. Retrieved from: https://www.washingtonpost.com/archive/politics/2001/04/10/they-die-piece-by-piece/f172dd3c-0383-49f8-b6d8-347e04b68da1/

Zanette, L.Y., and Clinchy, M. "Ecology and Neurobiology of Fear in Free-living Wildlife." *Annual Review of Ecology, Evolution, and Systematics* 51, no. 1 (2020): 297–318. https://doi.org/10.1146/annurev-ecolsys-011720-124613

29

ECOMASCULINITIES, BOYHOODS AND CRITICAL ANIMAL PEDAGOGY

Richard Twine

If assessing the success of ecofeminist scholarship, one would have to include the range of fields it has helped to shape. These include critical animal studies and vegan studies, and to this we can now add ecomasculinities. Arguably long overdue, a field of critical inquiry specifically interested in the interrelationships between masculinities and the more-than-human is an important and necessary development in the urgent need to address the crises that we are now living through, if not experiencing equally. In this case the line of influence is rather clear. Several scholars with longstanding research locations within ecofeminism have also been influential in constructing the ecomasculinities field.[1] Moreover, several ecomasculinities authors have made clear the importance of ecofeminism to their project.[2] Indeed, the multidisciplinary field has emerged through overlaps between masculinities studies and ecofeminism, at first, since the late 90s, quite fragmentary, but more recently coalescing in clearer conceptualisations and collections of work spanning the humanities and social sciences.[3] Ecomasculinities research is both a critical and creative endeavour, interested in both the role of masculinities in creating the ecological crisis and in examining how gender contestation and reinvention are vital for societal transformation.

In this chapter I begin to map out some future directions for ecomasculinities research focused around two areas that might be in danger of being marginalised: human/animal relations and childhood. First it is worth mentioning that also due to its co-location within men and masculinities research, ecomasculinities work has relevance for a broad range of activism and scholarship in, for example, gender-based violence, childhood studies, ecocriticism, feminist philosophy and the environmental social sciences. Despite the influence of ecofeminism on the emerging ecomasculinities field, it is not inevitable that human/animal relations will be fully attended to. Ecofeminism has seen its own internal struggles around questions of animal ethics,[4] and it is now familiar to critical animal scholars to see environmental sub-fields routinely fail to give due attention to human/animal relations.[5] Whilst the eco in ecomasculinities must be inclusive of human/animal relations to be ontologically coherent, research to date has lacked this focus to an extent.[6] One way in which some ecomasculinities researchers have addressed this is to focus on the often-noted cultural masculinisation of animal consumption as demonstrative of hegemonic masculinity and

DOI: 10.4324/9781003273400-36

vegan masculinities as a possible practice of ecomasculinity.[7] Ecomasculinities can open a rich seam of social research which also exceeds these, albeit important, foci.

The recent work of Martin Hultman and Paul Pulé has been important for developing and bringing together new work on ecomasculinities.[8] They have made the useful distinction between industrial/breadwinner masculinities, ecomodern masculinities and ecological masculinities.[9] The first of these is employed 'interchangeably with malestream, patriarchal, hegemonic, and normative masculinities (which we apply primarily to men, but also to the masculinities adopted by some women and non-binary/genderqueer people as well)'.[10] The bracketed text is noteworthy, as it takes us beyond a simplistic conflation of masculinities with 'male bodies'. The shifting meanings, materialities and doings of masculinity shape the gendered landscape for all. For Hultman and Pulé, industrial/breadwinner masculinities refer to the interplay between capitalism and gender, in which, during the unfolding of the Capitalocene, capitalism has prized, rewarded, and relied upon gendered practices that dispassionately exploit both 'nature-associated humans' and the more-than-human, including nonhuman animal species. Sociologist of masculinities David Morgan noted 40 years ago that Max Weber's famous argument that a protestant work ethic acted as a synergy for the development of capitalism[11] could also be read retrospectively as a study of masculinity[12] with its pre-requisites of (self) control and 'rationality' for entrepreneurial success. Hultman and Pulé's notion of industrial masculinities underlines that capitalism *required* particular gender identities. They point to those involved in possessing and managing the means of production, and more specifically the managers of fossil fuel industries, bankers, and shareholders, as exemplars of this type of relationship between masculinity and ecology. Clearly it is not a homogenous category, for example, certainly white and Global North dominated, but bifurcated by social class differences. Industrial/breadwinner masculinities might in some contexts be thought of as a coalition across social class difference, wherein some working-class men may come to advocate for the same economic imaginaries as the managerial class. As Hultman and Pulé contend, 'Industrial/breadwinner masculinities are bound to the pursuits of industrial growth, since the two require each other to thrive'.[13] This can be seen to play out within anti-ecological sentiment from trade unions and male working-class conservatism. These masculinities embody an entrenched disavowal of care toward the more-than-human within a typically essentialist understanding of how men and boys should be. They are visible, for example, in the way that practices of climate denialism, wildlife crime and meat consumption are gendered.[14]

Their second category, ecomodern masculinities, are embodied by those who recognise the existence of environmental degradation but align themselves with attempts to make capitalism more sustainable. Hultman and Pulé intend the connection here with the broader thesis of ecological modernisation,[15] the move to green (and greenwash) capitalism in order to preserve its core features of exploitation, growth and consumerism. These are masculinities that pay lip service to 'environmental issues' but fail to make any meaningful transformative change in the face of contemporary crises. Exemplars may include institutional actors such as in government, education or the private sector involved in instigating shallow policies in terms of either their mitigative or transformative effect. Ecomodern masculinities are the equivalent of the 'new man' who acknowledged second-wave feminism but made little real attempt to contest the deeply gendered status quo. Ecomodern masculinities are attracted to technological solutions to environmental problems instead of also examining their root causes. Hultman and Pulé offer the figures of Elon Musk and Arnold Schwarzenegger as individuals who embody this approach in their ecomodern practices.

In further developing the ecomasculinities field, they seek to explore alternatives to these two approaches which can offer 'a path for modern Western men and masculinities that prioritises deeper, broader and wider responses to global social and environmental problems—responses that are both personal and political, individual and systemic and therefore truly transformational'.[16] This urgent and overdue task necessarily involves engaging with previous work on the relationship between constructions of masculinity vis-à-vis practices of care, compassion, justice and empathy. It implies social research attentive to the intersections of these constructions with social class, age, 'race'/ethnicity, sexuality and disability. It requires an acknowledgement of capitalism as an extra-economic form of human organisation and historical force which has become hegemonic: producing nature and being shaped by it. Counter to dualism, capitalism does not 'act on nature' but is thoroughly embroiled in the web of life.[17] In this way of thinking, which names our epoch the Capitalocene, the climate crisis is historicised in terms of colonialism, racism, patriarchy and anthropocentrism. Furthermore, it must be acknowledged that masculinities, especially the aforementioned industrial masculinities, are also embroiled in the web of life, partly shaping present-day ecologies and human/animal relations. How might the web of life be remade through innovation in our ideas of gender and our gendered practices? What would human entanglements with the rest of nature look like if masculinities were premised on compassion, care and flourishing? How can children, and perhaps boys specifically, be dissuaded from dispassionate disengagement with the more than human?

Those scanning early developments within ecomasculinities might be disappointed by the relative lack of research and theorisation around gender and nonhuman animals specifically. As implied previously, the cross-cultural masculinisation of animal consumption and vegan masculinities as a practice of ecomasculinity are relatively 'low-hanging fruit' when it comes to investigating practices surrounding masculinities and other animals, and there is already a research base here external to that explicitly defined as ecomasculinities.[18] There is a potentially interesting phenomena at play here. Many men are indeed drawn to environmentalism and academically, the environmental humanities and social sciences. An element of this might be the sort of white heroic protector saviourism which has historically very much been part of hegemonic masculinity narratives. Another element might be a personal search for 'deep meaning' and pseudo-spiritual self-discovery. Again, this might not trouble a deeply traditional understanding of masculinity in terms of the gendered figure of the philosopher. Yet the question of men–animal relations gets to the heart of practices associated with domination and hierarchy and may be more troubling for men to confront. This can be noted in forms of masculine ecologisation which have self-imagined male embeddedness in nature via male as predator hunting practices, as if tapping into a primal mythical masculine core. Using essentialist takes on gender and representing hunting as a 'spiritual asset and biological drive' have long been criticised by ecofeminists.[19] The interesting and disturbing phenomena at play here is that some men may be brought to the ecological table exactly by familiar tropes of hegemonic masculinity but be less willing to relinquish its privilege and power.

Nevertheless, a few contributors to recent developments in the ecomasculinities field have made a point of including the themes of veganism and animal consumption.[20] Aavik, for example, carried out interview research with Estonian and Finnish men, exploring rich narratives of engagement with care and compassion and adding to understandings of how men undergoing vegan transition may relate their experience to gender issues. Specifically, Aavik found that many of these men tended to regard veganism as related and interlinked to a broader theme of social justice, and even those who did not recognised the relationship between animal consumption and gender stereotypes of masculinity. Many of her sample were engaged in re-assessing the role of care in their lives, for self, ecology, human and nonhuman animal others. Such studies remind us of the importance of continued social research on meat, masculinities and veganism and not simply to take these associations as read. People are resisting these meanings and practices, and some men are acting as progressive agents for change. Ideally vegan transition for self-defined males opens the door to a parallel profeminist transition because it brings into view experientially the conformity of other men and boys with gender scripts of care disavowal. As has long been noted, men participating in veganism and/or environmental politics are not immune to practicing 'male dominance or the unwelcome re-surfacing of hegemonic masculinity',[21] but such 'heganism'[22] can be addressed by contesting veganism itself and making the case for its intersectional definition.

Ramos-Gay and Alonso-Recarte have also recently suggested specific research around a framing of 'zoomasculinities' to capture the history and diversity of relationships between men, masculinity and nonhuman animals—acknowledging but not reducing these relationships to systemic complicity in exploitation.[23] There need not be a tension within the development of the ecomasculinities field as long as it, contra some other areas of ecological specialisation in the humanities and social sciences, ontologically understands the 'eco' as inclusive of human/animal relations. The kinds of research directions for such as inclusive ecomasculinities can be elaborated by noting another possible risk of omission.

Here I refer to childhoods and boyhoods specifically. Ecomasculinities research can be very valuable in thinking about childhood development and learning and as interventions into the social reproduction of anthropocentric values and practice. Yet despite their groundbreaking importance, the two recent works on ecomasculinities contain no specific focus or chapters on childhood or boyhood.[24] Similarly, there is very little engagement with the field of childhood studies or the related boy studies. A gap needs to be closed here, not least because in childhood studies (and in early childhood studies, which focuses on the ages of 0–8) there has now been a noticeable engagement with environmental topics and sustainability.[25] There is increasing recognition over the marginalisation of children from environmental decision-making as the decisions made by a largely white, Western middle-aged wealthy male demographic increasingly compromise the viability of life for future generations of humans and other animals. It is also noteworthy that whilst the climate and biodiversity crises have informed a new politicisation of children and childhood, the emergence of a small number of children as political citizens has been quite gendered.

We might wonder why there have not been high-profile boys alongside the emergence of the likes of Severn Cullis-Suzuki, Kehkashan Basu or Greta Thunberg? I would suggest that an ecomasculinities focus on the political agency of children must be embedded within recent traditions of childhood studies that have argued for the voice and rights of children to be taken seriously. This dovetails with critical feminist pedagogy around the rights of children not to be subject to normative stereotypical notions of gender which reinforce inequalities and curtail opportunities and freedom of self-definition. Thus, we might surmise that the comparative lack of boys emerging as voices critical of the aforementioned crises is linked in some way to the gendering of care and the feminisation of environmentalism and animal advocacy.[26] The vegan climate activist Greta Thunberg has been particularly successful in converting individual activism into collective struggle via the school strike movement that she inspired,[27] so much so that she has provoked backlash responses from such exemplars of industrial masculinities as former U.S. President Donald Trump, former Australian Prime Minister Scott Morrison and Brexit funder Aaron Banks.

Turning to childhood studies work on children and climate change, we can note how the historical disempowerment of children has worked partly through their representation as human futures, their present-day voice as children largely silenced and trivialised. Thus, the development of childhood studies has sought to innovate methodologically to contest this representation and to work with children to give their voice a platform on a broad range of issues. As Lee outlines, 'identifying children with the future has tended unjustly to silence children as present-day members of society and to sanction their use as a resource instead of their inclusion as citizens with views and preferences of their own'.[28] The climate crisis brings cultures that have historically constructed children as innocent and in constant need of protection into conflict with the need to empower children about both the causes and likely trajectories of climate breakdown. If this empowerment via meaningful climate pedagogy is absent, if children are denied a role in making the future, they remain cast as 'passively awaiting their role as janitors of the future',[29] as inheritors of adult incompetence.

This brings us to a sound assertion: one cannot credibly empower children around contemporary crises without including a focus on human/animal relations. The global commodification of farmed animals is a significant contributor to both the climate and biodiversity crises due to its emissions and also its avaricious and profligate use of land.[30] It is a major player within the Capitalocenic remaking of the web of life. Although tragically effects of these crises are already rapidly unfolding, to avert catastrophic climate change urgent transformational change is needed. Whilst much of that change refers to infrastructures of energy and mobility and substituting the dominant carbon and growth-based economics, transforming the global food system also affords significant opportunities and co-benefits in the decommodification of animals, reducing the emissions that threaten all species and offering improved human health. The overall standards of living of Global North citizens have benefitted from these colonial and capitalist histories. From the perspective of a just transition, arguably there is a moral imperative that the countries of the Global North transition deeper and faster.[31] This should be translating now into novel pedagogies which allow children to critically engage with dominant anthropocentric ideologies and their related food practices.

Many such societies are living within an amplified 'fleischgeist', Standen and Wizansky's useful term to refer to the growing cultural attention to and anxieties around meat.[32] In some contexts, such as the UK, this is translating into the contestation of children's food practices, the emergence of vegan parenting and plant-based school meals (typically as an option rather than a replacement). The fleischgeist appears symptomatic of the troubling of the social norm of animal consumption, for years naturalised through a range of discursive strategies (humans need meat, men need meat, children need meat and so on) that have gradually come to lose some of their force.

Furthermore, critical animal studies work on childhood and education has argued that anthropocentric practices such as animal consumption rely upon their active reproduction by adult-dominated societal institutions such as the education system, the family and the media and in cultural products such as computer games.[33] Recent social psychological work might also suggest that children do not tend to want to 'naturally' advocate for the slaughterhouse. For example, McGuire et al. found that compared with adults and young adults, children (aged 9–11) demonstrated less speciesism and deemed eating meat and animal products less morally acceptable.[34] Wilks et al. found that children (aged 5–9) were less likely to morally prioritise humans over nonhuman animals, suggesting that human exceptionalism is learned later in development and so likely socially acquired.[35] What would be especially interesting is whether from an ecomasculinities perspectives that process of learning which typically and ultimately manifests in greater anthropocentrism by young adulthood is related to encounters with specific meanings of gender.

As noted by Myers, there is a discursive tradition in the West of valorising children, referring to this as the 'child of nature', which views children and animals as morally good, juxtaposed against a fallen or corrupt notion of adult culture or civilisation.[36] Shaped by romanticist thought, this has been a significant influence on the co-location of children and animals as both innocent and good. Notions of childhood innocence have typically been used as reasons for shielding children from the harsh realities of society and thus conflict with a desire to empower children into becoming active citizens. There is no need to explain the aforementioned social psychological research in terms of a falsely universal essential moral child. Some children (often but not only boys)[37] are capable of violence toward other animals, and if there is a propensity for younger children to overall resist anthropocentrism and practices of violence, it may be telling us more about the way cultures suffuse childhoods with representations of animals or the incongruency of the slaughterhouse with the contemporary moral sensibilities of most people of all ages rather than seeing children as symbols for an innate human goodness.

It is clear that in many cultures of the Global North and beyond there is a broadly successful process of child development at play in which young people learn selective and arguably inconsistent moral meanings toward other animals based on normative social categorisations such as 'pets' or 'farm animals'.[38] This is effectively imposed by adults on each new generation of children. The emergence of childhood studies has strived to democratise the hierarchical aspect of the child/adult binary and critique dominant narratives of child development through advocacy for children's voice and rights. Whilst this could imply a critical questioning of the routine social reproduction of animal consumption, a more recent focus in childhood studies may offer even more conceptual purchase in this direction.

As a more direct advocacy for children, work has coalesced around the idea of childism,[39] which Wall argues is 'the needed critical lens for deconstructing adultism across research and societies and reconstructing more age-inclusive scholarly and social imaginations'.[40] Here adultism refers to the way in which 'social understandings and practices have historically been dominated by adults and adult points of view'.[41] As a response to this, Wall envisages the aim of childism being 'to critically restructure historically engrained norms of adultism'.[42] Taking this idea into thinking the climate crisis, Biswas and Mattheis offer a positive appraisal of the school strike movement for its ability to provoke critical reflection on educational philosophy and to subvert adultist hierarchical norms about learning. Rather than punitive responses to children, adults ought to see this social movement as a chance to learn *from* children, to better understand the politics of intergenerational justice[43] and specifically to understand how contemporary practice norms and ways of organising society are compromising the future.

The same can be said for children challenging the adult norm of animal consumption, whether that is on climate or inclusive of other related grounds. It is here that such developments in childhood studies dovetail with those in critical animal studies, especially the recent formulation of critical animal pedagogy (CAP).[44] Dinker and Pedersen intend CAP as an 'an alternative education where students at all levels across the curriculum are invited to explore both a critical analytic and a radically transformative approach to animals and affect in education'.[45] In this way CAP, like a feminist pedagogy that empowers children to understand gender essentialism and resist gender stereotypes, is a form of childism in its questioning of adult-imposed social norms. CAP can equip learners with the reflexive knowledge to understand their own affective construction of other animals, to be self-critical and to better understand broader social and institutional influences upon their own moral categorisations. CAP is childist because it is an 'invitation for adults to let children contribute to their own ongoing formation'.[46] Furthermore, it is especially pertinent for the tradition of childhood studies that has contested universalistic and Western centric models of child development. CAP speaks to a posthuman turn in (early) childhood studies[47] which recognises that the politics of childhood development are also the politics of the human. Historically our dominant narratives of childhood development are where our cultures have sedimented assumptions of what it means to be human, playing out the usual confusions over the 'biological' and the 'social', the 'human' and the 'animal'.[48] Giving space to children to resist animal consumption allows them to perform the human differently, contesting narratives of the compulsory omnivorous human and ultimately to re-assess and transcend anthropocentric meat culture.

That CAP foregrounds affective transformation may be even more significant for boys, especially those who might have been castigated for demonstrating 'unmanly' emotional attachments to other animals. For teachers and parents, it then becomes important to avoid a gendered 'shutting down' of such attachments so that boys especially are not policed into a dispassionate and instrumental relation toward other animals. Indeed, children should be encouraged and given space to talk and reflect on human/animal relations and for caring values and practices to be nurtured. Ecomasculinites requires an intersectional lens here in order to better understand the dynamics in which some childhoods and boyhoods more than others are steered away from connectedness, compassion and caring. It also needs to systematically approach the cultural worlds of children. Focusing for example on children's and boys' media may be instructive for better understanding how intersections of gender and animals are circulated. Within vegan studies to date there has also been minimal work on childhoods, but there has been analysis of vegan and animal rights children's books, including some which foreground the caring practices of boys.[49]

Bringing together ecomasculinities, childhood studies and critical animal pedagogy can be a fruitful avenue for new research, for interventions and for transformative change. It can contribute to longstanding research around parenting, familial dynamics, teacher training and humane education. It means listening to children and boys about their relations with other animals as well as critically questioning the adult imposition of violent practices such as meat eating and hunting as traditionally masculine and 'human' rites of passage. If boyhoods can be premised upon caring relations with the more-than-human, then it can help produce men who are up to the task of questioning anthropocentrism and acting in solidarity with others as climate leaders. Situating oneself in the rest of nature is also bound up in knowledge of one's own body. Teaching children about their bodies as an integral part of their agential capacities and ecological interdependency is also then key for young people in understanding the links between self-care, respect for the bodies of other animals and ecological flourishing.

Appropriately, Hultman and Pulé have helped to form the Starfish collective,[50] based in Sweden but increasingly international, that seeks to foreground the role of destructive social constructions of masculinities in the ecological crisis and aims to work with and change organisations by redefining masculinity towards ethics of care. Yet all such initiatives and attempts at transformative change in this direction must contend with the persistent synergy alluded to previously between industrial masculinity and capitalism. The heart of this synergy can be found in what ecofeminist philosopher Val Plumwood termed a sado-dispassionate and institutionalised form of rationality,[51] which nullifies empathy, attempts to extricate the human from nature and acts in denial of our ecological interdependencies. Re-imagined childhoods, boyhoods and human/animal relations are integral ingredients for ecological masculinities, but they are equally fundamental for transition beyond the Capitalocene itself.

Notes

1 For example, Greta Gaard and Sherilyn MacGregor.
2 P.M. Pulé and M. Hultman, eds., *Men, Masculinities, and Earth: Contending with the (m)Anthropocene* (Springer Nature, 2021).
3 Key texts include M. Allister, *Eco-man—New Perspectives on Masculinity and Nature* (Charlottesville: University of Virginia Press, 2004); S. MacGregor and N. Seymour, eds., *Men and Nature: Hegemonic Masculinities and Environmental Change* (RCC Perspectives, 2017). https://doi.org/10.5282/rcc/7977; R. Cenamor and S.L. Brandt, eds., *Ecomasculinities—Negotiating Male Gender Identity in U.S. Fiction* (New York: Lexington, 2019); M. Hultman and P.M. Pulé, *Ecological Masculinities: Theoretical Foundations and Practical Guidance* (London: Routledge, 2018); Pulé and Hultman, eds., *Men, Masculinities, and Earth: Contending with the (m)Anthropocene.*
4 For debates about veganism, see V. Plumwood, "Integrating Ethical Frameworks for Animals, Humans, and Nature: A Critical Feminist Eco-socialist Analysis," *Ethics & the Environment* 5, no. 2 (2000): 285–322; and C.J. Adams and L. Gruen. eds., *Ecofeminism: Feminist Intersections with Other Animals and the Earth*, 2nd ed. (Bloomsbury Publishing USA, 2021).
5 See I.M. Bergmann, "The Intersection of Animals and Global Sustainability—A Critical Studies Terrain for Better Policies?" *Proceedings* 73, no. 1 (2020): 12. https://doi.org/10.3390/IECA2020-08895
6 For critique on this, see R. Twine, "Masculinity, Nature, Ecofeminism, and the 'Anthropo'cene," in *Men, Masculinities, and Earth: Contending with the (m)Anthropocene*, eds. P.M. Pulé and M. Hultman (Cham: Springer International Publishing, 2021), 117–133.
7 K. Aavik, "Vegan men: Towards Greater Care for (Non)human Others, Earth and Self," in *Men, Masculinities, and Earth: Contending with the (m)Anthropocene*, eds. P.M. Pulé and M. Hultman (Cham: Springer International Publishing, 2021), 329–350; Twine, "Masculinity, Nature, Ecofeminism, and the 'Anthropo'cene."
8 Hultman and Pulé, *Ecological Masculinities: Theoretical Foundations and Practical Guidance*; Pulé and Hultman, eds., *Men, Masculinities, and Earth: Contending with the (m)Anthropocene.*
9 Ibid.
10 Hultman and Pulé, *Ecological Masculinities: Theoretical Foundations and Practical Guidance,* 40.
11 M. Weber, *The Protestant Ethic and the Spirit of Capitalism* (New York: Routledge, 2001 [1904])
12 D. Morgan, "Men, Masculinity and the Process of Sociological Enquiry," in *Doing Feminist Research*, ed. H. Roberts (London: Routledge, 1981), 83–113.
13 Hultman and Pulé, *Ecological Masculinities: Theoretical Foundations and Practical Guidance,* 42.
14 A.M. McCright and R.E. Dunlap, "Cool Dudes: The Denial of Climate Change among Conservative White Males in the United States," *Global Environmental Change* 21, no. 4 (2011): 1163–1172. https://doi.org/10.1016/j.gloenvcha.2011.06.003; F. Massé, N. Givá and E. Lunstrum, "A Feminist Political Ecology of Wildlife Crime: The Gendered Dimensions of a Poaching Economy and Its Impacts in Southern Africa," *Geoforum* 126 (2021): 205–214. https://doi.org/10.1016/j.geoforum.2021.07.031; R. Sollund, "Wildlife Crime: A Crime of Hegemonic Masculinity?" *Social Sciences* 9, no. 6 (2020): 93. https://doi.org/10.3390/socsci9060093; A. Peeters, G. Ouvrein, A. Dhoest and C. De Backer, "It's Not Just Meat, Mate! The Importance of Gender Differences in Meat Consumption," *Food, Culture and Society* (2023), https://doi.org/10.1080/15528014.2022.2125723; C.J. Adams, *The Sexual Politics of Meat: A Feminist-Vegetarian Critical Theory* (New York: Continuum, 1990).

15 Hultman and Pulé, *Ecological Masculinities: Theoretical Foundations and Practical Guidance.*
16 Ibid., 51.
17 J. Moore, *Capitalism in the Web of Life: Ecology and the Accumulation of Capital* (London: Verso Books, 2015)
18 J. Sobal, "Men, Meat, and Marriage: Models of Masculinity," *Food and Foodways* 13 (1–2) (2005): 135–158. https://doi.org/10.1080/07409710590915409; M.B. Ruby and S.J. Heine, "Meat, Morals, and Masculinity," *Appetite* 56, no. 2 (2011): 447–450. https://doi.org/10.1016/j.appet.2011.01.018; K.C. Sumpter, "Masculinity and Meat Consumption: An Analysis through the Theoretical Lens of Hegemonic Masculinity and Alternative Masculinity Theories," *Sociology Compass* 9, no. 2 (2015): 104–114.
19 M. Kheel, "The Killing Game: An Ecofeminist Critique of Hunting," *Journal of the Philosophy of Sport* 23, no. 1 (1996): 30–44. https://doi.org/10.1080/00948705.1996.9714529: 30.
20 Aavik, "Vegan men: Towards Greater Care for (Non)human Others, Earth and Self"; G. Gaard, "Queering the Climate," in *Men, Masculinities, and Earth: Contending with the (m)Anthropocene*, eds. P.M. Pulé and M. Hultman (Cham: Springer International Publishing, 2021), 515–536. B. Pease, "From Ecomasculinity to Profeminist Environmentalism: Recreating Men's Relationship with Nature," in *Men, Masculinities, and Earth: Contending with the (m)Anthropocene*, eds. P.M. Pulé and M. Hultman (Cham: Springer International Publishing, 2021), 537–558; Twine, "Masculinity, Nature, Ecofeminism, and the 'Anthropo'cene."
21 R. Twine, "Masculinity, Nature, Ecofeminism" (1997), retrieved from: http://richardtwine.com/ecofem/masc.pdf. The U.S. animal advocacy movement, for example, has been blighted by a series of cases of the male sexual harassment of women.
22 L. Wright, *The Vegan Studies Project: Food, Animals, and Gender in the Age of Terror* (Athens: University of Georgia Press, 2015)
23 I. Ramos-Gay and C. Alonso-Recarte, "Zoomasculinities: At the Intersection between Animals, Animality, and Masculinity," *Men and Masculinities* 23, no. 5 (2020): 807–813. https://doi.org/10.1177/1097184X20965125.
24 Hultman and Pulé, *Ecological Masculinities: Theoretical Foundations and Practical Guidance*; Pulé and Hultman, eds., 2021.
25 For example, J. Siraj-Blatchford, C. Mogharreban, and E. Park, eds., *International Research on Education for Sustainable Development in Early Childhood* (Heidelberg: Springer, 2016); K. Weldemariam, D. Boyd, N. Hirst, B.M. Sageidet, J.K. Browder, L. Grogan et al., "A Critical Analysis of Concepts Associated with Sustainability in Early Childhood Curriculum Frameworks across Five National Contexts," *International Journal of Early Childhood* 49, no. 3 (2017): 333–351. https://doi.org/10.1007/s13158-017-0202-8.
26 There could also be a desire in the media to promote notions of childhood innocence, which historically have been racialized and gendered as white and feminine.
27 See A. Holmberg and A. Alvinius, "Children's Protest in Relation to the Climate Emergency: A Qualitative Study on a New Form of Resistance Promoting Political and Social Change," *Childhood* 27, no. 1 (2020): 78–92. https://doi.org/10.1177/0907568219879970; Thunberg's veganism has arguably less of a focus in her coverage and influence.
28 N. Lee, *Childhood and Biopolitics: Climate Change, Life Processes and Human Futures* (Springer, 2013), 4.
29 Ibid., 130.
30 R. Twine, *The Climate Crisis and Other Animals* (Sydney: Sydney University Press, 2024)
31 However, co-benefits should be open to all. It is a mistake to think that poorer countries should be made to repeat the maldevelopment of the Global North.
32 A. Standen and S. Wizansky, "Hello from Meatpaper," *Meatpaper*, no. 1 (2007).
33 M. Cole and K. Stewart, *Our Children and Other Animals: The Cultural Construction of Human–animal Relations in Childhood* (London: Routledge, 2014)
34 L. McGuire, S.B. Palmer, and N.S. Faber, "The Development of Speciesism: Age-related Differences in the Moral View of Animals," *Social Psychological and Personality Science* (2023), https://doi.org/10.1177/19485506221086182.
35 M. Wilks, L. Caviola, G. Kahane, and P. Bloom, "Children Prioritize Humans over Animals Less Than Adults Do," *Psychological Science* 32, no. 1 (2021): 27–38. https://doi.org/10.1177/0956797620960398.

36 G. Myers, *The Significance of Children and Animals: Social Development and Our Connections to Other Species* (West Lafayette: Purdue University Press, 2007), 22–25.
37 L. Merz-Perez, K.M. Heide, and I.J. Silverman, "Childhood Cruelty to Animals and Subsequent Violence against Humans," *International Journal of Offender Therapy and Comparative Criminology* 45, no. 5 (2001): 556–573. https://doi.org/10.1177/0306624X01455003.
38 Cole and Stewart, *Our Children and Other Animals: The Cultural Construction of Human–animal Relations in Childhood.*
39 J. Wall, "From Childhood Studies to Childism: Reconstructing the Scholarly and Social Imaginations," *Children's Geographies* 20, no. 3 (2022): 257–270. https://doi.org/10.1080/14733285.2019.1668912; T. Biswas and N. Mattheis, "Strikingly Educational: A Childist Perspective on Children's Civil Disobedience for Climate Justice," *Educational Philosophy and Theory* 54, no. 2 (2022): 145–157. https://doi.org/10.1080/00131857.2021.1880390.
40 Wall, "From Childhood Studies to Childism: Reconstructing the Scholarly and Social Imaginations," 1.
41 Ibid., 4.
42 Ibid.
43 D.F. Lawson, K.T. Stevenson, M. Nils Peterson, S.J. Carrier, R. Strnad, and E. Seekamp, "Intergenerational Learning: Are Children Key in Spurring Climate Action?" *Global Environmental Change* 53 (2018): 204–208. https://doi.org/10.1016/j.gloenvcha.2018.10.002.
44 K.G. Dinker, "Critical Creatures: Children as Pioneers of Posthuman Pedagogies," *International Journal of Sociology and Social Policy* 41, no. 3/4 (2021): 391–406. https://doi.org/10.1108/IJSSP-10-2020-0464; K.G. Dinker and H. Pedersen, "Critical Animal Pedagogies: Re-learning Our Relations with Animal Others," in *The Palgrave International Handbook of Alternative Education*, eds. H.E. Lees and N. Noddings (London: Palgrave Macmillan UK, 2016), 415–430; K.G. Dinker and H. Pedersen, "Critical Animal Pedagogy: Explorations toward Reflective Practice," in *Education for Total Liberation: Critical Animal Pedagogy and Teaching against Speciesism,* eds. A. Nocella, C. Drew, and A.E. George (New York: Peter Lang U.S., 2019); J. Oakley, "What Can an Animal Liberation Perspective Contribute to Environmental Education?" in *Animals in Environmental Education,* eds. T. Lloro-Bidart and V.S. Banschbach (London: Palgrave, 2019), 19–34.
45 Dinker and Pedersen, "Critical Animal Pedagogies: Re-learning Our Relations with Animal Others," 418.
46 Biswas and Mattheis, "Strikingly Educational: A Childist Perspective on Children's Civil Disobedience for Climate Justice," 9.
47 K. Malone, M. Tesar, and S. Arndt, *Theorising Posthuman Childhood Studies* (Heidelberg: Springer, 2020)
48 E. Burman, *Deconstructing Developmental Psychology* (London: Routledge, 2007)
49 See M. Koljonen, "Thinking and Caring Boys Go Vegan: Two European Books That Introduce Vegan Identity to Children," *Bookbird: A Journal of International Children's Literature* 57, no. 3 (2019): 13–22.
50 See https://starfishcollective.se/?page_id=2
51 V. Plumwood, *Environmental Culture: The Ecological Crisis of Reason* (London: Routledge, 2002)

Bibliography

Aavik, K. "Vegan men: Towards Greater Care for (Non)human Others, Earth and Self." In *Men, Masculinities, and Earth: Contending with the (m)Anthropocene*, edited by P.M. Pulé and M. Hultman, 329–350. Cham: Springer International Publishing, 2021.

Adams, C.J., and Gruen, L., eds. *Ecofeminism: Feminist Intersections with Other Animals and the Earth.* 2nd ed. Bloomsbury Publishing USA, 2021.

Allister, M. *Eco-man—New Perspectives on Masculinity and Nature.* Charlottesville: University of Virginia Press, 2004.

Bergmann, I.M. "The Intersection of Animals and Global Sustainability—A Critical Studies Terrain for Better Policies?" *Proceedings* 73, no. 1 (2020): 12. https://doi.org/10.3390/IECA2020-08895.

Biswas, T., and Mattheis, N. "Strikingly Educational: A Childist Perspective on Children's Civil Disobedience for Climate Justice." *Educational Philosophy and Theory* 54, no. 2 (2022): 145–157. https://doi.org/10.1080/00131857.2021.1880390.

Burman, E. *Deconstructing Developmental Psychology*. London: Routledge, 2007.

Cenamor, R., and Brandt, S.L., eds. *Ecomasculinities—Negotiating Male Gender Identity in U.S. Fiction*. New York: Lexington, 2019.

Cole, M., and Stewart, K. *Our Children and Other Animals: The Cultural Construction of Human–animal Relations in Childhood*. London: Routledge, 2014.

Dinker, K.G. "Critical Creatures: Children as Pioneers of Posthuman Pedagogies." *International Journal of Sociology and Social Policy* 41, no. 3/4 (2021): 391–406. https://doi.org/10.1108/IJSSP-10-2020-0464

Dinker, K.G., and Pedersen, H. "Critical Animal Pedagogies: Re-learning Our Relations with Animal Others." In *The Palgrave International Handbook of Alternative Education*, edited by H.E. Lees and N. Noddings, 415–430. London: Palgrave Macmillan UK, 2016.

Dinker, K.G., and Pedersen, H. "Critical Animal Pedagogy: Explorations toward Reflective Practice." In *Education for Total Liberation: Critical Animal Pedagogy and Teaching against Speciesism*, edited by A. Nocella, C. Drew, and A.E. George, 47–64. New York: Peter Lang U.S., 2019.

Gaard, G. "Queering the Climate." In *Men, Masculinities, and Earth: Contending with the (m) Anthropocene*, edited by P.M. Pulé and M. Hultman, 515–536. Cham: Springer International Publishing, 2021.

Holmberg, A., and Alvinius, A. "Children's Protest in Relation to the Climate Emergency: A Qualitative Study on a New Form of Resistance Promoting Political and Social Change." *Childhood* 27, no. 1 (2020): 78–92. https://doi.org/10.1177/0907568219879970.

Hultman, M., and Pulé, P.M. *Ecological Masculinities: Theoretical Foundations and Practical Guidance*. London: Routledge, 2018.

Kheel, M. "The Killing Game: An Ecofeminist Critique of Hunting." *Journal of the Philosophy of Sport* 23, no. 1 (1996): 30–44. https://doi.org/10.1080/00948705.1996.9714529.

Koljonen, M. "Thinking and Caring Boys Go Vegan: Two European Books That Introduce Vegan Identity to Children." *Bookbird: A Journal of International Children's Literature* 57, no. 3 (2019): 13–22.

Lawson, D.F., Stevenson, K.T., Nils Peterson, M., Carrier, S.J., Strnad, R., and Seekamp, E. "Intergenerational Learning: Are Children Key in Spurring Climate Action?" *Global Environmental Change* 53 (2018): 204–208. https://doi.org/10.1016/j.gloenvcha.2018.10.002.

Lee, N. *Childhood and Biopolitics: Climate Change, Life Processes and Human Futures*. Springer, 2013.

MacGregor, S., and Seymour, N., eds. *Men and Nature: Hegemonic Masculinities and Environmental Change*. RCC Perspectives, 2017. https://doi.org/10.5282/rcc/7977.

Malone, K., Tesar, M., and Arndt, S. *Theorising Posthuman Childhood Studies*. Heidelberg: Springer, 2020.

McCright, A.M., and Dunlap, R.E. "Cool Dudes: The Denial of Climate Change among Conservative White Males in the United States." *Global Environmental Change* 21, no. 4 (2011): 1163–1172. https://doi.org/10.1016/j.gloenvcha.2011.06.003.

McGuire, L., Palmer, S.B., and Faber, N.S. "The Development of Speciesism: Age-related Differences in the Moral View of Animals." *Social Psychological and Personality Science* 14, no. 2 (2023): 228–237. https://doi.org/10.1177/19485506221086182.

Massé, F., Givá, N., and Lunstrum, E. "A Feminist Political Ecology of Wildlife Crime: The Gendered Dimensions of a Poaching Economy and Its Impacts in Southern Africa." *Geoforum* 126 (2021): 205–214. https://doi.org/10.1016/j.geoforum.2021.07.031.

Merz-Perez, L., Heide, K.M., and Silverman, I.J. "Childhood Cruelty to Animals and Subsequent Violence against Humans." *International Journal of Offender Therapy and Comparative Criminology* 45, no. 5 (2001): 556–573. https://doi.org/10.1177/0306624X01455003.

Moore, J. *Capitalism in the Web of Life: Ecology and the Accumulation of Capital*. London: Verso Books, 2015.

Morgan, D. "Men, Masculinity and the Process of Sociological Enquiry." In *Doing Feminist Research*, edited by H. Roberts, 83–113. London: Routledge, 1981.

Myers, G. *The Significance of Children and Animals: Social Development and Our Connections to Other Species*. West Lafayette: Purdue University Press, 2007.
Nocella, A., Drew, C., and George, A.E., eds. *Education for Total Liberation: Critical Animal Pedagogy and Teaching against Speciesism*. New York: Peter Lang U.S., 2019.
Oakley, J. "What Can an Animal Liberation Perspective Contribute to Environmental Education?" In *Animals in Environmental Education*, edited by T. Lloro-Bidart and V.S. Banschbach, 19–34. London: Palgrave, 2019.
Pease, B. "From Ecomasculinity to Profeminist Environmentalism: Recreating Men's Relationship with Nature." In *Men, Masculinities, and Earth: Contending with the (m)Anthropocene*, edited by P.M. Pulé and M. Hultman, 537–558. Cham: Springer International Publishing, 2021.
Peeters, A., Ouvrein, G., Dhoest, A., and De Backer, C. "It's Not Just Meat, Mate! The Importance of Gender Differences in Meat Consumption." *Food, Culture and Society* 25, no. 5 (2023): 1193–1214. https://doi.org/10.1080/15528014.2022.2125723.
Plumwood, V. "Integrating Ethical Frameworks for Animals, Humans, and Nature: A Critical Feminist Eco-socialist Analysis." *Ethics & the Environment* 5, no. 2 (2000): 285–322.
Plumwood, V. *Environmental Culture: The Ecological Crisis of Reason*. London: Routledge, 2002.
Pulé, P.M., and Hultman, M., eds. *Men, Masculinities, and Earth: Contending with the (m)Anthropocene*. Springer Nature, 2021.
Ramos-Gay, I., and Alonso-Recarte, C. "Zoomasculinities: At the Intersection between Animals, Animality, and Masculinity." *Men and Masculinities* 23, no. 5 (2020): 807–813. https://doi.org/10.1177/1097184X20965125.
Ruby, M.B., and Heine, S.J. "Meat, Morals, and Masculinity." *Appetite* 56, no. 2 (2011): 447–450. https://doi.org/10.1016/j.appet.2011.01.018.
Siraj-Blatchford, J., Mogharreban, C., and Park, E., eds. *International Research on Education for Sustainable Development in Early Childhood*. Heidelberg: Springer, 2016.
Sobal, J. "Men, Meat, and Marriage: Models of Masculinity." *Food and Foodways* 13, no. 1–2 (2005): 135–158. https://doi.org/10.1080/07409710590915409.
Sollund, R. "Wildlife Crime: A Crime of Hegemonic Masculinity?" *Social Sciences* 9, no. 6 (2020): 93. https://doi.org/10.3390/socsci9060093.
Standen, A., and Wizansky, S. "Hello from Meatpaper." *Meatpaper*, no. 1 (2007).
Sumpter, K.C. "Masculinity and Meat Consumption: An Analysis through the Theoretical Lens of Hegemonic Masculinity and Alternative Masculinity Theories." *Sociology Compass* 9, no. 2 (2015): 104–114.
Twine, R. "Masculinity, Nature, Ecofeminism." (1997). Retrieved from: http://richardtwine.com/ecofem/masc.pdf
Twine, R. "Masculinity, Nature, Ecofeminism, and the 'Anthropo'cene." In *Men, Masculinities, and Earth: Contending with the (m)Anthropocene*, edited by P.M. Pulé and M. Hultman, 117–133. Cham: Springer International Publishing, 2021.
Twine, R. *The Climate Crisis and Other Animals*. Sydney: Sydney University Press, 2024.
Wall, J. "From Childhood Studies to Childism: Reconstructing the Scholarly and Social Imaginations." *Children's Geographies* 20, no. 3 (2022): 257–270. https://doi.org/10.1080/14733285.2019.1668912.
Weber, M. *The Protestant Ethic and the Spirit of Capitalism*. New York: Routledge, 2001 [1904].
Weldemariam, K., Boyd, D., Hirst, N., Sageidet, B.M., Browder, J.K., Grogan, L. et al. "A Critical Analysis of Concepts Associated with Sustainability in Early Childhood Curriculum Frameworks across Five National Contexts." *International Journal of Early Childhood* 49, no. 3 (2017): 333–351. https://doi.org/10.1007/s13158-017-0202-8.
Wilks, M., Caviola, L., Kahane, G., and Bloom, P. "Children prioritize Humans over Animals Less Than Adults Do." *Psychological Science* 32, no. 1 (2021): 27–38. https://doi.org/10.1177/0956797620960398.
Wright, L. *The Vegan Studies Project: Food, Animals, and Gender in the Age of Terror*. Athens: University of Georgia Press, 2015.

30

GENDER AND ANIMALS UNDER BRAHMINICAL PATRIARCHY

Rama Ganesan

The way we think about meat-eating in the west is very different from the way we think about it in the Indian subcontinent and among the south Asian diaspora. A deeper understanding of what meat and meat-eating means in the South Asian context and the politics of meat and beef in India will go a long way towards expanding our framework for animals' rights. In this chapter, I will attempt to outline how South Asian vegetarianism makes us rethink some of our favorite ideas about meat in our diet. The crux of the issue is caste, the inherited, inborn, hierarchically ordered status of a human being. Caste determines the cultural milieu of the human, from what they eat, where they live, to what jobs they do, from birth through death and, ostensibly, beyond. This chapter argues that it is important to ensure that our theorizing is inclusive of eastern understanding of animals in our world.

The form that patriarchy takes under the caste system has been termed "Brahminical patriarchy,"[1] where Brahminism is defined as the root of what has evolved into Hinduism. Understanding Brahmanical patriarchy will cause us to question our notions of how meat is related to masculinity and what masculine power and patriarchy really are. While brute physical strength is deployed within Brahminical patriarchal power, it is matched and often superseded by the authority of sanctity. As contemporary social scientists have noted and demonstrated, social power is split between the martial and the clerical functions. Masculinity is associated, superficially, with physical power and assertion, but I argue that sanctity or divinity is a neglected dimension of patriarchy. Purity or sanctity is vitally important to Brahminical patriarchy. It designates what is "sacred" and what is "polluted." In Brahminical patriarchy, vegetarianism is a proxy for upper-caste purity. Theologian Arvind Theodore muses[2] about the power wielded by upper-caste men who might not be "masculine" in the physical sense:

> men not only derive the power to control and dominate other bodies through their maleness or muscularity but also by their caste affiliations . . . in the process of "othering the other"—in the monitoring and disciplining of bodies of women and Dalits—dominant-caste men construct a "pure" self.

DOI: 10.4324/9781003273400-37

The codes of Brahminical patriarchy[3] comprise explicit rules on the subjugation of women and the lower castes. But also included within its hierarchical ordering are non-human animals. Animals hold great importance in Hindu myth-making. It is fair to say that among existing major religions, Hinduism makes other animal species an integral part of its worship. Yet this elevation of other mythologized nonhumans coexists with the dehumanization of both women and oppressed caste people. How are we to reconcile this with rather simplistic notions of "animalization" as a form of dehumanization? In what follows, I critically evaluate the western idea of animalization and argue that contemporary animal rights theory confounds dehumanization with animalization. Including caste-based dehumanization within our framework expands our understanding of the true place of other animals in human culture.

Ternary Societies and Patriarchal Dimensions

In *Capital and Ideology*,[4] Thomas Piketty describes a fundamental distinction between social groups into three roles, what he terms as the "ternary society": the clergy, the nobility, and the third estate. This tripartite social system is, he argues, a general form that is replicated in many premodern societies around the world, with some minor variations. The clergy is the spiritual leadership, the nobility or warrior class provides protection and security, and the third estate is the common people who do the day-to-day work. Such partitioning of society is seen in premodern Europe, as well as in Hinduism, Islam, and much of the far east. In these societies, the clergy and the nobility generally owned the majority of the land and thus held power over the remaining people.

Piketty notes that the Hindu varna system is more explicitly a quaternary system of Brahmins (priests), Kshatriyas (warriors and nobility), Vaishyas (land-owning farmers and craftspeople), and Shudras (mostly landless serfs). However, even in Europe, the laboring class is in practice divided into two subgroups, the land-owning and landless workers, or free peasants and serfs. So, this is a minor classificatory variation on the fundamental ternary scheme.

The Vedic scriptures describe further categories below the four varnas, called "avarna" or not part of the varna classification. These people lived outside of society, in tribes deprecated as "uncivilized" and "untouchable." Today these people are referred to as Dalit ("broken people") or Adivasi (indigenous people). Each of the varnas was held to be born of a different part of the godhead.

In comparison with other premodern social systems, a unique feature of Hinduism is that social hierarchy and class inequality is based on an ethic of purity: "Among its distinctive features were an emphasis on ritual and dietary purity, strong endogamy within jatis,[5] and separation and exclusion dividing upper from lower classes (untouchables)" (p 319).

Vegetarianism Does Not Challenge Patriarchy

Western analyses of masculinity, especially within the animal rights framework, have observed that eating meat is the cornerstone of masculine power, male dominance, and patriarchy. By far, the work that is most often cited to connect patriarchy and meat-eating is Carol J Adams' 1990 book, *The Sexual Politics of Meat*.[6] This work is clearly not intended to represent Brahminical patriarchy, which, as discussed, is closely associated with vegetarianism. This is not to deny the association of some types of patriarchal power with meat-eating, or to say that Adams' work may not accurately describe some forms of patriarchy. Nevertheless, within the Indian context, the clergy held patriarchal power, and in Brahminical patriarchy, this is expressed through the purity of vegetarianism.

For anyone living under Brahminical patriarchy, statements cited in *The Sexual Politics of Meat* such as "meat is for brain workers" or "meat is a symbol of patriarchy" ring as inaccurate. Through history, Brahmin males are those who entrusted themselves as custodians of the scriptures, to preside over important ceremonies, and to intercede between humans and the gods. Education was the prerogative of Brahmins (and other privileged castes) but was denied to Shudras and Dalits. To this day, Brahmins hold a disproportionate number of positions as knowledge workers—in administration,[7] academia,[8] media,[9] and other professional posts.

Elsewhere in her book, Adams draws attention to the language used to describe women and their bodies—as "pieces of meat." She connects meat-eating culture to violence against women's bodies. But despite the hegemony of vegetarianism, India is still a place of violence against women, and all forms of violence including rape threats are a common way to silence and control women. Given the prevalence of victim-blaming, we have every reason to suspect that the reported rates of rape (88 a day in 2019) are underestimations.[10]

Adams' thesis in *The Sexual Politics of Meat* is that the oppression of women in patriarchy is linked to the oppression of animals and the consumption of meat: that by freeing yourself from one, you can free yourself from the other. It also holds that vegetarianism and veganism challenge patriarchy. From the vantage of Brahminical patriarchy, though, we can see that this thesis is misleading. Vegetarianism does not challenge patriarchy; on the contrary, it is one of the very pillars of Brahminical patriarchy. In *The Sexual Politics of Meat*, Adams is tacitly focusing on white, western patriarchy, whereas the Indian context demonstrates that patriarchal oppression of women can also be virulently oppressive when it is vegetarian.

The idea that meat-eating is an expression of male supremacy is influential in critical animal studies and continues to be referred to widely without the type of nuance that I attempt in this chapter.[11] Recently Adams has engaged in activism in the context of the #MeToo movement, to speak out against the sexism, sexual harassment, and sexual assault of women within the larger animal rights community.[12] What these types of gendered violence make clear is that, contrary to the views expressed in *The Sexual Politics of Meat*, veganism/vegetarianism, even when combined with a concern for animal welfare, is not sufficient to overthrow patriarchal attitudes and behaviors among some male leaders in the animal rights movement.

How Did Vegetarianism Become Associated with Brahminism?

Let us for a moment reflect upon some of the nuances in who does or does not eat meat in India today and who eats what type of meat. As we can see from Table 30.1, by far it is the Brahmins, only 5% of the population, who have the highest reported rates of vegetarianism, at 65%. Other Forward Castes and Other Backward Castes are two categories that consist of the remaining varnas (Kshatriya, Vaishya, and Shudra) of varying privilege. For both sets of groups, the incidence of vegetarianism is about a third. The lowest rates of vegetarianism are seen among the Scheduled Castes and Scheduled Tribes, which are Dalits and Adivasis, respectively.

Table 30.1 IHDS Data 2011–2012[13]

Classification	*% of Population (Source: IHDS)*	*% Vegetarian (Source: Natrajan & Jacob, 2018)*[14]
Forward Caste—Brahmin only	5	65
Other Forward Caste (predominantly Kshatriya and Vaishya)	15	32
Other Backward Caste (predominantly Shudra)	36	31
Scheduled Castes (SC, Dalit)	22	13
Scheduled Tribes (ST, Adivasi)	8	8
Muslims other than SC/ST	11	1
Christians other than SC/ST	2	28

Informants in this survey classified as Jains are less than 1% of the sample, and they are 98% vegetarian.

It is critical to note that even when privileged castes eat meat, as a rule the diet will not include beef. It is traditionally only the Dalits and the Adivasis who are beef-eaters. Outside of the Hindu fold, Muslims and Christians also report higher levels of beef-eating.[15] Overall, and at first glance closer to Adams' thesis, women are more likely to be vegetarian in India than men (in one survey, 29% among women vs 20% among men). Across all caste categories, women report 10 percentage points higher incidence of vegetarianism than men. This indicates that in addition to caste-based practices, gender norms also influence reported incidence of meat-eating. Within a patriarchal system, it is women of the household who are most involved in food preparation and therefore uphold caste divisions (see subsequently).[16] Taken overall, no more than 30% of Indians eat a vegetarian diet, and in practice it might be closer to 20%, given the stigma attached to meat-eating.[17] How did it come to be that only the "uppermost" caste shows a high incidence of vegetarianism? According to historical records, Brahmins did eat all kinds of animal flesh, including beef, but at a certain point in time, they relegated meat-eating to the lower castes and took up vegetarianism. B. R. Ambedkar described the theory that best fits historical facts and current customs, and I draw upon it in the following.[18]

The earliest scriptures of Brahmanism, the religion that would develop into Hinduism, include descriptions of animal sacrifices. Such practices were not always a part of the civilization in the Indus River Valley. Rather, they were introduced by Aryans, nomadic pastoralists who migrated into the subcontinent from the Eurasian Steppe in the second millennium BCE.[19] They instituted a varna system that attested to the superiority of the Aryans over the native inhabitants of the area. Horses and cows, as well as other animals like goats, were killed in ritual sacrifices that were presided over by Aryan priests, who came to be known as Brahmins. Because the cow was central to the pastoralist's way of life, it was the main sacrificial animal. But in the agricultural society of the Indus River Valley, sacrificing a "useful" animal is necessarily expensive, and together with the inherent violence of sacrifices, there rose up a resistance to Brahminism. Buddhism and Jainism were part of the resistance, and these disciplines were preaching against both the needless slaughter as well as the assumed superiority, spiritual and otherwise, of the Brahmins. In order to counter these objections but still remain in a position of power, Brahmins adopted vegetarianism in a stroke of one-upmanship. At the same time, those outcaste communities who lived on the margins of society by scavenging dead cow carcasses became further discriminated against with an additional burden, that of untouchability.[20]

Purity

The concept of purity in moral psychology has received a great deal of attention recently,[21] perhaps because it taps into moral compulsion that is quite distinct from doing/not doing direct physical or emotional harm to an individual. There is a clear evolutionary basis for a mechanism to avoid disease contagion. It is possible that avoiding contact with dead bodies and contaminated food has a strong selection basis. What is not clear, however, is whether all types of moral condemnation can be related to disgust mechanisms. For instance, Brahminism allows and encourages marriages between a man and his sister's daughter, but this would be considered incestuous and immoral in other cultures.

Conceptualizations of purity in the literature have been varied, but the immorality of certain acts can only be explained by appeals to natural order, sanctity, and defilement in a primarily religious context.[22] For our purposes, the four-fold varna system describes a divine natural order, the contravention of which is an immoral act. A patriarchal culture built on the foundation of purity necessarily manifests its concerns through restrictions on women's bodies. Two of the primary ways in which Brahminical patriarchy imposes purity rules on women's bodies are through restrictions on their sexuality and their diet.

Endogamy is the means by which caste is maintained over time.[23] If a group means to split itself off from others, it can only do so by restricting members' marriage partners to within the group. Without enforced endogamy, members may marry individuals from outside of the group, thereby destroying group integrity. This endogamy is maintained primarily by a strict control of women's sexuality. Indeed, the religious texts exhort us to keep a stringent sexual control of women by projecting upon them an innately promiscuous nature. The opening lines spoken by Arjuna in the *Bhagavad Gita* state that the immorality of women leads to mixing of castes, leading to a hell on earth where all family values are destroyed.

> When irreligion is prominent in the family, O Krsna, the women of the family become corrupt, and from the degradation of womanhood, O descendant of Vrsni, comes unwanted progeny.
>
> When there is increase of unwanted population, a hellish situation is created both for the family and for those who destroy the family tradition. In such corrupt families, there is no offering of oblations of food and water to the ancestors.
>
> Due to the evil deeds of the destroyers of family tradition, all kinds of community projects and family welfare activities are devastated.[24]

As the Brahmin caste is most concerned with maintaining purity of its lineage, it does so by imposing the strictest prohibitions of sexual expression among its women. Ambedkar speculates as to the reasons behind the Brahminical practices of child marriages and prohibition of widow remarriage.[25] Early betrothal of children is intended to bond the couple together even before any temptation to stray outside caste can arise. Widows were kept from remarrying through the infamous practice of "sati," or burning widows at the deceased husbands' funeral pyre. Another more mundane method is through social degradation of the widow to enforce a compulsory widowhood upon her. A Brahmin widow undergoes a social death, thus becoming a type of outcaste herself. Similar rules did not apply to the lower-caste widow, who was encouraged to remarry to maintain numbers among laboring castes.

As noted previously, women report a higher incidence of vegetarianism within each caste group.[26] It is no stretch to note that this is related to their position as upholders of patriarchy by embodying its purity rules, not just in sexual relations but also in diet. Thus, although women are more likely than men to be vegetarian in India, this does not undermine the link between vegetarianism and Brahminical patriarchy. As the anthropologist Leela Dube has noted, food is a critical element in the ritual idiom of purity and pollution. Women, who are in charge of procuring meals, are therefore in a prime position to play a critical role in the hierarchical ordering of castes.[27]

Animalization or Dehumanization?

As human societies subjugate both humans and non-human animals, the relationship between these two types of oppressions is an important question for study. In the American context, animal rights activists often compare animal use to human slavery, in not always a sensitive or considerate manner.[28] However, some influential scholars have theorized this relationship not just to promote animal rights but to analyze and understand race relations in the west.

Let's take, for example, Claire Jean Kim's thesis in *Dangerous Crossings*.[29] Kim writes that the conventional wisdom is that people in power have used animalization to "dehumanize" less powerful groups. Kim states that some human groups, specifically people of African descent, Native Americans, and Chinese people, were considered animal-like, animality being integral to the production of human racial difference. She describes these racialized groups as inhabiting the "borderlands" between human (defined as the white race) and the animal. People of African descent were likened to apes, imagined as the "missing link" between humans and other primates. Slavery was said to "civilize" people of African descent by rendering them more tractable, similar to domesticated animals. Native Americans, on the other hand, were likened to wild creatures, the wolves in the forests and the buffalo on the plains, giving white colonists license to appropriate their land. The Chinese laborers entering the U.S. were likened to swarming pestilential beasts or insects, or dragon-like monsters, needing to be exterminated.

To take a slightly different example let us consider Aph and Syl Ko's nuanced ideas in their book, *Aphro-ism*.[30] They suggest that Eurocentric models view "human" and "animal" as a binary and that African Americans are placed somewhere in between them. As they write:

> A human being is fundamentally opposite to animals. With these poles set in place—the former as extreme superiority and the latter as extreme inferiority—those who authored this system placed themselves in the former position and from there divided humanity along a spectrum that went all the way "down" to "the animal.[31]

In *Racism as Zoological Witchcraft*, Aph Ko takes this argument a step further, describing animal oppression as a form of racism: "The zoologo-racial order is the true foundation of white supremacy."[32]

These authors elaborate on their position to suggest that just as some humans are animalized, some animals are "racialized." Pit bull dogs, for example, are raced as Black and discriminated against, similar to racialized humans.[33]

Despite minor variations in these and other similar theories, it is evident that scholars in the field of critical animal studies have held that animalization was an integral part of racial discrimination, especially in the context of white supremacy in the USA.[34] By considering caste divisions, however, we can see quite a different mechanism at work. It is much less obvious that discriminated groups of humans are placed in a "borderland" between humans and other animals. No obvious dimensionality between human and animal existed in ancient India when the varna system was devised. In fact, animals are classified along with humans into different varna groups. For this part of my thesis, I rely on the scholarship of Brian K. Smith in the book *Classifying the Universe* (1994),[35] which examines the Vedas—religious texts composed, broadly, in the millennium before the common era—to ascertain the roots of the Indian classification system.

The Vedic varna divisions operate not just among humans but also among different species of non-human animals, among plants, among elements, and everything else in the physical and non-physical universe. Smith explains, "varna might be regarded as the "root metaphor" or "master narrative" of Vedic thought. The varnas functioned as supercategories which cut across the boundaries of the species or discrete classes. Varna thus describes the classificatory scheme in which components were vertically or hierarchically ordered and interconnected as well as horizontally categorized, linking together components from different species.[36]

In the Vedic literature, each varna group was associated with certain animals. Some humans were identified with certain animals in order to classify those people as altogether uncivilized and outside the realm of proper humanity; other humans were associated with other kinds of animals in order to reinforce the divisions of the ideal civilized society and to extend the varna system to the animal kingdom.

Animal sacrifices were the key sacred rites of Vedic Brahminism. Indeed, according to mythology, the very cosmos was created from a primordial sacrifice performed by the creator. The distinction between domesticated "village" animals and the free-roaming "jungle" animals is very important, and scriptures detail that the most auspicious animals for sacrifice are domesticated animals. The very function of religious rituals and sacrifice was to impose order on the universe, and the sacrifice of jungle animals was believed to lead to chaos. Jungle animals might well be hunted, but ritual sacrifice could only be performed on domesticated animals.

Humans are also considered domesticated animals. The ritual power of the Brahmin associated him with the preeminent sacrificial animals, the goat and the cow. The Kshatriya, the man of physical power and strength, was associated with the horse and the bull. Vaishya are associated with sheep, as it is a fecund animal always found in large herds. Vaishya are also associated with the cow—but here not because of the cow's sacrificial quality but because cows are food animals raised by Vaishya. A lower-class or Shudra animal is the ass, as the "least of animals," a bearer of others. The humans occupying the margins, the outcastes, were associated with undomesticated wild animals. "However, some humans are assimilated to the non-sacrificial, inedible wild animals of the jungle, beyond the pale of "proper" humanity altogether" (p. 255). The wild counterpart of the domesticated (cultured) human is termed a pseudo-human, or barbarian of the jungle. They are "empty shells" of the real human beings.[37]

The Aryans saw themselves as sacred beings and their non-Aryan neighbors not as non-human animals but as a lower order of men, a nonsacred counterpart, a degraded being.[38] We can see that this is quite different from the way western scholars have theorized Black and Native dehumanization, which is predominantly in terms of animalization.

One might dismiss this distinction as merely different ways in which groups are othered and discriminated against, but I suggest that it tells us something fundamental about the ways in which dehumanization works and doesn't work. Recent work is questioning the assumed causal relationship between dehumanization and animalization, that humans are devalued via being compared to other animals.[39] Harriet Over has delineated several challenges for the dehumanization hypothesis, especially via a direct animalization process.[40] Racialized groups may be ascribed negative characteristics by comparison with animals, but at the same time, the oppressor groups may ascribe positive characteristics to themselves in the same manner. For example, even as the Nazis compared Jews to rats or lice, they compared themselves to predators like wolves and eagles. Even the same animal can be used to denote positive characteristics in one context and negative characteristics in another. The snake is often used to signify treachery, but during the American Revolution, the Gadsden flag depicted the American people as a snake to indicate their own defiance against the British.[41] Animal comparisons may be used to substantiate or justify contempt towards marginalized groups, but ultimately animalization may not be the causal factor in their marginalization. The use of animals in othering groups tells us more about the language of metaphors in general, rather than about a specific animalization process.

It is important to note at this juncture that the oppressed group has often chosen to express its resistance by associating with an animal. Dalits themselves have chosen the buffalo as their symbol, as a display of their own endurance, and to claim their history and inheritance.[42] The Indo-European migrants brought with them the cow, but the buffalo is an indigenous domesticated animal that predates the cow on the subcontinent.[43]

A unique implication of the Vedic classification, the remnants of which we still witness today, is that some animals are elevated above some humans. Despite beef being standard fare for Dalits and Muslims, the butchering of cows has been banned in most Indian states over the last decade.[44] While this is ostensibly seen as protecting the cow, it is really about advancing the ruling party's Hindu-nationalist project. This has given rise to vigilante groups who attack Dalits and Muslims on the suspicion of eating or trading in beef. These vigilantes state that a cow's life is more important than that of a human.[45] Victims' families are deterred from pursuing the matter, as they are threatened with counter-charges of cow slaughter.

It is self-evident that subjugation of human groups by an oppressor group is a process of casting the oppressed as inferior relative to the oppressor. But it is highly questionable whether animalization has a direct causal role in human discrimination. Syl Ko has reinterpreted and refined her ideas more recently[46] to state that "to fail to be a Human [that is, white or Brahmin] is not to be a nonhuman animal." There are two reasons for this. First, because the subjugated group cannot embody the failure to be fully human if they are a non-human animal. Second, to fully cement the oppressor's project, the oppressed need to internalize their inferiority, and this can only happen if both oppressor and oppressed are of the same broad category. It is noteworthy that Syl Ko's analysis mirrors that of Brian Smith's interpretation of Vedic classification of outcaste groups as "pseudo-men" or "empty shells of real human beings."[47]

To summarize, my study of the varna system's oppressive tactics indicates that we need to look beyond animalization to understand how human groups are devalued.[48] The same conclusion has been reached independently by other authors,[49] who have surveyed at least two fields of study that are different from the caste system.

Conclusion

This chapter has enumerated a few of the important ways in which critical animal studies has neglected to consider the Indian context and how Brahminical patriarchy challenges common arguments within critical animal studies. Western thought has tended to focus on "masculinity" as a force of physical power. Considering Brahminical patriarchy, I have suggested that we expand our conception of patriarchal power by including sanctity/purity as an important dimension that is separable from physical power. Western thought has tended to view humans and nonhuman animals as occupying either ends of a bipolar continuum. In contrast, the Brahminical varna system categorizes and ranks both human beings and other animals, sometimes elevating certain animals over oppressed human communities. This is not to suggest, at least at this stage, that Brahminical patriarchy follows radically different principles from patriarchy in the west. Rather, these are variations on a theme that reveal the underlying machinations of a controlling power. The clergy has wielded patriarchal power in the west in premodern times, and we can still see vestiges of its reach today.[50] That clerical power became associated with purity and vegetarianism in Brahminical patriarchy is a context-dependent variation. Dehumanization is a process by which those in power have rationalized the subjugation of other humans, and "animalization" is just one example of the ways in which dehumanization has been accomplished. White supremacy exacerbates ill health and shortens life spans of a population by imposing a meat- and dairy-heavy diet, and Brahminism does the same by imposing vegetarianism on a largely undernourished population.[51] In both these cases (and others), groups marginalized by gender and species have been subject to a master narrative that justifies their oppression.

Notes

1 Uma Chakravarti, "Conceptualizing Brahminical Patriarchy in Early India," *Economic and Political Weekly* 28, no. 14 (1993): 579–585.
2 Arvind Theodore, "The Violent Materialization of Caste in Indian Matchmaking and Sairat," *Justice Collective* (2021), retrieved from: https://justicecollective.in/2021/10/19/the-violent-materialization-of-caste-in-indian-matchmaking/, accessed September 2, 2022.
3 Uma Chakravarti, *Gendering Caste: Through a Feminist Lens* (Sage, 2018)
4 Thomas Piketty, *Capital and Ideology* (Cambridge, MA: Harvard University Press, 2020).
5 I will use "caste" to mean varna classifications in this chapter. The four-fold varna system is a scriptural construct, and in practice social categories are lot more complex, numerous, and sometimes more porous. There are a large number of occupational "jatis" or professional guilds, each forming a regional endogamous group. Although the details are intricate, both sociological and genetic studies indicate that we are justified in clustering jati groups around the four varna classifications—especially at either end of the stratification, Brahmins and Dalits, as shown in

recent genomic analysis by Vagheesh Narasimhan, Nick Patterson, Priya Moorjani, Nadin Rohland, Rebecca Bernardos, Mallick Swapan, and Lazaridis Iosif, "The Formation of Human Populations in South and Central Asia," *Science* 365 (2019), retrieved from: https://www.ncbi.nlm.nih.gov/pmc/articles/PMC6822619/.

6 Carol J. Adams, *The Sexual Politics of Meat* (Bloomsbury Academic, 2010)

7 Namit Saxena, "Disproportionate Representation at the Supreme Court: A Perspective Based on Caste and Religion of Judges," *Bar and Bench* (2021), retrieved from: https://www.barandbench.com/columns/disproportionate-representation-supreme-court-caste-and-religion-of-judges.

8 Ajantha Subramanian, *The Caste of Merit: Engineering Education in India* (Cambridge, MA: Harvard University Press, 2019).

9 Haripriya Suresh, "Dominant Castes Occupy Most Leadership Positions in Indian Newsrooms: Report," *The Newsminute* (2019), retrieved from: https://www.thenewsminute.com/article/dominant-castes-occupy-most-leadership-positions-indian-newsrooms-report-106622.

10 Times of India, "India Sees 88 Rape Cases a Day; Conviction Rate below 30%," (2020). https://timesofindia.indiatimes.com/india/india-sees-88-rape-cases-a-day-but-conviction-rate-below-30/articleshow/78526440.cms.

11 Carol J. Adams, "The Oxford Union Debate On 'This House Would Move Beyond Meat'," *caroljadams.com* (2022), retrieved from: https://caroljadams.com/carol-adams-blog/the-oxford-union-debate-on-this-house-would-move-beyond-meat, accessed June 7, 2023.

12 Carol J. Adams, "No Means No: A Call to Boycott ARNC2021," *caroljadams.com* (2021), retrieved from: https://caroljadams.com/carol-adams-blog/p3rwt3z85t6136xt0zsqtgv1eupf70, accessed June 7, 2023.

13 India Human Development Survey, *India Human Development Survey* (2012), retrieved from: https://ihds.umd.edu/social-groups, accessed September 2, 2022.

14 Balmurli Natrajan and Suraj Jacob, "'Provincialising' Vegetarianism," *Economic & Political Weekly* LIII, no. 9 (2018): 54–64.

15 Ibid.

16 Leela Dube, "Seed and Earth: The Symbolism of Biological Reproduction and Sexual Relations of Production in Visibility and Power," in *Essays on Women in Society and Development*, eds. Leela Dube, Eleanor Leacock, and Ardener Shirley (Delhi: Oxford University Press, 1986).

17 Balmurli Natrajan and Suraj Jacob, "'Provincialising' Vegetarianism," *Economic & Political Weekly* LIII, no. 9 (2018): 54–64.

18 Bhimarao Ramji Ambedkar, *The Untouchables: Who They Were and Why They Became Untouchables* (New Delhi: Kalpaz Publications, 1948); Alex George and S Anand, *Beef, Brahmins & Broken Men, an Annotated Critical Selection from the Untouchables by B. R. Ambedkar* (New York: Columbia University Press, 2020).

19 Tony Joseph, *Early Indians* (New Delhi: Juggernaut Books, 2018); David Reich, *Who We Are and How We Got Here: Ancient Dna and the New Science of the Human Past* (New York: Pantheon Books, 2018).

20 Kancha Ilaiah Shepherd, "Freedom to Eat," *Caravan Magazine* (2019).

21 Jonathan Haidt, *The Righteous Mind* (Harlow, England: Penguin Books, 2013); Oliver Scott Curry, "What's Wrong with Moral Foundations Theory, and How to Get Moral Psychology Right," *Behavioral Scientist* (2019), retrieved from: https://behavioralscientist.org/whats-wrong-with-moral-foundations-theory-and-how-to-get-moral-psychology-right/, accessed September 2, 2022.

22 Kurt Gray, Nicholas DiMaggio, Chelsea Schein, and Frank Kachanoff, "The Problem of Purity in Moral Psychology," *Personality and Social Psychology Review* (2022), retrieved from: https://psyarxiv.com/vfyut/, accessed September 2, 2022.

23 Bhimarao Ramji Ambedkar, "Castes in India: Their Mechanism, Genesis and Development," *Columbia.edu* (1916), retrieved from: http://www.columbia.edu/itc/mealac/pritchett/00ambedkar/txt_ambedkar_castes.html, accessed September 2, 2022.

24 Bhaktivedanta Swami Prabhupada, *Bhagavad Gita, Chapter 1: Observing the Armies on the Battlefield of Kuruksetra* (trans. 1968), retrieved from: https://asitis.com/1, accessed September 2, 2022.

25 Bhimarao Ramji Ambedkar, "Castes in India: Their Mechanism, Genesis and Development," *Columbia.edu* (1916), retrieved from: http://www.columbia.edu/itc/mealac/pritchett/00ambedkar/txt_ambedkar_castes.html, accessed September 2, 2022.

26 Balmurli Natrajan and Suraj Jacob, "'Provincialising' Vegetarianism," *Economic & Political Weekly* LIII, no. 9 (2018): 54–64.
27 Leela Dube, "Seed and Earth: The Symbolism of Biological Reproduction and Sexual Relations of Production in Visibility and Power," in *Essays on Women in Society and Development*, eds. Leela Dube, Eleanor Leacock, and Ardener Shirley (Delhi: Oxford University Press, 1986).
28 A. Breeze Harper, *Sistah Vegan* (New York: Lantern Books ed. 2010).
29 Claire Jean Kim, *Dangerous Crossings* (New York: Cambridge University Press, 2015).
30 Aph Ko and Syl Ko, *Aphro-ism* (New York: Lantern Books, 2017).
31 Ibid., 67.
32 Aph Ko, *Racism as Zoological Witchcraft* (New York: Lantern Books, 2019), 58.
33 Kim, *Dangerous Crossings;* see also Claire Jean Kim, "Murder and Mattering in Harambe's House," *Politics and Animals* 3 (2017) for further elucidation of her ideas on race and animalization (C.K. Kim 2017); Ko and Ko, *Aphro-ism*
34 The animalization theory for race is often misinterpreted in lay advocacy and in the media to suggest that dismantling the human–animal boundary will automatically annihilate racial discrimination. Street advocates claim that raising children to respect animals will eradicate racism (Earthling Ed, 2016 https://youtu.be/ZXu74W8A8BQ), and newspaper headlines claim going vegan will combat racial violence (The Cornell Daily Sun 2016, retrieved from: https://cornellsun.com/2018/02/23/veganism-a-way-to-combat-race-based-violence-says-columbia-lecturer/). These are simplistic arguments not borne out by observations; see Donaldson and Kymlicka, 2014).
35 Brian Smith, *Classifying the Universe* (New York, NY: Oxford University Press, 1994); See also Embodiment of Dharma in Animals by Andrea Guttierrez in The Oxford History of Hinduism: Hindu Law: A New History of Dharmasastra (2017)
36 Smith, *Classifying the Universe,* 12–13.
37 Ibid.
38 Ibid.
39 Harriet Over, "Seven Challenges for the Dehumanization Hypothesis," *Perspectives on Psychological Science* 16, no. 1 (2021): 3–13.
40 Ibid.
41 Ibid.
42 Pramod Ranjan, ed., *Mahishasur: The People's Hero* (New Delhi: Forward Press Books, 2017)
43 Kancha Ilaiah Shepherd in Buffalo Nationalism (2004) further speculated the light color of cows in contrast with the dark buffalo further plays into the divide, based on the distinction between light-skinned Indo-Aryan settlers and darker-skinned native Dravidians.
44 Rama Ganesan, "The Humane Myth of Ahimsa," *Vegan Feminist Network* (2021). retrieved: https://veganfeministnetwork.com/the-humane-myth-of-ahimsa/.
45 Surinder S. Jodhka and Dhar Murali, "Cow, Caste and Communal Politics: Dalit Killings in Jhajjar," *Economic and Political Weekly* 38, no. 3 (2003): 174–176.
46 Syl Ko, "An Interview with Syl Ko," *Tierautonomie* 6, no. 1 (2019): 1–22. retrieved: https://simorgh.de/tierautonomie/JG6_2019_1.pdf.
47 Smith, *Classifying the Universe.*
48 I am not overlooking the idea that in a fundamental way humans hold all non-human animals in an altogether different category of "lesser" being. That is, the varna system categorization might be superimposed on a more essential distinction where non-human animals are held beneath humans of all castes. Even though cow vigilantes state cows are more important than (some) humans, we know that the value of the cow is merely symbolic. Cows are still mistreated in the dairy industry, in slaughter for export, allowed to die on the streets, and so on, which belies the claim that cows' lives are valuable. Exploring the idea of superimposed categories is, however, beyond the scope of this chapter.
49 Over, "Seven Challenges for the Dehumanization Hypothesis."; Ko, "An Interview with Syl Ko."
50 Piketty, *Capital and Ideology.*
51 Kajori Banerjee, "Disparity in Childhood Stunting in India: Relative Importance of Community-level Nutrition and Sanitary Practices," *PLOS ONE* (2020), retrieved from: https://www.ncbi.nlm.nih.gov/pmc/articles/PMC7462311/.

Bibliography

Adams, Carol J. *The Sexual Politics of Meat*. Bloomsbury Academic, 2010.
Adams, Carol J. "No Means No: A Call to Boycott ARNC2021." *caroljadams.com*, April 20, 2021. Retrieved from: https://caroljadams.com/carol-adams-blog/p3rwt3z85t6136xt0zsqtgv1eupf70.
Adams, Carol J. "The Oxford Union Debate on 'This House Would Move Beyond Meat'." *caroljadams.com*, December 22, 2022. Retrieved from: https://caroljadams.com/carol-adams-blog/the-oxford-union-debate-on-this-house-would-move-beyond-meat.
Ambedkar, Bhimarao Ramji. "Castes in India: Their Mechanism, Genesis and Development." *Columbia.edu*, May 9, 1916. Retrieved from: http://www.columbia.edu/itc/mealac/pritchett/00ambedkar/txt_ambedkar_castes.html.
Ambedkar, Bhimarao Ramji. *The Untouchables: Who They Were and Why They Became Untouchables*. Gyran Books reprint, 2017 (1948).
Banerjee, Kajori. "Disparity in Childhood Stunting in India: Relative Importance of Community-level Nutrition and Sanitary Practices." *PLOS ONE* (2020). Retrieved from: https://www.ncbi.nlm.nih.gov/pmc/articles/PMC7462311/.
Chakravarti, Uma. "Conceptualizing Brahminical Patriarchy in Early India." *Economic and Political Weekly* 28, no. 14 (1993): 579–585.
Chakravarti, Uma. *Gendering Caste: through a Feminist Lens*. Sage, 2018.
Curry, Oliver Scott. "What's Wrong with Moral Foundations Theory, and How to get Moral Psychology Right." *Behavioral Scientist*, March 26, 2019. Retrieved from: https://behavioralscientist.org/whats-wrong-with-moral-foundations-theory-and-how-to-get-moral-psychology-right/.
Dube, Leela. "Seed and Earth: The Symbolism of Biological Reproduction and Sexual Relations of Production in Visibility and Power." In *Essays on Women In Society and Development*, edited by Leela in Dube, Eleanor Leacock and Ardener Shirley. Delhi: Oxford University Press, 1986.
Ganesan, Rama. "The Humane Myth of Ahimsa." *Vegan Feminist Network*, September 26, 2021. Retrieved from: https://veganfeministnetwork.com/the-humane-myth-of-ahimsa/.
George, Alex, and Anand, S. *Beef, Brahmins & Broken Men, An Annotated Critical Selection from the Untouchables by B. R. Ambedkar*. Columbia University Press, 2020.
Gray, Kurt, DiMaggio, Nicholas, Schein, Chelsea, and Kachanoff, Frank. "The Problem of Purity in Moral Psychology." *Personality and Social Psychology Review* (2022). Retrieved from: https://psyarxiv.com/vfyut/.
Haidt, Jonathan. *The Righteous Mind*. Harlow, England: Penguin Books, 2013.
Harper, A. Breeze, ed. *Sistah Vegan*. New York: Lantern Books, 2010.
India Human Development Survey. *India Human Development Survey* (2012). Retrieved from: https://ihds.umd.edu/social-groups, accessed September 2, 2022.
Jodhka, Surinder S., and Dhar, Murali. "Cow, Caste and Communal Politics: Dalit Killings in Jhajjar." *Economic and Political Weekly* 38, no. 3 (2003): 174–176.
Joseph, Tony. *Early Indians*. New Delhi: Juggernaut Books, 2018.
Kim, Claire Jean. *Dangerous Crossings*. New York: Cambridge University Press, 2015.
Kim, Claire Kim. "Murder and Mattering in Harambe's House." *Politics and Animals* 3 (2017).
Ko, Aph. *Racism as Zoological Witchcraft*. New York: Lantern Books, 2019.
Ko, Aph, and Ko, Syl. *Aphro-ism*. New York: Lantern Books, 2017.
Ko, Syl. "An Interview with Syl Ko." *Tierautonomie* 6, no. 1 (2019): 1–22. Retrieved from: https://simorgh.de/tierautonomie/JG6_2019_1.pdf.
Narasimhan, Vagheesh, Patterson, Nick, Moorjani, Priya, Rohland, Nadin, Bernardos, Rebecca, Swapan, Mallick, and Iosif, Lazaridis. "The Formation of Human Populations in South and Central Asia." *Science* 365, no. 6457 (2019). Retrieved from: https://www.ncbi.nlm.nih.gov/pmc/articles/PMC6822619/.
Natrajan, Balmurli, and Jacob, Suraj. "'Provincialising' Vegetarianism." *Economic & Political Weekly* LIII, no. 9 (2018): 54–64.
Over, Harriet. "Seven Challenges for the Dehumanization Hypothesis." *Perspectives on Psychological Science* 16, no. 1 (2021): 3–13.
Piketty, Thomas. *Capital and Ideology*. Cambridge, MA: Harvard University Press, 2020.
Prabhupada, Bhaktivedanta Swami, trans. *Bhagavad Gita, Chapter 1: Observing the Armies on the Battlefield of Kuruksetra*. 1968. Retrieved from: https://asitis.com/1.

Ranjan, Pramod, ed. *Mahishasur: The People's Hero*. New Delhi: Forward Press Books, 2017.
Reich, David. *Who We Are and How We Got Here: Ancient DNA and the New Science of the Human Past*. New York: Pantheon Books, 2018.
Saxena, Namit. "Disproportionate Representation at the Supreme Court: A Perspective Based on Caste and Religion of Judges." *Bar and Bench*, May 22, 2021. Retrieved from: https://www.barandbench.com/columns/disproportionate-representation-supreme-court-caste-and-religion-of-judges.
Shepherd, Kancha Ilaiah. "Freedom to Eat." *Caravan Magazine*, November 1, 2019.
Smith, Brian. *Classifying the Universe*. New York, NY: Oxford University Press, 1994.
Subramanian, Ajantha. *The Caste of Merit: Engineering Education in India*. Cambridge, MA: Harvard University Press, 2019.
Suresh, Haripriya. "Dominant Castes Occupy Most Leadership Positions in Indian Newsrooms: Report." *The Newsminute*, August 3, 2019. Retrieved from: https://www.thenewsminute.com/article/dominant-castes-occupy-most-leadership-positions-indian-newsrooms-report-106622.
Theodore, Arvind. "The Violent Materialization of Caste in Indian Matchmaking and Sairat." *Justice Collective*, October 19, 2021. Retrieved from: https://justicecollective.in/2021/10/19/the-violent-materialization-of-caste-in-indian-matchmaking/.
Times of India. "India Sees 88 Rape Cases a Day; Conviction Rate below 30%." October 7, 2020. Retrieved from: https://timesofindia.indiatimes.com/india/india-sees-88-rape-cases-a-day-but-conviction-rate-below-30/articleshow/78526440.cms.

PART VII

Gender and Animals in Folklore and Fiction

31

FROM FOLKLORE TO FACTORY FARMS

Gender, Sex and Chickenkind

Annie Potts

For centuries, human assumptions about masculinity and femininity dominated depictions of chickenkind. Across different cultures, roosters (or cockerels) were portrayed as brave and vigilant leaders of their flocks and hens as the epitome of maternal devotion and domesticity, willing to sacrifice their own lives to save their chicks. Today, despite billions of chickens being killed globally for various forms of consumption, most people—and especially those living in Western societies—seldom think about these birds, and if they do, they are more likely to be trivialized as stupid or cowardly. This chapter explores how chickens have shifted from being respected and even feared in pre-industrial human cultures to becoming the most abused and disregarded land creatures on the planet today. Emphasis is placed on how stereotypes about gender, as well as efforts to exploit the reproductive bodies of chickens, have aided the devastating downfall of these birds.

I begin by discussing how traditional beliefs about 'proper' masculine and feminine behaviours have influenced the symbolism of roosters and hens in mythological depictions of chickens. I will also explore how chickens who transgress human ideals of gender may appear as ominous or unlikeable figures in folklore, while real-life birds who breach such norms—such as hens who crow or roosters who sit on eggs—have faced lethal consequences. I move on to examine two examples of extreme exploitation occurring today related to the gendering and sexing of roosters and hens, namely the practice of cockfighting as a masculinist competition between men and the business of egg farming as the ultimate manipulation of hens for human profit and consumption.

Gendering Chickens in Mythology and Folklore

While it is not possible to cover all cultural variations of chickens' gendered depictions in this chapter, I have selected a small range from across the globe. One positive quality of roosters shared across cultures and throughout history relates to their dutiful vigilance, particularly their reliable pronouncement of a new day at the first sign of the sun's rays. In ancient Iran (2000–700 BC), for example, the rooster was actually considered the most sacred of birds due to his relationship with the Persian sun god, Mithras, his crowing at dawn dispelling the demonic spirits of night-time. Fourteenth-century BC Egyptian pharaoh

DOI: 10.4324/9781003273400-39

Akhenaten also referred to the wonder of roosters in his 'Hymn to the Sun', while the Jewish Talmud praises God for giving "understanding to the cock to distinguish between night and day", and the Hadith states: "When you hear a cock crow, ask for Allah's blessings for he has seen an angel".[1]

The ancient Greek myth of Alectryon offers an explanation for how the rooster came to noisily welcome each day. Charged with guarding the lovers Ares and Aphrodite as they conducted their illicit affair, Alectryon fell asleep at his post, missing the rising of Helios the sun and allowing their discovery. For his incompetence, Alectryon is turned into a rooster so he might forevermore remember to proclaim the coming of dawn. For early Christians the crowing of roosters at sunrise was a reminder of the Resurrection and Christ's triumph over darkness, sin and evil. This is one reason why roosters have been common symbols on church steeples.[2]

Representations of roosters as solar entities can be contrasted with other religious or mystical understandings linking these birds to darkness and the moon. In Ancient Greece, for instance, black roosters were special familiars of Hecate, the powerful goddess of the night, magic and spell, while on Rapa Nui (Easter Island), white roosters were feared for their links to sorcery.[3]

While traditional masculinist portrayals of roosters tend to focus on their abilities to watch, guard and defend, gendered assumptions about hens relate to ideals about maternity and domesticity. Because such qualities are devalued in patriarchal societies, hens are less likely to be viewed as heroic or extraordinary—rather, they represent the mundane. However, one creation narrative from West Africa exalts the life-giving qualities of a white hen who, having been placed on a sandy bank by the young god Obatala, commences to carve out the world's mountains and valleys with her scritching.[4] Sixteenth-century Italian ornithologist Ulisse Aldrovandi wrote in the first treatise on all matters relating to chickens that he greatly admired hens' "noble example of love of their offspring", especially because they often "offer [the chicks] food they have collected and neglect [their] own hunger".[5] Thus, in Aldrovandi's opinion at least, hens were ideal mothers because they happily made sacrifices for their young.

In patriarchal societies, 'femininity' may be viewed as mysterious, dangerous, untrustworthy and ambiguous. When women and chickens appear in close relationships or as hybrid figures in mythology or customary tales, there may be negative or even sinister connotations. The Slavic female entity known as Baba Yaga is an ambiguous character, sometimes depicted as a malevolent witch who rides a mortar and pestle (or broomstick) and devours children and sometimes as a nurturing feminine figure who loves animals and the home.[6] She is so complex because of her different interpretations that one expert proclaims Baba Yaga "the most feminist character in folklore",[7] largely because—whether evil or kind—she eschews the conventions of marriage, motherhood and family duties by living alone in the forest, doing exactly what she wants.

Baba Yaga's windowless house is perched on chickens' legs, which dance about the forest and obey the command to stand up and turn around when access to a secret door is required. Hence there is a direct association between Baba Yaga and chickenkind. It has been speculated that the house's chicken legs link Baba Yaga to Slavic tropes of domesticity and healing related to hens and chicks. For instance, in Russian tradition, women who are pregnant may be warned away from collecting eggs laid in henhouses for fear their heir babies develop 'hen sickness' (constant crying). If children become affected by this malady, they are taken to the chicken coop, which is asked to restore health to the ailing child "by giving her or him a 'human life' instead of a 'chicken life'."[8]

Another fascinating figure of Slavic origin is the cheeky house-spirit known as Kikimora, who resembles a peasant woman with hen's legs and feet and a long beak for her nose. Kikimora links women and chickens to domesticity in her obsession with inspecting family homes to check they are clean. If she is disappointed with the mess or filth of a house, she ensures its inhabitants suffer from insomnia by whistling in their ears or tickling their feet throughout the night. If she is pleased with what she finds, Kikimora retires to the hen-house.[9] For protection from Kikimora, Eastern European hens and roosters have their own champion, the chicken god Kirinyi Bog.[10]

Crowing Hens and Laying Roosters

Traditionally, as discussed, roosters have been masculinized as vigilant defenders of their flocks, while hens have been feminized as 'ideal mothers'. When chickens have been observed transgressing 'correct' gender roles and behaviours, there have been grave consequences for the actual birds. In this section, gendered mythologies, folklore and superstitions lead to lethal outcomes for roosters accused of laying and hens heard crowing.

In her 1913 article on the folklore of Oxfordshire from 1840 until 1900, Angelina Parker discusses some of the customs and superstitions she learned of while living in the English village of Long Handborough. One is the practice of farmers' wives declaring "A whistling woman and a crowing hen are neither good for God nor men" as they chop off the heads of hens heard to be crowing. Such incidents were considered very unlucky, but interestingly, according to Parker, the compulsory destruction of the transgressive hen was not so much about reversing any ill luck as it was a punishment "to protest against the hen's usurping the privileges of the cock".[11]

Seventy years later, in her analysis of intersex figures in folklore, Ursula Mittwoch exposes a similar yet more sinister proverb historically recited in German villages: "Den Mädchen die da pfeifen und den Hühnern, die da krähen, denen muss man bei Zeiten den Hals umdrehen" ('whistling maids and crowing hens should have their necks wrung without delay').[12] Singing or whistling maids were thought to be witches whose tunes or whistles were forms of communication with the devil, while crowing hens foretold unexpected death.[13] The crowing hen as a bad omen goes back centuries, a reference occurring, for instance, in Latin playwright Terence's comedy *Phormio* from 161 BC.[14] In nineteenth-century Bohemia (Czechoslovakia), the colour of the crowing hen's plumage provided more detail on the type of ill luck to be expected: a white crowing hen meant a death in the family, while red and black crowing hens foretold of fire or theft, respectively. To prevent the manifestation of these misfortunes, an accused hen's feet were broken, and she was thrown alive into water.[15]

I can vouch, after living with chickens for over thirty years, that it is actually quite common for hens to crow, especially if they reside in rooster-less flocks. Rather than a scary breach of gender normativity, hens who crow are cleverly attempting to resettle the pecking order, usually after some change has occurred. It is very important for chickens to recognize their own and others' positions within their carefully negotiated social structures and to behave accordingly. Knowing one's place in a flock is 'chicken manners'.

While crowing hens have been feared harbingers of misfortune, roosters who disrupt culturally determined expectations of correct masculine behaviour have also been vilified and punished. In the Middle Ages in Europe it was believed that certain roosters produced eggs, a notion possibly derived from people finding tiny yolkless eggs which they presumed belonged to male chickens.[16] Such eggs were apparently coveted by sorcerers who credited them with special powers for 'black' magic. Roosters accused of laying eggs were sometimes brought to trial, especially in middle parts of Europe, where animals could be persecuted for conspiring with witches. As folklorist L.F. Newman points out, many ordinary back-yard cockerels were condemned as heretics and burnt at the stake for the unnatural practice of laying an egg (hens who crowed also met this fate). At one such trial in 1474, the magistrate of Basle in Switzerland sentenced to death a rooster charged with laying an egg, and the historical record of this event contends that, upon cutting open the rooster, the executioner found three more.[17]

Two and a half centuries later, in 1730, a similar trial and execution took place, again in Switzerland—this time an elderly rooster was accused of laying an egg containing a much-feared monster.[18] Laying roosters transgressed nature, specifically the God-given rightful roles of males and females, but people of the time were also terrified of what they believed hatched from these eggs if they were incubated—a creature know as a cockatrice. The unusual eggs containing cockatrices were believed to come from very old roosters and to have been incubated over many years by serpents or toads.[19] A similar monster known as a basilisk existed in Ancient Greece; however, he emerged from eggs laid by toads or snakes and incubated by roosters. The cockatrice resembled a snake with a rooster's comb, while the more ancient basilisk retained the serpent's form but was also depicted standing on chicken's legs and wearing a crown, for he was also known as the King of Reptiles. The basilisk was dangerous because not just his venom, but also his smell, glance and bite were considered poisonous: merely glancing at a basilisk resulted in death. For protection, travellers were advised to carry with them 'proud' cockerels in case they met a basilisk on the road. This was because basilisks (hatched from roosters' eggs) could be killed if confronted by correct masculinity in the form of a crowing rooster.[20]

In East and Central Africa there also exist tales of creatures half-snake, half-rooster. However gender conformity, rather than gender transgression, characterizes these legendary serpents. Both male and female crowing crested cobras are said to look like snakes and sport bright red comb-like crests, with the male also possessing wattles like a chicken. They don't pose much danger to people, as they almost exclusively live off maggots contained in carrion. In accordance with their gendered conventionality, the male is said "to crow like a rooster when advancing on his mate, while the female cobra responds in turn by clucking seductively".[21]

Extreme Exploitation of Roosters and Hens

As discussed, chickens have historically held prominent places across different cultural traditions—as vigilant guardians and exemplary parents, and even as helpers in the world's creation. Affirmative representations of chickens tend to be associated with an admiration for the way they fulfil gendered expectations; negative or fearful narratives may derive from the observation that such expectations have been transgressed or reversed, as in the cases of crowing hens and basilisks incubated from roosters' eggs. In this section, I turn to examine two tragic ways in which roosters and hens are currently exploited based on ideas about gender and/or knowledge of their reproductive bodies. The first example, cockfighting, has a very long history, and, while it is now illegal in many countries, it is still considered a legitimate traditional 'sport' for men in others. The second example focuses on the commercial egg industry, which, through the increasing manipulation and control of hens' reproductive bodies over the past century or so, has not only reduced the hen, once admired for her dedication to motherhood, to a 'machine' of capitalism but has also removed any natural connection between mother hens and their eggs and chicks. In the process of turning hens into 'egg machines', humans have become effectively distanced from any meaningful connection to chickenkind's prior narratives and symbolism, reverent or monstrous.

Cockfighting and Heteronormative Masculine Performance

In those regions of the world where chickens were first domesticated, such as South-east Asia, India, China and Iran, roosters were originally valued as 'natural alarm clocks' who reliably greeted the new day, and as 'natural alarm systems' who alerted people to disturbances or intruders. They were also important symbolically in religious discourse, and known for their magical or mystical properties in fortune-telling and other forms of divination; this latter role is thought to have influenced the beginning of their use by humans in competitive bird fighting around 3,500 years ago. While cockfighting therefore has ancient origins, spreading, as chickens did through domestication processes, from South-east Asia into Europe and the Americas, it is no longer legal in many countries (although it continues to be practised underground) and is of particular significance in the Philippines (where it is known as sabong), Bali and Puerto Rico.[22]

Enforced cockfighting involves two carefully bred and trained roosters typically fighting each other to the death while imprisoned in a circular pit surrounded by men who have gambled on a winner. In nature, roosters confront each other to establish a pecking order and territorial boundaries for them and their flocks. They rarely kill each other during these stand-offs, since it quickly becomes apparent who the dominant bird is once they start their combat (comb length also provides clues for roosters regarding hierarchy and status). This allows the other rooster to relinquish the fight and leave the situation before any serious injury.[23] In competitive cockfighting, there is no escape for either bird. Each rooster is fitted with steel spurs (called gaffs or slashers) comprising sharp curving blades of 1–3 inches in length; in the Philippines sharp knives (tari) are attached to a rooster's left foot. Roosters' combs are cut down prior to fighting so neither 'competitor' can 'read' details from the other's comb as they would do in a natural environment.[24]

Cockfighting is an obvious example of heteronormative macho masculinity being imposed upon male birds both symbolically and physically. The human owners of fighting cockerels have a metonymic relationship with their birds: a victorious rooster confirms his owner's masculinity, while the losing bird emasculates his owner (men who compete through their roosters are known as 'cockers'). In his exploration of cockfighting, folklorist Alan Dundes provided some poignant insights of this intensively masculinist 'sport', albeit from a Eurocentric psychoanalytic approach.[25] He contends that several aspects of cockfighting reveal its function as phallic combat between men. One relates to the phallic association of roosters with 'cocks' (penises), which he argues can also be observed in the positioning of roosters on patriarchal (and penile) architecture, such as the steeples of churches. Dundes stays on a religious theme when he claims that the Christian idea of 'resurrection'—a miraculous rising from the dead—is expressed in routine practices of cockfighting undertaken by cockers to revive or 'resurrect' (or, as Dundes puts it, 're-erect') their birds when they appear to be losing a fight or near death. Such 'revivals' involve cockers inserting hot chili or red pepper into the cloaca (anuses) of the injured or weakened roosters, which provides such a painful shock to the birds that they are immediately alert again. In some cases, the cockers blow their own breath down the throats or up the cloaca of exhausted or fatally injured roosters. Another phallic ritual identified by Dundes as part of cockfighting relates to the tendency of cockers to bounce their birds while rubbing and stroking their necks, an activity resembling men's masturbation.[26] Indeed, the American expression "to choke the chicken" is a standard euphemism for penile masturbation.[27]

Since cockfighting is a violent form of heterosexual masculine competition, homophobia and misogyny are also evident in its discourse and practices. For example, in Malay, two roosters are said to be betrothed prior to their fight, which determines who will be valorized as the groom (the dominant one) and who will be denigrated as the bride (the deceased weaker bird). In Brazil, cockfighters shame the dying bird by jeering "the mother's blood is showing", and in Mexico, the term gallo-gallina (rooster-hen) is an expletive implying someone is cowardly or homosexual.[28]

Hens, Eggs and Chicks as Commodities

The well-known story of the little red hen follows a hard-working hen as she discovers a wheat seed and decides to grow it for the benefit of her chicks and others living on a farm. In one version of this story, the little red hen asks a cat, a pig and a rat to help her plant and harvest the seed. They all refuse. When she later asks them to help her bake the wheat flour into bread, they refuse again. However, when the hen, out of habit, asks these animals to help her eat the bread she has sowed, harvested and baked herself, the pig, cat and rat enthusiastically agree to share it with her. At this point in the story, the little red hen comes to her senses and refuses them even a crumb. Instead she and her chicks enjoy the food she alone prepared.[29] The accepted lessons of this children's story include the benefits of working hard, helping out and sharing the load. In her reading of this story, cultural theorist Susan Merill Squier suggests that the figure of the little red hen also symbolizes the historical place of women in agriculture and stresses how chickens are liminal creatures, placed between "the 'egg money'" of the wife on traditional farms and the massive profits of the new industrialized world of intensive meat and egg production.[30]

In contemporary western societies, women have more or less lost the more personalized experiences of backyard flocks of egg-laying hens referred to by Squier, as well as the small income these provided, for reasons elaborated upon in the following. Women and chickens remain connected to each other, however, through newer disrespectful tropes and images associated with agriculture and its links to patriarchal anthropocentrism. Slang sayings such as 'chicken ranch' (a brothel) and 'chicken dinner' (a woman deemed attractive by men) point to such common disparagement of women and hens.[31] Women and chickens (hens predominantly) are also objectified and consumed in popular cultural texts, including meat, dairy and egg advertising. As Carol J. Adams has shown in *The Pornography of Meat* (2003), promotions for animal products tend to target a heteronormative masculinist audience, with common usage of salacious images and slogans that effectively animalize women and feminize/sexualize hens.[32]

I found a particularly disturbing example of this on-line in a recent visual advertisement for chicken seasoning, which depicts the naked and bronzed carcass of a cooked hen posed on a plate in an unmistakably pornographic posture. Her bottom is in the air and she is gazing straight at the viewer in a suggestive way complicit with heterosexual porn: her head is intact (not usually the case when people purchase chicken meat) and human eyes have replaced hens' eyes, complete with elongated eyelashes that have been photoshopped onto her face. A similarly shocking KFC advertisement from the United States promotes the 'Hillary' special, which comprises "2 fat thighs with small breasts and a left wing",[33] a blatantly misogynistic animalization of a prominent woman politician. These two examples show how meat culture[34] both shapes and is shaped by the objectification and consumption of women and chickens: in the first case, the hen's carcass has retained her head and is posed in a pornographic way in order to appear more sexualized; in the second case, Hillary Clinton has been dismembered in order to be consumed like any one of the billions of birds slaughtered yearly by KFC.

The hen referred to in both the KFC and chicken seasoning ads hatches into and perishes within the chicken meat industry. Many people remain unaware of the distinction between the chicks killed for their meat and the 'layer hens' exploited for their eggs. Each type of bird has its own history of increasing manipulation and complex exploitation resulting from developments in agriscience and technology over the past century or so, but it is beyond the scope of this chapter to provide a detailed overview.[35] The 'chicks' mentioned previously are just that: they are male and female baby birds housed post-hatching in cramped sheds of up to 50,000 birds until they are swept up during the night by massive collecting machines and taken to slaughter at around 4–6 weeks of age. By the time they are killed for consumption they are still cheeping for their mothers and sporting baby feathers. Meat chicks (also known as broilers) are unhealthy and disabled birds, often unable to stand by the time they are slaughtered since their skeletons can no longer hold the weight of rapidly developing muscle on their breasts and thighs, an outcome of 'advances' in selective breeding, 'chicken nutrition' and tactics to ensure these birds remain inactive and grow bigger faster. Even worse, because meat chicks are bred to become so big in their brief lives, they are more readily trivialized or disparaged in any representations of them, since they appear to look 'fat' and 'greedy'.[36]

While the hens of the egg industry may live longer—sometimes up until eighteen months before they are considered 'spent' and sent to slaughter—their lives are every bit as miserable. In nature, a wild hen may live well past ten years and lay a couple of clutches of eggs each spring or summer; however, the reproductive systems of commercial layer hens have been engineered to such an extent that they have no option but to produce almost an egg a day (around 300 per year). Hens who are rescued from battery farms only live on average three years, as they develop diseases of their reproductive organs, including prolapses, cancers and egg peritonitis, a very common terminal illness which occurs when a hen is no longer able, due to a heavily fatigued reproductive system, to push out an egg. The egg breaks inside her and she develops a kind of sepsis from which she cannot recover.[37]

In order to create such diseased and disabled birds[38] and profit from their ova, every biological feature of hens was first investigated and experimented on. The qualities pre-industrial societies had admired in hens were no longer of value; instead, ways had to be found to foil the natural reproductive lives of hens. For example, the seasonal laying of hens—in response to longer or shorter days—limited profit, so egg farmers introduced artificial lighting in sheds to confuse the birds into producing eggs all the time.[39] The invention of the incubator and the colony brooder meant that chicks could be separated from their mothers prior to hatching, while the confinement of thousands of hens in cages and automated sheds led to greater control of birds and less need for human labour. Meanwhile scientists worked to create more lucrative breeds of egg-laying hens, the commercial sensitivity of such endeavours requiring breeds to be numbered and birds to become intellectual property.[40]

While they will never experience hatching and raising chicks, hens on battery farms still attempt to prepare nests before laying eggs, as they would do in nature. The egg industry calls these futile efforts 'vacuum nesting' since no material is supplied to hens to help them build a nest.[41] For economic reasons, it is imperative in both the egg and chicken meat industries that mothers be separated from chicks and that both be kept away from the natural environment—the only time a hen or a broiler chick might see the light of day or feel rain on their feathers is when they are being transported from shed to slaughter.[42]

A Plea for Chickenkind

In 1996 U.S. television personality Dick Clark told the following (now well-known) 'joke':

> Where did Paul Simon get the idea to write "Mother and Child Reunion?"
> From a chicken-and-egg dish at a Chinese restaurant.[43]

Clark's 'witticism' typifies the devaluation of chickens following industrialization. Here a slaughtered hen (who appears as chicken meat in this dish) is ridiculed for finally getting to meet her offspring, the egg, just when they are both about to be eaten. This 'joke' demonstrates misogyny, racism and speciesism in its supposed punchline, and it also indicates just how far we have come from any kind of esteem for chickens.

These birds are now so distanced from Western consumers' lives that they are not even regarded as sentient beings: they are viewed merely as 'food on legs' even while still alive. Promotional material tempting new farmers into the poultry industry blatantly acknowledges this when it claims: "When you choose a career in the poultry industry you may not see a chicken or an egg—except at mealtime".[44]

It is my contention that, despite Clark's reference to mothers and offspring, chickens have actually been *degendered*, especially in Western societies. Roosters are no longer honoured and hens no long admired. Since gender is a human construct, this might have been beneficial for chickens, if it were not for the more important point that, as the previous promo enticing new poultry farmers shows, these birds have now also become *deanimalized*. We routinely objectify them and trivialize their lives and deaths. I hope this chapter has made it obvious that we urgently need to develop a more respectful relationship with chickenkind—one that does not deanimalize, invisibilize and deindividualize these birds in the name of consumption and profit, and also one that does not impose our cultural beliefs about masculinity and femininity onto roosters and hens.

Acknowledgements

My heartfelt thanks to Katya Krylova for her help translating and explaining material in Russian on Baba Yaga.

Notes

1 Annie Potts, *Chicken* (London: Reaktion, 2012), 75.
2 Ibid.
3 Ibid.
4 Donna Rosenberg, *World Mythology* (Chicago: NTC Publishing Group, 1994), 404.
5 Ulisse Aldrovandi, *Aldrovandi on Chickens: The Ornithology of Ulisse Aldrovandi*, Vol. II, Book XIV [1600], trans. L.R. Lind (Norman, OK, 1963), 143.
6 Katya Krylova, personal communication, December 2022.

7 David Barnett, *Baba Yaga: The Greatest Witch of All?* (BBC Culture, 2022), retrieved from: https://www.bbc.com/culture/article/20221118-baba-yaga-the-greatest-wicked-witch-of-all?fbclid=IwAR3P_62WC-uRklvauP416FwU65G7MPtAz5vQzUNGWjo8ZLzKMC5LqX9xbUU, accessed December 22, 2022; also Christina Hendry and Lindy Ryan, eds., *Into the Forest: Tales of the Baba Yaga* (Black Spot Books, 2022).
8 Andreas Johns, *Baba Yaga* (New York, 2004), 166.
9 Potts, *Chicken.*
10 Ibid.
11 Angelina Parker, "Oxfordshire Village Folklore (1840–1900)," *Folklore* 24, no. 1 (1913): 90.
12 Ursula Mittwoch, "Whistling Maids and Crowing Hens—Hermaphroditism in Folklore and Biology," *Perspectives in Biology and Medicine* 24, no. 4 (1981): 595.
13 Ibid.
14 Ibid.
15 Ibid.
16 L.F. Newman, "Notes on the Folklore of Poultry," *Folklore* 53, no. 2 (1942): 104–111; yolkless eggs are actually first-time eggs laid by very young hens called pullets.
17 Ibid.
18 Ibid.
19 Carey Miller, *Dictionary of Monsters and Mysterious Beasts* (London, 1987).
20 Potts, *Chicken*; similar figures exist in folklore of the Philippines (where the creature that manifests from the egg is a green lizard) and Portugal (where the hatchling is a scorpion). See Benedict Anderson, "The Rooster's Egg," *New Left Review* 2 (2000): 47.
21 Ibid.
22 Ibid.
23 Ibid.
24 Alan Dundes, *Cockfighting: A Casebook* (Madison, WI, 1994).
25 Ibid.
26 Ibid.
27 Potts, *Chicken*, 71.
28 Dundes, *Cockfighting*, 259.
29 Susan Merrill Squier, *Poultry Science, Chicken Culture: A Partial Alphabet* (Rutgers University Press, 2011).
30 Ibid., 138.
31 Carol J. Adams, *The Pornography of Meat* (New York, 2004).
32 Ibid.
33 Ibid., 78.
34 Annie Potts, "What Is Meat Culture?" in *Meat Culture*, ed. A. Potts (Boston and Leiden: Brill, 2016).
35 Please see Hattie Ellis's *Planet Chicken: The Shameful Story of the Bird on Your Plate* (Sceptre, 2007); Annie Potts' *Chicken* (Chapter 6, pp. 139–173); Roger Horowitz's, *Putting Meat on the American Table: Taste, Technology, Transformation* (Baltimore: The Johns Hopkins University Press, 2006); and William Boyd's "Making Meat: Science, Technology and American Poultry Production, *Technology and Culture* XLII (2001), for histories of chicken farming—from famly flocks to intensive farms. I also recommend Clare Druce's *Chicken's Lib* (Bluemoose, 2013) and Barbara J. King's *Personalities on the Plate: The Lives and Minds of Animals We Eat* (Chapter 4) (Chicago: University of Chicago Press, 2017).
36 Potts, *Chicken.*
37 I.F. Keymer, "Disorders of the Avian Female Reproductive System," *Avian Pathology* 9, no. 3 (1980): 405–419.
38 Sunaura Taylor, *Beasts of Burden: Animal and Disability Liberation* (New York and London: The New Press, 2017).
39 Steve Striffler, *Chicken: The Dangerous Transformation of America's Favorite Food* (New Haven and London: Yale University Press, 2005); and Potts, *Chicken* (Chapter 6).
40 John Steele Gordon, "The Chicken Story," *American Heritage* 47, no. 5 (1996), retrieved from: https://www.americanheritage.com/chicken-story, accessed December 22, 2022.

41 S.F. Smith, M.C. Appleby, and B.O. Hughes, "Nesting and Dust Bathing by Hens in Cages: Matching and Mis-matching between Behaviour and Environment," *British Poultry Science* 34, no. 1 (1993): 21–33.
42 If the reader is wondering where male chickens fit into the egg industry: they don't. Males are killed by a method known as 'instantaneous fragmentation' (they are macerated en masse within hours of hatching), as they are dispensable within a system focused on expoiting female reproductive bodies. A few are 'saved' for breeding purposes, and their lives are even worse than battery hens' lives. See Erik Marcus, *Meat Market* (Boston, MA, 2005).
43 Cited in Carol J. Adams, *Pornography of Meat*, p. 150.
44 Karen Davis, *Prisoned Chickens, Poisoned Eggs* (Summertown, TN, 1996), pp. 85–86.

Bibliography

Adams, Carol J. *The Pornography of Meat*. New York, 2004.
Aldrovandi, Ulisse. *Aldrovandi on Chickens: The Ornithology of Ulisse Aldrovandi*, Vol. II, Book XIV [1600]. Translated by L.R. Lind. Norman, OK, 1963.
Anderson, Benedict. "The Rooster's Egg." *New Left Review* 2 (2000): 47.
Barnett, David. "Baba Yaga: The Greatest Witch of All?" *BBC Culture*, 2022. Retrieved from: https://www.bbc.com/culture/article/20221118-baba-yaga-the-greatest-wicked-witch-of-all?fbclid=IwAR3P_62WC-uRklvauP416FwU65G7MPtAz5vQzUNGWjo8ZLzKMC5LqX9xbUU, accessed December 22, 2022.
Boyd, William. "Making Meat: Science, Technology and American Poultry Production." *Technology and Culture* XLII (2001).
Davis, Karen. *Prisoned Chickens, Poisoned Eggs*. Summertown, TN, 1996.
Druce, Clare. *Chicken's Lib*. Hebden Bridge, UK: Bluemoose, 2013.
Dundes, Alan. *Cockfighting: A Casebook*. Madison, WI: University of Wisconsin Press, 1994.
Ellis, Hattie. *Planet Chicken: The Shameful Story of the Bird on Your Plate*. London: Sceptre, 2007.
Gordon, John Steele. "The Chicken Story." *American Heritage* 47, no. 5 (1996). Retrieved from: https://www.americanheritage.com/chicken-story, accessed December 22, 2022.
Hendry, Christina, and Ryan, Lindy, eds. *Into the Forest: Tales of the Baba Yaga*. London: Black Spot Books, 2022.
Horowitz, Roger. *Putting Meat on the American Table: Taste, Technology, Transformation*. Baltimore: The Johns Hopkins University Press, 2006.
Johns, Andreas. *Baba Yaga: The Ambiguous Mother and Witch of the Russian Folktale*. New York: Peter Lang, 2004.
Keymer, I.F. "Disorders of the Avian Female Reproductive System." *Avian Pathology* 9, no. 3 (1980): 405–419.
King, Barbara J. *Personalities on the Plate: The Lives and Minds of Animals We Eat*. (Chapter 4) Chicago: University of Chicago Press, 2017.
Marcus, Erik. *Meat Market*. Boston, MA: Brio Press, 2005.
Miller, Carey. *Dictionary of Monsters and Mysterious Beasts*. London: Olympia, 1987.
Mittwoch, Ursula. "Whistling Maids and Crowing Hens—Hermaphroditism in Folklore and Biology." *Perspectives in Biology and Medicine* 24, no. 4 (1981).
Newman, L.F. "Notes on the Folklore of Poultry." *Folklore* 53, no. 2 (1942): 104–111.
Parker, Angelina. "Oxfordshire Village Folklore (1840–1900)." *Folklore* 24, no. 1 (1913).
Potts, Annie. *Chicken*. London: Reaktion, 2012.
Potts, Annie. "What is Meat Culture?" In *Meat Culture*, edited by A. Potts. Boston and Leiden: Brill, 2016.
Rosenberg, Donna. *World Mythology*. Chicago: NTC Publishing Group, 1994.
Smith, S.F., Appleby, M.C., and Hughes, B.O. "Nesting and Dust Bathing by Hens in Cages: Matching and Mis-matching between Behaviour and Environment." *British Poultry Science* 34, no. 1 (1993): 21–33.
Squier, Susan Merrill. *Poultry Science, Chicken Culture: A Partial Alphabet*. New Brunswick, NJ: Rutgers University Press, 2011.
Striffler, Steve. *Chicken: The Dangerous Transformation of America's Favorite Food*. New Haven and London: Yale University Press, 2005.
Taylor, Sunaura. *Beasts of Burden: Animal and Disability Liberation*. New York and London: The New Press, 2017.

32
GOTHIC SNAKES
Snake Handling, Snake Women and a Post-Secular Serpentine Practice

Sue Hall Pyke

Introduction

This chapter is part of an ongoing study of snake–human relations that is reorienting me from the harms of ophidiophobia. All through my childhood I was terrified of snakes. This culturally inherited fright led me to lead others to snake killings. Now, committed to cohabitating ethically with animals other than humans, despite my invasive body, unsettled on land sovereign to the Eastern Maar Nation, a 'stray', as Barbara Creed would have it,[1] I find myself ethically bound to seek new ways to live with the tiger snakes and brown snakes around this place, families of snakes that go back at least as far as the human families who have always been and will always be of this earth in ancestrally material ways that far exceed my own white settler presence.

The snake vilification that shaped me in my formative years is grounded in the masculinist white church that bred me. I might go to a rainbow ecofeminist church now, when I'm not making up my own church, but the underpinning patriarchal structures of Christianity remain intact in the broader Western culture I inhabit. This is exhibited in the fact that it is legal for me to kill a snake who enters my home, no matter that this house is built on snake territory. This, under the act that 'protects' animals. It is no accident that a legal system that does little about the killing of a snake for being a snake where it was born also is inept at reducing the rate of women being murdered in their own homes. In Australia, a 'femicide' happens once a week.[2] Most likely, snakes die from 'shovel disease' more often than this, but there are only statistics around snakebites. About a third of these snakebites are caused by deliberately handling a snake, often with the intention to kill, and two thirds of these deliberate handlers are adult men.[3]

I do not conflate the killing of snakes with the killing of women to reduce the snakeness of snakes or the humanness of humans. Instead, I am pointing to the assumption of a right to harm that is also implicit in the righteous masculinist Christian culture of judgement. The logic of control exercised by men handling both snakes and women in ways that suit them is the same logic that draws snakes and women together as an evil that needs to be regulated by men. While feminist theology is complex and emergent in theological practice, and hermeneutics around speciesism are developing in generative ways, in many

DOI: 10.4324/9781003273400-40

conservative Christian churches there is little to disturb the centuries-old signification of snakes and women as standing for sin, treachery and temptation, in a holy trinity where Wisdom Sophia is made masculine in the run of fathersonandholyspirit.

The Christian church's contempt for snakes, and its erasure of women, does not sit well with the central Christian edict of loving one's neighbours as oneself as I understand it now, but I don't remember seeing this contradiction as a young schoolgirl, singing that bloodthirsty line from Julia Ward Howe's Battle Hymn of the Republic: 'Let the Hero, born of woman, crush the serpent with his heel'.[4] I heard this rip-roaring swashbuckling song again, not so long ago, in St Paul's Cathedral, in the centre of Naarm/Melbourne. Distracted from my task of lighting a candle to mark the death of a friend's father who adhered to the Anglican church, I stared at these words. Crushing a serpent with my heel? I wanted no part of that judgement.

The inspiration for Howe's hymn, apart from a civil war, comes from Genesis 3:15, a verse close to the beginning of the Old Testament. Disobeyed, the God figure curses the sneaky snake of knowledge: 'I will put enmity between thee and the woman, and between thy seed and her seed; it shall bruise thy head, and thou shalt bruise his heel'.[5] I quote the King James version, not only because it was the version that Howe read. The King James version was given to me by my grandmother, a strong influence in the familial Christian culture that informs my writing position here. She went to church each Sunday until her dying days.

Singing up bad relations between (and because of) 'women' and 'snakes' goes further back than Howe, right into the dark ages of Christian theological tradition, where the godly right to crush snakes, by their heads or other parts of their body, was extended to the right to trample women into patriarchal line. The sixth-century hymn composer, St Romanos the Melodos, was one among many when he described Eve as 'a snake more dangerous and snakier than the snake'.[6] This symbolic snake/woman relationship that entwines ideas of Eve and evil, the snake and sin, the corruptive snake and the snaky female, was unlikely to have been intended by Howe, an abolitionist, suffragist, pacifist and social activist. Yet her hymn's symbolic language, with its ready contempt for the life of a snake, holds an acceptance of violence that works to also harm women. It may be the case that privileged women are experiencing a shift of power that began with the shift to state control from Church rule, and it is true that snakes are protected under the law, but such changes are hindered by an influential old-school Christian culture that weakens opportunities for real change. Women and snakes are still getting killed.

This deadly influence came to life, for me, when I was teaching Gothic literature. The early movements of this churchy genre are marked by devious snakes and impious women, evil villains at worst, wild beasts at best, in need of containment or, even better, death.

Gothic Handlings of Literary Snakes

John Milton's *Paradise Lost* shaped the early beginnings of Gothic literature, perhaps because this epic poem represented the Bible's Eve and serpent in unbound ways that reflect the Gothic genre, which is nothing if not uncontained.[7] Milton's poem builds, perhaps unintentionally, on Gnostic readings of the Bible, where the snake can be both agent and victim of the Fall.[8] In this theological corner, Milton's Christ, the serpentine Satan, knowledge and redemption all can be read as one.[9] Milton expands the cultural room Gnosticism creates for snakes in his doubling and mirroring story of trust lost and wisdom gained, creating questions around the masculinist certainty that women, like snakes, are harbingers of evil.[10]

Despite the ambivalence in Milton's text, where good is not set against a binary of evil, feminist critique of *Paradise Lost* goes back almost as far as the text itself. In Milton's Garden of Eden, the snake's encounter with Eve is as follows:

> . . . replete with guile
> Into her heart too easy entrance won:
> Fixed on the fruit she gazed, which to behold
> Might tempt alone, and in her ears the sound
> Yet rung of persuasive words, impregned
> With reason, to her seeming, and with truth[11]

Eve's apparent willingness to be persuaded, her 'easy entrance' to corruption, rightly set off the feminist sensibilities of Mary Wollstonecraft, who derided Milton for disempowering Eve in his narrative of the 'frail first mother'.[12] Eve's agency continues to be questioned in contemporary feminist scholarship. Alison Bare's evaluation of claims for egalitarianism in Milton's text, through the 'crux of the narrative in the temptation scenes,' finds little that is liberating in Milton's ophidian Satan, who dominates Eve to serve 'his own attempts at diminishing God's power'.[13] Similarly, for literary scholar Alison Milbark, Milton's Satan cannot be equated with the good associated with the Christ figure, because evildoing is this serpent's 'sole delight'.[14]

Shannon Miller finds more agency in Milton's Eve, reading beyond the 'popular and longstanding tradition' that aligns Eve either with the snake and therefore evil, or as a victim of a 'seducing force that must be resisted'.[15] Sarah Morrison also provides a more complicated reading, in her work with the 'perplexingly indeterminate nature of the Serpent'.[16] For Morrison, Milton's snake signifies both death and life, sexual productivity and transgression, wisdom and trickery, beneficence and evil.[17] Morrison's reading becomes refreshingly zoocentric when she argues that Milton's poem refuses false separations between humans and the rest of nature.

Morrison's expansion from anthropocentric readings nudges me to think again about sublime metamorphosis, a rich aspect of Gothic literature since its inception.[18] What might shift for a reader when a nonhuman speaks through a human text?[19] The serpent is speaking in words that Eve understands. So, is the serpent a human-speaking snake, or is Eve a snake-speaking human, or are both these creatures becoming something else altogether, together, creepy crawling in that Garden of Eden?

Metamorphosis can be understood as a way of seeing each body as always and already the body of the other. As philosopher Emanuele Coccia puts it, a body (snake or human, or any other body) is 'the gateway to an infinity of other worlds'.[20] Coccia's materialist perspective on metamorphosis invites me to take Milton's description of Satan personally. Where is the snake in my infinity? Seeing the body of a snake in my own body creates a reading of Milton's verses interrupted by strikethrough and parenthesis.

> ~~His~~ (My) Visage drawn ~~he~~ (I) felt to sharp and spare,
> ~~His~~ (My) Armes clung to ~~his~~ (my) Ribs, ~~his~~ (my) Leggs entwining
> Each other, till supplanted down ~~he~~ (I) fell
> A monstrous Serpent on ~~his~~ (my) Belly prone[21]

My revisions draw something reptilian, if not straight out snaky, in the gauntness of my aging woman's face, my collagen-rejecting skin folding closer to the structure of my skull that will, in the end, be recognised first as animal, then as human. Arms, harder to lift each year, closer to the ribs, weak-kneed legs folding into one other. Fall by monstrous fall, the ground will have me prone in the end.

My metamorphic imagining, which lays together the im/possibility of becoming-serpent together with my being of the same earth as the snakes that live alongside me, stretches towards the 'sympoiesis' or 'making-with' described by Donna Haraway. In this play, snakes become my 'oddkin', my 'godkin'.[22] The radical potential of intentional metamorphosis, felt in my human body, where the commonality of the animal takes primacy over an all-too-human focus on species, offers up a powerful embodied affect.

In my playing-snake with Milton's poem, the gateway suggested by Coccia opens into an imaginative encounter that fosters more empathetic relations with the snakes around me.[23] This potential is not so present in the metamorphic inclinations of three Gothic Romantic writers who responded to Milton's text.

Monk Lewis's early Gothic novel, *The Monk: A Romance* sets the tone for harmful metamorphic trajectories in Gothic representations, because in this text the serpent stands, unambiguously, for serpentine supernatural womanly evil. Ambrosio, the head of the Catholic church's fanciest monastery, is corrupted by the bite of a snake, who may well be Matilda, a woman dressed as a man who is the devil who becomes a serpent, a demonic being whose hair writhes in the retribution of her fall. These uncovered tresses are 'supplied by living snakes' who sound out 'frightful hissings'.[24] Ambrosio, along with Matilda, is banished to this hell of final judgement.

Samuel Taylor Coleridge was outraged by the 'libidinous minuteness' of *The Monk*, worried about the harms that might come to sons and daughters from this novel's depravity.[25] Yet Coleridge writes with spice enough in his own Gothic version of serpentine evil. In his narrative ballad "Christobel", the gormless titular character is seduced by snaky Geraldine, not only through her hooded singular 'snake's small eye' with its phallic 'malice' that 'blinks dull and shy' but also through her voice, her words and the womanly hiss of her voice.[26] This queer Geraldine does not change from woman to snake on the page, but the snake is present in her body, mirroring Lewis's depiction of Matilda as a metamorphic metaphoric unholy snake.

John Keat's "Lamia" depicts a corruptive snaky woman that bears comparison to Lewis's Matilda and Coleridge's Geraldine. Lamia is held inside a 'serpent prison' of masculinist constructions of beauty, her scaled skin 'all crimson barr'd/And full of silver moons'.[27] This 'palpitating snake' of great splendour is betrayed by the man she loves, then conquered by a spear-like male gaze that sees her as a snake. Her personhood lost, Lamia must vanish or be killed.[28] This is a fate that has been shared by snakes in my neighbourhood for the past two centuries.

My readings of these three works are not the only readings possible. Milbark has noted elsewhere the 'creative theological work' of the Gothic, where its open possibilities allow a broad range of interpretations.[29] However, Angela Carter's postmodern Gothic novel *Heroes and Villains* does, in part, parody the conflation of bad woman and bad snake that I find in these three interlinked texts.[30] Carter's Marianne, the novel's central nonconformist character, becomes enemy of the people when a snake, as quick as 'variegated lightning,' bites her for not looking where she's going.[31] Scandalously, Marianne lives instead of dying. Next she meets a decaying snake behind 'beribboned bars,' lifelessly playing the part of a symbol needed in 'crafting a new religion'.[32] This lesson is well-learned by a young child who tells Marianne she should be kept 'in a cage, like that snake', all the better to be prodded with a stick through the bars.[33] Marianne's third encounter with a snake is through the tattoo across the back of a man who rapes her. She reads Adam, Eve and the tree of knowledge and the snake as a 'relic of the survival of Judeo-Christian iconography'.[34] Carter's pastiche of the fearful 'infernal serpent' undermines the early Romantic Gothic texts that worked so hard to enclose snakes and women.[35]

The Southern Gothic of the Appalachian Snake Handlers

The Romantic Gothic snake, evil, just like a bad woman, is different to what I would call, following the mischief of Carter, the crafted new religion of snake-handling. The cultural space occupied by these snake handling practices, high in Tennessee's Appalachian hills, where snakes are used to test the power of God, has been the focus of a good number of movies and documentaries. One such work, *Alabama Snake*, was recently glossed as a 'Southern Gothic portrait', a genre netting I cast wider to catch the church as well.[36] Like most Gothic narratives *Alabama Snake* is a tale of 'fear and love'.[37] Films follow the same narrative. In *Them That Follow*, men hold snakes tight and women tighter.[38] It is a Gothic frame replicated in the trinity of Lewis, Coleridge and Keats.

Ethnographers describe snake handling as a 'private, localized community tradition'.[39] From this perspective, snake handling is a 'complex traditional religious belief', not a 'socially dangerous deviant civic practice'.[40] As a scholar of the Gothic, I recognise these ideas of danger and deviance. Sociologists consider snake handling as a pathological expression of repressed trauma, or explain the practice as compensatory escapism from an arduous life of impoverishment. Again, Gothic works can be read the same way. But these literary, ethnographic and sociological interpretations hold little meaning for those of the faith. The Appalachian appellants see themselves as led by an edict in the Gospel of Mark to take up serpents so they may be cleansed of sin and readied for resurrection. The elected of God will demonstrate their faith by taking up serpents, as well as casting out demons, speaking in tongues, drinking poison and healing with their hands.[41] As the King James Bible puts it:

> 17 And these signs shall follow them that believe; In my name shall they cast out devils; they shall speak with new tongues; 18 They shall take up serpents; and if they drink any deadly thing, it shall not hurt them; they shall lay hands on the sick, and they shall recover.[42]

For the 100 or so churches that 'take up serpents', snake handling is a Christian duty.[43] Snake handling allows 'convincing demonstrations of God's power', and by extension, demonstrates the power given to humans by God, because God's spirit moves the handler to handle them.[44] With typical Gothic slippage, snakes are the satanic enemy that must be overpowered, and, at the same time, the holy divine inhabits both the snake and the handler.[45]

Snake handling worship communities are dismissed as sects by the larger Church of God, but from where I stand, these churches' fundamentalist interpretation of the Bible brings them into proximity with other fundamentalist Christian churches.[46] I have not attended a snake handling service, but the footage reminds me of the fundamentalist churches I have entered in my time, where men speak and women listen and sin is present in all who do not see the Bible as 'supernaturally preserved' by a God who will save only those of a 'genuine faith'.[47] Women, like the snakes in the snake handling churches, are closely controlled in these communities, their identity 'subjugated' through ideas of submission and male authority.[48]

Yet for those in the snake handling churches, the spirit is non-hierarchical, moving anyone at all to handle the snakes through an 'intense emotionality' created together by the laity.[49] Music is made, and the spirit of God is waited upon, until one of the worshippers in the gathering is called by the spirit to open the box.[50] The released snakes are 'passed amongst the obedient' as part of a two-hour service that is then followed by prayer and testimony.[51] Hopes for resurrection relate directly to the 'death in that box'.[52] Together with the potential for death, the handling of the snake brings 'joy, blessedness, gladness, goodness, peace, victory, wonder and greatness'; in the evocative words of one female participant, snakes are handled for the 'bubbling of my soul'.[53] It is the bubbling joy that comes with handling the snake, not the bite, that provides succour to this faith.[54]

As the previous quote implies, while popular cultural texts depict snake-handling churches as full of marginalised, poor, uneducated white men who have inherited this practice from their fathers, there is more to it than that. It is true that some families have been handling snakes for four generations, but some have never touched a snake before, some are educated, some are rich and both men and women attend the services.[55] While women do not preach in these churches, they most certainly do testify.[56] Estimates suggest that nearly 50% of the snake handling is by women.[57] Interestingly, those who die are more often men than women, and of these, the male preachers are no small percentage.[58] Perhaps these fatalities reflect gendered leadership, where men are more active in handling the snakes. The video footage does suggest this. Alternatively, the death of a man, and particularly a snake-handling preacher, may be deemed more newsworthy. It could be that women handle snakes in less threatening ways or at different points in the service. No matter who is bitten, or who dies, the bites show that handling snakes is part of a harmful utilitarianist relationship that privileges the will of humans above the will of snakes. Women are part of this process. Complicit, like younger me, screaming when I saw a snake near the house.

There are laws in Tennessee to stop snakes being handled against their will, yet nothing yet has stopped the human will represented by human-made snake boxes and human godling hands that wrap around the bodies of snakes. This physical oppression of snakes is not the focus of the participants, nor do studies of snake handling churches spend much time on how it might feel to be a snake detained for these services. It is an instructive silence. Nothing about the fact that a timber rattlesnake would rather move around a territory between twenty and forty hectares.[59] Nothing about the 'forested habitat' of copperheads, who prefer 'wooded hillsides with abundant logs, leaf litter, or rocks for cover'.[60] The snake-holding boxes may well be of legal size, but they are so much smaller than the hectares needed to meet the behavioural requirements of a snake. When they are not in the ceremonial boxes brought into the worship service, the snakes are kept in other jail-like structures. Perhaps a snake pit, a larger box, a hessian bag. These containments say all that needs to be said about the will of the snakes. It is a strategic silencing, like the one Morrison notes, in her discussion of *Paradise Lost*: 'the power of the serpent as sign in the Christian exegetical tradition does not so much answer as overshadow troubling questions about the fate of the naturalistic serpent'.[61] In the human-centred context of sacred snake handling, the signs that are watched for do not include looking for captivity stress. There are no reports detailing snakes pushing against their enclosures, being off their food, fleeing, flattening their body, hiding their head, puffing up their body, hissing, fighting with other snakes, stiffening, playing dead, mouth-breathing or vomiting. There is, however, a measure of snake agency, if not will, in the numbers of human handlers who have been bitten. Snakes only bite creatures as large as humans when they are in fear of their life.

Conclusion

The Christian church's long history of violent relations with women and snakes shaped my early responses to snakes. I screamed and men killed. Decades later I see such killings as murder, see myself as an accomplice, informer, as a part of that violence. I now also understand the holding of a snake as an expression of violence and power.[62] Just as women will escape domestic violence once it is safe to do so, snakes will leave their enclosures, including the hands that hold them, contain them, enclose them, imprison them, just as soon as they can. If a snake does not make its way into one's hands, then surely taking them up, into one's hands, can be read as an act against their will, the antithesis of loving one's neighbour.

Yet giving snakes the widest berth possible seems a bare counter to the cultural hatred that kills snakes, just as avoiding men does not keep women safe where they live. Perhaps my need to 'do good', to *love* snakes, expresses, once again, the Christian acculturation embedded in my body. The Christian church is an institution that centralises ideas of loving justice in ways that justify enclosing snakes and women in subjugated positions in the name of love. My idea of love is shaped by the same culture of judgement that shapes the snake-handlers in the Appalachian hills and the Romancing Gothic writers and hymn writers like Howe. We have held the same Bible. These cultural shackles are likely to keep both snakes and women within hurtful confines no matter my ecofeminist intentions of care. Yet the ambivalent Milton and rebellious Carter and the materialist Coccia also form part of my cultural inheritance, giving me room to move.

Feeling my body as snake through Milton, refusing the tropes of my tradition like Carter, I find the cellular level outlined by Coccia offers a post-secular escape from Christian ideologies that hold snakes captive. The snakes in my immediate vicinity, perhaps under my feet and the floor as I write, have hearts that my own heart might well be connecting with, at this very moment. We are one body beating in old animalistic patterns.[63] The ripple of heartbeat and muscle that is both of me, and of the snakes in my life, demonstrates a shared force of life that feels divine. Just as the food from my garden is, in part, the snakes who have lived and died on Eastern Maar Country through millennia, stray as I am, an uninvited and invasive white settler, my ash or humus or whatever is left of me after what can be reused is reused will move into new life here, after my death. I may become microbe or worm, mouse or snake. And already, and always, the soil that is my body is also the soil that is the body of the snakes around here. I shed my cells like skin.

Attending to this fact alone might be enough, but it is not enough for me.[64] In the last year or so I have been turning devoutly to the earth that is life that is me, attending to a practice I call Snake Church. This regular worship, that happens without pre-planned intent, does not demand the presence of a snake. The moment I am called by a kookaburra, I become a Snake Church attendee.[65] This church began with the simple act of being so quiet, so contemplative, so still, that it felt a lot like worship. But in the process of developing this chapter, I have found myself playing snake. Snake Church is broadening, opening me towards a performative metamorphosis.[66] Finding faith in the animal of my body, inspired by Milton's text, I newly devote myself to my snake neighbours, using my body to refuse literary reductions and literal incarcerations. Like Milton's serpent, Lewis's Matilda, Coleridge's Geraldine, Keat's Lamia and Carter's tiger (snake) Marianne, I play still, I play bask, I play vibrations under my body, I play being alert enough to slither away from human harm. In part, this is the transformative 'spiritual engagement', or 'dance with animacy' suggested by Teja Brooks Pribac.[67] It is a mode of divine contemplation, with a ludic tendency towards a radical and fluid metamorphic force of change, that makes no demands on the snakes that I live alongside. I would like to think I am performing the partial uncommon worlding of hope outlined in Petra Tschakert's theorising of new relations with 'Unknown Others'.[68] Staying still, becoming-with snake, every time I am called to do so by a kookaburra, is an active reminder of the animal connection I have with my neighbours.

In this godly becoming-with, alongside and within the multitude of forms that are always and already my body, the snakes I barely know are imaginatively enfleshed in this body writing these words. Unless it is snakes themselves writing me towards a more respectful trusting faith in the snake aspects of my body, snakes themselves creating this emergent embodied game of snake, snakes co-creating the rules emerging from the earth of this place. I hope so. That's my kind of church.

Notes

1 Barbara Creed, *Stray: Human–Animal Ethics in the Anthropocene* (Power Publications. 2017).

2 Australia's National Research Organisation for Women's Safety. 'Intimate partner violence homicides 2010–2018. Australian Domestic and Family Violence Death Review Network. *Data Report* (2022). Jenny Mouzos, "Femicide: An Overview of Major Findings," *Trends and Issues in Crime and Criminal Justice*. Australian Institute of Criminology. No. 124 (1999). See also Clifton P. Flynn, "Examining the Links between Animal Abuse and Human Violence," *Crime, Law and Social Change* 55 (2011): 453–468. This collection's focus is on gender and speciesism, and so I have focused on the 90% of humans who suffer domestic violence. This is not to discount the 10% of people who identify with male or non-binary genders.

3 Dennis K. Wasko and Stephan G. Bullard, "An Analysis of Media-Reported Venomous Snakebites in the United States, 2011–2013," *Wilderness and Environmental Medicine* 27, no. 2 (2016): 219–226. https://doi.org/10.1016/j.wem.2016.01.004

4 Julia Ward Howe, "The Battle Hymn of the Republic," *The Atlantic Monthly* 9, no. 52 (1862), retrieved from: https://www.theatlantic.com/magazine/archive/1862/02/the-battle-hymn-of-the-republic/308052/

5 https://www.kingjamesbibleonline.org/Genesis-Chapter-3/#15

6 Feminist theologian Eva Catafygioti Topping writes scathingly that the vitriol of St Romanos was not unusual; he was 'far from being a lone voice'. Eva Catafygioti Topping, "Reflections of an Orthodox Feminist," *Greek American Review* 41, 42, scholarship.tricolib.brynmawr.edu/bitstream/handle/10066/10614/Catafygiotu-Topping_44_536.pdf?sequence=1&isAllowed=y Extract from Holy Mothers of *Orthodoxy—Women and the Church*. However, emerging feminist scholarship positions St Romanos as a voice that gave presence to bold New Testament women as sites of potential transformation previously unseen. Erin Galgay Walsh, "Sanctifying Boldness: New Testament Women in Narsai, Jacob of Serugh, and Romanos Melodos" (PhD thesis, Duke University, Graduate Program in Religion, 2019). See page 9. https://dukespace.lib.duke.edu/dspace/bitstream/handle/10161/19813/Walsh_duke_0066D_15282.pdf?sequence=1

7 It is notoriously difficult to define the Gothic, but in Clive Bloom's thorough historical and cultural review, the Gothic is described as 'a mode of cultural production that pertains to the exploration of otherness and uncanny familiarity' with a focus on 'the construction and dissolution of boundaries' that reflect social fears, and in all this, 'the Gothic does not stand still,' it is 'a hybrid mode' that 'often appears in collaboration with other literary forms, modes and genres.' Clive Bloom,

"Introduction to the Gothic Handbook Series: Welcome to Hell," in *The Palgrave Handbook of Contemporary Gothic,* ed. Clive Bloom (Palgrave Macmillan, 2020), See pages 8, 13 and 23.

8 Paul Robertson and Gabrielle Scott, "Gnostic Thought in Milton's *Paradise Lost,*" *Gnosis: Journal of Gnostic Studies* 7, no. 2 (2022): 171–223. doi.org/10.1163/2451859X-00702003. See also, Neil Forsyth, "The Sign of the Dove and Serpent," *Milton Quarterly* 34, no. 2 (2000): 57–65.

9 Forsyth, "The Sign of the Dove and Serpent," 57–65.

10 Neil Forsyth, *The Satanic Epic* (Princeton University Press, 2002). See page 311. The whole quote is: 'Behold, I send you forth as sheep in the midst of wolves: be ye therefore wise as serpents, and harmless as doves' (Matthew 10:16, King James version, retrieved from: https://www.kingjamesbibleonline.org/Matthew-10-16/). For more on gnosticism in this context see Whimont's argument that the Gnostics saw the serpent to be 'no other than the Savior himself who initiated the world of salvation through urging men to consciousness'. E.C. Whitmont, *The Symbolic Quest* (G.P. Putnam's Sons, 1969). See page 255.

11 John Milton, *Paradise Lost* (IX, 1–6) (1674), retrieved from: www.bartleby.com/360/4/155.html

12 Mary Wollstonecraft, *A Vindication of the Rights of Woman: with Strictures on Political and Moral Subjects'* (Project Gutenberg, 1792)

13 Alison Bates, "Feminism Regained: Exposing the Objectification of Eve in John Milton's *Paradise Lost,*" *English Studies: A Journal of English Language and Literature* 99, no. 2 (2018): 93–112. https://doi.org/10.1080/0013838X.2017.1405324

14 As Alison Milbank puts it, Milton demonstrated the 'aesthetic appeal of the Catholic liturgy and architecture as an instructive mode of spiritual education'. Alison Milbank, *God and the Gothic: Religion, Romance, and Reality in the English Literary Tradition* (Oxford University Press, 2018). See page 54.

15 Shannon Miller, "Serpentine Eve: Milton and the Seventeenth-Century Debate over Women," *Milton Quarterly* 42, no. 1 (2008): 48–49.

16 Sarah Morrison, "The Accommodating Serpent and God's Grace in Paradise Lost," *SEL Studies in English Literature 1500–1900* 49, no. 1 (2009): 173–195. doi 10.1353/sel.0.0048. See pages 175–176.

17 Shannon Miller, "Serpentine Eve: Milton and the Seventeenth-Century Debate over Women," *Milton Quarterly* 42, no. 1 (2008): 44–68. See page 57.

18 Susan Pyke, "Creaturely Shifts: Contemporary Animal Crossings through the Alluring Trace of the Romantic Sublime," *TEXT Journal of Writing and Writing Courses* 41 (2017): 1–19.

19 For classics scholar Kalina Allendorf, Milton's Eve expresses wonder on two levels. The first is 'the physical capacity of animals to express human language,' which was, as Allendorf notes, an object of speculation in Milton's time. The way that the voice of the snakes in Gothic texts echo Milton's work (and, indeed, the Bible) is of much interest to Allendorf who shows how, in *Paradise Lost*, the lead to the Fall is accompanied by the serpent's voice, speaking directly to Eve. Allendorf goes on to argue Milton's fall is the 'misplaced wonder' that lies in 'attributing human cognition to animals.' Kalina Allendorf, "Lucretian Subversion: Animal Speech and Misplaced Wonder in *Paradise Lost,*" *Milton Quarterly* 52, no. 1 (2018). See page 44. There is potentially a slightly different perspective that could be applied here, in the light of critical literary animal studies' thinking about the radical potential of anthropomorphism and metamorphosis.

20 Emanelle Coccia, *Metamorphoses* (Polity Press, 2022). Se pages 121–122.

21 Milton, *Paradise Lost* (X, 511–516).

22 Donna Haraway, *Staying with the Trouble: Making Kin in the Chthulucene* (Duke University Press, 2016). See pages 2 and 5.

23 Claire Parkinson makes a good argument for 'the mediated encounter' in literary works which break down the barriers between species, implicitly suggesting that the metamorphic moment might be a 'meeting point' for readerly affect and nonhuman agency. Claire Parkinson, *Animals, Anthropomorphism and Mediated Encounters (*Routledge, 2020). See page 4. Brian Massumi has theorised this beautifully in his work about ludic play. Brian Massumi, *What Animals Teach Us About Politics* (Duke University Press, 1984). See page 86. The 'politics of fabulation' has been further extended by Erin Manning. Erin Manning, *For a Pragmatics of the Useless* (Duke University Press, 2020). See page 139.

24 Mathew Lewis, "Monk," 1796 in *The Monk* (New York: Oxford University Press, 2008). See page 433.

25 Samuel Taylor Coleridge, *The Collected Works of Samuel Taylor Coleridge, Volume 11: Shorter Works and Fragments: Volume II* (Princeton University Press, 2019). See page 61.
26 Samuel Taylor Coleridge, 1816 (1797). "Cristobel," retrieved from: www.poetryfoundation.org/poems-and-poets/poems/detail/43971. See lines 571 and 574.
27 John Keats, "Lamia," (1820), retrieved from: www.online-literature.com/keats/2055/. See line 204.
28 Ibid. See line 45.
29 Alison Milbank, *God and the Gothic: Religion, Romance, and Reality in the English Literary Tradition* (Oxford University Press, 2018). See page 3.
30 Jeremy Chow offers a strong literary outline of the 'queer eco-Gothic' that is present in the 'serpentine appetites' of Matilda, Geraldine and Lamia. Linking these three queer works together highlights the power of this trinity in their long cultural moment. Jeremy Chow, "Snaking into the Gothic: Serpentine Sensuousness in Lewis and Coleridge," *Humanities* 10, no. 1 (2021). See pages 3 and 15. doi.org/10.3390/h10010052
31 Angela Carter, *Heroes and Villains* (Penguin, 2011) (1969). See pages 38 and 32.
32 Ibid. See pages 117 and 70.
33 Ibid. See page 71. Sometimes the snake forms a 'phallic cult' and sometimes it does not, depending on the mood of the community's spiritual leader. See page 34.
34 Ibid. See page 135.
35 There is metamorphosis in the novel as well, although Marianne becomes a tiger, not a snake. This gives me joy. Where I live, a tiger *is* a snake.
36 Jan Read, "Thomas G. Burton, MA'58, PhD'66, Serpents and Stories," *Vanderbilt News* (2021), news.vanderbilt.edu/2021/04/22/thomas-g-burton-ma58-phd66-serpents-and-stories/
37 *Alabama Snake*. Directed by Theo Love. Written by Thomas G. Burton, Theo Love and Bryan Storkel. HBO. For trailer, see: www.youtube.com/watch?v=Rdh8ZIpQyP0
38 The trailer for *Them That Follow* begins its story of sexuality, faith and power with four snakes, writhing captive in a deep pit. 'They're so beautiful' says a young woman, trapped under the gaze of a man. The scenes escalate, the tension builds, a man's preacherly voice with the tenor of conviction intones, 'The serpent will purify you, the serpent will cleanse you'. The trailer ends with a woman's lips close to the dangle of a snake's head. *Them That Follow*. Directed and written by Britt Poulton and Dan Madison Savage, 2019. For trailer, see: https://www.youtube.com/watch?v=nN1x1ZG869Y
39 Jamie Sarafan, "The Women and the Word: Serpent Handling and the Women of the Church of God, 1914–1935" (PhD thesis, University of Colorado, 2016). See page 137.
40 William Glass, "Review," *Tennessee Historical Quarterly* 52, no. 4 (1993): 266–267. Review of: Burton, Thomas. *Serpent-Handing Believers* (University of Tennessee Press, 1993), See page 137.
41 This practice aligns with the ideas of God-blessed enmity discussed earlier, where the 'power to tread on serpents,' is justified by notions of human stewardship drawn from Genesis. *King James Bible*. Luke: 10:19, retrieved from: www.kingjamesbibleonline.org
42 *King James Bible*. Mark 16:17–18, retrieved from: www.kingjamesbibleonline.org
43 ABC, "Snake-Handling Pentecostal Pastor Dies from Snake Bite," https://abcnews.go.com/US/snake-handling-pentecostal-pastor-dies-snake-bite/story?id=22551754#:~:text=It's%20estimated%20that%20125%20churches,more%20than%20a%20century%20ago.
44 Ralph Wood and Paul Williamson, *Them That Believe: The Power and Meaning of the Christian Serpent-Handling Tradition* (University of California Press, 2018). See page xv.
45 Snake handling also enacts, on a slant, the biblical actions of Apostle Paul, bitten by a snake when gathering sticks for a fire. The King James version of Acts 28:3–6 goes like this: '3. And when Paul had gathered a bundle of sticks, and laid them on the fire, there came a viper out of the heat, and fastened on his hand. 4 And when the barbarians saw the venomous beast hang on his hand, they said among themselves, No doubt this man is a murderer, whom, though he hath escaped the sea, yet vengeance suffereth not to live. 5 And he shook off the beast into the fire, and felt no harm. 6 Howbeit they looked when he should have swollen, or fallen down dead suddenly: but after they had looked a great while, and saw no harm come to him, they changed their minds, and said that he was a god.' Others have argued that the sign of a true follower of God requires accidental bite, as experienced by Paul, rather than a testing of snakes to make the sign visible, as suggested in the Book of Mark, but for the snake-handlers, the congregations who follow the edict of the Gospel of

Mark, true believers will be known by the taking up of snakes. Preachers who are bitten continue to preach, if they survive. According to the narrator of this chronicle, Paul survived by the power of his faith. However, for snake handling churches, being killed by a snake does not mean one is ungodly. The Bible is a living text because it is read by those who live according to the mores of their times and places. https://www.kingjamesbibleonline.org

46 The snake handling churches originated from the Church of God, a radical non-segregated church. It was only when the movement bifurcated in 1914, and the egalitarian and largely black congregation stayed where the church began, that the conservative nature of this church prospered. In Tennessee, break-away segregating churches became more conservative, finding a measure of their faith through the body of snakes trapped to perform to their reading of the Bible. Ralph Wood, and Paul Williamson, *Them That Believe: The Power and Meaning of the Christian Serpent-Handling Tradition* (University of California Press, 2018).

47 Appalachian Church, "What We Believe," last updated 2022, appchurch.org/index.php/what-we-believe/.

48 Erica Appelros, "Gender within Christian Fundamentalism: A Philosophical Analysis of Conceptual Oppression," *International Journal of Philosophy and Theology* 75, no. 5 (2014): 460–473. See page 469, https://doi.org/10.1080/21692327.2015.1036906

49 Wood and Williamson. *Them That Believe: The Power and Meaning of the Christian Serpent-Handling Tradition.* See page 16.

50 Some of this powerful footage can be found in 'A Family Tradition.' *Snake Salvation.* Produced by Jerry Decker and Matthew Testa. National Geographic Television (2013), retrieved from: https://www.youtube.com/watch?v=k1-BhaX5GSE.

51 Wood and Williamson, *Them That Believe: The Power and Meaning of the Christian Serpent-Handling Tradition*, 5.

52 Paul Williamson and Howard R. Pollio, "The Phenomenology of Religious Serpent Handling: A Rationale and Thematic Study of Extemporaneous Sermons," *Journal for the Scientific Study of Religion* 38, no. 2 (1999): 203–218, 209.

53 Ibid.

54 Wood and Williamson. *Them That Believe: The Power and Meaning of the Christian Serpent-Handling Tradition:* See page 37.

55 Ibid.

56 While snake handling is 'almost exclusively associated with men', in its earlier manifestations women 'accepted and promoted the controversial sign', according to Jamie Sarafan, who notes that the 'institutionally-sanctioned, powerful, female-led and run fundraising arm' helped the 'transition away from their near-exclusive reliance on the signs that follow for religious authority.' Jamie Sarafan, "The Women and the Word: Serpent Handling and the Women of the Church of God, 1914–1935" (PhD thesis, University of Colorado, 2016). See pages 2, 9, 132 and 136, retrieved from: scholar.colorado.edu/downloads/9k41zf775.

57 Williamson and Pollio, "The Phenomenology of Religious Serpent Handling: A Rationale and Thematic Study of Extemporaneous Sermons." See page 216.

58 This has not been formally quantified, but indicative numbers have been anonymously gathered. Of the 87 snakebite mortalities detailed from a survey of newspaper reports, 25 are from snake handling. Of these, 4 who died were women, 15 were men, and 6 of these men were preachers. "List of Fatal Snake Bites in the United States," *Wikipedia.* https://en.wikipedia.org/wiki/List_of_fatal_snake_bites_in_the_United_States. Note that these numbers are based on media reports. Note also that it is not considered godly to seek medical assistance when bitten, so the number of bites has not been formally quantified. Media commentators find it very newsworthy when the leaders of these churches die, and much has been made of the fact that the founder of this practice, George Went Hensley, died from a snake bite.

59 Clifford Warwick, Phillip Arena, and Catrina Steedman, "Spatial Considerations for Captive Snakes," *Journal of Veterinary Behaviour* 30 (2019): 37–48. See page 40.

60 Tennessee Wildlife Resources Agency, "Copperhead," Tennessee Wildlife Resources Agency, retrieved from: https://www.tn.gov/twra/wildlife/reptiles/snakes/copperhead.html

61 Sarah Morrison, "The Accommodating Serpent and God's Grace in *Paradise Lost,*" *SEL Studies in English Literature 1500–1900* 49 (2009): 173–195. https://doi.org/10.1353/sel.0.0048. See page 174. This is not to dismiss the important readings of the snake as symbol, most interestingly

told by Harding Pitt, who paints the serpent as part of Milton's project to depict 'pagan beliefs as belated fallacies' and 'fallen angels' as an 'elaborate genealogy of pagan deities' in a comment on 'sacred histories'. In this careful reading of the literary allusions in *Paradise Lost,* the idea of a metamorphosis is present in Ovid's Cadmus who 'calls upon the gods to change him into a serpent' and 'Aesculapius' and his 'serpent-entwined caduceus' and Jove, who 'took serpent form' to father 'with human mothers' two 'conquerors, 'Alexander the Great' and 'Scipio Africanus'. Here, 'the Satan-possessed serpent' is the backsliding route to pagan rites, working on Eve's vainglorious tendencies towards 'self-deification' as the Greeks would have humans be gods. In this reading the serpent is less a snake than a 'subtext of all allusions'. Pitt Harding, "Milton's Serpent and the Birth of Pagan Error," *Studies in English Literature 1500–1900* (Winter 2007). See pages 164, 165, 167–8, 173 and 174. Morrison also considers the way Milton's Eve's internal shape has the same 'color Serpentine' as the tempter snake. The womb as space of the snake within might be read within these lines. This is old knowledge that goes back to matriarchal understandings of the body. There is not room in this chapter to look to the cultural role of snakes in other religions, but it is worth noting that globally, the snake is a symbol for eternal life, wisdom and fertility. In sympathy with Carl Jung, who made much of this fact, Ruth-Inge Heinze considers snakes across cultural symbology, noting in particular the ways in which fertility is associated with the serpent, activating examples from Mexico, Cambodia, Kenya and South India. She finds that the snake is both 'an archetype', and a power who 'opens new doors of perception'. Ruth-Inge Heinze, "Symbols and Signs, Myths and Archetypes: A Cross Cultural Survey of the Serpent," *Shaman* 10, no. 1–2 (2002). See pages 33, 41, 41, 43 and 51. There is good room for more ecofeminist interruptions in these spaces, by those culturally positioned to work on such analyses.

62 It must be noted, however, that I began to turn my fear to love by holding a captive snake. Susan Pyke, "Citizen Snake: Uncoiling Human Bindings for Life," in *The Materiality of Love: Essays on Affection and Cultural Practice* (Routledge, 2017).

63 Quantum of shared DNA might be evoked here, but more specific data is more interesting. Both humans and rattlesnakes share the ability to sense noxious chemicals. Helix, "This Gene Helps Snakes 'See' Heat—and Helps Humans Steer Clear of Harmful Substances," *Helix* (2022). https://blog.helix.com/2018/03/infrared-sensing-weekly-gene-trpa1/

64 I am not alone in looking for ways to express faith in ways that approach nonhuman animals differently. For example, David Cough has called to the Christian church to consider non-human animals more equitably. His premise is that all creatures (surely, snakes included) are 'good creatures of God', are 'beloved of God' and 'praise and delight God in their particular modes of flourishing', meaning the appropriate Christian response to snakes (amongst other species) must involve a 'love and respect' that seeks 'their flourishing'. David L. Clough, *On Animals: Volume Two: Theological Ethics* (T&T Clark, 2018). See page 243.The first of Clough's two-volume work *On Animals* was lauded as 'indisputably the most important and comprehensive theological treatment of animals to have appeared in any language at any time in the Christian tradition'. Brian Brock, "Review: David L. Clough, *On Animals. Volume 1: Systematic Theology*," *International Journal of Systematic Theology* 17, no. 3 (2015): 357–360. Ellen Grace Lesser and Christopher Southgate deemed the second volume 'a more extensive and stronger book'. Ellen Grace Lesser and Christopher Southgate, "On Animals: An Extended Review of David Clough's Two-Volume Work," *Studies in Christian Ethics* 34, no. 1 (2000): 2–21, 88–98.

65 Sue Hall Pyke, "Snake Church," *Animal Studies Journal* 11, no. 1 (2022): 102–120, https://doi.org/10.3390/rel13070597

66 I learn here from the workshop led by Terry Hurtado and Joana Formosinho, where participants played cow in a way I still feel in my body. 'Imagining Cow Being.' *Minding Animals Conference (MAC4)* Mexico City, 2018.

67 Teja Brooks Pribac, "Narrating Animals, between Fear and Resilience," *Religions* 13, no. 597 (2022): 1–14. https://doi.org/10.3390/rel13070597. See also Teja Brooks Pribac, "Spiritual Animal," in *Enter the Animal* (Sydney University Press, 2021). See pages 151–154

68 Petra Tschakert, "More-Than-Human Solidarity and Multispecies Justice in the Climate Crisis," *Environmental Politics* 31, no. 2 (2022): 277–296. https://doi.org/10.1080/09644016.2020.1853448

Bibliography

ABC. "Snake-Handling Pentecostal Pastor Dies from Snake Bite." (2014). Retrieved from: https://abcnews.go.com/US/snake-handling-pentecostal-pastor-dies-snake-bite/story?id=22551754#:~:text=It's%20estimated%20that%20125%20churches,more%20than%20a%20century%20ago.
Allendorf, Kalina. "Lucretian Subversion: Animal Speech and Misplaced Wonder in Paradise Lost." *Milton Quarterly* 52, no. 1 (2018).
Appalachian Church. "What We Believe." (Last updated 2022). Retrieved from: appchurch.org/index.php/what-we-believe/.
Appelros, Erica. "Gender within Christian Fundamentalism: A Philosophical Analysis of Conceptual Oppression." *International Journal of Philosophy and Theology* 75, no. 5 (2014): 460–473. https://doi.org/10.1080/21692327.2015.1036906
Australia's National Research Organisation for Women's Safety. "Intimate Partner Violence Homicides 2010–2018. Australian Domestic and Family Violence Death Review Network." *Data Report* (2022).
Bates, Alison. "Feminism Regained: Exposing the Objectification of Eve in John Milton's Paradise Lost." *English Studies: A Journal of English Language and Literature* 99, no. 2 (2018): 93–112. https://doi.org/10.1080/0013838X.2017.1405324
Bloom, Clive. "Introduction to the Gothic Handbook Series: Welcome to Hell." In *The Palgrave Handbook of Contemporary Gothic*, edited by Clive Bloom. Palgrave Macmillan, 2020.
Brock, Brian. "Review: David L. Clough, On Animals. Volume 1: Systematic Theology.' *International Journal of Systematic Theology* 17, no. 3 (2015): 357–360.
Carter, Angela. *Heroes and Villains*. Penguin, 2011 (1969).
Chow, Jeremy. "Snaking into the Gothic: Serpentine Sensuousness in Lewis and Coleridge." *Humanities* 10, no. 1 (2021).
Coccia, Emanelle. *Metamorphoses*. Polity Press, 2022.
Coleridge, Samuel Taylor. "Cristobel." 1816 (1797). Retrieved from: www.poetryfoundation.org/poems-and-poets/poems/detail/43971.
Coleridge, Samuel Taylor. *The Collected Works of Samuel Taylor Coleridge, Volume 11: Shorter Works and Fragments: Volume II*. Princeton University Press, 2019.
Creed, Barbara. *Stray: Human–Animal Ethics in the Anthropocene*. Power Publications, 2017.
Directed and written by Poulton, Britt, and Savage, Dan Madison. Them that Follow (2019).
Directed by Theo Love. Written by Thomas G. Burton, Theo Love and Bryan Storkel. Alabama Snake. HBO.
Flynn, Clifton P. "Examining the Links between Animal Abuse and Human Violence." *Crime, Law and Social Change* 55 (2011): 453–468.
Forsyth, Neil. "The Sign of the Dove and Serpent." *Milton Quarterly* 34, no. 2 (2000): 57–65.
Forsyth, Neil. *The Satanic Epic*. Princeton University Press, 2002.
Glass, William. "Review." *Tennessee Historical Quarterly* 52, no. 4 (1993): 266–267.
Haraway, Donna. *Staying with the Trouble: Making Kin in the Chthulucene*. Duke University Press, 2016.
Harding, Pitt. "Milton's Serpent and the Birth of Pagan Error." *Studies in English Literature 1500–1900* 47, no. 1 (2007): 161–177.
Heinze, Ruth-Inge. "Symbols and Signs, Myths and Archetypes: A Cross Cultural Survey of the Serpent." *Shaman* 10, no. 1–2 (2002).
Helix. "This Gene Helps Snakes 'See' Heat—and Helps Humans Steer Clear of Harmful Substances." *Helix*. 2022. Retrieved from: https://blog.helix.com/2018/03/infrared-sensing-weekly-gene-trpa1/
Howe, Julia Ward. "The Battle Hymn of the Republic." *The Atlantic Monthly* 9, no. 52 (1862). Retrieved from: https://www.theatlantic.com/magazine/archive/1862/02/the-battle-hymn-of-the-republic/308052/
Hurtado, Terry, and Formosinho, Joana. "Imagining Cow Being." *Minding Animals Conference (MAC4)*, Mexico City, 2018.
Keats, John. "Lamia." (1820). Retrieved from: www.online-literature.com/keats/2055/.
King James Bible.
Lesser, Ellen Grace, and Southgate, Christopher. "On Animals: An Extended Review of David Clough's Two-Volume Work." *Studies in Christian Ethics* 34, no. 1, 2–21 (2021): 88–98.

Lewis, Mathew 'Monk.' In *The Monk*. New York: Oxford University Press, 1796 (2008).
Manning, Erin. *For a Pragmatics of the Useless*. Duke University Press, 2020.
Massumi, Brian. *What Animals Teach Us About Politics*. Duke University Press, 1984.
Milbank, Alison. *God and the Gothic: Religion, Romance, and Reality in the English Literary Tradition*. Oxford University Press, 2018.
Miller, Shannon. "Serpentine Eve: Milton and the Seventeenth-Century Debate over Women." *Milton Quarterly* 42, no. 1 (2008).
Milton, John. *Paradise Lost*. 1674. Retrieved from: www.bartleby.com/360/4/155.html
Morrison, Sarah. "The Accommodating Serpent and God's Grace in Paradise Lost." *SEL Studies in English Literature 1500–1900* 49, no. 1 (2009): 173–195. doi 10.1353/sel.0.0048.
Mouzos, Jenny. "Femicide: An Overview of Major Findings." *Trends and Issues in Crime and Criminal Justice*. Australian Institute of Criminology, no. 124 (1999).
National Geographic Television. "A Family Tradition." *Snake Salvation*. Produced by Jerry Decker and Matthew Testa (2013). Retrieved from: https://www.youtube.com/watch?v=k1-BhaX5GSE.
Parkinson, Claire. *Animals, Anthropomorphism and Mediated Encounters*. Routledge, 2020.
Pribac, Teja Brooks. "Spiritual Animal." In *Enter the Animal*, by Teja Brooks Pibac. Sydney University Press, 2021.
Pribac, Teja Brooks. "Narrating Animals, between Fear and Resilience". *Religions* 13, no. 597 (2022): 1–14. https://doi.org/10.3390/rel13070597
Pyke, Sue Hall. "Snake Church." *Animal Studies Journal* 11, no. 1 (2022): 102–120. https://doi.org/10.3390/rel13070597
Pyke, Susan. "Citizen Snake: Uncoiling Human Bindings for Life." In *The Materiality of Love: Essays on Affection and Cultural Practice*, edited by Anna Malinowska and Michael Gratzke. Routledge, 2017a.
Pyke, Susan. "Creaturely Shifts: Contemporary Animal Crossings through the Alluring Trace of the Romantic Sublime." *TEXT Journal of Writing and Writing Courses* 41 (2017b): 1–19.
Read, Jan. "Thomas G. Burton, MA'58, PhD'66, Serpents and Stories." *Vanderbilt News*, 2021. Retrieved from: news.vanderbilt.edu/2021/04/22/thomas-g-burton-ma58-phd66-serpents-and-stories/
Robertson, Paul, and Scott, Gabrielle. "Gnostic Thought in Milton's Paradise Lost." *Gnosis: Journal of Gnostic Studies* 7, no. 2 (2022): 171–223. https://doi.org/10.1163/2451859X-00702003
Sarafan, Jamie. "The Women and the Word: Serpent Handling and the Women of the Church of God, 1914–1935." PhD thesis, University of Colorado, 2016.
Tschakert, Petra. "More-Than-Human Solidarity and Multispecies Justice in the Climate Crisis." *Environmental Politics* 31, no. 2 (2022): 277–296. https://doi.org/10.1080/09644016.2020.1853448
Tennessee Wildlife Resources Agency. "Copperhead." Tennessee Wildlife Resources Agency. Retrieved from: https://www.tn.gov/twra/wildlife/reptiles/snakes/copperhead.html
Topping, Eva Catafygioti. "Reflections of an Orthodox Feminist." *Greek American Review* (1991). Retrieved from: scholarship.tricolib.brynmawr.edu/bitstream/handle/10066/10614/Catafygiotu-Topping_44_536.pdf?sequence=1&isAllowed=y
Walsh, Erin Galgay. "Sanctifying Boldness: New Testament Women in Narsai, Jacob of Serugh, and Romanos Melodos." PhD thesis, Duke University, Graduate Program in Religion, 2019. Retrieved from: https://dukespace.lib.duke.edu/dspace/bitstream/handle/10161/19813/Walsh_duke_0066D_15282.pdf?sequence=1
Warwick, Clifford, Arena, Phillip, and Steedman, Catrina. "Spatial Considerations for Captive Snakes." *Journal of Veterinary Behaviour* 30 (2019): 37–48.
Wasko, Dennis K., and Bullard, Stephan G. "An Analysis of Media-Reported Venomous Snakebites in the United States, 2011–2013." *Wilderness and Environmental Medicine* 27, no. 2 (2016): 219–226. https://doi.org/10.1016/j.wem.2016.01.004
Whitmont, E.C. *The Symbolic Quest*. G.P. Putnam's Sons, 1969.
Wikipedia. "List of Fatal Snake Bites in the United States." *Wikipedia*. Retrieved from: https://en.wikipedia.org/wiki/List_of_fatal_snake_bites_in_the_United_States.

Williamson, Paul, and Pollio, Howard R. "The Phenomenology of Religious Serpent Handling: A Rationale and Thematic Study of Extemporaneous Sermons." *Journal for the Scientific Study of Religion* 38, no. 2 (1999): 203–218.
Wollstonecraft, Mary. *A Vindication of the Rights of Woman: With Strictures on Political and Moral Subjects'*. Project Gutenberg, 1792.
Wood, Ralph, and Williamson, Paul. *Them That Believe: The Power and Meaning of the Christian Serpent-Handling Tradition*. University of California Press, 2018.

33

MASCULINITY AND MULTISPECIES LABOUR

A Feminist Animal Studies Reading of *In the Skin of the Lion*

Tessa Wotherspoon

Introduction

In this chapter I offer a feminist animal studies reading of Michael Ondaatje's 1987 novel, *In the Skin of a Lion*.[1] While scholarly attention has explored the text's engagement with performative masculinities,[2] the role of the ever-present non-humans in the text have not been recognised. In the opening chapter of the novel, Ondaatje begins signalling to the reader that *In the Skin of a Lion* is not only a book about oppressed masculine labourers but also the oppression of non-human animals. Here we see the protagonist, Patrick Lewis, as a boy watching loggers on their way to work before the sun had risen, passing a field of cattle. Ondaatje writes:

> Sometimes the men put their hands on the warm flanks of these animals and receive their heat as they pass. They put their thin-gloved hands on these black and white creatures, who are barely discernible in the last of the night's darkness. They must do this gently, without any sense of attack or right. They do not own this land as the owner of the cows does.[3]

This scene foreshadows human and non-human entanglements of exploitation and violence, an entanglement that is central to the disruption of dominant forms of masculinity in the book. Moving beyond representations of oppression, I argue that *In the Skin of a Lion* offers a transformative approach to this interconnection between masculine and non-human labourers. Thus, *In the Skin of a Lion* re-historicises the relationship between male and non-human labourers, presenting a masculinity in early 20th-century Toronto that subverts dominant Western constructs.

Sri Lankan-born writer Michael Ondaatje is known for producing novels that intermingle history with fiction and poetry. *In the Skin of a Lion*[4] is no exception. Centred on the life of Patrick Lewis, the text is set largely in Toronto, Ontario, in the 1920s and 30s. The novel begins in Patrick's adolescence, when he is being raised by his father, Hazen Lewis, in a small community in Ontario. Patrick eventually moves to Toronto, where he enters a series

DOI: 10.4324/9781003273400-41

of dangerous and dirty jobs, including the construction of a tunnel under Lake Ontario and working in a leather tannery. Culturally, Ondaatje describes North America at the time as "without language", with "gestures and work and bloodlines"[5] as the primary means of producing hierarchy and meaning, hierarchies which subordinate the marginal labourers in the text. Ondaatje intertwines Patrick's life with that of many other masculine, largely immigrant, and low-paid labourers, including Nicholas Temelcoff, an immigrant labourer from Macedonia whose masculinity I use as a point of analysis. Other characters include Clara Dickens, Alice Gull, and Alice's daughter Hana, three women who, in addition to animals, become central figures in the transformation of Patrick and Nicholas' masculine performances. Despite the text's fictional narrative, it is nonetheless based on the invisible histories of marginal workers in early 20th-century Toronto, signalling Ondaatje's intent to uncover the hidden truths behind the history of the city.

This chapter approaches sex and gender as social constructs. Exploring the performative nature of masculinity, Kadri Aavik notes: "just as masculinities are not natural or fixed, the category 'man' is not to be taken for granted either".[6] While there is no universal construct of masculinity, there are normative performances within many cultural contexts. In the Western world, hegemonic masculinities and their connections to violence, virality, and dominance produce negative relationships with the environment, non-human animals, women, and men who perform gender non-normatively.[7] In this chapter, I use the term "dominant masculinity" to denote these performances of Western masculinity.

Representations of Masculinity

Scholars have attended to gendered performance as a central theme of *In the Skin of a Lion*.[8] Through a character analysis of Patrick, for instance, Burkitt asserts that Ondaatje "destabilize[s] binary constructions of gender by foregrounding the performance that is involved in any demonstration of manhood".[9] Overbye[10] agrees that Ondaatje's text is subversive of dominant expressions of masculinity, with Patrick's body becoming "the site of resistance in the social construction of the self".[11] According to Overbye, through Patrick's character Ondaatje reconciles the relationship between the self and the body—a relationship often fractured by dominant embodiments of masculinity. Extending these arguments, I explore the relationship between such masculine gender performance and animals in *In the Skin of a Lion*.

The greatest influence on Patrick's performance of masculinity in his adolescence is his father, Hazen Lewis. Hazen is described as skilled, isolated, and quiet mannered, with a dedication to his work as a dynamiter for a logging company—work that eventually kills him. Growing up Patrick rarely even hears the voice of his father, and when he does, his dialect is described as an "unemotional tongue".[12] Reflecting the isolation of his workplace, Ondaatje notes that Hazen made himself "as self-sufficient, as invisible as possible".[13] Hazen's lack of emotion appears to extend to his relationship to animals, with Ondaatje noting that Hazen "would step up to his horse and assume it, as if it were a train, as if flesh and blood did not exist".[14] As Aavik[15] explores, the dominant masculine relationship to animals is largely shaped by cultural context and material conditions. In much of the Western colonial world, dominant masculinities are commonly associated with an unfeeling and qualitative domination of the non-human.[16] Yet, despite Hazen's commitment to self-sufficiency and reserved demeanour, there are moments in the text where he strays from this attitude, demonstrating a fracture in his normative masculine performance. For instance, with the help of Patrick, Hazen saves a cow from a freezing lake, putting both himself and Patrick at risk due to the extreme temperatures. This act of compassion towards the cow even catches Patrick off guard, as he notes his father's obsession with "not wasting things",[17] especially in relation to rope—rope which they used to save this cow's life.

Nicholas, a worker constructing the Bloor Street Viaduct, also embodies a dominant performance of masculinity at the start of the text. When Nicholas first immigrated to Canada, he worked nights in a bakery, studying English obsessively during the day in recognition that "if he did not learn the language he would be lost".[18] Eventually he finds work in the construction of the Bloor Viaduct, a bridge which will connect Eastern and Western Toronto, the country to the city.[19] In his work on the viaduct, Nicholas is described as a secluded and fearless daredevil, taking on some of the most dangerous tasks of the project. Due to his willingness to risk his life for this labour, Nicholas's work is more highly valued than the average labourer, resulting in a higher wage that the rest of the construction workers. Nicholas is almost completely devoid of human connection, not even looking at the face of his supervisor while he is being given instructions, "as if he must hear the orders nakedly without seeing a face around the words".[20] Instead, his "eyes hook objects" and do not even dare look in pleasure at food that he is going to consume, "watching instead a man attaching a pulley to the elevated railings or studying the expensive leather on the shoes of the architects".[21] Nicholas "has no portrait of himself"[22] and appears boy-like, hyper-fixated and fanatical. Essentially, Nicholas lives and breathes his labour.

Upon moving to Toronto, Patrick's masculinity begins to be influenced by the materiality of his new life, including the attitudes and behaviours of the men around him. While Patrick is described during his adolescence as soft and delicate, a lover of literature with deep and complex emotions, he eventually takes up some aspects of dominant masculinity. Much as Nicholas is emotionally disconnected from his work, Patrick's body becomes mechanical in its labouring. His body is shown as physically interconnected with his shovel working on the tunnel under Lake Ontario, for instance, with "each blow against the shale wall jar[ring] up from the palms into the shoulders as if the body is hit".[23] Following in his father's footsteps, Patrick becomes a dynamiter, which Ondaatje notes is for Patrick "the only ease in this terrible place where he feels banished from the world".[24]

Moreover, Patrick's isolation from women for much of his young life leads him to fantasise about stereotypical tropes of femininity. When Patrick recalls his childhood growing up on a farm in the countryside, his mind immediately wanders to "his first seduction in a hay bed, the angry girl slapping him when both were full and guilty".[25] Before meeting Clara Dickens, Patrick primarily thought of women as they were depicted in the books he had read, being "rescued from runaway horses, from frozen pond accidents".[26]

In juxtaposition to the marginal status of labourers, Rowland Harris, the commissioner for public works, functions as the ideal vision of bourgeois dominant masculinity in the text—a performance that both Nicholas and Patrick come to subvert. Commissioner Harris leads two of the most significant projects in the book, the construction of the Bloor Street Viaduct and the tunnel under Lake Ontario. He is obsessed with dominating and transforming nature, referring to his waterworks project as his "palace of water".[27] Harris goes to great lengths to ensure his success, regardless of the impact on human and non-human labourers. It is in part the impacts of Harris' dominating embodiment of masculinity that I explore in the next section of this chapter.

More than human entanglements in dirty and violent labour

Throughout the narrative, Ondaatje reminds readers that multiple species played a role in the physical labour of 20th-century Toronto. The human labourer's body is marked through work in multiple ways: illness, smell, dismemberment, and death. In turn, these marginalised, often immigrant workers, the lines between the body as the vessel of an individual subject and the body as a mechanism of labour are blurred. One of the violent and dirty jobs Patrick takes on is at Wickett and Craig's leather tannery as a "pilot man", where he cuts the skin of animals into leather. Wickett and Craig was founded in 1867 and is still in production today, marketed as of 2022 as "a world premier" tannery requiring "a labour-intensive method".[28] However, the original building that housed the tannery stayed open until 1990 and was described by Toronto photographer, Peter MacCallum, who captured photos of the tannery over the years up to its demise, as "very dirty" and "a terrible place to work".[29] MacCallum further comments on the smell that came from the building, which was "foul smelling downstairs; sweet smelling upstairs"; according to MacCallum, he has not "smelled anything like it since".[30]

The smell described by MacCallum, while later in the building's history, is much like the description in *In the Skin of a Lion*. While Patrick himself develops a certain smell from his work with animal skin at the tannery, he can't help but compare it to that of the dyers, whose foul smell never leaves, regardless of the length of their shower. The dyers were paid one dollar each day, and "nobody could last in that job more than six months and only the desperate took it".[31] Yet, it is not only the smell that plagues the dyers, as the chemical in the dye causes them to develop health conditions such as tuberculosis—the dire impacts of which may not become visible until later in life. While the dyers are placed at the greatest risk, each job at the tannery presents workplace hazards.

The killing floor in particular presents challenges to both mental and physical well-being, given the inherently violent nature of the job. Ondaatje describes workers moving "among the bellowing cattle stunning them towards death with sledge hammers, the dead eyes still flickering while their skins were removed".[32] Unsurprisingly given the violence of such a scene, recent research has explored the psychological and criminal impacts of slaughterhouse labour, including the influence of the job on interpersonal relationships.[33] Fitzgerald et al., for example, have explored the impact of the presence of the slaughterhouse on the surrounding community.[34] The theoretical findings of this research indicate that the presence of slaughterhouses increases arrest rates, including arrests for violent crimes such as sex offences.

The human labourers in *In a Skin of a Lion* are brutally exploited, however they do not fully mirror the experiences of non-human animals in the text, who suffer violence at generally far quicker speeds, and are often perpetually tied to their labour. For instance, the mules and pit horses who aid in the construction of the tunnel under Lake Ontario live in the tunnel where they forcibly labour, thus providing no separation between life and work. In *Animals, work, and the promise of interspecies solidarity*,[35] Kendra Coulter argues that in labour which involves both humans and non-human animals, the quality of human working conditions impacts the treatment of non-human animals. From this perspective, poor conditions and pay for human workers corresponds to a worse yet environment for animals. Additionally demonstrating intersections of human–animal labour, Coulter compares the labour of animals to that of women's domestic labour; animal labour and women's domestic labour are similar in that they are unpaid, undervalued, and often not considered "work" at all. Additionally, as Coulter notes, while there are varying levels of agency and social valuations of these labourers, these largely exploitative forms of work contribute significantly to "people, economies, societies and corporate interests everywhere".[36] As we see in Ondaatje's text, the labour that constructs the tunnel under Lake Ontario does not just serve the purpose of supplying waterworks for the city of Toronto, but also an aesthetic function in the mind of Commissioner Harris. Harris daydreams, for instance, of the "marble walls, the copper-banded roofs" and "the brass railings curved up three flights like an immaculate fiction" in the waterworks building.[37]

The construction of a tunnel under Lake Ontario serves as one of the strongest examples of multispecies labour in the text. Both men and non-human animals work in the construction of the tunnel, dominated by the vision of Commissioner Harris. Despite the devaluation of animal labourers in the text, the scene of mules and pit horses being lowered into the tunnel sticks in Patrick's and the other labourer's minds: "the teeth of the animals distinct, that screaming, the feet bound so they wouldn't slash out and break themselves, lowered forty feet down and remaining there until they died or the tunnel reached the selected mark under the lake".[38] Echoing Coulter's call to bring to the fore both human and animals at labour,[39] Ondaatje reminds us that despite their difference in treatment, there may not be much of a distinction between humans and animals at work: "The brain of the mule no more and no less knowledgeable than the body of a man who dug into a clay wall in front of him".[40]

Through his saturation in such violence, it is apparent to Patrick that the history of such violent labour ought to be remembered. However, he learns that in the eyes of the broader public, this violence is invisible. In the latter half of the text Patrick discovers that while his and his fellow labourer's lives and stories are entangled within the web of society as "part of a mural",[41] they are at the same time actively restricted from being told. When Patrick seeks out traces of the history of common labourers' work on the Bloor Street Viaduct, he quickly finds that nothing has been documented. Instead, Patrick finds recorded "every detail about the soil, the wood, the weight of concrete".[42] Even during the construction of the bridge, the labour remained practically out of sight from the eyes of the general population, with the primary existing photographs captured from afar in "time-lapse evolution",[43] or centred on the work of the engineers, architects, and commissioners. Ondaatje notes how it is often only the smell that calls attention to the construction that is taking place, in the darkness of the night or the early hours of the morning. Essentially, the workers, both human and non-human, have been visually removed from the construction of the viaduct, their piece of the mosaic replaced by statistics and material.

Given the marginal status of those working in construction during this time, it becomes clear why this invisibility persisted. As Ondaatje illustrates, the labourers in *In the Skin of a Lion* experience racism and xenophobia, which influenced their agency both materially and socially. Although 19th- to mid-20th- century Toronto was a large and ever-developing city, it nonetheless contained deep-seated British Protestantism, with the social and economic power laying in the hands of the Anglo-Protestant elite.[44] These power dynamics played out in Canadian immigration policy, which remained largely racially, ethnically, and nationally discriminatory, with a preference for white and English-speaking migrants.[45] In fact, it was nearly impossible for certain ethnic and racial groups to even enter the country legally.

Therefore, prejudices and discrimination were reproduced both socially and in government policy and were strongly tied to the economy and labour value. For instance, the 1920s saw an increase in xenophobia, with the labour of immigrants strongly devalued.[46] As a result, and especially after the financial crisis of 1929, immigration was sharply decreased.[47] This left little room for immigrant and working class economic and social mobility, especially for those who were unable to speak English.[48] As in the case of Nicholas, even when non-English-speaking immigrants learned the language, it was still nearly impossible for them to make their way up the social ladder. These individuals were left to the "shadows" of urban Toronto, often forming distinct communities, forced to take on low paid, dangerous, and dirty work.[49]

Like the human labourers, many of the animals, such as mules and cattle, in *In the Skin of a Lion* were also the product of labour migration. Mules, a cross-breed between donkeys and horses, were the result of intentional human breeding.[50] As a result of colonial expansionism, mules were first introduced to North America in the 15th century and have since been used by humans primarily as pack animals, contributing to mining, the construction of infrastructure, and even wars. While cattle were similarly introduced to North America in the 15th century, the rising presence of extractive industries beginning in the 18th century increased the presence of cattle in Canada.[51] The settlement of these industries by colonial forces increased the demand for meat, and in addition, it was hoped that cattle would aid in agricultural settlements[52] and cement the mission of colonialism and economic imperialism. By the early 20th century, cattle production was growing rapidly in Canada.

As argued by Brubaker, one of the core requirements for diaspora is boundary maintenance, which may be maintained by social exclusion or alienation by the host nation.[53] For both the humans and non-human animals in *In the Skin of a Lion*, this phenomenon plays out with these diasporic labourers being isolated to the countryside, within their workplace, or in the darkness of the night. Building upon this phenomenon, Muller demonstrates how animals as well as human labourers can experience a similar sense of zombified "social death" in the context of the slaughterhouse.[54] As *In the Skin of a Lion* demonstrates, this social death may well extend beyond the slaughterhouse to other forms of dirty and violent labour involving humans and non-human animals. Both the human and the non-human animal's suffering is invisible to the common eye within the text; they often work in darkness, sometimes underground or on the outskirts of society, and therefore sleep during the day; they are immigrants, poor, othered, or disconnected from the greater society.

Transformation

Beyond depicting the oppression and domination of humans and non-human animals, *In the Skin of a Lion* is "a novel about the wearing and the removal of masks; the shedding of skin, the transformations and translations of identity".[55] A prominent collective moment of such transformation occurs in the leather tannery where Patrick works. Ondaatje writes:

> Circular pools had been cut into the stone—into which the men leapt waist-deep within the reds and ochres and greens, leapt in embracing the skins of recently slaughtered animals. In the round wells four-foot in diameter they heaved and stomped, ensuring the dye went solidly into the pores of the skins that had been part of a live animal the previous day. The men stepped out in colours up to their necks, pulling wet hides out after them so it appeared they had removed the skin from their own bodies. They had leapt into different colours as if into different countries.[56]

Furthermore, while the dyers shower off the dye at the end of the day: "the colour disrobed itself from the body, fell in one piece to their ankles, and they stepped out, in the erotica of being made free".[57] While the toxic chemicals which the dyers leapt into will lead to their slow death, this scene embodies a symbolic transformation of both the masculine labourers as well as the animals whose skin is being dyed. Ondaatje acknowledges that it is the "skins *of* recently slaughtered animals" that are being dyed, restoring visibility to these beings. Simultaneously, by focusing on the aesthetics of the dyed animal's flesh, the bodies of the masculine labourers are entangled within that of the animal's flesh: an embodied merging of two oppressed groups. While indeed the impacts of these actions are mere moments within lives of brutality and violence, the power of this scene reminds the reader of the possibilities that can arise through a regard for non-humans.

Similarly, both Patrick and Nicholas experience a kind of transformation in the text. During his work on the construction of the Bloor Street Viaduct, Nicholas saves a nun who fell from a bridge, an encounter that changes the trajectory of his life. Later in the text we learn that the nun's name is Alice Gull, and she will save Nicholas in turn, rescuing him from the bonds of dominant masculinity. It is the experience of connection with Alice that provides Nicholas the path to enlightenment, and it is Alice who gives him a curiosity about beauty, the desire to seek out human connection, to pay attention to reality outside of his labour, and to abandon his dangerous work for a quiet life as a baker. Yet this experience also provides Alice the ability to transform herself, and she cuts away her nun's habit with Nicholas's shears following the incident on the bridge.

While Patrick takes up traits of dominant masculinity in the text, it becomes clear in the conclusion of the text that his performance of gender remains subversive. Most notably, Patrick consistently shows a sensitivity to animals, a sense which is heightened following the traumas of his life. Noticing cages of dogs on the street, for instance, Patrick can't help but compare the dog's position to that of his own when he was held captive in prison: "He got closer to the cages, looked into the eyes which saw nothing, the way his own face in prison had looked in a metal mirror".[58] In tandem with this heightened sensitivity, Patrick moves away from his formerly masculine-centred kinship structure, adopting Alice's daughter, Hana, following Alice's death.

Conclusion

In the Skin of a Lion is more than an account of marginal human labourers in early 20th-century Toronto. It offers a transformative approach to the entanglement of human and non-human labourers, re-historicising the role of animals in male-dominated workplaces. In the final scene of the novel, Patrick hears from Clara, and as he and Alice are on their way to meet her, he allows Alice to take the wheel, "pretending to luxuriate in the passenger seat, making animal-like noises of satisfaction".[59] At the same time that Ondaatje presents subversive masculinities in the conclusion of *In the Skin of a Lion*, he draws attention to the non-human world which surrounds them. As Patrick and Hana are on their way to meet Clara, Ondaatje mentions non-human animals two final times, describing "A dog's chain hung off a step railing" and "racoons pausing on steps seemingly tamed as if owning the territory of the porch".[60] In this way, in concluding *In the Skin of a Lion*, Ondaatje invokes both the persistent oppression of animals and the possibilities of their agency.

Notes

1 Michael Ondaatje, *In the Skin of a Lion*, 1st ed. (Random House of Canada, 1996).
2 See Karen Overbye, "Re-Membering the Body: Constructing the Self as Hero in In the Skin of a Lion," *Studies in Canadian Literature* 17, no. 2 (June 6, 1992). See also Katharine Burkitt, "In Their Fathers' Footsteps: Performing Masculinity and Fatherhood in the Work of Les Murray and Michael Ondaatje," in *Performing Masculinities*, eds. Rainer Emig and Antony Rowland (Palgrave Macmillan, 2010), 130–148.
3 Ondaatje, *In the Skin of a Lion*, 7.
4 Ibid.
5 Ibid., 43.
6 Kadri Aavik, *Contesting Anthropocentric Masculinities through Veganism: Lived Experiences of Vegan Men* (Cham: Springer International Publishing AG, 2023). https://doi.org/10.1007/978-3-031-19507-5.:11
7 Gwen Hunnicutt, *Gender Violence in Ecofeminist Perspective: Intersections of Animal Oppression, Patriarchy and Domination of the Earth*. Taylor & Francis Group. 1st ed. (London: Routledge, 2019). https://doi.org/10.4324/9781351026222.
8 Katharine Burkitt, "In Their Fathers' Footsteps: Performing Masculinity and Fatherhood in the Work of Les Murray and Michael Ondaatje," in *Performing Masculinities*, eds. Rainer Emig and Antony Rowland (Palgrave Macmillan, 2010), 132.
9 Katharine Burkitt, "In Their Fathers' Footsteps: Performing Masculinity and Fatherhood in the Work of Les Murray and Michael Ondaatje," Essay, in *Performing Masculinities*, eds. Rainer Emig and Antony Rowland (Palgrave Macmillan, 2010), 130–148.
10 Overbye, *Re-Membering the Body: Constructing the Self as Hero in In the Skin of a Lion.*
11 Ibid., n.p.
12 Ondaatje, *In the Skin of a Lion*, 19
13 Ibid., 18
14 Ibid., 15
15 Aavik, *Contesting Anthropocentric Masculinities through Veganism*
16 Hunnicutt, *Gender Violence in Ecofeminist Perspective*
17 Ondaatje, *In the Skin of a Lion*, 14
18 Ibid., 46
19 Ann Marie Murnaghan, "The City, the Country, and Toronto's Bloor Viaduct, 1897–1919," *Urban History Review* 42, no. 1 (2013): pp. 41–50, https://doi.org/10.3138/uhr.42.01.03.
20 Ondaatje, *In the Skin of a Lion*, 42
21 Ibid., 42.
22 Ibid.
23 Ibid., 105.
24 Ibid., 107.
25 Ibid., 53.
26 Ibid., 61.
27 Ibid., 221.
28 "The Wickett & Craig Story," Wickett & Craig, retrieved from: https://wickett-craig.com/the-wickett-craig-story/, accessed July 14, 2022.
29 Peter MacCallum via Christopher Hume, "Capturing Toronto's Grimy Past," *Toronto Star* (2008), retrieved from: https://www.thestar.com/life/homes/2008/02/12/capturing_torontos_grimy_past.html.
30 Peter MacCallum via Christopher Hume, "Capturing Toronto's Grimy Past."
31 Ondaatje, *In the Skin of a Lion*, 131.
32 Ibid.
33 Jessica H. Leibler, Patricia A. Janulewicz, and Melissa J. Perry, "Prevalence of Serious Psychological Distress among Slaughterhouse Workers at a United States Beef Packing Plant," *Work* 57, no. 1 (July 2017): 105–109, https://doi.org/10.3233/wor-172543.
Kelly Struthers Montford and Tessa Wotherspoon, "The Contagion of Slow Violence: The Slaughterhouse and Covid-19," *Animal Studies Journal* 10, no. 1 (2021): pp. 80–113, https://doi.org/10.14453/asj.v10i1.6.

Stephanie Marek Muller, "Zombification, Social Death, and the Slaughterhouse: U.S. Industrial Practices of Livestock Slaughter," *American Studies* 57, no. 3 (2018): 81–101, https://doi.org/10.1353/ams.2018.0048.

34 Amy J. Fitzgerald, Linda Kalof, and Thomas Dietz, "Slaughterhouses and Increased Crime Rates," *Organization & Environment* 22, no. 2 (2009): 158–184, https://doi.org/10.1177/1086026609338164.

35 Kendra Coulter, *Animals, Work, and the Promise of Interspecies Solidarity* (New York, NY: Palgrave Macmillan, 2016).

36 Ibid., 77

37 Ondaatje, *In the Skin of a Lion*, 109.

38 Ibid., 108.

39 Coulter, *Animals, Work, and the Promise of Interspecies Solidarity*

40 Ondaatje, *In the Skin of a Lion*, 108

41 Ibid., 145

42 Ibid.

43 Ibid., 26.

44 Harold Troper, "Becoming an Immigrant City: A History of Immigration into Toronto since the Second World War," in *The World in a City*, eds. Paul Anisef and Michael Lanphier (University of Toronto Press, 2003), 19–62, retrieved from: https://www.jstor.org/stable/10.3138/9781442670259.

45 Ibid.

46 Ibid.

47 Ibid.

48 Ibid.

49 Ibid.

50 Dave Babb, "History of the Mule," American Mule Museum (2013), retrieved from: https://www.mulemuseum.org/history-of-the-mule.html#:~:text=In%201495%2C%20Christopher%20Columbus%20brought,exploration%20into%20the%20American%20mainland.

51 Ian MacLachlan, "The Historical Development of Cattle Production in Canada," *The Historical Development of Cattle Production in Canada*, retrieved from: https://opus.uleth.ca/handle/10133/303, accessed July 25, 2022.

52 MacLachlan, "The Historical Development of Cattle Production in Canada."

53 Roger Brubaker, "The 'Diaspora' Diaspora," *Ethnic and Racial Studies* 28, no. 1 (2005): 1–19.

54 Muller, "Zombification, Social Death, and the Slaughterhouse."

55 "Michael Ondaatje," British Council, retrieved from: https://literature.britishcouncil.org/writer/michael-ondaatje, accessed May 15, 2023.

56 Ondaatje, *In the Skin of a Lion*, 130.

57 Ibid., 132.

58 Ibid., 210.

59 Ibid., 244.

60 Ibid.

Bibliography

Aavik, Kadri. *Contesting Anthropocentric Masculinities through Veganism: Lived Experiences of Vegan Men*. Cham: Springer International Publishing AG, 2023. https://doi.org/10.1007/978-3-031-19507-5.

Babb, Dave. "History of the Mule." American Mule Museum, 2013. Retrieved from: https://www.mulemuseum.org/history-of-the-mule.html#:~:text=In%201495%2C%20Christopher%20Columbus%20brought,exploration%20into%20the%20American%20mainland.

Brubaker, Roger. "The 'Diaspora' Diaspora." *Ethnic and Racial Studies* 28, no. 1 (2005): 1–19.

Burkitt, Katharine. "In Their Fathers' Footsteps: Performing Masculinity and Fatherhood in the Work of Les Murray and Michael Ondaatje." In *Performing Masculinities*, edited by Rainer Emig and Antony Rowland, 130–148. Palgrave Macmillan, 2010.

City of Toronto, Carl Benn. "The First Half of the 20th Century, 1901–51." City of Toronto, December 11, 2017. Retrieved from: https://www.toronto.ca/explore-enjoy/history-art-culture/museums/virtual-exhibits/history-of-toronto/the-first-half-of-the-20th-century-1901-51/.

Coulter, Kendra. *Animals, Work, and the Promise of Interspecies Solidarity*. New York, NY: Palgrave Macmillan, 2016.
Cudworth, Erika, McKie, Ruth E., and Turgoose, Di. "Introduction: Locating Feminist Animal Studies." In *Feminist Animal Studies*, 1st ed., 1–15. London: Routledge, 2022. https://doi.org/10.4324/9781003222620.
Fitzgerald, Amy J., Kalof, Linda, and Dietz, Thomas. "Slaughterhouses and Increased Crime Rates." *Organization & Environment* 22, no. 2 (2009): 158–184. https://doi.org/10.1177/1086026609338164.
Hume, Christopher. "Capturing Toronto's Grimy Past." *Toronto Star*, February 12, 2008. Retrieved from: https://www.thestar.com/life/homes/2008/02/12/capturing_torontos_grimy_past.html.
Hunnicutt, Gwen. *Gender Violence in Ecofeminist Perspective: Intersections of Animal Oppression, Patriarchy and Domination of the Earth*. 1st ed. Taylor & Francis Group. London: Routledge, 2019. https://doi.org/10.4324/9781351026222.
Kemp, Tom. *Historical Patterns of Industrialization*. Taylor & Francis Group. 2nd ed. London: Routledge, 1993. https://doi.org/10.4324/9781315844527.
Leibler, Jessica H., Janulewicz, Patricia A., and Perry, Melissa J. "Prevalence of Serious Psychological Distress among Slaughterhouse Workers at a United States Beef Packing Plant." *Work* 57, no. 1 (2017): 105–109. https://doi.org/10.3233/wor-172543.
MacLachlan, Ian. Ms. "The Historical Development of Cattle Production in Canada." Retrieved from: https://opus.uleth.ca/handle/10133/303.
"Michael Ondaatje." British Council. Retrieved from: https://literature.britishcouncil.org/writer/michael-ondaatje.
Muller, Stephanie Marek. "Zombification, Social Death, and the Slaughterhouse: U.S. Industrial Practices of Livestock Slaughter." *American Studies* 57, no. 3 (2018): 81–101. https://doi.org/10.1353/ams.2018.0048.
Murnaghan, Ann Marie. "The City, the Country, and Toronto's Bloor Viaduct, 1897–1919." *Urban History Review* 42, no. 1 (2013): 41–50. https://doi.org/10.3138/uhr.42.01.03.
Ondaatje, Michael. *In the Skin of a Lion*. 1st ed. Random House of Canada, 1996.
Overbye, Karen. "Re-Membering the Body: Constructing the Self as Hero in In the Skin of a Lion." *Studies in Canadian Literature* 17, no. 2 (June 6, 1992).
Struthers Montford, Kelly, and Wotherspoon, Tessa. "The Contagion of Slow Violence: The Slaughterhouse and Covid-19." *Animal Studies Journal* 10, no. 1 (2021): 80–113. https://doi.org/10.14453/asj.v10i1.6.
Troper, Harold. "Becoming an Immigrant City: A History of Immigration into Toronto since the Second World War." In *The World in a City*, edited by Paul Anisef and Michael Lanphier, 19–62. University of Toronto Press, 2003. Retrieved from: https://www.jstor.org/stable/10.3138/9781442670259.
"The Wickett & Craig Story." Wickett & Craig. Retrieved from: https://wickett-craig.com/the-wickett-craig-story/.

34
MARGARET ATWOOD'S DAIRYSCAPE

Emelia Quinn

Introduction

This chapter calls attention to the proliferation of dairy in the fiction of Margaret Atwood: from the consistency with which the characters across her prodigious literary output reach for a glass of milk and a cheese sandwich; to her liberal use of the adjectives "cheesy", "milky", and "buttery"; and to her employment of dairy imagery to signal everything from female oppression to the end of the world.

In what follows I offer a vegan theoretical reading of Atwood's use of dairy as camp spectacle. This reading of dairy as camp opens a space for deconstructing the myriad of meanings foisted upon dairy in the Western imagination through a vegan aesthetic lens that finds reparative possibilities in the co-mingling of pleasure, excess, and complicity. In order to advance such a reading, I first contextualise my study in relation to the wide range of research published over recent years in interdisciplinary milk studies and outline the various ways in which gender, animals, and eating have previously been explored in scholarship on Atwood. I then offer an accretion of examples from across what I am calling Atwood's "dairyscape" in order to establish dairy as a recurrent and excessive motif across her work.

In a different context, Jonathan Culler identifies what he terms as the "vealism" at the heart of Gustave Flaubert's *Madame Bovary* (1856). For Culler, the strange and surprising bovine motifs found throughout *Madame Bovary* (not least in Emma Bovary's name itself) demonstrate Flaubert's postmodern challenge to literary representation. Culler argues that "the proliferation of bovine elements undermines their representational quality, making them elements which refer to one another and to the mechanical process that produces them rather than a theme".[1] Found in meals and character and locations names, Culler notes how "In each of these cases we have a minor, insignificant detail, apparently working, through its very triviality, to connote the real; yet where we expect the real, we get more veal".[2] Bovine language is thus seen to function in Flaubert's writing as a prescient postmodernist sentiment that draws attention to the limits of literary representation. For Culler, Flaubert takes a parodic stance towards the representational where his "vealism puts us in the position of trying to interpret—without adequate means—the collision and collusion of the representational and antirepresentational".[3]

DOI: 10.4324/9781003273400-42

I suggest that something similar to "vealism" is happening in Atwood's work. In the proliferation and repetition of dairy language across Atwood's oeuvre, dairy repeatedly draws attention to itself as an overdetermined literary construction. Important from a vegan theoretical perspective is the way in which this playful dairy imaginary opens a space from which to question our deeply held cultural investments in milk and cheese.

Milk's Meanings[4]

Recent years have seen the publication of a plethora of histories of milk-drinking that establish the surprisingly recent emergence of milk-drinking in Western culture.[5] Many other works explore the wide-ranging cultural meanings of milk.[6] Mathilde Cohen and Yoriko Otomo, for instance, note the gendered dynamics of milk cultures, where "lactating animals and human mothers are devalued and commodified while the recipients of the milk . . . are elevated into a framework of greater value, socially and economically". In addition, Carol J. Adams has coined the term *feminized proteins* to refer to milk and eggs, a term that recognises "that female animals are doubly oppressed, in their living *and* in their dying".[7]

However, rather than exposing the institutionalised cruelty of nonhuman animals within the modern milk industry—from the enforced impregnation of dairy cows to the use of their calves in the veal industry—work in milk studies tends to focus on the symbolic and cultural meanings that attach themselves to milk. The designation of milk as "nature's perfect food", for instance, creates an association between milk and a pastoral ideal.[8] Work in milk studies explores the paradoxes of such an association, with fresh milk-drinking established as an invention of industrial modernity. While butter and cheese consumption were relatively common across human history, fresh-milk drinking was rare, becoming widespread only in cities in the mid-nineteenth century when urbanisation put new demands on working mothers which increased the need for alternatives to breastmilk.[9] Milk's relation to industrial modernity is also highlighted in such work through its imbrication with practices of insemination, purification, pasteurisation, refrigeration, and transportation that were required to make widespread milk-drinking possible.

In a North American context, the history of the institutionalisation of milk-drinking further disrupts milk's image as natural and universal. As E. Melanie DuPuis details, our conception of cow's milk as essential to children's dietary health emerged from the economic concerns and targeted propaganda campaigns of the U.S. National Dairy Council (NDC) in the first half of the twentieth century. According to DuPuis, the publications of the NDC became central to "defining the healthy child's body and behavior in America".[10] Milk production and consumption increased rapidly over the first half of the twentieth century, encouraged by milk programs for children and milk subsidies for those supplying milk to schools. The Second World War also provided a further boost to milk production and consumption, with milk linked to the war effort through a nationalist rhetoric focused on maintaining strong troops and a healthy population.[11]

In 1923, then President Henry Hoover declared that "the very growth and virility of the white race" was dependent on the U.S. dairy industry,[12] just one of many examples of what Tobias Linné and Ally McCrow-Young note as "a racialized politics of milk" that "manifests in discourses about social perfection and white racial superiority" in the late nineteenth and early twentieth centuries.[13] The normalisation of milk-drinking as a key component of a healthy diet and the construction of bodies as "lactose intolerant" has been noted by many as a form of Eurocentrism that negates the fact that a large percentage of the world's population are biologically incapable of digesting the lactose found in milk beyond a few years of age.[14] For Vasile Stanescu, the promotion of milk-drinking and lactose tolerance in contemporary alt-right discourses (particularly in the U.S.) must be understood in relation to a longer history of milk and race. Stanescu cites, for instance, a 1930s agricultural history of New York that asserted a causative link between dairy consumption and white racial superiority:

> A casual look at the races of people seems to show that those using much milk are the strongest physically and mentally, and the most enduring of the people of the world. Of all races, the Aryans seem to have been the heaviest drinkers of milk and the greatest users of butter and cheese, a fact that may in part account for the quick and high development of this division of human beings.

Milk here emerges as "a commodity fetish seemingly unifying 'intelligence' and 'race' ".[15]

Overdetermined milk discourses—which imbue milk with deeply held beliefs about gender, race, the nation, and human health and progress—feed into our understanding of Atwood's dairyscape by revealing dairy's ability to attract polymorphous and often competing structures of meaning. Attention to the work of milk studies might also suggest that Atwood's dairy imaginary is not so unique or surprising, given both Atwood's longstanding interest in the dynamics of gendered power and oppression and the saturation of our everyday language with dairy and its metaphors. As Melanie Jackson and Esther Leslie note,

> Milk helps us form our first words and catalyzes a language that is quickly adapted to a language of cogitation and communication, forming a social and cultural matrix of metaphor: skim, condense, homogenize, express, churn, curdle, culture, sour, combine, separate.

They argue that milk possesses an "expressability" that manifests as a "capacity to be images, to seep into language and be made metaphorical".[16] However, I argue that the use of dairy in Atwood is so excessive and abundant that it draws attention to itself as a textual construction and offers therefore a site from which to critically reflect on the ways in which we project gendered, racial, and national values onto dairy products.

Animals, Gender, and Eating

Atwood's work is perhaps best known for its feminist themes, with her novels exploring issues of female empowerment, disempowerment, and complicity. Sarah Sceats has demonstrated the ways in which such explorations are intimately tied to questions of eating, with Atwood making "extensive symbolic use of food and eating to highlight themes such as the commodification of women, the duplicity of sexual predation or the negative power of the victim".[17] Others have focused on the specific role of meat and animal products within this gustatory imaginary and over the past few decades Atwood's work has become a staple of animal studies scholarship: from Adams's analysis of *The Edible Woman* in *The Sexual Politics of Meat,* to a wealth of articles and book chapters considering the intersections between Atwood's treatment of gender and animals,[18] meat-eating,[19] and veganism.[20]

Atwood's texts demonstrate overt engagement with the "absent referent" animal first expounded by Adams.[21] Examples of instances in which absent referent animals become visible abound, such as when the protagonist of *Cat's Eye* describes the turkey on her family dining table as having "thrown off its disguise as a meal and revealed itself . . . for what it is, a large dead bird" or when Marian of *The Edible Woman* recognises her steak as "part of a real cow that once moved and ate and was killed".[22] However, Atwood's novels resist the vegetarian politics Adams sees as resulting from the exposure of the logic of the absent referent, with Atwood remaining sceptical of vegetarianism and veganism, presented most often in her novels as naïve projections of innocence that fail to acknowledge our implication in a world of inevitable consumption.

In *The Edible Woman,* Marian's increasing identification with dead and dismembered nonhuman animals leads her to the rejection of various animal foods. Her rejection of red meat, for example, comes to signal her own increasing identification of cows and women, as in her description of her female friend's expression of a desire to become a mother as reminiscent "of a farmer discussing cattle-breeding".[23] For Sceats, Marian's disgust at eggs functions as a metaphor for her "disgust with mature femaleness and the generation of life",[24] connecting her unconscious veganism to reproductive politics. However, Marian's veganism comes to an end at the close of the novel as she triumphantly proclaims that "[she] had steak for lunch",[25] before gorging on an effigy of herself made from cake. These are two seemingly necessary steps for her to re-enter civilisation and abandon what is presented in the novel, according to Chloë Taylor, as a "neurotic and self-deceived" victim complex.[26] Marian re-enters the world as a consumer and recognises, through the literal consumption of the cake-based image of herself she has constructed, her own collusion in women's commodification at the hands of men.

This reading of *The Edible Woman* is just one example of the ways in which meat and animal products offer metaphoric support to broader questions of female identity and power in Atwood's fiction. As will be detailed in the following, dairy accompanies similar explorations of gendered power dynamics but has been overlooked in existing criticism, appearing only as part of broader food-based critiques.[27] Dairy is though worth considering as its own distinct symbolic matrix. While meat and eggs offer a relatively stable set of symbolic markers in Atwood's fiction—a projection of shared victimhood in the case of meat and a symbol of maternal reproductivity in the case of eggs[28]—dairy finds itself tethered to an endlessly proliferating series of meanings and repeatedly appears as a site of metatextual reflection, drawing attention to its status in the novels as an overdetermined textual construction.

Atwood's Dairyscape

Cheese sandwiches and glasses of milk abound in Atwood's diegetic world(s), appearing, on the surface, as part of a background realism. However, at other points dairy language seeps into increasingly incongruous similes, as, for example, in the descriptions of the weather in *The Blind Assassin,* where the afternoon light is "like melted butter" and the mist "swirl[s] like skim milk in the air".[29] The following examples are grouped into loose categories that highlight the dominant metaphors around which dairy coalesces and demonstrate just some of the ways in which such imagery unravels into a series of endless textual deferrals. I offer only brief summaries of just some of the dairy-based references to be found across

Atwood's work since a comprehensive catalogue is beyond the scope of this chapter. *Cat's Eye,* for example, to choose a particularly dairy-laden text, features sixty-two instances of dairy as a noun (whether cheese, milk, yoghurt, or cream); eight mentions of milking as a verb or adjectives such as cheesy or milky; four references to curdling, condensing, churning, and skimming; and nineteen further etymological derivatives such as milkweed plants and butterflies.[30] While not therefore aiming for an exhaustive survey, the following chronicling of Atwood's dairyscape offers a select few examples of dairy's function in her texts. Despite this narrowing, the following remains nonetheless lengthy and dense, designed to simulate the experience of reading Atwood, in which dairy accumulation happens at an insistent and often overwhelming rate.

Dairy and Gender

Dairy appears insistently in Atwood's writing as an overt metaphor for female commodification. Women's bodies are draped in dairy-clothing, as with the milk-protein fabric targeted at female consumers in *The Blind Assassin,*[31] and the milk-silk harvested from goat and spider splices in *Oryx & Crake.*[32] Women are also frequently described *as* cheese. In *The Heart Goes Last,* Charmaine is asked to "Play the piece of cheese to Ed's rat", and in *Life Before Man,* Nate's desire to feel protected by women sees the narrative suggest his desire to be a mouse, with his estranged wife Elizabeth envisioned as a cage with a heart of "pure cheese".[33] In these latter two examples, the conventional patriarchal notion of women as irresistible temptations and sirens luring unsuspecting men into danger is situated in relation to the long-standing cultural myth connecting mice with a love for cheese. It is worth noting here that the relation between mice and cheese is a cultural fiction, an association that gained cultural dominance in North America by the success of cheese-loving mice in the U.S. cartoon *Tom and Jerry.*[34] The allusion in *The Heart Goes Last* and *Life Before Man* to a cartoonish depiction of mice and cheese as a metaphor for female oppression thus invites a reflection on a particular type of commodification of women, one that unravels not to any pure referent but to a metaphor of already impure origins that implies the threat of an encroaching U.S. cultural imperialism.

Atwood also offers overt metatextual reflections on the textual origins of dairy-based misogyny in *Lady Oracle, Bodily Harm,* and *The Blind Assassin*: three novels that each feature female writers as their protagonists. In the costume gothics that Joan of *Lady Oracle* writes, for instance, women are described as "whey-faced chit[s]" with "milky throat[s]".[35] In *Bodily Harm,* the murder mystery novels Rennie avidly consumes to avoid political engagement sees the heroine characterised by her "skin like clotted cream".[36] And in *The Blind Assassin,* the science fiction stories found in protagonist Iris's autobiographical novel describe young girls "dressed in cheesecloth trousers and little metal brassieres like two funnels joined by a chain".[37] Here, the fiction either consumed or produced by women reiterates a logic of dairy-based objectification. That this mirrors Atwood's own obsessive use of dairy descriptors renders her complicit, as a female author, in dairy-based objectifications. Reflecting on the dairy imagery of the low-brow literatures within Atwood's diegetic worlds offers a wry reflection on her own positionality as an author of generically hybrid fiction immersed in dairy-based imagery that associates milk and women as consumable objects. Here dairy offers a reflection on a shared writerly and readerly complicity in the language of female commodification.

The accoutrements of cheese-making are also repeatedly associated with female commodification through the spectacle of young women on stage.[38] In "Salome was a Dancer", Atwood's twelve-year-old Salome appears in a school play wearing only "[s]even layers of cheesecloth".[39] Here dairy unravels not to any solid referent but back to another textual referent: Oscar Wilde's 1891 play *Salome* which first described Salome's dance as "the dance of the seven veils". Wilde's description positioned the dance as "a pivotal display of deviant desire" that "invoked and contributed to the evocative mythology surrounding the story", positioning Salome as a *femme fatale* and read by many as a precedent of modern striptease.[40] Wilde's text, originally written in French and censored in England until 1931, drew on a long history of re-writings and imaginings of the New Testament story of the execution of John the Baptist and as well as French re-writings, most notably Flaubert's 1877 *Hérodias*. Atwood's reference to "[s]even layers of cheesecloth" thus inserts dairy into an assemblage of prior textual references. The cheesecloth here becomes a parodic reference whereby, as with Culler's outline of Flaubert's vealism, dairy draws attention to itself as textual invention. The sinister narratorial voice of Atwood's short story places blame on the precocious child Salome—"All those middle-aged dads sitting with their legs crossed. Oh, she knew what she was doing!"[41]—and positions the sexual and physical violence that later befalls her as "all the mother's fault".[42] Here the cheesecloth layers might be read as the accretion of layers of historic literary constructions of female sexuality that have worked to denigrate female agency.

Female sexuality is also entangled with the role of women as producers of milk through their maternal function. In *Cat's Eye,* the protagonist Elaine's art school curriculum sees a proliferation of "milk-fed and pinky-white" depictions of the Virgin Mary.[43] Elaine notes how the breastfeeding aspect of the Virgin Mary's representation is entirely omitted from discussion, with the professor and students made uncomfortable by those images that appear with "a hand pressing the nipple and even real milk".[44] These paintings prompt an assertion by the women of their desire to bottle-feed, which is seen as "more sanitary".[45] This latter statement signals an internalised disgust directed towards the female body and its generativity. Here milk appears at the juncture of cultural constructions of ideal womanhood, and the role of patriarchal authorities in dictating the terms of our engagement with women's bodies and the contradictory valuation of virginity and maternity while condemning sexuality and embodiment.

Similarly, just as dairy emphasises female desirability, dairy metaphors also spoil, curdle, and sour to reveal a putrid and disgusting female body. This is aptly demonstrated in the acknowledgement by Offred, in *The Handmaid's Tale,* a novel in which daily milk drinking serves a metonymic function in relation to the Handmaids' "role as infant nurturers",[46] that her use of butter as a moisturiser grants only a temporary illusion of desirability since "The butter is greasy and it will go rancid and I will smell like an old cheese".[47] Similarly, in *The Blind Assassin,* the remains of Iris's fridge reflect her own ageing body: "An end of cheese, wrapped in greasy paper and hard and translucent as toenails".[48] Iris's mother also "smelled of milk" in the lead-up to her death.[49] Such references position the desirability of the female body as, like fresh milk, highly perishable.

In addition to being associated with the aging female body, the disgust-inducing properties of cheese are linked to dysfunctional female bodies that resist patriarchal visions of buxom maternity. In *The Handmaid's Tale,* the commander's wife refuses to acknowledge Offred's presence "although she knows I'm there. I can tell she knows, it's like a smell, her knowledge; something gone sour, like old milk".[50] Here it is not Offred herself who is the soured milk but her presence that forces a reflection on the wife's sense of her own bodily insufficiency: the fear of being infertile in a world which imbues pregnancy with Divine purpose. In *Bodily Harm,* Rennie, following a partial mastectomy due to a tumour in her breast, becomes convinced that her boyfriend can smell her sickness through her bandages: "there was a faint odour of decay seeping through the binding, like an off cheese".[51] This latter example draws attention to the gulf between idealised images of women and the pernicious violence of such images. It also mirrors the novel's exploration of the contrast between the picture-perfect beauty of the fictionalised postcolonial islands Rennie visits as a tourist and the continuing violence and injustice beneath the surface as perpetuated by neo-colonial policies.

Dairy and Nationalism

Atwood's dairyscape also engages with the changing shape of the dairy industry in the first half of the twentieth century and can be accorded the function of historical realism, tracing the significance of dairy to post-war North America. In *Cat's Eye,* Elaine lives a nomadic life in early childhood, where during the Second World War, "Meat and cheese are scarce, they are rationed".[52] Later, her and her family gorge on "hunks of cheese" since "There's more cheese, now that the war is over".[53] Milk then becomes cemented as part of childhood health, with the students at the school "issued small bottles of milk".[54] We hear also of the detritus left behind by milk consumption where her brother avidly collects milk-bottle tops "from dozens of dairies; he carried sheafs of them around in his pockets . . . and stood them up against walls and threw other milk-bottle tops at them to win more".[55] This mundane description of a childhood fad plays into the novel's broader nostalgia for a time in which milk was hand-delivered from local dairies. Indeed, the milk-bottle top collection is a fad that passes: "Then it was pop-bottle tops, then cigarette cards, then sightings of licence plates from different provinces and states".[56] Here there is a movement outwards beyond the local to the national and international. Dairy here signals a distinct Canadian anxiety about U.S. influence in the movement from cars arriving from Canadian provinces to U.S. states.

Post-war increases in dairy are treated in Atwood's works with suspicion and a sense of loss for an imagined ideal of a rural Canadian past. In *The Blind Assassin,* the end of the war sees an abundance of meat and "mountains of yellow butter". This takes on a biting tone as the narrator reflects that "Everyone ate and ate. They stuffed themselves full of technicolour meat and all the technicolour food they could get".[57] The yellow of the butter here is associated with the technicolour, and we can read it as vibrant but also tinged by its association with the technological, the filmic, and the illusory. In *Lady Oracle,* Joan's visits to the Canadian National Exhibition with her aunt, a site for the nationalistic display of agricultural development, sees them "head for the Pure Foods first every year to see the cow made of real butter; one year they made the Queen instead".[58] Here the image of the Queen carved from mountains of butter offers an ironic undercutting of dairy's abundance, with the purity of the "Pure Foods" logo tainted by the colonial undertones of the royal tribute.

The decline of localised milk production is also linked to cultural and economic colonisation in *Surfacing*. The narrator's neighbour Paul, living in rural Quebec, sees his milking shed transformed into a garage, with his cow metaphorically "killed by the milkbottle".[59] *Bodily Harm* also offers an overt critique of neo-colonial agricultural tariffs and trade agreements as Rennie—a tourist on a fictionalised Caribbean Island—finds herself unable to order yoghurt at her hotel. The waitress explains that "They got Pioneer industries for dairy now" and that "Dairy don't make no yoghurt. Yoghurt need powder milk. Powder milk outlaw, you can't buy it".[60] While associated through the name "Pioneer industries" with American neo-imperialism, *Bodily Harm* encourages a questioning of the false innocence of the "sweet Canadians" (an ironic refrain that recurs throughout the novel) who are equally implicated in the political unrest and poverty of the islands. In her life as a journalist prior to arriving on the islands, Rennie writes a profile as part of *Pandora* magazine's "Woman of Achievement Series", a series that includes "a female executive from a cheese food company".[61] This reconstituted cheese product comes to have more sinister consequences through the subtle references to the collapse of the local dairy industry on the islands and functions as a metaphor for the superficiality of Rennie's life in Toronto.

Dairy further signals a changing Canada in the description, in *Cat's Eye,* of the disappearance of the Union Jack from the Canadian flag, with the Canadian Red Ensign replaced in 1965 with the flag still in use today. Elaine struggles to acknowledge the new flag as an authentic vision of Canada, with the maple leaf at its centre appearing "like a trademark for margarine of the cheaper variety",[62] signalling a distinction between the real butter of the past and a new, cheaper, and lower-quality alternative.[63] Here again dairy appears as a signifier of artifice and an anxiety about the meaning of national identity in the wake of British colonialism. European influences also come under scrutiny by Elaine, with the cappuccino she is served by an "imitation Italian waiter" representing the unwelcome and alienating gentrification of an urban Toronto she no longer recognises.[64] Cappuccinos are also a sign of gentrification and alienation in *The Blind Assassin.* The introduction of cappuccinos in Iris's local diner sees her quip that "*I don't need that fluff on my coffee. Looks like shaving cream. One swallow and you're foaming at the mouth*".[65] Associated with rabies, as an incurable taint threatening to infect the population, dairy here becomes associated with the threat of alternative European cultures to a fixed idea of pre-war Canadian life.

Dairy and Nostalgia

The industrialisation of the dairy industry lingers as a threatening presence throughout Atwood's novels and functions as a locus point for anxieties about the fake and the real. Fake cheese abounds in the dystopic pre-apocalyptic world of her *MaddAddam* trilogy, from "marbled hunks of cheesefood" to "cheese food in a tube",[66] butter and powdered milk substitutes,[67] and "some sort of quasi-cheese product".[68] Such anxieties about cheesefood links to Jovian Parry's assessment of the nostalgia for "real" meat found in *Oryx and Crake* where

> The continued prestige enjoyed in the novel by "real" meat . . . is due to a longing for something authentic and natural in a world where consumers are so fundamentally divorced from the process of production, where pastoral interaction with the natural world is a fond but distant memory, and clever facsimiles of "real" food products abound.[69]

Similarly, Katarina Labudova has noted the role of "real" cheese as a sign of freedom in *The Handmaid's Tale.* For Labudova, "Ersatz Gileadean cheese . . . becomes the epitome of everything that is disgusting in the dystopian regime of fake food, fake relationships and fake religion".[70]

Even while Atwood's novels demonstrate a pervasive sense of nostalgia for an imagined time of subsistence dairying, such nostalgia is also undercut at various points by the acknowledgement of the ways in which such rural ideals are bound to patriarchal structures of control. Walking into town with a fellow handmaid, Offred of *The Handmaid's Tale* notes that "We must look good from a distance: picturesque, like Dutch milkmaids on a wallpaper frieze".[71] Here, the milkmaid represents an idealised image of femininity—a static image that represents a certain ideal of purity and a perceived antidote to urban life—derived from a patriarchal society that rigidly controls the women's clothing, movements, and diet. Laura in *The Blind Assassin*—the paradigm of exploited innocence—also finds herself framed as a rural milkmaid, joining a volunteer group who visit local hospitals "dressed up in dairy-maid pinafores with tulips appliqued on their bibs".[72] Framed here in relation to an image bespeaking female virtue and maternal care, this construction of Laura is later suggested by Iris as an enticing image for older, predatory men.[73]

Atwood's cautionary approach to nostalgia is also closely implicated with images of rural dairy production in *Cat's Eye*. Elaine recounts a memory from early childhood where a boy placed frozen horse-dung on the freshly washed white sheets hanging out to dry in their garden. The dung has come from "the milk-wagon horse" and leads Elaine to the reflection that "all sheets are white, all milk comes from horses".[74] Full understanding of the significance of this formulation, which demonstrates a state of ignorance through which the cow becomes an absent referent of the milking process, only comes later in the novel, with several encounters of ruined white sheets: the "oval splotch of blood" on the bed following Elaine's mother's miscarriage and the "great splotch of red blood" on the bedsheet following the botched abortion of a young art student.[75] Here the reader's knowledge that all milk certainly does not come from horses encourages a reflection on the concealment of violent gendered realities signalled by the white sheets. Behind the surface of the romanticised picture of the milk-wagon horse, and the implication of virginity and innocence in white sheets, is the reality of sexual oppression and violence.

Deconstructing Dairy, or Dairy Camp?

In the previous, dairy accumulates into a variety of structures of meaning. Dairy functions as a signifier for Atwood's interest in female disempowerment, with the commodification of bovine reproductivity providing a metaphorical linkage to female commodification and reproductive rights, which is further supported by Atwood's frequent comparison of women to cows.[76] The visceral properties of dairy also simultaneously highlight the contradictions of patriarchal culture: an apt symbol of the circulation of female bodies as something both desirable and disgusting. Dairy also appears as a generic signifier for an imagined ideal of Canadian life under threat from urbanisation and foreign influence. Technological interventions, from skimming to cheesefoods, also provide a fertile terrain from which Atwood can stage her interest in the distinctions between appearance and reality. In its passing and casual references, the assumption is that such dairy meanings are so familiar to the reader as to render them a helpful metaphorical shorthand for Atwood's broader interest in the interplay between innocence and complicity. However, I suggest we might also read dairy's relentless appearance in a different manner.

My own interest in Atwood's dairyscape has stretched over the past seven years, consisting of obsessive noting and cataloguing of her dairy references. Such an obsession though has failed at each point to result in a cohesive argument, finding myself unable to discern a logical explanation for its recurrence. Atwood's dairyscape has led to paranoic tendencies in my reading practices: to the creation of vast spreadsheets in the hope of being able to untangle some form of pattern, to allow some form of meaning to coalesce. However, I want to now turn away from the paranoic fantasy of revealing the hidden web of sense behind her dairyscape. Instead, I want to draw from Eve Kosofsky Segwick's work[77] to suggest the reparative possibilities of accepting the possibility that Atwood's dairy imagery, through accumulating *too much* meaning, comes to mean nothing at all. Extending my existing work on vegan camp, I propose that dairy proliferates in Atwood's novels to such an extent that we can read it, through a vegan lens, as a form of absurdity or campy playfulness that draws attention to its status as an empty signifier upon which so much cultural baggage is attached.

For Jackson and Leslie, "milk is polymorphic with an inclination for promiscuous collaboration—whether it be with bacteria, with cartoon avatars, with economics, pornography, racial politics, or genetic re-calibration".[78] Colliding in Atwood are the endless human meanings of dairy that resist a clear or fixed referent. The polymorphic inclination is made clear through the self-referentiality of dairy in Atwood's texts: aware at each point of its textual collaborations and deferrals. As with Culler's analysis of Flaubert's "vealism", Atwood's dairyscape draws our attention to a writerly tic. Such an authorial intrusion also implicates Atwood in the language of costume gothics, science fiction stories, and murder mysteries, as argued previously, embodying a postmodern breakdown of traditional divisions between high and low fiction. The intrusion of dairy metaphors at points of social critique signals Atwood's authorial intrusion into the text and therefore attracts a consideration of the perhaps inescapable role of literary texts in supporting and colluding with oppressive systems.[79]

While a vegan reading practice might be expected to take issue with Culler's silence as to the lived bovine reality behind the veal references that he reads as symbols of the anti-representational, I take a vegan theoretical approach to Atwood's dairyscape that repeats this absenting by reading dairy as pure linguistic construct. As I have argued in my previous work on vegan camp, there is a productivity to turning against the imperative to recover the absent referent animal and to dwell on the surface of language. Such a dwelling allows us to better acknowledge the symbolic investments we place upon dairy; the human meaning foisted upon such surfaces.

Vegan camp is an aesthetic category that seeks to rehabilitate vegan pleasures. As I have previously written:

> vegan camp uses the structures in which it is implicated to reimagine a relation to the material world, offering a possible survival strategy for vegans: the ability to revel in the instability of human attachments to meat, in the paradoxical nature of the desire to consume and understand nonhuman animals, and to accept the impossibility of a pure or complete veganism. In refusing to look beyond the surface, vegan camp laughs at the suggestion that dead animal bodies could constitute a position so central to notions of human identity.[80]

I want to suggest here the vegan potentialities of reading the proliferation of surface significations in Atwood's dairyscape as a parody of the representational faith placed in dairy. A disruption therefore to human meanings of dairy, the camp spectacle of Atwood's literary dairyscape is to be found in the accretion of referents through which anxiety is generated around "the inability to attach clear meaning to slaughtered animal remains".[81] In a vegan reader's recognition of a certain complicity derived from our ability to make sense of this increasingly bizarre and absurd proliferation of dairy we find ourselves returning to Atwood's broader thematic concern with individual complicity in violence and her scepticism of veganism and its association with moral purity.

Atwood herself seems anxious in her writing to return to a world of dairy stability, as is evident in her overt disdain for cheesefoods. However, this is an anxiety that is itself undermined by the enmeshment of such longing with misogyny, neo-imperialism, and false projections of innocence. The desire for return to a fixed dairy referent is simultaneously undermined by the deconstructive tendencies of her work, in which the referent for such a pastoral ideal unravels only to reveal further textual referents. A vegan camp reading seeks to draw satirical pleasure from this longing and to parody our reliance on the exploitation of nonhuman animals to make sense of ourselves within a white, Western, and capitalist frame.

A vegan camp reading of Atwood's dairyscape thus provides a sense of spectacle and pleasure for vegan readers by embracing rather than seek to negate veganism's often uncomfortable associations with purity politics. Such a reading turns away from the revelation of the absent referent animal to remain on the surface of dairy metaphors and to acknowledge the absurd level of investment we place in animal products—whether around race, gender, or national identity—whilst simultaneously acknowledging the ways in which vegans, through their position in an inescapable system of animal exploitation, are implicated in such absurdity. A vegan camp reading of Atwood's dairyscape also enacts a form of utopian aspiration for a vegan future, not through a return to an imagined past, but through the promotion of creative ways of detaching ourselves from the cultural weight of dairy. It is indeed arguably only through such creative detachment that we can begin to envision a world without dairy.

Acknowledgements

For their feedback on earlier drafts of this chapter, I would like to thank Daniel Ibrahim Abdalla and Laura Wright.

Notes

1 Jonathan Culler, "The Uses of *Madame Bovary*," in *Flaubert and Postmodernism*, eds. Naomi Schor and Henry F. Majewski (Lincoln; London: University of Nebraska Press, 1984), 7.
2 Ibid.
3 Ibid., 9.
4 When referring to "milk" in this chapter, I am predominantly referring to cow's milk.
5 Peter Atkins, *Liquid Materialities: A History of Milk, Science and the Law* (Abingdon: Routledge, 2016); E. Melanie DuPuis, *Nature's Perfect Food: How Milk Became America's Drink* (New York: New York University Press, 2002); Mark Kurlansky, *Milk! A 10,000-Year Food Fracas* (New York: Bloomsbury, 2018); Ann Mendelson, *The Surprising Story of Milk through the Ages* (New York: Knopf, 2008); Kendra Smith-Howard, *Pure and Modern Milk: An Environmental History since 1900* (Oxford: Oxford University Press, 2013); Deborah Valenze, *Milk: A Local and Global History* (New Haven: Yale University Press, 2011); Hannah Velten, *Milk: A Global History* (London: Reaktion Books, 2010)
6 Mathilde Cohen and Yoriko Otomo, eds., *Making Milk: The Past, Present and Future of our Primary Food* (London: Bloomsbury Academic, 2017); Richie Nimmo, *Milk, Modernity and the*

Making of the Human: Purifying the Social (Abingdon: Routledge, 2012); Andrea S. Wiley, *Cultures of Milk: The Biology and Meaning of Dairy Products in the United States and India* (Cambridge, MA: Harvard University Press, 2014); Andrea S. Wiley, *Re-imagining Milk: Cultural and Biological Perspectives* (Abingdon: Routledge, 2015).

7 Cohen and Otomo, introduction to *Making Milk: The Past, Present and Future of our Primary Food* (London: Bloomsbury Academic, 2017), 1; Carol J. Adams, *The Sexual Politics of Meat: A Feminist-Vegetarian Critical Theory* (London: Bloomsbury Academic, 2015), 113, emphasis in original.

8 DuPuis, *Nature's Perfect Food.*

9 Ibid., 46–66; Wiley, *Cultures of Milk,* 35–36.

10 DuPuis, *Nature's Perfect Food,* 111.

11 Wiley, *Cultures of Milk,* 40.

12 Cited in Mathilde Cohen, "Of Milk and the Constitution," *Harvard Journal of Law & Gender,* 40 (2017): 148.

13 Tobias Linné and Ally McCrow-Young, "Plant Milk: From Obscurity to Visions of a Post-Dairy Society," in *Making Milk: The Past, Present and Future of our Primary Food,* eds. Mathilde Cohen and Yoriko Otomo (London: Bloomsbury Academic, 2017), 200.

14 See Wiley, *Cultures of Milk,* 106–107.

15 Vasile Stanescu, " 'White Power Milk': Milk, Dietary Racism, and the 'Alt-Right'," *Animal Studies Journal* 7, no. 2 (2018): 108.

16 Melanie Jackson and Esther Leslie, "Unreliable Matriarchs," in *Making Milk: The Past, Present and Future of our Primary Food,* eds. Mathilde Cohen and Yoriko Otomo (London: Bloomsbury Academic, 2017), 63.

17 Sarah Sceats, *Food, Consumption, and the Body in Contemporary Women's Fiction* (Cambridge: Cambridge University Press, 2000), 4.

18 Tania Aguila-Way, "Beyond the Logic of Solidarity as Sameness: The Critique of Animal Instrumentalization in Margaret Atwood's *Surfacing* and Marian Engel's Bear," *ISLE: Interdisciplinary Studies in Literature and Environment* 23, no. 1 (2016): 5–29; Robert McKay, "Identifying with the Animals: Language, Subjectivity and the Animal Politics of Atwood's *Surfacing,*" in *Figuring Animals: Essays on Animal Images in Art Literature, Philosophy and Popular Culture,* eds. Mary Sanders Pollock and Catherine Rainwater (Basingstoke: Palgrave Macmillan, 2005), 207–227.

19 Susan McHugh, "Real Artificial: Tissue-cultured Meat, Genetically Modified Farm Animals, and Fictions," *Configurations* 18, no. 1 (2010): 181–197; Jovian Parry, "*Oryx and Crake* and the New Nostalgia for Meat," *Society and Animals,* no. 17 (2009): 241–256.

20 Tatiana Prorokova-Konrad, "Veganism, Ecoethics, and Climate Change in Margaret Atwood's 'MaddAddam' Trilogy," in *The Routledge Handbook of Vegan Studies,* ed. Laura Wright (Abingdon: Routledge, 2021), 76–88; Emelia Quinn, "Margaret Atwood and Monstrous Vegan Words," in *Reading Veganism: The Monstrous Vegan, 1818 to Present,* 89–115 (Oxford: Oxford University Press, 2021); Chloë Taylor, "Abnormal Appetites: Foucault, Atwood, and the Normalization of an Animal-Based Diet," *Journal for Critical Animal Studies* 10, no. 4 (2012): 130–148; Laura Wright, "Vegan Zombies of the Apocalypse: McCarthy's *The Road* and Atwood's *The Year of the Flood,*" in *The Vegan Studies Project: Food, Animals, and Gender in the Age of Terror* (Athens: University of Georgia Press, 2015), 68–88.

21 For Adams, animals become absent referents in three key ways: "One is literally: . . . through meat eating they are literally absent because they are dead. Another is definitionally: when we eat animals we change the way we talk about them, for instance, we no longer talk about baby animals but about veal or lamb. . . . The third way is metaphorical. Animals become metaphors for describing people's experiences" (21).

22 Margaret Atwood, *Cat's Eye* (1988; London: Virago, 1990), 131; Margaret Atwood, *The Edible Woman* (1969; London: Virago, 1980), 151.

23 Atwood, *Edible Woman,* 42.

24 Sceats, *Food, Consumption, and the Body,* 97.

25 Atwood, *Edible Woman,* 280.

26 Taylor, "Abnormal Appetites," 134.

27 See Maria Christou, *Eating Otherwise: The Philosophy of Food in Twentieth-Century Literature* (Cambridge: Cambridge University Press, 2017). Christou briefly mentions the milk-drinking

in *The Handmaid's Tale* as part of her analysis of the figurative auto-cannibalism of the novel, "whereby one becomes who one is through 'eating' oneself" (135). Here dairy is just one of numerous food products that symbolizes the handmaids' "incorporation of their role as infant nurturers": (134).

28 Atwood herself elaborates on Canadian animal stories as symptoms of a national victim complex in *Survival: A Thematic Guide to Canadian Literature* (Toronto: Anansi, 1972). For more on the link between meat and female oppression in Atwood see Adams, *The Sexual Politics*. For work on the reproductive resonance of eggs in Atwood's fiction, see Maria Christou's excellent work in *Eating Otherwise*. The role of eggs as metonyms of wombs is also detailed in Katarina Labudova, "Testimonies in *The Testaments* by Margaret Atwood: Images of Food in Gilead," *ELOPE* 17, no. 1 (2020): 97–110, and Karen Stein, "Margaret Atwood's Modest Proposal: *The Handmaid's Tale*," *Canadian Literature* 148 (1996): 57–72.

29 Margaret Atwood, *The Blind Assassin* (Toronto: McClelland & Stewart, 2000), 17, 150.

30 Butterflies and milkweed plants are terms that draw on the language of dairy. As Greta Gaard has noted, terms such as milkweed "extend cultural conceptions of milk" where "To see this milk everywhere in 'milky' plants and fruits is to articulate not just recognition but yearning—and an invitation to appropriate and use the 'milky' substance for human purposes" ("Toward a Feminist Postcolonial Milk Studies," *American Quarterly* 65, no. 3 (2013): 226, 227).

31 Atwood, *Blind Assassin,* 225.

32 Margaret Atwood, *Oryx and Crake* (Toronto: McClelland & Stewart, 2003), 199.

33 Margaret Atwood, *The Heart Goes Last* (London: Bloomsbury, 2015), 237; Margaret Atwood, *Life Before Man* (1979; London: Virago, 1982), 162.

34 See, for example, a 2006 study claiming that mice are more likely to turn away from cheese due to its strong smell: http://news.bbc.co.uk/1/hi/england/manchester/5319210.stm.

35 Margaret Atwood, *Lady Oracle* (1976; London: Virago, 2009), 138, 183.

36 Margaret Atwood, *Bodily Harm* (1981; London: Vintage, 1996), 246.

37 Atwood, *Blind Assassin,* 153.

38 See, for example Atwood, *Life Before Man,* 71 and Atwood, *Lady Oracle,* 44.

39 Margaret Atwood, "Four Short Pieces: King Log in Exile, Post-Colonial, Salome Was a Dancer & Take Charge," *Daedalus* 134, no. 2 (2005): 121.

40 Tony W. Garland, "Deviant Desires and Dance: the *Femme Fatale* Status of Salome and the Dance of the Seven Veils," in *Refiguring Oscar Wilde's* Salome, ed. Michael Y. Bennett (Amsterdam: Rodopi, 2011), 125, 126.

41 Atwood, "Four Short," 121.

42 Ibid., 122.

43 Atwood, *Cat's Eye,* 283.

44 Ibid., 283.

45 Ibid., 284.

46 Christou, *Eating Otherwise,* 134.

47 Margaret Atwood, *The Handmaid's Tale* (1985; Boston: Houghton Mifflin Company, 1986), 97.

48 Atwood, *Blind Assassin,* 56.

49 Ibid., 93.

50 Atwood, *Handmaid's Tale,* 46.

51 Atwood, *Bodily Harm,* 49.

52 Atwood, *Cat's Eye,* 22.

53 Ibid., 28.

54 Ibid., 45.

55 Ibid., 60.

56 Ibid.

57 Atwood, *The Blind Assassin,* 506.

58 Ibid., 92. For an account of the emergence of butter sculptures as a key piece of propaganda for a newly industrialized dairy industry in the early twentieth century see Pamela H. Simpson, "Butter Cows and Butter Buildings: A History of an Unconventional Sculptural Medium," *Winterthur Portfolio* 41, no. 1 (2007): 1–20. Simpson's article explicitly references the success of the butter-sculpture of the Queen at the 1952 Toronto National Exhibition and its controversy in Britain for the sensational quantities of butter available to waste.

59 Margaret Atwood, *Surfacing* (1972; London: Virago, 1996), 13.
60 Atwood, *Bodily Harm,* 62.
61 Ibid., 66.
62 Atwood, *Cat's Eye,* 311.
63 See Simpson, "Butter Cows," for more on the threat of oleomargarine to the Noth American dairy industry. Other disparagements of margarine can be found across Atwood's oeuvre as when in *Lady Oracle*, just as Joan is trying to share her true self with husband Arthur, an advert comes on the television where "A famous figure-skater praised margarine, unconvincingly" (299).
64 Atwood, *Cat's Eye,* 151.
65 Atwood, *Blind Assassin,* 44, italics in original.
66 Atwood, *Oryx and Crake,* 107, 320.
67 Atwood, *Year of the Flood,* 150, 317.
68 Margaret Atwood, *MaddAddam* (2013; London: Virago, 2014), 102.
69 Parry, "New Nostalgia," 248.
70 Labudova, "Testimonies," 107.
71 Atwood, *Handmaid's Tale,* 212.
72 Atwood, *Blind Assassin,* 421.
73 Ibid., 486. A further example of the employment of the milkmaid at a site of ambivalent escapism can also be found in Rennie's experience in the squalid conditions of an island prison in *Bodily Harm,* where she braids her cellmate Lora's hair to make her resemble a German milkmaid (282).
74 Atwood, *Cat's Eye,* 122.
75 Ibid., 166, 320.
76 For a particularly explicit example, see *The Handmaid's Tale,* where the Aunts, the older women responsible for indoctrinating and policing the behaviour of new handmaids, enforce their control through "electric cattle prods slung on thongs from their leather belts" (4) while tattoos on the women's ankles resemble "a cattle-brand" (254).
77 Eve Kosofsky Sedgwick, "Paranoid Reading and Reparative Reading, or, You're So Paranoid, You Probably Think This Essay Is About You," in *Touching Feeling: Affect, Pedagogy, Performativity,* ed. Adam Frank (Durham, NC: Duke University Press, 2003), 123–151.
78 Jackson and Leslie, "Unreliable Matriarchs," 63.
79 See Quinn, *Reading Veganism,* where I explore moments in Atwood's work in which she explores the idea that "the act of writing is a mode of carnivorous appetite in itself, a violence that erodes agency and difference by seeking to speak for the other" (95).
80 Emelia Quinn, "Notes on Vegan Camp," *PMLA* 135, no. 5 (2020): 927.
81 Ibid., 924.

Bibliography

Adams, Carol J. The Sexual Politics of Meat: A Feminist-Vegetarian Critical Theory. Bloomsbury Academic, 2015.
Aguila-Way, Tania. "Beyond the Logic of Solidarity as Sameness: The Critique of Animal Instrumentalization in Margaret Atwood's Surfacing and Marian Engel's Bear." ISLE: Interdisciplinary Studies in Literature and Environment 23, no. 1 (2016): 5–29.
Atkins, Peter. Liquid Materialities: A History of Milk, Science and the Law. Abingdon: Routledge, 2016.
Atwood, Margaret. Survival: A Thematic Guide to Canadian Literature. Toronto: Anansi, 1972.
Atwood, Margaret. The Edible Woman. London: Virago, 1980. First published 1969.
Atwood, Margaret. Life before Man. London: Virago, 1982. First published 1979.
Atwood, Margaret. Cat's Eye. Toronto: McClelland & Stewart, 1988.
Atwood, Margaret. The Handmaid's Tale. Boston: Houghton Mifflin Company, 1986. First published 1985.
Atwood, Margaret. Bodily Harm. London: Vintage, 1996a. First published 1981.
Atwood, Margaret. Surfacing. London: Virago, 1996b. First published 1972.
Atwood, Margaret. The Blind Assassin. Toronto: McClelland & Stewart, 2000.
Atwood, Margaret. Oryx and Crake. Toronto: McClelland & Stewart, 2003.

Atwood, Margaret. "Four Short Pieces: King Log in Exile, Post-Colonial, Salome Was a Dancer & Take Charge." Daedalus 134, no. 2 (2005): 119–123.
Atwood, Margaret. Lady Oracle. London: Virago, 2009. First published 1976.
Atwood, Margaret. The Heart Goes Last. London: Bloomsbury, 2015.
Christou, Maria. Eating Otherwise: The Philosophy of Food in Twentieth-Century Literature. Cambridge: Cambridge University Press, 2017.
Cohen, Mathilde. "Of Milk and the Constitution." Harvard Journal of Law & Gender 40 (2017): 115–182.
Cohen, Mathilde, and Otomo, Yoriko, eds. Making Milk: The Past, Present and Future of our Primary Food. London: Bloomsbury Academic, 2017.
Culler, Jonathan. "The Uses of Madame Bovary." In Flaubert and Postmodernism, edited by Naomi Schor and Henry F. Majewski, 1–12. Lincoln and London: University of Nebraska Press, 1984.
DuPuis, E. Melanie. Nature's Perfect Food: How Milk Became America's Drink. New York: New York University Press, 2002.
Gaard, Greta. "Toward a Feminist Postcolonial Milk Studies." American Quarterly 65, no. 3 (2013): 595–617.
Garland, Tony W. "Deviant Desires and Dance: the Femme Fatale Status of Salome and the Dance of the Seven Veils." In Refiguring Oscar Wilde's Salome, edited by Michael Y. Bennett, 125–143. Amsterdam: Rodopi, 2011.
Jackson, Melanie and Esther Leslie. "Unreliable Matriarchs." In Making Milk: The Past, Present and Future of our Primary Food, edited by Mathilde Cohen and Yoriko Otomo, 63–80. London: Bloomsbury Academic, 2017.
Kurlansky, Mark. Milk! A 10,000-Year Food Fracas. New York: Bloomsbury, 2018.
Labudova, Katarina. "Testimonies in The Testaments by Margaret Atwood: Images of Food in Gilead." ELOPE 17, no. 1 (2020): 97–110.
Linné, Tobias and Ally McCrow-Young. "Plant Milk: From Obscurity to Visions of a Post-Dairy Society." In Making Milk: The Past, Present and Future of our Primary Food, edited by Mathilde Cohen and Yoriko Otomo, 195–212. London: Bloomsbury Academic, 2017.
McHugh, Susan. "Real Artificial: Tissue-cultured Meat, Genetically Modified Farm Animals, and Fictions." Configurations 18, no. 1 (2010): 181–197.
McKay, Robert. "Identifying with the Animals: Language, Subjectivity and the Animal Politics of Atwood's Surfacing." In Figuring Animals: Essays on Animal Images in Art Literature, Philosophy and Popular Culture, edited by Mary Sanders Pollock and Catherine Rainwater, 207–227. Basingstoke: Palgrave Macmillan, 2005.
Mendelson, Ann. The Surprising Story of Milk through the Ages. New York: Knopf, 2008.
Nimmo, Richie. Milk, Modernity and the Making of the Human: Purifying the Social. Abingdon: Routledge, 2012.
Parry, Jovian. "Oryx and Crake and the New Nostalgia for Meat." Society and Animals, no. 17 (2009): 241–256.
Prorokova-Konrad, Tatiana. "Veganism, Ecoethics, and Climate Change in Margaret Atwood's 'MaddAddam' Trilogy." In The Routledge Handbook of Vegan Studies, edited by Laura Wright, 76–88. Abingdon: Routledge, 2021.
Quinn, Emelia. "Notes on Vegan Camp." PMLA 135, no. 5 (2020): 914–930.
Quinn, Emelia. Reading Veganism: The Monstrous Vegan, 1818 to Present. Oxford: Oxford University Press, 2021.
Sceats, Sarah. Food, Consumption, and the Body in Contemporary Women's Fiction. Cambridge: Cambridge University Press, 2000.
Sedgwick, Eve Kosofsky. "Paranoid Reading and Reparative Reading, or, You're So Paranoid, You Probably Think This Essay Is About You." In Touching Feeling: Affect, Pedagogy, Performativity, edited by Adam Frank, 123–151. Durham, NC: Duke University Press, 2003.
Simpson, Pamela H. "Butter Cows and Butter Buildings: A History of an Unconventional Sculptural Medium." Winterthur Portfolio 41, no. 1 (2007): 1–20.
Smith-Howard, Kendra. Pure and Modern Milk: An Environmental History since 1900. Oxford: Oxford University Press, 2013.
Stanescu, Vasile. "'White Power Milk': Milk, Dietary Racism, and the 'Alt-Right'." Animal Studies Journal 7, no. 2 (2018): 103–128.

Stein, Karen. "Margaret Atwood's Modest Proposal: The Handmaid's Tale." Canadian Literature 148 (1996): 57–72.
Taylor, Chloë. "Abnormal Appetites: Foucault, Atwood, and the Normalization of an Animal-Based Diet." Journal for Critical Animal Studies 10, no. 4 (2012): 130–148.
Valenze, Deborah. Milk: A Local and Global History. New Haven: Yale University Press, 2011.
Velten, Hannah. Milk: A Global History. London: Reaktion Books, 2010.
Wiley, Andrea S. Cultures of Milk: The Biology and Meaning of Dairy Products in the United States and India. Cambridge, MA.: Harvard University Press, 2014.
Wiley, Andrea S. Re-imagining Milk: Cultural and Biological Perspectives. Abingdon: Routledge, 2015.
Wright, Laura. "Vegan Zombies of the Apocalypse: McCarthy's The Road and Atwood's The Year of the Flood." In The Vegan Studies Project: Food, Animals, and Gender in the Age of Terror, edited by Laura Wright, 68–88. Athens: University of Georgia Press, 2015.

35
A HUT OF HER OWN

Deborah Slicer

Elizabeth Costello is in the doldrums. Two men—the one who fathered her, the other the father's namesake—are omnipresent in her life. And now she wants a new one.

Bio

Readers in animal studies are most certainly familiar with Elizabeth's famous 1999 lectures at Appleton College (later published as *The Lives of Animals*), the ones in which she pins Reason—that "vast tautology"—to the mat, offending the philosophers; compares industrial animal agriculture to the Holocaust, offending everybody else; refuses to make nice at a faculty dinner, where she is barraged with questions about her vegetarianism; frustrates her audiences with long digressions and complicated non-answers to questions, causing her son, John, a professor of physics, to wish she'd just stayed home with her cats. After her visits her grandchildren pick at their food, asking, Is this veal? Free-range? Dolphin safe? So now they are no longer allowed to eat or even be alone with their stooped-shouldered, fleshy, white-haired, tired, old, shabbily dressed grandmother, who smells of cold cream.[1]

The award ceremony at Altona College in Pennsylvania in 1995 (published in *Elizabeth Costello*) is less famous than the Appleton event. Again, John accompanies (and narrates for) her. She is still old and tired. Her hair is greasy, lifeless. She wears white shoes that make her look like Daisy Duck, he observes.[2]

Elizabeth's critique of rationalism, her case on behalf of the literary and moral imagination in service of the heart, joyful embodiment, even her misanthropic outbursts are spot on and the bases for her moral ontology and epistemology. But, sadly, what Elizabeth *says* and what Elizabeth *does* and *is* are very different, at least in *Lives*.

Elizabeth isn't writing lyric poetry. She's engaged in intellectual cock fighting with philosophers because "[t]here is no position outside of reason where you can stand and lecture about reason and pass judgment on reason," as her daughter-in-law, the philosopher, says.[3] Except the madhouse, she adds.

And she's hardly an example of joyful embodiment, "fullness of being," but rather a prime instance of her nemesis', Descartes', "soul as a pea in a shell."[4]

 DOI: 10.4324/9781003273400-43

As for her sympathetic imagination, she claims to know what it's like to be an oyster, a bat, a corpse, claims that if she can imagine this, others can too. And once we know what "it's like" we know what we ought to do. Being alive to the world is to be in sympathy with the world. Fair enough. (But, really?! An oyster? A corpse?) Elizabeth is certainly alive to the atrocities but insensible to joy.

Is anything to be done?

On the trip to Altona when an admirer asks her if she sees herself as challenging Joyce in her famous novel about Molly Bloom, Elizabeth says no. "But certain books are so prodigally inventive that there is plenty of material left in the end, material that almost invites you to take it over and use it to build something of your own."[5]

Say no more.

Elizabeth Costello boards a bus.

A Hut of Her Own

January 10 at half-past noon, the Rimrock Stagebus stops at Clearwater and Highway 200. Elizabeth Costello steps down from the heated vestibule into metallic cold, light blowing snow, windy silence. Clearwater runs north, the direction taken by the warm bus, 200 lopes south. Her ride, someone whose face she's never seen before, coming from a long way, someone who is a long time coming, if they come, will come from the west.

Here stands Elizabeth Costello—the internationally acclaimed Australian writer, feminist icon, and the author of a late 20th-century animal studies manifesto—in the middle of winter in the middle of Montana, waiting along this drifted, empty highway. She wears an expensive Nordic blue coat, mouse-gray gloves, scarf, and fat boots, grips a rolling suitcase. Improbable, even fantastic, she thinks, and she feels plenty foolish.

Elizabeth watches seventeen magpies strung along the branches of a stubby winter tree. She's fond of these intelligent tricksters, though some detractors call them flying skunks. Behind them a herd of seven cow bison face plow deep trenches through deep drifts. At eight hundred pounds these ladies can push a lot of snow, make good time. A cow with a nose full of snow *Kacks.*

Elizabeth Costello has never experienced such quiet. Companionable.

At half-past one a small woman in pressed green coveralls, Elizabeth's age, shows up in an old, yellow Chevy truck, apologizing profusely. Delays. Weather. Elizabeth of course reassures her no harm was done. And by dark she's settled in the Hut, a twelve by fourteen-foot cabin, mostly windows on stilts, south facing, primitive necessities, furnished with a cot for serious sleeping and an over-sized red chair and ottoman for afternoon naps. Meals arrive three times daily—lemony cashew sauces, macadamia-nut cheeses, saffron polentas, lavender sorbet. Tea and calissons at four. "How do you like the Hut? the woman asks Elizabeth the morning after she arrives. Elizabeth likes it, likes it very much, decides to stay a while.

That first morning, when Elizabeth opens her eyes and sees the four horses lunging skyward, snapping at snowflakes, everything, just as it is, nameless and already perfectly loved, suddenly steps nearer.

What Is It Like to Be a Horse?

Elizabeth's first acquaintance is a sixteen-hand, all black, thirteen-year-old quarter horse, who lives in a large field near the woman's copper-roofed barn. Asa. He drinks from a creek that runs through the back field, where he naps under big cottonwoods, often with the steer, Chuck McGraw, who he sometimes chases, bullies, because moving other four-leggeds feels good, and because generally Asa thinks highly of himself. Or sometimes just because. His best friend of nine years, a little good-time goat, Dusty Miles, smells rank and sounds like a gagging crow. But Dusty is nimble and always game for head wrestling, butt biting. And, more recently, there's Max.

A seventeen-hand, dark bay, Irish-born thoroughbred, worldly, confident, the top of Elizabeth's head falls an inch short of his shoulder. Max is a semi-retired eventer with international papers and prize money from the recent Nationals in Denver. He's killing time here while his rider decides between college and him. Asa and Max are the Terrible Two. When they got bored chasing each other with rubber feed pans in their mouths, they chased the feeder, a timid two-legged, who started waving a long-flagged stick when they ran tight-bucking circles around her. After they played chicken with the tractor, and one day the tractor gunned it and turned on them like a bull, they found the fire hoses in a large utility box that someone had left in their field, unsecured. Grey, flat, like a pair of fifteen-foot shoe laces with heavy brass headers, unrolling the things and then dragging them out into the field probably took a coordinated effort.

When a responsible party found them swinging hoses at each other's rear ends, or, when in range, at the other guy's head, Max had a bloody ear, and Asa was side-stepping the beating Max was aiming at his rump. After each successful *Whomp* they galloped screaming circles side-to-side and lunged their swan-like necks at the people fetched to stop them.

Truly, breaking it up was risky. But the real reason the spoilers held off was because this game of just-for-the-hell-of-it-bad-ass-slapstick, well. It was joyful. Because one ton is a lot of élan, and who wants to stand in the way of that?

Certainly not Elizabeth. Elizabeth might know what it's like to be an oyster, a bat, a corpse, and various people who never even existed, like Molly Bloom. But she cannot fathom, not yet, what it's like to be a horse or even a human being who is comfortable around horses. But the woman who keeps these horses, while she will always think mostly like a woman, which mostly means she thinks too much, has, over five decades with equines, developed a sense, one that moves her body mostly the right ways before she stops to think. She senses when an equine is going to bite, or pull back at the tie rail, jig, bolt, if an equine will kick out, or swing a thirteen-pound head NO, senses all these things, and why, nanoseconds before any of them happens.

She also intuits the herd vibe, whether relaxed or fearful or bored or tired or hungry, playful or joyful. It is an energetic thing, and a kinesthetic thing, and a physiognomic thing, and what they have taught her, she says, by patience and forgiveness and generosity and exchanged kindnesses.

After the Appleton lectures, more than one friendly critic was disappointed that Elizabeth made no reference to her personal experiences with other animals, her embodied, heart-felt experiences, the very ones she claims ground our moral relations with them.[6] In truth, she hasn't lived with many other animals and had always thought, only half-jokingly, that raising two children came close enough to living with puppies and piglets. Still, sometimes she worries she's a faker. Here she'll leave less to the imagination and immerse herself in the lives of real animals, though she didn't sign on for this feed-time spectacle. Elizabeth is sure that the two she was asked to feed delight in making her an object of comic torture. Yesterday the pirouetting bullies caused her to slip in the snow, and then, while she lay there, deftly nibbled at the hay that had fallen in her hair. One even nibbled her ear, gently, maybe even, she thought, apologetically.

Elizabeth and the woman who owns this place, a friend of her daughter's, are comfortably uncurious about each other. They're on friendly terms, but not overly. They speak, but seldom and briefly. Tending her tiny black woodstove and feeding the four-leggeds twice daily are her only responsibilities.

One day she cuts her hair into white, short, soft spikes. Cuts it just because, and notices that dropping the extra weight between her ears feels good.

For the first few weeks Elizabeth mostly watches the winter weather move across the mountain tops, northwest to southeast, spectacular storms, Stark Mountain's snows and melts.

Before coming here Elizabeth felt like a bead in a handheld galaxy game, that miniature sort of pinball distraction. Will she land inside angelic Saturn, or the angry red planet, or Earth? Or will she rattle around deep space forever because the hand is inept?

*

By late March the horses tease Elizabeth affectionately but have stopped terrorizing her for laughs. This morning the woman takes a pencil out of Elizabeth's hand and passes her a pair of muddy boots. She calls for Elizabeth to follow her, and a voluptuous black mare, Lu, and a fox-orange thoroughbred mare, Indie, to the creek, which is deep, fast running, noisy. The woman sits bareback on top of Lu mid-creek. Just off the track, Indie has probably never even seen a creek, much less crossed one. She paws and paces along the bank behind Lu for a good hour. The woman explains how often an unconfident horse will not enter turbid water—a creek, even a large puddle, whether six inches deep or abyssal—even a dark matt in the barn doorway, if she cannot see bottom. The horse may jump or at best clumsily bolt through, or, most likely, refuse to move forward. But a more experienced horse, like Lu, will show her how it's done, will enter first, leaving the reluctant horse behind to watch. If a typical human being leads the reluctant horse she's dragging or on top of, then she rubs and praises in that chirping, affected human sort of way, inevitably rushing, envisioning a mile over that way rather than being right here, losing patience, correcting, over-thinking the whole process. While a lead horse might sigh. Lick. Chew. Manure into the abyss, nonchalantly. Stand, snatch grass along the bank, or drink, or play-splash, tail-swart flies. Or do nothing at all. Doze in the delicious lap-lap of cool water on her belly. Close her eyes. Fording this way can take five minutes, or all day, or weeks of coming back to the same place, starting over.

In another half-hour Indie is shin deep, wary but standing calmly behind Lu. A brave effort. Tomorrow, the woman says, they'll come back to this place. Do it all over again.

*

A cloud of April midges flies ahead of her like a discombobulated soul trying to find its way down the muddy path. Grasses' wild sugars climb calf-high along fence grids both sides of the wet track. Boot prints, hoof prints, red mud slowly works its skin smooth again. And the river ahead, running spring rain, snow melt, a child's pink bicycle, forever flocks of geese in perfect pitch.

Elizabeth most loves these early morning walks when heavy dew strains even pine boughs under the weight of water gathered from the atmosphere at dew point. At six a.m., bird songs, like a child whistling through sugar. And so many smells and more intensely when the forest is wet—musky pissed patches, vanilla-scented Ponderosa pine bark, fish skank off the pond, grape-jelly smelling lupine. Two or three hours roaming and she returns to her hut-home drenched, even chilled. Strips, towels, calls it a bath. This morning she finds a strange handwritten note with her coffee and toasts, a set of instructions for a game.

While she's aware it sounds inane, she does wonder which of the Two left this note and why.

Running

Players: Generally three or more (a small herd). The very occasional two-legged. In which case equine handicapping is optional.

Playing Field: A five-to-seven-acre field. Tree stands conducive to running screaming circles. A narrow creek, preferably year-round flow, which players may leap or, during summer months, pummel until drenched. One hill with no more than 30-degree grade, dusty, low ankle-biting scrub. No loose rocks. NO prairie dog holes. Killdeer heroically defending ground nests.

Rules: No stopping to defecate, eat grass, startle at squirrels, boulders, shadows. Players may stop to blow, scream, shadowbox, buck. When possible, avoid trampling heroic killdeer and chicks.

Cardinal Rule: Stay out of kicking range. Remember, a typical 1200 lb. player kicks back 5 feet, delivers sidekicks at 4 feet, a mantis-style front punch from 6 feet. A swinging 13 lb. head stops a 2-legged dead, drop dead, if she catches it in the face. Better players dance on the head of a pin, turn east into northwest in a half-breath, let fly.

To Begin Play: Any player with the vibe, regardless of pecking order, may display socially contagious behavior. Rabbit ears, eye rolls, shadow box, fart. Lip tremble, tail flag (mandatory if Arab descent), jig, scream, fart. Lower center of gravity and propel four feet vertically, bolt, head toss, fart.

Object of the Game: Run

The running game is not about life or death. Or about beauty or courage. Or about stamina, speed, or strength. It isn't "about."

It is running.

*

Yesterday Elizabeth was running, trying too hard and too successfully to keep up, when Asa kicked back and caught her left breast. Roughed up like any other running horse, but a hundred-pound woman, no horse, an alto sort of ache reverberated round and around her chest, like sound inside a bell. And her heart quieted. Then fluttered back to life again.

*

Three days ago, when you brought home the new horse, Chief, I watched something that perhaps you can explain.

I don't know why I smell strawberries, when there probably isn't a strawberry in the entire valley in August. Bear season, huckleberry season, but not strawberries.

Elizabeth usually ignores this woman's irrelevant prattling and chooses to ignore this too. The woman is bent over, cleaning mud out of Lu's right, front foot. Lu's ears swivel forward toward the horses in the field in front of her and then back toward Elizabeth when she speaks. So at least the mare is listening.

I saw the new horse, Chief, the leggy paint, put into a field with five other horses, one of whom was Asa. I was told that everyone knew Chief, as he'd spent a previous summer with this group. But that was a year ago. For whatever reason yesterday he was treated as an interloper. Asa led the charge and the herd, screaming, chased Chief for a good twenty minutes through the creek, several times around a run-in shed, finally pinning him at the east fence, where he began pacing and calling pathetically to whomever he'd left at home. So long as he was on the fence, the herd didn't molest him. But whenever he crossed some invisible line west, they ran him against the fence again, where he paced and clearly considered jumping. This went on for most of an afternoon. Except for Zimba.

What about Zimba?

Zimba, that low-ranking light grey of some exotic breed, the youngest, infrequently handled, watches Chief. Intently. Everyone else is watching too, heads down, mouths full.

But Zimba isn't eating?

Right, he's studying the back and forth, and his ears fan out whenever Chief whinnies. He watches. Then, at four o'clock, he leaves the herd and slowly trots east to Chief, apparently doing what he knows how to do to be non-threatening. Chief watches Zimba approaching. And when he arrives noses sniff. Zimba puts his long neck on top of Chief's neck, rubs. Lays his head on Chief's back, rubs. The two walk away from the fence together, together they graze all that evening and into the next day, together. No one bothers them when they're together. They went their separate ways today. No one bothered.

Sweet Zimba's soul sleeps with its windows wide open.

Meaning?

He's low ranking, been there. He could have moved himself a notch up the pecking order if he'd chased Chief too. What you witnessed was kindness. Is your chest bruised?

Pardon?

Your running game.

The size of a dime.

He could have killed you, Elizabeth, if he'd aimed to.

In the morning, with breakfast, a novel on the Hut porch. *The Mare*, by Mary Gaitskill, a highlighted passage.

"That night I dreamed of horses running together like they were water with a brain that could decide where to go. Except you could see their faces and their feet and tails coming out and then going back into the water of themselves."[7]

What Is It Like to Be?

Elizabeth wakes from an odd dream at dawn. A grey tabby cat, Lassie, who has come into the Hut during the night, is sleeping temple-to-temple with Elizabeth. Lassie's front paws knead the pillow, her back legs gently kick, her nose twitches, and her mouth is slightly open. Elizabeth dreamed she was near the horses' run-in shed, and small animals moved around the paddock. A skunk, mice, various other long, furred, grey four-footed creatures she can't identify go about their business in the silent half-dark. And their squeaks and flutters. When she wakes up she feels both giddy and empty like a wagging balloon.

These cats, Sophie and Lassie, who sometimes sleep with her and lounge in her sheets in the heat of the afternoon, they bring in tics and other crawly creatures that climb off their fur into Elizabeth's bed. All this insect life bites, sometimes tickles her at night, crawling along a shoulder or across her lips, and make her itch.

Sometimes one of the horses, a dun, bred a mustang-thoroughbred cross grooms her head and back with his lips. She's watched those lips jitterbug after fine hay leavings across dirt, dexterous and gentle.

Often, she spends all day at the pond about a mile's walk from the Hut, sits among the late-summer fireweed watching coots and grebes glide across the dimpling face of the water. Magpies, a large mischief, whose squawking she associates with the fallen angels' heckling at the gates of Hell, mob song birds and once even a bald eagle mid-air. Frequently the eagle sits on a snag opposite her, eyeing the pond, looking for fish, she assumes. But maybe she shouldn't assume. Maybe he enjoys the same pastoral view that brings Elizabeth to the pond, the unexpectant quiet, the same just-because-sleepy surrender to three in the afternoon. And the mountain lion once. A tail as long and thick as her arm, feet large as catcher's mitts. Three or four yards to her east, the lion had come for a drink, and surely she must have scented Elizabeth there. Elizabeth has been told to be quiet like the forest or else sing softly to a lion. She is quieter than the forest, so joyful, she forgets to breathe.

Before walking home, Elizabeth might take off her clothes, lie on her back in the pond, allow the water to hold her up, admire her bare breasts as they float just above the water's surface, like water lilies. Or she might watch the little fish swim through reflections of passing clouds around her submerged feet, wider, flatter, surer than they used to be, carrying her for some seventy years.

Leaping Poetry

"Ill-mannered rodent!" Elizabeth shakes her fist at the large gray squirrel who picks pine cones high up a Ponderosa pine, drops them on her metal rooftop. The incessant *plop-pop, plop-pop* has driven her out of the Hut. October, and she's aware that this is just another animal with more sense than she has, preparing for the immanent winter. Shorter days, and the horses' coats lose summer hair and grow wooly-dense. The wild turkeys pulse themselves straight up into the Hawthorne bushes, sounding like umbrellas opening and closing, to eat berries that make them winter-fat. Elizabeth Costello must take stock too.

*

The horse, the dun, the one the woman calls "Lips," lips Elizabeth's eyebrows, like mumbling bees. Solemn and serious, and not. The teaser.

He's messing with you, Elizabeth. He knows he's breaking the rules. He be excess, be expatiatic, exsusciationtic, being precocious, promiscuous, prooemionic, proteustic. He be bad.

I'm not altogether comfortable with this. He has an ear.

He be nibbling your ears, neck, nipping. Is it lovemaking or combat? Enter the land of -Esque, Elizabeth, where we co-exist, co-mingle, collapse—yield more, a little something different and more, the zone of improvisation. You're a poet, Elizabeth. Innovative, improvisatory, fast, leaping associations. The bottomless feeling one experiences as the further a poem or play gets from its initial worldly circumstances without breaking the thread. Our contact with being, joy.[8]

I worry about the breaking part. He's switched from lips to teeth. Can I trust I won't lose an ear?

Oh, well, things could fly off the rails, a nibble could always become a bite. Do you trust him?

No.

Ah! A double-charge of reality. Isn't it positively aesthetic, Elizabeth?

No.

A nibble and not-nibble. Bite hard but not too hard. A suspension of the logic of excluded middle. Tension that positively swells with possibility. Joy![9]

And one possibility is that I lose an ear.

Not likely. Though I wouldn't struggle just now. And, Elizabeth, he knows that. He is enacting an abstraction from both lovemaking and fighting. Lovemaking and not, fighting and not. And he knows that you know the difference. Don't you, Elizabeth?

No! Oh rocks! When is he going to let go!?

Bite him back, Elizabeth. See what comes.

How can I possibly? He has me by the ear.

Dear Elizabeth, "you are expecting things instead of being. . . . Being is complete. . . . Remember the age of six? Stay there. Stay at the age of six. . . . And do with the horse what you would do, and then see what is coming. . . . You can drag the horse, you can do a lot of things, but the horse from the inside will not come."[10]

*

Elizabeth is walking creekside in November in an early-coming dark, when, incredibly, a 1200-pound animal is suddenly, silently beside her, walking too, walking with his head down so that he and Elizabeth are eye to eye. Asa moves a little closer to her right. She moves a little closer to his left. With his right eye he can watch the equines he's left behind, while his left, a soft eye, a curious eye, kindly watches her. He has chosen to walk with Elizabeth. She chooses to walk with Asa. They enjoy this walk together in an early-coming dark.

Dear, John

John is once again chauffeuring his mother to an airport. Recently she came from Montana to this city to attend a writers' conference, where she had top billing. He's surprised to see her waiting on the curb along with another attendee, a man approximately her age, wiry, tidy-looking. John Coetzee. Introductions are rather complicated. A confused quiet follows, each man assessing the other's face. "Don't worry," Elizabeth says, "he's quite harmless." Both men raise a paw and shake.

The two pile in, Elizabeth in the front seat, Coetzee in the back. All this is rather awkward because John has brought his two children along. Their grandmother is of course delighted. She insists on stashing their car seats in the trunk along with the luggage, and holds both children on her lap.

"Mother, this is illegal, irresponsible. We'll be royally fined if we're stopped, and we'd deserve it."

She ignores him, tells the children about wombats and Koalas in Australia, far-fetched stories about horses in Montana.

Forlornly, four-year old Charlie tells her his parents have forbidden the boys a dog. Elizabeth demonically starts chanting, "Dog!Dog!Dog!Dog!Dog!Dog!" until the children and Elizabeth literally howl with laugher, and then laugh some more. So unlike her!

The hair is a nice touch, Coetzee eventually says from the back. And it's good to hear your laughter. You're a little old for the Converse high-tops you're wearing, though.

To which Elizabeth replies, Nonsense!

You've been cherry picking the text, he says.

Have I? Well, it feels more like pulling carrots and digging after potatoes. There's something Persephonous about your Elizabeth and her Johns.

Elizabeth does acknowledge that not only women like Molly Bloom and Penelope but men like Leopold Bloom and Ulysses are prisoners of cultural myths, of gender, for example, and Humanism, he says.[11]

Let's just remember that some prisoners get better treatment than others and leave it at that.

Elizabeth says that once men start redeeming and liberating masculine stereotyped characters it will be "a grand spectacle" "But seriously," she says, "we can't go on parasitizing the classics forever. I am not excluding myself from the charge. We've got to start doing some inventing of our own.[12]

A "grand spectacle." No doubt men will make it so. "Inventing." Well, we don't have to start ex nihilo, like gods. There are teachers, individuals who are easy out of bounds, who have already crossed the blood-brain barrier, she says somewhat cryptically, hoping he'll occupy himself with that remark.

He smiles, nods his head. Even closes his eyes.

But, then:

John does say he loves his mother. He's protective of her, proud, Coetzee reminds her in a soft, placating voice.

Her son is frenetically tapping his thumbs against the steering wheel. He knows what's coming next.

Oh, yes? Well, he also says his mother is a cat. "One of those large cats that pause as they eviscerate their victim and, across the torn-open belly, give you a cold yellow stare."[13] Not quite Derrida's kitty, though they have the same effect: existential disorientation, emasculation, the threat of annihilation.

Both men exchange accusatory glances in the rearview mirror.

Ho—and this? As she is sleeping John imagines his mother's gullet, "pink and ugly, contracting as it swallows like a python drawing things down to the pear-shaped belly sac. . . . No he says to himself, that is not where I came from, that is not it."[14] Very pretty. The uterine-shaped belly sack, contracting, pulling him back in.

And that business about knowing what it's like to be a bat, an oyster, even a corpse?? First, why go to such extremes to make the obvious point that human beings are morally lazy, especially in the moral imagination department? Second, while it makes sense to say a bat has a perspective, experiences "battiness" (a morally salient capacity among other salient capacities and contexts), I don't have to know what it's like to be batty in order to sympathize with a bat. Third, even an imagination marinated in the intimate life world of another animal, pick an easy one, say a horse, can only best-guess a hazy version of "what-it's-like." I concede Thomas Nagel's general point about the limits of the human imagination, humbly and in awe of the mystery of bat-being.[15]

She hears something like cackling from the back. Cackling or whatever, it's certainly not contrite. More thumb thrumming in the front.

Elizabeth comments it's a nice December afternoon and she'd like to open her window. John asks her please don't. She does it anyway, half-way. Then decides it's too noisy. Powers the thing back up. Down. Up. Half-down. Down.

John gives her his archbishoprical look.

And why in God's name do you want to give it a happy ending? Denouement? the genius in the backseat shouts over road noise.

She looks hard at her son. She glances at the John at her back. She begins organizing her personals: passport, visa, money, her notebooks of prose and sketches from out west. Elizabeth is considering France, where her daughter lives in a coastal village in Brittany. There, Elizabeth is welcome to stay for as long as she likes. She's not yet sure what she likes. Probably, she likes water. She will go to Brittany, see what comes.

I'm reviving the happy ending. Joie, hope. It could come back into vogue, like cold cream, she finally answers.

Six-year-old Max runs his hands through his grandmother's white hair until it sticks straight up in ridges.

You look like T-Rex! he screeches.

Everyone turns to look at Elizabeth's new upright head.

Giddy snowflakes begin to blow through her open window.

Both children scramble around their grandmother's lap, stick their arms outside. Four small hands catching at falling snow.

Ridiculous, John says under his breath.

Elizabeth reaches for him, clasps his wrist, leans close, smiles: There, there, John she says. There, there. This will soon be over.

Notes

1 J.M. Coetzee, *The Lives of Animals* (Princeton: Princeton University Press, 1999).
2 J.M. Coetzee, *Elizabeth Costello* (New York: Penguin, 2003).
3 Ibid., 48.
4 Ibid., 34.
5 Coetzee, *Elizabeth Costello*, 12.
6 For example, see Barbara Smuts, commentary in *The Lives of Animals* (Princeton, NJ: Princeton University Press, 1999), 107–108.
7 Mary Gaitskill, *The Mare* (New York: Pantheon, 2015), 148.
8 Robert Bly, *Leaping Poetry* (Boston, 1975). Bly coined the term "leaping poetry" and developed the idea.
9 Brain Massumi, *What Animals Teach Us About Politics* (Durham: University of North Carolina Press, 2014). This section is very much inspired by Massumi.
10 Klause Hempfling, *The Path of the Horse* (Stormy May Productions, 2008). I quote Hempfling.
11 Coetzee, *Elizabeth Costello*, 13–14.
12 Ibid., 14.
13 Ibid., 5.
14 Ibid., 32.
15 Ed Yong's new book *An Immense World* (New York: Random House, 2022), about other animals' sensual *Umwelten*, greatly enhances my sympathy for Nagel.

Bibliography

Bly, Robert. *Leaping Poetry: An Idea with Poems and Translations*. Boston: Pitt Poetry, 1975.
Coetzee, J.M. *The Lives of Animals*. Princeton: Princeton University Press, 1999.

Coetzee, J.M. *Elizabeth Costello*. New York.: Penguin, 2003.
Gaitskill, Mary. *The Mare*. New York: Pantheon, 2015.
Hempfling, Klause. *The Path of the Horse*. Stormy May Productions, 2008.
Massumi, Brain. *What Animals Teach Us About Politics*. Durham: University of North Carolina Press, 2014.
Smuts, Barbara. *The Lives of Animals*. Princeton, N.J.: Princeton University Press, 1999.
Yong, Ed. *An Immense World*. New York: Random House, 2022.

PART VIII

Activism and Advocacy

36
ANIMAL RESISTANCE

Tim Reijsoo

Introduction

Animal resistance is important because it is a phenomenon necessarily associated with oppression and brings to light the perspectives of the animals themselves. When animals resist their enslavement and oppression, they challenge the socially constructed position of human privilege both by drawing attention to the contingency of the categories to which they are assigned (e.g. zoo animals, livestock animals, etc.) and by fighting their material domination.[1]

In order to understand animal oppression, we need to attend to their standpoints. Animal resistance shows the depth of animals' agency, and recognition of animal agency is essential to deconstruct the social category of the human. Human innocence, a psychological disposition of cultivated or wilful ignorance, characterises hierarchical systems and is essential in maintaining and perpetuating the unequal and exploitative societal group relations.[2] This disposition explains why most humans do not recognise animal resistance.

Animals have always resisted humans and their systems of domination.[3] When animals resist they challenge both the material and symbolic domination of humans. Foregrounding animals in their bid for freedom, namely as advocates who make humans aware of the unjust relations that exist between species and who challenge the socially constructed position of unexamined human privilege, is important if we want to overcome a narrow human-centred perspective and build a just multispecies society.

In order to analyse animal oppression, it is essential to discuss and look at the concept of resistance, and how such a concept operates when it transcends the species border. This involves questions such as: How do animals resist their oppression? Can humans and animals form alliances in a multi-species oppressive structure? The chapter is structured as follows. First, I discuss how to conceptualise and historise animal resistance. Second, I look at examples of animal resistance and the strategies of animal enterprises downplaying their resistance. Third, I discuss ways in which nonhuman animals have put up resistance leading to changes in the public's perception. Finally, I conclude the chapter with a case study.

DOI: 10.4324/9781003273400-45

Conceptualising and Historicising Animal Resistance

According to Fahim Amir[4] and Dinesh Wadiwel,[5] resisting one's domination is in itself political. These scholars argue that the space of resistant politics is characterised by a continuum of forms of resistance, not by a winner-takes-all situation where, on the one hand, everything that is human is political and, on the other, everything that is not human is devoid of the political. As such, animals are part of this continuum, and although this does not mean placing them on a par with humans in every way and in every aspect, it does mean working out partial connections.[6] Following Amir and Wadiwel, I see animals as political agents capable of resistance. In fact, animals engage in resistance across a whole spectrum of rebellious actions. Nonhuman animals hide from crowds when they are expected to perform, deliberately perform poorly, ignore commands, stop working, and bite their exploiters.[7] When animals refuse to cooperate or try to escape from their enclosure, they communicate their dissent with respect to the situation humans have created for them. The French philosopher Michel Foucault understood critique to mean refusal to be governed in some way.[8] As the examples in this chapter will demonstrate, animals often refuse to be governed by humans. Thus, in conceptualising animal resistance, the ethical question "can they suffer?" gives way to the political question: where and how do animals put up resistance? Once it is acknowledged that animals are political agents capable of resistance, we can start to view the history of animal domestication as a continual process of power dynamics between different species, instead of viewing domestication as a one-sided engineering endeavour led by the human species.

The hegemonic narrative in society views the process of animal domestication as beneficial to both humans and animals. Fortunately, this narrative has been contested and challenged by a number of historians.[9] Animals have forcefully been integrated into human dominated political, cultural, and economic systems, in most of history. The reason some animal species were integrated into society in the first place was due to their ability for human social interaction. Nevertheless, this does not mean the process of domestication was benign; quite the opposite. Domesticating animals requires suppressing their attempts at resistance and their desire to break free from human enclosures and remain autonomous. Humans, such as farmers, continuously need to find new ways to adapt their systems precisely because animals resist.[10] From the cruel ancient practices of inventing devices that mutilate female animal udders to prevent their own children from consuming breastmilk, all the way to modern-day applications of advanced surveillance technologies monitoring animals in zoos, all are proof that animals do not cooperate.[11]

Even among domesticated animals, we find numerous acts of revolt. As Patterson[12] explains, the process of domestication necessarily required suppressing resistance by eliminating defiant animals to prevent them from reproducing. However, despite these selective breeding practices which aim to get rid of autonomous or "difficult" animal individuals and their traits, humans have not succeeded in eliminating domestic animals' resistance. For example, to this day, farmers fear that resistant animals will teach other captive individuals ways to resist.[13] Sheep farmers in New Zealand, for example, are afraid that the lambs who flee from the ranch by opening the fences can teach others to do so as well. For this reason, "farmers will shoot them so they can't pass on their knowledge."[14] Nicole Shukin argues that, "the rendering of animal capital is surely first contested by animals themselves, who neither live *unhistorically* nor live with the historic passivity regularly attributed to them."[15] For humans to ignore animal individuality and agency, they thus have to actively cultivate a form of wilful ignorance, which I call human innocence. To illustrate this ambivalence of the human–animal condition, and the entanglement with inter-human systems of oppression, we can look at the horse.

Already in *Capital*, Karl Marx[16] pointed out how animals with a mind of their own have caused significant problems for the industrial production process: "Of all great motors handed down from the manufacturing period, horse power is the worst, partly because a horse has a head of its [sic] own, partly because he is costly, and the extent to which he is applicable in factories is very restricted."[17] In this statement, animal agency is implicitly recognised whereas the broader context of submission to humans remains uncontested. The animal labour scholar Kendra Coulter[18] explains why horses turned out to pose a particular obstacle for the capitalist production process. Horses are livings beings who decide whether to cooperate or to disobey commands. Their moods can change due to a variety of factors. The examples of horse behaviour show the inherent ambivalence of the power dynamics at work. On the one hand, animal agency is appropriated for capital gain and poses an obstacle in their instrumentalisation; on the other hand, animal agency has to be denied because of the moral consequences this could have regarding the position of horses in human dominated societies. The cultural archive contains more subtle references to horse resistance.

The word that characterises the neoliberal era par excellence, to *manage*, is itself derived from putting down horse resistance. The term "manager" as the leading figure for organisation, administration, and supervision, is derived from the verb "to manage."[19] It first appears in the sixteenth century in the sense of horse riding or dressage, inspired by the French *manège* (riding school). Etymologically the word means to control or to condition. Recently managers have even been refining their social skills in "equine-assisted leadership training," extending the notion of getting a horse to do what you want to doing the same with your employees. A horse who totally resists dressage, a being who is completely unmanageable, is sometimes called a *crazy horse*.

The label *crazy horse* captures the intersectional nature of oppression through the structural similarities and commonalities between animal and human oppression because this term is used to appropriate and eliminate both human and animal resistance. Indigenous leaders who put up fierce resistance against the colonising U.S. army were mockingly called Crazy Horse, just as horses who put up fierce resistance against their jockeys (or *managers*) are labelled. By refusing to call them each what they really are, namely remarkable displays of defiance, strength, and courage, and instead replace this acknowledgement with the demeaning label of *crazy*, a violent epistemic and discursive move to eliminate the resistance is in fact performed.[20] Hence, the silencing of animal subjects, their struggles, and cultures, reminds us in structurally relevant ways of how so called "dominant races" silenced the cultural histories of the colonised.[21] The cultural archive serves as a repository of memory for us to excavate the manifold ways animals have put up resistance and how their oppression is both discursively and materially interconnected with human suffering throughout history.

Before turning to the next section, I would like to briefly mention one more way to conceptualise animal resistance. Some scholars suggest that under certain circumstances suicide can count as a final act of resistance.[22] They argue that when the only way to escape one's exploitation is self-destruction, this is an act of resistance. For example, mother sows in factory farms are for the most part kept in so-called farrowing crates where they can barely move. Mother sows in farrowing crates have repeatedly been reported to smack their heads against the metal struts, sometimes until they succumb to their injuries. Although Blanchette reports on this phenomenon, he terms this a mental tic on the part of the sows,[23] however Amir questions the human–animal boundary about this contested final act of bodily and (mental) autonomy.[24] I share Amir's critique of Blanchette for never exploring whether this could be a form of animal resistance; the last resort of the mother sow: a suicide attempt.

Pigs are not the only animals whose suicide attempts have been observed. Numerous captive animals refuse to eat to the point of starvation. For example, cetaceans and other marine mammals seem to consciously prefer death over life in marine parks and scientific experiment facilities.[25] They have been witnessed to consciously stop breathing and end their life as a result. Although humans continue to deny that animals have the cognitive capacities to commit suicide, systems of advanced animal exploitation are carefully designed to take away from animals even the right to end their own lives.

Whether suicide is committed as a deliberate act from a clear state of mind, or whether it is committed almost automatically by a mind and body suffering from severe mental distresses, does not matter for our purposes. Resistance doesn't always have to be intentional. For example, in *Abnormal*, Michel Foucault writes of the demonic possession of religious women as a "resistance effect" to confession,[26] while in *Psychiatric Power*, he names hysteria as involuntary resistance to medical power on the part of severely medicalised women.[27] Both types of resistance involved "convulsions of the flesh"[28] in which bodies refused to behave according to the logics of domination to which they were subjected, and are characterised by Foucault as acts of resistance. As in the cases of animals resisting their domination in factory farms, in neither case did these acts of resistance liberate woman from their institutionalised captivity or qualify as deliberate plans to revolt.[29] Nonetheless, they show that when bodies and minds, human or animal, are pushed to their very limits, they will resist, even when resistance is pointless because there is no possible way to escape.

Industry Response to Animal Resistance

Having discussed the importance of acknowledging animal resistance to change how humans view animals, I now turn to instances of animals intentionally resisting their oppression and how their oppressors try to silence their dissent. In order for human privilege to function, animal resistance needs to remain a myth. Besides physical suppression of their resistance, this means that institutions also need to epistemically and discursively marginalise animals. To deconstruct human supremacy, it is important to understand the institutional processes playing a significant role in oppressing animals. One key institution in silencing animal resistance is the zoo. The zoo was originally constituted as a site where human and animal oppression were interconnected. I will briefly discuss the entanglement of colonialism with speciesism in the zoo before continuing with the zoo's role in silencing animal resistance.

Zoos exhibited indigenous people and people of colour before they became exclusive live animal exhibitions, or rather animal humiliations. In fact, zoos were a direct materialisation of and tribute to the colonial empires. The very reason behind the zoo's existence was to showcase stolen exotic treasures and individuals (both human and animal) from overseas to Europe's metropolitan populations. By zoos displaying these exotic items, animals, and even humans, the imperial and colonial ideology materialised.[30] Zoos conveyed to the European metropolitan population the notion that these overseas territories, animals, and peoples were now conquered and brought under the colonisers' authority and control.[31] This colonial institution thus simultaneously exploited humans and other animals.

To show how zoos silence animal resistance, I turn to the work of Jason Hribal. According to the animal entertainment industry, animals rarely refuse to cooperate, despite overwhelming evidence to the contrary. Orangutans, for example, have developed a variety of creative means to overcome detention technologies and their captors. They learned how to sabotage locking mechanisms, and some even used wood or rubber material to withstand the electric wires surrounding their enclosures when trying to escape.[32] Chimpanzees likewise have a notorious reputation in both test facilities and zoos. At the Toledo Zoo in Ohio, for example, a chimpanzee would hold milk in her mouth until it got sour and then spit it over her trainers. Trainers started wearing protecting gear as a result.[33] Tigers have broken out to attack visitors who threw stones at them. In their pursuit they would leave the rest of the zoo visitors undisturbed and would only harm their abusers.[34] Another example concerns their sex lives. Marine mammals are infamous for refusing to reproduce in captivity. Almost none of the hundreds of marine parks worldwide have managed to successfully breed orcas in captivity. Even if employees were successful in causing a successful pregnancy, the mothers would simply neglect to nurse their newborns. As a result, aquariums often go through a whole series of whales before just one of them made it into adolescence. In fact, many animal species refuse to reproduce in captivity and with partners selected for them by humans.[35]

Moreover, circus and zoos representatives will emphasise that animals act on the basis of 'instinct,' that they are 'wild' by nature and that attacks against employees and the public are mere 'accidents.' This is the narrative institutions transmit to silence animal agency and intentionality. Animals know exactly which behaviours are rewarded and which ones are punished. Due to the regime in zoos and circuses, they understand there will be consequences for incorrect actions. The consequences for refusing to perform or attacking a trainer range from being beaten, to having reduced food rations, or being placed in solitary confinement. However, they still carry out such actions, which means they resist intentionally. Yet those who refer to animal agency are often accused of anthropomorphism. When humans ascribe complex emotions and cognitions to animals, such as revenge, they will be met with ridicule, as agency is exclusively reserved for members of the human species.[36] Not only do animals have a history, they also co-created history, and their resistance has led directly to historical change more than once. When stories of animal defiance make it into the news headlines, a public debate about the social/moral status of animals usually follows, which can support the emancipation process of animals in society. Likewise, elephants also played a major role in abolishing elephant exploitation in circuses due to their many acts of refusal to perform and retaliation against the humans who were exploiting them.[37]

Finally, in response to an escape, a zoo or circus will fortify its enclosures. When human casualties result from animal resistance, a zoo or circus punishes the animal to convince the public extreme physical and mental conditioning prevents any chance of similar future events. Another response is to simply send the animal somewhere else (where they usually await a similar if not worse fate) to re-establish trust among their clientele and the general public. Despite zoos and circuses denying animal agency, they will occasionally admit the relationship between animal handlers and elephants is primarily antagonistic, coercive, or even violent.[38] For example, the entertainment industry appropriated and capitalised on elephant resistance by creating the narrative of the "mad elephant" and by portraying elephant retaliation as part of the show's spectacle.[39] However, as animal welfare concerns grew, the industry "removed" the most rebellious elephants. Punishing elephants was done exclusively backstage. The secrecy surrounding animal resistance has everything to do with keeping intact the public's perception the hierarchical situation is perfectly normal and natural. Moreover, human innocence can distort perception. Instead of becoming aware of their privilege when witnessing intelligent creatures perform ridiculous tricks in captivity, humans conclude the animals in these entertainment industries are privileged individuals, because they have many professional human caretakers overseeing them and no need to provide food for themselves.

Consequences Animal Resistance on Public Perception

The reason the industry is keen on hiding the phenomenon of animal resistance is because of the impacts it can have on public perception. The consequences of animal resistance reaching the general public are not straightforwardly positive with respect to animal liberation. In this section, I discuss a variety of responses to animal resistance and how these responses relate to human privilege. I also discuss some commonalities between injustice against animals and injustice against marginalised humans. I end with a critical note on how oppressive conditions severely limit the possibilities for resistance given its effect on animal psychology.

A term we can borrow for our purpose is Timothy Pachirat's[40] notion of the 'politics of sight'. What he means by this is the fine-tuned and interconnected system of politics, media, law, culture, and industry, which cooperatively ensure and decide what the general public is allowed to see and in which way they are allowed to perceive it. The *politics of sight* is a useful way to understand how the human gaze is managed regarding animal resistance. When animals escape and humans encounter runaway animals in the public sphere as a result, these animals come to be recognised as having their own unique consciousness and are viewed as individuals. When animals escape, they cause a physical and conceptual rupture by residing in places where they are not supposed to be. The 'politics of sight' thus encompasses both material and discursive policing. Moreover, as Tim Cresswell writes, border transgressions enable us to "realize that a boundary even existed."[41] In crossing these borders, animals create a social impact on the world around them. These human-constructed borders maintain a system that enables sentient beings to be turned into products and which is specifically designed to shelter those who consume these animal-derived products or services from feeling discomfort about their origins.[42] These borders keep the oppression of animals hidden from both actual view and the symbolic or imaginative view. The 'politics of sight' can explain why successful acts of resistance put up by individual animals rarely lead to structural changes concerning the exploitation and moral status of nonhuman animals in society. While runaway agricultural animals are able to win the hearts of humans, for instance, the vast majority of animals are exploited and sent to slaughterhouses every day without public outcry.

When public concern does go out to a particular animal, it can even result in fortifying and strengthening the oppressive structure by reinforcing or borrowing from another system of oppression and its corresponding epistemic and discursive violence.[43] Colling[44] describes instances of agricultural animals, such as cows and sheep, who escaped on their way to slaughter. When local media reported on this, occasionally members of the public recognised their individuality and raised money to ensure this particular individual would not end up in a slaughterhouse but instead would live out the rest of his or her life in an animal sanctuary. However, when the public agrees a defiant animal individual deserves not to be killed and instead should be provided with amnesty given his or her courageous attempt to escape slaughter, this is usually ascribed to the animal's outstanding individual qualities and strength. Consequently, it is not a result of the captive and exploitative conditions causing the animal to escape, but it is in large part due to their own effort and merit accounting for this rare occurrence. The animal must have been an exceptional individual and different from the rest of their kind. Instead of acknowledging that the oppressive conditions that animals are forced to live in are responsible for attempts at escape, the focus and praise are placed on the merit of the heroic qualities pertaining to the particular animal individual. This celebration of escaped animals as special in comparison to other less fortunate individuals resembles other discourses of exceptionalism, such as international refugee discourse: a select few can be provided citizenship, while the majority of migrants and their suffering remain wilfully ignored. To counter this narrative, it is essential that animal advocates remind the public that all captive animals would escape if they could.

Besides these ideological obstacles jeopardising the acknowledgment of animal resistance, the harsh reality is that misery and suffering do not necessarily lead to resistance. As with humans, animals process experiences of trauma and pain differently. When animals do not resist, this does not mean they are content with their condition or that their social role is appropriate. Even if animals resisted at first, they might get tired after numerous rounds of horrific torment and give up. In addition, many animals are unable to resist because the condition of their exploitation is so extreme, including situations where a cage is barely bigger than the animal. Moreover, in virtue of domestication, farmed animals have been subject to intensive breeding practices culling the most resistant tendencies and thereby engineering an artificial type of extremely obedient and docile animal species.

The public reaction to animal resistance varies because of a combination of factors. Human innocence, epistemic violence, and industry motives jeopardise the emancipatory process for animals. It is essential that human allies amplify animals' own voices and prevent their heroic resistance against their oppressors from having been in vain. In the following case study, I look at the resistance of orcas. Orcas are particularly infamous for their resistance, both in history and today. By zooming in on the history of orca resistance, the conceptual work on animal resistance discussed previously becomes tangible and their agency obvious.

Case Study: Orca Resistance

The institutions paying the highest cost, in terms of human casualties, for exploiting "their animal performers" are likely to be the so-called marine amusement parks and aquaria, such as SeaWorld. Numerous trainers have been killed by orcas taking revenge. Tilikum's revenge, for example, lasted for over half an hour.[45] The credibility of marine amusement parks has declined considerably over the years, with more and more countries banning the cruel practice. Complete abolition remains to be seen.

Orca culture is effectively destroyed in captivity because these animals have highly evolved and cohesive matriarchal cultures which are made impossible in captivity. Human-mediated captivity brings different nations of orcas, who speak different dialects and languages, into tiny, artificial, and unhealthy environments.[46] Fierce resistance to this fate occurs not only on the part of orcas already held captive but also during the process of capturing these wild animals. Orcas resist being captured and entire pods have gone to great lengths to prevent capture and ensure rescue of their fellow group and family members. For centuries before being captured to serve the needs of the entertainment industry, these social mammals living in the ocean were already hunted and at war with humans.

The hunting of sperm whales began in the early 18th century to fuel the city lights of urbanising Europe and North America. The oil and ambergris obtained from their dead bodies became a profitable, yet dangerous, business for the humans. During hunts, a whale hunting ship would launch several small harpooner boats. Sperm whales resisted these ambushes by attacking these boats upon sight. They also took more intentional revenge against the hunters by attacking the ones who had been seen harming members of their pod.[47] Many of these calculated retaliations of destroying hunting ships by sperm whales were recorded and inspired works of fiction, both novels and movies. In addition, recent research by leading whale scientists reveals whales not only physically attacked whaling ships. With time, they developed diversion tactics and learned other defensive behaviours to lure the hunting ships away from the pods. Whales who experienced and survived a hunt passed on those strategies to other populations, who adopted the same. In other words, whales use cultural transmission to respond to a threat, and this kind of learning turns out to be very rapid.[48] In doing so, sperm whales not only developed signalling calls for these particular lethal threats but entire strategies to fool the sailors and their crews.[49] This behaviour means the way humans interact with animals (peacefully or violently) also has consequences for our perception of them, as they respond accordingly and adjust their behaviour to that of humans.

Conclusion

Animal resistance is of significant social importance in their struggle for liberation because the animals themselves challenge the material, conceptual, and psychological borders responsible for their ontological and subjected status. If humans want to offset their species privilege, they need to join the animals in their resistance. However, we need to make sure the way in which we conceptualise our solidarity with nonhuman animals does justice to their agency and dignity. When nonhuman animals are represented as mere helpless victims utterly dependent on human saviours, animal rights advocacy reinforces and perpetuates the narrative that it aims to dismantle.[50] When we claim to stand in solidarity with the animals, we therefore need to proceed from the activity of the affected themselves; otherwise the word solidarity would lose its meaning. The animals themselves are their own best advocates when it comes to communicating the arbitrariness of speciesism to humans. Besides humans assisting animals in their bids for freedom, many animal species free themselves and each other.[51] Humans should try to work towards building multispecies alliances. In addition to conceptually tearing down the human constructed boundaries between "us" and "them," humans need to find ways to dissolve physical boundaries separating human and animal worlds. We must reinvent the different ways (species-specific) life can be integrated in mutually acknowledging ways. Sanctuaries function as the avant-garde in this transition to more compassionate and just multispecies communities.[52]

Furthermore, animal resistance allows us to rewrite history from a multi-species perspective. Constructing narratives of history where animal agency is central in turn can assist in overcoming and challenging the systems of epistemic violence rooted deep within institutional structures. Deconstructing narratives of human supremacy by replacing them with an account that acknowledges animals' agency and resistance can put a hold to the tendency of oppressive systems to entrench and solidify. In fact, most oppressive systems, because of the cumulative effect of social power, are perceived as the natural order of things. For oppression to remain socially acceptable, it paradoxically needs to remain socially invisible, in the sense that the hierarchy is perceived as "natural" and not as a collective social process inherently subject to change. A historical narrative from a multispecies perspective acknowledging animal resistance refutes the mainstream narrative of animal domestication as a benign process, and it highlights the contingency of the institutional domination of humans over animals.

Notes

1 S. Colling, *Animal Resistance in the Global Capitalist Era,* in The Animal Turn series (Michigan State University Press, 2021); systems of oppression necessarily entail privileges for certain groups and individuals: when some are oppressed (Group X), others are privileged as a result (Group Y) (M. Joy, *The Vegan Matrix: Understanding and Discussing Privilege Among Vegans to Build a More Inclusive and Empowered Movement* (Brooklyn, NY: Lantern Publishing & Media, 2020). Just as white privilege belongs to institutional racism, speciesism logically also has its associated privilege: privilege by virtue of species membership. Fortunately, we can expose privilege by showing its inherent arbitrary foundation.

2 I extend Gloria Wekker's notion of *White Innocence* (G. Wekker, *White Innocence: Paradoxes of Colonialism and Race* (Duke University Press, 2016) to analyse the domination of humans over animals. Wilful ignorance is found among many humans who participate in animal oppression by choosing to consume animal products and supporting zoos, circuses, and animal testing. While most humans say they care about animals, when confronted with their exploitation, they actively choose to look away from these injustices that are required to maintain their current way of life. This phenomenon is also known as the *meat paradox*; that is, people like to consume animal products but do not want to be reminded of the nonhuman animal individual whose body is eaten (E. Aaltola, "The Meat Paradox, Omnivore's Akrasia, and Animal Ethics," *Animals* 9, no. 12 (2019): 1125.

3 C. Patterson, *Eternal Treblinka: Our Treatment of Animals and the Holocaust* (Lantern Books: New York, 2002); D. Wadiwel, *The War against Animals* (Koninklijke Brill Rodopi NV, Leiden: The Netherlands, 2015).

4 F. Amir, *Being and Swine: The End of Nature (As We Knew It)* Originally published in German as *Schwein und Zeit: Tiere, Politik, Revolte* (Edition: Nautilus, Hamburg, Germany, 2018) First published in English (Between the Lines: Toronto Canada, 2020), trans. Geoffrey C. Howes and Corvin Russell.

5 Wadiwel, *The War against Animals.*

6 M. Strathern, *Partial Connections* (Rowman Altamira, 2005).

7 J. Hribal, *Fear of the Animal Planet: The Hidden History of Animal Resistance* (CounterPunch and AK Press, 2010).

8 M. Foucault, *History of Sexuality*, Vol. I (Vintage Reissue, 1976).

9 D. Nibert, *Animal Oppression and Human Violence: Domesecration, Capitalism and Global Conflict* (Colombia University Press, 2013); Patterson, *Eternal Treblinka: Our Treatment of Animals and the Holocaust;* J. Mason, *An Unnatural Order: The Roots of Our Destruction of Nature* (Continuum International Publishing Group, 1998).

10 Patterson, *Eternal Treblinka: Our Treatment of Animals and the Holocaust.*

11 Ibid.; Wadiwel, *The War against Animals.*

12 Patterson, *Eternal Treblinka: Our Treatment of Animals and the Holocaust*

13 Colling, *Animal Resistance in the Global Capitalist Era*

14 Jeffry M. Masson, *The Pig Who Sang to the Moon: The Emotional World of Farm Animals* (New York: Ballentine Books, 2003), 103

15 N. Shukin, *Animal Capital: Rendering Life in Biopolitical Times* (Minneapolis: University of Minnesota Press, 2009), 130

16 K. Marx and F. Engels, "Capital: Critique of Political Economy," in *Marx & Engels Collected Works Vol 35* (London: Lawrence & Wishart, 1996).

17 Ibid., 379.

18 K. Coulter, "Horse Power: Gender, Work, and Wealth in Canadian Show Jumping," in *Gender and Equestrian Sport: Riding around the World*, eds. Miriam Adelman and Jorge Knijnik (Dordrecht: Springer, 2013).

19 Amir, *Being and Swine: The End of Nature (As We Knew It)*.
20 "Crazy Horse" also shows the intersection of sanism and speciesism. Although 'crazy' is used to pathologise both the Indigenous leader and the defiant horse, 'crazy' as an insult or pejorative label also reflects an instance of sanism (S. Leblanc and E.A. Kinsella, "Toward Epistemic Justice: A Critically Reflexive Examination of 'Sanism' and Implications for Knowledge Generation," *Studies in Social Justice* 10, no. 1 (2016): 59–78.). For a study of the intersections of animal oppression and sanism, see Hallie Abelman, "Cripping Mad Cow Disease," in *Disability and Animality: Crip Perspectives in Critical Animal Studies*, eds. Stephanie Jenkins, Kelly Struthers Montford, and Chloë Taylor (Routledge, 2020), 212–220.
21 C. Spivak, "Can the Subaltern Speak?" in *Marxism and the Interpretation of Culture*, eds. Cary Nelson and Lawrence Grossberg (Urbana: University of Illinois Press, 1988), 271–313; D. Chakrabarty, *Provincializing Europe: Postcolonial thought and historical difference* (Princeton University Press, 2000).
22 Amir, *Being and Swine: The End of Nature (As We Knew It)*; Wadiwel, *The War against Animals*.
23 A. Blanchette, "Herding Species: Biosecurity, Posthuman Labor, and the American Industrial Pig," *Cultural Anthropology* 30, no. 4 (2015): 640–669.
24 Amir, *Being and Swine: The End of Nature (As We Knew It)*
25 Another example of animal suicide is documented among bears: "On bile farms, moon bears are kept for up to twenty years in small wire cages and have their gall bladders milked daily, a painful procedure that leads to infections and disease. Because of the atrocious conditions, bears may try to kill themselves by punching their own stomachs so they are fitted with iron vests. One day a bear managed to escape her tiny prison after hearing her cub in distress. But the mother and child had nowhere to escape to, and to avoid a lifetime of torture, the bear hugged and strangled her cub to death and killed herself by running into a wall." (S. Colling, S. Parson, and A. Arrigoni, "Until All Are Free: Total Liberation through Revolutionary Decolonization, Groundless Solidarity, and a Relationship Framework," in *Defining Critical Animal Studies: An Intersectional Social Justice Approach for Liberation*, ed. Anthony J. Nocella (Peter Lang, 2014), 65).
26 M. Foucault, "Abnormal," in *Lectures at the Collège de France* (New York: Picador, 1974–1975) (2004)
27 M. Foucault, "Psychiatric Power," in *Lectures at the Collège de France* (New York: Picador. 1973–1974) (2006).
28 Foucault, "Abnormal."
29 C. Taylor, "Foucault and Critical Animal Studies: Genealogies of Agricultural Power," *Philosophy Compass* 8, no. 6 (2013): 539–551.
30 Nibert, *Animal Oppression and Human Violence: Domesecration, Capitalism and Global Conflict;* J. Sanbonmatsu, ed., *Critical Theory and Animal Liberation* (Rowman & Littlefield Publishers, 2011).
31 "The Congolese man Ota Benga was exhibited in the Bronx Zoo in a cage with an orangutan" (C. Kim, *Dangerous Crossings: Race, Species, and Nature in a Multicultural Age* (Cambridge: Cambridge University Press, 2015), 43))
32 Hribal, *Fear of the Animal Planet: The Hidden History of Animal Resistance*
33 Ibid.
34 Ibid.
35 R. Hermes et al., "Assisted Reproduction in Female Rhinoceros and Elephants—Current Status and Future Perspective," *Reproduction in Domestic Animals* 42.s2 (2007): 33–44.
36 Hribal, *Fear of the Animal Planet: The Hidden History of Animal Resistance*.
37 Colling, *Animal Resistance in the Global Capitalist Era*.
38 Hribal, *Fear of the Animal Planet: The Hidden History of Animal Resistance*.
39 Colling, *Animal Resistance in the Global Capitalist Era*
40 T. Pachirat, *Every Twelve Seconds: Industrialized Slaughter and the Politics of Sight* (Yale University Press, 2011).
41 T. Cresswell, *In Place/Out of Place: Geography, Ideology, and Transgression* (University of Minnesota Press, 1996), 22
42 Aaltola, "The Meat Paradox, Omnivore's Akrasia, and Animal Ethics"; M. Joy, *Why We Love Dogs, Eat Pigs and Wear Cows: An Introduction to Carnism* (Conari Press/Red Wheel, 2010).

43 A.J. Nocella, J. Sorenson, K.E Socha, and A. Matsuoka, eds., *Defining Critical Animal Studies: An Intersectional Social Justice Approach for Liberation* (Nova Iorque: Peter Lang, 2014)
44 Colling, *Animal Resistance in the Global Capitalist Era.*
45 Hribal, *Fear of the Animal Planet: The Hidden History of Animal Resistance.*
46 Colling, *Animal Resistance in the Global Capitalist Era.*
47 Ibid.
48 H. Whitehead, T.D. Smith, and L. Rendell, "Adaptation of Sperm Whales to Open-boat Whalers: Rapid Social Learning on a Large Scale?" *Biology Letters* 17 (2021).
49 Researchers conducted extensive data analysis from historical records, and what they found is that as hunts continued over time, fewer and fewer kills would be associated with the number of times whales would be spotted. This led them to conclude that the diminishing return rate of whaling expeditions must be ascribed to efforts on the part of the whales and their cunning tactics of deception (Whitehead, Smith and Rendell, "Adaptation of Sperm Whales to Open-boat Whalers: Rapid Social Learning on a Large Scale?").
50 Amir, *Being and Swine: The End of Nature (As We Knew It).*
51 "News that there was an escaped cow living in the forest spread quickly. Neighbours caught glimpses of her before she disappeared into the bush. Weeks went by and people began to wonder how the calf was surviving on her own. It turned out she was in good company. Cameras set up by hunters caught her on film, and to the surprise of many, she wasn't alone. Bonnie was travelling, foraging, and grazing with a herd of deer. The deer offered her the companionship she needed after losing her family and showed her how to survive in the wild" (Colling, *Animal Resistance in the Global Capitalist Era,* 70).
52 E. Abrell, "Sanctuary-Making as Rural Political Action," *Journal for the Anthropology of North America* 22, no. 2 (2019): 109–111; S. Donaldson and W. Kymlicka, "Farmed Animal Sanctuaries: The Heart of the Movement? A Socio-Political Perspective," *Politics and Animals* no. 1: (2015): 50–74; E. Meijer, "Sanctuary Politics and the Borders of the Demos: A Comparison of Human and Nonhuman Animal Sanctuaries," *Krisis, Journal for Contemporary Philosophy* 41, no. 2 (2021): 35–48.

Bibliography

Aaltola, E. "The Meat Paradox, Omnivore's Akrasia, and Animal Ethics." *Animals* 9, no. 12 (2019): 1125.

Abrell, E. "Sanctuary-Making as Rural Political Action." *Journal for the Anthropology of North America* 22, no. 2 (2019): 109–111.

Amir, F. *Being and Swine: The End of Nature (As We Knew It).* Originally published in German in 2018 as Schwein und Zeit: Tiere, Politik, Revolte by Edition Nautilus, Hamburg, Germany. First published in English in 2020 by Between the Lines, Toronto Canada. Translated by Geoffrey C. Howes and Corvin Russell, 2020.

Blanchette, A. "Herding Species: Biosecurity, Posthuman Labor, and the American Industrial Pig." *Cultural Anthropology* 30, no. 4 (2015): 640–669.

Chakrabarty, D. *Provincializing Europe: Postcolonial Thought and Historical Difference.* Princeton University Press, 2000.

Colling, S. *Animal Resistance in the Global Capitalist Era.* Michigan State University Press, 2021.

Colling, S., Parson, S., and Arrigoni, A. "Until All Are Free: Total Liberation through Revolutionary Decolonization, Groundless Solidarity, and a Relationship Framework." In *Defining Critical Animal Studies: An Intersectional Social Justice Approach for Liberation,* edited by Anthony J. Nocella. Peter Lang, 2014.

Coulter, K. "Horse Power: Gender, Work, and Wealth in Canadian Show Jumping." In *Gender and Equestrian Sport: Riding around the World,* edited by Miriam Adelman and Jorge Knijnik. Dordrecht: Springer, 2013.

Cresswell, T. *In Place/Out of Place: Geography, Ideology, and Transgression.* University of Minnesota Press, 1996.

Donaldson, S., and Kymlicka, W. "Farmed Animal Sanctuaries: The Heart of the Movement? A Sociopolitical Perspective." *Politics and Animals* 1, no. 1 (2015): 50–74.

Foucault, M. *Psychiatric Power: Lectures at the Collège de France*. New York: Picador, 1973–1974 (2006).
Foucault, M. *Abnormal: Lectures at the Collège de France*. New York: Picador, 1974–1975 (2004).
Foucault, M. *History of Sexuality*. Vol. I. Éditions Gallmard, 1976.
Hermes, R et al. "Assisted Reproduction in Female Rhinoceros and Elephants—Current Status and Future Perspective." *Reproduction in Domestic Animals* 42.s2 (2007): 33–44.
Hribal, J. *Fear of the Animal Planet: The Hidden History of Animal Resistance*. CounterPunch and AK Press, 2010.
Joy, M. *Why We Love Dogs, Eat Pigs and Wear Cows: An Introduction to Carnism*. Conari Press/Red Wheel, 2010.
Joy, M. *The Vegan Matrix: Understanding and Discussing Privilege Among Vegans to Build a More Inclusive and Empowered Movement*. Brooklyn, NY: Lantern Publishing & Media, 2020.
Kim, C. *Dangerous Crossings: Race, Species, and Nature in a Multicultural Age*. Cambridge: Cambridge University Press, 2015.
Leblanc, S., and Kinsella, E.A. "Toward Epistemic Justice: A Critically Reflexive Examination of 'Sanism' and Implications for Knowledge Generation." *Studies in Social Justice* 10, no. 1 (2016): 59–78.
Macho, T. *Das Leben nehmen: Suizid in der Moderne*. Berlin: Suhrkamp, 2017.
Marx, K., and Engels, F. "Capital: Critique of Political Economy." In *Marx & Engels Collected Works*, Vol. 35. London: Lawrence & Wishart, 1996.
Mason, J. *An Unnatural Order: The Roots of Our Destruction of Nature*. Continuum International Publishing Group, 1998.
Masson, Jeffry M. *The Pig Who Sang to the Moon: The Emotional World of Farm Animals*. New York: Ballentine Books, 2003.
Meijer, E. "Sanctuary Politics and the Borders of the Demos: A Comparison of Human and Nonhuman Animal Sanctuaries." *Krisis, Journal for Contemporary Philosophy* 41, no. 2 (2021): 35–48.
Nibert, D. *Animal Oppression and Human Violence: Domesecration, Capitalism and Global Conflict*. Colombia University Press, 2013.
Nocella II, A.J., Sorenson, J., Socha, K. e Matsuoka, A., eds. *Defining Critical Animal Studies: An Intersectional Social Justice Approach for Liberation*. Nova Iorque: Peter Lang, 2014.
Pachirat, T. *Every Twelve Seconds: Industrialized Slaughter and the Politics of Sight*. Yale University Press, 2011.
Patterson, C. *Eternal Treblinka: Our Treatment of Animals and the Holocaust*. New York: Lantern Books, 2002.
Sanbonmatsu, J., ed. *Critical Theory and Animal Liberation*. Rowman & Littlefield Publishers, 2011.
Shukin, N. *Animal Capital: Rendering Life in Biopolitical Times*. Minneapolis: University of Minnesota Press, 2009.
Spivak, C. "Can the Subaltern Speak?" In *Marxism and the Interpretation of Culture*, edited by Cary Nelson and Lawrence Grossberg, 271–313. Urbana: University of Illinois Press, 1988.
Strathern, M. *Partial Connections*. Rowman Altamira, 2005.
Taylor, C. "Foucault and Critical Animal Studies: Genealogies of Agricultural Power." *Philosophy Compass* 8, no. 6 (2013): 539–551.
Wadiwel, D. *The War against Animals*. The Netherlands: Koninklijke Brill Rodopi NV, Leiden, 2015.
Waggoner, J. "Race, Gender, and Sanism: Remapping Mad Feminist Genealogies." *Signs: Journal of Women in Culture and Society* 47, no. 4 (2022): 885–904.
Wekker, G. *White Innocence: Paradoxes of Colonialism and Race*. Duke University Press, 2016.
Whitehead, H., Smith, T.D., and Rendell, L. "Adaptation of Sperm Whales to Open-boat Whalers: Rapid Social Learning on a Large Scale?" *Biology Letters* 17 (2021).

37
WOMEN'S CONTRIBUTIONS TO THE MOVEMENTS FOR ANIMAL PROTECTION AND RIGHTS

Katja M. Guenther

Volunteer at an animal shelter, join a group of people rescuing injured seabirds, sign up for training in wildlife rehabilitation, gather to bear witness to the final moments of pigs' lives as they are driven through the gates of a slaughterhouse, or attend a rally against the fur trade and you're sure to notice that most of the people around you appear to be women. Across different forms of advocacy and volunteerism centered on improving animals' life chances, and in different regions of the world, women are overrepresented.[1] This has been mostly true since the inception of animal welfare activism in the United States and Europe in the mid-1800s and remains true even as animal advocacy has become more diverse in terms of the backgrounds of participants.

Women have always been at the heart of animal-focused activism, whether advancing welfarist approaches, which work towards reducing animal suffering without necessarily changing the underlying structural conditions that support human exploitation of other-than-human animals, or liberationist approaches, which work to dismantle the ideologies and practices that reproduce human domination over other-than-human animals. In the United States and other countries, women are overrepresented both among animal rights activists and among volunteers who dedicate unpaid labor to caring for animals in need, such as unhoused companion animals, injured or orphaned free-roaming animals, and farmed animals living in sanctuary settings. The preponderance of women in these arenas has been a subject of curiosity and inquiry. In fact, the causes of the gender gap in animal advocacy and activism is something of a perennial question.[2] Here, I focus less on *why* women in the U.S. appear to be so much more likely to take action for animal rights and welfare than do men and more on *what* women have contributed to the struggle for animal welfare and rights. This is important because women's experiences—like those of animals—are too often erased from history and their contributions unrecognized. Further, women in the U.S. have brought a specific set of perspectives to animal protectionism and animal rights activism that have reflected and at times resisted dominant norms about gender, womanhood, and motherhood. Importantly, I also address how the feminization of animal-centered activism has both helped and hindered the progress of this movement.

I start this chapter by providing a brief historical overview of women's activism on behalf of animals in the U.S. before homing in on some of the key constraints women have

DOI: 10.4324/9781003273400-46

faced specifically *as women* trying to eventuate gains in animal welfare and animal rights. In the final section of the chapter, I also attend to the contributions women have made as animal advocates, illuminate that women brought particular perspectives based in their own experiences *as women* to this activism and show how women's standpoints have in many ways come to define the contemporary field of animal welfare and animal rights activism in the United States.

A Brief History of Women's Animal Advocacy in the United States

Women's activism in pursuit of animal rights and welfare dates at least to the early 19th century with the rise of the animal protection movement. Anti-cruelty movements spread globally during the latter portion of the 1800s and into the 1900s, contributing to growing public awareness of animal protection claims and to the passage of animal welfare laws in numerous countries. In the early period of organized activism (roughly 1860–1915), the emphasis in animal protection advocacy centered on stopping cruelty to working animals (especially horses), ending or at least regulating blood sports, improving conditions for farmed animals, helping unhoused companion animals, and reducing or ending the emergent practice of vivisection, which became increasingly integrated into medical training and laboratory research during the 19th and early 20th centuries. In the United States, these areas of emphasis reflected women's concerns that the public's exposure to violent spectacle such as the beating of carriage horses and the culling of stray dogs would undermine morality, that animal abuse and blood sports like cockfighting and dogfighting were linked to men's use of alcohol and thus were also tied to men's abuse of women, and that since women and animals shared a subordinate place in society women should resist men's violence against members of both groups.

Women's efforts to establish certain behaviors as abusive towards animals and to promote a compassionate worldview rested on a complex set of beliefs and values, some newer to the United States. The influence of white femininity was powerful in the development of the commitment among humane advocates to center the purported moral threat animal cruelty posed to men and children exposed to it and to the effort to propagate humane education programs. Historian Bernard Unti (2002) contextualizes the centrality of women's roles as socializers of children, particularly boys, who emphasized compassion for animals:

> One explanation for the pervasiveness of the kindness-to-animals ethic lies in its consonance with the republican gender ideology of the post-Revolutionary United States. Early American society assumed a set of paternalistic relationships both within and outside of the family, emphasizing the importance of a virtuous citizenry devoted to republican principles of governance. This made education of the boy especially critical, since as a man he would take his place of leadership over family, chattel, property, and social institutions. Responsibility for educating the child for his leadership role rested with women, who were assumed to be the repositories of gentle virtue, compassionate feeling, and deep devotion—a buffering presence against the heartless struggles of the business world and the masculine public sphere. Humane education provided one important means for insulating boys against the tyrannical tendencies that might undermine civic life were they to go unchecked. Animals were nicely suited for instruction that impressed upon the child their helplessness and dependence upon him, and his considerable power over them.[3]

Beyond consideration of just the moral socialization of children (especially boys, who were understood as lacking the intrinsic gentleness and compassion bestowed upon girls), Americans—especially white Americans—saw themselves as the "moral stewards" of all those whom they viewed as unable to shepherd themselves, including Black Americans. The United States also took on the role of moral stewardship outside of the United States. As historian Janet Davis[4] documents, the "gospel of kindness" became a core American value that animal activists worked to export to parts of the world the United States was incorporating into its empire. Some animal activists opposed the expansion of American empire through military means—during the Progressive Era (1890s–1920s) in the United States, movements focused on peace and anti-militarism, prison reform, temperance, woman suffrage, anti-lynching, disability advocacy, and child welfare often overlapped with animal protectionist activism. At the same time, animal protectionists did support a moral empire in which emergent American ideas about how animals should be treated prevailed.

This gospel of kindness was thus not only about animals, but also about race and colonialism, both within and outside of the United States. Historian Paula Tarankow (2022) writes that

> Pervasive turn-of-the [20th] century evolutionary theory that linked cruelty with savagery and barbarism and humane behavior to civilization stabilized [animal protection] reformers' assumptions not only that pain and civilization were antithetical to each other but also that sympathy with animal suffering was a marker of racial difference.[5]

Humane treatment of animals emerged as part of nativity and whiteness, while brutality and cruelty to animals was associated with immigrant status (especially Irish and Italian) and Blackness. This was true even though many Blacks were involved in and supportive of efforts to improve the conditions of animals[6] and often mobilized African Americans' participation in humane advocacy as evidence of their own humanity.

The racist, classist, and xenophobic undertones (and overtones!) of animal advocacy doesn't mean that white and wealthy humane advocates spared the upper classes from their efforts at reform. For example, Henry Bergh, who founded the American Society for the Prevention of Cruelty to Animals (ASPCA) and was himself part of the white American elite, both expressed anti-immigrant views (especially against Irish Americans) and also agitated around some of the ways wealthy whites abused animals, such as through hunting. Still, activities the public associated with people of color and low-income people, like dog- and cock-fighting and especially the use of animals for labor, became clarion calls for the need for change to protect animals.

Animal protectionists used variable tactics, including efforts at policing and criminalizing how humans treated animals and developing services for animals, as well as efforts at humane education intended to teach people—and especially children—how to engage with animals in ways that did not cause harm. The emphasis on humane education extended beyond the private sphere of the household, where women were to socialize children to be compassionate, into women's activism to expose children to the ethic of kindness to animals in public schools and other venues. British animal advocate Catherine Smithies, for instance, developed the Bands of Mercy, local organizations of children and teenagers focused on the humane treatment of animals. Transplanted to the U.S. from Britain by George Angell, founder of the Massachusetts Society for the Prevention of Cruelty to

Animals (MSPCA), and Reverend Thomas Timmins, the Bands of Mercy existed across the United States. Among other activities, the Bands of Mercy sang songs written by Sarah J. Eddy and Charlotte Farrington, hymns that emphasize that God wants children to be kind to all living beings, no matter how small. The songs are a call to action, imploring children to protect animals and others who are vulnerable as acts of compassion and goodness. While men were at the helm of the national leadership, women were much of the rank-and-file adult presence in Bands of Mercy chapters across the nation, including in the American South, where white and Black children participated in racially segregated chapters.

Women produced much of the educational material intended to enhance children's sense of care for animals, including novels in which cruelty to animals featured centrally, such as Anna Sewell's (1877) children's classic, *Black Beauty: The Autobiography of a Horse* and numerous other "autobiographies" of animals. The American Humane Education Society (AHES) produced a series of texts targeting young readers, most of which were written by women. In their roles as teachers and educational reformers, women also promoted the incorporation of kindness to animals into education.

Educational and religious institutions, along with the family, were acceptable arenas for women's activism. But animal activism also opened opportunities for women to be involved in public advocacy in domains from which they had mostly been excluded and where their presence was more novel and sometimes controversial. The antivivisection movement became an especially central site for women's activism for animals in the United States in the late 19th and early 20th centuries, for instance. Here, women came into direct contact—and conflict—with scientists, doctors, and policymakers. Although the antivivisection movement initially included men, within the first decade of its mobilization, the movement was peopled almost entirely by women, in part because of the movement's vocal criticism of male-dominated professions like medicine and science. The antivivisection movement was also closely allied with women's organizing for temperance in the U.S., where women activists were "inclined to perceive the brutalized vivisector as cousin to the cruel drunkard."[7] By the turn of the 20th century, antivivisectionists were also closely allied with suffragists in the United States, and women's political events, such as major conferences, typically included women with commitments to suffrage, ending vivisection, temperance, and other issues.

Vivisection was one arena in which more affluent white women activists did stand up specifically to men of their race and class, challenging emergent practices of science and suggesting that vivisection would place white masculinity at risk for contagion of immorality—immorality already linked to people of color and poor people. The antivivisection movement employed many frames and rationales in its work, including especially concerns that that the practice of vivisection was immoral and brutal in ways that undermined core Christian values and that the practice of vivisection supported an ethos of violence and callousness against animals that would corrupt men. But the American Medical Association (AMA) countered the claims and framings anti-vivisectionists employed, ultimately succeeding in delegitimating the claims anti-vivisections were making as anti-science and anti-progress.[8] The AMA, which touted the necessity of medical research in pursuit of prevention and treatment for public health epidemics like tuberculosis and influenza, appealed to a growing public interest in medical breakthroughs that, during this time period, substantially advanced public health and helped develop vaccines and treatments for some of the most feared diseases. The rabies vaccine, for instance, became an international sensation and was developed by French scientist Louis Pasteur, who tested on rabbits in developing the vaccine. Physicians and scientists asserted that anti-vivisectionists were making false claims about the practice.[9] Defenders of vivisection successfully branded anti-vivisectionists as sentimental, ignorant, and short-sighted women.

The anti-vivisection movement struggled to frame their campaign in a way that resonated with a public eager for medical and scientific advances and generally accepting that some degree of cruelty to animals, whether in the context of medical research or meat farming, was necessary to maintain human health and happiness. Sometimes this resulted in seeming back-pedaling: the American Anti-Vivisection Society (AAVS, which was founded by Caroline Earle White, a co-founder of the Pennsylvania Society for the Prevention of Cruelty to Animals and arguably the most influential animal protectionist of the era), for instance, moved between an emphasis on some regulation to total abolition.[10] As animal-based research became the norm within medical research, anti-vivisectionists continued to struggle with their messaging, a struggle that continues to this day.

Animal rights advocates faced formidable foes in the late 19th century and throughout the 20th century: science and medicine, the animal agriculture industry, and other political and economic groups and institutions that relied on human exploitation of other-than-human animals. Animal rights activists were a source of opposition to the exploitation of animals, but many of their more radical impulses abated in the early part of the 20th century until well after the Second World War. Instead, during the early-to-mid 1900s, animal rights activism homed in on a narrower set of issues with a particular emphasis on companion animal welfare, the effort to provide companion animals in shelters more humane deaths, and to reduce the overall rate of shelter killing.

In response to previous experiences, broader cultural resonance, and structural and environmental conditions after the Second World War, activism on behalf of animals in the United States incorporated shifting issues during the second half of the 20th century. With mechanization, humans in the United States had almost entirely stopped using animals to plow fields and for transport, but animals continued to labor in a range of contexts, including as entertainers in zoos, circuses, and theme parks, and as producers of milk, flesh, skin, and feathers, among others. Working to improve the conditions for working and farmed animals and campaigns to abolish animal labor reemerged as core arenas of animal activism.

Energized by engagement with cross-over movements advocating for the rights of marginalized groups and with the modern environmental movement, the animal rights movement reemerged and expanded in its range of strategies and tactics in the late 20th century. This included confrontational protests that sought shock value alongside more traditional state-centered strategies that attempted to use legislative means to improve the lives and deaths of animals. The later part of the 20th century also ushered in radical animal liberationist groups (who sometimes overlapped with radical environmentalists) that used tactics of disruption and economic damage to challenge laboratory use of animals and the horrific conditions in factory farms. Diverse bedfellows came into contact with one another, including those committed to the abolition of human exploitation of animals; land and species conservationists who often had their roots in hunting and/or science; radical and more moderate tacticians; and activists affiliated with the state, non-profit-agencies, and/or who were acting as part of loose networks. A notable success in the U.S. was the use of public education and boycotting to push for the end of so-called animaltainment, or animals used for entertainment. The closing of Ringling Bros. and Barnum & Bailey Circus in 2007, which the company said was in part because of pressure from animal rights activists, was a major victory for animal rights proponents. People for the Ethical Treatment of Animals (PETA), founded and helmed by Ingrid Newkirk, spearheaded efforts to end the use of animals in circuses. Success with legislation and voter initiatives helped set new minimum standards of care for farmed animals in many states, even as the animal agriculture industry maintained firm political and cultural control. Although Americans remain the second-highest per capita consumers of meat in the world, the animal rights movement also contributed to the mainstreaming of ethical vegetarianism and veganism and to general public awareness of animal issues. Overall, the movement has had many victories, but its overarching goal of substantially reducing or even eliminating human exploitation of animals remains very far from realized.

Animal Activism and the Reproduction of Gender Inequality

Dating back to its inception in the United States, women carved out animal protectionism and rights as spaces in which women could be involved in community and political issues. The animal protection and rights movements have also been sites for the maintenance and thus the reproduction of gender inequality. This is not to suggest that *all* animal advocacy organizations and campaigns have failed to address sexism or to promote and foster the activism and full inclusion of women. Rather, when looking at the arena of animal rights activism as a whole, it's notable that in its early history, men often sought deliberately to exclude women, and that in its later history, men have benefitted from the so-called glass escalator, or the invisible path of rapid advancement and unearned reward that men who enter women-dominated spaces routinely experience.[11] This has been true throughout the movement's history: while some women have held leadership positions, men have been and continue to be grossly overrepresented among leaders. ASPCA founder Henry Bergh staunchly objected to women serving on the board of directors of that organization, viewing them as unsuitable for such work, and no woman held that role for decades after this death. While not all men involved in the early animal protection movement held such views, the animal protection movement often limited women's influence and sidelined them in non-leadership positions, even as it relied on them as the core of activists.

Today, barriers to women in leadership and the sexist cultures that endure in many animal welfare and rights organizations are less often about men excluding women from positions of power outright and in total (although there are contexts in which this has happened),[12] and yet many animal-focused organizations reinforce existing gender inequalities through organizational practices ranging from failure to accommodate the needs of women workers to tolerating sexual harassment of women to placing men on glass escalators. Masculine bodies and voices carry with them political legitimacy, challenge the stereotype that animal advocates and activists are "soft" and "overly emotional" (i.e., feminine) and hold power and authority over women, such that the broader social structure also helps uphold gender inequalities within animal advocacy groups and organizations. In spite of the intensive involvement of women, few animal welfare and rights organizations have taken concrete steps to integrate feminist values into their organizational cultures and practices.

Through an in-depth assessment and analysis of women's participation in animal rights activism in the United States, Emily Gaarder (2011)[13] highlights both the causes of the prevalence of women within the movement, as well as how gender is used within the movement both as a tool and a barrier. In reflecting on her findings over a decade later, it is dispiriting that the gender inequalities she identifies persist. More recently, Lisa Kemmerer (2023)[14] both reflects on her decades of work in the animal liberation movement and analyzes findings from an open survey of women involved in animal advocacy. She finds troubling consistency in the narratives of women in the movement about overt sexism they encountered within organizations for which they have worked in paid or unpaid roles. These accounts echo the narratives that continue to circulate within animal rights circles about unrepentantly sexist men colleagues and mentors.

Additionally, animal advocacy organizations have often fallen back onto sexist tropes and/or the exploitation of women's bodies to further their causes. Various campaigns by People for the Ethical Treatment of Animals are the most widely known examples of this, such as their infamous "I'd Rather Go Naked Than Wear Fur" campaign. Among other controversial elements, PETA's campaigns have also involved inappropriate and ahistorical metaphors that fail to acknowledge the distinctive traumas and legacies of slavery and the Holocaust. Corey Lee Wrenn's contribution to this volume details another such case. Campaigns that rely on sexist representations of women and/or use women as absent referents create a movement culture that many women find alienating and troubling. While some women accept the idea that "sex sells," others report how these campaigns leave them feeling shut out, demeaned, and belittled.[15]

That the gender imbalance among animal welfarists and liberationists persists is itself, in my view, not problematic. What is problematic is that this imbalance reflects sexist practices and beliefs within the movement for a better world for animals, upholds the cultural belief that connecting with animals is a feminine activity, contributes to the marginalization of animal rights positions in political life, supports the lack of mainstreaming of animal studies into many academic disciplines where it clearly belongs (such as sociology, political science, and public policy, to name a few; see for instance Irvine 2008 for an engagement with the marginalization of animal studies and Gaard 2012 and Probyn-Rapsey, O'Sullivan, and Watt 2019 for engagements with the marginalization of feminist animal studies),[16] *and* helps explain the continued political and cultural marginalization of animal rights perspectives in the U.S. As Gaarder (2011) attests, "A movement dominated by women struggles for legitimacy."[17]

Some of the Key Consequences of Women's Overrepresentation in Animal Advocacy

The consequences of women's overrepresentation among activists and advocates for other-than-human animals have been manifold. Throughout the 150 years of women's involvement in animal rights activism in the United States, women's standpoints have shaped their work. In invoking standpoint, I refer to the work of feminist sociologist Dorothy Smith (1974)[18] and others, who show how women's positions within the broader relations of power in society contribute to women having a particular view of both everyday life and the social structures that so often constrain them. Women's standpoints of course vary across nation, class, race, ethnicity dis/ability, and other dimensions, but women as a group do share a set of analogous experiences in that they are generally the expected caregivers in families and the reproducers of culture (including morality) within families, routinely encounter an overlapping set of social beliefs about women's "natural" roles in society (such as mothering), and experience oppression through structural and individual discrimination and/or violence.

A women's standpoint brings values like compassion and care into their interactions with, and advocacy for, other-than-human animals. Their expected role as caregivers and moral stewards within families offers them moral bases for expressing concern for animals; as discussed previously, morality has long been central to those seeking to protect animals and improve their life conditions. As those generally responsible for socializing children within families, women have used that role to promote kindness to animals, and have extended their role within the family to serve as agents of change on behalf of animals in the various spheres, including through humane education.

Women spearheaded and dominated the branch of the early animal protection movement focused on humane education. They were also less likely than men to promote carceral responses to animal cruelty and animal management (such as licensing, impoundment, etc.). The women of the Pennsylvania SPCA, for instance, objected to licensing fees that the government of Philadelphia imposed and routinely paid the fees for working-class and poor people who wanted to get their animals back from the pound. While ASPCA founder Henry Bergh often turned to the courts for remediation, and believed that cruelty to animals "could only be stopped with police power, not poems,"[19] women animal protectionists centered humane education—through books, music and song, newsletters, and direct outreach—in their efforts.

Women also occupied—and continue to occupy—a key position in the family that has enabled them to expand the concept of family to include animals. Advocates for animals, especially in the United States and Europe in the late 20th century and beyond, have pushed for the expansion of how humans conceptualize family to include companion animals, and they assert that farmed animals and animals used for labor (including as laboratory test subjects) have their own familial bonds that should not be disrupted. While people who are not women can and do make claims about families and animals, women in these cultural contexts have a particular authority to do so as those primarily responsible for caring for families and for companion animals. The companion animal care industry has gratefully supported and facilitated the idea that animals are like family members to humans, and as all aspects of social life have become increasingly commodified, so, too, have companion animals. For critical feminist women (and others), this commodification detracts from feminist goals, but for many supporters of animal rights, this is a welcome social change.

Assertions about the familial bonds of farmed animals have been slower to gain cultural traction, likely because they would demand a response that includes reassessing how humans break these bonds and potentially radically changing farming practices. Yet women remain uniquely compelling carriers of this message because of their social standing as arbiters of compassion and family boundaries. Companion animal protection has been the low-hanging fruit of the animal rights movement (although there is still much progress to be made in animal sheltering in the United States and in the welfare of companion animals elsewhere).[20] Substantial inroads against animaltainment may reflect women's concern with the representational exploitation of women's and animal's bodies.

Although the animal protection and rights movement certainly has contributed to a sea change in public views of many aspects of human–animal relations, it is a movement whose work is largely unfinished. The preponderance of women in animal rights movements is one of the challenges this movement has faced in the United States. Certainly, women's substantial participation in animal activism and advocacy seems to have contributed to the sidelining of many of the demands of such activists and advocates. This is true for the early period of women's involvement in animal protectionism, when women did not have the right to vote and had limited power in electoral politics, as well as later in animal welfare and liberation movements, where women's involvement was often used as a basis for dismissing these causes. The broader social belief that women are more sensitive and emotional than men has served to support the idea that people who care about animals are "soft" and that animal rights and suffering aren't politically relevant issues. Even with the substantial involvement of people who are not women, caring about animals still is seen as something that women mostly do.

In their analyses of the resistance to the field of animal studies within academia, Susan Fraiman (2012) and Fiona Proby-Rapsey, Siobhan O'Sullivan, and Yvette Watt (2019) have identified and critiqued what they call the "pussy panic" that animal studies as a field of study triggers. Their analyses can be readily extended to make sense of how and why animal-centered activism is dismissed as sentimental and what the consequences of the feminization of animal rights activism are. As Probyn-Rapsey, O'Sullivan, and Watt (2019)[21] discuss, pussy panic refers to two different but connected concerns about animal studies as an area of academic study. First, a pussy panic is about a fear of being seen as doing something ridiculous, like "siding with animals" in a sentimental way[22] and in a way that is also associated with women. Second, as they engage with Fraiman, Probyn-Rapsey, O'Sullivan, and Watt observe that she "extrapolates from the first concern a pussy panic in the field of animal studies itself, a masculinist turn that seeks institutional acceptance and credibility by turning away from the work of women."[23] Fraiman argues that the field of animal studies only became "legitimate" in the academy when its adherents began to point to Jacques Derrida as the founding thinker of the area—when, in fact, several feminist scholars, including Carol J. Adams, Donna Haraway, Vicki Hearne, and Harriet Ritvo, among others, were the pioneers of animal studies. Shifting the idea of pussy panic into the context of the animal welfare and rights movement, animal advocacy struggles for legitimacy both because the movement represents the interests of a marginalized and exploited group (animals) and because its adherents are members of a marginalized and exploited group (women). Men in the contemporary movement benefit from the glass escalator because, among other things, their presence legitimizes the claims of the animal rights movement.

This pussy panic about animal rights also serves as an easy cover for the animal industry: painting animal rights supporters and activists as a group of hysterical women is a cheap shot but an effective and enduring one. I want to stress here that while numerous feminist scholars have written about the importance and significance of drawing on emotions in advocating for animals and theorizing human relations with them,[24] many women committed to animal rights and protection have long stressed empiricism. Dating all the way back to the inception of the animal protection movement, Caroline Earle White and her women colleagues in the Women's Branch of the Pennsylvania SPCA worked studiously to compare different methods of killing shelter animals to try to identify and propagate the most humane method. Countless women scientists, perhaps most notably Jane Goodall, are committed to animal rights. In his analysis of how a local group of animal activists in the United States uses emotion and science, Julian Allister Groves (2001)[25] finds that that the mostly professional middle-to-upper-class women involved use scientific and rational language to talk about their work and make claims for animal protection. That is, they do not rely solely, or even heavily, on the earlier rhetoric of compassion, kindness, and morality that defined animal protection between 1860 and 1930, nor the emotional discourses commonly associated with animal-centered activism. Instead, they seek to distance themselves from activists they see as overly emotional (and notably link this with radicalism) and work instead to emphasize scientific and philosophical reasons for supporting animal rights.

Today's animal rights movement is diverse in its tactics and strategies, including its engagement with emotions and its practices of gender. Many of the influential animal rights organizations in the United States are highly professionalized and bureaucratized institutions that rely on science and data to guide their work, and they also use deeply emotional advertising/outreach centered on compassion to appeal for support and donations. Men hold many key leadership positions, even as women continue to dominate the rank-and-file of animal advocacy. Efforts to acknowledge the contributions of women and members of other marginalized groups to the movement have emerged in the 2020s, including We Animals Media's Unbound Project (2023),[26] which highlights the work of women advocates for animals. In spite of the overlaps in the early animal protection movement with abolition, child protection, and temperance, among other causes, the later connections between the animal rights movement, environmentalism, and feminism (especially ecofeminism) and the contemporary push among radical animal rights groups to attend to and address diverse justice projects,[27] the contemporary animal rights movement in the United States largely remains a single-issue movement. Great potential remains for interconnecting commitments to diverse forms of justice—animal, disability, economic, environmental, food, gender, Indigenous, racial—which would advance what David Naguib Pellow (2014)[28] terms total liberation, or an ethic and practice of justice that is inclusive of all humans, nonhumans, and ecosystems. Feminist theorizing has been at the forefront of intersectional thinking and action for decades, and holds particular promise for guiding such efforts.

Notes

1 Emily Gaarder, *Women and the Animal Rights Movement* (Newark: Rutgers University Press, 2011); Lisa Kemmerer, *Oppressive Liberation: Sexism in Animal Activism* (London and New York: Palgrave Macmillan, 2023) (forthcoming); Marie Mika, "Framing the Issue: Religion, Secular Ethics and the Case of Animal Rights Mobilization," *Social Forces* 85, no. 2 (2006): 915–941.

2 For example, Corwin R. Kruse, "Gender, Views of Nature, and Support for Animal Rights," *Society and Animals* 7, no. 3 (1999): 179–198; Charles W. Peek, Nancy J. Bell, and Charlotte C. Dunham, "Gender, Gender Ideology, and Animal Rights Advocacy," *Gender & Society* 10, no. 4 (1997): 464–478; Gaarder, *Women and the Animal Rights Movement*.

3 Bernard Oreste Unti, "The Quality of Mercy: Organized Animal Protection in the United States 1866–1930" (Doctoral thesis submitted to the Department of History, American University, 2002): 52–53.

4 Janet M. Davis, *The Gospel of Kindness: Animal Welfare and the Making of Modern America* (Oxford and New York: Oxford University Press, 2016); Janet M. Davis, "Cockfight Nationalism: Blood Sport and the Moral Politics of American Empire and Nation Building," *American Quarterly* 65, no. 3 (2013): 549–574. https://doi.org/10.1353/aq.2013.0035.

5 Paula Tarankow, "The Historical Roots of Humane Carceral Logics in the United States," in *Carceral Logics: Human Incarceration and Animal Cruelty*, eds. Lori Gruen and Justin Marceau, 38, 15–35 (Cambridge: Cambridge University Press, 2022), 27.

6 Paula Tarankow, "Loyal Animals, Faithful Slaves: Animal Advocacy, Race, and the Memory of Slavery" (Doctoral thesis submitted to the Department of History, Indiana University, 2019)

7 Craig Buettinger, "Women and Antivivisection in Late-Nineteenth Century America," *Journal of Social History* 30, no. 4 (1997): 860

8 Stephen R. Hausmann, " 'We Must Perform Experiments on Some Living Body': Antivivisection and American Medicine: 1850–1915," *The Journal of the Gilded Age and Progressive Era* 16, no. 3 (2017): 264–283. https://doi.org/10.1017/S1537781417000196.

9 Buettinger, "Women and Antivivisection in Late-Nineteenth Century America."

10 Unti, "The Quality of Mercy: Organized Animal Protection in the United States 1866–1930."

11 Christine L. Williams, "The Glass Escalator: Hidden Advantages for Men in the 'Female' Professions," *Social Problems* 39, no. 3 (1992): 253–267. https://doi.org/10.2307/3096961
12 See: Kemmerer, *Oppressive Liberation: Sexism in Animal Activism.*
13 Gaarder, *Women and the Animal Rights Movement*
14 Kemmerer, *Oppressive Liberation: Sexism in Animal Activism.*
15 Maneesha Deckha, "Disturbing Images: PETA and the Feminist Ethics of Animal Advocacy," *Ethics & the Environment* 13, no. 2 (2008): 35–76. https://doi.org/10.2979/ete.2008.13.2.35; Gaarder, *Women and the Animal Rights Movement*; Renata Bongiorno, Paul G. Bain, and Nick Haslam, "When Sex Doesn't Sell: Using Sexualized Images of Women Reduces Support for Ethical Campaigns," *PLoS ONE* 8, no. 12 (2013). https://doi.org/10.1371/journal.pone.0083311.
16 Greta Gaard, "Feminist Animal Studies in the U.S.: Bodies Matter," *Deportate, Esuli, Profughe: Rivista Tematica Di Studi Sulla Memoria Femminile* 20 (2012): 14–21, https://doi.org/http://dx.doi.org/10.1353/een.2010.0056; Leslie Irvine, "Animals and Sociology," *Sociology Compass* 2, no. 6 (2008): 1954–1971, https://doi.org/https://doi.org/10.1111/j.1751-9020.2008.00163.x. Fiona Probyn-Rapsey, Siobhan O'Sullivan, and Yvette Watt, " 'Pussy Panic' and Glass Elevators: How Gender is Shaping the Field of Animal Studies," *Australian Feminist Studies* 34, no. 100 (2019): 198–215, https://doi.org/10.1080/08164649.2019.1644605.
17 Gaarder, *Women and the Animal Rights Movement,* 11.
18 Dorothy Smith, "Women's Perspective as a Radical Critique of Sociology," *Sociological Inquiry* 44, no. 1 (1974): 7–13.
19 Ernest Freeberg, *A Traitor To His Species: Henry Bergh and the Birth of the Animal Rights Movement* (New York: Basic Books, 2020)
20 Katja M. Guenther, *The Lives and Deaths of Shelter Animals* (Stanford: Stanford University Press, 2020).
21 Probyn-Rapsey, O'Sullivan, and Watt, " 'Pussy Panic' and Glass Elevators: How Gender is Shaping the Field of Animal Studies."
22 Ibid., 199.
23 Ibid.
24 For example, Josephine Donovan, "Feminism and the Treatment of Animals: From Care to Dialogue," *Signs: Journal of Women in Culture and Society* 31, no. 2 (2006): 305–329. https://doi.org/10.1086/491750; Kathryn Gillespie, "Witnessing Animal Others: Bearing Witness, Grief, and the Political Function of Emotion," *Hypatia* 31, no. 3 (2016): 572–588. https://doi.org/10.1111/hypa.12261; Lori Gruen, *Entangled Empathy: An Alternative Ethic for Our Relationships with Animals* (New York: Lantern Books, 2015)
25 Julian McAllister Groves, "Animal Rights and the Politics of Emotion: Folk Constructs of Emotions in the Animal Rights Movement," in *Passionate Politics: Emotions and Social Movements*, eds. Jeff Goodwin, James M. Jasper, and Francesca Polletta (Chicago and London: University of Chicago Press, 2001), 212–229
26 We Animals Media, "The Unbound Project," (2023), retrieved from: https://unboundproject.org/, accessed May 7, 2023.
27 David Naguib Pellow, *Total Liberation: The Power and Promise of Animal Rights and the Radical Earth Movement* (University of Minneapolis Press, 2014)
28 Pellow, *Total Liberation: The Power and Promise of Animal Rights and the Radical Earth Movement.*

Bibliography

Bongiorno, Renata, Bain, Paul G., and Haslam, Nick. "When Sex Doesn't Sell: Using Sexualized Images of Women Reduces Support for Ethical Campaigns." *PLoS ONE* 8, no. 12 (2013). https://doi.org/10.1371/journal.pone.0083311.

Buettinger, Craig. "Women and Antivivisection in Late-Nineteenth Century America." *Journal of Social History* 30, no. 4 (1997): 857–872.

Davis, Janet M. "Cockfight Nationalism: Blood Sport and the Moral Politics of American Empire and Nation Building." *American Quarterly* 65, no. 3 (2013): 549–574. https://doi.org/10.1353/aq.2013.0035.

Davis, Janet M. *The Gospel of Kindness: Animal Welfare and the Making of Modern America*. Oxford and New York: Oxford University Press, 2016.

Deckha, Maneesha. "Disturbing Images: PETA and the Feminist Ethics of Animal Advocacy." *Ethics & the Environment* 13, no. 2 (2008): 35–76. https://doi.org/10.2979/ete.2008.13.2.35.

Donovan, Josephine. "Feminism and the Treatment of Animals: From Care to Dialogue." *Signs: Journal of Women in Culture and Society* 31, no. 2 (2006): 305–329. https://doi.org/10.1086/491750.

Freeberg, Ernest. A *Traitor To His Species: Henry Bergh and the Birth of the Animal Rights Movement*. New York: Basic Books, 2020.

Gaard, Greta. "Feminist Animal Studies in the U.S.: Bodies Matter." *Deportate, Esuli, Profughe: Rivista Tematica Di Studi Sulla Memoria Femminile* 20 (2012): 14–21. https://doi.org/http://dx.doi.org/10.1353/een.2010.0056.

Gaarder, Emily. *Women and the Animal Rights Movement*. Newark: Rutgers University Press, 2011.

Gillespie, Kathryn. "Witnessing Animal Others: Bearing Witness, Grief, and the Political Function of Emotion." *Hypatia* 31, no. 3 (2016): 572–588. https://doi.org/10.1111/hypa.12261.

Groves, Julian McAllister. "Animal Rights and the Politics of Emotion: Folk Constructs of Emotions in the Animal Rights Movement." In *Passionate Politics: Emotions and Social Movements*, edited by Jeff Goodwin, James M. Jasper, and Francesca Polletta, 212–229. Chicago and London: University of Chicago Press, 2001.

Gruen, Lori. *Entangled Empathy: An Alternative Ethic for Our Relationships with Animals*. New York: Lantern Books, 2015.

Guenther, Katja M. *The Lives and Deaths of Shelter Animals*. Stanford: Stanford University Press, 2020.

Hausmann, Stephen R. "'We Must Perform Experiments on Some Living Body': Antivivisection and American Medicine: 1850–1915." *The Journal of the Gilded Age and Progressive Era* 16, no. 3 (2017): 264–283. https://doi.org/10.1017/S1537781417000196.

Irvine, Leslie. "Animals and Sociology." *Sociology Compass* 2, no. 6 (2008): 1954–1971. https://doi.org/10.1111/j.1751-9020.2008.00163.x.

Kemmerer, Lisa. *Oppressive Liberation: Sexism in Animal Activism*. London and New York: Palgrave Macmillan, 2023, forthcoming.

Kruse, Corwin R. "Gender, Views of Nature, and Support for Animal Rights." *Society and Animals* 7, no. 3 (1999): 179–198.

Mika, Marie. "Framing the Issue: Religion, Secular Ethics and the Case of Animal Rights Mobilization." *Social Forces* 85, no. 2 (2006): 915–941.

Peek, Charles W., Bell, Nancy J., and Dunham, Charlotte C. "Gender, Gender Ideology, and Animal Rights Advocacy." *Gender & Society* 10, no. 4 (1997): 464–478.

Pellow, David Naguib. *Total Liberation: The Power and Promise of Animal Rights and the Radical Earth Movement*. University of Minneapolis, 2014.

Probyn-Rapsey, Fiona, O'Sullivan, Siobhan, and Watt, Yvette. "'Pussy Panic' and Glass Elevators: How Gender Is Shaping the Field of Animal Studies." *Australian Feminist Studies* 34, no. 100 (2019): 198–215. https://doi.org/10.1080/08164649.2019.1644605.

Smith, Dorothy. "Women's Perspective as a Radical Critique of Sociology." *Sociological Inquiry* 44, no. 1 (1974): 7–13.

Tarankow, Paula. "Loyal Animals, Faithful Slaves: Animal Advocacy, Race, and the Memory of Slavery." Doctoral thesis submitted to the Department of History, Indiana University, 2019.

Tarankow, Paula. "The Historical Roots of Humane Carceral Logics in the United States." In *Carceral Logics: Human Incarceration and Animal Cruelty*, edited by Lori Gruen and Justin Marceau, 38, 15–35. Cambridge: Cambridge University Press, 2022.

Unti, Bernard Oreste. "The Quality of Mercy: Organized Animal Protection in the United States 1866–1930." Doctoral thesis submitted to the Department of History, American University, 2002.

We Animals Media. "The Unbound Project." (2023). Retrieved from: https://unboundproject.org/.

Williams, Christine L. "The Glass Escalator: Hidden Advantages for Men in the 'Female' Professions." *Social Problems* 39, no. 3 (1992): 253–267. https://doi.org/10.2307/3096961

38

THE REPRESSION OF ANIMAL ACTIVISM

An Interview with Tayler Zavitz

Chloë Taylor (CT): Thank you, Tayler, for agreeing to do this interview with me. Can you start by saying a bit about your work both as an activist and as a scholar of activism?

Tayler Zavitz (TZ): Thank you so much for inviting me to be a part of this interview. It's a pleasure to be speaking with you. I'm a PhD candidate and instructor in the Department of Sociology at the University of Victoria. My background in academia was mainly in critical sociology with a specific focus in critical animal studies. But I also do have a background in political science and public law as well. My activism started about 15 years ago and I've been active in a variety of grassroots activism campaigns, doing local outreach and advocacy, going to protests, doing speaking engagements, and that sort of thing. I've also worked with a number of large animal organizations doing organizing and taking part in slaughterhouse vigils, traveling to set up organization chapters in other places in the world, like in the U.S., for example, working on undercover investigations. So I've kind of had my hands in a lot of different areas of activism over the years.

My academic career as a scholar of activism really kind of followed from getting involved with activism myself and really learning about activist repression that was taking place, specifically at that time it was going on in the U.S., through things like ag–gag legislation and the Animal Enterprise Terrorism Act, two types of legislation that specifically criminalize animal activists. Ag-gag legislation criminalizes animal activists for exposing animal suffering on factory farms, and the Animal Enterprise Terrorism Act makes it not only a criminal offense, but "terrorism" to cause an animal enterprise a loss of profit, or, conspiring to do so. I really started to dive into that area of research and then wanted to uncover what was happening in Canada in that respect. And from there, that's really where my PhD work comes in now, and that work continues to have that critical animal studies foundation

DOI: 10.4324/9781003273400-47

because it's an area of academia that, while definitely expanding and growing with each year, it's still often widely ignored or almost diminished as a legitimate field of study, despite the fact that these issues, the large scale animal suffering and exploitation that we're seeing, while obviously important on their own, also intersects with so many other social justice issues.

I honestly believe that a great deal of people are really just ignorant to what's actually happening to animals, and I really do see this through my teaching all the time. People often just have no real knowledge about animal industries and the reality of what's happening within them. And so for me, a big part of my decision to pursue my academic path was twofold. One, simply to educate, to create that space within academic discussions, to highlight the reality animals are facing within the animal industrial complex, and then, second, to work within the system to help legitimize the field of critical animal studies and push for a more truly inclusive education system.

CT: What are some of the ways in which animal activism has been criminalized and repressed that you examine in your research?

TZ: I look at my research in two different ways. I look at the criminalization in terms of the legal criminalization, so actual legislation, actual criminalization through the criminal justice system. So that would be things like the ag-gag laws that have now come up in Canada as well, as well as ecoterrorism labeling of activists, as terrorists under the legal system or the use of, we don't have ecoterrorist legislation currently in Canada, like the Animal Enterprise Terrorism Act in the United States, for example, but that rhetoric is still being pushed throughout the criminal justice system here as well. So I look at these sorts of cases of activists being legitimately charged and convicted, if that's the case. Like the example of Anita [Kranjc] back in 2015, is one case that I look at. Looking at how in that specific example she was giving water to a pig on a slaughter truck, and she was charged with criminal mischief and faced heavy fines and jail time for doing so. She was acquitted of that. But that was a long legal process that she had to go through. And so there's a number of cases like that in Canada that I've looked at where activists have been criminalized under the current legislation. So whether that's things like criminal mischief or trespassing, those sorts of laws that we had on the books at the time that were being used to try and silence or deter activists from taking part in those sorts of activities. Most cases that I've looked at have been acquitted or haven't actually gone forward. One that is different is the case of the Excelsior Four, which went forward, and Amy Soranno and Nick Schafer were both charged and convicted. And that was really the first time that we've seen activists in Canada who have actually been charged, convicted and sentenced. And in that case, it was specific to a Meat the Victims event that they took part in here in British Columbia. And so that's kind of been one side of my research is looking at how the legal system and the criminal justice system is moving forward in terms of criminalizing animal activists to try and silence them, to try and deter them from taking part in these sorts of activities.

But the other side of that is the ideological side that I'm looking at in terms of the criminalization. And that's just through things like education systems, through media, through these ideological ways of talking about animal activists. And that really comes down to things like language and the rhetoric that's used to talk about animal activists and how that specific sort of criminalization, just through ideology and through our understandings of animal activists, are criminalizing them far before they even get to the criminal justice system. And so this conversation about how media is representing animal activists, for example, and how that then perpetuates a specific idea about who they are, the actions that they do. And so it's an interesting kind of dynamic because I think a lot of the time we're looking at or when we're talking about these issues of criminalization, it's often talking about that specific legal system because it is so problematic in terms of things like ag-gag legislation and something that obviously we have to be concerned about. But I think that tends to be the focus because of those strict sort of punishments that are being implemented. But I think an important conversation that isn't often seen or talked about as often is that ideological conversation around animal activists that has created the situation for these sorts of legal actions to be taken in the first place. And so it's this notion of ecoterrorism, this fear mongering that's created this moral panic that's created through this sort of rhetoric. Really the ideology is just to protect the animal industrial complex, right? It's a specific idea that's promoted, specific narrative that's told in order to create this sort of panic, to silence the critics of this industry.

CT: And do you have personal experiences of activists being criminalized, either yourself or people you know.

TZ: Yes, ideologically, absolutely, right? Everyone that I know that's involved really in the movement. But yes, I have had two family members of mine who have been kind of personally criminalized. One was my sister, who was protesting a Ministry of Natural Resources–conducted deer hunt in our local town. And she was standing in front of a driveway where the ministry trucks were all trying to come out and people were protesting and standing with signs in front of the driveway. And this specific ministry worker didn't want to wait anymore and ended up using his vehicle to push her across the street out of the way. And it was all on film. Luckily, she was not overly physically harmed in that situation, which we're all very grateful for. But in that situation, when there was talks afterwards of charges being laid against that specific ministry worker, the police in that situation really pushed forward how my sister could be charged for criminal mischief for standing in front of the driveway. And so it was just this interesting kind of conversation around something as serious as, like vehicular assault versus somebody standing with a sign, much like you would see with a labor picket, right? Like standing in front of a driveway, just blocking access. And that sort of being the main focus in that specific case. And in this conversation, I will say, that's not me advocating for the use of the criminal justice system, like for punishment. I believe in decarceralization and moving away from the current criminal justice system. But I think my point to that is just to highlight the inconsistencies and the discrepancy of how the system is used, who's criminalized and who is not, right? Somebody who is just openly willing to hit somebody with a vehicle, that's seen as fine and somebody standing with a sign is not, right?

And then in the same event or the same protest that's taken place, my father was actually arrested and charged with assault of a Ministry of Natural Resources worker despite the fact that the entire thing was on film and he was nowhere near the individual when he says that this took place. And so in that situation, my father, who is somebody who is well known and very well respected in his community and has served his community in healthcare and community work for decades, had to go be fingerprinted and put in a jail cell while he awaited these charges to talk to somebody about what was going on. And I think for me that was something that, a time that,

I was probably the most angry at the system, but I was also the most proud of him for standing up against an industry that will do whatever they can to try and silence those who are standing up against what they're doing.

And so I think in those specific situations and those are just two directly related to me, but those just speak volumes in terms of what evidence you can have, what you can show to have happened or not have happened, and the things that will still go forward despite that, based on the power of these sorts of industries. And I mean, luckily the charges against my father were eventually dropped, but same sort of idea. Like it's creating that fear, wasting time, wasting money for these individuals and wrapping them up in a system that is very scary.

CT: The type of animal activism that tends to be criminalized as ecoterrorism is usually direct action activism that breaks the law, for instance, by trespassing onto farms, although in the examples you just gave, your sister and your father weren't doing anything illegal. Your research also shows the legislation is so broad that people could potentially be targeted just by happening to witness cruelty to animals that is protected under ag-gag laws. But there are forms of animal advocacy that do not even break the law, and in these cases, animal advocates tend to be stigmatized in other ways as crazy or hysterical, for instance, and it tends to be sort of feminized. So what role do you see gender playing in these different forms of activism and the ways they are undermined?

TZ: Yes, this reminds me of a situation at a circus protest a number of years ago that I was at, where I was protesting with a group of individuals at this circus that comes through town like once a year. And as I was standing there, a man was yelling at me for protesting the event and told me that I should be at home cooking dinner for my husband. And I remember just that stopping and thinking, what? Why is that the automatic sort of response? And I think gender definitely plays a role in, like you're saying, the way animal activism is stigmatized as being overly sentimental, or overly emotional or weak. And it is feminized, right? It's seen as lesser than because of those things. And of course, women care about animals because we're too emotional and need to toughen up about the reality of the world. And so this sort of rhetoric is consistent in terms of talking about animal activism. I think because of that, it's often not taken seriously or seen as irrational. I think it's kind of passed off as a bleeding heart movement that doesn't actually hold any sort of legitimacy.

CT: Animal activism has been critiqued for using tactics, images, and comparisons that are experienced as racist, sexist, fat phobic, and otherwise oppressive to minoritized groups of humans. So I'm thinking especially of PETA ads that compare animal oppression to the Holocaust and human enslavement, the "Save the Whales, Lose the Blubber, Go Vegetarian" PETA ad that obviously exploits discrimination against fat people to advance their cause, the recent ad that appropriates a symbol of Black Lives Matter by showing animals taking a knee, and then the frequent use of female nudity to promote veganism, also in PETA ads. Animal activists have also been critiqued for targeting the practices of minoritized groups, the eating of shark fin soup, for instance. What are your thoughts on these forms of animal activism and these kinds of charges of racism and sexism within animal activism?

TZ: I think the movement has often done itself a serious disservice by taking part in some of these sorts of campaigns. I think they can show a real disregard and an insensitivity for the suffering and oppression that human beings have gone through and continue to go through. And some of them, like these blatantly fat phobic or sexist or racist ads, in my opinion, have no place in a movement that's based on compassion and equality. The whole idea behind veganism and the animal rights movement is this idea of having compassion, having that idea of equality, having that moving forward of ideology to be progressive. And these sorts of campaigns are taking us back, right? They're not doing anything other than further oppressing others to try and promote not oppressing animals, it doesn't add up.

There are campaigns like you're talking about—the Holocaust campaigns, or slavery campaigns—which are fairly popular with PETA, and I think those people often struggle with seeing why they're problematic because there *are* connections between oppressions, right? And I think people want to try and get people to understand those connections. But I think there's other ways of doing that that aren't problematic and aren't overtly potentially offensive to those who have struggled themselves. And especially when we see examples of white activists doing street outreach and going up to a person of color and saying, "Well, animals deal with things like slavery," and having no understanding of what that actually means to that individual, right? And so there is this sort of insensitivity that's really, I think, prevalent in the movement. And I think that also comes from a place of ignorance and a place of not understanding those connections or understanding other social justice movements. So as a movement, we, the animal rights movement, should and need to be allies of other social justice movements. We should be helping to fight alongside them and raising our voices in support of social equality, not being the ones contributing to further inequality.

CT: Although it's easy to criticize PETA ads such as these, in an interview for Plant Powered Radio, you mentioned that it was a PETA video that made you go vegetarian. And you also said that it was the film *Earthlings* that made you go from a passive vegetarian to an angry and active vegan, promising to do everything in your power to improve the lives of animals. So both of those experiences show the impact that media can have. And in your article "Animal Oppression and Solidarity: Examining Representations of Animals and Their Allies in Twenty-First Century Media", you explore representations of the Animal Liberation Front in Media. What are your thoughts on the Animal Liberation Front in particular and how they have been depicted in the media, both negatively and more rarely, positively? And maybe first you can say a little bit about the Animal Liberation Front and the sorts of tactics they use.

TZ: Yes, it is easy to criticize and critique PETA for the problematic things that they do and I also think it's important to do so. That's how we grow and evolve as a movement and become more effective. But at the same time, I can also acknowledge that they have done important work throughout their history as an organization. And some of that work, like you said, directly impacted me and made me, at the young age of twelve, say that I would never eat animals again. I do believe media is very impactful and that can be in a good way or a bad way, and we can definitely see that specifically around activists like those in the ALF.

The ALF is a global, decentralized, and anonymous network of dedicated, underground individuals who take direct action to liberate animals from oppression. The ALF target all animal exploitation industries and use direct action and illegal tactics, like breaking into buildings to liberate animals, property damage, and arson. Personally, I support direct action work. I have a great deal of respect for those who put their lives and freedom on the line to save the lives of animals and disrupt these horrifically violent industries. I know they're often seen as controversial, and some argue that they make the movement as a whole look bad or create a specific idea about the movement as a whole. But I think that in itself really speaks to the power of media because that's how media has framed the ALF.

I've done a great deal of archival research for my dissertation and looking at specifically newspaper clippings of what was going on with animal activists throughout its history in Canada and throughout the 80s and 90s in Canada, there was a number of ALF actions that were taking place and the media really had like a field day with them. You saw language like "lunatics" or "dangerous radicals", "thieves", "criminals", "militants", "extremists", "kidnappers", "terrorists", right? This is the sort of common rhetoric that we were seeing. And usually that was alongside the photo of an activist wearing a balaclava to create this very kind of scary picture. One actual example that comes to mind is in the mid-80s, like 1985, there was a monkey who was liberated by the ALF from a laboratory at the University of Western Ontario. And after the liberation took place, the laboratory came out and said that the monkey could be infected with a herpes virus that was deadly to humans. And so the narrative then that was prevalent throughout the media was this fear mongering about the ALF now putting the public in danger and spreading this disease. So they weren't just criminals anymore, but now they were a public health risk, right? So most media representation of the ALF is not positive. It often highlights things like theft and breaking and entering and having to get an increase of security and people feeling threatened or unsafe. And there's typically a general disregard for the real issue at hand, which is the animals that were liberated in those situations, right? Like what about the ethics of purposefully giving a monkey herpes! Why is there not a public outcry about that? So there is typically a general disregard for the animals, but, I can't say that's always the case. There are some articles that I found where there was somewhat sympathetic kind of tone to the issue in terms of the animals, but they were still pretty harsh in terms of the activists themselves.

So I think when it came to these two specific examples of media that we were talking about in the article, *Okja*, the film and the book, *We Are All Completely Beside Ourselves,* I think they stood out to us so much because it was like a complete flip of the narrative around the ALF, which you just don't commonly see. In *Okja*, for example, there's a full scene where ALF activists are explicitly outlining their philosophies and their codes that they follow. And really, it was taken directly, almost word for word, from ALF press literature. And so for the first time, you really had this actual understanding of who the ALF are in their own words, as opposed to what the media was creating for them. And in Fowler's book, we also see these accurate depictions of how and why the ALF operates. She talks about real life historical context, talking about different liberations that have taken place throughout the ALF's history, and also outlines descriptions of horrendous treatment of animals within these animal industries. So, again, creating a very different narrative and a very accurate narrative of what's actually taking place.

So these two examples really kind of set the stage for this undeniable rethinking of the ALF and who they are and what they do. And there was also a clear highlighting in both of the ALF and how they're deemed as terrorists and really the absurdity of those labels as well. As we say in the article, those that are engaging with the film or reading the book, they're ultimately exposed to the reality that animal activists aren't these violent criminals or terrorists, but rather they're compassionate individuals who are motivated by things like empathy and justice. And in a world where we're seeing through the mainstream media that activists are typically villainized or called these specific words to create this sort of fear around them, these media sources do the opposite. They humanize them. They make them relatable, they highlight these sorts of inconsistencies in this unfounded characterization and expose the reality of who these people are.

CT: In the same article, you drawn the work of Timothy Pachirat, who spent six months working in a slaughterhouse in order to write a book on the topic *Every Twelve Seconds: Industrialized Slaughter and the Politics of Sight*. Animal advocates working for PETA have also gone undercover in slaughterhouses with hidden cameras to expose the horrors of these places. What do you think of these forms of critical animal studies research and activism that involve going undercover in slaughterhouses or into other sites of animal oppression and even participating in the killing of animals with the objective of ultimately exposing that violence to the public.

TZ: Similar to how I feel about direct action in terms of the ALF, I'm somebody who believes that a variety of tactics is necessary in order to really push the movement forward. I think some things that some will find problematic or they don't necessarily agree with, like direct action for example, because it's breaking the law, others will see that and that will be what impacts them the most and changes their ideology about these sorts of issues. And so I think there's room for all sorts of tactics in the movement as long as those tactics are not oppressing somebody else by doing so, like these ads that we're talking about.

In terms of undercover work, I worked for an organization who did undercover investigations. I saw the impact that those investigations had on people when they came out into the mainstream media. I think it's something that's important in terms of seeing something that is so hidden. It's something that we do not typically have the availability to see. We don't have the opportunity to have these sorts of conversations from the actual animals' point of view, seeing it from what their actual experience is. We hear about it kind of on a periphery when we talk about these sorts of issues oftentimes. And we can tell somebody about these sorts of things. We can tell people about the suffering that's taking place or what animals are going through. But I think seeing it is a completely different situation for people. And like I said, that's what changed me. And having gone to farms and gone to slaughterhouses in the U.S. and Canada and there were times where I would have my phone recording or live streaming at the time and I saw the impact that that had on people that I knew, just in my own kind of small circle of people who I had been talking about veganism to for years, and who never really paid attention to it. I think there's something very powerful in bearing witness to those animals in that time in their life.

I think it's got its pros and cons. Obviously taking part in the killing of animals is an issue, but it's also something that I know some who have worked in those specific situations doing undercover work, who I've talked to at length about these sorts of issues have said like: "It's going to happen whether I'm there or not". And in some respect, they preferred that they were the one to do it because they would be kinder. And so, it's a situation where we have to ask, is it important to have that footage? I think so. And is it important to be able to be in those spaces? I think it is. But obviously that's going to come with problems on its own.

CT: A number of feminist critical animal studies scholars have pointed out that since the 19th-century anti-vivisection and vegetarian movements, more women than men have been engaged in animal advocacy and are vegetarian or vegan. At the same time, men are overrepresented as the leaders of the movement, with Peter Singer dubbed as the father. So many animal rights organizations are headed by men, even while women do most of the work, although PETA is an exception. What do you think, or why do you think it is that more women than men are vegan and advocate for animals? And why do you think men nonetheless lead the movement, or at least head many of its organizations and are the most influential authors? Has this been your own experience in animal advocacy and scholarship?

TZ: Yes. I find it interesting, or ironic rather, that even in a movement that, as we talked about earlier, is often feminized, seen as emotional or irrational or weak because of the fact that it's mostly women, that men still hold the powerful positions. And I think if that's not the patriarchy at work, I don't know what is. It's just I think that overarching system that we can see play out.

I think socialization obviously plays a role here. If we look back to childhood socialization all the way up to current socialization within society, boys and men are socialized to be tough, uncaring, unfeeling, aggressive, strong, take part in violence to be tough. This sort of rhetoric of eating animals is masculine, it's tough, killing animals, like hunting, that makes you tough. This sort of thing is perpetuated consistently within society. And then on the flip side, obviously we come back to this idea that girls and women are socialized to be inferior or caregivers, maternal. Eating plants is weak, it's feminine. So this sort of socialization obviously plays a role in why more women than men are vegan and take part in animal activism.

I think in terms of if we look at the men still leading the movement or being heads of organization or getting the most attention in terms of their activist work, as I said, I do think the patriarchal system is still at play here. Just in general, there's still an undertone of women not being good enough to hold these positions. There's still a lot of men in powerful positions who use these positions to exploit and target women. And we've seen that play out in our movement as well. And I think that's another reason why it's so important for our movement to understand and align our values with other social justice movements to try and eliminate some of this that's taking place.

And I also think that it's often because women aren't generally promoting themselves the way that men do. I think currently in a time of a technological age, in a time of social media, we see a lot of activist work online. And I think a lot of those women who are doing the work aren't the ones that are plastering their faces on social media. They're not sitting back and spending hours editing videos of themselves to promote themselves. They're typically behind the scenes. They're doing the hard labor, they're doing the work, the difficult work that is often unrecognized and is often then used by these individuals to promote themselves. And so it's so common that folks in the mainstream animal rights movement look at this history of the movement as only including like you're saying, like Peter Singer or James Aspey and Earthling Ed because they're the ones that have the most media attention.

And really this is an absolute disservice to the movement and to those who have done the work before them. I mean, not only is it just overtly problematic, because it's like a blatant inaccurate and flawed representation of our movement and one that obviously lacks serious insight and important pivotal moments, but it also completely erases the incredible work that's been done and continues to be done by so many women and women of color specifically. Just as a starting point, Jill Phipps, A. Breeze Harper, Carol J. Adams, Liz White, Lesli Bisgould, Maneesha Deckha, Twyla Francois, Aph Ko, Lori Gruen, pattrice jones, Jo-Anne McArthur. This is just to name a few, and they're the ones that you don't typically hear about because they're the ones that are just busy doing the work.

I think a lot of what this does is simply upholds this male saviorism, and usually white male saviorism, that is already so rampant in this movement. And not to mention that most of these men and these higher positions often still promote oppressive views and problematic individuals. That these individuals or ideas that are using things like these PETA campaigns, these really problematic, overtly sexist or racist campaigns. And so that creates a problem in itself. And so there needs to be this focus on that intersection of ideology and social justice movements. And I think our actual history of this movement has so much depth to it, so many crucial individuals and turning points and promoting it in such a simplistic way is not only offensive to all of those who came before us and who created and shaped the movement to what it is today, but also who, just because they fought tirelessly and still do, to try and push the movement forward.

CT: Since the 19th century many women and animal advocates have been feminists and have connected their advocacy to their animal advocacy to their feminism. Is your own animal advocacy connected to your feminism, or is your feminism connected to your animal advocacy?

TZ: Yes, I would say my feminism really emerged because of my animal activism and learning about the connection between the oppression of animals and the oppression of women. I think becoming involved in animal advocacy work and then having the opportunity at the time to be at a university that had one of the only critical animal studies programs in Canada, I was suddenly emerged in learning about the world of inequality and oppression and violence and, at the same time, the world of social justice work. And so having that opportunity to learn about and engage with these intersecting issues like reading the work of Carol J. Adams and Maneesha Deckha and those who are working in these fields, having also faced oppression myself based on being a woman, I was then really easily able to see that connection. And I would say that that's where my feminism really flourished.

CT: Although there's a strong historical and conceptual relationship between feminism and animal advocacy, in my own life, I've never experienced so much pushback against my veganism and animal advocacy as I have from feminists. Although much less discussed, I think these conversations about animals within feminism might parallel the fraught discussions of race with white feminists, where a group that is used to identifying as the social justice warriors and the oppressed or the allies of the oppressed is very uncomfortable finding themselves situated on the side of the oppressors and the privileged in terms of this issue. What are your experiences as an animal activist in feminist academic spaces?

TZ: Yes, I would say it's very similar, for those feminists that are outside of the critical animal studies space, obviously. I think for me as well, I came from, like I said, Brock University that had this critical animal studies program. I came from a department where even those who weren't engaged with critical animal studies understood it, and it was something that was just accepted, it was normalized. And I think leaving that kind of little bubble was a bit of a rude awakening for me when I moved and got into a new department. And I was very surprised that those who were the most dismissive or those that I had the most butting of heads with about issues weren't the people that I thought they would be. I had a number of professors who had zero real interest in animal issues or didn't know much about animal issues, who were very supportive of the work that I was doing, whereas feminist scholars were the opposite. And it was very confronting for me.

I was put in a situation where I had multiple well known feminist theorists who were extremely dismissive and condescending about the work that I was doing, to the point that one came to do a guest lecture in our feminist theory class, and is very well-known feminist theorist, and I asked her a question about how she navigated. She was talking about how she created new language within her department to talk about women's issues at the time, many years previously, and how difficult it was and how she struggled with it and all of these issues. And I was really thinking at the time of how similar it was in terms of how I had to approach departments, for example, and how I had to have these conversations and try and recreate a specific language around animals. And I asked a question about how did you overcome that? What was your way through that? How did you navigate that? Because this is something similar to what I'm doing. And the response that I got still to this day kind of really shocks me in that it was just completely condescending. I was basically told the research that I was doing was invalid, that it didn't matter. And I just couldn't believe how there was such clear parallels between the work that they did and the struggles that they had and the work and struggles that I had and that there was nothing, there was no conversation that took place about it, or about how we can move this forward. It was just very much: what you're doing is irrelevant.

I think it does stem from this feeling of being called out. And not even me actually verbally calling out, but just my presence as an animal advocate and a vegan, called out the fact that there were things that they were taking part in that could be seen as hypocritical with what they're promoting or what they're advocating for in their own work. And I think similar to how defensive people can get when you talk to them about eating animals, whether they're feminist scholar or not, I think it just is confronting to people because they have to acknowledge their own practices. They have to acknowledge the things that they're doing that's contributing. And I think especially, as you said, those who consider themselves these very strong allies and those who are dedicated to social justice to have to then deal with that dilemma of, okay, well, now I'm the one that's actually taking part in that, I think that's very confronting for people, and I think it's something that makes people very uncomfortable. And I think that tends to be why you get that kind of feedback from those whom you would maybe least expect it from.

39
SEXISM IN ANIMAL ACTIVISM
The Foie Gras Campaigns

Corey Lee Wrenn

Foie gras (which translates to "fat liver" in French) consists of liver taken from force-fed and force-fattened ducks and, to a lesser extent, gooses[1] (gooses being more expensive and time intensive to exploit). While foie gras has been in production for hundreds, if not thousands, of years, its industrialization in the 1960s dramatically increased the number of animals impacted.[2] Approximately 40 million individuals each year are currently used, abused, and killed for this specialty product, the majority of whom live and die in southwest France.[3] Adams[4] has argued that nonhuman animal agriculture is a deeply gendered industry, whereby nonhuman animals are routinely feminized in order to facilitate their objectification, butchering, and consumption. An analysis of foie gras production expands this observation by underscoring the gendered and institutional elements of nonhuman animal agriculture beyond biological sex, as the vast majority of foie gras victims are male.

This twist further proves itself particularly relevant to understanding efforts to *resist* foie gras. As of this writing, foie gras production (and, in some cases, its importation and sale) is banned in many European countries, parts of the United States, the United Kingdom, Australia, Argentina, India, and elsewhere. It remains one of the longest-pursued campaigns in the modern nonhuman animal rights movement. It also provides a revealing case study in the gender politics of anti-speciesism, notably the persistent inability of the movement to transgress sexist scripts in its effort to challenge speciesist cultural constructions. Foie gras comes from male ducks, yet nonhuman animal rights mobilization, specifically that associated with People for the Ethical Treatment of Animals (PETA), maintains tactics that both villainize and victimize women.

Using vegan feminist theory, this chapter critically analyzes PETA's anti-foie gras campaigning (predominantly that which transpires in the United States, United Kingdom, and Europe) and its cultural implications for understanding both gender and species relations in the wider public and within activist spaces. Foie gras production aligns with gendered roles of male domination and female subservience, and anti-speciesism activists have attempted to accentuate this relationship using female activists as foie gras victims. However, they are also known to target women as perpetrators. Anti-foie gras campaigns, furthermore, tend to rely on sexist (and often violent) imagery and ideas about women. This is a tendency, I argue, that is deeply problematic in a society that is as patriarchal as it is human

DOI: 10.4324/9781003273400-48

supremacist. Gender scripting in anti-speciesism campaigning must be carefully employed (if employed at all) to avoid intersectional failure when drawing comparisons between sexism and speciesism.

Sex and Gender in Foie Gras Production

The making of "meat"[5] is a deeply patriarchal affair. Vegan feminist scholars have noted that speciesist agriculture tends to follow sexist scripts whereby "farmers" (culturally envisioned as male) own, control, physically manipulate, and financially exploit farmed animals.[6] Male farmers may be outnumbered by their female counterparts in some areas, but nonhuman animal agriculture remains male dominated and largely male owned.[7] Women, Black women in particular, have been disproportionately assigned to less prestigious, lower paying, more dangerous and psychologically damaging jobs in the American poultry industry,[8] suggesting similar gender inequalities in foie gras production may be present. The United States, United Kingdom, and Western Europe, for that matter, also rely heavily on vulnerable immigrant laborers to undertake difficult and loathsome agricultural work. Migrant women in this industry are especially subject to sexual violence and slavery.[9] These gender dynamics are obviously not outwardly promoted by nonhuman animal agricultural corporations, and how this transpires in foie gras production is not well documented. However, an idealized image of femininity has been key to its status as a wholesome delicacy. In France, women have been important to the branding and customer service elements of foie gras sales, a role some women farmers have celebrated as vital to the preservation and transmission of cultural heritage.[10]

Gender inequality is also found in the treatment of the nonhuman products themselves. While I will argue that all farmed animals exploited in the food system are feminized by way of their subservient position, it is also the case that female bodies are disproportionately exploited given their capacity to offer their own bodies for capitalist gain as well as their sexual productions (namely eggs, breastmilk, and children to repopulate the system).[11] Furthermore, male bodies, in industries such as that of dairy and egg production, are rendered low value or even worthless; they are frequently killed immediately or very soon after birth. Male chicks, for instance, are ground alive in an industrial grinder just minutes after hatching if not dumped in large trash receptacles where they will be suffocated or crushed under thousands of their brothers. In foie gras production, however, the product is the liver not the egg, and thus male bodies are prized, as they tend to grow larger than females. Female duck and goose hatchlings are killed immediately after birth, as are rooster chicks in the egg industry. Indeed, reproduction has a much more marginal role in foie gras production; the animals used are hybrid "mules" created from two different subspecies and are unable to reproduce themselves.

The sex of the victims in foie gras is key to production, but gender relations, vegan feminism emphasizes, are also highly relevant. Nonhuman animal agriculture, in general, tends to be male owned and male benefiting, but its very structure is patriarchal in design. Consider, for instance, that it is generally considered a masculine role in patriarchal cultures to provide food and shelter, but this "protection" and "provision" comes at the cost of entrapment in the domestic setting. Here, ducks have been literally caged and isolated, preventing movement and comradery with other victims. The "feeding" practice, in particular, is highly sexualized. "Farmers," who may be male or female, take on a masculine role by restraining their victims and forcing a long pipe or funnel into their mouths and down their throats to fill their stomachs close to (and sometimes beyond) bursting (Figure 39.1). In traditional methods, the bird's body will be fully restrained by the "farmer's" hands, legs, and feet while the violation occurs. In modern methods, birds remain encased in their cages and their faces and beaks are manipulated by an employee who is aided by a pneumatic pump rather than gravity. The technological shift seems to further masculinize the process, removing even the human touch involved in handling victims, replacing it with cold, rationalized machinery. The vulnerability of on-the-ground employees in many foie gras facilities (as is typical in the "meat" industry) adds another layer of domination.[12] Although these workers take on a masculine role in their manipulation of incarcerated ducks and gooses, they themselves are, if to a lesser extent, also victims of patriarchal power relations given the exploitative conditions that frequently characterize their work.

Nonhuman animal rights campaigners have rightly detected the sexualized nature of foie gras production. A Humane League protest against the London Grill in Philadelphia, for instance, reportedly shamed management with the chant "How many ducks have you raped today?"[13] (Caro 2009). United Poultry Concerns director Karen Davis[14] has also argued that the exploitation of farmed animals is a form of rape:

> The rape of farmed animals is an ancient practice, not only because these animals have always been readily available for sexual assault on the farm, but because farmed-animal production is based on physically manipulating and controlling animals' sex

Figure 39.1 Forced feeding of a restrained duck.

> lives and reproductive organs. Sexually abusive in essence, animal farming invites crude conduct and attitudes toward the animals on the part of producers and consumers alike.

To the first point, we might typically think of this in the context of layer hens and dairy cows, but ducks killed for foie gras are sexually manipulated as well. Ducks must be "artificially inseminated" to reproduce, for instance. Furthermore, incubation temperature will be adjusted to encourage a genetic predisposition to increasing liver weight for the industry.[15] Baby birds just one to three weeks old are then subjected to several weeks of "preparation" (i.e. grooming) for their future force-feeding. This preparation involves modified feeding practices that facilitate rapid growth, manipulate their throats to be able to withstand a lifetime of force-feeding, and acclimation to the psychical constraint of their bodies necessary for that feeding.[16] The gag reflex proves to be a major difficulty in grooming young birds for the industry, for instance, and here the intersections with women's sexual exploitation is perhaps at its clearest.

A number of injustices are imposed on foie gras victims that are unique to ducks and gooses, however. In many cases, these victims are kept in darkness except for feeding periods in order to reduce panic and fear. Birds frequently exhibit psychological and physical signs of distress and injury, including feather-picking, pacing, panting, efforts to flee "farmers" at feeding times, difficulty standing and walking due to distended livers, fighting with other victims, bodily injuries gleaned from the feeding practices and cage designs, digestive difficulties, cannibalism, and so on.[17] These violations of birds' dignity and bodily integrity have become routinized because humans find their sick and disabled organs a delicacy (although it is important to acknowledge that it is predominantly men who control the food industry and influence the diets of all genders). Hundreds of thousands of ducks and gooses each year do not survive long enough to reach their final destination at the slaughterhouse.

Although Davis does not explicitly acknowledge the sexual acts farmed animals must endure, as noted, it is not difficult to draw parallels. In the case of dairy, for instance, farmers (almost always male) are the ones who impregnate cows, not bulls. Men restrain cows, glove their arm in what amounts to an extended condom (designed as it is to ease penetration and protect the arm), and then insert their arm into each victim's rectum. From here, he is able to manipulate her reproductive organs and use his other hand to penetrate her vulva with an instrument containing sperm. In foie gras production, sexual organs might not be immediately violated in the process of force-feeding, but the process is sexualized nonetheless as the throats of ducks and gooses are penetrated with feeding tubes that ejaculate fatty meal. In all cases—the rape of women, the forced impregnation of dairy cows, and the forced feeding of ducks—a patriarchal industry violates the bodily autonomy of vulnerable groups for profit or pleasure. In all cases, it is an exaction of power.

Davis's observation that the control of birds' bodies in food production is inherently sexually abusive is also exemplified in the biosecurity practices of the foie gras industry. The intensive conditions most birds live under facilitate repeated zoonotic diseases, particularly for those who are living in "free range" facilities as they have greater contact with free-living animals and other environmental contagions.[18] Indoor "farming" allows for the full control (Figure 39.2) not only of birds' bodies but of nature itself. Although "free-range" alternatives do more to serve consumer-enticing myths about happy, healthy animals who willingly give up their lives to please humans than they do for *actually* increasing the welfare conditions for victims, the industry's preference for indoor factory farming does speak

Figure 39.2 Confinement conditions that typify the experience of foie gras victims.

to the level of control sought in foie gras production. There can be no consent in the making of foie gras when eating, drinking, reproduction, movement, longevity, and ability to maintain bodily integrity (few animals eat themselves to the point of sickness outside of human institutions) are strictly controlled. No duck or goose willingly walks to slaughter, for that matter.

Sex and Gender in Foie Gras Protest

Beyond the mechanics of corporal oppression in "farming," research has noted that body politics have been central to the foie gras debates as well.[19] Advocates for foie gras sometimes insist that free-living ducks and gooses overeat to prepare for winter months just as their force-fed domesticated relatives would; industrial force-feeding is akin to helping them along with natural behavior. Advocates for the ducks and gooses, however, point to their capacity for physical and psychological suffering. As is often the case with producers and consumers of pornography, those who profit from the foie gras industry or indulge in its products as consumers ascribe to scripts of denial that frame the duck victims as happy, well-fed, and treated to the good life. Adams[20] has noted that, as in pornography, the inherent violence in the objectification of marginalized bodies for privileged consumption is made possible with sexualization, humor, fragmentation, and ultimately the disconnection of the final product from the person who was exploited to create it. In the case of foie gras, the trappings of fine cuisine (including exoticism, elaborate culinary displays, and upscale prices) ensure that the consumer focuses on the food as fare and sensual experience, not as a decomposing internal organ of a person who lived and died in horrific conditions. Activists are understandably compelled to penetrate the fantasy and romance of foie gras consumption to accentuate the crass power dynamics that make the product possible. The engagement with gender politics, it seems, serves as a particularly relevant tactic, or at least one that stands out as most available in a deeply gendered culture.

Femininity as a gender role is consistently stereotyped as a performance of servility and a site of masculine control and violence. In this way, foie gras production displays the hallmarks of gender as a system of power relations. Foie gras, however, disrupts the conflation of sex and gender that is typical in Western cultures. *Male* ducks in the foie gras industry are feminized as they are dominated by "farmers," and their bodies, from birth to death, are sites of extreme violence. This poses a bit of a conundrum for the nonhuman animal rights movement, whether this is consciously acknowledged or not, as, traditionally, campaigners have sought to challenge speciesism by drawing on sexism as a poignant analogy. This tactic could be criticized for conflating oppressions despite distinctive qualities of these oppression that women and other animals independently experience. For all the injustices women face, for instance, they are ultimately human and considerably more privileged than nonhuman animals. Nonhuman animals, meanwhile, lacking literacy in human gender politics, would not place relevance on many aspects of sexism. Analogy-making of this kind can also reinforce stereotypes about women and reify gender essentialism. Not all women are equally or inherently weak, vulnerable to violence, and attuned to nature and other animals; not all men are compelled to engage with that violence, eat other animals, use women, and so on.

I argue that this tendency to analogize speciesism with sexism may simply be a reflection of the larger number of female activists available and the prevalence of sexism in the male-ruled movement, sexism that condones the exploitation of female activists.[21] Although foie gras does not align with conventionally understood gendered social relations in the nonhuman animal rights movement (male farmers oppressing female nonhumans), the movement attempts to wedge this conventional gender script into traditional campaigning styles which normally utilize women as proxies of other animals. Perhaps this awkward analogy reflects activist ignorance of the sexual politics of foie gras production (a distinct possibility, as few persons outside of the industry are aware of the particulars beyond the infamously gruesome practice of force-feeding). More likely, the strategy reflects sexist assumptions about female victimization in farming and female objectification in social movement campaigning. Campaigners may be unable to fathom the possibility of male activists serving as metaphorical recipients of male violence like that experienced by male ducks and gooses used, abused, and killed for foie gras.

A similar theme seems to permeate modern foie gras campaigning. Although both women *and* men consume foie gras, women are also used in some PETA campaigns as representations of the guilty diner, while men in these campaigns are more likely to be represented as a sort of disembodied or indirect conduit. Female consumers, that is, are more likely to be vilified for participating in a patriarchal, male-benefiting industry than men. In several PETA foie gras (Figures 39.3 and 39.4), women are used to analogize the feminized, oppressed nonhuman *and* the evils of frivolous consumption. Street demonstrations replicate these images in real life with women bound and strapped to dining chairs or forced to their knees. Demonstrations of this kind feature women choking and gagging on a feeding tube manipulated by another, usually male, activist. Some protests include oozing blood

Figure 39.3 PETA foie gras campaign depicting the victim as female with pornographic undertones.

Figure 39.4 PETA foie gras campaign depicting both the victim and perpetrator as female with undertones of punishment and pornography.

painted on women's mouths and may even feature women passed out in their own bloody vomit. The regurgitation is sometimes riddled with grains used in force-feeding ducks and gooses. In other protests, women cower on their knees below their male counterpart with expressions of pain and fear, wholly reminiscent of gonzo pornography and other misogynistic hardcore tropes reflecting the cultural legacy of the 1972 film *Deep Throat*.[22]

As the foie gras industry took hold in the United States at the turn of the 20th century, protest gathered momentum with the efforts of organizations such as In Defense of Animals, Animal Protection and Rescue League,[23] the Humane League, Farm Sanctuary, Viva! USA,[24] and PETA.[25] Given the complexity, diffusiveness, and relative spontaneity of activist networks, it is not possible to know the extent to which PETA's aforementioned sexist strategy is utilized in foie gras protest and, to be fair, this is not the only tactic it deploys in opposition to the industry. However, images of street demonstrations involving men's simulated assault on women diners hosted by PETA US, PETA UK, and PETA France can be found with a simple internet search of news coverage and stock photo websites, suggesting its persistent use in an international repertoire. One stunt even features the founder Ingrid Newkirk playing the victim.[26]

Foie Gras Protest Without the Sexism

Though I problematize the highly sexualized and misogynistic tactics commonly employed in PETA's foie gras campaigning, not all activists and organizations have taken this approach. Michelin-starred chef Alexis Gauthier, for instance, claims he was motivated to drop foie gras from his restaurant (which traded in 20 kilos each week) in favor of a veganized analog ("faux gras") following a protest outside his restaurant.[27] He collaborated with Animal Equality UK to deliver a quarter of a million signatures to the British government in support of a ban as part of an ongoing petition. The media utilized by Animal Equality UK to mobilize change relies on graphic footage taken from foie gras production sites. In these street protests, activists hold placards depicting duck and goose victims themselves rather than relying on female activists to impersonate them through the lens of human gender roles. PETA's 1991 undercover footage documentary *Victims of Indulgence* also had a sensational impact, encouraging significant legislative change in the United States.[28] The Animal Protection and Rescue League also utilized this tactic to rally support with *Delicacy of Despair* in 2003. This campaign included the open rescue of several ducks and the creation of an informative website, GourmetCruelty.com. A year later, Farm Sanctuary levied its website NoFoieGras.org and undercover footage for the successful removal of foie gras from Wolfgang Puck restaurants. This campaign also solicited pledges to keep foie gras off the menu in over 1,000 other restaurants. The Humane League personally met with restaurateurs in its home city of Philadelphia, providing information and images to good effect. By way of another example, PETA enlisted celebrities to describe the experiences of foie gras victims for the successful passing of California's ban in 2004.[29] Sexism, in other words, is not necessary to achieve anti-speciesist goals. Images of ducks and gooses can be a powerful means of raising critical awareness about a system of violence largely hidden from society. For many consumers, the *magnitude* and *complexity* of violence inherent to animal-based food production is, for various sociological and psychological reasons, generally obscured from their awareness beyond the basic understanding that these animals must be manipulated and killed in some way.

Consumer awareness, however, is not enough to create institutional change. Morally shocking images have certain limitations in a sexist society saturated with objectifying, violent, and pornographic imagery. For consumers who *have* been reached by anti-speciesist outreach efforts, cognitive dissonance (and confirmation bias) can encourage them to fall back on industry propaganda that promises high welfare and a "good life" for its victims. Sometimes this humane-washing is not even necessary to secure foie gras as a delicacy. One study finds that about half of foie gras consumers are aware of how the product is created, but they continue to consume it and many would consume *more* if it were more affordable.[30] What this could suggest is that the normalcy of owning and oppressing others for personal pleasure is the core problem. Vegan feminism recognizes the interlocking nature of speciesism and sexism, asserting that patriarchal norms that make violence and control ubiquitous must be disrupted to achieve a vegan world. Successful campaigning against foie gras must therefore be contextualized within a larger feminist framework. At the time of this writing for instance, the British government has reneged on its plan to ban imports of foie gras. The reversal is championed as a matter of consumer choice and individual freedom.[31] Such a narrative is not unlike that used in defense of pornography. In framing privilege and participation in inequality as a matter of personal consumer choice, the unjust system itself remains unexamined and the interests of the privileged consumer are protected as paramount.

The foie gras industry is horrific and must be abolished. To accomplish this, however, activists will need to think strategically about the efficacy of their tactics. The sexism that saturates society has created a problematic common sense that "sex sells." Exceedingly little research is available to assist campaigners in effectively negotiating with sexual objectification and rape scripts. What empirical research *has* been conducted suggests that sexually imbrued tactics trigger a critical feminist response from their audience, a response that is generally dismissive of anti-speciesism and uninterested in offering solidarity.[32] One study of male Australian university students found that sexualized PETA campaigns generated less support than non-sexualized comparison campaigns.[33] The dehumanization of women in the sexualized campaign "was the only significant mediator" accounting for the participants' disapproval. There is an immediate imperative, then, to perform additional research regarding the efficacy of sex, rape, and misogynistic scripts in anti-speciesist campaigning. At the very least, the Bongiorno et al. study does suggest that some men (specifically those enrolled in university who are more educated and liberal than the general public) will resist the nonhuman animal rights movement's stoking of sexist culture.

Conclusion

The production of "foie gras" is highly gendered, reflecting patriarchal norms of institutionalized male dominance, control, and consumption of feminized, objectified beings. In this case, the foie gras industry is predominantly male owned and highly masculinized, particularly with its rationalized, emotion- and pain-denying, commodifying, throat-stuffing approach to speciesism. This is only amplified by its reliance on feminized immigrant and precarious labor, which entangles the domination of vulnerable humans and nonhumans alike. However, although their livers are sexualized as a decadent, sensual, and almost taboo treat for the privileged diner to lustfully consume, the ducks and gooses who are victimized in this industry tend to be male. Anti-speciesist activists have typically relied on female activists to act as proxies for nonhuman victims in campaigning, presumably because it is often female animals who are abused in speciesist systems (such as diary and egg industries). In this case, choosing women's bodies does not align with the reality of the speciesist industry in question: the victims are usually male, and, for that matter, foie gras production in the West often employs women to conduct the force-feeding. This is particularly true of preindustrial France, where force-feeding ducks was assigned to women such that it has become an iconic image in the foie gras industry's nationalistic propaganda.[34] Activists are also unclear on the patriarchal source of oppression. For instance, while some campaigns depict men (sometimes disembodied with only their hands and arms visible) holding down or choking women who are supposed to represent ducks and gooses, women are simultaneously depicted as the guilty diners responsible for supporting foie gras through their self-centered consumption.

Women's self-indulgence and a failure to practice selflessness for others has long been considered a violation of the feminine gender role. Consider, for instance, the state's history of force-feeding imprisoned suffragettes, which was not only a physical violation but also a symbolic act of sexual violence and political silencing. Although it has been suggested that the force-feeding of feminists at the turn of the 20th century was an instance of intersectional oppression between women and other animals (both groups were subjected to scientific torture in the name of medicine),[35] I think it is safe to say that the message to women in the audience of a foie gras protest is anything but a call to solidarity. Women, it seems, are simulated as being tortured and punished in the public sphere via street demonstrations or in print campaigns for consuming foie gras. The punishment depicted in this protest imagery is designed to fit the crime by sentencing the women to meet the same fate as the ducks and gooses. Men's responsibility for creating, steering, and sustaining the industry goes largely unexamined.

Foie gras campaigning provides powerful insight into the sexist assumptions that permeate anti-speciesist repertoires and activist culture. Using women as proxies for other animals is not done to demonstrate to the public how things "really are." Rather, sexualized violence against women's bodies is a familiar script to activists and the public; the aim is to align protest frames with widespread cultural understandings. Blaming and shaming women for the ills of the world is a trope that activists seem to exploit with hopes of resonance in a sexist society (and this tactic likely reflects sexist beliefs within the movement as well). Furthermore, this sexist imagery is frequently sexualized, predictably so given the increasing pornification of modern society and the marketplace. This adoption of sexist scripts in anti-speciesism is a clear case of intersectional failure. If the sexualized objectification of nonhuman animals is to be challenged, this cannot be done successfully if the nonhuman animal rights movement uncritically persists in the vilification and sexualized objectification of women. Patriarchy and capitalism must be disrupted, not reinforced. Foie gras ducks and gooses cannot be liberated so long as the same ideological mechanisms are bulwarked in advocacy spaces.

Notes

1 Mass terms such as "geese" are avoided so as to respect the personhood of individuals victimized in the system.
2 M. Caro, *The Foie Gras Wars* (London: Simon & Schuster, 2009)
3 M. Delpont, V. Blondel, L. Robertet, H. Duret, J. Guerin, J. Vaillancourt, and M. Paul, "Biosecurity Practices on Foie Gras Duck Farms, Southwest France," *Preventative Veterinary Medicine* 158 (2018): 78–88.

4 C. Adams, *The Pornography of Meat* (New York: Continuum Publishing, 2003)
5 Euphemistic terms that mask human violence against animals are denoted with quotation marks.
6 K. Davis, *#MeToo for the Voiceless: Why We Can't Ignore the Animal Victims of Human Sexual Assault* (2018), retrieved from: https://www.alternet.org/2018/03/metoo-voiceless-why-we-cant-ignore-animal-victims-human-sexual-assault/, accessed May 11, 2022; K. Ducey, "The Chicken-Industrial Complex and Elite White Men," in *Animal Oppression and Capitalism*, ed. D. Nibert (Santa Barbara: Praeger, 2018), 1–17.
7 D. Mombauer and V. Wijenayake, "Feminist Viewpoints are Vital to Getting Livestock Policies Right," *Forest Cover* 66 (2021): 3–6.
8 L. Gray, *We Just Keep Running the Line: Black Southern Women and the Poultry Processing Industry* (Baton Rouge: Louisiana State University Press, 2014).
9 L. Palumbo and A. Sciurba, *The Vulnerability to Exploitation of Women Migrant Workers in Agriculture in the EU* (Brussels: European Union, 2018).
10 N. Mainet-Delair, "Les 'Dames-Fermières' du Foie Gras, dans le Salignacois, en Périgord Noir," *Acta Geographica* 134 (2003): 47–61.
11 C. Wrenn, "Toward a Vegan Feminist Theory of the State," in *Animal Oppression and Capitalism*, ed. D. Nibert (Santa Barbara: Praeger Press, 2017), 201–230.
12 J. Joyce, J. Nevins, and J. Schneiderman, "Commodification, Violence, and the Making of Workers and Ducks at Hudson Valley Foie Gras," in *Critical Animal Geographies*, eds. K. Gillespie and R. Collard (London: Routledge, 2017), 93–107.
13 M. Caro, *The Foie Gras Wars: How a 5,000-Year-Old Delicacy Inspired the World's Fiercest Food Fight* (New York: Simon & Schuster, 2009).
14 K. Davis, *Blurring the Boundary Between Humans and Other Animals* (2022), retrieved from: https://www.upc-online.org/welfare/220120_blurring_the_boundary_between_humans_and_other_animals.html, accessed February 7, 2022.
15 W. Massimino, S. Davail, M. Bernadet, T. Pioche, A. Tavernier, K. Ricaud, K. Contier, C. Bonnefont, H. Mnase, M. Morisson, B. Fauconneau, A. Collin, S. Panserat, and M. Houssier, "Positive Impact of Thermal Manipulation During Embryogenesis on Foie Gras Production in Mule Ducks," *Frontiers in Physiology* 10 (2019): 1–12, https://doi.org/10.3389/fphys.2019.01495.
16 P. Le Neindre, P. Willeberg, P. Jensen, D. Broom, J. Harting, R. Dantzer, D. Morton, P. Bénard, M. Verga, J. Faure, B. Nicks, and I. Estevez, "Welfare Aspects of the Production of Foie Gras in Ducks and Geese," *Report of the Scientific Committee on Animal Health and Animal Welfare*. European Commission (1998).
17 Ibid.
18 Delpont et al., "Biosecurity Practices on Foie Gras Duck Farms, Southwest France."
19 R. Youatt, "Power, Pain, and the Interspecies Politics of Foie Gras," *Political Research Quarterly* 62, no. 2 (2012): 346–358.
20 Adams, *The Pornography of Meat*.
21 C. Wrenn, "The Role of Professionalization Regarding Female Exploitation Nonhuman Animal Rights Movement," *Journal of Gender Studies* 24, no. 2 (2015): 131–146.
22 C. Itzin, "Pornography and the Construction of Misogyny," *The Journal of Sexual Aggression* 8, no. 3 (2002): 4–42.
23 H. Canavan, *The Decade-Long Foie Gras Fight*, Explained (2015), retrieved from: https://www.eater.com/2015/1/9/7513743/foie-gras-ban-california-history-appeal-peta-aldf, accessed February 16, 2023.
24 Caro, *The Foie Gras Wars*.
25 Priya S, *PETA's Foie Gras Campaign Highlights from Over the Years* (2018), retrieved from: https://www.peta.org.uk/blog/petas-foie-gras-campaign-highlights-from-over-the-years/, accessed February 16, 2023.
26 M. Kretzer, *PETA President Bound and Force-Fed During Protest* (2013), retrieved from: https://www.peta.org/blog/peta-president-bound-force-fed-protest/, accessed February 16, 2023.
27 A. Gauthier, "Why This Michelin-Starred Chef Turned His Back on Foie Gras for Good," *Plant Based News* (2021), retrieved from: https://plantbasednews.org/opinion/opinion-piece/michelin-starred-chef-foie-gras, accessed November 9, 2021.
28 Caro, *The Foie Gras Wars*.

29 Ibid.
30 A. Czibolya and E. Lendvai, "Examination of Foie Gras Consumption Habits," *Analecta Technico Szegedinensia* 9, no. 1 (2015): 18–24.
31 H. Horton, *Tory MPs Plan Revolt Over U-Turn on Fur and Foie Gras Import Ban* (2022), retrieved from: https://www.theguardian.com/world/2022/feb/21/tory-mps-plan-revolt-over-u-turn-fur-foie-gras-import-ban, accessed February 22, 2022.
32 C. Wrenn, *A Rational Approach to Animal Rights* (London: Palgrave, 2016).
33 R. Bongiorno, P. Bain, and N. Haslam, "When Sex Doesn't Sell: Using Sexualized Images of Women Reduces Support for Ethical Campaigns," *PLoS ONE* 8, no. 12 (2013): e83311.
34 M. DeSoucey, *Contested Tastes: Foie Gras and the Politics of Food* (Princeton: Princeton University Press, 2016).
35 I. Miller, "Necessary Torture? Vivisection, Suffragette Force-Feeding, and Responses to Scientific Medicine in Britain c. 1970–1920," *Journal of the History of Medicine and Allied Sciences* 64, no. 3 (2009): 333–372.

Bibliography

Adams, C. *The Pornography of Meat*. New York: Continuum Publishing, 2003.
Bongiorno, R., Bain, P., and Haslam, N. "When Sex Doesn't Sell: Using Sexualized Images of Women Reduces Support for Ethical Campaigns." *PLoS ONE* 8, no. 12 (2013): e83311.
Canavan, H. "The Decade-Long Foie Gras Fight." *Explained* (2015). Retrieved from: https://www.eater.com/2015/1/9/7513743/foie-gras-ban-california-history-appeal-peta-aldf.
Caro, M. *The Foie Gras Wars*. London: Simon & Schuster, 2009.
Czibolya, A., and Lendvai, E. "Examination of Foie Gras Consumption Habits." *Analecta Technico Szegedinensia* 9, no. 1 (2015): 18–24.
Davis, K. "#MeToo for the Voiceless: Why We Can't Ignore the Animal Victims of Human Sexual Assault." *Alternet,* 2018. Retrieved from: https://www.alternet.org/2018/03/metoo-voiceless-why-we-cant-ignore-animal-victims-human-sexual-assault/
Davis, K. *Blurring the Boundary Between Humans and Other Animals* (2022). Retrieved from: https://www.upc-online.org/welfare/220120_blurring_the_boundary_between_humans_and_other_animals.html.
Delpont, M., Blondel, V., Robertet, L., Duret, H., Guerin, J., Vaillancourt J., and Paul, M. "Biosecurity Practices on Foie Gras Duck Farms, Southwest France." *Preventative Veterinary Medicine* 158 (2018): 78–88.
DeSoucey, M. *Contested Tastes: Foie Gras and the Politics of Food*. Princeton: Princeton University Press, 2016.
Ducey, K. "The Chicken-Industrial Complex and Elite White Men." In *Animal Oppression and Capitalism*, edited by D. Nibert, 1–17. Santa Barbara: Praeger, 2018.
Gauthier, A. "Why This Michelin-Starred Chef Turned His Back on Foie Gras for Good." *Plant Based News,* 2021. Retrieved from: https://plantbasednews.org/opinion/opinion-piece/michelin-starred-chef-foie-gras.
Gray, L. *We Just Keep Running the Line: Black Southern Women and the Poultry Processing Industry*. Baton Rouge: Louisiana State University Press, 2014.
Horton, H. "Tory MPs Plan Revolt Over U-Turn on Fur and Foie Gras Import Ban." *The Guardian,* 2022. Retrieved from: https://www.theguardian.com/world/2022/feb/21/tory-mps-plan-revolt-over-u-turn-fur-foie-gras-import-ban.
Itzin, C. "Pornography and the Construction of Misogyny." *The Journal of Sexual Aggression* 8, no. 3 (2002): 4–42.
Joyce, J.J. Nevins, and Schneiderman, J. "Commodification, Violence, and the Making of Workers and Ducks at Hudson Valley Foie Gras." In *Critical Animal Geographies*, edited by K. Gillespie and R. Collard, 93–107. London: Routledge, 2017.
Kretzer, M. "PETA President Bound and Force-Fed During Protest." *Peta,* 2013. Retrieved from: https://www.peta.org/blog/peta-president-bound-force-fed-protest/.
Le Neindre, P., Willeberg, P., Jensen, P., Broom, D., Harting, J., Dantzer, R., Morton, D., Bénard, P., Verga, M., Faure, J., Nicks, B., and Estevez, I. "Welfare Aspects of the Production of Foie Gras

in Ducks and Geese." *Report of the Scientific Committee on Animal Health and Animal Welfare*. European Commission, 1998.
Mainet-Delair, N. "Les 'Dames-Fermières' du Foie Gras, dans le Salignacois, en Périgord Noir." *Acta Geographica* 134 (2003): 47–61.
Massimino, W., Davail, S., Bernadet, M., Pioche, T., Tavernier, A., Ricaud, K., Contier, K., Bonnefont, C., Mnase, H., Morisson, M., Fauconneau, B., Collin, A., Panserat, S., and Houssier, M. "Positive Impact of Thermal Manipulation During Embryogenesis on Foie Gras Production in Mule Ducks." *Frontiers in Physiology* (2019). https://doi.org/10.3389/fphys.2019.01495.
Miller, I. "Necessary Torture? Vivisection, Suffragette Force-Feeding, and Responses to Scientific Medicine in Britain c. 1970–1920." *Journal of the History of Medicine and Allied Sciences* 64, no. 3 (2009): 333–372.
Mombauer, D., and Wijenayake, V. "Feminist Viewpoints are Vital to Getting Livestock Policies Right." *Forest Cover* 66 (December 2021): 3–6.
Palumbo, L., and Sciurba, A. *The Vulnerability to Exploitation of Women Migrant Workers in Agriculture in the EU*. Brussels: European Union, 2018.
S, Priya. "PETA's Foie Gras Campaign Highlights from Over the Years." *Peta*, 2018. Retrieved from: https://www.peta.org.uk/blog/petas-foie-gras-campaign-highlights-from-over-the-years/.
Willsher, K. " 'We Love Foie Gras': French Outrage at UK Plan to Ban Imports of 'Cruel' Delicacy." *The Guardian*, 2021. Retrieved from: https://www.theguardian.com/environment/2021/apr/17/we-love-foie-gras-french-outrage-uk-plan-import-ban-delicacy.
Wrenn, C. "The Role of Professionalization Regarding Female Exploitation Nonhuman Animal Rights Movement." *Journal of Gender Studies* 24, no. 2 (2015): 131–146.
Wrenn, C. *A Rational Approach to Animal Rights*. London: Palgrave, 2016.
Wrenn, C. "Toward a Vegan Feminist Theory of the State." In *Animal Oppression and Capitalism*, edited by D. Nibert, 201–230. Santa Barbara: Praeger Press, 2017.
Youatt, R. "Power, Pain, and the Interspecies Politics of Foie Gras." *Political Research Quarterly* 62, no. 2 (2012): 346–358.

40

RACISM AND ANTI-RACISM IN ANIMAL ADVOCACY

Darren Chang

In 2019, online media platform BuzzFeed published the first of a series of detailed reports entitled "WWF's Secret War," which accused Switzerland-based international non-governmental conservation organization World Wildlife Fund (WWF) of being complicit in systemic violence carried out by forest and park rangers across several countries (Nepal, India, Cameroon, Democratic Republic of Congo, Republic of Congo, and the Central African Republic).[1] With "more than 100 interviews and thousands of pages of documents, including confidential memos, internal budgets, and emails discussing weapons purchases," investigators revealed that, for at least two decades, anti-poaching units supported by WWF (through lobbying, funding, training, and supplying of equipment) engaged in human rights violations, such as torture, murders, sexual assaults, and beatings.[2] Although a year later an independent panel of experts cleared WWF of complicity in the aforementioned abuses, the panel's report noted that WWF was inconsistent in their approach to upholding human rights and that the organization failed to "assess the human rights risks or develop an effective plan to prevent and respond to abuses" despite being aware of "the potential for human rights abuses by ecoguards."[3] Critics of WWF were less generous in response to the panel's findings; human rights organization Survival International asserted that the panel and WWF have shifted the blame towards government rangers and refused to take responsibility,[4] while another conservation organization, Rainforest Foundation UK, "accused the WWF of a 'lack of contrition' over the independent review," failing to issue sincere apologies to victims of human rights abuses connected to WWF campaigns.[5]

Throughout the timeframe that WWF was allegedly embroiled in human rights violations, popular support for militarized rangers appeared to be on the rise, especially in the West and the Global North. This trend has been reflected in the overwhelmingly positive reception of the 2014 British documentary film *Virunga*: advertised with the tagline "Conservation Is War," the film follows the stories of rangers defending mountain gorillas and the bio-diverse environment in eastern Congo amidst an armed conflict, poaching, and resource extraction by Global North corporations.[6] Similarly, Akashinga, an armed, (Black) all-women anti-poaching unit in Zimbabwe led by a (white) former Australian special forces soldier and founder of International Anti-Poaching Foundation, has been

DOI: 10.4324/9781003273400-49

prominently featured in mainstream media, including a short film promoted by National Geographic entitled *Akashinga: The Brave Ones*.[7]

Oppressed peoples in nation-states that continue to suffer from historical injustices of Western imperialism and colonialism are thus interpellated both as criminals in animal trafficking industries deserving of a militarized punitive response and as glorified courageous animal protectors who enforce animal protection legislations in ways that could jeopardize human rights. This is, I suggest, one example of Zakiyyah Iman Jackson's analysis that "black(ened) people . . . are cast as sub, supra, and human *simultaneously*."[8] It is not simply the historically produced structural forces that reproduce carceral and militarized violence against black(ened) people, but also "that blackness is produced as sub/super/human at once . . . at the register of ontology."[9] Furthermore, these ontological conceptions of blackness have direct legal and material consequences. Jackson emphasizes the fact that while dispossessed and impoverished peoples in primarily Global South states that continue to be afflicted by histories of colonization and global capitalist financial institutions (e.g. the IMF and World Bank) might participate in the multibillion-dollar illegal wildlife trade, "the captured animals and the wealth generated from their labor spiral upward to the West—*but not the criminal prosecution*."[10] All the while, Global North animal advocates seeking to overcome anthropocentrism and speciesism through strategies grounded in liberal humanism often "do not subject the very humanity they want to decenter and/or expand to sufficient interrogation," resulting in their reliance on authorizing "the violence of the state, one that protects, criminalizes, enforces, and prosecutes differentially based on race, class, gender, sexuality, national origin, religion, ability, and immigration status."[11]

I begin this chapter with Jackson's critical analyses alongside the WWF example to illustrate the various scales and dimensions on which racism is an inescapable injustice that animal advocates must confront. From the level of the individuals experiencing the ongoing harms of enduring colonial-capitalism and imperialism to the institutions they are situated in, from the socioeconomic structures that shape power hierarchies and financial disparities in animal advocacy organizations to the ideological and epistemological foundations that shape animal advocacy strategies, racism is ever present. Following the aim of this volume to serve as a companion to studying the connections between gender and animals, this chapter focuses on highlighting three interconnected areas of animal advocacy where theorists and activists influenced by feminist and Black feminist thought (e.g. Kimberlé Crenshaw's concept of intersectionality) have intervened to make animal advocacy more responsive to racial justice. The three areas are (1) racism internal to a white-dominated animal movement, (2) racism as reflected in animal advocacy strategies and tactics, and (3) racism in the worldviews and knowledges that shape our understandings of the root problems creating injustices for humans and nonhumans. The next section of the chapter discusses the problematic reciprocal relations between of the first two areas of concern, and by way of conclusion, the final section of this chapter considers the possibilities of radically transforming animal advocacy in relation to the third point.

The Legacy of White Dominance in the Animal Movements

In her seminal work on the concept of privilege, feminist and antiracist educator Peggy McIntosh defines privilege as sets of unearned and hidden advantages systemically conferred to certain groups to uphold their dominance in society.[12] McIntosh describes how coming to understand oppressions as interlocking helped her to recognize the ways in which men are unwilling to acknowledge male privilege, much as white people are taught to deny the existence of white privilege.[13] Coming from a different standpoint, Kimberlé Crenshaw's concept of intersectionality makes visible the experiences of people who are forced to confront structural and cultural oppressions because their identities put them at the intersection of multiple forms of oppression (e.g. women of color simultaneously experiencing racism and sexism in qualitatively different ways from how men of color experience racism and how white women experience sexism).[14] As Rita Dhamoon points out, these early theoretical groundworks by feminist and critical race theory scholars and activists resulted in the contemporary mainstreamed comprehension of how "subjectivity is differently and differentially constituted through relations of privilege and penalty, with real material effects," and how different forms of oppression are "mutually reinforcing and relational, thereby challenging the notion that some forms of difference are intrinsically separable and more significant than others."[15] The critical analysis that oppressions are inseparable and socially constructed helps us to grasp how Western, Eurocentric animal advocacy organizations have sustained whiteness (inclusive of all the privileges across race, gender, sexuality, class, ability, species, and nationality) since the inception of their movement.

The tendency of animal advocates to reinforce rather than challenge hierarchical social orders and dominant assumptions about social classes in their efforts to help animals was already apparent shortly after the creation of the earliest animal protection organizations. Historian Harriet Ritvo has shown that members of the Royal Society for the Prevention of Cruelty to Animals (RSPCA) during the Victorian era had disproportionately utilized legal instruments to prosecute cruel treatment of animals perpetrated by the lower classes in England, while often disregarding if not actively supporting and participating in upper-class activities such as horseracing and fox hunting.[16] For the RSPCA, promoting animal welfare became not only a crusade to better the lives of animals but also a moralizing project to identify "cruelty as a lower-class propensity" that threatened the Victorian social order, which required correction through education and discipline.[17]

Across the Atlantic Ocean, similar relations and dynamics played out as some of the first humane societies and SPCAs were created in the United States during the same era. Paula Tarankow's research on the historical roots of humane carceral logics in the U.S. details the process by which white supremacy combined with "sentimentalism, liberalism, paternalism, and Christian theology" to produce a surge of humanitarianism that served as a moral benchmark "for national belonging, assimilation, and readiness for citizenship."[18] Tarankow notes that the recourse to legal measures in animal protection had "significant gendered components," where "[w]hite male reformers espoused a type of Christian manhood motivated by a sense of injustice rather than mere sentimental love for animals."[19] This merging of concern for animals with state power "coincided with the dissemination of crime statistics that explained criminal behaviour based on racial difference" led by "white social scientists, social reformers, journalists, law enforcement officials, and politicians of the day," which construed Black people, immigrants, and colonized peoples as prone to criminality and unfit for citizenship.[20] The production of this benchmark that measured white humanity against Black humanity produced a type of respectability politics that pressured Black reformers and Black animal protectionists who believed in the inseparability of antiracism and animal advocacy to establish their own animal welfare organizations; however, Tarankow highlighted in a footnote that

> Black animal protectionists appear not to have held any executive leadership positions within white-controlled anticruelty societies in either the North or the South, although some Black reformers were affiliated with or employed by the American Humane Education Society under the aegis of the Massachusetts Society for the Prevention of Cruelty to Animals.[21]

As with other segments of society, contemporary animal movements in the West and the Global North inherited these enduring racial injustices, now more deeply entrenched and expanded across an international scale. White supremacy's hold on the animal movement is reflected in the lack of racial diversity in the prominent mainstream animal organizations that prompted activist responses within the movement to directly address this issue. Resistance to white supremacy in animal advocacy is not a new phenomenon, as people of color have always had a presence and have demanded better representation and inclusion. More recent examples include A. Breeze Harper's edited volume *Sistah Vegan: Black Female Vegans Speak on Food, Identity, Healthy, and Society*, first published in 2010, which featured diverse contributors that discussed their relations to veganism, a movement and lifeway that shares profound overlaps with animal rights/liberation.[22] Vegan animal liberation and food justice organizations led by women of color, such as Food Empowerment Project, have fought for animals and exploited humans (e.g. migrant farm workers) as interconnected injustices within food production systems since its creation in 2007.[23]

Such antiracist and social justice efforts in the animal movement are parts of larger intergenerational anti-oppression struggles that have taken place over many decades, and are examples of how social movements build on shared momentum to generate the undercurrents of change. This is one of the reasons the Black Lives Matter movement's sustained push to foreground racial injustices and white supremacy in the public consciousness in light of police violence and murders of Black people in the U.S. since 2015 has culminated in forcing the animal movement to wrestle with its own overwhelming whiteness over the past several years. One example of this labor was undertaken by Encompass, a women-of-color-led organization and network formed in 2020 (now disbanded as of mid-2022)[24] that provided resources and facilitated racial equity workshops for animal advocates; part of Encompass' work has been documented in *Antiracism in Animal Advocacy: Igniting Cultural Transformation*, a collection of essays by Black, Indigenous, and People of the Global Majority (BIPGM) animal advocates and white allies reflecting on the representational and material inequities and injustices perpetuating in the movement.[25] Similarly, founded in 2014 in the U.S., the Multicultural Veterinary Medical Association (MCVMA) emerged with the recognition that veterinary medicine (an often overlooked, but vital part of animal protection work) is "one of the least culturally, racially, and ethnically diverse professions in America."[26]

One of the key lessons from the resistance against a legacy of white dominance in animal advocacy is that, not only is racism an inherent injustice that reinforces other forms of oppression and therefore requires dismantling in order to achieve justice and liberation for all humans and nonhumans, a failure to address racism has strategic consequences as well. In an edited volume that critically appraises Effective Altruism (EA), Christopher Sebastian discusses in his chapter how the overwhelmingly white membership of EA and the movement's ignorance towards racial injustice and anti-Blackness not only hurt their cause, but bring about more harms in the world.[27] EA is "a philosophical and social movement based on reason and evidence" that aspires to "do the *most* good in an impartial manner," inspired by utilitarian philosophers like William MacAskill and Peter Singer, and has substantial stakes in how to make animal activism more effective.[28] As Sebastian argues, given that

> the worst outcomes for people of color globally are experienced by people who are Black, EA's refusal to explicitly contest anti-Blackness and be more informed by Black experiences and insights conflicts with the movement's goal of "doing the 'most good, most effectively," because it reveals a poverty of political and philosophical thought.[29]

Many of EA's strategies and tactics exemplify classic white saviorism through charity; Effective Altruists focus almost exclusively on assessing the impacts of charitable organizations in order to solicit, encourage, and direct wealthy donors into funding projects and organizations that EA charity evaluating bodies recommend as the most effective based on their quantifiable metrics. As Sebastian puts it, in a global context where systemic injustices are historically rooted and institutionally engrained, there is "something fundamentally condescending and dishonest in believing that the solutions" to structurally produced harms could be "conceptualized by people who largely have profited from" historical injustices.[30] It is also highly concerning that a mostly white, wealth-accumulating minority beneficiary of historical injustices possess an excess amount of power and wealth to influence which organizations and what types of campaigns deserve backing with their financial and material resources.

To illustrate the magnitude of this concern, consider that People for the Ethical Treatment of Animals (PETA), one of the major international animal organizations with $66 million in revenues in 2020 alone, has repeatedly implemented campaigns that are racially insensitive at best and outright racist at worst.[31] Sebastian examines several recent examples of PETA's racist campaigns: in 2014, social media users criticized PETA for what some described as "veganism by extortion" and disrespecting basic human rights when PETA offered to help poor residents in Detroit, USA (a city where 78% of the population identified as Black) pay for their water bills if the residents were willing to go vegan for a month after the residents had their water shut off due to nonpayment; in 2015, PETA congratulated and celebrated Maricopa Country, Arizona (USA) sheriff Joe Arpaio for "serving vegetarian meals to immigrant detainees in an open-air jail" despite Arpaio's transparent racial profiling and public comments proudly comparing his jail to concentration camps; in 2020, PETA continued their historical pattern of appropriating symbols of Black resistance when they released an animated Super Bowl advertisement depicting animals taking a knee during the U.S. national anthem, a gesture "popularized by Black U.S. football player Colin Kaepernick, who knelt during the anthem before games as an act of peaceful resistance to disproportionate police violence in U.S. law enforcement."[32]

As white-dominated Global North animal advocacy organizations continue to accumulate the most funding and resources that enable them to dictate the strategies and directions of advocacy efforts, their campaigns often replicated racist colonial and imperial power relations that fail to account for existing and ongoing injustices. Elsewhere, I presented a critical look from an anticolonial framework at how major international animal and environmental non-governmental organizations (International Fund for Animal Welfare,

Greenpeace, Humane Society International, and the Sea Shepherd Conservation Society) exacerbated the economic hardships that a history of colonialism has imposed on the Inuit communities through campaigns that sought to end the commercial seal hunt.[33] In *Dangerous Crossings: Race, Species, and Nature in a Multicultural Age*, Claire Jean Kim investigated three

> impassioned disputes: the battle over the live animal markets in San Francisco's Chinatown, the uproar over the conviction of NFL superstar Michael Vick on dogfighting charges, and the firestorm over the Makah tribe's decision to resume whaling in the Pacific Northwest after a hiatus of more than seventy years.[34]

While Kim urged readers to move beyond the oversimplified single-optic narratives where racialized animal exploiters and antiracists activists accused mostly white animal advocates of being racist cultural imperialists while animal advocates claimed racial innocence and colorblindness in their campaigns targeting racialized minorities, each case nonetheless depicted elements of racism within white animal activism. Jacqueline Dalziell and Dinesh Wadiwel offer another example in their critique of the imperialist, nationalist, and Orientalist discourses in Australian animal advocacy campaigns against live exporting farmed animals to Middle Eastern and Asian nations, which discursively produced these export nations "as savage and regressive . . . uncivilized Oriental other[s]" while naturalizing Australians as "more compassionate" and "more properly developed" in their moral sensibilities and superiority.[35]

It is important to note that as whiteness and white supremacy are socially constructed and institutionally upheld ideologies and structures, their arrangements and formations are not exclusive to the West and the Global North, and are often manifested in postcolonial nation-states in the form of ultranationalism and fascism. Yamini Narayanan has examined how Hindu-led cow protectionism in India, particularly through cow "sanctuaries" known as gaushalas, have been used to mobilize bovine bodies as "symbolic capital in the intersections of anthropatriarchy and gendered ultranationalist Hindu patriarchy" to advance Hindu extremism.[36]

In sum, examples of how racism, imperialism, colonialism, and other enmeshed forms of violence problematize animal advocacy will be endlessly reproduced so long as animal movements either refuse or fail to dismantle these oppressive ideologies and structures. In the final section of this chapter, I consider what transformations might need to occur at the epistemological and ontological levels of our conceptions of human–nonhuman relations if the animal movements were to radically disrupt and move beyond supremacist logics.

Towards Transformative Conceptions of Humanity/Animality in Collective Liberation

Recent theoretical development at the confluence of race, species/animality, and gender have produced more congruent and coherent analyses that accept these socially constructed categories and subjectivities to be "dynamically interconstituted *all the way down*."[37] Much of this theorizing has been advanced by Black, Indigenous, and People of Color thinkers engaging critically with the concept of intersectionality. On one level, these conversations grappled with which theoretical and epistemological frameworks might better capture the complex realities and relations between different forms of oppression. For example, Kim argues for "taxonomies" as the locus of power that produces hierarchical orderings of life.[38] Meanwhile, Aph Ko and Syl Ko have offered Afrofuturism, Black Veganism, and multidimensionality as more favorable approaches to attaining human and nonhuman liberation.[39]

On another level, the conversations deal with what Jackson, à la Sylvia Wynter, describes as a "praxis of being."[40] Quite literally, topics in this area treat "human" and "animal" as verbs, and question both the meaning of being human/animal and how to be human/animal. In *Nature's Wild: Love, Sex, and Law in the Caribbean*, Andil Gosine presents an alternative account of queer desires in the Caribbean by engaging with colonialism, humanism, and animality.[41] Recognizing how Caribbean people have been animalized by colonizers as a means to justify committing racialized violence against them, Gosine proposes that those whose lives have been historically threatened by animalization ought to "embrace being marked 'animal' or, at the very least, refuse to heed the call to prove" that they are not animal.[42] Jackson, however, takes a different position and offers an "interrogation of humanism by identifying our shared being with the nonhuman without suggesting that some members of humanity bear the burden of 'the animal'."[43]

I end this chapter by underscoring the previous theoretical developments, as these are, in my view, some of the undertheorized areas that hold tremendous potential for readers researching in fields related to gender and animal studies. In addition to the shifting activist practices in making the animal movements more just, diverse, and equitable, transforming and revolutionizing our worldviews on human-nonhuman relations and subjectivities might be what is necessary to truly end racism and other injustices within and beyond the animal movements.

Notes

1 Katie J.M. Baker and Tom Warren, "WWF Funds Guards Who Have Tortured And Killed People," *BuzzFeed* (March 4, 2019), retrieved from: https://www.buzzfeednews.com/article/tomwarren/wwf-world-wide-fund-nature-parks-torture-death
2 Ibid.
3 Navi Pillay, John H. Knox, and Kathy MacKinnon, "Embedding Human Rights in Nature Conservation: From Intent to Action," *World Wildlife Fund* (2020), retrieved from: https://wwfint.awsassets.panda.org/downloads/independent_review___independent_panel_of_experts_final_report_24_nov_2020.pdf
4 Peter Beaumont, "Report Clears WWF of Complicity in Violent Abuses by Conservation Rangers," *The Guardian* (November 25, 2020), retrieved from: https://www.theguardian.com/environment/2020/nov/25/report-clears-wwf-of-complicity-in-violent-abuses-by-conservation-rangers
5 BBC News, "WWF Vows to 'Do More' after Human Rights Abuse Reports," *BBC News* (November 24, 2020), retrieved from: https://www.bbc.com/news/world-55069926
6 Virunga, "About," *Virunga,* retrieved from: https://virungamovie.com/#about
7 National Geographic, "Akashinga: The Brave Ones," *National Geographic*, retrieved from: https://films.nationalgeographic.com/akashinga, accessed January 11, 2023
8 Jackson, *Becoming Human*, 35, emphasis in original.
9 Ibid., 3.
10 Ibid., 16, emphasis in original.
11 Ibid., 15–16.
12 Peggy McIntosh, "White Privilege: Unpacking the Invisible Knapsack," *Freedom Magazine,* July/August 1989: 10–12.
13 Ibid.
14 Kimberle Crenshaw, "Mapping the Margins: Intersectionality, Identity Politics, and Violence Against Women of Color," *Stanford Law Review* 43, no. 6 (1991): 1241–1299.
15 Rita Kaur Dhamoon, "Considerations on Mainstreaming Intersectionality," *Political Research Quarterly* 64, no. 1 (2011): 240.

16 Harriet Ritvo, *The Animal Estate: the English and Other Creatures in the Victorian Age* (Cambridge, MA: Harvard University Press, 1987), 133–135.
17 Ibid., 135.
18 Paula Tarankow, "Saved: The Historical Roots of Humane Carceral Logics in the United States," in *Carceral Logics: Human Incarceration and Animal Captivity*, eds. Lori Gruen and Justin Marceau (Cambridge University Press, 2022), 20–21.
19 Ibid., 25.
20 Ibid., 26.
21 Ibid., 22–23.
22 A. Breeze Harper, *Sistah Vegan: Black Female Vegans Speak on Food, Identity, Health, and Society* (New York: Lantern Books, 2010).
23 Food Empowerment Project, "Missions and Values," *Food Empowerment Project*, retrieved from: https://foodispower.org/mission-and-values/, accessed April 23, 2023
24 Some of Encompass' work is being carried forward by Apex Advocacy, a Black-led organization empowering BIPOC animal activists (see: https://www.apexadvocacy.org/about)
25 Jasmin Singer, Aryenish Birdie, and Michelle Rojas-Soto, *Antiracism in Animal Advocacy: Igniting Cultural Transformation* (New York: Lantern Publishing & Media, 2021)
26 Multicultural Veterinary Medical Association, "About MCVMA," *Multicultural Veterinary Medical Association*, retrieved from: https://mcvma.org/about-mcvma/, accessed April 23, 2023
27 Christopher Sebastian, "Anti-Blackness and the Effective Altruist," in *The Good It Promises, the Harm It Does: Critical Essays on Effective Altruism*, eds. Carol J. Adams, Alice Crary, and Lori Gruen (New York: Oxford University Press, 2023), 26–40.
28 Ibid., 27, emphasis in original.
29 Ibid., 29.
30 Ibid., 28.
31 Ibid., 34.
32 Ibid.
33 Darren Chang, "Tensions in Contemporary Indigenous and Animal Advocacy Struggles: The Commercial Seal Hunt as a Case Study," in *Colonialism and Animality: Anti-Colonial Perspectives in Critical Animal Studies*, eds. Kelly Struthers Montford and Chloë Taylor (Milton: Routledge, 2020), 29–49.
34 Claire Jean Kim, *Dangerous Crossings: Race, Species, and Nature in a Multicultural Age* (New York, NY: Cambridge University Press, 2015).
35 Jacqueline Dalziell and Dinesh Joseph Wadiwel, "Live Exports, Animal Advocacy, Race and 'Animal Nationalism'," in *Meat Culture*, ed. Annie Potts (Leiden and Boston, MA: Brill, 2016), 84.
36 Yamini Narayanan, " 'Cow Is a Mother, Mothers Can Do Anything for Their Children!' Gaushalas as Landscapes of Anthropatriarchy and Hindu Patriarchy," *Hypatia* 34, no. 2 (2019): 197–198.
37 Claire Jean Kim, "Murder and Mattering in Harambe's House," *Politics and Animals* 3 (2017): 10, emphasis in original.
38 Kim, *Dangerous Crossing*.
39 Aph Ko and Syl Ko, *Aphro-Ism: Essays on Pop Culture, Feminism, and Black Veganism from Two Sisters* (New York: Lantern Books, 2017); Aph Ko, *Racism as Zoological Witchcraft: A Guide for Getting Out* (Brooklyn: Lantern Books, 2019).
40 Jackson, *Becoming Human*, 19.
41 Andil Gosine, *Nature's Wild: Love, Sex, and Law in the Caribbean* (Durham: Duke University Press, 2021).
42 Ibid., 8.
43 Jackson, *Becoming Animal*, 12.

Bibliography

Baker, Katie J.M., and Warren, Tom. "WWF Funds Guards Who Have Tortured And Killed People." *BuzzFeed*, March 4, 2019. Retrieved from: https://www.buzzfeednews.com/article/tomwarren/wwf-world-wide-fund-nature-parks-torture-death

BBC News. "WWF Vows to 'Do More' after Human Rights Abuse Reports." *BBC News*, November 24, 2020. Retrieved from: https://www.bbc.com/news/world-55069926

Beaumont, Peter. "Report Clears WWF of Complicity in Violent Abuses by Conservation Rangers." *The Guardian*, November 25, 2020. Retrieved from: https://www.theguardian.com/environment/2020/nov/25/report-clears-wwf-of-complicity-in-violent-abuses-by-conservation-rangers

Chang, Darren. "Tensions in Contemporary Indigenous and Animal Advocacy Struggles: The Commercial Seal Hunt as a Case Study." In *Colonialism and Animality: Anti-Colonial Perspectives in Critical Animal Studies*, edited by Kelly Struthers Montford and Chloë Taylor, 29–49. Milton: Routledge, 2020.

Crenshaw, Kimberle. "Mapping the Margins: Intersectionality, Identity Politics, and Violence Against Women of Color." *Stanford Law Review* 43, no. 6 (1991): 1241–1299.

Dalziell, Jacqueline, and Wadiwel, Dinesh Joseph. "Live Exports, Animal Advocacy, Race and 'Animal Nationalism.'" In *Meat Culture*, edited by Annie Potts, 73–89. Leiden and Boston, MA: Brill, 2016.

Dhamoon, Rita Kaur. "Considerations on Mainstreaming Intersectionality." *Political Research Quarterly* 64, no. 1 (2011): 230–243.

Food Empowerment Project. "Missions and Values." *Food Empowerment Project*, Retrieved from: https://foodispower.org/mission-and-values/

Gosine, Andil. *Nature's Wild: Love, Sex, and Law in the Caribbean*. Durham: Duke University Press, 2021.

Harper, Breeze A. *Sistah Vegan: Black Female Vegans Speak on Food, Identity, Health, and Society*. New York: Lantern Books, 2010.

Jackson, Zakiyyah Iman. *Becoming Human: Matter and Meaning in an Antiblack World*. New York, NY: New York University Press, 2020.

Kim, Claire Jean. *Dangerous Crossings: Race, Species, and Nature in a Multicultural Age*. New York, NY: Cambridge University Press, 2015.

Kim, Claire Jean. "Murder and Mattering in Harambe's House." *Politics and Animals* 3 (2017): 1–15.

Ko, Aph. *Racism as Zoological Witchcraft: A Guide for Getting Out*. Brooklyn: Lantern Books, 2019.

Ko, Aph, and Ko, Syl. *Aphro-Ism: Essays on Pop Culture, Feminism, and Black Veganism from Two Sisters*. New York: Lantern Books, 2017.

McIntosh, Peggy. "White Privilege: Unpacking the Invisible Knapsack." *Freedom Magazine*, July/August, 1989: 10–12.

Multicultural Veterinary Medical Association, "About MCVMA." *Multicultural Veterinary Medical Association*. Retrieved from: https://mcvma.org/about-mcvma/

Narayanan, Yamini. "'Cow Is a Mother, Mothers Can Do Anything for Their Children!' Gaushalas as Landscapes of Anthropatriarchy and Hindu Patriarchy." *Hypatia* 34, no. 2 (2019): 195–221.

National Geographic. "Akashinga: The Brave Ones." *National Geographic*, 2020. Retrieved from: https://films.nationalgeographic.com/akashinga

Pillay, Navi, Knox, John H., and MacKinnon, Kathy. "Embedding Human Rights in Nature Conservation: From Intent to Action." *World Wildlife Fund*, 2020. Retrieved from: https://wwfint.awsassets.panda.org/downloads/independent_review___independent_panel_of_experts__final_report_24_nov_2020.pdf

Ritvo, Harriet. *The Animal Estate: the English and Other Creatures in the Victorian Age*. Cambridge, Mass: Harvard University Press, 1987.

Sebastian, Christopher. "Anti-Blackness and the Effective Altruist." In *The Good It Promises, the Harm It Does: Critical Essays on Effective Altruism*, edited by Carol J. Adams, Alice Crary, and Lori Gruen, 26–40. New York: Oxford University Press, 2023.

Singer, Jasmin, Birdie, Aryenish, and Rojas-Soto, Michelle. *Antiracism in Animal Advocacy: Igniting Cultural Transformation*. New York: Lantern Publishing & Media, 2021.

Tarankow, Paula. "Saved: The Historical Roots of Humane Carceral Logics in the United States." In *Carceral Logics: Human Incarceration and Animal Captivity*, edited by Lori Gruen and Justin Marceau, 15–36. Cambridge University Press, 2022.

Virunga. "About." *Virunga*. Retrieved from: https://virungamovie.com/#about

PART IX

Multispecies Justice

41

A GRATEFUL ACKNOWLEDGEMENT

Gender Theory and Multispecies Justice

Danielle Celermajer and Darren Chang

Introduction

Multispecies justice, as we understand it, does not propose a substantive, thick, or strong theory of justice.[1] It is rather a field of thought and praxis where the justice concerns of all earth beings, including (*inter alia*) animals other than humans and all humans, plants, rivers, mountains, soils, and ecological systems, *and* their relationships, might be brought and thought together in the context of the polycrisis that is unfolding in the twenty-first century.[2] However, as numerous groups of women and nonbinary people have continuously reminded putatively inclusive gender justice movements that nevertheless render invisible or distort distinctive experiences of injustice and aspirations for justice, thinking together does not mean thinking justice in a newly expansive but nevertheless uniform manner.[3] Multispecies justice is rather an aspirational movement where the different but intersecting justice claims and concerns of all earth beings might engage each other with a view to imagining, creating, and sustaining institutions that are more alive to encoded exclusions and hence sensitive to the differences not only amongst claimants but between the logics and communication styles of their claims.[4] In this regard, and consistent with feminists' insistence on the grounded, embodied character of theory production, multispecies justice arose at a distinctive historical moment and as a response to what was occurring both on the theoretical plane and in a world where all earth beings are, albeit highly differentially and on different time scales, straining under the impacts neoliberal capitalism and extractivism are having on the conditions of liveability.[5]

Multispecies justice has been offered up as a lens on ethics and institutional transformation in the hope that it will assist differently placed thinkers and justice movements in the urgent task of addressing the intensifying violence and suffering driven by ecocidal forms of human life.[6] It arose in, and speaks to, a moment in time, albeit one whose pathologies and threats are likely to extend beyond the short term, into medium and long-term futures. At the same time, it is critical to acknowledge the streams of thought and activism that have flowed into, informed, and nourished this contemporary justice frame and the ways in which its foundational commitments and ideas draw from the work of other justice movements. Critically, as Māori political theorist Christine Winter has argued, multispecies

DOI: 10.4324/9781003273400-51

justice must not repeat the epistemic erasure Zoe Todd identified had been committed by western philosophy's so-called novel ontological turn:[7] a failure to acknowledge and learn from Indigenous peoples' long held and articulated knowledges, practices, and protocols, in this case in relation to justice beyond the human.[8]

Elsewhere, one of us was a co-author of an early article on multispecies justice that sought, amongst other objectives, to trace some of the roots of the idea, both in theory and in social movements.[9] In that article, the formative role of various forms of feminism, in particular ecofeminism, new materialist, and posthumanist feminisms, was acknowledged and named, but given the survey nature of the original article, this was done in a fairly epigrammatic manner. While a fully adequate treatment of the influence of various types of feminist thought and action on multispecies justice is beyond the scope of even a dedicated chapter, here, we seek to give flesh to the earlier fleeting recognition.

In doing so, we organise our ideas around three linked themes. First, and in the next section, we introduce the idea of multispecies justice by explicitly tracing how it arose as a theoretical response to the experience that a range of earth beings were having of violence, oppression, and exploitation and to the failure of the hegemonic theories or dominant frames of justice to adequately theorise or address these experiences or to offer sufficiently transformative possibilities. In other words, while theories of multispecies justice have primarily been articulated in academic contexts, their impetus lies in the experience of, and outrage at, injustice. Feminist theory is certainly not unique in this regard, and indeed one might argue that all critical justice theories emerge from a similar constellation of experience and theoretical shortcomings. Nevertheless, as a dynamic body comprising theoretical moves that consistently and self-consciously cleave back to bodies that are both harmed and aspire to live otherwise, feminisms constitute an exemplar for how multispecies justice arose and what it might emulate.

Second, we consider a constellation of multispecies justice's key conceptual commitments, comprising its relational, anti-dualist, and non-hierarchical ontology; its rejection of human exceptionalism; and its insistence on the entanglement of humans in their more-than-human worlds. Here, we trace the strong continuities between multispecies justice and the radical theoretical moves ecofeminists were already making from the 1970s on. Indeed, the more we return to the writings of early ecofeminists, the more we are struck by both the prescience of theories that would later gain the accolades of the academy and the injustice of their being refused such recognition in their own time.

The third theme around which we organise our thinking concerns multispecies justice's insistence on attending to the margins where exclusion and silencing occurs, and a commitment to justice with respect to the dynamics of voice. Importantly, this goes further than ensuring that those who have historically been relegated to the other side of exclusion are now (merely) recognised as subjects of justice, to actively combatting the institutional structures that have disabled their participation in the articulation of the meaning and conditions of justice. Thus, and again drawing on feminisms' attentiveness to epistemic injustice and the historical constitution of exclusionary logics and structures, multispecies justice is curious about those earth beings or relations that have been excluded from hegemonic justice theories and must also be committed to a radical honesty about the forms of exclusion that will inevitably occur under its own moniker. We conclude this chapter with some thoughts regarding politicised communication across the radical differences of multispecies relations.

The Emergence of Multispecies Justice and the Experience of Confinement

By the early twenty-first century, it was becoming increasingly difficult to justify the continued exclusions from the scope of justice theory of the grave harms that *all* earth beings were experiencing as a result of ever-expanding industrialisation and development, the persistent legacies of colonialism, and intensifying capitalist extractivism. No doubt the exclusions had always been harmful and voices had long been raised against them, but three principal trends illuminated both their impacts and their injustice. The first was the intensification of harms committed against others, including animals, forests, rivers, soils, and ecological systems. The second was the erosion of the empirical basis that had been relied upon to legitimate the various points of differentiation between humans and all others. From within western systems themselves, research was increasingly uncovering the complex cognitive, affective, relational, and communicative capacities not only of other animals but also of the plant world and beyond.[10] Third, the entanglement of the violence and harm dominant systems wrought on the most vulnerable humans, other animals, and ecological systems was becoming increasingly evident. Examples abound: the relationship between agribusiness, deforestation, the destruction of Indigenous peoples' lifeworlds, biodiversity loss, and climate change in places like the Amazon or West Papua and Malaysia;[11] the multidimensional forms of violence and degradation associated with industrial animal agriculture;[12] or the devastation climate change–induced wildfires cause for terrestrial and ocean ecosystems as the capacity of the animals for whom they form homes to inhabit them becomes ever diminished.[13] In these cases and so many others, what stands out is the violence across human–animal–plant–earth boundaries. No doubt tracking historical forms of power and privilege, we acknowledge the impacts are differential, as is the capacity to adapt or develop forms of protection, but we are now in the age of omnicide, the killing of everyone and everything.[14]

On the theoretical level, frustration with the boundaries of 'traditional' fields of justice theory is also increasingly felt. By the second decade of the twenty-first century, certain dynamics or trends seemed to be occurring amongst theorists working within the 'fields' of intra-human justice, environmental or ecological justice, and animal justice. Many (although by no means all) were (most often without acknowledgment) picking up on the insights ecofeminists already laid out in the 1970s (to which we return later), that the logics or sources of the gender-based injustices were also the logics or sources of injustices against beings who had been, and largely remained, outside most social movements' official circle of concern. As poignantly summarised by Kim Erno, the wages of "the idolatry of global capitalism"—exclusion, exploitation, expulsion, and extermination—were demanded across gendered and raced human bodies and the earth.[15] Theorists such as Claire Jean Kim and Joshua Bennett were theorising these connections explicitly with respect to race and animality,[16] as feminists like Gruen continued to do with respect to gender,[17] and in the environmental justice space, feminist theorists like Sherilyn MacGregor and Giovanna De Chiro sought to keep gender alive in a field where it continued to be largely marginalised.[18] At the same time, Indigenous theorists such as Zoe Todd, Kim TallBear, Christine Winter, and Kyle Powys Whyte were rendering explicit the mutually enforcing dynamics of colonial violence, environmental destruction, and hegemonic western theories of justice.[19]

Further, western theory was finally catching up with ecological realities well understood by First Nations peoples and described by Arturo Escobar as 'a decolonial view of nature', where the flourishing of all beings depends on the maintenance of relationships and flows.[20] Thus, at least in some circles, it was now beginning to be appreciated that insofar as they were conceptualised as properties of the bounded and human individual, many of the prized pillars of hegemonic western justice theory—specifically autonomy, agency, and freedom—had been deployed to justify the truncation of ecological relationships and flows and hence were complicit with the systematic production of injustice.[21] As more and more disciplines, from biology and physics through social theory, embraced ontologies of entanglement, an ethics of entanglement became more pressing.[22] Multispecies justice seeks to articulate such an ethic.

As noted in the introduction, multispecies justice arose as a space into which theorists and activists who might otherwise identify themselves within the fields separated out by speciesist, hierarchical, and atomising logics might come together to think through what justice for all earth beings would look like in the era where they all, and especially some amongst them, face extermination, violence, and injury. In doing so, however, and critical to its approach, it neither assumes nor seeks out a higher-order principle, or, as expressed by feminist geographer Gibson-Graham, it does not look for 'strong theory', rather embracing weak theory that "powerfully attends to nuance, diversity, and overdetermined interaction. Weak theory does not elaborate and confirm what we already know; it observes, interprets, and yields to emerging knowledge."[23]

In fact, that an irreducible multiplicity of perspectives and concerns can find a home in and be welcomed to the field must be amongst the core commitments of multispecies justice if it is to operate with integrity according to principles intrinsic to the multispecies ethic. In this regard, one can trace a strong line between certain feminist theories and social movements and a commitment to epistemic and ontological pluralism, which, following Isabel Stengers or Marisol De la Cadena, one might call cosmopolitical or many-worlds approaches.[24] At the same time, and again, standing on the shoulders of decades of feminist thought, multispecies justice embraces the principle that a radical commitment to pluralism only heightens the importance of vigilance to the dynamics of power and structural injustices,[25] including in (though not limited to) forms of epistemic injustice and violence.[26] In the case of multispecies justice, the demands for epistemic and communicative pluralism become still more radical and continue to come up against assumptions foundational to western thought about the types of capacities required to make claims that are worthy of and intelligible to the adjudication of justice.[27]

Finally, the use of the word "justice" in multispecies justice merits particular comment. As Val Plumwood astutely observed, in the Kantian tradition, adopted even by biocentric philosophers like Paul Taylor, because the realm or definition of morality or ethics within which justice is located has been established "as distance from emotion and 'particular fondness,' morality is then seen as the domain of reason and its touchstone, belief."[28] The effect is to treat "care, viewed as 'inclination' or 'desire,' as irrelevant to morality."[29]

Looking to Ecofeminism and the Centrality of Relational Logics

As noted, multispecies justice has at its core a constellation of commitments to an ontology of entanglement and an ethics of relationality. In this regard, ecofeminism constitutes a critical inspiration. Two edited volumes of ecofeminist writings spanning nearly two decades, *Ecofeminism: Women, Animals, Nature* (1993) and *Ecofeminism: Feminist Intersections with Other Animals and the Earth* (2022), both emphasise interconnectedness, interdependence, and relationality as defining features of how ecofeminists conceptualise human–nonhuman relations.[30] Further, and consistent with the comments made earlier about an understanding of justice and ethics, the emphasis on relationality also entails a move away from the dualism of reason and emotion and an active embrace of embodied and affective relations as a mode of ethical understanding. Ecofeminist theories and practices entail affective engagements with more-than-humans, as well as thinking contextually on issues and conflicts within and across species.[31] In recognising affect as a guide for action, ecofeminists allow for the relevance of the entire spectrum of feelings and emotions, from care, compassion, empathy, desire, and joy to grief, pain, rage, and a sense of loss.[32] Together, affect-guided actions and contextual thinking form the basis of holistic and critical approaches ecofeminists embody and perform in challenging oppressions on all relational scales, from interpersonal to institutional. For instance, Greta Gaard argues that "the insights of ecology, feminism, and socialism" shape "ecofeminism's basic premise that the ideology which authorizes oppressions such as those based on race, class, gender, sexuality, physical abilities, and species is the same ideology which sanctions the oppression of nature."[33] Carol J. Adams and Lori Gruen similarly state that ecofeminism "addresses the various ways that misogyny, heteronormativity, white supremacy, colonialism, and ableism are informed by and support pernicious anthropocentrism, and how analysing the ways these forces interconnect can produce less violent, more just practices."[34]

Central to ecofeminist philosophy is a rejection of value dualism, which upholds hierarchical and exclusionary relations through "fictitious binaries," such as "culture/nature; male/female; mind/body; master/slave; reason/emotion; and human/nonhuman."[35] Beyond demarcating and perpetuating difference, these oppositional binaries also "mark those with power and those available to be exploited," rationalising the dominance of one and the subordination of the other.[36] Dismantling oppressive binary logics is one of the key ecofeminist legacies and learnings instructing the work of multispecies justice. The starting point of multispecies justice entails a radical suspicion of any assumed hierarchy in relations across species.

The rejection of such logics has some critical implications for the ways in which multispecies justice seeks to operate. First, like Gruen, we are "critical of 'extensionist accounts' that move out from a human center, usually assumed to be rather homogenous, to more distant others based on their likeness to those in the center."[37] A classic example of such extensionist or expansionist accounts within the field of animal ethics would be Peter Singer's argument that altruism originates within a community of close kin, and it is the human capacity for moral reasoning that continually expands our circle of moral consideration and concern to include the interests of others, both humans and nonhumans.[38] Not only does Singer's method elevate the capacity for reason above emotional and affective ways of ethical relating, his utilitarianism is exclusionary in its grounding in sameness/likeness, applying only to those proven sentient or to possess the capacity to suffer.[39]

Another example is the move to extend human rights or personhood rights to other animals in the legal sphere, exemplified by the strategies of Nonhuman Rights Project and the Great Ape Project.[40] Maneesha Deckha points to the danger that such strategies may further entrench the historically exclusionary legal category of liberal humanism's concept of personhood and its binary logics, which separate the humanness of personhood from the concept of animality that personhood is perpetually defined against.[41] Adopting the critical frameworks developed by "feminist, postcolonial, and queer theory scholars," Deckha points out that the exclusionary function within "masculinist and colonial and humanist logic of legal personhood and rights" is emblematic of rather than exceptional to those concepts.[42] In citing Deckha's concerns, we are not taking an absolute position against strategies that involve recognising personhood of the more than human but rather pointing to the risks of assimilative logics. In other systems of meaning, in particular those of First Nations peoples, personhood already challenges the humanist and individualist ontology underpinning the western conception, and in this sense, anticolonial worldviews can dismantle and enrich those ontologies and the concept of personhood itself. Moreover, where personhood is occupied by different types of beings, the concept, which is after all a socio-historical one, may be transformed.

The larger point is that multispecies justice does not look to expand the reach of justice by 'allowing' the affordance of moral value or bequeathing valued status as a subject of justice on the basis of the possession of a particular property or set of properties, as if there exists a circle of moral considerability emanating out from the moral exemplar of the putative rational human and encompassing others on the basis of proximity. Rather, and again, on the basis of its relational ontological commitments, justice is understood as a property of relations, where the good of different types of beings is relationally constituted.[43] While this does not dissolve dilemmas concerning the reach of justice or about the relative claims of different types of beings, it casts a very different light on the demarcation problem—who is in and who is out?—that continues to haunt attempts to expand justice beyond the human.

A second critical implication of the rejection of dualistic and hierarchical logics for multispecies justice concerns what follows for institutional and more specifically political transformation. As Sherilyn MacGregor observes, at the heart of ecofeminist Val Plumwood's critiques of the network of dualisms lies a political strategy aimed at deconstructing the architecture of dominant liberal political systems: "the appropriation of reproductive labour and material nature as private property and disavowed dependency as a condition of masculinist citizenship."[44] Thus, multispecies justice entails not simply a politics of inclusion but one of exposing the exploitative and extractivist processes that have been the conditions of possibility for the participation of the traditionally valourised and included.

Postcolonial Feminist and Queer Theories: Multispecies Justice and Attention to the Margins

The tendency to always consider those who might be intentionally or unintentionally excluded and marginalised by dominant norms and institutions or might be deliberately silenced and erased, as well as the power dynamics in these processes, can be found in the works of postcolonial feminist and queer theorists concerning race, gender, and sexuality, particularly in critical investigations of erasures and silencing within archives or historical narratives more generally. One example is Anjali Arondekar's critical analysis of Indian colonial archives, where Arondekar notes that the ways in which sexuality studies continually discover queer subjectivities in the process of " 'queering' pasts" comes with trade-offs, whereby "the recovery of the hidden documents of homosexuality surrenders presence, but only to reinstate its archival liminality."[45] In other words, despite homosexuality having been "naturalized as a frequent phenomenon" in colonial India, the colonial archives structurally enforce the absence of homosexuality by concealing it as an open secret that has been sparsely documented, therefore maintaining the oppressive epistemological systems that require a successful self-fashioning of queer subjects to rely on constant disclosure and discovery.[46]

Hil Malatino's critical look at the disparities between *transnormative* subjects ("subjects who, save for their status as trans, are otherwise highly assimilable—gender normative, heterosexual, middle class, well educated, racialized as white") versus nontransnormative subjects in relation to various systems and institutions is yet another example of postcolonial feminist and queer attentiveness to how some subjects become invisiblised, even within already marginalised communities.[47] Malatino examines the ways in which transnormative subjects have come to be the ones occupying heavy presences within "medical archives of transsexuality" and how they "have the least mitigated access to medical technologies of gender transition—hormones, surgery, and continued care," while nontransnormative subjects face routine and systematic "institutional and interpersonal violence . . . including death—by homicide and suicide . . . [and] also by lack of access to quality, affordable, trans-competent health care."[48]

Taking after such postcolonial feminist and queer interventions to critically ensure that differences are present and represented on their own terms, multispecies justice applies similar politicised ethics, always questioning and attending to subjects who may have been erased, silenced, ignored, or simply excluded from or denied subjecthood entirely. In this regard, Chloë Taylor and James Stanescu's application of Judith Butler's ideas around the precarity and grievablility of life to other animals are consistent with these interventions in the multispecies justice context.[49] Thinking about the subjects and relations that may have been left without a seat at the table is both a crucial precursor and concurrent step to imagining and putting into practice communicative interactions with more-than-human others.

Conclusion

Beyond the recognition of beings other than humans as subjects of justice, multispecies justice demands an inclusion of the agency and voices of other species in the processes of political decision-making that shape our collective existence in our shared worlds. Against the idea that other animals are "voiceless," Lauren Corman has called for conceptualising the voice metaphor in animal advocacy in ways that "highlight difference" and "indicate non-unitary subjectivity, experience, relationality, and resistance" in the lives of nonhuman animals.[50] Similarly, Eva Meijer has instructed us to consider radically different forms of communication for imagining interspecies democracies.[51] Following Josephine Donovan, these types of cross-species political engagements involve moving from care to dialogue with more-than-humans[52] In an overtly political tone, Catriona Sandiland's calls for applying to nature "the principles of equality, self-determination and even subjectivity" as "as co-conspirator in human activity and as an interdependent actor fully included in the process of democratization."[53]

To return, though, to the omnicidal context in which multispecies justice has emerged, the political processes for transformation will need to be Janus faced: attentive to both the conditions of epistemic and structural violence made manifest in historical exclusions and marginalisations and to the possibilities of forging different cleavages of solidarity and formations of community. As pattrice jones suggests within the context of trauma recovery involving eco/feminist psychology, the "intrapsychic and interpersonal cleavages" formed by traumatic alienations, as well as our "estrangement from nature, other animals, and one's own animality," require embodied and inclusive processes that bring better harmony and attunement between "our animal bodies . . . and the systems in which our bodies are situated."[54] Learning about the lifeways and desires of more-than-humans through dialogue and communication must therefore transcend the limitations of anthropocentric languages through our embodied experiences and relationalities with other beings and entities.

One future direction or project for multispecies justice must then involve implementing means to creatively and respectfully institutionalise the lessons and guidance from the theoretical and activist traditions we have traced in this chapter so as to facilitate the political and politicised communication necessary to pursue justice across radical differences.

Notes

1 On strong theory, see Julie K. Gibson-Graham, "Rethinking the Economy with Thick Description and Weak Theory," *Current Anthropology* 55, no. S9 (2014): S147–S153.

2 The term polycrisis has been coined to describe the interlocking and multidimensional political and environmental crises of the twenty-first century; see Thomas Homer-Dixon, Ortwin Renn, Johan Rockstrom, Jonathan F. Donges, and Scott Janzwood, "A Call for an International Research Program on the Risk of a Global Polycrisis," *Available at SSRN 4058592* (2021). See also Thomas Homer-Dixon and Johan Rockström, "What Happens When a Cascade of Crises Occur?" *New York Times* (November 13, 2022).

3 Amongst the large and diverse body of literature, see, for example, Benita Roth, *Separate Roads to Feminism: Black, Chicana, and White Feminist Movements in America's Second Wave* (Cambridge University Press, 2004); Anita Ghai, "Disabled Women: An Excluded Agenda of Indian Feminism," *Hypatia* 17, no. 3 (2002): 49–66; Aileen Moreton-Robinson, *Talkin'up to the White Woman: Indigenous Women and Feminism* (University of Minnesota Press, 2021); Ien Ang, "I'm a Feminist But . . . 'Other' Women and Postnational Feminism," in *Transitions* (Routledge, 2020), 57–73.

4 See Danielle Celermajer, Sria Chatterjee, Alasdair Cochrane, Stefanie Fishel, Astrida Neimanis, Anne O'Brien, Susan Reid et al., "Justice through a Multispecies Lens," *Contemporary Political Theory* 19, no. 3 (2020): 475–512.

5 Janae Davis, Alex A Moulton, Levi Van Sant, and Brian Williams, "Anthropocene, Capitalocene, . . . Plantationocene? A Manifesto for Ecological Justice in an Age of Global Crises," *Geography Compass* 13, no. 5 (2019): e12438.

6 On ecocide, see Polly Higgins, Damien Short, and Nigel South, "Protecting the Planet: A Proposal for a Law of Ecocide," *Crime, Law and Social Change* 59, no. 3 (2013): 251–266. On the connections between ecocide, genocide, capitalism, and colonialism, see also Martin Crook, Damien Short, and Nigel South, "Ecocide, Genocide, Capitalism and Colonialism: Consequences for Indigenous Peoples and Glocal Ecosystems Environments," *Theoretical Criminology* 22, no. 3 (2018): 298–317.

7 Zoe Todd, "An Indigenous Feminist's Take on the Ontological Turn: 'Ontology' is Just Another Word for Colonialism," *Journal of Historical Sociology* 29, no. 1 (2016): 4–22.

8 Christine J. Winter, "Introduction: What's the Value of Multispecies Justice?" *Environmental Politics* 31, no. 2 (2022): 251–257.

9 Danielle Celermajer, David Schlosberg, Lauren Rickards, Makere Stewart-Harawira, Mathias Thaler, Petra Tschakert, Blanche Verlie, and Christine Winter, "Multispecies Justice: Theories, Challenges, and a Research Agenda for Environmental Politics," *Environmental Politics* 30, no. 1–2 (2021): 119–140.

10 Monica Gagliano, *Thus Spoke the Plant: A Remarkable Journey of Groundbreaking Scientific Discoveries and Personal Encounters with Plants* (North Atlantic Books, 2018); M.A. Gorzelak, A.K. Asay, B.J. Pickles, and S.W. Simard, "Inter-plant Communication through Mycorrhizal Networks Mediates Complex Adaptive Behaviour in Plant Communities," *AoB Plants* (2015 May 15): 7: plv050, https://doi.org/10.1093/aobpla/plv050. PMID: 25979966; PMCID: PMC4497361.

11 Kerry W. Bowman, Samuel A. Dale, Sumana Dhanani, Jevithen Nehru, and Benjamin T. Rabishaw, "Environmental Degradation of Indigenous Protected Areas of the Amazon as a Slow Onset Event," *Current Opinion in Environmental Sustainability* 50 (2021): 260–271; Oliver Pye, "Palm

oil as a Transnational Crisis in South-East Asia," *ASEAS-Austrian Journal of South-East Asian Studies* 2, no. 2 (2009): 81–101.

12 Aurora Moses and Paige Tomaselli, "Industrial Animal Agriculture in the United States: Concentrated Animal Feeding Operations (CAFOs)," in *International Farm Animal, Wildlife and Food Safety Law* (Cham: Springer, 2017), 185–214.

13 Danielle Celermajer, Rosemary Lyster, Glenda M. Wardle, Rachel Walmsley, and Ed Couzens, "The Australian Bushfire Disaster: How to Avoid Repeating This Catastrophe for Biodiversity," *Wiley Interdisciplinary Reviews: Climate Change* 12, no. 3 (2021): e704.

14 Danielle Celermajer, *Summertime* (Random House Australia, 2021).

15 Kim Erno, "Update: Global Capitalism's Attack on Mother Earth and Her Indigenous Daughters," in *Global Femicide*, eds. Brenda Anderson; et al. (Open Educational Resources Publishing Program Regina, 2021).

16 Claire Jean Kim, *Dangerous Crossings* (Cambridge University Press, 2015); Joshua Bennett, *Being Property Once Myself: Blackness and the End of Man* (Harvard University Press, 2020).

17 Lori Gruen, "Dismantling Oppression: An Analysis of the Connection between Women and Animals," in *Living with Contradictions* (Routledge, 2018), 537–548.

18 Sherilyn MacGregor, *Routledge Handbook of Gender and Environment* (Taylor & Francis, 2017); Giovanna Di Chiro, "Welcome to the White (M) Anthropocene?: A Feminist-Environmentalist Critique," in *Routledge Handbook of Gender and Environment*, ed. Sherilyn MacGregor (Routledge, 2017), 487–505.

19 Kim TallBear, "Why Interspecies Thinking Needs Indigenous Standpoints," *Cultural Anthropology* 24 (2011); Zoe Todd, "Fish Pluralities: Human–animal Relations and Sites of Engagement in Paulatuuq, Arctic Canada," *Études/inuit/studies* 38, no. 1–2 (2014); Kyle Whyte, "Settler Colonialism, Ecology, and Environmental Injustice," *Environment and Society* 9, no. 1 (2018); Christine J Winter, *Subjects of Intergenerational Justice: Indigenous Philosophy, the Environment and Relationships* (Routledge, 2021).

20 Such a view, as Escobar puts it, "calls for seeing the interrelatedness of ecological, economic, and cultural processes that come to produce what humans call nature." Arturo Escobar, "Territories of Difference," in *Territories of Difference* (Duke University Press, 2008), 154. See also a similar analysis of the 'environmentalism of the poor' in Ramachandra Guha and Joan Martínez Alier. *Varieties of Environmentalism: Essays North and South* (Routledge, 2013).

21 David Schlosberg, "Ecological Justice and the Anthropocene," in *Political Animals and Animal Politics*, eds. M. Wissenburg and D. Schlosberg (London: Palgrave Macmillan, 2014).

22 Karen Barad, *Meeting the Universe Halfway: Quantum Physics and the Entanglement of Matter and Meaning* (Duke University Press, 2007); Eduardo Kohn, *How Forests Think: Toward an Anthropology beyond the Human* (University of California Press, 2013); Donna J Haraway, *Staying with the Trouble: Making Kin in the Chthulucene* (Duke University Press, 2016).

23 Gibson-Graham, "Rethinking the Economy with Thick Description and Weak Theory," S149.

24 Isabelle Stengers, "Including Nonhumans in Political Theory: Opening Pandora's Box," in *Political Matter: Technoscience, Democracy, and Public Life*, eds. Bruce Braun and Sarah J Whatmore (Minneapolis: University of Minnesota Press, 2010); Marisol De la Cadena and Mario Blaser, *A World of Many Worlds* (Duke University Press, 2018).

25 Iris Marion Young, *Responsibility for Justice* (Oxford University Press, 2010). Note the claim here is not that feminist theory has an unbroken line to commitments to the recognition of differentials in power and it has only been the insistence of women of colour, trans women, and women with disability that has exposed the limits of feminist power contentions.

26 Astrida Neimanis, *Bodies of Water: Posthuman Feminist Phenomenology* (Bloomsbury Publishing, 2019); José Medina, *The Epistemology of Resistance: Gender and Racial Oppression, Epistemic Injustice, and the Social Imagination* (Oxford University Press, 2012).

27 See, for example, Neimanis and Chatterjee's contribution in Danielle Celermajer et al., "Justice through a Multispecies Lens," *Contemporary Political Theory* (2020).

28 Val Plumwood, "Nature, Self, and Gender: Feminism, Environmental Philosophy, and the Critique of Rationalism," *Hypatia* 6, no. 1 (1991): 3–27, 4–5.

29 Lori Gruen articulates a similar position in, Lori Gruen, *Entangled Empathy: An Alternative Ethic for Our Relationships with Animals* (New York: Lantern Books, 2015).

30 Carol J. Adams and Lori Gruen. *Ecofeminism: Feminist Intersections with Other Animals and the Earth*, 2nd ed. (New York: Bloomsbury Academic, 2022); Greta Claire Gaard. *Ecofeminism: Women, Animals, Nature* (Philadelphia: Temple University Press, 1993).
31 Adams and Gruen, *Ecofeminism*.
32 Ibid.
33 Gaard, *Ecofeminism*, 1.
34 Adams and Gruen, *Ecofeminism*, xxi.
35 Carol J. Adams and Lori Gruen, "Ecofeminist Footings," in *Ecofeminism: Feminist Intersections with Other Animals and the Earth*, 2nd ed., eds. Carol J. Adams and Lori Gruen (New York: Bloomsbury Academic, 2022), 3–4. See also I. Diamond and G. Orenstein, *Reweaving the World: The Emergence of Ecofeminism* (San Francisco: Sierra Club Books, 1990) and critically, Val Plumwood, *Feminism and the Mastery of Nature* (Routledge, 2002).
36 Adams and Gruen, "Ecofeminist Footings," 3.
37 Lori Gruen, "Expressing Entangled Empathy: A Reply," *Hypatia* 32, no. 2 (2017): 452.
38 Peter Singer, *The Expanding Circle: Ethics and Sociobiology*, 1st ed. (New York: Farrar, Straus & Giroux, 1981).
39 Peter Singer, *Animal Liberation: The Definitive Classic of the Animal Movement*. 40th anniversary ed. (New York: Open Road Media, 2009).
40 Maneesha Deckha, *Animals as Legal Beings: Contesting Anthropocentric Legal Orders* (Toronto: University of Toronto Press, 2021), 88.
41 Ibid., 87–94.
42 Ibid., 92.
43 Critical here is the difference between intrinsic value understood as the difference between means and ends and intrinsic value understood as relational and non-relational. See Christine Korsgaard, "Two Distinctions in Goodness," *Philosophical Review* 92 (1983): 169–195. Critical to our argument is the point made by Karen Green that intrinsic values (values as ends) can nevertheless derive from the relational properties of a being. See Karen Green, "Two Distinctions in Environmental Goodness," *Environmental Values* 5 (1996): 31–46
44 Sherilyn MacGregor, "Making Matter Great Again? Ecofeminism, New Materialism and the Everyday Turn in Environmental Politics," *Environmental Politics* 30, no. 1–2 (2021): 41–60. The relevant text of Plumwood's is Val Plumwood, "Has Democracy Failed Ecology? An Ecofeminist Perspective," *Environmental Politics* 4, no. 4 (1995): 134–168.
45 Anjali R. Arondekar, *For the Record: On Sexuality and the Colonial Archive in India* (Durham: Duke University Press, 2009), 7.
46 Ibid., 7, 9.
47 Hil Malatino, *Queer Embodiment: Monstrosity, Medical Violence, and Intersex Experience* (Lincoln: University of Nebraska Press, 2019), 110.
48 Ibid.
49 Chloë Taylor, "The Precarious Lives of Animals: Butler, Coetzee, and Animal Ethics," *Philosophy Today (Celina)* 52, no. 1 (2008): 60–72; James Stanescu, "Species Trouble: Judith Butler, Mourning, and the Precarious Lives of Animals," *Hypatia* 27, no. 3 (2012): 567–582.
50 Lauren Corman, "The Ventriloquist's Burden: Animal Advocacy and the Problem of Speaking for Others," in *Animal Subjects 2.0*, eds. Jodey Castricano and Lauren Corman (Waterloo, Ontario: Wilfrid Laurier University Press, 2016), 501.
51 Eva Meijer, *When Animals Speak: Toward an Interspecies Democracy*, Vol. 1 (NYU Press, 2019).
52 Josephine Donovan, "Feminism and the Treatment of Animals: From Care to Dialogue," *Signs: Journal of Women in Culture and Society* 31, no. 2 (2006): 305–329.
53 Catriona Sandilands, *The Good-Natured Feminist: Ecofeminism and the Quest for Democracy* (University of Minnesota Press, 1999), 195.
54 Pattrice Jones, "Roosters, Hawks and Dawgs: Toward an Inclusive, Embodied Eco/feminist Psychology," *Feminism & Psychology* 20, no. 3 (2010): 376.

Bibliography

Adams, Carol J., and Gruen, Lori. *Ecofeminism: Feminist Intersections with Other Animals and the Earth*. 2nd ed. New York: Bloomsbury Academic, 2022a.

Adams, Carol J., and Gruen, Lori. "Ecofeminist Footings." In *Ecofeminism: Feminist Intersections with Other Animals and the Earth*, 2nd ed., edited by Carol J. Adams and Lori Gruen, 1–43. New York: Bloomsbury Academic, 2022b.
Arondekar, Anjali R. *For the Record: on Sexuality and the Colonial Archive in India*. Durham: Duke University Press, 2009.
Barad, Karen. *Meeting the Universe Halfway: Quantum Physics and the Entanglement of Matter and Meaning*. Duke University Press, 2007.
Celermajer, Danielle. *Summertime*. Random House Australia, 2021.
Celermajer, Danielle, Chatterjee, Sria, Cochrane, Alasdair, Fishel, Stefanie, Neimanis, Astrida, O'Brien, Anne, Reid, Susan *et al.* "Justice through a Multispecies Lens." *Contemporary Political Theory* (2020): 1–38.
Corman, Lauren. "The Ventriloquist's Burden: Animal Advocacy and the Problem of Speaking for Others." In *Animal Subjects 2.0*, edited by Jodey Castricano and Lauren Corman, Waterloo, 473–512. Ontario: Wilfrid Laurier University Press, 2016.
Deckha, Maneesha. *Animals as Legal Beings: Contesting Anthropocentric Legal Orders*. Toronto: University of Toronto Press, 2021.
De la Cadena, Marisol, and Blaser, Mario. *A World of Many Worlds*. Duke University Press, 2018.
Donovan, Josephine. "Feminism and the Treatment of Animals: From Care to Dialogue." *Signs: Journal of Women in Culture and Society* 31, no. 2 (2006): 305–329.
Erno, Kim. "Update: Global Capitalism's Attack on Mother Earth and Her Indigenous Daughters." In *Global Femicide*, edited by Brenda Anderson;, Shauneen Pete;, Wendee Kubik, and Mary Rucklos-Hampton, 62–70. Open Educational Resources Publishing Program Regina, 2021.
Escobar, Arturo. *Territories of Difference: Place, Movements, Life, Redes*. Durham: Duke University Press, 2008.
Gaard, Greta Claire. *Ecofeminism: Women, Animals, Nature*. Philadelphia: Temple University Press, 1993.
Gruen, Lori. "Expressing Entangled Empathy: A Reply." *Hypatia* 32, no. 2 (2017): 452–462.
Haraway, Donna J. *Staying with the Trouble: Making Kin in the Chthulucene*. Duke University Press, 2016.
Jones, Pattrice. "Roosters, Hawks and Dawgs: Toward an Inclusive, Embodied Eco/feminist Psychology." *Feminism & Psychology* 20, no. 3 (2010): 365–380.
Kohn, Eduardo. *How Forests Think: Toward an Anthropology Beyond the Human*. Univ of California Press, 2013.
Malatino, Hil. *Queer Embodiment: Monstrosity, Medical Violence, and Intersex Experience*. Lincoln: University of Nebraska Press, 2019.
Medina, José. *The Epistemology of Resistance: Gender and Racial Oppression, Epistemic Injustice, and the Social Imagination*. Oxford University Press, 2012.
Meijer, Eva. *When Animals Speak: Toward an Interspecies Democracy*. Vol. 1. NYU Press, 2019.
Neimanis, Astrida. *Bodies of Water: Posthuman Feminist Phenomenology*. Bloomsbury Publishing, 2019.
Schlosberg, David. "Ecological Justice and the Anthropocene." In *Political Animals and Animal Politics*, edited by M. Wissenburg and D. Schlosberg, 75–89. London: Palgrave MacMillan, 2014.
Singer, Peter. *The Expanding Circle: Ethics and Sociobiology*. 1st ed. New York: Farrar, Straus & Giroux, 1981.
Singer, Peter. *Animal Liberation: The Definitive Classic of the Animal Movement*. 40th anniversary ed. New York: Open Road Media, 2009.
Stanescu, James. "Species Trouble: Judith Butler, Mourning, and the Precarious Lives of Animals." *Hypatia* 27, no. 3 (2012): 567–582.
Stengers, Isabelle. "Including Nonhumans in Political Theory: Opening Pandora's Box." In *Political Matter: Technoscience, Democracy, and Public Life*, edited by Bruce Braun and Sarah J Whatmore. Minneapolis: University of Minnesota Press, 2010.
TallBear, Kim. "Why Interspecies Thinking Needs Indigenous Standpoints." *Cultural Anthropology* 24 (2011): 1–8.
Taylor, Chloë. "The Precarious Lives of Animals: Butler, Coetzee, and Animal Ethics." *Philosophy Today (Celina)* 52, no. 1 (2008): 60–72.
Todd, Zoe. "Fish Pluralities: Human–animal Relations and Sites of Engagement in Paulatuuq, Arctic Canada." *Études/inuit/studies* 38, no. 1–2 (2014): 217–238.

Whyte, Kyle. "Settler Colonialism, Ecology, and Environmental Injustice." *Environment and Society* 9, no. 1 (2018): 125–144.

Winter, Christine J. *Subjects of Intergenerational Justice: Indigenous Philosophy, the Environment and Relationships*. Routledge, 2021.

Young, Iris Marion. *Responsibility for Justice*. Oxford University Press, 2010.

42

TWO-SPIRIT FEMINISM AND INDIGENOUS ECOLOGICAL GOVERNANCE

Margaret Robinson

Introduction

In this chapter, I examine how relations between humans and other animals are expressed in land protection activism, in which Indigenous women and Two-Spirit people are highly visible. I approach this work as a Two-Spirit woman and a member of Lennox Island First Nation. My own activism has focussed on queer liberation rather than land protection, which is a subject I approach as a scholar rather than as a participant.

Models of Land Protection

Settler occupation of Indigenous homelands has been devastating to the environment, to the animals that rely upon it, and to Indigenous ways of relating to land and other animals. Land protection is therefore a central concern for Indigenous nations. Due to our long pre-colonial histories, what Indigenous peoples mean by 'land protection' may differ from how others frame the concept. In Canada, I note that settler approaches to land protection often follow a Christian stewardship model in which humans manage natural resources on behalf of the Creator. For some, this manifests in a desire to save animals, plants, and water from the evils of pollution. Such ecologically minded settlers can be found in most land protection movements. However, a stewardship approach in which humans claim God-like dominion over all other animals is also found in land-management approaches that favour extractive industries such as oil and gas. Profit-driven stewardship demands that humans gain economically from nature. This expectation may be informed by the Parable of the Talents, found in the *Gospel of Matthew* (25:14–30) and the *Gospel of Luke* (19: 11–27). In that story, three servants are entrusted with varying amounts of currency by their master, who gives them no additional directions concerning the funds. Two of the servants increase their wealth through business dealings, while the third hides the money to keep it secure, earning nothing. When the master returns, the servants who pursued profit are rewarded while the servant that did not is punished. As I interpret the tale, savvy managing of economic investments are a metaphor for religious faithfulness, tacitly sacralizing the pursuit of profit. The secularization of Europe and North America since the 1960s has stripped

DOI: 10.4324/9781003273400-52

religious language from much environmental and resource development talk, but the view of humans as powerful decision-makers pursuing profit remains and has reached unfettered heights under multinational capitalism. The United Nations Economic Commission for Europe, for example considers land as "a resource from both an environmental and an economic perspective" and argues that all nations are responsible to ensure "environmentally sound and sustainable development worldwide."[1]

The vision underlying sustainable management of resources is constant profit. That mandate reflects settler capitalist values such as privatization of property, self-interest, and free enterprise but fails to embody traditional Indigenous economic values such as connection to place, relationality, reciprocity, or long-term planning.[2]

Lawyer Camille Boulianne notes that under Canadian law, "environmental stewardship is often referred to as "resource management,"[3] which seems to fit the stewardship-as-profit-seeking model. Protecting land from development does not seem to be a viable option for resource managers, and indeed, First Nations leaders report being pressured to accept development projects that they and their communities do not want.[4] Although the stewardship model emerges from settler culture, it is not without Indigenous adherents. Some Indigenous communities have developed models of stewardship that combine settler values of control and profit with Indigenous knowledge. For example, Chief Matthew Kakekaspan, of Fort Severn Cree Nation, along with colleagues in Thunder Bay,[5] has outlined an Indigenous stewardship model for managing polar bear populations. The model includes Cree traditions, such as hunting polar bears, protecting polar bear dens, and passing on traditional knowledge about polar bears but also stresses economic benefits anticipated by the sale of polar bear hide, crafts, and other "polar bear products," as well as the development of a tourism industry to view the bears.[6] Cree values can be seen in the emphasis on hunting as part of a subsistence lifestyle affirmed by treaty, as well as on equitable distribution of economic benefit. To this is added the developmental goal of "[p]romoting economic self-sufficiency" through "ecologically sustainable micro-enterprise development."[7] Indigenous stewardship models such as this hybrid model may frame animals as a solution to unemployment and poverty in Indigenous communities.

The role of animals is an area where land management often differs from land protection. The settler stewardship model considers humans and other animals as essentially separate from the land on which they rely. The definition of land offered by the United Nations Economic Commission for Europe, for example, excludes "[o]bjects that are not attached to the soil, such as motor cars, animals and human beings" but notes that such objects "will be subject to the rights [of the landholder] that control the use of the space that they occupy."[8] This view posits humans and other animals as objects of landowner control and as transferrable to other places, rather than as part of an interconnected ecosystem that evolved in symbiosis. In stewardship models, the challenge of land protection is to manage and distribute resources to achieve profit, with animals, water, and soil treated as objects of exchange. Animals and their essential habitats are framed as resources that can be used or sold when the price is right. If the needs of other animals are considered at all, it is often at the level of population, with each species expected to justify their existence by supporting human profit. Against the stewardship model, I would describe Indigenous approaches to land as taking a holistic perspective in which land and animals are essentially interconnected.[9] Management strategies in an Indigenous model lean toward preservation of habitat and attempt to balance the needs of all beings sharing the land. This is generally beneficial for animals when compared to the settler stewardship model, even if we consider that the needs of animals in Indigenous land protection must be balanced against the needs of plants, humans, and other beings.

To discuss Indigenous models of land protection, however, I must acknowledge the overgeneralization of labelling any approach as 'Indigenous.' The concept of 'Indigenous people' combines so many culturally and politically distinct nations that the term only makes sense under colonialism, which is a relatively recent development. My own nation, the Mi'kmaq, has endured colonialism longer than most, yet our engagement with settlers makes up only 500 years of our 13,500-year history.[10] Being Mi'kmaq is larger than our experience of occupation. Without colonialism, I would not identify as Indigenous but as Mi'kmaw. Back even further, before we became friends to newcomers, I would be an L'nu (one of the human beings). In other First Nations as well, self-identity labels often translate as 'the people' or 'the human beings.' I take this pattern to indicate that our primary relations are with other animals in our homeland rather than with neighbouring humans, and it is this centring of relation between humans and other animals that informs my work.

Approaches to land protection I have encountered in Indigenous circles differ from the settler stewardship model in three significant ways: 1) by viewing land (or parts of it) and other animals as having spirit and personality, and as capable of relation; 2) by viewing humans and other animals as connected to land at a fundamental level; and 3) by viewing other animals as subjects (or persons) similar to humans, especially in their rights to land and water.[11] The first of these differences—a view of the land and other animals as beings with spirit—is rooted in what Mi'kmaw Elder Jane Meade describes as a philosophy that centres "the connectiveness of all of creation," requiring us to behave "as a sacred being living among other sacred beings."[12] To explain what this worldview means for the human–animal relation requires understanding that Indigenous taxonomies draw different lines when distinguishing living being from non-living beings. The Mi'kmaw language, for example, refers to beings that possess a spirit as animate and those that do not as inanimate. The animate includes humans and other animals but also beings such as the sun, stars, and rivers.[13] In a Mi'kmaw worldview, all these beings must be considered in land protection decision-making.

An example of how land protection differs from stewardship can be seen in the case of Kluskap's Mountain. Mikmaw knowledge keeper Clifford Paul notes that our people describe Kluskap's Mountain as "the centre of the universe" since it contains the cave where the Mi'kmaw hero Kluskap left this world for another.[14] Paul reports that some Mi'kmaq go to the cave as a spiritual pilgrimage. Mi'kmaw activist Elizabeth Marshall identifies the mountain as home to Wi'klatmu'jk (the little people), small trickster spirits that live in wooded areas.[15] Whether one believes in Wi'klatmu'jk or Kluskap, the mountain holds special importance to the Mi'kmaq. In 2017 Mi'kmaw people and our settler allies protested an attempt by the Mining Association of Nova Scotia to remove the mountain from the Nova Scotia Wilderness Areas Protection Act.[16] The Mining Association proposed "adding a 'land swap' mechanism to the Protection Act that would permit resource extraction companies "to access protected land by purchasing land of at least equal size and ecological value outside of the protected areas and arranging for it to be protected instead."[17] This land swap proposal rejects the individuality of land and animals, considers their cultural and spiritual relationality insignificance, and ignores the land as home to specific animals, instead framing land as interchangeable and considering animals at the level of species, if at all.

A second way the Indigenous land protection movements I have encountered differ from stewardship is by asserting that humans and other animals have an essential connection to land as opposed to being transferable to other places. Mi'kmaw historian and archaeologist Roger Lewis explains how the verb weji-sqalia'timk (to sprout from the land) "expresses the Mi'kmaw understanding of their being rooted here . . . they sprouted from this land much like a plant sprouts from it."[18] Perhaps such self-understanding does not emerge among a people that has moved, or been displaced, or has lived in several regions in quick succession. To the Mikmaq, however, Mi'kma'ki is not simply a place to live; it is our home, the homeland of our ancestors, and the place we hope our descendants will live until the end of time. In a relationship this long-term, the other partner is not interchangeable with one of similar size and value. For many Mi'kmaw, their relationship with land is a spiritual one. A study about conservation and land management that interviewed Mi'kmaw community members in Eskasoni, Membertou, Potlotek, Wagmatcook, and We'koqma'q emphasized the need to centre respect for animals, land, water, and culture. "The land is not sacred," the study reported, "it is the relationship to the land that is sacred."[19] That a relationship to land remains important for me as a secular atheist suggests that a sense of place taps deeply into who we are as a people. Land protection activism is similarly an expression of Indigenous self-identity, not simply of political perspective.

A third feature of Indigenous land protection is the affirmation of animal personhood.[20] This concept acknowledges humans as animals and considers other animals similar to humans in important ways, particularly in relation to needs and rights. As Mi'kmaw Elder Albert Marshall explains, "no one being is greater than the next . . . we are equal."[21] This viewpoint is expressed in the phrase m'sit no'kmaq, meaning 'all my relations.' Included as relations are "animals, plants, inorganic matter, such as rocks and the land itself."[22] Without a sanctified profit motive undergirding land talk, the Indigenous approaches I know prioritize living in balance with other animals on the land. As Elder Marshall adds, "each one of us has a responsibility to the balance of the system."[23] Achieving balance may require rejecting opportunities to profit at the expense of other animals, and the value of doing minimal harm to others seems foundational to Mi'kmaw economics. The Mi'kmaw concept of netukulimk, for example, has been translated as 'avoiding not having enough,'[24] which is important departure from profit accumulation.[25] In use, the term seems flexible. Mi'kmaw linguist Bernie Francis describes netukulimk to include "anything from hunting large or small animals and birds, fishing, attending a blueberry and potato harvest."[26]

The Unama'ki Institute of Natural Resources, a non-profit resource governance organization headed by Mi'kmaw Chiefs in the Unama'ki region, defines netukulimk as "achieving adequate standards of community nutrition and economic well-being without jeopardizing the integrity, diversity, or productivity of our environment."[27] To me, references to productivity evoke a stewardship model. However, the Institute affirms that a Mi'kmaw approach to managing resources "includes a spiritual element that ties together people, plants, animals, and the environment,"[28] suggesting their model may be hybrid. Others have highlighted the relational aspect of netukulimk as well, describing it as "a relationship of mutual aid and interconnectedness between humans and the natural world,"[29] and as incorporating "respect, responsibility, relationship, and reciprocity into every aspect of . . . life."[30] Regardless of one's definition of netukulimk, the value of 'avoiding not having enough' can be applied to all beings, not just to humans.

For many Indigenous land and water protectors, the issue is not solely one of protecting land and animals, but also of protecting the sacred. This may entail protecting specific places for spiritual reasons or protecting sacred relations with the land and other animals. Such work engages issues of Indigenous sovereignty by shaping our relations with occupying settler states and raises issues of First Nations governance as we question how decisions are made and by whom. Issues of sovereignty and governance become more urgent as environmental degradation spurs changes to climate that affect the liveability of our homeland.

Governance in an Apocalypse

Indigenous nations in North America governed their homelands effectively long before Europeans arrived, but the devastating impact of settlers on the environment has caused land protection to become a central struggle. Brooklyn Leo, a nonbinary Cherokee, describes our current time period as "post-apocalyptic" due to the devastation wrought by colonialism.[31] Leo describes how the impacts of colonialism have compounded, with European diseases killing up to 90% of Indigenous people, and settler land theft causing "a cascade of extinction events for animal, vegetative, and geological life" as interdependent relations among humans and animals were ruptured.[32] In addition to epidemiological and ecological catastrophes, Indigenous nations face ongoing genocide through settler efforts to reduce or eliminate Indigenous populations (through sterilization or murder), and to control or assimilate us by destroying our cultures, languages, and religions (through institutions such as residential schools). This has impacted women and Two-Spirit people particularly heavily, as Leo argues:

> Traditional medicine makers, matrifocal leaders, tribal knowledge-keepers, and Two Spirit peoples were systematically targeted for routine humiliation, bodily injury, and extermination, activating a wholesale restructuring of Indigenous social life.[33]

Land protection in an apocalypse unites Indigenous peoples and other animals in our homelands because we face threats to extinction under settler colonialism. Women and Two-Spirit people are active in grassroots land protection, perhaps because we do not mirror settler male authority as easily as Indigenous men and are co-opted less easily by political systems that treat land and animals as disposable for the right price.

Indigenous ecofeminism positions land protection as an extension of women's traditional leadership. Many Indigenous nations traditionally accorded women a significant role in governance. Cree/Métis scholar Kim Anderson offers the example of Haudenosaunee society, which is matrilineal, in which clan mothers have a traditional role in selecting leaders and negotiating treaties.[34] While the traditional role of women varies by Indigenous nation, women's involvement in traditional governance is a common theme. Cherokee poet Marilou Awiakta reports that when Cherokee Chief Attakullakulla met with Europeans to negotiate a treaty he asked, "Where are your women?" Awiakta interprets the chief's question as indicating his suspicion at the lack of gender balance, a suspicion she extends to US politics of her own time. "I look to the Congress," Awiakta writes, "the Joint Chiefs of Staff, the Nuclear Regulatory Commission . . . to the hierarchies of my church, my university, my city, my children's school. 'Where are your women?' I ask."[35]

Sadly, Awiakta's observation now applies to Indigenous governments as well, as women were pushed from leadership roles. The Dominion of Canada formed in 1867 and by 1869 had passed the Gradual Enfranchisement Act, decreeing that women classed as "Indian" by the settler government lose that classification (termed "Indian status") if they marry a man without such status.[36] This removed Indigenous women from positions of power by declaring them non-Indigenous. Soon after, Canada's 1876 Indian Act imposed an electoral process to replace existing Indigenous governments and removed long-standing rights for Indigenous women, including the right to vote, to speak at public meetings, and to run for political office.[37] The Indian Act consolidated political power in male chiefs and councils elected by male community members, forming Indigenous governments that mirrored the colonial governments occupying our territories.[38] First Nations women in Canada would not regain the right to vote in their own communities or run for office until the 1950s and could not vote in Canadian elections until the 1960s.

Dr. Angele Alook, a member of Bigstone Cree Nation, notes that "colonial structures broke, and continue to break, traditional governing practices . . . by undermining the roles of women."[39] In 2019 Indigenous women formed only 27% of First Nations councillors and only 19% of chiefs.[40] Male domination of governments means women are often excluded from "discussions and decisions about water management," and as a result women's knowledge about the land and other animals "has not necessarily been brought to bear on the development of protocols and practices" of land management.[41] This power imbalance directly impacts land protection, as research with legislators has found women are significantly more likely to endorse environmental protection than men.[42] The experience and knowledge of Indigenous women is essential to making decisions about 'resource governance.' This is especially true if our decisions about land, water, and other animals are to be informed by an awareness of the harms done by colonial systems and with an awareness of opportunities for resistance and transformation.[43]

While part of the apocalypse described by Brooklyn Leo includes the political exclusion of Indigenous women, a second part is the disempowerment of Two-Spirit people. Since the cultural resurgence that began in the 1960s, Indigenous people who do gender differently have been developing new terms to express their identities (such as Two-Spirit, Indigiqueer or QueerNDN) and asserting our right to participate fully in our nations. While some Two-Spirit people are elected to office, data on Two-Spirit leadership is not collected. My experience suggests that many Two-Spirit people are leaders in their communities, but often in service roles lacking decision-making authority. In an article co-authored with Bird,[44] we argued that Two-Spirit people must be empowered to direct and inform work 'development' work, but we also noted that such work is often shaped by settler expectations and imposed upon First Nations.

I agree with Michi Saagiig Nishnaabeg scholar Dr. Leanne Betasamoke Simpson that Canada is built on heteropatriarchy, which Simson defines as "the structural privileging of . . . white, heterosexual, masculine control."[45] Heteropatriarchal governance is a problem that cannot be solved by simply electing more women and Two-Spirit people to roles in government. I agree with Dene scholar Dr. Glen Coulthard,[46] who argues that recognition within the settler state fails to address structures of oppression. Representation alone cannot prevent us from imitating the masculinist domination of settler governance structures or internalizing profit-driven colonial ambitions. Participation in colonial governance structures may even reinforce the oppression inherent in hierarchical governance systems.[47] To offer governance alternatives, we must support the decision-making of Indigenous women and Two-Spirit people outside of settler-controlled governance structures, and we must continually challenge the centring of profit as a value over-riding all others. As Alook and Bidder write, "For Indigenous nations to live, capitalism must die."[48]

Women and Two-Spirit People as Leaders in Land Protection

Excluded from governance, women and Two-Spirit people often express their agency through land protection. The 2016–2017 movement at Standing Rock, for example, to protect water from the Dakota Access Pipeline was led by Indigenous women. Joye Braun, a Cheyenne River Sioux, raised the first teepee at what later became known as Oceti Sakowin Oyate (Seven Council Fire) by Mnisose (also called the Missouri River).[49] Women and their supports joined together in prayer and ceremony to protect local burial grounds and preserve safe water for future generations.

Land protection is dangerous work. Protestors are routinely assaulted by security forces representing corporate and state interests. Protestors at Standing Rock report being gassed, hit by water cannons, and having dogs set upon them.[50] In 2017, under the presidency of Donald Trump, the US National Guard forcibly removed the Standing Rock land protectors, and Dakota Access (a company owned by Texas-based Energy Transfer) installed their pipeline through the Lakota homeland. Extractive projects, such as pipelines, are often accompanied by additional gendered and sexualized violence when settler men installing or maintaining extractive equipment commit recreational sexual violence against Indigenous girls, women, and Two-Spirit people.[51] Dr. Simpson argues that such gender violence is a tactic of settler land theft and occupation that aims to keep women and Two-Spirit people too busy surviving to organize politically.[52]

The leadership of women in Indigenous land protection movements has been noted by the Royal Canadian Mounted Police (RCMP), the federal and national police body in Canada. The Aboriginal Television Network (APTN) obtained an internal RCMP report called Project Sitka[53] that detailed police surveillance of Indigenous land protectors, primarily those who spoke out against pipelines and shale gas. The RCMP had compiled a list of 89 Indigenous activists they believed "pose a criminal threat."[54] Included on the list was Suzanne Patles, a mother from Eskasoni First Nation. Although names were redacted in the RCMP report, Patles was identified through report details, including her membership in the Mi'kmaq Warrior Society and participation in a national speaking tour. "I guess you wear it like a badge of pride," Patles told APTN, "You know you are doing your job right if you are on a government watch list."[55]

Physical violence against land protectors by settler police and private security is an extension of the state violence of land theft and often indistinguishable from the violence against the land and other animals inherent to extraction industries. Dr. Simpson, for example, describes how gender violence is used to "remove agency from the plant and animal worlds," enabling settlers to reframe them as "natural resources" for their own use.[56] However, surviving violence may reinforce the need to protect land. Many of those targeted by state surveillance and whose land protection work is criminalized are already targeted by colonial violence due to their Indigeneity, gender, sexuality, and/or economic class. Excluded from governance by Indian Act structures and targeted by colonial authorities, women and Two-Spirit people may feel they have less to lose by confronting land exploitation. The land protectors I know personally are often young women in poverty who nonetheless risk beatings and arrest. In the following, I discuss two examples of successful land and water protection by women and Two-Spirit people and their allies in my own homeland: the first to protect water against fracking in Elsipogtog and the second to protect the Sipekne'katik River.

Elsipogtog

I first learned of the anti-fracking protests near Elsipogtog First Nation in 2013, from my cousin, Jim Robinson, who was then working for the Aboriginal Peoples Television Network as a camera operator. Jim was covering the action in Elsipogtog, where the police were attacking water protectors. The settler government of New Brunswick had allowed Southwestern Energy Resources, a Texas-based company, to explore for shale gas in the province, using a process known as fracking.[57] Studies find that fracking increases air pollution,[58] rates of cancer and cardiac and respiratory disease, and mortality[59] and reduces soil fertility by changing its density, acidity, and distribution of organic matter.[60] In a letter to the New Brunswick Hydraulic Fracturing Commission, the Grand Council of the Wolastokewiyik warned that their oral traditions report flooding and earthquakes in the area,[61] increasing the danger posed by fracking and seismic ground tests.

Southwestern Energy Resources began seismic blasting to test for gas deposits, and the Wolastokewiyik (People of the Beautiful River) and the Mi'kmaq, neither of whom had given permission for seismic testing on their homelands, began to protest. They blocked access to the trucks and equipment used for testing. Mikmaw Elder Doris Copage preferred to be called a protector rather than a protestor. "We are here to protect our water, our land," she said. "We have a river. It's a beautiful river, we love it and we respect it."[62]

While attending the 36th International Two-Spirit Gathering in 2023, I got to discussing Elsipogtog with some of the attendees. Three had been at the protests, helping to rebuild generators, renting and erecting tents, and bringing food to the Mi'kmaq Warriors Society, who maintained an encampment on the site where the fracking trucks were parked. Land protectors had the support of Indigenous government. On October 1, 2013, Chief Arron Sock of Elsipogtog First Nation announced his Council had issued an eviction notice to Southwestern Energy Resources, instructing them to remove their equipment from the territory.[63] On October 9, Sock announced the eviction notice had been ignored and stated, "we have been compelled to act to save our water, land and animals from ruin."[64]

On October 17, the RCMP moved in to arrest land protectors as they slept. Video soon emerged of police brutality against Elders and women. Suzanne Patles was struck in the head by an assault rifle, pulled from her vehicle, then hit several more times before being arrested.[65] Doris Copage, a 66-year-old Mi'kmaq Elder from Elsipogtog, was pepper-sprayed in the face.[66] Amanda Polchies, a 28-year-old mother from Elsipogtog, witnessed the attack on Elder Copage. "So many people got hurt," Polchies said. "I prayed for the women that were in pain, I prayed for my people, I prayed for the RCMP officers."[67] A photograph of Polchies kneeling on the highway, as 21 police officers in riot gear advance, came to represent the role of women in land protection.[68] The police are heavily armed, and Polchies holds only an eagle feather, a symbol of connection to the Creator and our animal relations.

Women reported that the RCMP used tear gas, pepper spray, rubber bullets, and dogs against the land protectors, eventually arresting 40 people.[69] The Civilian Review and Complaints Commission, an agency of the settler government, acknowledged the RCMP had broken Canadian law but still declared their actions against the Indigenous protestors "necessary and proportional."[70] Although the RCMP were not disciplined, Suzanne Patles was charged three times. None of the charges against Patles went to trial, although male members of the Mi'kmaw Warriors Society were sentenced to 15 months each.[71] Ultimately, the province's fracking plans did not go ahead, and Southwestern Energy Resources withdrew their equipment.

Sipekne'katik River Protests

Another successful defence of waterways in Mi'kma'ki took place near Spiekne'katik. Led by Mi'kmaw Grandmothers, this protest opposed a proposal by Alton Gas to store high-pressure hydrocarbon gas in salt caverns on the Sipekne'katik River, which would have poisoned the fish and other river animals with saline.[72] Cheryl Maloney, president of the Nova Scotia Native Women's Association, explained that "Indigenous women are traditionally responsible for water."[73] Mi'kmaw water protectors, including the Mi'kmaw Warrior Society, connected their protection of the river to their treaty rights by establishing a trading post on the river and emphasizing the animals in the river as a food source. Water protectors traded earrings, set eel traps, and shared teachings about traditional eel fishing in the river. As one activist explains, "we did it symbolically but purposefully . . .[to] exercise our right of trade" and "show traditional land use."[74]

While hunting and fishing were traditionally overseen by district chiefs, Mi'kmaw knowledge keeper Muin'iskwoq noted that a council of grandmothers was traditionally consulted.[75] Without such councils incorporated into elected government, women's voices are diminished. The chief of Spiekne'katik First Nation, Rufus Copage, supported the water protectors, but Millbrook Chief Bob Gloade expressed interest in the project provided it was under Mi'kmaw control.[76] Faced with ongoing opposition, the Alton Gas project too was shelved.

Sovereignty and Leadership

I agree with Brooklyn Leo that "The revitalization of Indigenous ways of knowing and being with land is central to addressing the devastating impacts of climate change."[77] Effective leadership in the post-apocalypse will require restoring the political power of women and Two-Spirit people. As Jaye Simpson,[78] an Oji-Cree Saulteaux trans woman, notes, "To return the land is one thing, but who the land goes to and who it is governed by is important." Removing women from positions of leadership and decision-making has been a tactic of colonial land theft. Criminalizing of land protection efforts is an additional level of state violence and one that can impact the employability of land protectors for the rest of their lives. Despite the risks, Indigenous women and Two-Spirit people continue to embrace relational worldviews that place animals on an equal footing with humans and to express that commitment through activism.

Notes

1 United Nations Economic Commission for Europe, "Land Administration Guideline with Special Reference to Countries in Transition," (Geneva, 1996). Retrieved from: https://unece.org/DAM/hlm/documents/Publications/land.administration.guidelines.e.pdf, 56, 108.

2 C.A. Hilton, *Indigenomics: Taking a Seat at the Economic Table* (New Society Publishers, 2021)

3 C. Boulianne, *Fostering Netukulimk: The Mi'kmaq Right to Fisheries Conservation and Co-management* (McGill University, Canada, 2022), 18.

4 N. Bird and M. Robinson, "Two-Spirit Issues in Development," in *The Routledge Handbook of Indigenous Development*, ed. N. Postero (Routledge, 2023)

5 M. Kakekaspan, B. Walmark, R.H. Lemelin, M. Dowsley, and D. Mowbray, "Developing a Polar Bear Co-management Strategy in Ontario through the Indigenous Stewardship Model," *Polar Record* 49, no. 3 (2013).

6 Ibid., 235.

7 Ibid.
8 United Nations Economic Commission for Europe, "Land Administration Guideline with Special Reference to Countries in Transition," 10.
9 Hilton, *Indigenomics: Taking a Seat at the Economic Table.*
10 R. Lewis, "Mi'kma'ki At 13,500," *Atlantic Books #83, Features History* (2021) retrieved from: https://atlanticbooks.ca/stories/mikmaki-at-13500/
11 M. Robinson, "The Role of Animals in Mi'kmaw Spirituality," *Concilium: An International Journal of Theology* 4 (2022): 131–145. https://concilium.hymnsam.co.uk/issues/20224-animals-and-theologies/
12 Nova Scotia Curriculum, "Netukulimk, " (2020): 3. Retrieved from: https://curriculum.novascotia.ca/sites/default/files/documents/resource-files/Netukulimk_ENG.pdf
13 P. Paul-Martin, "Animate and Inanimate Objects in Mi'kmaq," *Mi'kmaq Talking Posters, Smith/Francis Orthography. Eastern Woodland Print Communications* (2021). Retrieved from: https://firstnationhelp.com/ali/posters/pdf/animate_inanimate.pdf
14 Mi'kmawey Debert Cultural Centre, "Kluskap's Sacred Cave," (2014): 00:50, retrieved from: https://www.mikmaweydebert.ca/ancestors-live-here/kluskaps-mountain/kluskaps-sacred-cave/
15 M. Googoo, "Rally Held to Protect Sacred Kluscap Mountain from Mining," *Kukuwes.com* (2017), Retrieved from: http://kukukwes.com/2017/12/18/rally-held-to-protect-sacred-kluscap-mountain-from-mining/: para 19
16 Ibid.
17 Ibid., para 11–12.
18 Lewis, "Mi'kma'ki At 13,500," para 5.
19 Unama'ki Institute of Natural Resources, "Tan Teloltí'k: How We Are Doing Now. Final Project report," *Environment and Climate Change Canada* (2020): 18, retrieved from: https://www.uinr.ca/wp-content/uploads/2020/09/IPCA-Report-2020.pdf
20 Robinson, "The Role of Animals in Mi'kmaw Spirituality."
21 A. Marshall, "Albert Marshall," *Tepi'ketuek: Mi'kmaw Archives* (2013): para 4. Retrieved from: https://mikmawarchives.ca/authors/albert-marshall
22 Hilton, *Indigenomics: Taking a Seat at the Economic Table,* 49.
23 Marshall, "Albert Marshall," para 4.
24 Boulianne, *Fostering Netukulimk: The Mi'kmaq Right to Fisheries Conservation and Co-management*
25 Robinson, "The Role of Animals in Mi'kmaw Spirituality."
26 T. Sable and B. Francis, *The Language of This Land, Mi'kma'ki* (Cape Breton University Press, 2012), 105
27 Unama'ki Institute of Natural Resources, "Tan Teloltí'k: How We Are Doing Now. Final Project report," 9.
28 Ibid.
29 Boulianne, *Fostering Netukulimk: The Mi'kmaq Right to Fisheries Conservation and Co-management,* 10, 19.
30 Nova Scotia Curriculum, "Netukulimk," 1.
31 B. Leo*, "Ancestral Lands and Genders: A Queer Indigenous Critique of Settler Climate Change and Post-Apocalyptic Narratives," *Radical Philosophy Review* 26, no. 1 (2023): 22.
32 Ibid., 21.
33 Ibid., 22.
34 K. Anderson, *A Recognition of Being: Reconstructing Native Womanhood* (Sumach Press, 2016), 66
35 M. Awiakta, *Selu: Seeking the Corn-mother's Wisdom* (Fulcrum Publishing, 1993), 92
36 Anderson, *A Recognition of Being: Reconstructing Native Womanhood*
37 Ibid., 68–69
38 M. Cannon, "Revisiting Histories of Legal Assimilation, Racialized Injustice, and the Future of Indian Status in Canada," Aboriginal Policy Research Consortium International 97 (2007). Retrieved from: https://ir.lib.uwo.ca/cgi/viewcontent.cgi?article=1347&context=aprci
39 A. Alook and H. Bidder, "Why Environmental Activism Needs Indigenous Feminism. Feminist Dispatches," York Centre for Feminist Research (n.d.), retrieved from: https://www.yorku.ca/cfr/why-environmental-activism-needs-indigenous-feminism/: para 9

40 Statistics Canada, "Gender Results Framework: Data Table on the Representation of Men and Women in First Nations Band Councils and Chiefs in First Nations Communities in Canada, 2019," *The Daily* (2021), retrieved from: https://www150.statcan.gc.ca/n1/daily-quotidien/210413/dq210413f-eng.htm
41 Alook and Bidder, "Why Environmental Activism Needs Indigenous Feminism. Feminist Dispatches," para 9.
42 L. Ramstetter and F. Habersack, "Do Women Make a Difference? Analysing Environmental Attitudes and Actions of Members of the European Parliament," *Environmental Politics* (2019): 1063–1084. https://doi.org/10.1080/09644016.2019.1609156
43 Alook and Bidder, "Why Environmental Activism Needs Indigenous Feminism. Feminist Dispatches," para 12.
44 Bird and Robinson, "Two-Spirit Issues in Development."
45 L.B. Simpson, *As We Have Always Done: Indigenous Freedom through Radical Resistance* (Indigenous Americas: University of Minnesota Press, 2017), 91
46 G. Coulthard, *Red Skins, White Masks: Rejecting the Colonial Politics of Recognition* (University of Minnesota Press: Minneapolis, MN, 2014)
47 Bird and Robinson, "Two-Spirit Issues in Development."
48 Alook and Bidder, "Why Environmental Activism Needs Indigenous Feminism. Feminist Dispatches."
49 K. Kickingwoman, "Joye Braun, the Firestorm, Dies at 53," *Indian Country Today* (2022). Retrieved from: https://ictnews.org/news/joye-braun-the-firestorm-dies-at-53
50 N. Estes and J. Dhillon, eds., *Standing with Standing Rock: Voices from the# NoDAPL Movement* (University of Minnesota Press, 2019)
51 Eg. A. Condes, "Man Camps and Bad Men: Litigating Violence against American Indian Women," *Northwestern University Law Review* 116, no. 2 (2021): 515–560.
52 L.B. Simpson, "Not Murdered & Not Missing: Rebelling Against Colonial Gender Violence. Rabble.ca It Ends Here: Rebelling Against Colonial Gender Violence," (2014). Retrieved from: https://rabble.ca/feminism/itendshere-rebelling-against-colonial-gender-violence/; Simpson, *As We Have Always Done: Indigenous Freedom through Radical Resistance*
53 Royal Canadian Mounted Police, "Project Sitka: Serious Criminality Associated to Large Public Order Events with National Implications. National Intelligence Coordination Centre, National Tactical Intelligence Priority (January 2014–January 2015)," File #20131509123. Processed by CSIS under the provisions of the Privacy Act and Access to Information Act (2015). Retrieved from: https://warriorpublications.files.wordpress.com/2016/11/project-sitka-report.pdf
54 Ibid., 8.
55 J. Barrera, "Identities of Two Mi'kmaq Warriors on RCMP 'Threat' List Revealed," (2016): para 6, retrieved from: https://www.aptnnews.ca/national-news/identities-of-two-mikmaq-warriors-on-rcmp-threat-list-revealed/
56 Simpson, "Not Murdered & Not Missing: Rebelling Against Colonial Gender Violence. Rabble.ca It Ends Here: Rebelling Against Colonial Gender Violence," para 23.
57 Fault Lines, "Elsipogtog: The Fire Over Water," (2013), retrieved from: https://www.aljazeera.com/program/fault-lines/2013/12/6/elsipogtog-the-fire-over-water
58 S.E. Wilde, J.R. Hopkins, A.C. Lewis, R.E. Dunmore, G. Allen, J.R. Pitt, R.S. Ward, and R.M. Purvis, "The Air Quality Impacts of Pre-operational Hydraulic Fracturing Activities," *Science of the Total Environment* 858 (2023): 1–13. https://doi.org/10.1016/j.scitotenv.2022.159702
59 N. Apergis, G. Mustafa, and S.G. Dastidar, "An Analysis of the Impact of Unconventional Oil and Gas Activities on Public Health: New Evidence across Oklahoma Counties," *Energy Economics* 97 (2021): 105223.
60 N. Birkhimer, T.M DeSutter, K. Jore, J. Staricka, and M. Meehan, "Effects of Pipeline and Wellpad Reclamation on Topsoil Properties. A Meta-analysis," *Agrosystems, Geosciences & Environment* 6, no. 2 (2023): e20387.
61 Maliseet Grand Council, "Letter to the New Brunswick Hydraulic Fracturing Commission," (2015). Retrieved from: https://nben.ca/en/component/tags/tag/fracking.html
62 K. Kalliber, "Mi'kmaq Anti-Fracking Protest Brings Women to the Front Lines to Fight for Water," *Tulalip News* (2013), retrieved from: https://www.tulalipnews.com/wp/tag/elsipogtog-first-nation/

63 CBC News, "Elsipogtog Chief Arren Sock Says Council Is Reclaiming Unoccupied Reserve Lands," (2013), retrieved from: https://www.cbc.ca/news/canada/new-brunswick/first-nations-chief-issues-eviction-notice-to-swn-resources-1.1874870
64 S. da Silva, ""FRACK OFF!" Elsipogtog First Nation Announces Major Land Reclamation in Ongoing Anti-fracking Struggle," *Two Row Times* (2013). Retrieved from: https://tworowtimes.com/news/national/frack-elsipogtog-first-nation-announces-major-land-reclamation-ongoing-anti-fracking-struggle/
65 Fault Lines, "Elsipogtog: The Fire Over Water."
66 Kalliber, "Mi'kmaq Anti-Fracking Protest Brings Women to the Front Lines to Fight for Water."
67 Ibid., Para 2.
68 CBC News, "Ossie Michelin on His Iconic Fracking Protest Image," (2014), retrieved from: https://www.cbc.ca/news/canada/newfoundland-labrador/ossie-michelin-on-his-iconic-fracking-protest-image-1.2502891
69 Kalliber, "Mi'kmaq Anti-Fracking Protest Brings Women to the Front Lines to Fight for Water."
70 B. Forester, "Watchdog Has 'Serious Concerns' about RCMP's Response to Findings of Elsipogtog Investigation," *APTN News* (2020), retrieved from: https://www.aptnnews.ca/national-news/watchdog-has-serious-concerns-about-rcmps-response-to-findings-of-elsipogtog-investigation/
71 T. Roach, "2 Mi'kmaq Warriors Sentenced to 15 Months over Elsipogtog Fracking Fight," *APTN News* (2014), retrieved from: https://www.aptnnews.ca/national-news/2-mikmaq-warriors-sentenced-15-months-elsipogtog-fracking-fight/
72 S. Pictou, "Wolastoqiyik and Mi'kmaq Grandmothers-Land/Water Defenders Sharing and Learning Circle: Generating Knowledge for Action," (2021), retrieved from: https://www.kairoscanada.org/wp-content/uploads/2021/09/Grandmothers_Land_Defense_Report_Pictou_2021.pdf, 2
73 Kalliber, "Mi'kmaq Anti-Fracking Protest Brings Women to the Front Lines to Fight for Water."
74 Pictou, "Wolastoqiyik and Mi'kmaq Grandmothers-Land/Water Defenders Sharing and Learning Circle: Generating Knowledge for Action," 22.
75 Jean Augustine-McIsaac (Kaqtukwasisip Muin'iskw), "Mi'kmaw Daily Life-Organization," *Mi'kmaw Spirit* (2016), retrieved from: http://www.muiniskw.org/pgCulture1b.htm
76 P. Withers, "Millbrook Chief Wants Indigenous Monitors to Be Able to Shut Down Alton Gas Project," *CBC News* (2019), retrieved from: https://www.cbc.ca/news/canada/nova-scotia/alton-gas-storage-project-shubenacadie-river-bob-gloade-1.5018576
77 Leo*, "Ancestral Lands and Genders: A Queer Indigenous Critique of Settler Climate Change and Post-Apocalyptic Narratives," 22.
78 J. Simpson, "Land Back Means Protecting Black and Indigenous Trans Women," *Briarpatch Magazine* (2020). Retrieved from: https://briarpatchmagazine.com/articles/view/land-back-meansprotecting-black-and-indigenous-trans-women

Bibliography

Alook, A., and Bidder, H. "Why Environmental Activism Needs Indigenous Feminism. Feminist Dispatches." *York Centre for Feminist Research*. Retrieved from: https://www.yorku.ca/cfr/why-environmental-activism-needs-indigenous-feminism/

Anderson, K. *A Recognition of Being: Reconstructing Native Womanhood*. Sumach Press, 2016.

Apergis, N., Mustafa, G., and Dastidar, S.G. "An Analysis of the Impact of Unconventional Oil and Gas Activities on Public Health: New Evidence across Oklahoma Counties." *Energy Economics* 97 (2021): 105223.

Augustine-McIsaac (Kaqtukwasisip Muin'iskw), Jean. "Mi'kmaw Daily Life-Organization." *Mi'kmaw Spirit*, March 2016. Retrieved from: http://www.muiniskw.org/pgCulture1b.htm

Awiakta, M. *Selu: Seeking the Corn-mother's Wisdom*. Fulcrum Publishing, 1993.

Barrera, J. "Identities of Two Mi'kmaq Warriors on RCMP 'Threat' List Revealed." *Apt News*, 2016. Retrieved from: https://www.aptnnews.ca/national-news/identities-of-two-mikmaq-warriors-on-rcmp-threat-list-revealed/

Bird, N., and Robinson, M. "Two-Spirit Issues in Development." In *The Routledge Handbook of Indigenous Development*, edited by N Postero, 56–64. Routledge, 2023.

Birkhimer, N., DeSutter, T.M., Jore, K., Staricka, J., and Meehan, M. "Effects of Pipeline and Well-pad Reclamation on Topsoil Properties: A Meta-analysis." *Agrosystems, Geosciences & Environment* 6, no. 2 (2023): e20387.

Boulianne, C. *Fostering Netukulimk: The Mi'kmaq Right to Fisheries Conservation and Co-management*. McGill University (Canada), 2022. Retrieved from: https://search.proquest.com/openview/a178a3c484f0611bf791c4797348256a/1?pq-origsite=gscholar&cbl=18750&diss=y&casa_token=w725z3rF4oAAAAAA:5DKXOHSwpEXfiSYwd5dtFLykWyHu4e5t_412w4fkUB4MbUw_UdvCjxSuNR0BsQyoTUZrryDu3c9p

Cannon, M. "Revisiting Histories of Legal Assimilation, Racialized Injustice, and the Future of Indian Status in Canada." *Aboriginal Policy Research Consortium International* 97 (2007). Retrieved from: https://ir.lib.uwo.ca/cgi/viewcontent.cgi?article=1347&context=aprci

CBC News. "Elsipogtog Chief Arren Sock Says Council Is Reclaiming Unoccupied Reserve Lands." *CBC News*, 2013. Retrieved from: https://www.cbc.ca/news/canada/new-brunswick/first-nations-chief-issues-eviction-notice-to-swn-resources-1.1874870

CBC News. "Ossie Michelin on His Iconic Fracking Protest Image." *CBC News*, 2014. Retrieved from: https://www.cbc.ca/news/canada/newfoundland-labryzador/ossie-michelin-on-his-iconic-fracking-protest-image-1.2502891

Condes, A. "Man Camps and Bad Men: Litigating Violence against American Indian Women." *Northwestern University Law Review* 116, no. 2 (2021): 515–560.

Coulthard, G. *Red Skins, White Masks: Rejecting the Colonial Politics of Recognition*. University of Minnesota Press: Minneapolis, MN, 2014.

da Silva, S. ""FRACK OFF!" Elsipogtog First Nation Announces Major Land Reclamation in Ongoing Anti-fracking Struggle." *Two Row Times*, 2013. Retrieved from: https://tworowtimes.com/news/national/frack-elsipogtog-first-nation-announces-major-land-reclamation-ongoing-anti-fracking-struggle/

Estes, N., and Dhillon, J., eds. *Standing with Standing Rock: Voices from the# NoDAPL Movement*. U of Minnesota Press, 2019.

Fault Lines. "Elsipogtog: The Fire Over Water." *Al Jazeera*, 2013. Retrieved from: https://www.aljazeera.com/program/fault-lines/2013/12/6/elsipogtog-the-fire-over-water

Forester, B. "Watchdog Has 'Serious Concerns' about RCMP's Response to Findings of Elsipogtog Investigation." *APTN News*, 2022. Retrieved from: https://www.aptnnews.ca/national-news/watchdog-has-serious-concerns-about-rcmps-response-to-findings-of-elsipogtog-investigation/

Googoo, M. "Rally Held to Protect Sacred Kluscap Mountain from Mining." *Kukuwes.com*, 2017. Retrieved from: http://kukukwes.com/2017/12/18/rally-held-to-protect-sacred-kluscap-mountain-from-mining/

Hilton, C.A. *Indigenomics: Taking a Seat at the Economic Table*. New Society Publishers, 2021.

Kakekaspan, M., Walmark, B., Lemelin, R.H., Dowsley, M., and Mowbray, D. "Developing a Polar Bear Co-management Strategy in Ontario through the Indigenous Stewardship Model." *Polar Record* 49, no. 3 (2013): 230–236.

Kalliber, K. "Mi'kmaq Anti-Fracking Protest Brings Women to the Front Lines to Fight for Water." *Tulalip News*, 2013. Retrieved from: https://www.tulalipnews.com/wp/tag/elsipogtog-first-nation/

Kickingwoman, K. "Joye Braun, the Firestorm, Dies at 53." *Indian Country Today*, 2022. Retrieved from: https://ictnews.org/news/joye-braun-the-firestorm-dies-at-53

Leo*, B. "Ancestral Lands and Genders: A Queer Indigenous Critique of Settler Climate Change and Post-Apocalyptic Narratives." *Radical Philosophy Review* 26, no. 1 (2023): 21–40.

Lewis, R. "Mi'kma'ki At 13,500." *Atlantic Books #83, Features History*, 2021. Retrieved from: https://atlanticbooks.ca/stories/mikmaki-at-13500/

Maliseet Grand Council. "Letter to the New Brunswick Hydraulic Fracturing Commission." *New Brunswick Environmental Network*, 2015. Retrieved from: https://nben.ca/en/component/tags/tag/fracking.html

Marshall, A. "Albert Marshall." *Tepi'ketuek: Mi'kmaw Archives*, 2013. Retrieved from: https://mikmawarchives.ca/authors/albert-marshall

Mi'kmawey Debert Cultural Centre. "Kluskap's Sacred Cave." (2014). Retrieved from: https://www.mikmaweydebert.ca/ancestors-live-here/kluskaps-mountain/kluskaps-sacred-cave/

Nova Scotia Curriculum. "Netukulimk." (2020). Retrieved from: https://curriculum.novascotia.ca/sites/default/files/documents/resource-files/Netukulimk_ENG.pdf

Paul-Martin, P. "Animate and Inanimate Objects in Mi'kmaq." *Mi'kmaq Talking Posters, Smith/Francis Orthography. Eastern Woodland Print Communications*, 2021. Retrieved from: https://firstnationhelp.com/ali/posters/pdf/animate_inanimate.pdf

Pictou, S. "Wolastoqiyik and Mi'kmaq Grandmothers-Land/Water Defenders Sharing and Learning Circle: Generating Knowledge for Action." *Kairoscanda*, 2021. Retrieved from: https://www.kairoscanada.org/wp-content/uploads/2021/09/Grandmothers_Land_Defense_Report_Pictou_2021.pdf

Ramstetter, L., and Habersack, F. "Do Women Make a Difference? Analysing Environmental Attitudes and Actions of Members of the European Parliament." *Environmental Politics* (2019): 1063–1084. https://doi.org/10.1080/09644016.2019.1609156

Roach, T. "2 Mi'kmaq Warriors Sentenced to 15 Months over Elsipogtog Fracking Fight." *APTN News*, 2014. Retrieved from: https://www.aptnnews.ca/national-news/2-mikmaq-warriors-sentenced-15-months-elsipogtog-fracking-fight/

Robinson, M. "The Role of Animals in Mi'kmaw Spirituality." *Concilium: An International Journal of Theology* 4 (2022): 131–145. Retrieved from: https://concilium.hymnsam.co.uk/issues/20224-animals-and-theologies/

Royal Canadian Mounted Police. "Project Sitka: Serious Criminality Associated to Large Public Order Events with National Implications. National Intelligence Coordination Centre, National Tactical Intelligence Priority (January 2014–January 2015), File #20131509123. Processed by CSIS under the Provisions of the Privacy Act and Access to Information Act." *Warrior Publications*, 2015. Retrieved from: https://warriorpublications.files.wordpress.com/2016/11/project-sitka-report.pdf

Sable, T., and Francis, B. *The Language of This Land, Mi'kma'ki*. Cape Breton University Press, 2012.

Simpson, J. "Land Back Means Protecting Black and Indigenous Trans Women." *Briarpatch Magazine*, 2020. Retrieved from: https://briarpatchmagazine.com/articles/view/land-back-meansprotecting-black-and-indigenous-trans-women

Simpson, L.B. "Not Murdered & Not Missing: Rebelling Against Colonial Gender Violence. Rabble.ca It Ends Here: Rebelling Against Colonial Gender Violence." *Rabble*, 2014. Retrieved from: https://rabble.ca/feminism/itendshere-rebelling-against-colonial-gender-violence/

Simpson, L.B. *As We Have Always Done: Indigenous Freedom through Radical Resistance (Indigenous Americas)*. University of Minnesota Press, 2017.

Statistics Canada. "Gender Results Framework: Data Table on the Representation of Men and Women in First Nations Band Councils and Chiefs in First Nations communities in Canada, 2019." *The Daily*, 2021. Retrieved from: https://www150.statcan.gc.ca/n1/daily-quotidien/210413/dq210413f-eng.htm

Unama'ki Institute of Natural Resources. "Tan Telolti'k: How We Are Doing Now. Final Project Report." *Environment and Climate Change Canada*, 2020. Retrieved from: https://www.uinr.ca/wp-content/uploads/2020/09/IPCA-Report-2020.pdf

United Nations Economic Commission for Europe. "Land Administration Guideline with Special Reference to Countries in Transition." *Geneva*,(1996). Retrieved from: https://unece.org/DAM/hlm/documents/Publications/land.administration.guidelines.e.pdf

Wilde, S.E., Hopkins, J.R., Lewis, A.C., Dunmore, R.E., Allen, G., Pitt, J.R., Ward, R.S., and Purvis, R.M. "The Air Quality Impacts of Pre-operational Hydraulic Fracturing Activities." *Science of the Total Environment* 858 (2023): 1–13. https://doi.org/10.1016/j.scitotenv.2022.159702

Withers, P. "Millbrook Chief Wants Indigenous Monitors to Be Able to Shut Down Alton Gas Project." *CBC News*, 2019. Retrieved from: https://www.cbc.ca/news/canada/nova-scotia/alton-gas-storage-project-shubenacadie-river-bob-gloade-1.5018576

43

PRONOUNS IN MORE-THAN-HUMAN WORLDS AND INDIGENOUS AND COLONIAL LAWS[1]

e Campbell[2]

Phoenix

Phoenix was born in Quw'utsun territory at the now-closed Rescue and Sanctuary for Threatened Animals and now resides at Home for Hooves Sanctuary. When I met them, they were just a baby and could fit entirely in my hand. Phoenix would sit with me as I stared down my computer screen looking for grant opportunities and would delight in spreading their feet out on my warm keyboard on cold, drizzly days, pecking away at the keys under the protection of the stoop where we would do our work. One day, to my surprise, Phoenix wrote out "seed" on my computer (amongst other random characters). I was so excited, I exclaimed to another volunteer "Look at what she just typed out on my computer!" "Oh cool," they replied, "but, also, just so you know, Phoenix is a *he*!" Unbeknownst to me, who had just used she/her pronouns for Phoenix from (perhaps?)[3] out of nowhere, updates from RASTA's page had referred to Phoenix as having the combined singular pronoun of "he/she" or "she/he." RASTA's Instagram account shared,

> I'm still not sure whether Phoenix is a boy or girl (or both) and our followers seem pretty evenly divided on what the gender is so for simplicity sake I'm just going to call Phoenix a "he" for now. :/[4]

The she/him seed conversation has been etched into my brain ever since. While I had stopped volunteering in person at the Sanctuary at that time, I stammered when, on April 30, 2020, the Sanctuary posted that Phoenix had started to hang out in a nest[5] and, by May 1, 2020, the Sanctuary shared, using "they/them" pronouns,

> Phoenix, although having many of the physical traits of a male turkey, appears to be a hermaphrodite just like momma Gerry!! At first we noticed Phoenix was nesting, which was new behaviour, and then one day they laid an egg. :o[6]

DOI: 10.4324/9781003273400-53

Figure 43.1 Phoenix. e Campbell.

Of course, this news had me waving my enby flag, having come out as nonbinary myself just months earlier, using the pronouns they/them/theirs, and that summer heading into a gender-affirming surgery. But, in a broader way, this experience had me questioning my use of pronouns in my relationships with lands, waters, and the "two-legged, four-legged, winged, finned, rooted, and flowing":[7] How do pronouns function as imposed language on more-than-human worlds? Importantly, for my work as someone who engages in questions of law through analyzing Indigenous stories that, for the most part, have been translated into English, how does English, as a colonial language used within what is currently known as Canada, impose colonial genders on more-than-human worlds and facilitate oppressive legal relationships with more-than-humans? How do animals relate to sex and gender and, if so, how does that relate to their pronouns? And, finally, what are the ways to move forward in a way that advances multi-species justice?[8]

Language and Law

Language can shape thought (linguistic relativity) and our worldview (linguistic determinism).[9] In other words, different languages create different social realities and distinct worlds.[10] As bell hooks states, "Language is also a place of struggle."[11] Language has the power to maintain and sustain power relations, and failures to account for and attend to oppression based in language can reproduce violence.[12] For example, Leanne Betasamosake Simpson highlights how the colonial imposition of English and legal and political regimes that accompanied it superimposed colonial gender binaries and prevented the "gender variance encoded in our language [Nishnaabemowin]."[13] The colonial imposition of language can, and historically has also had material repercussions for the more-than-human world, including beliefs that animals do not have gender, the imposition of human genders on animals, and ignorance of the gender diverse experiences of animals.[14]

Some argue that language is solely a human tool.[15] However, members of more-than-human worlds speak in their own languages[16] and "human" languages.[17] Plants and fungi, too, have their own languages.[18] Many Indigenous epistemologies also point to how animals are not "voiceless,"[19] despite disavowal by anthropologists and others that Indigenous peoples' knowledge concerning relationships with animals is anthropomorphic[20] or is reduced to metaphor.[21] Some Indigenous legal orders hold language as being gifted or developed with or by animals and of animals understanding or using human languages to communicate.[22] For example, Thomas Highway, Cree author and playwright, notes, "our Elders like to say that there was a time eons ago when humans and animals spoke one language," highlighting how some theorists consider language to have come through and between interactions with animals.[23] Late to the party and still anchored within Aristotelian thought, Cartesian positivism, and human exceptionalism, Western thought is becoming increasingly attuned to theories around interspecies communication, affect attunement, and multispecies sociality.[24]

Human and nonhuman languages are institutions that embed law, whether written, vocalized, sung, braided, weaved, painted, danced, or otherwise. As Darcy Lindberg asserts, languages creates "building blocks for law to emerge, but also for the collective coordination of action of our relationships with lands, waters, and other non-human beings."[25] Like language, law is gendered,[26] is a "site of gender struggle,"[27] and can be anthropocentric,[28] sometimes in hidden ways.[29] Legal analyses failing to account for relations of power that inhere to gender and anthropocentrism reproduce violence[30] and reinforce colonialism.[31] While some legal scholars argue that law is solely a human tool,[32] others such as Lindberg locate law in more-than-human communities, citing treaties, agreements, and kinship relationships within and with more-than-human worlds; highlighting ways that these relationships have been upheld through obligations not to transgress animals, including through speech;[33] and problematizing how non-human voices have been removed from treaties through "[t]he imposition of Eurocentric legal reasoning."[34] Consequently, more-than-human beings can be and are participants in their social construction, have and are participants in legal orders, have their own languages, and can communicate with us, if only we develop or remember the literacies to be able to understand when it is happening.[35]

Note on Terminology

I use the terms "more-than-human worlds" and "more-than-human beings."[36] Other ways to refer to more-than-human worlds and beings in English in ways that have attempted to move beyond colonial, anthropocentric, and Cartesian ideologies have included nonhumans; beyond humans; other-than-humans;[37] all my relations;[38] and lands, waters, and the "two-legged, four-legged, winged, finned, rooted, and flowing."[39] One starting point is to consider how thought can be given to different approaches to referring to more-than-human beings and worlds: generalizing approaches (animals; more-than-humans; non-humans;

Figure 43.2 Beyond human. e Campbell.

other-than-humans); category/kind-based approaches (i.e. species); noun-centric (what a being is called, i.e. "Phoenix"); verb-based approaches (what a being is doing, i.e. "flowing"); state-based approaches (the state a being is in, i.e. rooted); mixed approaches; and so on.[40] Each approach carries with it different positives and negatives. Generalizing approaches may create hierarchies (i.e. "more than"); reinforce anthropocentrism (i.e. "animals" when referred to non-humans); or work to remediate power imbalances, speak back to anthropocentrism, or better articulate worldviews and legal conceptualizations of more-than-human beings (i.e. "more than"; "beyond"; "worlds"). They may also "other" beings (i.e. "non-"; "other-than"). Category-centric approaches can import ontologies around "inherent" or imposed qualities and associated narratives (e.g. scientific philosophies). Noun-centric approaches can bring with them the violence of naming (depending on who is doing the naming). Verb-centric and state-centric approaches appeal to what a being (or beings) are doing or have done rather than appealing to an inherent nature. While I have defaulted to more-than-human beings and more-than-human worlds in this chapter, the methodologies outlined in the end of this chapter may be better placed in determining more ethical ways to relate to discuss more-than-human worlds and beings in ways that can be language, legal order, and relationship specific.

Pronouns

Gender, anthropocentrism, colonialism, law, and language converge in the use of pronouns. The imposition of English pronouns "she," "he," and "it" has played a significant role in the erasure and imposition of genders on human and more-than-human worlds.

In English, popular singular pronouns have historically included "it," "she," "he," and "they." Neopronouns (a catch-all category for all other kinds of personal pronouns, some of which denote gender and some of which do not) are numerous and are developing rapidly.[41] Examples of neopronouns include ze/hir/hirs, xe/xem/xyr, fae/faer/faeself, noun-self pronouns and nominalization (for example, kitten/kittenself or Ellen/Ellenself), pronouns including emojis, and others.[42] Nonbinary and neopronouns have a long history in English.[43] The pronoun "they" has "been used as a nonbinary singular pronoun since 1375," and neopronouns such as "oh" and "(h)a" date back to the 1300s.[44] Importantly, however, regardless of this history or lack thereof for some pronouns, it is important to note neopronouns

> are real [pronouns] because they carry meaning and are understood by others. . . . Transgender and gender nonconforming people are the experts on our lives, and we invented these pronouns to make our gendered language more inclusive of us.[45]

Despite their history and continued importance, the use of "they" and neopronouns has received a significant amount of backlash.[46] Institutionally, the *Chicago Manual of Style* (the default legal style manual in the United States and Canada [where the McGill style guide does not apply]) historically has rejected the singular "they" in formal writing.[47] In its most recent edition, the *Chicago Manual* has recommend the use of "they" only when in reference to someone who has a "stated *preference* for a specific pronoun."[48]

Overwhelmingly, in English, the pronoun used to refer to the more-than-human beings in formal writing is "it."[49] The use of "it" has a colonial, anthropocentric, racist, ableist, and sexist history. Those rejecting the use of pronouns other than "it" to refer to more-than-human beings have argued the use of gendered pronouns is anthropomorphic[50] or discount their use as simple rhetorical devices.[51] This is despite the fact that, historically, the argument against anthropomorphism has resulted in worse, not better, understandings of more-than-human beings as a result of its commitments to biases and prejudices.[52] "It" has been and continues to be used in reference to human and more-than-human beings derogatorily as a pejorative or slur;[53] to indicate inanimateness or "thing-ness," objectify, dehumanize, or depersonify;[54] to indicate less-than-humanness;[55] to render intelligible to Western systems of thought;[56] to excuse abusive treatment;[57] to indicate lack of personal connection or detachment[58] for purposes of commodification or indication of ownership;[59] to impose a hierarchy (including, for example, religious hierarchies of humans above more-than-human beings or hierarchies between more-than-human beings);[60] or to avoid acknowledging similarity.[61] Many have criticized the use of "it" to describe more-than-human beings.[62] Joan Dunayer has argued for an interventionist approach to the use of "it" (as well as "that" and "which") in reference to more-than-human beings (to "consider inserting [sic] to mark this pronoun use as speciesist").[63] Despite this history, however, there are individuals who are reclaiming the pronoun "it."[64]

Unsurprisingly, "she" and "he" are the other two commonly used pronouns in reference to non-human nonhuman animals in English. In some cases, when referring to nonhuman animals, the use of gendered pronouns will correspond to their assigned sex. Cases in which a nonhuman's assigned sex is referenced in relation to their pronouns include when they are "owned"; when they are "domesticated"; when the speaker considers themselves to have a personal relationship with them; to indicate amount of empathy the speaker holds for them; when they are considered a "higher" animal (i.e. based on anthropocentric, cisheterosexist, or colonial criteria); when they are perceived as "active" rather than "passive"; when they are construed as having an ability to speak; when they are considered sentient; whether they are liked or perceived positively; or when they are considered useful to humans.[65] The use of gendered pronouns with regard to more-than-humans can also be done to highlight sentience, beingness, or kinship bonds within more-than-human worlds.[66]

In cases where gender identity or lack thereof is unknown, someone has not assigned sex, or it is judged as irrelevant,[67] pronouns often default to "it" (for reasons cited previously), "he," or "she." These ascriptions cannot often be understood outside of anthropocentric, colonial, or cisheterosexist ideologies.[68] Pronominalization is overwhelmingly male oriented, perhaps reminiscent of the historical use of "he" as a "nonbinary" pronoun.[69] Overwhelmingly, the pronoun "he" is used most commonly when an animal is considered aggressive, big, ugly, active, or wise and can also be used to reference specific types of nonhuman animals (such as birds).[70] The use of "he" as a generic pronoun has received heavy critique from feminists.[71] By comparison, the pronoun "she" is used most commonly when an animal is considered gentle, maternal, passive, defensive or aversive,[72] small, weak, or in need of protection.[73] The colonial[74] and cisheterosexist use of "she" as a pronoun is illustrated by the use of she in patriotic imagery of nations, countries, cities, towns, and territory more broadly (for example, controlling "her" borders or protecting the "motherland").[75] Importantly, "she" is also used to "other" beings[76] and to emphasize the view of women as closer to nature and nonhuman animals,[77] including in "mother earth" rhetoric.[78]

Gender, Genderlessness, and More-Than-Human Beings[79]

Is Gender Helpful?

One important question related to the use of pronouns in more-than-human worlds and Indigenous and colonial law is whether, and to what extent, gender can be a liberatory category for anyone, including more-than-human beings.[80] Given its situatedness within colonial discourses within Western epistemologies of the self[81] (and with it colonialism, ableism, racism, cisheterosexism, and so on), for some, gender may not be a category that has utility in truly advancing kinship relationships with more-than-human beings.[82] Of course, gender has many definitions: "causational" (i.e. based off of sex assigned at birth),[83] dualistic, performative,[84] societally constructed,[85] contextual,[86] individually mediated,[87] a site of colonial domination, and as a potential site of decolonization.[88]

Even if gender is a helpful category to some people and within some circumstances, there are questions as to the appropriateness of operationalizing it in reference to or in conversation with more-than-human beings.[89] Some examples of this would be the inherent violence of gendering another being;[90] fears of appropriating the experiences of women, and/or trans, nonbinary, agender peoples, among others;[91] and equating assigned sex with gender without problematizing sex, and processes of assigning sex, in the first place.[92] It is also informed by limitations in present-day capacities for language-sharing with more-than-human beings, interpreting the language used by more-than-humans, and issues of

Figure 43.3 I'm not gendering you. e Campbell.

consent.[93] This also opens up questions around the (mis)gendering of more-than-human worlds: What happens when we gender other beings? Can you misgender beings in more-than-human worlds (especially keeping in mind the specific lived experiences of Two-Spirit, trans, gender-nonconforming, third gender, genderqueer, agender, and other individuals who have experienced the often devastating and dysmorphic experience of being misgendered)?[94]

Operating From the Assumption That Gender Can Be Helpful

With all of the earlier being said, this paper operates from the assumption that gender can be a helpful category, particularly for some beings in particular contexts.[95] Importantly, interacting with the concept of more-than-human-beings and gender can assist the possibility of increasing gender knowledge[96] and gender literacy, and the act of "doing gender" that is negotiated in everyday interactions and relationships,[97] despite the invasive presence of colonial hegemonic cisheteromasculinity.[98]

Speaking to the prevalence of gender in more-than-human worlds, Anishinaabeg scholar Leanne Betasamosake Simpson states that the gender binary is colonial,[99] highlighting that "the land . . . provides endless examples of queerness and diverse sexualities and genders."[100] As Bruce Bagemihl points out, there are numerous examples of the diverse genders of nonhuman animals observed and documented extensively by Indigenous communities, charted through oral histories, ceremonies and rituals, experiences, poetry, festivals, dance, hunting protocols and practices, clothing, masks, performances, art, and other resources of law.[101] Western science continues to catch up with these insights (including in the study of intersex nonhuman animals, "gender-mixing" in bears, correlations between laterality and gender non-conforming beings, and so on), resulting in varying terminologies to refer to different genders in more-than-human worlds.[102]

English Pronouns Used to Describe More-Than-Human Beings in Colonial Canadian Law

My presumption going into this research would be that most style guides of various courts across Canada counsel would defer to the *Chicago Manual of Style*, as this is the default style guide to refer to if the *Canadian Guide to Uniform Legal Citation* is not dispositive of an issue.[103] The *Chicago Manual of Style* uses the term "it" to refer to nonhumans and recommends the use of the coreference "which" when referring to nonhuman animals.[104] A survey of English language decisions and statutes in Canada showed heavy reliance on the pronoun "it" (singular) and co-reference "which" (as opposed to who/whom) when referring to nonhuman animals.[105] Interestingly, however, some of these cases still use the term "mother" to describe nonhuman animals while still using the word "it" to describe them.[106]

One exception is the administrative tribunals, the British Columbia Farm Industry Review Board, which hears appeals of decisions of the British Columbia Society for the Prevention of Cruelty to Animals pursuant to section 20.3 of the *Prevention of Cruelty to Animals Act*.[107] This tribunal often uses "it" and gendered pronouns interchangeably within decisions (with respect to this point, in some cases, decisions around pronoun use appear to be informed by a mixture of reflecting the statutory language and the language of those who are giving evidence, however, other reasons such as those provided earlier could have had a hand in the pronoun mixing).[108]

Another exception to the use of "it" was in the key case *Reece v Edmonton*, the 2010 trial decision of the Court of Queen's Bench Alberta. This decision does not use gendered pronouns for Lucy the Elephant, instead using her name.[109] However, the appellate decision in the Court of Appeal of Alberta uses pronouns "she/her" in both the reasons for judgement written by the Honourable Mr. Justice Slatter (concurred with by Mr. Justice Costigan) and dissenting reasons for judgement by the Honourable Chief Justice Fraser, which also uses the co-reference "whom."[110]

The Use of English Pronouns to Describe the More-Than-Human World Within Indigenous Legal Orders

Statements of Indigenous laws in English often ascribe particular pronouns for different beings, including more-than-human beings. For the reasons described previously, statements of Indigenous law using English pronouns with regard to both humans and more-than-humans may result in distortion due to colonial lenses (both as a result of the English language itself and/or as a result of the imposition of Western ideologies and processes of colonization).[111] Gloria Anzaldúa speaks about the desexification and de-animalization of Coatlaopeuh, a serpent goddess, by the Spaniards and the Church.[112] Qwo-Li Driskill emphasizes the imposition of European genders onto Indigenous communities and the impacts that this had on anthropologists' accounting for stories or events.[113] Leanne Betasamosake Simpson highlights that, despite Two-Spirit/queer (2SQ) individuals being highly visible to anthropologists and missionaries,[114] "gendered divisions of labour" were "exaggerated by anthropologists."[115] Awanigiizhik Bruce also underscores the impact of "Eurocentric ideologies and assimilations/acculturations" on Anishinaabemowin, highlighting instead numerous terms that describe gender-diverse individuals in Anishinaabemowin,[116] pronouns that are not reflected in English translations of oral histories.

Various Indigenous scholars and others contend with the embeddedness of western gender binaries within Indigenous legal resources impacted by colonizing influences through augmented understandings of the use of pronouns. Alex Wilson of Opaskwayak Cree Nation shares,

> We call the moon grandmother and the earth mother in English but in Cree that isn't the case. What is important is the relational aspect acknowledging some kind of kinship. In Cree, the land (aski) is not gendered. . . . Same for water. It's not gendered but it has a spirit of life and it's fluid.[117]

In the context of nêhiyaw law, Leona Makokis, Kristina Kopp, Ralph Bodor, Ariel Veldhuisen, and Amanda Torres describe the limitations of gendered language in English:

> [A] fundamental difference between the Western world and the *nêhiyaw* world is that the *nêhiyaw* language does not have gender-based pronouns. Rather, *nêhiyaw* worldviews organize connection in terms of animate or inanimate, a difference that has fundamental implications for relationship roles and responsibilities. Position and connection are not based within or differentiated primarily by gender, rather it is the role that is provided in connection to the child that prioritizes the relationship. In other words, animate beings that create and provide love and caring for small traveling spirits are not distinguished so much by gender but more by role. English

> gender-based terms, such as aunt or uncle, would not be relevant or appropriate in this context. . . . The actual *nêhiyaw* kinship terms include the connection of that specific relationship role to the child, family, and community and gender is indicated within the description of that relationship. To emphasize, *nêhiyaw* relationship terms convey a job description and the focus is on the role, relationship, and connection–the action and the verb. . . . In addition, something that may be considered inanimate in a Western world may be understood as being animate in an Indigenous world.[118]

As other examples, the Ktunaxa *Qat'muk Declaration*, which concerns itself with Kławła Tukłułakʔis, the Grizzly Bear Spirit, uses the terms "it" and "itself,"[119] and the *Te Awa Tupua Act* uses "it" to describe the Whanganui river[120] (although some have brought up the possibility of assigning a different, perhaps gendered, pronoun to the Whanganui river)[121] in ways that may seem at odds with Ktunaxa and Māori views of these beings, especially when interpreted from a colonial legal imaginaries (though not, perhaps, if looked at from a critical discourse or worldedness approach, as discussed in the following sections). Consequently, there are overarching difficulties of translation and the corresponding conceptualizations of the pronouns "it," "he," and "she" in relation to Indigenous languages and legal orders. This is not to say there are not ways to attend to this difficulty, but just that it continues to persist and invite conversation. As one example of this, Leanne Betasamosake Simpson responds to this difficulty through varying pronoun usage, sharing that, for example, for one Michi Saagiig Nishnaabeg story: "I have chosen to gender the main character as a girl because I identify as a woman, but the story can be and should be told using all genders."[122]

Options

a. Prescriptive Models and Political Correctness Approaches: "She"/"He"/"They"

Prescriptive models and political correctness approaches have numerous shortcomings. For example, People for the Ethical Treatment of Animals (PETA) and the style guide developed by Debra Merskin have argued for prescriptive models to grammar: the idea that nonhuman animals should be understood as he or she when assigned sex or "gender" are known and "they" when they are not known.[123] Limiting pronouns to these three may (at least in their definitions of these terms heretofore) fall prey to biological determinist beliefs of sex and gender and the myth of biological sex.[124] This approach is situated within a cisheterosexist and ableist reading of sex and gender, failing to recognize the diverse ways that more-than-humans interact with their worlds, including through and beyond gender. This approach is also unresponsive to the inherent issues in any prescriptive approach to grammar.[125] Importantly, in addition, these approaches fail to reconcile the colonial underpinnings of pronoun use within the English language and the recognition of the operation of this language on stolen lands.

b. Defaulting to "They"

There is also an option to using, or even defaulting, to the pronoun "they" for more-than-human beings, drawing on a multiplicity of languages that do not encode gender into their personal pronouns.[126] While this option does not carry with it the history of "thing-ification," as does the pronoun "it," the approach may, in some cases, result in a missed opportunity to engage in more complex conversations around gender and more-than-human worlds, and, in doing so, miss out on the creation of richer relationships.

c. Nominalization and Noun-Self Pronouns

Nominalization and name-self or species-self (e.g. "Bellaself") pronouns is another option for describing more-than-human beings. Like the pronoun "they," nominalization and noun-self pronouns already have popular use within 2SLGBTTQQIA+[127] communities. Some difficulties arise with the use of noun-self pronouns and nominalization with both human and more-than-human beings. One, for example, is the process of naming, which for some beings is experienced as a violent process.[128] In the case of species-self language ("kittenself"), these may be understood within essentialized species positions that bring with them certain conceptualizations of who a species is and that may not inhere to certain legal orders.[129] Further, nominalization and noun-self pronouns may also fail to indicate gender where it would be helpful. Nominalization and noun-self pronouns could also be creatively expanded to interrupt English linguistic imperialism and reflect or encode inter-societal relations with more-than-human beings, for example, through using Indigenous placenames and names of other beings when combined with worldedness and critical discourse analysis methods described in the following.[130]

Critical Discourse Analysis

Critical discourse analysis (CDA) provides "an account of the role of language, language use, discourse or communicative events in the (re)production of dominance and inequality."[131] When looking at linguistic features, such as pronouns, CDA can "reveal hidden ideological assumptions on which discourse is based,"[132] particularly when those hidden ideological assumptions are based in "common sense" ideologies.[133] As Stibbe articulates,

> A discourse approach . . . [in comparison to a political correctness approach] would recognize that it is a particular *combination* of features that create models of the world. So the use of the pronoun *it* in industry texts may be part of a discourse that objectifies animals, but the same pronoun could be used as part of a quite different discourse of empathy and respectful distance.[134]

CDA facilitates processes of making clear when discourses become what Stibbe terms "*destructive discourses*": "discourses that potentially construct inhumane and ecologically damaging relationships between humans and animals."[135] As Carl Mika points out, however, one of the significant limitations of CDA: "CDA . . . is designed to disclose power relations. Yet, if it relies on a notion of language and concept of things that is unholistic, it will perform the same function as more obviously orthodox methods."[136]

Example: Reclaiming "It"

As Stibbe highlights, there are examples of the use of the pronoun "it" in law within a context that, from a critical discourse perspective, do not seem to deny the more-than-human world dignity and that may require caution around taking into account the broader context of any use of the pronoun's ("it's") before labelling particular wording as speciesist or deciding to intervene in someone's words to add, for example, a "[sic]," as Dunayer suggests. One example includes the work individuals are doing to reclaim "it" as a pronoun for themselves to speak back to depersonifying, colonialism, gender, and ableism.[137] Other examples are the *Qat'muk Declaration* and the *Te Awa Tupua Act* (discussed previously) when read from the perspective of Ktunaxa and Māori legal imaginaries. However, that has to be balanced with the relations of power inherent in the sustained history of the term (discussed previously) and may miss opportunities around engaging with gender in relationship with the more-than-human world, better representing Indigenous relationships with these beings, and interrupting English language imperialism.

Example: Interchangeable Pronoun Use and Pronoun Mixing

One example of what I would see as falling under CDA analysis that has been used by Indigenous and 2SLGBTTQQIA+ communities in dealing with the bounds of colonial gender norms has been to use pronouns interchangeably. This is a method used by numerous Elders across Indigenous communities,[138] including Hän elders, as Alexandra Winton highlights,[139] "There are no gender pronouns in the Hän language, so often, Tr'ondëk Hwëch'in elders use 'he' and 'she' interchangeably" when speaking in English."[140] Solomon Ratt also shares that Cree speakers also interchange pronoun use: "English has the personal pronouns 'he' and 'she' for the third person, but we only have the term 'wiya' for these, so it is not surprising to hear a Cree speaker using 'she' when talking about a male in English!"[141] I find this method a very persuasive way to interrupt English language imperialism and to take into account the presence of the fluidity of gender and other values such as proximity, kinship, and animacy, noted previously. Pronoun mixing and switching have also been personally helpful to me when I am introducing myself in Indigenous languages, often as a stand-in until I have a chance to have deeper conversations about how best to introduce myself as a nonbinary person within that language. This method is also perhaps helpful when engaging in intersocietal discussions around law, providing for the representation of multiple relationships and genderings within the more-than-human world, particularly when combined with worldedness methods, defined later.

Worldedness

Worldedness provides a further option for engaging with pronouns and more-than-human worlds. Worldedness is "a metaphysics and term analysis (method)"[142] that, briefly described, "explain[s] the fact of one thing's constitution by all others . . . , and it reflects the belief that all things are interconnected or, indeed, one."[143] Carl Mika draws on Arola to argue that this method is part of a drive for a "holistic ontology is beyond simply epistemic"; it is an "ethical responsibility."[144] Mika explains from a Māori worldedness perspective,

> If a term does nurture a deep view of all things being "worlded" or co-constitutional, then it may be said to sit well with that most fundamental Māori metaphysics and may then indeed simply *sit well*, not doing harm to the region it is located in or to the representing self. If, however, the term reveals a landscape of strict division and strong individual presence, then it is opposed to that Māori metaphysics, will

> therefore threaten a Māori ability to represent things in accordance with those initial principles, and will be "unhealthy" in a Māori sense.

Mika emphasizes worldedness's ability as a method to be able to recognize, for example, ideas as "evolving, ancestral entitles that need to be ethically represented in their own right as ancestral."[145]

This approach is perhaps most consistent with the cultivation of expressions and understandings of legal relations with more-than-human worlds[146] as based in webs of interrelations informed by more-than-human beings (seeing them as both language-holders and participants in legal orders), Indigenous peoples,[147] (nation-specific) queer communities and identities,[148] and place (particularly when on Indigenous lands and stolen lands). Such an approach would need to be informed by Indigenous communities' understandings and gender literacies respecting more-than-human worlds,[149] especially those informed by Two-Spirit, trans, agender, and queer members of those communities. Consequently, taking a step forward in this area means being in relationship with linguistic methods for revitalizing Indigenous laws and languages.[150] Methods and methodologies used in this sense could include those already used for interpreting language and law, including Two-Spirit research methodologies;[151] meta-principle methods, grammar as worldview method, word part methods, word clusters methods, and placenames methods;[152] sound-based methods (where people work with the meanings of each of the sounds of and, in doing so, craft pronouns, for humans and more-than-humans based on individualized relationships).[153] This approach, of course, would necessitate taking a deeper look at land-centered[154] (or beyond-human-world-centered) and gender literacies on personal societal, and intersocietal levels. It may also require taking hard looks at the ways that gender and anthropocentrism operate and pervade various spaces. Importantly, it also means not using a pan-Indigenous approach or romanticizing Indigenous languages and laws as panacea,[155] necessitating deep and rigorous engagement with and alongside Indigenous communities, likely in nation-, community-, and family-specific understandings of more-than-human beings.

Example: Ki/Kin

One example of a worldedness-based approach could be the pronoun "ki/kin." Robin Wall Kimmerer, botanist and member of the Citizen Potawatomi nation, argues for Potawatomi, an Anishinaabe language,[156] to inform a new pronoun in English of "ki" ("kin" plural) that could interrupt the linguistic imperialism and de-animism inherent to English based on the "Anishinaabe word for beings of the living Earth . . . Bemaadiziiaaki . . . to signify a being of the living Earth."[157]

Example: Nêhiyaw Law Informing Pronoun Use

Worldedness-based approaches based in the Nêhiyaw worldview may consider how Nêhiyaw wâhkôtowin, or the "relatedness of all things," which is "inseparably tied" to the "agency of non-human beings and things,"[158] may relate to more-than-human worlds, genders, and pronoun use, specifically how concepts of ohcînemowin (not to transgress more-than-human worlds through speech) and the importance of the precision of language in reference to legal relations[159] might relate pronoun use with more-than-humans in discussions of more-than-human beings. Given that nêhiyawêwin (Plains Cree–Y dialect) distinguishes from animate and inanimate objects and proximity, rather than using gendered pronouns[160] based on understandings of animacy and inspiritedness within the Nêhiyaw worldview,[161] it is possible that discussions of pronouns and words used to describe more-than-human beings and worlds in English, particularly when used within Nêhiyaw legal contexts, could shift to reflect concepts including animacy, inspiritedness, and ohcinêmowin in generative ways.

Conclusion[162]

As a result of their visibility within English language and critical and important shifts in mainstream culture, pronouns may receive disproportionate focus in comparison to other ways of relating to each other through language and otherwise. That being said, they do provide a helpful starting point. Relating to each other ethically when it comes to gender, genderlessness, and animacy is not just about pronouns; it extends past linguistic questions (concerning pronominalization, nominalization, adverb and adjective use, referents, and so on) into questions of social interaction and relation as part of a broader practice in engaging in multispecies justice. My hope is that this discussion provides avenues for fostering relationships with and between human and more-than-human worlds. I have attempted to open space for thinking about how English creates obstacles for and with the beyond human world and the importance of looking to other frameworks to provide that space. While I don't have the answers, I believe this path is necessary both as an ethical responsibility with respect to listening to the more-than-human beings we engage with, like Phoenix, on what it means to relate to them appropriately, specifically when we do so on Indigenous territories and on stolen lands.

Notes

1 I am extremely grateful that the thinking in this chapter was developed with the assistance of Darren Chang, Darcy Lindberg, Tara Williamson, Jessica Asch, and, of course, Phoenix. For transparency, I have made notes where they have provided assistance or thoughts. I am also very thankful to Chloë Taylor for her continued support in pursuing this topic and finalizing this chapter. The thinking in this chapter is also very much thanks to seeds planted by Jodey Castricano and Dan Irving. Content warnings: This chapter discusses the imposition of colonial genders on Indigenous peoples. This chapter also includes discussions of transantagonistic policies and protests. Also, please note, the comics in this article are the intellectual property of e Campbell. For permission to share or reproduce these images, please contact e at campbell.ellen@outlook.com.

2 My name is e/Ellen Campbell (they/them/theirs). I am a settler, nonbinary, and neurospicy artist, lawyer, and researcher living, sketching, and working as an uninvited guest on MÁLEXEŁ and Quw'utsun territories. I have Scottish, English, Norse, French, and German ancestry, and I grew up in Michi Saagiig Nishnaabeg territory in Keswick and Mississauga (Ontario) and syilx territory in Kelowna (British Columbia).

3 My use of she/her pronouns with Phoenix may be because of unconscious associations of those infantile and nonhuman with the pronouns "she." See Marina Teterina, "The Use of Gendered Pronouns with Animal Referents in English," *Limbaj şi Context* 4, no. 2 (2012): 1.

4 Rescue and Sanctuary for Threatened Animals, Instagram Post, August 9, 2019.

5 Rescue and Sanctuary for Threatened Animals, Instagram Post, April 30, 2020.

6 Rescue and Sanctuary for Threatened Animals, Instagram Post, May 1, 2020.

7 Kathleen M.K. Menke, *Spirits of the North—Two-legged, Four-legged, Winged, Finned, Rooted, and Flowing* (Crystal Images, 2010). See the section "Note on Terminology" concerning commentary on how more-than-human beings and worlds are referred to in this Chapter.

8 Conversation with Darren Chang 27 October 2022. On "multispecies justice," see Donna Haraway, *When Species Meet* (Minneapolis: University of Minnesota Press, 2008). It has been developed further in Eben Kirksey, Sophie Chao, Karin Bolender, eds, *The Promise of Multispecies Justice* (Duke U P, 2022).
9 Edward Sapir, "The Status of Linguistics as a Science," *Language* 4, no. 4 (1929): 207 at 209.
10 Ibid.
11 bell hooks, *Talking Back: Thinking Feminist, Thinking Black* (Boston: South End Press, 1999) at 28.
12 Norman Fairclough, *Language and Power* (London: Longman, 1989) at 84, online (pdf), retrieved from: *ResearchGate* <researchgate.net/publication/49551220_Language_and_Power>.
13 Leanne Betasamosake Simpson, *As We Have Always Done: Indigenous Freedom through Radical Resistance* (University of Minnesota Press), 127.
14 Ibid., 118. Barbara Noske, *Beyond Boundaries: Humans and Animals* (Montréal: Black Rose Books, 1997), 80 at 109–110.
15 René Descartes, "Animals are Machines," in *Environmental Ethics: Divergence and Convergence*, eds. S.J. Armstorn and R.G. Botzler (New York: McGraw-Hill, 1993), 281 at 281–282. On the uptake of this view within contemporary science, see Donald J. Barnes, "A Matter of Change," in *In Defense of Animals*, ed. Peter Singer (New York: Basil Blackwell, 1985), 157. For further discussion, see Jodey Castricano, "Introduction: Animal Subjects in a Posthuman World," in *Animal Subjects*, ed. Jodey Castricano (Wilfred Laurier University Press, 2008) at 9.
16 For examples of historicity and communication between species, see Barbara Noske's discussion of elephant and baboon memory and cultural transmission within and between species Barbara Noske, supra n 14 at 111–112. Also see Ulla Hedeager, "Is Language Unique to the Human Species," at 10, online (pdf), retrieved from: columbia.edu/~rmk7/HC/HC_Readings/AnimalComm.pdf.
17 National Geographic, "Watch Koko the Gorilla Use Sign Language in this 1981 Film" (June 22, 2018) at 06m:00s, retrieved from: youtube.com/watch?v=FqJf1mB5PjQ.
18 Will Sullivan, "Plants Make Noises When Stressed, Study Finds," *Smithsonian Magazine* (April 4, 2023), retrieved from: smithsonianmag.com/smart-news/plants-make-noises-when-stressed-study-finds-180981920/#:~:text=Stressed%20plants%20that%20have%20been,week%20in%20the%20journal%20Cell.; Suzanne Simard, "The foundational role of mycorrhizal networks in self-organization of interior Douglas-fir forests," retrieved from: sciencedirect.com/science/article/pii/S0378112709003351; Peter Wohlleben, *The Hidden Life of Trees* (Musqueam, Squamish, Tsleil-Waututh Lands: Greystone Books Ltd, 2016).
19 American Society for the Prevention of Cruelty to Animals, "A Voice for the Voiceless," retrieved from: secure.aspca.org/team/a-voice-for-the-voiceless, accessed April 25, 2023.
20 Lindsay Keegitah Borrows, "The Land Is Our Casebook" (University of Alberta LLM thesis, 2021): 42 at page 19 of Chapter 1 [unpublished].
21 Paul Nadasdy, "The Gift in the Animal: The Ontology of Hunting and Human–animal Sociality," *American Ethnologist* 34, no. 1 (2007): 25 at 26, online (pdf), retrieved from: anthropology.cornell.edu/sites/anthro/files/Nadasdy%202007%20Gift%20in%20the%20Animal.pdf.
22 Ibid at 34, 36. Also see Hadley Friedland, "The Wetiko (Windigo) Legal Principles: Responding to Harmful People in Cree, Anishinabek and Saulteaux Societies—Past, Present and Future Uses, with a Focus on Contemporary Violence and Child Victimization Concerns" (LLM thesis with the University of Alberta, 2009): 60 [unpublished].
23 Thomas Highway, *Laughing with the Trickster: On Sex, Death, and Accordions* (Toronto: House of Anansi Press Inc, 2022) at 7–8.
24 Cynthia Willett and Malini Suchak, "Sociality," in *Critical Terms for Animal Studies*, ed. Lori Gruen (Chicago: University of Chicago Press, 2018), 377. Victoria Gill, "Human and Wild Apes Share Common Language," *BBC News* (January 25, 2023), retrieved from: bbc.com/news/science-environment-64387401?fbclid=PAAaaC-0bL4B3t5n_3PvpaMuwK3ic9p-kTtO7xtRyL1x-cn_MosrhNKT9oHtU.
25 Darcy Lindberg, "Nêhiyaw Âskiy Wiyasiwêwina: Plains Cree Earth and Constitutional/Ecological Reconciliation" (Dissertation, Doctor of Philosophy (Faculty of Law), University of Victoria, 2020): 35, 122, 128, 204.

26 Carol Smart, *Law, Crime and Sexuality: Essays in Feminism* (London: Sage Publications Ltd, 1995); Emily Jane Snyder, "Representations of Women in Cree Legal Educational Materials: An Indigenous Feminist Legal Theoretical Analysis" (Doctor of Philosophy, Department of Sociology, University of Alberta, 2014): 15 [unpublished]; Darcy Lindberg, "kihcitwâw kîkway meskocipayi-win (Sacred Changes): Transforming Gendered Protocols in Cree Ceremonies through Cree Law" (Master of Laws thesis, Faculty of Law, University of Victoria, 2017): 19 [unpublished].

27 Carol Smart as paraphrased in Emily Snyder, Val Napoleon, and John Borrows, "Gender and Violence: Drawing on Indigenous Legal Resources," *UBC Law Review* 48, no. 2 (2015): 594 at 607; Darcy Lindberg, supra n 27 at 19 [unpublished].

28 Maneesha Deckha, "Unsettling Anthropocentric Legal Systems: Reconciliation, Indigenous Laws, and Animal Personhood," *Journal of Intercultural Studies* 41, no. 1 (2020): 77 at 78.

29 Carolyn Korsmeyer, *Gender and Aesthetics: An Introduction* (New York: Routledge, 2004), 3. Emily Snyder, "Representations of Women in Cree Legal Education Materials: An Indigenous Feminist Legal Theoretical Analysis" (DPhil, University of Alberta, 2014): 13, retrieved from: era.library.ualberta.ca/items/15997cd2–0909–4b8c-ad37-c6e1e9f513a0/view/201d1dbc-fb9b-4660–895d-443f26213059/Final-20Dissertation-20-20Emily-20Snyder.pdf.

30 See generally Emily Snyder, "Indigenous Feminist Legal Theory," *Canadian Journal of Women and the Law* 26, no. 2 (2014): 365; Darcy Lindberg, supra n 27 at iii.

31 Darcy Lindberg, supra n 27 at 29, 100.

32 See John Borrows, *Law's Indigenous Ethics* (Toronto: University of Toronto Press, 2019) at 22.

33 Darcy Lindberg, supra n 26 at x, 35, 161, 174.

34 Darcy Lindberg, "Transforming Buffalo: Plains Cree Constitutionalism and Food Sovereignty," in *Food Law and Policy in Canada* (No Place: Carswell, 2019), 37 at 48 [*Transforming Buffalo*]. Darcy Lindberg, "Excerpts from Nêhiyaw Âskiy Wiyasiwêwina: Plains Cree EarthLaw and Constitutional/Ecological Reconciliation," indigenous-law-association-at-mcgill.com/2021/06/03/excerpts-from-nehiyaw-askiy-wiyasiwewina-plains-cree-earthlaw-and-constitutional-ecological-reconciliation/, accessed October 30, 2022.

35 Victoria Gill, "Human and Wild Apes Share Common Language," *BBC News* (January 25, 2023), retrieved from: bbc.com/news/science-environment-64387401?fbclid=PAAaaC-0bL4B3t5n_3PvpaMuwK3ic9p-kTtO7xtRyL1x-cn_MosrhNKT9oHtU.

36 David Abram, *The Spell of the Sensuous: Perception and Language in a More-Than-Human World* (New York: Pantheon Books, 1996); Mark Dowie, "The Myth of a Wilderness Without Humans," *MIT Press*, online: <thereader.mitpress.mit.edu/the-myth-of-a-wilderness-without-humans/>.

37 Matthew Wildcat, "Wahkohtowin *in Action,*" retrieved from: ualberta.ca/wahkohtowin/media-library/data-lists-pdfs/wahkotowin-in-action.pdf; Colin Scott, "Ontology and Ethics in Cree Hunting: Animism, Totemism and Practical Knowledge," in *The Handbook of Contemporary Animism*, ed. Graham Harvey, at 159.

38 Chief Wayne Roan and Earle Waugh, "Meanings of Sacred Pipe" (2004), online: *Nature's Laws,* retrieved from: wayback.archive-it.org/2217/20101208172609/http://www.albertasource.ca/natureslaws/traditions/ritual_meanings_pipe.html [Webpage Archive Captured December 8, 2010].

39 Kathleen M.K. Menke, supra n 7.

40 For lack of better terminology, I have used the aforementioned categories to describe each of these types of terms as placeholders. There are likely better ways to describe each of these terms. There are also likely far more numerous ways to conceptualize different kinds of terms used. Further, distinctions between each of these categories are not necessarily set in stone and are somewhat artificially imposed.

41 Ezra Marcus, "A Guide to Neopronouns," *The New York Times* (April 8, 2021), retrieved from: nytimes.com/2021/04/08/style/neopronouns-nonbinary-explainer.html.

42 Pronoun Wiki, "Pronouns," retrieved from: pronoun.fandom.com/wiki/Category:Pronouns, accessed October 29, 2022. Also see Pronouny, "Pronouny," retrieved from: pronouny.xyz, accessed October 30, 2022.

43 For further history, see Dennis Baron, *What's Your Pronoun?* (New York: Liveright Publishing Corporation, 2020).

44 Dennis Baron, "A Brief History of Singular 'They'," Oxford English Dictionary (September 4, 2018), retrieved from: public.oed.com/blog/a-brief-history-of-singular-they/> [*A brief history*]; Carrd,

"What Are Neopronouns," retrieved from: neopronounss.carrd.co/#what, accessed October 30, 2022. Also see Michea B, "Neopronouns, They're Way Older Than You Think" (January 27, 2020), retrieved from: *Medium* themicheab.medium.com/neopronouns-theyre-way-older-than-you-think-ae0ec7030ac0; Dennis Baron, "Nonbinary Pronouns Are Older Than You Think" (October 13, 2018), retrieved from: *The Web of Language* <blogs.illinois.edu/view/25/705317>.

45 My Kid Is Gay, "Defining: Neopronouns" (2019), retrieved from: mykidisgay.com/blog/defining-neopronouns, accessed October 30, 2022.

46 Courtney Dickson and Andrew Kurjata, "Arrests Made as Thousands Attend Rallies for and against Teaching Gender Diversity in BC Schools" (September 19, 2023), retrieved from: cbc.ca/news/canada/british-columbia/sogi-marches-bc-1.6972195.

47 *Chicago Manual of Style* (16th ed.) s 5.46.

48 *Chicago Manual of Style* (17th ed.) s 5.48. Emphasis added to highlight what can be trans antagonistic beliefs around "preferences" for pronouns.

49 Marina Teterina, supra n 3 at 1.

50 For discussion of anthropomorphism and law, and the shortcomings of arguments against anthropomorphism, see Lindsay Borrows, supra n 21. On anthropomorphism more broadly, see Barbara Noske, supra n 14 at 89.

51 Marina Teterina and Elena Crestianicov, "Personification or Sexification of Countries in English?" (4:2) at 6–7, online (pdf), retrieved from: usarb.md/limbaj_context/volumes/v7/art/teterina%20crestianicov.pdf.

52 Barbara Noske, supra n 14 at 90; Jodey Castricano, supra n 15 at 9.

53 "It" has been used as a slur against trans people, gender non-conforming people, Black people, Indigenous people, People of Colour, and disabled people. See Susan Stryker, "My Words to Victor Frankenstein Above the Village of Chamounix: Performing Transgender Rage," in *The Transgender Studies Reader*, eds. Susan Stryker and Stephen Whittle (New York: Routledge, 2006) 244 at 246; Christopher Rufo, "Pronouns Unbound," *City Journal* (September 29, 2022), retrieved from: city-journal.org/san-francisco-public-schools-subvert-parental-rights.

54 Arran Stibbe, supra n 16 at 152. Also see Sunaura Taylor, *Beasts of Burden, Animal and Disability Liberation* (New York: The New Press, 2017) at 59–60.

55 Marina Teterina and Elena Crestianicov, supra n 51 at 7. Debra Merskin, "*She, He,* Not *It*: Language, Personal Pronouns, and Animal Advocacy," *Journal of World Languages* 8, no. 2 (2022): 391 at 397. Also see Sunaura Taylor, supra n 54 at 59–60.

56 Susan Bordo, "The Cartesian Masculinization of Thought," *Journal of Women in Culture and Society* 11, no. 3 (1986): 439 at 452.

57 Susan J Armstrong and Richard Boltzer, "Animal Ethics: A Sketch of How it Developed and Where It Is Now," in Richard Boltzer and Susan J Armstrong, eds, *The Animal Ethics Reader* (New York: Routledge, 2003) 1 at 3; Arran Stibbe, supra n 16 at 153.

58 Marina Teterina, supra n 3 at 1.

59 Arran Stibbe, *Animals Erased: Discourse, Ecology, and Reconnection with the Natural World* (2012, Wesleyan U P) at 5; Marina Teterina, supra n 3 at 4; Arran Stibbe, supra n 16 at 152, 153; Debra Merskin, "*She, he,* not *it*: Language, Personal Pronouns, and Animal Advocacy," *Journal of World Languages* 8, no. 2 (2022): 391 at 396.

60 Marina Teterina, supra n 3 at 2; Arran Stibbe, supra n 16 at 152.

61 Carol J. Adams, *The Sexual Politics of Meat: A Feminist-Vegetarian Critical Theory* (New York: Bloomsbury Academic, 2010) at 93. Arran Stibbe, supra n 16 at 151.

62 Robin Wall Kimmerer, *Nature Needs a New Pronoun*, cited previously. Also see the Animal Studies journal submission guidelines: University of Wollongong Australia, "Animal Studies Journal: Policies," retrieved from: ro.uow.edu.au/asj/policies.html, accessed October 30, 2022.

63 Joan Dunayer, "On Speciesist Language," *On the Issues* (Winter 1990), online: <https://ontheissuesmagazine.com/earth/on-speciesist-language/>; Also see Debra Merskin, "*She, he,* not *it*: Language, Personal Pronouns, and Animal Advocacy," *Journal of World Languages* 8, no. 3 (2022): 391 at 405.

64 Conversation with Darren Chang 22 October 2022. Also see Christopher Rufo, supra n 53.

65 See Marina Teterina, supra n 3.

66 See Maneesha Deckha, "Veganism, Dairy, and Decolonization," *Journal of Human Rights and the Environment* 11, no. 2 (2020): 244.

67 Carol J. Adams, supra n 61 at 93; Marina Teterina, supra n 3 at 7.
68 Marina Teterina, supra n 3 at 2. In some cases, people may carry over genders ascribed from other languages to describe animals. See Anke Frohlich's use of "she" with Flaco the owl here: Naaman Zhou, "Evenings in the Park with Flaco" (February 22, 2023) *The New Yorker*, online: <newyorker.com/news/our-local-correspondents/evenings-in-the-park-with-flaco>.
69 Carol J. Adams, supra n 61 at 93; Marina Teterina and Elena Crestianicov, supra n 51 at 2.
70 Barbara Noske, supra n 14 at 109. Marina Teterina, supra n 3 at 2.
71 Carol J. Adams, supra n 61 at 93.
72 Barbara Noske, supra n 14 at 109.
73 Gloria Anzaldúa, *Borderland/La Frontera: The New Mestiza* (San Fransisco: Aunt Lute Books, 1987) at 17.
74 Marina Teterina and Elena Crestianicov, supra n 51 at 4.
75 Ibid.
76 Susan Bordo, supra n 56 at 454.
77 Gloria Anzaldúa, supra n 73.
78 For example, Ellen Cronan Rose, "The Good Mother: From Gaia to Gilead," *Frontiers: A Journal of Women Studies* 12, no. 1 (1991): 91; Carolyn Merchant, *The Death of Nature: Women, Ecology, and the Scientific Revolution,* 2nd ed. (New York: Harper One, 1990) at xv and 4.
79 This is a comic I made based on an Instagram reel where Carrington Wentz is shooing away a turkey, who is following Carrington down the road. For the original reel, see Carrington Wentz (November 18, 2022), online: *Instagram,* retrieved from: instagram.com/p/ClGxnOasvDR/?utm_source=ig_embed&utm_campaign=embed_video_watch_again.
80 Conversation with Darren Chang 22 October 2022. For example, see Gloria Anzaldúa, supra n 73 at 26: "I know things older than Freud, older than gender." Also see summary of key work summarized in Letitia Meynell and Andrew Lopez, "Gendering Animals," *Synthese* 199 (2021): 4287–4311 at 4290.
81 Gloria Anzaldúa as paraphrased by Micha Cárdenas in Boellstorff et al, "Decolonizing Transgender: A Roundtable Discussion," *Transgender Studies Quarterly* 1, no. 3 (2014): 434 at 435, online (pdf): <ericastanleydotnet.files.wordpress.com/2012/05/tsq-2014-boellstorff-419-39.pdf>.
82 See for example, Billy-Ray Belcourt, "The Cree Word for a Body Like Mine is Weesageechak," in *This Wound Is a World: Poems* (Calgary: Frontenac House Ltd, 2017), 9 at 9.
83 Judith Butler, *Gender Trouble* at 46–54, cited previously.
84 Ibid.
85 Ibid.
86 Ibid at 12, 15.
87 Ibid at 46–54.
88 María Lugones, "Toward a Decolonial Feminism," *Hypatia* 25, no. 4 (2010): 743.
89 See summaries of these conversations in Letitia Meynell and Andrew Lopez, supra n 81.
90 Susan Stryker, "My Words to Victor Frankenstein Above the Village of Chamounix: Performing Transgender Rage," in *The Transgender Studies Reader*, eds. Susan Stryker and Stephen Whittle (New York: Routledge, 2006), 244 at 253.
91 See Donna Haraway as paraphrased in Barbara Noske, supra n 14 at 92, 110.
92 Letitia Meynell and Andrew Lopez, supra n 81.
93 See Darcy Lindberg, supra n 26 at 160, 180–183.
94 On misgendering, see A.K. O'Loughlin, "Gender-as-lived: The Coloniality of Gender in Schools as a Queer Teacher Listens in to Complicated Moments of Resistance," *Indo-Pacific Journal of Phenomenology* 19, no. 1 (2019): 49 at 4. Regarding animals and gender, see Letitia Meynell and Andrew Lopez, supra n 81 at 4303.
95 Kiera Ladner, "Gendering Decolonization, Decolonizing Gender," Paper Presented at the 80th Annual Conference of the Canadian Political Science Association UBC June 2008 (2008), retrieved from: https://cpsa-acsp.ca/papers-2008/Ladner.pdf, at 6; Also see Darcy Lindberg, supra n 27 at 9. Regarding Carl Mika, see Carl Mika, "A Māori Worlded Speculation on Terms 'Sex' and 'Gender'," in *Honouring Our Ancestors*, eds. Alison Green and Leonie Pihama (Wellington: Te Herenga Waka University Press, 2023) at e-book location 891 at n 2. Also see Letitia Meynell and Andrew Lopez, supra n 81 at 4303.

96 Marecek, Crawford, and Popp, "On the Construction of Gender, Sex, and Sexualities," in *The Psychology of Gender,* 2nd ed., eds. A.H. Eagly, A.E. Beall, and R.J. Sternberg (New York: Guilford Press, 2004), 192. Stacey J. Connell, "Gender and Power: Society, the Person, and Sexual Politics," *Contemporary Sociology* 17, no. 5 (1988).
97 C West and D H Zimmerman, "Doing Gender," in *Doing Gender, Doing Difference: Inequality, Power, and Institutional Change*, eds. S. Fenstermaker and C. West (New York: Routledge, 2002) 3.
98 Stacey, "Gender and Power," 595.
99 Leanne Betasamosake Simpson supra n 13 at 123.
100 Leanne Betasamosake Simpson includes a footnote here to Bruce Bagemihl, *Biological Exuberance: Animal Homosexuality and Natural Diversity* (New York: St Martin's Press, 1999) in ibid at 118.
101 Bruce Bagemihl, "Left-Handed Bears and Androgynous Cassowaries: Informing Biology with Indigenous Knowledge" and "Chimeras, Freemartins, and Gynandromorphs: The Scientific Reality of Indigenous 'Myths'," in *Biological Exuberance: Animal Homosexuality and Natural Diversity* (New York: St Martin's Press, 1999) np. Also see Joan Roughgarden, *Evolution's Rainbow: Diversity, Gender, and Sexuality in Nature and People* (London: University of California Press, 2004) at 37.
102 Ibid. Also see gender traitors in Letitia Meynell and Andrew Lopez, supra n 81 at 4300.
103 Nicolas Labbé-Corbin and Me Daniel Boyer, *Canadian Guide to Uniform Legal Citation*, 9th ed (2018) at E-3.
104 *Chicago Manual of Style* (17th ed.) ss 5.43, 5.46, and 5.47.
105 Here is a list of examples of cases from various jurisdictions on various topics (including animal welfare, hunting, labour relations) that use the pronoun "it" and/or co-referent "which" in reference to animals: *R v Zhou*, 2022 BCSC 362, paras 13–14. *Captive Wildlife Regulations*, RRS c W-13 Reg 13 s 15(1). *Manziak v Animal Protection Services of Saskatchewan*, 2022 SKQB 75, para 28. *Mokelky v Animal Protection Services of Saskatchewan Inc*, 2019 SKQB 80 (CanLII), para 24. *Rhéaume v Chief Animal Welfare Inspector*, 2022 ONCACRB 18 (CanLII), paras 48–51. *Flying E Ranche Ltd v Attorney General of Canada*, 2022 ONSC 601 (CanLII), paras 225–228. *R c McWueen*, 2022 QCCQ 2801 (CanLII), para 98. *Animal Protection Act*, SNS 2018 c 21, s 2(1)(a). *J.S. v Nova Scotia Society for the Prevention of Cruelty*, 2022 NSAWAB 3 (CanLII) at paras 80, 6, 62. *Safe Food for Canadians Regulations*, SOR 2018–108. *Health of Animals Act*, SC 1990 c 21, s 16(1). *Canada (Attorney General) v Denfield Livestock Sales Limited,* 2010 FCA 36 (CanLII), para 20. *Alsager v Canada* (Agriculture and Agri-Food), 2011 FC 1071 (CanLII), paras 32, 56. *Professional Institute of the Public Service of Canada v Canada Food Inspection Agency*, 2011 PSLRB 16 (CanLII) para 124.
106 *Mokelky v Animal Protection Services of Saskatchewan Inc*, 2019 SKQB 80 (CanLII), para 24. *R v Zhou*, 2022 BCSC 362, paras 13–14.
107 *Prevention of Cruelty to Animals* Act, RSBC 1996, c 374 s 20.3.
108 *Viitre v British Columbia Society for the Prevention of Cruelty to Animals*, BCFIRB, 2016, paras 4 "it," 5 "he," 17 "he," 18 "he," 23 "it," 32 "it," 40 "it." *Faye Parkinson v British Columbia Society for Prevention of Cruelty to Animals*, BCFIRB, 2013 paras 36 "he," 49 "it," 50 "it," 67 "it." *Kelpin v British Columbia for the Society of Prevention of Cruelty to Animals*, BCFIRB, 2015, paras 20 "it," 33 "it," 35 "he."
109 *Reece v Edmonton (City)*, 2010 ABQB 538, para 1.
110 *Reece v Edmonton (City)*, 2011 ABCA 238, paras 3, 39, 58.
111 Leanne Betasamosake Simpson, supra n 13 at 124. Qwo Li-Driskill, *Asegi Stories: Cherokee Queer and Two-Spirit Memory* (U of Arizona P, 2016) e-book location 526.
112 See Gloria Anzaldúa, supra n 73 at 27.
113 Qwo Li-Driskill, *Asegi Stories: Cherokee Queer and Two-Spirit Memory* (University of Arizona Press, 2016) e-book location 526.
114 Leanne Betasamosake Simpson, supra n 13 at 124.
115 Ibid., at 128.
116 Kai Pyle, Manidoo Ma'iingan, and Charles Lippert, and Awanigiizhik Bruce, "Anishinaabe Gender Terms" (December 13, 2021), online (photo): *Instagram*, retrieved from: instagram.com/p/CXbxiN2vkR1/?igshid=MzRlODBiNWFlZA%3D%3D.

117 Alex Wilson (Opaskwayak Cree Nation) quoted in Native Youth Sexual Health Network and Women's Earth Alliance, "Violence on the Land, Violence on Our Bodies: Buidling and Indigenous Response to Environmental Justice," at 5, online (pdf), retrieved from: landbodydefense.org/uploads/files/VLVBToolkit_2016.pdf.
118 Leona Makokis, Kristina Kopp, Ralph Bodor, Ariel Veldhuisen, and Amanda Torres, "Cree Relationship Mapping: *nêhiyaw kesi wâhkotohk—How We Are Related*," 15, no. 1 (2020) online (pdf): <erudit.org/en/journals/fpcfr/2020-v15-n1-fpcfr05200/1068362ar.pdf>. Thank you to Darcy Lindberg for recommending this paper to me.
119 Ktunaxa Nation, *Qat'muk Declaration* (November 15, 2010), online (pdf): <ktunaxa.org/who-we-are/qatmuk-declaration/>.
120 *Te Awa Tupua (Whanganui River Claims Settlement Act)* (2017), retrieved from: legislation.govt.nz/act/public/2017/0007/latest/whole.html.
121 Laura Walters, "If the Whanganui River is a person, is it just like you and me," *Stuff* (March 16, 2017), retrieved from: stuff.co.nz/national/90516475/if-the-whanganui-river-is-a-person-is-it-just-like-you-and-me.
122 Leanne Betasamosake Simpson, "Land as Pedagogy: Nishnaabeg Intelligence and Rebellious Transformation," *Decolonization: Indigeneity, Education & Society* 3, no. 2 (2014): 1 at 2. Please note that Leanne Betasamosake Simpson's approach to the story from her own community does not mean that outsiders have permission to change genders within oral histories they are reading.
123 Ingrid E. Newkirk, "Clearly, an Animal Isn't an 'It': PETA Wants Change to AP Stylebook" (March 24, 2022), retrieved from: *People for the Ethical Treatment of Animals* peta.org/media/news-releases/clearly-an-animal-isnt-an-it-peta-wants-change-to-ap-stylebook/. Debra Merskin, "*She*, *He*, Not *It*: Language, Personal Pronouns, and Animal Advocacy," *Journal of World Languages* 8, no. 3 (2022): 391 at 405.
124 Simón(é) D Sun, "Stop Using Phony Science to Justify Transphobia," *Scientific American* (June 13, 2019), retrieved from: blogs.scientificamerican.com/voices/stop-using-phony-science-to-justify-transphobia/.
125 Arran Stibbe, supra n 59 at 5.
126 For example, in Nêhiyawêwin and Hän, there is no gender included in personal pronouns. Solomon Ratt, *Mâci-nêhiyawêwin Beginning Cree* (Regina: University of Regina Press, 2016) at 73. Alexandra Winton, "Tr'ondëk Hwëch'in Governance, Law, and Cosmology" (April 2019) at 27 n 56. Also see on nsyilxcən, Jeanette Armstrong, speaking on nsilxcən, as interviewed by Kim Anderson, *A Recognition of Being: Reconstructing Native Womanhood* (Toronto: Sumach Press, 2000) at 130.
127 This is an iterative and constantly evolving acronym. The version I have used here stands for Two-Spirit, lesbian, gay, bisexual, transsexual, transgender, queer, questioning, intersex, and asexual folks, with the plus sign indicating folks who are not included in this acronym.
128 Jacques Derrida, *Of Grammatology*, trans. Gayatri Chakravorty Spivak (Baltimore: The Johns Hopkins University Press, 1974) at 112; James R. Dawes, "Language, Violence, and Human Rights Law," *Yale Journal of Law & the Humanities* 11 (1999): 215, retrieved from: openyls.law.yale.edu/bitstream/handle/20.500.13051/7280/11_11YaleJL_Human215_1999_.pdf?sequence=2&isAllowed=y.
129 Here I am thinking of Indigenous legal orders which recognize animals transforming into different kinds of animals (i.e. from orcas to humans, etc).
130 For example, on naming and place names within Hul'qumi'num legal tradition, see Sarah Morales, "Stl'ul nup: Legal Landscapes of the Hul'qumi'num Mustimuhw" (2016) 33 Windsor Y B Access Just 103 at 118, online: <canlii.org/en/commentary/doc/2016CanLIIDocs182#!fragment//BQCwhgziBcwMYgK4DsDWszIQewE4BUBTADwBdoByCgSgBpltTCIBFRQ3AT0otokLC4EbDtyp8BQkAGU8pAELcASgFEAMioBqAQQByAYRW1SYAEbRS2ONWpA>
131 Teun A. van Diijk, "Principles of critical discourse analysis," *Discourse & Society* 4, no. 2 (1993): 249 at 282.
132 Chilton and Schäffner as paraphrased in Arran Stibbe, supra n 16 at 149.
133 Norman Fairclough, supra n 12 at 149.
134 Arran Stibbe, supra n 59 at 5.
135 Ibid.
136 Carl Mika, supra n 96 at 736 n 2.

137 Christopher Rufo, supra n 53.
138 Conversations with Tara Williamson over Fall 2022.
139 Alexandra Winton, cited previously at 27 n 56.
140 Ibid.
141 Solomon Ratt, cited previously at 73.
142 Carl Mika, supra n 96 at 746.
143 Ibid at 746.
144 Ibid at 760.
145 Ibid.
146 Lindsay Keegitah Borrows, "The More-Than-Human World as Historical Precedent: Looking to Distant Time Stories People Tell About Their Homelands" (LLM thesis, University of Alberta): 25.
147 Craig Womack, *Drowning in Fire: Native American Literary Separatism* (Minneapolis: University of Minnesota Press, 1999), 273.
148 Ibid. Also see Leanne Betasamosake Simpson, supra n 13.
149 On Indigenous gender literacies, see, for example Willie Seymour as cited in Brian Thom, "Coast Salish Senses of Place: Dwelling, Meaning, Power, Property and Territory in the Coast Salish World" (PhD, McGill University, 2005) at 178, online (pdf): <web.uvic.ca/~bthom1/Media/pdfs/senses_of_place.pdf>.
150 Lindsay Keegitah Borrows, *Otter's Journey through Indigenous Language and Law* (Vancouver: UBC Press, 2018), xi.
151 See Sarah Hunt, Sandy Lambert, and Alex Wilson, "Body Sovereignty: A Collaborative Reflection on Two-Spirit Methodologies," in *Honouring Our Ancestors*, eds. Alison Green and Leonie Pihama (Wellington: Te Herenga Waka University Press, 2023) at e-book location 4836.
152 Naomi Metallic, "Five Linguistic Methods for Revitalizing Indigenous Laws" (May 17, 2022) *McGill L J*, online (pdf): papers.ssrn.com/sol3/papers.cfm?abstract_id=4099156.
153 Thanks to Tara Williamson for highlighting the possibility of constructing pronouns using the sound-based method. See Caroline Helen (Roy) Fuhst, *Understanding Anishinaabemowin, Students' Edition: Understanding all the Sounds That Are Heard* (2012).
154 See, for example, Jeff Corntassel, "Restorying Indigenous Landscapes: Community Regeneration and Resurgence," in *Plants, People, and Places*, ed. Nancy J. Turner (Ottawa: McGill-Queen's University Press, 2020) 350 at 351.
155 See Val Napoleon, "Indigenous Women Talking: The Work of Indigenous Feminisms in the World" (2020) at 16–17, online (pdf): <ilru.ca/wp-content/uploads/2020/08/Napoleon-Indigenous-Women-Talking-0072.pdf>.
156 Robin Wall Kimmerer, *Braiding Sweetgrass* (Minneapolis: Milkweed Editions, 2013) at 49.
157 Robin Wall Kimmerer, *Nature Needs a New Pronoun*, cited previously.
158 Darcy Lindberg, supra n 26 at 22. Also see Sylvia Saysewahum McAdam, *Nationhood Interrupted: Revitalizing Nêhiyaw Legal Systems* (Saskatoon: Purich Publishing, 2014) at 10. Also see Maria Campbell, "We Need to Return to the Principles of Wâhkôtowin," *Eagle Feather News* (November 2007), online: *M Gouldhawke Wordpress*, retrieved from: mgouldhawke.wordpress.com/2019/11/05/we-need-to-return-to-the-principles-of-wahkotowin-maria-campbell-2007/.
159 Darcy Lindberg, supra n 26 at x, 35, 161, 174.
160 Conversation with Darcy Lindberg, September 22, 2022.
161 Solomon Ratt, cited previously at 63.
162 Thank you to Jessica Asch for assisting me with this conclusion.

Bibliography

Adams, Carol. The Sexual Politics of Meat: A Feminist-Vegetarian Critical Theory. New York: Bloomsbury Academic, 2010.

American Society for the Prevention of Cruelty to Animals. "A Voice for the Voiceless." ASPCA. Retrieved from: secure.aspca.org/team/a-voice-for-the-voiceless

Anzaldúa, Gloria. Borderland/La Frontera: The New Mestiza. San Fransisco: Aunt Lute Books, 1987.

Armstrong, Susan J., and Boltzer, Richard. "Animal Ethics: A Sketch of How it Developed and Where it is Now." In The Animal Ethics Reader, edited by Richard Boltzer and Susan J Armstrong. New York: Routledge, 2003.

B, Michea. "Neopronouns, They're Way Older Than You Think." The Medium, January 27, 2020. Retrieved from: themicheab.medium.com/neopronouns-theyre-way-older-than-you-think-ae0ec7030ac
Bagemihl, Bruce. "Left-Handed Bears and Androgynous Cassowaries: Informing Biology with Indigenous Knowledge." and "Chimeras, Freemartins, and Gynandromorphs: The Scientific Reality of Indigenous 'Myths'." In Biological Exuberance: Animal Homosexuality and Natural Diversity. New York: St Martin's Press, 1999.
Barnes, Donald J. "A Matter of Change." In In Defense of Animals, edited by Peter Singer. New York: Basil Blackwell, 1985.
Baron, Dennis. "A Brief History of Singular 'They'." Oxford English Dictionary, September 4, 2018. Retrieved from: public.oed.com/blog/a-brief-history-of-singular-they/
Baron, Dennis. "Nonbinary Pronouns Are Older Than You Think." The Web of Language, October 13, 2018. Retrieved from: blogs.illinois.edu/view/25/705317
Baron, Dennis. What's Your Pronoun? New York: Liveright Publishing Corporation, 2020.
Belcourt, Billy-Ray. "Animal Bodies, Colonial Subjects: (Re)Locating Animality in Decolonial Thought." Societies 5 (2015).
Belcourt, Billy-Ray. "The Cree Word for a Body Like Mine is Weesageechak." In This Wound is a World: Poems, by Billy-Ray Belcourt. Calgary: Frontenac House Ltd, 2017.
Birke, Lynda. "Intimate Familiarities? Feminism and Human–animal Studies." Society & Animals 10, no. 4 (2000).
Bordo, Susan. "The Cartesian Masculinization of Thought." Journal of Women in Culture and Society 11, no. 3 (1986).
Borrows, John. Law's Indigenous Ethics. Toronto: University of Toronto Press, 2019.
Borrows, Lindsay Keegitah. Otter's Journey through Indigenous Language and Law. Vancouver: UBC Press, 2018.
Borrows, Lindsay Keegitah. "The Land Is Our Casebook." University of Alberta LLM thesis, 2021 (unpublished) 42 at page 19 of Chapter 1.
Borrows, Lindsay Keegitah. "The More-Than-Human World as Historical Precedent: Looking to Distant Time Stories People Tell About Their Homelands." LLM thesis, University of Alberta.
Briant, Louis de. "Q'est-ce que l'écritre Inclusive." Le Journal du Dimanche (19 February 2021). Retrieved from: lejdd.fr/Societe/quest-ce-que-lecriture-inclusive-4026119
Butler, Judith. Gender Trouble. New York: Routledge, 1990.
Campbell, Maria. "We Need to Return to the Principles of Wâhkôtowin." Eagle Feather News, November 2007. Retrieved from: mgouldhawke.wordpress.com/2019/11/05/we-need-to-return-to-the-principles-of-wahkotowin-maria-campbell-2007/
Carrd. "What Are Neopronouns." Retrieved from: neopronounss.carrd.co/#what, accessed October 30, 2022.
Castricano, Jodey. "Introduction: Animal Subjects in a Posthuman World." In Animal Subjects, edited by Jodey Castricano. Wilfred Laurier University Press, 2008.
Cavalieri, Paola. "The Animal Debate: A Reexamination." In In Defense of Animals: The Second Wave, edited by Peter Singer. Malden: Blackwell Publishing, 2006.
Cheng, Amy. "A French Dictionary Added a Gender-neutral Pronoun. Opponents say it's too 'woke'." The Washington Post, November 18, 2021. Retrieved from: washingtonpost.com/world/2021/11/18/iel-petit-robert-gender-neutral-woke/
Connell, Stacey J. "Gender and Power: Society, the Person, and Sexual Politics." Contemporary Sociology 17, no. 5 (1988).
Deckha, Maneesha. "Unsettling Anthropocentric Legal Systems: Reconciliation, Indigenous Laws, and Animal Personhood." Journal of Intercultural Studies 41, no. 1 (2020a): 77–97.
Deckha, Maneesha. "Veganism, Dairy, and Decolonization." Journal of Human Rights and the Environment 11, no. 2 (2020b).
Descartes, René. "Animals are Machines." In Environmental Ethics: Divergence and Convergence, edited by S J Armstorn and R G Botzler. New York: McGraw-Hill, 1993.
Dickson, Courtney and Andrew Kurjata, "Arrests Made as Thousands Attend Rallies for and against Teaching Gender Diversity in BC Schools." September 19, 2023. Retrieved from: cbc.ca/news/canada/british-columbia/sogi-marches-bc-1.6972195.
Fairclough, Norman. Language and Power. London: Longman, 1989.
Faye Parkinson v British Columbia Society for Prevention of Cruelty to Animals, BCFIRB, 2013.

Friere, Paulo. Pedagogy of the Oppressed. Translated by Myra Bergman Ramos. New York: Continuum International Publishing Group, 2005.
Fuhst, Caroline Helen (Roy). Understanding Anishinaabemowin, Students' Edition: Understanding All the Sounds That Are Heard (2012).
Gill, Victoria. "Human and Wild Apes Share Common Language." BBC News, January 25, 2023. Retrievedfrom:bbc.com/news/science-environment-64387401?fbclid=PAAaaC-0bL4B3t5n_3Pvpa MuwK3ic9p-kTtO7xtRyL1x-cn_MosrhNKT9oHtU
Haddad, Raphaël. Manuel d'Écriture Inclusive. (Mots-Clés, 2016). Retrieved from: univ-tlse3.fr/medias/fichier/manuel-decriture_1482308453426-pdf#:~:text=L'utilisation%20du%20point%20milieu,%2C%20sénior·e·s
Haraway, Donna. When Species Meet. Minneapolis: University of Minnesota Press, 2008.
Harding, Sandra. "Women's Standpoints on Nature: What Makes Them Possible?" Osiris 12 (1997).
Hedeager, Ulla. "Is Language Unique to the Human Species." Columbia University. Retrieved from: columbia.edu/~rmk7/HC/HC_Readings/AnimalComm.pdf
Highway, Thomas. Laughing with the Trickster: On Sex, Death, and Accordions. Toronto: House of Anansi Press Inc, 2022.
hooks, bell. Talking Back: Thinking Feminist, Thinking Black. Boston: South End Press, 1999.
Hunt, Sarah, Sandy Lambert, and Alex Wilson. "Body Sovereignty: A Collaborative Reflection on Two-Spirit Methodologies." Honouring Our Ancestors, eds Alison Green and Leonie Pihama (Wellington: Te Herenga Waka University Press, 2023)
Kelpin v British Columbia for the Society of Prevention of Cruelty to Animals, BCFIRB, 2015.
Kimmerer, Robin Wall. Braiding Sweetgrass. Minneapolis: Milkweed Editions, 2013.
Kimmerer, Robin Wall. "Nature Needs a New Pronoun: To Stop the Age of Extinction, Let's Start by Ditching It." March 30, 2015. Retrieved from: yesmagazine.org/issue/together-earth/2015/03/30/alternative-grammar-a-new-language-of-kinship
Kirk, Gwyn. "Ecofeminism and Environmental Justice: Bridges across Gender, Race and Class." Frontiers: A Journal of Women Studies 18, no. 2 (1997).
Kirksey, Eben, Chao, Sophie, and Bolender, Karin, eds. The Promise of Multispecies Justice. Duke University Press, 2022.
Korsmeyer, Carolyn. Gender and Aesthetics: An Introduction. New York: Routledge, 2004.
Labbé-Corbin, Nicolas, and Boyer, Daniel. Me. Canadian Guide to Uniform Legal Citation. 9th ed. McGill Law Journal, 2018.
Ladner, Kiera. "Gendering Decolonization, Decolonizing Gender." Paper Presented at the 80th Annual Conference of the Canadian Political Science Association UBC June 2008 (2008). Retrieved from: cpsa-acsp.ca/papers-2008/Ladner.pdf
Lesbian, Gay, Bisexual, Transgender, Queer Plus (LGBTQ+) Resource Centre. "Gender Pronouns." Retrieved from: uwm.edu/lgbtrc/support/gender-pronouns/, accessed October 30, 2022.
Li-Driskill, Qwo. Asegi Stories: Cherokee Queer and Two-Spirit Memory. University of Arizona Press, 2016.
Lindberg, Darcy. "kihcitwâw kîkway meskocipayiwin (Sacred Changes): Transforming Gendered Protocols in Cree Ceremonies through Cree Law." Master of Laws thesis, Faculty of Law, University of Victoria, 2017.
Lindberg, Darcy. "Transforming Buffalo: Plains Cree Constitutionalism and Food Sovereignty." In Food Law and Policy in Canada, by Heather McLeod-Kilmurray, Angela Lee and Nathalie Chalifour. Carswell, 2019.
Lindberg, Darcy. "Excerpts from Nêhiyaw Âskiy Wiyasiwêwina: Plains Cree EarthLaw and Constitutional/Ecological Reconciliation." Indigenous Law Association, 2021. Retrieved from: indigenous-law-association-at-mcgill.com/2021/06/03/excerpts-from-nehiyaw-askiy-wiyasiwewina-plains-cree earthlaw-and-constitutional-ecological-reconciliation/
Lugones, María. "Toward a Decolonial Feminism." Hypatia 25, no. 4 (2010).
Makokis, Leona, Kopp, Kristina, Bodor, Ralph, Veldhuisen, Ariel, and Torres, Amanda. "Cree Relationship Mapping: nêhiyaw kesi wâhkotohk—How We Are Related." 15, no. 1 (2020). Retrieved from: erudit.org/en/journals/fpcfr/2020-v15-n1-fpcfr05200/1068362ar.pdf
Marcus, Ezra. "A Guide to Neopronouns." The New York Times, April 8, 2021. Retrieved from: nytimes.com/2021/04/08/style/neopronouns-nonbinary-explainer.html

Marecek, Crawford, & Popp. "On the Construction of Gender, Sex, and Sexualities." In The Psychology of Gender, 2nd ed., edited by A.H. Eagly, A.E. Beall, and R.J. Sternberg. New York: Guilford Press, 2004.
McAdam, Sylvia Saysewahum. Nationhood Interrupted: Revitalizing Nêhiyaw Legal Systems. Saskatoon: Purich Publishing, 2014.
Merchant, Carolyn. The Death of Nature: Women, Ecology, and the Scientific Revolution. 2nd ed. New York: Harper One, 1990.
Merskin, Debra. "She, He, Not It: Language, Personal Pronouns, and Animal Advocacy." Journal of World Languages 8, no. 2 (2022).
Metallic, Naomi. "Five Linguistic Methods for Revitalizing Indigenous Laws." McGill L J, May 17, 2022. Retrieved from: papers.ssrn.com/sol3/papers.cfm?abstract_id=4099156
My Kid is Gay. "Defining: Neopronouns." My Kid Is Gay, 2019. Retrieved from: mykidisgay.com/blog/defining-neopronouns
Napoleon, Val. "Indigenous Women Talking: The Work of Indigenous Feminisms in the World" (2020) at 16–17. Retrieved from: ilru.ca/wp-content/uploads/2020/08/Napoleon-Indigenous-Women-Talking-0072.pdf.
National Geographic. "Watch Koko the Gorilla Use Sign Language in This 1981 Film." YouTube, June 22, 2018. Retrieved from: youtube.com/watch?v=FqJf1mB5PjQ
Native Youth Sexual Health Network and Women's Earth Alliance. "Violence on the Land, Violence on Our Bodies: Building an Indigenous Response to Environmental Justice." Retrieved from: landbodydefense.org/uploads/files/VLVBToolkit_2016.pdf
Newkirk, Ingrid E. "Clearly, an Animal Isn't an 'It': PETA Wants Change to AP Stylebook." People for the Ethical Treatment of Animals, March 24, 2022. Retrieved from: peta.org/media/news-releases/clearly-an-animal-isnt-an-it-peta-wants-change-to-ap-stylebook/
Noske, Barbara. Beyond Boundaries: Humans and Animals. Montréal: Black Rose Books, 1997.
O'Loughlin, A.K. "Gender-as-lived: The Coloniality of Gender in Schools as a Queer Teacher Listens in to Complicated Moments of Resistance." Indo-Pacific Journal of Phenomenology 19, no. 1 (2019).
"Personal Pronouns and Gender." Chicago Manual of Style s 5.43.
Prevention of Cruelty to Animals Act, RSBC 1996, c 374 s 20.3.
Pronoun Wiki. "Pronouns." Retrieved from: pronoun.fandom.com/wiki/Category:Pronouns, accessed October 29, 2022.
Pronouny. "Pronouny." Retrieved from: pronouny.xyz, accessed October 30, 2022.
Ratt, Solomon. Mâci-nêhiyawêwin Beginning Cree. Regina: University of Regina Press, 2016.
Reece v Edmonton (City). (2010). ABQB 538, para 1.
Rescue and Sanctuary for Threatened Animals. Instagram Post, August 9, 2019.
Rescue and Sanctuary for Threatened Animals. Instagram Post, April 30, 2020.
Rescue and Sanctuary for Threatened Animals. Instagram Post, May 1, 2020.
Roan, Chief Wayne, and Waugh, Earle. "Meanings of Sacred Pipe." Nature's Laws, 2004. Retrieved from: wayback.archive-it.org/2217/20101208172609/http://www.albertasource.ca/natureslaws/traditions/ritual_meanings_pipe.html
Rose, Ellen Cronan. "The Good Mother: From Gaia to Gilead." Frontiers: A Journal of Women Studies 12, no. 1 (1991).
Roughgarden, Joan. Evolution's Rainbow: Diversity, Gender, and Sexuality in Nature and People. London: University of California Press, 2004.
Rufo, Christopher. "Pronouns Unbound." City Journal (September 29, 2022). Retrieved from: cityjournal.org/san-francisco-public-schools-subvert-parental-rights
Sapir, Edward. "The Status of Linguistics as a Science." Language 4, no. 4 (1929).
Simard, Suzanne. "The Foundational Role of Mycorrhizal Networks in Self-organization of Interior Douglas-fir Forests." Forest Ecology and Management 258 (2009). Retrieved from: sciencedirect.com/science/article/pii/S0378112709003351
Simpson, Leanne Betasamosake. "Land as Pedagogy: Nishnaabeg Intelligence and Rebellious Transformation." Decolonization: Indigeneity, Education & Society 3, no. 2 (2014).
Simpson, Leanne Betasamosake. As We Have Always Done. Minneapolis: U of Minnesota P, 2017.

Simpson, Leanne Betasamosake. "Indigenous Queer Normativity." In As We Have Always Done: Indigenous Freedom through Radical Resistance, by Leanne Betasamosake Simpson. University of Minnesota Press, 2021.
Smart, Carol. Law, Crime and Sexuality: Essays in Feminism. London: Sage Publications Ltd, 1995.
Snyder, Emily Jane. Representations of Women in Cree Legal Educational Materials: An Indigenous Feminist Legal Theoretical Analysis. Doctor of Philosophy, Department of Sociology, University of Alberta, 2014a.
Snyder, Emily Jane. "Indigenous Feminist Legal Theory." Canadian Journal of Women and the Law 26, no. 2 (2014b).
Stibbe, Arran. "Language, Power and the Social Construction of Animals." Society & Animals 9, no. 2 (2001).
Stibbe, Arran. Animals Erased: Discourse, Ecology, and Reconnection with the Natural World. Wesleyan University Press, 2012.
Stryker, Susan. "My Words to Victor Frankenstein Above the Village of Chamounix: Performing Transgender Rage." In The Transgender Studies Reader, edited by Susan Stryker and Stephen Whittle. New York: Routledge, 2006.
Sullivan, Will. "Plants Make Noises When Stressed, Study Finds." Smithsonian Magazine, April 4, 2023. Retrieved from: smithsonianmag.com/smart-news/plants-make-noises-when-stressed-study-finds-180981920/#:~:text=Stressed%20plants%20that%20have%20been,week%20in%20the%20journal%20Cell
Sun, Simón(é) D. "Stop Using Phony Science to Justify Transphobia." Scientific American, June 13, 2019. Retrieved from: blogs.scientificamerican.com/voices/stop-using-phony-science-to-justify-transphobia/
Taylor, Sunaura. Beasts of Burden, Animal and Disability Liberation. New York: The New Press, 2017.
Teterina, Marina. "The Use of Gendered Pronouns with Animal Referents in English." Limbaj şi context 4, no. 2 (2012).
Teterina, Marina, and Crestianicov, Elena. "Personification or Sexification of Countries in English?" Online (pdf) Retrieved from: usarb.md/limbaj_context/volumes/v7/art/teterina%20crestianicov.pdf
Thom, Brian. Coast Salish Senses of Place: Dwelling, Meaning, Power, Property and Territory in the Coast Salish World. (PhD, McGill University, 2005).
Turner, Nancy J. Plants, People, and Places. Ottawa: McGill-Queen's University Press, 2020.
University of Wollongong Australia. "Animal Studies Journal: Policies." Retrieved from: ro.uow.edu.au/asj/policies.html, accessed October 30, 2022.
van Diijk, Teun A. "Principles of Critical Discourse Analysis." Discourse & Society 4, no. 2 (1993).
Viitre v British Columbia Society for the Prevention of Cruelty to Animals, BCFIRB, 2016.
Warren, Karren. Ecofeminist Philosophy: A Western Perspective on What It Is and Why It Matters. Lanham: Rowman and Littlefield, 2000.
Wentz, Carrington. Instagram, November 18, 2022. Retrieved from: instagram.com/p/ClGxnOasvDR/?utm_source=ig_embed&utm_campaign=embed_video_watch_again.
West, C., and Zimmerman, D.H. "Doing Gender." In Doing Gender, Doing Difference: Inequality, Power, and Institutional Change, edited by S. Fenstermaker and C. West. New York: Routledge, 2002.
Willett, Cynthia, and Suchak, Malini. "Sociality." In Critical Terms for Animal Studies, edited by Lori Gruen. Chicago: University of Chicago Press, 2018.
Winton, Alexandra. "Tr'ondëk Hwëch'in Governance, Law, and Cosmology." April 2019.
Wohlleben, Peter. The Hidden Life of Trees. Musqueam, Squamish, Tsleil-Waututh Lands: Greystone Books Ltd, 2016.
Womack, Craig. Drowning in Fire: Native American Literary Separatism. Minneapolis: University of Minnesota Press, 1999.
Zhou, Naaman. "Evenings in the Park with Flaco." The New Yorker, February 22, 2023. Retrieved from: newyorker.com/news/our-local-correspondents/evenings-in-the-park-with-flaco.

44

RACIAL JUSTICE, ANIMAL JUSTICE

An Interview with Claire Jean Kim

Chloë Taylor [CT]: Claire Jean Kim is professor of political science and Asian American studies at the University of California Irvine. She is the author of three monographs. Her first two monographs, *Bitter Fruit: The Politics of Black-Korean Conflict in New York City* (Yale University Press, 2000) and *Dangerous Crossings: Race, Species, and Nature in a Multicultural Age* (Cambridge University Press, 2015), were each the recipient of the Best Book Award from the American Political Science Association Section on Race, Ethnicity, and Politics, among other prestigious book awards. Her most recent monograph, *Asian Americans in an Anti-Black World*, was just published in October 2023. In addition to being a prolific writer of academic texts, Professor Kim is a public intellectual, having been a guest commentator on MSNBC, PBS, and NPR and having written articles for *The Los Angeles Times*, *The Nation*, *The American Scholar*, and *Ms. Magazine*, as well as being regularly interviewed on podcasts and for documentary films on topics involving race, animals, and environmental issues. Thank you, Claire, for agreeing to take the time to do this interview at what I know is a very busy time.

Claire Jean Kim [CJK]: You're very welcome, Chloë. I'm happy to do it.

CT: We are having this interview in October 2023, and your third monograph was published with Cambridge University Press last month: *Asian Americans in an Anti-Black World*. The topic of this book is broader in scope but returns to the terrain of your first monograph, *Bitter Fruit: The Politics of Black-Korean Conflict in New York City*, published with Yale University Press in 2000. Last month also saw the *Students for Fair Admissions v*

DOI: 10.4324/9781003273400-54

Harvard Supreme Court decision that gutted affirmative action in U.S. universities. The suit was filed in 2013 and claimed Asian Americans were discriminated against by Harvard's race-conscious undergraduate admission process in favour of Black applicants. Given your longstanding interest and expertise in anti-Asian and anti-Black racism in the United States, and the sometimes conflictual relationship between the two, this case was clearly of concern to you and indeed you published an important article on the topic in 2018 in the *DuBois Review*, "Are Asians the New Blacks? Affirmative Action, Anti-Blackness and the 'Sociometry' of Race." Just a few days ago you also published an online article, "Better Asians Than Blacks," for *The American Scholar* on the Supreme Court decision. Can you start out by saying a bit about your position on the Supreme Court decision, how you are currently responding to it, and how it relates to your scholarship?

CJK: My book *Asian Americans in an Anti-Black World* talks about this case. Even though the book came out when I didn't yet know the outcome of the case, I guessed, of course, with most people, what it would be. So people who support affirmative action have long been concerned that the U.S. Supreme Court would strike the policy down because the Court's has been deeply divided on the issue. In the three prior cases that the court looked at on affirmative action in higher education, affirmative action survived by a hair in each case. In the first case, *[Regents of the University of California v] Bakke* (1978), [it survived] only when the court shifted the rationale for affirmative action from remediation for past discrimination to diversity. This was a fateful shift and I think predicted the outcome we see in the cases involving Harvard and University of North Carolina.

What we see in U.S. politics today is a strong, organized, sustained push from the right to roll back civil rights reforms. Some of the same people who brought down affirmative action also eviscerated the Voting Rights Act of 1965 in a 2013 U.S. Supreme Court case. It's a very organized, concerted effort by the right to take us back to an earlier era, which really is an era of segregation and even greater racial and economic inequality than we have now. This is disturbing although not surprising given that Trump stacked the court recently with three right-wing appointees after blocking the appointment of Merrick Garland under President Obama. This case, along with the *Dobbs* decision on abortion, raises the question of whether and how much the Supreme Court cares about its legitimacy as an institution. At this point, it seems to care most about advancing a right-wing agenda.

CT: In 2021 you published an essay in *Ms. Magazine* that also draws attention to how anti-Asian racism and anti-Blackness function differently and are sometimes in conflict in the U.S. In this essay, you draw attention to how the Covid-19 the Hate Crimes Act was passed by Congress with strong support (it was only nearly unanimous in the Senate, not in the House) and got a great deal of media attention, the function of this bill being to give more power and funds to the police and criminal legal system to punish anti-Asian hate crimes in the context of the pandemic. As you note, there is little evidence that policing and criminal law are effective ways to address hate crimes and so what this bill essentially did was put more money and power into institutions known to brutalize Black people, and this at a time when the Black Lives Matter movement was arguing for defunding the police. Meanwhile, you observe that a human rights report on police violence issued by the International Commission of Inquiry on Systemic Racist Police Violence Against People of African Descent in the United State and the George Floyd Justice in Policing Act were ignored by the media and the George Floyd bill failed to get through the Republican-controlled Senate. Can you say a bit about what prompted you to write this essay and also your decision to publish in a popular feminist magazine?

CJK: This story encapsulates the larger story I try to tell in the book, which is the way that Asian Americans are positioned above Black people in the U.S. racial order, and the way that the powers that be weaponize Asian Americans in order to reproduce a racial order that is anti-Black, and how Asian American at times help to weaponize themselves—which is what we see in the affirmative action case so strikingly. It's really, at this point, a conservative white-Asian collusion to reproduce structural anti-Blackness. In both cases, in the article and in the book, what I'm interested in is taking what people see is happening to Asian Americans and putting it in the larger context of structural anti-Blackness. How do we revisit or rethink Asian American history by putting it in this context? How do we think about the Covid-19 Hate Crimes Act and the general rise in hate against Asian Americans during Covid in relation to anti-Blackness?

Why is it that this international commission on anti-Black police violence has gotten no traction? There's almost no mention of it in the U.S. media. I think I saw one article in the *Guardian*. And then the Justice for George Floyd Policing Act, which was introduced into Congress after the murder of George Floyd, died in the Senate, where all racial justice bills go to die. Both of those important efforts addressing anti-Black violence are ignored or killed, and instead what we get is a pro-carceral solution to hate crimes against Asian Americans, which, as you suggested, doesn't really help Asian Americans. It doesn't address the racial positioning that sets Asian Americans up as targets of xenophobic resentment. It's just throwing money at the police and saying, "we've done something for Asian Americans." This is a cynical gesture in the middle of a Black Lives Matter resurgence after the murder of George Floyd which only appears to help Asian Americans and also ignores what's happening to Black communities. This is political theatre which conveys which lives matter more and which less.

I had the good fortune to get to know Professor Michele Goodwin, who's on the faculty at UCI [University of California at Irvine] Law, and she's just a remarkable scholar and activist, and she is an editor at *Ms. Magazine*. She heard me present on this topic and asked if I would appear on her podcast and also publish a piece in *Ms. Magazine*, which is what I decided to do.

On the issue of publishing for a more popular audience, it's unfortunately something that you only get the privilege to do later on in your career, because the way that the academy is set up is you are rewarded for publishing with academic presses, not with popular presses. You are rewarded for publishing in academic journals, not for publishing in the *LA Times* or *The Nation*. In fact, you get no credit for doing popular writing so it's only when you have reached a certain point in your career that you can contemplate these other kinds of writing, and I think it's so important to reach out to a broader audience. I'm certainly trying to do that with the recent book, to have more of a public presence and public voice. The point of working on these issues is, of course, to be at the table and have a voice in some of these debates and conversations.

CT: While we have started out by talking about your first and third monographs, there are close connections between these and your middle monograph, *Dangerous Crossings: Race, Species, and Nature in a Multicultural Age*, published in 2015 with Cambridge University Press. While your second monograph is about race and species, it relates to your other two monographs in that you carefully analyse differences in the ways that anti-Asian racism, Anti-Black racism and also racism against Indigenous people have functioned in the U.S. context in terms of another hierarchical way of taxonomizing living beings that is species. So this is a two-part question. I am hoping, first, that you can speak to the relationship between this middle book, which has been so significant to the field of critical animal studies, and your earlier and more recent work. And second, I am hoping that you can say a bit about how you have analysed what you describe as "synergies of race and species" in *Dangerous Crossings*. Why as a critical race scholar did it make sense for you to write this book on species as well as race?

CJK: I think you're the first person to notice that the three books really are connected in the sense that they are all looking at differentiations within and/or across taxonomies of power. Some of my colleagues who are race scholars were not that open to the second book, by which I mean, for example, they would ask me, "how's the animal book going?" A book about race and animals was just "the animal book," right? And it was almost like, "when is Claire going to finish that book so she can get back to writing about race?" Even though it was about race as well as animals.

There's still a certain lack of receptiveness [to animal studies] in the academy. Although animal studies has grown in a very exciting way, in many parts of the academy, in more traditional disciplines, even ethnic studies, there is still resistance to it. I think one still gets labelled "the person working on animal issues," and so I'm yet to feel there's a real integration of these issues for me, because in the world, they're not integrated. We still have movements that are separate. We still have intellectual formations that are separate.

I was looking at real-life conflicts between the white majority and racialized or Indigenous groups and their use of animals. And I'm looking at those conflicts and thinking there are ways that race and species are playing out here that are not being really adequately teased out. On a theoretical level, it was interesting to me. Also on a personal level, the timing of the book was when I had recently become vegetarian and later vegan. I had recently gotten interested in animal liberation issues. I was experiencing changes in my personal life that were political changes that made me more interested in writing about animals, and so rather than simply throw aside race and just write about animals, I thought, there really is a connection here. I wanted to do something to integrate the two.

CT: I think readers of this interview will be more aware of the ways that racialized groups of humans have been dehumanized or animalized than the inverse, the ways in which certain species or categories of animals are racialized, but in *Dangerous Crossings* you argue that "certain nonhuman species get racialized or imbued with negative meanings associated with despised human groups." Katja M. Guenther has also discussed this phenomenon in her 2020 book, *The Lives and Deaths of Shelter Animals*, where she describes attempts on the part of white female animal shelter volunteers to "whiten" and "feminize" pit bull–type dogs who are associated with Black masculinity in ways that can be deadly for them. In a 2019 lecture, "Who Is the Human and Who Is the Animal in the Human–Animal Divide," philosopher of race and animal ethicist Syl Ko also argues that there has been both an animalization of racialized humans and a racialization of nonhuman animals. Can you explain what is meant by nonhuman species being racialized? What would be some examples of this?

CJK: I think the classic example is the example Katja used, and I talk about in *Dangerous Crossings* too with regard to the Michael Vick case, and that is the Blackening or whitening of pit bulls. Another example I can think of is when in the Midwest, in the Great Lakes area, they had what they called "the invasion of Asian carp," and they were clogging the waterways. They were killing off other kinds of fish. They were interfering with recreational boating and tourism. And there was an emphasis on them having Asian origins, right? And this was playing into old tropes about invading Asiatic hordes. Racializing the carp heightened white American anxiety about the arrival of these fish and the threat that they posed.

I think the way that this happens, this sort of exchange of meaning across the animal–human barrier, is that no matter which ethnological framework you're looking at that draws upon that Chain of Being idea of the graduated difference in a hierarchy, the lowest human is always right above the highest animal, right? Animality gets associated with the lowest rung of humanity in these frameworks, which is Blackness.

But I think you more often see the animalization of marginalized human groups. And the reason for that would be that even the lowliest human still stands above other animals, right? When you use an animal to describe a lowly human, you're emphasizing the lowliness of the human and you're pulling that human almost down through the species barrier. When you racialize an animal with a human reference, it doesn't have quite the same impact.

CT: In *Dangerous Crossings* you not only explore the relationship between race and species, or how these are entangled taxonomies, but also how there are ongoing conflicts around culturally specific ways that racialized groups of people treat animals—Indigenous hunting, live animal markets in Chinatowns, and African American men engaging in dog fighting being the examples you focus on in the book. Although you have shown interconnections between race and species, these examples put social movements *against* racism and speciesism in conflict with one another, and so your book is as much about tensions between racial and species oppression as it is about the connections between these oppressions. Can you say a bit about this topic and whether you are seeing any growing sensitivity to race on the part of animal activists?

CJK: It depends on what you mean by growing sensitivity. I've been doing this kind of work for maybe 20 years and certainly it seems *de rigueur* now for animal activists, if they're planning a big conference, for example, or an anthology, it's pretty likely that they'll feel some need to include race in some way, right? Maybe they'll include something on Indigeneity out of a sense that there's more of a convergence between issues around indigenous ontologies and epistemologies and animal and environmental issues than there is with race and the latter. The animal movement needs to be more diverse. It needs to present a more diverse face to the public. It needs to be able to address communities that are not white.

But I don't think there's a deeper recognition of the ways that the animal movement needs to engage with structural anti-Blackness. In the first place, I don't think most animal advocates, like most people, have been educated in this U.S. educational system to understand what structural anti-Blackness means. The right wing is increasingly, as you know, going after critical race theory, saying we can't teach it in schools. The latest I've heard is that the governor in Florida is saying middle schools have to emphasize that slavery had positive aspects, for example, teaching slaves trades that after emancipation they could use to support themselves. We see where this is going. It's a whitewashing of structural anti-Blackness past and present. It's disabling Americans from understanding the persistence of anti-Blackness and what it means, how profoundly and comprehensively it structures society. And so when you ask animal advocates to pay attention to race, they don't realize necessarily that it means digging down to the foundations of animal studies and the animal movement and really questioning how do these projects, like other projects that don't acknowledge anti-Blackness, how do they end up reproducing it?

It's a real dilemma and this is why I wrote the most recent piece, "The Fight for Animals in a Moment of Danger." It seems to me that there are at least two paths. One path would be to keep on doing what the animal movement is doing and saying we are separate from anti-Blackness. We support Black liberation in theory and we may draw upon it as a symbol, but we are not going to engage and really take that on as our fight. And that road, I think, leads ultimately to giving succour to fascism because I think that's what many parts of the world are sliding toward, especially in the context of climate crisis. And the other road would say we recognize Black liberation is a struggle for all beings and a better world and that our struggle has to be part of that struggle. Think about it, do we really want to create a world where animals are free and Black people remain abjected, subordinated, and in many ways still enslaved? I think that's the kind of question we, as animal advocates and animal scholars, need to ask ourselves.

CT: You published two other pieces that I am aware of on race and species after *Dangerous Crossings*, your 2017 article, "Murder and Mattering in Harambe's House," and your 2018 "Abolition" chapter for Lori Gruen's *Critical Terms for Animal Studies*. Can you say a bit about how each of these pieces extended the arguments you made in *Dangerous Crossings*?

CJK: The abolition article was my first attempt to really look at how the animal movement is inadvertently appropriating from the Black freedom struggle in a way that is problematic. Of course, I'm not alone in having talked about this. A. Breeze Harper has written about this, and other Black scholars and non-Black scholars have done so. I wanted to look specifically at the choice of the word "abolitionism" to describe an animal liberationist perspective, because that is, of course, the phrase that Stephen Best and Gary Francione and others have used. I am highly critical of the idea of abolitionism [in animal liberation] as a new freedom struggle that replaces the original abolitionism because that one has succeeded. If you are a student of race and U.S.

history, you know that that original abolitionist struggle is far from over. This is not just a mistaken reading of history, it's a disturbing disavowal of the way that anti-Blackness continues to structure our lives today.

In the Harambe piece. I wanted to get at what I think are the two main referents for building the idea of the Human in Western culture, the Black—and I use that deliberately to mean an ungendered Black figure—and the animal. In Black studies, they emphasize the Human as constructed vis à vis the Black. In animal studies, they emphasize the Human as constructed vis à vis the animal. What I'm trying to talk about in my work is that Blackness and animality are themselves co-constituted, because these taxonomies of power are deeply entangled with one another. So there's something about these two ideas, Blackness and animality, closely linked, that serve together as a foil for the idea of the Human, and that's what I try to get at in that piece by looking at the murder or the killing of Harambe in the Cincinnati Zoo in 2016.

CT: As you mentioned earlier, you have a forthcoming essay, "The Fight for Animals in a Moment of Danger," in which you return to the issue of species and race. You make three points in the article. First, you reiterate your earlier arguments about the race as a species construct, or race and species as two dimensions of a single taxonomy. You return to Harambe in this part of the paper, arguing that the zookeepers' decision to kill the gorilla to save the Black child momentarily elevated a Black person to the level of a human being and hence above the animal, when white supremacy is premised on the equation of Blackness and animality. You argue that the online white fury over this act functioned to restore order, dragging Blackness back down to the level of the animal.

In the second part of the paper you reflect on the ongoing ways in which animal activism erases anti-Blackness and indeed this erasure is foundational to animal liberation movements, which often premise their arguments on racism being a thing of the past, as you were talking about regarding the use of the term "abolition" in animal liberation movements. As you show, from Jeremy Bentham in the early 19th century to contemporary authors such as Peter Singer and Gary Francione, we are told that just as human slavery was abolished, so too must animal slavery be abolished, or just as we overcame racial prejudice, now we must overcome species prejudice. In each case racism is spoken of as if it were a thing of the past, and there is no investment in knowing about, let alone doing anything about, ongoing racial oppression.

Why, in this article, do you argue that it is so critical to think about race and species together in this particular moment?

CJK: *Dangerous Crossings* was my first attempt to grapple with the race and species nexus, and as I was finishing that book, I started to read in Black studies more about Frantz Fanon's work and more about structural anti-Blackness. I want to revisit Black studies as a field because that's how I educated myself and how I continue to educate myself. Structural anti-Blackness means that the world is structured around the phobic avoidance of Blackness. Ideas about Blackness that arose during the period of slavery, that were used to justify the institution, have not died. Once slavery was formally ended, we saw it continue institutionally in a variety of ways, including the convict leasing system and mass incarceration today. The phobic ideas about Blackness remain, which is why Black people remain hyper-segregated and much more so than any other group of colour, and that is because of this continual distancing from Blackness. When we look at the police and the kind of animus and hatred they express for Black people, they are expressing it for the rest of society, because it's there in the fabric of American society. The police are not an aberration.

Today, I think it's important for us to relate animal issues to the larger political context and the issue of ecofascism, because under Trump, we saw the movement toward fascism in the U.S., which I think continues in some ways, for instance in the attacks on voting rights and the attacks on history curricula. These are from the fascist playbook, these kinds of politics. We see fascism resurging in other parts of the world, in Brazil, India, Hungary, Italy, Germany, and France. And my concern is, as things get more dire and shortages become more acute and hundreds of millions of people become climate refugees, as the world heats up, we're going to see more societies closing up, as they've already done in Europe, for example, to North African migrants, saying, "This is ours. You can't come in here" and watching the hundreds of millions die at the gates or in the sea, wherever they might be. This is the moment of danger I'm referring to.

Ecofascism is an ideology coming out of German thought, where the idea of whiteness is connected with the idea of protecting nature and the land and animals. Hitler, of course, had certain kinds of eco-fascist ideas himself. Environmentalism and animal liberation can be co-opted by ecofascists. For the left, the question is what do we have to say to ecofascism? It's not enough to just say we want animals liberated, because we need to have ideas about racial justice and racial liberation incorporated into that agenda in order to meet the challenge that ecofascism poses.

CT: In "The Fight for Animals in a Moment of Danger," you discuss the cover art by Sue Coe for Gary Francione's and Ann Charlton's (2015) book, *Animal Rights: The Abolitionist Approach*, a particularly glaring example of the erasure of anti-Blackness within animal liberation movements. Can you explain your argument about Sue Coe's artwork for this book cover?

CJK: I want to preface this by saying I love Sue Coe's artwork. I think it's incredibly powerful and impactful. I am not criticizing her artwork per se. I want to comment on what she is doing in this front and back piece because it is symptomatic of what the animal movement is doing.

The front cover shows animals in subjection. They're downcast, they're chained, everything's dark, and the word "Abolition" is on a banner, and there's a kneeling male slave in chains, and then a standing woman who may also be a slave based on how she's dressed. Then you go to the back cover, and the back cover says "Vegan," and there's a sun rising, and all the shackles have disappeared, and the animals are dancing. They're holding hands and they're dancing, and it's like the after, right? There's the before picture, and then there's the after picture, after abolition. And then you wonder, where did the two slaves go? And they are nowhere to be found. It's easy to miss this because you're focusing on the change in the status of the animals. But once you start thinking about it, why are there no free Black people? You can see how it's impossible to put the freed slaves in, given the way the [animal liberation] movement thinks now, because what are we going to say? Animal abolitionism freed Black people from slavery and now they're liberated? We know that's not true. Are we going to say that they're still in chains? Are we going to depict them still as slaves? No, because that doesn't make veganism and abolition look morally sound. We're just going to eliminate them from the picture and therefore reveal that we've been using them as props. We want them in the first picture because it helps us get a sense of what animals are going through. And once Black people are done being used as the props, we can remove them.

Again, I don't think this is the thought process of Sue Coe or the authors of the book at all. They would describe themselves, I'm sure, as advocates of racial justice, but it is symptomatic, I think, again, of the movement, and here I'm talking about the abolitionist liberationist aspect of the movement, that it's not asking the hard questions. How do these two liberation struggles articulate with one another? How should they be articulating with one another? Can we have animal liberation without Black liberation? Do we want to really have a world where Black people are still in chains and animals are running free?

CT: Many scholars in critical animal studies argue for including species within intersectionalist studies, and that critical animal studies needs to be intersectional in order to think about animals in relation to race, gender, sexuality, disability, and class. At the end of your forthcoming anthology chapter, you argue, however, that simply adding animals into pre-existing intersectionality frameworks isn't adequate to the task that you're describing, and I think you've argued this elsewhere too. Can you say a bit about why you think that intersectionality is not the right theoretical framework here?

CJK: Intersectionality is a very powerful framework, and I understand its great appeal and how much it improves upon other frameworks which don't seem to allow attention to more than one issue at a time. I get that. If we look back to early ecofeminists, like Val Plumwood, for example, she's talking about multiple dimensions of oppression and how we have to see them as all connected. This kind of work has really laid the groundwork for being able to theorize more about these issues today.

But how much can 'intersectionality' help us think through how these different forms of oppression are related to one another? How do we think through which forms might be more foundational than others, which might shape the others? There's a limit to how much this concept can help us here.

CT: Although *Dangerous Crossings* has been taken up as a trailblazing and extremely important book within critical animal studies academic circles, you've encountered resistance to your ideas on the part of animal and environmental activists more generally. Can you tell us about the kind of reactions you've experienced when you try to bring considerations of anti-Blackness into activist debates about animals? Have you encountered similar kinds of resistance on the part of anti-racism activists and critical race scholars when you try to bring consideration of species oppression into discussions of race?

CJK: I think I've experienced more of the first, that is, animal and environmental activists expressing some resistance to thinking about anti-Blackness, than I have the second, which is race scholars expressing resistance to talking about species and environmental issues. And that's probably just an artifact of the fact that I have been invited to do more of the first kind of talk than the second. In other words, I've generally gotten more interest in the work from people who are in animal and environmental circles saying "come talk to us about race" than I have from race circles saying "come talk to us about animals." It's making me think of John Sanbonmatsu, who said progressives and leftists should consider animal oppression as one of the important oppressions that we should talk about. It shouldn't be excluded from the issues that leftists are concerned with.

During Covid I did a few Zoom workshops or conferences where I spoke about anti-Blackness to animal and environmental studies scholars. In the first one, there was, I remember, one white woman scholar who was getting very worked up about the indifference to animals that some Black writers show, and so we got into it a little bit around that.

I gave a talk recently in Paris, and it was almost entirely silent during the Q and A. I think there was one person who asked a question, and that was it. And afterwards I asked the organizers, what is going on? The one guy who had asked the question wrote to me and said, there were some translation issues, there was fatigue at the end of the conference, but also maybe they just hadn't thought about anti-Blackness very much, so when they heard me speak about it, they didn't know what to do with it. And maybe there was even a little bit of "We already devote our lives to helping animals. Why do we have to then take on this whole other social issue?"

I think that the reaction in terms of bringing anti-Blackness in has been variable. I would say it's ranged from just being polite and hearing me out to some resistance. On the one hand, this is about the privilege of not being Black. On the other hand, it's a practical challenge at some level. Even if people are open to learning about anti-Blackness, they don't know what to do, and so I think doing reading groups would be a great idea for animal activists. Like the way they did in the 60s when they had groups like the Black Panthers. They would sit together and [discuss] Mao and Marx and Lenin and they would teach each other, teach themselves about these things, and if animal people got together and had these reading groups about anti-Blackness, I think that's one way to move forward.

CT: Since the publication of *Dangerous Crossings* there has been a flurry of other books published in critical animal studies considering the imbrications of race and species. I've already mentioned Syl Ko's work, and there is also the work that Syl Ko has done with her sister Aph Ko, collected in the 2017 book, *Aphro-ism*, and Aph Ko's 2018 monograph, *Racism as Zoological Witchcraft*, Lindgren Johnson's 2018 book *Race Matters, Animal Matters*, Bénédicte Boisseron's 2018 *Afro-Dog: Blackness and the Animal Question*, Jasmin Singer's 2021 edited collection, *Antiracism in Animal Advocacy*, Joshua Bennett's 2022 book, *Being Property Once Myself: Blackness and the End of Man*, and the 2023 edited volume, *Animals and Race*. There are other books—like Katja M. Guenther's monograph on shelter animals that I mentioned—that are not specifically about race but provide intersectional feminist analyses of topics related to animal oppression. You engage with some of these works—especially the work of Aph Ko and Joshua Bennett—in the forthcoming article we've just been discussing. Can you say a bit about the ways the topics of race and species have been explored by other authors, and how your own work has been taken up over the last 8 years within critical animal studies?

CJK: Great titles on that list. I wrote the foreword for Aph Ko's single authored book [*Racism as Zoological Witchcraft: A Guide to Getting Out*] and I think her work and Syl Ko's work combines theoretical acumen with this way of putting things that people who are not academics can understand. It's incredibly important work. It's exciting to see the work that is coming out now in animal studies. Boisseron's historical analysis, Bennett's literary analysis—this is as good as writing gets. My hope is that all of this work will prompt a paradigm shift in the field, and what I mean by this is invigorated and innovative thinking about how to do animal studies in an anti-Black world. That's what I'm writing toward.

45
TRANSFORMATIVE JUSTICE FOR ANIMALS
Lessons from Anti-Carceral Feminism

Kelly Struthers Montford, Darren Chang and Selingul Yalcin

Introduction

Much like second-wave feminists have advocated for legal reform and criminalization to elevate the status of victims, and to remedy gender-based inequalities and violence against women, there is now a 'carceral turn' in animal protection[1] in both Canada[2] and the United States.[3] At the same time that the legal denunciation of animal abuse is occurring, so too is the animal protection movement's interest in restorative justice (RJ).[4] RJ is often taken to be the antithesis of 'tough-on-crime' and its application appropriate only in minor cases. Yet both responses to harm are similar in their focus on individual responsibility to the exclusion of social structures that make both animals and marginalized humans vulnerable. Increasingly, RJ is also now administered by the criminal punishment system (CPS) and has institutional and financial relationships with the CPS rather than operating outside of and independently from it, as do transformative justice (TJ) initiatives. In this chapter, we argue that the failures of carceral feminism can illustrate how the criminal legal system is an unlikely solution to animal cruelty, as it often enables harms against those whom it purportedly protects. We instead suggest that TJ approaches are better positioned to address the structural, community, and interpersonal factors shaping harms to animals. To do so, we first show that in the context of animal advocacy, the criminal legal system by and large prioritizes protecting industrial and corporate interests over animals and animal advocates. Following that, we assess the use of RJ in cases of animal cruelty to examine its alternative potential and limitations. We then review the development of TJ practices and consider the U.S. Humane Society's "Pets for Life" program as a TJ approach to humans and their companion animals. Finally, we evaluate the potential for RJ and TJ approaches to address the structural violence of animal agriculture and argue that the goals of both approaches align with a feminist animal care ethics. We do so by drawing on farmed animal sanctuary practices and relationships to suggest that the application of some RJ principles in TJ praxis has the potential to achieve forms of interpersonal and structural justice for farmed animals.

DOI: 10.4324/9781003273400-55

The Harms of Legal Systems on Farmed Animals and Animal Advocates

At the same time that the carceral turn in animal protection is underway,[5] the law continues to enable systemic harm against animals and activists. For farmed animals, their dual status of being legal property and economic commodity means that by default, legal institutions and society at large will tend to be biased against farmed animals when their well-being comes into conflict with economic interests. If animal rescuers find escaped farmed animals, they can be legally forced to return the animals to their 'owners,' sometimes to the exact sites and unsafe conditions in which they were found—such as the side of a highway without food, water, or access to veterinary care, as was the case in the aftermath of Hurricane Florence.[6] This reality for activists is also reflected in the recent case concerning four animal activists who were charged with 21 indictable offenses after obtaining and publicizing video footage depicting animal suffering inside Excelsior Hog Farm in Abbotsford, British Columbia, Canada. Two of the activists, Amy Soranno and Nick Shafer, have since been convicted and sentenced to jail time for exposing systemic violence against animals.[7] In other instances, farmed animals have been 'rescued' from cruelty only to be harmed by their 'adopters.' Molly the pig, confiscated and rehomed by the BC SPCA due to abuse, was later eaten by her guardians, with the BC SPCA saying they had no recourse given extant laws.[8]

Lessons from carceral feminism and the victim's rights movement are instructive in understanding how animals have become the latest subject of this movement. Feminist legal theorist Aya Gruber, for her part, calls the trajectory of animal protection "familiar territory."[9] Recent 'tough on animal crime' laws, which seek to institute prison sentences or to increase possible sentence lengths, have been justified explicitly on punitive grounds, even when the reasons for harm are "so readily traced to economic precarity and social marginality."[10] Victims' rights movements and organizations emerged in the 1970s with the goal of ensuring better treatment by prosecutors. Advocates were largely middle-class white women, who remain disproportionately unlikely to be *victims of crime* and tended to be pro-law enforcement and supportive of the court system and the death penalty.[11] This demographic represented many aspects of the ideal victim: white, legally innocent, vulnerable, and largely performing hegemonic femininity[12] in their deference to and support of by-design patriarchal legal systems.[13] Victims' rights are not irreparably yoked to carceral responses. In the U.S., however, they have been deployed in such a way—and this way has been "zero-sum: the more the defendant suffered, the more the victims healed."[14]

The timing of the movement paralleled the rise of neoliberalism, on which carceral feminism capitalized. Mimi Kim's interrogation of carceral feminism shows that in the U.S., the 1970s marked the beginning of the rise of neoliberalism that began to shape both the political agenda and public opinion.[15] With this, there was increased skepticism in state programs to alleviate poverty or redistribute wealth. As a consequence, social benefits were significantly reduced, and those dependent on state assistance were increasingly demonized. The belief that the government's role was to improve inequitable social conditions was replaced by the belief that its role was to solve and/or control crime. As a result, as the welfare state receded, and funds previously used for equity-based programs and services were reallocated to the criminal legal system, thus resulting in its continued expansion. This shift to 'crime' also meant a shift from groups to the individual, thereby attempting to depoliticize structural issues.[16] What this meant for feminist anti-violence efforts, which were usually explicitly anti-racist, was that they would get more political support and traction *if* they began to talk about sexual assault and domestic violence *as crime control issues*. Sexual and domestic violence were no longer framed as social and political issues, but instead issues of crime. In defining social and political issues as crime, the cause is not rooted in racist, colonial, and heteropatriarchal systemic structures but as a lack of police intervention and punishment of offenders.[17] Hence, carceral feminists pushed for the mandatory arrest and prosecution of domestic violence. Gender-based justice became synonymous with tough-on-crime policies and punishment, and in the U.S. context, mainstream feminism had essentially married crime control.[18]

Talking about 'crime' is a coded way to talk about race.[19] The outcomes of tough-on-crime 'feminist' justice produced negative and retraumatizing effects for all women, but structurally speaking, women who are non-white, Indigenous, poor, queer, trans, unhoused, or disabled or have histories of mental health diagnoses or criminalization are those disproportionately impacted. Men's arrests increased but were eclipsed by the arrest rates of women, especially Black women, engaging in self-defense. Arrests turned out to benefit white women, as arrest records established a paper trail that saw them receive favorable divorce, custody, alimony, and child support decisions. Employed white men took notice, and assault rates decreased. For impoverished and marginalized families, this had no material benefit; instead, it invited other state services such as child protection and immigration enforcement. Forms of material support advocated for by racialized advocates, such as financial support, shelters, services such as child care, and employment opportunities that would actually help women leave abusive situations, were absent from victims' rights organizing and narratives, as they did not resonate with the needs articulated by the core of the movement.[20] The carceral response then represents a very small subset of victims, who are taken as universal by way of their whiteness and economic status.

We suggest that companion animals—on which recent tough-on-animal-crime laws are premised—are a parallel iteration of the ideal victimhood of the victims' rights movement, this time encapsulated through mainstream animal protection organizations. Current animal advocacy narratives often portray animals as the *most* innocent of victims. They are attributed the capacity to feel and experience positive human emotions but not those of guilt, legal responsibility, or malicious intent. Those seeking carceral responses project upon animal victims the same narratives of the victims' rights movement begun in the 1970s, such as total devastation by way of victimization and vengefulness. Some of the worst cases are sensationalized to pass harsh laws, and it is positioned as a taken for granted that animals would want justice served through vengeful punishment, including the caging of their human abusers. Gruber, however, suggests that there are limits to what we can infer and that it might very well be the case that, like human victims of interpersonal harm, animals could desire an improvement in their treatment, namely that the abuse stop, and that, having experienced caging and captivity, they would be hesitant to support such initiatives, and they would prefer to remain with their human companions should conditions improve.[21] Carceral responses then fail to center animals as aggrieved parties, with the logic often being that being tough on animal crime is required to promote human safety, a position that makes expanding the carceral system palatable to those far beyond the animal protection movement.[22]

In the case of companion animals, for whom 'animal carceralism' could arguably be best designed, the outcomes for humans and animals have been detrimental despite narratives of status elevation and mattering. Prosecutorial approaches to crimes against animals frame criminal charges as "send[ing] a strong message to both the offender and society as a whole that the proper and humane treatment of animals matters (whether wild, livestock, or pet)" in addition to being a way to incapacitate the offender.[23] This is especially the case for unhoused persons who have been charged with animal neglect for failing to provide veterinary care, even when the cost is completely out of their reach, and have seen their companion animals removed and euthanized without their consent. A punitive approach such as this does not change the material conditions that cause animal suffering nor prevent it in the future. Instead, carceral animal protectionism might then have the effect of criminalizing poverty[24] rather than *preventing* harms to animals. Punishments such as incarceration or fines can create a legacy of poverty.[25] Unpaid fines and fees can result in arrest warrants, leading to cycles of rearrest and criminalization, a phenomenon that sociologists refer to as permanent punishment.[26] While there has been an increasing move to criminalize animal abuse, there is a concurrent movement among legal scholars, professionals, advocates and animal protectionists to apply RJ practices in cases of harms against animals.[27]

Restorative Justice and Animal Victimization

RJ, influenced by and borrowing from Indigenous legal traditions, is meant to be a holistic response to offending that seeks to *repair* the harm endured by victims and survivors of crime. Proponents of RJ argue that it does not share the criminal legal system's focus on retribution; it moves beyond traditional legal actors and centers offenders, victims, and stakeholders by focusing not on the violation of state laws but on "crime [as] a violation of people and relationships."[28] Dialogue between the victim, offender, and broader community is foundational to the process, and offenders must admit to having perpetrated the offense prior to the process beginning. That the offender take responsibility, be honest, and be accountable for their actions is central to the process, as is a focus on their rehabilitation. Victims are meant to feel empowered, receive a form of restitution, get information about events and the motivations of the offender, and have a space to speak the truth of their experience. RJ processes are also meant to understand the complexity of offenders and the factors leading to the event(s) in question, and remedies are the result of a collaborative effort between all parties.

Brittany Hill explores how RJ can be applied in cases of animal cruelty and positions it as able to recognize animals as victims, take crimes against them seriously, address underlying issues driving the harm, and potentially address other detrimental behaviors. Animal interests can be represented in RJ's dialogue-centric process by surrogate victims, which can include the harmed animal's guardian if they were not the offender or veterinarians as they can attest to the physiological and psychological impacts of the offense. For Hill, justice for animal victims is realized by "protecting future animal victims," which RJ is better positioned to achieve than retributive approaches.[29] RJ approaches are most often used to remedy individual-to-individual cases, with state actors and institutions often thinking it is only appropriate to do so in less serious cases of harm, and this has been the case in the use of RJ in harms against animals as well.[30]

As it currently operates, RJ exists in relation to the criminal legal system, not as its alternative. Programs often require referrals from the judicial authorities, are run as state programs, do not result in the dismissal or sealing of charges, and in some cases are integrated into regular criminal punishment processes and/or inform sentencing outcomes.[31] RJ is now more a state diversion program that compels participation, as failing to take (full) responsibility will stream one back into the court process.

Crucially, members of INCITE! Women of Color Against Violence have articulated that RJ "models often depend on a romanticized notion of 'community' that seldom exists in practice," and with the absence of such ideal communities, RJ measures often fail to "actually hold perpetrators of gender violence accountable."[32] This critical perspective is consistent with reflections by members of Philly Stands Up (PSU), a transformative justice and community accountability (TJ/CA) group that initially emerged as a response to sexual violence that occurred in Philadelphia's anarchist community. Regarding RJ, PSU members questioned

> *what is it that we are restoring?* Would these efforts lead us to the same troubled, problematic world plagued with patriarchy, homophobia, fatphobia, insecurity, heterosexism, racism, anxiety, depression, ableism, and all of the other conditions that feed into sexualized violence in the first place?[33]

PSU members recognized that any communities or relations prior to a given occurrence of harm may well have been unjust and structurally facilitated the harm in the first place; therefore, uncritical attempts to simply restore damaged relations back to a status quo would fail to fundamentally transform the structural causes of the harms. Although the use of RJ in responding to animal abuse is nascent and few cases of animal harm have been addressed in this manner, it is the case that so far, the status quo largely remains intact.

In a now-sealed 2020 case in Vermont, an unhoused individual experiencing mental illness was charged for animal abuse and neglect for having an elderly dog—also his companion—outside who was very ill.[34] Though he knew the dog was unwell, he could not afford adequate veterinary care. The police seized the dog and euthanized him without this human's consent. RJ processes set out that he had to take responsibility for what he 'had done' and through various meetings had to establish trust with a veterinarian to whom he could reach out in the future when he had another dog. The goal to prevent future harm by connecting him to services to the benefit of the dog, which should prevent his future criminalization, is laudable and important. However, this process squarely focused on the 'offender's' responsibility, with the weight of the court looming over the process. The police and veterinarian who euthanized the dog without his permission or his presence, to our knowledge, did not have to take responsibility for their actions or sowing of distrust among the community. It is not evident that role of the broader community in enabling this harm was addressed by this process. Should the offender, however, not take responsibility and participate as the prosecutor deemed necessary, his case would revert to the courts. If he successfully completed the RJ process, as determined by the prosecutor, his record would be sealed. The anthropocentric, ableist, and classist structures shaping the marginalization of this man and the dog in this case remain largely unchallenged, although future harm in the context of him having a future companion animal may be mitigated. With these limitations in mind, we propose that TJ can be a more effective way to address interpersonal and structural harms facing animals, as it is community specific, separate from the state, and takes a shared and capacious approach to accountability and responsibility.

Transformative Justice: Principles and Praxis

Whereas RJ has increasingly reflected the criminal legal system's goals (blame, retribution, and punishment) and is administered by the state carceral system and/or contains state agents as members, TJ is an approach to responding to harm that operates outside of such logics and institutions. It is an agile, context-specific, and community-based approach to addressing harm. TJ also divorces punishment from ideas of justice[35] and takes power from the state and brings it back to the survivor and community as arbiters of what 'matters.' TJ was largely born out of the organizing work of on-the-ground community groups who experienced and/or understood the interrelated nature of interpersonal, community, and state violence and recognized that positioning the state as the solution to violence would both be ineffective and invite more harm into their communities. That is, they knew that they were better equipped to keep each other safe than were criminal punishment responses and that calling them to respond to harm would run the risk of deportation, police harassment and abuse, or criminalization and/or cause offenders or survivors to be shunned from their communities.

Addressing not only the interpersonal nature of harm, TJ is focused on the community's role in allowing harm to occur, with some collectives clear that their processes are ones of TJ and community accountability.[36] A survivor-centric approach is foundational and means that the TJ intervention is designed with input from the survivor and is premised on the best ways they can be supported by the community, including meeting immediate needs for safety; supporting survivors' self-determination according to what they identify as their own needs and desires; and/or validating, reducing, and/or sharing responsibility for ending the violence.[37] TJ initiatives have also been successfully developed to address serious forms of harms for which state institutions do not allow RJ approaches to be used, such as sexual assault and child sexual abuse.[38] Feminists who are skeptical or critical of anti-carceral approaches have also tended to appeal to sexual violence against children as cases that are far more challenging or nearly impossible for TJ and prison abolitionism to address.[39] However, as Chloë Taylor has noted, not only do "law-and-order responses to crimes of sexual and gendered violence revictimize and fail complainants," prisons are themselves institutions that uphold and reproduce misogyny, sexual violence, and other forms of oppression by normalizing a culture of rape.[40] Meanwhile, organizations that practice TJ/CA, such as Oakland-based GenerationFIVE, work to help communities understand how the "root causes of common responses to reports of child sexual abuse, such as denial, minimization, horror, victim blaming, vigilantism, and pity" further entrench power imbalances that sustain the harms, while empowering children and youth "to believe that their ideas, experiences, desires and opinions are important, that they are worthy of respect, and that they are capable of self-determination" as a means to end the conditions that make them vulnerable to sexual abuse.[41] How might TJ, which centers upon survivor-identified needs and demands for safety, be applied to situations of animal harm? And, relatedly, how might some of the community capacity initiatives be applied in a non-anthropocentric manner to work towards the elimination of violence? In the next section of this chapter, we consider how TJ/CA can address interpersonal and structural harms experienced by animals and how this aligns with feminist animal care theory. We also suggest that some RJ principles and practices can be incorporated in such praxis aimed at improving human–animal relations and achieving justice more broadly.

Animals as Victims and Survivors of Interpersonal and Structural Violences

We propose that the Humane Society of the United States' Pets for Life (PFL) program is an anti-carceral animal justice approach that embodies TJ praxis. PFL seeks to keep companion animals in homes in a way that does not stigmatize or punish their human companions. It does so by facilitating and/or funding access to services, whether paying for care and supplies, training classes, or transporting animals to veterinary clinics. PFL focuses on some of the structural conditions (racialization, poverty, geographical isolation) that allow harm to occur, and their interventions seek to change these material conditions in the immediate present while working towards community self-sufficiency. Although white, middle-class people in urban centers with access to resources could also perpetrate violence against, or neglect animals, PFL prioritizes dispossessed and marginalized communities precisely because they are overpoliced and disproportionately criminalized. Individuals and communities identify their needs, and support does not come with conditions attached, such as repayment. Clients who could otherwise face criminalization for the condition of their companion animals are not punished but supported, and these interventions provide sustainable alternatives to the carceral system as a solution to animal suffering.[42]

In line with previous arguments for adopting TJ/CA praxis as a justice-seeking route that refuses to further legitimize law enforcement and criminal punishment systems that reproduce harm towards both victims and members of their communities, there are similarly compelling reasons for applying TJ/CA praxis in addressing structural violence against farmed animals.

Common practices of care at sanctuaries include rehabilitating residents from the abuse and abandonment that they have suffered at the hands of individual humans or the animal agriculture industry.[43] Beyond rehabilitation, sanctuaries aim to ensure that residents can flourish according to their needs, behaviors, and activities typical to their species. They also refrain from activities constituting animal exploitation, and many sanctuaries run educational programs to advocate on behalf of animals, raising awareness around the structural violence of animal agriculture and getting the public to recognize, respect, and relate to each animal as unique individual subjects with their own personalities.[44] These care practices at sanctuaries are all intended to radically reshape human relations with farmed animals against a relation in which animals are understood as property. In many ways, these standard sanctuary activities already embody the basic principles of TJ, from the educational efforts that engage broader communities beyond the sanctuary to target the structural causes of systemic violence against farmed animals to the victim-centered rehabilitation process that also seeks to empower the animals themselves to explore new possibilities in a life outside of agricultural exploitation.

Given that proposed RJ practices in cases of animal harm[45] include victim–offender dialogues, long-term sanctuary staff who have developed deep bonds and high levels of trust with animal residents could fulfill this surrogate role. Talking circles are also used in RJ and, when divorced from state punishment, align with TJ praxis. This is where processes of RJ expand to include more individuals from communities at large to share resources and knowledge to address systemic causes of harms and prevent their future occurrences: *again, transforming rather than restoring the original conditions that perpetuate harm.* Sanctuaries are also well equipped to host talking circles with members of their surrounding communities, either within the sanctuary itself if the safety of the animals can be ensured or elsewhere outside the sanctuary. Sanctuary staff could also sit on community restorative boards (CRBs) composed of trained members who discuss the harms and consequences with the offender, provide education, and propose appropriate sanctions as a practice of accountability rather than punishment. We would suggest that the outcomes of such approaches could also align with TJ's foundational aspect of changing the material conditions shaping farmed animal exploitation.

Josephine Donovan proposes that a feminist animal care approach takes animals as "the epistemological source of theorizing" about themselves.[46] This is fundamentally about dialogue, "listening to animals, paying emotional attention, taking seriously—*caring about*—what they are telling us."[47] To be able to listen and morally respond, Donovan suggests that feminist standpoint theory can be extended to animals in the sense that it takes "objectification" and "bodily experience" as sources of knowledge.[48] For animals, she argues that their often complete bodily exploitation and commodification make them appropriate subjects to articulate their standpoints. However, the question then becomes: how do we *listen* and engage in subject-to-subject dialogue? She suggests, as do proponents of RJ, that "human advocates are required to articulate the standpoint of the animals—gleaned . . . in dialogue with them—to wit, that they do not wish to be slaughtered and treated in painful and exploitative ways,"[49] for example. As well, TJ's survivor-centric approach can be understood as the praxis of standpoint feminist theory in which experiential accounts and emotional responses to perpetration and victimization are not dismissed but affirmed as knowledge of the events in question. A care ethics of this sort also entails a political and contextual assessment of the situation, thus sharing with TJ and feminism a commitment to analyzing and improving the material conditions that allow a situation of harm to occur in the first place. It also squarely places accountability on broader structures, communities and the individuals involved and requires an ethical response based on what the animal(s) in question desire and/or communicate. Perhaps then, TJ is a politicized feminist ethics of care in its commitment to improving the material conditions faced by oppressed groups and its liberatory praxis.

Conclusion

As with carceral feminism, carceral animal protectionism will have similar results: the conditions giving rise to harm will be exacerbated and worsened. While the animal protection movement is of course not monolithic, those committed to anti-speciesist work cannot respond to harms against nonhuman animals by appealing to the criminal punishment system and the patriarchal, racist, and speciesist values upon which it is premised and upholds. We should also be cautious about turning to RJ. While the current interest in the movement is laudable, feminist anti-violence organizing is instructive and necessarily raises scrutiny as to how such initiatives currently operate and their potential to change the very structures of anthropocentrism that not only permit, but in the context of racial capitalism, incentivize wide-scale harm against animals in animal agriculture, biomedical, and entertainment industries. These relations cannot be beneficial to animals and must not be restored. Instead, certain RJ principles can be applied to TJ approaches that together have the potential to transform the relations problematized herein. Indeed, the socially transformative aspects of RJ align with TJ praxis, both of which reflect earlier calls of feminist animal care theory to listen, politically assess, and ethically respond to interpersonal and structural harms against animals. The settler state continues to in part reproduce itself based on the subjugation of animal 'resources.' Put otherwise, it has a vested interest in enabling harm against animals. Animal justice can then only occur outside of and against the state and its institutions.

Notes

1 Justin Marceau, *Beyond Cages: Animal Law and Criminal Punishment* (Cambridge, Cambridge University Press, 2022), https://doi.org/10.1017/9781108277877; Justin Marceau, "Carceral Logics beyond Incarceration," in *Animal Law in Context: The Limits of Carceral Strategies* (Cambridge University Press, 2022); Aya Gruber, "Humanizing Animals, Dehumanizing Humans," in *Carceral Logics: Human Incarceration and Animal Captivity* eds. Justin

Marceau and Lori Gruen (Cambridge: Cambridge University Press, 2022), 158–186, https://doi.org/10.1017/9781108919210.012.

2 Edmonton Police Service, "Animal Cruelty Investigation Unit (ACIU)," *Crime Prevention Unit* (2019), retrieved from: https://www.edmontonpolice.ca/CrimePrevention/PersonalFamilySafety/AnimalCrueltyInvestigationUnit, accessed July 26, 2022; Department of Justice Government of Canada, "Bestiality and Animal Fighting (Bill C-84)" (October 18, 2018), retrieved from: https://www.justice.gc.ca/eng/csj-sjc/pl/baf-bca/index.html. See Criminal Code sections, 444.1(1)(b), 447 (cruelty to animals), and 160 (bestiality), for example; Ontario SPCA and Humane Society, "2019 Annual Report," Ontario SPCA and Humane Society (2019), retrieved from: https://ontariospca.ca/2019-annual-report/. Per their report, The Ontario Society for Prevention of Cruelty to Animals and Humane Society provided enforcement training to the Cornwall Police to "share resources and knowledge of animal well-being and behavior, as well as tips on navigating complex animal welfare legislation."

3 Kelly Montford and Eva Kasprzycka, "The 'Carceral Enjoyments' of Animal Protection," *Building Abolition: Decarceration and Social Justice*, eds. Chloë Taylor and Kelly Montford (Routledge, 2021).

4 Hill, "Animal Victims and Restorative Justice"; Vermont Law and Graduate School and University of San Francisco Law, "Animals and Restorative Justice Symposium" (July 29, 2022), retrieved from: https://www.vermontlaw.edu/news-and-events/event/animals-and-restorative-justice-symposium; Brittany Hill, "Restoring Justice for Animal Victims," *Journal of Animal & Natural Resource Law* 17 (2021): 217–248; Vermont Law and Graduate School and University of San Francisco Law, "Animals and Restorative Justice Symposium."

5 Justin Marceau, *Beyond Cages: Animal Law and Criminal Punishment* (Cambridge: Cambridge University Press, 2019), https://doi.org/10.1017/9781108277877.

6 Stephanie Eccles and Elisabeth Stoddard, "Hurricane Florence's Impact: Policies on Animals Living in Confined Animal Feeding Operations in Eastern North Carolina," *World Animal Protection*, May 3, 2021, retrieved from: https://www.worldanimalprotection.us/blogs/has-industry-learned-anything-hurricane-florence

7 CBC News, "Animal rights activists sentenced to time in jail for 2019 protest at B.C. hog farm," *CBC News* (October 14, 2022), retrieved from: https://www.cbc.ca/news/canada/british-columbia/b-c-animal-rights-activists-get-jail-time-1.6614762

8 Amy Judd, "'I Do Promise That Molly Died Humanely': Owner who killed adopted pig apologizes," *Global News* (2018), retrieved from: https://globalnews.ca/news/4047050/molly-pig-bc-spca-owner-killed-eaten/.

9 Aya Gruber, "Humanizing Animals, Dehumanizing Humans," in *Carceral Logics: Human Incarceration and Animal Captivity*, eds. Justin Marceau and Lori Gruen (Cambridge: Cambridge University Press, 2022), 167, https://doi.org/10.1017/9781108919210.012.

10 Ibid.

11 Ibid.

12 Jennifer M. Kilty and Sylvie Frigon, *The Enigma of a Violent Woman: A Critical Examination of the Case of Karla Homolka* (Abingdon and New York: Routledge, 2016).

13 Josephine Donovan, "Feminism and the Treatment of Animals: From Care to Dialogue," *Signs: Journal of Women in Culture and Society* 31, no. 2 (January 2006): 305–329, https://doi.org/10.1086/491750.

14 Gruber, "Humanizing Animals, Dehumanizing Humans," 172.

15 Mimi E. Kim, "From Carceral Feminism to Transformative Justice: Women-of-Color Feminism and Alternatives to Incarceration," *Journal of Ethnic & Cultural Diversity in Social Work* 27, no. 3 (July 3, 2018): 219–233, https://doi.org/10.1080/15313204.2018.1474827.

16 Janine Brodie, "We Are All Equal Now: Contemporary Gender Politics in Canada," *Feminist Theory* 9, no. 2 (August 1, 2008): 145–164, https://doi.org/10.1177/1464700108090408.

17 Gruber, "Humanizing Animals, Dehumanizing Humans."

18 Kim, "From Carceral Feminism to Transformative Justice."

19 Ibid.

20 Gruber, "Humanizing Animals, Dehumanizing Humans."

21 Ibid.

22 Ibid.; Edmonton Police Service, "Animal Cruelty Investigation Unit (ACIU)."

23 Ashley N. Beck, "Giving a Voice to the Voiceless: A Prosecutor's Efforts to Combat Animal Cruelty," in *Carceral Logics: Human Incarceration and Animal Captivity*, eds. Justin Marceau and Lori Gruen (Cambridge: Cambridge University Press, 2022), 55, https://doi.org/10.1017/9781108919210.005.
24 The American Society for the Prevention of Cruelty to Animals (ASPCA) states that: "When pet owners with incomes lower than $50,000 were asked which service might have helped them the most, the majority indicated free or low-cost veterinary care (40%) . . . free or low-cost pet food (30%), free or low-cost temporary pet care or boarding (30%) and assistance in paying pet deposits for housing (17%) "More than 1 Million Households Forced to Give up Their Beloved Pet Each Year, ASPCA Research Reveals," ASPCA, retrieved from: https://www.aspca.org/about-us/press-releases/more-1-million-households-forced-give-their-beloved-pet-each-year-aspca, accessed July 30, 2022. The difficulty affording services such as veterinary care and goods like food and medicine for low-income guardians provides the conditions for understanding these persons as neglectful rather than disproportionately affected by a privatized animal care industry.
25 Devah Pager et al., "Criminalizing Poverty: The Consequences of Court Fees in a Randomized Experiment," *American Sociological Review* 87, no. 3 (June 1, 2022): 529–553.
26 Ibid.
27 Hill, "Animal Victims and Restorative Justice."
28 Howard Zehr, *The Little Book of Restorative Justice: Revised and Updated* (Good Books, 2015), 19, retrieved from: http://ebookcentral.proquest.com/lib/ryerson/detail.action?docID=1922319.
29 Hill, "Restoring Justice for Animal Victims."
30 Hill, "Animal Victims and Restorative Justice"; Sally Adams, "Case Studies in Restorative Justice," Animals and Restorative Justice Symposium (July 29, 2022), retrieved from: https://www.vermontlaw.edu/news-and-events/event/animals-and-restorative-justice-symposium.
31 Hill, "Animal Victims and Restorative Justice"; Sally Adams, "Case Studies in Restorative Justice."
32 INCITE! Women of Color Against Violence, INCITE, *Color of Violence: The INCITE! Anthology* (Durham: Duke University Press, 2016), 8.
33 Esteban Lance Kelly, "Philly Stands Up: Inside the Politics and Poetics of Transformative Justice and Community Accountability in Sexual Assault Situations," *Social Justice (San Francisco, Calif.)* 37, no. 4 (2012): 49.
34 Adams, "Case Studies in Restorative Justice."
35 Kai Cheng Thom, "What to Do When You Have Been Abusive," *Truthout*, retrieved from: https://truthout.org/articles/what-to-do-when-you-have-been-abusive/, accessed August 26, 2022.
36 Bay Area Transformative Justice Collective, "Transformative Justice and Community Accountability," 2013, retrieved from: https://trueleappress.files.wordpress.com/2020/02/162.-bay-area-transformative-justice-collective-batjc-transformative-justice-and-community-accountability.pdf.BATJC 2013
37 Ibid.
38 Kelly, "Philly Stands Up," 44–57.
39 Chloë Taylor, "Anti-Carceral Feminism and Sexual Assault—A Defense: A Critique of the Critique of the Critique of Carceral Feminism," *Social Philosophy Today* 34 (2018): 29–49.
40 Ibid., 30, 32.
41 Ibid., 44–45.
42 The Humane Society of the United States, "Keeping Pets for Life," The Humane Society of the United States, 2022, retrieved from: https://www.humanesociety.org/issues/keeping-pets-life; Danielle Tepper, "Hand to Hand: Pets for Life Delivers Peace of Mind to Philly Pet Owners," The Humane Society of the United States, Support the Humane Society of the United States (December 10, 2021), retrieved from: https://secured.humanesociety.org/page/78105/subscribe/1?locale=en-US; Brianna Grant, "'No Lack of Love': Minnesota Reservation Joins the Pets for Life Mentorship Program," The Humane Society of the United States (April 1, 2022), retrieved from: https://www.humanesociety.org/news/no-lack-love; Amanda Arrington and Micheal Markarian, " Serving Pets in Poverty: A New Frontier for the Animal Welfare Movement," The Sustainable Development Law & Policy Brief (Washington, DC: American University Washington College of Law, 2018), retrieved from: https://www.humanesociety.org/sites/default/files/docs/American%20University%20PFL%20brief.pdf.
43 Sue Donaldson and Will Kymlicka, "Farmed Animal Sanctuaries: The Heart of the Movement? A Socio-political Perspective," *Politics and Animals* 1 (2015): 50–74.
44 Ibid., 51.

45 Brittany Hill, "Restoring Justice for Animal Victims," *Journal of Animal & Natural Resource Law* 17 (2021): 217–248.
46 Donovan, "Feminism and the Treatment of Animals," 306.
47 Ibid.
48 Ibid., 319.
49 Ibid., 320.

Bibliography

Arrington, Amanda K., and Markarian, Michael. "Serving Pets in Poverty: A New Frontier for the Animal Welfare Movement." *Sustainable Development Law and Policy* 18, no. 1 (January 1, 2018): 11. Retrieved from: https://digitalcommons.wcl.american.edu/cgi/viewcontent.cgi?article=1605&context=sdlp.

ASPCA. "More Than 1 Million Households Forced to Give up Their Beloved Pet Each Year, ASPCA Research Reveals." Retrieved from: https://www.aspca.org/about-us/press-releases/more-1-million-households-forced-give-their-beloved-pet-each-year-aspca, accessed July 30, 2023.

Bay Area Transformative Justice Collective. "Transformative Justice and Community Accountability." (2013). Retrieved from: https://trueleappress.files.wordpress.com/2020/02/162.-bay-area-transformative-justice-collective-batjc-transformative-justice-and-community-accountability.pdf.

Beck, Ashley N. "Giving a Voice to the Voiceless: A Prosecutor's Efforts to Combat Animal Cruelty." In *Carceral Logics: Human Incarceration and Animal Captivity*, edited by Justin Marceau and Lori Gruen, 53–69. Cambridge: Cambridge University Press, 2022. https://doi.org/10.1017/9781108919210.005

Brodie, Janine. "We Are All Equal Now: Contemporary Gender Politics in Canada." *Feminist Theory* 9, no. 2 (August 1, 2008): 145–164. https://doi.org/10.1177/1464700108090408.

CBC News. "Animal Rights Activists Sentenced to Time in Jail for 2019 Protest at B.C. Hog Farm." October 13, 2022. Retrieved from: https://www.cbc.ca/news/canada/british-columbia/b-c-animal-rights-activists-get-jail-time-1.6614762.

Donaldson, Sue, and Kymlicka, Will. "Farmed Animal Sanctuaries: The Heart of the Movement?." *Politics and Animals* 1, no. 1 (2015): 50–74. Retrieved from: http://journals.lub.lu.se/index.php/pa/article/view/15045/14797.

Donovan, Josephine. "Feminism and the Treatment of Animals: From Care to Dialogue." *Signs: Journal of Women in Culture and Society* 31, no. 2 (January 1, 2006): 305–329. https://doi.org/10.1086/491750.

Eccles, Stephanie, and Stoddard, Dr Elisabeth. *Hurricane Florence's Impact: Policies on Animals Living in Confined Animal Feeding Operations in Eastern North Carolina*. World Animal Protection, May 2021. Retrieved from: https://dkt6rvnu67rqj.cloudfront.net/cdn/ff/PYBUDryP7nQFgb1G24YcyvuMNvRGR-VeXQxxJ_PTAlQ/1620145317/public/media/World_Animal_Protection-Impact_of_Hurricane_Florence_on_CAFOs_in_North_Carolina%28May2021%29_0.pdf.

Edmonton Police Service. "Animal Cruelty Investigation Unit (ACIU)." (n.d.). Retrieved from: https://www.edmontonpolice.ca/CrimePrevention/PersonalFamilySafety/AnimalCrueltyInvestigationUnit.

Government of Canada, Department of Justice, Electronic Communications. "Bestiality and Animal Fighting (Bill C-84)." September 1, 2021. Retrieved from: https://www.justice.gc.ca/eng/csj-sjc/pl/baf-bca/index.html#:~:text=Bill%20C%2D84%20represents%20a,practices%2C%20including%20Indigenous%20harvesting%20rights.

Grant, Brianna. "'No Lack of Love': Minnesota Reservation Joins the Pets for Life Mentorship Program." *The Humane Society of the United States*, April 1, 2022. Retrieved from: https://www.humanesociety.org/news/no-lack-love.

Gruber, Aya. "Humanizing Animals, Dehumanizing Humans." In *Carceral Logics: Human Incarceration and Animal Captivity*, edited by Justin Marceau and Lori Gruen, 158–186. Cambridge: Cambridge University Press, 2022. https://doi.org/10.1017/9781108919210.012.

Hill, Brittany. "Restoring Justice for Animal Victims." *Animal & Natural Resource Law Review* 17 (2021): 217–248. Retrieved from: https://www.animallaw.info/sites/default/files/ANRLR%20Vol%2017.pdf.

Hill, Brittany. "Redefining Animal Cruelty Responses Through Restorative Justice." *Presented at the Animals and Restorative Justice Symposium,* July 29, 2022. Retrieved from: https://www.vermontlaw.edu/news-and-events/event/animals-and-restorative-justice-symposium.

The Humane Society of the United States. "Keeping Pets for Life." (2022). Retrieved from: https://www.humanesociety.org/issues/keeping-pets-life.

INCITE! Women of Color Against Violence, ed. *Color of Violence: The INCITE! Anthology*. Duke University Press, 2016. https://doi.org/10.2307/j.ctv1220mvs.

Judd, Amy. " 'I Do Promise That Molly Died Humanely': Owner Who Killed Adopted Pig Apologizes." *Global News*, February 25, 2018. Retrieved from: https://globalnews.ca/news/4047050/molly-pig-bc-spca-owner-killed-eaten/.Retrieved from:

Kelly, Esteban Lance. "Philly Stands Up: Inside the Politics and Poetics of Transformative Justice and Community Accountability in Sexual Assault Situations." *Social Justice* 37, no. 4 (2010): 44–57. Retrieved from: https://www.questia.com/library/journal/1G1-290112392/philly-stands-up-inside-the-politics-and-poetics.

Kilty, Jennifer M., and Frigon, Sylvie. *The Enigma of a Violent Woman: A Critical Examination of the Case of Karla Homolka*. Routledge, 2016. Retrieved from: https://www.routledge.com/The-Enigma-of-a-Violent-Woman-A-Critical-Examination-of-the-Case-of-Karla/Kilty-Frigon/p/book/9780367596750.

Kim, Mimi E. "From Carceral Feminism to Transformative Justice: Women-of-Color Feminism and Alternatives to Incarceration." *Journal of Ethnic & Cultural Diversity in Social Work* 27, no. 3 (May 30, 2018): 219–233. https://doi.org/10.1080/15313204.2018.1474827.

Marceau, Justin F. *Beyond Cages: Animal Law and Criminal Punishment*. Cambridge University Press, 2019. https://doi.org/10.1017/9781108277877.

Montford, Kelly Struthers, and Eva Kasprzycka. "The 'Carceral Enjoyments' of Animal Protection." In *Building Abolition*, edited by Kelly Struthers Montford and Chloë Taylor, 227–247. Abingdon and New York: Routledge, 2021.

Ontario SPCA & Humane Society. *Annual Report 2019*. Toronto: Ontario SPCA, 2019. Retrieved from: https://ontariospca.ca/wp-content/uploads/2020/06/AnnualReport-2019-Online.pdf.

Pager, Devah, Goldstein, Rebecca, Ho, Helen, and Western, Bruce. "Criminalizing Poverty: The Consequences of Court Fees in a Randomized Experiment." *American Sociological Review* 87, no. 3 (February 20, 2022): 529–553. https://doi.org/10.1177/00031224221075783.

Taylor, Chloë. "Anti-Carceral Feminism and Sexual Assault—A Defense." *Social Philosophy Today* 34 (January 1, 2018): 29–49. https://doi.org/10.5840/socphiltoday201862656.

Tepper, Danielle. "Hand to Hand: Pets for Life Delivers Peace of Mind to Philly Pet Owners." *The Humane Society of the United States,* December 10, 2021. Retrieved from: https://secured.humanesociety.org/page/78105/subscribe/1?locale=en-U.S.

Thom, Kai Cheng. "What to Do When You Have Been Abusive." *Truthout*, January 26, 2020. Retrieved from: https://truthout.org/articles/what-to-do-when-you-have-been-abusive/.

Vermont Law and Graduate School, & Adams, Sally. "Animals and Restorative Justice Symposium 2022." *YouTube,* August 1, 2022. Retrieved from: https://www.youtube.com/watch?v=VjZBFEJ7qu0.

Vermont Law and Graduate School and University of San Francisco Law. "Animals and Restorative Justice Symposium." *Vermont Law School,* July 29, 2022. Retrieved from: https://www.vermontlaw.edu/news-and-events/event/animals-and-restorative-justice-symposium

Zehr, Howard. *The Little Book of Restorative Justice: Revised and Updated*. Good Books, 2015. Retrieved from: http://ebookcentral.proquest.com/lib/ryerson/detail.action?docID=1922319.

PART X

Multispecies Futures

46
QUEER FUTURES AND MUTUAL AID

Pierre Cloutier de Repentigny

1. Introduction

Referring to the radical act of self-immolation performed, in all of its finality, by David S Buckel—a prominent human rights lawyer, gay and trans rights advocate, and environmentalist—in reaction to our societies' toxic dependence on fossil fuel, Tremblay and Swarbrick state that "Buckel's self-annihilation points to a contradiction that environmental politics too often straightens out: that to live today is to accelerate extinction."[1] This refers to the footprint Western lifestyles leave on the Earth and its impact on species past, present, and future extinction, including our own.[2] We are undeniably within the sixth mass extinction event.[3] This gloomy prospect can lead to a future without humans, for better or for worse. Nonetheless, without diminishing the radical act of Bucknel, a future where we form part of the multispecies makeup of the biosphere is possible. Imagining this future and living it will not be easy tasks. These tasks will be many, but among them, especially when thinking about preserving the "multi" of species and thinking against extinction, is challenging the dominant extractive conception of other-than-human animals as feed for the furnaces of the capitalist machine. Imagining this future is a very queer task. As Muñoz affirms, "[q]ueerness is essentially about the rejection of a here and now and an insistence on potentiality or concrete possibility for another world."[4] In this context, queerness is about the rejection of extinction and insistence on the possibility of a multispecies future.

It is beyond the scope of this chapter to offer a roadmap to this multispecies future. Instead, it adopts the more modest goal of exploring what queer theory and the practice of mutual aid can offer for the construction of this future. It first gathers lessons from the teachings of queer theory on otherness, violence, and liberation and how these concepts can inform visions of multispecies future. It then shifts towards praxis and offers the framework of mutual aid as a means of achieving a future in line with queer teachings. The chapter concludes by highlighting the importance of committing to a queer multispecies future.

DOI: 10.4324/9781003273400-57

2. Queer Liberation and Animals

Queerness, by its very nature, is a difficult concept to define. Something founded on the transgression of social norms cannot be easily captured. Nonetheless, this idea of existing outside and as counter to social norms offers enough grounding for queerness to lead to a coherent theory. As Romero states, "queer theorists materialized as a radical challenge to a constellation of liberal and legal assumptions . . . about human subjectivity, especially those concerning gender and sexuality."[5] While challenging human subjectivity is often viewed within the confines of human/social relations, the work of queer ecologies has expanded the ambit of queer theory to the other-than-human world.[6] As Chen notes, queer, in the scholarly world, has come to mean a methodology rather than a sexuality or gender, a study of queerness as "an array of subjectivities, intimacies, beings, and spaces located outside of the heteronormative"[7] or a study of the *other* than heteronormative. Indeed, my choice of nomenclature for the world external to the dominant view of humanity, the *other*-than-human world (as opposed to the often-used non-human or more-than-human), is purposefully aligned with this vision of queer theory. Not only does it reflect the rejection of hierarchies that is core to queer theory (non- and more-than reflecting a positive or negative positioning vis-à-vis the anthropos), but it directly characterises as *other* the part of the world anthropocentric cultures, like Western cultures, typically externalises.[8] Otherness refers here to a state of opposition to social norms or, as Ahmed reminds us of bell hooks' characterisation of queer, of "being at odds".[9] Queerness or otherness is standing outside of western social norms while bearing the brunt of the oppression they foster. The *other*-world is a queer world.

Other-than-human animals can be understood as the ultimate other, always at odds with anthropocentric norms, always outside of their hegemonic structures. The *other*-animal does not willingly or consciously participate in the re/production of social norms. They do not benefit from these norms. As occupying the wrong end of the human/animal socially constructed duality, they in fact suffer greatly from the enactment of social norms. Law makes this othering process quite transparent by categorising human animals as persons and legal subjects (currently as Black, Indigenous and other people have been denied personhood in the past by Western legal systems) and other-than-human animals as property and legal objects.[10] This binary allows the instrumentalisation and exploitation of the object by the subject based on anthropocentric social norms.[11] That is, Western societies view animals as devoid of the moral character needed to participate in society (i.e., to gain personhood) without actually reflecting about the value underlying and the origin of these moral characteristics. From a queer perspective these constructed "moral characteristics" attached to personhood are particularly suspicious given the history and use of morality as means to exclude "deviant" sexualities and genders from society. This legal characterisation is but a manifestation of the larger cultural othering of other-than-human animals. They are either valuable by participating within capitalist production and consumption (e.g., livestock, pets, living resources) or are relegated to exotic externalities that exist outside society (wild animals) and, when in conflict with it, need to be managed, often to the point of extinction.[12] They have no place of their own; they are not allowed to exist for their own sake. Other-than-human animals are at odds within our hegemonic ways of living.

The exclusion and devaluation inherent in the process of othering result in a power dynamic where a normative group, through oppression, maintains power over those unlike them. Oppressive systems and the violence they produce, including the oppression of other-than-human animals, are often interrelated in various ways.[13] For examples, animality is often used to diminish or characterise black people as less than human. Naturality, and the behaviours of other-animals, is often mobilised to attack non-normative sexualities and genders. I could go on. Obviously, this implicates a particular social construction of nature, animal, human, race, and so on, but that is the point. These arguments are fabricated to maintain a discourse where the power of the dominant group is legitimised. The attempt is to make these power relations seem intrinsic to our world. But, as queer theory keeps pointing out, they are not.[14] What is real is the violence and other negative consequences these oppressive systems impose on queer beings and the benefits they procure to the same group at the top of the imperialist white supremacist capitalist patriarchy pyramid.[15] This interconnection of violence and its underlying hegemony allow for a more holistic picture of oppression, the sharing of insights between different perspectives/fields (e.g., animal and queer studies), and to build bridges between movements fighting for liberation.

The violence faced by other-than-human animals is multiple and well documented: extinction or speciescide; death; treatment akin to torture; mutilation; forced reproduction; captivity; and loss of habitat/home, food sources, and kin, to name a few.[16] This violence is often framed in terms of crisis, especially the ecological crisis and the collapse of biodiversity. It is undeniable that species extinction and the continuum of (other-than-human) animal death and suffering are embedded in the current ecological crisis (as victim and/or cause). This crisis is firmly entrenched within the treadmill of production:[17] industrial agriculture for its constant abuse of "live stocks" and its large-scale destruction of habitats (directly through land-use change and indirectly through various pollutants like pesticides and greenhouse gases); industrial fisheries and aquaculture for its direct extraction and laboratory-like production of fish; and countless other industries like mining, forestry, oil and gas, and housing development that result in large-scale direct and indirect annihilation of other-than-human animals. The treadmill of production supports, maintains, and benefits from imperialist white supremacist capitalist patriarchy and its legal apparatus.[18] The result is everyday violence against the other-than-human,[19] an assemblage of harm perpetuated by humans or a subset embedded in imperialist white supremacist capitalist patriarchy.[20] This violence against other-than-human animals, like the violence against queer human animals, is pervasive, omnipresent, and intertwined with the state and the ideology it represents, what Stanley calls an atmosphere of violence.[21] Atmosphere is an apt descriptor, as it possesses, as Ahuja puts it, "a double valence: it signals both the interspecies intimacy structured by geophysical forces of the earth and the ambient senses of crisis, withering, and extermination that intensify as the underside of neoliberal freedom."[22] The queer subject, whatever and whoever they might be, is entangled in a world that, despite needing interdependency to exist, is bent on fostering death and isolation. The tragedy of violence against the other is its inescapability in the immediate.

How does one liberate the other from the atmosphere of violence? Queer theory has long contended with this question. Queerness rejects the now, rejects this atmosphere, with the hope, desire, and commitment to build a life worth living.[23] "The present [atmosphere] is not enough. It is impoverished and toxic for queers and other people who do not feel the privilege of majoritarian belonging, normative tastes, and 'rational' expectations."[24] The queer project has always been utopian and forward looking, as "[t]o inherit the past in this world for queers would be to inherit one's own disappearance."[25] This does not mean forgetting the past and ignoring the present. Rather it entails understanding the present non-queer condition to "participate in a hermeneutic that wishes to describe a collective futurity, a notion of futurity that functions as a historical materialist critique."[26] The situation of other-than-human animals, breathing in the toxic atmosphere, may appear hopeless, but the future does not have to be. It holds promise: the promise of liberation.

In detaching ourselves from the present and rejecting the political impasses that characterise it, we are automatically rejecting a path towards the future where "liberation" (if it can even be called that) is premised on including the other within the normative. This is exemplified by the concept of homonormativity, when gay people adopt or assimilate into an otherwise heteronormative lifestyle,[27] which is incompatible with a queer future.[28] Plumwood highlights the pitfall of normative integration through the application of what is valued in a dualism (nature/culture or human/other-animal) to the other, what used to be devalued, rather than rejecting and rethinking the value system altogether.[29] Thus, attempting to integrate or valorise other-than-human animals within society (often based on similarities to the characteristic used to justify oppression) continues to operate within the hegemonies they condemn (at least on the surface) and will therefore not lead to liberation. These acts simply reproduce and reconfigure human normativity. As Luciano and Chen affirm, "queer theory has long been suspicious of the politics of rehabilitation and inclusion to which liberal-humanist values lead, and because 'full humanity' has never been the only horizon for queer becoming."[30] The history of queer oppression and the inefficacy of inclusion or equality measures in liberating queer people demonstrate the severe limitations of these approaches. In other words, animalnormativity—bringing the animal into the human—will not lead to other-animal liberation.

Instead of tinkering with the machine spewing this atmosphere of violence to make the air slightly less toxic, we need to go beyond reversing the exclusion of other-than-human animals and towards an "active, deliberate and reflective positioning" of ourselves *with* the other-than-human world "against destructive and dualizing form of culture."[31] Thinking with the other-than-human world requires understanding what unites us with all others; we are finite, interdependent, embodied, sensory, vulnerable, and bound by the inevitability and omnipresence of death.[32] Our interdependency or entanglement with the other-than-human world goes beyond other-animals and includes what western cultures view as non-living, like mountains, rivers, objects, and so on.[33] After all, the ontology of the world we inhabit forms "a nontotalizable, open-ended concatenation of interrelations that blur and confound boundaries at practically any level."[34] The question of liberation then becomes: what does it mean to be part of the entangled mesh of our lifeword(s) and to exist within it in a way that will not unravel the other-than-human world?

3. Mutual Aid for Multispecies Liberation

In thinking about the futurity of multispecies liberation, there is a level of uncertainty about queer futures and how to achieve them, but this uncertainty is inherent to the process of imagining and embodying a different world, and in fact, this uncertainty relates to the idea that a queer future must be flexible and mutable, as a just future is not singular or hegemonic.[35] Detailed prescriptions or a roadmap for a queer multispecies future are not feasible within the confines of a chapter (nor should they be proclaimed by one person). Multispecies liberation is an iterative process; one that is accomplished every day by a constellation of queer acts. What is feasible is formulating some basic principles and strategies for these acts. The overarching framework I propose is that of mutual aid. "Mutual aid is a form of political participation in which people take responsibility for caring for one another and changing political conditions . . . by actually building new social relations that are more survivable."[36] Mutual aid, as a bottom-up strategy of change anchored in resisting the atmosphere of violence, offers a perfect methodology for building a queer future as a way to connect all the queer acts of liberation. Its principles are easily adaptable and applicable to queer movements entangled with the other-than-human world. But, as Ahuja warns, taking interspecies entanglement seriously is but a starting point and not an ethical end in and of itself.[37] Mutual aid is already imbued by strong ethics of care and liberation. One need only extend this ethic—which is already linked to the other-than-human given the unity, so to speak, of the hegemony and violence it opposes—to all others.

Mutual aid provides means to create coordinated collective care as radical education and alternatives that counter the extractive and hegemonic imperialist white supremacist capitalist patriarchy.[38] Spade outlines three kinds of work within the mutual aid framework:

> (a) work to dismantle existing harmful systems and/or beat back their expansion, (b) work to directly provide for people targeted by such systems and institutions, and (c) work to build an alternative infrastructure through which people can get their needs met.[39]

He adds that these acts should also demonstrate the failures of the current system and how the status of the other and its stigma are a product of the system rather than a personal fault, politicise the work and build solidarity through it, and teach skills both for the work itself and for collective and antiauthoritarian governance. These types of work neatly capture both the queer necessity of dealing with the violent present and the utopian desire to build a future free from it. Spade also outlines what mutual aid is not. It is not charity and does not make the aid conditional on some moralising idea of deservingness. It does not replicate saviourism and paternalism. He mentions the concern that mutual aid projects are being co-opted by neoliberalism. Ensuring that these projects and movements remain bottom up and governed by the participants themselves (based on horizontal, participatory, and consensus-based decision making) and maintaining autonomy from elite institutions, government, and the corporate world can help prevent co-optation. Ultimately, "[w]e might understand mutual aid projects as frontline work in a war over who will control social relations and how survival will be reproduced, especially in the face of worsening crises."[40]

One thing is conspicuously absent from mutual aid work: state-based measures (i.e., reforms). Spade summarises the issues as follows: "[r]eforms often merely tinker with existing harmful conditions, failing to reach the root causes."[41] This has been noted by many queer theorists, who have long pointed out the futility of petitioning or reforming the western state-based legal orders for better queer life.[42] Reforms may grant to some, those who assimilate, reprieve from the violence or even a hand in its propagation to *others*, but they will not liberate the queer subject from the hegemony that relies on this violence to maintain itself. The history of gay rights demonstrates that reforms allowing gay people fuller participation in normative social orders are to the detriment of more marginalised queer people (Black, Indigenous, trans, disabled, etc.) and often render those assimilating complicit in their oppression.[43] Reform offers us two equally violent options: literal death or metaphysical death through assimilation.[44] Stanley argues convincingly that "racialized anti-trans/queer violence is a necessary expression of the liberal state."[45] Violence towards the other, in its multiplicity, is a necessary precondition of the state form modernity brought forth. Treating other-than-human animals with respect and recognising their agency is not the norm for the liberal state. What is normal is their exploitation for the self-preservation of the state form. The state is an othering space for other-than-human animals; the atmosphere it produces, the one we are forced to inhabit, is violent. Reforms can alleviate some violence, but they will not stop our societies' extractivist relationship with other-animals. The state-capitalist machine won't liberate us, won't liberate *them*, the others, just because we gave it a shiny new coat of paint and tinkered with it so its violence would be more "humane."

The queer future we can build through mutual aid will thus be outside the state. As articulated by Benjamin, the end of the violence cycle embedded in the legal form and the hegemonic forces that maintain it will come with the abolition of state power.[46] The concept of abolition is quite apt for queer futurity. In a way, we are trying to abolish the present condition to create conditions for a new liveable future. The current order perpetuating imperialist white supremacist capitalist patriarchy, represented by the state form, simply cannot continue if we are to end the atmosphere of violence that is so foundational to its existence.[47] Abolishing the state is also a precondition to the emergence of a different way or organising ourselves *with* the other-than-human world instead of separate from it and at odds with it. Mutual aid projects can start building this alternative and demonstrate how a different life, a queer life, is not only possible but desirable. Ultimately, the state must die for the queer to live. The state must be abolished for the other-animals to be liberated.

Mutual aid offers instead a practical way forward to reform ethical relationships with other-than-human animals as part of the larger project of liberation. While discussions of the ethical norms of queer theory are often fraught, I agree with Zanghellini that queer ethics offers something more than the rejection of social normativity; its political commitments, some outlined in the previous section of this chapter, make it quite clear.[48] In fact, we can make the distinction between social norms (morality) and the reflective practice of relating in a positive way to others (ethics).[49] Queer ethics could thus be qualified as a form of iterative virtue ethics guided by broad principles of openness, mutuality, care, autonomy, and non-oppression. What this body of principles refers to in its essence is the ability of individuals to exist fully without fear of judgement and live without oppression, on the one hand, and, on the other, to relate to others in a non-violent matter, taking into account our mutual ethical responsibilities, that is, the need to care for each other and support each other in solidarity for the betterment of all and with intellectual openness.[50] The imprecise contour of this approach is very much in line with the unwritten nature of queer futurity; how could we write precise codes of conduct for a utopia that does not yet exist? It is through the practice of queer ethics and the framework of mutual aid that we can develop a broader and more detailed understanding of multispecies queer ethics. This ethic can be "organically" developed through the communities of care and mutual aid that are formed both through daily practice[51] and through respectful and transparent practices of conflict resolution.[52]

With these general and non-exhaustive guidelines and principles in mind, what would multispecies mutual aid look like? There is obviously not one model, and the context of each "mutual aid project" (whether it is a specific project or the ways of life of a particular community) will greatly influence how multispecies mutual aid can be operationalised. I therefore try to offer two broad examples to canvas the praxis of multispecies mutual aid in two different contexts (situated in the land we now call Canada): one urban, one rural (both imagined in the St. Lawrence valley, as it is the land I am most familiar with). These contexts are also anchored in communities that are predominantly settler-colonial (again due to my positionality). Before going into the two examples, I will start with some overarching issues and end with a few additional relevant reflections to assist in thinking and developing the type of multispecies future I have painted throughout this chapter.

Other-than-human animal participants in a mutual aid project are entering in a mutually beneficial (or reciprocal) system of cooperation. They are thus not instruments for the realisation of the project (e.g., as a source of food or as forced labour) but are considered part of the queer community that is formed on an equal footing with all other participants (human or otherwise). The other-animal must therefore be a willing participant and be allowed to exist as they are rather than conforming to our expectations (in line with the ethics of openness and autonomy). While it is more difficult to communicate with other-animals, anyone who spends any amount of time with an other-animal companion knows that is it far from impossible and often becomes easier with time and attention. We may not be able to get a cat's consent, for example, but their desire to stay with us and return to us is a clear sign that they wish to continue the multispecies relationship that has formed.[53] In fact, the caring nature of mutual aid partnership (no excessive work, respecting boundaries, showing love, having all the necessities, etc.) ensures, in a way, that the other-animal partners have a desire to stay and continue to contribute to the project. In many respects, this will be established on a case-by-case basis (as it always is when human animals partner up), but it is clear that certain ways of relating are excluded due to their violation of the core ideals of queer ethics. Some examples include involuntary and unnecessary medical or other treatments (e.g., experimenting on animals), motivating the labour of other-animals through violence (e.g., whipping a horse to make them move or obey), confining other-animals (e.g., keeping pigs inside a small barn at all time where they can barely move), and capturing wild other-animals (there are exceptions when, for example, an individual is injured and cannot be returned to the wild after treatment).

In a rural setting, one way of building a multispecies mutual aid project could be through food-based initiatives like a small farm collective that establishes good relations with other-than-human animal participants. Given the large-scale habitat destruction (particularly wetlands in the St Lawrence Valley) in the name of agriculture, such a project should not require the creation of new arable land. The work of solitary and liberation embedded in mutual aid does not stop at some defined project border but should take into account its larger impact, here on wild other-animals that will likely share the space occupied by the project. If we were to treat other-human partners with the greatest care and respect but do so through wanton destruction of wildlife, then we are simply reproducing the atmosphere of violence so prevalent today. Within ecologically sustainable practices (e.g., not using pesticides and chemical fertilisers, not destroying habitats, using the appropriate plant species for the environment, etc.), a farming collective can emerge where other-animals have a crucial role to play. Some of the relationships will be more direct, as the partners will directly inhabit the space shared with humans. For example, pigs serve as a great way to compost food, as they are able to eat pretty much any organic matter that will not be used for other purposes (e.g., part of plants that are not edible for humans) and in return produce fertilisers for the next crop. The pigs get a source of food and shelter, in addition to mutual companionship, as part of the contribution. Goats can participate by offering milk and controlling overgrown vegetation (goats are, after all, the ultimate lawnmower). Sheep can offer wool to make clothing. Horses can assist with mobility across the land and carrying equipment, stocks, and so on to where they are needed. Bees can be kept in hives to ensure the pollination of the crops and offer honey and wax. An example of indirect contribution from other animals could be restoring a nearby bat habitat, which in return could lead to increased pollination and better insect population control. The bats get a stable home and a source of food in return for their participation in the collective. Another example includes maintaining a healthy soil to ensure a stable insect population both as a source of food for other species and as key players in a healthy plant ecosystem, which ultimately contributes to the collective. These relationships are more distant, but they are still mutually beneficial and caring and participate in a broader understanding of multispecies solidarity. Ultimately, the project creates a (relatively) self-sufficient community of care where the food produced would assist with the needs of all involved and others who are in need (potentially in partnership with an urban project that can in return aid the collective). It would create non-violent and alternative infrastructures of food production and care.

This form of mutual aid project may not appear as obvious in an urban setting, but they are equally feasible in that context. One important aspect, even more so than in a rural setting given population density, is living sustainably. While this aspect bears repeating in a western-settler context, the idea of living in harmony or in balance with the rest of the living and non-living world is far from new and is deeply integrated into different worldviews such as Indigenous and Buddhist cosmovisions.[54] It might be harder for a mutual aid project to be self-sufficient in a city given the greater need to rely on products and services that are outside of the project's control, but there are ways to mitigate our ecological footprints even within state systems, and the connection and solidarity work between projects can also assist in greatly reducing our reliance on unsustainable goods and services. In fact, this interconnection of mutual aid communities contributes to a multispecies future by relying on, for example, the food produced by the collective briefly described previously rather than industrial agriculture and all of its violence. The urban project can provide, in support, services and goods that are not as easily attainable in a rural setting, such as healthcare for all participants, the production of various tools, and access to larger populations for produces originating in the rural community. By ensuring a thriving rural mutual aid community, the urban one participates in fostering queer ethics and good multispecies relations. Given the stressful nature of urban life, a key contribution of other-animal participants like cats and dogs is their ability to improve our general wellbeing. Mutual aid participation extends beyond the material realm by providing, for example, emotional support and valuable teaching on governance and life itself, a theme especially present within Indigenous cosmovisions.[55] Other-animals can be great teachers for those who are willing to "listen." Additionally, as much as humans provide care labour to other-animals, they clearly return the favour in improving our own mental health.[56] As such, mutual support should not be conceptualised as only material support (including for human animals). In exchange for this, we care for other-animals, including by providing them with food and shelter. In addition to what are traditionally viewed as pets, humans can create these bonds with wild other-animals in urban spaces too. For example, a person started to build in their backyard small structures meant for other-animals (frogs in this case).[57] While they only had one species in mind at first, this small effort ended up recreating a small ecosystem, which undoubtedly contributes to the ecological health of the urban space and the support of other-than-human animals, with whom we share spaces, even urban ones, in addition to the wellbeing of the person. Other examples that people are already developing include keeping chickens in urban gardens to help control insect population and have access to eggs and keeping beehives and planting native flowers for pollination, honey, and wax. Finally, urban mutual aid projects can also bear a greater burden of the solidarity work between projects and movements, what could be called the "purely" political work of mutual aid. Through this support, they assist in the greater work of liberation and provide greater access for humans to queer models of multispecies relationships.

In all projects, it should be noted that the complexity of human relationships often leads to conflict and that it would be naïve to believe that none involving other-animals will arise. Mutual aid is built on the knowledge that disputes will happen and that consequently a conflict resolution system must exist. This process is built on collective, transparent, and non-hierarchical decision-making.[58] The process will vary depending on the community involved, but the limited oral communication skills of other-animals (at least from our human perspective) could make things difficult. Nonetheless, if all human participants in a multispecies mutual aid project take the collective responsibility of caring for other-animal participants and advocating for their wellbeing and pay attention to the behaviour and interaction of the other-animal participants, the conflict resolution process will have a much greater chance of fully and fairly considering the interests of other-animals. This is not to say it will be easy, but human conflicts are not either. The fact that some participants are not human is not an excuse to avoid a fair process built on queer ethics.

While I have relied in part on Indigenous knowledge to construct this queer multispecies futurity, I have deliberately avoided expressing views on how Indigenous peoples should relate to other-than-human animals. This is not only because it is not my place as a white settler but also because Indigenous peoples already have systems of multispecies relationships. The issue is that they have been disrupted to various degrees by settler colonialism.[59] The work in this context is thus more one of decolonisation to ensure their ability to live according to their rich and long history of mutually beneficial relationships between all living (and non-living) things.[60] There is obviously solidarity that can and should be built between Indigenous and queer understandings of multispecies futures,[61] but we (as in settlers) must do so on the terms of Indigenous nations and communities.

Overall, mutual aid projects for a multispecies future allow for human animals to learn how to form meaningful and respectful (queer) relationships with other-animals, education being a central tenet of mutual aid. This is not only beneficial for the individuals themselves, but it also serves as a behavioural model for others, demonstrating the possibility of a queer multispecies future. If we increase projects like the ones described, even for those that are not as obvious, and if we ensure that other-than-human animals are considered equally in them,[62] we can, brick by brick, year by year, slowly create a viable alternative to the *state* of the atmosphere of violence.

4. Conclusion

The oppression that comes with being othered often makes the future appear hopeless. The omnipresence of violence engendered by oppression with the tacit or explicit involvement of the state, the institution supposed to represent the public interest, creates the illusion that opposition to the machine that chews the queer subject up for the benefit of non-othered others is pointless. In the case of other-than-human animals, otherness is at a level where agency and the ability to feel physical and emotional pain of many species are still being questioned or doubted by those who benefit from the exploitation of those species. When the system is built to extract from other-animals in a multitude of ways, to the point of creating an atmosphere of violence, and when reforming the system is presented as the only legitimate option, one can wonder if other-than-human animal liberation is even possible. In this bleakness, the queer gaze helps to disturb, disorient, and disorganise the normative order of things.[63] If there is one overarching teaching to get from queer ecologies, it is that this order—the status quo, social normativity—has nothing natural to it. This opens the door then to other possibilities, queer possibilities of a different life. Learning from the past and its impact on the present, we can—no, we should—imagine and start living towards a future free from the violent imperialist white supremacist capitalist patriarchal machine, a queer multispecies future. This utopian vision is not as far out of reach as it may appear. The mutual aid movement lights the way forward by demonstrating how new ways of living can be constructed outside the state, in opposition to its hegemony, and with care for all involved. There is nothing stopping us from adapting these tactics to integrate other-than-human animals and start building a multispecies future, in solidarity and complementarity to other movements seeking to liberate the others.[64] "A queer politics does involve a commitment to a certain way of inhabiting the world,"[65] and we must start inhabiting this world in a queer way to liberate other-than-human animals, to liberate ourselves.

Notes

1 Jean-Thomas Tremblay and Steven Swarbrick, "Destructive Environmentalism: The Queer Impossibility of *First Reformed*," *Discourse* 43, no. 1 (2021): 4.
2 I add the qualifier Western to highlight that it is not all human life that impacts the biosphere in an extinction-inducing manner (see Davis and Todd, 2017).
3 Gerardo Ceballos et al., "Accelerated Modern Human–Induced Species Losses: Entering the Sixth Mass Extinction," *Science Advances* 1, no. 5 (2015): e1400253.
4 José Estéban Muñoz, *Cruising Utopia: The There and Then of Queer Futurity* (New York University Press, 2019), 1.
5 Adam Romero, "Methodological Descriptions: 'Feminist' and 'Queer' Legal Theories," in *Feminist and Queer Legal Theory: Intimate Encounters, Uncomfortable Conversation*, eds. Martha Albertson Fineman, Jack E. Jackson, and Adam P. Romero (Ashgate Publishing, 2009), 190.
6 Dana Luciano and Mel Y. Chen, "Has the Queer Ever Been Human?" *GLQ* 21, no. 2–3 (2015): 183. Catriona Sandilands, "Queer Ecology," in *Keywords for Environmental Studies*, eds. Joni Adamson, William A. Gleason and David N. Pellow (New York: NYU Press, 2016), 169.
7 Mel Y. Chen, *Animacies: Biopolitics, Racial Mattering, and Queer Affect* (Duke University Press, 2012), 104.
8 Christine Bauhardt, "Rethinking Gender and Nature from a Material(ist) Perspective: Feminist Economics, Queer Ecologies, and Resource Politics," *European Journal of Women's Studies* 20, no. 4 (2013): 361. Greta Gaard, "Toward a Queer Ecofeminism," *Hypatia* 12, no. 1 (1997): 137. Kim TallBear, "Beyond the Life/Not-Life Binary: A Feminist-Indigenous Reading of Cryopreservation, Interspecies Thinking, and the New Materialisms," in *Cryopolitics: Frozen Life in a Melting World*, eds. Joanna Radin and Emma Kowal (MIT Press, 2017), 179.
9 Sara Ahmed, "Feminism as Lifework: A Dedication to bell hooks," online (blog): *feministkilljoys* https://feministkilljoys.com/2022/09/20/feminism-as-lifework-a-dedication-to-bell-hooks/, accessed September 20, 2022
10 Wendy A. Adams, "Human Subjects and Animal Objects: Animals as 'Other' in the Law," *Journal of Animal Law and Ethics* 3 (2009): 29
11 This should not be read as viewing subjectification as always good and desirable and objectification as always bad and demeaning (e.g., Gayle Rubin, "Misguided, Dangerous, and Wrong: An Analysis of AntiPornography Politics," in *Bad Girls and Dirty Pictures: The Challenge to Reclaim Feminism*, eds. Allison Assiter and Avedon Carol (Pluto Press, 1993), 18; Catriona Sandilands, "Eco Homo: Queering the Ecological Body Politic," *Social Philosophy Today* 19 (2003)). The binary of subject/object is the problem, not the categories themselves. We should critique the

power dynamic resulting from the hierarchy produced by the binary as much as we ought to critique, as I am doing here, the material effects of the application of the binary.

12 See: David Nibert, ed., *Animal Oppression and Capitalism*, Vol. 1 & 2 (Praeger Publishers, 2017).

13 Chen, *Animacies: Biopolitics, Racial Mattering, and Queer Affect*. Eli Clare, *Exile and Pride: Disability, Queerness, and Liberation* (Duke University Press, 2015). Gaard, "Toward a Queer Ecofeminism." Anna Grear, "Deconstructing *Anthropos*: A Critical Legal Reflection on 'Anthropocentric' Law and Anthropocene 'Humanity'" *Law and Critique* 26 (2015): 225. Karen J Warren, *Ecofeminist Philosophy: A Western Perspective on What It Is and Why It Matters* (Rowman & Littlefield, 2000).

14 Gaard, "Toward a Queer Ecofeminism." Sandilands, "Eco Homo: Queering the Ecological Body Politic," 17. Catriona Mortimer-Sandilands and Bruce Erickson, "Introduction: A genealogy of Queer Ecologies," in *Queer Ecologies: Sex, Nature, Politics, Desire*, eds. Catriona Mortimer-Sandilands and Bruce Erickson (Indiana University Press, 2010), 1.

15 bell hooks, *Killing Rage: Ending Racism* (Holt, 1995). bell hooks, *Writing Beyond Race: Living Theory and Practice* (Routledge, 2013)

16 Victoria Braithwaite, *Do Fish Feel Pain?* (Oxford University Press, 2010). Gerardo Ceballos, Paul R Ehrlich and Rodolfo Dirzo, "Biological Annihilation via the Ongoing Sixth Mass Extinction Signaled by Vertebrate Population Losses and Declines," *PNAS* 114, no. 30 (2017): E6089. Sandra Diaz et al, *Summary for Policymakers of the Global Assessment Report on Biodiversity and Ecosystem Services of the Intergovernmental Science-Policy Platform on Biodiversity and Ecosystem Services* (IPBES, 2019). Rodolfo Dirzo et al, "Defaunation in the Anthropocene," *Science* 345, no. 6195 (2014): 401. Paul Eggleton, "The State of the World's Insects," *Annual Review of Environment and Resources* 45, no. 61 (2020). Gail A. Eisnitz, *Slaughterhouse: The Shocking Story of Greed, Neglect, And Inhumane Treatment Inside the U.S. Meat Industry*, 2nd ed (Prometheus, 2006). Philip Lymbery, *Farmageddon: The True Cost of Cheap Meat* (Bloomsbury, 2014)

17 Kenneth A. Gould, David N. Pellow, and Allan Schnaiberg, *The Treadmill of Production: Injustice & Unsustainability in the Global Economy* (Routledge, 2016).

18 Gould, Pellow and Schnaiberg, *The Treadmill of Production: Injustice & Unsustainability in the Global Economy*. Michael M'Gonigle and Louise Takeda, "The Liberal Limits of Environmental Law: A Green Legal Critique," *Pace Environmental Law Review* 30, no. 3 (2013): 1005. Eric A. Stanley, *Atmospheres of Violence: Structuring Antagonism and the TransQueer Ungovernable* (Duke University Press, 2021).

19 Neel Ahuja, "Intimate Atmospheres: Queer Theory in a Time of Extinctions," *GLQ* 21, no. 2–3 (2015): 365

20 Pierre Cloutier de Repentigny, "Responsibility in End Time: Environmental Harm and the Role of Law in the Anthropocene," in *Green Criminology and the Law*, eds. James Gacek and Richard Jochelson (Palgrave Macmillan, 2022), 235. Grear, "Deconstructing *Anthropos*: A Critical Legal Reflection on 'Anthropocentric' Law and Anthropocene 'Humanity'."

21 Stanley, *Atmospheres of Violence: Structuring Antagonism and the TransQueer Ungovernable*.

22 Ahuja, "Intimate Atmospheres: Queer Theory in a Time of Extinctions," 367.

23 Sara Ahmed, *Queer Phenomenology: Orientations, Objects, Others* (Duke University Press, 2006). Muñoz, *Cruising Utopia: The There and Then of Queer Futurity*

24 Muñoz, *Cruising Utopia: The There and Then of Queer Futurity*, 27

25 Ahmed, *Queer Phenomenology: Orientations, Objects, Others*, 179.

26 Muñoz, *Cruising Utopia: The There and Then of Queer Futurity*, 26.

27 Lisa Duggan, "The New Homonormativity: The Sexual Politics of Neoliberalism," in *Materializing Democracy*, eds. Russ Castronovo and Dana D Nelson (Duke University Press, 2002), 175.

28 Muñoz, *Cruising Utopia: The There and Then of Queer Futurity*.

29 Val Plumwood, "Feminism and Ecofeminism: Beyond the Dualistic Assumptions of Women, Men and Nature," *The Ecologist* 22, no. 1 (1992): 8

30 Luciano and Chen, "Has the Queer Ever Been Human?" 188

31 Plumwood, "Feminism and Ecofeminism: Beyond the Dualistic Assumptions of Women, Men and Nature," 13. See also: Gaard, "Toward a Queer Ecofeminism." James Stanescu, "Species Trouble: Judith Butler, Mourning, and the Precarious Lives of Animals," *Hypatia* 27, no. 3 (2012): 567

32 Stanescu, "Species Trouble: Judith Butler, Mourning, and the Precarious Lives of Animals."

33 Chen, *Animacies: Biopolitics, Racial Mattering, and Queer Affect*. TallBear, "Beyond the Life/Not-Life Binary: A Feminist-Indigenous Reading of Cryopreservation, Interspecies Thinking, and the New Materialisms"
34 Timothy Morton, "Queer Ecology," *PMLA* 125, no. 2 (2010): 275.
35 Alexandre Martins and Caia Maria Coelho, "Notes on the (Im)possibilities of an Anti-colonial Queer Abolition of the (Carceral) World," *GLQ* 28, no. 2 (2022): 207
36 Dean Spade, "Solidarity Not Charity: Mutual Aid for Mobilization and Survival," *Social Text* 38, no. 1 (2020): 136
37 Ahuja, "Intimate Atmospheres: Queer Theory in a Time of Extinctions."
38 Spade, "Solidarity Not Charity: Mutual Aid for Mobilization and Survival"
39 Ibid., 134.
40 Ibid., 147.
41 Ibid., 132.
42 Florence Ashley, "Don't Be So Hateful: The Insufficiency of Anti-Discrimination and Hate Crime Laws in Improving Trans Well-Being," *University of Toronto Law Journal* 68, no. 1 (2018): 1. Dean Spade, *Normal Life: Administrative Violence, Critical Trans Politics, and the Limits of Law* (Duke University Press, 2015). Stanley, *Atmospheres of Violence: Structuring Antagonism and the TransQueer Ungovernable*.
43 Stanley, *Atmospheres of Violence: Structuring Antagonism and the TransQueer Ungovernable*.
44 Ahmed, *Queer Phenomenology: Orientations, Objects, Others*
45 Stanley, *Atmospheres of Violence: Structuring Antagonism and the TransQueer Ungovernable*, 10
46 Walter Benjamin, "Critique of Violence," in *Walter Benjamin: Selected Writings Volume 1, 1912–1926*, eds. Marcus Bullock and Michael W Jennings (Harvard University Press, 1996), 236.
47 Stanley, *Atmospheres of Violence: Structuring Antagonism and the TransQueer Ungovernable*. Marquis Bey and Jesse A. Goldberg, "Queer as in Abolition Now!" *GLQ* 28, no. 2 (2022): 159. Martins and Coelho, "Notes on the (Im)possibilities of an Anti-colonial Queer Abolition of the (Carceral) World."
48 Aleardo Zanghellini, "Queer, Antinormativity, Counter-Normativity and Abjection," *Griffith Law Review* 18, no. 1 (2009): 1.
49 Carlos A. Ball, "Sexual Ethics and Postmodernism in Gay Rights Philosophy," *North Carolina Law Review* 80 (2002): 371.
50 See: Ball, "Sexual Ethics and Postmodernism in Gay Rights Philosophy." Spade, "Solidarity Not Charity: Mutual Aid for Mobilization and Survival." Hil Malatino, *Trans Care* (University of Minnesota Press, 2020). Kim Tallbear, "Close Encounters of the Colonial Kind," *American Indian Culture and Research Journal* 45, no. 1 (2021): 157.
51 Billy-Ray Belcourt and Lindsay Nixon, "What Do We Mean by Queer Indigenous Ethics?" online: *canadianart*, retrieved from: https://canadianart.ca/features/what-do-we-mean-by-queerindigenousethics/, accessed May 28, 2018
52 Spade, "Solidarity Not Charity: Mutual Aid for Mobilization and Survival."
53 One only has to browse YouTube for a while to see how diverse, dynamic, and complex multispecies relations are, including species often believed to be incapable of complex emotions like insects or fish.
54 Pierre Cloutier de Repentigny, "Responsibility in End Time: Environmental Harm and the Role of Law in the Anthropocene." Heather Davis and Zoe Todd, "On the Importance of a Date, or, Decolonizing the Anthropocene," *ACME* 16, no. 4 (2017): 761. Pragati Sahni, *Environmental Ethics in Buddhism: A Virtues Approach* (Routledge, 2008).
55 See: Rochelle Baker, "I Swam with the Salmon—They Taught Me about Dignity and Strength," online: *National Observer*, retrieved from: https://www.nationalobserver.com/2022/10/06/news/i-swam-salmon-they-taught-me-about-dignity-and-strength, accessed October 6, 2022. John Borrows, "Living between Water and Rocks: First Nations, Environmental Planning and Democracy," *The University of Toronto Law Journal* 47, no. 4 (1997): 417. Alexis Pauline Gumbs, *Undrowned: Black Feminist Lessons from Marine Mammals* (AK Press, 2020). Zoe Todd, "Fish Pluralities: Human–animal Relations and Sites of Engagement in Paulatuuq, Arctic Canada," *Études/Inuit/Studies* 38, no. 1–2 (2014): 217.
56 For example, Andrea Beetz et al., "Psychosocial and Psychophysiological Effects of Human–animal Interactions: The Possible Role of Oxytocin," *Frontiers in Psychology, Art* 3 (2012): 234

57 See https://www.tumblr.com/take-a-dip-in-the-deadpool/710384161188429824/if-you-like-frogs-or-possums-or-cool-builds-or
58 Spade, "Solidarity Not Charity: Mutual Aid for Mobilization and Survival"
59 Davis and Todd, "On the Importance of a Date, or, Decolonizing the Anthropocene." Michael Simpson, "The Anthropocene as Colonial Discourse," *Environment and Planning D: Society and Space* 38, no. 1 (2020): 53. Ruba Salih and Olaf Corry, "Displacing the Anthropocene: Colonisation, Extinction and the Unruliness of Nature in Palestine," Environment and Planning E: Nature and Space 5, no. 1 (2022): 381, https://doi.org/10.1177/2514848620982834
60 See: Borrows, "Living between Water and Rocks: First Nations, Environmental Planning and Democracy." Todd, "Fish Pluralities: Human–animal Relations and Sites of Engagement in Paulatuuq, Arctic Canada." TallBear, "Beyond the Life/Not-Life Binary: A Feminist-Indigenous Reading of Cryopreservation, Interspecies Thinking, and the New Materialisms." Kim TallBear, "Making Love and Relations Beyond Settler Sex and Family," in *Making Kin Not Population*, eds. Adele E. Clarke and Donna Haraway (Prickly Paradigm Press, 2018), 145.
61 Driskill, Qwo-Li, *Asegi Stories* (University of Arizona Press, 2016).
62 For example, prison abolitionist movements may not be an obvious site, but Struthers Montford's chapter demonstrates that there is a link. Additionally, many movements and projects involving Indigenous peoples will have embedded in them considerations for the other-than-human world (TallBear, 2017).
63 Ahmed, *Queer Phenomenology: Orientations, Objects, Others*
64 Maneesha Deckha, "Toward a Postcolonial, Posthumanist Feminist Theory: Centralizing Race and Culture in Feminist Work on Nonhuman Animals," *Hypatia* 27, no. 3 (2012): 527. Spade, "Solidarity Not Charity: Mutual Aid for Mobilization and Survival."
65 Ahmed, *Queer Phenomenology: Orientations, Objects, Others*, 161.

Bibliography

Adams, Wendy A. "Human Subjects and Animal Objects: Animals as 'Other' in the Law." *Journal of Animal Law and Ethics* 3, no. 29 (2009).

Ahmed, Sara. *Queer Phenomenology: Orientations, Objects, Others*. Duke University Press, 2006.

Ahmed, Sara. "Feminism as Lifework: A Dedication to bell hooks." *feministkilljoys* online (blog), September 20, 2022. Retrieved from: https://feministkilljoys.com/2022/09/20/feminism-as-lifework-a-dedication-to-bell-hooks/

Ahuja, Neel. "Intimate Atmospheres: Queer Theory in a Time of Extinctions." *GLQ* 21, no. 2–3 (2015): 365.

Ashley, Florence. "Don't Be So Hateful: The Insufficiency of Anti-Discrimination and Hate Crime Laws in Improving Trans Well-Being." *University of Toronto Law Journal* 68, no. 1 (2018).

Baker, Rochelle. "I Swam with the Salmon—They Taught Me about Dignity and Strength." *National Observer*, October 6, 2022. Retrieved from: https://www.nationalobserver.com/2022/10/06/news/i-swam-salmon-they-taught-me-about-dignity-and-strength

Ball, Carlos A. "Sexual Ethics and Postmodernism in Gay Rights Philosophy." *Carolina Law Review*, 80 North (2002).

Bauhardt, Christine, "Rethinking Gender and Nature from a Material(ist) Perspective: Feminist Economics, Queer Ecologies, and Resource Politics." *European Journal of Women's Studies* 20, no. 4 (2013).

Beetz, Andrea et al. "Psychosocial and Psychophysiological Effects of Human–animal Interactions: The Possible Role of Oxytocin." *Frontiers in Psychology* 3 (2012).

Belcourt, Billy-Ray, and Nixon, Lindsay. "What Do We Mean by Queer Indigenous Ethics?" *canadianart* online, May 28, 2018. Retrieved from: https://canadianart.ca/features/what-do-we-mean-by-queerindigenousethics/

Benjamin, Walter. "Critique of Violence." In *Walter Benjamin: Selected Writings Volume 1, 1912–1926*, edited by Marcus Bullock and Michael W. Jennings. Harvard University Press, 1996.

Bey, Marquis, and Goldberg, Jesse A. "Queer as in Abolition Now!" *GLQ* 28, no. 2 (2022).

Borrows, John. "Living between Water and Rocks: First Nations, Environmental Planning and Democracy." *The University of Toronto Law Journal* 47, no. 4 (1997).

Braithwaite, Victoria. *Do Fish Feel Pain?* Oxford University Press, 2010.
Ceballos, Gerardo et al. "Accelerated Modern Human–Induced Species Losses: Entering the Sixth Mass Extinction." *Science Advances* 1, no. 5 (2015).
Ceballos, Gerardo, Ehrlich, Paul R., and Dirzo, Rodolfo. "Biological Annihilation via the Ongoing Sixth Mass Extinction Signaled by Vertebrate Population Losses and Declines." *PNAS* 114, no. 30 (2017).
Chen, Mel Y. *Animacies: Biopolitics, Racial Mattering, and Queer Affect*. Duke University Press, 2012.
Clare, Eli. *Exile and Pride: Disability, Queerness, and Liberation*. Duke University Press, 2015.
Cloutier de Repentigny, Pierre. "Responsibility in End Time: Environmental Harm and the Role of Law in the Anthropocene." In *Green Criminology and the Law*, James Gacek and Richard Jochelson. Palgrave Macmillan, 2022.
Davis, Heather, and Todd, Zoe. "On the Importance of a Date, or, Decolonizing the Anthropocene." *ACME* 16, no. 4 (2017).
Deckha, Maneesha. "Toward a Postcolonial, Posthumanist Feminist Theory: Centralizing Race and Culture in Feminist Work on Nonhuman Animals." *Hypatia* 27, no. 3 (2012).
Diaz, Sandra et al. *Summary for Policymakers of the Global Assessment Report on Biodiversity and Ecosystem Services of the Intergovernmental Science-Policy Platform on Biodiversity and Ecosystem Services*. IPBES, 2019.
Dirzo, Rodolfo et al. "Defaunation in the Anthropocene." *Science* 345, no. 6195 (2014).
Driskill, Qwo-Li. *Asegi Stories*. University of Arizona Press, 2016.
Duggan, Lisa. "The New Homonormativity: The Sexual Politics of Neoliberalism." In *Materializing Democracy*, edited by Russ Castronovo and Dana D. Nelson. Duke University Press, 2002.
Eggleton, Paul. "The State of the World's Insects." *Annual Review of Environment and Resources* 45 (2020).
Eisnitz, Gail A. *Slaughterhouse: The Shocking Story of Greed, Neglect, And Inhumane Treatment Inside the U.S. Meat Industry*. 2nd ed. Prometheus, 2006.
Gaard, Greta. "Toward a Queer Ecofeminism." *Hypatia* 12, no. 1 (1997).
Gould, Kenneth A., Pellow, David N., and Schnaiberg, Allan. *The Treadmill of Production: Injustice & Unsustainability in the Global Economy*. Routledge, 2016.
Grear, Anna. "Deconstructing *Anthropos*: A Critical Legal Reflection on 'Anthropocentric' Law and Anthropocene 'Humanity.'" *Law and Critique* 26 (2015).
Gumbs, Alexis Pauline. *Undrowned: Black Feminist Lessons from Marine Mammals*. AK Press, 2020.
hooks, bell. *Killing Rage: Ending Racism*. Holt, 1995.
hooks, bell. *Writing Beyond Race: Living Theory and Practice*. Routledge, 2013.
Luciano, Dana, and Chen, Mel Y. "Has the Queer Ever Been Human?." *GLQ* 21, no. 2–3 (2015).
Lymbery, Philip. *Farmageddon: The True Cost of Cheap Meat*. Bloomsbury, 2014.
Malatino, Hil. *Trans Care*. University of Minnesota Press, 2020.
Martins, Alexandre, and Coelho, Caia Maria. "Notes on the (Im)possibilities of an Anti-colonial Queer Abolition of the (Carceral) World." *GLQ* 28, no. 2 (2022).
M'Gonigle, Michael, and Takeda, Louise. "The Liberal Limits of Environmental Law: A Green Legal Critique." *Pace Environmental Law Review* 30, no. 3 (2013).
Mortimer-Sandilands, Catriona, and Erickson, Bruce. "Introduction: A Genealogy of Queer Ecologies." In *Queer Ecologies: Sex, Nature, Politics, Desire*, edited by Catriona Mortimer-Sandilands and Bruce Erickson. Indiana University Press, 2010.
Morton, Timothy. "Queer Ecology." *PMLA* 125, no. 2 (2010).
Muñoz, José Estéban. *Cruising Utopia: The There and Then of Queer Futurity*. New York University Press, 2019.
Nibert, David, ed. *Animal Oppression and Capitalism*. Vol. 1 & 2. Praeger Publishers, 2017.
Plumwood, Val. "Feminism and Ecofeminism: Beyond the Dualistic Assumptions of Women, Men and Nature." *The Ecologist* 22, no. 1 (1992).
Romero, Adam. "Methodological Descriptions: 'Feminist' and 'Queer' Legal Theories." In *Feminist and Queer Legal Theory: Intimate Encounters, Uncomfortable Conversation*, edited by Martha Albertson Fineman, Jack E. Jackson, and Adam P. Romero. Ashgate Publishing, 2009.

Rubin, Gayle. "Misguided, Dangerous, and Wrong: An Analysis of AntiPornography Politics." In *Bad Girls and Dirty Pictures: The Challenge to Reclaim Feminism*, edited by Allison Assiter and Avedon Carol. Pluto Press, 1993.
Sahni, Pragati. *Environmental Ethics in Buddhism: A Virtues Approach*. Routledge, 2008.
Salih, Ruba, and Corry, Olaf. "Displacing the Anthropocene: Colonisation, Extinction and the Unruliness of Nature in Palestine." *Environment and Planning: Nature and Space* 5, no. 1 (2022).
Sandilands, Catriona. "Eco Homo: Queering the Ecological Body Politic." *Social Philosophy Today* 19 (2003).
Sandilands, Catriona. "Queer Ecology." In *Keywords for Environmental Studies*, edited by Joni Adamson, William A. Gleason, and David N. Pellow. New York: NYU Press, 2016.
Simpson, Michael. "The Anthropocene as Colonial Discourse." *Environment and Planning: Society and Space* 38, no. 1 (2020).
Spade, Dean. *Normal Life: Administrative Violence, Critical Trans Politics, and the Limits of Law*. Duke University Press, 2015.
Spade, Dean. "Solidarity Not Charity: Mutual Aid for Mobilization and Survival." *Social Text* 38, no. 1 (2020).
Stanescu, James. "Species Trouble: Judith Butler, Mourning, and the Precarious Lives of Animals." *Hypatia,* 27, no. 3 (2012).
Stanley, Eric A. *Atmospheres of Violence: Structuring Antagonism and the TransQueer Ungovernable*. Duke University Press, 2021.
TallBear, Kim. "Beyond the Life/Not-Life Binary: A Feminist-Indigenous Reading of Cryopreservation, Interspecies Thinking, and the New Materialisms." In *Cryopolitics: Frozen Life in a Melting World*, edited by Joanna Radin and Emma Kowal. MIT Press, 2017.
TallBear, Kim. "Making Love and Relations Beyond Settler Sex and Family." In *Making Kin Not Population,* edited by Adele E. Clarke and Donna Haraway. Prickly Paradigm Press, 2018.
Tallbear, Kim. "Close Encounters of the Colonial Kind." *American Indian Culture and Research Journal* 45, no. 1 (2021).
Todd, Zoe. "Fish Pluralities: Human–animal Relations and Sites of Engagement in Paulatuuq, Arctic Canada." *Études/Inuit/Studies* 38, no. 1–2 (2014).
Tremblay, Jean-Thomas, and Swarbrick, Steven. "Destructive Environmentalism: The Queer Impossibility of *First Reformed*." *Discourse* 43, no. 1 (2021).
Warren, Karen J. *Ecofeminist Philosophy: A Western Perspective on What It Is and Why It Matters*. Rowman & Littlefield, 2000.
Zanghellini, Aleardo. "Queer, Antinormativity, Counter-Normativity and Abjection."*Griffith Law Review* 18, no. 1 (2009).

47
ZOONOSIS

Tessa Laird

Dangerous Liaisons and Archival Bodies

Since the global COVID-19 pandemic, the once-specialist term "zoonosis" has become something of a household word. From the Greek *zōon* (animal) and *nosos* (disease), a zoonosis is, according to the Merriam-Webster dictionary definition, "an infection or disease that is transmissible from animals to humans under natural conditions".[1]

This of course presupposes that humans are not animals, or rather, we are such unique animals that we must be cordoned off from those vast multitudes which are done the violence of being corralled under the singular category "animal".[2] This violence of categorisation and the domination it inspires is a key driver of ecofeminist critique, which questions the West's idealisation of "rationality" and distance from nature. As unlikely as it may seem that zoonotic diseases can be read through the lens of gender studies, Val Plumwood's cogent analysis of the ways in which oppressed peoples are "often both feminised and naturalised",[3] while the natural world is consistently cast as female, should leave us with no doubt that human, more-than-human, and microbial naturecultures[4] are utterly entangled and that they all need to be considered within a continuum of intersectional and ecofeminist praxis.

Canadian epidemiologist David Waltner-Toews has the decency to refer to zoonoses as being spread to humans by "*other* animals" (my emphasis), as a reminder that "people are animals—exceptional animals, to be sure, but animals nonetheless" and that, for better or worse, "the sharing of microbes among species is normal".[5] Rather than inhabiting a world of cordons and categories, we are all—that is, animals, plants, fungi, and lichens—the result of what Lynn Margulis coined "hypersex", which occurs most notably among bacteria, doing away with the need for sexual reproduction and existing entirely outside the gender binaries humans are so keen to impose upon the natural world.[6] If bacteria can simply enter other bacteria, which grow and reproduce inside each other, thus leading to new life forms, then our revulsion towards the "dangerous interspecies liaisons"[7] which result in zoonoses may itself be a kind of disorder: a pathological rejection of permeable boundaries and of the endless variation this engenders, whether at the level of species, race, gender and sexuality, neurodiversity, or ability.

 DOI: 10.4324/9781003273400-58

It is true that bacterial hypersex, as thrilling as it sounds, can begin as an infection (then again, how many times does romantic literature portray love as a malady?).[8] Paradoxically, microbes that fail to destroy their hosts survive, while "perfect pathogens" kill off their hosts and therefore die. Imperfect pathogens eventually become new, permanently infected bodies, like ours, whose cells house "a promiscuous past, a long record of hypersex—permanent bacterial matings".[9] Viruses are smaller micro-organisms than bacteria, true parasites that require living cells to reproduce.[10] As they move from host to host, they mutate, creating genetic archives.[11] Numerous diseases that have sculpted the landscape of human history have non-human animal origins: hepatitis from our primate ancestors; anthrax from hunting and butchering animals; and, since the agricultural revolution around 10,000 years ago, a smorgasbord of ailments, including "stomach ulcers from sheep, tuberculosis from goats, whooping cough from pigs, measles from cattle, glanders from horses, smallpox from water buffalo, typhoid fever from chickens, and influenza from ducks".[12] Today, we might update the list to include COVID-19 from bats, if indeed we could be sure of its still-contested origins.[13] Even so, such an interspecies litany, rather than eliciting disgust, simply underscores our entangled histories and the fact that our reliance upon, and exploitation of, other animals comes with costs as well as benefits.

If our bodies are archives of microbial exchanges between our ancestors, cousins, and neighbours,[14] then "staying alive" for any and all species, is a matter of "livable collaborations", as Anna Tsing puts it. This means "working across difference", which inevitably "leads to contaminations" yet without which "we all die."[15] Tsing's equation of contamination with collaboration acknowledges that our body is full of benign bacteria and viruses that make us who we are and that we are all "contaminated by our encounters; they change who we are as we make way for others".[16]

Better Narratives: Carrier Bags and Cosmopolitics

If we have co-evolved with diseases, if they are our collaborators, as Tsing suggests, then we need "better narratives" than war, in which viruses are imagined as enemy forces to be overthrown by our combined "defences".[17] Perhaps we should consult storytellers, such as Ursula Le Guin, whose ecofeminist "Carrier Bag Theory of Fiction" imagines a mode of storytelling which highlights micro actions and attention to detail (for example, the gathering of oats) rather than the grand narrative of the hunt.[18] This is akin to Waltner-Toes' proposition of politics over war as a better metaphor for thinking with and through infectious diseases, while Isabelle Stengers calls her proposition for ethical living *cosmopolitics*, in which the polis is no longer a city state inhabited only by humans but a cosmos of interconnections requiring constant negotiations.[19]

Cosmopolitical diplomats, be they epidemiologists or shamans, study patterns. Staying with Western science for a moment (and discussing shamanic modes later), such patterns might involve an awareness of distinctions between direct, cyclo-, meta- and saprozoonoses.[20] Direct, as the name suggests, is a zoonosis passed straight from a vertebrate animal species and humans, without any other mediating host, via bites and scratches or contaminated food.[21] Cyclozoonoses require more than one vertebrate species to make the jump but no invertebrates. As with the epistemological violence of the all-encompassing "animal", when it comes to discussing zoonoses, it seems that whether you wear your skeleton on the inside or the outside is the next great taxonomic split in a knowledge system that loves to divide and conquer.[22] Metazoonoses require a combination of vertebrates and invertebrates (such as fleas or mosquitoes), to complete their cross-over into the world of humans. Saprozoonoses require other ingredients altogether to perfect their viral cocktails, such as soil, plants, or water.[23] When Waltner-Toes was writing the second edition of his popular *On Pandemics*, he noted that these categories were proving highly unstable, that everything was inextricably connected, and that "natural history and biology, social relationships, economics, and ethics" all form part of the fabric of zoonotic diseases.[24]

Milkmaids and Medical His-Stories

Tying gender, animals, and zoonotic diseases together, the origin of the modern vaccine is reflected in its etymology. From the Latin *vaccinus*, "of or pertaining to cows",[25] vaccines were first developed by English physician Edward Jenner, who had noted that milkmaids' immunity to smallpox seemed to be predicated on their exposure to cowpox.[26] Jenner, however, was capitalising on existing folk wisdom shared among the milkmaids, as well as a nascent form of inoculation brought into England from Turkey by Lady Mary Montagu, herself fascinated by the folk medicine practiced by local women.[27] In fact, forms of smallpox inoculation had been practiced for centuries in India and even had genealogies in Vedic scriptures, and although this information was accessible, Britons continued to die from smallpox until Jenner's "mastery" of vaccination, because his masculinist empiricism was seen in opposition to the feminine, folk, and foreign ways of "scratch" inoculation.[28] In her article "Contracting Xenophobia", Rajani Sudan argues this "distinction between inoculation and vaccination was predicated not as much upon empirical evidence as on ideologies of gender, race, and nationalism".[29] Both the milkmaid Sarah Nelms and Blossom, the cow she caught cowpox from, have been inducted into the medical hall of fame for their respective roles in the emergence of the smallpox vaccine, but as unconscious feminine actors in a medical his-story. Not only that, they become the wholesome face of English innovation in a blatant coverup of the Eastern origins of vaccines. It is no wonder, then, that the title of Sudan's article suggests that the real disease of Victorian England was xenophobia.

Zoognosis: Transmitting Knowledges

The definition of zoonosis that I put forth here is not intended as a biology primer, nor indeed to capitalise on the panic or horror that is automatically evoked by the utterance of certain names—bubonic plague, Ebola, HIV, bird flu, swine flu[30]—but rather as a reminder, if we even needed one, of our absolute imbrication in the Gaian forces that co-constitute the Earth and, while not wanting to fall into biological determinism or essentialist feminism, as a reminder, too, that those concepts (Gaia and Earth) are traditionally coded as feminine. To re-mind is perhaps not what we need in a world which has been ravaged by the Cartesian mind-body split, placing the earthly, and the feminine, as the always-submissive bottom of a binary power relationship, so instead I'll use the Gnostic term *anamnesis*, an epiphanic unforgetting, an influx of hitherto occluded knowledge. Elsewhere I have suggested that adding a silent "g" to zoonosis creates a mutation with its own viral vectors; *zoognosis* is the transmission of *knowledges* from animals to humans via infectious diseases. It then becomes a matter of translation—what are the animals trying to tell us?[31] At the conclusion of this chapter, and utilising a Black feminist theoretical framework proposed by Alexis Pauline Gumbs, I will attempt to "translate" my own experience with the currently ubiquitous zoonosis, COVID-19.[32] Gumbs' Black feminism can be equated with ecofeminism in that it promotes identification *with*, rather than identification *of*, animal species, in a refusal of the segregationist approach implicit in western science.[33]

Contracting Homophobia

Such an approach doesn't intend to minimise the real cost of zoonotic outbreaks in human populations, which too often take their toll on those already suffering from the depredations of capitalism, colonialism, racism, patriarchy, and, in the case of HIV AIDS, homophobia. The human immunodeficient virus (HIV) can lead to stage three acquired immunodeficiency syndrome (AIDS), which turns the body into a tabula rasa for every illness imaginable. In the 1980s and 1990s, the acquisition of this particular zoonosis saw afflicted humans treated as animals. Nowhere is the disgust and horror associated with the unravelling of the human body from HIV more apparent than in David Cronenberg's 1986 film *The Fly*, in which the suppurating figure of scientist Seth Brundle vomits, sheds, and disintegrates as he transforms into something wholly inhuman.[34] Such an abject spectacle, it could be argued, mirrors the homophobia already manifest in society; a demonisation of the queer body as animalistic "other".[35]

At the time of writing, a global outbreak of monkeypox outside of its endemic West and Central Africa, is predominantly affecting gay and bisexual men. Despite Tedros Adhanom Ghebreyesus declaring a global health emergency, the dissemination of vaccines is slow enough to indicate to certain communities that their health is not a priority. In so-called Australia where this was written, the federal government was spending more money and bureaucratic energy on ramping up border security to stop foot-and-mouth disease spreading from Indonesia via airline sandwiches and dirty flip-flops brought back by tourists.[36] If this agricultural disease takes hold, it will mean a mass "cull" of animals whose fate was already to be murdered for food. Interestingly, when these deaths in the millions take place *without* the resulting sandwiches, meat-eaters can experience a powerful grief. Thinking of these events through the lens of *zoognosis*, it seems clear that non-human animals are sending messages about the hazardous nature and ethical quagmire of industrial food production to their human–animal kin.

Human Vectors and Microfauna Extinctions

One of the clues to the human exceptionalism evident in our reaction to zoonoses as exceptional events, is the idea of "reverse zoonoses" (like "reverse racism", the term screams bias). Humans are animals, and unsurprisingly, we spread diseases to other animals in a process technically known as zooanthroponosis. After the outbreak of COVID-19, civet owners and bat carers alike wore protective equipment so as not to infect the animals under their supervision. Zookeepers around the world found big cats were coming down with the virus, reporting sluggish lions and coughing tigers, while even some house cats tested positive for COVID-19, although no deaths were reported. Dogs and cats were far less lucky in the Great Plague of Europe, but their mass deaths are rarely considered in the history books. And while bats have borne the brunt of the blame for supposedly being the origin of COVID-19, white nose syndrome, responsible for the deaths of millions of North American bats, is likely spread by human cavers and hikers using boots and equipment from expeditions in other parts of the world where bat species with resistance to certain diseases unwittingly end up spreading those same diseases to their cousins via human intermediaries.

Microorganisms may create havoc, but they also create the very conditions for life itself; after all, it was cyanobacteria who, billions of years ago, introduced oxygen into Earth's atmosphere as a by-product of photosynthesis.[37] Increasingly, viruses are being seen as key drivers of genetic diversity and evolutionary complexity.[38] We don't bestow the catch-all moniker "animal" upon bacteria or viruses, yet, like animals, they are animate and propelled by what the 17th-century philosopher Spinoza called *conatus*, a will to live. Waltner-Toes refers to such organisms as "microfauna", noting that even the most environmentally friendly humans rarely spare a thought for these microscopic ecologies. Indeed, according to Timothy Morton, "viruses, and virulence, are shunned in environmental ideology".[39] Yet, the 6th Great Extinction event for animal species affects them too, for, as their hosts die, so do "the trillions of viruses, yeasts, fungi, and bacteria for whom these disappearing animals are home".[40] Then again, new/old viruses are emerging from the melting permafrost,[41] and global warming is increasing the range of many diseases once confined to the tropics. The future is sure to be teeming with microbial possibilities: inflorescences as well as die-offs.

"Unnatural" Alliances

If bacteria created the conditions for life and diseases made us who we are today, it is no surprise some theorists, such as Tsing, imagine contagion as a positive transformative process. Deleuze and Guattari's oft-quoted chapter on becomings from *A Thousand Plateaus* figures becoming-animal as a contagion which can sweep up whole communities, turning individual subjects into roaming packs. Preferring the unruly collective to the sovereign subject, Deleuze and Guattari see the metamorphic capacities of contagion as desirable because they disrupt the calcification of socio-political structures (of course, this has been critiqued vis-à-vis the retrenchment of privilege versus precarity during the COVID-19 pandemic, including glaring gender disparity).[42] The heterogenous assemblages of contagion, or the "animal peopling of the human being",[43] create "unnatural" alliances between kingdoms, which the philosophers see as the purview of wizards and shamans and preferrable to the reproduction of mimicry or heredity. To be "unnatural" is desirable, since it breaks with convention and creates new possibilities. Margaret Atwood describes a zoonotic disease as having "got into you and changed things inside you. It rearranged you, cell by cell".[44] While this might result in drastic symptoms, including death, in micro doses, such internal recalibrations recall Guattari's term "molecular revolution" to signify social changes emerging from a "micro-politics of desire"[45] rather than the clumsy machinations of macro-political processes.

Politics aside, there is no doubt that molecular realignment takes place as a disease takes hold of a body. In Francis Ford Coppola's AIDS-era adaptation of *Dracula* (1992), the screen between scenes fills with blood, its swarming molecules signifying both sexual excitation and contagion. Deleuze and Guattari name each chapter or "plateau" in *A Thousand Plateaus* after a significant year; their chapter on becomings is prefaced by 1730, which, they declare, was peak season for European vampires. Not caring for metaphor, they claim to "passionately believe" in vampires,[46] figures which cross-contaminate becomings-animal with sexually transmitted disease, aristocratic abuse of power, and xenophobia, not to mention chiropterophobia, or fear of bats.[47] The implied sexuality of shapeshifting vampires, neither human nor animal, hints at bestiality, the unspeakable crime often invoked at the onset of a zoonotic outbreak; witness the common myth of AIDS being spread by humans having sex with monkeys.

Jack Halberstam suggests we rethink our dissociation with bestiality, given our intensely intimate relationships with companion animals and considering queer practices and re-imaginings of what constitutes a sex act.[48] Donna Haraway's intimate relationship with her beloved dog Cayenne, with whom she imagines a viral affinity, is a case in point, "I bet if you checked our DNA, you'd find some potent transfections between us. Her saliva must have the viral vectors".[49] Women who shower their affections upon animals rather than men are often regarded with suspicion, still carrying the taint of Inquisitors' reports from mid–14th-century Europe in which heretics were accused of animal worship, most notably the infamous *bacium sub cauda* or "kiss under the tail" of the Devil incarnate in animal form.[50] Associations between the Devil, witches, and animals might be interpreted as a fear of zoonotic diseases, fuelled by the immense death toll of the bubonic plague. An Inquisitorial document described "a powder made from the body of a cat stuffed with herbs, grain and fruit, which is then hurled down from mountaintops in order to cause plague".[51] Such tales led to reprisal killing of cats, which presumably only made the plague spread further and faster by allowing the rat population to increase.[52] This bears relevance to today's scapegoating of bats at almost every zoonotic outbreak, disregarding the fact that by eating literally tonnes of mosquitoes every night, bats help minimise the potential for malarial outbreaks, while the fruit-eaters and pollinators are responsible for re-foresting and therefore repairing the lungs of our damaged planet.

Animaladies

Haraway's use of the term "transfection" (the introduction of genetic material at the cellular level), along with her zealous ardour for Cayenne, makes me wonder about the connections between affection and infection, hinted at by the wonderful hybrid term *animaladies*. Coined by Fiona Probyn-Rapsey to highlight the dis-ease of human–animal relationships, the term also knowingly embraces the conflation of women with madness as well as women with animals, and sometimes all three together, for example, the figure of the "crazy cat lady" (CCL).[53] Probyn-Rapsey designates the "CCL" as a kind of "folk devil" reflecting anxieties around both femininity and animality, a "figure of gendered dysfunction" whose "love of cats is fundamentally misanthropic".[54] But the CCL as "animal hoarder" can also be a smokescreen for the "industrial scale hoarding" of factory farming[55] while trivialising the good work of (predominantly female) animal rescuers and advocates.[56] The use of the CCL in this respect is a deadly combination of misogyny with anthropocentrism, making animal hoarding "a symptom of gender rather than a broader issue of human arrogance".[57] The difference between the CCL and the factory farm, according to Probyn-Rapsey, is one of degree rather than kind,[58] where concerned citizens can respond to animal hoarders as exceptional rather than a reflection of structural, societal disregard for animal life.[59]

In the figure of the CCL, the "malady" is her "bestial" affections, though there may also be, as Haraway surmises, viral vectors in play. In fact, schizophrenia researcher E. Fuller Torrey is convinced of the connections between psychosis and toxoplasmosis (carried by cats and occasionally transmitted to their humans). Torrey traces the rise of "madness" in Britain to the increasing favour bestowed on domesticated cats by artists, writers, and the aristocracy after feline associations with witches began to wain.[60] He gathers a cast of odd characters, particularly 19th-century painters and poets, renowned for their love of cats but surprisingly doesn't mention Victorian postcard artist Louis Wain, whose proto-psychedelic felines have led many to speculate on his mental health.[61] Those who prefer feline company to human society are often regarded with suspicion at best, derision at worse, a feedback loop which makes feline friends all the more desirable. In such a situation, whether a human is infected by toxoplasmosis and its ensuing effects on mind and mood or not, cats operate as pharmakon, both poison *and* cure, and this may be seen as a predecessor to the concept of the vaccine, where exposure to minute doses of a virus can paradoxically keep the disease at bay.[62]

Supernatural Pathogens and Cannibal Metaphysics

It may seem strange that in discussing zoonoses, I have touched on vampires and witches, but such folktales often encode folk wisdom. Rather than perceiving such beliefs as antiquated, French anthropologist Frédéric Keck suggests public health officials should imagine emerging pathogens as "supernatural" beings.[63] Ritual prohibitions and attunement to invisible forces can serve real-world purposes of protection. James Frazer's classic of early anthropology, *The Golden Bough*, traces global practices of sympathetic magic, including an ancient rite to cure jaundice, in which the yellow pallor of the invalid is ritualistically "shooed" into a yellow bird tied to the end of the sickbed.[64] While I don't know of any "real world" correlation to health in this instance, sympathetic magic makes Western taxonomies, which thrive on segregation, irrelevant. Instead, a world of complex ecological interconnections emerges, with all its concomitant care and awareness.[65]

Eduardo Viveiros de Castro weaponises Western misunderstandings of Amazonian Amerindian ontologies, using the term "cannibal metaphysics" to indicate a world view which is, contrary to the Western imagination of cannibalism, far from extractive, and in which who eats whom, when and how, is governed by shamanic rituals of diplomacy. This seems particularly relevant when considering the supposed alimentary origins of many zoonotic diseases, including COVID-19. For Waltner-Toes, the "most intimate contact of all" is not sexual intercourse but "eating another species".[66] Yet we are reluctant to accept that such intimacy comes with a cost. Brazilian anthropologist Els Lagrou notes that the Huni Kuin, as well as many other Amerindian peoples, attribute diseases to the eating of animals, who take revenge with *nisun*, a "headache or vertigo that can result in sickness or death".[67] That the Huni Kuin are known as the "bat people" by their enemies makes their *zoognosis* all the more redolent in the era of COVID-19, where the "bat people" can be seen to be taking their revenge for being displaced, victimised, and eaten, although, at the time of writing, other animals were being implicated in the emergence of the virus.[68]

"We Are All Cannibals"

Regardless of where COVID-19 actually originated, there is no doubt that it frequently surfaced in spaces where human bodies collided with dead animals, such as abattoirs, bringing to mind bovine spongiform encephalopathy (BSE), better known as "mad cow disease". The outbreak of this zoonotic disease in the UK in the 1990s was caused by forcing herbivores to eat the "recycled" carcasses of other cows—effectively transforming them into not just carnivores but cannibals and prompting the eminent anthropologist Claude Lévi-Strauss to write the essay "We Are All Cannibals".[69] Like Viveiros de Castro, Lévi-Strauss was interested in challenging the hysteria that accompanies the term, and similarly, thinking across scales to include diseases and parasites, Waltner-Toes quotes W.H. Auden in order to take the stigma out of humanity's greatest fear: becoming meat, "The slogan of Hell is to eat or be eaten; the slogan of heaven is to eat and be eaten".[70] Auden's words are not necessarily about being swallowed by a tiger or crocodile, though famously, Plumwood's awakening to the radical *un*exceptionalism of human beings came as she was in the jaws of a crocodile.[71] To eat as we are being eaten is an acknowledgement that we are all being consumed by micro-organisms, in life and after death, and that if we better understand these processes, "we might even begin to understand ourselves".[72] This quest for self-knowledge as radically integrated rather than separate from nature might be termed "ecodelia",[73] a kind of cosmic *mis-en-abyme*, as, according to Lynn Margulis, spirochetes, or spiral bacteria, helped create the rods and cones which make up animal eyes, so that "when we look through a microscope at spirochetes, we are nothing more or less than bacteria looking at bacteria, trying to understand ourselves".[74]

The Virus Is Capitalism

If our bodies are archives of encounters with disease, so too, for better or worse, are our social systems. Waltner-Toes takes the example of bubonic plague, which led to the collapse of feudalism in western Europe, and this in turn "led to the rise of the middle class, capitalism, shopping malls, and evangelical consumer cults".[75] In an (avian-flu-infected) egg-or-chicken conundrum, we might wonder whether zoonoses cause capitalism or vice versa.[76] Uruguayan philosopher Sandino Núñez writes that COVID-19 showed us once and for all that

> the liberal market is totally incapable of organising a health programme; that drugs, medical technologies and systems shouldn't become commodities; that the transport system shouldn't be in the hands of businessmen whose only motivation is revenue and profit; that food and hygienic supplies shouldn't solely benefit those who own a car and a trolley and fight in supermarkets to buy things the prices of which are hiked every ten minutes; that the media can't keep earning fortunes thanks to the generalised medieval panic that they've been installing for decades in the masses.[77]

He concludes that the real disease is capital itself. In forcing us to acknowledge this, Jean-Luc Nancy says, the coronavirus is therefore a "communovirus", which asks us to come together to "dethrone" the crowned virus (and perhaps behead capitalism while we are at it).[78]

Other commentators pointed out that new understandings of community brought about by COVID-19 ought necessarily to be multispecies, and sometimes these were articulated in the mode of gentle storytelling as promulgated by Le Guin's "Carrier Bag Theory of Fiction". Camila Marambio and Nina Lykke wrote a fairy story in which a "bat, a pangolin and a virus decided that they would give the Earth a break". The unlikely trio collaborate to teach humans "a lesson about kinship and shared vulnerability."[79] Sabrina Orah Mark wrote of a "Fairy Tale Virus" in which the "Virus With a Crown on Its Head" reigned the land. In this fable, the virus's mother is a bat, and Mark embellishes Thomas Nagel's eternal philosophical question to ask not just "What is it like to be a bat?" but what it is like "to be the bat that started this cavalcade of coughing that shook a whole entire planet"? She decides there is no need to wonder, because in fact, *everything* is a bat! "I am a bat. You are a bat. . . . Every wildflower we ever picked is a bat. Sex is a bat and the soup you'll eat tonight is a bat. . . . And God, who created all the bats, is also a bat. And not believing in God is a bat, too".[80] In Mark's tale, which I consider ecofeminist as much as poetic and fanciful, everything is connected. You cannot make scapegoats of bats, or of goats, for that matter—for blame is like echolocation—it bounces back! You cannot project it onto something else and think that it will never return. Call it karma. Call it chickens coming home to roost with an avian flu made of secret herbs and spices and growth hormones.

What Is It Like to Be a Bat?

Perhaps the answer to "What is it like to be a bat?" is to catch COVID-19 and really *feel* it, absorb its messages. Laura Jean McKay's *The Animals in that Country* imagines a "zooflu" which enables the afflicted to understand the languages of animals. Part of that hallucinatory novel's power surely comes from the fact that McKay wrote much of the first draft while battling chikungunya fever.[81] How appropriate, then, that I am writing the closing part of this chapter on zoonosis while sick with COVID-19. So, here, I too will attempt to attune to the *gnosis*, wisdom, coded within the zoonosis.

I already spend a lot of time thinking of/with/for and through bats, but COVID-19's supposed origin makes me feel even closer to these magical beings. China may be the land this zoonosis emerged from, but it is also the land where bats are revered as harbingers of luck. Because the words for *bat* and *luck* are homophonic, the phrase "bats come down from the sky" is interpreted as "let good fortune come down upon you".[82] So, using the magical thinking of Marks, if we imagine ourselves as *one* with bats and, amending Margulis's self-reflexive bacteria, become bats looking at bats, what fortunate wisdom do we as people-bats understand as emanating from the bat-people? First of all, they remind us to breathe. Alexis Pauline Gumbs interprets the wisdom of marine mammals primarily in relation to the fact that they can breathe in "unbreathable circumstances", and this is "what we do every day in the chokehold of racial gendered ableist capitalism".[83] Gumbs reminds us that "the scale of our breathing is planetary",[84] and COVID-19 reminds me, as my breath becomes more laboured, that I share my breath with every other living creature. Whether we have lungs, gills, or stomata, we know what it's like to expand as we take in the good, sweet stuff and breathe out that which no longer serves us. We also know what it's like when something impedes that breath: a weight on the chest, a blockage of the nose, a constriction in the throat. We know what it's like to labour for something, just as we know what it is to take something for granted. The best way to be reminded of something's preciousness is to have it taken away. So, as some people lost their abilities to breathe without machines, I am grateful for this vital mechanism that most of us enjoy every day without so much as a thank you.

Like many people, I've also lost some of my capacity to smell, which leads to a reflection on the importance of the senses, although of course each being's sensory apparatus is wondrously unique. Having less of one sense can mean a refinement or intensification of another. The senses are indivisible, an entangled ecology: think of the nose of a horseshoe bat, finely tuned for echolocation, not just scent. Echolocation operates as a mapping system; it is essential for microbats to see.[85] Imagine that—a nose that can hear, see, and smell, all in one! As my nose sniffs but doesn't smell, as my hearing fails due to blocked ears, and my eyes can't seem to focus for long, I'm humbled by the bat's extraordinary capacities.

I ache all over but especially at the joints—these pliable parts of the body that make everything possible. This reminds me to give thanks for walking, crawling, waving, swimming, dancing, flying, whatever it is that you (oh creature of planet Earth) do, that makes you, you. My knees are inhabited by the ghosts of other knees—bat knees are in reverse to human knees. Perhaps that is why they ache so much—I have another skeleton inside me, just as my embryo once had a tail which got swallowed up as I developed in utero; there are ghostly echoes of all my more-than-human ancestors inside me. Right now, these bat-knees-ghosting-my-own-knees ache, like growing pains . . . as I morph into another being—a human-plus-COVID-19 being.

And speaking of things that hurt . . . I have a newfound propensity to cry . . . at everything! To feel the pain of others. . . . No longer is listening to the news an experience of abstracted suffering. Each item hits home, and my eyes fog up and my face crumples like a wrinkle-faced bat, whose visage is accordioned with cavities and holes, the better to resonate sympathetically with the world. Like other echolocating bats, the wrinkle-faced bat cries its heart out because it is alive, and each of those cries bounces back, as the world joins in, creating a chorus, no, a *symphony* of cries.[86] We cry like newborn babies—we are alive! Gumbs brilliantly parses the echolocation cries of the Indus river dolphin, whose numbers are finally coming back, as saying "Here. Here. Here". A survival against the odds.

> Here is all of me. And here we are. Here. Inside this blinding presence. Here. A constant call in a moving world. Here. All of it. Here. Here. Humbly listening towards home. And here. And here. Right here. My poem for you. My offered presence. This turbid life. Yes. Here you go.[87]

Stop the World, Open the Door!

But the most enduring message from COVID-19 is: slow down. Everything takes longer. Bats have fast metabolisms, but those in cold climes know how to hibernate, know how to slow their heartbeats; they even know how to postpone their babies' due dates until the time is right. This is the best wisdom for these hypercapitalist times. As Paul Preciado put it at the start of the great pandemic pause, this moment was akin to a global shamanic abstention, a refusal to take part in all the extraction and consumption that makes the world "go round" (until it's giddy and vomits). When the world won't stop turning in the wrong direction, the only option is to "stop the world".[88]

Michael Taussig writes that the virus "opens the doors of perception"[89] whether one is in its physical grip or in its wider socio-cultural ambit. Arundhati Roy refers to COVID-19 as a portal, "a gateway between one world and the next", where we are given the option to "imagine another world".[90] Thinking with and through COVID-19 has been a valuable opportunity to take stock, to see otherwise, and particularly to experience time differently. Just as Plumwood had her epiphany in the jaws of a crocodile, sometimes it takes a dramatic alteration of the senses, a rearrangement of molecules, to think differently and to radically alter the way we live. As the pessimists (or perhaps they are just realists?) remind us, more pandemics are on the way. Will we be open to the messages that animals are sending us through viral vectors?

Notes

1 Merriam-Webster, "Zoonosis," retrieved from: https://www.merriam-webster.com/dictionary/zoonosis.
2 Jacques Derrida, "The Animal That Therefore I Am (More to Follow)". Jacques Derrida, *The Animal That Therefore I Am*, ed. Marie-Louise Mallet, trans. David Wills (New York: Fordham University Press, 2008), 24 and passim.
3 Val Plumwood, *Feminism and the Mastery of Nature* (London: Routledge, 1993), 18.
4 Donna Haraway, *The Companion Species Manifesto: Dogs, People and Significant Otherness* (Chicago: Prickly Paradigm Press, 2003), and subsequent Haraway publications.
5 David Waltner-Toes, *On Pandemics: Deadly Diseases from Bubonic Plague to Coronavirus* (Carlton, VIC: Schwartz Books, 2020), 38.
6 Lynn Margulis and Dorion Sagan. *What is Sex?* (New York: Simon & Schuster, 1998) 73.
7 Ibid.
8 There are too many examples to list here, but a classic among them is Gabriel García Márquez's *Love in the Time of Cholera*, in which the protagonist's symptoms of lovesickness are so extreme, it is assumed he has the titular disease. Márquez, *Love in the Time of Cholera,* trans. Edith Grossman (New York: Penguin Books, 1999).
9 Margulis and Sagan, *What Is Sex?* 74.
10 Vinh-Kim Nguyen, "Of What Are Epidemics the Symptom? Speed, Interlinkage, and Infrastructure in Molecular Anthropology," in *The Anthropology of Epidemics*, eds. Ann H. Kelly, Frédéric Keck, and Christos Lynteris (London: Routledge, 2019), 157.
11 Margulis and Sagan, *What Is Sex?* 172–173.
12 E. Fuller Torrey, *Parasites, Pussycats and Psychosis: The Unknown Dangers of Human Toxoplasmosis* (Cham, Switzerland: Springer, 2022), 1.
13 In this chapter, I will be repeating the oft-quoted view that COVID-19 has its origins in horseshoe bats, but with the caveat that studies are still inconclusive. Moreover, this is not the first time bats have been scapegoated as viral reservoirs or vectors without substantive proof. Bat scientist and advocate Merlin Tuttle has made it his life's work to set the record straight for bats. See his article "A Viral Witch Hunt," *Issues in Science and Technology* (27 March, 2020), retrieved from: https://issues.org/a-viral-witch-hunt-bats/
14 Vinh-Kim Nguyen, "Of What Are Epidemics the Symptom? Speed, Interlinkage, and Infrastructure in Molecular Anthropology," 154.
15 Anna Lowenhaupt Tsing, *Mushroom at the End of the World: On the Possibility of Life in Capitalist Ruins* (Princeton, NJ: Princeton University Press, 2012), 28.
16 Ibid., 27.
17 Waltner-Toes, *On Pandemics: Deadly Diseases from Bubonic Plague to Coronavirus,* 15. See also "A Triptych of Viral Tales," by Nina Lykke and Camila Marambio, in *Kerb: Journal of Landscape Architecture*, #28, "Decentre—Designing for Coexistence in a Time of Crisis," 2020, retrieved from: https://kerb-journal.com/, which argues for the "new capacities" that viruses give us. See also Marambio's concept, following Octavia Butler's *Xenogenesis* trilogy, of "cancer as talent": https://breathablefutures.wordpress.com/2020/06/04/interview-with-camila-marambio/
18 Ursula K. Le Guin, "The Carrier Bag Theory of Fiction," in *Ecocriticism: The Essential Reader*, ed. Ken Hiltner (New York: Routledge, 2014).
19 Isabelle Stengers, *Cosmopolitics,* trans. Robert Bononno (Minneapolis: University of Minnesota Press, 2010).
20 Waltner-Toes, *On Pandemics: Deadly Diseases from Bubonic Plague to Coronavirus,* 40. Here, Waltner-Toes is following veterinary epidemiologist Calvin Schwabe's formula.
21 Ibid. Here, Waltner-Toes again reflects on the ways in which scientific language, when used uncritically, reinforces habitual thinking: "If the reservoir is a single animal, it's called a reservoir host (which should mean, if life were fair, that we would call the microbe a guest, which we don't; we denigrate it by calling it a parasite or pathogen)".
22 Apalech scholar Tyson Yunkaporta has a lot to say about the ways in which Western taxonomies insist on division rather than relation, finding evidence of this in the scars left on natural landscapes by violent naming practices. For Yunkaporta and countless peoples, the so-called "Great

Dividing Range" is "the body of the Rainbow Serpent" which "divides nothing . . . but connects systems along a massive songline". Running parallel to the range is the Great Barrier Reef, which is, he says, "another serpent in carpet snake form, which is a barrier to nothing . . . but another infinitely connective story". Tyson Yunkaporta, *Sand Talk: How Indigenous Thinking Can Save the World* (Melbourne: Text Publishing, 2019), 53.

23 Waltner-Toes, *On Pandemics: Deadly Diseases from Bubonic Plague to Coronavirus,* 47.

24 Ibid., 46.

25 Online Etymology Dictionary, "Vaccination," retrieved from: https://www.etymonline.com/word/vaccination.

26 Centers for Disease Control and Prevention, "History of Smallpox," retrieved from: https://www.cdc.gov/smallpox/history/history.html#:~:text=The%20basis%20for%20vaccination%20began,used%20to%20protect%20against%20smallpox.

27 Sanjoy Datta, Neerja Bhatla, Margaret Burgess, Matti Lehtinen, and Hans Bock, "Women and Vaccinations: From Smallpox to the Future, a Tribute to a Partnership Benefiting Humanity for over 200 Years," *Human Vaccines* 5, no. 7 (2009): 450–451.

28 Rajani Sudan, "Contracting Xenophobia: Etiology, Inoculation, and the Limits of British Imperialism," in *Fear and Loathing in Victorian England*, eds. Maria Bachman, Heidi Kaufman, and Marlene Tromp (Columbus: Ohio State University Press, 2013), 88–89.

29 Ibid., 91.

30 There is something about zoonotic diseases in particular which inspires artists, authors, and filmmakers to new depths of abjection, although this can also be strangely comedic. Richard Preston's 1994 non-fiction bestseller *The Hot Zone* contains hideous passages in which victims of Marburg disease vomit blood, and yet the author, when visiting the site of supposed contamination, comments on the bat guano found there as "a spinach-green paste speckled with grey blobs, which reminded me of Oysters Rockefeller" (*The Hot Zone*, Sydney, 1994, 359). In a completely fictional account of a zoonotic pandemic (though one concocted in a lab), Margaret Atwood writes the disease ran through a character "like shit through a goose. It was like watching pink sorbet on a barbecue—instant meltdown" (*Oryx and Crake* (New York: Nan A. Talese, 2003), 253). Hollywood diseasesploitation generally has less facility with language: the film *Contagion* proffers the unforgettably forgettable line "Somewhere in the world the wrong pig met up with the wrong bat" (dir. Steven Soderbergh, 2011).

31 Tessa Laird, "Zoognosis: When Animal Knowledges Go Viral. Laura Jean McKay's *The Animals in That Country*, Contagion, Becoming-Animal, and the Politics of Predation," *Animal Studies Journal* 10, no. 1 (2021): 30–56.

32 I do so as an incontrovertibly white feminist, but as Gumbs herself says, her book is for everyone, because "a world where queer Black feminine folks are living their most abundant, expressed, and loving lives is a world where everyone is free". Alexis Pauline Gumbs, *Undrowned: Black Feminist Lessons from Marine Mammals* (Chico, CA: AK Press, 2020), 13. In a live interview with adrienne maree brown, where brown asks Gumbs regarding *Undrowned*, "What makes this Black feminist wisdom?" Gumbs recounts a reviewer's response to the book that "this could apply to anyone". Gumbs gleefully declares, "That's the whole point—everything about Black feminism applies to everyone. . . . Is there anything that's not Black Feminism?" Charis Circle, "Undrowned—Black Feminist Lessons from Marine Mammals—Alexis Pauline Gumbs and Adrienne Maree Brown" (November 20, 2020), retrieved from: https://www.youtube.com/watch?v=-3_GUGaZ0rI

33 Gumbs differentiates between these two forms of identification by deftly italicising the Black feminist, expansive, empathetic kind (8).

34 "As if evoking the liminal dream-state of Franz Kafka's 'Metamorphosis', Seth concludes that he is 'an insect who dreamt he was a man, and loved it. But now the dream is over, and the insect is awake' ". Yves Saint-Cyr, "Desire, Disease, Death, and David Cronenberg: The Operatic Anxieties of The Fly," *The Canadian Review of Comparative Literature* 38, no. 4 (2011): 457. Sometimes illness is the best way to become-animal. In Deborah Bird Rose's final book, she writes of the cancer which plays with her human proprioception—so that she becomes-ant: *Shimmer: Flying Fox Exuberance in Worlds of Peril* (Edinburgh: Edinburgh University Press, 2022), 107.

35 Granted, Cronenberg's *The Fly* presents viewers with a heterosexual love story gone wrong, but there is no doubt that its affective power at the time of release drew on deep-seated fears of the AIDS virus, which was disproportionately affecting queer communities.

36 Amelia Dunn, "Why You Should Throw Out the Thongs You Wore in Bali," SBS News (15 July, 2022), retrieved from: https://www.sbs.com.au/news/article/why-you-should-throw-out-the-thongs-you-wore-in-bali/h70zxfw1v. Waltner-Toes points out that a more effective way to present such pandemics spreading in intensive animal farming would be to require farms to give workers paid sick leave and health insurance (30). Or we could just stop intensive animal farming.
37 Dorian Sagan and Eric D. Schneider. *Into the Cool: Energy Flow, Thermodynamics and Life* (Chicago: The University of Chicago Press, 2005), 282.
38 There is more recent research that supports this view, but Luis P. Villarreal's "Can Viruses Make Us Human?" is an early example of such a "viro-centric" view, *Proceedings of the American Philosophical Society* 148, no. 3 (2004): 296–323.
39 Timothy Morton, *The Ecological Thought* (Cambridge, MA: Harvard University Press, 2010), 2.
40 Waltner-Toes, *On Pandemics: Deadly Diseases from Bubonic Plague to Coronavirus,* 15.
41 See Linda Stupart's speculative fiction, "Thawing, Dissolving, Disappearing, Bleeding Etc.: A Scientific Study of Trauma, Time Travel and the Melting Polar Ice Caps," *Art + Australia, Multinaturalism* 57, no. 1 (2021), ed. Tessa Laird, retrieved from: https://artandaustralia.com/online/thawing-dissolving-disappearing-bleeding-etc-scientific-study-trauma-time-travel-and-melting-polar-0.html
42 See United Nations, "Policy Brief: The Impact of COVID-19 on Women," (2020), retrieved from: https://www.unwomen.org/en/digital-library/publications/2020/04/policy-brief-the-impact-of-covid-19-on-women
43 Gilles Deleuze and Félix Guattari, *A Thousand Plateaus: Capitalism and Schizophrenia,* trans. Brian Massumi (University of Minnesota Press: Minneapolis, 2009), 242. See also Laird, "Zoognosis: When Animal Knowledges Go Viral," 38.
44 Atwood, *Oryx and Crake*, 20–21.
45 It is possible Guattari detourned the term "molecular revolution" from Timothy Leary, who gave a lecture with this title in 1966 (published in Timothy Leary, *The Politics of Ecstasy* (London: Paladin, 1970), 269–293). Or it is just as likely that Guattari came to the term himself, through his and Deleuze's opposition of the molecular to the molar and his investment in a "micro-politics of desire": Félix Guattari, *Molecular Revolution: Psychiatry and Politics*, trans. Rosemary Sheed (Middlesex: Penguin, 1984), 82–107.
46 Deleuze and Guattari, *A Thousand Plateaus*, 275. They also casually refer to themselves as sorcerers.
47 Tessa Laird, *Bat* (London: Reaktion, 2018).
48 Jack Halberstam, *Wild Things: The Disorder of Desire* (Durham: Duke University Press, 2020), 155. Halberstam also makes the case that zombies are a more important "contagion" to think-with than vampires in the 21st century.
49 Donna J. Haraway, *When Species Meet* (Minneapolis: University of Minnesota Press, 2007), 15.
50 Silvia Federici, *Caliban and the Witch* (New York: Autonomedia, 2004), 40.
51 Torrey, *Parasites, Pussycats and Psychosis*, p. 33.
52 Ibid.
53 Lori Gruen and Fiona Probyn-Rapsey, "Distillations," in *Animaladies: Gender, Animals, and Madness,* eds. Lori Gruen and Fiona Probyn-Rapsey (New York: Bloomsbury Academic, 2019), 7.
54 Fiona Probyn-Rapsey, "The 'Crazy Cat Lady'," in *Animaladies*, eds. Gruen and Probyn-Rapsey (New York: Bloomsbury Academic, 2019), 175–177.
55 Ibid., 175.
56 Ibid., 183.
57 Ibid., 179.
58 Ibid., 181.
59 Ibid., 182.
60 Torrey, *Parasites, Pussycats and Psychosis*, 71–109.
61 Sadly, the fascinating life story of Louis Wain has recently been turned into an unnecessarily sentimental biopic, starring none other than Benedict Cumberbatch in the lead role: *The Electrical Life of Louis Wain* (dir. Will Sharpe), 2021.
62 For a lengthy exposition on the potentials and pitfalls of Plato's pharmakon, see: Jacques Derrida, *Dissemination,* trans. Barbara Johnson (Chicago: University of Chicago Press, 1981).
63 Frédéric Keck, *Avian Reservoirs: Virus Hunters and Bird Watchers in Chinese Sentinel Posts* (Durham: Duke University Press, 2020), 22.

64 James George Frazer, *The Golden Bough: A Study in Magic and Religion* (London: Macmillan, 1974), 20–21.
65 While some readers may consider a bird being tied to a bed as hardly indicative of care, let's not forget the billions of animals that are systematically incarcerated, abused, and murdered for the advancement, and sometimes simply the entrenchment, of Western medicine.
66 Waltner-Toes, *On Pandemics: Deadly Diseases from Bubonic Plague to Coronavirus,* 42.
67 Els Lagrou, "Nisun: The Vengeance of the Bat People or What It Can Teach Us about the New Coronavirus," *Voices of Amerikua*, trans. Julien Bismuth (2020), retrieved from: https://voicesofamerikua.net/nisun-the-vengeance-of-the-bat-people-or-what-it-can-teach-usabout-the-new-coronavirus/
68 Victoria Gill, "Covid origin studies say evidence points to Wuhan market," *BBC News* (July 26, 2022), retrieved from: https://www.bbc.com/news/science-environment-62307383
69 Claude Lévi-Strauss, "We Are All Cannibals," in *We Are All Cannibals and Other Essays,* trans. Jane Marie Todd (New York: Columbia University Press, 2016)
70 Waltner-Toes, *On Pandemics: Deadly Diseases from Bubonic Plague to Coronavirus,* 17.
71 Val Plumwood, "Being Prey," in *The New Earth Reader: The Best of Terra Nova*, eds. David Rothenberg and Marta Ulvaeus (Cambridge, MA: The MIT Press, 1999)
72 Waltner-Toes, *On Pandemics: Deadly Diseases from Bubonic Plague to Coronavirus,* 18.
73 Stacy Alaimo, "Your Shell on Acid: Material Immersion, Anthropocene Dissolves," in *Exposed: Environmental Politics and Pleasures in Posthuman Times* (Minneapolis: University of Minnesota Press, 2016)
74 Waltner-Toes, *On Pandemics: Deadly Diseases from Bubonic Plague to Coronavirus,* 81.
75 Ibid., 67.
76 The book *Neoliberal Ebola* argues that latent ebola pathogens only ever flourish in the landscapes of multi-national agribusinesses in West Africa, which denude the landscape and impoverish the local people. Wallace et al. note the origins of the capital investments which lead to such outbreaks and asks if we might characterise New York, London, and Hong Kong as disease "hot spots" instead of the West African townships those financial centres devastate. Robert G. Wallace, Richard Kock, Luke Bergmann, Marius Gilbert, Lenny Hogerwerf, Claudia Pittiglio, Raffaele Mattioli, and Rodrick Wallace, "Did Neoliberalizing West Africa's Forests Produce a Vaccine-Resistant Ebola?" *Neoliberal Ebola: Modeling Disease Emergence from Finance to Forest and Farm*, eds. Robert G. Wallace and Rodrick Wallace (Cham: Springer International, 2016), 61.
77 Sandino Núñez, "Virus Virus," in *Plague Proportions*, Melbourne School of Continental Philosophy, trans. Reynaldo Young, retrieved from: https://mscp.org.au/plague-proportions/virus-virus-2
78 Jean-Luc Nancy, "Communovirus," *Libéracion* (March 24, 2020), retrieved from: https://www.liberation.fr/debats/2020/03/24/communovirus_1782922/
79 Lykke and Marambio, "A Tryptich of Viral Tales," 106.
80 Sabrina Orah Mark, "The Fairy-Tale Virus," *The Paris Review* (April 6, 2020), retrieved from: https://www.theparisreview.org/blog/2020/04/06/the-fairytale-virus/. See also Thomas Nagel, "What Is It Like to Be a Bat?" *The Philosophical Review* 83, no. 4 (1974): 435–450.
81 Kelly Burke, "Laura Jean McKay Wins $100,000 Victorian Literature Prize for the Animals in That Country," *The Guardian* (1 February, 2021), retrieved from: https://www.theguardian.com/culture/2021/feb/01/laura-jean-mckay-wins-100000-victorian-literature-prize-for-the-animals-in-that-country
82 Muhammad Asif and Majid Ali, "Chinese Traditions Folk Art, Festivals and Symbolism," *International Journal of Research* 6, no. 1 (January 2019): 4.
83 Gumbs, *Undrowned,* 2.
84 Ibid.
85 See: Tessa Laird, with Liang Luscombe and Joel Stern, "Locating Echoes: Reverberations between Bats and Humans," as part of *The Art of Possibly Animate Things*, in *You Can't Trust Music*, curated by Xenia Benivolski, e-flux, (2022), retrieved from: https://yctm.e-flux.com/the-art-of-possibly-animate-things
86 See Jakob von Uexküll's charming concept of Nature's "symphony" in *A Foray into the World of Animals and Humans,* trans. Joseph D. O'Neill (Minneapolis: University of Minnesota Press, 2010), 208. In keeping with the mode of storytelling I propose in this chapter, von Uexküll's value as a biologist is precisely the fact that his writing is "fabulous, fabricated and child-like", see Stephen Loo and Undine Sellbach, "A Picture Book of Invisible Worlds: Semblances of Insects and Humans in Jakob von Uexküll's Laboratory," *Angelaki: Journal of the Theoretical Humanities* 18, no. 1 (2013): 46.

87 Gumbs, *Undrowned*, 68–69.
88 Paul B. Preciado, "On the Verge," *Artforum International* (July/August 2020), retrieved from: https://www.artforum.com/print/202006/paul-b-preciado-on-revolution-83286
89 Michael Taussig, "Would a Shaman Help?" *Critical Inquiry: In the Moment* (March 30, 2020), retrieved from: https://critinq.wordpress.com/2020/03/30/would-a-shaman-help/
90 Arundhati Roy, "The Pandemic Is a Portal," *Financial Times* (April 4, 2020), retrieved from: https://www.ft.com/content/10d8f5e8–74eb-11ea-95fe-fcd274e920ca

Bibliography

Alaimo, Stacy. "Your Shell on Acid: Material Immersion, Anthropocene Dissolves." In *Exposed: Environmental Politics and Pleasures in Posthuman Times*. Minneapolis: University of Minnesota Press, 2016.

Asif, Muhammad Asif, and Ali, Majid. "Chinese Traditions Folk Art, Festivals and Symbolism." *International Journal of Research* 6, no. 1 (January 2019).

Atwood, Margaret. *Oryx and Crake*. New York: Nan A. Talese, 2003.

Burke, Kelly. "Laura Jean McKay Wins $100,000 Victorian Literature Prize for The Animals in That Country." *The Guardian*, February 1, 2021. Retrieved from: https://www.theguardian.com/culture/2021/feb/01/laura-jean-mckay-wins-100000-victorian-literature-prize-for-the-animals-in-that-country

Centers for Disease Control and Prevention. "History of Smallpox." Retrieved from: https://www.cdc.gov/smallpox/history/history.html#:~:text=The%20basis%20for%20vaccination%20began,used%20to%20protect%20against%20smallpox.

Charis Circle. "Undrowned—Black Feminist Lessons from Marine Mammals—Alexis Pauline Gumbs and Adrienne Maree Brown." *YouTube*, November 20, 2020. Retrieved from: https://www.youtube.com/watch?v=-3_GUGaZ0rI

Datta, Sanjoy, Bhatla, Neerja, Burgess, Margaret, Lehtinen, Matti, and Bock, Hans. "Women and Vaccinations: From Smallpox to the Future, a Tribute to a Partnership Benefiting Humanity for over 200 Years." *Human Vaccines* 5, no. 7 (2009): 450–451.

Deleuze, Gilles, and Guattari, Félix. *A Thousand Plateaus: Capitalism and Schizophrenia*. Translated by Brian Massumi. University of Minnesota Press: Minneapolis, 2009.

Derrida, Jacques. *Dissemination*. Translated by Barbara Johnson. Chicago: University of Chicago Press, 1981.

Derrida, Jacques. *The Animal That Therefore I Am*. Edited by Marie-Louise Mallet, Translated by David Wills. New York: Fordham University Press, 2008.

Dunn, Amelia. "Why You Should Throw Out the Thongs You Wore in Bali." *SBS News*, July 15, 2022. Retrieved from: https://www.sbs.com.au/news/article/why-you-should-throw-out-the-thongs-you-wore-in-bali/h70zxfw1v.

Federici, Silvia. *Caliban and the Witch*. New York: Autonomedia, 2004.

Frazer, James George. *The Golden Bough: A Study in Magic and Religion*. London: Macmillan, 1974.

Gill, Victoria. "Covid Origin Studies Say Evidence Points to Wuhan Market." *BBC News*, July 26, 2022. Retrieved from: https://www.bbc.com/news/science-environment-62307383

Gruen, Lori, and Probyn-Rapsey, Fiona. "Distillations." In *Animaladies: Gender, Animals, and Madness*, edited by Lori Gruen and Fiona Probyn-Rapsey. New York: Bloomsbury Academic, 2019.

Guattari, Félix. *Molecular Revolution: Psychiatry and Politics*, Translated by Rosemary Sheed. Middlesex: Penguin, 1984.

Gumbs, Alexis Pauline. *Undrowned: Black Feminist Lessons from Marine Mammals*. Chico, CA: AK Press, 2020.

Halberstam, Jack. *Wild Things: The Disorder of Desire*. Durham: Duke University Press, 2020.

Haraway, Donna. *The Companion Species Manifesto: Dogs, People and Significant Otherness*. Chicago: Prickly Paradigm Press, 2003.

Haraway, Donna J. *When Species Meet*. Minneapolis: University of Minnesota Press, 2007.

Keck, Frédéric. *Avian Reservoirs: Virus Hunters and Bird Watchers in Chinese Sentinel Posts*. Durham: Duke University Press, 2020.

Lagrou, Els. "Nisun: The Vengeance of the Bat People or What It Can Teach Us about the New Coronavirus." *Voices of Amerikua*, trans. Julien Bismuth (2020). Retrieved from: https://voicesofamerikua.net/nisun-the-vengeance-of-the-bat-people-or-what-it-can-teach-usabout-the-new-coronavirus/

Laird, Tessa, with Liang Luscombe and Joel Stern. "Locating Echoes: Reverberations between Bats and Humans." As part of *The Art of Possibly Animate Things*, in *You Can't Trust Music*, curated by Xenia Benivolski, e-flux. (2022). Retrieved from: https://yctm.e-flux.com/the-art-of-possibly-animate-things

Laird, Tessa. *Bat*. London: Reaktion, 2018.

Laird, Tessa. "Zoognosis: When Animal Knowledges Go Viral. Laura Jean McKay's *The Animals in That Country*, Contagion, Becoming-Animal, and the Politics of Predation." *Animal Studies Journal* 10, no. 1 (2021): 30–56.

Le Guin, Ursula K. "The Carrier Bag Theory of Fiction." In *Ecocriticism: The Essential Reader*, edited by Ken Hiltner. New York: Routledge, 2014.

Leary, Timothy. "Molecular Revolution." In *The Politics of Ecstasy*, by Timothy Leary, 269–293. London: Paladin, 1970.

Lévi-Strauss, Claude. "We Are All Cannibals." In *We Are All Cannibals and Other Essays*. Translated by Jane Marie Todd. New York: Columbia University Press, 2016.

Loo, Stephen, and Sellbach, Undine. "A Picture Book of Invisible Worlds: Semblances of Insects and Humans in Jakob von Uexküll's Laboratory." *Angelaki: Journal of the Theoretical Humanities* 18, no. 1 (2013).

Lykke, Nina, and Marambio, Camila. "A Triptych of Viral Tales." *Kerb: Journal of Landscape Architecture*, #28, "Decentre—Designing for Coexistence in a Time of Crisis." (2020). Retrieved from: https://kerb-journal.com/

Margulis, Lynn, and Sagan, Dorion. *What is Sex?* New York: Simon & Schuster, 1998.

Mark, Sabrina Orah. "The Fairy-Tale Virus." *The Paris Review*, April 6, 2020. Retrieved from: https://www.theparisreview.org/blog/2020/04/06/the-fairytale-virus/.

Márquez, Gabriel García. *Love in the Time of Cholera*. Translated by Edith Grossman. New York: Penguin Books, 1999.

Morton, Timothy. *The Ecological Thought*. Cambridge, Mass: Harvard University Press, 2010.

Nagel, Thomas. "What Is It Like to Be a Bat?" *The Philosophical Review* 83, no. 4 (1974): 435–450.

Nancy, Jean-Luc. "Communovirus." *Libéracion*, March 24, 2020. Retrieved from: https://www.liberation.fr/debats/2020/03/24/communovirus_1782922/

Nguyen, Vinh-Kim. "Of What Are Epidemics the Symptom? Speed, Interlinkage, and Infrastructure in Molecular Anthropology." In *The Anthropology of Epidemics*, edited by Ann H. Kelly, Frédéric Keck, and Christos Lynteris. London: Routledge, 2019.

Núñez, Sandino. "Virus Virus." *Plague Proportions*, Melbourne School of Continental Philosophy, Translated by Reynaldo Young. Retrieved from: https://mscp.org.au/plague-proportions/virus-virus-2

Online Etymology Dictionary. "Vaccination." Retrieved from: https://www.etymonline.com/word/vaccination.

Plumwood, Val. *Feminism and the Mastery of Nature*. London: Routledge, 1993.

Plumwood, Val. "Being Prey." In *The New Earth Reader: The Best of Terra Nova*, edited by David Rothenberg and Marta Ulvaeus. Cambridge, MA: The MIT Press, 1999.

Preciado, Paul B. "On the Verge." *Artforum International*, July/August 2020. Retrieved from: https://www.artforum.com/print/202006/paul-b-preciado-on-revolution-83286

Preston, Richard. *The Hot Zone*. Sydney, 1994.

Probyn-Rapsey, Fiona. "The 'Crazy Cat Lady'." In *Animaladies: Gender, Animals, and Madness*, edited by Lori Gruen and Fiona Probyn-Rapsey, 175–177. New York: Bloomsbury Academic, 2019.

Rose, Deborah Bird. *Shimmer: Flying Fox Exuberance in Worlds of Peril*. Edinburgh: Edinburgh University Press, 2022.

Roy, Arundhati. "The Pandemic Is a Portal." *Financial Times*, April 4, 2020. Retrieved from: https://www.ft.com/content/10d8f5e8-74eb-11ea-95fe-fcd274e920ca

Sagan, Dorian, and Schneider, Eric D. *Into the Cool: Energy Flow, Thermodynamics and Life*. Chicago: The University of Chicago Press, 2005.

Saint-Cyr, Yves. "Desire, Disease, Death, and David Cronenberg: The Operatic Anxieties of The Fly." *The Canadian Review of Comparative Literature* 38, no. 4 (2011).

Sharpe, Will, dir. *The Electrical Life of Louis Wain*. 2021.

Soderbergh, Steven, dir. *Contagion*. 2011.

Stengers, Isabelle. *Cosmopolitics*. Translated by Robert Bononno. Minneapolis: University of Minnesota Press, 2010.

Stupart, Linda. "Thawing, Dissolving, Disappearing, Bleeding Etc.: A Scientific Study of Trauma, Time Travel and the Melting Polar Ice Caps." *Art + Australia*, "Multinaturalism", Vol. 57 (1) (2021), edited by Tessa Laird. Retrieved from: https://artandaustralia.com/online/thawing-dissolving-disappearing-bleeding-etc-scientific-study-trauma-time-travel-and-melting-polar-0.html

Sudan, Rajani. "Contracting Xenophobia: Etiology, Inoculation, and the Limits of British Imperialism." In *Fear and Loathing in Victorian England,* edited by Maria Bachman, Heidi Kaufman, and Marlene Tromp. Columbus: Ohio State University Press, 2013.

Taussig, Michael. "Would a Shaman Help?." *Critical Inquiry: In the Moment*, March 30, 2020. Retrieved from: https://critinq.wordpress.com/2020/03/30/would-a-shaman-help/

Torrey, E. Fuller. *Parasites, Pussycats and Psychosis: The Unknown Dangers of Human Toxoplasmosis*. Cham, Switzerland: Springer, 2022.

Tsing, Anna Lowenhaupt. *Mushroom at the End of the World: On the Possibility of Life in Capitalist Ruins*. New Jersey: Princeton University Press, 2012.

Tuttle, Merlin. "A Viral Witch Hunt." *Issues in Science and Technology*, March 27, 2020. Retrieved from: https://issues.org/a-viral-witch-hunt-bats/

Uexküll, Jakob von. "Nature's "Symphony." In *A Foray into the World of Animals and Humans*, translated by Joseph D. O'Neill. Minneapolis: University of Minnesota Press, 2010.

United Nations. "Policy Brief: The Impact of COVID-19 on Women." (2020). Retrieved from: https://www.unwomen.org/en/digital-library/publications/2020/04/policy-brief-the-impact-of-covid-19-on-women

Villarreal, Luis P. "Can Viruses Make Us Human?" *Proceedings of the American Philosophical Society* 148, no. 3 (2004): 296–323.

Wallace, Robert G., Kock, Richard, Bergmann, Luke, Gilbert, Marius, Hogerwerf, Lenny, Pittiglio, Claudia, Mattioli, Raffaele, and Wallace, Rodrick. "Did Neoliberalizing West Africa's Forests Produce a Vaccine-Resistant Ebola?" In *Neoliberal Ebola: Modeling Disease Emergence from Finance to Forest and Farm*, edited by Robert G. Wallace and Rodrick Wallace. Cham: Springer International, 2016.

Waltner-Toes, David. *On Pandemics: Deadly Diseases from Bubonic Plague to Coronavirus*. Carlton, VIC: Schwartz Books, 2020.

Yunkaporta, Tyson. *Sand Talk: how Indigenous Thinking Can Save the World*. Melbourne: Text Publishing, 2019.

48

ELACHISTOCENE VS. ANTHROPO-SCENES

Inheriting an Epoch Defined by Mass Extinction

Dylan Hall

A Cold and Broken Hallelujah: The Anthropocene and Its Discontents

A wide range of scientists, journalists, and artists have used the Anthropocene—Epoch of Humans—as a shorthand for humanity's alarming impacts, but the term is also promoted by a very specific group of scientists who urge the adoption of the Anthropocene as the official name for a new geological epoch. Atmospheric chemist Paul Crutzen, ecologist Eugene Stoermer, geologist Jan Zalaciewicz, earth system scientists Johan Rockström and Will Steffen, environmental historian John McNeill, and geographer Erle Ellis—all white men in rich countries—are among but some of the most vocal and prominent. In their versions of the story, an often undifferentiated 'humanity,' or the rather business-like 'human enterprise,' is 'overwhelming the great forces of nature,'[1] pumping carbon dioxide into the atmosphere and acidifying the oceans while filling them with plastic—driving a universal 'us,' in the words of Rockström and his colleagues, beyond 'a safe operating space for humanity.[2] Importantly, the earth systems scientists tell us that we are shifting out of the stability of the Holocene, the last 10,000 years of extremely unusual stability in climatic temperatures and precipitation patterns that made the rise of cities and agriculture possible, and this is both an immense threat to modern civilization, and why we are in a new epoch. The geological argument layers on top of this the observation that the vast scale of human activity will be observable in rock for millions of years, through sediments that reveal the immense increase in plastic and plutonium and chicken bones, and the abrupt end of fossils that reveal once extensive biodiversity.

Critics of the Anthropocene have engaged in proposing alternative names, playfully and seriously, to highlight drivers of destruction beyond a simple idea of 'the human enterprise.' Better the Eurocene or Anglocene, to point at the lost languages, the natural-cultural impacts of European colonization, the increasing domination of English in global business and science, including most conversations about the Anthropocene (like this one).[3] Or maybe the Capitalocene, advocated by Jason Moore and colleagues to point at our ravenous economy—the unequal and exploitative organization system for a huge number of anthropos—fundamentally based on debt, profit maximization, inequality, and infinite

DOI: 10.4324/9781003273400-59

exponential growth.[4] With the Manthropocene and Northropocene, Kate Raworth points out that the majority of Anthropocene scientists are men in the Global North, mansplaining ecological crisis as a problem of 'human nature.' This ignores feminists who have long shown how patriarchal agricultural empires and the exploitation of women by men are linked to hierarchies that also justify the exploitation of animals, waters, and minerals. This also ignores a host of decolonial thinkers who have long argued that states and corporations of the Global North are rich precisely because they extract wealth from the states, peoples, lands, and waters of the Global South, leaving pollution and poverty in the wake.[5] The Plantationocene identifies the global spread of the plantation, enslavement and domination of humans and animals in industrialized food production, and the plantation as the leading driver of extinctions over the last decades.[6] Simultaneously, Alf Hornborg's Technocene points to accelerating frankenstein technologies that enable and encourage increasingly bigger open-pit mines, mega-dams, fishing trawlers, nuclear bombs, and drone-driven tractors.[7] Donna Haraway's Cthulucene asks us to remember that the human has never only been human, that we are all bound up in tentacular and microbial and gossamer agencies and thus might yet make compostable allegiances with the living world that birthed us.[8]

I appreciate the insights offered by all these names (and the many others that have been proposed) but I also believe that conservative geologists in the Anthropocene Working Group of the International Commission on Stratigraphy will never seriously consider any of these names. There is only one name that I have stumbled upon which would fit with the geological naming scheme that has thus far determined the geological time scale, proposed by geologist Jill Schneidermann in a mere footnote to her 2015 essay 'Naming the Anthropocene.' Schneiderman proposed the *Elachistocene*—literally 'the time of least new life'—to center the magnitude of the Sixth Extinction at the heart of this epoch. Sympathetic to protests that the Anthropocene homogenizes humanity, Schneiderman asked: 'why not label the new epoch with a name that acknowledges the much less contested sixth extinction and increased diminishment of species on earth in this epoch?'[9]

Schneiderman's argument in sum goes like this: The geological time scale is an immense achievement describing the deep multi-billion year history of life on Earth based on records left by ancient fossils. The standard geologic time scale grew 'unsystematically in fits and starts' from relative dating based on chronostratigraphy—comparing sedimentary layers—to more recent radiometric dating that has revealed previously unimaginable time scales, which Schneiderman describes as having 'changed the way humans view our place in the fullness of time.'[10] One of the most extraordinary things revealed by deep time is that the gradual unfolding of evolution is punctuated by catastrophic events when life cannot keep up with the scale and rate of change. It is precisely extinction events and the life that follows that have defined *how* the time scale has been divided thus far. Mass extinction is exceptionally rare, colossal, and irreversible: there have only been five that have occurred in the 540 million years since eukaryotic life emerged from the oceans, defined as greater than 75% of species gone. These divide some of the major Eras and Ages of the GTS, with other smaller extinction events pockmarked between them dividing epochs and periods. The first era defined by extinctions is the Paleozoic (Ancient Life), which ended with the largest mass extinction event in Earth's history, known as the end-Permian or 'The Great Dying,' when ~96% of species went extinct about 252 million years ago, in as little as 160,000 years. It took millions of years to recover from this mass extinction. All life on earth today descends from the four percent of species that survived. It is humbling and terrifying to learn that, according to leading theories, approximately 252 million years ago, there was a massive release of carbon into the air and 'temperatures soared—the seas warmed by as much as eighteen degrees—and the chemistry of the oceans went haywire.'[11] Dissolved oxygen dropped so low that many organisms suffocated, and in approximately 200,000 years, over 90% of life was obliterated.

The Mesozoic—Middle Life—Era contains the famous Triassic, Jurassic, and Cretaceous Periods when dinosaurs were the dominant vertebrates. The Mesozoic begins with the Great Dying 252 million years ago and ends with a meteorite that eliminated ~75% of all species, including all the dinosaurs that couldn't fly. This marked the beginning of the Cenozoic—New Life—Era that began 66 million years ago. The 'cene' in Anthropocene, Holocene, and Pleistocene refers to the 'new' life in this Era. The niches opened by the end cretaceous mass extinction allowed flowering plants, tropical and temperate forests, birds (descended from avian dinosaurs), and mammals to thrive. The epochs of the periods of the Cenozoic Era are 'named to indicate the proportion of present-day . . . organisms in the fossil record since the beginning of the era.'[12] Thus, Paleocene means 'ancient new life'; Eocene means 'dawn' of new life; Oligocene means 'scanty' new life; Miocene means 'less' new life; Pliocene means 'more' new life; Pleistocene means 'most' new life; and Holocene, the epoch characterizing the last 10,000 years, means 'whole' or 'entirely' new life. The Pleistocene epoch, which lasted from 2.5 million years ago to about 12,000 years ago, is named—'most new'—to reflect that there was an explosion of new life represented in fossil records. Schneiderman explains that the Elachistocene is the antithesis of this—derived from the suffix *elachistos*, the antonym of *pleistos*—and thus literally meaning *the time of least new life*.

I understand that the Anthropocene is here to stay, as it has already proliferated through our culture, but in this chapter, I playfully and seriously consider the value of the Elachistocene as a thought experiment. I imagine the Elachistocene as the underbelly of the Anthropocene, revealing ghosts in a time of monsters by focusing on understanding the direct and indirect drivers of wild animal extinctions and population annihilations. Critical animal studies has long been rightly suspicious of the category *species*, and I argue that looking closely at global patterns driving extinctions helps us to understand how structural and systemic violence against the lives of industrially farmed animals is driving the extinctions of wild animals. The Elachistocene is supported by the lessons learned from the Capitalocene, Eurocene, Manthropocene, Plantationocene, Homogenocene, and other such scenes of inequality and injustice that describe deep drivers of intersecting natural-cultural crises while grounding these other diagnoses of socio-economic illnesses and inequality with a call to listen for our animal kin whose symphony is being obliterated from the orchestra of life and replaced by the tortured cries of broiler chickens and cows in feedlots. The Anthropocene might claim to bring together the divided worlds of the social and natural sciences, but it is still anthropocentric—after Eileen Crist and Leonard Cohen, 'cold and broken though it be, it's still a hallelujah.'[13] The Elachistocene is especially useful as an intervention in the thinking of scientists, because it follows the naming systems already in place, calls on earth scientists to become aware of the anthropocentrism baked into the long-troubled history of science, and asks scientists to be inspired by that rare and noble strain of science that asks us not to ignore the human but to decenter the human and expand ethical consideration well beyond us to include the rest of life—the living planet—that we are birthed from, enmeshed in, sustained by, and will die on.

The Sixth Extinction: Burning the Tapestry of Life

Death is a necessary part of organic life; death feeds life and allows it to continue. What is now occurring is not the same. Extinction eviscerates the possibility of species and lineages continuing that have persisted for millions of years. Diving into the science of global biodiversity crisis, it has been far harder for me to attempt to understand than I thought: what *is* happening to life on earth? Understanding the sixth extinction seems initially simple—an immense number of species must be going extinct. And yet, the more I delved, the harder it was to understand what to focus on, both in terms of drivers and the maze of scientific terms and categories. What makes a *mass* extinction by the standards of geology is if over 75% of species go extinct, and according to the International Union for the Convention of Nature (IUCN), only ~2% of species are currently listed as extinct.[14] A simple calculation of extinct species would seem to suggest that we are far from a sixth mass extinction event but it turns out this is deeply misleading, failing to consider *how* mass extinctions happen, disguising the scale of the eradication of animal populations, the evisceration of ecosystems, spreading oceanic dead zones, the destructive synergies *between* the drivers of extinction, and the trajectory we are on. The *least new life* of the Elachistocene contains much more than the category of species.

The *mass* part is the heavy part of this extinction event. Everywhere we look there are massive problems with *scale and rate of growth*. Mass consumption and mass production; mass destruction and mass distraction; mass unemployment and mass layoffs; mass hysteria and mass media; mass shootings, mass incarceration, mass graves, and mass extinction. The masses weigh heavy on my mind/heart/body, as does the literal mass of humans and farmed mammals weigh heavy on earth: global biomass of wild mammals has fallen by ~82% since 1492, and today if you weighed all the large mammals on earth, just 3% are living in the wild. The rest is some 30% of global mammal biomass in human bodies, and 67% is composed of the domestic animals that feed us.[15] When my grandmother was born in 1925, the world population reached 2 billion. When she celebrated her 97th birthday in 2022, the population had reached 8 billion. A quadrupling in one woman's lifetime. This massification of populations at a category called the global is *deeply problematic* in what it obscures and the ways that it has been used by xenophobic anti-immigration politicians, but as Haraway insists, that doesn't mean these numbers don't describe something real.[16]

The human population is not the main problem but it is still a problem. Consumption and inequality is horrific and important, and I will spend some time here—according to the global ecological footprint calculator (one such troubled method of counting), the average Canadian (like me) consumes 8.08 hectares of Earth, just above Americans at 8.04 and well above Colombians at 1.91, while the average human in Chad consumes 0.01 hectares. And, as multiple studies have shown, within wealthy countries—the greater one's income, the greater one's impact. Thus, rich people of the world (which includes the middle classes in Canada) have the disproportionately highest impact. In 2018, Donna Haraway writes how pointing out the problem of population is always dangerous because White families have been enculturated to believe—'White nuclear families in suburban homes: The American Dream. Dangerous populations of color everywhere: The American nightmare, including population bombs.' The irony of this, of course, is that the real consumptive, earth-eating population bomb was the very image of the American Dream: the post WWII baby boom.

Haraway invented an idiom for understanding matterings on earth that I have not been able to shake—The Born and The Disappeared. She engages in these constructed categories of global populations with care, insisting that the reader understand that entertaining any category of Big Numbers means engaging in dangerous universalizing inherent in the genre. She explains that the Born Ones include 'the almost unimaginable (but countable in deeply flawed data sets and globalizing modeling operations) multi-billions of human beings, industrial food animals, and companion pets enterprised up to bloated consumer status.'[17] Conversely, the Disappeared include:

> human resisters to criminal nation states, the imprisoned, missing generations of the Indigenous and other oppressed people and peoples, unruly women . . . disposable young people of every race or ethnicity, migrants, refugees and displaced people, stateless people, human beings subject to ethnic cleansing and genocide, and already about 50% of all vertebrate wildlife that were living on earth's lands and oceans less than 50 years ago, plus 76% of fresh water species.[18]

In 2006, Deborah Bird Rose called extinction 'double death,' the termination of entire patterns of living and dying on earth. Double-death has produced many of The Disappeared, whose 'conjoined twin,' in Haraway's language, The Born Ones, is thus produced by double-birth. Double birth is 'forced life for economist value production,'[19] the perversion of forced birthing, or hatching or germinating humans and other beings, especially industrial food animals and plants.[20] As a child of the children of post–World War II booming and consuming Canadian white middle-class affluence, I too, with Haraway, am one of The Born Ones. If Anthropocene is aligned with The Born Ones, then Elachistocene is the inverse, aligned with the Disappeared and the vanishing.

If there were to be a prophet of the Elachistocene, it would be Eileen Crist. In her book *Abundant Earth: Towards an Ecological Civilization,* Crist delivers a more succinct, comprehensive, and devastating summary of the sixth extinction than I have read elsewhere, and she contains it within the first chapter: 'Unraveling Earth's Biodiversity.'[21] Crist navigates the genre of Big Numbers while still managing to convey the emotional magnitude of ongoing 'biological annihilation' through her close attention to language and her oscillations between planetary and local calamity. She argues that only seeing the immensity of the crisis will allow us to dream of the scale of transformation needed to reverse the horror of present trends: this is why we need the Elachistocene—to not just see Old Man Anthropos.[22] I use Crist's work in my attempt to understand the ecology of the sixth extinction at the heart of this epoch.

The Elachistocene's focus on 'least new life' is useful because it focuses on a broad category—life. The fabric of life on earth is understood through numerous languages and cultures that have existed long before science and English were globalized through colonialism, many of which are far more reverent and respectful of animal life than science, but none of which I wish to romanticize nor appropriate as a non-Indigenous person. Thus, while acknowledging the reductive and colonial history and assumptions baked into the universalizations of science and the many other human voices worth listening to, I will nonetheless use the language of ecology and geology in my attempt to understand the rippling web of life, while doing my best not to reduce the living world simply to a resource.

While using the language of science, it is especially important to understand that science often will inherently minimize the moral magnitude of what is happening. Eileen

Crist identifies 'entitled instrumentalism' as one of the worst manifestations of anthropocentrism in our society and shows how it emerges in the language of engineers, politicians, economists, and even ecologists to facilitate and justify our trajectory towards ever more extinctions. Crist points out the way that a human-centered, entitled instrumentalism is baked into the scientific and managerial language of biologists, economists, engineers, planners, and politicians: the abundance of nature becomes resources—warped to be understood through a lens of utility for 'us:' fisheries, fish stock, livestock, timber, weeds, vermin, pests. As Audra Mitchell demonstrates, to Indigenous peoples, salmon, bison, water, and land are kin, relatives, and family: peoples with agency and social communities of their own.[23]

To understand what is happening to life, first I needed to understand what life is. I especially appreciated this definition from Eileen Crist:

> Earth's being is a cosmos that self-creates itself through the resonances of its innumerable members, who, barring rare and large-scale catastrophes, keep swelling into a plenum of diversified kinds, abundant numbers, different ways of life, and exquisitely convoluted relationships—all unfurling as a slow-motion upsurge of biodiversity over geological time.[24]

Life's richness exists at the entangled levels of species, subspecies, populations, genes, behaviors, minds, and ecologies; this richness is 'self-perpetuating and builds more of itself over time.'[25] Right now, this richness is burning. The concepts of extinction and biodiversity rely on the concept of species and while species does refer to actual biological entities, the term is also an abstraction. In the real world, a species is not a discrete entity but consists of a dynamic web of living beings and their relationships. Crist demonstrates this with the example of wolves in North America. It is estimated that pre-contact, there were approximately 500,000 wolves in North America. They roamed in distinct populations (metapopulations) and contained at least 23 subspecies. Wolves today occupy merely 5% of their historic range. Large numbers and wide ranges allowed for immense variety—in populations, phenotypes, and genetics. A huge degree of genetic variation actually occurs *within* populations, and this variety underpins resilience and evolutionary change over geological time.[26] Biodiversity also includes Earth's *biodisperity,* or ecological variety in different places. This illuminates continents and islands separated hundreds of millions of years ago that have allowed the evolution of interdependent communities of plants, animals, fungi, and insects in specific regions, climates, and altitudes—*biomes*—including wet and dry tropical forests, temperate and boreal forests, deserts, alpine meadows, tundra, grasslands, estuaries, and ocean depths.

Biodiversity has steadily and slowly increased over time, beginning about 3.8 billion years ago and, after billions of years of microscopic, bacterial, and algae evolution, accelerating rapidly after the Cambrian explosion 550 million years ago. Biodiversity gradually swells because new species form faster than the background (or naturally occurring) extinction of species. This process is interrupted only during episodes of mass extinction, when catastrophic events—such as immense volcanic eruptions or meteorite impacts—obliterate an immense amount of life. Crist urges us to understand that mass extinction events and background extinctions are *categorically* different—unlike background extinctions or even large extinction events, mass extinctions are extremely rare, global in scope, affect the majority of life-forms, demolish biodiversity at large, involve immense climatic and chemical changes to the atmosphere and ocean, and occur relatively rapidly in geological time, obliterating entire lineages of beings. Mass extinctions in the past have opened up space for new species to proliferate, but it has always taken multiple millions of years for the biosphere to generate a new symphony, a timeline that is meaningless for our children, their children, and their imaginably distant descendants.

The Sixth Extinction: An Unnatural History, by Elizabeth Kolbert, is arguably one of the best-selling books on extinction—a *New York Times* bestseller and winner of the Pulitzer Prize. Kolbert travels the world, interviewing scientists and ecologists and explaining the ongoing devastation of life through stories about specific animals that are long extinct, recently extinct, or threatened with extinction—mastodons, great auks, ichthyosaurs, graptolites, coral reefs, Amazon rainforest brown bats, Hawaiian crows, and Panamanian golden frogs—revealing the pattern of extinctions of species and ecosystems sweeping the globe. She explains ecological and paleo-geological concepts in engaging ways, often through a historical discussion of the European scientists who discovered and debated ideas of evolution, extinction, and geological time through the 18th and 19th centuries. However, Kolbert's impressive communication of paleontological, ecological, geological, and scientific intellectual history is regrettably contained within a discussion of the science and the scientists and thus falls prey to many trappings of professional ecologists, geologists, and scientists of the Anthropocene: 1) a fatalistic, misanthropic, and depressing discussion of the root causes of the calamity as simply the human species, without a sustained examination of which humans and which human systems are responsible, and 2) immensely inadequate responses to the violent annihilation of species contained within the heroic but desperate work of conservation scientists in zoos and preserves. This is especially egregious given that Kolbert has the privileged mobility to travel to research stations around the world on the dime of *The New Yorker*. Thus, the Elachistocene would not, in and of itself, be an antidote to this tendency to blame 'one weedy species,'[27] and would need to be grounded by other names.

Elizabeth Kolbert's work in *The Sixth Extinction* inspired and infuriated Ashley Dawson, who was compelled to write *Extinction: A Radical History* as a critical response to Kolbert's explanation that it is a 'Madness Gene' within *Homo sapiens* that explains the destruction. Dawson sketches a brief radical history of ecocide to demonstrate that, though especially agricultural empires like the Babylonian and Romans have long engaged in land and animal abuse, 'it is only with the expansion of Europe and the development of modern capitalism that ecocide has taken on a truly global extent and planet-consuming destructiveness.'[28] Today, free-trade agreements and debilitating debt regimes imposed on poor countries by wealthy countries drives the violence against both the wild animals of the world and the many people of the world who have been displaced, destroyed, and recruited by the international system of nation states and global capitalism constructed after WWII. Dawson writes a historical materialist account of the Sixth Extinction and shows that European settlers in the Americas, Asia, Oceania, and Africa introduced an understanding of nature as property, as dead matter, as a resource to be exploited for human gain; anthropocentrism was imposed along with the expansion of capitalism, militarism, and eventually, the modern nation state. In British, French, German, Dutch, and Portuguese law, life is stripped of its agency, imposing a completely different vision of human relations to land, water, forests, birds, animals, insects and the rest of nature. Dawson dissects the inadequacy of global conservation schemes and makes arguments for responses that specifically target transforming the immensely unequal relationships between poor countries and wealthy countries that drive the accelerating destruction of biodiversity hotspots, especially in the remaining tropical rainforests of South America, Africa, and Asia.

The present extinction event is not a metaphor and not new. We are already in an impoverished world, where much of the tapestry of wildlife has been destroyed by the last 500 years of European expansion, ecological imperialism, technological ascendance, and the globalization of modernity. Crist writes that there is 'colossal public ignorance' of the ongoing unraveling of biodiversity. Many of us are distracted by the cities and screens that surround us, and even those of us who do go out of our way to pay attention to the forests and the streams take the world we grew up in as the baseline from which to measure change. Those of us who are settlers like myself have also been raised in a culture intent on burying and forgetting the ecological and colonial crimes of the past. In this way, each successive generation has a 'shifting ecological baseline' from which to apprehend an increasingly diminished world, such that we do not know what has been lost.

The destruction of life's diversity, complexity, and abundance over the last centuries means that it is increasingly difficult to fully hear the immense cacophony of life's orchestra. Only a few centuries ago, millions of whales sustained entire ecosystems with their carcasses and feces. One hundred thousand tigers roamed from Siberia to India, Sumatra, and Java. Immense rainbows of seething coral reefs stretched past teeming millions of sharks, swordfish, marlin, tuna, and other gigantic fish. Massive herds of large herbivores—bison, antelopes, aurochs, elephants, rhinos—plowed and fertilized grasslands, which in turn fed the herds and their numerous attendants. Above it all flew vast flocks of birds: raptors and scavengers with immense wingspans, tiny colorful and drab songbirds, migrating, wading, and nocturnal birds.

'Direct Drivers' of Extinction

Both Eileen Crist and Elizabeth Kolbert explore specific examples of what scientists call the direct drivers of extinction and highlight the synergies between them: habitat destruction and fragmentation, direct killing, pollution, nonnative species, and climate change/ocean acidification. Habitat destruction and fragmentation are the leading driver of biodiversity decline, accounting for 70% of plants and animals facing extinction.[29] Tropical forests are being decimated, and where they are not cleared for oil palm, soy, and cattle plantations, they are fragmented by roads, fences, power lines and pipelines. The main predictor of local extirpations is 'small population size,' and global extinctions are always made up of local and regional extirpations first. 'Islands' of habitat are too small to support big animals; isolated populations are weakened by lack of genetic exchange and increasingly vulnerable to fire, disease, and hunting.

It is an ugly truth to learn that 'direct killing by guns, snares, poison, hooks, and nets' is eliminating immense numbers of wild animals globally, and this is the primary threat to the remaining largest vertebrate species.[30] There are various motives: food, protecting livestock, traditional medicine, profit (think tusks and the ivory trade), or fear. Industrial fishing occurs at unfathomable scales; spreading gill nets up to 50 miles long and 50 feet deep act as literal 'walls of death' and have proliferated across the high seas. Due to consistent under-reporting, no one knows the proportion of animals killed as 'bycatch' a.k.a. killing and tossing unwanted species. The bushmeat market has exploded, with hunting transforming from a small-scale subsistence activity to a multi-billion dollar international trade, constituting 'the most significant threat to wildlife (especially mammals) in Africa,

Asia, and South America.'[31] Poachers are now moving into reserves and parks, which, unless heavily guarded, have become death traps. As the last remaining fragments of habitat, poachers know exactly where to find the remaining animals. Kolbert reports the tragedy that, 'with the exception of humans, all the great apes today are facing oblivion . . . chimpanzee and mountain gorilla numbers are half of what they were fifty years ago, while lowland gorilla populations have fallen by sixty percent in just two decades.'[32]

Invasive species, non-native species, the Columbian exchange, and the 'homogenization of Earth's biota' all describe 'an evolutionary experiment like no other in Earth's history.'[3334] Two hundred million years of increasingly separate evolution since the supercontinent of Pangea broke into the hemispheres of the world ended with the European 'Age of Exploration' (also the age of exploitation and extinction). In his global voyages, Darwin was fuzzily beginning to apprehend this intercontinental reshuffling when he observed that South America, Africa, and Australia were entirely similar in climate and topography, but they were populated by 'entirely *dis*similar flora and fauna.'[35] A species in a new spot has left many of its predators and competitors behind, and many of its potential prey have never needed to defend themselves against that type of organism. This shaking off of evolutionary history has caused ecological chaos for many species. Kolbert provides a few tragic examples. The rosy wolf-snail was introduced in Hawaii in the late 1950s to control another introduced species, the giant African snail, which had become an agricultural pest. Rather than preying on the giant African snail, the wolf-snail focused instead on Hawaii's small, colorful native snails. Today, of the more than 700 species of endemic snails, 'something like ninety percent are now extinct.'[36] The brown tree snake, *Boiga irregularis,* from northern Australia, was released in Guam in the 1940s. Guam's fauna was completely unprepared, and the snake multiplied rapidly. At the peak of its 'irruption' population densities 'were as high as forty snakes per acre.'[37] The snakes have destroyed the Guam flycatcher, Guam rail, Mariana fruit-dove, and two species of bat. The remaining bat—Marianas flying fox—remains but is highly endangered. Bats around the world are also being devastated by white-nose syndrome, *Geomyces destructans.* Many other species have been obliterated by epidemic waves of pathogens—Kolbert tells the story of the American chestnut, a tree which was dominant in deciduous eastern forests of America in the 1800s until chestnut blight, *Cryphonectria parasitica,* was introduced around the turn of the 20th century. By 1950, chestnuts had been obliterated—some 4 billion trees.[38] Non-native species, especially the introduction of rats, goats, and pigs on islands, have been 'solely responsible for 20 percent of extinctions since 1600 and partly responsible for half of them.'[39] The hyper-connected modern world has only accelerated these trends; if highways, clear-cuts, and palm oil plantations create 'islands' where none before existed, global trade and travel 'deny even the remotest islands their remoteness.'[40]

Pollution of water, soil, and air, especially with agricultural fertilizers, insecticides, fungicides, herbicides, and pesticides is one of the least discussed direct drivers of extinctions and extirpations—Kolbert never mentions it—but it is another essential aspect to understand. There are 400 estuaries worldwide that are classified as 'dead zones,' oxygen starved from agricultural run-off. Chemical pollution and noise pollution contributed to the recent extinction of the Baiji, the freshwater Yangtze dolphin. China's lakes and rivers are especially polluted by agricultural and chemical waste, such that '80 percent of the 50,000 kilometers of major channels can no longer support fish of any kind.'[41] Far closer to home for me, in the St. Lawrence estuary, a 2014 article reported that 'one in four of the St. Lawrence whales are dying of cancer, mostly intestinal cancer.'[42] Toxic pollutants have infected even the mammals of the high arctic, where polychlorinated biphenyls (PCBs) are found in dolphins, seals, orcas, and polar bears, compromising their reproductive systems. Plastic is filling the seas—forming the swirling 'great Pacific garbage patch' and other trash islands—simultaneously killing sea turtles, fish, and sea birds. Albatross chicks are found dead in their nests, decomposed bodies exhibiting colorful plastic surrounded by downy feathers. A sperm whale found in 2012 was found to have in his stomach 'thirty square meters of tarpaulin, over four meters of long hose, a nine-meter plastic rope, and two flowerpots.'[43] Microplastics also enter the bodies of almost all marine life (and increasingly human bodies) with as-yet-unknown consequences. There are ominous reports that insect life is decreasing even in protected areas—up to 78% in biomass in a rigorous 27-year study from natural areas in Germany—likely a result of the increasing toxicity of the global environment from the wholesale use of novel pesticides, herbicides, and insecticides.[44]

Climate change and ocean acidification are already causing trouble for many species and ecosystems, and this trouble is accelerating. For two and a half million years through the large climatic swings of the Pleistocene, there's been no evolutionary advantage in being able to deal with extra heat, since temperatures never got much warmer than they are now. Global carbon levels (and thus coming temperatures) have not been higher than they are today since the mid-Miocene, 15 million years ago. Under a business-as-usual scenario, CO_2 levels could reach a level not seen since 'the Antarctic Palms of the Eocene, some fifty million years ago.'[45]

The ocean absorbs an immense amount of carbon dioxide from the atmosphere—roughly one-third of the CO_2 humans have put into the air, 'a stunning 150 billion metric tons.'[46] The pH of the oceans have already dropped from 8.2 to 8.1, which means they are 'thirty percent more acidic than they were in 1800.'[47] Another way to conceive of this: seas are more acidic than they have been in 800,000 years, and a pH of 7.8 is expected by 2100—a tipping point for most species. Acidification effects 'such basic processes as metabolism, enzyme activity, and protein functions . . . alters the availability of key nutrients . . . [and] promotes the growth of toxic algae.'[48] Most horrifying is the impact on calcifers—those organisms that build a shell or scaffold out of the mineral calcium carbonate—echinoderms like starfish and urchins, mollusks like clams and oysters, crustaceans like barnacles, many types of seaweed, and all coral species. These species are not isolated from other oceanic life, and thus acidification threatens the ocean's entire food web. In healthy oceans, coccolithophores (algae that dwell in tiny shells) are food for copepods, which feed small fish and krill, which feed whales, dolphins, tuna, seabirds, and all sorts of other marine life—if the coccolithophores go, huge segments of the oceanic food web are likely to collapse.[49]

Kolbert gives a beautiful description of just how diverse and majestic coral reefs are before crushingly explaining why warming and acidifying oceans are killing them. Thousands, if not millions, of species rely on coral reefs and the living structures created by coral polyps and zooxanthellae. The Great Barrier Reef is one such epic ecosystem that extends for more than 1500 miles, and in some spots is 500 feet thick—91% is currently bleached. It is with immense sadness that I learned 'it is likely that reefs will be the first major ecosystem in the modern era to become ecologically extinct.'[50] 'Reef gaps' have appeared in the geological record during extinction events, and in each case—after the end Permian, late Devonian, and late Triassic extinctions—it took millions of years for reef construction to emerge again.

Species have 'climate envelopes'—a range of temperatures and moistures that they can live within. When climate changes, species can either adapt, move, or perish. In ice ages throughout the Pleistocene, most species survived by moving, but today life is severely constrained. Amphibians emerged roughly 300 million years ago and have survived three mass extinction events. In this century, Amphibians contend with the combined struggles of habitat destruction, climate disruption, pesticide pollution, competition with nonnative species, and the dreaded chytrid fungus. Amphibians are the class of animalia with the dubious distinction of highest number of critically endangered and extinct species, quickly followed by freshwater vertebrates. Crist and Kolbert agree on this at least—though each of these drivers in isolation is immense, it is *destructive synergies* between these drivers that make the present a mass extinction event.

Synthesizing Drivers: Plantationocene, Capitalocene, Eurocene

Crist does the work of considering what specific industrial activities are causing destruction most directly, highlighting 1) industrial food production and 2) international trade (to some, captured in the Plantationocene and the Capitalocene). Food production, on land and at sea, is by far the 'most ecologically devastating activity.'[51] Crop cultivation covers an equivalent area of earth the size of South America, while livestock grazing is allotted an area the size of Africa—together 40% of Earth's ice-free land. A full 98% of land where rice, wheat, and corn can be grown has been converted.[52] Wild and free life has been transformed and squeezed out of existence by billions of farmed animals and vast expanses of monoculture crops. Human societies globally are implementing two accelerating trends in the production and consumption of industrial food: the mass production and consumption of meat and feedstock, and the immense consumption of processed and packaged food and drinks, largely filled with the staple stuffers of modern monocrops—corn and palm oil and sugarcane—trucked, shipped, and flown around the world, producing immense amounts of plastic trash.

The ethical price of industrial animal agriculture is also horrific. There is a 'debasement of human integrity' in Crist's words 'in the maltreatment of once life-filled terrestrial and marine ecologies, as well as in the abuse of farm animals who are put through an industrial production system and mass slaughtered without being cared for or ever thanked.'[53] Meat production has a double impact: in developed nations, 'up to two-thirds of total grain production is devoted to feed,' while 'a quarter of the world's fish catch is fed to livestock and carnivorous fish in aquaculture operations.'[54] Livestock in the Global South *tripled* between 1980 and 2002. This has all been socially and economically engineered by political and corporate interests largely from wealthy nations targeting debt-crippled poor nations with 'development opportunities' and a growing global consumer population with cheap animal products—all of which is 'devastating wildlife, cruel to farmed animals, and unhealthy for people.'[55]

A study published in *Nature* in 2012 drew the direct lines between 25,000 endangered species and 15,000 export commodities produced in 187 countries. Extirpations of populations was directly linked to a global economy that prioritizes export–import relationships. Wealthy countries, like Canada, the U.S., Australia, Japan, and the European Union, had the greatest ecological impact from importing commodities produced elsewhere. Poorer countries across South and Central America, Africa, and all of Asia destroyed lands and waters within their borders for the sake of producing exports. The commodities most responsible are animal feed; food staples; and other grown staples like rubber, cotton, and lumber.

Extinction has been outsourced to the poorer areas of earth as animal feed and staple foods, and even tulips and roses are increasingly globalized commodities planted around the globe. Planted: 'such an innocent-seeming word,' observes Haraway.[56] She argues that the Plantationocene begins in 1492 because the Columbian exchange marks the origin of the plantation system and the unprecedented homogenization of the earth's wild and cultivated biota, what she calls the Great Simplification and Charles Mann calls the Homogenocene.[57] Expanding at the increasing pace of the Great Acceleration, the explosion of petro-fueled, biocidal, soil-mining, forest-razing, water-draining agriculture around the world after WWII is directly linked to simultaneous human population increases, dispossession and displacement of humans and other animals from their home ecologies, and multi-species extinctions.

In 'Solidarity with Animals,' an opening essay for the 2023 Great Transition Institute Forum, Eileen Crist argues that the relationship between modern humans and animals is central to understanding and responding to the crises of extinction, biodiversity collapse, and climate change. Crist argues that 'the violence pervading the animal economy is unleashing pandemonium across the planet,' and 'dominating animals has become coextensive with global environmental destruction.' Most particularly:

> large-scale animal agriculture (particularly CAFOs and feed monocultures) and defaunation (notably, industrial fishing and bushmeat for remote markets) are lead causes of tropical deforestation, species extinctions, ecological impoverishment, agrochemical and factory-farm pollution, pollinator declines, rapid climate change, freshwater depletion, soil degradation, and infectious epidemic disease.[58]

Size and acceleration are key to understanding both The Born Ones of the Anthropocene, Capitalocene, and Plantationocene and The Disappeared of the Elachistocene. The specters of ever-eating empire and ever-ravenous capital have desecrated, desolated, and homogenized life. Plantations have re-produced across the world, disappearing the homes of humans and other animals, relationships to land, and ancient lineages of species. Cows, pigs, chickens, soy, coffee, cacao, palm oil, rubber, cotton, mangoes, coconuts, jojoba for biofuel, quinoa for health fads, roses for romance, or some other commodity for export produced by biocidal industrial agriculture have destroyed and replaced every single last individual Pyrenean ibex (2000), Baiji river dolphin (2007), Pinta giant tortoise (2012), poo-uli (2019), Spix's macaw (2019), and splendid poison frog (2020)—to name just a few of the species whose double-death coroner's report includes habitat destruction for agricultural development as a cause. Huge swathes of forests and wetlands and prairies have been turned into industrialized monocropped agricultural lands and ranchlands where violence is often committed to domesticated animals to feed The Born Ones and their pets in horrendously wasteful and destructive patterns of globalized trade. Like Nick Cave in 2004, I've got the abattoir blues.

The End of Expansion

The Elachistocene leaves many questions. When would the Elachistocene begin? How to reckon with the controversial and convoluted debates about the late-Pleistocene extinctions 12,000–60,000 years ago? Or the extinctions on Pacific Islands? Or the deforestation caused by Ancient Rome and Sumer that largely led to the collapse of those civilizations? How to center extinctions without engendering paralyzing fatalism and myopic misanthropy? Audra Mitchell asks us to seriously consider—what is the relationship between white tears and extinction?[59] How to mobilize grief when grief is often targeted for commodification and funding for global environmental non-governmental organizations (ENGOs)? How to consider specific animal lives in extinctions while also trying to speak about the global? How to encourage scientists and ecologists to recognize how unhelpful misanthropy is, and the importance of seeing the individual animals connected in the webs?

I do not want to engender fatalism by suggesting mass extinction is inevitable in naming this epoch to center mass extinction, and perhaps this is the ultimate danger of the Elachistocene. Crist, Dawson, and Eisenstein all argue that the expansion and growth of human systems needs to be halted and remaining wild places need to be protected at massive scale, and much of their understanding of the need for transformation comes from seeing the value of life in and of itself. The Anthropocene narrative needs to be flipped upside down and inside out. For many proponents of the Anthropocene, the main urgency of our ecological crisis is that it 'could undermine the health, security, and sustainability of our civilization.'[60] The crisis in the Anthropocene, and in much environmental rhetoric, is a crisis for modernity, for the globalized industrial-agricultural-militaristic-colonial world-system that we live in and have become dependent upon. The crisis is for Anthropos, for Capital, for Plantations and the Born Ones, and the subsequent plea is to save these systems. But wildlife is valuable and sacred in its own right, not only in its utility to us. The Elachistocene calls on us to see the biodiversity crisis, the animal crisis, as caused by the very systems most apocalypse and Anthropocene writing calls us to save. If globalized capitalism persists in its destruction of the living membrane, and technologies of lab-grown meat are able to insulate and buffer the wealthiest from the coming storms and deserts, the crisis for the rest of life will still be horrifyingly real.

Mushrooms persist in capitalist ruins, allies abound in unexpected places, and I must return to the garden to keep my sanity. Perhaps in the future, if we are lucky, the majority of humans will spend less time digging up ancient fossils and sitting on computers puzzling with words and more time growing food in regenerative ways that allow for the persistence of local and global life. I have reckoned with failure throughout writing this—failure to say it all, failure to acknowledge all intersections, failure to maintain my own health and resist my addictions, failure to show up to my human and animal relations as I would like—and I cannot help but see the failure of conservationists and global biodiversity and climate agreements to stem the gushing tide of extinctions. I feel sick and scared and sad, and perhaps centering the loss of the Elachistocene is about facing up to this failure and all the losses of the last 500 years and realizing there isn't an immediate or easy escape. This doesn't mean that we can't acknowledge and celebrate how much good work has been done and how much worse it would be without it.

The Elachistocene, in centering the loss of life, has encouraged me to turn to those authors who have taken the scale of crisis seriously and who imagine the degrowing, debt-repaying, regenerative farming, and shrinking societies of Crist's 2019 *Ecological Civilization*. One interpretation of the Anthropocene that I observe in the attitudes of my colleagues is hopelessness, fear, and misanthropy. We are in need of a term that encourages sincere thought beyond our own species instead of focusing on tired stories of triumphant but tragic and violent humans. The Elachistocene shifts the narrative to the rest of life on earth, rather than calling to the fear and self-hatred of the Anthropocene, calling instead to our grief and our love for the vanishing. Yet, the Elachistocene is not an inspiring term or a call to action. For me, it is a muted scream. Using the dispassionate language of science can only ever go so far, but the magnitude it gestures too is the underbelly of the Anthropocene, the wriggling and slimy ghosts of The Disappeared, their hoots and shrieks harder to hear as they disappear. The Elachistocene offers a haunting that whispers in caves and forests. Far from the loud proclamations of Anthropos as geological agent, it tempts us to search for those squawks, hisses, and growls that tell stories of how abundant life both was and is, how old, and how persistent. The extinction event unfolding now, however horrific, is not the equivalent of an immense meteorite strike. Not yet.

Notes

1 Will Steffen, Paul J. Crutzen, and John R. McNeill, "The Anthropocene: Are Humans Now Overwhelming the Great Forces of Nature?," *Ambio* 36, no. 8 (2007): 614–621.
2 Johan Rockström et al., "A Safe Operating Space for Humanity," *Nature* 461, no. 7263 (September 2009): 472–475, https://doi.org/10.1038/461472a.
3 For excellent critical discussion of anglocentrism and eurocentrism in Anthropocene discourse, see Kathleen D. Morrison, "Provincializing the Anthropocene: Eurocentrism in the Earth System," in *At Nature's Edge: The Global Present and Long-Term History*, eds. Gunnel Cederlöf and Mahesh Rangarajan (New York, New York: Oxford University Press, 2018); Elizabeth Chatterjee, "The Asian Anthropocene: Electricity and Fossil Developmentalism," *Journal of Asian Studies* 79, no. 1 (February 1, 2020): 3–24.
4 Jason W. Moore et al., *Anthropocene or Capitalocene?: Nature, History, and the Crisis of Capitalism* (Oakland: PM Press, 2016).
5 Kate Raworth, "Must the Anthropocene Be a Manthropocene?," *The Guardian* (October 20, 2014). For discussion on the linkages between the exploitation of women and land, there are a wealth of sources. One discussion of classic work and development since is given in Carolyn Merchant, "The Scientific Revolution and the Death of Nature," *ISIS* 97, no. 3 (September 2006): 513–533. Another by the sisters Sunaura and Astra Taylor in "Our Animals, Ourselves," *Lux Magazine*, 3, 2024.
6 Janae Davis et al., "Anthropocene, Capitalocene, . . . Plantationocene?: A Manifesto for Ecological Justice in an Age of Global Crises," *Geography Compass* 13, no. 5 (2019): e12438, https://doi.org/10.1111/gec3.12438.
7 Alf Hornborg, "The Political Ecology of the Technocene: Uncovering Ecologically Unequal Exchange in the World-System," in *The Anthropocene and the Global Environmental Crisis*, eds. Christophe Bonneuil, François Gemenne, and Clive Hamilton (New York, New York: Routledge, 2015).
8 Donna Haraway and Frédéric Neyrat, "Anthropocene, Capitalocene, Plantationocene, Chthulucene: Making Kin," *Multitudes, Thinking Matters* 65, no. 4 (February 2, 2017): 75–81, https://doi.org/10.3917/mult.065.0075.
9 Jill Schneiderman, "Naming the Anthropocene," *PhiloSOPHIA* 5, no. 2 (2015): 199.
10 Ibid., 189.
11 Elizabeth Kolbert, *The Sixth Extinction: An Unnatural History* (Picador, 2014), 103.
12 Schneidermann, "Naming the Anthropocene," 200.
13 Eileen Crist, "On the Poverty of Our Nomenclature," *Environmental Humanities* 3, no. 1 (May 1, 2013): 129–147, https://doi.org/10.1215/22011919-3611266.
14 Listing a species as extinct with the International Union for Conservation of Nature requires meeting very stringent criteria, and many species that are likely extinct are still listed as critically endangered. For example, the Eskimo curlew (*Numenius borealis*) was last sighted in 1963, and Bachman's Warbler (*Vermivora bachmani)* was last seen in 1988, despite ongoing targeted searches. Both are still listed as critically endangered by the IUCN because of the slim chance they might persist. The desire is to avoid the 'Romeo Error,' where protective measures and funding might be removed from a species in need. However, this means that the onus to prove extinction is incredibly high, which Cowie et al. argue results in underestimating the true extinction rate and directing limited funding at lost causes. The IUCN 'Red List' is also highly biased towards mammals and birds.
15 Simon Lewis and Mark Maslin, *The Human Planet: How We Created the Anthropocene* (Yale University Press, 2018), 272.

16 Donna Haraway, "Making Kin in the Cthulucene: Reproducing Multispecies Justice," in *Making Kin Not Population*, eds. Adele E. Clarke and Donna Haraway (Chicago, Illinois: Prickly Paradigm Press, 2018).
17 Haraway, "Making Kin in the Cthulucene: Reproducing Multispecies Justice," 73.
18 Ibid.
19 Ibid., 70.
20 Ibid., 69.
21 Eileen Crist, *Abundant Earth: Toward An Ecological Civilization* (University of Chicago Press, 2019).
22 A playful moniker originally used by Steve Mentz in his book *Break Up the Anthropocene* (Minneapolis: University of Minnesota Press, 2019).
23 Audra Mitchell, "Revitalizing Laws, (Re)-Making Treaties, Dismantling Violence: Indigenous Resurgence against 'the Sixth Mass Extinction,'" *Social & Cultural Geography* 21, no. 7 (September 1, 2020): 909–924, https://doi.org/10.1080/14649365.2018.1528628.
24 Crist, *Abundant Earth*, 14.
25 Ibid., 12.
26 Ibid., 16.
27 Elizabeth Kolbert, *The Sixth Extinction: An Unnatural History* (Picador, 2014), 13.
28 Ashley Dawson, *Extinction: A Radical History* (OR Books, 2016), 41.
29 Crist, *Abundant Earth: Toward An Ecological Civilization,* 21
30 Ibid., 23.
31 Ibid., 24.
32 Kolbert, *The Sixth Extinction: An Unnatural History*, 181.
33 Lewis and Maslin, *The Human Planet: How We Created the Anthropocene*, 274.
34 Invasive species is the language used by ecologists, so I will use the term too, but I first want to acknowledge that it is a regrettably xenophobic and militaristic metaphor.
35 Kolbert, *The Sixth Extinction: An Unnatural History,* 196.
36 Ibid., 203.
37 Ibid.
38 Ibid., 204.
39 Crist, *Abundant Earth: Toward An Ecological Civilization*, 25.
40 Kolbert, *The Sixth Extinction: An Unnatural History,* 198.
41 Crist, *Abundant Earth: Toward An Ecological Civilization*, 25.
42 Ibid., 25.
43 Ibid., 26.
44 Dave Goulson, "The Insect Apocalypse, and Why It Matters," *Current Biology* 29, no. 19 (October 2019): R967–R971, https://doi.org/10.1016/j.cub.2019.06.069.
45 Kolbert, *The Sixth Extinction: An Unnatural History,* 172.
46 Ibid., 123.
47 Ibid., 114.
48 Ibid., 120–121.
49 Crist, *Abundant Earth: Toward An Ecological Civilization*, 26.
50 Kolbert, *The Sixth Extinction: An Unnatural History*, 130.
51 Crist, *Abundant Earth: Toward An Ecological Civilization*, 31.
52 Ibid., 32.
53 Ibid., 34.
54 Ibid., 35.
55 Ibid.
56 Haraway, "Making Kin in the Cthulucene: Reproducing Multispecies Justice."
57 Charles C. Mann, "The Dawn of the Homogenocene," *Orion Magazine* (April 2011), retrieved from: https://orionmagazine.org/article/the-dawn-of-the-homogenocene/.
58 Crist, "Solidarity with Animals: Opening essay for a GTI Forum."
59 Audra Mitchell, "Decolonizing against Extinction, Part III: White Tears and Mourning," *Worldly* (December 14, 2017), retrieved from: https://worldlyir.wordpress.com/2017/12/14/decolonizing-against-extinction-part-iii-white-tears-and-mourning/.
60 Steffen cited in Crist, "Solidarity with Animals: Opening Essay for a GTI Forum," 243.

Bibliography

Chatterjee, Elizabeth. "The Asian Anthropocene: Electricity and Fossil Developmentalism." *Journal of Asian Studies* 79, no. 1 (February 1, 2020): 3–24. https://doi.org/10.1017/S0021911819000573.
Crist, Eileen. "On the Poverty of Our Nomenclature." *Environmental Humanities* 3, no. 1 (May 1, 2013): 129–147. https://doi.org/10.1215/22011919-3611266.
Crist, Eileen. *Abundant Earth: Toward an Ecological Civilization*. University of Chicago Press, 2019.
Crist, Eileen. "Solidarity with Animals: Opening Essay for a GTI Forum." *Great Transition Initiative*.(February, 2023). Retrieved from: https://greattransition.org/gti-forum/solidarity-animals-crist#endnote_25
Davis, Janae, Moulton, Alex A., Van Sant, Levi, and Williams, Brian. "Anthropocene, Capitalocene, . . . Plantationocene?: A Manifesto for Ecological Justice in an Age of Global Crises." *Geography Compass* 13, no. 5 (2019): e12438. https://doi.org/10.1111/gec3.12438.
Dawson, Ashley. *Extinction: A Radical History*. OR Books, 2016.
Goulson, Dave. "The Insect Apocalypse, and Why It Matters." *Current Biology* 29, no. 19 (October 2019): R967–R971. https://doi.org/10.1016/j.cub.2019.06.069.
Haraway, Donna. "Making Kin in the Cthulucene: Reproducing Multispecies Justice." In *Making Kin Not Population*, edited by Adele E. Clarke and Donna Haraway. Prickly Paradigm Press, 2018.
Haraway, Donna, and Neyrat, Frédéric. "Anthropocene, Capitalocene, Plantationocene, Chthulucene: Making Kin." *Multitudes, Thinking Matters* 65, no. 4 (February 2, 2017): 75–81. https://doi.org/10.3917/mult.065.0075.
Kolbert, Elizabeth. *The Sixth Extinction: An Unnatural History*. Picador, 2014.
Lewis, Simon, and Maslin, Mark. *The Human Planet: How We Created the Anthropocene*. Yale University Press, 2018.
Mann, Charles C. "The Dawn of the Homogenocene." *Orion Magazine*, April 2011. Retrieved from: https://orionmagazine.org/article/the-dawn-of-the-homogenocene/.
Mentz, Steve. *Break Up the Anthropocene*. Minneapolis: University of Minnesota Press, 2019.
Merchant, Carolyn. "The Scientific Revolution and the Death of Nature." *Isis* 97, no. 3 (September 2006): 513–533. https://doi.org/10.1086/508090.
Mitchell, Audra. "Decolonizing against Extinction, Part III: White Tears and Mourning." *Worldly*, December 14, 2017. Retrieved from: https://worldlyir.wordpress.com/2017/12/14/decolonizing-against-extinction-part-iii-white-tears-and-mourning/.
Mitchell, Audra. "Revitalizing Laws, (Re)-Making Treaties, Dismantling Violence: Indigenous Resurgence against 'the Sixth Mass Extinction.'" *Social & Cultural Geography* 21, no. 7 (September 1, 2020): 909–924. https://doi.org/10.1080/14649365.2018.1528628.
Moore, Jason W., Altvater, Elmar, Crist, Eileen C., Haraway, Donna J., Hartley, Daniel, Parenti, Christian, and McBrien, Justin. *Anthropocene or Capitalocene?: Nature, History, and the Crisis of Capitalism*. Oakland: PM Press, 2016.
Morrison, Kathleen D. "Provincializing the Anthropocene: Eurocentrism in the Earth System." In *At Nature's Edge: The Global Present and Long-Term History*, edited by Gunnel Cederlöf and Mahesh Rangarajan. New York, New York: Oxford University Press, 2018. https://doi.org/10.1093/oso/9780199489077.003.0001.
Raworth, Kate. "Must the Anthropocene Be a Manthropocene?" *The Guardian*, October 20, 2014. Retrieved from: https://www.theguardian.com/commentisfree/2014/oct/20/anthropocene-working-group-science-gender-bias.
Rockström, Johan, Steffen, Will, Noone, Kevin, Persson, Åsa, Chapin, Stuart, Lambin, Eric F., Lenton, Timothy M., et al. "A Safe Operating Space for Humanity." *Nature* 461, no. 7263 (September 2009): 472–475. https://doi.org/10.1038/461472a.
Schneiderman, Jill. "Naming the Anthropocene." *PhiloSOPHIA* 5, no. 2 (2015).
Steffen, Will, Crutzen, Paul J., and McNeill, John R. "The Anthropocene: Are Humans Now Overwhelming the Great Forces of Nature?" *Ambio* 36, no. 8 (2007): 614–621.
Taylor, Sunaura, and Taylor, Astra, "Our Animals, Ourselves." *Lux Magazine*, 3, 2024. Retrieved from: https://lux-magazine.com/article/our-animals-ourselves/.

49
DE-EXTINCTION

jessie l. beier

1. De-Extinction Programming: Jurassic Parks and Colossal Problems

For over three decades now, audiences have witnessed de-extinct dinosaurs wreaking havoc on human civilization through the *Jurassic Park*, and later *Jurassic World*, science fiction media franchise.[1] The premise of the films, a now well-trampled narrative path, goes something like this: scientists, backed by a flamboyant venture capitalist and a board of detached investors, extract ancient dinosaur DNA from mosquitoes encased in fossilized amber and, with the aid of powerful computers and cutting-edge advancements in genetic engineering, "fill in the gaps" in the genome using frog DNA so as to revive extinct dinosaurs for the purpose of, at first, creating a theme park. As the film series plays out, the purpose of this de-extinction programming mutates from one founded in creating a (for-profit) world of simulated "nature" and prehistoric wonder to one aimed at fully "unlocking the power of the genome" in order to apply this discovery to the world of humans. With its ongoing renderings of de-extinct dinosaur destruction, *Jurassic Park* and its sequels—*The Lost World: Jurassic Park, Jurassic Park III, Jurassic World, Jurassic World: Fallen Kingdom* and *Jurassic World: Dominion*[2]—is often credited for offering a social critique of the dangers of Man-made technological intervention backed by nefarious economic interests, or, as the characters in the film often put it, the perils of "playing god" via genetic engineering.[3] This critical reading is, however, questionable given the film's own interpolations in the libidinal infrastructure of global capitalism, which, in this case, not only involve the use of costly filmic technologies to imagine the hazards of anthropogenic technologies run amok but also ongoing expansion and proliferation of the media franchise, which now includes six feature length films, two short films, numerous television spinoffs, video games, toys, comics, exhibitions and, of course, theme parks. In this way, the *Jurassic Park* films and their attendant media ecologies are prime examples of the expansion of cinematic commodification through the creation of a globally branded cultural phenomenon, one that follows in a longer line of the Hollywood monster-movie/disaster spectacle genre, this time shocking and delighting audiences through innovations in computer-generated effects and giant dinosaur animatronics. Taking this critical ambivalence as a starting point, this short investigation aims to think with *Jurassic Park* as just one text that echoes contemporary

DOI: 10.4324/9781003273400-60

critical eco-political discourses, which are, not unlike the film, entangled and enmeshed within the systems of subjugation and violence they (cl)aim to denounce. Specifically, this chapter digs into contemporary examples of *de-extinction programming* as they unfold on and off the screen and positions these examples as a weird site to ask some perhaps unconventional questions about the status and implication of both animals and gender in an era of extinction.

Jurassic Park: Dominion, the final film in the series, opens with an expository overview of its cinematic world in the form of a "NowThis News" segment titled "Dinosaurs in Our World." In this fabulated and web-optimized news update, the audience is dropped into a universe where the de-extinction of prehistoric species has led to a "frightening reality" characterized by giant sea creatures (a Mosasaurus, to be exact) taking down fishing vessels, revivified Pterodactyls interrupting wedding ceremonies and baby Velociraptors tormenting children in their back yards. Alongside these clips of now-invasive species interrupting all manner of human activities, the audience is informed that the pointedly named genetics company—Biosyn—has been tasked by the U.S. government with housing the displaced animals in their sanctuary and research facility in the Dolomite Mountain range of Italy. It is here where 20 of the prehistoric species now live in refuge but are also part of a broader Biosyn research project wherein the company uses the dinosaurs to study their prehistoric immune systems in the hopes of finding better treatments for human ailments such as cancer, Alzheimer's and autoimmune diseases. As Biosyn's CEO, Lewis Dodgson, states "at Biosyn, we are dedicated to the idea that *dinosaurs can teach us more about ourselves.*"[4] The video segment proceeds by articulating that, while the public is skeptical of Biosyn's efforts, specifically how such ventures have led to spikes in the company's profit margins, the genetics company is adamant that "we can handle bioengineering responsibly."[5] After a brief mention of rumors of a mysterious human clone that has led to a worldwide search, a plot point that was central to the previous film, the segment ends with a series of questions that fuel the story that follows: "Now that we've brought these animals back from extinction, can we face the consequences? Are we responsible for them? Or should they be left to fend for themselves?"[6]

Within this opening segment, which is offered as both a recap and narrative framing device, the film seems to take a critical position on the science of de-extinction asserting that (spoiler alert!) bringing back de-extinct animals has unexpected and, as the final film makes clear, potentially disastrous consequences. As enigmatic chaostician Ian Malcolm/Jeff Goldblum (indeed, it's hard to distinguish the line between the character and the actor playing him) orates in a particularly telling lecture to Biosyn employees in *Dominion*:

> Yep. You . . . you control the future of our survival on planet earth. According to you the solution is genetic power, but that same power could devastate the food supply, create new diseases, alter the climate even further. Unforeseen consequences occur and every time, every single time we all act surprised, because deep down, I don't think that any of us actually believe that these dangers are real.[7]

Here, Malcolm, who acts as a kind of ironic commentator in both the original books and throughout the film series, serves as a mouthpiece who reminds the audience that "[w]e not only lack dominion over nature, we're subordinate to it."[8] As foretold in each of the films, albeit in different ways, human ability to control and manage the genetic miracles roaming the park, and later the world, ultimately proves futile: technological accidents, natural disasters and human greed ultimately lead, time and time again, to a spectacle of carnage and destruction. But while each film presents episodes of chaos and calamity brought on by the revival of dinosaurs, in the end these upheavals are always followed by the restoration of order and security. In *Dominion*, for instance, even after the revelation of wide-reaching corporate corruption, the near-devastation of global food sources due to an experimental genetic program gone awry and a fiery inferno that appears to engulf an entire eco-system of de-extinct animals, order is restored through both scientific "fixing" (i.e. even more "revolutionary" genetic engineering) and government regulation (i.e. a U.N. decree).

And so, at the same time that characters like Malcolm provide direct commentary on the limits and dangers of human intervention and scenes of disaster drive this point home through dazzling cinematic spectacle, the film's underlying narrative is one wherein dynamic, heroic individuals (and their governments) are ultimately and inevitably able to restore order and control, a narrative that echoes pervading political imaginaries wherein the ingenuity and gumption of individual humans can and will overcome even the most perilous of planetary trajectories. In this way, *Jurassic Park* can be read as a key text, a primer of sorts, for the (Good) Anthropocene,[9] the current geological era, where, through its activities and its growing population, the human species is positioned as a geological force that has not only altered the stratigraphic record but, for the same reason, is capable of overcoming ecological degradation though (more) technological innovation and (continued) human intervention. With its promise of an ultimately "good" ending for "us" humans, *Jurassic Park* echoes this narrative, offering yet another epic tale wherein humanity—the *anthropos* of the Anthropocene—is redeemed and relaunched, in turn reifying a new set of grand narratives and naive universalisms that remain faithful to the legacies and logics of human dominance and exceptionalism.[10] It is in this sense that *Jurassic Park* offers an allegory for contemporary ecological politics, one wherein human mastery and dominion over the earth and its inhabitants are treated as both threat *and* salvation.

This political imaginary is evidenced in our own world's de-extinction programming, which is not just the stuff of science fiction but has gained significant public and scientific interest (and funding)[11] in recent years. De-extinction, also called resurrection biology or species revivalism, is the process through which extinct species can be brought back into existence through methods of selective breeding, cloning and/or genetic engineering often with the goal of reintroducing species to the wild and restoring ecosystems.[12] Despite being introduced through the fictional world of *Jurassic Park* over thirty years ago, the term de-extinction is relatively new, first gaining significant public interest in March of 2013 thanks to both a series of live-streamed TED Talks and a cover story in *National Geographic* magazine.[13] Since then, de-extinction projects have grown in both number and scope, with early flagship projects such as "The Great Passenger Pigeon Comeback"[14] and the regeneration of the world's last Pyrenean ibex now accompanied by initiatives aimed at bringing back the thylacine, aurochs and even the woolly mammoth. While definitions of de-extinction focus on the process of restoring species that have died out or have gone extinct, this language is somewhat misleading, as the goal is not exactly to revive or resurrect species for which no viable members remain but instead to create a *proxy*[15] of an extinct species or, as the literature describes it, a "functional equivalent" capable of restoring ecological functions or habitats that might have been lost as a result of species extinction.[16]

For Colossal Laboratories & Biosciences, the company behind the aforementioned woolly mammoth project, this proxy creation involves altering the genetic code of endangered Asian elephants in order to resurrect a "cold-resistant elephant with all of the core biological traits of the woolly mammoth."[17] Founded in September 2021 by Harvard genetics professor George Church and tech entrepreneur Ben Lamm, the project is framed as a pioneering initiative committed to developing a practical, working model of de-extinction by operating at the "leading edge of genetic engineering and restorative biology."[18] The goal of seeing the "Woolly Mammoth thunder upon the tundra once again"[19] is just a starting point for Colossal, which aims to develop a "de-extinction library" that will house genetic DNA and embryos from endangered animals so as to address and combat the "colossal problem" of extinction, a problem the company explicitly identifies as being *anthropogenic* in nature. As the Colossal website outlines, "of all the root causes of modern extinction, the most concerning and hyperactive factor is that of *human causation*."[20] Situating their revolutionary project within what scientists have called the Sixth Mass Extinction, the company asserts that "[e]xtinction is a colossal [human-made] problem facing the world. And Colossal is the company that's going to fix it."[21] Here, Colossal not only makes reference to the current era of "biological annihilation,"[22] one wherein species extinction now occurs at 1000 times the "background rate," but also draws attention to its solutionist ethos, one that aims to "fix" ecological issues by, in this case, "combin[ing] the science of genetics" with "the business of discovery," in order to "jumpstart nature's ancestral heartbeat."[23] Not unlike *Jurassic Park's* (Good) Anthropocene vibes, Colossal's de-extinction programming approaches the current era of extinction as that which merely requires a "technical fix," one that can and should be enacted by reasonable, responsible humans. Or, as one of Colossal's key investors, Tom Chi, puts it, "[w]e'll soon have the choice of moving from the role of blind destroyer of the world's species toward being the thoughtful kin of all the life around us."[24] In both the case of *Jurassic Park* and Colossal, then, de-extinction programming positions human intervention as both the cause of *and* solution to current planetary predicaments. As such, extinction is treated as a threat, but also, through the corollary project of de-extinction, an innovative fulcrum for reimagining humanity's unquestioned capacity for redemption and salvation even, or perhaps especially, in the face of unparalleled planetary suffering.

2. Re-Wiring Life For "Us": Ancient Animals and Data Dominion

In the fictional world of *Jurassic Park*, de-extinct animals not only offer a proxy for the unexpected and, as it turns out, destructive, consequences of anthropogenic intervention, but also a site for *humanity to see itself anew*. Once again, the film's supposed critique of scientific intervention gone too far is ultimately counter-actualized when the heroic humans of the film take control over their unruly de-extinct progeny and restore order through more careful—or rather, better-managed, better-governed—*dominion*. A similar affirmation of all-too-human salvation is at work in Colossal's woolly mammoth project, which not only aims to advance the economies of biology and genetics but also to "make humanity more human."[25] In both examples, animality is always-already conceived from the vantage of the anthropocentric culture it is made to serve, and thus animals become a figure against which humanity is able to define and, ultimately, redeem itself. This mode of defining the human is, of course, not new, marking yet another instance in a long line of thinking wherein animals are used as a metaphorical figure to construct and re-construct human identity and purpose by symbolizing what humans are and are not. The assumption of human exceptionalism, dominion and superiority over animals shows up, for instance, in the legacy of Western philosophy (i.e. through the Greeks), which proposed a highly ordered and anthropocentric view of the universe that led, in part, to the deliberate forgetting of animals, including the animality of humans, in Western philosophical thought.[26] The deep roots of Judeo-Christianity, which subtend broader (Good) Anthropocene discourses, are likewise entangled with the creation of demarcative boundaries that attribute the soul to Man alone and thus dominion over other creations as "divine sanction for the exploitation of animals."[27] Importantly, the conceptualization of animals and animality as counter to the human is one that is itself in constant flux, never static but instead open to modulation based on the modes of alterity and identity necessitated by and for human self-definition. As such, animals continue to be appropriated for various rhetorical purposes, occupying, at various moments, "a Procrustean bed of order and disorder, innocence and depravity, violence and peacefulness, masculinity and femininity, monarchy and democracy, wisdom and ignorance, as well as sacrifice, sexuality (both innocent and depraved), gluttony, drunkenness, and other seemingly disparate qualities."[28] In short, within this "psychical domestication" of animals, one that now suffuses the cultural imaginary of the West, animals are always-already for "us."[29] As such, animality in its various forms continues to be offered up as metaphorical figure that allows humanity to see itself as a differentiated, distinct and exceptional form of life.

This mode of domestication is alive and well within the world of *Jurassic Park*, but also the parallel world of Colossal, where de-extinction programming is made possible through stranger and stranger forms of abstraction and metaphorization of animality. Where de-extinction operates under the auspices of "saving" nature (and thus saving ourselves), it not only works to affirm, exacerbate and reify illusory nature/culture divides but necessitates the thoroughgoing division, manipulation, extraction and commodification of "nature" into smaller and smaller packets of information. As just one instance of today's asymmetrical information economies, wherein platform capitalism,[30] or something worse,[31] necessitates the ongoing creation of more and more data points, wherein data sensing, extraction and analysis have established a new regime of algorithmic intelligibility ultimately run by machines,[32] wherein, to put it bluntly, "data is the new oil,"[33] animals, including extinct ones, become yet another scene for the computational stratification of all things. In the example of *Jurassic Park's* de-extinction programming, but also real-world initiatives like Colossal's, this stratification is made possible by treating or, better, coding animals as mere information-processing systems. Within the science of de-extinction, animals are most often conceptualized in terms of bundles of genetic information and dynamic codes, all of which can be extracted, collected, analyzed and stored on computers and then instrumentalized in order to regenerate, through proxy creation, that which has become extinct.

Key to this instrumental understanding of life is a conceptualization of DNA as autotelic, that is, "as an entity working toward its own self-contained end."[34] This metaphorization is famously dramatized in *Jurassic Park* through the first film's public-facing explainer of the cloning (or proxy-creation) process, which features Mr. DNA™ talking about the dinosaur "blueprints" that have been left behind by mosquitos in fossilized tree sap before "filling in the gaps" with his own two hands. The result of this metaphorization is a "surplus of meaning associated with the 'master molecule'" (a.k.a. Mr. DNA™), an excess that highlights a "key conundrum" within the code-script metaphor.[35] That is, while the metaphorization of DNA has led to assumptions about the immutable, but also infinitely adaptable, nature of DNA, setting the stage for innovations in the bioinformatics economy in the process, this same code-script idea renders DNA, or "genetic text," as a "signification without a referent."[36] This is not just a philosophical or linguistic issue but has profound social consequences for the knowledge claims that result from such (faulty) representations. In the example of *Jurassic Park,* for instance, the untenability of cloning (for, after all, a true clone can never be created due to the "gaps" in the genome) is less a problem of possibility and more a problem of time and resources. That is, in *Jurassic Park,* but perhaps also in the case of Colossal's "de-extinction library," resurrecting extinct species is a matter of stockpiling the right resources (i.e. data, computers, funding) and buying enough time from project investors in order to eventually apply, and profit from, those resources. De-extinction, in this way, becomes a form of bioprospecting and, as such, offers yet another "humanist tale of the tendency of capitalist commodity production to exploit resources and labor, engage in short-range planning, and emphasize conspicuous consumption."[37] The metaphorization of DNA, which extends far beyond cartoon animations, conceptualizes animals, in this case, de-extinct ones, as "raw material" to be mined, engineered, cloned and, ultimately, monetized. Through de-extinction, the ancient lives of animals become yet another site of extraction; after mining, selling and combusting fossilized remains to the point of planetary ecocatastrophe, we now look to these ancient stores once again so as to *fix* the human-caused disaster and, in so doing, *fix the future.*

At issue here is not just the moral ramifications of "playing god" through genetic engineering, as *Jurassic Park* likes to draw attention to (in more or less ironic ways), nor is it the discouraging realization that investments in de-extinction actually draw attention and resources away from other conservation efforts.[38] Indeed, the concerns raised by de-extinction go beyond questions of scientific viability, beyond debates about the risks of invasiveness and unforeseen species interactions,[39] beyond issues of animal welfare (i.e. the various forms of violence and suffering caused by captive rearing, embryo implantation, ecosystem reintroduction and the health problems associated with cloning), which are, notably, left out of much of the de-extinction literature.[40] The issue, at least in the context of this chapter, is that de-extinction relies on a specific conceptualization of *life itself*, one that is always-already *given* to the ever-evolving demands and protocols of global capitalism. This is to say that the techniques of genetic engineering developed through de-extinction not only "confer godlike capacities" on human scientists (and tech entrepreneurs) to restore extinct species but also work to "*rewire life* to conform to the dictates of corporate profit."[41] Under the regime of biocapitalism, underwritten as it is by computational stratification and the surplus meanings established through the metaphorization of DNA, living organisms are increasingly viewed, as Colossal CEO puts it, as "programmable manufacturing systems."[42] Where animals, including human ones, can be reduced to a series of numbers and sequences of letters, they can be easily transposed into computer codes. From this view, the problem of extinction can be solved by collecting more information so as to provide the data necessary for regenesis and, ultimately, de-extinction.

What is being saved here, then, is not only de-extinct or endangered animals but a vision of humanity as both in need and worthy of saving. The example of de-extinction programming highlights the way that world-ending scenarios, such as the extinction of a species, can and should be overcome through the collection and instrumentalization, or *dominion*, of data via calculative measures that (cl)aim to bring about particular trajectories of and for the future. Through this mutated logic of dominion, animals are not only relegated to an inferior racial world and domesticated as a site of extraction and exploitation (i.e. for "unlocking the genome") but are also summoned to save us from our own demise.[43] From this vantage, the disastrous event of something like biodiversity loss or mass extinction does not actually put human supremacy into question but affirms it further. By projecting humanity past the extinction, albeit through fantasies of some pristine, pre-historic time, de-extinction programming envisions a world wherein humans find themselves flourishing *after* annihilation, and thus the future becomes the territory for vanquishing existential threats through the affirmation of an ultimately redeemable and redeeming anthropos.

3. Un-Fixing the Future: Queer Clones and Non-Reproductive Futurity

Where de-extinction programming affirms narratives of human redemption and salvation even, or perhaps especially, in times of planetary ecocatastrophe, it contributes to broader fantasies of *fixing* the future. This "fixing" is not just aimed at solving the environmental crises on the horizon or that are, in many cases, already here, but "fixing," through ongoing reproduction and regeneration, the systemic contradictions of capitalism. There is, perhaps, "no clearer example of the tendency of capital accumulation to destroy its own conditions of reproduction than the sixth extinction."[44] This is evidenced by the case of Colossal, whose self-proclaimed "business of discovery" is involved in much more than just "rewilding" efforts. Through offers of an exciting new financial venture and the setting of proprietary legal precedents, the bioengineering company is also dedicated to a range of speculative monetization schemes that harness both the power of genetic engineering and patent law so as to guarantee even further financial payoffs in the future. It is in this sense that companies like Colossal not only aim to "fix" pressing ecological issues but "fix" the future by offering "a seductive but dangerously deluding techno-fix for an environmental crisis generated by the systemic contradictions of capitalism."[45] In this example, today's extinction events are not simply denied, ignored or downplayed, but instead positioned as rationale, or *reason,* to push forward solutions that actually work to perpetuate—or *fix in place*—present omnicidal trajectories.

This same reasoning is key to broader (Good) Anthropocene fantasies wherein threats of ecological devastation, such as mass extinction, are mutated into *opportunities* for humanity to reaffirm its proper mode, which is, in many tellings of this story, characterized by a very specific human at the center. The anthropos of the (Good) Anthropocene is not only correlated and reduced to an ideal subjective form characterized by bounded individualism, self-regulation and a distinct separation from the ecologies it inhabits but has been developed as an undifferentiated and depoliticized category where the supposedly universal "human" is always "white, Western, modern, able-bodied and heterosexual man."[46] Indeed, the story of the Anthropocene is one that is deeply gendered, underscored by messianic narratives that not only reproduce conservative Christian eschatological tropes of dominion and salvation but also rely on masculinist-solutionist ambitions wherein climate change is seen as merely requiring a "technical fix," and thus anthropos himself is also fully fixable.[47] In the *Jurassic Park* saga, a gendered narrative is also at play, particularly in the way that "science" is pitted against other forms of knowledge and knowing. A prime example of this antagonism takes the form of the "genius," but excluded, scientist, who acts as a proxy for a masculinist form of instrumental rationality and modernist technological worldview more generally.[48] In the film, this rationality is seemingly critiqued and challenged through scenes of dinosaur-driven chaos wherein "Man," especially "scientific Man," is pitted against the technological innovations he creates. Indeed, in almost every instance, the "bad" scientists who become corrupted by either power or money are eaten by the dinosaurs they had a hand in creating. Of note here is how the de-extinct dinosaurs, "cyborgs"[49] born of equal parts technology and "nature," are coded female, and throughout the film series are often depicted as having alliances with women and children. At one level, then, the film's approach to gender seems to offer an eco-feminist critique counter to dominant (Good) Anthropocene narrations, one that highlights how masculinist-techno-solutionism proceeds through the domination of women and animals which, in turn, makes the world "a docile prey to technological ingenuity and know-how."[50] The critique, if we follow this line of thinking, comes, in part, when this "technological know-how" quite literally bites back. Or, as Malcolm puts it on his very first journey around the park: "God creates dinosaurs. God destroys dinosaurs. God creates Man. Man destroys God. Man creates dinosaurs," to which a female paleontologist replies: "[d]inosaurs eat man. Woman inherits earth."[51]

Upon closer look, however, the warnings the film seems to offer about the perils of human intervention and the disastrous consequences of usurping God are not just critiques of "science" or "Man" or even instrumental rationality but instead offer a moralistic telling of the dangers of "bad" or "unnatural" reproduction. In this way, *Jurassic Park* can be read as "profound[ly] anti-feminist" in the way that it reinscribes some very conservative gender meanings, particularly when it comes to its depiction of genetic technologies.[52] Within the world of the film, genetic technology, which offers a sort of stand-in for "science" more generally, is often represented as something done by and for Men and contrasted with the more "natural" dispositions and behaviors of women. These spurious correlations and faulty essentialisms are especially evident in *Jurassic Park's* representation of genetic manipulation, which is seen as "unnatural," not to mention highly dangerous. From the very get-go, the de-extinct dinosaurs central to the film are seen as "bad" because they are unnatural; the very circumstances of their creation—that is, genetic engineering backed by multinational corporations—ensure that they are and always will be "bad." Once again, and as the films assert in various ways, when humans mess with "nature," they pay. While there may be some validity to such claims, particularly in an era of anthropogenic climate disaster, this narrative works to reconfigure the meaning of "nature" in reactionary ways, thus reinscribing the nature/culture divide that subtends possibilities for earthly dominion in the first place. In the world of the film, DNA (and its engineering) is not only metaphorized as a way to rewire life itself but seen as a sign of male science, which is always counterpoised against a "natural, not-scientific motherhood."[53] Throughout the film series, and especially in the original books, the "unnatural" cloning of dinosaurs is consistently juxtaposed against the "natural" reproduction characteristic of the "normal," or "white nuclear family," highlighting how the movies seem intent on the making and re-making of heteronormative families.[54] This intention carries through to the final scenes of the final film of the series, which ends with human clone Maisie finally, and triumphantly, being accepted into and accepting of her very own "normal" family. Underlying all of the spectacle, then, *Jurassic Park* presents a story about both the threat to and salvation of human continuance (of a particular kind) into the future.

Jurassic Park insists that genetic technologies are "bad" or wrong not because of the way they fix financial speculation; for the unexpected consequences they might introduce within ecosystems; or for the risks they bring to real, live animals (including human ones) but because they threaten the "natural" or "pure" family, which is asserted, in the world of the film, but also in our own timeline, as "a bulwark against frightening possible futures."[55] A prime example of this reproductive anxiety came with the release of *Jurassic Park: Dominion,* which was hailed a feminist turn for the franchise—woman can "fix" the world too!—while also inspiring outrage in certain corners of the (conservative) internet, where critics argued that the film was subliminally normalizing "the world's current gender confusion."[56] Referencing the cloning plot line that comes to full fruition in the final film wherein the same cloning technology used to revive ancient species is applied to human genetics, one critic asserted that:

> [I]t's contrived, unnatural, inorganic, tampered with, . . . Mother Nature is smarter than humans. She's the birthing person of humans. There's always retribution for human fiddling foolishness. Let's hope "Jurassic World" remains only a movie concept.[57]

This moral outrage not only points to the deeply gendered way in which the planet itself is often conflated with "natural" reproduction (i.e. Mother Earth), and thus the limits of (some) eco-feminist critiques,[58] but also highlights a deep seated commitment to *reproductive futurism.* Reproductive futurism refers to the tendency to place hope for the future upon children and thus relies on the belief that the fundamental driver for creating a "better" future is to protect and advance the next generation.[59] As such, reproductive futurism not only relies on strictly heterosexual and heteronormative forms of family and relationality but also organizes political discourse and the social imaginary around a projected fantasy of continuance. Importantly, reproduction here does not simply refer to biological reproduction but a certain kind of social reproduction. Beyond the reactions to *Jurassic Park's* cloning narratives, commitments to reproductive futurism, and a concomitant fear about non-reproductive futurity, remain central to contemporary political discourses. The figure of the child continues to symbolize visions of hope for the future, deployed by so-called "pro-life" advocates and anti-GMO activists alike,[60] leading to a social consensus that is impossible to refuse because, as Edelman[61] famously asserted, it is politically impossible to be against the child. That is, unless that child is a clone.

The clone points to a form of queerness that is not only aligned with a rejection of heteronormative coupling but also the refusal of social imperatives of biological reproduction. The clone therefore aligns itself with negativity, with a refusal to participate in the symbolic or social order. With this in mind, the clones of *Jurassic Park*, be it de-extinct (proxy) animals or the human ones, actually work to mutate the narratives of heroic salvation and human redemption that seemingly define the film, in turn upending assumptions about the social imaginary as the projected fantasy of continuance and a "good" future for "us." Indeed, *Jurassic Park's* cloning narrative foretells a future that is increasingly characterized by a range of queer progeny who may not even be human, much less our biological offspring. *Jurassic Park* can, in this way, be resituated as a posthumanist tale of "the commodity value of disembodied genetic information, which lives on long after the 'original' body itself, along with all of its descendants."[62] Beyond tales of de-extinction, which are becoming more "real" by the day, signs of our increasingly posthumanist, non-reproductive future are unfolding all around, where more common examples of our queer progeny include things like plastic, which has seeped into every aspect of our lives in quite material ways. As plastic enters into water streams, but also blood streams, it contributes to very real cases of non-reproductivity (i.e. causing mutations and inhibiting sexual reproduction) while at the same time "birthing a future of strange new life forms adapted to deal with these chemicals."[63] Plastic can, in this way, be understood as just one instance of "non-filial human progeny" or "a bastard child that will most certainly outlive us,"[64] in turn raising questions about how to think through increasingly queer futurities, ones that threaten dominant social orders and interrupt relentless fantasies of (human) continuance.

This line of questioning can also be posed in relation to de-extinction programming, which must not only be scrutinized in terms of moral, economic, social and ecological ramifications but treated as an experimental site to ask questions about what kinds of allegiances and attachments, but also resistances and refusals, might be made with the queer, non-reproductive forces that increasingly define life on planet earth. Where de-extinction, in both the fictional and real-world examples offered here, provides a space for interrogating the way in which life itself is, at the same time, re-wired for capitalist reproduction *and* fissured from reproductive logic, it becomes a test case for the creation of "queer attachments," attachments that both celebrate the "excess of life" that characterizes a world that is increasingly toxic while also "politiciz[ing] the sites at which this excess is eradicated."[65] With this in mind, de-extinction offers more than just a critical example of human exceptionalism and dominion, raising numerous questions for critical, feminist thought as it navigates increasingly non-reproductive futures. After all, where representations of "nature" have become "a special effect of planetary surveillance technologies and the capitalist entertainment complex"[66] and, further, where our world is increasingly characterized by abstraction, virtuality and complexity, critical approaches must adopt an ethos that is at the same time "vehemently anti-naturalist"[67] *and* dedicated to caring for *this* world, in all of its un/natural glory. With this in mind, and by way of conclusion, the examples of de-extinction programming offered here highlight the importance of analytical modes founded in partial perspectives that are both "adamantly synthetic"[68] *and* capable of embracing negativity. This synthetic mode of negation is just one gesture that might be enacted in order to unfix a future seemingly fixed in place; we might embrace and deploy negativity "[n]ot in the hope of forging . . . some more perfect social order, . . . but rather to refuse the insistence of hope itself as affirmation, which is always affirmation of an order whose refusal will register as unthinkable, irresponsible, inhumane."[69] Returning to the world of *Jurassic Park* once more, and specifically back to the "NowThis News" segment that sets up the final film, the question is not how to "fix" the future but how to practice modes of collaborative survival that involve living with, and not in spite of, the ecological ruination and queer progeny that increasingly characterize life in an era of mass extinction.

Notes

1 The *Jurassic Park* films are based on the Michael Crichton science fiction novels *Jurassic Park* (1990) and *The Lost World* (1995). This chapter focuses on the film adaptations, along with the media ecologies inspired by the films, with specific focus on the final film in the series *Jurassic World: Dominion* (2022).

2 For the purpose of this chapter, I will refer to the world of these films simply as *Jurassic Park*.

3 M.J. Lacy, "Cinema and Ecopolitics: Existence in the Jurassic Park," *Millennium: Journal of International Studies* 30, no. 3 (2001): 635–645.

4 C. Trevorrow (Director). *Jurassic World: Dominion* [Film]. Amblin Entertainment; The Kennedy/Marshall Company; Perfect World Pictures (2022) (my italics)

5 Ibid.

6 Ibid.

7 Ibid.

8 Ibid.

9 The idea of the "Good Anthropocene" originated with Erle Ellis, a landscape ecologist and senior fellow at the Breakthrough Institute, who champions what he calls "postnatural environmentalism" and the assertion that we should "forget Mother Nature" and recognize that "this is a world of our making" ("Forget Mother Nature: This Is a World of Our Making," *New*

Scientist (2011), retrieved from https://www.newscientist.com/article/mg21028165-700-forget-mother-nature-this-is-a-world-of-our-making/).

10 For more on the Anthropocene's all-too-human narrations, see H. Davis and Z. Todd, "On the Importance of a Date, or Decolonizing the Anthropocene," *ACME: An International E-Journal for Critical Geographies* 16, no. 4 (2017); D. Haraway, *Staying with the Trouble: Making Kin in the Chthulucene* (Durham, NC: Duke University Press, 2016); N. Mirzoeff, "It's Not the Anthropocene, It's the White Supremacy Scene; or, The Geological Color Line," in *After Extinction*, ed. R. Grusin (Minneapolis; London: University of Minnesota Press, 2018); K. Yusoff, *A Billion Black Anthropoceness or None* (Minneapolis, MN, USA: University of Minnesota Press, 2018); J. Zylinska, *The End of Man: A Feminist Counterapocalypse* (University of Minnesota Press, 2018).

11 As of the summer of 2023, the Colossal project has raised over $75 million dollars, $60 million of which was raised in March 2022 through an oversubscribed Series A financing round led by Thomas Tull at One Ventures, one of Australia's leading venture capital firms.

12 S. Cohen, "The Ethics of De-extinction," *NanoEthics* 8, no. 2 (2014): 165–178; B.J. Novak, "De-extinction," *Gene* 9, no. 1 (2018): 548, https://doi.org/10.3390/genes9110548; B. Shapiro, "Pathways to De-extinction: How Close Can We Get to Resurrection of an Extinct Species?" *Functional Ecology* 31, no. 5 (2017): 996–1002.

13 Novak, "De-extinction."

14 "The Passenger Pigeon Comeback" is a project led by California-based nonprofit biotechnology company Revive & Restore and is one of many initiatives aimed at bringing biotechnological innovation to conversation biology with the mission of enhancing biodiversity through "genetic rescue" of endangered and extinct animals.

15 As the the International Union for the Conservation of Nature (IUCN) outlines, the use of proxy is meant to address the fact that the original extinct species' lineage can never be fully recovered, and therefore only an altered version of a living species can substitute (or proxy) for the extinct species.

16 IUCN/SSC. *IUCN SSC Guiding Principles on Creating Proxies of Extinct Species for Conservation Benefit (Verson 1.0)* (Gland, Switzerland: IUCN Species Survival Commission. 2016)

17 Colossal, "De-extinction Projects, Facts & Figures," *Colossal.com* (2022). https://colossal.com/de-extinction/

18 Ibid.

19 Ibid.

20 Ibid (my italics).

21 Ibid.

22 G. Ceballos, P.R. Ehrlich, and R. Dirzo, "Biological Annihilation via the Ongoing Sixth Mass Extinction Signaled by Vertebrate Population Losses and Declines," *PNAS* 114, no. 30 (2017): https://doi.org/10.1073/pnas.1704949114.

23 Colossal, "De-extinction Projects, Facts & Figures."

24 B. Swanger, "Colossal Biosciences, the Business Looking to Bring Back Woolly Mammoths, Raises $60 Million," *D Magazine* (2022), retrieved from https://www.dmagazine.com/business-economy/2022/03/colossal-biosciences-the-business-looking-to-bring-back-woolly-mammoths-raises-60-million/, accessed January 17, 2023.

25 Colossal, "De-extinction Projects, Facts, & Figures."

26 J. Derrida, *The Animal That Therefore I Am* (New York: Fordham University Press, 2008)

27 N. Batra, "Dominion, Empathy, and Symbiosis: Gender and Anthropocentrism in Romanticism," *ISLE: Interdisciplinary Studies in Literature and Environment* 3, no. 2 (1996): 116.

28 Ibid., 101

29 J.J. Wallin, "Catch 'em all and let man sort 'em out: Animals and Extinction in the World of Pokémon GO," in *Interrogating the Anthropocene: Ecology, Aesthetics, Pedagogy, and the Future in Question*, ed. J. Jagodzinski (Palgrave Macmillan, 2018), 158.

30 N. Srnicek, *Platform Capitalism* (Cambridge, UK/Malden, MA: Polity, 2017).

31 M. Wark, *Capital Is Dead: Is This Something Worse?* (London and New York: Verso Books, 2019).

32 L. Parisi, "Computation Turn," in *Posthuman Glossary*, eds. R. Braidotti and M. Hlavajova (London and New York: Bloomsbury Press, 2018), 88–91.

33 This memetic phrase has been forwarded by technology caputalists and proponents of Big Data such as UK mathematician and architect of Tesco's Clubcard, Clive Humby (2006), (widely credited as the first to coin the phrase) who writes: "Data is the new oil. It's valuable, but if unrefined it cannot really be used. It has to be changed into gas, plastic, chemicals, etc to create a valuable entity that drives profitable activity; so must data be broken down, analyzed for it to have value."
34 S.S. Turner, "Jurassic Park Technology in the Bioinformatics Economy: How Cloning Narratives Negotiate the Telos of DNA," *American Literature* 74, no. 4 (2002): 888
35 Ibid., 890
36 L. Kay, "Cybernetics, Information, Life: The Emergence of Scriptural Representations of Heredity," *Configurations* 5 (1997).
37 Turner, "Jurassic Park Technology in the Bioinformatics Economy: How Cloning Narratives Negotiate the Telos of DNA," 894.
38 J.R Bennett, R.F Maloney, T.E Steeves, J. Brazill-Boast, H.P Possingham, and P.J. Seddon, "Spending Limited Resources on De-extinction Could Lead to Net Biodiversity Loss," *Nature Ecology & Evolution* 1, no. 4 (2017).
39 C.N. Davis and M.D. Moran, "An Argument Supporting De-extinction and a Call for Field Research," *Frontiers of Biogeography* 8, no. (3) (2016): e28431.
40 H. Browning, "Won't Somebody Please Think of the Mammoths? De-extinction and Animal Welfare," *Journal of Agricultural and Environmental Ethics, 31* (2019).
41 A. Dawson, "Biocapitalism and de-extinction," in *After Extinction*, ed. R. Grusin (Minnesota: University of Minnesota Press, 2018), 176 my italics
42 Ibid., 180
43 Wallin, "Catch 'Em All and Let Man Sort 'Em Out: Animals and Extinction in the World of Pokémon GO."
44 A. Dawson, *Extinction: A Radical History* (New York and London: Or Books, 2016), 14
45 Ibid., 79.
46 C. Colebrook, "What Is the Anthropo-political?" in *Twilight of the Anthropocene Idols*, eds. T. Cohen, C. Colebrook, and J.H Miller (London: Open Humanities Press, 2016), 91.
47 Zylinska, *The End of Man: A Feminist Counterapocalypse*, 18
48 Lacy, "Cinema and Ecopolitics: Existence in the Jurassic Park."
49 D. Haraway, "A Cyborg Manifesto: Science, Technology, and Socialist-feminism in the Late Twentieth Century," in *Simians, Cyborgs, and Women: The Reinvention of Nature* (New York: Routledge, 1991).
50 Lacy, "Cinema and Ecopolitics: Existence in the Jurassic Park," 640.
51 S. Spielberg (Director), *Jurassic Park* [Film] (Universal Pictures/Amblin Entertainment, 1993)
52 L. Briggs and J.I. Kelber-Kaye, "There Is No Unauthorized Breeding in Jurassic Park: Gender and the Uses of Genetics," *NWSA Journal* 12, no. 3 (2000): 93
53 Ibid., 111
54 Ibid.
55 Ibid., 112
56 M. Jackson, "Film Review: 'Jurassic World: Dominion': Normalizing America's Gender Agenda," *The Epoch Times* (2022), retrieved from https://www.theepochtimes.com/film-review-jurassic-world-dominion-normalizing-americas-gender-agenda_4523686.html
57 Ibid.
58 J. Oksala, "Feminism, Capitalism, and Ecology," *Hypatia: A Journal of Feminist Philosophy* 33, no. (2) (2018).
59 L. Edelman, *No Future: Queer Theory and the Death Drive* (Duke University Press, 2004)
60 N. Seymour, *Strange Natures: Futurity, Empathy and the Queer Ecological Imagination* (Chicago: University of Illinois Press, 2013)
61 Edelman, *No Future: Queer Theory and the Death Drive.*
62 Turner, "Jurassic Park Technology in the Bioinformatics Economy: How Cloning Narratives Negotiate the Telos of DNA," 894.
63 H. Davis, "Toxic Progeny: The Plastisphere and Other Queer Futures," *philoSOPHIA* 5, no. 2 (2015): 232.
64 Ibid.

65 C. Mortimer-Sandilands and B. Erickson, "Introduction: A Geneaology of Queer Ecologies," in *Queer Ecologies: Sex, Nature, Politics, Desire*, eds. C. Mortimer-Sandilands and B. Erickson (Indianapolis: Indiana University Press, 2010), 37
66 Wallin, "Catch 'Em All and Let Man Sort 'Em Out: Animals and Extinction in the World of Pokémon GO," 168.
67 Laboria Cuboniks. *Xenofeminism: A Politics for Alienation* (2018), retrieved from: http://www.laboriacuboniks.net/
68 Ibid.
69 Edelman, *No Future: Queer Theory and the Death Drive,* 4.

Bibliography

Batra, N. "Dominion, Empathy, and Symbiosis: Gender and Anthropocentrism in Romanticism." *ISLE: Interdisciplinary Studies in Literature and Environment* 3, no. 2 (1996): 101–120.

Bennett, J.R., Maloney, R.F., Steeves, T.E., Brazill-Boast, J., Possingham, H.P., and Seddon, P.J. "Spending Limited Resources on De-extinction Could Lead to Net Biodiversity Loss." *Nature Ecology & Evolution* 1, no. 4 (2017): 0053.

Briggs, L., and Kelber-Kaye, J.I. "There Is No Unauthorized Breeding in Jurassic Park: Gender and the Uses of Genetics." *NWSA Journal* 12, no. 3 (2000): 92–113.

Browning, H. "Won't Somebody Please Think of the Mammoths? De-extinction and Animal Welfare." *Journal of Agricultural and Environmental Ethics* 31 (2019): 785–803.

Ceballos, G., Ehrlich, P.R., and Dirzo, R. "Biological Annihilation via the Ongoing Sixth Mass Extinction Signaled by Vertebrate Population Losses and Declines." *PNAS* 114, no. 30 (2017). https://doi.org/10.1073/pnas.1704949114.

Cohen, S. "The Ethics of De-extinction." *NanoEthics* 8, no. 2 (2014): 165–178.

Colebrook, C. "What is the Anthropo-political?" In *Twilight of the Anthropocene Idols,* edited by T. Cohen, C. Colebrook, and J.H. Miller, 81–125. London: Open Humanities Press, 2016.

Colossal. *De-extinction Projects, Facts & Figures*. Colossal.com, 2022. Retrieved from: https://colossal.com/de-extinction/

Davis, C.N., and Moran, M.D. "An Argument Supporting De-extinction and a Call for Field Research." *Frontiers of Biogeography* 8, no. 3 (2016): e28431.

Davis, H. "Toxic Progeny: The Plastisphere and Other Queer Futures." *philoSOPHIA* 5, no. 2 (2015): 231–250.

Davis, H., and Todd, Z. "On the Importance of a Date, or Decolonizing the Anthropocene." *ACME: An International E-Journal for Critical Geographies* 16, no. 4 (2017): 761–780.

Dawson, A. *Extinction: A Radical History*. New York/London: Or Books, 2016.

Dawson, A. "Biocapitalism and de-extinction." In *After Extinction*, edited by R. Grusin. Minnesota: University of Minnesota Press, 2018.

Derrida, J. *The Animal That Therefore I Am*. New York: Fordham University Press, 2008.

Edelman L. *No Future: Queer Theory and the Death Drive*. Duke University Press, 2004.

Ellis, E. "Forget Mother Nature: This Is a World of Our Making." *New Scientist* (2011). Retrieved from: https://www.newscientist.com/article/mg21028165-700-forget-mother-nature-this-is-a-world-of-our-making/

Haraway, D. "A Cyborg Manifesto: Science, Technology, and Socialist-feminism in the Late Twentieth Century." *Simians, Cyborgs, and Women: The Reinvention of Nature*. New York: Routledge, 1991.

Haraway, D. *Staying with the Trouble: Making kin in the Chthulucene*. Durham, NC: Duke University Press, 2016.

IUCN/SSC. *IUCN SSC Guiding Principles on Creating Proxies of Extinct Species for Conservation Benefit (Version 1.0)*. Gland, Switzerland: IUCN Species Survival Commission, 2016.

Jackson, M. "Film Review: 'Jurassic World: Dominion': Normalizing America's Gender Agenda." *The Epoch Times*, 2022. Retrieved from: https://www.theepochtimes.com/film-review-jurassic-world-dominion-normalizing-americas-gender-agenda_4523686.html

Kay, L. "Cybernetics, Information, Life: The Emergence of Scriptural Representations of Heredity." *Configurations* 5 (1997): 23–91.

Laboria, Cuboniks. *Xenofeminism: A Politics for Alienation*. 2018. Retrieved from: http://www.laboriacuboniks.net/
Lacy, M.J. "Cinema and Ecopolitics: Existence in the Jurassic Park." *Millennium: Journal of International Studies* 30, no. 3 (2001): 635–645.
Mirzoeff, N. "It's not the Anthropocene, It's the White Supremacy Scene; or, The Geological Color Line." In *After Extinction*, edited by R. Grusin. Minneapolis and London: University of Minnesota Press, 2018.
Mortimer-Sandilands, C., and Erickson, B. "Introduction: A Genealogy of Queer Ecologies." In *Queer Ecologies: Sex, Nature, Politics, Desire*, edited by C. Mortimer-Sandilands and B. Erickson. Indianapolis: Indiana University Press, 2010.
Novak, B.J. "De-Extinction." *Genes* 9, no. 11 (2018): 548. https://doi.org/10.3390/genes9110548
Oksala, J. "Feminism, Capitalism, and Ecology." *Hypatia: A Journal of Feminist Philosophy* 33, no. 2 (2018): 216–134.
Seymour, N. *Strange Natures: Futurity, Empathy and the Queer Ecological Imagination*. Chicago: University of Illinois Press, 2013.
Shapiro, B. "Pathways to De-extinction: How Close Can We Get to Resurrection of an Extinct Species?" *Functional Ecology* 31, no. 5 (2017): 996–1002.
Spielberg, S. (Director). *Jurassic Park* [Film]. Universal Pictures/Amblin Entertainment, 1993.
Srnicek, N. *Platform Capitalism*. Cambridge, UK/Malden, MA: Polity, 2017.
Swanger, B. "Colossal Biosciences, the Business Looking to Bring Back Woolly Mammoths, Raises $60 Million." *Dallas Magazine,* 2022. Retrieved January 17, 2023 from https://www.dmagazine.com/business-economy/2022/03/colossal-biosciences-the-business-looking-to-bring-back-woolly-mammoths-raises-60-million/
Trevorrow, C. (Director). *Jurassic World: Dominion* [Film]. Amblin Entertainment; The Kennedy/Marshall Company; Perfect World Pictures, 2022.
Turner, S.S. "Jurassic Park Technology in the Bioinformatics Economy: How Cloning Narratives Negotiate the Telos of DNA." *American Literature* 74, no. 4 (2002): 887–909.
Wallin, J.J. "Catch 'em all and let man sort 'em out: Animals and Extinction in the World of Pokémon GO." In *Interrogating the Anthropocene: Ecology, Aesthetics, Pedagogy, and the Future in Question*, edited by J. Jagodzinski, 155–174. Palgrave Macmillan, 2018.
Wark, M. *Capital Is Dead: Is This Something Worse?* London and New York: Verso Books, 2019.
Yusoff, K. *A Billion Black Anthropoceness or None*. Minneapolis, MN, USA: University of Minnesota Press, 2018.
Zylinska, J. *The End of Man: A Feminist Counterapocalypse*. University of Minnesota Press, 2018.

50
COMPOSTING LIFE-DEATH TIME

Danika Jorgensen-Skakum

Introduction[1]

This chapter is an intervention in human/nonhuman animal relations at the temporal interstices between life and death—especially in the context of the so-called Anthropocene, or this geological epoch marked by climate change and other (capitalist) ecological ills. Taking up Donna Haraway's invocation of compost,[2] I use human composting—namely the Recompose process first launched in Seattle, Washington, and the failed Infinity (mushroom) Suit—to illustrate the interspecies muddling that complicates dominant conceptions of "the end" as humans face the prospect of temporally complex mass death in a warming world.

In recent years, there has been renewed interest in environmentally friendly human body disposal. Usually these plans—for "green burial," mushroom suits or urns that convert remains into trees—stem from an expressed desire to *return to the Earth*. Death scholar Tony Walter attributes this desire to death's re-emergence in Western social life—a sort of cultural trend capitalizing on other greening initiatives.[3] I think this desire is more indicative of a temporal matter: a desire, perhaps, to enter into ecological or planetary time once our own anthropogenic time elapses. Possibly this deathly environmentalism even reflects green guilt, as if we might leave our bodies as a parting gift or apology to the Earth when our attempts to divert climate change fail. Either way, the supporting belief is that our bodies are good, natural, and quite possibly nutritious for the earth we call "Other."[4] We see ourselves as *Anthropos* until we have absolutely no choice; I argue that we must instead understand our position as *already* environment/earth/bios.

Thus, I forward a temporal shift that might help us think otherwise, addressing what it means to become, and especially what it means to become Other when death and life are fundamentally temporal encounters that stretch across multiple times, species, and generations: when they are not seen as transitory or diametrically opposed. There is no true "return to Earth," no Anthropos—only the messy, often fecund, and re/decomposing timescape of the compost pile.

DOI: 10.4324/9781003273400-61

Composting Life-Death Time

Time is popularly conceived as belonging to Anthropos, emerging solely in the relationship between human animals and their limited understanding of finite existence. To think of time as a resonance of human experience is thus to think of time as an articulation of human history, or a path from past to present and (hetero)futurism wherein the (white, patriarchal, colonizing) dominant subject takes precedence. This view is woefully inadequate to the challenge of understanding the significance of time (particularly in *this* epochal time), "as it is articulated both in the emergence of a planet from out of stardust and then in the emergence upon that planet of life and so of those reefs of limestone from which a mountain range was in part built."[5] Thus, negotiating life-death time requires alternative temporal formations capacious enough to account for timescales that have nothing to do with the traditionally dominant subject. Following James Hatley, we might think of time as relational rather than linear, perhaps as emerging through Aristotle's *phusis,* "in which my life *is given* in its very body through the manifold bodies of others who ultimately are of different kinds than myself."[6] In this temporal frame, we recognize that we are born into many (generational) times, "inextricably interwoven with and so always in gratitude to the innumerable lives of others for the very body that is one's own"[7]—not only the lives of humans but "the entirety of life in all its variations and through all its zoogenetic transformations."[8] Under a similar temporal analysis, Deborah Bird Rose "address[es] the gift of life as a multispecies offering at the intersection of sequential and synchronous time."[9]

Thinking with life-death time is therefore a turn to the timeliness, the temporality, of the body—or bod*ies*, to be more precise—resisting the impulse to romanticize a single timeline. It requires shedding the old desire to transcend mortality "through a distancing from the body, and, ultimately, through death."[10] It means inhabiting a transmortality dedicated neither to finitude nor infinitude but holding both at once—understanding that there is no immortality extracted beyond the body. There is no afterlife in the compost pile, only those lives-that-come-after, and the messy complications of engaged fleshy embodiment.

Time thus becomes a question of other subjects, ontological premise and shedding the classic mind-body dualism that has captured the attention of posthumanists, ecofeminists, and those in the environmental humanities—particularly as the distinction between mind and body easily becomes the distinction between human and nature. Elizabeth Grosz argues that "time is not merely the attribute of a subject, imposed by us on the world: it is a condition of what is living, of matter, of the real, of the universe itself."[11] However, Grosz seeks to rectify this temporal (mis)understanding by suggesting that all living beings exist "in a single temporality"—a resolution incompatible with compost.[12] We should aim not to bring others into our time but to dilate the possibilities of the temporal realm in relationships with other kin.

Donna Haraway forwards compost as divestment from the human and "the detumescing project of a self-making and planet-destroying CEO" alongside the many long, durational and quickened crises of the so-called Anthropocene.[13] With compost, we turn instead to collective world-making and the compost pile's sweaty, entangled relationships; affirming compost is thus refusing autopoiesis or self-making—a profound re-cognition of how we become who we become together across time and space.[14] Of course, thinking of time relationally is to embrace precarity, which

> is the condition of being vulnerable to others. Unpredictable encounters transform us; we are not in control, even of ourselves. Unable to rely on a stable structure of

> community, we are thrown into shifting assemblages, which remake us as well as our others. . . . A precarious world is a world without teleology. Interdeterminancy, the unplanned nature of time, is frightening, but thinking through precarity makes it evident that interdeterminancy also makes life possible.[15]

Such precarity is the antithesis of progress, which is "embedded, too, in widely accepted assumptions about what it means to be human": to "look forward—while other species, which live day to day, are thus dependent on us."[16] Indeed, "Without that driving beat, we might notice other temporal patterns. Each living thing remakes the world through seasonal pulses of growth, lifetime reproductive patterns, and geographies of expansion."[17] Tsing argues that even our preoccupation with *the end* (conceived of through the Anthropocene or otherwise) is an extension of progress, which "still controls us even in tales of ruination."[18]

The limits of thinking within mortality—the finitude of the end—is also why Heather Davis incorporates Povinelli's *extinguishment*, which

> recognizes that things live and die, re-composing in a different form, but without the drama of *the end*. Particular configurations of matter, politics, ideas, and organisms obviously cease to exist, while others come into being. However, extinguishment abandons the teleological impulse by recognizing the circularity and fecundity of living systems. *This* civilization may die, but within that death is the possibility for a reconfiguration with what may be left. Humanity will most certainly one day die off, and it wouldn't be that great a surprise if that happened in the relatively near future, but that doesn't mean that species won't evolve or mutate, or that our descendants, even if primarily bacterial, won't inherit the world we leave behind.[19]

Of course, conceiving of time as a series of multispecies and multigenerational encounters is nothing new, particularly in terms of Indigenous cosmologies. This is why Mark Rifkin proposes that "Rather than approaching time as an abstract, homogeneous measure of universal movement along a singular axis, we can think of it as plural, less as a temporality than *temporalities*."[20] In this way also, time is about becoming.[21] By enfolding Indigenous cosmologies in this discussion of time and death, my aim is not only to represent myself as a Métis scholar caught between (or among) worldviews but also to respond to critiques levelled by those like Zoe Todd. Todd warns against the gentrifying impetus of those who theorize about time on ecological scales—particularly in the epoch popularized as the Anthropocene—and posthumanism more broadly "for its erasure of non-European ontologies."[22] In other cases, Todd notes, posthumanist thinkers (like Donna Haraway and, indeed, Deborah Bird Rose) "borrow" from Indigenous theories to establish themselves and their own (white) intellectual traditions.[23] Helpfully, Todd's redress ("Indigenizing the Anthropocene") is both temporal and relational, drawn from Dwayne Donald's "ethic of historical consciousness:"

> This ethic holds that the past occurs simultaneously in the present and influences how we conceptualize the future. It requires that we see ourselves related to, and implicated in, the lives of those yet to come. It is an ethical imperative to recognize the significance of the relationships we have with others, how our histories and experiences are layered and position us in relation to each other, and how our futures as

> people similarly are tied together. It is also an ethical imperative to see that, despite our varied place-based cultures and knowledge systems, we live in the world together with others and must constantly think and act with reference to these relationships. Any knowledge we gain about the world interweaves us more deeply with these relationships, and gives us life.[24]

An ethic of historical consciousness thereby implicates us in life-death time. The temporalities of Indigenous kin transcend the dominant (white-heteropatriarchal) subject timeline discussed previously, revealing the centrality of relationships to the nature of time. In this conceptualization, no subject weighs more heavily than any other because each being is oriented toward other lives—including the bacterial. Death is not necessarily the end of these relationships, since ancestral ties and the promise of successive generations keep communities moving in multiple timelines. Moreover, the connection to place and place-based knowledge secures the multispecies legacies so integral to life-death time. As I indicated, "Life and death do not take place in isolation from others; they are thoroughly relational affairs for fleshy, mortal creatures."[25] Death and life are not opposed but extensions of each other held together by relationships. Thom Van Dooren, whose book *Flight Ways* focuses on avian encounters,

> emphasizes the "embodied temporality" of species, asking us to pay attention to species as evolving "ways of life" that are shared, produced, and nurtured in the world through the work of successive generations of living beings. Thinking in this way requires us to work across entirely different temporal horizons: to think about species in a way that acknowledges that they are vast evolutionary lineages stretched across millions of years, while not losing sight of the fleeting and fragile individual birds whose lives and labors both constitute and enable the continuity of this larger species.[26]

Work, or labor, is necessary to bring forth generational time that must be held in tension between the kin existing now.[27] By embracing an ethic of historical consciousness, I aim not to gentrify or supplant Indigenous knowledge but to uphold the relational nature of knowing and learning that fold me into composted and ethical theory. This theory makes it possible to consider what it means to become in life-death time and offers new directions for considering life-death in the Anthropocene.

Recomposing and Losing Infinity

In Seattle, Washington, architect Katrina Spade recently launched the first human composting facility. The original design for the project featured a massive rooftop garden supported in part by decomposing human tissue. The bereaved would lay their loved ones to rest beneath the surface, and, in time, they would be able to take some of the garden soil with them to memorialize the deceased. Surprisingly (or perhaps unsurprisingly), Spade received considerable resistance to this initial vision, particularly after delivering a much-viewed TED Talk on the subject. Critics objected to the collective treatment of remains and the loss of individual memorial space, so Spade altered her design. Recompose now uses individual composting pods, fit together like a giant indoor honeycomb. Adopting the term *recomposition*, Spade seeks to allay consumer fears about becoming nothing in the decomposition process. The "re-" of recomposition signals a renewed process of becoming, emphasizing the conversion to soil and/or a return to nature. The honeycomb-like pods ensure that those who receive memorial soil assume to receive the bodily contributions of their individual loved one.

I first learned about the Recompose project at the 2017 Death Salon—an international gathering of death professionals and laypeople interested in the "Death Positivity Movement." A fellow attendee—Leah Leighton, forensic anatomy specialist—pointed out immediately that recomposition would never be as individualized as Spade proposed. After all, Spade notes in her TED Talk that bacterial species are necessary in the recomposition process. These bacteria are unlikely to stay loyal to one honeycomb pod. There is also, of course, the question of the materials used to expedite the composting process—who and where and from which communities the materials derive. Further, what does it mean to conceive of dirt as an individualized substance of one, when in fact dirt is composed of many substances, organisms, and beings (both living and dead, or on their own trajectories of life-death time)? What does it mean to become something or nothing in death? To *become* earth?

Soil, by its very definition, is full of life—both macro and micro.[28] And, as Ladelle McWhorter notes, "That's the trouble with dirt. Dirt has no integrity. Dirt isn't a particular, identifiable thing. And yet it acts. . . . Dirt perpetuates itself."[29] Indeed, the designation "dirt" implies that it has gone somewhere it shouldn't, becoming more nuisance than fertile ground.[30] Yet, "dead" dirt can become soil even as soil can become dirt, traversing life and death through a series of micro-relations. Spade addresses the human-dirt relationship as follows:

> What's magical is that we cease to be human during this [recomposition] process. . . . Our molecules are rearranged into other molecules, and in fact what's created is not human remains. To give someone back the soil that is created from just their person would be purely symbolic. If what we're trying to do is reconnect with the fact that we're all part of this grand natural world, let's say ok, we really are part of this system that's greater than ourselves.[31]

Curiously, even as Spade distinguishes soil from human remains, she obscures the micro-relations that make recomposition possible. "Symbolic" memorial soil isn't just not-human, it's decidedly more-than-human. Haraway uses compost to expand this biotic-abiotic relationality: "[H]uman beings are not in a separate compost pile. We are humus, not homo . . . human beings are with and of the earth, and the other biotic and abiotic powers of this earth are the main story."[32] Kim Q. Hall proposes a similar "metaphysics of the compost," writing that "compost is not a singular, fixed thing. It is a process of decomposition, a process of becoming. Compost is simultaneously a materialization of decay and life. It teems with many varied organisms."[33] Indeed, becoming is always relational, a process more than a fixed destination.

In 2016, funeral professional Melissa N. Unfred was approached by the creators of the Infinity Burial Suit—a shroud "infused with mushroom mycelium promis[ing] to eat the toxins produced by a decomposing body and leave the soil in better condition."[34] The creator, artist Jae Rhim Lee, gained fame via a popular TED Talk and was eager to try out her shroud with Unfred's help and the assistance of a university researcher.[35] Unfred found a donor and assisted in shrouding the body with the specialized material the evening before the funeral. The Suit looked and smelled different than she expected—beige, not black, reeking of mushroom. However, once the shroud was installed beneath ground, it was months before Unfred heard from the team again. Eventually, she was contacted by the researcher, who reported that there had been no mycelium growth. The Infinity Suit did not work as hoped, and Unfred alleges the team has not responded to her inquiries since.

In her TED Talk on the Infinity project, Lee warns the audience about the toxicity of their human bodies—containers of "219 toxic pollutants" (0:56) that "return to the environment" upon death (1:27). The Infinity Mushroom on the suit she creates would theoretically consume the toxic human body, purifying in the process. She adds, "I realize this is not the kind of relationship we usually aspire to have with our food. We want to eat, not be eaten, by our food, right?" (3:49) This statement, of course, reflects the anthropocentric impulse to separate eating and being eaten, particularly as the distinction encapsulates the human/animal binary. Aaron Bell observes the violence made possible in this ontological distinction, as does Zipporah Weisberg.[36] For Val Plumwood, recognizing our edibility is key to recognizing our co-constitution with other beings. It is a consequence of individualism and "Human exceptionalism [which] positions us as the eaters of others who are never themselves eaten."[37]

Plumwood argues that "[O]ur worldview denies the most basic feature of animal existence on planet earth—that we are food and that through death we nourish others."[38] Of course, the concept of "nourishing" others requires some unpacking. How is it possible to uphold the generational labor and care of multispecies time—life-death time—while rejecting the premise that human animal bodies are innately nourishing or good for the E/earth? I prefer a politics of contamination. Most bodies, by virtue of being in this world, are contaminated, and this contamination introduces its own temporal relationships. According to Stacy Alaimo,

> As a particularly vivid example of trans-corporeal space, toxic bodies insist that environmentalism, human health, and social justice cannot be severed. They encourage us to imagine ourselves in constant interchange with the environment and, paradoxically perhaps, to imagine an epistemological space that allows for both the unpredictable becomings of other creatures and the limits of human knowledge. Toxic bodies may provoke material, trans-corporeal ethics that turn from the disembodied values and ideals of bounded individuals toward an attention to situated, evolving practices that have far-reaching and often unforeseen consequences for multiple peoples, species, and ecologies.[39]

Within the logic of compost, life-death time resists individualism wherever possible. The contaminants that circulate on this planet ensure this interconnection:

> Chemicals, plastics, living beings and their lineages, commitments, relationships, and much more are brought together here in a tangled web of interactions. Temporalities converge in this meeting of bodies, each carrying histories and presaging futures inscribed in them by evolutionary inheritances and/or the processes of their design and manufacture.[40]

Therefore, Heather Davis argues

> we must find ways of living without the categories and fantasies of containment, either in relation to time or in relation to matter. We must recognize the porousness of our bodies and thoughts that leach into economics and materials, that transfer our wastes across the planet and into the deep future.[41]

Contamination brings us together . . . for better or worse. Mycelia operate in different worlds—different timescales—than humans:

> Fungi are famous for changing shape in relation to their encounters and environments. Many are "potentially immortal," meaning they die from disease, injury, or lack of resources, but not from old age. Even this little fact can alert us to how much our thoughts about knowledge and existence just assume determinate life form and old age.[42]

Tsing asks, therefore,

> What if our indeterminate life form was not the shape of our bodies but rather the shape of our motions over time? Such interdeterminancy expands our concept of human life, showing us how we are transformed by encounter. Humans and fungi share such here-and-now transformations through encounter.[43]

The intermediacy of encounter is why it ultimately doesn't matter if the Infinity Suit works or not. It doesn't matter whether we are composted in one big pile or in our own honeycomb pods. Citing Alaimo's concept of trans-corporeality while discussing the Infinity Suit, Salome Rodeck notes, "Inextricably entangled with the world, bodies are always already part of the planetary assemblage, toxic and otherwise."[44] Thus, she argues "the infinity burial project is a symbol of an understanding of death as part of the constant renewal of interconnected webs of life."[45] Rodeck does argue, however, that "Infinity burial uncovers that even within a post-transcendental society, existence itself can be understood as the endless transformation of matter."[46] But remembering that there is no immortality without the body and labors of life-death time, I contend that what it means to transform matter is not necessarily for our flesh to turn into fungi, to become dirt, or to be eaten by nonhuman animal kin but for our temporal understanding of life-death to make room for compost.

Conclusion

The truth is, we cannot transition into ecological or planetary time; we have already arrived. We are here, contaminants and contaminated, alongside all of the other species with their own life-death times. Like Haraway notes, we are "with and of the earth," thereby implicated in all of the associated temporal formations and ethics.[47] Our desire to find more ethical means of body disposal should be born of this re-cognition that we are always becoming-*with*. We are eaten before death. Life and death are not two oppositional points but interstitial meeting places where we relate and labor in generational time. Thus, according to Donald's ethic of historical consciousness, the more we learn from each other and come into relationship, the more responsibility we amass and the greater the entanglement. Our ontologies become messier upon every new connection—which is not to say that compost is beautiful. On the contrary, compost is mostly shit, or what is excreted upon consumption. Other elements of compost include the wreckage of insect birth, the threat of bacterial exposure, the seepage of metals, and the toxicity of plastics. But compost *is* fertile, making room for life-death times that may result in yet-unthought assemblages, new collaborations, and reconnections with kin across all timelines—relationships we need now more than ever to face the so-called Anthropocene and other multitemporal crises.

Notes

1 Author's Note: Sections of this chapter have appeared in my master's thesis: D. Jorgensen-Skakum, "Death Positivity and Death Justice in the Anthropocene" (Master's thesis, 2018). https://era.library.ualberta.ca/items/9bf8c4f6-c83f-4898-9e01-267f71fe5d45
2 See Donna Haraway, "Staying with the Trouble: Anthropocene, Capitalocene, Chthulucene," in *Anthropocene or Capitalocene? Nature, History, and the Crisis of Capitalism*, ed. Jason Moore (PM Press, 2016), 34–77. Also Donna Haraway, *Staying with the Trouble: Making Kin in the Chthulucene* (Duke University Press, 2016).
3 Tony Walter, *What Death Means Now: Thinking Critically About Dying and Grieving* (Bristol University Press, 2017).
4 The distinction between Earth and earth is a conscious one, intended to reflect the difference between thinking morally of our ecology as the entity *Earth* and thinking materially in terms of human and nonhuman bodily (de)composition.
5 James Hatley, "The Virtue of Temporal Discernment: Rethinking the Extent and Coherence of the Good in a Time of Mass Species Extinction," *Environmental Philosophy* 9, no. 1 (2012): 2.
6 Ibid., 10.
7 Ibid.
8 Ibid., 11.
9 Deborah Bird Rose, "Multispecies Knots of Ethical Time," *Environmental Philosophy* 9, no. 1 (2012): 128.
10 Elizabeth Grosz, *The Nick of Time: Politics, Evolution, and the Untimely* (Duke University Press, 2004), 6.
11 Ibid., 4–5.
12 Ibid., 5.
13 Donna Haraway, *Staying with the Trouble: Making Kin in the Chthulucene,* 32.
14 Ibid., 33.
15 Anna Tsing, *The Mushroom at the End of the World* (Princeton University Press, 2015), 20.
16 Ibid., 21
17 Ibid.
18 Ibid.
19 Heather Davis, "Life and Death in the Anthropocene: A Short History of Plastic," in *Art in the Anthropocene: Environments and Epistemologies*, eds. Heather Davis and Etienne Turpin (Open Humanities Press, 2013), 347–358, retrieved from: http://openhumanitiespress.org/books/download/Davis-Turpin_2015_Art-in-the-Anthropocene.pdf, 355.
20 Mark Rifkin, *Beyond Settler Time: Temporal Sovereignty and Indigenous Self-Determination* (Duke University Press, 2017), 2.
21 Ibid., 2.
22 Zoe Todd, "Indigenizing the Anthropocene," in *Art in the Anthropocene* (Open Humanities Press, 2013), 244–245.
23 Ibid., 246.
24 Donald as cited by Todd, "Indigenizing the Anthropocene," 251.
25 Thom Van Dooren, *Flight Ways: Life and Loss at the Edge of Extinction* (Columbia University Press, 2016), 4.
26 Ibid., 22.
27 Though, to be fair, notions of the "present" are complex and unlikely to be carried between species—an idea to be discussed elsewhere—the kin who came before, and the kin who are still to come.
28 Colby Moorberg, " 'Soil' vs. 'Dirt'," Blog, *Colby Digs Soil* (blog) (December 2011), retrieved from: https://colbydigssoil.com/2011/12/28/soil-vs-dirt/.
29 Ladelle McWhorter, *Bodies and Pleasures: Foucault and the Politics of Sexual Normalization* (Indiana University Press, 1999), 162.
30 Ella Wilson, "Soil vs. Dirt: You'll Be Surprised to Know the Difference," Blog, *Tiny Plantation* (blog) (August 4, 2017), retrieved from: https://www.tinyplantation.com/soil-fertilizers/soil-vs-dirt.
31 *When I Die, Recompose Me*, TEDxOrcasIsland (2016), retrieved from: https://www.ted.com/talks/katrina_spade_when_i_die_recompose_me.
32 Donna Haraway, "Staying With the Trouble: Anthropocene, Capitalocene, Chthulucene," 59.

33 Kim Q. Hall, "Toward a Queer Crip Feminist Politics of Food," *PhiloSOPHIA* 4, no. 2 (2014): 177–196, 190–191.
34 Melissa Unfred, *Four Years Ago Today, 2016, I Met up with the Creator of the Mushroom Burial Shroud at a Dingy Crematory* [Photograph]. Instagram. [@mod_mortician], *My Mushroom Burial Suit* (YouTube, 2011), retrieved from: https://www.youtube.com/watch?v=_7rS_d1fiUc.
35 *My Mushroom Burial Suit,* YouTube (2011), retrieved from: https://www.youtube.com/watch?v=_7rS_d1fiUc.
36 Aaron Bell, "The Dialectic of Anthropocentrism," in *Critical Theory and Animal Liberation*, ed. John Sanbonmatsu (Rowman & Littlefield Publishers, Inc., 2011): 171. See also Zipporah Weisberg, "Speciesism as Pathology," in *Critical Theory and Animal Liberation*, ed. John Sanbonmatsu (Rowman & Littlefield Publishers, Inc., 2011): 177.
37 Val Plumwood, "Tasteless: Towards a Food-Based Approach to Death," *The Forum on Religion and Ecology Newsletter* 1, no. 2 (n.d.): 1–7, 1. See also ames Stanescu, "Species Trouble: Judith Butler, Mourning, and the Precarious Lives of Animals," *Hypatia* 27, no. 3 (2012): 567–582, https://doi.org/10.1111/j.1527-2001.2012.01280.x.
38 Plumwood, "Tasteless," 1.
39 Stacy Alaimo, *Bodily Natures: Science, Environment, and the Material Self* (Indiana University Press, 2010), 22.
40 Van Dooren, *Flight Ways,* 32.
41 Davis, "Life and Death in the Anthropocene," 356.
42 Tsing, *The Mushroom at the End of the World,* 47.
43 Ibid.
44 Salome Rodeck, "Dying with 'Infinity Mushrooms': Mortuary Rituals, Mycroremediation and Multispecies Legacies," *Women, Gender & Research* 28, no. 3–4 (2019): 62–74, 66, retrieved from: https://tidsskrift.dk/KKF/article/view/116309.
45 Rodeck, "Dying with 'Infinity Mushrooms'," 68.
46 Ibid., 71.
47 Haraway, "Staying with the Trouble," 59.

Bibliography

Alaimo, Stacy. *Bodily Natures: Science, Environment, and the Material Self.* Indiana University Press, 2010.

Bell, Aaron. "The Dialectic of Anthropocentrism." In *Critical Theory and Animal Liberation*, edited by John Sanbonmatsu. Rowman & Littlefield Publishers, Inc., 2011.

Davis, Heather. "Life and Death in the Anthropocene: A Short History of Plastic." In *Art in the Anthropocene: Environments and Epistemologies*, edited by Heather Davis and Etienne Turpin, 347–358. Open Humanities Press, 2013. Retrieved from: http://openhumanitiespress.org/books/download/Davis-Turpin_2015_Art-in-the-Anthropocene.pdf

Grosz, Elizabeth. *The Nick of Time: Politics, Evolution, and the Untimely.* Duke University Press, 2004.

Hall, Kim Q. "Toward a Queer Crip Feminist Politics of Food." *PhiloSOPHIA* 4, no. 2 (2014).

Haraway, Donna. *Staying with the Trouble: Making Kin in the Chthulucene.* Duke University Press, 2016a.

Haraway, Donna. "Staying With the Trouble: Anthropocene, Capitalocene, Chthulucene." In *Anthropocene or Capitalocene? Nature, History, and the Crisis of Capitalism*, edited by Jason Moore, 34–77. PM Press, 2016b.

Hatley, James. "The Virtue of Temporal Discernment: Rethinking the Extent and Coherence of the Good in a Time of Mass Species Extinction." *Environmental Philosophy* 9, no. 1 (2012).

Jorgensen-Skakum, D. "Death Positivity and Death Justice in the Anthropocene." Master's thesis, 2018. Retrieved from: https://era.library.ualberta.ca/items/9bf8c4f6-c83f-4898-9e01-267f71fe5d45

McWhorter, Ladelle. *Bodies and Pleasures: Foucault and the Politics of Sexual Normalization.* Indiana University Press, 1999.

Moorberg, Colby. " 'Soil' vs. 'Dirt.' " *Colby Digs Soil* (blog), December 2011. Retrieved from: https://colbydigssoil.com/2011/12/28/soil-vs-dirt/.

My Mushroom Burial Suit. YouTube, 2011. Retrieved from: https://www.youtube.com/watch?v=_7rS_d1fiUc.

Plumwood, Val. "Tasteless: Towards a Food-Based Approach to Death." *The Forum on Religion and Ecology Newsletter* 1, no. 2 (n.d.): 1–7.

Rifkin, Mark. *Beyond Settler Time: Temporal Sovereignty and Indigenous Self-Determination*. Duke University Press, 2017.

Rodeck, Salome. "Dying with 'Infinity Mushrooms': Mortuary Rituals, Mycroremediation and Multispecies Legacies." *Women, Gender & Research* 28, no. 3–4 (2019): 62–74. Retrieved from: https://tidsskrift.dk/KKF/article/view/116309.

Rose, Deborah Bird. "Multispecies Knots of Ethical Time." *Environmental Philosophy* 9, no. 1 (2012).

Stanescu, James. "Species Trouble: Judith Butler, Mourning, and the Precarious Lives of Animals." *Hypatia* 27, no. 3 (2012): 567–582. https://doi.org/10.1111/j.1527-2001.2012.01280.x.

Todd, Zoe. "Indigenizing the Anthropocene." In *Art in the Anthropocene: Encounters Among Aesthetics, Politics, Environments and Epistemologies,* edited by Heather Davis and Etienne Turpin. Open Humanities Press, 2013.

Tsing, Anna. *The Mushroom at the End of the World*. Princeton University Press, 2015.

Unfred, Melissa. *Four Years Ago Today, 2016, I Met Up with the Creator of the Mushroom Burial Shroud at a Dingy Crematory* [Photograph]. Instagram. [@mod_mortician].

Van Dooren, Thom. *Flight Ways: Life and Loss at the Edge of Extinction*. Columbia University Press, 2016.

Walter, Tony. *What Death Means Now: Thinking Critically About Dying and Grieving*. Bristol University Press, 2017.

Weisberg, Zipporah. "Speciesism as Pathology." In *Critical Theory and Animal Liberation*, edited by John Sanbonmatsu. Rowman & Littlefield Publishers, Inc., 2011.

When I Die, Recompose Me, TEDxOrcasIsland, 2016. Retrieved from: https://www.ted.com/talks/katrina_spade_when_i_die_recompose_me.

Wilson, Ella. "Soil vs. Dirt: You'll Be Surprised to Know the Difference." *Tiny Plantation* (blog), August 4, 2017. Retrieved from: https://www.tinyplantation.com/soil-fertilizers/soil-vs-dirt

INDEX

For Product Safety Concerns and Information please contact our EU representative GPSR@taylorandfrancis.com Taylor & Francis Verlag GmbH, Kaufingerstraße 24, 80331 München, Germany

Batch number: 10399613

Printed by Printforce, the Netherlands